The Kidney
From Normal Development to Congenital Disease

The Kidney

From Normal Development to Congenital Disease

Edited by

Peter D. Vize

University of Calgary, Calgary, Alberta, Canada

Adrian S. Woolf

Institute of Child Health, London, United Kingdom

Johnathan B. L. Bard

Edinburgh Unversity, Edinburgh, United Kingdom

ACADEMIC PRESS

An imprint of Elsevier Science

Amsterdam   Boston   London   New York   Oxford   Paris
San Diego   San Francisco   Singapore   Sydney   Tokyo

An imprint of Elsevier Science.
525 B Street, Suite 1900, San Diego, California 92101-4495, USA
http://www.academicpress.com

Academic Press
 84 Theobalds Road, London WC1X 8RR, UK
http://www.academicpress.com

Library of Congress Catalog Card Number: 2002107395

International Standard Book Number: 0-12-722441-6

PRINTED IN CHINA
02   03   04   05   06   07   CTP   9   8   7   6   5   4   3   2   1

Contents

v

Section III: Congenital Disease

Contributors

Numbers in parenthesis indicate page numbers on which authors contributions begin.

Dale R. Abrahamson (221)
Department of Anatomy and Cell Biology, The University of Kansas Medical Center, Kansas City, Kansas, United States of America.

Jonathan Bard (139, 181)
Department of Biomedical Sciences, Edinburgh University, Edinburgh, United Kingdom.

Hallgrímur Benediktsson (149)
Department of Pathology and Laboratory Medicine, University of Calgary Medical School, Calgary, Alberta, Canada.

Nicholas Bockett (411)
Cancer Genetics Laboratory, University of Otago, Dunedin, New Zealand.

Cathy Boucher (433)
Cambridge Institute for Medical Research, Addenbrokes Hospital, Cambridge, United Kingdom.

Pierre Bouchet (87)
Department Informatique, Université Henri Poincaré, Vandoeuvre lès Nancy, France.

Thomas J. Carroll (19, 343)
Harvard University, Boston, Massachusetts, United States of America.

Eun Ah Cho (195)
University of Michigan, Ann Arbor, Michigan, United States of America.

Jamie Davies (165)
Department of Anatomy, Edinburgh University, Edinburgh, United Kingdom.

Igor B. Dawid (119)
National Institute of Health, Bethesda, Maryland, United States of America.

Gregory R. Dressler (195)
Department of Pathology, University of Michigan, Ann Arbor, Michigan, United States of America.

Iain Drummond (61)
Massachusetts General Hospital, Charleston, Massachusetts, United States of America.

Michael Eccles (411)
Cancer Genetics Laboratory, University of Otago, Dunedin, New Zealand.

Marie Claire Gubler (395)
Hospital Necker Enfants Malade, Paris, France.

Benedikt Hallgrímsson (149)
Department of Cell Biology and Anatomy, University of Calgary Medical School, Calgary, Alberta, Canada.

Monika H. Hermanns (377)
Nephro-Urology Unit, Institute of Child Health, University College London, London, United Kingdom.

Doris Herzlinger (51)
Departments of Physiology and Urology, Cornell University Medical College, New York, New York, United States of America.

Christer Holmberg (475)
Hospital for Sick Children and Adolescents, University of Helsinki, Helsinki, Finland.

Neil A. Hukriede (119)
National Institute of Health, Bethesda, Maryland, United States of America.

Hannu Jalanko (475)
Hospital for Sick Children and Adolescents, University of Helsinki, Helsinki, Finland.

Richard G. James (51)
Beth Israel Deaconess Medical Center, Boston, Massachusetts, United States of America.

Cécile Jeanpierre (395)
    Hospital Necker-Enfants Malade, Paris, France.
Elizabeth A. Jones (93)
    University of Warwick, Coventry, United Kingdom.
Sharon Karp (211)
    Division of Nephrology, Indiana University Medical
    Center, Indianapolis, Indiana, United States of America.
Anzhelika Listopadova (51)
    Cornell University Medical College, New York,
    New York, United States of America.
Arindam Majumdar (61)
    Massachusetts General Hospital, Charleston,
    Massachusetts, United States of America.
Andrew P. McMahon (343)
    Department of CMB, Harvard University, Cambridge,
    Massachusetts, United States of America.
Bruce Molitoris (211)
    Division of Nephrology, Indiana University Medical
    Center, Indianapolis, Indiana, United States of America.
Sharon Mulroy (433)
    Cambridge Institute for Medical Research, Addenbrokes
    Hospital, Cambridge, United Kingdom.
Kirsi Sainio (75, 181, 327)
    Program of Developmental Biology, University of
    Helsinki, Finland.
Richard Sandford (433)
    Cambridge Institute for Medical Research, Addenbrokes
    Hospital, Cambridge, United Kingdom.
Hannu Sariola (181)
    Institute of Biomedicine, University of Helsinki,
    Finland.
Lisa M. Satlin (267)
    Pediatric Nephrology, Mount Sinai School of Medicine,
    New York, New York, United States of America.
Lauri Saxén (xi)
    University of Helsinki, Helsinki, Finland.
Thomas M. Schultheiss (51)
    Beth Israel Deaconess Medical Center, Boston,
    Massachusetts, United States of America.
George J. Schwartz (267)
    Pediatric Nephrology, University of Rochester School of
    Medicine, Rochester, New York, United States of
    America.
Helen Skaer (7)
    Department of Zoology, Cambridge University,
    Cambridge, United Kingdom.

Cherie Stayner (411)
    Harvard Institues of Medicine, Boston, Massachusetts,
    United States of America.
Karl Tryggvason (475)
    Division of Matrix Biology, MBB Kavolinska Institute,
    Stockholm, Sweden.
William van't Hoff (461)
    Institute of Child Health, University College London
    Medical School, London, United Kingdom.
Marie D. Vazquez (87)
    Department de Cytologie, Histologie et Embryologie
    Faculte de Medicine, Vandoeuvre lès Nancy, France.
Peter D. Vize (1, 19, 87, 149)
    Department of Biological Sciences, University of Calgary,
    Calgary, Alberta, Canada.
Cheryl Walker (451)
    Department of Carcinogenesis, University of Texas/MD
    Anderson Cancer Center, Smithville, Texas, United States
    of America.
John B. Wallingford (19)
    Department of Molecular and Cell Biology, University of
    California, Berkeley, Berkeley, California, United States of
    America.
Ruixue Wang (221)
    Department of Anatomy and Cell Biology, University of
    Kansas Medical Center, Kansas City, Kansas, United
    States of America.
Brant M. Weinstein (119)
    National Institute of Health, Bethesda, Maryland, United
    States of America.
Simon J.M. Welham (377)
    Nephro-Urology Unit, Institute of Child Health, University
    College London, London, United Kingdom.
Paul J.D. Winyard (377, 433)
    Nephro-Urology Unit, Institute of Child Health, University
    College London, London, United Kingdom.
Craig B. Woda (267)
    Mount Sinai School of Medicine, New York, New York,
    United States of America.
Adrian S. Woolf (251, 377, 487)
    Nephro-Urology Unit, Institute of Child Health, University
    College London, London, United Kingdom.
Hai T. Yuan (251)
    Nephro-Urology Unit, Institute of Child Health, University
    College London, London, United Kingdom.

Foreword

The vertebrate kidney is a complex organ constructed from four synchronously developing cell lineages; the nephric mesenchyme, the pronephric-duct-derived epithelial collecting-duct system, and the vascular and neuronal systems. Its developmental richness has resulted in nephrogenesis becoming an excellent model system for exploring such key issues as: the problem of transitory and vestigial organs, programmed cell death (apoptosis), early determination (protodifferentiation), directed movements of cells and cell clusters (nephric duct), epithelial branching morphogenesis (ureter tree), cell aggregation and how a mesenchyme-epithelial transition can lead to the production of complex tubular structures with polarized epithelial cells(nephron formation). The underlying control mechanisms include the morphogenetic interactions between the various cell lineages and the characterization of the signalling pathways that regulate kidney development. Indeed, almost the entire set of genomic and postgenomic control mechanisms can be exploited in this model system.

The history of nephrogenesis is, as for many biological systems, the history of its methodology. The first years of the last century witnessed a thorough light microscopic description of the chain of kidney development, and this morphological analysis was completed in the mid-century by electron microscopy. Over the last two decades, the localization of developmentally regulated genes using in situ hybridisation and immunochemistry has given us complementary molecular anatomy (Davies and Brandli, 2002).

Experimental manipulations to investigate the development of the excretory system were introduced in the late 1930s by Gruenwald (Gruenwald, 1937), work that followed Rienhoff's avian kidney culture studies in the early 1920s (Rienhoff, 1922). The next technical breakthrough came in the 1950s when Grobstein managed to separate the main tissue components of the metanephric kidney, the ureter-derived epithelial bud and the mesenchymal blastema, and

was able to study their inductive interactions using his trans-filter technique (Grobstein, 1956). Today we can build on this technology to understand gene function by using blocking antibodies and antisense technology, and can integrate such studies with analysis of animals carrying targeted mutations that lead to gene loss and overexpression. Such animals can provide models of pediatric kidney disorders and so help develop treatments for them.

When I reviewed our knowledge of nephrogenesis and its control mechanism in the 1980s (Saxén, 1987), little was known of the molecular basis of kidney development. Over the last 15 or so years, our molecular understanding has increased beyond measure, and it is clear that the number of genes participating in just nephron formation and epithelial branching is far greater than we could ever have imagined (Davies and Brandli, 2002). An overall schema describing the many aspects of nephrogenesis does however remain to be constructed and this requires us to integrate our knowledge across the spectrum of kidney research. The present volume is thus most welcome and will undoubtedly stimulate further work on nephrogenesis at both the basic and applied levels.

Lauri Saxén
University of Helsinki

References

Davies, J. A., and Brandli, A. W. The Kidney Development Database, http://golgi.ana.ed.ac.uk./kidhome.html.

Gruenwald, P. (1937). Zur Entwicklungsmechanik der Urogeital-systems Beim Huhn Roux. Arch 136:786–813.

Reinhoff, W. F. (1922). Development and growth of the metanephros or permanent kidney in chick embryos. Johns Hopkins Hosp. Bull. 33:392–406.

Grobstein, C. (1956). Trans-fliler induction of tubules in mouse metanephrogenic mesenchyme. Exp. Cell Res. 10:424–440.

Saxén, L. (1987). Organogenesis of the Kidney. Cambridge University Press, Cambridge, UK.

Preface

In 1987, Lauri Saxén of Helsinki University published a short book entitled Organogenesis of the Kidney which summarized what was then known about kidney development, much of it based on the work of the author and his collaborators. In the last 15 years, with many other European and American laboratories making major contributions to the study of kidney development, our knowledge of the subject has increased almost beyond measure, and for two complementary reasons. The first, of course, is the molecular revolution: we now know of many hundreds of genes expressed during kidney development and are beginning to understand their function. The second is the realization that the different sorts of kidneys and the abnormalities that can arise during metanephric development are interrelated, and knowledge of any one aspect of organogenesis helps understand others.

The result is that anyone now interested in the kidney, be it for medical or basic research reasons, needs to know a great deal about its molecular, developmental, anatomical and functional bases, far more than he or she can hope to acquire from the primary literature. Hence this book, whose main purpose is to summarise such knowledge as we have in 2002 on many aspects of normal kidney development and how abnormal development can lead to congenital disease. There is however a second purpose: editing this book has made us aware of how much remains to be discovered about the kidney and we hope that those unsolved problems raised in each chapter will stimulate new research, produce novel solutions and lead to progress in curing or at least ameliorating congenital kidney disease.

We are grateful to all the authors and for their enthusiasm in fulfilling the difficult task asked of them; we also appreciate the generosity of our colleagues and their publishers in allowing us to use many original pictures – the book would have been far duller without these and also without the new drawings of our artists, George Kelvin and Kathy Stern. We would also like to thank Lauri Saxén both for his pioneering work on kidney development and for his Forward.

Peter Vize
Adrian Woolf
Jonathan Bard

1

Introduction: Embryonic Kidneys and Other Nephrogenic Models

Peter D. Vize

I am a reformed lover of mesoderm induction. My association with the pronephros began for opportunistic reasons with the original plan being to exploit the expression of pronephric genes as markers of the patterning and establishment of the intermediate mesoderm. However, after finding such markers and following their expression in forming pronephroi, I became more interested in how these genes contributed to the regulation of kidney morphogenesis than in simply using them as markers of earlier events. Upon exploring what was known about embryonic kidney development (very little) and what could be learned using modern molecular embryology (an enormous amount), my future research directions were established. The embryonic kidneys are an ideal system in which to explore cell signaling, specification, adhesion, shape change, morphogenesis, and of course organogenesis. In addition to being a wonderful intellectual problem, the analysis of embryonic kidney development has many advantages in terms of the availability of techniques with which to dissect the process. Some of the organisms with the most extensive and well developed embryonic kidneys are also those with the most highly advanced genetic and embryological tools—a perfect match. Finally, similar genetic networks regulate the development of all nephric organs so data gleaned from embryonic systems are as relevant to human congenital disease as they are to the understanding of a quaint model.

This first section of "The Kidney" covers the development of the embryonic kidneys, the pro- and mesonephroi, in a depth never before attempted in a text on kidney development and function. For those who no longer recall their undergraduate developmental biology course,

even the names of these organs may be unfamiliar. After all, some mammals (including humans) can survive until birth without any kidneys, so of what interest are transient organs that some would posit are nothing more than evolutionary artifacts? In this introduction some of the reasons for refraining from such an opinion will be explored, as will the renaissance of research into the use of embryonic kidneys as model systems for the analysis of organogenesis. The following chapters provide a detailed description of the anatomy, development, function, and molecular biology of the transient embryonic kidneys as a resource for those willing to accept my arguments regarding relevancy. Similar arguments can be made supporting the relevance of invertebrate models of nephrogenesis, and Chapter 2 opens with a review of Malpighian tubule morphogenesis in the fruit fly, *Drosophila melanogaster*.

The embryonic kidney of amphibians and fish is known as the pronephros, head kidney, or vorniere, while their adult kidney is known as a mesonephros, Wolffian body, or urniere. In some instances, fish and frog permanent mesonephroi are unnecessarily referred to as opisthonephroi, a term used to distinguish them from the transient mesonephroi of amniotes, but which results in more confusion than clarification. To begin the description of the development of vertebrate embryonic kidneys, a brief description of the occurrence of pro- and mesonephroi is appropriate. All vertebrates have distinct embryonic and adult kidneys (Goodrich, 1930; Burns, 1955; Saxén, 1987). Upon development of the adult kidney, the embryonic kidney usually either degenerates or becomes a part of the male reproductive system (Burns, 1955; Balinsky, 1970). In some

1

instances the embryonic kidney switches to a new role as a lymphoid organ (Balfour, 1882).

Well-developed, functional pronephroi are found in all fish, including dipnoids (e.g., lungfish), ganoids (e.g., sturgeon), and teleosts (e.g., zebrafish), and in all amphibians. Reptiles vary in the degree to which pronephroi form, with the more primitive reptiles having the most advanced pronephroi (Chapter 3). Birds have only a poorly developed pronephros, as do most mammals. In organisms with aquatic larvae, pronephroi are absolutely essential for survival. The pronephroi excrete copious amounts of dilute urine that allows such animals to maintain water balance. If the pronephroi are not functional, aquatic larvae die rapidly from oedema (Chapter 3).

Pronephric kidneys are very simple and form within a day or two of fertilization. They usually contain a single nephron with an external glomerulus or glomus (Fig. 1.1). This glomus filters blood in an identical manner to standard glomeruli, except that the filtrate is deposited into a cavity rather than into Bowman's space. In some instances, this cavity is the coelom, in others, a dorsal subcompartment of the coelom known as the nephrocoel, and in yet others into the pericardial cavity. The glomeral filtrate is collected from the receptive cavity by ciliated tubules known as nephrostomes. The nephrostomes in turn are linked to the pronephric tubules. These tubules have distinct proximal and distal segments. As with a classical mammalian nephron, the proximal segment functions in solute resorption and waste excretion, whereas the distal segment resorbs water. From the distal tubule urine passes down the pronephric duct to the cloaca. The entire pronephros is in essence a single large nephron. This section uses the term

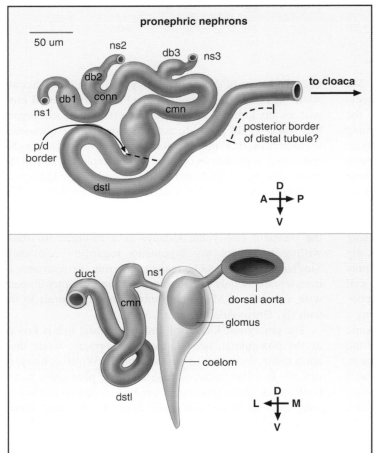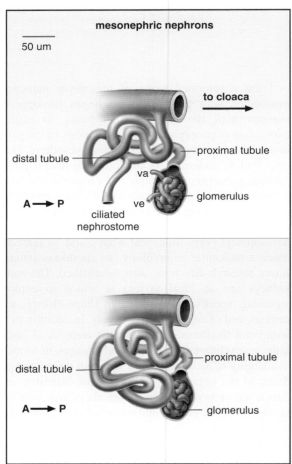

Figure 1.1 Embryonic kidney nephrons. (Left) Lateral and anterior views of a frog pronephric nephron at around the onset of function are illustrated. The anterior border of the distal tubule is marked in the lateral view. The posterior border of this segment has not yet been defined, but the transition region is indicated. (Right) Two common forms of mesonephric nephron are illustrated. In each case, a glomerulus projects into the tip of the proximal segment. In the upper example the nephron branches into a peritoneal funnel that links the proximal tubule to the coelom. This type of nephron receives fluids from two sources: the glomerulus via filtration and the coelom via ciliary action (Chapter 3). ns1, ns2, ns3, nephrostomes 1 through 3; db1, db2, db3; dorsal branches 1 through 3; cmn, common (or broad) tubule; dstl, distal tubule; duct, nephric duct; p/d border, border between proximal and distal tubule zones; va, vas afferens; ve, vas efferens.

pronephros to describe an embryonic kidney that either utilizes an external glomus or is anatomically distinct from the mesonephric kidney in the same organism. Details of pronephric anatomy and complete bibliographies are provided by Chapters 3 through 5.

Mesonephric kidneys are more complex in organization and consist of a linear sequence of nephrons (Fig. 1.2) linked to the nephric duct (Fig. 1.1). Mesonephric nephrons contain internal (or integrated) glomeruli, and in some instances, particularly in anterior mesonephric tubules, also link to the coelom via ciliated tubules called peritoneal funnels. Such funnels are sometimes referred to as nephrostomes, which they resemble very closely, but the correct nomenclature of the two structures allows one to specify whether the funnel links the coelom to the glomerulus or the glomerulus to the tubule. Nephrostomes are also sometimes present in mesonephroi so the distinction is important.

The mesonephros is first functional at around 7.5 days of development in the frog *Xenopus* (Nieuwkoop and Faber, 1994) and continues to grow along with the animal. In organisms in which the mesonephros is transient, the complexity of this organ is extremely variable, ranging from almost no nephrons in rodents to 34 in humans and 80 in pigs (Felix, 1912; Bremer, 1916; Table 1.1). The anatomy of a human mesonephros is illustrated in Fig. 1.3.

In animals in which the mesonephros is the terminal kidney, such as amphibians and fish, the final organ is very complex, containing a large number of nephrons, most of which have an internal glomerulus. In the example of the frog *Rana*, an adult mesonephros contains around 2000 nephrons (Richards, 1929) whereas an adult toad *(Bufo)* contains around 3000 (Møbjerg *et al.*, 1998). The general anatomy of the pro- and mesonephroi of the frog and the transition between the two types of kidney are illustrated in Fig. 1.2.

In amniotes the degree of development of the mesonephros, and even the presence of glomeruli, is linked to the form of placental development. Different amniotes have very different placentas. In some organisms the fetal and maternal tissues are opposed epithelia (e.g., the pig), whereas in others the maternal epithelium breaks down directly, bathing the intervillous spaces of the fetal epithelium directly, with blood, allowing for a more efficient supply of nutrients and removal of wastes (e.g., rodents, primates). Animals with the former type of placenta have large well-developed mesonephroi that remain until the adult kidney is functional, whereas those with the later have less complex embryonic kidneys that often degenerate prior to the formation of the metanephroi (Bremer, 1916; Witschi, 1956).

Metanephroi develop in all amniotes from reptiles through humans and are the most complex of kidneys (Chapter 10). Instead of the linear organization of nephrons found in mesonephroi, metanephroi have a branched architecture with arborized networks of nephrons. The development of metanephroi is covered in detail elsewhere in this book.

A final point worth discussing before presenting the embryonic kidneys and model systems in detail is the similarity between the genetic hierarchies that regulate the development of all three different forms of kidney. Evolution does not reinvent complex processes—it fine tunes existing systems to variations in the environment. As molecular biology began to identify genes and determine their function in embryonic development, it soon became clear that the developmental roles of individual genes were highly conserved between species. The human orthologue of a fly gene often performs a closely related function even though the development of these two organisms differs extraordinarily. In some instances, a mammalian gene can act as a functional substitute in an insect and rescue the animal from a loss of function phenotype (Leuzinger, 1998; Nagao, 1998). Given such conservation of function between species, it was not surprising to find that the different kidney forms utilize similar sets of genes to regulate their development (Vize *et al.*, 1997). Genes demonstrated to regulate key developmental steps in mammalian metanephric kidneys in targeted ablation experiments are expressed in the embryonic kidneys in patterns that imply similar activities. Also, in the few instances where gene function has been tested in embryonic kidneys, the results implied conserved activities in most instances. As the embryonic kidneys are obviously very different than those of the adult, there must be differences in gene expression—and some have been noted (Carroll and Vize, 1996; Chapter 3). Two of the model systems present in part 1, and hopefully three in the near future (Bronchain, 1999) are amenable to genetic screens on a scale impossible in a mammalian system. The embryos of some of these models are also excellent for the experimental embryology and microinjection approaches that have provided a wealth of information on the regulation of developmental processes in the recent past.

It is hoped that the following chapters provide a useful resource for those willing to explore the many advantages of the simple kidneys as model organogenesis systems.

Table 1.1   The Mesonephric Nephron Number

	Embryo length		
	6–10 mm	11–16 mm	21–40 mm
Guinea pig	0	14	0
Human	34	34	12
Rabbit	40	42	34
Cat	20	26	30
Sheep	20 (+6)	20 (+50)	20 (+50)
Pig	54	60	60

After Bremer (1916).

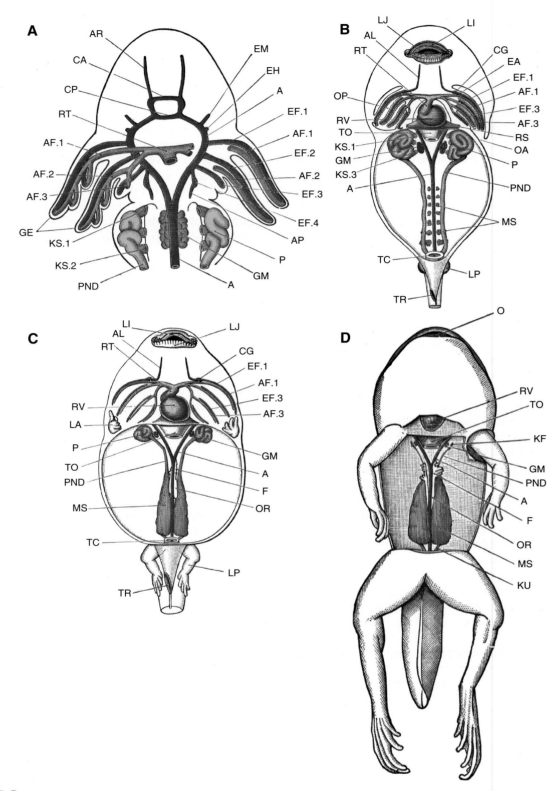

Figure 1.2 Transition between pro- and mesonephroi in the frog, *Rana temporaria,* ventral view (after Marshall, 1902). The arterial system is colored red, the venous system blue, and the pronephric glomus (GM) purple. Pronephric (P) and mesonephric (MS) tubules are in green and the nephric duct (PND) is in yellow. Tadpoles of 6.5 mm (A), 12 mm (B), 40 mm (C), and a metamorph (D). Additional labeled structures correspond to A, dorsal aorta; AF, afferent branchial vessels; AL, lingual artery; AP, pulmonary artery; AR, anterior cerebral artery, CA, anterior commissural artery; CG, carotid gland; CP, posterior commissural artery, EF, efferent branchial vessels; EH, efferent hyoidean vessel; EM, efferent mandibular vessel, GE, gill; GM, glomus; KS, nephrostome; KU, ureter; MS, mesonephros/mesonephric tubules; OR, genital ridge; PND, nephric duct; P, pronephros; KS, nephrostomes; RT, truncus arteriosus; RS, sinus venosus, RV, ventricle; TC, cloaca; TO, oesophagus, cut short; TR, rectal sprout.

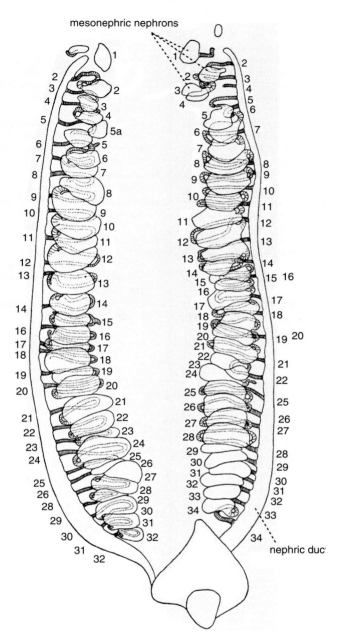

Figure 1.3 Human mesonephros (9.5 mm). Anterior nephrons are undergoing degeneration. Each nephron has an S-shaped tubule linking the glomerulus to the nephric duct. There is some variation in the spacing of the mesonephric tubules, and some glomeruli share a common collecting duct (e.g., glomeruli 15 and 16 and glomeruli 19 and 20 in the left mesonephros). After Felix (1912).

References

Balfour, F. M. (1882). On the nature of the organ in adult teleosteans and ganoids, which is usually regarded as the head-kidney or pronephros. *Quart. J. Micr. Sci.* **22**, Reprinted in "Works of Balfour", Eds. M. Foster and A. Sedgwick. Macmillan, London, 1885. pp. 848–853.

Balinsky, B. I. (1970). "An introduction to embryology." Saunders, Philadelphia.

Bremer, J. L. (1916). The interrelations of the mesonephros, kidney and placenta in different classes of animals. *Am. J. Anat.* **19**, 179–209.

Bronchain, O. J., Hartley, K. O., and Amaya, E. (1999). A gene trap approach in *Xenopus. Curr. Biol.* **21**, 1195–1198.

Burns, R. (1955). Urogenital system. *In* "Analysis of development" (B. Willier, P. Weiss, and V. Hamburger, Eds.). W.B. Saunders, Philadephia.

Carroll, T. J., and Vize, P. D. (1996). Wilms' tumor suppressor gene is involved in the development of disparate kidney forms: evidence from expression in the *Xenopus pronephros. Dev. Dyn.* **206**, 131–138.

Felix, W. (1912). The development of the urogenital organs. *In* "Manual of human embryology" (F. Kiebel and F. P. Mall, Eds.), pp. 752–979. J.B.Lippincott Co., Philadelphia.

Goodrich, E. S. (1930). "Studies on the structure and development of vertebrates." Macmillan and Co., London.

Leuzinger, S., Hirth, F., Gerlich, D., Acampora, D., Simeone, A., Gehring, W. J., Finkelstein, R., Furukubo-Tokunaga, K., and Reichert, H. (1998). Equivalence of the fly orthodenticle gene and the human OTX genes in embryonic brain development of *Drosophila. Development* **125**, 1703–1710.

Marshall, A. M. (1902). "The frog: an introduction to anatomy, histology, and embryology." Macmillan and Co., New York.

Møbjerg, N., Larsen, E. H., and Jespersen, Å. (2000). Morphology of the kidney in larvae of *Bufo viridis* (Amphibia, Anura, Bufonidae). *J. Morph.* **245**, 177–195.

Nagao, T., Leuzinger, S., Acampora, D., Simeone, A., Finkelstein, R., Reichert, H., and Furukubo-Tokunaga, K. (1998). Developmental rescue of *Drosophila cephalic* defects by the human Otx genes. *Proc. Natl. Acad. Sci. USA* **95**, 3737–3742.

Nieuwkoop, P. D., and Faber, J. (1994). "Normal table of *Xenopus laevis* (Daudin)." Garland, New York.

Richards, A. N. (1929). "Methods and results of direct investigations of the function of the kidney." Williams and Wilkins Company, Baltimore.

Saxén, L. (1987). "Organogenesis of the kidney." Cambridge University Press, Cambridge.

Vize, P. D., Seufert, D. W., Carroll, T. J., and Wallingford, J. B. (1997). Model systems for the study of kidney development: use of the pronephros in the analysis of organ induction and patterning. *Dev. Biol.* **188**, 189–204.

Witschi, E. (1956). "Development of Vertebrates." W.B. Saunders Company, Philadelphia.

2

Development of Malpighian Tubules in *Drosophila Melanogaster*

Helen Skaer

I. Introduction

A. Insect Renal Tubules

In the first part of the last century, Wigglesworth and Ramsay established the Malpighian tubules, the primary excretory epithelium of insects, as a classical physiological model for the study of transporting epithelia (Wigglesworth, 1939; Ramsay, 1954, 1955, 1958). Subsequent studies have shown that despite their apparently simple structure, these renal tubules perform many of the functions carried out by their more complex vertebrate counterparts. In addition, the activity of Malpighian tubules is highly regulated; e.g., the secretion of primary urine is under the control of a network of regulators, including stimulation by the cyclic AMP pathway, a leukokinin-mediated PKC/calcium-regulated pathway, and the nitric oxide/cGMP signaling pathway (Fig. 2.1C; reviewed in Dow *et al.*, 1998).

Insect Malpighian tubules are made up of a single layer of epithelial cells, which show a pronounced apico-basal polarity. Proximo-distal polarity is also apparent both structurally (Fig. 2.1A) and functionally (Maddrell, 1981; Wessing and Eichelberg, 1978); primary urine is secreted into the lumen of the distal region of the tubule through the activity of proton-linked ion transporters and is modified by selective reabsorption and regulated water movement as it flows through the proximal region of the tubules, the ureters, and into the hindgut (Figs. 2.1A and 2.1B). As the distal tubule epithelium is relatively leaky, both essential and potentially toxic compounds pass into the primary secreted fluid. Essential compounds are reabsorbed by specific transporters, whereas toxins are excreted by default. Metabolic waste products are transported actively into the urine: nitrogenous products as urate (which precipitates in the acid conditions of the lumen as uric acid) and other metabolic breakdown products, such as *p*-aminohippuric acid and ethereal sulfates. The tubule epithelium also has transporters for a range of complex organic toxins such as alkaloids and cardiac glycosides that may be encountered during feeding (Fig. 2.1B). Malpighian tubules therefore represent the major site of ionic and osmotic regulation, as well as the organ critical for clearing the insect blood, or hemolymph, of toxins.

B. *Drosophila* Malpighian Tubules

There are four Malpighian tubules in *Drosophila;* two pairs, each joined by a common ureter, which empties into the hindgut (Fig. 2.1A). Each tubule of the longer, anterior pair is made up of 144 (+10) cells and follows a regular

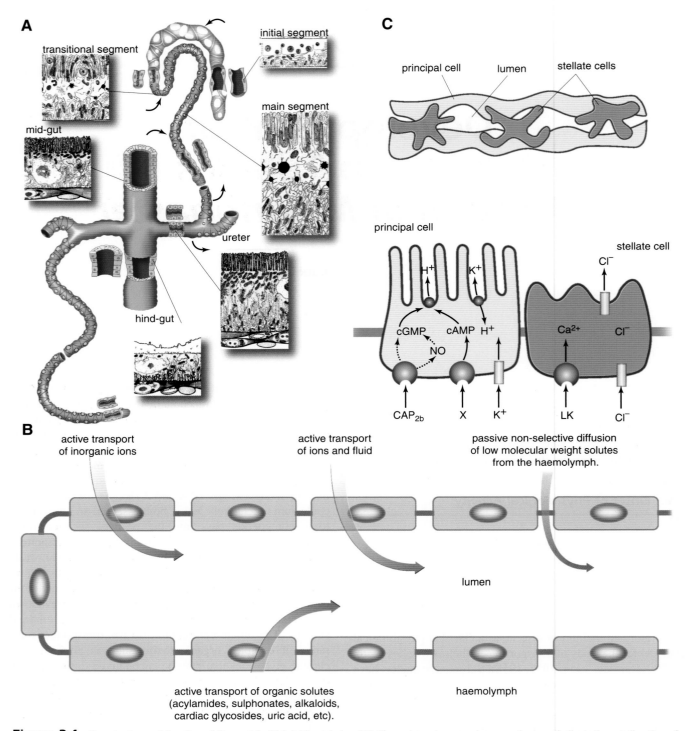

Figure 2.1 Organization and function of *Drosophila* Malpighian tubules. (A) The main regions are shown, and arrows indicate the net direction of transport; from hemolymph to lumen in the distal regions and from lumen to hemolymph in the proximal region, the adjacent ureter and hindgut. Modified from Wessing and Eichelberg (1978). (B) Diagram showing the various transport processes involved in tubule function. Modified from Maddrell and O'Donnell (1992). (C) Summary of the regulation and activity of transporters in the stellate and principal cells that underlie secretion. CAP2b; cardioacceleratory peptide 2b, LK; leukokinin, NO; nitric oxide. X indicates an as yet unknown agonist that stimulates the elevation of intracellular cyclic AMP levels. Drawings modified from O'Donnell *et al.* (1996).

course through the hemolymph, one on either side of the midgut, whereas each of the shorter posterior tubules, of 103 (+8) cells, runs back along the hindgut (Janning *et al.,* 1986; Skaer and Martinez Arias, 1992). The tubules can be divided into four domains—initial, transitional, main (secretory), and proximal (reapsorbtive) regions—both on structural grounds and through their patterns of gene expression (Fig. 2.1A) (Sozen *et al.,* 1997; Wessing and Eichelberg, 1978). Within these domains there are rather few different cell types: type I (principal) cells are found along the whole tubule length, type II (stellate) cells in all but the proximal region, and "tiny" (possibly neurosecretory) cells in just this proximal region. Principal and stellate cells of the main segment have been demonstrated to have specific physiological activities (Fig. 2.1C; O'Donnell *et al.,* 1996).

The substantial and rapid growth that characterizes insect development must be matched by the excretory capacity of the Malpighian tubules. Different insect groups achieve this by increasing either the number or the size of their tubules (Wigglesworth, 1939). In *Drosophila,* the four tubules laid down during embryogenesis increase in size, not through continuous cell division but by cell growth, involving endoreplicative amplification of their DNA. As the tubules persist through metamorphosis into the adult, the number of cells established during embryogenesis represents the mature number, whose growth underpins adult excretion. Both the allocation of cells instructed to adopt tubule fate and the proliferation of these cells during early development must be tightly regulated processes.

During embryonic development, a single cell, the tip cell, is specified and subsequently differentiates at the distal end of each tubule (Fig. 2.2). These cells are distinctive in their pattern of gene expression, their morphology, and their position; because each locates in a precise position in the mature embryo, they may well influence the three-dimensional organization of the tubules. Thus the tubules are structurally simple but display a number of interesting features: they contain a very reproducible number of cells, a regular pattern of different cell types, and take a stereotyped course through the hemolymph.

II. Tubule Development and the Genes That Regulate It

The tubules are ectodermal in origin, arising from the hindgut primordium close to its junction with the endodermal, posterior midgut. The tubules first appear as four buds that push out from the hindgut (Fig. 2.3A) and grow during the next few hours by cell proliferation (Figs. 2.3B and 2.3C). During this phase, a unique cell lineage is specified in each tubule, resulting in the production of the tip cell and its sibling (Figs. 2.2 and 2.5), cells that play an important role in regulating tubule development. Half-way through embryogenesis, the mature number of tubule cells has been established, and the short cylindrical structures (Fig. 2.3C) then undergo a dramatic convergent–extension movement that transforms them into elongated tubules, whose three-dimensional arrangement in the body cavity is regular and highly reproducible (Figs. 2.3D and 2.3E). Shortly after the completion of these morphogenetic movements, the physiological activity of the tubules becomes apparent by the appearance of crystals of uric acid in the tubule lumen, indicating that the differentiation of particular transport functions is complete (Fig. 2.3F). By the end of embryogenesis, the tubule cells have differentiated and mature excretory function is evident as urine secretion starts after hatching (Skaer *et al.,* 1990).

The apico-basal polarity of cells is established early in embryogenesis and is maintained throughout tubule development. Tubule growth thus contrasts with the condensation of nephric mesenchyme to form nephrons in the vertebrate metanephric kidney and is more akin to the extension of the ureteric bud in which cells maintain their epithelial character. However, the establishment of proximo-distal polarity and the patterning of cell differentiation, to produce the different tubule regions, unfold during embryogenesis.

A. Allocation of Malpighian Tubule Cells and Formation of Primordia

1. Signaling across Posterior Gut Boundaries Specifies Tubule Cells and Leads to Eversion of Tubule Primordia from the Hindgut

The tubules arise from the posterior gut, which is specified maternally as part of the embryonic terminalia. Activation of the receptor tyrosine kinase Torso leads through MAP kinase to the expression of two zygotic transcription factors, encoded by *tailless* and *huckebein* (reviewed in Duffy and Perrimon, 1994). The activities of these two genes and their targets define the posterior gut and subdivide it into two domains, the ectodermal hindgut and the endodermal posterior midgut, which later grows out and fuses with its anterior counterpart to form the complete alimentary canal (reviewed in Skaer, 1993). The Malpighian tubules

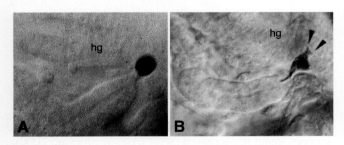

Figure 2.2 Malpighian tubule tip cells, A. stained with an antibody to Krüppel during the mitogenic phase and B. with 22C10, an antibody to Futsch, after cell division is completed. The tip cell differentiates an apical extension into the tubule lumen and forms processes (arrowheads) that will make contact with a nerve running up the hindgut (hg).

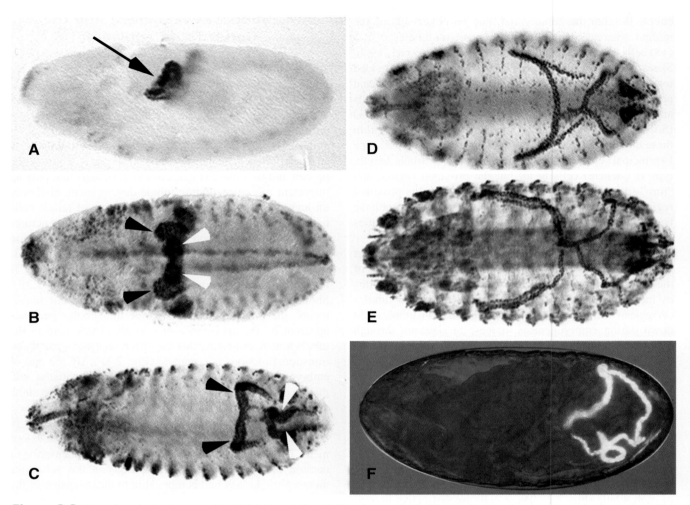

Figure 2.3 The embryonic development of the Malpighian tubules. (A–E) embryos stained with an antibody against the transcription factor, Cut, which labels the tubules and elements of the nervous system. The tubule primordia evert from the hindgut at the extended germ band stage, arrow in (A), and grow principally by cell division during germ band retraction (B, C) and thereafter by elongation, through stereotypic convergent-extension movements (D), which result in a reproducible arrangement of the mature tubules (E). Black arrowheads; anterior tubules. White arrowheads; posterior tubules. F. Urates, secreted into the tubule lumen of the posterior tubules crystallise by the end of embryogenesis, revealed as a bright deposit under polarised light. A, lateral view; B–E, dorsal views; anterior to the left.

segregate from the hindgut but paradoxically their specification also depends on the presence of the posterior midgut. Posterior midgut identity is conferred by the combined action of *huckebein* and a downstream target, the GATA factor encoded by *serpent*. In embryos mutant for these two genes, the posterior midgut does not form and although he hindgut develops, tubule cells fail to be specified (Ainsworth *et al.,* 2000).

This finding suggests that signaling between the two posterior gut domains patterns the hindgut primordium, allocating a subset of cells to the tubule fate. Signaling of this kind may act via the Wnt pathway, as attenuation of the pathway activator *armadillo* (β-catenin) leads to a reduction in the number of tubule cells specified, whereas loss of the Wnt pathway repressor *zeste-white3* (glycogen synthase kinase3) results in overrepresentation of tubule cells (Ainsworth *et al.,* 2000).

Tubule cells first become apparent in the hindgut as the expression of the zinc finger transcription factor, encoded by *Krüppel,* refines in the posterior gut to mark out a band of cells—those that will push out to form the tubule primordia. A precondition of *Krüppel* expression is the activity of a *Drosophila* HNF protein Forkhead (Gaul and Weigel, 1990), which is expressed throughout the gut. However, it is signaling from the posterior midgut that is critical in subdividing the shared hindgut/tubule primordium by maintaining localized *Krüppel* expression in the tubule cells, which leads to the expression of an immediate target, *cut,* a homeodomain transcription factor (Ainsworth *et al.,* 2000).

Analysis of mutants indicates that the coexpression of *Krüppel* and *cut* in the tubule cells initiates tubule eversion from the hindgut. In embryos mutant for *Krüppel,* the tubules fail to form (Gloor, 1950), and in those lacking the function of *cut,* the tubule cells form a multilayered blister on the surface of the hindgut and tubules never develop (Liu *et al.,* 1991). Strikingly, if the overexpression of *Krüppel* (and therefore also of *cut*) is engineered in the hindgut, supernumerary cells evert from the hindgut, enlarging the tubule primordia (Ainsworth *et al.,* 2000). Thus it appears that some of the targets of *Krüppel* and *cut* in the tubule cells are likely to be genes that modulate the cytoskeleton and/or modify cell adhesion and so bring about the first morphogenetic movement that characterizes this tissue. One candidate for this function might be *walrus.* The *walrus* mutant phenotype includes abnormalities in tubule eversion so that *cut*-expressing cells remain in the hindgut (Liu *et al.,* 1999). Although the nature of the *walrus*-predicted product, an electron transfer flavoprotein, has been established, its mode of action remains obscure (Flybase, http://fly.ebi.ac.uk:7081/).

2. A Switch to Malpighian Tubule Fate or Gradual Evolution of Tubule Differentiation?

The absence of tubules in *Krüppel* mutant embryos raises the possibility that *Krüppel* acts as a master switch, propelling cells into a tubule vs hindgut fate (Harbecke and Janning, 1989). However, cells in the enlarged mutant hindgut still express a tubule cell marker [an enhancer trap in Fasciclin 2 (Ghysen and O'Kane, 1989)] and later uric acid is deposited in the hindgut lumen (Skaer, 1993). Thus, in *Krüppel* mutant

embryos, cells remaining in the hindgut clearly differentiate some tubule characteristics, such as the ability to transport urates. Intriguingly, they also express hindgut markers (Liu and Jack, 1992), suggesting that they are partially transformed toward a hindgut fate. These findings suggest (1) that the principal activity of *Krüppel* is to promote the movement of tubule cells out of the hindgut and (2) that subsequent tubule cell differentiation depends on other factors, one of which might be that tubule cells escape the influence of signals in the hindgut, either from neighboring cells in the epithelium or from the visceral mesoderm that ensheathes the gut but never invests the tubules.

III. Generating Cells: Regulation of Cell Proliferation in the Tubule Primordia

A. Intercellular Signaling Dictates the Pattern of Cell Division

Early divisions of the zygotic nuclei in *Drosophila* are synchronous and are not accompanied by cytokinesis until the 14th mitotic cycle. At this stage, cellularization of the embryonic nuclei occurs and cell division becomes regulated spatially (Foe, 1989). The first division of this type (cycle 14) in the posterior gut encompasses the hindgut, before the tubule cells are specified (Foe, 1989). Once the tubule primordia are formed, the pattern of cell division becomes

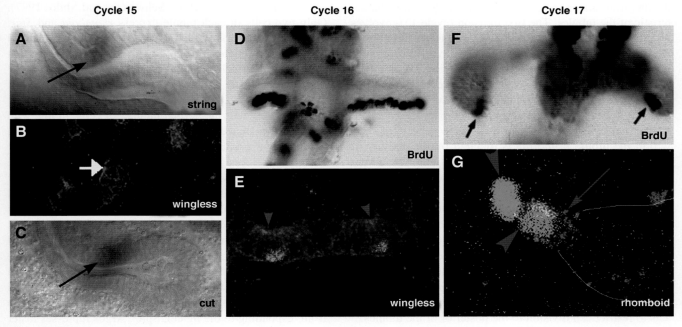

Figure 2.4 Patterning of cell cycles in the tubules and of genes that regulate them. (A–C) Cycle 15. G2 is marked by the expression of *string,* the homologue of cdc25 (A), in the tubule primordium (shown by the expression of *cut,* C). wingless is expressed in the same domain (B). (D and E) Cycle 16. The incorporation of bromodeoxyuridine (BrdU) reveals that cells only on the posterior side of the tubules cycle (D). At this stage, *wingless* (red) is expressed only in the posterior domain (E). Tip mother cells are selected in this posterior region (green). (F and G) Cycle 17 onward. BrdU is incorporated (arrows) only by cycling cells in the distal tubules (F), close to the tip cells (out of focus). Both the tip cell and its sibling (green) express *rhomboid* (red) and therefore secrete the EGF homologue, Spitz (G).

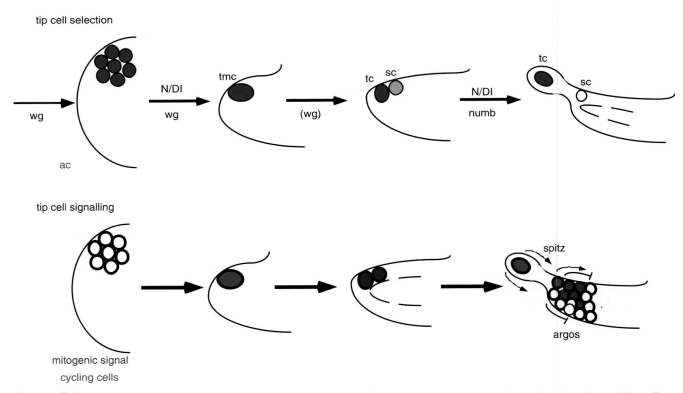

Figure 2.5 Summary of the selection and activity of the tip cell lineage in Malpighian tubules. tmc, tip mother cell; tc, tip cell; sc, sibling cell; ac, achaete; wg, wingless; N, Notch; Dl, Delta.

distinct from that in the gut epithelium and is highly characteristic of the tubules. The first division (cycle 15) involves all the tubule cells (Figs. 2.4A–2.4C) and is synchronous in all four primordia, whereas the second division (cycle 16) involves only a subset of cells in the posterior of each tubule (Figs. 2.4D and 2.4E). Subsequent cycles (cycle 17 onward) are restricted to the distal region of the developing tubules (Figs. 2.4F and 2.4G) and are nonsynchronous.

The patterning of cell proliferation to subsets of cells within the tubule primordia is dictated by the activity of two signaling pathways: (a) signaling by wingless, a Wnt homologue, and (b) through the *Drosophila* epidermal growth factor receptor (EGFr) pathway.

In embryos mutant for *wingless*, the tubule cells arrest in the first tubule-specific cycle (cycle 15), whereas in mutants lacking the activity of the EGFr pathway, they arrest in the first asynchronous cell division cycle (cycle 17). *wingless* is expressed initially throughout the tubule primordia but becomes restricted to the posterior domain after they evert (Figs. 2.4B and 2.4E). It is therefore possible that *wingless* acts to restrict cell division to the posterior of the tubules.

The ligand of the EGFr pathway, Spitz, an EGF-like molecule, is expressed ubiquitously as an inactive precursor, but is processed and secreted only from those cells that express *rhomboid* and *Star*, which encode proteins thought to be involved in the generation of the active Spitz ligand

(reviewed in Freeman, 1997; Schweitzer and Shilo, 1997). Only two cells in each tubule express *rhomboid* and *Star*, the tip cell and its sibling, located at the extreme distal end of each tubule (Fig. 2.4G). These cells are born at M16 and direct cell division in their neighbors from M17 onward. If they are ablated, cell proliferation ceases (Skaer, 1989; Hoch *et al.*, 1994), resulting in tubules containing approximately half the normal number of cells. In line with this, Spitz-mediated activation of the EGF receptor leads to upregulation of a target gene, the COUP transcription factor encoded by *seven up*, and to the activation of cell cycle regulators (Kerber *et al.*, 1998).

Thus signaling from the tip cell lineage at the distal end of the tubules establishes a late domain of proliferating cells (Fig. 2.4F). However, this domain is small and additional evidence suggests that attenuation of the mitogenic signal from the tip cell is not the only factor that delimits the zone of dividing cells. In other tissues, one of the targets of the EGF receptor is an inhibitor of the pathway, *argos*. Argos is secreted from activated cells and inhibits neighboring cells from responding to the ligand, Spitz. Argos represses faster than Spitz activates, possibly because it diffuses more rapidly so that the domain of EGF receptor activation is restricted by an inhibitory feedback loop. As removal of *argos* results in a slight increase in the number of tubule cells (Kerber *et al.*, 1998) and overexpressing *argos* in the tubules results in a significant reduction in number (Eckhardt and Skaer, unpublished results), this feeback loop also appears to act in the developing tubules, to limit the distal domain of dividing cells (see Fig. 2.5).

B. Roles of Tip Cell Lineage: Cell Specification, Lateral Inhibition, Asymmetric Cell Division, and Their Relationship to Mitogenic Signaling

Clearly establishing the tip cell lineage is critical in generating the normal complement of cells in the tubules (summarized in Fig. 2.5). This lineage consists of a precursor cell, which divides once to generate two daughters, one of which, the tip cell, differentiates a striking morphology, protruding from the distal end of the tubule (Fig. 2.2). The other daughter, the sibling cell, remains fully incorporated in the tubule epithelium (Figs. 2.4G and 2.5). The precursor cell is established in each tubule primordium from an equivalence group, a cluster of 8–12 cells expressing the basic helix-loop-helix family of proneural genes (such as *achaete*), which confers on them the potential to adopt the tip cell fate. Lateral inhibition, mediated by the neurogenic genes (including the ligand *Delta* and the receptor *Notch*), ensures that only one cell in each cluster continues to express the proneural genes and thus retains tip cell fate. This cell, the tip mother cell (tmc), is the progenitor, dividing once to generate daughters, both of which express the proneural genes, as well as *Notch* and *Delta*. Once again these cells signal through the neurogenic pathway to establish their fates. The tip cell continues to express *achaete* and its target *Krüppel*, whereas the other, repressed by Notch activation, loses *achaete* and *Krüppel* expression and becomes the sibling cell. Thus the tip cell is established in two stages: by the selection of a progenitor, the tip mother cell, from a group of equivalent cells and then by a competitive interaction between the sibling progeny of this precursor cell (Wan *et al.*, 2000).

Even more players are involved in establishing this lineage. In embryos mutant for *wingless,* tip cells fail to appear, resulting from defects during the first stage. The equivalence clusters are not established normally and *achaete* expression is lost from the tmc so that neither the tip cell nor its sibling is specified (Wan *et al.*, 2000). Two genes, *numb* and *osa* (also known as *eyelid*), are required at the second stage to establish the normal allocation of tip cell and sibling cell fates. *Numb* inhibits signaling through *Notch* (Guo *et al.*, 1996; Zhong *et al.*, 1996) and is asymmetrically segregated to the tip cell when the tip mother cell divides. This ensures deafness to *Notch* activation in the tip cell. In contrast, the absence of *Numb* in the sibling cell allows Delta-mediated activation of Notch leading to the repression of proneural gene expression and, as a result, of

their targets. In embryos mutant for *osa,* two tip cells appear and the expression of *Krüppel* persists in both daughters of the tip mother cell (Carrera *et al.*, 1998). Osa is a DNA-binding protein that forms part of the Brahma complex, whose activity remodels chromatin strucure, thereby regulating the accessibility of specific promotors to transcriptional regulators (Collins *et al.*, 1999). Thus the activity of Osa appears to be a prerequisite for the repression of tip cell genes in the tip cell lineage.

When tubules develop without the segregation of the tip cell lineage, e.g., in embryos carrying a deficiency that lacks proneural genes, cell division arrests in cycle 17 and the tubules contain approximately half the normal number of cells. In *wingless* mutants, there are no tip cells but the tubules contain even fewer cells, as cell proliferation is arrested earlier, during cycle 15. Somewhat surprisingly, in embryos mutant for *numb,* in which there are no tip cells but two sibling cells, cell division in the tubules is normal and the full complement of cells is produced. The same is true of tubules in which *numb* is overexpressed so that there are two tip cells but no sibling cell. Thus the tip cell and its sibling are equivalent in their ability to promote cell division in their neighbors. In line with this finding, both the tip cell and the sibling express *rhomboid* and therefore process Spitz (Sudarsan *et al.*, 2002; Fig. 2.4G). Thus the crucial event for the maintenance of cell division in the tubules is specification of the tip mother cell. Establishing distinct tip and sibling cell identity must therefore be significant for some other aspect of tubule development or function (see later).

C. Endoreplication and Cell Growth

Once cell division is complete in the tubules, all the cells enter S phase in the first round of endoreplication (Orr-Weaver, 1994; Smith and Orr-Weaver, 1991). The Malpighian tubule cells go through repeated rounds of endoreplication so that the DNA content of the adult tubule cells is 256C. Little is known about the timing of these cycles after embryogenesis in *Drosophila,* but a study in the blood-sucking insect *Rhodnius prolixus* has shown that the DNA content of the tubule cells doubles at every molt, that the increase in cell size follows the increase in DNA content, and that endoreplication appears to be regulated hormonally (Maddrell *et al.*, 1985).

IV. Morphogenetic Movements

A. Convergent–Extension Movements Result in Tubule Elongation

Soon after cell division in the tubules is complete, they start to elongate. This results from the interdigitation of neighboring cells, producing a classical convergent–extension movement (Figs. 2.3C–2.3E). The tubule diameter narrows as they elongate, and the number of cells around the circumference reduces from 6–10 immediately before elongation to just 2 cells afterward (Skaer, 1992, 1993). Screens for mutants that affect tubule morphogenesis have identified genes that are required for this process (Harbecke and Lengyel, 1995; Jack and Myette, 1999; Liu et al., 1999). Those that appear to be directly involved include genes such as *ribbon*, *raw*, and *zipper* (which encodes a non-muscle myosin heavy chain) (Young et al., 1993; Blake et al., 1999), whose activity is thought to be in regulating the actin cytoskeleton.

B. Organizing the Stereotypical Arrangement of Tubules in the Body Cavity

It is striking that by the time elongation is complete, each tubule has taken up a precise three-dimensional configuration in the body cavity and each tip cell is always located in exactly the same place; e.g., the two posterior tip cells contact nerve branches that run up each side of the hindgut visceral mesoderm. It appears that both the tip cell and its sibling are necessary for this stereotyped arrangement; in *numb* mutants (two sibling cells, no tip cell) or when *numb* is overexpressed (two tip cells, no sibling cell) the precise configuration is lost and the tubules lose their way so that the distal tips are positioned randomly (Wan et al., 2000). Mutants in *thick veins* (a receptor for the *Drosophila* BMP, *decapentaplegic*) and *schnurri* (a transcription factor activated by the BMP pathway) show a similar phenotype (Jack and Myette, 1999), indicating that intercellular signaling must play a role in guiding tubules to their correct locations.

The tubules also lose their way in embryos mutant for *myoblast city* (Ainsworth et al., 2000), a member of the gene family that includes human *DOCK-18* and *Caenorhabditis elegans ced-5* (Wu and Horwitz, 1998). These genes encode regulators of Rac (Klyokawa et al., 1998; Nolan et al., 1998), a Rho-GTPase that is active in reorganizing the cytoskeleton (Hall, 1998; Ridley et al., 1992), and are expressed in cells such as migratory cells and phagocytes that throw out membrane extensions.

This finding highlights a very intriguing parallel with the behavior of the distal tip cell in the gonad of the nematode *C. elegans*. This cell initially signals to its neighbors to maintain them in mitotic division (Kimble and White, 1981; Austin et al., 1989; Henderson et al., 1994), but later plays a critical role in guiding morphogenesis of the enlarging gonad arms (Blelloch and Kimble, 1999). In animals mutant for *ced-5*, distal tip cell migration and gonad morphogenesis are defective (Wu and Horwitz, 1998). Another gene essential for this distal tip cell-dependent morphogenesis has been identified in the nematode on the basis of the gonad

phenotype. It encodes a matrix metalloproteinase capable of modifying the extracellular matrix (Blelloch and Kimble, 1999). Members of this family of proteins are expressed in kidney interstitial mesenchyme and play an important role in kidney tubulogenesis (Lelongt et al., 1997) but although metalloproteinases have been identified in *Drosophila* (Flybase, http://fly.ebi.ac.uk:7081/), it is not yet known whether their activity is required for fly tubule morphogenesis.

The challenge now is to correlate the requirement for specialized cells at the distal tip of elongating tubules with the patterns of gene expression that allow them to guide and regulate tubule morphogenesis.

V. Onset of Physiological Activity

Malpighian tubules must be physiologically competent and responsive by the end of embryogenesis, as the hatchling larva starts feeding almost immediately. The transport of solutes, and urates in particular, is established before the end of embryogenesis, presumably as the digestion and metabolism of yolk result in the buildup of nitrogenous waste. However, little is known about the differentiation of specific cellular features that underlie these developments in *Drosophila*. In the hemipteran *Rhodnius prolixus*, the early onset of solute transport is established as soon as the formation of mature intercellular junctions between tubue cells allows the enlargement of the lumen, whereas fluid transport starts only with the full elaboration of apico-basal membrane architecture (Skaer et al., 1990).

The tip cell persists through all the larval instars and expresses many genes that are characteristic of neuronal cells, including cell adhesion molecules [such as Neuromusculin (Kania et al., 1993), microtubule-associated proteins (such as Futsch, (Hummel et al., 2000)] and synaptic proteins [such as Synaptotagmin, Synaptobrevin, Cysteine-string protein (Hoch et al., 1994)] and a novel neurotransmitter transporter (Johnson et al., 1999). These observations, as well as the association of tip cells with nerve branches on the visceral mesoderm, suggest a physiological role for tip cells in regulating tubule activity, either as neurosecretory cells or as sensors.

Sozen et al. (1997) isolated a panel of markers that allows them to catalogue the differentiated cell types of mature tubules. These include cells of as yet unknown function in the initial and transitional segments, cells that express a vATPase and transport K^+ into the lumen in a proton-dependent fashion and a smaller population of cells in the same segment that transport Cl^- and support the resulting water flux, via *Drip*, a *Drosophila* intergral protein/aquaporin homologue (Dow et al., 1995, 1998) (see figure 2.1C). The reapsorbtive region of the tubules as well as the ureter also contain specialised cells, including a small neurosecretory cell type (Sozen *et al.*, 1997). At present very little is known about the specification and differentiation of specialised tubule cells. However the markers that define these cell populations should themselves enable us to chart the separation of cell types and dissect apart the networks that define and regulate their differentiation.

A. *Drosophila* Malpighian tubules as a model for tubulogenesis?

Clearly there are profound differences between the development of Malpighian tubules and the nephrogenesis of vertebrate kidney, which involves the integration of vascular, neural and nephric tissues. The complex architecture of the glomerulus and use of filtration as the predominant mechanism for the formation of primary urine are absent in lower organisms. Despite this, there are many parallels between renal and Malpighian tubules in terms of their cellular structure and functional activity. Further, to generate either tissue, cells must be recruited, divide in a regulated fashion, arrange themselves to form an elongated, single-cell layered tubule with a specific 3-dimensional architecture and finally undergo patterned differentiation within the epithelium. Inductive interactions underlie cell recruitment in both meso- and metanephric kidney development and further cell interactions regulate tubule extension and branching (reviewed by Bard *et al.*, 1994; Kuure *et al.*, 2000; Lechner and Dressler, 1997) and see (Obara-Ishihara *et al.*, 1999). Similarly in *Drosophila* cells are recruited to form Malpighian tubules by inductive signalling in the posterior gut and cell interactions regulate tubule growth and extension. In both tissues tubulogenesis depends on the influence of specific groups of cells: in nephric tubules, in addition to the reciprocal interactions between the ureteric bud and nephrogenic mesenchyme, a subset of the mesenchyme, the stromal cells, promote branching of the ureteric bud, by regulating the expresssion of the GDNF receptor, c-ret (Mendelsohn *et al.*, 1999). In the Malpighian tubules the tip cell lineage regulates tubule growth by signalling through a different growth factor receptor, the EGFR.

It is tempting to draw molecular parallels between the factors that regulate fly and vertebrate renal tubulogenesis. For example the expression of Zn finger (*Krüppel* and *WT-1*, Pritchard-Jones *et al.*, 1990) and homeodomain (*Cut* and *Cux1*, Vanden Heuvel *et al.*, 1996) transcription factors play important roles early in nephrogenesis. Signalling through the Wnt pathway plays important roles in the development of both *Drosophila* and vertebrate renal tubules (Herzlinger *et al.*, 1994; Kispert *et al.*, 1998; Stark *et al.*, 1994). Furthermore, evidence indicates that a vertebrate EGF, *amphiregulin*, is expressed in the developing kidney and is able to stimulate branching of the ureteric bud in organ culture (Lee *et al.*, 1999). However, our state of knowledge is still too

sketchy to make correlations that are informative rather than simply intriguing. The study of tubulogenesis in an organism as genetically manipulable as *Drosophila* undoubtedly has an important role to play in uncovering the networks that regulate cellular processes common to the generation of both simple or more complex renal systems. As these networks are filled in, and with the information available from genomic analysis, it will become easier to identify common strategies and chart parallels in the deployment of conserved molecular pathways and to test their significance.

Acknowledgments

I am grateful to Vikram Sudarsan for help in preparing the figures and to the members of my laboratory for stimulating discussions and perceptive comments on the manuscript. This work is supported by the Welllcome Trust.

References

Ainsworth, C., Wan, S., and Skaer, H. (2000). Coordinating cell fate and morphogenesis in *Drosophila* renal tubules. *Phil. Trans. Roy. Soc. B* **355**, 931–937.

Austin, J., Maine, E., and Kimble, J. (1989). Genetics of intercellular signalling in *C. elegans*. *Development (Suppl.)*, 53–57.

Bard, J. B. L., McConnell, J. E., and Davies, J. A. (1994). Towards a genetic basis for kidney development. *Mech. Dev.* **48**, 3–11.

Blake, K. J., Myette, G., and Jack, J. (1999). The products of *ribbon* and *raw* are necessary for proper cell shape and cellular localization of nonmuscle myosin in *Drosophila*. *Dev. Biol.* **203**, 177–188.

Blelloch, R., and Kimble, J. (1999). Control of organ shape by a secreted metalloprotease in the nematode *Caenorhabditis elegans*. *Nature* **399**, 586–590.

Carrera, P., Abrell, S., Kerber, B., Walldorf, U., Preiss, A., Hoch, M., and Jäckle, H. (1998). A modifier screen in the eye reveals control genes for *Krüppel* activity in the *Drosophila* embryo. *Proc. Natl. Acad. Sci. USA* **95**, 10779–10784.

Collins, R. T., Furukawa, T., Tanese, N., and Treisman, J. E. (1999). Osa associates with the Brahma chromatin remodelling complex and promotes the activation of some target genes. *EMBO J.* **18**, 7029–7040.

Dow, J. A. T., Davies, S. A., and Sozen, M. A. (1998). Fluid secretion by the *Drosophila* Malpighian tubule. *Am. Zool.* **38**, 450–460.

Dow, J. A. T., Kelly, D. C., Davies, S. A., Maddrell, S. H. P., and Brown, D. (1995). A novel member of the major intrinsic protein family in *Drosophila*: Are aquaporins involved in insect Malpighian (renal) tubule fluid secretion? *J. Physiol.* **489**, 110P–111P.

Duffy, J., and Perrimon, N. (1994). The Torso pathway in *Drosophila*: Lessons on receptor tyrosine kinase signaling and pattern formation. *Dev. Biol.* **166**, 380–395.

Foe, V. E. (1989). Mitotic domains reveal early commitment of cells in *Drosophila* embryos. *Development* **107**, 1–22.

Freeman, M. (1997). Cell determination strategies in the *Drosophila* eye. *Development* **124**, 261–270.

Gaul, U., and Weigel, D. (1990). Regulation of Krüppel expression in the anlage of the Malpighian tubules in the *Drosophila* embryo. *Mech. Dev.* **33**, 57–67.

Ghysen, A., and O'Kane, C. (1989). Neural enhancer-like elements as specific cell markers in *Drosophila*. *Development* **105**, 35–52.

Gloor, H. (1950). Schädigungsmuster eines letalfaktors (Kr) von Drosphila melanogaster. *Arch. Julius Klaus-Stift. Vererbungsforsch. Sozialanthropol. Rassenhyg.* **25**, 38–44.

Guo, M., Jan, L. Y., and Jan, Y. N. (1996). Control of daughter cell fates during asymmetric division: Interaction of Numb and Notch. *Neuron* **17**, 27–41.

Hall, A. (1998). Rho GTPases and the actin cytoskeleton. *Science* **279**, 509–514.

Harbecke, R., and Janning, W. (1989). The segmentation gene *Krüppel* of *Drosophila* melanogaster has homeotic properties. *Genes Dev.* **3**, 114–122.

Harbecke, R., and Lengyel, J. (1995). Genes controlling posterior gut development in the *Drosophila* embryo. *Roux's Arch. Dev. Biol.* **204**, 308–329.

Henderson, S., Gao, S., Lambie, E., and Kimble, J. (1994). *lag-2* may encode a signaling ligand for the GLP-1 and LIN-12 receptors of *C. elegans*. *Development* **120**, 2913–2924.

Herzlinger, D., Qiao, J., Cohen, D., Ramakrishna, N., and Brown, A. M. C. (1994). Induction of kidney epithelial morphogenesis by cells expressing *Wnt-1*. *Dev. Biol.* **166**, 815–818.

Hoch, M., Broadie, K., Jäckle, H., and Skaer, H. (1994). Sequential fates in a single cell are established by the neurogenic cascade in the Malpighian tubules of *Drosophila*. *Development* **120**, 3439–3450.

Hummel, T., Krukkert, K., Roos, J., Davis, G., and Klambt, C. (2000). *Drosophila* Futsch/22C10 is a MAP1B-like protein required for dendritic and axonal development. *Neuron* **26**, 357–370.

Jack, J., and Myette, G. (1999). Mutations that alter the morphology of the Malpighian tubules in *Drosophila*. *Dev. Genes Evol.* **209**, 546–554.

Janning, W., Lutz, A., and Wissen, D. (1986). Clonal analysis of the blastodertm anlage of the Malpighian tubules in *Drosophila melanogaster*. *Roux's Arch. Dev. Biol.* **195**, 22–32.

Kania, A., Han, P. L., Kim, Y. T., and Bellen, H. (1993). Neuromusculin, a *Drosophila* gene expressed in peripheral neuronal precursors and muscles, encodes a cell adhesion molecule. *Neuron* **11**, 673–687.

Kerber, B., Fellert, S., and Hoch, M. (1998). Seven-up, the *Drosophila* homolog of the COUP-TF orphan receptors, controls cell proliferation in the insect kidney. *Genes Dev.* **12**, 1781–1786.

Kimble, J., and White, J. (1981). On the control of germ cell development in *Caenorhabditis elegans*. *Dev. Biol.* **81**, 208–219.

Kispert, A., Vainio, S., and McMahon, A. P. (1998). Wnt-4 is a mesenchymal signal for epithelial transformation of metanephric mesenchyme in the developing kidney. *Development* **125**, 4225–4234.

Klyokawa, E., Hashimoto, Y., Kobayashi, S., Sugimura, H., Kutata, T., and Matsuda, M. (1998). Activation of Rac1 by a Crk SH3-binding protein, DOCK180. *Genes Dev.* **12**, 3331–3336.

Kuure, S., Vuolteenaho, R., and Vanio, S. (2000). Kidney morphogenesis: Cellular and molecular regulation. *Mech. Dev.* **92**, 31–46.

Lechner, M. S., and Dressler, G. R. (1997). The molecular basis of kidney development. *Mech. Dev.* **62**, 105–120.

Lee, S, Huang, K., Palmer, R., Truong, V., Herzlinger, D., Kolquist, K., Wong, J., Paulding, C., Yoon, S., Gerald, W., Oliner, J., and Haber, D. (1999). The Wilms tumor suppressor *WT1* encodes a transcriptional activator of amphiregulin. *Cell* **98**, 663–673.

Lelongt, B., Trugnan, G., Murphy, G., and Ronco, P. (1997). Matrix metalloproteinases MMP2 and MMP9 are produced in early stages of kidney morphogenesis but only MMP9 is required for renal organogenesis in vitro. *J. Cell Biol.* **136**, 1363–1373.

Liu, S., and Jack, J. (1992). Regulatory interactions and role in cell type specification of the Malpighian tubules by *cut*, *Krüppel* and *caudal* genes of *Drosophila*. *Dev. Biol.* **150**, 133–143.

Liu, S., McLoed, E., and Jack, J. (1991). Four distinct regulatory regions of the *cut* locus and their effect on cell type specification in *Drosophila*. *Genetics* **127**, 151–159.

Liu, X., Kiss, K., and Lengyel, J. A. (1999). Identification of genes controlling Malpighian tubule and other epithelial morphogenesis in *Drosophila melanogaster*. *Genetics* **151**, 685–695.

Maddrell, S. H. P. (1981). The functional design of the insect excretory system. *J. Exp. Biol.* **90**, 1–15.

Maddrell, S. H. P., Lane, N. J., Harrison, J. B., and Gardiner, B. O. C. (1985). DNA replication in binucleate cells of the Malpighian tubules of Hemipteran insects. *Chromosoma* **91**, 201–209.

Maddrell, S. H. P., and O'Donnell, M. (1992). Insect Malpighian tubules V-ATPase action in ion and fluid transport. *J. Exp. Biol.* **172**, 417–429.

Mendelsohn, C., Batourina, E., Fung, S., Gilbert, T., and Dodd, J. (1999). Stromal cells mediate retinoid-dependent functions essential for renal development. *Development* **126**, 1139–1148.

Nolan, K., Barrett, K., Lu, Y., Hu, K., Vincent, S., and Settleman, J. (1998). Myoblast city, a *Drosophila* homolog of DOCK180/CED-5, is required in a rac signaling pathway utilized for multiple developmental processes. *Genes Dev.* **12**, 3337–3342.

O'Donnell, M. J., Dow, J. A. T., Huesmann, G. R., Tublitz, N. J., and Maddrell, S. H. P. (1996). Separate control of anion and cation transport in Malpighian tubules of *Drosophila melanogaster*. *J. Exp. Biol.* **199**, 1163–1175.

Obara-Ishihara, T., Kuhlman, J., Niswander, L., and Hertzlinger, D. (1999). The surface ectoderm is essential for nephric duct formation in intermediate mesoderm. *Development* **126**, 1103–1108.

Orr-Weaver, T. L. (1994). Developmental modification of the *Drosophila* cell cycle. *Trends Genet.* **10**, 321–327.

Pritchard-Jones, K., Fleming, S., Davidson, D., Bickmore, W., Porteous, D., Bard, J., Buckler, A., Pelletier, J., Housman, D., van Heyningen, V., and Hastie, N. (1990). The candidate Wilm's tumour gene is involved in genitourinary development. *Nature* **346**, 194–197.

Ramsay, J. A. (1954). Active transport of water by the Malpighian tubules of the stick insect, *Dixippus morosus* (Orthoptera, Phasmidae). *J. Exp. Biol.* **31**, 104–113.

Ramsay, J. A. (1955). The excretion of sodium, potassium and water by the Malpighain tubules of the stick insect, *Dixippus morosus* (Orthoptera, Phasmidae). *J. Exp. Biol.* **32**, 200–216.

Ramsay, J. A. (1958). Excretion by the Malpighian tubules of the stick insect *Dixippus morosus* (Orthoptera, Phasmidae): Amino acids, sugars and urea. *J. Exp. Biol.* **35**, 871–891.

Ridley, A. J., Paterson, H. F., Johnston, C. L., Diekmann, D., and Hall, A. (1992). The small GTP-binding Rac regulates growth factor-induced membrane ruffling. *Cell* **70**, 389–399.

Schweitzer, R., and Shilo, B. Z. (1997). A thousand and one roles for the *Drosophila* EGF receptor. *Trends Genet.* **13**, 191–196.

Skaer, H. (1989). Cell division in Malpighian tubule development in *Drosophila melanogaster* is regulated by a single tip cell. *Nature* **342**, 566–569.

Skaer, H. (1992). Cell proliferation and rearrangement in the development of the hemipteran, *Rhodnius prolixus*. *Dev. Biol.* **150**, 372–380.

Skaer, H. (1993). The alimentary canal. *In* "The Development of *Drosophila melanogaster*" (M. Bate and A. Martinez-Arias, eds.), pp. 941–1012. Cold Spring Harbor Laboratory Press, Cold Spring Harbor, NY.

Skaer, H., Harrison, J. B., and Maddrell, S. H. P. (1990). Physiological and structural maturation of a polarised epithelium: The Malpighian tubules of a blood-sucking insect, *Rhodnius prolixus*. *J. Cell Sci.* **96**, 537–547.

Skaer, H., and Martinez Arias, A. (1992). The *wingless* product is required for cell proliferation in the Malpighian tubules of *Drosophila melanogaster*. *Development* **116**, 745–754.

Smith, A. V., and Orr-Weaver, T. L. (1991). The regulation of the cell cycle during *Drosophila* embryogenesis: The transition to polyteny. *Development* **112**, 997–1008.

Sozen, M. A., Armstrong, J. D., Yang, M.-Y., Kaiser, K., and Dow, J. A. T. (1997). Functional compartments are specified to single-cell resolution in a *Drosophila* epithelium. *Proc. Natl. Acad Sc. USA* **94**, 5207–5212.

Stark, K., Vainio, S., Vassileva, G., and McMahon, A. P. (1994). Epithelial transformation of metanephric mesenchyme in the developing kidney regulated by *Wnt-4. Nature* **372**, 679–683.

Sudarsan, V., Pasalodos-Sanchez, S., Wan, S., Gampel, A., and Skaer, H. (2002). A genetic hierarcy establishes mitogenci signalling and mitotic competence in the renal tubules of *Drosophila. Development* **129**, 935–944.

Vanden Heuvel, G. B., Bodmer, R., McConnell, K. R., Nagami, G. T., and Igarashi, P. (1996). Expression of a cut-related homeobox gene in developing and polycystic mouse kidney. *Kidney Int.* **50**, 453–461.

Wan, S., Cato, A.-M., and Skaer, H. (2000). Multiple signalling pathways establish cell fate and cell number in *Drosophila* Malpighian tubules. *Dev. Biol.* **217**, 153–165.

Wessing, A., and Eichelberg, D. (1978). Malpighian tubules, rectal papillae and excretion. *In* "The Genetics and Biology of *Drosophila*" (M. Ashburner and T. R. F. Wright, eds.), Vol. 2c, pp. 1–42. Academic Press, London.

Wigglesworth, V. B. (1939). "The Principles of Insect Physiology." Methuen, London.

Wu, Y. C., and Horwitz, H. R. (1998). *C. elegans* phagocytosis and cell-migration protein CED-5 is similar to DOCK180. *Nature* **392**, 501–504.

Young, P. E., Richman, A. M., Ketchum, A. S., and Kiehart, D. P. (1993). Morphogenesis in *Drosophila* requires non-muscle myosin heavy chain function. *Genes Dev.* **7**, 29–41.

Zhong, W., Feder, J., Jiang, M., Jan, L., and Jan, Y. N. (1996). Asymmetric localisation of mammalian Numb homolog during mouse cortical neurogenesis. *Neuron* **17**, 43–53.

3

Induction, Development, and Physiology of the Pronephric Tubules

Peter D. Vize, Thomas J. Carroll, and John B. Wallingford

I. Introduction

The general anatomy and function of the vertebrate embryonic kidney was explored in Chapter 1. The present chapter is the first of three exploring the biology of the vertebrate pronephros in much greater depth. Each section covers one of the functional components: the *tubules*, which perform the resorptive and excretory functions of the pronephros; the *glomus*, its filtration unit; and the *duct*, which disposes of the urine, may participate in the regulation of acid–base balance, and plays a key role in the induction of the adult kidneys. Each of these chapters focuses on the biology and physiology of the various pronephric structures rather than the molecular processes that underpin them. The molecular regulation of their development is covered in depth in Chapter 8, and the experimental techniques used to study them in Chapter 9.

II. Tubule Fate and Origins

Before the processes of pronephric induction, specification, and morphogenesis are described, it is first worth discussing the spatial origins of the pronephros. This consideration plays an important part in understanding how this organ forms, as one must know where the organ anlage is located at the time of its specification before the inductive signals that regulate its formation can be mapped and characterized. On a practical level, one must also know which part of an embryo will later form the organ of interest in order to manipulate the anlage itself or its gene expression profile experimentally.

The past, or history, of a cell tissue, or organ describes its path through embryogenesis. This cannot be determined experimentally and can only be inferred through our knowledge of developmental processes such as gastrulation. The fate of a cell, however, can be determined experimentally.

A. Fate Mapping

Cell fate is different from cell specification. Fate refers to the future pathway of development of a cell should

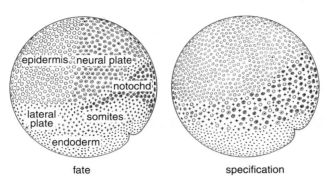

fate   specification

Figure 3.1 Fate versus specification. The embryo on the left illustrates the future fate of various regions of amphibian early gastrulae. If left undisturbed, each of these regions will give rise to the structures noted. The embryo on the right illustrates the state of *specification* of the same regions at this stage of development, assayed by explanting (removing) small pieces of embryonic tissue, allowing it to differentiate in a simple saline solution, and assaying differentiation histologically. Note that the dorsal cells of the animal pole will form the nervous system if left undisturbed (left), but at this point of development they remain specified as epidermis. After Holtfreter (1936) and Holtfreter and Hamburger (1955).

embryogenesis proceed undisturbed. It does not imply that a cell knows anything about its developmental future or that it is capable of adopting the appropriate course of differentiation without additional direction from embryonic patterning systems. Specification refers to a cell state in which enough information has already been received to direct a specific path of differentiation in a neutral environment.

Embryologists have been studying pronephric cell fate since the early 1920s, pioneered by workers such as Vogt (1929), Pasteels (1942), and Keller (1976). The concept of these experiments is very simple: an indelible mark is made on an embryo with a nontoxic vital dye, the exact position of the mark is recorded, and the fate of the dyed region is followed over time. The reality of such experiments is much less simple, as making very small, long-lasting, nontoxic marks in clearly identifiable positions and tracing their progress as they move through the embryo during gastrulation are technically very demanding. More recently, microinjection of fluorescent tracers has been used for fate mapping, and details of this technology can be found in Chapter 9.

The investigation of cell specification (as opposed to cell fate) first became possible in the 1930s when Johannes Holtfreter formulated a synthetic medium in which explanted amphibian embryo fragments could develop and differentiate (Holtfreter, 1931). Because amphibian embryos contain maternal yolk stores in every cell, explants (excised fragments) can survive for long periods of time in this simple medium. This means that one can test the specification status of a group of cells simply by cutting them out of the embryo using microsurgery and culturing them in Holtfreter's saline. If a group of cells has already been specified to form kidney, it will do so under such

neutral conditions. However, if the cells of the future kidney had not yet received their developmental programming and they were explanted and grown in isolation, they would only be able to follow the developmental instructions that they had received prior to removal. This distinction was made beautifully by Holtfreter's comparison (Fig. 3.1) of early gastrula cell fate (what cells become if left undisturbed) and cell specification (what cells know about their future). An excellent example of this distinction in gastrulae is the dorsal animal pole. In pregastrular stages, this region is specified as ectoderm, but these cells are fated to form the nervous system (Holtfreter, 1936; Fig. 3.1).

B. Early Origins

The regions that will later go on to form the pronephric kidneys have been mapped in many amphibians, fish, and birds. Amphibians have been one of the most widely studied classes for historical reasons, experimental simplicity, and ease of analysis. Unlike mammals, birds, and, to a lesser extent, fish, amphibian cells tend to remain associated with their neighboring cells during embryogenesis. This means that if one labels a single cell in a four cell frog embryo with a vital dye and allows the embryo to develop, the labeled cells will go on to form a variety of different tissues but will all remain in the same quadrant of the embryo. Migratory cell populations and cells that enter the circulatory system are obvious exceptions to this generalization. In contrast, if one were to label one cell of a four cell stage mammalian embryo with a vital dye and investigate its later fate, every tissue in the embryo would contain descendants of the labeled cells due to extensive cell mixing.

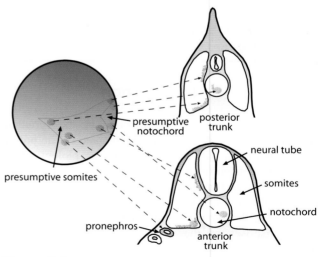

Figure 3.2 Mesodermal fate maps. Multiple colored marks made on the surface of a late blastula stage embryo were followed through morphogenesis to their final positions in a tailbud stage embryo. Note the later separation of marked presumptive notochord cells. After Pasteels (1942).

An example of a fate mapping experiment following presumptive pronephric, somitic, and notochord domains is illustrated in Fig. 3.2. In this experiment, groups of cells were marked with different colored dyes in a late blastula and the embryo was grown to early tailbud stage and then dissected to find the marked cells descendants (Pasteels, 1942). Gastrulation moves the initially superficial (surface) somitic mesoderm internally, resulting in marked cells being found in the medial somites adjacent to the neural tube and notochord. Marks that are initially very close to each other on the surface of the blastula end up far apart in the tadpole due to the lengthening of the embryo driven by convergence and extension movements during gastrulation. Similar experiments have been performed on earlier stage embryos, both by dye marking and by injecting fluorescent dyes.

The 32 cell stage pronephric fate maps of one salamander and two frog species are shown in Fig. 3.3 (Moody, 1987; Saint-Jeanet and Dawid, 1994; Delarue *et al.,* 1997). These maps were generated by injecting fluorescently labeled dyes into individual blastomeres at the 32 cell stage and scoring the frequency with which they contribute to pronephric tubules. The maps are all very similar, and in each case the pronephric tubules are most often derived from blastomere C3, whereas C4 and D3 also sometimes contribute.

In the teleost fish, *Danio rerio* (zebrafish), localization of the region fated to form the pronephros is only possible close to gastrulation; prior to this point, cell marking generates inconsistent results due to the lack of specification and a significant amount of cell mixing. Tracing at early stages must be performed with high molecular weight dyes, as cleavage stage cells are linked by cytoplasmic bridges that allow low molecular weight dyes to move from cell to cell (Kimmel and Law, 1985). Unlike amphibians that cleave in a holoblastic manner where the entire egg is

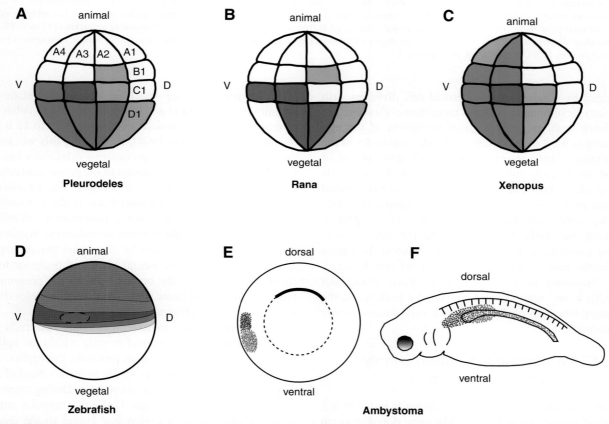

Figure 3.3 Pronephric fate maps. (A–C) Red indicates blastomeres that make a significant contribution to the pronephros, green blastomeres that sometimes contribute, and yellow blastomeres that rarely contribute. (A) A 32 cell pronephric fate map of the salamander, *Pleurodeles waltl* (after Delarue *et al.,* 1997). (B) A 32 cell pronephric fate map of the frog, *Rana pipiens* (after Saint-Jeanet and Dawid, 1994). (C) A 32 cell pronephric fate map of the frog *Xenopus laevis* (after Moody, 1987). (D) Late blastula *Danio rerio* (after Langland and Kimmel, 1997). The location of the pronephric mesoderm within the deep cells is marked in red. Brown indicates presumptive ectoderm and neurectoderm, and blue indicates mesoderm and yellow endoderm. Border regions that contribute to both adjoining germ layers are marked in purple and green. (E) Early gastrula *Ambystoma*, with the left ventrolateral mesoderm marked with vital dyes, a blue more dorsal region, and a red more ventral region. Adapted from Pasteels (1942). (F) The fate of cells marked in the embryo illustrated in E at the tailbud stage. Somites and pronephros are outlined in black. Blue cells contribute to the dorsal region of the third branchial arch, pronephric tubules, and the ventral part of somites 3 through 5, whereas the red region contributes to the ventral parts of somites 5 through 7, some lateral plate, the posterior portion of the pronephric tubules, and the entire pronephric duct (after Pasteels, 1942).

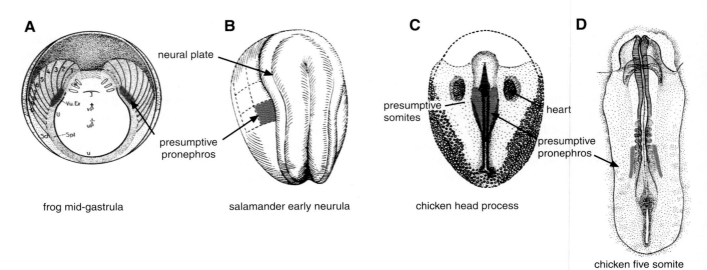

Figure 3.4 Later fate maps. Maps of pronephric regions in later development prior to overt pronephric morphogenesis. (A) Early/midgastrula fate map of the frog pronephric region. After Vogt (1929). Presumptive pronephric regions are marked in red. (B) Fate map of a stage 5 [head process, 19–22 h of development; Hamburger and Hamilton (1951)] chicken embryo (adapted from Rudnick, 1955). The presumptive nephric mesoderm is marked in red. (C) Presumptive pronephric region (red) in *Ambystoma* (axolotl) early neurula [based on the transplantation and extirpation experiments of Fales (1935) and their subsequent adaptation by Burns (1955)]. (D) Early stage 8 (5 somite) chicken embryo. The location of the presumptive nephric mesoderm is marked in red (after Nelson, 1953), beginning lateral to somite 5.

subdivided by a series of symmetrical cell divisions, fish (and bird) development utilizes meroblastic cleavage. In meroblastic embryos, only a small proportion of the egg cytoplasm is used to make cells, with the remainder of the egg forming a yolk reserve used to fuel development of the embryo. In early fish embryos, a cap of dividing cells sits atop a large yolk cell. This cap of cells has a superficial layer, the enveloping layer, and a mass of deep cells. Only the deep cells will contribute to the embryo proper. Following the early period of extensive cell mixing, marking experiments using injected fluorescent dyes have mapped the fate of deep cells (Langeland and Kimmel, 1997), including those of the presumptive pronephros (Fig. 3.3). If one ignores the large vegetal yolk mass, the pronephric fate map of the fish looks very similar to that of amphibians.

C. Pronephric Mesoderm Shortly before Specification

Later in development, when gastrulation is initiated, the pronephric precursors lie in a slightly more dorsal position (Fig. 3.3; Vogt 1929; Pasteels, 1942). At this stage of development a distinction can be made between regions fated to form the anteriodorso pronephric tubules and the more posterioventral distal tubule and pronephric duct, with the presumptive duct lying ventral to the prospective tubules (Pasteels, 1942). In Fig. 3.3, original data for both tubule and duct have been superimposed to correctly illustrate the relative positions of these two regions.

Pronephric cells are specified during gastrulation in amphibians (Brennan *et al.*, 1998), but the first visible sign of pronephric morphogenesis occurs some 10 to 12 h later (Nieuwkoop and Faber, 1994). It is during this window of time that patterning of the pronephros subdivides the pronephric region into the precursors of the pronephric tubules, duct, and glomus (Carroll *et al.*, 1999). The mesodermal derivatives of the C3 blastomeres that contribute extensively to the pronephros involute during gastrulation and migrate extensively. In amphibian early neurulae the somites lie below the neural plate, and by this stage the pronephric anlage lies lateral to the neural plate and posterior to the restriction that marks the border between the presumptive hindbrain and spinal chord (Fig. 3.4) in both urodeles (Fales, 1935) and anurans (Brennan *et al.*, 1998).

In birds the nephric mesoderm cells are, like cells with other mesodermal fates, spread broadly within the epiblast (Witschi, 1956). As gastrulation proceeds, the nephric cells migrate through the primitive streak to form a band of cells ventral to the paraxial somitic mesoderm. During regression of the primitive streak (stage 5) the presumptive nephric cells are localized to a region just lateral to the anterior streak and ventral to the presumptive spinal chord (Rudnick *et al.*, 1955; Fig. 3.4). As the posterior axis extends, this group of cells will become localized to a column of cells, the intermediate mesoderm, that lies between the somitic mesoderm and the lateral plate. The anterior extent of the pronephric mesoderm is somite 5.

Figure 3.4 also compares the position of presumptive pronephric cells in amphibians and birds. This stage of development immediately precedes the acquisition of pronephric specification. Removing presumptive pronephric cells from the vicinity of the overlying somites will block pronephric differentiation in both amphibians (Fales, 1935; Brennan *et al.,* 1998) and birds (Mauch *et al.,* 2000).

III. Pronephric Induction

The pronephros forms between the paraxial mesoderm (presumptive somites) and the lateral plate, just behind the head. If mesoderm from the presumptive pronephric region is removed from early gastrula stage amphibian embryos and is cultured in a minimal saline solution (Brennan *et al.,* 1998) or if it is transplanted to a heterologous site far from its normal position (Fales, 1935), it will not form pronephric tubules. This indicates that at these early stages this region has not yet received all of the signals necessary to set it upon the path to pronephric development. By late gastrula stages in *Xenopus,* or midneurula stages in axolotls, similar explants develop into pronephric tubules in isolation (Fales, 1935; Tung, 1935; Brennan *et al.,* 1998). During the intervening period of time, the presumptive pronephric region has been further instructed as to its developmental destiny. When a cell is instructed of its developmental future by another cell, it is called induction (Spemann, 1938).

In vertebrates, all components of the kidney form from mesoderm. Although a considerable amount is known about the inductive signals that first induce the mesoderm and then subdivide it into dorsal, lateral, and ventral domains (reviewed by Kimelman and Griffin, 2000), much less is known about the later refinements of this patterning that establish organ primordia (in German, anlage). The location of one inductive signal required for pronephric development has been mapped, but there are probably at least two other such signals. Evidence for the existence of each of these signals, their potential sources, and their role in pronephric specification and maintenance will be outlined here. For details of tubule anatomy, please see Fig. 1.1.

A. Signal 1: Anterior Somites

The presumptive anterior somites are essential for normal pronephric development. If the anterior somites are removed by generating body plan phenotypes lacking them (Seufert *et al.,* 1999) or if they are separated surgically from the presumptive pronephros (Mauch *et al.,* 2000), pronephroi do not form. They are not only necessary, in some cases they are a source of inductive signals sufficient to impart pronephric fate on presumptive pronephric mesoderm. Anterior somites can induce pronephric tubules when cultured in contact with competent amphibian mesoderm, and experimental manipulations that increase the amount of somitic tissue increase the size of the pronephric tubules

(Seufert *et al.,* 1999; Mauch *et al.,* 2000). The source or timing of this signal has not yet been mapped in detail, but it is not present throughout the somitic mesoderm, as evidenced by ventralized embryos that contain competent mesoderm plus posterior somites failing to form pronephroi (Seufert *et al.,* 1999). Furthermore, pronephric anlage transplanted to positions ventral to more posterior somites do not develop into normally sized pronephroi (Fales, 1935), which once again implies a posterior limitation.

The anterior somite signal may be responsible for the nephric field described by Holtfreter (1933). This term describes a region in which transplanted ectoderm can be induced to form pronephric tubules. In urodeles, the pronephric field extends within the dorsal lateral plate from the gills back to the cloaca. The field is strongest, and induces tubules at the highest frequencies, in the vicinity of normal pronephros ventral to the anterior somites (Holtfreter, 1933; Burns, 1955).

Pronephros-inducing anterior somites are one component of the dorsal axial structures. Ectopic dorsal axial structures can be induced in vertebrate embryos, including amphibian (Spemann and Mangold, 1924), bird (Waddington and Schmidt, 1933), fish (Oppenheimer, 1936), and mammals by a variety of experimental and genetic techniques (Chapter 9). Incomplete secondary axes in manipulated frog embryos sometimes contain pronephroi, even though such axes usually contain somites and neural tissue, but no notochord, once again indicating a role for somites in pronephric induction. When primary and secondary axes are closely opposed in duplicated embryos, large fused pronephroi sometimes form between them, presumably in response to overlapping pronephric induction fields (Fig. 3.5).

Transcriptional targets of the anterior somite-derived signal are at the moment unknown, but we can make some informed guesses based on the patterns of expression of early pronephric genes. At the time at which the pronephric tubules are specified, stage 12.5 in *Xenopus laevis,* the gene encoding the paired box gene *Pax-8* is activated transcriptionally in a patch directly behind the head and under the anterior somites in *Xenopus* embryos (Fig. 3.6; Carroll and Vize, 1999). As discussed in later sections, this gene probably plays a role in the specification of the pronephric mesoderm (Chapter 8). The location and timing of *Pax-8* expression are consistent with its expression being a direct response to the anterior somite-derived inductive signal. The experimental modulation of *Pax-8* expression in response to the patterning perturbations that affect pronephric size, such as those described earlier, has not yet been described but is necessary if we are to link this response to the anterior somite derived signal.

The pronephric patterning signal from anterior somites is by far the best characterized component of pronephric induction. Although we lack strong embryological evidence for the existence of additional signals their presence can be

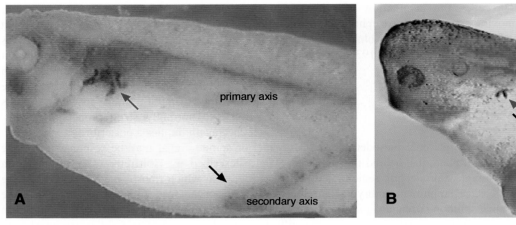

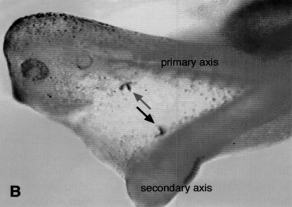

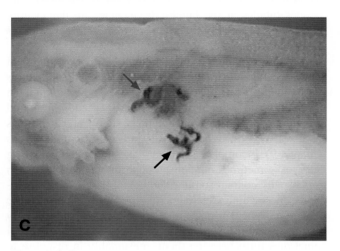

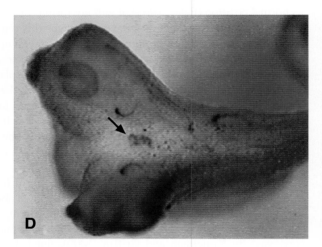

Figure 3.5 Pronephric kidney induction in secondary axes. The presence of pronephroi in secondary axes depends on the completeness of the axis or on its juxtaposition to the primary axis. (A) A *Xenopus laevis* embryo with a duplicated axis lacking a pronephros. The proximal tubules of the primary axis are indicated with an arrowhead. (B) Pronephroi present in both primary and secondary axes. In this instance, the secondary axis is more complete than that shown in A. (C) Well-developed pronephros in a secondary axis. In this instance, the secondary axis is very close to the primary axis. (D) An almost complete secondary axis. In this example, a single fused pronephros has formed midway between the two axes.

inferred from gene expression studies, where the activation and repression of subsets of genes within the pronephric anlage demonstrate the action of additional forces. This section describes the patterning events characterized by these modulations in gene expression. The role that these genes play in regulating kidney development is discussed in detail in Chapter 8.

B. Signal 2: *Xlim-1* Activator Signal

The second gene known to be activated in presumptive pronephric mesoderm coincident with the specification of this region is *lim-1*. Mammalian *lim-1* is absolutely essential for all stages of murine kidney development (Shawlot and Behringer, 1995), as it plays a critical role in the establishment of the intermediate mesoderm- precursor of the urogenital system (Tsang *et al.*, 2000). Although this gene is activated at the same time as the *Pax-8* gene, the spatial

expression patterns of these two genes are very distinct, implying that they are regulated by different signals. When the *Xlim-1* gene is first activated in the *Xenopus* presumptive pronephric mesoderm, its expression extends from the ventral border of the somites to the ventral belly mesoderm in a dorsoventral belt around the center of the embryo (Taira *et al.*, 1994, Carroll and Vize, 1999; Fig. 3.6). This belt has defined anterior and posterior boundaries and is the equivalent of approximately two to three future somite segments in width. Given the ventral extension of this band, plus its tight anterior and posterior delimitations, it is very unlikely that this gene is activated in response to the same signals as *Pax-8*. Over time, the ventral component of the *Xlim-1* expression domain is lost and its expression resolves to match that of *Pax-8* by *X. laevis* stage 23. This restriction indicates that *Xlim-1* requires a maintenance factor present only in the most dorsal portion of its expression domain. Candidates for this maintenance factor include the anterior somite-derived

XPax-8

Xlim-1

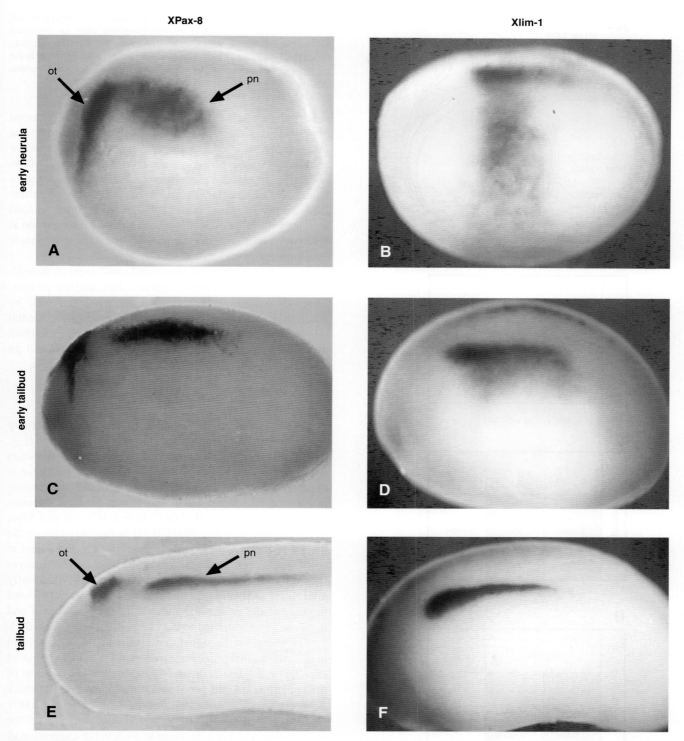

Figure 3.6 Early expression patterns of *Pax-8* and *lim-1* highlight genetic responses to early pronephric induction signals in *Xenopus laevis*. Expression of *XPax-8* is illustrated on the left and *Xlim-1* on the right. From Carroll and Vize (1999).

signal and the *Pax-8* gene that we have just postulated to mediate its effects. As to the signal that activates early *Xlim-1* expression, we can only speculate. The band of *Xlim-1* expression identifies cells with a similar anterioposterior position, and it is possible that the anterioposterior axis itself regulates expression of this gene in the intermediate mesoderm. Preliminary support for this model has been obtained in experiments investigating the effects of retinoids on pro-

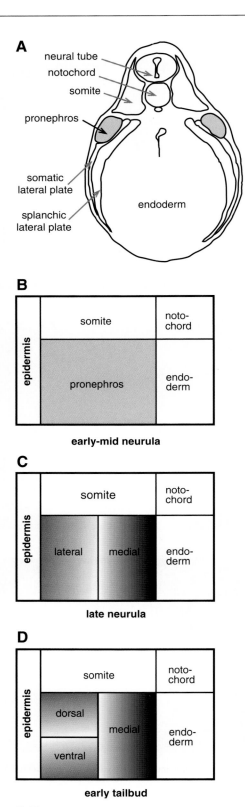

Figure 3.7 Patterning events that subdivide the *Xenopus* pronephric primordium. Hypothetical sequence of patterning events based on changes in gene expression within the forming pronephric primordia. (A) Transverse section through a NF stage 26 embryo (after Hausen and Riebesell, 1991) illustrating the relationship of the pronephric primordia to surrounding tissues. (B) Schematic representation of pronephric and surrounding tissues in a midlate neurula. Pronephric genes are expressed evenly throughout the primordia at this stage. (C) By stage 20 (late neurula/early tailbud), distinct lateral and medial domains exist within the pronephric mesoderm. (D) By stage 26 the lateral domain is further subdivided into dorsal and ventral domains.

nephric development (unpublished observations). Treatment of gastrula stage *Xenopus* embryos with retinoic acid results in the suppression of anteriodorsal properties, and anterior cells adopt more posterioventral fates than they would normally. In mildly posteriorized embryos, pronephric tubules are often enlarged, and in more severely posteriorized embryos, pronephric tubules sometimes extend anteriorly all of the way around the head. This expansion is consistent with anterior signals suppressing pronephric development.

The region defined by *Pax-8* and *lim-1* extends through precursors of the pronephric tubules, duct, and the glomus. This original pronephric domain is then subdivided by additional patterning signals that (1) establish a medial region that will later form the glomus and (2) establish distinct dorsolateral and ventrolateral domains that will go on to form the tubules and duct, respectively (Fig. 3.7).

C. Signal 3: Mediolateral Patterning

Genes expressed in response to the early patterning signals are activated throughout the pronephric anlage (Fig. 3.8). The first exception to this general pronephric expression is the Wilms' tumor gene-1 orthologue, *WT-1*. This gene is expressed only in the medial portion of the developing pronephric mesoderm and is the first indicator of mediolateral patterning in both frogs and fish (Carroll and Vize, 1996; Majumdar *et al.*, 2000). *WT-1* is first activated in the medial pronephric mesoderm at stage 18 in *X. laevis* and at some point prior to 24 h postfertilization in *Danio rerio* (Carroll and Vize, 1996; Drummond *et al.*, 1998). Shortly after *WT-1* activation, genes that were initially expressed throughout the pronephric mesoderm, such as *Pax-8* and *lim-1*, become restricted to the lateral pronephros (Fig. 3.8).

The source of the signal that results in *WT-1* expression in only the medial pronephric mesoderm is unknown, but an excellent candidate is the overlying epidermis. Removal of the epidermis inhibits pronephric development in the chick (Obara-Ishihara *et al.*, 1999, but see Mausch *et al.*, 2000). Obara-Ishihara *et al.*, (1999) found that removal of the epidermis resulted in a decrease in the level of expression of pronephric marker genes *Pax-2* and *Sim-1* and blocked subsequent kidney development. This tissue expresses *BMP-4*, a protein mediating many other inductive signals, and the effects of epidermal removal could be rescued with exogenous *BMP-4*. Epidermal *BMP-4* may well be the source of the mediolateral patterning signal. Its effects may be mediated through *WT-1* or localized *WT-1* expression

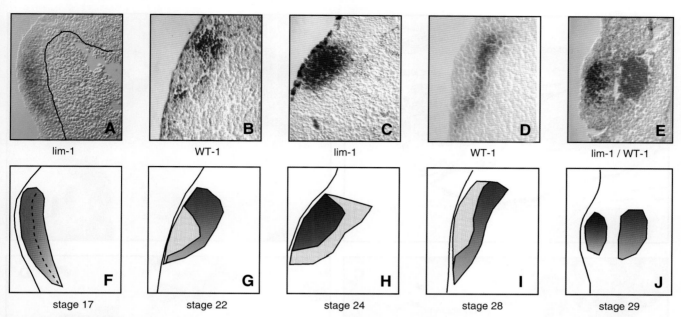

Figure 3.8 Gene expression patterns reveal the subdivision of the pronephros in *Xenopus*. (Top row) Transverse sections through embryos processed via *in situ* hybridization to reveal gene expression patterns of Xlim-1, WT-1, or both. (Bottom row) A graphic interpretation of what is immediately above. (A) *Xlim-1* expression at stage 17. The border between the mesoderm and the endoderm is marked with a black line. *Xlim-1* is expressed in both lateral and medial regions of the intermediate mesoderm. (B) Stage 22 expression of *xWT1* is restricted to just the medial portion of the pronephric primordium. (C) Stage 24 expression of *Xlim-1*. Transcription is now restricted to the lateral region. (D) Stage 28 *xWT1* expression in the medial domain. (E) Double *in situ* hybridization comparing *Xlim-1* and *WT1* expression. Note the absence of *Xlim-1* in the ventrolateral portion of the pronephros. This corresponds to the presumptive distal tubule.

may simply report the occurrence of the patterning. If epidermal *BMP-4* does directly result in restricted *WT-1* expression, it must do so by inhibiting *WT-1* expression in the lateral pronephric region. This is supported by the observation that in zebrafish lacking *Pax-2* (Majumdar *et al.*, 2000), *WT-1* expression expands laterally as one would expect if *Pax-2* mediates the *BMP-4* repressive signal. Another possible source of mediolateral patterning signals are the endoderm and notochord that lie medial to the primordia.

D. Signal 4: Dorsoventral Signal

The plasticity of the dorsoventral axis within the lateral mesoderm was demonstrated beautifully by Harrison's (1921) limb rotation experiments in urodeles. Harrison found that the future limb region can be removed and its dorsoventral axis inverted up to stage 32 without affecting the patterning of the resulting limb. Beyond this stage, the dorsoventral axis is fixed and rotation results in the development of a limb with an inverted dorsoventral orientation. Limbs develop relatively late in the scheme of amphibian development, and pronephric morphogenesis is well underway when the limb mesoderm overlying it has its dorsoventral axis fixed. The pronephric dorsoventral axis may be fixed quite early, as evidenced by 180° dorsoventral (D-V) rotations of midneurula ("widely open neural folds")

pronephric anlage in the frog *Discoglossus* generating D-V-inverted pronephroi (Tung, 1935). Such rotated pronephroi look relatively normal but have nephrostomes (Section IV) extending from the ventral region of the tubules rather than from the dorsal end as they would normally. As early as stage 12.5 (late gastrula) in the frog, embryos can be dissected into dorsal and ventral fragments that have different pronephric potentials, with dorsal fragments always forming pronephric tubules and ventral fragments forming minor amounts of tubule but a large amount of distal tubule and duct (Vize *et al.*, 1995). As there is little other than the lateral plate lying ventral to the pronephros, the most likely source of early dorsoventral patterning signals is the adjacent paraxial or more distant axial mesoderm, presumptive somite, or notochord, respectively.

Molecular markers of early dorsoventral patterning include wnt-4 (Carroll *et al.*, 1999) and components of the Notch/Delta pathway (McLaughlan *et al.*, 2000). *Wnt-4* is expressed initially throughout the pronephric primordium, but transcription becomes restricted to the dorsolateral region that will form the tubules by stage 26. *Serrate-1* also becomes localized to the dorsal compartment between stages 24 and 30 (McLaughlan *et al.*, 2000). These genes, plus other components of the Notch/Delta pathway, are expressed in broader patterns prior to this restriction. As the restriction in expression occurs considerably later than the time identified by Tung (1935) as the point at which the

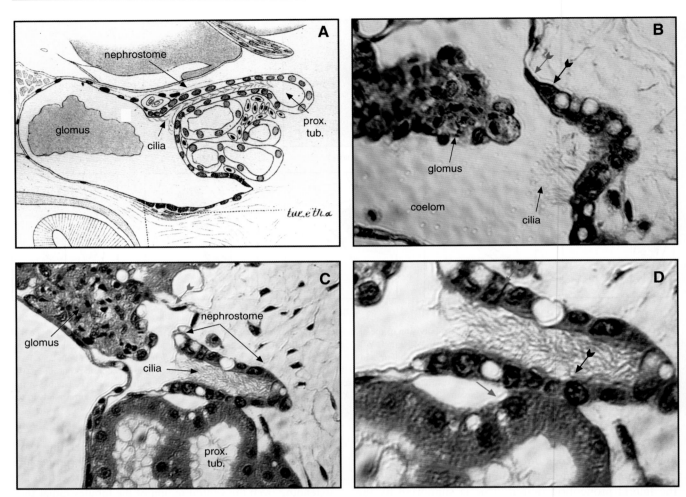

Figure 3.9 Anatomy of amphibian nephrostomes: *Ambystoma punctatum* (44-mm larvae) (A; Hall, 1904) and *Bombina orientalis* (B and C; sections made available by D.W. Seufert). (A) Transverse section of a mature urodele pronephros. An external glomus projects into the nephrocoel. The thin, ciliated nephrostome links the nephrocoel to the proximal tubules. Blood cells surround the tubules. (B) Transverse section through mature anuran pronephros (H&E stain). The nephrocoel is very small at this late stage of development, and the glomus is close to the opening on the nephrostome. Note the difference in morphology between the cuboidal nephrostomal epithelia and that of the high cuboidal proximal tubules and mesothelium of the coelom. Cilia within the nephrostome are stained pink. (C) Transverse section through opening of nephrostomal funnel. Cilia are present. The change in morphology from mesothelial (green arrow) to cuboidal (black arrow) epithelium is obvious. (D) Black arrow, nephrostomal cell. Red arrow, proximal tubule cell.

dorsoventral axis is fixed, if these genes are actually involved in dorsoventral patterning, they may do so during their early broad expression phases. *Wnt-4* is first activated in the *Xenopus* pronephros at stage 18, *Delta-1* at stage 19, *Notch-1* at stage 20, and *Serrate-1* at stage 22 (Carroll *et al.*, 1999; McLaughlan *et al.*, 2000).

E. Signal 5: Compartment Boundaries?

In order to avoid ambiguity in regions bordering cell populations with different developmental fates, boundaries are established that clearly demarcate to which group a cell belongs. The most common molecular mechanism used to establish such boundaries is the expression of Notch or Notch ligands such as Delta and Serrate. An elegant examination of the expression patterns of *Notch* and its ligands within the pronephros has provided evidence that this system may also act to establish compartments within the forming pronephros. McLaughlin *et al.* (2000) have demonstrated that *Notch* is normally expressed in the pronephric mesoderm and that Notch ligands are expressed in subdomains of this region. Artificial activation of the Notch pathway blocks duct development, whereas blocking Notch signaling enhances it, implying that this system functions to define a compartment of the developing pronephros as "not duct." Both proximal tubule and glomus marker gene expression are enhanced when Notch is activated. This pathway is also essential in adult kidneys, as defects in *Notch2* block glomerular development (McCright *et al.*, 2001).

F. Induction Summary

At least two overlapping signals act to establish the pronephric mesoderm in late gastrulae or early neurulae (Fig. 3.7). One of these signals is derived from the anterior somites and may act through *Pax-8*. Once established, the pronephric mesoderm is then subdivided into medial and proximal compartments. The proximal compartment then undergoes further subdivision to establish the precursors of the proximal tubules and the distal tubule/duct.

IV. Pronephric Anatomy

A. Pronephric Anatomy Around the Onset of Function

The anatomy of tubules is easiest to interpret shortly after each of its components have formed and before extension and looping lead to convolution and a more complex organization. In the axolotl, this corresponds to Harrison stage 35 (Fales, 1935), in *X. laevis* to Nieuwkoop and Faber (1994) stages 35–36, and in the zebrafish around 36 h postfertilization (Chapter 5). This is just prior to the pronephros first becoming functional, which is at *Xenopus* stage 38 (Nieuwkoop and Faber, 1994) and 40 to 48 h postfertilization in zebrafish (Drummond *et al.*, 1998). In order to best illustrate the anatomical features in whose development we are most interested in the clearest possible manner, the anatomy of pronephric tubules is described at this early transient stage and also at the time at which the final organ has formed.

Pronephric tubules are connected to the coelom on their rostrodorsal side by thin epithelial funnels called nephrostomes and to the cloaca on their caudal side by the nephric duct (Fig. 1.1). The tubules can be further subdivided into proximal tubules adjacent to the nephrostomes and a distal tubule linking the proximal tubules to the duct. Nephrostomes are quite distinct from the nephrotomes discussed in some of the older literature. While nephrostomes are ciliated tubules, nephrotomes are blocks of what were thought to be mesodermal segments that were the precursors of individual pronephric branches. The concept of nephrotomes is explored in Section V. Another important point of terminology is the distinction between a nephrostome and a peritoneal funnel (Kerr, 1919; Goodrich, 1930; Fig. 3.10). A nephrostome links the coelom to the nephric tubule, whereas a peritoneal funnel links the coelom to a glomerulus. Peritoneal funnels are also ciliated and could easily be mistaken for a nephrostome.

1. Nephrostomes/Nephrostomal Funnels

Amphibians, fish, and reptiles have between 1 and 12 nephrostomes linking pronephric tubules to the coelom. Nephrostomes are thin epithelial tubules lined with long cilia (Fig. 3.9). Cilia form at Harrison stages 35 to 36 in axoltols (Fales, 1935) and stage 33 in *X. laevis* (Nieuwkoop and Faber, 1994), initially facing outward toward the coelom but reorienting toward the interior of the pronephros at stage 37 in axolotls when fluid movement through the tubules begins. Cilia can be seen clearly in eosin-stained sections (Fig. 3.9) and can be stained specifically using antitubulin immunohistochemistry.

Excellent electron micrographs of the ciliary network have been published by Gendre (1969) and Møbjerg *et al.* (2000), among others. Cilia are very dense, and over 100 are observed in a single section through a *Rana* nephrostome, with each nephrostome containing approximately 500 cilia in total (Gendre, 1969). Each cilium contains a classical 9 + 2 arrangement of doublet and singlet microtubules (Gendre, 1969).

Because each nephrostome is linked to a branch of the proximal tubule, organisms with 10 nephrostomes have 10 proximal tubule branches, one linking each of the nephrostomes to the downstream tubule network. The transition from nephrostome to proximal tubule is characterized by a great reduction in the frequency of cilia and a change in epithelial morphology, from cuboidal or even low cylindrical, to a high cuboidal or columnar epithelium (Figs. 3.9 and 3.11). In addition to the difference in cell shape and size, nephrostomal tubules are narrower in overall diameter than proximal tubules (Fig. 3.9). The morphology of the nephrostomal epithelium is also distinct from that of the adjacent coelomic lining. Cuboidal cells with multiple cilia

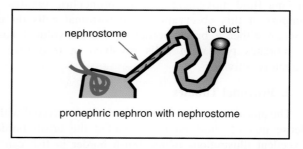

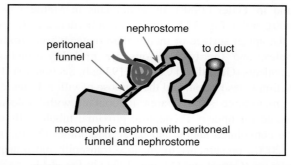

Figure 3.10 Nephrostomes and peritoneal funnels. (Top) A nephrostome links the coelom (or nephrocoel) to a pronephric proximal tubule. (Bottom) A peritoneal funnel links the coelom to the glomerulus in a mesonephric nephron. The glomerulus is linked to the mesonephric tubule by a nephrostome.

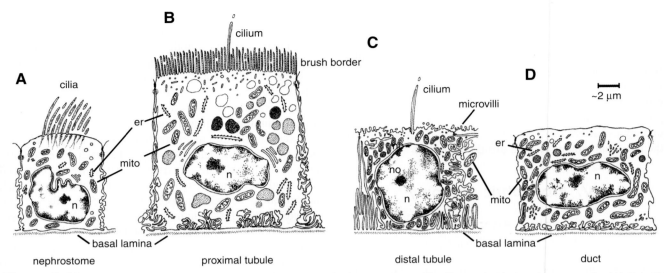

Figure 3.11 The four major epithelial cell morphologies observed in the pronephros. All cells are sketched at the same scale. (A) Cuboidal nephrostomal cell with multiple apical cilia. (B) High cuboidal proximal tubule cell with apical brush border and basal convolutions. (C) Cuboidal distal tubule cell with apical microvilli and extensively convoluted basal membrane. (D) Principal cells of the pronephric duct have low cylindrical morphology.The basal membrane is much less convoluted than that of distal tubule cells. er, endoplasmic reticulum. mito, mitochondria. Based on Møbjerg *et al.* (2000). Reprinted by permission of Wiley-Liss Inc., a subsidiary of John Wiley & Sons, Inc.

extend for a few cells each side of the nephrostomal tubule before the transition to the normal squamous morphology of the coelomic mesothelium is observed (Fig. 3.9; Wrobel and Süβ, 2000).

Nephrostomal funnels do not appear to be associated with blood vessels and their function is likely limited to driving fluid movement into the pronephric tubules. In support of this observation, nephrostomal cells do not contain either a brush border or the convoluted basal membranes typical of cells involved in resorptive or excretory activity.

2. Proximal Tubules

The apical surface of proximal tubules is covered with a dense mat of short microvilli called the brush border. Excellent illustrations of the brush border in fish can be found in Tytler (1988), in amphibians in Møbjerg *et al.* (2000), and in Fig. 3.11. Similar brush borders are observed on the apical surface of mammalian metanephric proximal tubules, and the two regions are probably functional equivalents (Gérard and Cordier, 1934a,b; Section VI). This function is resorption, and the purpose of villi is to increase the membrane surface area in contact with molecules targeted for uptake. Pronephric proximal tubule cells also have convoluted lateral and basal membranes (Møbjerg *et al.*, 2000), presumably to enhance transepithelial transport processes (Gérard and Cordier, 1934a,b; Fig. 3.11).

The proximal tubule has short branches on its dorsal aspect that link it to each of the nephrostomes (Fig. 1.1). The common region of the proximal tubules is sometimes referred to as the connecting tubule (Vize *et al.*, 1997), but

this name has also sometimes been used for the region linking the distal segment to the duct (see later). The connecting tubule (in our nomenclature) has a similar diameter to its dorsal branches and, as noted earlier, it is noticeably wider than nephrostomes. Fluids moving through the connecting proximal tubule drain into a single common, or broad, tubule that is of even greater diameter (at least initially). No histological differences between connecting and broad tubules have been reported to date: both stain with the pronephric apical membrane-specific antibody 3G8 and with a number of tubule-specific genes, including *Pax-8* (Carroll and Vize, 1999) and SMP-30 (Sato *et al.*, 2000).

At the caudal end of the broad tubule, the pronephric epithelium undergoes a constriction in diameter (Vize *et al.*, 1995). This constriction is also marked by a rapid decrease in 3G8 immunoreactivity, probably reflecting a reduction in brush border density. This transitional region extends only 20µm or so, and caudal to this point the tubule is of a uniform diameter over the remainder of its length. The point at which constriction of the broad tubule occurs and 3G8 immunoreactivity decreases is also the caudal border of expression of proximal tubule genes, including *Pax-8* and SMP-30. Organization of the proximal tubule systems of frogs is illustrated in Figs. 1.1 and 3.12.

Cells of the proximal tubule network express connexins, membrane proteins that form gap junctions. The Cx30 connexin gene is expressed throughout proximal tubule epithelia, but probably not in the distal tubule (Levin and Mercola, 2000; Levin personal communication). Active gap junctions would link ion transport processes in coupled cells.

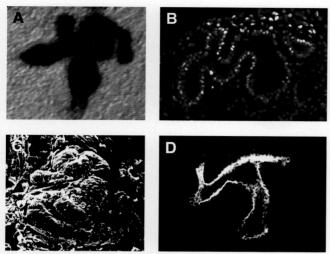

Figure 3.12 Pronephric proximal tubules of *Xenopus laevis* at around NF stage 35/36. All views are lateral, with dorsal up and anterior to the left. (A) Immunohistochemistry with monoclonal antibody 3G8. The antibody binds to an unidentified antigen localized to the apical surface of the proximal tubules. (B) Confocal optical section through an embryo stained with propidium iodide to visualize nuclei. The proximal tubule epithelia with its three dorsal branches and the wide common tubule are visible. (C) Scanning electron micrograph of an embryo that has had the overlying epidermis peeled away by microsurgery. One of the dorsal branches (the most posterior) has been lost in the process of sample preparation. (D) Confocal optical section using immunofluorescence to visualize 3G8 localization. The lining of the pronephric lumen is labeled with the antibody. The surrounding epithelium is also visible due to a faint auto-fluorescence. A and C are from Vize *et al.* (1995) and B and D are from Wallingford *et al.* (1998).

3. Distal Tubule

Following the *Xenopus* broad tubule, a narrower tubule loops in an "S" shape, with the top of the S linking to the broad tubule and the bottom of the S linking to the pronephric duct, which runs along the ventral border of the somites to the rectal diverticulum and the cloaca (Fig. 3.13). At least part of this S-shaped region, and probably all of it, corresponds to the future distal tubule, which will form a large percentage of the total tubule length in the mature pronephros (Møbjerg *et al.*, 2000). Previous descriptions of the early pronephric system have usually classified the epithelia as either a tubule or a duct without sufficient attention to the location of the distal segment (e.g., Vize *et al.*, 1995). Given the distinct functions of the distal tubule in pro-, meso-, and metanephroi (Gérard and Cordier, 1934a,b; Section VI), this is an oversight that needs to be addressed.

Pronephric distal tubule cells do not have an apical brush border (Fig. 3.11) but the basal membranes are folded extensively (Møbjerg *et al.*, 2000). With the exception of basal membrane convolutions present in distal cells, these cells are quite similar in overall appearance to those that follow them along the nephric system within the duct, and the transition between these two segments is not immediately obvious.

While the rostral end of the distal segment in *Xenopus* is marked by a slight widening in tubule diameter and the acquisition of 3G8 immunoreactivity (Fig. 3.12), the caudal end of this segment has not yet been defined accurately. In Fig. 3.13, immunohistochemistry of the *Xenopus* distal segment using antibody 4A6 is illustrated. At the onset of function, this antibody stains all nephric cells immediately caudal to the proximal tubules in a solid manner. After reaching the somites, the distal tubule/nephric duct turns sharply toward the cloaca. At this point, 4A6 staining becomes mosaic, with approximately 75% of the cells expressing the antigen in the anterior portion of the duct, decreasing to 5% or less of duct cells near the cloaca being 4A6 positive. The nephric system has been a complete epithelial tube for many hours by this point in time and this loss of expression reflects cell heterogeneity—not the end of the duct. It is quite possible that the solid region of 4A6 immunoreactivity defines the distal segment, whereas the mosaic segment corresponds to the nephric duct. The region expressing 4A6 in a solid manner also fails to express *Xlim-1* in early *Xenopus* tailbud stages, whereas both the proximal tubule and the duct are *Xlim-1* positive (Carroll *et al.*, 999). *Xiro3* (Bellefroid *et al.*, 1998) is also expressed in the distal segment, but has not yet been described in sufficient detail to draw conclusions as to its usefulness as a distal tubule marker. A more detailed analysis with additional molecular markers and further histology will be required to resolve this issue. A distal segment has not yet been mapped accurately in early fish pronephroi, but is indicated in some anatomical figures (e.g., Felix, 1897).

4. Duct

Details of duct development are covered in Chapter 4, but for comparative purposes, the general morphology of duct cells is worth mentioning briefly at this point. The duct contains a number of different cell types, each of which has the same general shape and size as distal tubule cells (Fig. 3.11). They do not, however, have the extensive basal membrane folding observed in distal cells (Møbjerg *et al.*, 2000), presumably reflecting the absence of transepithelial transport processes in this segment of the pronephric system.

The duct contains a mixture of at least two different cell types (Møbjerg *et al.*, 2000). Some of these cells stain with antibody 4A6 and probably correspond to principal cells. The remaining cells do not stain with 4A6 and probably correspond to the intercalating cells observed histologically. Pronephric intercalating cells are characterized by large numbers of mitochondria and are probably the functional equivalents of intercalating cells of the mammalian collecting duct, which function to control acid secretion and maintain acid/base balance.

The duct forms from pronephric mesoderm that migrates caudally, but also receives a contribution from the neural crest (Krotoski *et al.*, 1988; Collazo *et al.*, 1993), and trans-

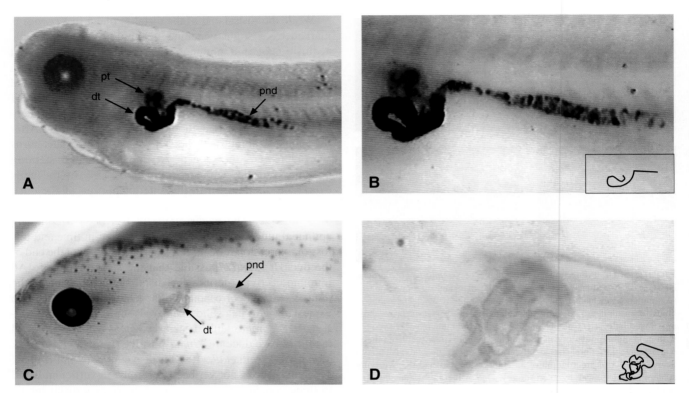

Figure 3.13 Distribution of 4A6 immunoreactivity in the frog pronephros. (A) Stage 39 *Xenopus* tadpole. Faint staining is observed in the proximal tubule, possibly artifactual. Solid staining is present within the distal tubule. Solid staining within a subset of individual cells is present in the pronephric duct. (B) Enlargement of A. Note the solid nature of reactivity within the distal tubule and mosaic staining in the duct. (C) Stage 41 *Xenopus* tadpole. 4A6 reactivity appears to be solid within both the distal tubule and the duct. (D) Enlargement of embryo shown in C. The distal tubule has grown in length and is now convoluted. No staining within the proximal tubule is observed. pt, proximal tubule; dt, distal tubule; pnd, pronephric duct.

planted crest cells can also migrate along the pronephric duct pathway (Zackson and Steinberg, 1986). The final pronephric fate of crest cells in this region has not yet been determined, but a molecular marker for this group of cells may have been identified: Gremlin (a member of the DAN family of BMP/nodal inhibitors). *Gremlin is* expressed in the frog neural crest, the distal tubule, and much of the pronephric duct (Hsu *et al.,* 1998). It is not expressed in the duct derived from the rectal diverticulum. The pronephric expression of *Gremlin* may mark duct neural crest derivatives, but this has not yet been demonstrated. The defined caudal boundary of *Gremlin* pronephric expression may indicate that crest cells stop migrating posteriorly when they reach the duct/diverticulum fusion point (Fig. 4.5). The posterior migration of crest cells may explain the observations of Cornish and Etkin (1993), who have described the presence of cells not derived from the anterior pronephric primordium in the duct.

5. Capsule

The pronephric tubules and their surrounding blood vessels are encased within an investing mesothelial capsule. This capsule develops from the ventrolateral edge of the somites and is first observed as a small cluster of cells between the overlying somite and the tubule primordium in frog early tailbud stages (Field, 1891). It gradually extends ventrally around the pronephric anlage to meet and fuse with the somatic plate. Morphogenesis of the capsule surrounding the somites begins at the same time and place, and the two capsules form a two-layered structure between these two organs.

B. Anatomy of the Mature Pronephros

The basic differences in pronephric anatomy that occur between the time at which this organ first becomes functional and its terminal morphology are illustrated in Fig. 1.2. These differences include a great expansion of pronephric tubule length, which continues to increase in frogs until degeneration during metamorphosis (Section VII). At stage 35, the frog proximal tubule is approximately 300μm in length from the nephrostome to the constriction at the proximal/distal boundary (Carroll and Vize, 1999). In a mature (8 mm) *Bufo* tadpole pronephros, the proximal tubule has grown to approximately 3.6 mm in length. The distal tubule undergoes even more growth, extending from a few hundred

Table 3.1   Pronephric Development Sources

Class	Order (or *subclass)	Genus	Common name	Key references
Amphibia	Anura	*Rana*	Frog	Field (1891); Fübringer (1878);
		Xenopus	Frog	Nieuwkoop and Faber (1994); Fox (1963); Vize *et al* (1995)
	Caudata (Urodela)	*Ambystoma*	Salamander	Gillespie and Armstrong (1985)
		Cryptobranchus	Giant salamander	Fox (1963)
	Gymnophiona	*Hypogeophis*	Caecilian	Brauer (1902); Wake (1970)
Cyclostomata	Petromyzontia*	*Petromyzon*	Lamprey	Wheeler (1900)
Pisces	Dipnoi*	*Neoceratodus*	Lungfish (Australian)	Fox (1960, 1962)
		Lepidosiren	Lungfish (South America)	Kerr (1919)
	Chondrostel	*Acipenser*	Sturgeon	Maschkowzeff (1926); Fraser (1928)
	Selachii	*Torpedo*	Electric ray	Rückert (1888)
		Squalus	Dogfish	Balfour (1881); Bates (1914)
	Teleostei	*Danio*	Zebrafish	Drummond (1998)
		Salmo	Trout, salmon	Felix (1897); Tytler (1988)
Reptilia	Crocodilia		Crocodile	Wiedersheim (1890)
	Lacertilia	*Lacerta*	Lizard	Kerens (1907)
	Ophidia	*Tropidontus*	Snake	Kerens (1907)
		Chrysemys	Turtle	Burlend (1913); Walsche (1929)
Aves	Galliformes	*Gallus*	Chicken	
Mammalia	Chiroptera	*Rhinolophus*	Bat	van der Stricht (1913); see Fraser (1920)
	Carnivora		Cat	Fraser (1920)
	Monotremata		Echidna	Keibel (1903)
	Polyprotodontia		Marmot	Janosik (1904)
	Insectivora		Mole	Kerens (1907)
	Rodentia		Rabbit	Kerens (1907)
	Primates		Human	Felix (1912)

micrometers to over 7 mm (Møbjerg *et al.*, 2000). The elongate distal tubule coils ventral to the proximal tubules (Fig. 3.13).

In addition to cilia lining the nephrostomes, pronephroi acquire additional ciliation in defined segments of the tubules in later stages of development. Such segments have been described in a variety of amphibians (e.g., McCurdy, 1931). Presumably, these regions maintain fluid movement rates along the tubule complex.

A number of authors have produced excellent three-dimensional reconstructions of mature pronephric tubules. Examples include those of *Rana* (frog) by Field (1891) and Marshall (1902), lamprey (jawless fish) by Wheeler (1900), *Ambystoma* (salamander) by Howland (1921), *Protopterus* (polypterinid fish) by Kerr (1919), *Triturus* (newt) by Mc-Curdy (1931), and *Bufo* (toad) by Møbjerg *et al.* (2000). The reconstructions of Marshall (1902) are presented in Fig. 1.2.

1. Vasculature of Mature Tubules

Solutes and water resorbed by pronephric tubules from the glomeral filtrate are returned to the circulation through a network of veins that surrounds the tubules known as the pronephric sinus. This sinus receives blood from multiple sources, including the vas efferentia of the glomus, the anterior and posterior cardinal veins, the branchial vein, and the external jugular vein (Fig. 3.14; Viertel and Richter, 1999). Blood exits the sinus via the duct of Cuvier (Fig. 3.14; Chapter 5).

C. Comparative Morphology

For detailed reviews on the comparative anatomy of pronephroi, the works of Balfour (1881), Forster and Sedgwick (1885), Kerr (1919), Goodrich (1930), Fraser (1950), and Fox (1963), plus those in Table 3.1, are recommended, and many of the older papers also provide thorough (if sometimes caustic) reviews of the field.

In general, all amphibians and fish have well-developed pronephroi, whereas amniotes have rudimentary or extremely transient pronephric kidneys. Exceptions to this generalization include poorly developed pronephroi in most sharks and rays (Balfour, 1881; Kerr, 1919) and well-developed and functional pronephroi in some reptilia, e.g., crocodiles (Wiedersheim, 1890). Even with this range in the degree of development and function, the anatomy of pronephric tubules in most vertebrates is very similar, at

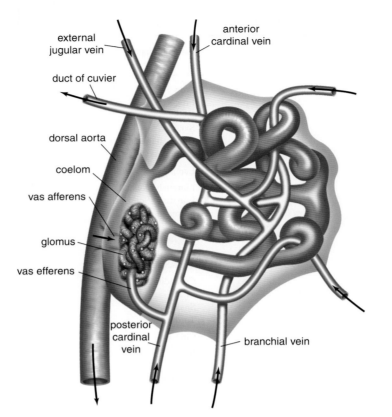

Figure 3.14 Pronephric vasculature. The pronephric glomus is supplied with blood by the dorsal aorta. Filtrate from the glomus is swept into the pronephric tubules through the ciliated nephrostomes. Solutes recovered from the filtrate are returned to the bloodstream by a network of vessels that weave through the tubules. The plexus of vessels receives input from multiple sources as indicated and is recovered via the duct of Cuvier.

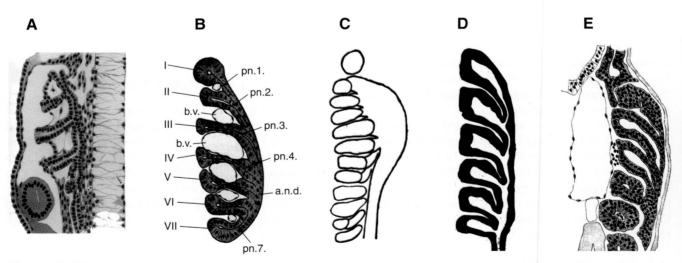

Figure 3.15 Frontal sections through the pronephroi of various organisms. (A) The lamprey (Wheeler, 1900). (B) The ray, *Torpedo*. After Rückert (1888). (C) The ray-finned fish, *Polypterus*. After Kerr (1919). (D) The caecilian, *Hypogeophis alternans*. After Brauer (1902). (E) The lizard, *Lacerta muralis*. After Kerens (1907).

least for the short period of time in which they exist. For example, Fig. 3.15 shows frontal sections through the tubules of a jawless fish, an electric ray, a lungfish, a legless amphibian, and a lizard, and all are remarkably similar. Details of variations of anatomy are discussed briefly later, and Table 1.1 provides examples of literature describing pronephric development in a wide range of taxa. The main difference observed among species is the number of dorsal proximal tubule branches. The general organization of pronephroi in both primitive and advanced amphibians and fish is illustrated in Fig. 3.16.

Even though amniotes only form transient pronephroi whose anatomy implies that they may not ever function as excretory organs, they may still play an important developmental role. The only real way to demonstrate that pronephric tubules have no function is to identify a viable mutant that lacks embryonic but not adult kidneys, and no such mutation has yet been identified. One possible developmental role for pronephric tubules is in the induction of the pronephric duct, without which neither the meso- nor the metanephric kidneys would form or function. Given that at least rudimentary pronephric tubules exist in all vertebrates, this is a distinct possibility. Ducts can form even when separated from tubules, but they do so poorly and may already have received inductive signals from the tubules prior to separation (Vize *et al.,* 1995). Even in species in which duct development slightly precedes that of the tubules, e.g., teleosts, presumptive tubules could still be the inducing tissue. For example, the axial mesoderm induces the overlying ectoderm to adopt a neural fate well before the axial mesoderm itself differentiates into notochord, and a similar delay in the timing of differentiation or morphogenesis of a responsive tissue is no reason to dismiss a

possible inductive role for the tubules. A second reason for abstaining from dismissing the developmental relevance of pronephroi prematurely is the very question of what genetic alterations led to the loss of advanced pronephric development and what changes have contributed to the variations in branch number and form observed in different animals. The transient kidneys may well be extremely useful in comparative molecular analyses linking gene expression patterns to changes in morphology. A brief discussion of the comparative anatomy of the pronephroi may therefore be of some use and is provided later. Figure 3.17 presents transverse sections illustrating the general anatomy of pronephroi in various taxa.

1. Amphibians

In anurans ("no tail," includes frogs and toads) and urodeles ("evident tail," includes newts and salamanders), general pronephric anatomy is essentially the same. Anuran pronephroi almost always have three dorsal branches, whereas most urodele pronephroi have two (Fig. 3.16). In more primitive salamanders, such as the giant salamanders of the genus *Cryptobranchus*, pronephroi have five branches, whereas congo eels (amphibians of the genus *Amphiuma*) also have more complex architecture with each of their three dorsal branches either splitting into two additional branches or each branch inducing two nephrostomes (Field, 1894). Caecilians (order Apoda or Gymnophiona) are legless, blind, worm-like amphibians that live their terrestrial adult lives burrowing in the soils of the tropics. Pronephroi of caecilians have been the subject of many studies, but as there is such wide variation in urogenital anatomy in this order, it is difficult to summarize briefly (for a review, see Wake, 1970). In general, caecilians have

| teleost | urodele | anuran | ganoid | gymnophiona |

Figure 3.16 Branching patterns in pronephroi. In general, more primitive species have more branching. Teleost fish (after Drummond *et al.,* 1998). Urodele (after Mangold, 1923). Anuran (after Field, 1891). Ganoid fish (after Fraser, 1928). Gymnophiona (Semon, 1890).

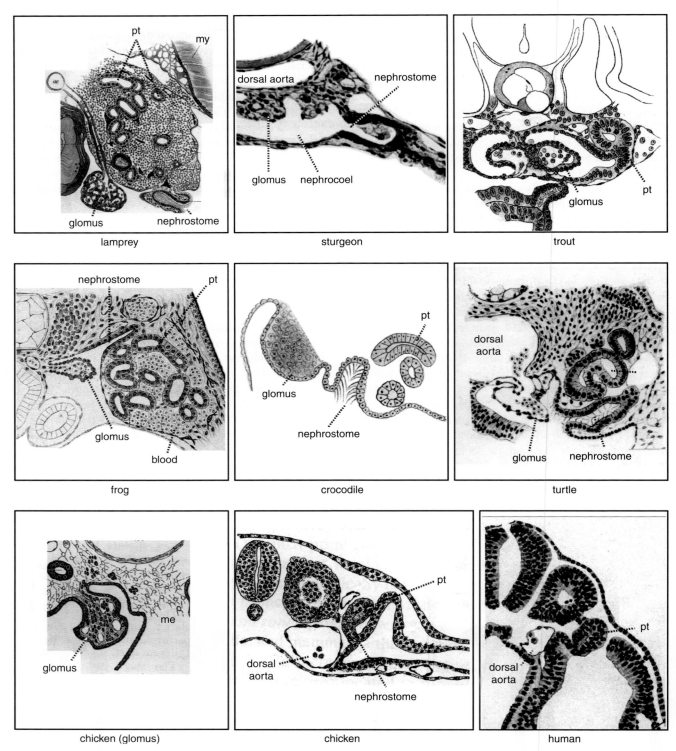

Figure 3.17 Comparative pronephric anatomy. Transverse sections through pronephroi of various species. The lamprey, after Wheeler (1900). The sturgeon fish *Acipenser rubicundus*, after Fraser (1928). Crocodile pronephros with well-developed external glomus and nephrostome. After Weidersheim (1890). Frog, *Rana temporaria*, pronephros. After Fürbringer (1878). Trout pronephros, after Felix (1897). Turtle pronephros, after Walsche (1929). The pronephric glomus of the chicken. After Balfour (1881). Chicken pronephric tubule, after Lillie (1919). Human pronephros, after Felix (1912). pt, proximal tubule.

complex pronephroi with up to 12 dorsal branches (Semon, 1890; Brauer, 1902; Wake, 1970). They often include an external glomus that extends over many body segments, but also have internal glomeruli in the more posterior region of the pronephros.

Most amphibian pronephroi have a single large glomus that projects into the coelom (Figs. 1.2 and 3.14). The glomus extends over multiple segments, but forms as a continuous, nonsegmented structure (Fig. 3.16). Details of glomeral development are provided in Chapter 5.

2. Fishes

Fishes, including dipnoi (lungfish), ganoids (e.g., sturgeon), gnathosomes (jawless fish), and teleosts (e.g., medaka, salmon, trout, zebrafish), both fresh and salt water, have well-developed pronephroi. With the exception of teleosts, fish pronephric anatomy is very similar to that in amphibians. An example of this conservation of anatomy is illustrated in Fig. 3.18 where the pronephric tubules of a newt are compared to those of a ganoid fish. Other than the number of dorsal branches and nephrostomes, the two organs are essentially identical. As mentioned earlier, more primitive amphibians have a larger number of dorsal branches and are even more similar to ganoid fish pronephroi (Fig. 3.16). Similarities are also obvious in the transverse section (Fig. 3.17).

Teleost pronephroi differ in a number of ways from those of other fish and other vertebrates. In this group, the vascular glomus develops as a dilation of the anterior tip of the nephric tubule rather than as separate infolding of the splanchnic intermediate mesoderm (Felix, 1897). The two glomera also migrate to the midline and fuse, generating a single midline glomus, feeding both lateral tubule complexes (Balfour, 1882; Felix, 1897; Tytler, 1988; Drummond *et al.*, 1998). With the intimate association of the glomus and the tubules, nephrostomes present in other pronephroi appear to be unnecessary, and the proximal

tubule is linked to the glomus via a short ciliated "neck" segment (Tytler, 1988). Superficially, the anatomy of a teleost pronephros resembles a mesonephric nephron with an internal glomerulus more closely than a pronephros. However, if one were to simply collapse the coelom of an amphibian and shorten the nephrostomes, its pronephros would appear much more teleostean.

The organization of pronephric tubules in elasmobranchs is quite variable with some authors (e.g., Balfour, 1881) claiming that no tubules were present, whereas others noted tubules with up to seven branches (e.g., Rückert, 1888; Bates, 1914; Fig. 3.15). This is probably due to the transient existence of the tubules, species differences, and the critical timing necessary to document their morphogenesis. Because some tubule primordia also survive longer than others, the number of tubule branches observed in a small sample size could vary considerably. No well-developed glomera form in either sharks or rays (Balfour, 1881; Kerr, 1919).

3. Reptiles

In reptiles, detailed descriptions of pronephric anatomy and morphogenesis have been available for many years. Excellent examples are listed in Table 3.1. Crocodiles have well-developed pronephroi with ciliated nephrostomes linking the tubules to the coelom (Wiedersheim, 1890; Fig. 3.17). This advanced degree of development led to the conclusion that pronephroi are very likely functional in this taxa, but some have questioned whether these organs may, in fact, be mesonephroi (see Wheeler, 1900). Given that they possess an external glomus and the tubules open into the duct, they appear to be both pronephric and functional by our standards.

Turtles also possess well-formed pronephric organs that include ciliated nephrostomes and large glomera (Wiedersheim, 1890; Gregory, 1900; Walsche, 1929; Fig. 3.17). Lizards and snakes also form pronephric tubules (six), but these may degenerate before becoming functional (Kerens, 1907; Kerr, 1919).

4. Birds

Pronephroi in birds have been described in considerable detail. One of the earliest detailed studies was by Balfour and Sedgwick (1879; Balfour, 1881). These authors illustrated the existence of a well-developed (if transitory) set of pronephric tubules and external glomera in addition to the permanent nephric duct (Fig. 3.17). In the chicken, approximately 12 rudimentary pronephric tubules form. By day 6 of development, the tubules have degenerated, but the glomera remain for another 2 days (Balfour and Sedgwick, 1879; Kerr, 1919).

5. Mammals

Development of the transient kidneys, both pro- and meso-nephroi, varies considerably among mammalian species

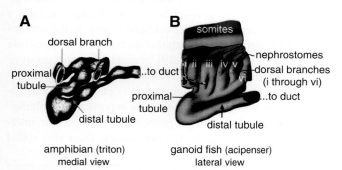

Figure 3.18 Comparison of amphibian and ganoid fish pronephric tubules. The urodele pronephros (A) has two dorsal branches (Mangold, 1924), whereas the fish (Fraser, 1928) pronephros (B) has six (Roman numerals). Other than the number of branches, the organs are very similar in anatomy.

(Bremer, 1916; Wintour and Moritz, 1997). In many mammals, the embryonic kidneys, and even the adult kidneys, are not essential for waste disposal or osmotic regulation prior to birth—as is evidenced by the live birth of humans completely lacking kidneys. As mentioned in Chapter 1, this is due to the evolution of efficient placental systems that provide the services otherwise rendered by the embryonic kidney. The development of a yolk sac may have contributed in a similar manner to the reduced physiological importance of pronephroi in reptiles and birds. However, pronephroi are present and their development has been described in many mammals, including bats, cats, echidnas, humans, mice, moles, and rabbits (Table 3.1).

In ruminants such as sheep and cattle, a large "giant" external glomerulus is present, but well-developed pronephric tubules, distinct from mesonephric tubules, do not form (Davies, 1951; Wintour and Moritz, 1997). The first nephrons of the mesonephros in these animals are linked to the coelom via peritoneal funnels and they may well function in the same manner as nephrostomes in amphibian–fish pronephroi—by cilia driving the movement of coelomic fluid into the proximal tubule. The distinction between the anterior-most tubules of the mesonephros and the posterior-most region of the pronephros is not always clear. Are the anterior-most tubules of the mesonephros actually pronephric tubules that contain an internal and external glomerulus or mesonephric tubules that contain a peritoneal funnel? As there is no reason why pronephric and mesonephric kidneys should not have anatomic features in common, or indeed that there should be intermediate type nephrons that fit into both classes, for most purposes such distinctions are unnecessary. Some authors have argued that the presence of peritoneal funnels argues for evolutionary relationship with aquatic species (Gaeth *et al.*, 1999), whereas others propose that their presence is required for the development of the male reproductive system (Wrobel and Süß, 2000). As a working definition, we will refer to kidneys as pronephric if they have an external glomus or glomerulus, or if two clearly distinct organs form in space or in time with the second organ being mesonephric in form.

Humans develop pronephric tubules and a rudimentary glomus—as extensive in fact as those of the reptiles (Fig. 3.17). The most complete description of human pronephric development is by Felix (1912), whose wonderful work on fish pronephroi was also mentioned earlier (Felix, 1897). Pronephric tubules are first seen in human embryos of 1.7 mm (4th week, 9 to 10 somites) and have undergone degeneration by 4.9 mm (Felix, 1912; Dodds, 1946). As more caudal tubules develop, more rostral ones degenerate, with an average tubule number being around six. The human glomus exists but is rudimentary, and no internal glomeruli are present in the pronephros. As with amphibian and bird pronephroi, the glomus persists longer than the tubules (Felix, 1912).

V. Morphogenesis

In the previous section, the anatomy of the pronephros was described at two discrete points: just prior to the onset of function and at the maximal level of pronephric complexity. This section explores the process by which the pronephroi obtain these two forms. The process of pronephric degeneration during metamorphosis is covered in Section VII.

A. Primordium Segregation

Approximately 10 h after the precursors of the *Xenopus* pronephric tubules are specified, they begin to segregate from the intermediate mesoderm. The process of mesodermal segregation into the pronephric primordium differs from the processes that form mesonephric and metanephric nephrons. In meso- and metanephroi, nephrons condense from a blastema of proliferating mesenchymal cells known as the nephric or nephrogenic blastema via a classical mesenchyme-to-epithelial transition (Chapter 12; Saxén, 1987). The pronephric nephron, however, forms more by a process of rearrangement, or segregation, than it does by condensation from a blastema. In fact, there is little evidence of the existence of a pronephric blastema in any vertebrate. Lateral views of pronephroi give the appearance of a sheet of mesenchymal cells (see later), but transverse sections indicate that this may be caused by the basal protrusion of columnar cells pushing outward.

As illustrated in Fig. 3.19, the first sign of pronephric morphogenesis in both fish and frogs is a change in cell shape in the somatic intermediate mesoderm.[1] Once there is a clear distinction between the nephric mesoderm and the intermediate mesoderm, this cell mass is sometimes referred to as a nephrotome. However, this may cause some confusion with nephrostome, a ciliated funnel that pushes fluids into the tubules, and it is probably useful to simply refer to such cell masses as the pronephric primordia, or anlagen. Following the adoption of a columnar morphology (*not* mesenchymal), pronephric cells reorganize to form a compact structure two cells wide just ventral to the somites (Fig. 3.19) prior to overt epithelialization. The transition from Fig. 3.19F to Fig. 3.19I presumably only requires further polarization of the cells, not any folding or mesenchymal migration. The process of early pronephric morphogenesis has been described in some reports as a "folding" (German: vornierenfalte = pronephric fold) of the intermediate mesoderm as though the somatic layer first rolled

---

[1]The term intermediate mesoderm is useful in distinguishing the most dorsal region of the lateral plate, even in organisms in which this region is not morphologically distinct, from the more ventral region that has a very different development fate. The intermediate mesoderm, lying just ventral to the somites, will form the urogenital system and much of the vasculature.

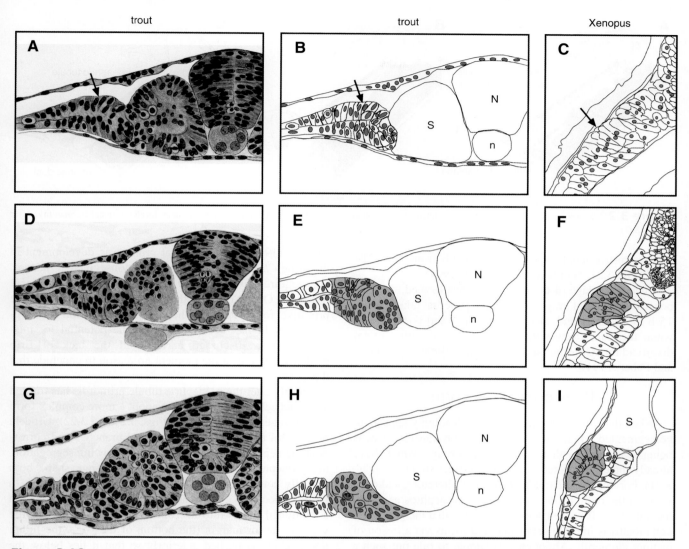

Figure 3.19 Early stages of pronephric morphogenesis in fishes and frogs. (A, D, and G) Progression of pronephric morphogenesis in the trout (Felix, 1897), and the frog, *Xenopus*. N, neural tube; n, notochord; S, somite. The pronephric primordia is shaded gray.

outward and then pinched off in a manner akin to past simple (and incorrect) models of neural tube formation. If one were to view Fig. 3.19I alone, one could image how this could be concluded, but when viewing the complete series, it is clearly incorrect. The entire process of pronephric segregation needs to be analyzed in a rigorous manner in a range of organisms before firm conclusions can be drawn as to its cellular or mechanical basis. It had been assumed for years that neural tube folding was a simple process, yet recent imaging advances have shown that it is actually a complex procedure involving medial migration, directed protrusive activity, cell intercalation, and convergent extension (Davidson and Keller, 1999). Pronephric segregation may be just as complex.

The process of primordia segregation begins with different portions of the pronephros in different organisms.

Segregation in teleost fish appears to begin with a more rostral portion of the pronephros than it does in amphibians. In the early stages depicted in Fig. 3.20, the pronephric duct (and possibly the yet to be defined teleost distal tubule) can be seen to be clearly segregated from the intermediate mesoderm, whereas the more rostral tubule precursors remain poorly defined condensations. The duct also segregates prior to the tubules in the chick (Jarzem and Meier, 1987), one of the many lines of evidence demonstrating that the claim that the duct forms by a "fusion" of the tubules is incorrect.[2] In amphibians, the tubule primordia forms first, with the duct precursors appearing a few hours later in an

---

[2]Other lines of evidence against this conclusion also exist (see Vize *et al.* 1995).

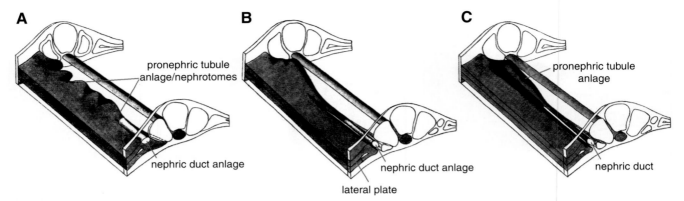

Figure 3.20 Morphogenesis of the pronephros in teleost fish. Three-dimensional reconstructions of trout (German: forelle) pronephric anatomy. After Felix (1897).

ordered rostrocaudal sequence that extends from stage 21+ to 27 (Fig. 3.21; Nieuwkoop and Faber, 1994).

Once the anlage of the pronephros has formed, the duct begins its caudal migration. This process takes about 15 h in *X. laevis* and 24 h in the axolotl and is completed when the duct fuses with the rostrally migrating rectal diverticulum (Gillespie and Armstrong, 1985; Nieuwkoop and Faber, 1994).

B. Acquisition of Form

When the anlage first appears, it misleadingly appears to be segmented when viewed laterally with clusters of cells being associated with the overlying somites, which have already undergone rotation and are distinct structures (Fig. 3.21). For this reason, much of the older literature, and even some of the more recent, refers to the pronephros as a segmental organ. It is even sometimes stated that the glomus in amphibians is initially segmental and forms via a fusion of distinct glomeruli. Although it is simple to rule out such a developmental mode for glomera by observing gene expression patterns in the splanchnic intermediate mesoderm throughout the process of glomeral development (e.g., Carroll and Vize, 1996), it is less easy to eliminate in the process of tubule development. As the somites are already segmented, their association with the pronephric anlage could well give the impression of segmentation that does not in fact exist. In sections through this early precursor structure in axolotls, it is evident that the majority of the structure appears to lack segmentation (Gillespie and Armstrong, 1985), and it is only in the regions of contact with the somites that any such organization appears to occur. Once the pronephric anlage has segregated, it shows no indication of any segmental organization (Fig. 3.21), but as development proceeds and nephrostomes develop under the somites, once again the appearance of a segmental-like organization is established. In the systems used most commonly to study pronephric development—frogs, salamanders, and fish—there is no strong evidence of a segmental mode of development.

Burlend (1913) also noted that pronephric development in the turtle *Chrysemys* is initiated in a nonsegmental manner, but acquires a metameric appearance as development proceeds. Of course, in other organisms, especially more primitive forms such as the hagfish, segmental development may in fact occur, but given the potential for misinterpretation discussed earlier and the lack of firm evidence, there is at the moment no reason to conclude that this is so.

Shortly after the pronephric tubule primordia has formed a distinct cell mass, it begins to adopt a more complex form that presages its final pattern (Figs. 3.21 and 3.22). Urodele pronephroi typically have two dorsal branches and corresponding nephrostomes (Section IV). Following segregation, the pronephric primordium begins to change shape both dorsally and ventroposteriorly. On the dorsal edge, a molding becomes evident as the anlage changes shape from a flattened oval into a rounded "Y" shape. On the ventroposterior margin (the bottom of the "Y"), the future distal tubule is pushed anteriorly so that it lies below the most anterior of the forming dorsal branches. This anterior loop of the pronephric system is present in amphibians (Mangold, 1923; Fig. 3.15), teleosts (Felix, 1897; Drummond *et al.,* 1998), ray-finned (Kerr, 1919), and ganoid fish (Fraser, 1928). It is possible that the posterior extension of the pronephric duct exerts an anterior force that buckles the anlage and pushes it anteriorly.

A detailed representation of the process of pronephric morphogenesis in anurans is illustrated in Fig. 3.23. Anuran pronephroi possess three dorsal branches, but the process is very similar in organisms with different branch numbers (Fig. 3.22). The reports on pronephroi with greater numbers of pronephric branches, e.g., ganoid fish, primitive salamanders, caecilians, and reptiles, are not yet complete enough to discuss further. The variation in pronephric anatomy among these species probably arises more by varying the number of common events than the form of events. In animals with three dorsal branches, three dorsal condensations appear.

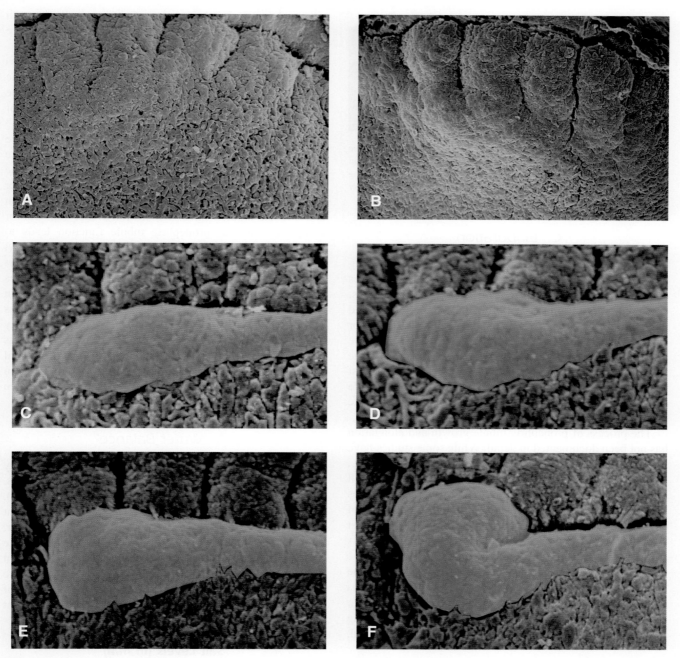

Figure 3.21 Pronephric morphogenesis in *Ambystoma*. Scanning electron micrographs of axolotl pronephroi. The pronephric primordium is shaded purple following its segregation. After Gillespie and Armstrong (1985). Images provided by J. Armstrong, University of Ottawa.

The molecular processes underlying dorsal branch formation are not known. The only two genes identified to date that are candidates in this process are *BMP-7* (Wang *et al.,* 1997) and a BMP-antagonist, *gremlin* (Hsu *et al.,* 1998). *BMP-7* is expressed in each of the three dorsal knobs of the *Xenopus* pronephric anlage at the time at which the dorsal branches begin to form. *Gremlin* is expressed in the presumptive distal tubule and duct region, presumably

limiting the range of action of the dorsally expressed BMP-7. Targeted disruption of the BMP-7 gene in mice results in abnormal kidney development, but kidneys do form and early nephric genes are activated normally (Dudley *et al.,* 1995). This indicates that this factor acts in the later stages of nephron formation rather than in their early specification, which is consistent with the expression observed in pronephroi.

Xenopus laevis   Ambystoma mexicanum

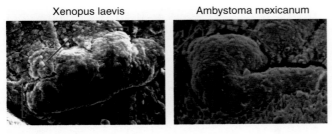

Figure 3.22 Comparison of dorsal branching in anurans (three branches) and urodeles (two branches). (Left) Scanning electron micrograph of *Xenopus* pronephros (Vize *et al.*, 1995). (Right) Scanning electron micro-graph of axolotl pronephros (16–17 somite stage, courtesy of J. Armstrong, University of Ottawa).

C. Elaboration

Morphogenesis of the pronephric tubules continues long after the onset of function. Following the processes described earlier (formation of the dorsal branches, epithelial polarization, formation of the tubule lumen, and the anterior looping of the distal segment), the major changes yet to occur include growth and extension of the proximal and distal segments and vascularization of the pronephric sinus. Nephrostomes also must form, being first observed at stage 28 in *Xenopus*, some 9.5 h after the tubule anlage first begins to segregate from the intermediate mesoderm (Nieuwkoop and Faber, 1994).

The process of pronephric tubule growth and elongation has not been documented in detail. While some studies have investigated proliferation rates in growing pronephroi (e.g., Chopra and Simnett, 1970), no specific zones of cell division have been observed. In pronephroi, genes implicated in pronephric specification have patterns of expression that later become restricted to the anterior-most region of the pronephric tubules. Their expression is strongest in a domain that could be either the nephrostome or in the dorsal branches that link the nephrostomes to the main proximal tubule. Examples of genes expressed in this manner include *lim-1, Pax-2,* and *wnt-4* (Carroll *et al.*, 1999). This may indicate that new tubule epithelium is being generated at these sites, but this remains to be demonstrated.

VI. Pronephric Function and Physiology

Frog mesonephroi were used in the first experimental demonstration of the filtration/resorption process, which is the basis of vertebrate kidney function (reviewed by Richards, 1929). In these experiments, microelectrodes were inserted into frog mesonephric tubules immediately adjacent to the glomerulus. Microanalysis of the glomerular filtrate demonstrated that it was essentially identical to blood, other than the absence of proteins and fats in the filtrate, and very different in composition to urine (Richards, 1929). Kidneys,

metanephric, mesonephric, and pronephric, actually do much more than remove nitrogenous wastes. Such functions are explored in this section.

The metanephric kidney has five main functions: (i) elimination of wastes, (ii) regulation of blood fluid volume and solute levels, (iii) control of blood pH, (iv) produce appropriate endocrine hormones, and (v) modification of some metabolites.

The process of blood filtration followed by selective resorption and excretion achieves the first three of these functions. Dye clearance studies have shown that the pronephric glomus also functions as a molecular filter (Gérard and Cordier, 1934; Jaffe, 1954), whereas extirpation and mutant studies have demonstrated that removal or compromization of pronephric tubule function leads to severe edema and death (Howland, 1921). Obviously, pronephroi are essential for normal water balance. As discussed later, they are also efficient at recovering solutes from the glomeral filtrate. The gills of fish and amphibian larvae play a major role in the excretion of nitrogenous wastes (Smith, 1929), and whereas the pronephric kidney may play a role in this process, it is not as critical as its osmoregulatory activity (Ultsch *et al.*, 1999). Circumstantial evidence shows that pronephroi contribute to the regulation of blood pH. At the moment there is no direct evidence indicating that pronephroi produce endocrine hormones.

A. Water Balance

The blood of frog and freshwater fish is strongly hyperosmotic relative to their surrounding environments, whereas the opposite is true of marine fish (Table 3.2). Given the ion gradient that exists between the environment and their body fluids, the permeability of their skin to water, and the considerable volumes of water ingested while feeding, tadpole and fish larvae are faced with a constant influx of water and must excrete large volumes of dilute urine to retain osmotic balance. Given the large volumes of urine that they produce, they must also be efficient at

Table 3.2   External and Internal Salt Levels (in m*M*)

	habitat [Na]	Na	K	Urea
Dogfish (adult)	470	287	5.4	354
Salmon (adult)	470	212	3.2	
Salmon (adult)	Freshwater	181	1.9	
Lamprey (adult)	470	120	3	
Frog (*Rana catesbeiana*)	Freshwater	90	4	
Frog (*R. cancrivora*)	0	153	5	
Frog (*R. cancrivora*)	470	265	6	

From Gordon and Tucker (1965) and Bentley (1971).

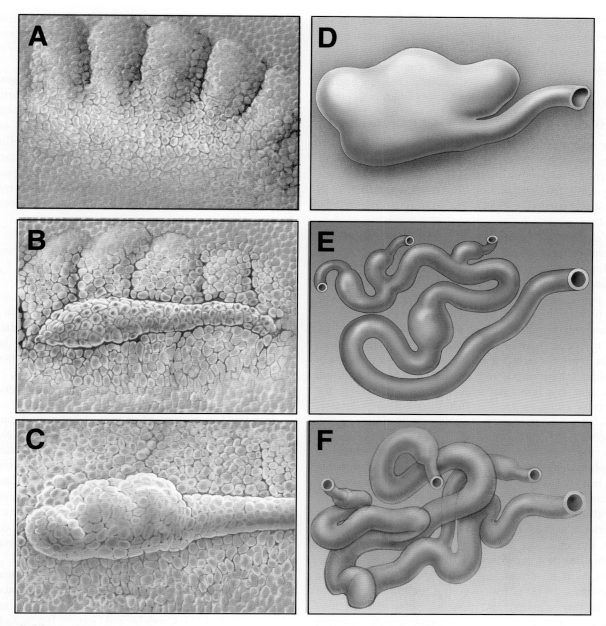

Figure 3.23 Morphogenesis of the amphibian pronephros. (A–C) Redrawn from scanning electron micrographs [after Vize *et al.* (1995) and originals of J. Armstrong, University of Ottawa]. (D and F) Histological reconstructions (Field, 1891). (E) From immunohistochemistry (Vize *et al.,* 1995).

recovering solutes from the glomeral filtrate or they would become depleted of key solutes very rapidly.

Amphibian tadpoles are good osmoregulators, and many species can survive in environments ranging from pure rain water to brackish puddles. Given the variable circumstances in which tadpoles find themselves growing, this adaptability is probably extremely useful. The tadpoles of one species, the crab-eating frog *Rana cancrivora*, are found in mangrove swamps and seawater, yet can be adapted to freshwater. Remarkably, the levels of sodium in the blood of *R. cancrivora* tadpoles changes only 2-fold when their environment is altered from fresh to salt water, over a 10-fold increase in osmotic pressure (Gordon and Tucker, 1965). The concentration of sodium in the urine, however, changes dramatically with such a change in environment. When in freshwater, the urine of *R.cancrivora* tadpoles has a much lower concentration of sodium than the blood, but when kept in salt water, the very small volumes of urine produced are essentially the same salinity as the blood (Gordon and Tucker, 1965). Frogs, like fish, cannot generate urine that is hypertonic, but can reduce water loss by generating an isotonic urine.

The pronephros is not the only organ responsible for controlling water balance, but it does play an essential role in this process. If the pronephroi are removed surgically (Howland, 1921) or their function is compromised by a genetic lesion (Drummond, 2000), animals become edemic rapidly. Both pronephroi must be removed to have this effect.

1. Hormonal Modulation of Water Excretion Rates

The regulation of water balance in amphibians, fish, and reptiles is largely mediated by controlling glomeral or glomerular filtration, as the rates of water resorption in the kidney are relatively constant (Bentley, 1971). This is in contrast to mammals, where water balance is regulated by controlling resorption rates.

A number of hormones can effect the glomeral/glomerular filtration rate in both pro- and mesonephroi, including vasotocin and oxytocin. Vasotocin is a peptide hormone stored in the neurohypophysis and has been found to be present at high levels in this organ in amphibians, birds, fish, and reptiles. Elevated levels of vasotocin reduce the glomeral filtration rate in pronephroi, which in turn leads to reduced urine flow and reduced sodium loss. Vasotocin effects on the pronephros have been reported in amphibians (Jorgensen *et al.,* 1946; Alvarado and Johnson, 1966), and this hormone is also known to regulate the glomerular filtration rate in fish mesonephroi (Bentley, 1971), but we are not aware of studies on its effects on fish pronephroi.

As mentioned earlier, water resorption rates in pronephroi are relatively constant, so this process probably plays less of a role in adapting to changes in environment than controlling the glomeral filtration rate. In terms of general homeostasis, however, the process of water resorption may be quite important. The distal tubule is the major site of water resorption in frog pro- and meso-nephroi (Bensley and Steen, 1928; Gérard and Cordier, 1934a), as it is in mammalian nephrons. This has been demonstrated in frogs by observing the concentration of dyes in this segment and also by demonstrating that uranium oxide-mediated tissue damage mostly occurred in this segment, presumably due to its concentration (Gérard and Cordier, 1934b).

B. Resorption

Given the large volumes of water excreted by the pronephroi, plus the hypoosmotic medium in which they live, amphibian and fish larvae would rapidly become ion depleted if efficient recovery systems were not in place in the pronephric tubules. The distribution of resorptive activity within the pronephric tubules has not been mapped as accurately as it has been in meso- and metanephric kidneys, but some studies have been performed. The most detailed are those of Gérard and Cordier [(1934a,b, also see

Ultsch *et al.* (1999)]. Gérard and Cordier (1934a,b) observed that these tubules can resorb injected dyes, such as carmine, within the proximal region and resorb water in the distal region (Figure 3.24).

In addition to the resorption of solutes, the proximal tubules of pronephroi can also resorb large particles. This activity increases along the proximal segment, with the highest activity being in the most posterior part of the proximal tubule. Cells in this region often contain yellowish inclusions that represent resorbed blood cholesterol. These cells also resorb colloids injected into the coelom, which are swept into the pronephric tubules by the nephrostomes (Gérard and Cordier, 1934b). This process is known as athrocytosis.

C. Elimination/Retention of Nitrogenous Waste

Pronephric kidneys play a far less important role in nitrogenous waste disposal in freshwater larvae than adult kidneys. The gills are the major site of ammonia and urea excretion and secretion in fish and in amphibian larvae, although in some instances the skin also plays a major role (Smith, 1929; Fanelli and Golstein, 1964; Balinsky, 1970). However, the pronephroi do participate in the excretion of nitrogenous wastes, so this area is reviewed briefly.

In general, animals that live in freshwater environments, such as amphibian larvae and freshwater fish, excrete nitrogen as ammonia, whereas terrestrial animals use urea for this purpose. It is thought that ammonia excretion may be a disadvantage under terrestrial conditions due to its high pH. In aquatic organisms, this is not a problem; as outlined earlier, the constant influx of water results in the excretion of copious amounts of dilute urine. This is not the case in terrestrial animals, so biochemical adaptations converting ammonia to urea are necessary.

Amphibians usually utilize both environments—water as larvae and land as adults. Their nitrogen metabolism changes during metamorphosis to adapt to this alteration (Munro, 1939, 1953; Candelas and Gomez, 1963). Tadpoles (with pronephric kidneys) excrete nitrogen primarily as ammonia (80–90%) but also as urea (10–20%). As the changeover from pro- to mesonephroi occurs during metamorphosis, this shifts so that most nitrogenous waste is now found as urea (85–90%). Note that this alteration does not depend on the kidneys—it is largely mediated by changes in liver metabolism—but the net result is that each form of kidney is confronted with a different form of nitrogenous waste (Balinsky, 1970).

It is interesting to note that in a species such as *X. laevis*, which is permanently aquatic, no such transition in nitrogen metabolism normally occurs, and the primary (80–87%) waste product of both larvae and adults is ammonia (Munro, 1953). Freshwater fish also simply continue to use ammonia

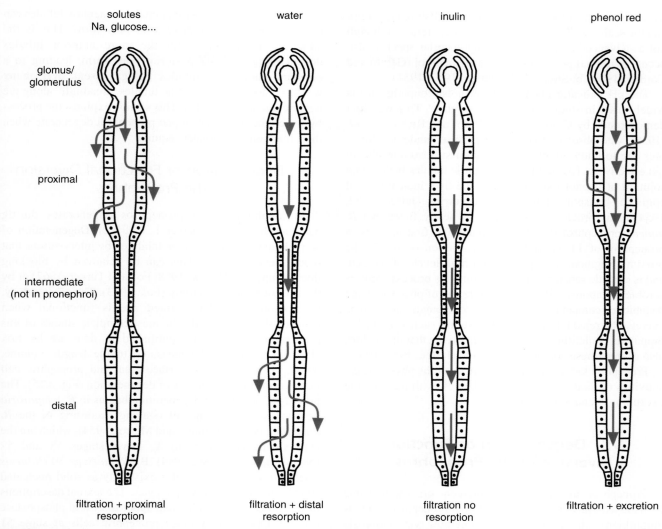

Figure 3.24 Variations on filtration–resorption in the kidney. Pathways of resorption or elimination of various molecules in metanephric nephrons are illustrated. Pronephroi utilize the same systems, but do not possess an intermediate segment. Inulin is a starch-like polysaccharide (MW 5200) that is not resorbed by tubules and is used as a standard to measure glomerular filtration in many physiological studies. After Smith (1943).

at all stages of development under normal circumstances. However, animals that encounter arid conditions periodically can temporarily switch over to urea production. Examples include the aquatic frog *X. laevis* (Balinsky, 1970), which under drought conditions can find itself trapped in dried-up pools. Under such circumstances, nitrogen metabolism adjusts to the situation by producing urea rather than ammonia and allowing the animal to minimize the production of urine and corresponding water loss. In aevistating lungfish, a similar switch in nitrogenous wastes occurs, but this may be regulated by shutting down ammonia production rather than increasing urea synthesis (Smith, 1930; Goldstein and Forster, 1970).

Under some circumstances, animals both generate and retain urea in order to increase osmotic pressure and reduce water loss. This situation is common in marine fishes such

as the elasmobranchs, where urea levels are so high that water flows inward through the gills even in a marine environment (Smith, 1953). Adults of *R.cancrivora* also use urea to raise their osmotic pressure and help prevent water loss in saline environments.

D. Excretion

In addition to the standard filtration/resorption process of waste disposal, adult kidney tubules also actively excrete some waste compounds into the proximal tubule (Fig. 3.24). In some animals, this mode of waste disposal has become so advanced that filtration/resorption and its associate cost in water loss have been disposed of entirely, as has the presence of the glomeruli (Marshall and Smith, 1930; Grafflin, 1937). Such aglomerular kidneys are found in salt water fish such as the midshipman, pufferfish (e.g., *Fugu*), and the seahorse. The process of tubular excretion, although not as advanced as that found in aglomerular species, also occurs in normal pro-, meso-, and metanephroi (Gérard and Cordier, 1934a,b; Smith, 1943; Jaffee, 1952, 1954).

The pH indicator phenol red (phenolsulfonphthalein) is widely used in studies of tubular excretion. This was first demonstrated by Goldring *et al.* (1936) and Richards *et al.* (1938), who observed that tubules perfused under too low a pressure to allow glomerular filtration still excrete phenol red effectively. Excised frog pronephric tubules bathed in a solution of phenol red accumulate the dye, indicating that epithelial transport of the dye is occurring (Jaffee, 1952, 1954). This process is not active in newly functional *R. pipiens* pronephroi (6-day-old larvae) and first appears in pronephroi of 11-day-old larvae. Excretion occurs in the proximal segment, is dependent on high levels of oxygen, and is cyanide sensitive, indicating that this process requires aerobic metabolism. Pronephric excretion of phenol red can be inhibited competitively with other compounds known to be excreted by renal tubules in other systems, such as *p*-amino-hippurate, penicillin, and uric acid, implying that pronephroi also excrete these compounds actively (Jaffee, 1952, 1954).

Pronephric kidneys therefore share all of the physiological activities of their more complex relatives, with the possible exception of endocrine function.

VII. Degeneration or Function Diversion of the Pronephros

Pronephroi are in general evanescent organs. In some animals with physiologically functional pronephroi, such as amphibians, the pronephros degenerates via apoptosis following the acquisition of function by the mesonephros. In other groups, such as fish, the hematopoietic function of the pronephros is retained and only the nephric components degenerate, leaving a lymphoid organ in place of the earlier excretory organ. In higher mammals, pronephric tubules can begin to degenerate shortly after they form. This is the situation in humans, with the more anterior tubules degenerating as more posterior ones form, leading to a constant number of tubules being present over many weeks of development, as the tubules "treadmill" along the nephric duct (Felix, 1912). This section explores the process by which the better developed pronephroi degenerate when their function is no longer required.

A. Degeneration or Functional Diversion of the Pronephros

In amphibians, the pronephros degenerates during metamorphosis (see Chapter 1, Fig. 1.2). Degeneration of amphibian pronephroi can be inhibited by interventions that block metamorphosis. This can be achieved by blocking thyroid function (Hurley, 1958; Fox and Turner, 1967) or by thyroid- or hypophysectomy (Fox, 1963). The adult kidney, the mesonephros, has formed and is functional when degeneration is initiated. The most complete studies of this pronephric degeneration published to date are by Fox (1962a,b,c, 1963), who measured tubule length, volume, lumen volume, internal surface area, and pronephric cell numbers during the window of degeneration (Fig. 3.25). The process of pronephric degeneration begins in *R. temporaria* between stages 47 and 49 (staging according to the *R. dalmatina* table of Cambar and Marrot, 1954), which are the approximate equivalent to *X. laevis* stages 55 and 57 (Nieuwkoop and Faber, 1994). By *Rana* stage 50 (*X. laevis* stage 59), pronephric tubules exist only as solid nucleated strands surrounded by lymphocytes. The textual descriptions of this process depict classical apoptosis. Acid phosphatase activity appears in *X. laevis* pronephric cells at stage 51 (Goldin and Fabian, 1971), prior to the histological detection of degeneration, and can be used as a marker.

The anterior portion of the nephric duct present between the pronephros and the mesonephros degenerates simul-

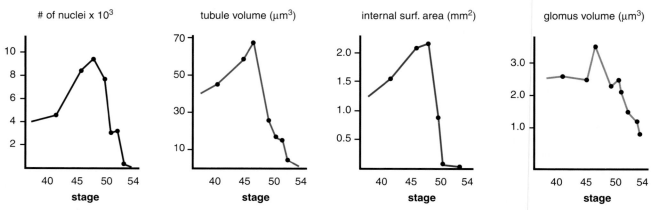

Figure 3.25 Quantitative changes in pronephric size and cell number during pronephric degeneration. After Fox (1962c).

taneously with the tubules. The pronephros and anterior duct have disappeared by *Rana* stage 54 (equivalent to *X.laevis* stage 66). Tahara *et al.* (1993) observed the first signs of degeneration in the tubules and duct of *X. laevis* at stage 56, and tubules and anterior duct had disappeared completely by stage 64.

The pronephric glomus begins to shrink at *Rana* stage 47 (Fig. 3.25, equivalent to *Xenopus* stage 55), but a vestigial glomus remains for a considerable period of time after degeneration of the pronephric tubules (Fox, 1962). Tahara *et al.* (1993) reported that the *X. laevis* glomus begins to atrophy at stage 44, around the time at which mesonephric vascularization begins (stage 47). However, this may simply reflect a compaction, as illustrated in Fig. 1.2, rather than degeneration.

In fish, both teleosts and ganoids, the pronephros does not degenerate, even in the adult. Rather, it becomes a lymphoid organ (Balfour, 1882) and plays a key role in hematopoiesis. Genes involved in immunoglobulin and T-cell receptor rearrangement can be used as markers to detect late teleost pronephroi (Kaatari and Irwin, 1985; Hansen and Kaattari, 1995, 1997). The adult pronephros does not contain any secretory tubules and does not play any role in the elimination of nitrogenous wastes in most fishes (Balfour, 1882). The transition point from an excretory to a lymphoid organ has not yet been characterized in the zebrafish, but is known to occur quite late, some time beyond 6 weeks of development (Drummond, 2000).

Amphibian pronephroi also function as a hematopoietic organ and are a major site of myeloid cell differentiation (Carpenter and Turpen, 1979). The hematopoietic tissue is distinct from the vascular sinus surrounding the tubules, being separated from it by an endothelial barrier. As with the excretion of nitrogenous wastes, this hematopoietic function is taken over by the mesonephros during metamorphosis.

In the chicken and in other organisms with evanescent pronephroi, degeneration of the tubules often begins shortly after they form. As with amphibians, chicken external glomera survive longer than tubules, but they also eventually disappear (Witschi, 1956).

VIII. Pronephric Tubules as a General Model of Tubulogenesis?

This chapter has demonstrated that there are many similarities among pronephric, mesonephric, and adult kidneys. As discussed in Chapters 1 and 8, almost every gene demonstrated to play an important role in mammalian kidney development by mutant or gene ablation studies has also been found to be expressed in the embryonic kidneys. Furthermore, in as far as the different anatomies of the kidneys allow, most of these genes are expressed in an analogous spatial pattern. Where differences in timing or distribution of a gene in different kidneys have been observed, they are extremely informative and may in fact lead to insights into gene function (Sainio *et al.*, 1997). It seems likely that the majority of genetic interactions governing nephrogenesis hold true for both complex and simple kidneys. However, a wide range of powerful embryological, genetic, and epigenetic techniques are available for investigating development of the embryonic kidneys that are unavailable (or vastly more time consuming) to studies of adult kidney development. Genetic screens have identified mutants in both tubule and vascular components of pronephric kidneys (Drummond *et al.*, 1998; Drummond, 2000) and provide wonderful starting points for investigating the very first steps in kidney specification. As screening techniques improve and can be targeted toward specific organs, e.g., by screening GFP-tagged transgenic lines (Chapter 9), the pronephric system may soon provide the foundation to uncover the genetic hierarchy regulating kidney development.

The simple kidneys also show promise as model systems for physiological investigation. Rather than the internal location of meso- and metanephroi, the pronephroi are present directly below the epidermis of amphibian larvae. The epidermis is transparent in fish and later amphibian larvae, and the kidney can be observed simply by looking at the embryos. It would be trivial to operate on these organs, insert electrodes, or perform a myriad of other forms of analysis. An additional advantage in doing such experiments on a pronephros is that the organ could first be manipulated genetically or embryologically via interventions in earlier developmental stages and the consequences on organogenesis and physiology determined only 2 days later. One could, for example, introduce a plasmid expressing a novel channel or express a channel in a novel segment and then investigate the effects on physiological function.

Studies since the early 1990s have generated the markers, tools, and understanding of the basic biology of simple kidney development. Genomics projects are providing whole genome equences, ESTs, and other resources at an extraordinay rate. The next 5 to 10 years should be interesting indeed.

References

Abdel-Malek, E. T. (1950). Early develoment of the urogenital system in the chick. *J. Morphol.* **86**, 599–626.

Alvarado, R. H., and Johnson, S. R. (1966). The effects of neurohypophysial hormones on water and sodium balance in larval and adult bullfrogs (Rana catesbeiana). Comp. Biochem. Physiol. **18**, 549–561.

Balfour, F. M. (1875). On the origin and history of the urogenital organs of vertebrates. *J. Anat. Physiol.* **10**. Reprinted in "Works of Balfour" (M. Foster and A. Sedgwick, eds.), pp. 135–167. Macmillan, London, 1885.

Balfour, F. M. (1881). "A Treatise on Comparative Embryology." Macmillan, London.

Balfour, F. M. (1882). On the nature of the organ in adult teleosteans and ganoids, which is usually regarded as the head-kidney or pronephros.

Quart. J. Micr. Sci. **22**. Reprinted in "Works of Balfour" (M. Foster and A. Sedgwick, eds.), pp. 848–853. Macmillan, London, 1885.

Balfour, F. M., and Sedgwick, A. (1879). On the existence of a head kidney in the embryo chick, and on certain points in the development of the Mullerian duct. *Quart. J. Micr. Sci.* **19**. Reprinted in "Works of Balfour" (M. Foster and A. Sedgwick, eds.), pp. 618–641. Macmillan, London, 1885.

Balinsky, J. B. (1970a). Nitrogen metabolism in amphibians. *In* "Comparative Biochemistry of Nitrogen Metablism" (J. W. Campbell, ed.), Vol. 2, pp. 519–637. Academic Press, London.

Balinsky, B. I. (1970b). "An Introduction to Embryology." Saunders, Philadelphia.

Balinsky, J. B., Choritz, E. L., Coe, C. G. L., and van der Schans, G. S. (1967). Urea cycle enzymes and urea excretion during the development and metamorphosis of *Xenopus laevis. Comp. Biochem. Physiol.* **22**, 53–57.

Bates, G. A. (1914). The pronephric duct in elasmobranchs. *J. Morphol.* **25**, 345–373.

Bellefroid, E. J., Kobbe, A., Gruss, P., Pieler, T., Gurdon, J. B., and Papalopulu, N. (1998). Xiro3 encodes a *Xenopus* homolog of the *Drosophila* Iroquois genes and functions in neural specification. *EMBO J.* **17**, 191–203.

Bensley, R. R., and Steen, W. B. (1928). The functions of the differentiated segments of the uriniferous tubule. *Am. J. Anat.* **41**, 75–96.

Bentley, P. J. (1971). "Endocrines and Osmoregulation: A Comparative Account of the Regulation of Water and Salt in Vertebrates." Springer-Verlag, New York.

Brauer, A. (1902). Beiträge zur kenntniss der entwicklung und anatomie der Gymnophionen. III. die entwicklung der excretionsorgane. *Zool. Jahrbüch.* **16**, 1–176.

Bremer, J. L. (1916). The interrelations of the mesonephros, kidney and placenta in different classes of animals. *Am. J. Anat.* **19**, 179–209.

Brennan, H. C., Nijjar, S., and Jones, E. A. (1998). The specification of the pronephric tubules and duct in *Xenopus laevis. Mech. Dev.* **75**, 127–137.

Bronchain, O. J., Hartley, K. O., and Amaya, E. (1999). A gene trap approach in *Xenopus. Curr. Biol.* **21**, 1195–1198.

Burlend, T. H. (1913). The pronephros of Chrysemys marginata. *Zool. Jahrbüch.* **26**, 1–90.

Cambar, R., and Marrot, B. (1954). Table chronologique du développement de la grenouille agile (*Rana dalmatina* Bon.). *Bull. Biol. France-Belgique* **88**, 168–177.

Carroll, T. J., and Vize, P. D. (1996). Wilms' tumor suppressor gene is involved in the development of disparate kidney forms: Evidence from expression in the *Xenopus* pronephros. *Dev. Dyn.* **206**, 131–138.

Carroll, T. J., and Vize, P. D. (1999). Synergism between Pax-8 and lim-1 in embryonic kidney development. *Dev. Biol.* **214**, 46–59.

Carroll, T. J., Wallingford, J. B., and Vize, P. D. (1999). Dynamic patterns of gene expression in the developing pronephros of *Xenopus laevis. Dev. Genet.* **24**, 199–207.

Chopra, D. P., and Simnett, J. D. (1970). Stimulation of cell division in pronephros of embryonic grafts following, partial nephrectomy in the host *(Xenopus laevis). J. Embryol. Exp. Morphol.* **24**, 525–533.

Collazo, A., Bronner-Fraser, M., and Fraser, S. E. (1993). Vital dye labelling of *Xenopus laevis* trunk neural crest reveals multipotency and novel pathways of migration. *Development* **118**, 363–376.

Davidson, L. A., and Keller, R. E. (1999). Neural tube closure in *Xenopus laevis* involves medial migration, directed protrusive activity, cell intercalation and convergent extension. *Development* **126**, 4547–4556.

Davies, J. (1951). Nephric development in the sheep with reference to the problem of the ruminant pronephros. *J. Anat.* **85**, 6–11.

Dodds, G. S. (1946). "The Essentials of Human Embryology." Wiley, New York.

Drummond, I. A. (2000). The zebrafish pronephros: A genetic system for studies of kidney development. *Pediatr. Nephrol.* **14**, 428–435.

Drummond, I. A., Majumdar, A., Hentschel, H., Elger, M., Solnica-Krezel,
L., Schier, A. F., Neuhauss, S. C., Stemple, D. L., Zwartkruis, F., Rangini, Z., Driever, W., and Fishman, M. C. (1998). Early development of the zebrafish pronephros and analysis of mutations affecting pronephric function. *Development* **125**, 4655–4667.

Dudley, A. T., Lyons, K. M., and Robertson, E. J. (1995). A requirement for bone morphogenetic protein-7 during development of the mammalian kidney and eye. *Genes Dev.* **9**, 2795–2807.

Fales, D. E. (1935). Experiments on the development of the pronephros of *Amblystoma punctatum. J. Exp. Zool.* **72**, 147–173.

Felix, W. (1897). Beiträge zur entwickelungsgeschichte ser salmoniden. *Anat. Hefte* **8**, 249–467.

Felix, W. (1912). The development of the urogenital organs. *In* "Manual of Human Embryology" (F. Kiebel and F. P. Mall, eds.), pp. 752–979. Lippincott, Philadelphia.

Field, H. H. (1891). The development of the pronephros and segmental duct in amphibia. *Bull. Museum Comp. Zool. Harvard* **21**, 201–340.

Field, H. H. (1894). Sur le développement des organs excréteurs chez l' *Amphiuma. C. R. Acad. Sci. Paris* **118**, 1221–1224.

Foster, M., and Sedgwick, A. (1885). "The Works of Francis Maitland Balfour." Macmillan, London.

Fox, H. (1960). Early pronephric growth in *Neoceratodus larvae. Proc. Zool. Soc. Lond.* **134**, 659–663.

Fox, H. (1961). Quantitative analysis of normal growth of the pronephric system in larval amphibia. *Proc. Zool. Soc. Lond.* **136**, 301–315.

Fox, H. (1962a). Degeneration of the remaining pronephros of *Rana temporaria* after unilateral pronephrectomy. *J. Embryol. Exp. Morphol.* **10**, 224–230.

Fox, H. (1962b). A study of the evolution of the amphibian and dipnoan pronephros by an analysis of its relationship with the anterior spinal nerves. *Proc. Zool. Soc. Lond.* **138**, 225–256.

Fox, H. (1962c). Growth and degeneration of the pronephric system of *Rana temporaria. J. Embryol. Exp. Morphol.* **10**, 130–114.

Fox, H. (1963). The amphibian pronephros. *Quart. Rev. Biol.* **38**, 1–25.

Fox, H., and Turner, S. C. (1967). A study of the relationship between the thyroid and larval growth in *Rana temporaria* and *Xenopus laevis. Arch. Biol.* **78**, 61–90.

Fraser, E. A. (1920). The pronephros and early development of the mesonephros in the cat. *J. Anat.* **54**, 287–304.

Fraser, E. A. (1928). Observations on the development of the pronephros of the sturgeon, *Acipenser rubicundus. Quart. J. Micr. Sci.* **71**, 75–112.

Fraser, E. A. (1950). The development of the vertebrate excretory system. *Biol. Rev. Camb. Phil. Soc.* **25**, 159–187.

Fürbringer, M. (1878). Zur vergleichenden anatomie und entwickslungsgeschichte der excretionsorgane der vertebraten. *Morpholog. Jahrbuch.* **4**, 1–111.

Gendre, P. (1969). Observations sur l'infrastructure et le fonctionnement du revêtement ciliare néphrostomial chez le pronéphros de la grenouille agile (Amphibien Anoure). *C. R. Soc. Biol.* **162**, 1743–1746.

Gérard, P., and Cordier, R. (1934a). Esquisse d'une histopathologie comparée du rein des vertebrates. *Biol. Rev. Camb. Phil. Soc.* **9**, 110–131.

Gérard, P., and Cordier, R. (1934b). Recherches d'histophysiologie comparée sur le pro- et le mésonéphros larvaires des anoures. *Zeitsch. Zellforschung Mikroskopische Anat.* **21**, 1–23.

Gillespie, L. L., and Armstrong, J. B. (1985). Formation of the pronephros and pronephric duct rudiment in the Mexican axolotl. *J. Morphol.* **185**, 217–222.

Goldin, G., and Fabian, B. (1971). An histochemical investigation of acid phosphatase activity in the pronephros of the developing *Xenopus laevis* tadpole. *Acta Embryol. Exp. (Palermo)* **1971**, 31–39.

Goldring, W., Clarke, R. W., and Smith, H. W. (1936). Phenol red clearance in normal man. *J. Clin. Invest.* **15**, 221–230.

Goldstein, L., and Forster, R. P. (1970). Nitrogen metabolism in fishes. *In* "Comparative Biochemistry of Nitrogen Metablism" (J. W. Campbell, ed.), Vol. 2, pp. 495–518. Academic Press, London.

Goodrich, E. S. (1930). "Studies on the Structure and Development of Vertebrates." Macmillan, London.

Gordon, M. S., and Tucker, V. A. (1965). Osmotic regulation in the tadpoles of the crab-eating frog (Rana cancrivora). J. Exp. Biol. **42**, 437–445.

Grafflin, A. L. (1937). Observations upon the aglomerular nature of certain teleostean kidneys. J. Morphol. **61**, 165–173.

Gregory, E. (1900). Observations on the development of the excretory system in turtles. Zool. Jahrbüch. **13**, 683–714.

Harrison, R. G. (1921). On the relation of symmetry in transplanted limbs. J. Exp. Zool. **32**, 1–136.

Heller, N., and Brandli, A. W. (1997). Xenopus Pax-2 displays multiple splice forms during embryogenesis and pronephric kidney development. Mech. Dev. **69**, 83–104.

Hochstetter, F. (1898). Ein beitrag zur vergleichenden anatomie des venensystems der edentaten. Morpholog. Jahrbuch. **25**, 362–376.

Holtfreter, J. (1931). Potenzprüfungen am amphibienkeim mit hilfe der isolationsmethode. Verh. d. deutschen Zool. Ges. 158–166.

Holtfreter, J. (1933). Der einfluss von wirtsalter und verschiedenen organbezirken auf die differenzierung von angelagertem gastrulaektoderm. Roux' Arch. Entw. Mech. **127**, 619–775.

Holtfreter, J. (1936). Regionale induktionen in xenoplastich zusammengesetzten explantaten. Roux' Arch. Entw. Mech. **134**, 466–550.

Howland, R. B. (1921). Experiments of the effect of removal of the pronephros of Amblystoma punctatum. J. Exp. Zool. **32**, 355–395.

Hsu, D. R., Economides, A. N., Wang, X., Eimon, P. M., and Harland, R. M. (1998). The Xenopus dorsalizing factor Gremlin identifies a novel family of secreted proteins that antagonize BMP activities. Mol. Cell **1**, 673–683.

Hurley, M. B. (1958). The role of the thyroid in kidney development in the tadpole. Growth **22**, 125–166.

Jaffee, O. C. (1952). "The Morphogenesis of the Pronephros of the Leopard Frog (Rana pipiens) and Studies on Pronephric Function. Ph.D. thesis, Department of Zoology, Indianna University.

Jaffee, O. C. (1954). Phenol red transport in the pronephros and mesonephros of the developing frog (Rana pipiens). J. Cell. Comp. Physiol. **44**, 347–363.

Janosik, J. (1904). Über die entwickelung der vorniere und des vornierenganges bei Säugern. Arch. Mikr. Anat. **64**. Referenced by Fraser (1920).

Jarzem, J., and Meier, S. P. (1987). A scanning electron microscope survey of the origin of the primordial pronephric duct cells in the avian embryo. Anat. Rec. **218**, 175–181.

Jorgensen, C. B., Levi, H., and Ussing, H. H. (1946). On the influence of neurohypophysial extracts on sodium metabolism in the axolotl Ambystoma mexicanum. Acta Physiol. Scand. **12**, 350–371.

Karsenty, G., Luo, G., Hofmann, C., and Bradley, A. (1996). BMP 7 is required for nephrogenesis, eye development, and skeletal patterning. Ann N. Y. Acad. Sci. **785**, 98–107.

Keibel, F. (1903). Ueber entwickelung des urogenitalapparates von echidna. Verh. Anat. Ges. **17**, 14–22.

Keller, R. E. (1976). Vital dye mapping of the gastrula and neurula of Xenopus laevis. II. Prospective areas and morphogenetic movements of the deep layer. Dev. Biol. **51**, 118–137.

Kerens, B. (1907). Recherches sur les premiéres phases du développement de l'appareil excréteur des amniotes. Arch. Biol. **22**, 493–648.

Kerr, J. G. (1919). "Text-book of Embryology." MacMillan, London.

Kimelman, D., and Griffin, K. J. (2000). Vertebrate mesendoderm induction and patterning. Curr. Opin. Genet. Dev. **10**, 350–356.

Kimmel, C. B., and Law, R. D. (1985). Cell lineage of zebrafish blastomeres. II. Formation of the yolk syncytial layer. Dev. Biol. **108**, 86–93.

Krotoski, D. M., Fraser, S. E., and Bronner-Fraser, M. (1988). Mapping of neural crest pathways in Xenopus laevis using inter- and intra-specific cell markers. Dev. Biol. **127**, 119–132.

Langeland, J., and Kimmel, C. B. (1997). The embryology of fish. In "Embryology: Constructing the Organism" (S. F. Gilbert and A. M. Raunio, eds.), pp. 383–407. Sinauer, Sunderland, MA.

Leuzinger, S., Hirth, F., Gerlich, D., Acampora, D., Simeone, A., Gehring, W. J., Finkelstein, R., Furukubo-Tokunaga, K., and Reichert, H. (1998). Equivalence of the fly orthodenticle gene and the human OTX genes in embryonic brain development of Drosophila. Development **125**, 1703–1710.

Levin, M., and Mercola, M. (2000). Expression of connexin 30 in Xenopus embryos and its involvement in hatching gland function. Dev. Dyn. **219**, 96–101.

Lillie, F. R. (1919). "The Development of the Chick." Henry Holt, New York.

Majumdar, A., Lun, K., Brand, M., and Drummond, I. A. (2000). Zebrafish no isthmus reveals a role for pax2.1 in tubule differentiation and patterning events in the pronephric primordia. Development **127**, 2089–2098.

Mangold, O. (1923). Transplantationsversuche zur frag der spezifität und der bildung der keimblätter bei Triton. Arch. Mikr. Anat. Entw. Mech. **100**, 198–301.

Marshall, A. M. (1902). "The Frog: An Introduction to Anatomy, Histology, and Embryology." Macmillan, New York.

Marshall, E. K., and Smith, H. W. (1930). The glomerular development of the vertebrate kidney in relation to habitat. Biol. Bull. **9**, 135–153.

Maschkowzeff, A. (1926). Zur phylogenie des urogenitalsystems der wirbeltiere auf grund der entwicklung des mesoderms, des pronephros, der analöffnung und der abdominalporen bei Acipenser stellatus. Zool. Jahrbüch. **48**, 201–272.

McCright, J., Gao, X., Shen, L., Lozier, J., Lan, Y., Maguire, M., Herzlinger, D., Weinmaster, G., Jiang, R., and Gridley, T. (2001). Defects in development of the kidney, heart and eye vasculature in mice homozygous for a hypomorphic Notch2 mutation. Development **128**, 491–502.

McCurdy, H. M. (1931). Development of the sex organs in Triturus torosus. Am. J. Anat. **47**, 367–401.

McLaughlin, K. A., Rones, M. S., and Mercola, M. (2000). Notch regulates cell fate in the developing pronephros. Dev. Biol. **227**, 567–580.

Møbjerg, N., Larsen, E. H., and Jespersen, Å. (1998). Morphology of the nephron in the mesonephros of Bufo bufo (Amphibia, Anura, Bufonidae). Acta Zool. (Stockholm) **79**, 31–50.

Møbjerg, N., Larsen, E. H., and Jespersen, Å. (2000). Morphology of the kidney in larvae of Bufo viridis (Amphibia, Anura, Bufonidae). J. Morphol. **245**, 177–195.

Munro, A. F. (1953). The ammonia and urea excretion of different species of amphibia during their development and metamorphosis. Biochem. J. **54**, 29–36.

Nagao, T., Leuzinger, S., Acampora, D., Simeone, A., Finkelstein, R., Reichert, H., and Furukubo-Tokunaga, K. (1998). Developmental rescue of Drosophila cephalic defects by the human Otx genes. Proc. Natl. Acad. Sci. USA **95**, 3737–3742.

Nieuwkoop, P. D., and Faber, J. (1994). "Normal Table of Xenopus laevis (Daudin)." Garland, New York.

Obara-Ishihara, T., Kuhlman, J., Niswander, L., and Herzlinger, D. (1999). The surface ectoderm is essential for nephric duct formation in intermediate mesoderm. Development **126**, 1103–1108.

Oppenheimer, J. M. (1936). Transplantation experiments in developing teleosts. J. Exp. Zool. **72.**

Pasteels, J. (1942). New observations concerning the maps of presumptive areas of the young amphibian gastrula (Amblystoma and Discoglossus). J. Exp. Zool. **89**, 255–281.

Poole, T. J. (1988). Cell rearrangement and directional migration in pronephric duct development. Scan. Microsc. **2**, 411–415.

Rappaport, R. (1955). The initiation of pronephric function in Rana pipiens. J. Exp. Zool. **128**, 481–487.

Richards, A. N. (1929). "Methods and Results of Direct Investigations of the Function of the Kidney." Williams and Wilkins C. Baltimore.

Richards, A. N., Bott, P. A., and Westfall, B. B. (1938). Experiments concerning the possibility that inulin is secreted by the renal tubules. Am. J. Physiol **123**, 281–295.

Rückert. (1888). Über die entstehung der exkretionsorgane bei Selachiern. *Arch. Anat. Entwick* Figure reproduced by Kerr (1919).

Sainio, K., Hellstedt, P., Kreidberg, J. A., Saxén, L., and Sariola, H. (1997). Differential regulation of two sets of mesonephric tubules by WT-1. *Development* **24**, 1293–1299.

Semon, R. (1890). Über die morphologische bedeutung der urniere in ihrem verhältnis zur vorniere und nebenniere und über ihre verbindung mit dem genitalsystem. *Anatomischer Anzeiger* **5**, 455–482.

Seufert, D. W., Brennan, H. C., DeGuire, J., Jones, E. A., and Vize, P. D. (1999). Developmental basis of pronephric defects in *Xenopus* body plan phenotypes. *Dev. Biol.* **215**, 233–242.

Shawlot, W., and Behringer, R. R. (1995). Requirement for Lim1 in head-organizer function. *Nature* **374**, 425–430.

Smith, H. W. (1929). The excretion of ammonia and urea by the gills of fish. *J. Biol. Chem.* **81**, 727–732.

Smith, H. W. (1930). Metabolism of the lung-fish. *J. Biol. Chem.* **88**, 97–130.

Smith, H. W. (1932). Water regulation and its evolution in the fishes. *Quart. Rev. Biol.* **7**, 1–25.

Smith, H. W. (1943). Newer methods of study of renal function in man. *In* "Lectures on the Kidney." University of Kansas, Lawrence.

Smith, H. W. (1953). "From Fish to Philosopher." Little Brown and Company, Boston.

Smith, A. M., Potter, M., and Merchant, E. B. (1967). Antibody-forming cells in the pronephros of the teleost *Lepomis macrochirus*. *J. Immunol.* **99**, 876–882.

Spemann, H., and Mangold, H. (1924). Über induktion von embryonalanlagen durch implantation artfremder organisatoren. *W. Roux' Arch. Entwicklungsmech. Organ.* **100**, 599–638.

Spemann, H. (1988). "Embryonic Development and Induction." Yale Univ. Press, New Haven.

Tahara, T., Ogawa, K., and Taniguchi, K. (1993). Ontogeny of the pronephros and mesonephros in the South African clawed frog, *Xenopus laevis* Daudin, with special reference to the appearance and movement of the renin-immunopositive cells. *Exp. Anim.* **42**, 601–610.

Taira, M., Otani, H., Jamrich, M., and Dawid, I. B. (1994). Expression of the LIM class homeobox gene Xlim-1 in pronephros and CNS cell lineages of *Xenopus* embryos is affected by retinoic acid and exogastrulation. *Development* **120**, 1525–1536.

Tsang, T. E., Shawlot, W., Kinder, S. J., Kobayashi, A., Kwan, K. M., Schughart, K., Kania, A., Jessell, T. M., Behringer, R. R., and Tam, P. P. (2000). Lim1 activity is required for intermediate mesoderm differentiation in the mouse embryo. *Dev. Biol.* **223**, 77–90.

Tung, T. C. (1935). On the time of determination of the dorso-ventral axis

of the pronephros in Discoglossus (Amphibian). *Peking Nat. Hist. Bull.* **10**, 115–116.

Tytler, P. (1988). Morphology of the pronephros of the juvenile brown trout, *Salmo trutta*. *J. Morphol.* **195**, 189–204.

Uchiyama, M., Murakami, T., Yoshizawa, H., and Wakasugi, C. (1990). Structure of the kidney in the crab-eating frog, *Rana cancrivora*. *J. Morphol.* **204**, 147–156.

Ultsch, G. R., Bradford, D. F., and Freda, J. (1999). Physiology: Coping with the environment. *In* "Tadpoles: The Biology of Anuran Larvae" (R. W. McDiarmid and R. Altig, eds.). University of Chicago Press, Chicago.

Vize, P. D., Jones, E. A., and Pfister, R. (1995). Development of the *Xenopus* pronephric system. *Dev. Biol.* **171**, 531–540.

Vize, P. D., Seufert, D. W., Carroll, T. J., and Wallingford, J. B. (1997). Model systems for the study of kidney development: use of the pronephros in the analysis of organ induction and patterning. *Dev. Biol.* **188**, 189–204.

Vogt, W. (1929). Gestaltungsanalyse am amphibienkeim mit örtlicher vitalfärbung. II. Teil. Gasrulation und mesodermbildung bei urodelen und anuran. *Roux' Arch. Entwick. Mech.* **120**, 384–706.

Waddington, C. H., and Schmidt, G. A. (1933). Inductions by heteroplastic grafts of the primitive streak in birds. *oux' Arch.* **128**, 521–563.

Wake, M. (1970). Evolutionary morphology of the caecilian urogenital system. II. the kidneys and urogenital ducts. *Acta. Anat.* **75**, 321–358.

Walsche, L. (1929). Etude sur le développement du pronéphros et du mésonéphros chez Chéloniens. *Arch. Biol.* **39**, 1–59 (with three additional plates).

Wang, S., Krinks, M., Kleinwaks, L., and Moos, M. (1997). A novel Xenopus homologue of bone morphogenetic protein-7 (BMP-7). *Genes Funct.* **1**, 59–271.

Wheeler, W. M. (1900). The development of the urogenital organs of the lamprey. *Zool. Jahrbüch.* **13**, 1–88.

Wiedersheim, R. (1890). Über die entwicklung des urogenitalapparates bei krokodilien und schildkröten. *Anat. Anzeiger* **5**, 337–344.

Wiedersheim, R. (1890). Über die entwicklung des urogenitalapparates bei krokodilien und schildkröten. *Arch. Mikr. Anat.* **36**, 410–468.

Witschi, E. (1956). "Development of Vertebrates." Saunders, Philadelphia.

Wrobel, K., and Süß, F. (2000). The significance of rudimentary nephrostomial tubules for the origin of the vertebrate gonad. *Anat. Embryol.* **201**, 273–290.

Yamada, T. (1940). Beeninflussung der differenzierungsleistung des isolierten mesoderms von molchkeimen durch zugefügtes chorda- und neural-material. *Okajimas Fol. Anat. Jap.* **19**, 131–197.

Zackson, S. L., and Steinberg, M. S. (1986). Cranial neural crest cells exhibit directed migration on the pronephric duct pathway: Further evidence for an *in vivo* adhesion gradient. *Dev. Biol.* **117**, 342–353.

4

Formation of the Nephric Duct

Thomas M. Schultheiss, Richard G. James,
Anzhelika Listopadova, and Doris Herzlinger

I. Introduction

During embryonic development, the intermediate mesoderm differentiates into tubular epithelial tissues of the kidney
and genital system. One of the first mesenchymal-to-epithelial
conversions that occurs results in formation of the nephric
duct (Saxén, 1987). This duct, which later in development is
called the mesonephric or Wolffian duct, differentiates into
portions of the male genital system and is required for all
further kidney development. It induces surrounding intermediate mesoderm to differentiate into nephrons of the mesonephric kidney, the excretory organ of a majority of vertebrate species and a developmental intermediary excretory
organ of birds and mammals (Gruenwald, 1942). In birds
and mammals the nephric duct issues a caudal diverticulum,
the ureteric bud, which is essential for the formation of their
adult excretory organ, the metanephros. The ureteric bud gives
rise to the water-conserving collecting system required for
land-dwelling animals and induces the surrounding caudal
intermediate mesoderm to differentiate into nephrons (Saxén,
1987). Despite the essential role the nephric duct and its derivatives play in the developing urogenital system, the tissue
interactions and molecules regulating nephric duct formation
in the early embryo remain poorly characterized.

A. Morphological Description of Nephric Duct Formation

The morphological transitions that occur in the intermediate mesoderm as it differentiates into the nephric duct
have been described in developing agnathostome, teleost,
amphibian, avian, and mammalian embryos (Price, 1897;
Brauer, 1902; Price, 1910; Torrey, 1943; Poole and Steinberg,
1981; Gillespie, *et al.,* 1985; Jarzem and Meier, 1987; Vize
et al., 1997). In the lamprey (Price, 1897, 1910) and in the
legless amphibian Gymnophonia (Brauer, 1902), the intermediate mesoderm is segmented, and the nephric duct is
formed through the fusion of portions of each nephric tubule.
In these organisms, the whole length of the duct is formed
by this type of segmental fusion. In contrast, in most other
amphibians and in birds and mammals, the nephric duct is
derived from a rudiment, which forms in the pronephric
region; during development, this rudiment migrates posteriorly to generate the duct of the more posterior regions of
the embryo. This chapter focuses on this latter mode of duct
formation, which is found in higher vertebrates. The reader
is referred to the studies referenced earlier for a detailed
description of the former mode of duct development [see
also Fraser (1950) for a review].

In *Xenopus laevis* and axolotl, the duct can first be detected at the caudal aspect of the pronephric rudiment (Poole and
Steinberg, 1981; Gillespie, *et al.,* 1985; Vize *et al.,* 1997).
The pronephric rudiment in these species can be visualized
by scanning electron microscopy (SEM) in stage 23 embryos
(12 and 5–7 somites, respectively). It appears as a slight
thickening in the somatopleuric mesoderm at the axial level

of the cervical somites and subsequently expands to form a distinct morphological structure located lateral to the somites between the surface ectoderm and the lateral plate mesoderm. As development proceeds, the nephric duct appears as a short extension of mesenchymal cells at the posterior end of the pronephric rudiment. At later stages of development, the distal tip of the duct can be detected at progressively more caudal axial levels joining the cloaca at stage 33/34 and 28–30, respectively.

Unlike amphibians and urodeles, the pronephric rudiment in birds cannot be visualized by scanning electron microscopy, although it can be demarcated by molecular marker analyses (see later discussion). The structural prominence of the pronephric rudiment in urodele and amphibian embryos as compared to birds most likely reflects the differing functional role the pronephros serves in these species (Torrey, 1965). The pronephros is the sole excretory organ of anuran and urodele-larvae, whereas its excretory function in the developing avian embryo is questionable; fully differentiated pronephric tubules have not been observed. Thus, the nephric duct is the first structure that can be detected by SEM in the developing avian urogenital system (Jarzem and Meier, 1987). At the 8–10 somite stage (HH St 9+ – 10), the duct appears as a short cord of mesenchymal cells located at the lateral edge of a groove that forms between the paraxial and the lateral mesoderm, extending from the axial levels of somites 6–10. As in urodeles and anurans, the avian duct projects caudally as development proceeds, fusing with the rectal diverticulum at the 33 somite stage (HH St 22). Differentiation of the extended cord of mesenchymal nephric duct progenitors into tubular epithelia begins at the 19 somite stage (HH St 13) and occurs in a rostral–caudal wave (Obara-Ishihara *et al.*, 1999).

As in birds, the mammalian pronephros is a rudimentary structure and probably does not possess any excretory function during embryonic development (Chapter 3; Torrey, 1943, 1965). In the rat, mesoderm destined to form the urogenital system or intermediate mesoderm can be visualized as a distinct population of cells located between the paraxial and the lateral plate mesoderm beginning at the 11 somite stage. A cord of mesenchymal cells appears to split directly off the dorsal-most aspect of the intermediate mesoderm at the 12 somite stage and projects caudally, fusing with the cloaca in embryos of 32–36 somites. Nephric duct epithelialization begins at the 30 somite stage and, as in the avian embryo, progresses in a rostral–caudal direction.

B. Gene Expression during Duct Morphogenesis

Transcripts expressed in the intermediate mesoderm as it differentiates into the nephric duct have been well described in the developing chick (Obara-Ishihara *et al.*, 1999; Mauch *et al.*, 2000; T. M. Schultheiss, unpublished observations). The pronephric rudiment of the chick can first be identified at the 6–7 somite stage as a population of *Pax-2/Lim-1/sim-1* expressing cells lying between the paraxial and the lateral plate mesoderm. By the 9–10 somite stage, *Lim-1* (Fujii *et al.*, 1994; Shawlot and Behringer, 1995) and *c-Sim1* (Pourquie *et al.*, 1996) become restricted to a dorsal subpopulation of the *Pax-2* expressing rudiment (Figs. 4.1A–F). These *Lim-1*, *c-Sim1* expressing cells exhibit a location consistent with the first morphologically identifiable duct structure, the cord of cells that can be visualized by SEM at the 9–10 somite stage. In contrast, *Pax-2* expressing cells ventral to the duct co-express mRNAs encoding *WT-1* (Pelletier *et al.*, 1991; Kreidberg *et al.*, 1993) (Fig. 4.1). Because WT-1 mRNA expression is restricted to nephron progenitors in developing murine renal tissues, it is likely that the ventral portion of the avian pronephric rudiment differentiates into rudimentary pronephric nephrons (Chapter 3). Later in development, the cord of mesenchymal duct progenitors projecting caudally is characterized by the expression of c-Sim-1, Lim-1, and Pax-2 mRNAs. By stage 17, mRNA encoding c-Ret, a tyrosine kinase receptor, which has been implicated in murine metanephric collecting system growth and branching, is upregulated in the differentiating duct (Schuchardt *et al.*, 1994, 1995). Transcripts encoding c-Sim-1, Lim-1, Pax-2, and c-Ret mRNAs are also expressed in the murine nephric duct, suggesting that several regulatory molecules mediating duct morphogenesis are conserved from birds to mammals (Dressler *et al.*, 1990; Fujii *et al.*, 1994; Schuchardt *et al.*, 1994; Fan *et al.*, 1996).

II. Nephric Duct Morphogenesis

From the morphological and molecular marker analyses described earlier, nephric duct morphogenesis can be subdivided into at least three stages; specification of duct progenitors, caudal elongation of the duct as a mesenchymal cord followed by its differentiation into an epithelial tubule.

A. Specification of Duct Progenitors

By "duct specification," we refer to the developmental events that determine which embryonic cells will differentiate as the nephric duct. This process begins well before there is morphological evidence of a duct. The following section on nephric duct specification discusses four issues: (a) the developmental history of duct precursors prior to the formation of the duct rudiment, (b) the timing of the developmental events that pattern the duct, (c) transcriptional regulation of duct specification, and (d) signals that regulate duct specification. Because, as outlined earlier, the duct arises in close association with the pronephros, the discussion of necessity includes some treatment of the specification of the pronephros itself. Indeed, as will be seen, much less is known

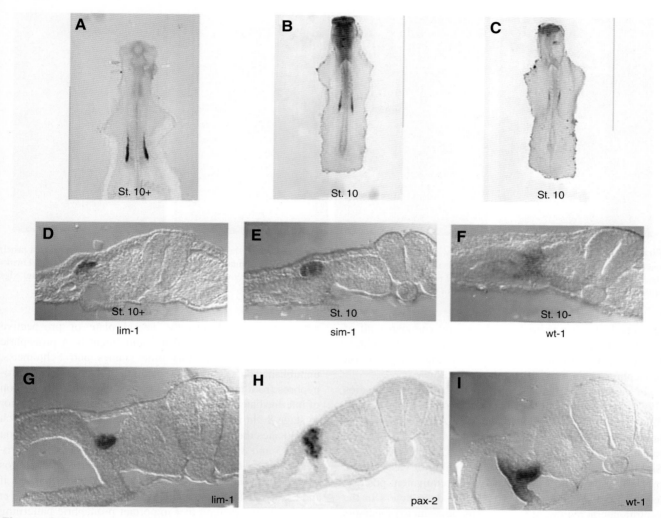

Figure 4.1 *In situ* hybridization for *lim-1*, *Pax-2*, *sim-1*, and *WT-1* at stages 10 (A–F) and 12 (G–I). Note that *lim-1* and *sim-1* expression is confined to the duct rudiment, *WT-1* is excluded from the duct, and *Pax-2* is expressed in both duct and nonductal tissues.

about duct specification than about specification of the prone-phros. Further discussion of pronephric specification can be found in Chapter 3.

1. Early Developmental History of Duct Precursors

Fate maps have been performed in several vertebrate species in order to locate pronephric precursors prior to the expression of duct genes or the formation of a morpholo-gically recognizable duct rudiment. These studies provide information regarding the environment in which duct precursors are located prior to their differentiation, and thus can suggest which tissues and signals might be patterning the developing duct. A general description of fate mapping is presented in Chapter 3, and experimental details of how such experiments are performed are discussed in Chapter 9.

Some of the most detailed pronephric fate maps have been performed in the chick. In some cases, resolution of the available fate maps is not fine enough to distinguish pro-spective duct rudiment from prospective pronephric tubules (but see Fig. 3.3). In the chick, however, because the pro-nephric tubules are rudimentary, the pronephros at the axial level of somites 6 to 10 is predominantly duct so the pronephric fate map is in large part a duct fate map. In the chick, the prospective pronephric intermediate mesoderm is located in the primitive streak at stage 5 at a location 60–75% of the length of the primitive streak (measured from the posterior end), posterior to the prospective somites, and anterior to the prospective lateral plate (Psychoyos and Stern, 1996). The prospective pronephros then migrates out of the primitive streak and into the mesodermal layer (Fig. 4.2; Psychoyos and Stern, 1996; James and Schultheiss, unpublished data). During its migration from the primitive streak to its final location, the prospective pronephros is in contact with the segmental plate and lateral plate mesoderm, as well as

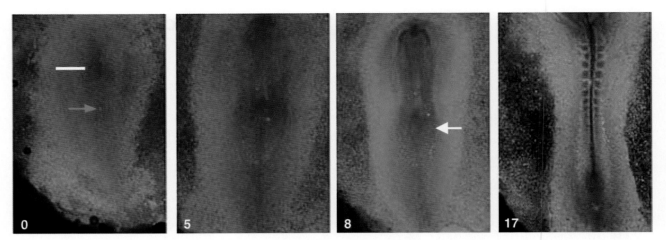

Figure 4.2 Fate map of the chick pronephros. The fluorescent dye DiI was injected into the stage 5 primitive streak (green arrows in the first panel) and the embryos were followed for 17 h. DiI labels the intermediate mesoderm posterior to the somite 5 axial level. The white line in the first panel marks Hensen's node. The white arrow in the third panel points out that during its migration, the prospective pronephros and duct are located at the lateral edge of the neural plate. Numbers indicate the number of hours after injection.

the neural plate, nonneural ectoderm, and endoderm, all of which could exert patterning influences on the pronephros. However, the influence of any of these tissues on pronephros or duct specification is not known.

Pronephric fate maps also exist for other species (Fig. 3.3). Injections of tracking dyes into individual cells of 32 cell embryos in *Xenopus* have localized the prospective pronephros to blastomeres C3 and C4, which also gives rise to parts of the somites and lateral plate (Chapter 3: Dale and Slack, 1987). In zebrafish, at the beginning of gastrulation, pronephric precursors are located in an equivalent position as in 32 cell *Xenopus* embryos, in the deep layers in the ventral–lateral regions of the embryos, adjacent to precursors of the anterior somites (Kimmel *et al.,* 1990). From there, the pronephric cells gastrulate and assume their final positions ventral to the anterior somites.

2. Timing of Duct Specification during Chick Embryogenesis

Transplant and explant studies have been performed in order to determine when during their developmental history the pronephros and the pronephric duct become patterned. In avian embryos, such experiments have determined that important events that pattern the pronephros take place between stages 5 and 8 (late gastrulation to the four somite stage). When prospective pronephros that lies within the chick primitive streak (at stage 5) is transplanted into noninter-mediate mesoderm regions of the streak, the transplanted tissue does not express pronephric genes, indicating that at this stage of development the chick pronephric precursors are not yet committed to an intermediate mesoderm fate (Garcia-Martinez and Schoenwolf, 1992; James and Schultheiss, unpublished data). By stage 8–9 (four to seven somites), prospective pronephros will express pronephric genes

when transplanted into the lateral plate or prospective somite region, indicating that commitment to a pronephric fate has occurred by this time (James and Schultheiss, unpublished data). Surgical separation of the prospective pronephros from more medial tissues results in the suppression of intermediate mesoderm gene expression if performed prior to stage 8 but not after stage 8 (Mauch *et al.,* 2000), which also locates the time of pronephric commitment to between stages 5 and 8, i.e., to the 6–8 h after the prospective pronephros exits the primitive streak. Prospective pronephros placed in tissue culture will express pronephric markers if the tissue is taken from stage 8 or older embryos (Mauch *et al.,* 2000), confirming that important pronephric patterning events are occurring prior to stage 8. These experiments do not distinguish between pronephric duct and tubule formation. Cultures of prospective *Xenopus* pronephros will express pronephric genes if placed in culture at stage 14 or later (neural plate stage), implying that pronephric specification in *Xenopus* takes place around stage 14 (Brennan *et al.,* 1998). While this study suggests that pronephric specification takes place somewhat earlier in *Xenopus* than in chick, it is not clear that equivalent rudiments were assayed in the chick and *Xenopus* studies.

3. Transcriptional Regulation of Duct Specification

As discussed earlier, the pronephric duct rudiment in chick is characterized by coexpression of the transcription factors *Lim-1, Pax-2,* and *c-Sim1.* Of these, *Lim-1* is the earliest to be expressed, being detectable in the pronephric region as early as stage 6 (R. G. James and T. M. Schultheiss, unpublished data). This early expression domain is broader than the actual prospective pronephros. By stage 8–9, when *Pax-2* and *c-Sim1* begin to be expressed, the *Lim-1* expression domain narrows down to the prospective duct rudiment

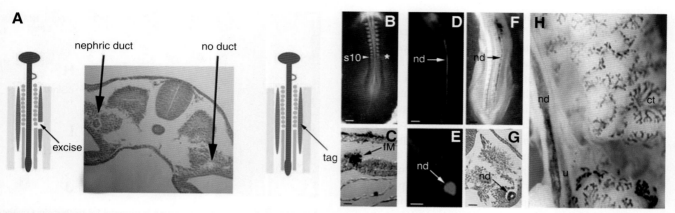

Figure 4.3 The nephric duct derives from a focal domain of the rostral intermediate mesoderm. The intermediate mesoderm located between the axial levels of somites 8–10 was excised from 10 somite chick embryos (HH 10). Eggs were resealed and incubated for an additional 24 h, and nephric duct morphogenesis was assayed by conventional histological techniques. **(A)** The nephric duct, an epithelial tubule composed of columnar epithelia, is present on the left, unoperated side of the embryo, whereas the duct is absent on the side of the embryo from which the intermediate mesoderm was excised. Ten somite chick embyos were injected with 1 nl Dil (B–E) or replication retrovirus encoding *lac-Z* (SNTZ) into the intermediate mesoderm between the axial levels of somites 8–10. Embryos were fixed immediately after injection (B and C), incubated for 24 h (D–G) or 14 days (H), and processed for morphological analyses. **(B)** Whole mount view of a representative embryo injected with 1 nl Dil. A 1-nl injection labels an area of approximately 100 μm². **(C)** Representative section of Dil injected embryo after photoconversion. IM, intermediate mesoderm is labeled with the brown, Dil reaction product. **(D)** Whole mount of embryo 24 h after targeted injection. Red fluorescent, Dil-labeled cells are present in a narrow band of tissue that projects caudally toward the cloaca. **(E)** Histological section of embryo 24 h after Dil injection. Dil-labeled cells comprise the nephric duct. **(F)** Whole mount of embryo 24 h after targeted injection of SNTZ. *lac-z*-expressing cells (blue) deriving from the intermediate mesoderm located between the axial levels of somites 8–10 are organized as a narrow band of tissue that projects caudally. **(G)** Histological section of embryo 24 h after SNTZ injection. Blue, *lac-z*-expressing cells present in the caudal aspect of the embryo are nephric duct epithelia. **(H)** Whole mount of embryo 13 days after targeted injection of SNTZ. Blue, *lac-z*-expressing cells deriving from the intermediate mesoderm located between the axial levels of somites 8–10 are located specifically in nephric duct (nd) or vas deferens, ureter (u), and metanephric-collecting tubules (ct).

(Fig. 4.1; R. G. James and T. M. Schultheiss, unpublished data). In *Xenopus laevis, lim-1* is detected at stage 12.5 in a broad region of the lateral mesoderm (Carroll and Vize, 1999; Carroll *et al.*, 1999). As in the chick, *Xenopus lim-1* is detected well before any morphological evidence of prone-phros development and in a broader domain that will even-tually give rise to the pronephros. *Pax-8* (a homologue of *Pax-2*) is also first detectable in *Xenopus* at about stage 12.5 in a domain that partially overlaps with *lim-1* (Carroll and Vize, 1999; Carroll *et al.*, 1999).

Targeted disruption of the *lim-1* gene in mice results in complete loss of duct and tubule formation, indicating that *lim-1* is required for duct formation (Shawlot and Behringer, 1995). The loss of tubules in *lim-1* mutant mice could be due to a requirement for *lim-1* in the tubules, a loss of induction by the duct, or both. Loss of *Pax-2* expression in mouse (Torres *et al.*, 1995) or zebrafish (Majumdar *et al.*, 2000), however, affects terminal stages of duct differentiation and nephron formation in both the meso- and the metanephric kidney. While this finding indicates that *Pax-2* activity is not required for initial phases of duct formation, the related gene *Pax-8* is also expressed in the pronephros and so it is still possible that some *Pax-2*-like activity is required (Plachov *et al.*, 1990). A *sim-1* mutant phenotype has not been reported.

Ectopic expression studies in *Xenopus* have investigated the role of *lim-1* and *Pax-8* in pronephros formation (Chapter 8). These studies found that ectopic expression of either *lim-1* or *Pax-8* generates ectopic ducts and tubules in the intermediate mesoderm and somite regions of the embryo (Carroll and Vize, 1999), an effect that is augmented when both genes are misexpressed together. Because both duct and tubular tissue are generated, the most likely interpretation of these experiments is that a combination of *lim-1* and *Pax-8* generates the pronephros itself, which then generates both duct and tubule under the influence of the tissue environment. The combined misexpression of *lim-1* and *Pax-8* generates ectopic pronephric tissue only in restricted areas of the embryo (the paraxial and intermediate mesoderm, which would normally give rise to somites and the urogenital system), implying that additional positive or negative signals regulate formation of the pronephros. Details of the molecular regulation of pronephric tubule specification are explored further in Chapter 8.

4. Signals that Specify the Duct

Several signaling systems have been implicated in the patterning of the pronephros and pronephric duct. Studies in *Xenopus* have found that kidney tubules can be induced with high frequency when animal caps are treated with a combination of activin and retinoic acid (Moriya *et al.*, 1993). Although the role of activin in pronephros formation *in vivo* is not clear, several lines of evidence suggest that retinoic acid may play a role in nephrogenesis *in vivo*. Ectopic application of retinoic acid can induce expression of ectopic *lim-1* in *Xenopus* embryo (Taira *et al.*, 1994), and mice deficient for the retinoic acid synthetic enzyme RALDH2 fail to form nephric ducts (Niederreither *et al.*, 1999). The failure to form nephric ducts in both *lim-1* and *RALDH2* mutant mice suggests that retinoic acid regulation of *lim-1* might be an important step in the initial steps of duct specification.

Work in *Xenopus* has implicated the Notch signaling pathway in regulating the allotment of pronephric cells into tubular or ductal fates (McLaughlin *et al.*, 2000). Components of the Notch signaling pathway are expressed in the pronephros beginning at stage 19. Experimental activation of the Notch pathway leads to repression of duct formation, whereas repression of Notch signaling leads to higher levels of expression of duct markers (although not ectopic duct formation). This suggests that Notch signaling may regulate allocation of cells in the pronephric rudiment between ductal and tubular fates.

BMP signaling in the ectoderm and lateral plate mesoderm has been found to be important for the maintenance of duct markers (Obara-Ishihara *et al.*, 1999), although a role for BMP signaling in the initial patterning of the duct or pronephros has not been determined (also see Chapter 3).

Several studies have investigated the influence of adjacent tissues on pronephros and duct patterning. Evidence shows that signals from dorsal embryonic structures can promote pronephros development. In *Xenopus*, dorsalized tissues can induce pronephric gene expression in tissue recombinations *in vitro* (Seufert *et al.*, 1999). In avian embryos, separation of the prospective pronephros from more dorsal structures prevents expression of pronephric genes (Mauch *et al.*, 2000). The molecular nature of this dorsal signal is not known.

One theory that has been presented in numerous older textbooks holds that the pronephric duct is formed from the fusion of the posterior portions of the pronephric tubules. In *Xenopus*, where the pronephros gives rise to both duct and tubules, it has been possible to study whether duct formation requires the presence of tubules. As early as stage 12.5, *Xenopus* embryos can be divided into dorsal and ventral halves, and in some of these embryos, the ventral half will give rise to duct and not tubule, whereas the dorsal part will give rise to tubules and not duct (Vize *et al.*, 1995, 1997). This implies that duct formation does not require the presence of tubular tissue and that any interaction between ductal and tubular precursors must take place prior to stage 12.5 (late gastrula, when *lim-1* is just beginning to be expressed in the prospective pronephros).

B. Nephric Duct Elongation

The second stage of nephric duct morphogenesis, the projection of mesenchymal duct progenitors toward the cloaca, begins at the 10–11 somite state in the avian embryo. The

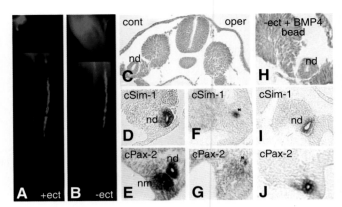

Figure 4.4 The surface ectoderm is essential for nephric duct epithelization, but not elongation, in the developing chick. The intermediate mesoderm adjacent to somite 10 was tagged with DiI at stage 10 and the surface ectoderm was left in place (A) or removed (B). Caudal migration of DiI-tagged nephric duct progenitors was assessed 24 h later by whole mount fluorescent microscopy. Caudal migration of nephric duct progenitors was identical in control (A) or operated (B) embryos, indicating that nephric duct elongation is independent of the overlying surface ectoderm. However, histological analyses of operated embryos demonstrate that the surface ectoderm is essential for nephric duct tubulogenesis (C). A representative section of urogenital system fixed 24 h after the surface ectoderm was removed from the left side of the embryo. On the control, unoperated side of the embryo, the nephric duct (nd) is characterized by tubular epithelia that express cSim-1 (D) and Pax-2 mRNAs (E). Pax-2 mRNA is also expressed in the nephrogenic intermediate mesoderm (nm) destined to form mesonephric nephrons. On the right, operated side of the embryo, although the surface ectoderm has regenerated, the differentiated duct is not visible by conventional morphological techniques (C). However, a cord of mesenchyme (*) expressing low levels of C-Sim-1 (F) and Pax-2 (G) is present. Pax-2 is expressed at very low levels in nephrogenic mesenchyme (nm) on the operated side of the embryo (G). Strikingly, a BMP-4-coated bead can restore nephric duct tubulogenesis when the surface ectoderm is removed. A representative embryo fixed 24 h after a BMP-4-soaked bead was placed on the intermediate mesoderm exposed by removing the surface ectoderm at stage 10. Hematoxylin and eosin stained section demonstrates that an epithelialized nephric duct (ND) is present (H). Moreover, a BMP-4 coated bead is sufficient to restore C-Sim-1 (I) and Pax-2 (J) expression in the nephric duct after surface ectoderm removal. In addition, normal levels of C-Pax-2 mRNA were detected in the nephrogenic intermediate mesoderm (NM) in the presence of a BMP-4-coated bead. Nephric duct tubulogenesis was not restored when a control BSA-coated bead was placed over the exposed intermediate mesoderm after surface ectoderm removal (data not shown).

rate of extension is 100–400 μm/h, and contact with the cloaca is established within a 24-h period (33 somite stage) (Bellairs *et al.*, 1995). Similar rates of extension have been documented in urodele embryos (Lynch and Fraser, 1990). Extirpiration and carbocyanine dye fate mapping experiments performed in urodele and anuran and embryos demonstrate conclusively that cells of the elongating duct derive from the rostral duct rudiment localized to the axial levels of the cervical somites (Poole and Steinberg, 1981; Lynch and Fraser, 1990). There is, however, evidence from chimeric *Xenopus* embryos that cells from the caudal mesoderm are incorporated into the duct as it fuses with the cloaca (Cornish and Etkin, 1993). In the chick, cells that give rise to the nephric duct are localized to rostral intermediate mesoderm between the axial levels of somites 8–10 (Fig. 4.3; Le Douarin and Fontaine, 1970; Obara-Ishihara et al., 1999).

Because cells comprising the duct derive from a rostral location, caudal duct elongation must be regulated by one or more of the following processes: cell proliferation, cell shape changes, and/or cell rearrangements. Proliferating cells have been detected in the urodele and avian nephric duct as it extends caudally. However, the rate of duct cell proliferation does not exceed that of cells in adjacent axial tissues over which the duct moves rapidly (Overton, 1959). Thus, it is unlikely that cell proliferation alone can account for the rapid rates of caudal duct elongation documented. Changes in cell shape have been observed during duct elongation in the developing chick but not in axoltl embryos (Poole and Steinberg, 1981; Jarzem and Meier, 1987). Thus it is likely that other mechanisms are also crucial for the elongation process. The rapid rate of duct elongation in all species examined is likely to be dependent on cell rearrangements. Carbocyanine dye studies in *Xenopus* demonstrate that cell rearrangements mediating duct elongation occur throughout the duct rudiment, whereas duct progenitors retain their position relative to each other (Lynch and Fraser, 1990). In contrast, duct elongation in the chick embryo is believed to be dependent on cell rearrangements localized solely at the caudal duct tip (Torrey, 1965). Data in support of this hypothesis are indirect and include the cessation of duct elongation when the caudal-most duct tip is severed (Jacob and Christ, 1984) and inhibition of duct extension when an obstacle is placed at its caudal-most aspect (Gruenwald, 1942). However, caudal duct extension still occurs when duct rudiment is rotated 180° (Jacob and Christ, 1984). This latter result strongly suggests that cell rearrangements throughout the duct rudiment mediate avian duct extension.

Molecular mechanisms regulating the cell rearrangements essential for duct extension remain elusive. Currently, the only experimental manipulations that have been used successfully to inhibit extension are rather nonspecific; caudal duct elongation is perturbed by the removal of phosphatidylinositol glycan-linked proteins or negative charges carried by polysialic acid residues associated with the cell adhesion molecule, NCAM (Zackson and Steinberg, 1989; Bellairs et al., 1995). Although this latter results suggests that NCAM is essential for duct elongation, the duct appears to form and elongate in mutant mice lacking NCAM gene expression (Cremer et al., 1994). Because duct progenitors appear to extend processes into an underlying basal lamina during the extension process, it is likely that migration along this substrate, in part, mediates cell rearrangements crucial for duct elongation. One report suggests that synthetic peptides that inhibit integrin–fibronectin interactions perturb duct elongation, whereas another study refutes this result (Jacob et al., 1991; Bellairs et al., 1995). Synthetic peptides that perturb integrin–laminin interactions do not impede duct elon-

gation, but instead appear to disrupt the direction of extension (Jacob et al., 1991).

When avian embryos are microinjected with peptides encoding the minimal sequence necessary for laminin-cell adhesion, the nephric duct extends but projects over the lateral plate mesoderm instead of following its normal pathway between somitic and lateral plate mesoderm (Jacob et al., 1991). Thus, it is likely that laminin isoforms present in the basal lamina between somitic and lateral plate mesoderms provide positional cues regulating the direction and location of duct extension. Furthermore, extensive studies in axoltl indicate that adhesive properties of the basal lamina underlying the extending duct support its extension (Gillespie and Armstrong, 1985; Zackson and Steinberg, 1987). Other positional cues regulating duct extension remain poorly understood. One report suggests that the surface ectoderm provides some of the directional cues regulating this process in developing axoltl embryos (Drawbridge et al., 1995). However, it is clear that the surface ectoderm is not required for caudal duct extension in either avian or *Xenopus* embryos (Figs. 4.4A and 4.4B; Obara-Ishihara et al., 1999; unpublished observations of H. Brennan and E. Jones).

C. The Rectal Diverticulum

As the nephric duct migrates posteriorly from the pronephric primordium, a separate group of duct precursors begins to migrate anteriorly from the cloaca. These cloacal outgrowths are known as the rectal diverticula or, in some instances, as cloacal horns. They first appear at approximately NF stage 32 in *X. laevis* as as small slits in the cloacal wall. They begin to extand anteriorly and meet with the nephric duct moving in the opposite direction at NF stage 37. The two anlage then fuse to create the complete duct primordia (Fig. 4.5). Useful molecular markers of the rectal diverticulum in *Xenopu*s include *Pax-2* (Fig. 4.5; Heller et al., 1997) and *claudin* (Brizuela et al., 2001), a tight junction encoding gene.

In amniotes, the duct also fuses with an anterior-projecting cloacal derivative. In birds, this is the urodeum (Burns, 1955). In mice, the migrating duct fuses with the urogenital sinus (Kaufman and Bard, 1999).

D. Nephric Duct Tubulogenesis

The third stage of nephric duct morphogenesis, the conversion of mesenchymal nephric duct progenitors into tubular epithelia, has been analyzed in the developing chick and shown to be dependent on proximate tissue interactions (Obara-Ishihara et al., 1999). Nephric duct tubulogenesis, but not elongation, is inhibited when the surface ectoderm overlying the differentiating duct is removed (Figs. 4.4A–4.4G). In addition, duct tubulogenesis is inhibited when an impermeable barrier is placed between the differentiating duct

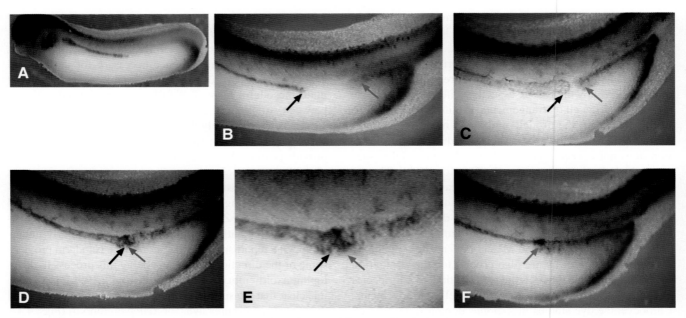

Figure 4.5 The duct is derived from both the pronephric rudiment and the rectal diverticulum in *Xenopus laevis. In situ* hybridization with a Pax-2 probe illustrates the posterior growth of the pronephric rudiment and anterior growth of the diverticulum. The two structures meet and fuse at NF stage 36/37. (A) Stage 28. (B) Stage 34/35. (C) Stage 35. (D) Stage 36. (E) Higher magnification of D. (F) Stage 37. All views are lateral. Images courtesy of Thomas J. Carroll and Peter D. Vize.

and the lateral plate mesoderm (unpublished observations, T. O. Ishihara and D. Herzlinger). Inhibition of duct tubulogenesis by these experimental manipulations results in a dramatic downregulation of c-Sim-1 and Pax-2 mRNAs, which are normally expressed abundantly in the duct as it epithelializes. These data suggest that factors secreted by tissues surrounding the differentiating duct are essential for mediating its conversion into an epithelial tubule. Thus the differentiation of the intermediate mesoderm into the nephric duct, like the differentiation of the caudal intermediate mesoderm into nephron epithelia later in development, is dependent on factors secreted by surrounding tissues. However, it is likely that some of the regulatory molecules mediating epithelial morphogenesis during duct and nephron formation differ. For example, *Pax-2* mRNA is upregulated in mesenchymal nephron progenitors prior to epithelialization and is required for this phenotypic conversion *in vitro* and *in vivo* (Dressler *et al.,* 1990; Rothenpieler and Dressler, 1993). Although *Pax-2* mRNA is expressed by mesenchymal duct progenitors early in urogenital development, *Pax-2* null mice form an epithelialized nephric duct, although duct differentiation is incomplete (Torres *et al.,* 1995).

One of the factors that has been implicated in regulating nephric duct tubulogenesis is BMP-4 (Obara-Ishihara *et al.,* 1999). Members of the BMP gene family mediate the differentiation of a variety of cell types, and *BMP-4* haploinsufficient urogenital defects have been observed in mice. *BMP-4* null mice die during gastrulation prior to the generation of the intermediate mesoderm (Winnier *et al.,* 1995;

Hogan, 1996; Dunn *et al.,* 1997). In the developing chick, the two tissues shown to be required for nephric duct tubulogenesis, the surface ectoderm and lateral plate mesoderm, both express BMP-4 mRNA (Pourquie *et al.,* 1996; Watanabe and Le Douarin, 1996). Furthermore, the surface ectoderm is required for the normal upregulation of BMP-4 transcripts in the lateral plate (Obara-Ishihara *et al.,* 1999). Thus, removal of the surface ectoderm dramatically decreases the levels of BMP-4 available to mesenchymal nephric duct progenitors as they transform into an epithelial tubule. Most importantly, a BMP-4-coated bead can restore normal levels of *Sim1* and *Pax-2* transcription and rescues nephric duct tubulogenesis in avian embryos from which the surface ectoderm was removed (Obara-Ishihara *et al.,* 1999; Figs. 4.4H–4.4J). Collectively, these data suggest that the surface ectoderm expresses and/or regulates the expression of BMP-4, or an unidentified member of the TGF-β family member, that is essential for nephric duct tubulogenesis. Currently, it is unclear if BMP-4 binds directly to mesenchymal duct progenitors mediating their differentiation into an epithelium or if BMPs regulate duct tubulogenesis indirectly by upregulating the expression of unidentified tubulogenic factors.

III. Conclusions

Clearly, much remains unknown about the regulation of nephric duct morphogenesis. We list a few unanswered questions here in order to highlight some of the large gaps in our understanding. What are the transcription factors and signaling molecules that specify duct rudiment formation? Why do duct precursors form only in a restricted area in the anterior region of the intermediate mesoderm (adjacent to somites 8–10 in the chick) while the intermediate mesoderm can give rise to nephrons along its entire length? What are the molecular cues regulating the direction and location of duct extension? Finally, what are cell biological processes that regulate the fusion of nephrons to the duct and what are the mechanisms and regulatory molecules mediating the fusion of duct to the cloaca?

Nephric duct extension appears to be dependent on the rearrangement of progenitors cells present in the duct rudiment that forms at the axial level of the cervical somites. The molecules that trigger duct extension and provide directional cues for this process remain poorly understood.

References

Bellairs, R., Lear, P., Yamada, K. M., Rutishauser, U., and Lash, J. W. (1995). Posterior extension of the chick nephric (Wolffian) duct: The role of fibronectin and NCAM polysialic acid. *Dev. Dyn.* **202**, 333–242.

Brandli, A. W. (1999). Towards a molecular anatomy of the Xenopus pronephric kidney. *Int. J. Dev. Biol.* **43**, 381–395.

Brauer, A. (1902). Beitrage zur kenntniss der entwicklung und anatomie der Gymnophionen. III. Die entwicklung der excretionsorgane. *Zool. Jb. Abt. 2 Anat. Ontog.* **16**, 1–176.

Brennan, H. C., Nijjar, S., and Jones, E. A. (1998). The specification of the pronephric tubules and duct in *Xenopus laevis*. *Mech. Dev.* **75**, 127–137.

Brizuela, B. J., Wessely, O., and De Robertis, E. M. (2001). Overexpression of the Xenopus tight-junction protein claudin causes randomization of the left-right body axis. *Dev. Biol.* **230**, 217–229.

Burns, R. K. (1955). Urogenital system. *In* "Analysis of Development" (B. H. Willier, P. A. Weiss, and V. Hamburger, eds.), pp. 462–491. Saunders, Philadelphia.

Carroll, T. J., and Vize, P. D. (1999). Synergism between Pax-8 and lim-1 in embryonic kidney development. *Dev. Biol.* **214**, 46–59.

Carroll, T. J., Wallingford, J. B., and Vize, P. D. (1999). Dynamic patterns of gene expression in the developing pro-nephros of *Xenopus laevis*. *Dev. Genet.* **24**, 199–207.

Cornish, J. A., and Etkin, L. D. (1993). The formation of the pronephric duct in Xenopus involves recruitment of pos-terior cells by migrating pronephric duct cells. *Dev. Biol.* **159**, 338–345.

Cremer, H., Lange, R., Christoph, A., Plomann, M., Vopper, G., Roes, J., Brown, R., Baldwin, S., Kraemer, P., Sheff, S., *et al.* (1994). Inactivation of the N-CAM gene in mice results in size reduction of the olfactory bulb and deficits in spatial learning. *Nature* **367**, 455–459.

Dale, L., and Jones, C. M. (1999). BMP signalling in early Xenopus development. *Bioessays* **21**, 751–760.

Drawbridge, J., Wolfe, A. E., Delgado, Y. L., and Steinberg, M. S. (1995). The epidermis is a source of directional information for the migrating pronephric duct in *Ambystoma mexicanum* embryos. *Dev. Biol.* **172**, 440–451.

Dressler, G. R., Deutsch U., Chowdhury, K., Nornes, H. O., and Gruss, P. (1990). Pax2, a new murine paired-box-containing gene and its expression in the developing excretory system. *Development* **109**, 787–795.

Dunn, N. R., Winnier, G. E., Hargett, L. K., Schrick, J. J., Fogo, A. B., and Hogan, B. L. (1997). Haploinsufficient phenotypes in Bmp4 heterozygous null mice and modification by mutations in Gli3 and Alx4. *Dev. Biol.* **188**, 235–247.

Fan, C. M., Kuwana, E., Bulfone, A., Fletcher, C. F., Copeland, N. G., Jenkins, N. A., Crews, S., Martinez, S., Puelles, L., Rubenstein, J. L., and Tessier-Lavigne, M. (1996). Expression patterns of two murine homologs of Drosophila single-minded suggest possible roles in embryonic patterning and in the pathogenesis of Down syndrome [published erratum appears in *Mol. Cell. Neurosci.* **7**(6), 519 (1996)]. *Mol. Cell. Neurosci.* **7**, 1–16.

Fraser, E. A. (1950). The development of the vertebrate excretory system. *Biol. Rev.* **25**, 159–187.

Fujii, T., Pichel, J. G., Taira, M., Toyama, R., Dawid, I. B., and Westphal, H. (1994). Expression patterns of the murine LIM class homeobox gene lim1 in the developing brain and excretory system. *Dev. Dyn.* **199**, 73–83.

Garcia-Martinez, V., and Schoenwolf, G. C. (1992). Positional control of mesoderm movement and fate during avian gastrulation and neurulation. *Dev. Dyn.* **193**, 249–256.

Gillespie, L. L., Armstrong, J. B., and Steinberg, M. S. (1985). Experimental evidence for a proteinaceous preseg-mental wave required for morphogenesis of axolotl meso-derm. *Dev. Biol.* **107**, 220–226.

Gruenwald, P. (1942). Experiments on the distrivution and activation of the nephrogenic potency in the embryonic mesenchyme. *Physiol. Zoo.* **15**, 396–409.

Heller, N., and Brandli, A. W. (1997). Xenopus Pax-2 displays multiple splice forms during embryogenesis and pronephric kidney development. *Mech. Dev.* **69**, 83–104.

Hogan, B. L. (1996). Bone morphogenetic proteins: Multi-functional regulators of vertebrate development. *Genes Dev.* **10**, 1580–1594.

Jacob, H. J., and Christ, B. (1984). Ultrastruckturelle unde experimentelle Untersuchungen zur Antwicklung des Wolffschen Ganes bie Huhnerembryonen. *Verb. Anta. Ges.* **78**, 291–292.

Jacob, M., Christ, B., Jacob, H. J., and Poelmann, R. E. (1991). The role of fibronectin and laminin in development and migration of the avian Wolffian duct with reference to somitogenesis. *Anat. Embryol.* **183**, 385–395.

James, R. G., and Schultheiss, T. M. (2002). Patterning of the Avian Intermediate Mesoderm by Lateral Plate and Axial Tissues. *Dev. Biol.* (in press).

Jarzem, J., and Meier, S. P. (1987). A scanning electron microscope survey of the origin of the primordial proneph-ric duct cells in the avian embryo. *Anat. Rec.* **218**, 175–181.

Kaufman, M. H., and Bard, J. B. L. (1999). "The Anatomical Basis of Mouse Development." Academic Press, San Diego.

Kimmel, C. B., Warga, R. M., and Schilling, T. F. (1990). Origin and organization of the zebrafish fate map. *Development* **108**, 581–594.

Kreidberg, J. A., Donovan, M. J., Goldstein, S. L., Rennke, H., Shepherd, K., Jones, R. C., and Jaenisch, R. (1996). Alpha 3 beta 1 integrin has a crucial role in kidney and lung organogenesis. *Development* **122**, 3537–3547.

Le Douarin, N. M., and Fontaine, J. (1970). Limites du territorie pronephritique caqpale de s'autodifferencier et der fournir l'ebauche primitive du canal de Wolff l'embryon de Poulet. *C. R. Acad. Sci.* **270**, 1708–1711.

Lynch, K., and Fraser, S. E. (1990). Cell migration in the formation of the pronephric duct in *Xenopus laevis*. *Dev. Biol.* **142**, 283–292.

Majumdar, A., Lun, K., Brand, M., and Drummond, I. A. (2000). Zebrafish no isthmus reveals a role for pax2.1 in tubule differentiation and patterning events in the pro-nephric primordia. *Development* **127**, 2089–2098.

Mauch, T. J., Yang, G., Wright, M., Smith, D., and Schoenwolf, G. C. (2000). Signals from trunk paraxial mesoderm induce pronephros formation in chick interme-diate mesoderm. *Dev. Biol.* **220**, 62–75.

McLaughlin, K. A., Rones, M. S., and Mercola, M. (2000). Notch regulates cell fate in the developing pro-nephros. *Dev. Biol.*

Moriya, N., Uchiyama, H., and Asashima, M. (1993). Induction of pronephric tubules by activin and retinoic acid in presumptive ectoderm of *Xenopus laevis*. *Dev. Growth Differ.* **35**, 123–128.

Niederreither, K., Subbarayan, V., Dolle, P., and Chambon, P. (1999). Embryonic retinoic acid synthesis is essential for early mouse post-implantation development. *Nature Genet* **21**, 444–448.

Obara-Ishihara, T., Kuhlman, J., Niswander, L., and Herzlinger, D. (1999). The surface ectoderm is essential for nephric duct formation in inter-mediate mesoderm. *Development* **126**, 1103–1108.

Overton, J. (1959). Mitotic pattern in the chick pro-nephric duct. *J. Embryol. Exp. Morphil.* **7**, 275–280.

Pelletier, J., Schalling, M., Buckler, A. J., Rogers, A., Haber, D. A., and Housman, D. (1991). Expression of the Wilms' tumor gene WT1 in the murine urogenital system. *Genes Dev.* **5**, 1345–1356.

Plachov, D., Chowdhury, K., Walther, C., Simon, D., Guenet, J. L., and Gruss, P. (1990). Pax8, a murine paired box gene expressed in the developing excretory system and thyroid gland. *Development* **110**, 643–651.

Poole, T. J., and Steinberg, M. S. (1981). Amphibian pro-nephric duct morphogenesis: Segregation, cell rearrangement and directed migration of the Ambystoma duct rudiment. *J. Embryol. Exp. Morphol.* **63**, 1–16.

Pourquie, O., Fan, C. M., Coltey, M., Hirsinger, E., Watanabe, Y., Breant, C., Francis-West, P., Brickell, P., Tessier-Lavigne, M., and Le Douarin, N. M. (1996). Lateral and axial signals involved in avian somite patterning: A role for BMP4. *Cell* **84**, 461–471.

Price, G. C. (1897). Development of the excretory organs of a myxinoid *Bdellostoma stouti* Lockington. *Zool. Jb. Abt. 2 Anat. Ontog.* **10**, 205–226.

Price, G. C. (1910). The structure and function of the adult head-kidney of *Bdellostoma stouti. J. Exp. Zool.* **9**, 849–864.

Psychoyos, D., and Stern, C. D. (1996). Fates and migra-tory routes of primitive streak cells in the chick embryo. *Development* **122**, 1523–1534.

Rothenpieler, U. W., and Dressler, G. R. (1993). Pax-2 is required for mesenchyme-to-epithelium conversion during kidney development. *Development* **119**, 711–720.

Saxén, L. (1987). "Organogenesis of the Kidney." Cambridge Univ. Press, Cambridge.

Schuchardt, A., D'Agati, V., Larsson-Blomberg, L., Costantini, F., and Pachnis, V. (1994). Defects in the kidney and enteric nervous system of mice lacking the tyrosine kinase receptor Ret. *Nature* **367**, 380–383.

Schuchardt, A., Srinivas, S., Pachnis, V., and Costantini, F. (1995). Isolation and characterization of a chicken homolog of the c-ret proto-oncogene. *Oncogene* **10**, 641–649.

Seufert, D. W., Brennan, H. C., DeGuire, J., Jones, E. A., and Vize, P. D. (1999). Developmental basis of pronephric defects in Xenopus body plan phenotypes. *Dev. Biol.* **215**, 233–242.

Shawlot, W., and Behringer, R. R. (1995). Requirement for Lim1 in head-organizer function. *Nature* **374**, 425–430.

Taira, M., Otani, H., Jamrich, M., and Dawid, I. B. (1994). Expression of the LIM class homeobox gene Xlim-1 in pronephros and CNS cell lineages of Xenopus embryos is affected by retinoic acid and exogastru-lation. *Development* **120**, 1525–1536.

Torres, M., Gomez-Pardo, E., Dressler, G. R., and Gruss, P. (1995). Pax-2 controls multiple steps of urogenital development. *Development* **121**, 4057–65.

Torrey, T. (1943). The development of the urinogenital system of the albino rat. *Am. J. Anat.* **72**, 37–58.

Torrey, T. W. (1965). Morphogenesis of the vertebrate kidney. *In* "Organo-genesis" (R. L. DeHahn, ed.), pp. 559–579. Holt, Reinhardt and Winston, New York.

Vize, P. D., Jones, E. A., and Pfister, R. (1995). Develop-ment of the Xenopus pronephric system. *Dev. Biol.* **171**, 531–540.

Vize, P. D., Seufert, D. W., Carroll, T. J., and Wallingford, J. B. (1997). Model systems for the study of kidney development: Use of the pronephros in the analysis of organ induction and patterning. *Dev. Biol.* **188**, 189–204.

Watanabe, Y., and Le Douarin, N. M. (1996). A role for BMP-4 in the development of subcutaneous cartilage. *Mech. Dev.* **57**, 69–78.

Winnier, G., Blessing, M., Labosky, P. A., and Hogan, B. L. (1995). Bone morphogenetic protein-4 is required for meso-derm formation and patterning in the mouse. *Genes Dev.* **9**, 2105–2116.

Zackson, S. L., and Steinberg, M. S. (1987). Chemotaxis or adhesion gradient? Pronephric duct elongation does not depend on distant sources of guidance information. *Dev. Biol.* **124**, 418–422.

Zackson, S. L., and Steinberg, M. S. (1989). Axolotl pronephric duct cell migration is sensitive to phosphatidy-linositol-specific phospholipase C. *Development* **105**, 1–7.

5

The Pronephric Glomus and Vasculature

Iain A. Drummond and Arindam Majumdar

I. Introduction

A. The Pronephros

Perhaps more than any other organ, the kidney has allowed the survival and evolution of organisms in new environments ranging from seawater to dry land (Smith, 1953). The universal demand for tissue fluid homeostasis and osmoregulation in diverse habitats has led to the development of a variety of kidney forms. In vertebrates with free-swimming larvae, including amphibians and teleost fish, the pronephros is the functional kidney of early life and is required for proper osmoregulation (Howland, 1921). Blood filtration is carried out by the pronephric glomerulus or glomus, which is supplied by capillaries branching from the dorsal aorta. These capillary tufts either hang in the body cavity as a multisegmental vascular structure [an "external" glomus as seen in amphibians, Vize et al. (1997)] or exist in a closed nephron, being surrounded by a nephrocoele or Bowman's space (an "internal" glomerulus as seen in teleosts; Tytler, 1988; Drummond et al., 1998, Fig. 5.1). This minor difference in pronephric structure between amphibians and teleosts is unlikely to be significant, as the developmental origins of the cells contributing to the glomerular components of the pronephros are the same (Goodrich, 1930; Armstrong, 1932; Tytler, 1988; Vize et al., 1997; Drummond et al., 1998).

The pronephric glomerulus is composed of cell types typical of kidneys of higher vertebrates, including fenestrated endothelial cells in capillary tufts and podocytes with extensive foot processes (Ellis and Youson, 1989; Hentschel and Elger, 1996; Drummond et al., 1998; Majumdar and Drummond, 1999). The pronephric nephron is completed by the pronephric tubules, which drain the nephrocoele via nephrostomes or ciliated openings of the tubules on the inner nephrocoele surface, and by the pronephric ducts, which maintain a caudal connection to the exterior at the cloaca (Vize et al., 1997; Drummond et al., 1998). Recovery of filtered metabolites is achieved by highly efficient vectorial transport processes in the pronephric tubules and ducts, with the recovery of ions and small molecules into the closely apposed venous circulation (Fig. 5.1).

B. Function of the Pronephros

The principal function of the kidney in freshwater vertebrates is to facilitate the elimination of water from the body fluids (Hickman and Trump, 1969; Chapter 3). Freshwater vertebrates maintain a body fluid osmolarity of 250–330 mOs, which is much higher than the water in which they swim (Hickman and Trump, 1969). This creates a constant threat of dilution and lethal edema, which is prevented by the excretion

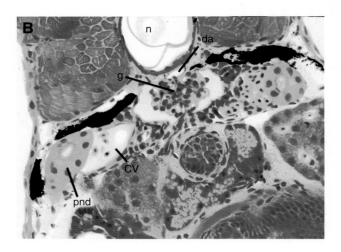

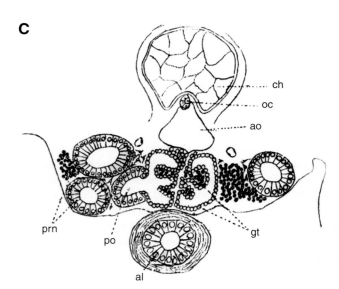

Figure 5.1 The pronephric kidney of teleost larvae. Larvae (A–C) sectioned just caudal to the pectoral fin (A). (B) A six day old zebrafish *Danio rerio* larva showing the "integrated" pronephric glomerulus (g) is positioned ventral to the notochord (n) and dorsal aorta (da). The pronephric ducts (pnd) lie laterally and are in close association with the cardinal veins (cv). (C) A ten day old trout larva showing the direct connection of the pronephric tubule (po) to the glomerulus (gl) ventral to the aorta (ao) (adapted from Balfour, 1880).

of a very dilute urine. The vital function of the pronephros was first shown by Howland (1921) and later by others (McClure, 1918; Swingle, 1919; Fales, 1935) using ablation of the pronephros in *A. punctatum* embryos and observing lethal edema at 8–12 days (bilateral ablation) or compensatory hypertrophy in the remaining pronephros (unilateral ablation). The initiation of pronephric osmoregulatory function in *Rana pipiens* embryos was shown to occur soon after the heart starts beating (Rappaport, 1955). The onset of glomerular filtration in the teleost pronephros has been demonstrated using injections of fluorescent small molecule tracers into the circulation. In the zebrafish, *Danio rerio,* glomerular filtration begins 40–48 h following fertilization and correlates with the ingrowth of capillaries from the dorsal aorta (Fig. 5.2; Drummond *et al.,* 1998). Several different zebrafish mutants affecting pronephric development suffer gross edema by day 5 of development and die shortly after (Drummond *et al.,* 1998). Although marine teleost larvae face quite a different problem regarding fluid homeostasis, having to maintain a blood osmolarity lower than the surrounding seawater, the onset of glomerular filtration in turbot and herring larvae similarly follows closely after the formation of the pronephric capillary tuft (Tytler *et al.,* 1996).

C. The Structure of the Pronephros

The pronephros forms from the most anterior nephrogenic mesoderm and is the first and most primitive of the three kidney types to develop during embryogenesis; the pro-, meso-, and metanephros (Goodrich, 1930; Marshall and Smith, 1930; Burns, 1955; Torrey, 1965; Saxén, 1987). It is found primarily in larval forms of lower vertebrates and is replaced later in development by the mesonephros in amphibians and in teleost fish. It is also sometimes referred to as an opisthonephros in these animals. The pronephros was first described by Müller in 1829, and the glomus was identified by Bidder as its vascular component in 1846 (reviewed in Adelmann, 1966). Formation of the pronephros

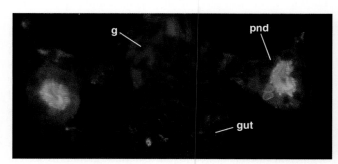

Figure 5.2 Glomerular filtration of flourescent dextran and reuptake by cells of the pronephric duct. Rhodamine dextran (10,000 MW) injected into the circulatory system in a 3-day zebrafish larvae passes the glomerular filter (g) and accumulates in pinocytotic vesicles in the pronephric duct epithelial cells (pnd). Counterstain: FITC wheat germ agglutinin.

in several amphibian species was described in detail more than a century ago by Field, who depicted the structure of the pronephric glomeral capillary tuft in great detail (Field, 1891). Studies since that time have focused primarily on the amphibian and teleost pronephros. The anatomy of amniote pronephroi is discussed in Chapter 3 (see Torrey, 1965; Saxén, 1987).

The hypothetical protovertebrate kidney, often referred to as the "archinephron," is believed to have been a segmentally reiterated structure derived from the intermediate mesoderm and consisting of a hollow nephrocoele surrounding a capillary tuft, connecting ventrally to the coelom and laterally, via segmental nephrostomes (coelomostomes) and tubules, to a common duct (Fig. 5.3A; Goodrich, 1930). Each segmental tubule was thought to have fused into the pronephric duct, which opened caudally into the cloaca (Goodrich, 1930). This general tissue organization of external glomera and reiterated nephrostomes and tubules persists in the amphibian and chicken pronephroi (Fig. 5.3B). In the course of evolution, the number of segments contributing to the pronephros has been reduced and the archinephron is thought to have lost its connection to the coelom with the nephrocoele becoming specialized as Bowman's capsule (Goodrich, 1930; Fox, 1963). With the closed integration of the glomerular capillary tuft and the pronephric tubule and duct comes the concept of the nephron as the functional unit of the kidney.

The anatomy of amphibian pronephroi has been reviewed extensively by Field (1891), Fox (1963), and, more recently, by Vize et al. (1995, 1997: Chapter 3). In *Xenopus laevis* and *Ambystoma punctatum,* the mature pronephros contains a glomus ventral to somites three through five (Nieuwkoop and Faber, 1994). The glomus is vascularized by capillaries ori-ginating from the dorsal aorta and projects into a nephrocoele space, which is open to the coelom. Light microscopy of the glomus in *X. laevis* embryos reveals a structure typical of glomeruli with podocytes arranged on a convoluted capillary tuft (Vize et al., 1995, 1997). Blood plasma filtered into the nephrocoele/ coelom is conveyed to pronephric tubules via openings in the coelomic lining, the nephrostomes (Field, 1891; Fox, 1963; Vize et al., 1997). The recovery of ions and metabolites occurs as a result of the transport properties of the tubule and duct epithelium (Chapter 3).

In the Gymnophionian (caecilian) *Ichthyophis koh-taoensis,* a large pronephric glomus extends over 10 body segments and consists of podocyte-lined, vascularized septa in an expansive nephrocoele (Bowman's space) (Wrobel and Süb, 2000). Communication between the coelom and the nephrocoele is intermittent along the length of the glomus, suggesting that these glomera are intermediate between external and internal glomeruli. The relative size of the Gymnophionan pronephros (large multisegmental glomus and 10 paired nephrostomes) vs the urodele and the anuran pronephros (smaller glomus; 2 or 3 nephrostomes respectively) illustrates Fox's suggestion (Fox, 1963) that, over the course of evolution, the number of body segments contributing to the pronephros has been reduced, with different classes of amphibians possessing pronephroi of various anteroposterior dimensions.

In agnathans (lamprey and hagfish), the pronephros persists as a functional excretory organ much longer than in amphibians and teleosts where the pronephros is replaced in adults by the mesonephros (Goodrich, 1930; Burns, 1955; Saxén, 1987; Ellis and Youson, 1989, 1990). In the lamprey *Petromyzon marinus,* paired, external pronephric glomera extend into the pericardial space, and nephrostomes are observed on the inner surface of the pericardium. Ultra-structurally, the glomera display typical podocytes and a well-developed mesangium (Ellis and Youson, 1989).

In freshwater teleosts (guppy, trout, and zebrafish), as

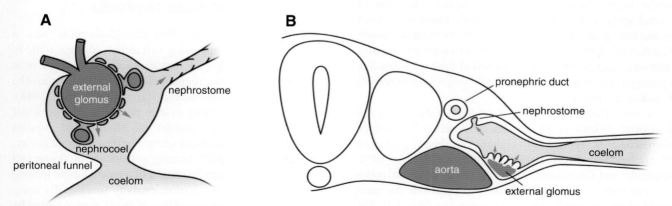

Figure 5.3 The archinephros and the persistence of the external glomus in the chick embryo. (A) Proposed structure for a segmentally repeated, ancestral archinephros composed of a vascular tuft within a nephrocoel, which remained open to the coelom. The nephrocoel is drained by the ciliated nephrostome and associated archinephric tubules and ducts. (B) The pronephros of the chick embryo consists of multiple "external" glomera positioned in a nephrocoel space that is open to the coelom. Filtered blood passes through the nephrocoel space and exits via multiple nephrostome openings on the dorsal surface of the nephrocoel to collect in the pronephric ducts.

well as marine teleosts (killifish, salmon, turbot, and herring), the pronephros consists of only two nephrons with glomeruli fused at the embryo midline just ventral to the dorsal aorta (Fig. 5.1B) (Balfour, 1880; Goodrich, 1930; Marshall and Smith, 1930; Armstrong, 1932; Newstead and Ford, 1960; Agarwal and John, 1988; Tytler, 1988; Hentschel and Elger, 1996; Tytler *et al.*, 1996; Drummond *et al.*, 1998; Drummond, 2000). Two pronephric tubules connect directly to the closed nephrocoele (Bowman's space), and paired bilateral pronephric ducts convey the altered blood filtrate outside the animal. The pronephric nephrons are typically located under somite 3. A connection between the nephrocoele and the coelom is lacking; peritoneal funnels and nephrostomes are not observed (Tytler, 1988; Drummond *et al.*, 1998). Presumably, nephrostomes in teleosts have become the neck of the pronephric tubule connecting to the nephrocoele (Bowman's space) and the nephron forms a closed system of blood filtration and tubular resorption.

As with amphibians, the teleost glomerulus is vascularized by capillaries derived from the dorsal aorta (Balfour, 1880; Tytler, 1988; Tytler *et al.*, 1996; Drummond *et al.*, 1998; Majumdar and Drummond, 1999; Drummond, 2000). However, the glomeruli are more medial and abut at the midline. A particularly close relationship of the glomerulus to the aorta is seen in zebrafish where the pronephric primordium appears to reside almost within the lumenal space of the aorta (Drummond *et al.*, 1998).

In the chick embryo, pronephric tubules have been demonstrated extending from somite 5 to 15 (Romanoff, 1960) and external glomera have been shown to form between somites 9 and 15 (Fig. 5.3B; Davies, 1950). Ultrastructurally, external glomera appear similar to mesonephric glomeruli (Jacob *et al.*, 1977), suggesting that in embryos that form a mesonephros, the distinction between pronephric and mesonephric nephrons may be subtle and graded. In fact, similar to pronephric glomera, mesonephric glomeruli of several species of freshwater vertebrates maintain a direct connection to the coelom via extended peritoneal funnels (Gaeth *et al.*, 1999). This feature of the elephant and manatee embryonic kidney has been interpreted as further evidence of a phylogenetic relationship between *Proboscidea* and *Sirenia* (Gaeth *et al.*, 1999).

At the ultrastructural level, the filtration barrier of the pronephric glomus or glomerulus as revealed by electron microscopy is typical of higher vertebrate glomeruli (Hentschel and Elger, 1996; Drummond *et al.*, 1998; Majumdar and Drummond, 1999). In zebrafish larvae, podocyte cell bodies with extensive foot processes line the outer aspect of the glomerular basement membrane (Fig. 5.4). The glomerular basement membrane itself is trilaminar with a central lamina densa bordered by lamina rarae on surfaces apposed to cellular processes. Capillary endothelia show fenestrations, which might be expected to allow a high rate of

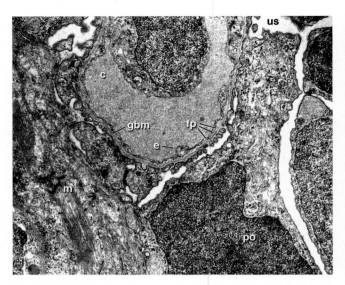

Figure 5.4 Ultrastructure of the pronephros in zebrafish. Glomerulus and filtration barrier of the wild-type pronephros at 3.5 days postfertilization. The filtration barrier is well developed, displaying an endothelium with open pores (without diaphragm), a thin trilaminar glomerular basement membrane (gbm), and podocytes (po) with primary processes and numerous interdigitating foot processes (fp). The filtration slits between the foot processes are bridged by one or sometimes two slit membranes. Mesangial cells (m) are also evident. us, urinary space; c, capillary space; e, endothelial cell. × 36,500.

glomerular filtration, facilitating the elimination of large amounts of fluid (Hickman and Trump, 1969; Majumdar and Drummond, 2000). Also present are cells resembling mesangial cells, which bridge the capillary branches. The glomerular capillary tuft is enclosed by a parietal epithelial cell layer (Bowman's capsule).

D. Relationship of the Pronephros to Invertebrate Nephridia

Despite earlier claims that the glomerulus was a vertebrate invention, appearing first as fish adapted to life in freshwater (Marshall and Smith, 1930), it is clear that segmental tubules and segmentally reiterated glomus-like structures exist in invertebrates (Ruppert and Smith, 1988). The invertebrate excretory system occurs in two principal forms. Protonephridia consist of blind-ended tubules leading from the ectodermal surface to the coelomic cavity, terminating in a flagella-bearing solenocyte or flame cell (Fig. 5.5A). Metanephridia consist of a patent tubule leading from the exterior to the coelomic cavity functionally associated with vascular outpouchings in the coelom covered by podocyte-like cells (Fig. 5.5B; Ruppert and Smith, 1988). These vascular specializations appear to be highly similar to the vertebrate glomus or "external" glomerulus. Both protonephridial flame cells and metanephridial podocytes have

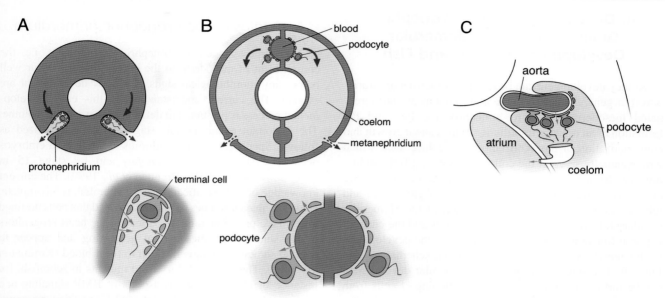

Figure 5.5 Invertebrate excretory systems and the cephalochordate kidney. (A) Invertebrate protonephridia consist of blind-ended tubules projecting from the ectoderm into connective tissue or the coelom. Terminal cells within the protonephridium are podocyte like in that they extend elaborate foot processes that contribute to a fluid filter involved in osmoregulation (Ruppert, 1994). Apical flagella and microvilli help drive filtered fluid through the pronephridium. (B) An invertebrate metanephridium is a ciliated pore or funnel that communicates between the coelom and the exterior. Metanephridia are thought to drain excess water from the coelom. Metanephridia are functionally paired with segmentally repeated outpouchings of the vasculature, which are covered on the coelomic surface with podocyte-like cells, creating a vascular filtration site similar to the vertebrate glomus. (C) Cephalochordate (Amphioxis) kidneys are segmentally repeated associations of podocyte-covered blood vessels suspended in the subchordal coelom and tubules that drain the blood filtrate into a common space surrounding the pharynx called the atrium (cephalochordates do not possess a pronephric duct). Much like the vertebrate glomus, cephalochordate kidneys are supplied by the aorta and are positioned in a nephrocoel-like space continuous with the coelom. Podocytes have extensive basal foot processes and bear apical flagella similar to the terminal cells of the invertebrate protonephridium. For a complete description see, Ruppert (1994).

extensive foot processes and form filtration slits on basement membranes (Ishii, 1980; Ruppert and Smith, 1988; Ruppert, 1994). Semper's original view that the segmental tubules of vertebrates were homologous with the metanephridia of Annelida was challenged by Goodrich, who maintained that the different germ layer origins of these epithelia (nephridia from ectoderm; coelomostomes/tubules from mesoderm) precluded homology (reviewed in Goodrich, 1930). Nonetheless, podocyte-like cells on blood vessels have been observed in several families of polychaete worms, and evidence for ultrafiltration of iron dextran (MW 4000–6000) by glomerular podocytes has been obtained in the acorn worm, *Saccoglossus kowalevskii* (Balser, 1985; Ruppert and Smith, 1988; Hansen, 1996). Based on detailed morphological and developmental observations, Ruppert (1994) argues that the vertebrates inherited the essential components of the archinephron, as well as the overall functional organization of the excretory system, from invertebrate ancestors.

A related primitive podocyte-like cell, the nephrocyte, is a component of the pericardium in *Drosophila* and is thought to regulate the composition of the hemolymph by selective fluid filtration based on molecular size and charge (Crossley, 1985; Rugendorff *et al.,* 1994). In conjunction with the Malpighian tubules (Chapter 2), nephrocytes have been proposed to

contribute to a filtration/excretory system for insects (Crossley, 1985). Although the excretory system of the nematode *Caenorhabditis elegans* has been well studied, it appears to be secretory in nature and lacks podocyte-like cells (Nelson *et al.,* 1983; Nelson and Riddle, 1984).

The excretory organs of the cephalochordate Amphioxis are paired, segmental, podocyte-covered blood vessels that protrude into the subchordal coelom, a cavity confluent with the coelom and similar to the nephrocoele (Fig. 5.5C). The podocyte-like cells in Amphioxis are unusual in that they bear a flagellum and, in some respects, resemble both protonephridial flame cells and metanephridial podocytes (Nakao, 1965; Moller and Ellis, 1974; Ruppert and Barnes, 1994). The blood filtrate is directed into so-called "nephridioducts," which open individually to the exterior atrium (a structure surrounding the gills), as Amphioxis lacks a common or pronephric duct (reviewed in Ruppert, 1994). The Amphioxis excretory unit thus may represent the closest link to the primitive, segmental "archinephron." In Ruppert's view, the indeterminate morphology of the podocytes in Amphioxis suggests that invertebrate filtration cells may be functionally interchangeable and could represent the ancestor of the vertebrate podocyte (Ruppert, 1994).

II. Development of the Pronephric Glomus: Stages of Glomerular Development in Frogs and Fish

Kidney development can be divided somewhat arbitrarily into four general stages: (1) the commitment of undifferentiated mesodermal cells to a nephrogenic fate and the formation of an organ primordium, (2) the caudal growth and epithelialization of the Wolffian or pronephric duct, which will eventually serve as the collecting system, (3) the induction and formation of the nephron and the cell patterning events associated with the differentiation of glomerular and tubular cell types, and (4) the formation of the glomerular capillary tuft by ingrowing endothelial cells and the onset of nephron function (Saxén, 1987). Stages in the development of the zebrafish pronephros are illustrated schematically in Fig. 5.6. The stages of pronephric glomerular development can be further subdivided as (1) a partitioning of cells from the intermediate mesoderm that will serve as podocyte progenitors, (2) the differentiation of podocytes with the elaboration of foot processes and the formation of a glomerular basement membrane, (3) the interaction of podocytes with nearby aortic endothelial cells to form the glomerular tuft, and finally (4) the integration of supporting mesangial cells into the glomerular architecture.

A. Origins of the pronephric primordium

The earliest stages of pronephric development, i.e. the commitment of mesoderm to the kidney lineage, occur well before any morphological signs of kidney development are evident. The current understanding of this early developmental process is presented in detail elsewhere in this volume. The important events of this stage can be summarized as follows. Specification of the glomus in amphibian embryos occurs by stage 12.5 in *Xenopus* and by stage 15 in *Ambystoma* (Fales, 1935; Brennan *et al.,* 1999). The current knowledge of cell specification in the zebrafish is incomplete. However, lineage studies have demonstrated that cells destined to form the pronephros lie just dorsal to the heart progenitors in the shield stage mesodermal germ ring and appear to overlap somewhat with cells fated to form blood (Kimmel *et al.,* 1990). The so-called dorsalized mutants in zebrafish, for instance *swirl*, affect various aspects of BMP signaling and show a reduction or elimination of pax2.1—positive, presumptive kidney intermediate mesoderm (Mullins *et al.,* 1996).

In *Ambystoma,* transplantation of presumptive pronephric tissue from stage 15 embryos into older hosts resulted in the differentiation of pronephric tubules and, in some cases, the formation of ectopic glomera (Fales, 1935). Although evidence suggests that the glomus and pronephric tubules are

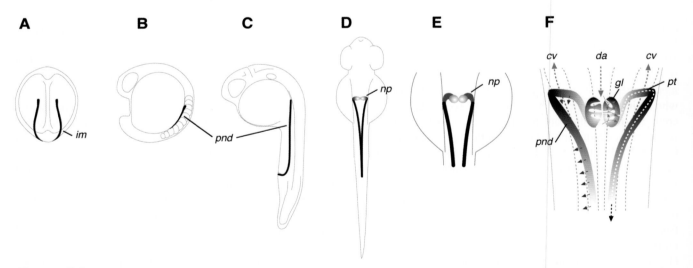

Figure 5.6 Developmental stages in formation of the zebrafish pronephros. (A) One to eight somite stage (10–13 hpf): Commitment of the intermediate mesoderm. Expression patterns of pax2.1 and lim-1 define the posterior regions of the intermediate mesoderm (im) and suggest that a nephrogenic field is established early in development. (B) Early somitogenesis to Prim-5 stage (12–24 hpf): Pronephric duct growth and formation of nephron primordia. Epithelialization of the pronephric duct (pnd) follows behind somitogenesis and is complete by 24 hpf (C). Nephron primordia (np) form at the anterior tips of the pronephric ducts by 24 hpf. (D and E) Prim-5 to High-Pec stage (24–42 hpf): Nephron primordia (np) ventral to somite three separate from the coelom and undergo integrated morphogenesis, giving rise to pronephric glomeruli (gl) and tubules (pt). (F) High-Pec stage to hatching (42–48 hpf): The onset of pronephric function. Vascularization of the pronephros by the ingrowth of capillaries from the dorsal aorta (da) starts at 40–42 hpf and continues over the next 24 h, correlating with the onset of glomerular filtration. Blood (red arrows) from the dorsal aorta (da) is filtered at the glomerulus (gl; arrows). The filtrate passes to the pronephric tubules (pt) and ducts (pnd) where it is modified by active transport processes and the recovery of small metabolites (blue arrows) to the cardinal veins (cv). Arrows in the vasculature indicate the direction of blood flow. Stages of zebrafish development refer to Kimmel *et al.* (1995).

specified at the same time (Brennan *et al.,* 1998, 1999), Howland (1921) and later Fales (1935) demonstrated that the glomus can develop after removal of the tubule and duct primordia, suggesting independent origins for these structures (Vize *et al.,* 1995). Nonetheless, interactions between tubular and glomerular components of the pronephric primordium can be inferred from the phenotype of the zebrafish mutant *no isthmus* (pax2.1), where absence of a functional tubule-specific gene, pax2.1, results in the absence of pronephric tubules and expanded expression of podocyte cell-specific markers (Majumdar *et al.,* 2000, see later).

In both amphibians and teleost fish, the first morphological sign of kidney organogenesis is the formation of the pronephric primordium as a mass of intermediate mesoderm lying ventral and lateral to the anterior somites (Armstrong, 1932; Nieuwkoop and Faber, 1994; Kimmel *et al.,* 1995; Vize *et al.,* 1995, 1997; Drummond *et al.,* 1998). The origins of the pronephric nephron are linked intimately to the formation of the coelom (Fig. 5.7). In the vertebrate embryo, the coelom is formed from the lateral plate mesoderm, a loose mesenchyme situated lateral to the somites. The splitting of the lateral plate into the somatic mesoderm underlying the body wall and the splanchnic mesoderm surrounding the gut is brought about by a mesenchyme to epithelial transition of cells in the center of the lateral plate, forming the coelomic cavity (Meier, 1980; Gilbert, 1991; Funayama *et al.,* 1999). The emergence of a primary pronephric fold between splanchnic and somatic mesodermal layers is first observed at stage 21 in *X. laevis,* at the 14 somite stage in *Fundulus,* and at the pharyngula stage (24 hpf) in zebrafish (Armstrong, 1932; Nieuwkoop and Faber, 1994; Drummond *et al.,* 1998, Chapter 3). The precursor of the glomus in amphibians forms at the medial edge

of the splanchnic mesoderm, in the intermediate mesoderm below somites 3 through 5 (Nieuwkoop and Faber, 1994; Vize *et al.,* 1997). In teleosts, including zebrafish, a primary pronephric fold pinches off completely from the coelom at the anterior extent of the formed pronephric ducts and ventral to the second somite, forming the pronephric nephron primordium (Armstrong, 1932; Tytler, 1988; Tytler *et al.,* 1996; Drummond *et al.,* 1998). The nephron primordia in teleosts undergo cell patterning events, resulting in differentiation of the pronephric glomerular and tubular components (see later) (Drummond *et al.,* 1998). Based on gene expression studies (see later), similar events are likely to occur in amphibian pronephric primordia (Carroll and Vize, 1996; Wallingford *et al.,* 1998; Carroll *et al.,* 1999).

B. Differentiation of Podocytes and Interactions with Blood Vessels

Vascularization of the glomerulus in teleosts occurs after the completion of pronephric duct and tubule development (Marshall and Smith, 1930; Armstrong, 1932; Tytler, 1988; Tytler *et al.,* 1996; Drummond *et al.,* 1998). In zebrafish, the bilateral glomerular primordia coalesce at 36–40 hpf ventral to the notochord, bringing the presumptive podocytes into intimate contact with capillary-forming endothelial cells of the dorsal aorta (Fig. 5.8A) (Drummond *et al.,* 1998). At this stage, podocytes begin to elaborate foot processes and show extensive interdigitation with the basement membrane supporting the overlying aortic endothelial cells (Fig. 5.8B). Further capillary growth and invasion of endothelial cells result in convolution of the glomerular basement membrane (Fig. 5.8C). Although the

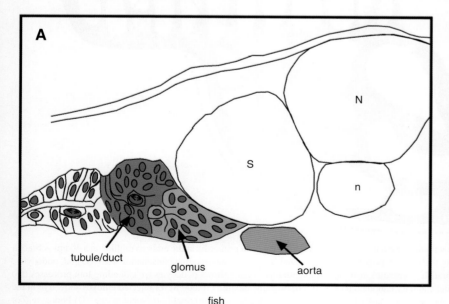

fish

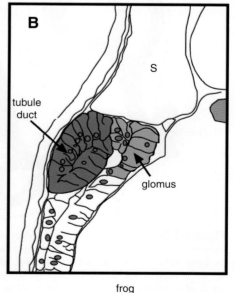

frog

Figure 5.7 Location of forming pronephric primordia in fish and frogs. In both the trout and *Xenopus* the tubule and duct primordia is colored blue, and the primordia of the glomus green. Note the forming coelom separating the two primordia in *Xenopus* (figure by Peter Vize).

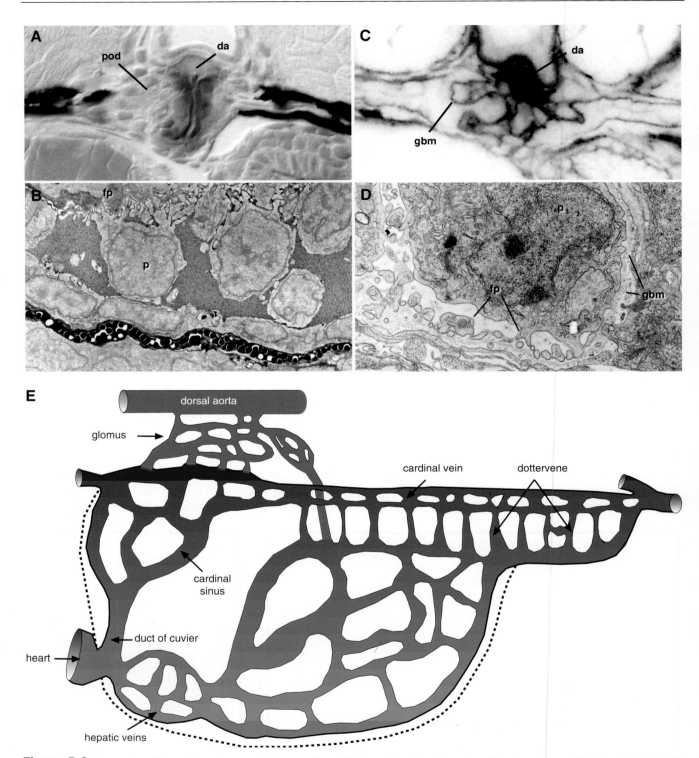

Figure 5.8 Interaction of pronephric podocytes with the vasculature. (A) Apposition of nephron primordia at the embryo midline in a 40 hpf zebrafish embryo. Aortic endothelial cells in the cleft separating the nephron primordia are visualized by endogenous alkaline phosphatase activity. pod, podocytes; da, dorsal aorta. (B) Ultrastructure of the forming zebrafish glomerulus at 40 hpf. A longitudinal section shows podocytes (p) extending foot processes (fp) in a dorsal direction and in close contact with overlying capillary endothelial cells. (C) Rhodamine dextran (10,000 MW)-injected embyos show dye in the dorsal aorta (da) and in the glomerular basement membrane (gbm) shown here graphically inverted from the original fluorescent image. (D) Podocyte foot process formation does not require signals from endothelial cells, as evidenced by the appearance of foot processes (fp) in cloche mutant embryos, which lack all vascular structures. gbm, glomerular basement membrane; p, podocyte cell body. (E) Diagram of the pronephric blood supply of the sturgeon.

timing of podocyte foot process formation correlates closely with interactions with the vasculature, the morphological differentiation of podocytes does not require any signals originating from endothelial cells. This can be inferred from the presence of foot processes on podocytes in the zebrafish mutant *cloche*, which lacks endothelial cells (Fig. 5.8D) (Majumdar and Drummond, 1999).

At the time of glomerular vascularization, the dorsal aorta is a well-formed vessel. In both zebrafish and *Xenopus*, endothelial cells branch out from the dorsal aorta as they invade the glomerular epithelium (Armstrong, 1932; Tytler, 1988; Tytler *et al.,* 1996; Cleaver *et al.,* 1997; Vize *et al.,* 1997; Drummond *et al.,* 1998; Majumdar and Drummond, 1999). This observation suggests that the glomerular capillaries form by sprouting, implying an angiogenic mechanism for glomerular formation. The zebrafish mutant *floating head* lacks a notochord and, as a consequence, fails to form a dorsal aorta (Fouquet *et al.,* 1997). Nonetheless, clusters of podocytes form in lateral positions and appear to recruit endothelial cells to form abnormal but functional glomeruli (Majumdar and Drummond, 2000). In this instance, endothelial cells that contribute to the glomerulus do not appear associated with any preformed vessel, suggesting that under some circumstances, vasculogenic mechanisms may also play a role in capillary tuft formation.

Formation of the cardinal vein and vascular sinus that will surround the pronephric tubules and eventually receive recovered small molecules from the glomerular filtrate occurs concurrently with pronephric duct growth and epithelialization. In teleosts and amphibians, cardinal vein and vascular sinus formation involve the condensation of mesenchyme and a conversion into endothelial cells (Armstrong, 1932; Cleaver *et al.,* 1997; Brandli, 1999). Thus, in contrast to the formation of the glomerular capillary tuft, blood sinus formation may be primarily a vasculogenic event. The mature pronephric vasculature of the sturgeon is very similar to the vasculature of the amphibian pronephros (Vize *et al.,* 1997; Brandli, 1999, Fig. 5.8E). The pronephric vasculature in other teleosts (zebrafish) differs in that there is no obvious connection of the glomerular capillary tuft to the veinous system and blood seems to percolate through the glomerulus and return to the dorsal aorta (I. A. Drummond, unpublished observations).

Mesangial cells have been observed in the zebrafish and lamprey pronephroi; however, there is currently little information on the mechanisms of their integration into the pronephric glomerulus.

III. Gene Expression and Function in Pronephric Glomerular Development

Conserved patterns of gene expression and gene function between the pronephros and the meso- and metanephric kidneys demonstrate that despite the differences in final organ morphology, cell differentiation in these three kidney types may employ similar molecular mechanisms.

The Wilms' tumor suppressor *wt1, lim-1, pax8,* and *pax2* genes are important regulators of vertebrate kidney development and perform important functions during multiple stages of nephrogenesis (reviewed in Lechner and Dressler, 1997). During early stages of pronephric development in *Xenopus* and zebrafish, *lim-1* and *pax8* are expressed in the nephrogenic intermediate mesoderm (Taira *et al.,* 1994; Toyama and Dawid, 1997; Pfeffer *et al.,* 1998; Carroll and Vize, 1999; Carroll *et al.,* 1999; Heller and Brandli, 1999). The results of mRNA injection studies in *Xenopus* embryos suggest that these two genes participate in defining the extent of mesoderm committed to form the pronephros (Carroll and Vize, 1999). Following expression of *lim-1* and *pax8, wt1* is expressed at stage 20 in *Xenopus* in a ridge on the dorsal side of the pronephric tubule and duct primordium (Carroll and Vize, 1996; Semba *et al.,* 1996). This staining corresponds to the splanchnic layer of the intermediate mesoderm where the glomus will eventually form. In zebrafish, bilateral bands of expression are observed in the intermediate mesoderm between somites 1 and 3 at the 10 somite stage. As the pronephric nephron primordia form and pinch off from the coelom (24 hpf), *wt1* staining becomes restricted to the primordia ventral to somite 3 and later to the glomerular epithelium as the primordia coalesce at the embryo midline (Fig. 5.9, Drummond *et al.,* 1998; Majumdar and Drummond, 1999, 2000; Majumdar *et al.,* 2000).

A. Vascularization of the Glomerulus

In addition to the expression of *wt1*, zebrafish pronephric podocytes express vascular endothelial growth factor (VEGF; Fig. 5.9E) (Majumdar and Drummond, 1999; Majumdar *et al.,* 2000), which is known to be important for blood vessel formation in a variety of vertebrate systems (Shalaby *et al.,* 1995; Carmeliet *et al.,* 1996; Ferrara *et al.,* 1996). In a complementary manner, capillary forming endothelial cells express flk-1, a VEGF receptor, and an early marker of the endothelial differentiation program (Figs. 5.9C and 5.9F) (Majumdar and Drummond, 1999). In zebrafish embryos at 40 hpf, *flk-1*-positive endothelial cells can be observed invading the glomerular epithelium (Fig. 5.9F). Endothelial cells also express alkaline phosphatase, which can be used in zebrafish to visualize the formation of blood vessels (Fig. 5.8A; Majumdar and Drummond, 1999; Drummond, 2000). In zebrafish *floating head* mutant embryos, the normal source of glomerular blood vessels, the dorsal aorta, is absent (Fouquet *et al.,* 1997) and the pronephric nephron primordia do not fuse at the midline, instead remaining at ectopic lateral positions in the embryo (Majumdar and

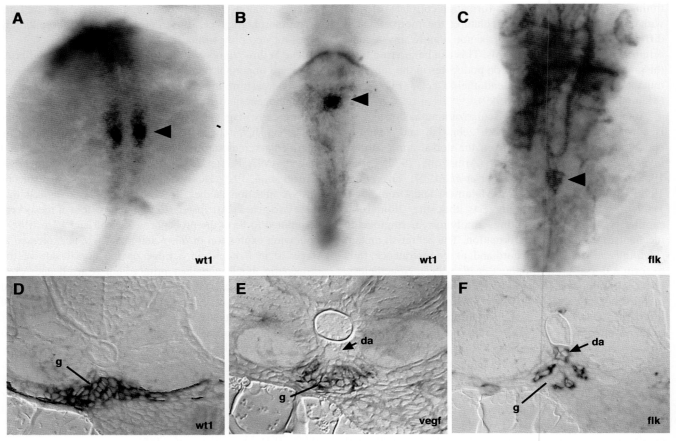

Figure 5.9   Gene expression in the developing pronephric glomerulus in zebrafish. (A) Whole-mount *in situ* hybridization of 24 hpf zebrafish embryos with a Wilms' tumor suppressor (wt1) probe reveals the progenitors of the podocytes (arrowhead) prior to nephron formation and (B) following midline fusion of the nephron primordia (arrowhead). (C) The receptor for vascular endothelial growth factor, flk, is expressed in the vascular endothelial cells throughout the embryo and in the forming glomerular capillaries (arrowhead). (D) Transverse section through the pronephros shows the restriction of wt1 expression to the forming podocytes. (E) vegf is expressed specifically in podocytes at the time of capillary ingrowth, whereas expression of its receptor, flk (F), can be observed on glomerular endothelial cells.

Drummond, 2000). These podocytes continue to express *wt1* and *vegf* and appear to recruit *flk-1*-positive endothelial cells to form a functional glomerulus (Majumdar and Drummond, 2000). These results support the idea that podocytes, by expressing *vegf,* play a primary role in attracting and assembling the glomerular capillary tuft. In *Xenopus, vegf* expression is observed in the forming glomus at stage 32, and *flk-1*-positive cells are seen in the forming blood sinus that will come to lie adjacent to the pronephric tubules and ducts as part of the veinous system (Cleaver *et al.,* 1997). Formation of the glomerular capillary tuft coincides with formation of the glomerular filtration barrier and the onset of glomerular filtration. Aside from the finding that antibodies raised against collagen type IV α1, 2 react positively with the zebrafish pronephric glomerular basement membrane (Majumdar and Drummond, 1999), little is known about gene expression required for pronephric glomerular basement membrane formation.

B. Nephron Patterning

In *Xenopus,* zebrafish, and chick embryos, the *pax2.1* gene is expressed first in the intermediate mesoderm prior to overt duct or tubule differentiation (Krauss *et al.,* 1991; Puschel *et al.,* 1992; Heller and Brandli, 1997; Pfeffer *et al.,* 1998; Carroll *et al.,* 1999; Heller and Brandli, 1999; Drummond, 2000; Majumdar *et al.,* 2000; Mauch *et al.,* 2000). In *Xenopus, pax2* is expressed simultaneously in the forming pronephric tubules and pronephric ducts (Heller and Brandli, 1997; Carroll *et al.,* 1999), whereas in zebrafish, duct expression precedes expression in the tubules, which form 6 h after duct development is completed (Fig. 5.10; Drummond *et al.,* 1998; Majumdar *et al.,* 2000). Several lines of evidence suggest that the pronephric glomerulus and tubule may develop under the influence of mutually antagonistic molecular signals involving *pax2* and *wt1.* Early ectopic expression of *Xenopus wt1* prevents the later expression of tubule and duct markers. The loss of marker expression and the inhibition of pronephric tubule

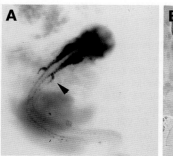
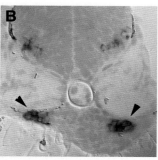

Figure 5.10 Expression of zebrafish pax2.1 in the forming pronephric tubules. (A) Whole-mount *in situ* hybridization reveals increased expression of pax2.1 at 30 hpf in the pronephric tubule progenitors (arrowhead), whereas transverse sections (B) show that the immediately adjacent cells in the nephron primordium ventral to the aorta (medial podocyte progenitors) are negative [arrowheads in (B); pax2.1 positive tubule progenitor cells].

development were shown to be a result of an inhibition of pronephric anlage formation (Wallingford *et al.,* 1998). The zebrafish mutant *no isthmus* lacks a functional *pax2.1* gene (Brand *et al.,* 1996). As might be expected from the expression pattern of this gene and its known function during mouse nephrogenesis (Torres *et al.,* 1995), *noi* embryos fail to form pronephric tubules (Majumdar *et al.,* 2000). However, in addition to the loss of tubules phenotype, *noi* embryos show an abnormal expanded expression of the podocyte markers *wt1* and *vegf* in cells of the pronephric duct (Majumdar *et al.,* 2000). The implication from these studies is that a normal function for *pax2.1* is to repress podocyte gene expression and restrict podocyte development to medial nephron primordia cells. Taken together, results from zebrafish and *Xenopus* suggest that *wt1* and *pax2.1* play important roles in demarcating the boundary between glomerular and tubular epithelium in the forming nephron.

IV. Summary: Future Prospects

For studies of kidney development, the pronephros offers a relevant model system that is amenable to both genetic and embryological approaches. Although there are clear differences in organ morphology among the pronephros, mesonephros, and metanephros, at the cellular level, many parallels exist that can be exploited to further our understanding of nephrogenesis. For example, in the development of all kidney forms, the conversion of mesenchyme to epithelia is a central event (Saxén, 1987). In the pronephros, the coelom and the nephrocoele form by a mesenchyme to epithelial transition (Goodrich, 1930; Meier, 1980; Funayama *et al.,* 1999), whereas in the metanephros, renal vesicles form reiteratively by a similar process (Saxén, 1987). Not unlike the nephrocoele, a renal vesicle may be thought of as an epithelial cavity that will receive vascular

ingrowth on one end while the other proliferates to become the proximal and distal tubule. Most all cell types found in the metanephros are also found in the pronephros, including endothelial cells, podocytes, mesangial cells, and polarized tubular epithelial cells. Similarities in gene expression and function also suggest that studies of the pronephros will have predictive value for studies of nephrogenesis in higher vertebrates.

The current developmental model systems most likely to yield further insights into pronephric development are the frog, *X. laevis*, the zebrafish, and the chick. In *Xenopus*, testing gene function by microinjecting mRNA has a long history and will no doubt see refinements as conditional expression systems and transgenesis become more widely employed (Kolm and Sive, 1995; Gammill and Sive, 1997). Transplantation and explant cultures are also likely to reveal the sources and identities of molecules that regulate pronephric cell differentiation. *Xenopus tropicalis,* a diploid with a reasonably short generation time, also offers the possibility of combining the rich embryology of *X. laevis* with the forward genetics approach to studies of gene function (Amaya *et al.,* 1998). The zebrafish offers advantages as a classical genetic system to pursue studies of organogenesis. Large-scale mutagenesis screens have yielded a wealth of mutants (Driever *et al.,* 1996), and current insertional approaches promise to speed the cloning of genes essential for pronephric development (Amsterdam *et al.,* 1999). As another fish model system for vertebrate genetics, the medaka *(Oryzias latipes)* is receiving increased attention (Ishikawa, 2000). Future mutagenesis screens using promoter/gene trap approaches (Friedrich and Soriano, 1991) or using fish engineered to express green fluorescent protein in a tissue-specific fashion (Long *et al.,* 1997) are likely to yield more insights into gene function in the developing kidney. The chick embryo enjoys the advantage of being manipulated readily by microdissection (Obara-Ishihara *et al.,* 1999) and by retrovirus-mediated gene expression (Laufer *et al.,* 1997; Logan and Tabin, 1998). Embryological studies in the chick clearly have demonstrated that somites are the source of a signal driving early pronephric development (Mauch *et al.,* 2000). Further studies of the chick pronephros promise to identify the molecules involved, opening the way to testing their general relevance to the development of higher kidney forms.

As we enter an era where a complete genome sequence is available for the human and a number of model organisms, the focus of developmental studies is shifting to understanding gene function and putting gene sequence and gene regulatory networks fully in the context of the developing organism. To this end, developmental model organisms will play an increasingly important role as readily manipulated systems that can yield relevant information on the process of nephrogenesis.

References

Adelmann, H. (1966). "Marcello Malpighi and the Evolution of Embryology." Cornell Univ. Press, Ithaca, NY.

Agarwal, S., and John, P. A. (1988). Studies on the development of the kidney of the guppy, *Lebistes reticulatus*. 1. The development of the pronephros. *J. Anim. Morphol. Physiol.* **35,** 17–24.

Amaya, E., Offield, M. F., and Grainger, R. M. (1998). Frog genetics: *Xenopus tropicalis* jumps into the future. *Trends Genet.* **14,** 253–255.

Amsterdam, A., Burgess, S., Golling, G., Chen, W., Sun, Z., Townsend, K., Farrington, S., Haldi, M., and Hopkins, N. (1999). A large-scale insertional mutagenesis screen in zebrafish. *Genes Dev.* **13,** 2713–2724.

Armstrong, P. B. (1932). The embryonic origin of function in the pronephros through differentiation and parenchyma-vascular association. *Am. J. Anat.* **51,** 157–188.

Balfour, F. M. (1880). "A Treatise on Comparative Embryology." Macmillan, London.

Balser, E. J. (1985). Ultrastructure and function of the proboscis complex of Saccoglossus (Enteropneusta). *Am. Zool.* **25,** 41.

Brand, M., Heisenberg, C. P., Jiang, Y. J., Beuchle, D., Lun, K., Furutani-Seiki, M., Granato, M., Haffter, P., Hammerschmidt, M., Kane, D. A., *et al.* (1996). Mutations in zebrafish genes affecting the formation of the boundary between midbrain and hindbrain. *Development* **123,** 179–190.

Brandli, A. W. (1999). Towards a molecular anatomy of the *Xenopus* pronephric kidney. *Int. J. Dev. Biol.* **43,** 381–395.

Brennan, H. C., Nijjar, S., and Jones, E. A. (1998). The specification of the pronephric tubules and duct in *Xenopus laevis. Mech. Dev.* **75,** 127–137.

Brennan, H. C., Nijjar, S., and Jones, E. A. (1999). The specification and growth factor inducibility of the pronephric glomus in *Xenopus laevis. Development* **126,** 5847–5856.

Burns, R. K. (1955). Urogenital system. *In* "Analysis of Development" (B. J. Willier, P. Weiss, and V. Hamburger, eds.), pp.462–491. Saunders, Philadelphia.

Carmeliet, P., Ferreira, V., Breier, G., Pollefeyt, S., Kieckens, L., Gertsenstein, M., Fahrig, M., Vandenhoeck, A., Harpal, K., Eberhardt, C., *et al.* (1996). Abnormal blood vessel development and lethality in embryos lacking a single VEGF allele. *Nature* **380,** 435–439.

Carroll, T. J., and Vize, P. D. (1996). Wilms' tumor suppressor gene is involved in the development of disparate kidney forms: Evidence from expression in the *Xenopus* pronephros. *Dev. Dyn.* **206,** 131–138.

Carroll, T. J., and Vize, P. D. (1999). Synergism between Pax-8 and lim-1 in embryonic kidney development. *Dev. Biol.* **214,** 46–59.

Carroll, T. J., Wallingford, J. B., and Vize, P. D. (1999). Dynamic patterns of gene expression in the developing pronephros of *Xenopus laevis. Dev. Genet.* **24,** 199–207.

Cleaver, O., Tonissen, K. F., Saha, M. S., and Krieg, P. A. (1997). Neovascularization of the *Xenopus* embryo. *Dev. Dyn.* **210,** 66–77.

Crossley, A. C. (1985). Nephrocytes and pericardial cells. *In* "Comprehensive Insect Physiology, Biochemistry and Pharmacology" (G. A. Kerkut and L. I. Gilbert, eds.). Pergamon Press, Oxford.

Davies, J. (1950). The pronephros and the early development of the mesonephros in the chick. *J. Anat.* **84,** 95–103.

Driever, W., Solnica-Krezel, L., Schier, A. F., Neuhauss, S. C., Malicki, J., Stemple, D. L., Stainier, D. Y., Zwartkruis, F., Abdelilah, S., Rangini, Z., *et al.* (1996). A genetic screen for mutations affecting embryogenesis in zebrafish. *Development* **123,** 37–46.

Drummond, I. A. (2000). The zebrafish pronephros: A genetic system for studies of kidney development. *Pediatr. Nephrol.* **14,** 428–435.

Drummond, I. A., Majumdar, A., Hentschel, H., Elger, M., Solnica-Krezel, L., Schier, A. F., Neuhauss, S. C., Stemple, D. L., Zwartkruis, F., Rangini, Z., *et al.* (1998). Early development of the zebrafish pronephros and analysis of mutations affecting pronephric function. *Development* **125,** 4655–4667.

Ellis, L. C., and Youson, J. H. (1989). Ultrastructure of the pronephric kidney in upstream migrant sea lamprey, *Petromyzon marinus* L. *Am. J. Anat.* **185,** 429–443.

Ellis, L. C., and Youson, J. H. (1990). Pronephric regression during larval life in the sea lamprey, *Petromyzon marinus* L: A histochemical and ultrastructural study. *Anat. Embryol. (Berl.)* **182,** 41–52.

Fales, D. E. (1935). Experiments on the development of the pronephros of *Ambystoma punctatum. J. Exp. Zool.* **72,** 147–173.

Ferrara, N., Carver-Moore, K., Chen, H., Dowd, M., Lu, L., O'Shea, K. S., Powell-Braxton, L., Hillan, K. J., and Moore, M. W. (1996). Heterozygous embryonic lethality induced by targeted inactivation of the VEGF gene. *Nature* **380,** 439–442.

Field, H. H. (1891). The development of the pronpehros and segmental duct in Amphibia. *Bull. Mus. Comp. Zoöl. Harvard* **21,** 201–340.

Fouquet, B., Weinstein, B. M., Serluca, F. C., and Fishman, M. C. (1997). Vessel patterning in the embryo of the zebrafish: Guidance by notochord. *Dev. Biol.* **183,** 37–48.

Fox, H. (1963). The amphibian pronephros. *Quart. Rev. Biol.* **38,** 1–25.

Friedrich, G., and Soriano, P. (1991). Promoter traps in embryonic stem cells: A genetic screen to identify and mutate developmental genes in mice. *Genes Dev.* **5,** 1513–1523.

Funayama, N., Sato, Y., Matsumoto, K., Ogura, T., and Takahashi, Y. (1999). Coelom formation: Binary decision of the lateral plate mesoderm is controlled by the ectoderm. *Development* **126,** 4129–4138.

Gaeth, A. P., Short, R. V., and Renfree, M. B. (1999). The developing renal, reproductive, and respiratory systems of the African elephant suggest an aquatic ancestry. *Proc. Natl. Acad. Sci. USA* **96,** 5555–5558.

Gammill, L. S., and Sive, H. (1997). Identification of otx2 target genes and restrictions in ectodermal competence during *Xenopus* cement gland formation. *Development* **124,** 471–481.

Gilbert, S. F. (1991). "Developmental Biology." Sinauer, Sunderland, MA.

Goodrich, E. S. (1930). "Studies on the Structure and Development of Vertebrates." Macmillan, London.

Hansen, U. (1996). Electron microscopic study of possible sites of ultrafiltration in *Lumbricus terrestris* (Annelida, Oligochaeta). *Tissue Cell* **28,** 195–203.

Heller, N., and Brandli, A. W. (1997). *Xenopus* Pax-2 displays multiple splice forms during embryogenesis and pronephric kidney development. *Mech. Dev.* **69,** 83–104.

Heller, N., and Brandli, A. W. (1999). *Xenopus* Pax-2/5/8 orthologues: Novel insights into Pax gene evolution and identification of Pax-8 as the earliest marker for otic and pronephric cell lineages. *Dev. Genet.* **24,** 208–219.

Hentschel, H., and Elger, M. (1996). Functional morphology of the developing pronephric kidney of zebrafish. *J. Am. Soc. Nephrol.* **7,** 1598.

Hickman, C. P., and Trump, B. F. (1969). The kidney. *In* "Fish Physiology" (Hoar and Randall, eds.), pp. 91–239. Academic Press, New York.

Howland, R. B. (1921). Experiments on the effect of the removal of the pronephros of *Ambystoma punctatum. J. Exp. Zool.* **32,** 355–384.

Ishii, S. (1980). The ultrastructure of the protonephridial flame cell of the freshwater planarian *Bdellocephala brunnea. Cell Tissue Res.* **206,** 441–449.

Ishikawa, Y. (2000). Medakafish as a model system for vertebrate developmental genetics. *Bioessays* **22,** 487–495.

Jacob, H. J., Jacob, M., and Christ, B. (1977). Die Ultrastruktur der externen Glomerula. Ein Beitrag zur Nierenentwicklung bie Hühnerembryonen. *Verh. Anat. Ges.* **71,** 909–912.

Kimmel, C. B., Ballard, W. W., Kimmel, S. R., Ullmann, B., and Schilling, T. F. (1995). Stages of embryonic development of the zebrafish. *Dev. Dyn.* **203,** 253–310.

Kimmel, C. B., Warga, R. M., and Schilling, T. F. (1990). Origin and organization of the zebrafish fate map. *Development* **108,** 581–594.

Kolm, P. J., and Sive, H. L. (1995). Efficient hormone-inducible protein function in *Xenopus laevis. Dev. Biol.* **171,** 267–272.

Krauss, S., Johansen, T., Korzh, V., and Fjose, A. (1991). Expression of the zebrafish paired box gene pax[zf-b] during early neurogenesis. *Development* **113**, 1193–1206.

Laufer, E., Dahn, R., Orozco, O. E., Yeo, C. Y., Pisenti, J., Henrique, D., Abbott, U. K., Fallon, J. F., and Tabin, C. (1997). Expression of Radical fringe in limb-bud ectoderm regulates apical ectodermal ridge formation [published erratum appears in *Nature[???]* 24;388(6640):400 (1997)]. *Nature* **386**, 366–373.

Lechner, M. S., and Dressler, G. R. (1997). The molecular basis of embryonic kidney development. *Mech. Dev.* **62**, 105–120.

Logan, M., and Tabin, C. (1998). Targeted gene misexpression in chick limb buds using avian replication-competent retroviruses. *Methods* **14**, 407–420.

Long, Q., Meng, A., Wang, H., Jessen, J. R., Farrell, M. J., and Lin, S. (1997). GATA-1 expression pattern can be recapitulated in living transgenic zebrafish using GFP reporter gene. *Development* **124**, 4105–4111.

Majumdar, A., and Drummond, I. A. (1999). Podocyte differentiation in the absence of endothelial cells as revealed in the zebrafish avascular mutant, cloche. *Dev. Genet.* **24**, 220–229.

Majumdar, A., and Drummond, I. A. (2000). The zebrafish floating head mutant demonstrates podocytes play an important role in directing glomerular differentiation. *Dev. Biol.* **222**, 147–157.

Majumdar, A., Lun, K., Brand, M., and Drummond, I. A. (2000). Zebrafish no isthmus reveals a role for pax2.1 in tubule differentiation and patterning events in the pronephric primordia. *Development* **127**, 2089–2098.

Marshall, E. K., and Smith, H. W. (1930). The glomerular development of the vertebrate kidney in relation to habitat. *Biol. Bull.* **59**, 135–153.

Mauch, T. J., Yang, G., Wright, M., Smith, D., and Schoenwolf, G. C. (2000). Signals from trunk paraxial mesoderm induce pronephros formation in chick intermediate mesoderm. *Dev. Biol.* **220**, 62–75.

McClure, C. F. W. (1918). On the experimental production of edema in larval and adult Anura. *J. Gen. Physiol.* **1**, 261–267.

Meier, S. (1980). Development of the chick embryo mesoblast: Pronephros, lateral plate, and early vasculature. *J. Embryol. Exp. Morphol.* **55**, 291–306.

Moller, P. C., and Ellis, R. A. (1974). Fine structure of the excretory system of Amphioxus (*Branchiostoma floridae*) and its response to osmotic stress. *Cell Tissue Res.* **148**, 1–9.

Müller, J. (1829). Ueber die Wolffschen Korper bei Embryonen der Frösche und Kröten. *Meckel's Arch. F. Anat. Physiol.* Jahrg. 1829, 65–70.

Mullins, M. C., Hammerschmidt, M., Kane, D. A., Odenthal, J., Brand, M., van Eeden, F. J., Furutani-Seiki, M., Granato, M., Haffter, P., Heisenberg, C. P., *et al.* (1996). Genes establishing dorsoventral pattern formation in the zebrafish embryo: The ventral specifying genes. *Development* **123**, 81–93.

Nakao, T. (1965). Excretory organs of Amphioxus (Branchiostoma) belcheri. *J. Ultrastruct. Res.* **12**, 1–12.

Nelson, F. K., Albert, P. S., and Riddle, D. L. (1983). Fine structure of the *Caenorhabditis elegans* secretory-excretory system. *J. Ultrastruct Res.* **82**, 156–171.

Nelson, F. K., and Riddle, D. L. (1984). Functional study of the *Caenorhabditis elegans* secretory-excretory system using laser microsurgery. *J. Exp. Zool.* **231**, 45–56.

Newstead, J. D., and Ford, P. (1960). Studies on the development of the kidney of the Pacific Salmon, *Oncorhynchus forbuscha* (Walbaum). 1. The development of the pronephros. *Can. J. Zool.* **36**, 15–21.

Nieuwkoop, P. D., and Faber, J. (1994). "Normal Table of *Xenopus laevis* (Daudin)." Garland, New York.

Obara-Ishihara, T., Kuhlman, J., Niswander, L., and Herzlinger, D. (1999). The surface ectoderm is essential for nephric duct formation in intermediate mesoderm. *Development* **126**, 1103–1108.

Pfeffer, P. L., Gerster, T., Lun, K., Brand, M., and Busslinger, M. (1998). Characterization of three novel members of the zebrafish Pax2/5/8 family: Dependency of Pax5 and Pax8 expression on the Pax2.1 (noi) function. *Development* **125**, 3063–3074.

Puschel, A. W., Westerfield, M., and Dressler, G. R. (1992). Comparative analysis of Pax-2 protein distributions during neurulation in mice and zebrafish. *Mech. Dev.* **38**, 197–208.

Rappaport, R. (1955). The initiation of pronephric function in *Rana pipiens*. *J. Exp. Zool.* **128**, 481–488.

Romanoff, A. L. (1960). The urogenital system. *In* "The Avian Embryo." Macmillan, New York.

Rugendorff, A., Younossi-Hartenstein, A., and Hartenstein, V. (1994). Embryonic origin and differentiation of the *Drosophila* heart. *Roux's Arch. Dev. Biol.* **203**, 266–280.

Ruppert, E. E. (1994). Evolutionary origin of the vertebrate nephron. *Am. Zool.* **34**, 542–553.

Ruppert, E. E., and Barnes, R. D. (1994). "Invertebrate Zoology." Saunders, Philadelphia.

Ruppert, E. E., and Smith, P. R. (1988). The functional organization of filtration nephridia. *Biol. Rev.* **63**, 231–258.

Saxén, L. (1987). "Organogenesis of the Kidney." Cambridge Univ. Press, Cambridge.

Semba, K., Saito-Ueno, R., Takayama, G., and Kondo, M. (1996). cDNA cloning and its pronephros-specific expression of the Wilms' tumor suppressor gene, WT1, from *Xenopus laevis*. *Gene* **175**, 167–172.

Shalaby, F., Rossant, J., Yamaguchi, T. P., Gertsenstein, M., Wu, X. F., Breitman, M. L., and Schuh, A. C. (1995). Failure of blood-island formation and vasculogenesis in Flk-1-deficient mice. *Nature* **376**, 62–66.

Smith, H. W. (1953). "From Fish to Philosopher." Little Brown, Boston.

Swingle, W. W. (1919). The experimental production of edema by nephrectomy. *J. Gen. Physiol.* **1**, 509–514.

Taira, M., Otani, H., Jamrich, M., and Dawid, I. B. (1994). Expression of the LIM class homeobox gene *Xlim-1* in pronephros and CNS cell lineages of *Xenopus* embryos is affected by retinoic acid and exogastrulation. *Development* **120**, 1525–1536.

Torres, M., Gomez-Pardo, E., Dressler, G. R., and Gruss, P. (1995). Pax-2 controls multiple steps of urogenital development. *Development* **121**, 4057–4065.

Torrey, R. W. (1965). Morphogenesis of the vertebrate kidney. *In* "Organogenesis" (R. L. DeHaan and H. Ursprung, eds.). Holt, Rinehart and Winston, New York.

Toyama, R., and Dawid, I. B. (1997). lim6, a novel LIM homeobox gene in the zebrafish: Comparison of its expression pattern with lim1. *Dev. Dyn.* **209**, 406–417.

Tytler, P. (1988). Morphology of the pronephros of the juvenile brown trout, *Salmo trutta*. *J. Morphol.* **195**, 189–204.

Tytler, P., Ireland, J., and Fitches, E. (1996). A study of the structure and function of the pronephros in the larvae of the turbot (*Scophthalmus maximus*) and the herring (*Clupea harengus*). *Mar. Fresh. Behav. Physiol.* **28**, 3–18.

Vize, P. D., Jones, E. A., and Pfister, R. (1995). Development of the *Xenopus* pronephric system. *Dev. Biol.* **171**, 531–540.

Vize, P. D., Seufert, D. W., Carroll, T. J., and Wallingford, J. B. (1997). Model systems for the study of kidney development: Use of the pronephros in the analysis of organ induction and patterning. *Dev. Biol.* **188**, 189–204.

Wallingford, J. B., Carroll, T. J., and Vize, P. D. (1998). Precocious expression of the Wilms' tumor gene xWT1 inhibits embryonic kidney development in *Xenopus laevis*. *Dev. Biol.* **202**, 103–112.

Wrobel, K. H., and Süb, F. (2000). The significance of rudimentary nephrostomial tubules for the origin of the vertebrate gonad. *Anat. Embryol. (Berl.)* **201**, 273–290.

6

Development of the Mesonephric Kidney

Kirsi Sainio

I. Introduction

The diversity of cell types and the complexity of anatomical structure increase during urogenital morphogenesis (Cunha, 1972; Saxén, 1987; Bard *et al.,* 1996). All stages of the developing urinary system are derivatives of the intermediate mesoderm. The nephric ridge is located dorsally, close to the neural tube and adjacent to the somites from cranial segments down to the lumbar region. In vertebrates, the second phase of the nephric development, that of the mesonephros, follows that of the pronephros (Chapters 3–5) and precedes the development of metanephros, the permanent kidney (Section II). Thus it is also called the intermediate kidney (German= urniere, Fig. 6.1).

The mesonephros serves as the excretory organ in adult amphibians and teleost fish. In amniotes, the mesonephros is a vestigial organ whose development has not gained as much attention as that of the metanephros, but this does not make it a less interesting target for research. On the contrary, this temporary organ would be a good model for those interested not only in epithelial differentiation, but also in the regulation of apoptosis (the regression phase), cell migration (between the developing gonad, adrenal gland, and mesonephric blastema), stem cell differentiation (hematopoietic and various other somatic cells), and hormonal determination of gene activity and developmental fate (sexual differences in gonadal duct system development). Furthermore, during the last two decades, an increasing amount of evidence suggests that the mesonephros is something else than purely a transitory organ with a short life span and minimal, if any, functional significance.

The purpose of this chapter is to give an overview of mesonephric development, using the murine species as models. Because the development and regression of the mesonephros vary tremendously from one species to another, the major differences between murine and other species are also discussed. The development of gonadal duct systems in both male and females is described, and an overview of the genetic regulation of mesonephric differentiation and regression is given. Finally, and partly as a future perspective, stem cell properties of the mesonephric area are described.

II. Mesonephric Development: An Anatomical Overview

The nephric blastemal ridge first forms a pronephros in the neck region. In all mammalian species, the pronephros

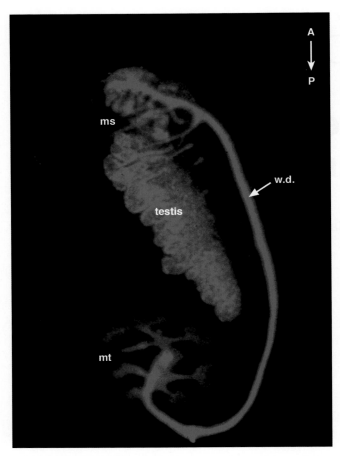

Figure 6.1 Three-dimensional composition of the urogenital region in a mouse embryo. Cranial mesonephric tubules (ms) are linked to the anterior portion of the mesonephric/Wolffian duct (w.d.) The metanephros (mt), with numerous ureteric branches, is linked to the posterior of the duct (lower left). The developing testis is visible between the two kidneys. Whole mount immunofluorescence of cytokeratin 8/18, mouse E11 urogenital block has grown *in vitro* for 3 days. Orientation of the anterior–posterior (A-P) axis is indicated in the upper right.

Fig. 6.2). More recently, a three-dimensional reconstruction of the mouse mesonephros (Vazquez *et al.,* 1998; Chapter 7) and whole mount immunohistochemistry (Sainio *et al.,* 1997a) and *in situ* hybridization techniques (Kume *et al.,* 2000) have given new insight into the anatomical relations between meso- and metanephric blastema, developing gonadal structures, and the Wolffian duct (Figs. 6.1, 7.1, and 7.2). The first, most cranial renal vesicles of the mesonephros develop in humans at day 25 of gestation, and in murine species at embryonic day 9 (E9; mouse) and embryonic day 11(E11; rat). The maximal number of mesonephric tubules in both mouse and rat is 18 to 26, and two distinct sets of tubules are observed throughout development. The first 4 to 6 tubular vesicles are connected to the common drain, the Wolffian duct, and have a short distal segment that might be even partly derived from the Wolffian duct (Sainio *et al.,* 1997a; Chapter 7). A similar segment has also been described in chicken (Croisille, 1976). The caudal tubules, which represent the majority of the mesonephric nephrons, are initially seen in close vicinity to the Wolffian duct but they never fuse with the duct (Figs. 6.3, 7.2, and 7.3). Thus, in murine species, only the most cranial tubules can be functional and the other tubules remain vestigial. Evolutionary, this is interesting, as a closely related species, the vole *Microtus rossiameridionalis,* has a completely different structure in its mesonephros (Fig. 6.4). Here, 30 to 40 well-developed tubules are all connected to the Wolffian duct and hence the organ could be functional. The two rodent lineages have diverged some 20 million years ago, quite recently in evolutionary time. The situation is different from murine species, in human (Martino and Zamboni, 1966; Figs. 1.3 and 7.1), in pig, and in sheep (Tiedemann 1976, 1979, 1983), which all develop from 30 to 50 mesonephric tubules and highly specified vasculature systems, suggesting that the mesonephros is physiologically functional in producing urine (Martino and Zamboni, 1966; Moore, 1977; Nistal and Panigua, 1984; Tiedemann, 1976, 1979, 1983; Tiedemann and Egerer, 1984).

The nephric duct is essential as both the common urinary drain and the terminal inducer of mesonephric tubules. The role of the duct in mesonephric induction is analogous to the interaction between the ureteric bud and the metanephric mesenchyme discussed in Chapters 12–14. In classic experiments by Boyden (1927), Gruenwald (1937) and Waddington (1938) showed that no mesonephric nephrons differentiate if the inductive interaction between the duct and the blastema is disrupted (Fig. 6.5). Indeed, these observations have been confirmed more recently in GATA-3 mutant mice that develop only short Wolffian ducts and hence only few mesonephric tubules (I. Pata, personal communication, Fig. 6.6). The inductive interaction performed by the Wolffian duct seems to involve similar, if not entirely the same, molecular mechanisms as the regulation of metanephric blastemal induction by the ureteric bud. Hence the

contains only a few nephrons that regress soon after they form (see Chapter 3). In mammals, the caudally descending nephric duct, initially called the pronephric duct, but also called the Wolffian or mesonephric duct at later stages, forms in the neck–trunk region from the pronephros (Toivonen, 1945; Smith and MacKay, 1991). Results from chicken suggest that the surface ectoderm next to the intermediate mesoderm is essential for nephric duct formation (Obara-Ishihara *et al.,* 1999; Chapter 4). The duct is permanent, but the pronephric tubules regress and this organ is eventually replaced by a new type of embryonic kidney, the mesonephros, in the middle part of the nephric ridge (Fig. 6.1).

The most detailed anatomical description of the mesonephros in various mammalian species has been performed in a series of classic histological and electron microscopy studies by Klaus Tiedemann (Tiedemann, 1976, 1979, 1983; Tiedemann and Egerer, 1984; Schiller and Tiedemann, 1981;

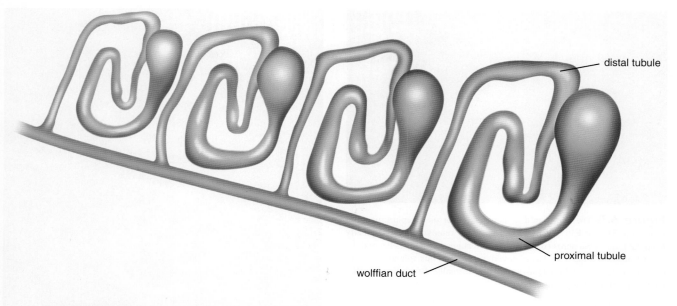

Figure 6.2 Organization of mesonephric nephrons in mammals. After Schiller and Tiedemann (1981.) w.d., Wolffian duct.

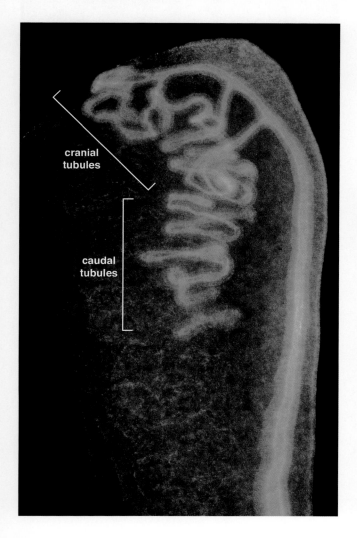

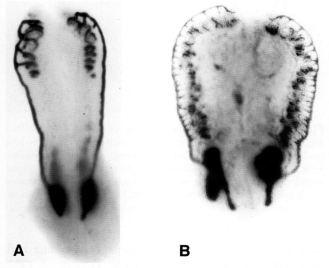

Figure 6.3 Only cranial mesonephric tubules are connected to the Wolffian duct in murine species. Caudal tubules remain unlinked and do not function in urine production (see Sainio *et al.*, 1997). Whole mount immunofluorescence of mouse E13 mesonephros with cytokeratin 8/18 antibodies.

Figure 6.4 Mouse (A) and the closely related species vole (B) have very different mesonephroi. The mouse has only rudimentary structures and 4 to 6 tubules connected to the duct, whereas the vole has approximately 40 mesonephric nephros all connected to the Wolffian duct. Whole mount *in situ* hybridization of paired-box gene pax2 at E11. Courtesy of Dr. Marjo Hytönen.

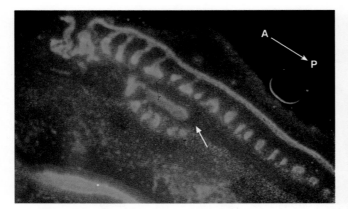

Figure 6.5 The Wolffian duct is essential for mesonephric nephron formation. The rat E11 Wolffian duct has been removed microsurgically from the left mesonephros (white arrow). No nephric structures beyond the point from which the duct was removed are seen on this side after 2 days of cultivation. The right side differentiates normally. Brush border (BB) whole mount immunofluorescence.

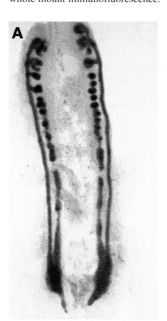
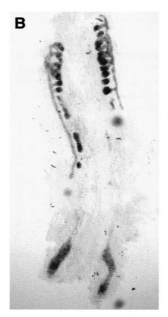

Figure 6.6 The GATA-3 transcription factor is essential for Wolffian duct elongation. Whole mount *in situ* hybridization of paired-box gene *pax2* shows the lack of mesonephric tubules in regions where Wolffian duct growth has stopped. In the wild-type urogenital ridge (A) at this stage, the nephric ridge is uniform. In the mutant urogenital area (B), *pax2* expression is still detected in the metanephric blastema, but the Wolffian duct is not elongating and the ureteric bud does not form. Both samples are mouse day E10. Courtesy of Dr. Illar Pata.

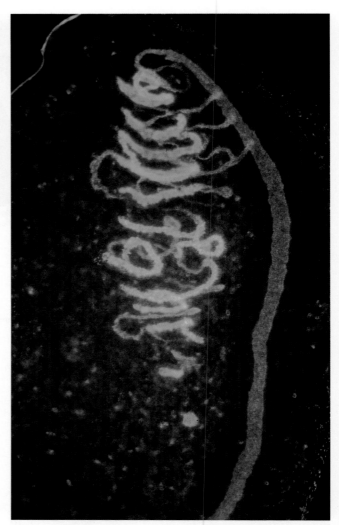

Figure 6.7 Mesonephric nephrons at mouse E13.5. The proximal segment can be recognized with antibodies against brush border epitopes. Mesonephric nephrons form short tubules that lack a loop of Henle and a juxtaglomerular apparatus, but have most of the other components of metanephric nephrons. Once again this example illustrates that only the cranial tubules are connected to the Wolffian duct. Mesonephric nephron formation and differentiation are more advanced in other mammalian species, including humans.

same heterologous inducers (e.g., spinal cord) induce both mesonephric and metanephric nephrons (Gruenwald, 1952; Chapter 13).

Morphogenesis of the tubular structures in the meso-nephros resembles that of the metanephros, but again there is

a great deal of variation in the size, structure, and functional maturity of the mesonephric tubules within species (Nelson *et al.*, 1992). Mesonephric mesenchyme first aggregates and forms a renal vesicle (Burns, 1955). Those vesicles that make the connection to the Wolffian duct seem to grow finger-like protrusions toward the basement membrane of the duct (Lawrence, 1992). The renal vesicle then differentiates into epithelial S-shaped structures that elongate and eventually form proximal tubules but no loop of Henle or juxtaglo-merular apparatus (Fig. 6.7). In murine species, small glomerular-like structures develop in the medial part of the tubule (Zamboni and Upadhyay, 1981; Smith and MacKay, 1991; Sainio *et al.,* 1997a) with primitive endothelial tufts (Smith and MacKay, 1991). In pig (Tiedemann and Egerer, 1984) and human (Martino and Zamboni, 1966) mesonephroi, both glomeruli and vasculature are extensive and well developed. The short distal tubular segment (Martino and Zamboni, 1966; Schiller and Tiedemann, 1981; Tiedemann and Egerer, 1984; Smith and MacKay, 1991; Croisille, 1976) closely resembles epithelia of the Wolffian duct (Martino and Zamboni, 1966; Croisille, 1976; Sainio *et al.,* 1997a), as it does in both pronephric and metanephric nephrons. The cranial mesonephric tubules frequently form branches that may reflect presumptive ependymal morphogenesis (Friebová- Zemanová and Goncharevskaya, 1982; Sainio *et al.,* 1997). In the chick, these "two-headed" nephrons were branched from the proximal segment, but in murine species, branching occurs in the distal segment. Branching within the proximal segment is also observed in pronephroi (Chapter 3).

In murine species, both mesonephric and metanephric tubules express characteristic markers for metanephric nephrons, i.e., brush border epitopes, cytokeratin-8/18, and the p75 neurotrophin receptor, although some stage differences appear (Sainio *et al.,* 1997a; Figs. 6.3 and 6.7). The glomerulus-like extension or mesonephroi expresses markers of metanephric glomeruli, such as podocalyxin (Sainio *et al.,* 1997a).

The mesonephric kidney regresses prenatally in mammals. Degradation is completed in rat by embryonic day 17 and in mouse by day 15 (reviewed by Saxén, 1987). Murine regression seems to start from the caudal mesonephric tubules and spread cranially (Smith and MacKay, 1991, Sainio *et al.,* 1997a), but in humans starts in the cranial region (Martino and Zamboni, 1966; Chapter 7) and in sheep from the middle region (Tiedemann, 1976). In all females, degradation is complete, but in males, remaining mesonephric tubules form part of the gonadal ducts (Moore, 1977; Orgebin-Crist, 1981; Nistal and Paniagua, 1984). The molecular mechanism regulating the mesonephric degradation is not known, but apoptosis has been described by morphological criteria in the mesonephric field already at early developmental stages, when pycnotic nuclei are observed frequently (Theiler, 1972; Smith and MacKay, 1991). However, it seems that regression of mesonephric tubules depends on local extrinsic stimuli rather than intrinsic factors. In the rat embryo, the regression phase takes place on days E17–E18. If the separated nonepigonadal mesonephric tubules of rat E12 embryo are grafted to the eye chamber of an adult rat, the tubules extend their normal life span at least sixfold (Runner, 1946). Moreover, such grafted tubules differentiate more extensively than normal in this environment. This result suggests that mesonephric tubule differentiation and regression are controlled by tissue surrounding the mesonephric tubules.

III. Molecular Basis of Mesonephric Development

Unlike in metanephric development, the regulatory molecules involved in mesonephric induction, differentiation, or regression are not well characterized. However, evidence suggests that no major differences exist between the characteristics of these nephric mesenchymes, and because the mesonephros differentiates in a similar fashion to the metanephros, it is not surprising that some key regulatory molecules in mesonephrogenesis are the same as in the permanent kidney (Chapter 20 and the kidney development database: http://golgi.ana.ed.ac.uk/kidhome.html). Morphologically, the main difference between these two kidneys is the lack of a collecting duct system in the mesonephros. Thus many of the molecules regulating ureteric budding and subsequent branching morphogenesis may not play as crucial role in mesonephric differentiation as they do in metanephrogenesis.

The three stages of kidney development appear during embryogenesis in a consecutive, partly overlapping spatiotemporal pattern. As in all the other parts of the embryo, Hox signaling probably regulates positional information during nephric ridge differentiation. Mouse mutations of the posterior abdominal B-type paralogue group of Hox genes (*Hoxa 11-Hoxd 11*) have some interesting urogenital features. First, mutation of *hoxa 11,* which is normally expressed in both metanephric and mesonephric mesenchyme, results in partial homeotic transformations in the mesonephric area. In females, the uterine endometrium is defective, and in male animals, the Wolffian duct–derived vas deferens seems to be transformed into epididymis and mutant testes fail to descend. Thus, both female and male mutant animals are infertile (Hsieh-Li *et al.,* 1995). The other posterior hox gene, *hoxd 11,* is also expressed both in meta- and mesonephric blastema. When combined with the *hoxa 11* mutant, double mutants show a severe, lethal urogenital phenotype with ureter branching morphogenesis defects (Davis *et al.,* 1995; Patterson *et al.,* 2001). The mesonephros is also altered in double mutants and again, homeotic transformation with the vas deferens transformed to the epididymis is observed (Davis *et al.,* 1995). The mesonephroi of these animals have not been analyzed in detail, but results from newborn animals suggest that the positional information provided by hox genes is essential for the normal specification and differentiation of the urogenital tissue.

Some regulatory molecules that are essential in mesonephrogenesis have been identified. Not surprisingly, the same molecules are again essential in metanephric development. The transcription factor paired box gene *Pax-2* (Dressler *et al.,* 1990; Deutsch and Gruss, 1991; Torres *et al.,* 1995) is expressed in pronephric and mesonephric tubules, in the Wolffian duct, in the early metanephric condensates, and in the ureteric bud and it is downregulated in mature metanephric epithelia (Dressler *et al.,* 1990; Dressler and Douglass, 1992; Phelps and Dressler, 1993; Figs. 6.4 and 6.6). *Pax-2* is necessary for the development of the excretory system, as targeted disruption of this gene results in the lack of meso- and metanephric kidneys, ureters, and the genital tract in homozygous mutant mice and in the reduction of the metanephric size in adult heterozygotes (Torres *et al.,* 1995). Thus, the primary defect in *Pax-2* null mutants is an early event associated with initial steps in the differentiation of the intermediate mesoderm. A signaling molecule from the transforming growth factor-β superfamily, bone morphogenetic protein-4 (BMP-4) secreted by the surface ectoderm, has been shown to regulate the expression of *Pax-2* by nephric duct progenitor cells of the intermediate mesoderm (Obata-Ishihara *et al.,* 1999). As shown by microsurgical tissue experiments (Gruenwald, 1937), Wolffian duct formation is essential for both meso- and metanephrogenesis (Figs. 6.5 and 6.6). Thus, BMP-4 is a candidate molecule to regulate early steps in the differentiation of this organ system. The other function of BMP-4 seems to be restriction of ureteric budding (Miyazaki *et al.,* 2000). Thus, BMP-4 is required in the mesonephric area for normal growth of the Wolffian duct and proper spatial budding of the ureter. The second growth factor crucial for ureteric budding, a stimulatory molecule glial cell line-derived neurotrophic factor (GDNF), is able to induce ectopic ureters in the caudal mesonephric area when expressed inappropriately (Sainio *et al.,* 1997b; Brophy *et al.,* 2001, Fig. 6.8).

The second transcriptional regulator crucial for urogenital development is Wilms' tumor gene-1, *WT-1* (Pritchard-Jones *et al.,* 1990; Kreidberg *et al.,* 1993). Targeted disruption of *WT1* (Kreidberg *et al.,* 1993) leads to embryonic lethality in homozygous mice. In these mice the ureteric bud does not form and the metanephric mesenchyme dies through apoptosis. The mesonephros in homozygous embryos still develops, although mesonephric tubules are not as numerous as in the wild-type littermates (Kreidberg *et al.,* 1993). Indeed, only the cranial mesonephric tubules develop in *WT-1 –/–* null mutants (Sainio *et al.,* 1997a; Fig. 6.9). Thus, development of the cranial mesonephric tubules is regulated differently from the more caudal ones.

A third interesting transcription factor that seems to regulate at least Wolffian duct maintenance is the homeobox gene *Emx-2,* a mouse homologue of a *Drosophila* head gap gene *empty spiracles* (*ems;* Yoshida *et al.,* 1997). The disruption of this gene in mice leads to severe defects in urogenital development (Miyamoto *et al.,* 1997; Chapter 20) with an apparently normal metanephric mesenchyme that fails to form tubules, possibly due to the failure of inductive interaction between the mesenchyme and the ureteric epithelium. Surprisingly, by histology, at least some mesonephric tubules are formed, but the Wolffian duct and the mesonephric area regress earlier than normal and no gonads or mesonephric-

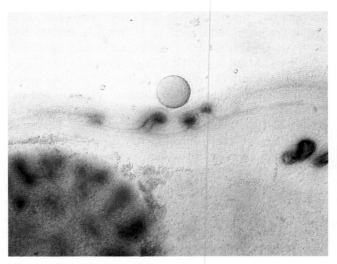

Figure 6.8 GDNF triggers budding from ectopic positions along the Wolffian duct. A heparin-coated bead was soaked with GDNF and placed adjacent to the Wolffian duct in between the metanephros (on the lower left) and caudal mesonephric tubules (on the right). Multiple ectopic buds were detected by whole mount *in situ* hybridization for GFRα1. Initially, the entire Wolffian duct expresses this coreceptor of GDNF, but expression is downregulated as development proceeds. Ectopic GDNF is able to reactivate α-receptor expression. It has been shown in *Xenopus* that GDNF signaling, especially α-receptor-mediated signaling, is needed for Wolffian duct elongation. Courtesy of Dr. Marjo Hytönen.

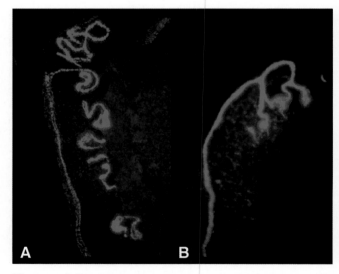

Figure 6.9 WT-1 regulates caudal mesonephric tubule formation. Whole mount immunofluorescence with anticytokeratin 8/18 shows normal mesonephric tubules in wild-type (A) mice and the lack of caudal nephrons in WT-1-deficient mice (B) (see Sainio *et al.,* 1997a).

derived gonadal ducts are formed. The molecular pathway under the control of *Emx-2* is not known.

Transcriptional regulators of the forkhead/winged helix family (presently known as the Fox family of transcription factors; Kaestner *et al.*, 2000) are essential for normal kidney development. A spontaneous mouse mutant *congenital hydrocephalus* (*Ch;* Grüneberg, 1943) is caused by a point mutation in the DNA-binding domain of Foxc1 (Kume *et al.*, 1998). Ureteric budding and branching morphogenesis defects, as well as the altered spatial–temporal segmentation of the metanephros–mesonephros, are similar in both *Ch* mice and in transgenic *Foxc1 –/–*mice (Green, 1970; Kume *et al.*, 2000). The downstream targets regulated by these factors are still unknown, but *Foxc1* mutant chondroblasts do not respond to TGFβ family members, TGFβ-1 and BMP-2 *in vitro* (Kume *et al.*, 1998). Mesonephric structures are observed within the metanephric area in *Ch* mutant kidneys, which may be a consequence of altered posterior migration of the metanephrogenic mesenchyme in this background. Another potentially important player in the *Foxc1* phenotype is *Eya1,* as both Eya1 and GDNF expression patterns are altered in a mutant background (Kume *et al.*, 2000). *Eya1,* a mouse homologue of the *Drosophila eyes absent (eya)* gene, which is essential for ommatidia formation in the fly eye (Bonini *et al.*, 1993), is an important transcriptional regulator of mammalian kidney morphogenesis. At least two of the four mouse *Eya* genes are expressed in the developing kidney (*Eya1* and *Eya2*). The *Eya1* gene is essential for metanephric development and null mutants fail to develop kidneys (Xu *et al.*, 1999). Eya target genes are activated when the molecule is translocated to the nucleus. For this, Eya needs another family of transcription modulators, Six [(mouse homologues to *Drosophila sine oculis* (so)]. In the *Drosophila* eye, cooperation of So, Eya, and Pax, together with retina-forming *dachshund* (mouse *Dac*), is needed for the complete development of functional ommatidia (Ohto *et al.*, 1999). In the kidney, *Eya1,* together with *Six2* and *Pax2* (Xu *et al.*, 1999; Ohto *et al.*, 1999) and possibly *Dac,* could regulate GDNF signaling and thus ureteric branching morphogenesis (Brophy *et al.*, 2001). The function of *Eya1* and *Six2,* and their relationship with Fox signaling, has not been characterized in the mesonephric kidney to date.

Maybe the most interesting regulatory molecule from the mesonephric point of view is the zinc finger transcription factor GATA-3. GATA-3 is expressed in the Wolffian duct and ureteric bud throughout nephrogenesis (George *et al.*, 1994), and the haploinsufficiency of GATA-3 causes human HDR syndrome (hypoparathyroidism, sensorineural deafness, renal anomaly syndrome, MIM 146255; Van Esch *et al.*, 2000). Mice with targeted disruption of the GATA-3 gene die at E11 or E12 of gestation due to massive internal bleeding (Pandolfi *et al.*, 1995). The urogenital phenotype of these mice is unique: they show very limited growth of the Wolffian duct and thus retarded mesonephrogenesis and no metanephrogenesis, probably due to the lack of a ureteric bud (I. Pata, personal communication, Fig. 6.6). Downstream molecules of GATA-3 still need to be elucidated, but an interaction between GATA-3 and a member of the Krüppel-like family of transcription factors, Klf6 (Fischer *et al.* 2001), has been suggested.

IV. Mesonephric Contribution to Gonadal Differentiation

The urinary organs and somatic structures of the gonads develop next to each other and have a common developmental origin, the intermediate mesoderm. Although functionally totally different when mature, these two organ systems have links during embryonic development (McLaren, 2000). During mammalian urogenital development, the long genital ridge faces the inner surface of the equally long mesonephros. Finally, some of the tubules of the mesonephros will remain as epididymal ducts of the adult male (Sainio *et al.*, 1997a; Fig. 6.10), whereas the Wolffian duct serves as the vas deferens and parts of it form the rete testis. In females, at least the rete ovarii, a group of anastomosing tubules (epididymis-like structure), and the epoöphron connecting them, are thought to be mesonephric derivatives (Kardong, 1995). Moreover, the Müllerian duct, the presumptive female ovarian duct, differentiates normally only in the presence of the Wolffian duct. This has been confirmed in both Emx-2- and GATA-3-deficient mice, where no Müllerian ducts form in the absence of a normal Wolffian duct.

In addition to the ductal system of the gonads, the mesonephros has been suggested to be important for the somatic cell differentiation of both the testis and the ovary (McLaren, 2000). Some studies even suggest that some of the somatic cells forming the sex cord are actually of mesonephric origin. Tissue cultures of urogenital ridges isolated from mice suggest that the mesonephros is necessary for proper differentiation of the seminiferous tubules of testis (Buehr *et al.*, 1993; Merchant-Larios *et al.*, 1993). This has

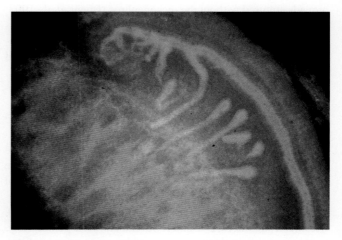

Figure 6.10 Double whole mount immunofluorescence of neural cell adhesion molecule L1 (green) and p75 neurotrophic receptor (red/orange) shows the relationship of the male gonad and the mesonephros. p75 has been shown to be important both for epididymal development and for myoid cell formation in the testical cord. Future epididymal structures in mouse are derivatives of the cranial mesonephric tubules seen in the vicinity of the cranial part of the testis. Caudal tubules are detiorated gradually. Rat E15 urogenital block.

been confirmed in mouse lines with abnormal testis cord differentiation (Albrecht *et al.,* 2000) and in a fetal gonad/ mesonephros coculture system that utilizes the tissues of transgenic mice expressing ubiquitous enhanced green fluorescent protein (Nishino *et al.,* 2000).

At least three types of cells migrate between the mesonephros and the developing gonad, especially toward the testis: endothelial, myoid, and fibroblast cells (Wartenberg, 1978; Martineau *et al.,* 1997). Angiogenic precursor cells proliferate and differentiate in the mesonephric area and subsequently invade the future gonad where they form the endothelial vascular network of the testis (Merchant-Larios *et al.,* 1993). Indeed, vascular endothelial growth factor (VEGF) activity, one of the major regulators of angiogenesis, is upregulated by testosterone (Sordello *et al.,* 1998) and might be a signaling molecule responsible for guiding migration of the mesonephros-derived endothelial cells into the developing testis. Moreover, the endothelial sialoglycoprotein podocalyxin (Schnabel *et al.,* 1989; Miettinen *et al.,* 1990) is highly expressed in the mesonephric area. Podocalyxin is expressed not only in the presumptive podocytes of the mesonephric tubules, but also in the mesonephric stroma surrounding the epithelial structures (Sainio *et al.,* 1997a) and might also be a marker for hematopoietic stem cells in the mesonephric stroma (Hara *et al.,* 1999).

The other cell types, myoid and fibroblast cells, probably shape the seminiferous cords by synthesizing basement membrane molecules (Merchant-Larios *et al.,* 1993). Myoid cells eventually form the peritubular smooth muscle layer around the testicular cords. The origin of testicular myoid and fibroblast cells from mesonephric stroma has been indicated, as these cells are absent in isolated impaired gonads cultured without the mesonephros (Buehr *et al.,* 1993; Merchant-Larios *et al.,* 1993). Interestingly, neurotrophins and their low- and high-affinity receptors (p75 and trk, respectively) have also been shown to affect testis cord myoid cell formation and epididymal development (Russo *et al.,* 1999; Campagnolo *et al.,* 2001; Fig. 6.10). Coculture of mesonephros and testis tissues has also shown that p75 expression in the testis does not occur without a mesonephric influence (Campagnolo *et al.,* 2001).

Mesonephric stromal cells have been shown to become testosterone-producing Leydig cells after a prolonged culture of mesonephric/gonadal chimeras between CD-1 mice genital ridges and ROSA26 mesonephroi expressing a β-galactosidase marker (Merchant-Larios and Morena-Mendoza, 1998). Results from other CD-1 and ROSA26 chimeric genital ridge and mesonephric tissue culture experiments suggested that cell migration from the mesonephros occurs exclusively into testis (Martineau *et al.,* 1997). The careful analysis of Wnt-4-deficient female embryos has shown that Leydig cell precursors derived from the mesonephros actually invade the genital ridge of both sexes (Vainio *et al.,* 1999). Wnt-4, a member of the Wnt-signaling molecule family that is

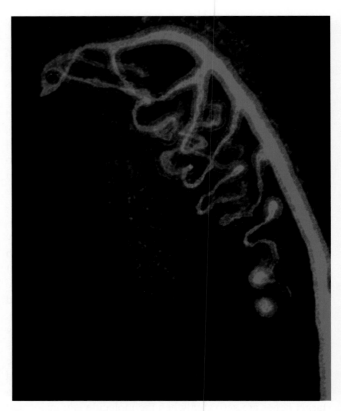

Figure 6.11 Mesonephric-derived ductal development in male gonads is dependent on the hormonal induction provided by the testis (Cunha *et al.,* 1983). The mouse E11 mesonephros has been cocultured with the embryonic testis for 7 days. Cranial tubular structures are well preserved, whereas few caudal structures remain. Without the contribution from the testis, all nephros would have regressed after 3–4 days in culture. Whole mount immunofluorescence with a pan-cytokeratin antibody.

expressed in the stromal compartment of the mesonephros but not in the mesonephric epithelium, suppresses the differentiation of ovarian Leydig cell precursors and is required to inhibit their testosterone production. The fetal ovary in *wnt-4*-deficient female mice embryos has Leydig cells that produce testosterone and consequently the female embryos are masculinized (Vainio *et al.,* 1999). The further cellular fate of the Leydig cell precursors that invade the female ovary is not known. Some of the mesonephric stromal cells are suggested to differentiate into granulosa cells of the ovary (reviewed by Wenzel and Odend'hal, 1985).

The origin of Sertoli cell precursors and the regulation of their development are more controversial. Several early studies, based on histological criteria, suggested that some or all Sertoli cells are derived from the mesonephros (Upadhyay *et al.,* 1981; Pelliniemi *et al.,* 1984; Wartenberg *et al.,* 1991). However, expression of the Y chromosomal male sex-determining factor *Sry* (Koopman *et al.,* 1990, Koopman (1991), which is observed in pre-Sertoli cells, is not found in recombination chimeras of mesonephros and the male genital ridge in which mature Sertoli cells form (Buehr *et* al., 1993; Merchant-Larios and Morena-Mendoza, 1998). Thus, either Sertoli cells differentiating in this system are not able to maintain their normal gene expression pattern or Sertoli cell precursors are derived from other mesodermal tissues than the mesonephros. New evidence from mesonephros and bipotent gonad recombinant tissue culture experiments shows that migrating mesonephric cells have an essential role in testicular cord formation, as well as in Sertoli cell differentiation (Tilmann and Capel, 1999).

Interestingly, both *Emx-2* and *WT-1*-deficient mutant mice lack not only the metanephros, but also the gonads (Miyamoto *et al.*, 1997; Kreidberg *et al.*, 1993). In both these lines, mutant individuals show abnormalities in mesonephric development in both sexes and the caudal tubules are completely absent (Miyamoto *et al.*, 1997; Sainio *et al.*, 1997a). *Emx-2* mutant mice have disrupted Wolffian duct regression that takes place much too early and in both sexes, whereas in *WT-1* mutants, only a few cranial mesonephric nephrons form and the caudal tubules are completely absent. The function of these two transcriptional regulators in gonadal differentiation is not known, but in mature testis, *WT-1* is expressed by Sertoli cells, arguing that the gonadal effect may be direct.

While a direct effect of the mesonephros on gonadal development has not yet been described, the regulatory action of the testis toward the mesonephros has been demonstrated. Cuhna *et al.* (1983) observed that mesonephric-derived ductal development in male gonads is in turn dependent on the hormonal induction provided by the developing testis. It is possible that this hormone dependence of mesonephric development in mice and rat embryos only functions in the cranial set of mesonephric tubules and requires testosterone.

V. Mesonephric Contribution to Other Organ Systems

The central location of mesonephros in the body cavity and the rather long existence of pro- and mesonephric organs during embryonic development render this region an optimal source for several cell lineages, not only for somatic cells of the sex glands, but also other cell types (Sainio and Raatikainen-Ahokas, 1999).

Hematopoiesis is a well-known model for stem cell differentiation. The initial hematopoietic activity was assigned to the yolk sac and then it was supposed to shift via fetal liver and spleen to bone marrow, where it remains in adults (Moore and Metcalf, 1970; Johnson and Moore, 1975). However, the limited differentiation repertoire of yolk sac-derived stem cells (reviewed by Dzierzak and Medvinsky, 1995) suggested that yolk sac hematopoiesis is transitory and has no derivatives in the adult animal, as had been suggested earlier in chicken (Dieterle-Lievre, 1975). A new

source for adult hematopoiesis was identified in the mouse embryo from a region including the dorsal aorta, genital ridge/gonads, and pro/mesonephros (aorta–gonad–mesonephros area [AGM]) (Medvinsky *et al.*, 1993; Medvinsky and Dzierzak, 1996). Hematopoietic precursors appear simultaneously in both the AGM region and the yolk sac at late embryonic day 9, but, unlike yolk sac precursors, stem cells from AGM are highly mitotic and can give rise to complete hematopoiesis in adult irradiated mice. Thus, definite hematopoiesis seems to be initiated at the AGM region, and fetal liver and bone marrow are later seeded by these hematopoietic stem cells (Medvinsky and Dzierzak, 1996; reviewed by Marshall and Thrasher, 2001). In accordance with experimental data, expression of erythropoietin, a regulator of red blood cell production, has been found in the developing bovine mesonephros (Wintour *et al.*, 1996). The pronephros is also a site of hematopoiesis (Chapter 3).

The hematopoietic transcription factor GATA-3 is expressed in placenta and T lymphocytes (Yang *et al.*, 1994), but also both by mouse (George *et al.*, 1994; Lakshmanan *et al.*, 1999) and human (Labastie *et al.*, 1995) mesonephric tubules, where it seems to regulate Wolffian duct growth (Fig. 6.6). GATA activators (Friend of GATAs; FOG) that are all involved in multiple steps of hematopoiesis are also expressed in the urogenital ridge, where they may regulate the hematopoietic differentiation of mesonephric stem cells. A further interesting possibility is that GATA-3 expression in mesonephros also regulates the differentiation of lymphocytes. The presumptive omentum (dorsal mesogastrium) of the E13 mouse embryo is a region between anterior limbs, foregut and mesonephros. Cells in the presumptive omentum may differentiate into Thy-1-positive lymphocytes, and thus omentum is a stem cell source for developing lymphocytes (Kubai and Auerbach, 1983). As the dorsal mesogastrium lies next to the mesonephros and is actually in contact with it, an interesting possibility is that the hematopoietic cell lineage of this tissue might be derived from the AGM. The other well-characterized regulatory gene involved in the development and maintenance of hematopoietic stem cells of the AGM region is transcription factor acute myeloid leukemia gene-1 (AML-1). Accordingly, AML-1-deficient mice have a dose-dependent reduction of hematopoietic stem cells derived from the paraaortic splanchnopleural and aorta–gonad–mesonephros region (Mukouyama *et al.*, 2000).

The origin of adrenal cortex cells in the mammalian embryo has not been fully elucidated. The adrenocortical tissue is of mesodermal origin, but the medullary cells are derivatives of the neural crest precursors. As with the gonads, the adrenal glands also develop in close contact with the nephric blastema and mesonephros. The expression of several marker molecules in these tissues has led Worbel and Süß (2000) to reevaluate adrenal cortical differentiation. In bovine fetuses, they have been able to show that the cortical cells originate from the mesonephros. Similar conclusions have been drawn by Upadhyay and Zamboni (1982) based on experiments in the sheep, suggesting that a nephric origin of adrenocortical cells is a general feature of all vertebrates.

VI. Summary

The mesonephric kidney is an embryonic organ that disappears in all mammalian species when the permanent kidney, the metanephros, is functional. In adult males, the epididymal ducts and in females, the rete ovarii are derived from the mesonephros.

The regulation of mesonephric differentiation is not well understood at the molecular level. However, some molecules regulating metanephric development, such as *WT-1* and *Pax-2,* are also important in mesonephric differentiation. In addition to epithelial tubules, the mesonephros also contains stromal cells. These stromal cells are suggested to be a source of multiple precursor cells. Mesonephric stromal cells invade the genital ridge of both sexes. In males, at least a portion of the endothelial and Leydig cells are derived from the mesonephros. Some pre-Sertoli cells, based on morphological criteria, may also be of mesonephric origin. Thus far there is no molecular evidence to support this assumption, but mesonephric cells are at least needed for the proper differentiation of Sertoli cells and testicular cord formation. Leydig cell precursors also seem to invade the presumptive female gonad, but there the differentiation and steroid production by these cells are suppressed by Wnt-4. The fate of pre-Leydig cells invading the female genital ridge is not clear. They may deteriorate apoptotically, but an interesting option is that these mesonephros-derived cells take another developmental pathway in the ovary.

Given the important role of the mesonephros in the development of many adult organs, including the metanephros, characterizing mesonephric defects may prove to be of value in studies investigating the failure of organogenesis in mutant lines.

References

Albrech, K. H., Capel, B., Washburn, L. L., and Eicher, E. M. (2000). Defective mesonephric cell migration is associated with abnormal testis cord development in C57BL(6J XY (mus domesticus) mice. *Dev. Biol.* **225,** 26–36.

Bard, J., Davies, J., Karavanova, I., Lehtonen, E., Sariola, H., and Vainio, S. (1996). Kidney development: The inductive interactions. *Sem. Cell Dev. Biol.* **7,** 195–202.

Bonini, N., Leiserson, W., and Benzer, S. (1993). The eyes absent gene: Genetic control of cell survival and differentiation in the developing Drosophila eye. *Cell* **72,** 379–395.

Boyden, E. A. (1927). Experimental obstruction of the mesonephric ducts. *Proc. Soc. Exp. Biol. Med.* **24,** 572–576.

Brophy, P. D, Ostrom, L., Lang, K. M., and Dressler G. R. (2001). Regulation of ureteric bud outgrowth by Pax2-dependent activation of the glial derived neurotrophic factor gene. *Development* **128,** 4747–4756

Buehr, M., Subin, G., and McLaren, A. (1993). Mesonephric contribution to testis differentiation in the fetal mouse. *Development* **117,** 273–281.

Burns, R. K. (1955). "Urogenital System: Analysis of Development" (B. J. Willier, P. Weiss, and V. Hamburger, eds.), pp. 462–491. Saunders, Philadelphia.

Campagnolo, L., Russo, M. A., Puglianiello, A., Favale, A., and Siracusa, G. (2001). Mesenchymal cell precursors of peritubular smooth muscle cells of the mouse testis can be identified by the presence of the p75 neurotrophin receptor. *Biol. Reprod.* **64,** 464–472.

Croisille, Y. (1976). On some recent contributions to the study of kidney tubulogenesis in mammals and birds. *In* "Tests of Teratogenicity *in Vitro*" (J. D. Ebert and M. Marois, eds.), pp. 149–170. North-Holland, Amsterdam.

Cunha, G. R. (1972). Tissue interactions between epithelium and mesenchyme of urogenital and integumental origin. *Anat. Rec.* **172,** 529–541.

Cunha, G. R., Chung, L. W. K., Shannon, J. M., Taguchi, O., and Fujii, H. (1983). Hormone-induced morphogenesis and growth, role of mesenchymal-epithelial interactions. *Rec. Prog. Horm. Res.* **39,** 559–597.

Davis, A. P., Witte, D. P., Hsieh-Li, H. M., Potter, S. S., and Capecchi, M. (1995) Absence of radius and ulna in mice lacking hoxa-11 and hoxd-11. *Nature* **375,** 791–795.

Deutsch, U., and Gruss, P. (1991). Murine paired domain proteins as regulatory factors of embryonic development. *Sem. Dev. Biol.* **2,** 413–424.

Dieterle-Lievre, F. (1975). On the origin of hematopoietic stem cells in the avian embryo: An experimental approach. *J. Embryol. Exp. Morphol.* **33,** 607–619.

Dressler, G., and Douglass, E. (1992). Pax-2 is a DNA-binding protein expressed in embryonic kidney and Wilms tumor. *Proc. Natl. Acad. Sci. USA* **89,** 1179–1183.

Dressler, G., Deutsch, U., Chowdhury, K., Nornes, H., and Gruss, P. (1990). Pax2, a new murine paired-box-containing gene and its expression in the developing excretory system. *Development* **109,** 787–795.

Dzierzak, E., and Medvinsky, A. (1995). Mouse embryonic hematopoiesis. *Trends Genet.* **11,** 359–366.

ischer, E. A., Verpont, M. C., Garrett-Sinha, L. A., Ronco, P. M., and Rossert, J. A. (2001) Klf6 is a zinc finger protein expressed in a cell-specific manner during kidney development. *J. Am. Soc. Nephr.* **12,** 726–735.

Friebová-Zemanová, Z., and Goncharevskaya, O. A. (1982). Formation of the chick mesonephros. *Anat. Embryol.* **165,** 125–139.

George, K. M., Leonard, M. W., Roth, M. E., Lieuw, K. H., Kioussis, D., Grosveld, F., and Engel, J. D. (1994). Embryonic expression and cloning of the murine GATA-3 gene. *Development* **120**(9), 2673–2686

Green, M. (1970). The developmental effects of congenital hydrocephalus (Ch) in the mouse. *Dev. Biol.* **23,** 585–608.

Gruenwald, P. (1937). Zür Entwicklungsmechanik der Urogenital-systems beim Huhn. *Wilhelm Roux Arch. Entw. Mech.* **136,** 786–813.

Gruenwald, P. (1952). Development of the excretory system. *Ann. N.Y. Acad. Sci.* **55,** 142–146.

Grüneberg, H. (1943). Congenital hydrocephalus in the mouse, a case of spurious pleiotropism. *J. Genet.* **45,** 1–21.

Grüneberg, H. (1953). Genetical studies on the skeleton of the mouse. VII. Congenital hydrocephalus. *J. Genet.* **51,** 327–358.

Hara, T., Nakano, Y., Tanaka, M., Tamura K., Sekiguchi, T., Minehata, K., Copeland, N. G., Jenkins, N. A., Okabe, M., Kogo, H., Mukouyama, Y., and Miyajima, A. (1999). Identification of podocalyxin-like protein 1 as a novel cell surface marker for hemangioblasts in the murine aorta-gonad-mesonephros region. *Immunity* **11,** 567–578.

Hsieh-Li, H. M., Witte, D. P., Weinstein, M., Branford, W., Li, H., Kersten, S., and Potter, S. S. (1995). Hoxa11 structure, extensive antisense transcription, and function in male and female fertility. *Development* **121,** 1373–1385.

Johnson, G. R., and Moore, M. A. S. (1975). Role of stem cell migration in initiation of mouse foetal liver haematopoiesis. *Nature* **258,** 726–728.

Kardong, K. (1995). The urogenital system: Embryonic development. *In* "Vertebrates: Comparative Anatomy, Function, Evolution" (M. J. Kemp and S. Dillon, eds.), pp. 534–538 Wm. C. Brown Dubuque, IA.

Kaestrner, K., Knochel, W., and Martinez, D. (2000). Unified nomenclature for the winged helix/forkhead transcription factors. *Genes Dev.* **14,** 142–149.

Koopman, P., Gubbay, J., Vivian, N., Goodfellow, P., and Lovell-Badge, R. (1991). Male development of chromosomally female mice transgenic for Sry. *Nature* **351,** 117–121.

Koopman, P., Munsterberg, A., Capel, B., Vivian, N., and Lovell-Badge, R. (1990). Expression of a candidate sex-determinating gene during mouse testis differentiation. *Nature* **348,** 450–452.

Kreidberg, J. A., Sariola, H., Loring, J., Maeda, M., Pelletier, J., Housman, D., and Jaenicsh, R. (1993). WT-1 is required for early kidney development. *Cell* **74,** 679–669.

Kubai, L., and Auerbach, R. (1983). A new source of embryonic lymphocytes in the mouse. *Nature* **301,** 154–156.

Kume, T., Deng, K. Y., Winfrey, V., Gould, D. B., Walter, M. A., and Hogan, B. L. (1998). The forkhead/winged helix gene Mf1 is disrupted in the pleiotrophic mouse mutation congenital hydrocephalus. *Cell* **93,** 985–996.

Kume, T., Deng, K., and Hogan, B. (2000) Murine forkhead/winged helix genes Foxc1(Mf1) and Foxc2 (Mfh1) are required for the early organogenesis of the kidney and urinary tract. *Development* **127,** 1387–1395.

Labastie, M. C., Catala, M., Gregoire, J. M., and Peault, B. (1995). The GATA-3 gene is expressed during human kidney embryogenesis. *Kidney Int.* **47**(6), 1597–1603.

Labastie, M. C., Cortes, F., Romeo, P.-H., Dulac, C., and Peault, B. (1998). Molecular identity of hematopoietic precursor cells emerging in the human embryo. *Blood* **92,** 3624–3635.

Lakshmanan, G., Lieuw, K. H., Lim, K.-C., Grosveld, F., Engel, J. D., and Karis, A. (1999). Localization of distant urogenital system-, central nervous system-, and endocardium-specific transcriptional regulatory elements in the GATA-3 locus. *Mol. Cell. Biol.* **19,** 1558–1568.

Lawrence, W. D., Whitaker, D., Sugimura, H., Cunha, G. R., Dickersin, G. R., and Robboy, S. J. (1992). An ultrastructural study of the developing urogenital tract in early human fetuses. *Am. J. Obset. Gynecol.* **167,** 185–193.

MacLaren, A. (2000). Germ and somatic cell lineages in the developing gonad. *Mol. Cell. Endocr.* **163,** 3–9.

Marshall, C. J., and Thrasher, A. J. (2001). The embryonic origins of human haematopoesis. *Br. J. Haematol.* **112,** 838–850.

Martineau, J., Nordqvist, K., Tilmann, C., Lovell-Badge, R., and Capel, B. (1997). Male specific cell migration into the developing gonad. *Curr. Biol.* **7,** 958–968.

Martino, C., and Zamboni, L. (1966). A morphologic study of the mesonephros of the human embryo. *J. Ultrastruct. Res.* **16,** 399–427.

Marshall, C. J., and Thrasher, A. (2001). The embryonic origins of the human haematopoiesis. *B. J. Hematol.* **112,** 838–850.

Medvinsky, A., and Dzierzak, E. (1996). Definitive hematopoiesis is autonomously initiated by the AGM region. *Cell* **86,** 897–906.

Merchant-Larios, H., and Moreno-Mendoza, N. (1998). Mesonephric stromal cells differentiate into Leydig cells in the mouse fetal testis. *Exp. Cell Res.* **244,** 230–238.

Merchant-Larios, H., Moreno-Mendoza, N., and Buehr, M. (1993). The role of the mesonephros in the cell differentiation and morphogenesis of the mouse fetal testis. *Int. J. Dev. Biol.* **37,** 407–415.

Miettinen, A., Dekan, G., and Farquhar, M. G. (1990). Monoclonal antibodies against membrane proteins of the rat glomerulus. *Am. J. Pathol.* **137,** 929–944.

Miyamoto, N., Yoshida, M., Kuratani S., Matsuo, I., and Aizawa S. (1997). Defects of urogenital development in mice lacking Emx-2. *Development* **124,** 1653–1664.

Miyazaki Y., Oshima K., Fogo A., Hogan B., and Ichikawa, I. (2000). Bone morphogenetic protein 4 regulates the budding site and elongation of the mouse ureter. *J. Clin. Invest.* **105,** 863–873.

Moore, K. L. (1977). The urogenital system. *In* "The developing Human: Clinically Oriented Embryology" (K. L. Moore, ed.), pp. 220–259. Saunders, Philadelphia.

Moore, M. A. S., and Metcalf, D. (1970). Ontogeny of the haematopoietic system: Yolk sac origin of *in vivo* and *in vitro* colony forming cells in the developing mouse embryo. *Br. J. Haematol.* **18,** 279–296.

Mukouyama, Y., Chiba, N., Okada, H., Ito, Y., Kanamaru, R., Miyajima, A., Satake, M., and Watanabe, T. (2000). The AML1 transcription factor functions to develop and maintain hematogenic precursor cells in the embryonic aorta-gonad-mesonephros region. *Dev. Biol.* **220,** 27–36.

Nelson, J. E., Yuemin, L., and Gemmell, R. T. (1992). Development of the urinary system of the marsupial native cat *Dasyurus hallucatus*. *Acta Anat.* **144,** 336–342.

Nishino, K., Kato, M., Yokouchi, K., Yamanouchi, K., Naito, K., and Tojo, H. (2000). Establishment of fetal gonad/mesonephros coculture system using EGFP transgenic mice. *J. Exp. Zool.* **286,** 320–327.

Nistal, M., and Paniagua, R. (1984). Development of the male genital tract. *In* "Testicular and Epidymal Pathology" (M. Nistal and R. Paniagua, eds.), pp. 1–13. Thieme-Stratton, New York.

Obara-Ishihara T., Kuhlman, J., Niswander, L., and Herzlinger, D.(1999). The surface ectoderm is essential for nephric duct formation in intermediate mesoderm. *Development* **126,** 1103–1108.

Ohto, H., Kamada, S., Tago, K., Tominaga, S-I., Ozaki, H., Sato, S., and Kawakami, K. (1999). Cooperation of Six and Eya in activation of their target genes through nuclear translocation of Eya. *Mol. Cell. Biol.* **19,** 6815–6824.

Orgebin-Crist, M. C. (1981). The influence of testicular function on related reproductive organs. *In* "The Testis" (H. Burger and D. de Kretser, eds.), pp. 239–253. Raven Press, New York.

Pandolfi, P. P., Roth, M. E., Karis, A., Leonard, M. W., Dzierzak, E., Grosveld, F., Engel, J. D., and Lindenbaum, M. H. (1995). Targeted disruption of the GATA-3 gene causes severe abnormalities in the nervous system and in fetal liver haematopoiesis. *Nature Genet.* **11,** 40–44.

Patterson, L., Pembaur, M., and Potter, S. S. (2001). Hoxa11 and Hoxd11 regulate branching morphogenesis of the ureteric bud in the developing kidney. *Development* **128,** 2153–2161.

Pelliniemi, L., Paranko, J., Grund, S. K., Fjordman, K., Foidart, J.-M., and Lakkala-Paranko, T. (1984). Morphological differentiation of Sertoli cells. *INSERM* **123,** 121–140.

Phelps, D. E., and Dressler G. R. (1993). Aberrant expression of Pax-2 in Danforth's short tail (Sd) mice. *Dev. Biol.* **157,** 251–258.

Pritchard-Jones, K., Fleming, S., Davidson, D., Bickmore, W., Porteous, D., Gosden, C., Bard, J., Buckler, A., Pelletier, J., Housman, D, van Heyningen, V., and Hastie, N. (1990). The candidate Wilms' tumour gene is involved in genitourinary development. *Nature* **345,** 194–197.

Runner, M. N. (1946) The development of the mesonephros of the albino rat in intraocular grafts. *J. Exp. Zool.* **103,** 305–319.

Russo, M. A., Giustizieri, M. L., Favale, A., Fantini, M. C., Campagnolo, L., Konda, D., Germano, F., Farini, D., Manna, C., and Siracusa, G. (1999). Spatiotemporal patterns of expression of neurotrophins and neurotrophin receptors in mice suggest functional roles in testicular and epididymal morphogenesis. *Biol. Reprod.* **61,** 1123–1132.

Sainio, K., Hellstedt, P., Kreidberg, J., Saxén, L., and Sariola, H. (1997a). Differential regulation of two sets of mesonephric tubules by WT-1. *Development* **124,** 1293–1299.

Sainio, K., and Raatikainen-Ahokas, A. (1999). Mesonephric kidney: A stem cell factory? *Int. J. Dev. Biol.* **43,** 435–439.

Sainio, K., Suvanto, P., Davies, J., Wartiovaara, J., Wartiovaara, K., Saarma, M., Arumäe, U., Meng, X., Lindahl, M., Pachnis, V., and Sariola, H. (1997b). Glial-cell-line-derived neurotrophic factor is required for bud initiation from ureteric epithelium. *Development* **124,** 4077–4087.

Saxén, L. (1987). "Organogenesis of the Kidney." Cambridge Univ. Press, Cambridge.

Schiller, A., and Tiedemann, K. (1981). The mature mesonephric nephron of the rabbit embryo. *Cell Tissue Rev.* **221**, 431–442.

Schnabel, E., Dekan, G., Miettinen, A., and Farquhar, M. G. (1989). Biogenesis of podocalyxin, the major glomerular sialoglycoprotein, in the newborn rat kidney. *Eur. J. Cell Biol.* **48**, 313–326.

Smith, C., and MacKay, S. (1991). Morphological development and fate of the mouse mesonephros. *J. Anat.* **174**, 171–184.

Sordello, S., Bertrand, N., and Plouet, J. (1998). Vascular endothelial growth factor is up-regulated in vitro and *in vivo* by androgens. *Biochem. Biophys. Res. Commun.* **251**(1), 287–290.

Theiler, K. (1989). "The House Mouse: Atlas of Embryonic Development." Springer-Verlag, New York.

Tiedemann, K. (1976). The mesonephros of cat and sheep: Comparative morphological and histochemical studies. *In* "Advances in Anatomy, Embryology and Cell Biology" (A. Brodal, W. Hild, J. Van Limborgh, R. Ortmann, T. H. Schiebler, G. Töndury, and E. Wolff, eds.), pp. 1–119. Springer-Verlag, Berlin.

Tiedemann, K. (1979). Architecture of the mesonephric nephron in pig and rabbit. *Anat. Embryol.* **157**, 105–112.

Tiedemann, K. (1983). The pig mesonephros. I. Enzyme histochemical observations on the segmentation of the nephron. *Anat. Embryol.* **137**, 113–123.

Tiedemann, K., and Egerer, G. (1984). Vascularization and glomerular ultrastructure in the pig mesonephros. *Cell Tissue Res.* **238**, 165–175.

Tilmann, C., and Capel, B. (1999). Mesonephric cell migration induces testis cord formation and Sertoli cell differentiation in the mammalian gonad. *Development* **126**, 2883–2890.

Toivonen, S. (1945). Über die Entwicklung der Vor- und Uriniere beim Kaninchen. *Ann. Acad. Aci. Fenn. Ser. A* **8**, 1–27.

Torres, M., Gómez-Pardo E., Dressler, G., and Gruss, P. (1995). Pax-2 controls multiple steps of urogenital development. *Development* **121**, 4057–4065.

Upadhyay, S., Luciani, J.-M., and Zamboni, L. (1981). The role of the mesonephros in the development of the mouse testis and its excurrent pathways. *In* "Development and Function of Reproductive Organs" (A. G. Byskov and H. Peters, eds.), pp. 18–27. Excerpta Medica, Amsterdam.

Vainio, S., Heikkilä, M., Kispert A., Chin, N., and McMahon, A.(1999). Female development in mammals is regulated by Wnt-4 signalling. *Nature* **397**, 405–409.

Van Esch, H., Groenen, P., Nesbit, M. A., Schuffenhauer, S., Lichtner, P., Vanderlinden, G., Harding, B., Beetz, R., Bilous, R. W., Holdaway, I,

Shaw, N. J., Fryns, J. P., Van de Ven, W., Thakker, R. V., and Devriendt, K. (2000). GATA3 haplo-insufficiency causes human HDR syndrome. *Nature* **406**, 419–422.

Vazquez, M. D., Bouchet, P., Mallet, J. L., Foliguet, B., Gerard, H., and Le Heup, B. (1998). 3D reconstruction of the mouse mesonephros. *Anat. Hist. Embryol.* **27**, 283–187.

Vize, P., Seufert, D. W., Carroll, T., and Wallingford, J. B. (1997). Model systems for the study of kidney development: Use of the pronephros in the analysis of organ induction and patterning. *Dev. Biol.* **188**, 189–204.

Waddington, C. H. (1938). The morphogenetic fuction of vestigial organ in the chick. *J. Exp. Biol.* **15**, 371–376.

Wartenberg, H., Kinsky, I., Viebahn, C., and Schmolke, C. (1991). Fine structural characteristics of testicular cord formation in the developing rabbit gonad. *J. Elect. Microsc. Techn.* **19**, 133–157.

Wartenberg, H. (1982). Development of the early human ovary and role of the mesonephros in the differentiation of the cortex. *Anat. Embryol.* **165**, 253–280.

Wenzel, J. G., and Odend'hal, S. (1985). The mammalian rete ovarii: A literature review. *Corn. Vet.* **75**, 411–425.

Wintour, E. M., Butkus, A., Earnest, L., and Pompolo, S. (1996). The erythropoietin gene is expressed strongly in the mammalian mesonephric kidney. *Blood* **88**, 3349–3353.

Wrobel, K.-H., and Süß, F. (1999). On the origin and prenatal development of the bovine adrenal gland. *Anat. Emryol.* **199**, 301–318.

Xu, P.-X., Adams, J., Peters, H., Brown, M. C., Heaney, S., and Maas, R. (1999). Eya1-deficient mice lacks ears and kidneys and show abnormal apoptosis of organ primordia. *Nature Genet.* **23**, 113–117.

Yang, Z., Gu, L., Romeo, P.-H., Bories, D., Motohashi, H., Yamamoto, M., and Engel, J. D.(1994). Human GATA-3 trans-activation, DNA-binding, and nuclear localization activities are organized into distinct structural domains. *Mol. Cell. Biol.* **14**, 2201–2212.

Yoshida, M., Suda, Y., Matsuo, I., Miyamoto, N., Takeda, N., Kuratani, S., and Aizawa, S (1997). Emx1 and Emx2 functions in development of dorsal telencephalon. *Development* **124**, 101–111

Zamboni, L., and Upadhyay, S. (1981). Ephemeral, rudimentary glomerular structures in the mesonephros of the mouse. *Anat. Rec.* **201**, 641–644.

Zamboni, L., and Upadhyay, S. (1982). The contribution of the mesonephros to the development of the sheep fetal testis. *Am. J. Anat.* **165**, 339–356.

7

Three-Dimensional Anatomy of Mammalian Mesonephroi

Marie D. Vazquez, Pierre Bouchet, and Peter D. Vize

I. Introduction

The aim of this work is to explore the organization and morphogenesis of mesonephric tubules during embryonic development. In humans, graphic representations of mesonephric tubules have only been reported for the first 12 weeks of development (Mac Callum, 1902; Shikinami, 1926; Altschule, 1930). Using similar methods, Bovy (1929), has described the development of the Wolffian body (mesonephros) in mice. With the advent of computer sciences, new approaches for the reconstruction of embryonic structures are now possible. Machin *et al.* (1996) performed the reconstruction of a human embryo of 5 weeks. In this reconstruction, mesonephroi are represented only by external contours, and the details of tubule organization are not described. In this chapter, 75 computer-generated models of murine and human mesonephroi are presented. Each was constructed from serially sectioned samples. The borders of mesonephric organs were determined manually and then three-dimensional (3-D) reconstruction software was used

to combine and segment the individual organs. The results are models of various stages of development that allow unique insights into the anatomy of this evanescent and poorly characterized organ.

II. Materials

A. Human

Three human embryos from voluntary abortions were obtained. The ages and stages of embryonic specimens, ascertained by crown–rump measurements, were according to O'Rhally and Gartner (1974): the smallest embryo was 7.4 mm and was classified as stage 15, 30 days postconception, and the two others were, respectively, 11 and 16 mm classified stage 17–42 days postconception and stage 19, 50 days postconception. Embryo sex was ascertained by gonadal morphology for 6- and 7-week embryos.

B. Murine

The embryos used for the models were obtained by natural mating of SPF outbred mice, OF1 Ico: OF1 (IOPS Caw) (stock maintained by the departmental colony of R. Janvier, Le Genest, France). Embryos were staged from the time point of vaginal plug observation, which was designated as 0.5 days postcoitum (pc). Embryos enclosed within their membrane were sexed by the chromatin test. For histology, E10.5, E11.5, E12.5, E13.5, E14.5, and E15.5 stages were utilized in this study.

III. Three-Dimensional Reconstruction

The usual approach for visualizing histological structures is based on surface-modeling algorithms (Huijsmans *et al.*, 1983). Because the automatic detection of tissue limits remains difficult, edges of histological structures were defined manually and input as a set of *x, y,* and *z* coordinates. By fastening points between contiguous contours with the shortest distance, the surface is obtained by the juxtaposition of gathered quadrilateral or triangular tiles. However, shortest path algorithms require a same number of constitutive points in each contour (West and Skytte, 1986), and in an embryo or organ, adjacent sections can be quite different, complicating this process. Moreover, anatomical structures often present multiple bifurcations, and it is sometimes difficult to perceive links within complex 3-D histological structures. To address these issues, an interpolation based on Boissonat's (1988) method was used that allows the construction of a polyhedral volume that can be intersected by planar contours and generate surface tiles at each branch (Lozanoff, 1991).

In our study, after searching for the mesonephric tubular zone with light microscopy, the external contours of distinct tubules were drawn manually onto acetate sheets from all serial sections. Each section was then scanned and its contour lines converted into polygonal lines. The surfaces were then rebuilt using GOCAD software and the previously obtained polygonal lines. A major advantage of the GOCAD program is the use of discrete smooth interpolation (DSI; Mallet, 1992). The DSI interpolator recomputes the *x, y,* and *z* coordinates of the free vertices. This smoothing procedure serves to reduce the presence of small irregularities. Small misalignments can be induced by compression or distorted thermal expansion during histological preparation or by microscopical procedures and interfere with model building if not removed by such an approach. DSI-generated vertices are considered to be sure positions, as opposed to the other vertices, which are regarded as free. Model generation has been described elsewhere (Vazquez *et al.*, 1998a,b).

From each of the GOCAD models generated using this approach, 72 images were generated every 10° and used to create computer animations, which are described in detail.

IV. Human Mesonephric Development

Three models of human mesonephroi at different embryonic ages are presented: 7-mm (4 weeks), 11-mm (6 weeks), and 16-mm (7 weeks) stages (Fig. 7.1). Animations illustrating these stages are available at the book web site. Representative views are also presented in Fig. 7.1.

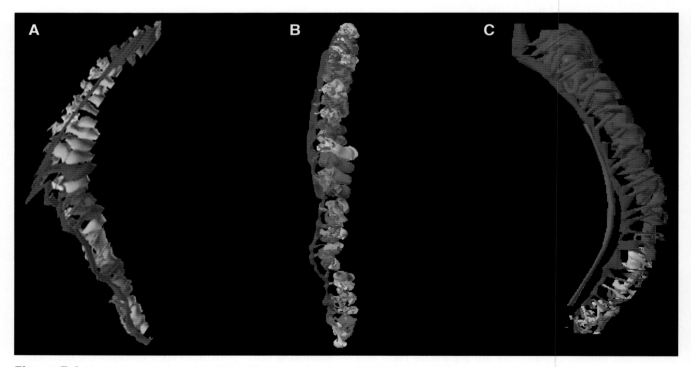

Figure 7.1 Human mesonephric models. (A) Model of the mesonephros of a 7.4-mm stage 15 (30 days pc) human embryo (unknown sex). There are 23 mesonephric tubules (red or pink) linked to the mesonephric duct (blue). A number of unlinked tubules (yellow or white) are also present that do not contact the duct. (B) Model of the mesonephros of an 11-mm stage 17 (42 days pc) male human embryo with 24 linked tubules. (C) Model of a 16-mm female stage 19 (50 days pc) mesonephros with 18 linked tubules. The mesonephric duct is colored blue and the Müllerian duct green.

At 4 and 6 weeks, the models show 23 or 24 tubules bound (Fig. 7.1, in red) to the Wolffian (mesonephric) duct. At the 7th week of development, the number has fallen to 18. These results are in accordance with Arey's description, which identified 17 tubules at 9 weeks of age. The models show that linked (bound) mesonephric tubules appear before unbound tubules. The bound tubules persist and differentiate and are relatively stable. Following the initial establishment of the mesonephros, a regressive process begins and the number of bound tubules is reduced from 24 to 18. This process is completed by the 7th week, and tubules are concentrated cranially, become short, and lose their aspect of a complete nephric unit. Isolated tubules (Fig. 7.1, in yellow), unbound to the mesonephric duct, are present at all three stages examined and have three characteristic shapes: a simple tubule, a tubule with a capsule, and a nephric unit with a proximal and distal tubule and a Bowman's capsule.

The great number of unbound tubules with distinct forms could be the result of unbound tubules developing separately or by unbound tubules forming by a regressive process from bound tubules.

V. Murine Mesonephric Development

Models were generated for male and female mice at 10.5, 11.5, 12.5, 13.5, and 15.5 days of development (see book website at http://www.academicpress.com/kidney for animated models). Twelve models were built for each stage, nine from males and three from females. Once again, animations are provided of all 60 of these reconstructions that allow viewing from all angles, and individual frames depicting particularly useful viewpoints are compiled in Fig. 7.2.

The male mesonephros at 10.5 days has a Wolffian body (Fig. 7.2, in blue) from which the Wolffian (mesonephric) duct separates laterally. Caudally, unlinked mesonephric tubules

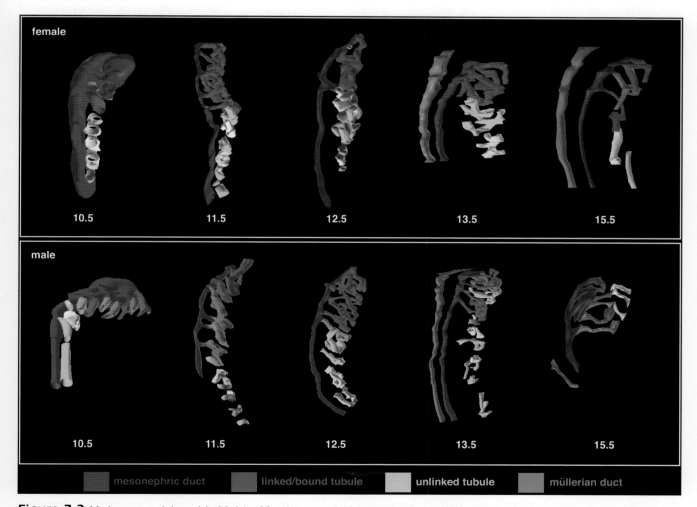

Figure 7.2 Murine mesonephric models. Models of female mesonephroi (top) and male mesonephroi (bottom). The stage of the embryo from which the model was created (in days pc) is noted below each image. Tubules linked to the mesonephric duct are shaded red or pink, whereas unlinked tubules are yellow or white. The mesonephric duct is colored blue and the Müllerian duct green.

(Fig. 7.2, in yellow or white) are present along the mesonephric duct. In the cranial part of the mesonephros, short mesonephric tubules (Fig. 7.2, in red or pink) are present linked to the duct. At 11.5 days the male mesonephric duct has four bifurcations at this point in time: the more cranial bifurcation gives two tubules with a downward rotation (relative to the caudal part of the embryo), the second and third bifurcations have two tubules with a downward rotation, and the more caudal bifurcation has two tubules. The more cranial tubule has an upward rotation relative to the cranial part of the embryo, whereas the most caudal tubule has a downward rotation relative to the caudal part of the embryo.

In the caudal portion of the female mesonephros at 11.5 days, the isolated tubules take the shape of a "J" then of an "S". Mesonephric tubules bound to the mesonephric duct all have a downward rotation. They are organized in a network, branching from the mesonephric duct at four points. From these four points of emergence, bifurcations occur at the origin of spiral tubules. This organization is unique to the female mesonephros and would explain why the volumes of some bound tubules in female mesonephros are bulkier than those of males.

At 12.5 days pc, tubules become more elongate, and in the mesonephroi of both males and females there is a moderate reduction in the number of linked tubules and a more cranial distribution.

At each stage models were generated from both male and female siblings (movies, litter A, litter B, etc.). In Fig. 7.3, mesonephroi of E13.5 siblings are illustrated. In male sibling mesonephroi, the cranial portion of both models contains circumvented cranial unlinked tubules, whereas in the previous model (day 12.5 siblings), they were only present in the caudal part of the mesonephros. Moreover, a number (11 and 17) of unlinked tubules are now present in the caudal region. The presence of unlinked tubules in the cranial part of both these mesonephroi could be the product of the regression of some cranial tubules. Cranial tubules bound to the mesonephric duct are short and rotate in the same direction as unbound tubules. Female sibling mesonephroi at 13.5 days also illustrate a novel feature: each contains three bifurcations with interconnected branches (Fig. 7.3). These connections give the cranial part of the female mesonephroi a bulky network organization. The subjacent tubule is short and straight, surrounded by isolated tubules. Mesonephroi therefore differ considerably between the sexes at this stage of development.

By day 14.5 of development, mesonephric regression has progressed and the majority of both linked and unlinked tubules are in the cranial region (Fig. 7.2).

At 15.5 days of development, the linked mesonephric tubules of the male mesonephros have anastomosed (Fig. 7.2, blue cranial area) to the mesonephric duct. This may represent the beginnings of formation of the rete testis (Buehr *et al.*, 1993). This model also allows visualization of the cranial concentration of the lateral extremities of the mesonephric

tubules. Unlinked tubules are rare and found only near the medial ends of the bound tubules.

In the female mesonephros at day 15.5 (Fig. 7.2), the linked tubules are linear, even rudimentary, and the isolated tubules are rare. Moreover, the mesonephric duct (Fig. 7.2, in blue) is now less bulky than the Müllerian duct (Fig. 7.2, in green).

VI. Conclusions

The data-processing protocol, based on the use of histological serial sections (Vazquez *et al.*, 1998a, b), allows the generation of 3-D mesonephric reconstructions that are more precise than those carried out by AUTOCAD (Machin *et al.*, 1996). Indeed, as these models illustrate, GOCAD can reconstruct complex junctions and objects.

The reconstructed mesonephroi confirm previous data on the form of the nephric units (see Chapter 6), the number of mesonephric tubules, and the cranio-caudal sequence of tubular appearance. Moreover, the female mesonephros of 7 weeks illustrates the existence of a multivesicular structure, and they efficiently display the alignment of linked and unlinked tubules to the mesonephric duct.

Murine models show distribution of the tubules during their development and allow them to be visualized from any angle. This is not the case with the histochemical technique *in toto* practiced on the mesonephric tubules of rats and mice (Sainio *et al.*, 1997; Chapter 6). Moreover, mobilization of the reconstructed objects allows researchers to detect elements that can be hidden by static views. The murine reconstructions improve upon the graphic representations carried out at the beginning of the last century by J. Bovy (1929). Moreover, GOCAD has a smoothing tool, DSI, improving the quality of the image and conferring on the 3-D reconstructions a representation closer to the real object. The program REYES, which was used to build reconstructions of the connection between the metanephric bud and the Wolffian duct of a *limb deformity* mutant mouse embryo at 12 days of development, did not correct the jagged aspect of the reconstruction (Maas *et al.*, 1994), and the difference between this model and those presented here is dramatic.

The GOCAD models allowed the characterization of several novel aspects of murine mesonephric development. First, no well-defined nephrostomes were observed, despite Brambells (1928) description of six nephrostomes (three cranial, three gonadic) being present at day 11. The reconstructions also show a distinction in the male–female organization of the tubules from 11.5 to 13.5 days of development and a network structure characteristic of female mesonephroi that is not present in males. The presented models contribute the first steps in building a 3-D atlas of the mesonephros. In addition to uncovering details of mesonephric anatomy, such models may serve as the scaffold upon which gene expression patterns can be projected and analyzed.

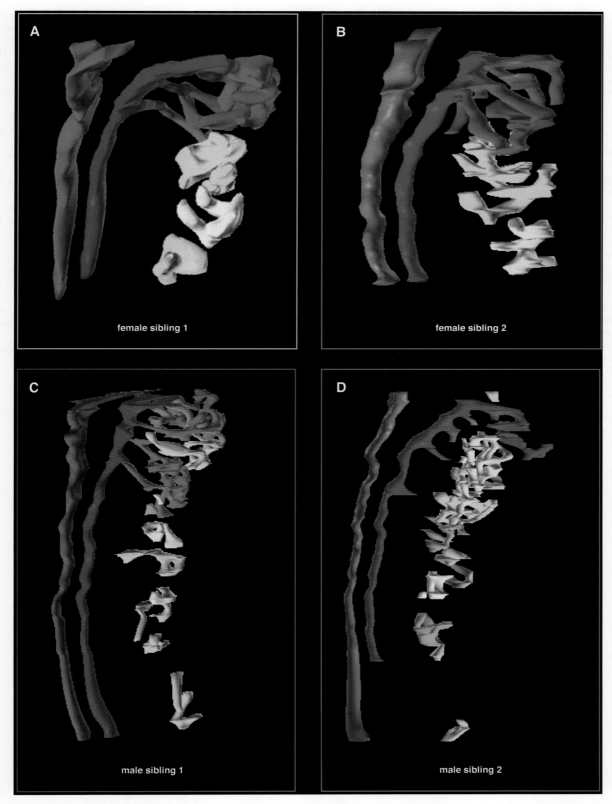

Figure 7.3 Variation in mesonephric anatomy between littermates. (A) E13.5 female mesonephric model. (B) Mesonephros of a female littermate of the embryo used to build the model shown in A. (C) E13.5 male mesonephric model. (D) Mesonephros of a male littermate of the embryo used to build the model shown in C. Color scheme is identical to that in Fig. 7.1 and 7.2. See book website for animated models of these models.

Acknowledgments

We are grateful to Dr. S. Magre (Laboratoire de Physiologie de la Reproduction, Universitie Pierre et Marie Curie, Paris, France) for valuable discussion and precious advice. We thank B. Cunin and M. Simonetti for their technical contributions to the work. This work was supported by AFFDU grants (Dorotee Leet).

References

Altschule, M. D. (1930). The changes in the mesonephric tubules of human embryos ten to twelve weeks old. *Anat. Rec.* **46,** 81–91.

Boissonnat, J. D. (1988). Shape reconstruction from planar cross-sections. *Comput. Vis. Graph. Image. Proc.* **132,** 240–259.

Bovy, J. (1929). Recherches sur le corps de Wolff et lorigine des connexions urogonitales chez la souris. *Arch. Biol.* **39,** 139–174.

Brambell, F. W. R. (1928). The development and morphology of the gonads of the mouse. II. The development of the Wolffian body and ducts. *Proc. Roy. Soc. Lond. [Biol.]* **102,** 206–221.

Buehr, M., Subin, G. U., and McLaren, A. (1993). Mesonephric contribution to testis differentiation in the fetal mouse. *Development* **117,** 273–281.

Huijsmans, D. P., Lamers, W. H., Los, L. A., and Strackee, J. (1986). Toward computerized morphometric facilities: A review of 58 software packages for computer-aided three-dimensional reconstruction, quantification, and picture generation from parallel serial sections. *Anat. Rec.* **216,** 449–470.

Karl, J., and Capel, B. (1995). Three-dimensional structure of the developing mouse genital ridge. *Phil. Trans. R. Soc. Lond. B* **350,** 235–242.

Lozanoff, S., and Deptuch, J. J. (1991). Implementing Boissonnats method for generating surface models of craniofacial cartilages. *Anat. Rec.* **229,** 556–564.

Maas, R. L., Zeller, R., Woychik, R. P., Vogt, T. F., and Lede, P. (1990). Disruption of formin-encoding transcripts in two mutant limb deformity alleles. *Nature* **346,** 853–855.

Mac Callum, J. B. (1902). Notes on the Wolffian body of higher mammals. *Am. J. Anat.* **1,** 245–328.

Machin, G. A., Sperber, G. H., Ongaro, I. and Murdoch, C. (1996). Computer graphic three-dimensional reconstruction of normal human embryo morphogenesis. *Anat. Embryol.* **194,** 439–444.

Mallet, J. L. (1992). Discrete smooth interpolation in geometric modelling. *Comput. Aided Des.* **24,** 178–191.

O'Rahilly, R., and Gartner, E. (1974). Developmental stages of the human embryo. *Bull. Ass. Anat.* **58,** 177–182.

Sainio, K., Hellstedt, P., Kreidberg, J. A., Saxon, L. and Sariola, H. (1997). Differential regulation of two sets of mesonephric tubules by WT-1. *Development* **124,** 1293–1299.

Shikinami, J. (1926). Detailed form of the Wolffian body in human embryos of the first eight weeks. *Contr. Embryol. Carneg. Instn.* **18,** 51–61.

Vazquez, M. D., Bouchet, P., Foliguet, B., Gerard, H., Mallet, J. L., and Leheup, B. (1998a). Differentiated aspect of female and male mouse mesonephroi. *Int. J. Dev. Biol.* **42,** 621–624.

Vazquez, M. D., Bouchet, P., Mallet, J. L., Foliguet, B., Gerard, H., and LeHeup, B. (1998b). 3D reconstruction of the mouse's mesonephros. *Anat. Histol. Embryol.* **27,** 283–287.

West, M. J., and Skytte, J. (1986). Anatomical modeling with computer-aided design. *Comput. Biomed. Res.* **19,** 535–542.

8

Molecular Control of Pronephric Development: An Overview

Elizabeth A. Jones

I. Introduction

Since the mid-1990s there has been a resurgence of interest in developmental studies aimed at the molecular dissection of pronephros development. The simplicity of the pronephric structure has allowed both experimental and genetic manipulation in a variety of vertebrate organisms to be carried out. Such experimentation can provide important advances in our knowledge of how the initial phases of kidney induction, specification, and patterning occur. Because these phases are repeated during the successive waves of kidney differentiation from the larval pronephros to the adult mesonephros or metanephros, it is likely that an understanding of the initial inductive and signaling events will also elucidate many of the later phases of kidney development and differentiation. Although there is a wealth of information available on the genetic inactivation of many of these genes, interpretation of the observed phenotypes is often highly complex and may not represent the primary lesion. Rather it may represent a secondary event, as the induction of the mesonephros is

dependent on the pronephros, and metanephric development is dependent on the mesonephros (Saxén, 1987). The molecular study of the initial events in kidney formation, the induction, patterning, and function of the pronephros specifically addresses the roles of these genes in the intiation of kidney organogenesis.

The aim of this review is not to list all of the available markers of pronephric cell lineages in vertebrate organisms, as these have been reviewed elsewhere (Vize *et al.,* 1997; Brändli, 1999), but rather to focus on those molecules, which in a variety of biological systems, have been shown to have conserved distributions and/or functions in vertebrate kidney forms. It focuses particularly on the pronephric studies that have been carried out and attempts to draw parallels in structure, developmental expression patterns, and function among the four developmental models for kidney development—mouse, chick, zebrafish, and *Xenopus*—with a view to establishing essential paradigms in kidney development. Many of the genes that have a role in kidney organogenesis have been identified by targeted mutagenesis studies. A brief review of these results has been included to allow developmental comparisons to be made. Where there is additional information on more primitive vertebrates and occasionally invertebrates, this has been included to give an evolutionary context. The review is divided into two main sections that focus on the roles of transcription factors and signaling molecules in pronephric development. There are several excellent reviews that cover other molecular players in pronephric development in amphibia and fish (Carroll *et al.,* 1999a; Brändli, 1999; Kuure *et al.,* 2000).

II. Transcription Factors Implicated in Development of the Pronephros

A. The LIM Family of Transcription Factors

The LIM domain family of transcriptional regulators controls the expression of genes that pattern the body and generate cell type specificity across a wide range of invertebrate and vertebrate organisms (Curtiss and Heilig, 1998; Dawid *et al.*, 1998). These molecules are characterized by the presence of two cysteine and histidine-rich LIM domains that occur in tandem at the N terminus of the protein, together with a homeodomain (Fig. 8.1). The homeodomain is capable of binding regulatory sequences in the promoters of target genes. LIM domains coordinate the binding of zinc ions, each forming two zinc finger-binding motifs. Family members have been identified in organisms from ascidians

to humans. All family members have been shown to have a functional role in the development of the nervous system, although many of the family members have discrete expression patterns in a variety of other tissues where these molecules are thought to play an equally important developmental role (Curtiss and Heilig, 1998).

Two major LIM-HD proteins have a defined role in vertebrate kidney development: *Lim-1* and *Lmx-1b*. In both of these cases, targeted mutagenesis experiments have been carried out in the mouse, which generate, in addition to a neural and a limb phenotype, respectively, a kidney phenotype (Shawlot and Behringer, 1995; Chen *et al.*, 1998). The *Lim-1* disruption results in mice lacking the pronephros, mesonephros, and the metanephros; however, the overall phenotype is so severe that an extensive analysis of the kidney phenotype is impossible. The mutant phenotype agrees well with the observed expression domains of *Lim-1* in the wild-

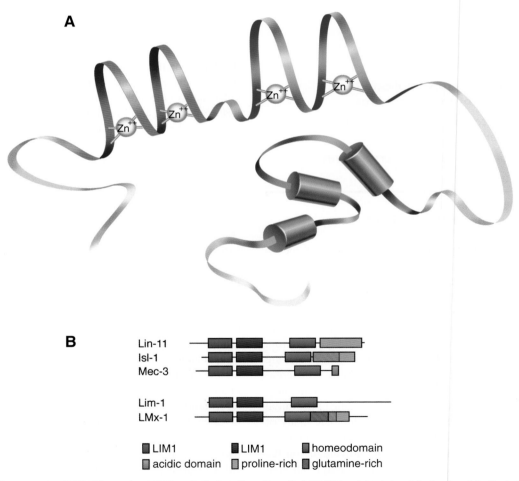

Figure 8.1 The structure of LIM-HD proteins. (A) Hypothetical configuration of a LIM-HD protein such as Lim1 or Lmx1 to illustrate the role of the LIM domains in preventing DNA binding of the homeodomain, thus preventing transcriptional activation. The two LIM domains, LIM1 and LIM2, are colored red and blue and the homeodomain is green. Breaks in the protein chains represent the variable lengths that different family members possess in this region. (B) Domain structure of the founder members of the group, Lin-11, Isl-1, and Mec-3, together with the two family members with roles in kidney development, Lim-1 and Lmx-1. Adapted from Curtiss and Heilig (1998).

type embryo, where it is initially expressed in the lateral plate of the 7 to 7.5 dpc embryo and then subsequently in the nephrogenic cords of the 9-day embryo (Barnes *et al.*, 1994). Targeted disruption of the *Lmx-1b* gene results in transgenic mice with a phenotype consistent with the role of this gene in dorsal limb specification. The dominant features of transgenic offspring include the absence of nails and patellae. In addition to the limb phenotype, mutant embryos exhibited a clear kidney phenotype. The kidneys contained abnormal deposits of periodic acid Schiff staining material in the distal convoluted tubules, consistent with aberrant glycoprotein expression. Ultrastructural analysis showed that the basement membrane associated with the glomerulae displayed prominent irregular thickenings and an occasional lack of continuity, suggesting an essential requirement for the *Lmx-1b* gene in embryonic metanephric kidney development. The phenotypes observed by the targeted disruption of the *Lmx-1b* gene are strikingly similar to those observed in nail patella syndrome patients (NPS). It has been shown that the *Lmx-1b* gene maps to the NPS locus in humans. Individual patients analyzed show mutations in the *Lmx-1b* gene, which are able to functionally disrupt the DNA-binding domain of the LIM-HD protein or are terminated prematurely to produce truncated proteins (Dreyer *et al.*, 1998). These data show that mutations in the *Lmx-1b* have profound effects on limb development and kidney function, perhaps focused on glomerular structure and/or function. There is, therefore, evidence from amniotes that LIM-HD proteins play a fundamental role in kidney development, although the exact functional roles are difficult to determine from mutant phenotypes, which are complex.

The most compelling evidence for the role of LIM proteins in the development of the pronephric kidney comes from the studies of *Xlim-1*, the *Xenopus Lim-1* orthologue, which is one of the earliest markers of pronephric specification in early development. *Xlim-1* was cloned by Taira and colleagues in 1992 using degenerate polymerase chain reaction (PCR) primers to conserved regions of the homeobox region of the *C. elegans* genes *mec-1*, *Isl-1*, *lin-11*, and *ceh-14*, which are all members of the LIM-HD family (Way and Chalfie, 1988; Karlsson *et al.*, 1990, Freyd *et al.*, 1990). *Xlim-1* has an expression pattern that marks cells contributing to three lineages, the notochord, the pronephros, and the nervous system; the expression is temporally biphasic. Initially, at early gastrula stages, expression is observed in the organizer region and later in the notochord. At later gastrula stages and early neurula stages, an additional phase of expression arises in the lateral plate mesoderm, forming a belt that extends around the whole embryo, ventral to the neural folds. This subsequently condenses into a teardrop-shaped domain by tailbud stages, which corresponds to the pronephric anlagen (Taira *et al.*, 1994). At this stage, subsets of cells within the central nervous system also start to express *Xlim-1*. As the pronephros differentiates *Xlim-1*, expression is retained in both the tubules and the duct of the pronephros. As the tubules

start to lumenize, *Xlim-1* expression becomes restricted to the tubules only, with the strongest expression being retained in the nephrostome tips (Chapter 3; Carroll *et al.*, 1999b).

The expression of *Xlim-1* within the pronephric lineage is dependent on the normal cell interactions that take place during gastrulation. In exogastrulae, where normal ectodermal/mesodermal interactions are disrupted, the expression of *Xlim-1* in the lateral plate mesoderm is reduced severely, whereas expression is retained in the notocord. This correlates with the finding that exogastrulae fail to develop either pronephric tubules or duct as assessed by immunostaining with the monoclonal antibody markers 3G8 and 4A6, respectively (Brennan and Jones, unpublished data; Vize *et al.*, 1995). Further circumstantial evidence supporting a role for *Xlim-1* in pronephric specification comes from growth factor and modifier studies. The expression of *Xlim-1* is upregulated by retinoic acid treatment in all of the three cell lineages in which it is expressed in the embryo, resulting in broader domains of expression. Furthermore, within these embryos the pronephros is expanded, probably as a consequence of additional recruitment of cells into the pronephric anlagen (Taira *et al.*, 1994). Treatment of animal caps derived from blastula stage embryos with either activin or RA results in the expression of *Xlim-1* in the animal cap tissue; treatment of the growth factor and modifier together results in a synergistic response (Taira *et al.*, 1994; Uochi and Asashima, 1996). It has been demonstrated by a number of laboratories that such treatment of animal caps with a wide range of concentrations of activin and RA results in the histological and immunohistological identification of pronephric tubules at high frequency (Fig. 8.2; Moriya *et al.*, 1993; Uochi and Asashima, 1996; Brennan *et al.*, 1999). Furthermore, a number of reports describe the formation of functional organs following *in vitro* stimulation under these conditions (reviewed in Ariizumi and Asashima, 2001). *Xlim-1* overexpression

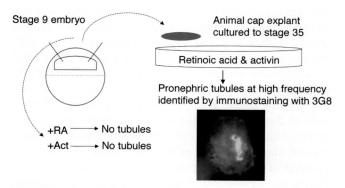

Figure 8.2 *In vitro* differentiation of kidney tubules. Animal caps are removed at stage 9 by manual dissection and are incubated in retinoic acid (RA), activin, or a combination of both. The caps were cultured until stage 35, fixed, and immunostained with the tubule-specific monoclonal antibody, 3G8 (Vize *et al.*, 1995). (Inset) Positive pronephric tubules in a cap treated with both RA and activin. Photograph supplied by Massé and Jones.

following the injection of synthetic mRNA into early embryos, however, is insufficient to cause a high frequency of ectopic kidney formation, although there is synergism between *Xlim-1* and *XPax-8* in kidney development following coinjection (Taira *et al.*, 1994; Carroll and Vize, 1999). Further light has been shed on the role of *Xlim-1* in pronephric development by Chan *et al.* (2000). These workers utilized the activin/RA induction system to study the effects of changing the timing of RA and activin treatment on pronephric induction, correlating this with the expression profile of *Xlim-1*. They established that *Xlim-1* was expressed 9 to 15 h after the activin/RA treatment at a time corresponding to pronephros differentiation. They also studied the effects of constitutively active or dominant-negative *Xlim-1* expression on pronephros formation in the animal cap system and in expression targeted to the pronephric primordium in whole embryos. They showed that injection of wild-type or constitutively active *Xlim-1*, with or without its binding partner *Xldb-1*, failed to induce pronephric tissue in either RA or activin-alone-treated caps. However, when caps were treated with both activin and RA, augmented pronephric formation was observed. In contrast, when a dominant-negative *Xlim-1-enR* was injected, this inhibited differentiation of the pronephros by 25–75%. Studies of targeted overexpression in whole tadpoles showed that a functional deficit in *Xlim-1* resulted in a failure of the pronephros to undergo tubulogenesis. Embryos formed the classic tor-shaped pronephros, which consisted of the three collecting tubules but failed to form the common tubule that linked these to the duct. Inhibition of duct formation was not observed. These observations led the authors to suggest that *Xlim-1*, although an important regulator of pronephric tubule development, is unable to induce pronephros independently, suggesting that there are other molecules expressed prior to *Xlim-1* and *XPax-8* that are involved in formation of the pronephric primordium.

Lim-1-related genes have also been cloned in zebrafish and in the sea urchin, *Hemicentrotus pulcherrimus*. The zebrafish gene, *lim1*, was cloned by degenerate PCR and was shown to be 88% identical at the predicted amino acid sequence with mouse and *Xenopus* (Toyama *et al.*, 1995). Whole-mount *in situ* hybridization shows that *lim1* mRNA is expressed around the entire margin of the embryo at 30% epiboly, with a region of more intense staining to one side. The staining pattern is refined such that at the shield stage, cells of the axial hypoblast within the shield are the only cells to retain expression. This expression pattern is similar but not identical to that of the early phase of *Xlim-1* expression in the organizer and subsequently the notochord. During the segmentation stages, *lim1* expression disappears from the shield and notochord and first becomes expressed at the 5 somite stage in the presumptive pronephric duct. *lim1* is then expressed in both the pronephric tubules and the pronephric duct; this pattern is retained until the 20 somite stage (Toyama and Dawid, 1997). There is, therefore, clear

conservation of expression patterns in the developing pronephric duct and tubules between zebrafish and amphibia. There is currently no functional analysis of the *lim1* gene to establish its role in zebrafish pronephric development.

Other LIM-HD molecules, such as Xlmx1b, have been shown to play a role in normal kidney, probably glomerular development, as described previously. Orthologues of this gene have been cloned in a variety of organisms, including *C.elegans* and *Xenopus* (Hobert *et al.*, 1999; Matsuda *et al.*, 2001; Haldin and Jones, unpublished data). The *C. elegans* orthologue, *lim-6*, is expressed in a small number of sensory, motor, and interneurons in the excretory gland cells of the excretory system and in the epithelium of the uterus. Its expression pattern was established by the fusion of genomic regions of the *lim-6* gene to green fluorescent protein (GFP) and is highly dynamic. The *C. elegans* excretory system is composed of four different cell types, one of which, the A-shaped excretory gland cell, expresses *lim-6* from late embryogenesis and throughout adulthood (Nelson *et al.*, 1983). Laser ablation of the excretory gland cell fails to have any significant effect on development of the *C. elegans* larva, and the developmental consequences of lack of expression of *lim-6* on excretory function have not yet been assessed (Hobert *et al.*, 1999). However, these authors propose that loss of function of *lim-6* affects GABAergic motor neurons involved in rhythmical enteric muscle contraction by preventing the normal formation of the normal axon morphology and also affecting the neuroendocrine outputs of the nervous system. Evidence from our laboratory also indicates a role for the orthologue of this gene isolated from *Xenopus* in neuron differentiation, and its expression is again retained in the excretory system, confined to the glomus (Haldin and Jones, submitted for publication). *Xlmx1b* and its orthologues therefore display a highly conserved expression pattern in the nervous system and the excretory system. We are currently in the process of exploring the functional role of this gene in pronephros development in *Xenopus laevis*.

Another LIM-only gene isolated from the trout, *Onchorhynchus mykiss*, also has restricted domains of expression found in the pronephros and branchial arches (Delalande and Rescan, 1998).

B. Role of the Pax Family of Transcription Factors

The Pax gene family of transcription factors has also been shown to play a major role in the development of the kidney in a variety of organisms. Their developmental roles, however, are spread across a variety of different organs. Pax proteins, like LIM proteins, are involved in the development and patterning of the brain (Dahl *et al.*, 1997; Dressler, 1999; Davies *et al.*, 1999). Pax genes were initially identified on the basis of homology with the *Drosophila* gene *paired*, which together with *gooseberry-proximal* and *gooseberry-*

distal contain the conserved paired box DNA motif, thus forming a subset of *Drosophila* homeobox-containing genes. There are now nine members of this group of regulatory genes identified in vertebrates (Dahl *et al.,* 1997).

The Pax gene family encodes transcription factors characterized by the presence of an N-terminal paired box, a DNA-binding domain forming a subset of homeobox genes, the octopeptide sequence, and in three of the four subgroups found in this family, either a partial or a complete additional homeobox located toward the C terminus, (Fig. 8.3). Genes within each group are very highly conserved and even in diverse species show conservation of spatial expression patterns, indicating that they play important conserved developmental functions. This contention is upheld by the fact that naturally occurring mutations in four Pax genes in both mouse and human result in profound congenital abnormalities. Three mutations at the *undulated* locus in the mouse have been shown to be due to mutations in the *Pax1* gene. Although there is not yet evidence that *PAX1* contributes to a human congenital disorder, it has been linked to spina bifida as a contributor to its pathogenesis (Wallin *et al.,* 1994; Dietrich and Gruss, 1995; Helwig *et al.,* 1995). *Pax3/PAX3* have been identified as the genes involved in *Splotch* mutations in the mouse and Waardenberg syndrome in the human. Both of these mutations result in neural defects: *PAX3* affects the neural crest derivatives involved in pigmentation, among other effects, whereas *Pax2* affects neural tube and muscle development (Epstein *et al.,* 1991; Hol *et al.,* 1996; Baldwin *et al.,* 1992; Tassabehji *et al.,* 1992). When mutated, *Pax6/PAX6* give rise to the *Small eye* phenotype in mice, and *PAX6* has been identified in aniridia patients as the genetic lesion (Macdonald and Wilson, 1996). The *Pax2/PAX2* gene is associated in the heterozygous state with kidney hypoplasia. In almost all of these cases, the phenotypes are dominant and are due to a haploinsufficiency, indicating that two copies of the wild-type gene are essential for normal development of the affected tissue.

The two Pax genes implicated in the control of kidney development are *Pax2* and *Pax8,* which fall into the Pax2, 5, 8 group. The *Pax2* gene is probably the best characterized and was characterized by Dressler *et al.,* (1990). Expression of this gene in the mouse is essentially restricted to the developing nervous and excretory systems where two distinct transcripts can be identified in all the positive tissues. *Pax2* can be detected initially in the mouse in the pronephric duct at around the level of the 10–12th somite. As the duct extends into the mesonephros, the mesonephric tubules also express *Pax2.* At the time of ureteric bud outgrowth into the metanephric mesenchyme, *Pax2* is concentrated in the condensing mesoderm at the ureteric bud tips and may be induced by activating molecules produced by tip outgrowth. Down-regulation of *Pax2* in the induced mesenchyme occurs following epithelialization in the S-shaped body, probably mediated by *WT1.* The mechanism by which *Pax2* expression is lost in the tubules is unknown. *Pax2* is therefore expressed in all the sequential forms of the mammalian kidney: pro-, meso-, and metanephros.

The importance of the expression of *Pax2* in the development of the excretory system is shown from a variety of experiments involving targeted mutagenesis removing endogenous function, deregulation of *Pax2* in transgenic gain-of-function experiments, and *in vitro* antisense oligonucleotide inhibition in organ cultures (Dressler *et al.,* 1992; Rothenpieler and Dressler, 1993; Torres *et al.,* 1995). Initial indications of the requirement of *Pax2* for the mesenchyme to epithelium conversion were obtained following the treatment of kidney organ cultures using phosphothiorate-substituted antisense oligonucleotides. Embryonic kidneys were excised from mouse embryos at E11.5 at which time the ureteric bud has just reached the metanephric mesenchyme and branched once symmetrically. These excised kidneys are capable of forming advanced kidney structures on culturing and are viable for up to 2 weeks. Incubation in the presence of antisense *Pax2* severely reduced the level of branching at the ureteric bud tips due to the inhibition of condensation and polarization of the metanephric mesenchyme following antisense treatment (Rothenpieler and Dressler, 1993). The mutant phenotype was established by Torres *et al.,* (1995). Homozygous mutant mice completely lacked kidneys ureters and genital tracts, suggesting that *Pax2* was essential in both ductal and mesenchymal components of the urinogenital system. The mesenchyme of the nephrogenic cord failed to epithelialize, and the Wolffian and Müllerian ducts developed only partially and degenerated during embryogenesis. Deregulated expression of the *Pax2* gene driven by a CMV promoter restricted the developmental potential of the renal

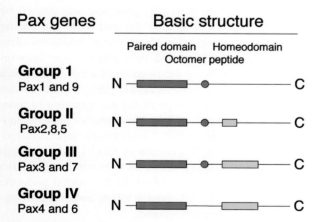

Figure 8.3 Basic domain structure of the Pax family of transcription factors. Pax genes have been divided into four groups based on their structure. They are made up of three structural elements: the paired box domain, the octopeptide, and the homeodomain shown in blue, red, and yellow, respectively. Group II Pax genes, *Pax2* and *Pax 8,* are implicated in kidney development and are characterized by a partial homeobox, which encodes only the first α helix of the homeodomain.

epithelial cells, indicating the importance of the correct regulation of *Pax2* for normal kidney development. Interestingly, functional equivalence of Pax2 and 5 has been shown by knock in experiments (Bouchard *et al.*, 2000). In these experiments, a Pax5 minigene was expressed under the control of the Pax2 locus. The Pax2[5ki] allele was able to rescue the majority of eye and kidney phenotypes characteristic of *Pax2* mutant embryos, suggesting that Pax2 and Pax5 have maintained their functional equivalence, despite their early evolutionary divergence.

Another report suggests that the level of *Pax2* expression may directly inflence the number of cells in the ureteric bud that enter apoptosis. High levels of Pax2 protect cells in the ureteric bud from programmed cell death (Torban *et al.*, 2000).

Pax2 has been isolated from a number of different organisms, including *Xenopus* (Heller and Brändli, 1997; Carroll and Vize, 1999), zebrafish (Krauss *et al.*, 1991, Puschel *et al.*, 1992), and chick (Mauch *et al.*, 2000). *XPax-2* displays multiple splice forms just as in the mouse, again with no tissue-restricted expression of the variants. Whole-mount *in situ* hybridization has shown that *XPax-2* transcripts are first detected soon after gastrulation and are detectable throughout gastrulation. The main tissues expressing *XPax-2* are the nervous system, the developing sensory system, the visceral arches, and the excretory system (Heller and Brändli, 1997).

Pax-2 transcripts are first detectable in the excretory system at Nieuwkoop and Faber stage 21 (Nieuwkoop and Faber, 1994). At this time the first indications of pronephric morphology become apparent as the somatic layer of intermediate mesoderm opposite somites 3 and 4 begins to thicken. Initial expression in a small extended stripe soon changes to an extended tear-drop by stage 24, with the future position of the pronephric duct being clearly decorated with transcripts. As the excre-tory system develops (Chapters 1 and 3) through differentiation of the pronephros into the nephrostomes, tubular caniculi, and the collecting tube, the elongation of the pronephric duct, and the outgrowth of the rectal diverticulum (Chapter 4), *XPax-2* expression is retained until at least stage 39. The formation of nephrostomes is accompanied by the increase in expression of *Pax-2* in these regions, coinciding with high levels of expression of both *Xlim1* and *Xwnt-4* (Carroll and Vize, 1999). Although the expression domain of *XPax-2* is consistent with a function in development and differen-tiation of the pronephros, it is expressed too late to have an early role in specification (Brennan *et al.*, 1998). It has been suggested that *XPax-2* may supercede *XPax-8* function in pronephric differentiation (Carroll and Vize 1999). The timing of expression of *XPax-2* coincides with the initiation of shape changes, which signal the start of pronephric differentiaton (Hausen and Reibesell, 1991; Vize *et al.*, 1997).

The zebrafish orthologue of the *Pax2* gene was initially isolated by Krauss *et al.* (1991) using a combined probe consisting of sequences from the paired box of *Pax1* and a cloned PCR-amplified product of the zebrafish, *pax[zf-b]*.

The zebrafish clone was highly homologous to *Pax2,* showing 96% identity at the amino acid level. The *pax[zf-b]* transcripts were located in four domains, the optic stalk, the otic vesicle, the developing pronephros, and the central nervous system, including both brain and spinal cord. The earliest transcripts that were detectable in the zebrafish appear in the rostral one-third of the embryo as two faint bands. This initial staining pattern then extends and ultimately forms an arrowhead-shaped zone of expression that coincides with particular morphological landmarks. The posterior border coincides with the furrow separating the midbrain from the hindbrain, whereas the anterior border coincides with the posterior border of the tectal ventricle; this region defines the isthmus or midhindbrain border (MHB). Transcripts were also detected in the pronephric primordium and the developing pronephric duct. Expression first appeared at about 12–14 h in a narrow band of mesodermal cells. This band expanded anteriorly by 15 h in a position corresponding in register to somite 3–5 and bifurcated posteriorly into a pattern that mapped onto the Wolffian duct and the ureteric bud, both of which form the excretory system. The transcript expression pattern was confirmed by studies of Puschel *et al.* (1992), who analyzed the expression of Pax2 protein by immunostaining in both mice and zebrafish. These authors confirmed that both the transient pronephros and the mesonephros expressed the protein. The role of the *pax[zf-b]* gene in both neural patterning and development of the excretory system was confirmed by studies aimed at the isolation of mutations in genes affecting formation of the boundary between the midbrain and the hindbrain (Brand *et al.*, 1996). The zebrafish mutation *no isthmus, noi,* had a clear defect in this region of the brain and has also been shown to lack both pronephric ducts and tubules, a similar phenoype to that observed in the mouse (Brand *et al.*, 1996). This mutation was shown to map to the *pax[zf-b]* gene and was caused by generation of a stop codon in the middle of the coding region, resulting in truncation of the protein. Additional silent mutations were also identified. *pax[zf-b]* has now been renamed *Pax2.1* following the isolation of a novel additional *Pax2* transcript, *Pax2.2* (Pfeffer *et al.*, 1998). Further studies have analyzed the phenotype of the *noi* mutation in more depth (Majumdar *et al.*, 2000). These studies have demonstrated a requirement for *Pax 2.1* in multiple aspects of pronephric development. Expression of *Pax2.1* in lateral cells of the pronephric primordium seems to be required to restrict the expression of *wt1* and *VEGF* to the medial podocyte progenitors and is also involved in tubule and duct differentiation and cloacal morphogenesis.

In a series of very elegant laser uncaging experiments, Serluca and Fishman (2001) have utilized the *Pax2.1* gene, together with *wt1* and *sim1*, to study the prepattern that exists in the pronephric field in zebrafish. They labeled small numbers of cells and followed their fates to show that there was a specific bounded region that gives rise to the pronephros.

Pax2.1-expressing cells gave rise to tubule and duct but not the podocyte population, which derived from the most anterior cells in the pronephric field.

Another member of the Pax gene family has been implicated in kidney development and, through recent, elegant functional work, has been shown to play an important role in the development of the pronephros in *Xenopus* (Carroll and Vize, 1999). *Pax8* was initially isolated from the mouse using a mixture of *Pax1, Pax2,* and *Pax3* paired box probes in a low-stringency screen. Comparisons of the paired box sequences of *Pax2* and *Pax8* showed that they were very similar, 71.9% at the amino acid level, and constituted two members of a distinct group of paired box genes (Plachov *et al.,* 1990). Northern analysis of the tissue distribution of this gene identified transcripts in adult kidney, but not in a variety of other adult organs. Unlike *Pax2,* multiple transcripts have not been identified. *In situ* analysis revealed expression at 10.5 dpc in the nephrogenic cord and in the anterior mesonephric tubules, but not in the nephric duct. In contrast to *Pax2, Pax8* is not expressed in the branching ureter, only in the condensing mesenchyme and in the epithelial structures forming from these condensations. These tissues represent the responding rather than the inducing tissues of the excretory system. At 16.5 dpc, *Pax8* is expressed strongly in the cortex of the metanephros, along with *Pax2,* at a time when morphogenesis is proceding.

The other major organ in which *Pax8* is expressed is the developing thyroid gland, where transcripts are first detected as the thyroid vesicles evaginate and bud off from the floor of the pharynx and are subsequently retained throughout thyroid development (Plachov *et al.,* 1990). Transient expression of *Pax8* also occurs in the myencephalon and through the entire length of the neural tube at 11.5 days of gestation, although this expression pattern is soon downregulated and is undetectable by 13.5 days. Surprisingly, the mutant mouse line generated by the *Pax8* gene deletion has no kidney phenotype, although it does display a lack of the thyroid gland (Mansouri *et al.,* 1998). This may be due to the genetic redundancy between *Pax2* and *Pax8,* which are expressed in overlapping domains in the mammalian kidney. However, such redundancy is not found in amphibia, where their expression patterns are both temporally and spatially distinct (Carroll and Vize, 1999).

XPax-8 was cloned by a combination of library screening and PCR strategies from whole stage 22 embryo cDNA. *In situ* analysis showed that *XPax-8* expression is detected at late gastrulation in both the otic vesicles and the presumptive pronephros, several hours prior to the expression of *XPax-2* transcripts in this region, and is maintained through late tadpole stages (Carroll and Vize, 1999). In contrast, there is no reported expression at the midhindbrain boundary, unlike *Pax8* in mouse and zebrafish. The timing of pronephric expresson of *XPax-8* coincides absolutely with the time at which the pronephros is specified. This was defined

as that time that isolated explants derived from the presumptive pronephric region will develop into pronephros within ectodermal wraps (Brennan *et al.,* 1998). The expression of *XPax-8* is consistent with its having a direct role in this process (Carroll and Vize, 1999).

Injection of synthetic *XPax-8* mRNA alone targeted into the C2 blastomere of the 32 cell embryo has been shown to lead to the development of either enlarged or ectopic pronephroi in 30% of the cases. Further experiments have shown dramatically that *Xlim-1* can synergize with *XPax-8* (and also *XPax-2*) to generate ectopic kidney (Carroll and Vize, 1999). In these experiments, targeted overexpression of single mRNAs for *XPax-8* or *Xlim-1* resulted in the enlargement of the pronephric tubules on the injected side as compared with the contralateral side. Three dorsal branches are found in wild-type pronephroi, whereas multiple branches were found in the overexpressing embryos, with these pronephroi occupying a much larger area than those on the uninjected side. However, many of the cells that inherited the injected mRNA failed to form pronephric tubules, although some small, ectopic, 3G8-staining material was found within the somites. Because LIM proteins exert their transcriptional control at least in part through their protein interactive LIM domains, the authors postulated that cofactors might be limiting in these experiments. Because *XPax-8* and *Xlim-1* colocalize in the pronephric anlagen, they coinjected the two mRNAs. The injected embryos developed up to five times more branches than control pronephoi, with the overall area of the pronephros being nearly four times greater. Embryos also demonstrated additional duct tissue, as illustrated by 4A6 immunostaining. *XPax-2* was also capable of causing the same effects as *XPax-8* when coinjected with *Xlim-1,* showing that *XPax-8* and *XPax-2* are functionally redundant in *Xenopus,* although their spatial and temporal separation *in vivo* suggests that this redundancy has no relevance during normal development.

The zebrafish orthologue of *Pax8* was also cloned along with the remaining members of the Pax 2/5/8 family (Pfeffer *et al.,* 1998). As in *Xenopus,* the *Pax8* orthologue is expressed very early in the pronephric and otic placode primordia prior to the expression of either *Pax2.1* or *2.2* and is the earliest developmental marker of these structures. Zebrafish *Pax8* was cloned by PCR using oligonucleotides homologous to the mammalian *Pax8* and subsequent screening of a 1-day-old cDNA library to obtain the full-length clone. Transcripts were detected in two domains. The anterior domain corresponds to the prospective mid/hindbrain boundary and the primordium of the otic vesicle. The posterior domain extends from the middle of the anterior posterior axis to its caudal end along the edges of the embryonic shield in the intermediate mesoderm. This expression domain is maintained in this region through 10 somite stage embryos and is later seen in embryos of 20 somites, 27 h, in the pronephric tubules and duct. This expression pattern is very similar to that of *Pax2.1* in the presumptive excretory system, except that *Pax8* is expressed earlier than *Pax2.1.* Indeed, *Pax2.1* is essential for the initiation of the *Pax5* and *Pax8* expression domains in the isthmus, as mutations in the *Pax2.1* gene, *noi,* which produces a truncated protein lacking any DNA-binding and transactivation domains, caused a failure in the transcription of these genes. In the pronephric field, *Pax2.1* does not activate *Pax8* expression, but is responsible for the maintenance of its expression. This dependence suggests a genetic interaction between these two family members.

The clear importance of the Pax2/5/8 family in both MHB patterning and excretory system development has led to the search for homologues of these genes in the lower chordates (Kozmic *et al.,* 1999). In *Amphioxus,* a single member of this family has been isolated, *AmphiPax2/5/8,* representing an ancestral gene existing prior to the duplication event leading to separate family members in vertebrates. Although the neural domain of expression is initially more posterior to that expected of a gene family involved in the MHB formation, the other domains of expression of *AmphiPax2/5/8* are very similar with transcripts detected in the eye, nephridium, thyroid-like structures, and pharyngeal gill slits, which constitute the sum of the domains of vertebrate *Pax2, Pax5,* and *Pax8* with the exception of the otic placode domain of expression. The nephridium is a thickening on the mesothelial wall of the first muscular somite on the left side and is the excretory organ of *Amphioxus.* No functional data are available for this organism.

C. Role of the Tumor Suppressor gene, WT-1

The Wilms' tumor suppressor gene WT-1 encodes a transcription factor, which, in all cases studied so far, has a major functional role in kidney organogenesis and differentiation (reviewed in Hastie, 1994; Englert, 1998; Little *et al.,* 1999). WT-1 has many properties characteristic of a transcription factor, the N terminus carries a glutamine/proline-rich region, the molecule carries both an activation and a repression domain (Wang *et al.,* 1993), and the C terminus carries four Cys2/His2 zinc finger domains (Haber *et al.,* 1991). The WT-1 protein is expressed as a protein of 52–65 kDa arising as a result of there being two translational initiation sites and two alternative spliced exons, which yields four identifiable isoforms, WT-IA-D (Fig 8.4). The alternative spliced products contain exon 5, which encodes a 17 amino acid stretch, or utilize an alternative splice donor sequence between exons 9 and 10, which results in a three amino acid insertion, KTS. This insertion disrupts the normal spacing between zinc fingers 3 and 4. Both KTS + and – splice forms bind to DNA, although with differing affinities and possibly different targets. The role of the alternative translational start point variant is not clear. DNA binding occurs through a GC-rich consensus-binding site.

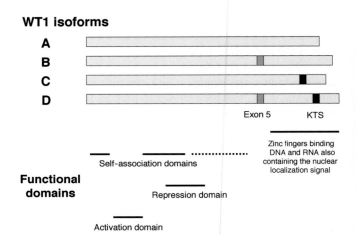

Figure 8.4 The structure of WT1 isoforms. WT1 genes encodes four different isoforms (A–D) as a result of the alternate splicing of two exons: exon 5, encoding 17 amino acids, and KTS, an alternate splice site at the end of exon 9 leading to a 9-bp insert. The major functional regions, including DNA and RNA binding, activation, repression, and WT1 protein self-association, are indicated. Figure not to scale.

A number of molecular targets of WT1 have been established both *in vivo* and *in vitro,* and the WT1 protein is clearly involved in both positive and negative regulation at the transcriptional level. The WT-1 protein has been shown to repress a number of genes, including growth factors and cognate receptor genes, by direct interaction with promoter sequences in *in vitro* transfection assays (Reddy and Licht, 1996). In addition, endogenous repression has been demonstrated in some growth factor receptors and growth factor transcriptional control, e.g., epidermal growth factor receptor, insulin-like growth factor receptor, and platelet-derived growth factor (Gashler *et al.,* 1992; Englert *et al.,* 1995; Werner *et al.,* 1995). *Pax-2* is also repressed by WT-1 by a direct interaction illustrating a direct link between a tumor suppressor gene and a developmental control gene (Ryan *et al.,* 1995). These repressive activities are entirely consistent with the role of WT-1 as a growth regulator via tumor suppression. In addition, positively regulated targets of WT-1 have now been identified, which include syndecan 1, a proteoglycan whose induction is coincident with epithelial differentiation during kidney development (Cook *et al.,* 1996), the anti-apoptotic factor bcl-2 (Mayo *et al.,* 1999), and insulin-like growth factor 2 (Werner *et al.,* 1995). These studies illustrate the pivotal role of WT-1 as a tissue-restricted growth regulator.

Furthermore, it has been shown that WT-1 can colocalize with RNA splicing machinery, suggesting that WT1 can play a direct role in posttranscriptional processing of RNA as well as in transcription itself (Larsson *et al.,* 1995). Interestingly, it is the most abundant +KTS isoform of WT1 that associates with the components of the splicing machinery rather than the –KTS form, which seems to be the higher affinity DNA-binding isoform, thus emphasizing the functional division between the different isoforms. This association with posttranscriptional events within the cell has been further studied by the identification, via structural modeling, of an RNA-binding motif in the WT-1 protein (Kennedy *et al.*, 1996). It has also been demonstrated directly that both KTS plus and minus isoforms bind to insulin-like growth factor 2 mRNA (Caricasole *et al.*, 1996).

Evidence as to the importance of the WT-1 gene in kidney development comes from studies of naturally occurring mutations causing disease in humans (reviewed in Hastie, 1994), from mutagenesis studies in the mouse (Kreidberg, 1993), and from overexpression studies in *Xenopus* (Wallingford *et al.*, 1998). It is interesting to note, however, that the precise role of WT-1 may well vary in embryonic and adult kidneys (Wallingford *et al.*, 1998).

The WT-1 protein is first expressed in the early 9 dpc mouse embryo in the intermediate mesoderm lateral to the coelomic cavity. Within hours, this expression domain extends to the mesoderm surrounding the whole coelomic cavity and in the urogenital ridge. By 12.5 dpc, expression has increased in the induced nephrogenic mesenchyme, particularly in the nephrogenic condensations. By 20 dpc, the majority of the expression domain has been downregulated with the exception of the glomeruli in the metanephros (Armstrong *et al.*, 1992). The human expression pattern is very similar to that of the mouse at appropriate stages of development, with the intermediate mesoderm being the major expressing tissue.

In humans, germline mutations in the *WT-1* gene are associated with both Wilms' tumors and urogenital malformations. The identification of chromosome deletions in the 11p13 region of Wilms' tumor sufferers allowed the isolation of the *WT-1* gene (Call *et al.*, 1990; Gessler *et al.*, 1990). Individuals who are heterozygous for null mutations at the *WT-1* locus include those with the WAGR syndrome (Wilms' tumor, aniridia, genitourinary malformations, and mental retardation). These individuals contain cytogenetically visible deletions of the 11p13 region that affect the structural integrity of the *WT-1* gene. Individuals carrying the Denys–Drash syndrome have a more serious phenotype. These individuals have streak gonads and ambiguous genitalia. They carry a dominant point mutation in the *WT-1* gene that appears to be responsible for the disorder. Both these syndromes result in predisposition to Wilms' tumor. This occurs on the loss of the remaining wild-type allele in the tumor tissue, thus defining *WT1* as a tumor suppressor gene.

In the mouse, the *WT-1* mutant phenotype was generated by targeted disruption of the *WT-1* gene in ES cells (Kreidberg *et al.*, 1993). Transgenic lines generating homozygous mutant embryos were embryonic lethal. At day 11 dpc, the metanephric blastema underwent apoptosis, the ureteric bud failed to grow out from the Wolffian duct, and the inductive events that normally lead to the formation of the metanephric kidney failed to occur. In addition, other organs, such as the heart, where *WT-1* is know to be expressed, showed mesothelial abnormalities. Interestingly, *Pax-2* was also absent from the nephrogenic blastema. It has been shown that *WT-1* transcriptionally upregulates the anti-apoptotic *bcl-2* gene through a direct interaction, thus indicating how disrupted *WT-1* expression can result in decreased apoptosis and increased oncogenic potential, leading to Wilms' tumor (Mayo *et al.*, 1999).

A remarkable finding has shown that the two splice variants have distinct functions in both kidney formation and in the sex determination pathway (Hammes *et al.*, 2001). Homozygous mutant mice have been generated deficient for one or another splice variant in an animal model of the Frasier mutation. Mouse lines lacking KTS demonstrate an increase in stromal tissue and a decrease in tubular epthelium; however, the glomerular number was unchanged. Glomerular tufts were, however, abnormal in size and glomerular podocytes were abnormal. Immunohistological expression patterns of both WT-1 and Pax2 were unchanged. In KTS homozygous animals, there was a much more extreme kidney phenotype with a dramatic reduction in kidney size with a reduction in both stromal components and glomerular number. Those glomeruli that did form were contracted dramatically. In these kidneys, the distribution of WT-1 was normal, but the distribution of Pax2, while normal in position, showed that the condensing blastema had a much looser appearance. These data show a remarkable functional role for the two isoforms of the WT-1 protein.

The role of the *WT-1* gene in pronephric development has been studied in amphibian systems. The *Xenopus WT-1* gene has been cloned and characterized (Carroll and Vize, 1996; Semba *et al.*, 1996). As in the mouse and human, the major site of expression is in the developing kidney; however, there is no expression in the pronephric tubules, only in the glomus, the vascular filtration device of the larval pronephros. This is the functional equivalent to the glomerulus, where late expression of *WT-1* persists in both mouse and human. This illustrates an interesting utilization of this transcription factor, despite the different embryonic origins of the glomerulus and the glomus. The *xWT1* gene is identical to the human gene over the zinc finger domains except for the first two amino acids, although the glutamine/proline region at the N terminus is not conserved. The *Xenopus* gene does express the alternatively spliced KTS, exon 9, although the other most common alternative spliced form, exon 5, is not present. This does support the importance of expressed alternative spliced products throughout the vertebrate species. RT-PCR analysis indicates that strong zygotic activation of *xWT1* is seen between stages 18 and 21, which is confirmed by *in situ* analysis that indicates staining to the developing glomus at stage 20. The staining does not extend to the developing tubules and duct in the pronephric anlagen. At stage 36 the heart also expresses *xWT1*. Expression of *xWT1* persists in adult tissues in both the kidney and the testis, although it is not expressed in the ovary (Semba *et al.*, 1996).

Functional studies in amphibia have been addressed via overexpression of mRNA targeted into the blastomeres fated to give rise to the pronephros (Wallingford *et al.*, 1998). Ectopic expression of xWT1 of both KTS plus and minus splice forms inhibited the development of pronephric tubules as identified by immunostaining with the monoclonal antibody marker 3G8 (Vize *et al.*, 1995). This phenotype was shown to be due to the failure to form a pronephric anlagen of the appropriate size rather than a defect in epithelialization. It has been suggested that the roles of xWT1 and WT-1 in the formation of the vascular components of the kidney may be more ancestral than their roles in the mesenchymal condensations, found in higher vertebrates only (Carroll *et al.*, 1999a). Further work in the highly manipulable amphibian embryo may throw light on this in the future.

Partial Wilms' tumor cDNA sequences have also been isolated from chick, zebrafish, and alligator (Kent *et al.*, 1995). Chick WT1, CWT1, shows 92 and 93% amino acid sequence identity to the mammalian mouse and human clones, respectively. The CWT1 clone, however, does not contain a 13 polyproline stretch found in the human sequence. The cDNA clone was incomplete, lacking the fourth zinc finger. *In situ* analysis in whole-mount embryos showed staining along the paired urogenital ridges at stage 16, and by stage 19, CWT1 was expressed in the first wave of tubule differentiation in the pronephric region. Diffuse expression is also seen on the ventral edge of the somites and the cells overlying the heart and the mesonephros. The allantois is strongly stained at this stage. Later in development the staining persists in the mesonephric glomeruli and in the genital ridge and there is a high level of expression in the condensing mesenchyme of the metanephric blastema. CWT1 also has domains of expression in the indifferent gonad and in mesothelia overlying the heart, liver, and gut. The expression of WT1 protein in both chick and quail shows that the protein is located in specific areas of the mesothelium adjacent to the nephric ducts (Carmona *et al.*, 2001).

A partial cDNA clone was also identified from alligator cDNA (Kent *et al.*, 1995). Expression of alligator *WT1*, *AWT1*, is also detected in the urogenital ridge, somites, and heart. At later stages of development, a novel site of expression is in the developing forelimb. During comma and S-shaped stages of tubule formation, *AWT1* is expressed throughout the nephron. No functional data currently exist for the role of *WT1* in either alligator or chick.

Wt1 has been cloned in zebrafish (Drummond *et al.*, 1998), and its expression pattern was compared with that of Pax2.1 in a study of the prepattern of the zebrafish pronephros (Serluca and Fishman, 2001). Its expression first appears as bilateral stripes at the 2 to 3 somite stage embryo,

which extend from the first to the fourth somite, its anterior limit extending further than that of Pax2.1. A later wave of expression appears in a region medial to the original expression pattern at the 11 somite stage and it is these cells that include those giving rise to the pronephric primordia. The initial striped expression pattern fades as somitogenesis proceeds, the medial cell pattern refines, and from 40 h postfertilization expression is limited to the podocytes in the glomerulus, a domain of expression characterized by the lack of Pax2.1 transcripts. There is therefore a highly consistent pattern of WT1 expression conserved throughout vertebrates.

D. Role of the HNF Family of Winged Helix Transcription Factors

The winged helix family of transcription factors, exemplified by the hepatocyte nuclear factor (HNF) or LFB, family of transcription factors, are implicated in kidney development. These transcription factors carry a characteristic winged helix structure, the HNF-3 domain, whose integrity is essential for DNA binding, and a homeodomain (reviewed in Kaufmann and Knochel, 1996). There are two closely related HNF-1 genes in all vertebrates except fish; HNF-1α (LF-B1) and HNF-1β (also called vHNF1 or LF-B3). They share a similar structure, with 75% sequence identity in the dimerisation domain, and 93% identity in the DNA-binding domain. They are most divergent in their transactivation domains (47%). Both molecules are able to act as transcriptional activators of the same targets, although the transactivation potential of HNF-1β is less. HNF-1α and HNF-1β can heterodimerize with each other through their N-terminal domains or with additional binding partners, e.g., DCoH. Binding to DCoH increases the transactivation capacity (reviewed in Ryffel, 2001).

In adult tissue the HNF-1α factor was initially identified as being highly represented in liver, with additional expression domains in kidney, pancreas, and stomach. The expression of HNF-1β is higher in the kidney and less high in the liver. Both factors are characterized by being expressed in highly polarized epithelial cells. During embryonic kidney development, HNF-1β is expressed in the condensing metanephric mesenchyme surrounding the tip of the ureteric bud and in the bud itself. HNF-1α is expressed somewhat later during the development of the S-shaped bodies.

In amphibia, HNF-1β transcripts are first detected during the midgastrula stage, stage 10.5 (Demartis *et al.*, 1994), in the internalized endoderm. At the start of neurulation, the neurectoderm starts to express, and by midneurula, stage 16, all three germ layers express transcripts. The intermediate mesoderm, from which the pronephros is derived, clearly expresses the transcripts, whereas the somites and lateral plate mesoderm do not. As the pronephric anlagen becomes distinct, pronephric specific expression is clearly detectable in both tubule and duct primordia. As a transcription factor with consensus binding sites in many tissue-specific genes, this expression pattern makes this an attractive candidate factor for involvement in the early phases of pronephric development. DcoH has also been identified in amphibia and its developmental profile has been mapped (Pogge et al., 1995). It is also expressed in the liver, gut, and pronephros where it is localized to the nuclei. HNF4, another member of the HNF family, has also been cloned in amphibia. This factor transactivates the HNF-1α gene via an activin responsive element in the promoter (Holewa et al., 1996; Weber et al., 1996)

The roles of the HNF 1 molecules in normal kidney development have been illustrated by studies on naturally occurring mutations in humans. In the heterozygous form, these cause a variety of renal defects in addition to the early onset of diabetes (the insulin gene carries a HNF1-binding site in the promoter region). These phenotypic studies are complemented by the study of mutant mouse lines, which clearly distinguish the different developmental potencies of the HNF-1α and HNF-1β genes. HNF-1α mutant mice fail to thrive and die around weaning, following a progressive wasting syndrome, with enlarged livers. Liver-specific genes tranisactivated by HNF-1α, e.g., albumin, α1-antitrypsin, and α and β fibrinogen and phenylalanine hydroxylase, are either reduced or silenced completely. The mice suffer from phenylketonuria. In addition to liver phenotypes, these mice suffer from severe renal Fanconi syndrome as a result of renal proximal tubule dysfunction, which causes huge urinary glucose loss and consequent energy and water loss (Pontoglio et al., 1996). This exteme phenotype is in contrast to HNF-1α mutant mice, generated by the Cre-loxP recombination system, which develop noninsulin-dependent diabetes and Laron dwarfism (Lee et al., 1998). These studies have established that the HNF1α has an essential role in postnatal development, although it is dispensible in embryonic development.

Targeted disruption of the HNF-1β (vHNF-1) has severe lethal embryonic consequences (Barbacci et al., 1999; Coffinier et al., 1999). Deficient mice develop normally to the blastocyst stage and start implantation, but die soon after due to a lack of normal extraembryonic visceral endoderm. These experiments suggest that HNF-1β (vHNF-1) occupies a fundamental position in the regulatory network that regulates visceral endoderm formation. Further embryonic development depends on visceral endoderm formation for the processes of gastrulation, anterior neural development, and posterior mesoderm development to proceed.

An elegant piece of work has established, in a heterologous system, the role of HNF-1β in pronephric development. Ryffel and colleagues show that when a mutated human HNF-1β gene, which is normally associated with renal agenesis, is expressed in transfection experiments, it behaves as a gain-of-function mutation with increased transactivation potential (Wild et al., 2000). Expression of this transcription factor in Xenopus embryos leads to defective development and agenesis of the pronephros, a phenotype very similar to the overexpression phenotype generated by HNF-1β mRNA injection into embryos. Use of an additional mutant, which in the human results in the reduction of nephron number, has little effect on the pronephros. This work establishes the conservation of function across humans and Xenopus and identifies the importance of this transcription factor in pronephros development, which cannot be investigated easily in mammalian models due to early embryonic lethal phenotypes (Barbacci et al., 1999; Coffinier et al., 1999).

III. Growth Factors in Pronephric Kidney Development

A. Role of the Wnt Family of Signaling Molecules

Members of the Wnt family are now recognized as one of the most important families of developmental regulators involved in key steps in the development of a variety of organisms and a number of biological systems. Family members have been identified in species as far apart as the nematode, C. elegans, and human. Wnt proteins form a family of secreted glycoproteins, although their solubility is very limited. They are usually between 350 and 400 amino acids in length and carry 23–24 conserved cysteine residues. They have been shown to have important roles in gastrulation, the dorso/ventral patterning of the limb, central nervous system development, segment polarity, axis formation, and kidney formation (reviewed in Lee et al., 1995; Cadigan and Nusse, 1997; 1998; Vainio et al., 1999). The Wnt signal is transduced through the frizzled family of seven-membrane-pass receptors, expressed on neighboring cells. Once secreted into the extracellular space, Wnt proteins are able to interact with a number of secreted molecules, which can modulate their activity, including glycosaminoglycans, which bind Wnts with high affinity, and Frps, which resemble the ligand-binding domain of the frizzled family of receptors. Once received by the frizzled receptor, the wingless signal is passed, via dishevelled, resulting in the inactivation of Zw3/GSK-3. Zw3/GSK-3 normally phosphorylates the armadillo/βcatenin complex, thus labeling it for ubiquitin degradation. The receipt of the Wnt signal prevents this and allows gene transcriptional activation via the TCF1/lef1/pangolin complex (reviewed in Nusse, 1999).

An excess of 15 Wnt genes have been identified in human and mouse, many of which have defined orthologues in Xenopus, chicken, and zebrafish. Following completion of the Drosophila and C. elegans genome projects, these two species have seven and five identifiable Wnt homologues in their genomes, respectively (http://www.stanford.edu/~rnusse/wntgenes/vertwnt.html). This chapter focuses only on those members of the Wnt family that have been shown by functional approaches to have a role in metanephric or meso-nephric kidney development or to have a protein or RNA distribution consistent with a conserved role in kidney development in other organisms. This will include Wnt 1, 4 7b, and 11 and their orthologues.

1. Potential Roles of Wnt-11 and Wnt-7b in Kidney Development

Wnt-11 is initially detected in E6.75 in the posterior part of the mouse embryo where the primitive streak has formed. By day E7, two zones of expression can be seen: one in the node and one in the allantois and extraembryonic mesoderm. At day 7.75, expression is in the forming heart tube, becoming confined to the myocardium 2 days later. At this time, E9.5 expression occurs in the developing somites, a unique area for Wnt expression, but one that may have considerable importance for the development of the urogenital system. It is still not entirely clear what the role of this gene is in kidney studies from mutant phenotypes in mice. Wnt-11 is first expressed in the developing excretory system in the mouse E9 embryo. At E10.5 there is a strong domain of expression in the epithelium of the Wolffian duct adjacent to the region from which the ureteric bud will form. Expression is then maintained at the tip of the ureteric bud throughout growth and branching such that each newly formed tip expresses Wnt11, whereas the region between the two newly formed branches is downregulated. Wnt-11 also has additional zones of expression. Initial homozygous Wnt-11 -/- mice appeared to have no phenotype, although studies carried out on another background have proved to be lethal. It remains to be seen whether there is a kidney phenotype associated with this lethality (McMahon, personal communication). It is interesting to note that the application of beads coated in GDNF, the ligand of c-Ret, can directly induce the transcriptional upregulation of Wnt-11. The GDNF/c-ret complex has been shown to have a clear effect on ureteric development (see later).

The metanephric mesenchyme is thought to be the source of the signal that promotes ureteric growth and branching. In Danforth's short-tail mutant mice, homozygous mutant embryos show severe tail truncation and have agenic kidneys. In these mice the relative proximity of the ureteric bud and the nephrogenic mesenchyme is disrupted, *Wnt-11* expression is also disrupted, and the characteristic branching morphology found in normal kidney development is not apparent (Kispert *et al.,* 1996). However, the fact that *Wnt-11* expression can be maintained by lung mesenchyme, albeit that epithelia are of a lung type, suggests that two types of signal regulate ureteric branching: (1) a nonspecific signal that promotes the growth and branching of the ureteric bud and (2) a specific signal that promotes kidney morphology. Kispert *et al.,* (1996) investigated whether proteoglycans might be the source of the nonspecific signal by disrupting the integrity and function of proteoglycans with sodium chlorate,

pentosan polysulfate, or suramin treatments (Davies and Garrod, 1995; Kispert *et al.,* 1996). *Wnt-11* expression was abolished rapidly, followed by the loss of *c-ret* expression, ampullae-like morphology, and growth and branching of the ureteric tips. These experiments indicated that sulfated proteoglycans might be important for the maintenance of *Wnt-11* expression. The potential importance of proteoglycans in pronephric development has also been demonstrated in our laboratory by the injection of suramin into *Xenopus* embryos at blastula and early gastrula stages (Nijjar and Jones, unpublished results). This treatment inhibits the formation of pronephric duct but not tubules. Wnt-11 has been isolated from the human and has similar expression patterns in the developing urogenital system to that of the mouse, although additional zones of expression can be identified in other tissues (Lako *et al.,* 1998).

Wnt-11 has been cloned from both avian and zebrafish sources. In both these cases, *Wnt-11* transcripts are detected in the mesoderm; however, initially no detection of transcripts has been reported in either of these cases in the developing urinogenital system (Eisenberg *et al.,* 1997; Makita *et al.,* 1998). This apparent deviation from the expression observed in mouse and human is somewhat surprising in view of the fact that *Xenopus Wnt-11, Xwnt-11,* has been reported to be expressed in the pronephric tubules but not in the pronephric duct (Ku and Melton, 1993). Another report suggests that *Cwnt-11* is expressed in the tips of the branching ureteric bud, a distribution more consistent with that in other vertebrates (Stark *et al.,* 2000). In *Xenopus,* there is also expression of *Xwnt-11* in the somite tissue. Experiments from our laboratory, in collaboration with Peter Vize, have shown the importance of somite tissue in signaling in the process of pronephric specification. Previously unspecified presumptive pronephric material isolated from early gastrula embryos can become specified to form tubule material when combined in Holtfreter sandwiches with dissected anterior somites (Seufert *et al.,* 1999). These data perhaps suggest that Wnt-11 may have a role in both induction and branching, a hypothesis that can be tested directly in the *Xenopus* system, but that this function may have been lost in the zebrafish, perhaps due to functional redundancy, with other members of the Wnt family. In the absence of clear functional data from the mouse knockout model and incomplete data from other organisms, the role of Wnt-11 in pronephric development must remain somewhat of an open question, although in the mouse a function in the metanephric kidney is suggested.

In the mouse, *Wnt-7b* is expressed in a highly characteristic pattern within the epithelia of the ureteric bud and collecting ducts of the metanephros, but not the mesonephros (Kispert *et al.,* 1996). However, it is expressed relatively late and the domain of expression does not extend to the ureteric bud tips. This perhaps makes *Wnt-7b* an unlikely candidate for a role in the earliest stages of metanephric development, perhaps suggesting a role in maintaining/inducing differentiation in the epithelia of the ureteric bud and collecting tubules. Unfortunately, due to the early embryonic lethality of the *Wnt-7b* mutant mouse, no additional information on the functional role of this gene is available (Carroll *et al.,* 2001).

2. Wnt-4, a Major Player

Perhaps the clearest evidence for the importance of Wnt-secreted proteins in kidney development come from studies on *Wnt-4* in mouse development. *Wnt-4* is expressed in the mesenchymal condensations surrounding the branching ureteric bud in positions corresponding to cells, giving rise to the first pretubular aggregates at day 11.5 dpc (Stark *et al.,* 1994). Later expression is detected in the comma-shaped bodies and then the descending S-shaped bodies where expression is restricted to the region where fusion was occurring with the collecting duct. When fusion is complete, expression of *Wnt-4* is lost, and expression is then seen only in the newly forming tubules on the periphery of the developing metanephric kidney (Stark *et al.,* 1994). Interestingly, like many other Wnts, *Wnt-4* is expressed in the neural tube. This location is compatible with the observation that the neural tube can act as a heterotypic inducer of kidney tubule in nephrogenic mesenchyme (Saxén, 1987).

Mutant *Wnt-4* mice were generated by the targeted disruption of the third exon of the *Wnt-4* gene with a selection cassette in ES cells. Adult mice heterozygous for this mutation showed no apparent phenotype. However, homozygous mutant mice died within 24 h of birth. The *Wnt-4* gene was therefore not essential for fetal survival. Examination of the mutant embryos from 18.5 dpc to neonates showed agenic kidneys that lacked nephrons. The kidneys consisted only of condensed nephrogenic mesenchyme and branched collecting duct structures, leading to kidney failure in neonates (Stark *et al.,* 1994). *Wnt-4* is thought to act as an autoactivator of the mesenchyme to epithelial transition in tubule formation and has been shown to be sufficient to trigger tubulogenesis in isolated metanephric mesenchyme (Kispert *et al.,* 1998). Tubulogenesis can be initiated in *in vitro* experiments that combine metanephric mesenchyme explants with cells expressing *Wnt-4*. Transfilter experiments demonstrate that cell contact is essential for induction to occur. Furthermore, *Wnt-4* is only required for the initiation of tubulogenesis, as *Wnt-4*-expressing cells can induce neighboring nephrogenic mesenchyme derived from null mutant mice to form tubules (Kispert *et al.,* 1998).

Wnt-4 has been cloned in the chicken (Holliday *et al.,* 1995; Tanda *et al.,* 1995). The functional embryonic kidney in the chicken is the mesonephros. In the mesonephros, expression of *CWnt-4* is also expressed throughout mesonephric tubulogenesis from the point of aggregating mesenchyme to the fusion of the epithelial tubules to the Wolffian duct. This suggests that *Wnt-4* orthologues have a conserved developmental function that may extend throughout all three

kidney forms. This is at least supported by the distributions of *Xwnt-4* and *zfwnt-4* in developing *Xenopus* and zebrafish pronephroi.

Xwnt-4 transcripts are first detectable by late gastrula stage *Xenopus* embryos, NF stage 12.5–13. Initial expression studies carried out by Northern analysis indicated that the transcripts were found mainly in anterior and dorsal dissected fragments. Treatment with LiCl resulted in an upregulation of *Xwnt-4* expression. Whole-mount *in situ* studies confirmed the distribution and indicated that transcripts were expressed in the neural ectoderm at stage 13 and ultimately after a dynamic pattern of gene expression become restricted to the midbrain, hindbrain, and floor plate of the neural tube by stage 23. *Xwnt-4* is also expressed in the region of the developing pronephros in tailbud embryos, although its precise localization was not defined clearly at this point (McGrew *et al.,* 1992). The localization of *Xwnt-4* has been reinvestigated by double *in situ* analysis with other important markers of pronephros; *Xlim-1, xWT-1,* and *XPax-2* (Carroll *et al.,* 1999b). At stage 18, just prior to the first morphological signs of pronephric development, *Xwnt-4* is expressed weakly throughout the pronephric mesoderm. During tailbud stages the expression is stronger in the tubule anlage, declining in the duct anlage. By stage 33, there is no expression in the duct, and expression in the tubule anlage is confined to the anterodorsal part in the developing nephrostomes (Carroll *et al.,* 1999b; Brändli, 1999). This localization of *Xwnt-4* to the developing tubules is highly suggestive of a conserved function for *Xwnt-4* in pronephric tubulogenesis, but there is as yet no published functional data to support this view.

Wnt-4 has been cloned from the zebrafish (Ungar *et al.,* 1995; Blader *et al.,* 1996). The expression pattern in these embryos was determined, with expression initiating in the forebrain and midbrain–hindbrain boundary at 11 h and then appearing in the posterior hindbrain. At about 13 h, expression is detected in the developing pronephros, although this expression has not been analyzed in detail and has been suggested to be ectodermal rather than mesodermal (Blader *et al.,* 1996). Expression of *zfwnt-4* is detected in the floor plate at 12.5 h and remains until between 24 and 36 h of development. Overexpression experiments have been carried out in both zebrafish and *Xenopus,* but no pronephric phenotype has been identified. Perhaps this should be revisited in view of the available molecular markers to identify the normal pronephric structure (Ungar *et al.,* 1995; Vize *et al.,* 1995).

Other members of the Wnt family, including *Wnt1,* have been shown to be able to induce tubulogenesis of metanephric mesenchyme *in vitro* (Hertzlinger *et al.,* 1994). However, these Wnts are not normally expressed in the developing urinogenital system, although they are expressed in the nervous system, which is capable of heterologous induction of tubules. This demonstrates the functional redundancy of this family of molecules.

3. Frizzled Receptor Proteins and Their Soluble Receptor Counterparts

All Wnt proteins, as members of the Wingless class of proteins, are secreted signaling proteins that exert their effects through binding to cell surface receptor molecules. Receptor proteins have been identified as seven-pass membrane proteins of the frizzled family. These receptors have been shown to bind the Wingless protein and, as a consequence of binding, elevate the expression of downstream proteins in the signaling pathway (Bhanot *et al.*, 1996). Members of this protein family have been identified in human, mouse, rat, chicken, *Xenopus,* and zebrafish, and related sequences are found in *C. elegans* (Wang *et al.*, 1996; Kawakami *et al.*, 2000; Shi *et al.*, 1997; Nasevicius *et al.*, 2000). *Frizzled-4* has been isolated in the chick and is expressed in all three kidney forms. *CFz-4* is expressed in the pronephros caudal to the third somite at stage 10 and ultimately becomes restricted to the newly forming glomeruli and tubules in the mesonephros and metanephros (Stark *et al.*, 2000). This distibution is consistent with *Cwnt-4* expression, suggesting that together they function in tubule formation, but not duct formation, although no direct functional data are available. More recently, functional experiments have directly identified a role for the Wnt receptor protein, frizzled-7, upstream of β-catenin in dorso/ventral patterning in *Xenopus* (Medina *et al.*, 2000; Sumanas *et al.*, 2000). These receptor proteins have varied patterns of expression in both embryonic and adult tissues consistent with a role in contributing to pattern formation (Wang *et al.*, 1996; Kawakami *et al.*, 2000; Shi *et al.*, 1997; Nasevicius *et al.*, 2000). One of these proteins has a reported distribution that includes the pronephric tubules, but not the pronephric duct, consistent with a role for *Xfz3* in pronephric development. The region of expression in the pronephric anlagen overlaps with the expression domains of both *Xwnt4* and *Xwnt11*. However, there is no direct evidence to link either ligand with this receptor. Overexpression experiments have been carried out by injecting mRNA into embryos at the two-cell stage. However the observed phenotypes were not specifically analyzed for any tubular defects in pronephros development, perhaps an area for further analysis (Shi *et al.*, 1997). A frizzled domain containing protein that carries a Wnt-binding region has also been identified in *Xenopus. Crescent* transcripts are located initally in the Spemann organizer and later in the pronephros at tailbud stages (stage 26) (Shibata *et al.*, 2000). This gene is upregulated synergistically by the expression of *Xlim, Ldb1,* and *Siamois* in animal caps, suggesting a role in pronephric development, which is, as yet, uninvestigated.

An additional level of control of Wnt signaling may prove to have some significance in the development of the kidney. Truncated soluble frizzled-related proteins have been identified in mammals, chickens, and *Xenopus* (Leyns *et al.*, 1997; Wang *et al.*, 1997; reviewed in Moon *et al.*, 1997; Lahder *et al.*, 2000). These proteins can bind to Wnt ligands extra-cellularly, thus preventing interactions with their cognate receptor. An increasing body of evidence shows that such interactions have important developmental consequences (Lescher *et al.*, 1998, Jaspard *et al.*, 2000) and that sFRP-2 is directly involved in metanephric kidney development via an interaction with Wnt 4. This protein shows an overlapping expression domain with Wnt-4 in mesodermal aggregates and simple epithelial bodies in metanephric development. Direct interaction of the Wnt4 protein with sFRP-2 has been shown by coprecipitation experiments. Hence, sFRP-2 is a target of the Wnt-4 signaling pathway in metanephric kidney development and may modulate Wnt-4 signaling (Lescher *et al.*, 1998). Increasing knowledge about the distributions of specific frizzled receptors and their related soluble proteins, SFRPs, will certainly add to our knowledge of the role and the control of Wnt protein signaling in pronephric development.

B. The BMP Family of Signaling Molecules

Bone morphogenetic proteins are one of the largest multifunctional families of developmental signaling molecules that regulate biological processes as diverse as cell differentiation, apoptosis, cell proliferation, cell fate determination, and morphogenesis (for a review, see Hogan, 1996). BMPs are members of the TGF-β superfamily and are all synthesized as precursor molecules that are activated by proteolytic cleavage to yield carboxy-terminal mature dimers. The protein chains are characterized by the presence of seven cysteines in the mature protein domain, six of which contribute to a cysteine knot. Several of the BMP family members, including BMP 4, 5, and 7, are expressed during renal development.

Only one member of the BMP family, BMP7, has been shown to have an essential role in kidney development in targeted mutagenesis experiments (Dudley *et al.*, 1995; Luo *et al.*, 1995). BMP-deficient mice were produced by stem cell technology, generating a null allele by targeting exons 6 and 7 in the mature protein domain. Heterozygous mice were viable at birth and phenotypically normal; however, crossing produced litters of normal size, but approximately a quarter of the pups had distinctive morphological features. They were significantly smaller in size than normal, exhibited polydactyly in the hindlimbs, and had either reduced or missing eyes. All of the abnormal pups died within 48 h of birth. Autopsy of the mutant offspring showed that they had small dysgenic kidneys with distended collecting ducts. The kidneys contained approximately 3% of the normal number of glomeruli as compared to wild type. The mutant kidneys showed no evidence of any metanephric mesenchyme nor glomerulus formation in the cortical region. These results indicate a critical role for BMP7 in kidney development (Luo *et al.*, 1995).

The basis of the embryological defect in homozygous mutant mice was established by a histological analysis of a timed embryological sequence. By day 12.5, mutant embryos already clearly show either a total or a partial lack of mesenchymal condensates surrounding the ureteric buds. At later stages, very few comma and S-shaped bodies and maturing glomeruli are observed in comparison to wild type. The absence of BMP7 affects the expression of molecular markers characteristic of the metanephric mesenchyme (WT1, Wnt-4) but not those of epithelial cells (Pax-2, laminin A). This loss of gene expression is thought to result from a loss of induced mesenchyme due to apoptosis.

There is considerable discussion in the literature as to whether BMP7 can act as an inducer as well as a survival factor (Godin et al., 1999). The distribution of BMP7 in the tips of the ureteric bud is certainly consistent with an inducing role, but the null mutation does not block nephrogenesis, suggesting that BMP7 is not an inducer of metanephric mesenchyme. It is reported that both antisense and oligonucleotides specific for BMP7 can block nephrogenesis in spinal cord/metanephric mesoderm recombinant cultures (Vukicevic et al., 1996). However, it has also been reported that similar recombinant cultures from BMP7 mutant mice can induce nephrogenesis, suggesting that BMP7 is not necessary for the induction (Dudley et al., 1999). These data would suggest that BMP7 is not necessary for the induction of metanephric mesenchyme.

Although there is strong evidence for the importance of BMP7 in metanephric kidney development, there is little information on its role in pronephric development. XBMP7 has been isolated from Xenopus and appears to have a dynamic and complex pattern of expression, which includes the pronephros. XBMP7 is expressed at stage 30 in the nephrostomes and tubules in similar domains to Xlim-1. Overexpression studies in both whole embryos and animal caps have been carried out to investigate the function of this gene in early embryogenesis. These experiments indicate that one of the major roles of this molecule is in hematopoiesis. The effect of overexpression on the pronephric phenotype has not been addressed specifically and might indicate a specific role in pronephric development. The role of BMP molecules and their receptors in pronephric development is still an area that has not been fully established, although it is likely to again illustrate the functional conservation of sigaling molecules throughout the vertebrates (Kuure et al., 2000). BMP7 has also been isolated from both zebrafish and chick. In the zebrafish, it has been established that the snailhouse mutant is caused by a mutation in the bmp7 gene. However, there have been no specific studies on the role of this gene in pronephric development, nor is it reported to be expressed in the developing pronephros (Dick et al., 2000; Schmid et al., 2000). Like many other BMP family members, the function of bmp7 in the zebrafish is to ventralize the embryo, possibly via heterodimer formation with bmp2b. The snailhouse mutation consequently has a strongly dorsalized mutant phenotype. BMP7 has also been studied in the chick, where it plays a role in patterning ventral midline cells where the coexpression of sonic hedghog with BMP7 appears to mediate the ability of prechordal

mesoderm to induce the ventral midline cells of the rostral diencephalon (Dale et al., 1999; Vesque et al., 2000).

C. The FGF Family of Growth Factors and the Pronephros

There is widespread evidence for the involvement of FGF signaling in kidney development, although much of this evidence is confined to studies on the mesonephric and metanephric forms rather than the pronephros. In in vitro mouse models of nephrogenesis, FGF-2 has been shown to be able to mediate the early inductive events in renal development (Perantoni et al., 1995). Incubation of explanted metanephric mesenchyme in the presence of FGF2 results in the characteristic condensation and prolonged survival seen when an inducing source, e.g., spinal cord, is present. In addition, transcriptional activation of WT1 occurred, although the later stages of tubule formation failed to occur. This deficiency can be overcome, however, by conditioned medium derived from a ureteric bud cell line (Karavanova et al., 1996), which, together with FGF, support complete differentiation of isolated mesenchyme into nephron epithelia and glomeruli-like structures. FGF7 has also been shown to play a role in modulating the growth of the ureteric bud and the number of nephrons. Overexpression of FGF7 under the control of the human apolipoprotein E promoter and enhancer in transgenic mice resulted in a kidney phenotype resembling polycystic kidney disease (Nguyen et al., 1996). Analysis of FGF7 null transgenic mice has shown that mice had markedly smaller ureteric buds and reduced nephron numbers than wild-type controls, a condition that could be improved by the ectopic addition of FGF7 to in vitro cultures (Qiao et al., 1999). The view is now that the BMP and FGF signaling pathways interact to promote growth, maintain mesenchymal competence, and inhibit tubulogenesis (Dudley et al., 1999). FGF receptors are also found within the developing kidney, and soluble dominant-negative FGF receptors introduced as a transgene driven by the MT1 promoter generated animals with either no kidney or a disorganized kidney severely reduced in size (Orr-Urtreger et al., 1993; Celli et al.,1998).

FGF family members have also been identified in the kidney in other vertebrates. FGF2 has been shown to be localized immunohistologically to the epithelial cells of both the pronephric tubules and the duct in the chick embryo. Its expression is maintained during mesonephric induction and differentiation where higher levels of protein are observed in the epithelia of the mesonephric tubules than the duct. Interestingly, during development of the mesonephros, the proteins undergo a profound change in subcellular location, becoming nuclear in the developing podocytes while remaining cytoplasmic in other cell types. FGF2 is also found in a nuclear location in the podocytes of the metanephros, suggesting an important, but yet undefined role for the localization of this protein (Dono and Zeller, 1994). Although dominant-negative FGF receptor experiments have been carried out in the chick, they have focused on the role of FGF signaling in cardiac myocyte proliferation and limb muscle differentiation. These experiments throw no light on the importance of FGF signaling in the chick kidney (Mima *et al.*, 1995; Itoh *et al.*, 1996; Flanagan-Street *et al.*, 2000). Dominant-negative FGF experiments have also been carried out in the zebrafish, but the pronephros was not studied, (Rodaway *et al.*, 1999).

FGF has been identified as an important developmental growth factor in amphibia for a number of years where its effects have been analyzed both *in vivo* and *in vitro* (for a review, see Isaacs, 1997). bFGF does not induce morphologically identifiable pronephric kidney in animal caps (Green *et al.*, 1990), but it does act with retinoic acid to induce *xWT1*, a marker of the pronephric glomus (Brennan *et al.*, 1999). Both maternal and zygotically expressed FGFs have been cloned in *Xenopus* and their expression patterns established (Tannahill *et al.*, 1992; Isaacs *et al.*, 1992; Isaacs, 1997), and specific FGF receptors have also been identified, one of which, PFR-4, is expressed in the pronephros among other tissues (Launay *et al.*, 1994). The role of FGF signaling in the development of the embryo past midblastula has been investigated by transgenesis (Kroll and Amaya, 1996). The resulting embryos had severe defects in the formation of the trunk and tail, although the head structures were fairly normal. Mesoderm induction occurred in these embryos, although the mesoderm failed to be maintained, resulting in embryos lacking axial mesoderm, including both notochord and somites. The formation of the pronephros was not determined in these experiments, although the absence of other mesodermal cell types would suggest that the pronephros would be absent. Transgenic experiments targeting *XFD* to the pronephros would allow the role of FGF signaling in the development of the pronephros to be determined unambiguously.

D. GDNF and Its Receptor Complex

Glial cell-derived neurotrophic factor (GDNF) is a potent neurotrophin of the TGF-β family, which has been shown by gene targeting to have a fundamental role in the induction of ureter branching in the developing metanephros in higher vertebrates. The GDNF signal is mediated through its interaction with the receptor tyrosine kinase, c-ret, which was originally identified as part of an oncogenic fusion protein resulting from a chromosomal translocation (for reviews, see Robertson and Mason, 1997; Sariola and Saarma, 1999; Fig. 8.5).

GDNF is expressed in a highly dynamic pattern throughout development, including the gastrointestinal tract, the developing urinogenital sytem, the limbs, and the head (Hellmich *et al.*, 1996). It is only expressed in tissues of mesenchymal origin and is never found expressed in epithelia, although it is expressed in a number of organs that develop via mesenchymal/epithelial interactions. It is first detected

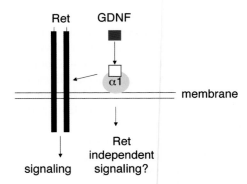

Figure 8.5 The GDNF signaling complex. The GDNF family of proteins binds to GPI-linked GFRα receptors, and the complex is delivered to the ret receptor. Ret dimerization and autophosphorylation trigger intracellular signaling events, leading to activation of transcription. There is currently some evidence that GDNF may also be able to signal via GFRα independently of ret.

in the developing excretory system in the mouse embryo at E11.5, just prior to the start of differentiation of the metanephros, at E12. GDNF transcripts are detected in the undifferentiated cortical mesenchyme adjacent to the mesenchyme condensing around the ureteric buds. The importance of this neurotrophic factor in kidney development was established in gene disruption experiments. Homozygous null mutant mice, generated by replacement of the third exon of *GDNF* with a *neo* cassette, were born at the expected frequency, but died within 24 h. Analysis of their internal morphology demonstrated that the kidneys were affected severely, being reduced in size or bi- or unilaterally missing (Pichel *et al.*, 1996; Moore *et al.*, 1996).

Indications that GDNF and the tyrosine kinase receptor c-ret might be functionally connected came from a realization that *ret* mutant mice had an almost identical phenotype to that for disrupted *GDNF*. These null mutants showed a failure of the ureteric buds to grow and branch appropriately in pups with small kidneys. The more extreme phenotype of severe dysgenesis seems to be a result of failure of the ureteric bud to form (Schuchardt *et al.*, 1994). *C-ret* expression is found frequently in tissues adjacent to those expressing GDNF (Pachnis *et al.*, 1993). Within the developing excretory system, *c-ret* is first expressed earlier than *GDNF* in the nephrotomes as early as E8.5 embryos. Subsequent expression is found in the pronephric and mesonephric duct at E9–10.5. In the metanephric kidney, *c-ret* is expressed in the epithelial cells of the branching ureteric bud and later in development in the outer nephrogenic zone, the region where the active differentiation of renal tubules and glomeruli is still occurring. The direct interaction of GDNF and c-ret was shown by Durbec *et al.*, (1996), among others, in a *Xenopus* oocyte bioassay and by direct ^{125}I-labeled ligand binding. This role is also supported by experiments in amphibia that indicate that GDNF and GFRα may play a role in the pronephric duct guidance. Signaling through these two molecules is sufficient to allow normal pronephric duct migration, as shown by the ectopic application of soluble GFRα to embryos lacking all GPI-linked proteins and by the demonstration that the excess addition of GFRα, which can compete out GDNF binding in normal embryos, prevents duct migration (Drawbridge *et al.,* 2000). It is now clear that the GDNF signal is actually transduced by binding to a receptor complex consisting of GPI-linked GFRα receptors and c-ret. GFRα receptors bind the GDNF ligand, with this complex binding to c-ret resulting in ret autophosphorylation and the activation of intracellular signaling cascades (reviewed in Sariola and Saarma, 1999). The role of GDNF in the initiation of the ureteric bud and the control of its branching is now clearly established (Pepicelli *et al.,* 1997; Sainio *et al.,* 1997).

All the members of the GDNF/c-ret/GFRα signaling complex have been cloned in the chick and their expression patterns established (Schuchardt *et al.,* 1995; Homma *et al.,* 2000) *GDNF* transcripts were found in the mesonephric and metanephric kidneys in stage 18 embryos, but were not reported in the pronephros. *Ret* is expressed somewhat earlier in the mesonephric duct at stage 13. *GFRα-1* is expressed even earlier in the pronephric intermediate mesoderm at stage 10 and in the ureteric duct at stage 18. The early localization of *GFRα-1* in pronephric tissue suggests that perhaps an alternative ligand of the GDNF family is active at this stage, with GDNF taking over a later role in duct development. Localization of the receptor and signaling molecules in adjacent tissues to those expressing ligand suggests a conserved role for these molecules in kidney development, although functional information is not yet available in the chick system. This view is also supported by the observations that the ret receptor has been identified in zebrafish and is also located in the pronephric ducts of 4–6 somite stage embryos expression persisting in the posterior duct of the embryonic and adult mesonephros (Marcos-Gutiérrez *et al.,* 1997). The zebrafish orthologues of GDNF and the ligand binding component of its receptor GFRα1 have also been cloned and their expression patterns established. GDNF is first detectable 14 h postfertilization in the ventral half of the anterior somites, where the pronephric duct condenses. Expression continues during the period when the duct forms but ceases thereafter, GFRα1 is expressed transiently in the developing pronephric ducts at 18 to 20 h postfertilization (Shepherd *et al.,* 2001). However, perturbation of GDNF expression by morpholino treatment failed to inhibit pronephric kidney formation, although it did prevent development of the enteric nervous system. This result perhaps suggests that GDNF is critical for metanephric but not for pronephric development. This is supported by the observation that GFRα1-deficient mice do not have any identifed pronephric or mesonephric abnormalities (Cacalano *et al.,* 1998) but then fails to reconcile the observations from amphibia (Drawbridge *et al.,* 2000).

IV. Conclusions and Future Perspectives

One of the most astounding aspects of developmental biology is the remarkable conservation of a molecular mechanism that exists between organisms as far apart as *Drosophila* and human. This conservation is displayed quite clearly in the organogenesis of the kidney, and particularly relevant for this chapter on the development and patterning of the pronephros. In practical terms, this means that any molecular information gleaned from the major systems discussed in this chapter, chick, zebrafish, amphibian, or lower vertebrate, will be of relevance to all the other model systems and even to higher vertebrates. For closely related organisms, the parallels are remarkable, whereas for those further apart, the principles often hold, although the details may be different. This places studies on the molecular control of pronephric development in a central position within the field of kidney development. The pronephros is sufficiently simple in the aforementioned organisms to be experimentally tractable, whether it be by dissection, expression of ectopically introduced genes, modification of the expression of ectopically introduced genes, mutation analysis, or for the detection of previously unidentified genetic markers at early stages of pronephric development. All these methods have yielded and will continue to yield temporal, spatial, and, most importantly, functional data, which are often harder to achieve in higher vertebrates.

Xenopus embryos have classically provided a system for investigation by dissection. This system has been utilized to great effect in pronephric studies, despite the relatively late developmental appearance of this organ. Small pieces of presumptive pronephric material have been cultured to establish the time of specification of each of the three pronephric components (Brennan *et al.,* 1998, 1999), to establish the position of duct and tubule primordia within the anlagen (Seufert *et al.,* 1999), and to study the effects of growth factors on pronephric tissue formation in animal caps (Taira *et al.,* 1994; Uochi and Asashima, 1996; Brennan *et al.,* 1999). Dissection in the chick has also provided considerable information, including that on the molecular cues involved in duct guidance (Obara-Ishihara *et al.,* 1999; Chapter 4).

The identification of *in vitro* conditions for the production of ectopic kidney is proving to be a powerful starting point for the identification of new markers of the pronephric kidney using a variety of subtractive hybridisation approaches (Peng *et al.,* 2000; Sato *et al,* 2000; Seville *et al.,* 2001; Massé and Jones, unpublished data). The ease with which chick, zebrafish, and amphibian embryos can be used to establish spatial and temporal domains of expression by *in situ* analysis has ensured that expression patterns can be established accurately from the earliest time points of pronephros development (Carroll *et al.,* 1999; Drummond *et al.,* 1998; Seville *et al.,* 2002; chapters 3, 4, and 5). The identification of new marker genes and investigations of their functional roles will add to the molecular knowledge of this organ. It is hoped that experiments aimed at establishing functional heirarchies will place all the pieces of the molecular jigsaw together and ultimately answer some of the currently unanswered questions relating to its induction, patterning, and function.

The relatively large size of amphibian embryos allows the microinjection of mRNAs into blastomeres of the early embryo to allow expression of the ectopically introduced genes. This technique has been used to considerable effect in establishing the role of those genes identified as being expressed in the pronephros in *Xenopus*. Injection of *Xlim1* and *WT1* and coinjection of *Xlim1* and *XPax8* (Wallingford *et al.*, 1998; Carroll and Vize, 1999) give rise to pronephric-

specific abnormalities, indicating direct functions in pronephric development. Furthermore, the modification of ectopically introduced genes by the generation of constitutively active, dominant negative, or hormone-inducible constructs has allowed the levels or timing of pronephros-specific effects to be established or controlled (Chan *et al.*, 2000; McLaughlin *et al.*, 2000). Table 8.1 summarizes the key experiments that have been carried out.

Xenopus laevis has the advantages of ease of manipulation and hence functional analysis, but due to the pseudo-tetraploid nature of its genome is not suitable for mutation analysis (Chapter 9). This may, to an extent, be alleviated by

Table 8.1   Summary of Overexpression Experiments of Genes Implicated in Pronephric Development Carried out in *Xenopus* Indicating the Resulting Phenotype.

Gene expressed	Material injected	Position of injection	Phenotype in pronephros	Reference
Xlim-1	mRNA transcribed *in vitro*	Injected into the C tier at the 32 cell stage	15% enlarged and 19% ectopic tubules. No abnormality was found in the duct	Carroll and Vize (1999)
Xlim-1, Xlim-1enR Xlim1-VP16 +/− Xldb-1	mRNA transcribed *in vitro*	Injected into the animal pole of embryos and then dissected as animal caps and incubated in combinations of RA and activin	Constitutively active or wild-type Lim-1 failed to induce pronephros with RA or activin treatment alone, but augmented pronephros when treated with both. The dominant-negative Xlim-1enR inhibited kidney formation	Chan *et al.* (2000)
Xlim-1enR	mRNA transcribed *in vitro*	Injected into the C tier at the 32 cell stage	A functional deficit of Xlim1 prevents the process of tubulo-, genesis with embryos failing to progress pass the tor stage	Chan *et al.* (2000)
Xpax8/2	mRNA transcribed *in vitro*	Injected into the C tier at the 32 cell stage	7% enlarged and 23% ectopic tubules. No abnormality was found in the duct. A similar effect was observed with Pax-2	Carroll and Vize (1999)
HNF-1β	mRNA transcribed from a human *HNF-1β* gene normally associated with renal agenesis or wild-type mRNA from *Xenopus*	Injected unilaterally into the 2 cell embryo	In both cases, defective development and pronephric agenesis were observed in 95% of the injected animals	Wild *et al.* (2000)
Xpax 8+ Xlim-1	mRNA transcribed *in vitro*	Injected into the C tier at the 32 cell stage	19% enlarged and 47% with ectopic tubules with multiple branches. No abnormality was found in the duct	Carroll and Vize (1999)
xWT1	mRNA transcribed *in vitro* from both KTS plus and minus forms	Injected into the C tier at the 32 cell stage	Reduced size of pronephric tubules due to failure to form a pronephric anlagen of the appropriate size	Wallingford *et al.* (1998)
GR-Su(H)VP16 GR-Su(H)^{DBM} GR-notch ICD	mRNA transcribed *in vitro*	Injected into one ventrovegetal blastomere and expression induced with dexamethasone at stage 20	Obliteration of duct markers following the activation by dexamethasone of GR-Su(H) VP16 or GR-Su (H)VP16 or GR-notch ICD as a result of increased notch signaling. Dominant-negative inhibition of notch signaling increased duct formation and expression of glomus markers	McLaughlin *et al.* (2000)

the adoption of *Xenopus tropicalis* as an alternative model amphibian. The zebrafish provides an established genetic system for the identification and characterization of mutations that affect pronephros form and/or function (Chapter 9). Mutants such as *noi* have proved invaluable in establishing the role of pax2.1 in the developing pronephros (Brand *et al.*, 1996; Majumdar *et al.*, 2000). Studies by Drummond *et al.* (1998) have identified 15 recessive mutants that result in the development of cysts instead of normal pronephric tubules. Studies such as these can only help our understanding of pronephric development and should feed across species as they become fully characterized in the zebrafish.

A. A Model Based on *Xenopus*

So what are the key elements in our current understanding of the molecular players in pronephric induction patterning and differentiation? Figure 8.6 summarizes the temporal distributions of candidate molecules in the three components of the *Xenopus* pronephros. The earliest marker of pronephric differentiation is *Xlim-1*, which is expressed at the late gastrula stage, concomitant with the time at which the specification of tubules is occurring. While it is clear that this molecule plays a major role in formation of the pronephros, its expression is not sufficient to cause significant development of ectopic tubules unless it is coinjected with *XPax-8* (Carroll and Vize, 1999). Furthermore, in *in vitro* animal cap systems, constitutively active forms failed to induce tubules in activin- or RA-treated tissue, but augmented tubule formation when presumptive tubule tissue was already present following activin and RA treatment (Chan *et al.*, 2000). These studies suggest that *Xlim-1* is not necessary to induce pronephros, but dominant-negative forms suggest that it is essential for tubulogenesis. This is somewhat different from the story in the mouse where there is no kidney formation in *Lim1*-deficient animals (Shawlot and Behringer, 1995). Perhaps this difference is due to the difficulty of observing mutant mice at stages when pronephric development is taking place. The zebrafish model supports an early role for *lim-1* where it is expressed initially in the duct and then in the tubules (Toyoma and Dawid, 1997). Either the identification of a mutant or functional experiments involving morpholino-impaired expression would help confirm the observations in *Xenopus*.

Xlim-1 is normally expressed in a domain that overlaps with *XPax-8*, and the coexpression experiments of Carroll and Vize (1999) would suggest that *XPax-8* acts with *Xlim-1* in tubule development. Because these experiments also showed an increase in duct immunostaining with 4A6 following injection of both *Xlim-1* and *XPax-8*, they may also play a role in duct development. These experiments suggest that *Xlim-1* and *XPax-8* are certainly important regulators of tubulogenesis, but raise the question of what molecules cause their expression in the late gastrula. The identification of those

molecules that induce the expression of *Xlim-1* and *XPax-8* in the pronephric field must be a major goal in understanding development of the pronephros.

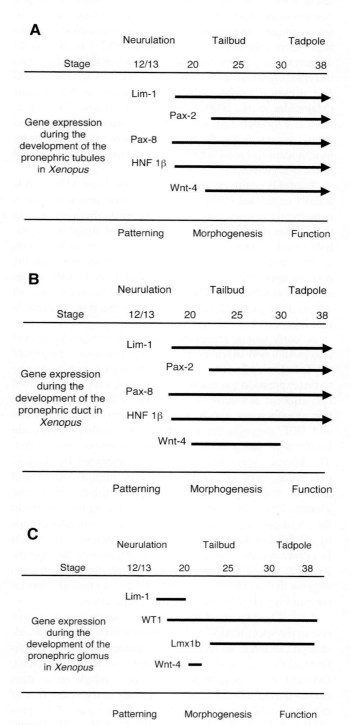

Figure 8.6 Gene expression patterns during pronephros patterning, morphogenesis, and function in *Xenopus*. (A) Temporal gene expression profiles in pronephric tubules. (B) Temporal gene expression profiles in the pronephric duct. (C) Temporal gene expression profiles in the glomus.

XPax-2 has a later expression pattern in the pronephric anlagen of the tailbud embryo and clearly has redundant functions that it shares with *XPax-8*. Like *XPax-8*, it is expressed in both the tubule and the duct, but its major expression domain in pronephroi at late stages of morphogenesis is in nephrostomes. The major role of *XPax-2* expression seems to be to restrict the expression domain of *xWT-1* to the medial domain of the pronephric anlagen, a function that is conserved in zebrafish where zebrafish *pax2.1* restricts *wt-1* and *VEG F* expression to the podocyte precursors in the glomus (Majumdar *et al.*, 2000; Serluca and Fishman, 2001). It is unclear why such high levels of expression of *XPax-2,8* and *Xlim-1* are retained in the nephrostomes or tubule tips during morphogenesis (see Chapter 3), but *Xwnt-4* is also localized to these regions.

Xwnt-4 is positively regulated by *Xlim-1* and *XPax-8*. It is initally expressed in the whole pronephric precursor, but is subseqently restricted to the lateral portion that gives rise to the tubules (Carroll *et al.*, 1999b). *Xwnt-4* signaling is probably mediated by the *xfz-3* receptor, which colocalizes with it. The restriction of *Xwnt-4* expression coincides with the relative upregulation of *XPax-2,8* and *Xlim-1* in the nephrostomes or tubule tips. Downregulation of *Xwnt-4* in the duct occurs during tailbud stages, suggesting that it plays a role in tubulogenesis but not in duct formation.

At early tailbud stages the pronephric anlagen is divided into medial and lateral portions by the expression pattern of xWT-1 (Fig. 3.7). The nature of the signal that gives rise to thisexpression has not been identified, but may arise from the adjacent ectoderm or endoderm (Chapter 3). The fact that glomus tissue can be induced by a combination of RA and FGF suggests the possibility that perhaps FGFs are involved in this process. This has yet to be investigated. The lateral expression of *xWT-1* is restricted by *XPax-2* expression, and *xWT-1* itself acts to inhibit *Xlim-1* and *Xwnt-4*, thus separa-ting presumptive tubules from the glomus. *XLmx-1b* is also expressed in late tailbud stages in the presumptive glomus. It is induced *in vitro* in animal caps by *xWT-1*, but does not induce *xWT1*, which suggests that it lies downstream of the earliest differentiation of the glomus. Its exact role is not yet clear, but morpholino experiments suggest that its presence is essential for normal glomerular morphology (Haldin and Jones, in preparation).

Less is known about the molecular interactions that give rise to the specification and differentiation of duct tissue, although candidate molecules in the process are being identified and the regulatory hierarchies have yet to be established. Division of the anlagen into presumptive tubule versus duct occurs during tailbud stages and results in the restriction of tubule-specific genes to the dorsal-anterior part of the pronephric anlagen (Vize *et al.*, 1995; Carroll *et al.*, 1999b; Chapter 3). It has been suggested that somites may be responsible for this signal, and indeed somites have been shown to be able to contribute to pronephric development in explants.

It is possible that *Xwnt-11,* which is found in somites, is responsible for this effect although this has yet to be tested. The Notch signaling pathway has also been shown to affect the development of duct (McLaughlin *et al.*, 2000). These experiments suggest that endogenous Notch signaling functions to inhibit duct formation in the dorsoanterior portion of the anlagen; overexpression, therefore, changes the allocation of the anlagen, resulting in the reduction of expression of duct markers. Relatively low levels of *XPax-2,8* and *Xlim-1* are present in the duct. *Xlim-1* is restricted finally to the posterior extending tip, whereas *XPax2* and *XPax8* expression is maintained in the duct—XPax-2 for some time but *XPax-8* only briefly. A postulated role for *XPax-2* in duct development is supported by the study of *zfpax2.1* mutants, which show that *Pax2.1* is required to support both duct and tubule development.

The BMP antagonist *gremlin* has been shown to be expressed in the duct (Hsu *et al.*, 1998), suggesting that inhibition of a BMP signal might be involved in duct development. It has been suggested that this could act by antagonizing the BMP repression of FGF activity in the Shh/FGF4 feedback loop as in limb development (Lappin *et al.*, 2001). There is evidence of *BMP7* expression in the developing tubules, perhaps interference with *BMP* signaling by *gremlin* is an essential prerequisite for duct development. Perhaps overexpression of *gremlin* in the tubule primordia might address this question.

The question of the involvement of the GDNF/c-ret/GFRα signaling complex in the development of the pronephric duct in *Xenopus* is still an open question. Studies from both chick and zebrafish suggest that it has a role in duct formation in these organisms, although the morpholino experiments of Shepherd *et al.* (2001) suggest that this may not be an essential role. Clearly the cloning of these components in *Xenopus* and functional studies here will add to our knowledge of duct development.

B. The Way Forward

There are three ways in which major advances are likely to be made in understanding the molecular control of pronephros development and differentiation. The first is the continued search for new genes that are expressed early in the developing pronephros, at the time of specification. An understanding of their expression domains, together with a knowledge of the genetic heirarchies within which they sit, should allow progress to be made on the molecules that induce and pattern all the elements of the pronephros. It should also identify those genes with particular physiological functions as the pronephros becomes an active excretory organ (see Chapter 3).

The second major way forward is made possible by the development of morpholino antisense technology that can be used to specifically inhibit the translation of mRNAs in early stages of development (reviewed in Corey and Abrams 2001; Ekker and Larson 2001). The morpholino oligonucleotide (MO) is so called because it contains a nonionic morpholine backbone rather than the standard sugar backbone of either DNA or RNA. Morpholinos inhibit translation due to the double-stranded mRNA/MO hybrid inhibiting the scanning of the mRNA by the 40s ribosomal subunit, thereby inhibiting translation (Summerton, 1999). The MO can target both maternal and zygotic transcripts and has sufficient longevity to be useful in pronephric organogenesis. We have used this technology successfully to inhibit the translation of Xanx-4 in the pronephros and have established a clear pronephric phenotype that can be rescued by wild-type mRNA (Seville *et al.,* 2002). This approach could be used with respect to all the major players identified to have a role in pronephric development and is an extremely powerful tool to reduce the expression of target genes. The use of RNAi as a means of reducing expression appears unreliable in vertebrates, although very powerful in invertebrates (see Chapter 9 for a review). Partial interference with *Xlim1* has been reported by RNAi, but no kidney phenotype was reported (Nakano *et al.,* 2000). We have attempted RNAi interference assays with *Xanx-4* and other genes, but have failed to observe any specific interference in embryos (Seville and Jones, unpublished results).

The final approach that may yield new data will be to fully exploit transgenics and targeted expression. As yet there have been no pronephric promoters published, although we are currently engaged in the analysis of the *Xanx-4* promoter (Seville, Massé, Collins and Jones, unpublished results). The ability to drive gene expression behind endogenous/pronephric-specific promoters will allow expression to be targeted precisely to the tissue of interest and will allow the disentangling of effects of the gene under study in tissues other than the pronephros. This is particulary relevant to genes such as *Xlim1*, *Pax2/8*, and *Xlmx1b*, which have clear domains of expression in the nervous system, as well as important roles in pronephric development. Experiments of this type could be carried out by transgenesis or by injection of plasmid clones to targeted sites, thus minimizing the effects of mosaicism.

The pronephros in lower vertebrates forms a highly manipulable system to study organogenesis. We have already identified some of the key players and their interactions to generate a patterned kidney. We have the technical tools to establish further levels of control and we are in the position to take advantage of functional genomics. The next few years will prove an interesting period in the study of pronephric induction patterning and differentiation.

Acknowledgments

I am endebted to all past and current members of my laboratory, particularly Hannah Brennan, Rachel Seville, Caroline Haldin, Robert Collins, Qian Chen, Surinder Bhamra, Robert Taylor, Sarbjit Nijjar, Kaine Massé, and Mark Barnett for all the efforts they have put into the pronephric projects in my laboratory. My research has been funded by the BBSRC.

References

Ariizumi, T., and Asashima, M. (2001). *In vitro* induction systems for analyses of amphibian organogenesis and body patterning. *Int. J. Dev. Biol.* **45,** 273–279.

Armstrong, J. F., Pritchard-Jones, K., Bickmore, W., Hastie, N. D., and Bard, J. B. L. (1992). The expression of the Wilms' tumor gene, *WT1* in the developing mammalian embryo. *Mech. Dev.* **40,** 85–97.

Baldwin, C. T., Hoth, C. F., Amos, J. A., Da-Silva, E. O., and Milunsky, A. (1992). An exonic mutation in the *HuP2* paired domain gene causes Waardenburg's syndrome. *Nature* **355,** 637–638.

Barbacci, E., Reber, M., Ott, M.-O., Breillat, C., and Huetz, F. (1999). Variant hepatocyte nuclear factor 1 is required for visceral endoderm specification. *Development* **126,** 4795–4805.

Barnes, J. D., Crosby, J. L. Jones, C. M. Wright, C. V. E., and Hogan, B. L. (1994). Embryonic expression of *Lim-1*, the mouse homologue of *Xenopus Lim-1*, suggests a role in lateral mesoderm differentiation and development. *Dev. Biol.* **161,** 168–178.

Bhanot, P., Brink, M., Samos, C. H., Hsieh, J.-C., Wang, Y., Macke, J. P., Andrew, D., Nathans, J., and Nusse, R. (1996). A new member of the *frizzled* family from Drosophila functions as a wingless receptor. *Nature* **382,** 225–230.

Blader, P., Strähle, U., and Ingham, P. W. (1996). Three *Wnt* genes expressed in a wide variety of tissues during development of the zebrafish, *Danio rerio*: Developmental and evolutionary perspective. *Genes Evol.* **206,** 3–13.

Bouchard, M., Pfeffer, P., and Busslinger, M. (2000). Functional equivalence of the transcription factors Pax2 and Pax5 in mouse development. *Development* **127,** 3703–3713.

Brand, M., Heisenberg, C.-P., Jiang, Y-J., Beuchle, D., Lun, K., Furutani-Seiki, M., Granato, M., Mullins, M., Odenthal, J., van Eeden, F. J. M., and Nusslein-Volhard (1996). Mutations in the zebrafish genes affecting the formation of the boundary between midbrain and hindbrain. *Development* **123,** 179–190.

Brändli, A. W. (1999). Towards a molecular anatomy of the *Xenopus* pronephric kidney. *Int. J. Dev. Biol.* **43,** 381–395.

Brennan, H. C., Nijjar, S., and Jones, E. A. (1998). The specification of the pronephric tubules and duct in *Xenopus laevis*. *Mech. Dev.* **75,** 127–137.

Brennan H. C., Nijjar, S., and Jones, E. A. (1999). The specification and growth factor inducibility of the pronephric glomus in *Xenopus laevis*. *Development* **126,** 5847–5856.

Cadigan, K. M., and Nusse, R. (1997). Wnt signaling: A common theme in animal development. *Genes Dev.* **11,** 3287–3305.

Calcalano, G., Farinas, I., Wang, L. C., Hagler, K., Forgie, A., Moore, M., Armanini, M., Phillips, H., Ryan, A. M., Reichardt, L. F., Hynes, M., Davies, A., and Rosenthal, A. (1998). GRFalpha 1 is an essential receptor component for GDNF in the developing nervous system and kidney. *Neuron* **21,** 53–62.

Call, C. M., Glaser, T., Ito, C. Y., Buckler, A. J., Pelletier, J., Haber, D. A., Rose, E. A., Krai, A., Yeger, H., Lewes, W. H.., Jones, C., and Housman, D. E. (1990). Isolation and characterisation of a zinc finger polypeptide gene at the human chromosome 11 Wilms' tumor locus. *Cell* **60,** 509–520.

Caricasole, A., Duarte, A., Larsson, S. H., Hastie, N. D., Little, M., Holmes, G., Todorov, I., and Ward, A. (1996). RNA binding by the Wilms' tumor suppressor zinc finger proteins. *Proc. Natl. Acad. Sci. USA* **93,** 7562–7566.

Carmona R., M. Gonzalez-Iriate, M., Perez-Pomares, J. M., and Munoz-Chapuli, R. (2001). Localisation of the Wilm's tumour protein WT1 in avian embryos. *Cell Tissue Res.* **303,** 173–186.

Carroll, T. J., and Vize, P. D. (1996). Wilms tumor suppressor gene is involved in the development of disparate kidney forms: Evidence from expression in the *Xenopus* pronephros. *Dev. Dyn.* **206**, 131–138.

Carroll, T. J., and Vize, P. D. (1999). Synergism between *Pax-8* and *lim-1* in embryonic kidney. *Dev. Biol.* **214**, 46–59.

Carroll, T., Wallingford, J., Seufert, D., and Vize, P. D. (1999a). Molecular regulation of pronephric development. *Curr. Top. Dev. Biol* **44**, 67–100.

Carroll, T. J., Wallingford, J. B., and Vize, P. D. (1999b). Dynamic patterns of gene expression in the developing pronephros of *Xenopus laevis*. *Dev. Genet.* **24**, 199–207.

Carroll, T. J., Ishibashi, M., Par, B. A., and McMahon, A. P. (2001). The role of Wnts in the development of the kidney collecting ducts. *Dev. Biol.* **235**, 205–206.

Celli, G., LaRochelle, J. W., Mackem, S., Sharp, R., and Merlino, G. (1998). Soluble dominant negative receptor uncovers essential role for fibr-blast growth factors in multiorgan induction and patterning. *EMBO J.* **17**, 1642–1655.

Chan, T.-C., Takahashi, S., and Asashima, M. (2000). A role for *Xlim-1* in pronephros development in *Xenopus laevis*. *Dev. Biol.* **228**, 256–269.

Chen, H., Lun, Y., Ovchinnikov, D., Kokubu, H., Oberg, K. C., Pepicelli, C. V., Gan, L., Lee, B., and Johnson R. (1998). Limb and kidney defects in *Lmx1b* mutant mice suggest an involvement of LMX1B in human nail patella syndrome. *Nature Genet.* **19**, 51–55.

Coffinier, C., Thépot, D., Babinet, C., Yaniv, M., and Barra, J. (1999). Essential role for the homeoprotein vHNF1/HNF1β in visceral endoderm differentiation. *Development* **126**, 4785–4794.

Cook, D. M., Hinkes, M. T., Bernfied, M., and Rauscher, F. J., III (1996). Transcriptional activation of the *Syndecan-1* promoter by the Wilms' tumor protein WT1. *Oncogene* **13**, 1789–1799.

Corey, D. R., and Abrams, J. M. (2001). Morpholino antisense oligonucleotides tools for investigating vertebrate development. *Genome Biol.* **2**, 1015.1–1015.5.

Curtiss, J., and Heilig, J. S. (1998). DeLIMiting development. *Bioassays* **20**, 58–69.

Dahl, E., Koseki, H., and Balling R. (1997). *Pax* genes and organogenesis. *Bioessays* **19**, 755–765.

Dale, K., Sattar, N., Heemskerk, J., Clarke, J. D. W., Placzec, M., and Dodd, J. (1999). Differential patterning of ventral midline cells by axial mesoderm is regulated by BMP7 and chordin. *Development* **126**, 397–408.

Davies, J. A., and Garrod, D. R. (1995). Induction of early stages of kidney tubule differentiation by lithium ions. *Dev. Biol.* **167**, 50–60.

Davies, J. A., Perera, A. D., and Walker, C. L. (1999). Mechanisms of epithelial development and neoplasia in the metanephric kidney. *Int J. Dev. Biol.* **43**, 473–478.

Dawid, I. B., Breen, J. J., and Toyama, R. (1998). LIM domains: Multiple roles as adapters and functional modifiers in protein interactions. *TIG* **14**, 156–161.

Delalande, J. M., and Rescan, P. Y. (1998). Expression of a cysteine-rich protein (CRP) encoding gene during early development of the trout. *Mech. Dev.* **76**, 179–183.

Demartis, A., Maffei, M., Vignali, R., Barsacchi, G., and De Simone, V. (1994). Cloning and developmental expression of *LFB3/HNF1β* transcription factor in *Xenopus laevis*. *Mech. Dev.* **47**, 19–28.

Dick, A., Hild, M., Bauer, H., Imai, Y., Maifeld, H., Schier, A. F., Talbot, W. S., Bouwmeester, T., and Hammerschmid, M. (2000). Essential role of BMP7 (*snailhouse*) and its prodomain in dorsoventral patterning of the zebrafish embryo. *Development* **127**, 343–354.

Dietrich, S., and Gruss, P. (1995). Undulated phenotypes suggest a role of *Pax-1* for the development of vertebral and extravertebral structures. *Dev. Biol.* **167**, 529–548.

Dono, R., and Zeller, R. (1994). Cell-type-specific nuclear translocation of fibroblast growth factor-2 isoforms during chicken kidney and limb morphogenesis. *Dev. Biol.* **163**, 316–330.

Drawbridge, J., Meighan, C. M., and Mitchell, E. A. (2000). GDNF and GFRα-1 are components of the axolotl pronephric duct guidance system. *Dev. Biol.* **228**, 116–124.

Dressler, G. R. (1999). Kidney development branches out. *Dev. Genet.* **24**, 189–193.

Dressler, G. R. Deutsch, U., Chowdhury, K., Nornes, H. O., and Gruss, P. (1990). *Pax2*, a new murine paired-box-containing gene and its expression in the developing excretory system. *Development* **109**, 787–795.

Dressler, G. R., Wilkinson, J. E., Rothenpieler, U. W., Patterson, L. T., Williams-Simons, L., and Westphal, H. (1993). Deregulation of *Pax-2* in transgenic mice generates severe kidney abnormalities. *Nature* **362**, 65–67.

Dreyer, S. D., Zhou, G., Baldini, A., Winterpatch, A., Zabel, B., Coles, W., Johnson, R. L., and Lee, B. (1998). Mutations in *LMX1B* cause abnormal and renal dysplasia in nail patella syndrome. *Nature Genet.* **19**, 47–50.

Dudley, A., Lyons, K., and Robertson, E. J. (1995). A requirement for bone morphogenetic protein-7 during development of the mammalian kidney and eye. *Genes Dev.* **9**, 2795–2807.

Dudley, A. T., Godin, R. E., and Roberson, E. J. (1999). Interaction between FGF and BMP signaling pathways regulates development of metanephric mesenchyme. *Genes Dev.* **13**, 1601–1613.

Durbec, P., Marcos-Gutierrez, C. V., Kilkeeny, C., Grigoriou, M., Wartiowaara, K., Suvanto, P., Smith, D., Ponder, B., Costantini, F., Saarma, M., Sariola, H., and Pachnis, V. (1996). GDNF signalling through the Ret receptor tyrosine kinase. *Nature* **381**, 789–793.

Ekker, S. C., and Larson, J. D. (2001). Morphant technology in model developmental systems. *Genesis* **30**, 89–93.

Eisenberg, C. A., Gourdie, R. G., and Eisenberg, L. M. (1997). Wnt-11 is expressed in early avian mesoderm and required for the differentiation of the quail mesoderm cell line QCE-6. *Development* **124**, 525–536.

Englert, C. (1998). WT1:-More than a transcription factor. *TIBS* **23**, 389–393.

Englert, C., Hou, X., Maheswaran, S, Bennett, P., Ngwu, C., Re, G. G. Garvin, A. J., Rosner, M. R., and Haber, D. A. (1995). *WT1* suppresses the synthesis of the epidermal growth factor receptor and induces apoptosis. *EMBO J.* **14**, 4662–4675.

Epstein, D., Vekemans, M., and Gros, P. (1991). *Splotch (Sp²ᴴ)*, a mutation affecting development of the mouse neural tube, shows a deletion within the paired homeodomain of Pax-3. *Cell* **67**, 767–774.

Flanagan-Street, H., Hannon, K., McAvoy, M. J., Hulinger, R., and Olwin, B. B. (2000). Loss of FGF receptor 1 signalling reduces skeletal muscle mass and disrupts myofibre organisation in the developing limb. *Dev. Biol.* **218**, 21–37.

Freyd, G., Kim, S. K., and Horvitz, H. R. (1990). Novel cystein-rich motif and homeodomain in the product of the *Caenorhabditis elegans* cell lineage gene *lin-11*. *Nature* **344**, 876–879.

Gashler, A., Bonthron, D., Madden, S., Rauscher, F. J., III, Collins, T., and Sukhatme, V. (1992). Human platelet-derived growth factor A chain is transcriptionally repressed by the Wilms' tumor suppressor *WT1*. *Proc. Natl. Acad. Sci. USA* **88**, 9618–9622, 10984–10988.

Gessler, M., Poustka, A., Cavanee, W., Neve, R. L., Orkin, S. H., and Bruns, G. A. P. (1990). Homozygous deletion in Wilms' tumours of a zinc finger gene identified by chromosome jumping. *Nature* **343**, 775–778.

Godin, R. E., Robertson, E. J., and Dudley, A. T. (1999). Role of BMP family members during kidney development. *Int. J. Dev. Biol.* **43**, 405–411.

Green, J. B. A., Howes, G., Symes, K., Cooke, J., and Smith, J. C. (1990). The biological effects of XTC-MIF: Quantitative comparison with *Xenopus* bFGF. *Development* **108**, 173–183.

Haber, D. A., Sohn, R. L., Buckler, A. J., Pelletier, J., Call, K. M., and Housman, D. E. (1991). Alternative splicing and genomic structure of the Wilms' tumor gene *WT1*. *Proc. Natl. Acad. Sci. USA* **88**, 9618–9622.

Hammes, A. Guo, J.-K., Lutsch, G., Leheste, J.-R., Landrock, D., Ziegler, U., Gubler, M.-C., and Schedl, A. (2001). Two splice variants of the Wilms' tumour 1 gene have distinct functions during sex determination and nephron formation. *Cell* **106**, 319–329.

Hastie, N. D. (1994). The genetics of Wilms' tumor: A case of disrupted development. *Annu. Rev. Genet.* **28**, 523–558.

Hausen, P., and Reibesell, M. (1991). "The Early Development of *Xenopus laevis*." Springer-Verlag, Berlin.

Heller, N., and Brändli, A. W. (1997). *Xenopus Pax-2* displays multiple splice forms during embryogenesis and pronephric kidney development. *Mech. Dev.* **69**, 83–104.

Hellmich, H. L., Kos, L., Cho, E. S., Mahon, K. A., and Zimmer, A. (1996). Embryonic expression of glial cell line derived neurotrophic factor (GDNF) suggests multiple developmental roles in neural differentiation and epithelial-mesenchymal interactions. *Mech. Dev.* **54**, 95–105.

Helwig, U., Imai, K., Schmahl, W., Thomas, B. E. Varnum, D. S., Nadeau, J. H., and Balling, R. (1995). Interaction between *undulated* and *Patch* leads to an extreme form of spinabifida in double mutant mice. *Nature Genet.* **11**, 60–63.

Hertzlinger, D., Qiao, J., Cohen, D., Ramakrishna, N., and Brown, A. M. C. (1994). Induction of kidney epithelial morphogenesis by cells expressing *Wnt-1*. *Dev. Biol.* **166**, 815–818.

Hobert, H., Tessmar, K., and Ruvkun, G. (1999). The *Caenorhabditis elegans lim-6* LIM homeobox gene regulates neurite outgrowth and function of particular GABAergic neurons. *Development* **126**, 1547–1562.

Hogan, B. L. M. (1996). Bone morphogenetic proteins multifunctional regulators of vertebrate development. *Gene Dev.* **10**, 1580–1594.

Hol, F. A., Guerds, M. P., Chatkup, S., Shugart, Y. Y., Balling, R., Shirander-Stumpel, C. T., Johnson, W. G., Hamel, B. C., and Mariman, E. C. (1996). *Pax* genes and human neural tube defects: An amino acid substitution in PAX1 in a patient with spina bifida. *J. Med. Genet.* **33**, 655–660.

Holewa, B., Strandmann, E., Zapp, D., Lorenz, P., and Ryffel, G. U. (1996). Transcriptional hierarchy in *Xenopus* embryogenesis: HNF4 a maternal factor involved in the developmental activation of the gene encoding the tissue specific transcription factor HNF1α (LFB1). *Mech Dev.* **54**, 45–57.

Holliday, M., McMahon, J. A., and McMahon, A. P. (1995). Wnt expression patterns in chick embryo nervous system. *Mech. Dev.* **52**, 9–25.

Homma, S., Oppenheim, R. W., Yaginuma, H., and Kimura, S. (2000). Expression pattern of GDNF, c-ret, and GFRαs suggests novel roles for GDNF ligands during early organogenesis in the chick embryo. *Dev. Biol.* **217**, 121–137.

Hsu, D. R., Economides, A. N., Wang, X., *et al.* (1998). The *Xenopus* dorsalising factor gremlin identifies a novel famift of secreted proteins than antagonise BMP activities. *Mol. Cell* **1**, 673–683.

Isaacs, H. V. (1997). New perspectives on the role of fibroblast growth factor family in amphibian development. *Cell. Mol. Life Sci.* **53**, 350–361.

Isaacs, H. V., Tannahill, D., and Slack, J. M. W. (1992). Expression of a novel FGF in the *Xenopus* embryo: A new candidate inducing factor for mesoderm formation and anteroposterior specification. *Development* **114**, 711–720.

Itoh, N., Mima, T., and Mikawa, T. (1996). Loss of fibroblast growth factor receptors is necessary for terminal differentiation of embryonic limb muscle. *Development* **122**, 291–300.

Jaspard, B., Couffinhal, T., Dufourcq, P., Moreau, C., and Duplaa, C. (2000). Expression pattern of mouse cFRP-1 and mWnt-8 during heart morphogenesis. *Mech. Dev.* **90**, 263–267.

Karavanova, I. D., Dove, L. F., Resau, J. H., and Perantoni, A. O. (1996). Conditioned medium from a rat ureteric bud cell line in combination with bFGF induces complete differentiation of isolated metanephric mesenchyme. *Development* **122**, 4159–4167.

Karlsson, O., Thor, S., Norberg, T., Ohlsson, H., and Edlund, T. (1990). Insulin gene enhancer binding protein, *Isl-1* is a member of a novel class of proteins containing both a homeo and a Cys-His domain. *Nature* **344**, 879–882.

Kaufmann, E., and Knochel, W. (1996). Five years on the wings of fork head. *Mech. Dev.* **57**, 3–20.

Kawakami, Y., Wada, N., Nishimatsu, S., Komaguchi, C., Noji, S., and Nohno, T. (2000). Identification of the chick *frizzled-10* expressed in the devloping limb and central nervous system. *Mech. Dev.* **92**, 375–378.

Kennedy, D., Ramsdale, T., Mattick, J., and Little, M. (1996). An RNA recognition motif in the Wilms' tumor protein (WT1) revealed by structural modelling. *Nature Genet.* **12**, 329–332.

Kent, J., Coriat, A.-M., Sharpe, P. T., Hastie, N. D., and van Heyningen, V. (1995). The evolution of *WT1* sequence and expression pattern in the vertebrates. *Oncogene* **11**, 1781–1792.

Kispert, A., Vainio, S., Shen, L., Rowitch, D. H., and McMahon, A. P. (1996). Proteoglycans are required for the maintenance of Wnt-11 expresion in the ureter tips. *Development* **122**, 3627–3637.

Kispert, A., Vainio, S., and McMahon, A. P. (1998). Wnt-4 is a mesenchymal signal for epithelial transformation of metanephric mesenchyme in the developing kidney. *Development* **125**, 4225–4234.

Kozmik, Z., Holland, N. D., Kalousova, A., Paces, J., Schubert, M., and Holland, L. Z. (1999). Characterisation of an amphioxus paired box gene, *AmphiPax2/5/8:* Developmental expression patterns in optic support cells, nephridium, thyroid-like structures and pharyngeal gill slits, but not in the midbrain-hindbrain boundary region. *Development* **126**, 1295–1304.

Krauss, S., Johansen, T., Korzh, V., and Fjose, A. (1991). Expression of the zebrafish paired box gene *pax [zf-b]* during early neurogenesis. *Development* **113**, 1193–1206.

Kreidberg,, J. A., Sariola, H., Loring, J. M., Maeda, M, Pelletier, J., Housman, D., and Jaenisch, R. (1993). Wt1 is required for early kidney development. *Cell* **74**, 679–691.

Kroll, K. L., and Amaya, E. (1996). Transgenic *Xenopus* embryos from sperm nuclear transplantations reveal FGF signalling requirement during gastrulation. *Development* **122**, 3173–3183.

Ku, M., and Melton, D. A. (1993). *XWnt-11:* A maternally expressed *Xenopus wnt* gene. *Development* **119**, 1161–1173.

Kuure, S., Vuolteenaho, R., And Vainio, S. (2000). Kidney morphogenesis: Cellular and molecular regulation. *Mech. Dev.* **92**, 31–45

Lahder, R. K., Church, V. L., Allen, S., Robson, L., Abdelfattah, A., Brown, N. A., Hattersley, G., Rosen, V., Luyten, F. P., Dale, L., and Francis-West, P. H. (2000). Cloning and expression of the Wnt antagonists *Sfrp-2* and *Frzb* during chick development. *Dev. Biol.* **218**, 183–198.

Lako, M., Strachan, T., Bullen, P., Wilson, D. I., Robson, S. C., and Lindsay, S. (1998). Isolation, characterisation and embryonic expression of *WNT11*, a gene which maps to 11q13.5 and has possible roles in the development of skeleton, kidney and lung. *Gene* **219**, 101–110.

Lappin, D. W. P., Hensey, C., McMahon, R., Godson, C., and Brady, H. R. (2000). Gremlins, glomeruli and diabetic nephropathy. *Curr. Opin. Nephrol. Hyper.* **9**, 469–472

Larsson, S. H., Charlieu, J.-P., Miyagawa, K., Engelkamp, D., Rassoulzagdegan, M., Ross, A., Cuzin, F., van Heyningen, V., and Hastie, N. D. (1995). Subnuclear localisation of WT1 in splicing or transcription factor domains is regulated by alternative splicing. *Cell* **81**, 391–401.

Launay, C., Fromentoux, V., Thery, C., Shi, D.-L., and Boucaut, J.-C. (1994). Comparative analysis of the tissue distribution of three fibroblast growth factor receptor mRNAs during amphibian morphogenesis. *Differentiation* **58**, 101–111.

Lee, S. M., Dickenson, M. E., Parr, B. A., Vainio, S., and McMahon, A. P. (1995). Molecular genetic analysis of Wnt signals in mouse development. *Semin. Dev. Biol.* **6**, 267–274.

Lee, Y.-H., Sauer, B., and Gonzalez, F. J. (1998). Laron dwarfism and non-insulin-dependent diabetes mellitis in the *Hnf-1α* knockout mouse. *Mol. Cell. Biol.* **18**, 3059–3068.

Lescher, B., Haenig, B., and Kispert, A. (1998). s-FRP-2 is a target of the Wnt-4 signaling pathway in the developing metanephric kidney. *Dev. Dyn.* **213**, 440–451.

Leyns, L., Bouwmeester, T., Kim, S.-H., Piccolo, S., and De Robertis, E. M. (1997). Frzb-1 is a secreted antagonist of Wnt signaling expressed in the Spemann organiser. *Cell* **88**, 747–756.

Little, M., Holmes, G., and Walsh, P. (1999). Wt1: What has the last decade told us? *Bioessays* **21**, 191–202.

Luo, G., Hofmann, C., Bronckers, A. L. J. J., Sohocki, M., Bradley, A., and Karsenty, G. (1995). BMP-7 is an inducer of nephrogenesis and is also required for eye development and skeletal patterning. *Genes Dev.* **9**, 2808–2820.

Majumdar, A., Lun, K., Brand, M., and Drummond, I. A. (2000). Zebrafish *no isthmus* reveals a role for *pax2.1* in tubule differentiation and patterning events in thr pronephric primordium. *Development* **127**, 2089–2098.

Makita, R., Mizuno, T., Koshida, S., Kuroiwa, A., and Takeda, H. (1998). Zebrafish *wnt11:* Pattern and regulation of the expression by the yolk cell and No tail activity. *Mech. Dev.* **71**, 165–176.

Mansouri, A., Chowdhury, K., and Gruss, P. (1998). Follicular cells of the thyroid gland require *Pax8* gene function. *Nature Genet.* **19**, 87–90.

Matsuda, H., Yokoyama, H., Endo, T., Tamura, K., and Ide, H.(2001). An epidermal signal regulates *Lmx-1* expression and dorsal-ventral pattern during *Xenopus* limb regeneration. *Dev. Biol.* **229**, 351–362.

Marcos-Gutiérrez, C. M., Wilson, S. W., Holder, N., and Pachnis, V. (1997). The zebrafish homologue of the ret receptor and its pattern of expression during embryogenesis. *Oncogene* **14**, 879–889.

Mauch, T. J., Yang, G., Wright, M., Smith, D., and Schoenwolf, G. C. (2000). Signals from trunk paraxial mesoderm induce pronephros formation in chick intermediate mesoderm. *Dev.Biol.* **220**, 62–75.

Mayo, M. W., Wang, C.-Y., Druin, S. S., Madrid, L. V., Marshall, A. F., Reed, J. C., Weissman, B., and Baldwin, A. S. (1999). WT1 modulates apoptosis by transcriptionally upregulating the bcl-2 proto-oncogene. *EMBO J.* **14**, 3990–4003.

McDonald, R., and Wilson, S. W. (1996). Pax proteins and eye development. *Curr. Opin. Neurobiol.* **6**, 49–56.

McGrew, L., Orre, A., and Moon, R. T. (1992). Analysis of *Xwnt-4* in embryos of *Xenopus laevis:* A *Wnt* family member expressed in the brain and floor plate. *Development* **115**, 463–473.

McLaughlin, K. A., Rones, M. S., and Mercola, M. (2000). Notch regulates cell fate in the developing pronephros. *Dev. Biol.* **227**, 567–580.

Medina, A., Reintsch, W., and Steinbesser, H. (2000). *Xenopus frizzled* can act in canonical and non-canonical *Wnt* signalling pathways: Implications on early patterning and morphogenesis. *Mech. Dev.* **92**, 227–237.

Mima, T., Ueno, H., Fischman, D. A., and Mikawa, T. (1995). Fibroblast growth factor receptor is required for *in vivo* cardiac myocyte proliferation at early embryonic stages of heart development. *Proc. Natl. Acad. Sci. USA* **92**, 467–471.

Moon, R. T., Brown, J. D, Yang-Snyder, J. A., and Miller, J. R. (1997). Structurally related receptors and antagonists compete for secreted Wnt ligands. *Cell* **88**, 725–728.

Moore, M. W., Klein, R. D., Farinas, I., Sauer, H., Armanini, M., Phillips, H., Reichardt, L. F., Ryan, A. M., Carver-Moore, K., and Rosenthal, A. (1996). Renal and neuronal abnormalities in mice lacking *GDNF*. *Nature* **382**, 76–79.

Moriya, N., Uchiyama, H., and Asashima, M. (1993). Induction of pronephric tubules by activin and retinoic acid in presumptive ectoderm of *Xenopus laevis. Dev. Growth Differ.* **35**, 123–128.

Nasevicius, A., Hyatt, T. M., Hermanson, S. B., and Ekker, S. C. (2000). Sequence expression and location of zebrafish *frizzled 10. Mech. Dev.* **92**, 311–314.

Nelson, F. K., Albert, P. S., and Riddle, D. L. (1983). Fine structure of the *Caenorhabditis elegans* secretory-excretory system. *J. Ultrastruct. Res.* **82**, 156–171.

Nguyen, H. Q., Danilenko, D. M., Bucay, N., DeRose, M. L., Van, G. Y., Thomason, A., and Simonet, W. S. (1996). Expression of keratinocyte growth factor in embryonic liver of transgenic mice causes changes in epithelial growth and differentiation resulting in polycystic kidneys and other organ malformations. *Oncogene* **12**, 2109–2119.

Nieuwkoop, P. D., and Faber, J. (1994). "Normal Table of *Xenopus laevis* (Daudin)." Garland, New York.

Nusse, R. (1999). Wnt targets: Repression and activation. *TIG* **15**, 1–3.

Obara-Ishihara, T., Kuhlman, J., Niswander, L., and Herzlinger, D. (1999). The surface ectoderm is essential for nephric duct formation in intermediate mesoderm. *Development* **126**, 1103–1108.

Orr-Urtreger, A., Bedford, M. T., Burakova, T., Arman, E., Zimmer, Y., Yayon, A., Givol, D., and Lonai, P. (1993). Developmental localisation of the splicing alternatives of fibroblast growth factor receptor-2 (FFGFR-2). *Dev. Biol.* **158**, 475–486.

Pachnis, V., Mankoo, B., and Costantini, F. (1993). Expression of the *c-ret* proto-oncogene during mouse development. *Development* **119**, 1005–1017.

Pepicelli, C. V., Kispert, A., Rowitch, D. H., and McMahon, A. P. (1997). GDNF induces branching and increased cell proliferation in the ureter of the mouse. *Dev. Biol.* **192**, 193–198.

Peng, Y., Kok, K. H., Xu, R.-H., Kwok, K. H. H., Tay, D., Fung, P. C. W., Kung, H.-F., and Lin, C. M. (2000). Maternal cold inducible RNA binding protein is required for embryonic kidney formation in *Xenopus laevis. FEBS Lett.* **482**, 37–43.

Perantoni, A. O., Love, L. F., and Karavanova, I. (1995). Basic fibroblast growth factor can mediate the ealy inductive events in renal development. *Proc. Natl. Acad. Sci. USA* **92**, 4696–4700.

Pfeffer, P. L., Gerster, T., Lun, K., Brand, M., and Busslinger, M. (1998). Characterisation of three novel members of the zebrafish *Pax2/5/8* family: Dependency of *Pax5* and *Pax8* expression on the *Pax2.1* (noi) function. *Development* **125**, 3063–3074.

Pichel, J. G., Shen, L., Sheng, H. Z., Granholm, A.-E., Drago, J., Grinberg, A., Lee, E. J., Huang, S. P., Saarma, M., Hoffer, B. J., Sariola, H., and Westphal, H. (1996). Defects in enteric innnervation and kidney development in mice lacking *GDNF. Nature* **382**, 73–79.

Plachov, D., Chowdhury, K., Walther, C., Simon, D., Guenet, J.-L., and Gruss, P. (1990). *Pax8,* a murine paired box gene expressed in the developing excretory system and thyroid gland. *Development* **110**, 643–651.

Pogge, V., Strandmann, E., and Ryffel, G. U. (1995). Developmental expression of the maternal protein XDcoH, the dimerisation cofactor of the homeoprotein LFB1 (HNF1). *Development* **121**, 1217–1226.

Pontoglio, M., Barra, J., Hadchouel, M., Doyen, A., Kress, C., Bach, J. P., Babinet, C., and Yaniv, M. (1996). Hepatocyte nuclear factor 1 inactivation results in hepatic dysfunction, phenylketonuria, and renal Fanconi syndrome. *Cell* **84**, 575–585.

Puschel, A. W., Westerfield, M., and Dressler, G. R. (1992). Comparative analysis of Pax-2 protein distributions during neurulation in mice and zebrafish. *Mech. Dev.* **38**, 197–208.

Qiao, J., Uzzo, R., Obara-Ishihara, T., Degenstein, L., and Fuchs, E. (1999). FGF-7 modulates ureteric bud growth and nephron number in the developing kidney. *Development* **126**, 547–554.

Reddy, J. C., and Licht, J. D. (1996). The WT1 Wilms' tumor suppressor: How much do we really know? *Biochim. Biophys. Acta* **1287**, 1–28.

Robertson, K., and Mason, I. (1997). The GDNF-RET signalling partnership. *TIG* **13**, 1–3.

Rodaway, A., Takeda, H., Koshida, S., Broadbent, J., Price, B., Smith, J. C., and Holder, N. (1999). Induction of the mesendoderm in the zebrafish germ ring by yolk cell-derived TGFβ family signals and discrimination of mesoderm and endoderm by FGF. *Development* **126**, 3067–3078.

Rothenpieler, U. W., and Dressler, G. R. (1993). *Pax-2* is required for mesenchymal-to-epithelium conversion during kidney development. *Development* **119**, 711–720.

Ryan, G., Steele-Perkins, V., Morris, J. F., Rausher, F. J., III, and Dressler, G. (1995). Repression of *Pax-2* by *WT1* during normal kidney development. *Development* **121**, 867–875.

Ryffel, G. U. (2001). Mutations in the human genes encoding the transcription factors of the hepatocyte nuclear factor (HNF)1 and HNF4 families: Functional and pathological consequences. *J. Mol. Endocrinol.* **27**, 11–29.

Sainio, K., Suvanto, P., Davies, J., Wartiovaara, J., Wartiovaara, K., Saarma, M., Aumäe, U., Meng, X., Lindahm, M., Pachnis, V., and Sariola, H.

(1997). Glial-cell-derived neurotrophic factor is required for bud initiation from ureteric epithelium. *Development* 124, 4077–4087.

Sariola, H., and Saarma, M. (1999). GDNF and its receptors in the regulation of the ureteric branching. *Int. J. Dev. Biol.* 43, 413–418.

Sato, A., Asashima, M., Yokota, T., and Nishinakamara, R. (2000). Cloning and expression pattern of a pronephros-specific gene *XSMP-30*. *Mech. Dev.* 92, 273–275.

Saxen, L. (1987). "Organogenesis of the Kidney." Cambridge Univ. Press, Cambridge.

Schmid, B., Furthauer, M., Connors, S. A., Trout, J., Thisse, B., Thisse, C., and Mullins, M. C. (2000). Equivalent genetic role for *bmp7/snailhouse* and *bmp2b/swirl* in dorsoventral pattern formation. *Development* 127, 957–967.

Schuchardt, A., D'Agati, V., Larsson-Blomberg, L., Costantini, F., and Pachnis, V. (1994). Defects in the kidney and the enteric nervous system of mice lacking the tyrosine kinase receptor RET. *Nature* 367, 380–383.

Schuchardt , A., Srinivas, S., Pachnis, V., and Costantini, F. (1995). Isolation and characterisation of a chicken homolog of the *c-ret* protooncogene. *Oncogene* 10, 641–649.

Semba, K., Saito-Ueno, R., Takayama, G., and Kondo, M. (1996). cDNA cloning and its pronephros specific expression of the Wilms' tumor suppressor gene, *WT1*, from *Xenopus laevis*. *Gene* 175, 167–172.

Serluca, F. C., and Fishman, M. C. (2001). Prepattern in the pronephric kidney field of zebrafish. *Development* 128, 2233–2241.

Seufert, D. W., Deguire, J, Brennan, H. C., Jones, E. A., and Vize, P. D. (1999). The developmental basis of pronephric defects in *Xenopus* body plan phenotypes. *Dev. Biol.* 215, 233–242.

Seville, R. A., Nijjar, S. A., Barnett, M. W. Massé, K., and Jones E. A. (2002). Annexin-4 *(Xanx-4)* has a functional role in the formation of the pronephric tubules. *Development* 129, 1693–1704.

Shawlot, W., and Behringer, R. R. (1995). Requirement for *Lim1* in head organiser function. *Nature* 374, 425–430.

Shepherd, I. T., Beattie, C. E., and Raible, D. W. (2001). Functional analysis of zebrafish *GDNF*. *Dev. Biol.* 231, 420–435.

Shi, D.-L., Goisset, C., and Boucaut, J.-C. (1997). Expression of *Xfz3*, a *Xenopus* frizzled family member, is restricted to the early nervous system. *Mech. Dev.* 70, 35–47.

Shibata, M., Ono, H., Hisaka, H., Shinga, J., and Taira, M. (2000). *Xenopus crescent* encoding a Frizzled-like domain is expressed in the Spemann organiser and the pronephros. *Mech. Dev.* 96, 243–246.

Stark, K., Vainio, S., Vassileva, G., and McMahon, A. P. (1994). Epithelial transformation of metanephric mesenchyme in the developing kidney regulated by *Wnt-4*. *Nature* 372, 679–683.

Stark, M. R., Rao, M. H., Schoenwolf, G. C., Yang, G., Smith, D., and Mauch, T. J. (2000). *Frizzle-4* expression during chick kidney development. *Mech. Dev.* 98, 121–125.

Sumanas, S., Strege, P., Heasman, J., and Ekker, S. C. (2000). The putative Wnt receptor *Xenopus frizzled-7* functions upstream of β-catenin in vertebrate dorsoventral mesoderm patterning. *Development* 127, 1981–1990.

Summerton, J., and Weller, D. (1997). Morpholino antisense oligomers: Design, preparation and properties. *Antisense Nucleic Acid Drug Dev.* 7, 187–195.

Taira, M., Jamrich, M., Good, P. L., and Dawid, I. B. (1992). The LIM domain-containing homeobox gene *Xlim-1 is expressed* specifically in the organiser region of *Xenopus* gastrula embryos. *Genes Dev.* 6, 356–366.

Taira, M., Otani, H., Jamrich, M., and Dawid, I. B. (1994). Expression of the LIM homeodomain protein Xlim-1 in pronephros and CNS cell lineages of *Xenopus* embryos is affected by retinoic acid and exogastrulation. *Development* 120, 1525–1536.

Tanda, N., Kawakami, Y., Sait, T., Noji, S., and Nohno, T. (1995). Cloning and characterisation of *Wnt-4* and *Wnt-11* cDNAs from the chick embryo. *DNA* 5, 277–281.

Tannahill, D., Isaacs, H. V., Close, M. J., Peters, G., and Slack, J. M. W. (1992). Developmental expression of the *Xenopus int-2 (FGF-3)* gene: Activation by mesodermal and neural induction. *Development* 115, 695–702.

Tassabehji, M., Read, A. P., Newton, V. V. E., Harris, R., Balling, R., Gruss, P., and Strachan, T. (1992). Waardenburg's syndrome patients have mutations in the human homologue of the *Pax-3* paired-box gene. *Nature* 355, 635–636.

Torban, E., Eccles, M. R., Favor, J., and Goodyer, P. R. (2000). PAX2 suppresses apoptosis in renal collecting duct cells. *Am. J. Pathol.* 157, 833–842.

Torres, M., Gomez-Pardo, E., Dressler, G. R., and Gruss, P. (1995). *Pax-2* controls multiple steps of urogenital development. *Development* 121, 4057–4065.

Toyama, R., and Dawid, I. B. (1997). *lim6*, a novel LIM homeobox gene in the zebrafish: Comparison of its expresion pattern with *lim1*. *Dev. Dyn.* 209, 406–417.

Toyama, R., O'Connell, M. L., Wright, C. V. E., Kuehn, M. R., and Dawid, I. B. (1995). *nodal* induces ectopic *goosecoid* and *lim1* expression and axis duplication in zebrafish. *Development* 121, 383–391.

Ungar, A. R., Kelly, G. M., and Moon, R. T. (1995). *Wnt4* affects morphogenesis when misexpressed in the zebrafish embryo. *Mech. Dev.* 52, 153–164.

Uochi, T., and Asashima, M. (1996). Sequential gene expression during pronephric tubule formation *In vitro* in *Xenopus* ectoderm. *Dev. Growth Differ.* 38, 625–634.

Vainio, S. J., Itaränta, P. V., Peräsaari, J. P., and Uusitalo, M. S. (1999). Wnts as kidney tubule inducing factors. *Int. J. Dev. Biol.* 43, 419–423.

Vesque, C., Ellis, S., Lee, A., Szabo, M., Thomas, P., Beddington, R., and Placzec, M. (2000). Development of chick axial mesoderm: Specification of prechordal mesoderm by anterior endoderm-derived TGFβ family signalling. *Development* 127, 2795–2809.

Vize, P. D., Jones, E. A., and Pfister, R. (1995). Development of the *Xenopus* pronephros. *Dev. Biol.* 171, 531–540.

Vize, P. D., Seufert, D. W., Carroll, T. J., and Wallingford, J. B. (1997). Model systems for the study of kidney development: Use of the pronephros in the analysis of organ induction and patterning. *Dev. Biol.* 188, 189–204.

Vukicevic, S., Kopp, J. B. , Luyten, F. P., and Sampath, T. K. (1996). Induction of nephrogenic mesenchyme by osteogenic protein 1 (BMP-7). *Proc. Natl. Acad. Sci. USA* 93, 9021–9026.

Wallin, J., Wilting, J., Koseki, H., Christ, B., and Balling, R. (1994). The role of *Pax-1* in axial skeleton development. *Development* 120, 1109–1121.

Wallingford, J. B. Carroll, T. J., and Vize, P. D. (1998). Precocious expression of the Wilms' tumor gene *xWT1* inhibits embryonic kidney development in *Xenopus laevis*. *Dev. Biol.* 202, 103–112.

Wang, Z. Y., Madden, S. L., Deuel, T. F., and Rauscher, F. J., III (1992). The Wilms' tumor product, WT1, represses transcription of the platelet derived growth factor A-chain gene. *J. Biol. Chem.* 267, 21999–22002.

Wang, Y., Macke, J. P. Abella, B. S., Andreasson, K., Worley, P., Gilbert, D. J., Copeland, N. G., Jenkins, N. A., and Nathans, J. (1996). A large family of putative transmembrane receptors homologous to the product of the *Drosophila* tissue polarity gene *frizzled*. *J. Biol. Chem.* 271, 4468–4476.

Wang, S., Krinks, M., Kleinwaks, L., and Moos, M., Jr. (1997). A novel Xenopus homologue of bone morphogenetic protein-7 (BMP-7). *Genes Funct.* 1, 259–271.

Wang, S., Krinks, M., Lin, K., Luyten, F. P., and Moos, M., Jr. (1997). Frzb, a secreted protein expressed in the Spemann organiser, bind and inhibits Wnt-8. *Cell* **88,** 757–766.

Way, W. C., and Chalfie, M. (1998). *mec-3,* a homeobox-containing gene that specifies differentiation of the touch receptor neurons in *C. elegans. Cell* **54,** 5–16.

Weber, H., Holewa, B., Jones, E. A., and Ryffel, G. U. (1996). Mesoderm and endoderm differentiation in animal cap explants: Identification of the HNF4-binding site as an activin responsive element in the *Xenopus HNF1α* promoter. *Development* **122,** 1975–1984.

Werner, H., Shen-Orr, Z., Rausher, F. J., III, Morris, F. J., Roberts, C. T., Jr., and Le Roith, D. (1995). Inhibition of cellular proliferation by the Wilms' tumor suppressor *WT1* is associated with the suppression of *Insulin-Like Growth Factor 1 receptor* gene expression. *Mol. Cell. Biol.* **15,** 3517–3522.

Wild, W., Pogge, V., Strandmann, E., Nastos, A., Senkel, S., Lingott-Frieg, A., Bulman, M., Bingham, C., Ellard, S., Tattersley, A. T., and Ryffel, G. U. (2000). The mutated human gene encoding hepatocyte nuclear factor 1β inhibits kidney formation in developing Xenopus embryos. *Proc. Natl. Acad. Sci. USA* **97,** 4695–4700.

9

Embryological, Genetic, and Molecular Tools for Investigating Embryonic Kidney Development

Neil A. Hukriede, Brant M. Weinstein, and Igor B. Dawid

I. Introduction
II. Molecular Embryology
III. Cellular Embryology
IV. Transgenic Methods
V. Classical Genetic Methods: Mutant Screens
References

I. Introduction

Lower vertebrates provide highly accessible models for the observation and experimental manipulation of embryonic development. This chapter focuses on some of the molecular, cellular, and genetic methods available for studying developmental processes in two commonly used lower vertebrate model systems, zebrafish *(Danio rerio)* and *Xenopus laevis,* although many of these methods are also potentially applicable to other lower vertebrates. Zebrafish and *Xenopus* eggs are both fertilized externally, and their embryos develop outside the mother from the single cell stage on. The yolk-laden frog embryo lends itself readily to culturing and microsurgical manipulation of blocks of tissue and is particularly useful for tissue grafting and explant culture. The optically clear zebrafish embryo is especially useful for manipulating single or small groups of cells and for *in vivo* microscopic imaging. This chapter discusses experimental methods used in zebrafish and *Xenopus* and highlight the useful properties of each

of these vertebrates for studying pronephric development. Due to the broad scope of this chapter and space constraints, we provide only a brief overview of most techniques. For a more comprehensive description of the methods surveyed and detailed protocols for their use, we refer the reader to a list of laboratory manuals and other resource materials provided at the end of this chapter.

II. Molecular Embryology

Ultimately, development can be viewed as the result of the concerted and orchestrated action of intracellular and extracellular molecules that regulate and carry out various morphogenetic processes. Fundamental understanding of developmental processes therefore requires detailed knowledge of the molecular processes underlying them. This, in turn, requires the use of methods for assaying and interpreting the activities of different types of genes and gene products *in vivo*. This section describes methods for detecting gene expression *in vivo* and for testing the functional roles of genes by altering, enhancing, interfering with, or eliminating their functions from the embryo.

A. Detecting Gene Expression

In order to analyze gene function within embryos, it is first necessary to have methods for detecting the temporal and spatial patterns of expression of specific genes. This can

Figure 9.1 *In situ* hybridization with markers focusing on the pronephros in zebrafish (A–D) and *Xenopus* (E–H). The *in situ* expression patterns of *Lim1* (A,C,E,G) and *Pax2.1* (B,D,F,H) are shown. A and B are early segmentation period embryos, whereas C and D are midlate segmentation period embryos. E and F are early tailbud stage embryos, whereas G and H are tailbud stage embryos. Arrows indicate pronephric staining. *Xenopus* pictures provided by Peter Vize, University of Texas, Austin.

be accomplished in a number of ways. Whole-mount *in situ* hybridization and immunohistochemistry can elucidate both the location and the time of expression of specific mRNAs or proteins; in the case of *in situ* hybridization (Fig. 9.1), they can even generate a semiquantitative estimate of relative levels of gene expression. Northern blot analysis, quantitative RT-PCR, and Western blot analysis offer more precise measures of mRNA and protein expression levels, although these do not provide the spatial resolution of the former two methods. All five of the aforementioned techniques are used in conjunction with methods described in this chapter for manipulating gene expression *in vivo*. A variety of pronephros-specific genes have been cloned in both *Xenopus* and zebrafish that serve as markers to study pronephric development and detect changes in the pronephros in experimentally manipulated embryos (Carroll *et al.*, 1999a,b). The function of most of these genes is still under investigation, and they also serve as substrates for the *in vivo* perturbation experiments described later. Some of these genes and their functional activities are discussed in chapter 9.

B. Methods for Ectopic Gene Expression in Lower Vertebrates

One of the first courses of action when studying the function of a gene of interest is to express it ectopically within the embryo. In this and the immediately following sections of this chapter we focus on transient expression. Stable expression in the form of transgenics is discussed in

a later section. When studying gene function in *Xenopus* and zebrafish, the most effective method for ectopically expressing a gene is by injecting synthetic messenger RNA (mRNA) into cleavage stage embryos (Krieg and Melton, 1987). Injections can be targeted effectively into specific blastomeres in *Xenopus;* the dorsal/ventral polarity of the embryo is clearly apparent at the 4 cell stage. This is more difficult in zebrafish because dorsal/ventral polarity is not apparent in embryos until much later and the blastomeres are linked cytoplasmically in a single syncytium until the 16 cell stage. Synthetic mRNAs are usually translated directly after injection, resulting in protein expression that is widespread and usually at very high levels. There are advantages and disadvantages to such high exogenous levels of protein. One advantage is that enough protein is likely present to either enhance or antagonize endogenous signaling systems, thereby generating detectable phenotypes. In addition, high protein levels are an advantage for studying the pronephric kidney, as it is not specified until stage 12.5 in *Xenopus* or during segmentation in zebrafish (Brennan *et al.*, 1998, 1999; Drummond *et al.*, 1998). The advantage of high levels of protein expression can also be a potential disadvantage, as artificially high expression levels can lead to artifactual results.

An additional complication that can arise in using an early embryonic injection of mRNA to study the function of genes involved in pronephric development is that many of these genes are also involved in other, earlier aspects of development. The *Lim1* gene, for example, is expressed in the dorsal organizer, notochord, brain, and sensory neurons in addition to the kidney. The four *Lim1* knockout mice that developed to term displayed a loss of nephric tissues, although the mice were also affected in many other aspects, including loss of all tissue anterior to rhombomere three (Shawlot and Behringer, 1995). In *Xenopus, Xlim1* has also established effects on early kidney development (Carroll and Vize, 1999). However, an early injection of modified forms of Xlim1 causes severe malformations in injected embryos long before the pronephros forms, potentially masking a pronephric defect (Kodjabachian *et al.,* 2000). Despite problems such as these, synthetic RNA injections are a valuable resource for studying the pronephric function of genes important for early development, particularly when used in conjunction with animal cap assays, explant studies, or injections targeted to specific blastomeres (see later). In addition, there are genes that do not play major developmental roles prior to pronephros formation, such as xWT1 or XSMP-30 (Carroll and Vize, 1996; Sato *et al.,* 2000; Wallingford *et al.,* 1998). With genes such as these, it may be possible to inject early cleavage stage embryos and obtain pronephros-specific effects.

Injection of linear DNA constructs provides a possible way to circumvent early effects of synthetic mRNA injections. *In vivo* transcription of mRNA from a linear DNA template will not begin before the midblastula transition (MBT), the point at which embryos switch from using maternal RNAs to transcribing zygotic RNAs (Almouzni and Wolffe, 1995; Etkin, 1985; Shiokawa *et al.*, 1994). A further delay and/or spatial restriction of the transcription of an injected gene of interest can be affected by the choice of a specific promoter. Additionally, DNA is relatively more stable than mRNA *in vivo* and will persist later into embryonic development. A disadvantage of DNA injections is that unlike injected mRNA, which distributes itself relatively uniformly after injection, injected DNA has a highly mosaic distribution, i.e., only a subset of the cells arising from the injected blastomere actually contain significant amounts of injected DNA (Sargent and Mathers, 1991; Vize *et al.*, 1991). Another disadvantage of DNA injections is that only low amounts of DNA can be injected into early cleavage stage embryos due to the relatively high toxicity of DNA compared to RNA.

C. Modifying the Function of Ectopically Introduced Genes

A transient injection of DNA or RNA encoding modified forms of a gene of interest can also be performed to either enhance or antagonize activity *in vivo*. The nature of these modifications depends on the type of gene, but generally involves direct modification of the protein sequence and/or the addition of repressing or enhancing domains. Direct modifications can take a variety of forms. Receptors and ligands can often be truncated to make dominant-negative molecules. Three examples of this are the FGF and BMP receptors and the Delta ligands. For the FGF receptor, removing the cytoplasmic domain generates a dominant negative molecule, and this acts by upon binding of the ligand, the dominant-negative receptor dimerizes with wild type receptors and blocks signal transduction by impeding downstream phosphorylation events (Amaya *et al.*, 1991). In addition, cytoplasmically truncated forms of the BMP receptor act in a dominant-negative fashion by out-competing endogenous receptors for ligand binding, whereas truncated Delta ligands are thought to function by outcompeting endogenous ligands by binding to the Notch receptor (Chitnis *et al.*, 1995; Graff *et al.*, 1994; Maeno *et al.*, 1994; Suzuki *et al.*, 1994). Creating and introducing modified cytoplasmic proteins that titrate out molecules that act either upstream or downstream of the protein is often an effective strategy. One example of this is the dominant-negative GSK3 kinase, which can mimic down-regulation of the Wnt pathway on the dorsal side of the early gastrula embryo (He *et al.*, 1995; Pierce and Kimelman, 1995). The specificity or activity of transcription factors can also be altered by deleting or changing DNA or protein-binding domains. This has been used to test the *in vivo* function of numerous molecules. Examples include the modification of homeodomains and deletion of basic domains from transcriptional activators so that they can no longer

bind to DNA but still form hetero- and homodimers (Muller *et al.*, 1988).

It has become popular to modify transcription factors to effectively change them into neomorphs by adding exogenous domains that either activate or repress transcription. A transcriptional repressor can be turned into an activator by addition of the strong transcriptional activation domain of the herpesvirus protein VP16 (Denecke *et al.*, 1993; Xu *et al.*, 1993). Conversely, transcriptional activators can be turned into repressors by the addition of a strong repression domain, such as that from the *Drosophila melanogaster* Engrailed (Eng) protein (Jaynes and O'Farrell, 1991). An example of the usefulness of the Eng repression domain is the *Xbra* study in *Xenopus* and the *siamois* and *Mix.1* studies where both Eng and VP16 chimeras were used successfully (Conlon *et al.*, 1996; Kessler, 1997; Lemaire *et al.*, 1998).

D. Controlling the Spatial and Temporal Expression of Ectopically Introduced Genes

Enhancing or antagonizing gene function can provide a plethora of useful data, but most of the molecular techniques thus far described were developed with the idea of addressing early developmental effects, such as dorsal/ventral patterning. As mentioned previously, the function of a gene during pronephric development can be masked by its involvement in earlier developmental events. One way to overcome this problem is to control the temporal and/or spatial expression of an exogenously introduced gene. There are several methods available for doing this. These include the use of linear DNA, inducible and tissue-specific promoters (their use in stable transgenic lines is addressed in Section IV), a steroid hormone-inducible system, targeted blastomere injections in *Xenopus,* and finally explant cultures.

As mentioned previously, DNA injections provide some degree of temporal control in that DNA is not transcribed until the MBT. Inducible promoters can provide much greater control over temporal expression. The Hsp70 heat shock promoter has been isolated from both zebrafish and *Xenopus* and has been used effectively to control the expression of injected genes in both systems. The expression of introduced genes fused to this promoter can be controlled by simply heating injected embryos for a relatively short period of time (Halloran *et al.*, Lele *et al.*, 1997; Wheeler *et al.*, 2000). This allows expression to be restricted to a specific temporal window (a highly advantageous feature for studying pronephric development), but does not provide any spatial regulation. Furthermore, as noted earlier, DNA injections are highly mosaic, and transient injections of linear DNA under the control of an inducible promoter generate expression in only a restricted subset of cells. The latter problem can be overcome using transgenics. Transgenic heat shock-inducible lines have been used successfully to achieve both widespread and spatially restricted expression in both zebrafish and *Xenopus* (Halloran *et al.,* 2000; Wheeler *et al.,* 2000). Their use is discussed later in this chapter.

Another method to control temporal expression is by using a steroid hormone-inducible system. The system consists of fusing the hormone-binding domain of steroid receptors to a gene of choice (Mattioni *et al.,* 1994). In the absence of hormone, the chimera remains in an inactive state, but upon the addition of hormone, rapid activation of the chimeric protein occurs (Kolm and Sive, 1995; Scherrer *et al.,* 1993; Tsai and O'Malley, 1994). Because the regulation is at the level of the protein, RNA can be injected into early cleavage stage embryos, and the chimeric protein is distributed relatively ubiquitously. Thus, relatively high and ubiquitous (nonmosaic) levels of gene activity can be achieved upon induction (Kolm and Sive, 1995). In *Xenopus,* this type of system has been used successfully to study *Brachyury, Otx2* and *MyoD* and *Notch* (Gammill and Sive, 1997; Kolm and Sive, 1995; Tada *et al.,* 1997, McLaughlin *et al.,* 2000). A potential disadvantage of steroid hormone induction is that the system has been reported to be leaky. Injection of a steroid hormone chimera with a protein that has a potent early developmental effect could cause early defects and again mask its later pronephric function.

Blastomere-targeted injection is a technique that can be used for the spatial restriction of gene products or in conjunction with the temporal methods mentioned earlier. Fate maps of 16 and 32 cell stage *Xenopus* embryos have been prepared, providing descriptions of the descendants of each blastomere (Dale and Slack, 1987; Moody, 1987a,b). At the 32 cell stage, a majority of the pronephric tissue arises from a single pair of blastomeres (the "C3" blastomeres), although other blastomeres also possess the ability to produce pronephric tissue (Dale and Slack, 1987). Thus, injection of C3 blastomeres provides some degree of restriction of effects to the pronephros. Thirty-two cell injections require some additional experience. Embryos do not always develop as in textbook pictures, and identification of the correct blastomere to target can be challenging. Furthermore, 32 cell blastomeres are much smaller than those at the 2 or 4 cell stages, and only smaller amounts of mRNA can be injected. C3-targeted injections have proven useful for addressing questions of pronephric development. By injecting *XPax-8* and *Xlim1* into the C tier of 32 cell stage embryos, it was shown that these genes synergize to produce expanded and ectopic pronephroi (Carroll and Vize, 1999). It was essential to perform this experiment with targeted injections because, as mentioned earlier, earlier injections of *Xlim1* cause global morphological changes in the embryo.

An alternative method for performing "spatially restricted" experiments is provided by explant cultures. Explanted tissues can be removed from specific regions and at specific times from a donor embryo and then cultured in the presence or absence of growth factors to study the role of these molecules

(Brennan *et al.,* 1998, 1999). Animal cap explants, in culture, give rise to ectoderm derivatives. However, the cap has the ability to respond to inductive signals such as activin and FGF (Cooke *et al.,* 1987; Cornell and Kimelman, 1994; Green *et al.,* 1990, 1992). In response to activin or FGF, animal caps form primitive mesoderm, and if retinoic acid is added, pronephric tissue results (Brennan *et al.,* 1999; Moriya *et al.,* 1993; Uochi and Asashima, 1996). In addition, other regions of the embryo can also be used to study the effects of inductive molecules on pronephric tissue, such as the lateral marginal zone, which gives rise to the presumptive pronephric mesoderm. For example, tissue from pronephric-competent lateral mesoderm in *Xenopus* at stage 11.5 can be cultured in ectodermal wraps to address issues of specification of pronephric tissue (Seufert *et al.,* 1999).

E. Interfering with Endogenous Gene Expression

Although it is useful to introduce ectopic positively or negatively acting versions of the gene of interest, it is frequently desirable to entirely remove the normal gene or its expression from embryos, generating embryos that are "hypomorphic" or "null" for the gene. Classical genetic methods for generating and recovering mutants in developmentally important genes are discussed in Section V. In mice, targeted disruption of genes by homologous recombination provides an effective means for removing a gene of interest from either all or a subset of the tissues of a mouse embryo. Comparable methods for performing homologous recombination and selecting for recombinants are currently unavailable in lower vertebrates. However, a number of alternative methods are available for interfering with the expression of endogenous genes.

One possible alternative is the use of double-stranded RNA (RNAi) injections to antagonize gene function. RNAi molecules are thought to antagonize gene function by causing degradation of the target RNA (Bass, 2000). RNAi was developed in *Caenorhabditis elegans* and has also been used successfully in *Drosophila,* planarians, trypanosomes, and plants (Fire *et al.,* 1998; Kennerdell and Carthew, 1998; Ngo *et al.,* 1998; Sanchez Alvarado and Newmark, 1999; Waterhouse *et al.,* 1998). To date, very few instances of successful RNAi experiments have been reported in vertebrates. There have been a few reports of successful interference with gene expression and function in zebrafish, *Xenopus,* and mice (Li *et al.,* 2000; Nakano *et al.,* 2000; Wargelius *et al.,* 1999; Wianny and Zernicka-Goetz, 2000). It has been reported that these sorts of RNAi injections cause nonspecific reduction in many RNA species, that many of the phenotypes reported in other studies can also be generated by nonspecific RNAi molecules, and that RNAi injection causes reduction in the levels of many different mRNA species simultaneously (Oates *et al.,* 2000). Thus, the question of whether RNAi can generate specific effects in vertebrates is still undecided.

Morpholine oligonucleotides ("morpholinos") provide a promising method for interefering with specific genes in zebrafish and *Xenopus*. Morpholinos are antisense oligonucleotides in which the five-member ribose ring has been replaced by a six-member morpholine ring (Summerton and Weller, 1997). One major advantage of this modification is that while the half-life of a normal antisense oligonuclotide is less than 5 min in an embryo, morpholinos can be stable for several days (Hudziak *et al.*, 1996). Morpholinos act by binding to the target transcript and inhibiting translation by up to 90%. They are highly specific for the target transcript in that four changes to a 25 member morpholino destroys the effects. One potential drawback may be the amount of morpholino needed to inhibit translation. Because morpholinos act in a stoichiometric manner by binding irreversibly to their target transcript, highly abundant transcripts may not be inhibited. Nevertheless, morpholinos have been used successfully in *Xenopus* to inhibit catenin function and may be a useful method for eliminating the expression of proteins that affect pronephric development injections (Heasman *et al.*, 2000), not only by injecting early into an embryo, but also by targeted blastomere.

Three additional methods that may be applicable to interfering with endogenous gene function are antisense oligonucleotides, modified antisense oligonucleotides, and antibodies. Antisense oligonucleotide technology has proven useful for depleting maternally transcribed gene products (Heasman *et al.*, 1994). As stated earlier, one difficulty in using such technology for zygotic genes is that oligonucleotides are not stable and, when injected into the blastula, they are degraded rapidly. To circumvent this problem, the oligonucleotide can be modified so that it is not degraded readily in the embryo. One way this is accomplished is by modifying the phosphodiester bonds of the oligonucleotide in order to make them inaccessible to nucleases (Dagle *et al.*, 2000). The modified antisense oligonucleotides can be injected into embryos to target the gene transcript for RNase H-dependent degradation. Finally, injection of antibodies into early cleavage stage embryos has resulted in blocking protein function, such as was done for XlHbox1 (Vize *et al.*, 1991; Wright *et al.*, 1989). Antibodies have been shown to be stable in *Xenopus*, lasting to at least stage 32 for anti-XlHbox1 (Cho *et al.*, 1988). As with morpholinos, modified antisense oligonucleotides and antibodies may prove useful for perturbing gene function in the developing pronephros, but to date very little has been accomplished.

III. Cellular Embryology

One of the great strengths of the lower vertebrate developmental models is the accessibility of their embryos to obser-

vation and experimental manipulation, and the opportunities this accessibility affords for *in vivo* analysis of the behavior and interactions of different cell and tissue types. This section reviews some of the methods available in *Xenopus* and zebrafish for examining and experimentally manipulating cells and tissues *in vivo*.

A. Tools of the Trade

The techniques described in this section involve the use of a small armamentarium of tools, some common in most laboratories and some fairly esoteric and expensive. Most of these tools are either microscopes or are peripheral to microscopes used to image living embryos. Dissecting stereomicroscopes with good optics are a necessity. These are used for basic observations, screening, data documentation at low magnifications, and performing some of the procedures described in this section. They should allow for magnifications of at least 50X, preferably up to 100X. "Zoom" magnification changing is desirable, as is an epifluorescence attachment for visualizing fluorescent samples conveniently at low magnification.

It is also necessary to have at least one good compound microscope equipped for both Nomarski/DIC optics and epifluorescence. The quality of DIC or Nomarski optics and epifluorescence sensitivity are two of the more important criteria in judging microscopes and objective lenses used for *in vivo* imaging. Higher numerical aperture (NA) water immersion optics provide the best DIC. For three-dimensional (3D) and 4D (3D time-lapse) analysis, it is necessary to have a computer-controlled motorized focus motor to capture images at defined depth "steps" (Z steps). A computer-controlled motorized stage able to provide automated X–Y movements is also useful for performing time-lapse analysis on multiple specimens simultaneously. Microscopes can be fitted with computer-controllable electronic shutters in order to minimize illumination of the specimen with either transmitted or epifluorescent light and to rapidly and conveniently switch between these two light sources in the course of an experiment. A motorized fluorescence filter turret allows automation of the process of changing between different fluorescence filters, useful for time-lapse experiments in which multiple fluorophores are employed. Motorized filter wheels are also available for applications requiring rapid switching between different fluorophores (e.g., near-simultaneous fluorescence ratio imaging). Most microscope manufacturers now provide the devices just described as built-in options on their microscope stands. A variety of third-party suppliers also produce them.

In order to be useful for data collection, microscopes must also be outfitted with cameras to capture images. For most imaging applications, photographic film cameras have largely been replaced by a variety of sensitive, high-resolution electronic cameras. The choice of camera is very highly dependent on the particular application(s) envisaged. A variety of considerations need to be taken into account. Is the camera to be used for high light level applications (such as DIC/Nomarski), low light level applications (fluorescence detection), or both? What level of camera or pixel resolution is required? Is color detection necessary or are black and white images sufficient? Are the samples to be imaged motionless, undergoing slow motion, or moving rapidly? These are just a few of the questions that inform the choice of camera to be used. Often a single camera will not in fact be ideal for all of the applications desired, and it is not uncommon to outfit imaging microscopes with two or more different cameras. For many imaging applications, such as time-lapse microscopy, cameras, automated devices, and attachments all need to be centrally computer controlled using software that is capable of collecting image data, driving the devices, and activating them in the desired temporal sequence. A variety of commercial packages are available for this purpose, as well as some quite good free software, notably "NIH Image" (http://rsb.info.nih.gov/nih-image/Default.html). A variety of stand-alone software packages are also available for image processing and data handling, although in many cases these functions are also integrated into commercial image collection and automation software.

Confocal microscopy has become a very important tool for fluorescence imaging in developmental biology (Paddock, 1994; Tadrous, 2000). Confocal microscopes "optically section" fluorescent specimens, greatly reducing out of the plane of focus fluorescence and generating images that are less complex, but with greater preservation of fine details, than those obtained via conventional fluorescence microscopy. In addition, a series of confocal optical sections can be recombined digitally to give an accurate spatial representation of *in vivo* fluorescent structures. A new refinement to confocal microscopy called two-photon confocal microscopy has also become available. Two-photon confocal microscopes use lasers that output light at double the wavelength required to excite the particular fluorophore being imaged. When a fluorophore molecule is hit simultaneously by two photons of double the excitation wavelength, their energies sum and the fluorophore becomes excited. This only occurs at the focal point of the objective lens where photon energy densities are sufficiently elevated to make simultaneous collision a probable event. In practice, this means that two-photon imaging delivers a thinner, tighter optical section than conventional confocal microscopes, improving the "confocality" of the system and reducing photobleaching due to out-of-plane fluorescence. The longer wavelength light used for excitation also penetrates tissues more effectively, making it possible to image more opaque and/or deeper tissues.

The dizzying array of choices of equipment available for microscopic imaging and the rapid pace of new technical developments and product changes make it difficult to make specific recommendations. The best approach is to begin by thinking carefully but broadly about the desired applications, drawing up a list of detailed requirements for an envisaged imaging system. Once these have been determined, contact sales representatives for the microscope suppliers and allow them come up with specific product recommendations. Imaging equipment is quite expensive. You should expect to be shown that the imaging system performs as needed for your desired application *before* they are purchased.

B. *In Vivo* Labeling Methods

In order to study the behavior of cells in living animals, it is often necessary to begin by vitally "marking" the cells or tissues to be studied so that they and their progeny can be detected at later stages. A variety of methods have been employed for this. In classical embryological studies such as amphibian grafting experiments, "intrinsic" markers were often used. These are cytological differences between the cells of two related species, which allow the cells of one species to be recognized when grafted into the other. Pigmentation differences are one common example of this (see Spemann, 1938). Differences in the appearance of the nucleoli in nuclei of chick and quail embryos served as the basis for the well-characterized chick–quail chimera system used to great effect for many developmental studies (LeDouarin *et al.,* 1975). In *Xenopus* embryos, induction of changes in the ploidy of embryos used as donors has also been used to mark these cells. In addition to these "intrinsic" markers, regions to be studied or fate mapped were often labeled extrinsically by application of carbon granules or vital dyes to the surface of the tissue in order to visually mark it (Vogt, 1929), (fig. 3.2).

In most recent studies, however, marking of cells and tissues has been accomplished using fluorescent dye labeling. Lipophilic dyes such as diI and diO can be applied to the cell surface in a manner analogous to the classical vital dyes to mark cell membranes (Honig and Hume, 1986). These dyes have been widely used for tracing cell populations (e.g., Austin, 1995). Their bright fluorescence and uniform cell surface labeling make it possible to detect even fine processes extending from labeled cells. This has made these particular dyes also very useful for studying neuronal axon projection *in vivo*. Intracellular injection of fluorescent dyes has been the most widely used method for labeling cells for fate mapping (e.g., Dale and Slack, 1987; Kimmel *et al.,* 1990). Caged fluorescence activation is also useful for labeling cells for fate mapping studies (Kozlowski *et al.,* 1997) (Fig. 9.2). Caged fluorescent dyes are fluorescent molecules that have been modified covalently such that the fluorescence is quenched and they are rendered nonfluorescent. These modifications are designed in such a way that they can be cleaved off from the fluorescent molecule by exposure to ultraviolet irradiation, restoring fluorescence to the molecule. These dyes have been used extensively to study cytoskeletal dynamics and have also

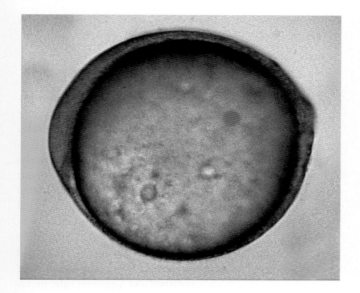

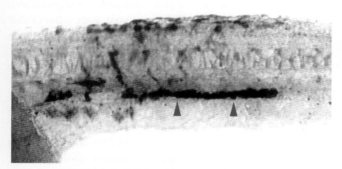

Figure 9.2 Labeling cells in living zebrafish embryos using fluorescent dye uncaging. A nonfluorescent, dextran-conjugated caged dye is injected into single cell embryos and the embryo is allowed to develop further. At later stages, adjacent groups of cells or even single cells can be made fluorescent by "uncaging" the dye using ultraviolet light provided by either epifluorescence illumination or a laser microbeam. (Top) a 2 somite stage embryo immediately after uncaging of a small patch of lateral cells by epifluorescence illumination. Lateral view; anterior is to the left. (Bottom) Labeled cell populations from a similar embryo at the 30 somite stage. In this embryo, the labeled group of cell included a large number of lateral mesodermal progenitors of the pronephric tubule, which is labeled prominently (arrows). Lateral view of the trunk; anterior is to the left.

found a use for temporally and spatially specific cell labeling in zebrafish embryos, which are ideal for this because of their optical clarity. Finally, innately fluorescent transgenic animals are potentially very useful for tracking specific cells *in vivo* (see later). Few such transgenics have yet been generated, however, and there are no published reports yet of experiments performed using these animals to follow cellular behaviors *in vivo*.

C. Fate Mapping and Cell Lineage Analysis

Fate maps have provided a very useful tool for developmental biologists, showing what regions and what cells in

early "undifferentiated" embryos contribute to different tissues and structures in later "differentiated" embryos. Fate maps provide a foundation for experimental study of the origins of these tissues and structures–their morphogenesis, differentiation, and inductive interactions that give rise to them. A fate map is prepared by first vitally labeling a cell or small group of cells in an embryo at a particular stage of development. The embryo is then allowed to develop to a later stage when differentiation of a variety of tissues has taken place and assessed to determine which of these tissues received contributions from the labeled cell population. By labeling many embryos, at many different locations, it is possible to generate an overlapping "map" of how each region of the early embryo contributes to different later tissues and structures.

General fate maps have been prepared for blastoderm stages of *Xenopus* (Dale and Slack, 1987; Moody, 1987b) and zebrafish (Kimmel *et al.,* 1990). The intermixing of cells during amphibian embryogenesis and gastrulation movements is less pronounced than that in zebrafish embryos, making for more consistent, clearly delineated, and less overlapping fate maps, at least at early stages (Wetts and Fraser, 1989; Wilson *et al.,* 1995). Although general fate maps have been constructed for the *Xenopus* and zebrafish, more specific fate maps have also been prepared for particular tissues or organs, e.g., the gut, vasculature, and neural plate in *Xenopus* (Chalmers and Slack, 2000; Keller *et al.,* 1992; Mills *et al.,* 1999) or the nervous system and embryonic shield (organizer) in zebrafish (Woo, 1995; Melby *et al.,* 1996). Specific fate maps have not yet been prepared for pronephric tissue in either amphibians or fish, although pronephric domains have been mapped in relation to other tissues as part of the general fate maps described earlier (Dale and Slack, 1987). More detailed maps of prepronephric domains would be useful for addressing a variety of questions. Are "subdomains" of the pronephros set aside at earlier stages of development? What tissue progenitors are adjacent to pronephric-fated domains at earlier stages, which could potentially be providing inductive signals for pronephric development? Do nephric and non-nephric tissues share common progenitors at various stages of development, and how is this important for the specification of the nephric lineage?

Although the latter question can be addressed at a crude level by fate mapping, a technique called lineage tracing can give a much higher resolution picture of the relationships between different tissues (see Warga and Kimmel, 1990). As for fate mapping, a single cell or small group of cells is fluorescently lineage labeled at a defined stage of development. However, rather than simply examining the disposition of their progeny at a single time point, time-lapse videomicroscopy is used to track the cells continuously as they develop, migrate, and undergo morphogenesis into differentiated tissues. This method allows determination of the precise lineage relationships between the descendents of the labeled cells and when and where they are specified to different tissue types. The technique is most applicable to the optically clear zebrafish embryo, where single cells and their progenitors can be imaged noninvasively even deep within an embryo. Time-lapse methods can also be used to study migration pathways and morphogenetic movements and the changing spatial relationships between different tissues as development proceeds. The knowledge of where, when, and in proximity to what other tissues lineage decisions occur can provide valuable insight into possible sources of inductive signals. Once fate maps and/or cell lineage relationships are completed, the significance of these can be examined experimentally using some of the methods that are described immediately here or molecularly and genetically using methods described in other sections of this chapter. Maps representing our present knowledge of pronephric fate are available in Chapter 3.

D. Microsurgical Manipulation and Explant Culture

Explant culture and microsurgical manipulations have been used extensively on amphibian embryos to analyze the specification and determination of cells to different differentiated fates and the inductive interactions that guide these fate decisions. The cells of amphibian embryos are generally large, yolky, and robust, and different tissues are relatively easily isolated intact. Explant culture involves removing groups of cells from an embryo and culturing them in a media either free from inductive factors or containing defined exogenously added factors. Culturing explants in the absence of inductive factors tests lineage commitment or specification, whereas tissue transplantation and explant coculture experiments assay inductive interactions between different tissues and tests their state of determination. Although the explant and transplant assays can provide similar results, the transplant assay is a more stringent test of commitment to a given cell fate, as transplanted tissues can receive additional inductive signals in their new surroundings that may alter the cell fate they would otherwise adopt.

Explants have been used in *Xenopus* to determine the stages at which pronephric tissue is specified. Presumptive pronephric mesoderm was removed from donor embryos at specific stages and was placed between two stage nine animal caps, forming an ectodermal wrap (Brennan *et al.,* 1998, 1999). Ectodermal wraps are then cultured to later stages of development (stages 33–38) and can be tested by either RT-PCR or immunostaining to assay for the presence of pronephric tissue. Using this technology, the stage at which glomus, tubule, and duct specification occur was determined. Explant culture is also feasible in zebrafish (Grinblat *et al.,* 1999). Although pronephric explant studies have not been performed in zebrafish, shield (the zebrafish dorsal organizer) explants have been used to study neural induction, and fore-

brain determination was also studied using explant techniques (Grinblat *et al.,* 1998; Sagerstrom *et al.,* 1996).

E. Cell Transplantation and Mosaic Analysis

Genetic mosaic analysis has been used as a tool for understanding gene function in a variety of species, including *Drosophila,* nematodes, mice, and zebrafish. Genetic mosaics are embryos with a mixture of cells of different genotypes. Most frequently, these consist of "wild-type" cells and cells with mutations in a particular gene or genetic locus. These mosaics can be used to test the "cell autonomy" of a mutated gene, i.e., whether the gene is required within cells affected phenotypically by the mutation (cell autonomous) or exogenously to those cells (cell nonautonomous). This can be extremely valuable information for assessing the functional role of a mutated gene and for determining when and in what tissue this function is accomplished. As a general rule, mutations in genes that act as receptors, intracellular signal transducers, transcriptional regulators, or have other intracellular roles will often cause cell autonomous defects within those cells. In contrast, mutations in genes encoding extracellular signaling molecules will often cause cell nonautonomous defects in other cells that are the signaling targets. Transplantation mosaics could be used to examine the function of genes necessary for pronephric development. What tissue do they function in? When are they required? A variety of methods are used to generate mosaics. In *Drosophila,* mosaics are produced by using ionizing radiation or, most recently, site-specific recombination (Xu and Rubin, 1993) to induce clones of cells with a deletion that uncovers a heterozygous mutation of interest. In mice, chimeras are made by aggregating cells from two different embryos (Gardner and Davies, 2000).

In zebrafish, cell transplantation is used to introduce genetically distinct donor cells into host embryos, generally during blastula or gastrula stages. This method has been described in detail elsewhere (Moens and Fritz, 1999). Briefly, one to eight cell donor embryos are lineage labeled, generally using dextrans conjugated with fluorescent dyes such as rhodamine. For histological detection and/or detection after *in situ* hybridization procedures, donor embryos can also be labeled with biotin-conjugated dextran. Additionally, donor embryos can be coinjected with RNA, DNA, or proteins if it is desired to test their affects in mosaic embryos. Host embryos are not labeled. Donor and host embryos are allowed to continue their development until they have reached an appropriate stage for transplantation. Cells are then removed from a labeled donor embryo using a sharpened, beveled micropipette and are transferred into one or more unlabeled host embryos. Host and donor embryos are both retained and allowed to develop until they have reached a predetermined stage for assaying the consequences of the transplant by microscopic observation and/or histochemical or *in situ* hybridization methods.

F. Confocal Microangiography

Confocal microangiography is a method that has not yet found wide use, but which is very useful for studying the vascular system and the vascularization of internal organs (such as the pronephros) in the zebrafish (Fig. 9.3). The small size and high degree of optical clarity make it possible to image through the entire depth of a living embryo. Confocal microangiography has a number of advantages over "traditional" methods used to delineate the vasculature, such as injection of India ink, colored dyes, or plastic resins. The technique does not kill or even significantly harm the specimens, and images are collected from living animals with an active, fully inflated circulation. The resolution of the method is limited only by the optical resolution of the confocal microscope used. Vessels that might be obscured by surrounding vessels when using alternative "opaque" methods are imaged readily using confocal microangiography. Furthermore, images collected using confocal angiography can be rendered directly and immediately into three-dimensional representations. The technique has already been used to derive a complete, 3D atlas of the vascular anatomy of the zebrafish embryo and early larva, the first available for any vertebrate species (Isogai *et al.,* submitted for publication). Kidney morphogenesis is intimately intertwined with the formation and differentiation of the vasculature that

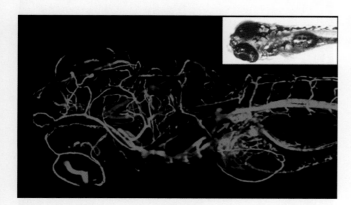

Figure 9.3 Visualization of the detailed anatomy of the vasculature using confocal microangiography. Confocal microangiography (Weinstein *et al.,* 1995). is performed by injecting a fluorescent dye into the circulation of a zebrafish embryo or larva and then collecting optical sections through the animal using a confocal microscope. This method is uniquely well suited to the optically clear embryo or larva of the zebrafish. The image shown is a confocal microangioram of the anterior half of a 6-day-old zebrafish embryo. Dorsal–lateral view, anterior to the left. (Inset) A comparable view of a 6-day-old embryo photographed with incident light. In the microangiogram, vascularization of the pronephric glomus (arrow) is readily apparent.

supplies this organ, making examination of kidney vascularization an important research area (Majumdar and Drummond, 1999).

G. Additional Methods

In addition to those described earlier, a large variety of additional methods are available for studying living animals. Cell borders, particular intracellular structures or organelles, and various specialized cell types can all be labeled specifically *in vivo* using a variety of different fluorescent probes (many of which are available commercially through Molecular Probes, Inc., Eugene, OR). These labeled cells or structures can then be examined via time-lapse videomicroscopy on either conventional or confocal microscopes to follow their behavior in real time in living embryos. Transgenic embryos that endogenously express fluorescent molecules such as GFP, YFP, BFP, or RFP (see later) provide a powerful new tool for marking and studying specific cell populations *in vivo*. Cells and tissues labeled by any of the methods discussed can be isolated by fluorescence-activated cell sorting (FACS) to isolate specific cell types or progenitor cells for molecular analysis or culture *in vitro*.

IV. Transgenic Methods

The ability to reintroduce a gene back into the embryo and control the spatial and temporal expression of the transgene has been one of the major recent advances in developmental biology (Rubin and Spradling, 1982; Spradling and Rubin, 1982). Both zebrafish and *Xenopus* are amenable to transgenesis and a variety of techniques can be applied (Halloran *et al.,* 2000; Higashijima *et al.,* 1997; Huang *et al.,* 1999; Ivics *et al.,* 1993, 1997; Kroll and Amaya, 1996; Long *et al.,* 1997; Meng *et al.,* 1999; Offield *et al.,* 2000; Raz *et al.,* 1998; Stuart *et al.,* 1988). A great advantage for kidney development is that genes affecting later developmental events can be studied without fear of influencing early developmental events. In addition, as addressed in Section V, transgenesis can be used as a form of mutagen.

A. Transgenesis in *Xenopus*

Transgenic lines have been established in both *Xenopus laevis* and *Xenopus tropicalis* (Kroll and Amaya, 1996; Marsh-Armstrong *et al.,* 1999). *X. tropicalis* offers some advantages over *X. laevis* for genetic analyses because it is diploid rather than tetraploid and because its smaller size, shorter generation time, and larger number of progeny make it more amenable to large-scale analysis (Amaya and Kroll, 1999). A technique developed by Kroll and Amaya (1996) allows stable germline integrants to be established in the F1 generation. Briefly, decondensed sperm nuclei are digested lightly with an infrequently cutting restriction endonuclease, such as *NotI*, and are then incubated with transgene DNA cut with the same enzyme. Sperm nuclei containing incorporated transgene DNA are then injected into oocytes. Because integration of transgene DNA occurs at the level of the sperm DNA rather than in an embryo, every cell of the resulting animal contains the transgene, and germline transmission to the next generation is guaranteed. Thus, transgenic lines can be established in a relatively short period of time and with minimal effort. Furthermore, the original fertilized transgenic embryos can be used directly for functional assays (Huang and Brown, 2000a,b; Huang *et al.,* 1999; Kroll and Amaya, 1996; Offield *et al.,* 2000).

B. Transgenesis in Zebrafish

Establishing transgenic lines in zebrafish has been accomplished by a variety of methods. The simplest is directly injecting linear DNA into early zebrafish embryos (Bayer and Campos-Ortega, 1992; Culp *et al.,* 1991; Lin *et al.,* 1994; Stuart *et al.,* 1988, 1990). The germline integration rates with this method are relatively high if the injection occurs within the first 15 min after fertilization (Higashijima *et al.,* 1997). The advantage of this method is the simplicity of preparation. Constructs of various sizes can be made and then simply linearized by restriction enzyme digestion and injected. One disadvantage is that linear DNA fragments seldom integrate as single copies, but instead as large concatamers, although some instances of integration of single or double copies have been noted (Bayer and Campos-Ortega, 1992; Gibbs *et al.,* 1994). Although integration of multiple copies of the exogenous gene could possibly allow for a strong induction, there are frequently negative effects on the region of integration and/or silencing of the genes(s) in these large concatamers via eukaryotic silencing mechanisms (Henikoff, 1998). Many of the problems with linear DNA can potentially be overcome by injecting large linear DNA fragments (such as BAC or PAC clones), which have been reported to integrate with a lower copy number and yield more stable integrants that are less prone to chromatin silencing.

Transgenes can also be integrated into the genome using transposable elements or retroviral vectors. A Tc3 element from *Caenorhabditis elegans* and a reconstituted zebrafish Tc1 element have both been shown to function in zebrafish (Ivics *et al.,* 1997; Raz *et al.,* 1998). The major advantage of transposon-mediated integration is that greater than 99% of integrates are single copies (Ivics *et al.,* 1997, 1999). One possible disadvantage is the limited size of transgenes that can be incorporated into the transposable element, although the upper limits of this size have not yet been fully explored (Raz *et al.,* 1998). Retroviral elements appear to be highly effective for insertional mutagenesis (see later), but their use

as vectors for introducing exogenous genes into the zebrafish germline has not yet been explored. However, there are known to be strong constraints on the size of inserts that can be introduced into retroviruses, which might limit their usefulness as vectors for introducing exogenous genes.

C. Controlling Gene Expression in Transgenic Animals

Although efficient methods are available for introducing exogenous genes into both *Xenopus* and zebrafish embryos and for integrating them into the germline, there is a need for effective methods to control the temporal and spatial expression of the introduced genes. There are a number of methods available for doing this, some of which were already mentioned in the context of transient DNA injections (see Section II.D.) The spatial and temporal expression of a gene can be restricted effectively by the choice of promoter used to drive expression of the gene (Fig. 9.4). A variety of different promoter sequences have been cloned from zebrafish and *Xenopus* that drive tightly spatially and temporally

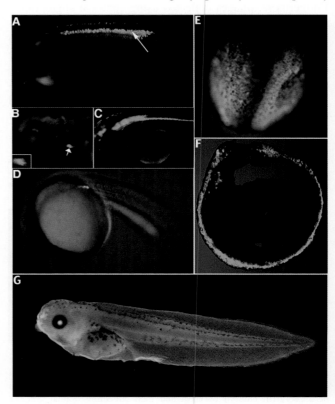

Figure 9.4 Transgenic zebrafish and *Xenopus*. (A–D) Zebrafish transgenic embryos showing GFP expression driven by various promoters. (A) Blood, (B) thymus, (C) neurons, and (D) pancreas. (E–G) *Xenopus tropicalis* embryos displaying GFP expression driven by the *Dlx3* promoter. (E) Stage 20 embryo showing epidermal expression. (F) Cross section of E. (G) Tadpole stage embryo. Zebrafish pictures provided by Shuo Lin, Medical College of Georgia, and *Xenopus* pictures provided by Tom Sargent, National Institutes of Health.

controlled transgene expression in blood cells (Long *et al.,* 1997; Meng *et al.,* 1999), muscle (Higashijima *et al.,* 1997), and the eye (Knox *et al.,* 1998; Offield *et al.,* 2000). At present there are no published pronephros promoters established as transgenic lines. Pronephric specific transgenes will be useful not only for driving spatially restricted expression, but also for genetic screens.

The usefulness of specific promoters can be increased greatly by using them in binary or two-component transgenesis. This method, first developed in *Drosophila melanogaster,* involves constructing two types of transgenic lines (Brand and Perrimon, 1993; Phelps and Brand, 1998). An "activator" line contains a specific promoter driving expression of a specific transcriptional activator (yeast GAL4, which has no endogenous targets in vertebrates, is often employed). A "reporter" line contains the target sequence for the transcriptional activator (such as the GAL4 UAS sequence) fused to a gene of interest. Breeding the activator and reporter lines together results in embryos that contain both transgenic integrants, thus activating expression of the "reporter" gene. Binary transgenesis has a number of advantages over standard transgenesis. The "reporter" or test gene is normally silent in the reporter line, an important feature when expressing genes with toxic or strong developmental effects. Because any activator line can potentially be bred to any reported line, this method also allows any activator promoter to be used to drive the expression of any reporter gene without the need for preparing new constructs or deriving new transgenic animals, exponentially amplifying the usefulness of both. The effectiveness of the GAL4-UAS binary transgenesis system was demonstrated in the zebrafish and shows strong potential for controlling the expression of transgenic target genes (Scheer and Campos-Ortega, 1999).

A variety of inducible promoters are also available for the temporal control of gene expression, including tetracycline-, ecdysone-, and heat shock-inducible promoters. Germline transgenic *Xenopus* or zebrafish have been prepared with gene expression driven by their respective heat shock promoters (Halloran *et al.,* 2000; Krone *et al.,* 1997; Lele *et al.,* 1997; Wheeler *et al.,* 2000). These animals allow fully inducible, ubiquitous expression of genes of interest. Spatial control of heat shock promoter-driven expression in these transgenic animals can also be achieved, using laser induction. This method was first established in *C. elegans* and *D. melanogaster* (Halfon *et al.,* 1997; Harris *et al.,* 1996; Stringham and Candido, 1993), but has been adapted for use in the zebrafish, where the optical clarity of the embryo makes this a very effective technique. Laser induction was used to express Sema3A1, a repulsive semaphorin, specifically in zebrafish motor axons and demonstrate that axon extension is retarded as a result (Halloran *et al.,* 2000). In a similar manner, individual cells or groups of cells thought to be involved in pronephric formation could be laser induced in transgenic animals to express specific gene products.

D. Applications of Transgenesis

What can we do with transgenic animals? One important application is driving the tissue-specific expression of "test" genes. This is a particularly useful feature for studies of the pronephros or other later-developing organs, where it is highly desirable that a gene not be expressed in other tissues or at earlier stages of development. Trangenic animals are also highly useful for "marking" specific tissues endogenously in living embryos. The use of fluorescent proteins such as GFP, BFP, and RFP in transgenic constructs makes it possible to follow, in real time, the intraembryonic cell movements and cell behaviors of specific cells giving rise to particular tissues. These "marked" cells can also be isolated for transplantation, explant, or culture studies. Cells can be FACS sorted and used to prepare tissue-specific libraries or enriched cell populations for culture.

Finally, "marked" transgenic lines are very useful for screening for mutational defects in the marked tissues or monitoring phenotypic changes in mutants. In zebrafish, with the transparent nature of the embryos, mutagenized lines can utilize a promoter expressed in specific tissues, such as the pronephros, and a simple visual screen can be used to monitor for changes in the fluorescent protein expression. There are also a wide variety of transient experiments that can utilize reporter constructs. From monitoring overexpression assays to cell sorting, the only limits are what the experimenter can conjure.

V. Classical Genetic Methods: Mutant Screens

The *sine qua non* of developmental biology has been, and continues to be, the isolation and study of mutations causing specific defects in developmentally important genes. Although perturbation of gene expression using exogenously introduced molecules provides valuable information about the role of a gene in the developing organism, as discussed earlier, overexpression of molecules or synthetic modifications can sometimes give misleading results when trying to address the true role of the gene. While *X. tropicalis* shows promise as a genetically tractable organism, large-scale genetic screens have yet to be performed using this organism. The zebrafish, however, has already demonstrated its usefulness in a number of large-scale genetic screens and is the main focus of this section (Amsterdam *et al.,* 1999; Driever *et al.,* 1996; Haffter *et al.,* 1996).

A number of characteristics of the zebrafish make it a favorable vertebrate to carry out mutation screens. Zebrafish are small, allowing large numbers of animals to be maintained in a small space. They breed frequently, giving large numbers of progeny, and have a generation time of only 3 months. In addition, zebrafish embryos are externally fertilized and optically clear, facilitating the isolation of mutants in genes with essential early roles in both externally and internally apparent developmental processes (Driever *et al.*, 1994; Haffter and Nusslein-Volhard, 1996). Zebrafish mutant screens have been conducted using a number of different mutagenic methods, including γ irradiation, ethylnitrosourea (ENU), psoralen, and retroviral insertion (Amsterdam *et al.*, 1999; Ando and Mishina, 1998; Driever *et al.*, 1996; Fritz *et al.*, 1996; Haffter *et al.*, 1996). Mutations have been recovered from mutagenized animals using a number of different methods for detecting mutants and a number of different screening strategies, some of which rely on zebrafish genetic "tricks." The remainder of this chapter reviews briefly some of these.

A. Mutagens

One of the most important considerations in choosing a mutagen is the types of mutants desired. If point mutations are desired, which can result in not only null mutants but also hyper- and hypomorphic mutants, ENU is the mutagen of choice (Driever *et al.*, 1996; Grunwald and Streisinger, 1992; Haffter *et al.*, 1996). If deletions are desired, either ionizing radiation or trimethylpsoralen (TMP) has been used successfully (Ando and Mishina, 1998; Chakrabarti *et al.*, 1983; Fritz *et al.*, 1996; Walker and Streisinger, 1983). Ionizing radiation such as γ rays and X-rays can also produce point mutations and deletions, ranging in size from a few base pairs to many megabases. In general, genomes disrupted by radiation are often rearranged extensively, which can make it difficult to isolate the affected locus (Fritz *et al.*, 1996). The mutagen TMP was originally used in *Escherichia coli* and *C. elegans* to produce small deletions (Piette and Hearst, 1985; Sladek *et al.*, 1989; Yandell *et al.*, 1994). TMP also efficiently induces mutations in the zebrafish, although the structure of the mutations generated has yet to be examined (Ando and Mishina, 1998).

One disadvantage of both irradiation and chemical mutagens is that it is difficult to isolate the defective gene from a mutant by positional cloning. Major advances have been made in creating and increasing the density of meiotic and radiation hybrid maps in zebrafish, and it is likely that in the future (particularly with completion of the sequencing of the zebrafish genome) finding point mutations will become much easier (Fornzler *et al.*, 1998; Gates *et al.*, 1999; Geisler *et al.*, 1999; Hukriede *et al.*, 1999; Johnson *et al.*, 1995; Kelly *et al.*, 2000; Knapik *et al.*, 1996; Postlethwait *et al.*, 1994; Shimoda *et al.*, 1999). However, at the present time, positional cloning is still an arduous task (Talbot and Schier, 1999). An alternative to using chemical and irradiation techniques is insertional mutagenesis using DNA microinjection, transposable elements, or retroviral integrases (Allende *et al.*, 1996; Amsterdam *et al.*, 1999; Culp *et al.*, 1991; Ivics *et al.*, 1993, 1999; Stuart *et al.*, 1988). To date, the most successful insertional mutagenesis has been accomplished using retroviruses, and a large-scale screen is currently under way using a pseudo-typed retroviral insertional vector (Allende *et al.*, 1996; Amsterdam *et al.*, 1999).

B. Detecting Mutants

Mutants can be detected by a variety of methods. The simplest, most widely used, and, in most cases, the most effective method in the zebrafish is to use direct visual inspection. Because of the optical clarity of the zebrafish embryo, dissecting microscopes can be used to noninvasively observe all internal regions of the embryo and assay for the proper formation of most internal organs. Most of the large-scale screens conducted thus far have relied exclusively or primarily on this method to detect mutants.

Noninvasive visual inspection has been supplemented by a variety of other methods for screening for mutants. Staining with heart chamber-specific antibodies was used successfully in a screen for mutants with defects in heart patterning (Yelon and Stainier, 1999). diI and diO labeling was used to screen for mutants with defective projection of optic neurons (Karlstrom *et al.*, 1996; Trowe *et al.*, 1996). Behavioral screening for the optokinetic reflex in zebrafish embryos was also used to screen for mutants with functionally defective visual systems (Brockerhoff *et al.*, 1995). Screening for aberrant gene expression patterns by *in situ* hybridization has also been used to look for mutants affecting particular developmental pathways (Moens *et al.*, 1996). The ease and reproducibility of *in situ* procedures and optical clarity of the zebrafish embryo make rapid *in situ* screening of large numbers of embryos relatively straightforward. The use of transgenic animals expressing GFP in a tissue-specific fashion is a potentially very powerful method for screening for mutants in particular tissues that is only beginning to be exploited. In all of these screens, ENU was used as the mutagen, although there is no *a priori* reason why other mutagens could not have been employed.

In addition to these "random" screening methods, "pseudo reverse genetic" approaches are also available for screening for defects in particular genes. A PCR-based detection scheme can be used to look for deletions in particular genes of choice (Fritz *et al.*, 1996). Because the absence of a portion of a gene locus is required, the mutagen is usually limited to either ionizing radiation or TMP. The great advantage of mutants isolated by a PCR-based screen is that the affected locus is already known, eliminating the need to map and/or clone it. The major disadvantage is that the mutants are often large deletions or rearrangements, and determining that an observed phenotype is due only to the deletion of a particular gene may require screening for point mutant alleles (see Section V,C). Mutations in predetermined genes can also be isolated by other strategies, such as enzyme assays or transgenic reporter assays, although these other methods have the potential to identify mutants in upstream or interacting genes in addition to those in the targeted gene (often a good thing!).

C. Screening Strategies

Beyond the choice of mutagen and method for mutant detection, a strategy for uncovering mutants must be chosen. A number of different strategies have been used to identify interesting mutants in the zebrafish, including "classical" F3 screening, haploid screening, early pressure ("EP") screening, allele screening, and suppressor and enhancer screening. We review briefly the first three methods mentioned (diagrammed in Fig. 9.5). Suppressor and enhancer screens have not yet been reported in the zebrafish, although they have been used widely and productively in nonvertebrates. Allele screens have been performed in zebrafish, but are not covered in this chapter (van Eeden *et al.,* 1999).

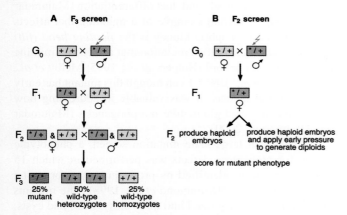

Figure 9.5 Mutational screening strategies employed in the zebrafish. (A) Classical F3 screens. Mutagenized "G0" male zebrafish (filled in purple) are crossed to wild-type females in order to generate F1 families. Individual F1 fish are then crossed either to wild-type fish (shown here) or to other F1 progeny of a different mutagenized G0 (not shown here) to generate F2 families. Intercrossing different F1 fish instead of outcrossing doubles the number of genomes screened per F2 family. Sibling fish from F2 families are intercrossed to produce F3 mutant offspring. A given recessive mutation should be present in a heterozygous, phenotypically "silent" state in one-half of the fish in an F2 family. The other 50% are geno-typically wild type (filled in yellow). Therefore, $1/2 \times 1/2$ or 25% of the F2 sibling intercrosses will produce homozygous mutant embryos that display the mutant phenotype. In these crosses, 25% of the embryos will be mutants (red box) whereas 75% will be phenotypically wild type. (B) Haploid or early pressure F2 screens. F1 families are generated as for classical F3 screens. However, instead of outcrossing or intercrossing F1 fish to generate normal diploid F2 families, F1 females are used to generate either haploid F2 progeny or early pressure ("EP") diploid F2 progeny. Fifty percent of haploid F2 embryos should display a given mutant phenotype, although this number is often substantially lower because the mutation is mosaic in the germline of the F1 animal. Fifty percent of diploid EP F2 embryos should display a given mutant phenotype if the mutation is tightly centromere linked. This frequency drops to zero for mutations unlinked to the centromere because of the meiotic recombination in EP animals. As for haploid F2 progeny, F1 mosaicism can further reduce the frequencies at which mutations are detected.

D. Classical F3 Screen

The classical three-generation breeding scheme for driving mutations to homozygosity (Fig. 9.5A) has generally been proven the most reliable strategy for isolating recessive lethal mutations. Two large-scale mutagenesis screens using this strategy have been performed successfully in zebrafish, producing a large array of mutations in multiple developmental pathways (Driever *et al.,* 1996; Haffter and Nusslein-Volhard, 1996). The advantage of this method of screening is that mutations are detected in "normal" diploid organisms, rather than in the developmentally impaired animals scored in haploid screens (see later). Furthermore, the stable Mendelian inheritance patterns of recessive mutants obtained from F2 families and multiplicity of mutant animals in these families make mutant identification, verification, and subsequent recovery much simpler. One disadvantage of F3 screening is that it is quite labor- and space-intensive. Housing and screening the large number of F2 families generated in a screen with even a reasonably moderate throughput can be beyond the capacity of most small laboratories. F3 screens can be performed on a smaller scale by reducing the number of genomes screened per week, but this will extend the length of time needed to perform the screen considerably.

E. Haploid F2 Screen

One of the genetic tricks available in the zebrafish is the ability to create embryos that contain a haploid rather than diploid genome. Gynogenetic haploids are produced by *in vitro* fertilizing eggs with UV-irradiated sperm, which can still activate the egg but not contribute DNA, whereas androgenetic haploids are produced by using normal sperm to fertilize γ-ray, X-ray, or UV-irradiated eggs (Walker, 1999). The resulting haploid embryos from either procedure look very similar and can be reared for up to about 3 days to screen for mutations that affect early embryonic patterning, segmentation, organogenesis, or other processes. The haploid screening strategy (Fig. 9.5B) has already been used in conjunction with visual inspection, *in situ* hybridization, and PCR detection of mutants (Cheng and Moore, 1997). *Valentino,* a zebrafish mutation that alters krox20 expression in rhombomere 5, was first identified by *in situ* haploid screening (Moens *et al.,* 1996). In addition, haploid screening is the strategy of choice for PCR-based detection of mutants (Fritz *et al.,* 1996). The main advantage of a haploid F2 screen is that an equivalent number of genomes can be screened on a very much smaller scale than an F3 screen. Because only one filial generation is needed, a large number of F2 families do not need to be reared and a much smaller number of fish tanks are needed. A disadvantage is that haploid embryos develop with a variety of abnormalities, such as duplicated otic vesicles, absence of otoliths, and malformed and/or incorrectly projecting Mauthner neurons (Walker, 1999). The usefulness of haploid embryos for screening for defects in a particular developmental process depends on whether the spectrum of haploid defects impinges on this process. A large mutational load in the F1 parent can also pose difficulties in scoring particular phenotypes. In general, defects in earlier developmental events are detected more reliably in haploid embryos. Another potential disadvantage of this screening strategy is that mutants can be more difficult to detect, verify, and recover from F1 animals than from F2 families. Only a single F1 animal will carry a given mutation, and in practice the mutation is frequently mosaic in the germline of this animal so that only a fraction of its progeny (<50%) will carry the mutation. F1 animals are also notoriously unhealthy and can die before out-crossing can be done or fail to breed.

F. Early Pressure F2 Screen

A number of methods are available for converting haploid (1n) zebrafish into diploid (2n) animals by preventing either the second meiotic division or the first zygotic division. These include early pressure (high hydrostatic pressure inhibition of second meiotic division), late pressure (high hydrostatic pressure inhibition of first zygotic division), or heat shock (inhibiting the second meiotic division). Early pressure has been used most often for mutant screens. Early pressure (EP) genetic screens (Fig. 9.5B) are similar to haploid screens in that F2 offspring of F1 generation females are scored. In effect, in EP embryos, the same genome is being screened as is in haploid embryos, only as a diploid rather than a haploid.

EP screens, like haploid screens, offer the advantage of smaller scale F2 screening rather than F3 screening. They also allow the assessment of phenotypes in the background of a "normal" diploid rather than haploid embryo, eliminating the developmental defects that hamper haploid screening. There are, however, a number of problems associated with EP screening. EP adds additional complexity to a screen. Furthermore, unless EP procedures are done carefully, a large number of embryos can remain haploid. A more fundamental problem is that EP does not result in a completely homozygous genome. Because EP results in inhibition of the second meiotic division, the genome has already been subjected to the first round of meiosis and recombination events have occurred (Johnson et al., 1995; Streisinger et al., 1986). Thus, the fraction of embryos homozygous for a mutation in a gene will be reduced in relationship to the distance of that gene from the centromere (Beattie et al., 1999). One way around the loss of homozygosity from EP is to use heat shock (HS) to make homozygous diploid embryos. HS results in the first mitotic division being inhibited, thus the genome remains homozygous not only at proximal regions of the chromosome, but also at the most distal points (Streisinger et al., 1981). While HS would seem to be a better situation for screening homozygous diploids, the major disadvantage is that only 10–20% of the offspring survive the process (Beattie et al., 1999).

G. Pronephric Kidney Mutants

Even though many genes that affect pronephric development are known to play multiple roles during developmental patterning, specific mutations that affect these genes have proven invaluable. An example from zebrafish is the no isthmus (noi) mutation. noi is a mutation in the pax2.1 gene and has developmental defects in axon pathway formation, midbrain–hindbrain boundary, and the visual system in addition to the kidney (Brand et al., 1996; Lun and Brand, 1998; Macdonald et al., 1997). Even though multiple defects are associated with the noi mutation, a study of noi mutants focusing on pronephric-specific defects revealed that the mutation affected tubule and duct differentiation (Majumdar et al., 2000). A second example of a mutation that affects more than the pronephric kidney is the floating head (flh) mutation. flh is a zebrafish mutation that affects the midline and lacks the notochord (Halpern et al., 1995; Melby et al., 1996; Talbot et al., 1995). Even though this mutant has early developmental defects, it was valuable in addressing how the midline affected glomerular morphogenesis (Majumdar and Drummond, 2000).

As part of a large-scale mutation screen, a phenotypic screen for pronephric defects was performed in which 15 mutated loci were identified by pronephric cyst formation and general edema (Drummond et al., 1998). The mutants identified were categorized into three classes. All had cysts, but in addition, two of the groups also had body curvature and one group had eye degeneration. One specific mutation from this screen is the double bubble (dbb) mutation (Fig. 9.6). In the dbb mutation, the glomerulus becomes distended when blood filtration is being established, and it is thought that the primary defect is in the formation of the glomerulus rather than a secondary defect due to the inability to withstand filtration pressure (Drummond et al., 1998). The ability to study cystic mutations, such as dbb in zebrafish, especially early in development, is very important. In human cystic diseases, developmental or cellular defects in epithelia caused by cystic mutations are often obscured by compensatory tissue responses, making it difficult to define the primary cause of cystic disease (Drummond, 2000). Although this screen has provided valuable inroads to pronephric mutant screening, using techniques we have addressed earlier, more advanced screens could be performed. Using a reporter for pronephric development, by in situ hybridization of pronephros-specific genes or GFP expression driven by a pronephros specific promoter, questions pertaining to early stages of pronephric development can be the focus. In this manner, all sorts of screens can be developed, which not only

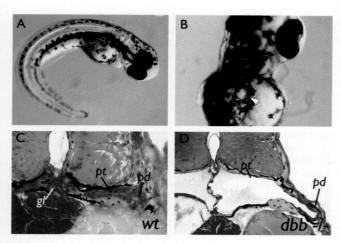

Figure 9.6 Zebrafish pronephros mutant *double bubble (dbb)*. (A) The mutant *dbbm153* emerges from its chorion with a ventrally curved body axis, and within 3–4 h, bilateral pronephric cysts (arrow in B) are evident. (C) Cross section of wild-type pronephros at 3.5 days past fertilization. (D) Section of *dbbm468* showing a grossly distended pronephric cyst in place of the pronephric tubule and a glomerulus reduced to a flattened septum at the midline. *gl*, glomerulus; *pt*, pronephric tubule; *pd*, pronephric duct. Reproduced with permission from Iain Drummond, Massachusettes General Hospital.

focus on general pronephric development, but on specific structures of the pronephros.

References for Protocols

"Early Development of *Xenopus laevis:* A Laboratory Manual" (2000). Eds. Hazel L. Sive, Robert M. Grainger, and Richard M. Harland. Cold Spring Harbor Laboratory Press, Cold Spring Harbor, NY.

"Developmental Biology Protocols," Vol. I (2000). Eds. Rocky S. Tuan and Cecilia W. Lo. (Methods in Molecular Biology, Vol. 135). Humana Press, Totowa, NJ.

"Developmental Biology Protocols," Vol. II (2000). Eds. Rocky S. Tuan and Cecilia W. Lo. (Methods in Molecular Biology, Vol. 136). Humana Press, Totowa, NJ.

"Developmental Biology Protocols," Vol. III (2000). Eds. Rocky S. Tuan and Cecilia W. Lo. (Methods in Molecular Biology, Vol. 137). Humana Press, Totowa, NJ.

"The Zebrafish: Biology" (1999). Eds. H. William Detrich III, Monte Westerfield, and Leonard I. Zon. (Methods in Cell Biology, Vol. 59). Academic Press, San Diego, CA.

"The Zebrafish: Genetics and Genomics" (1999). Eds. H. William Detrich III, Monte Westerfield, and Leonard I. Zon (Methods in Cell Biology, Vol. 60). Academic Press, San Diego, CA.

"Molecular Embryology: Methods and Protocols" (1999). Eds. Paul T. Sharpe and Ivor Mason (Methods in Molecular Biology, Vol. 97). Humana Press, Totowa, NJ.

"Molecular Methods in Developmental Biology: *Xenopus* and Zebrafish" (1999). Ed. Matthew Guille (Methods in Molecular Biology, Vol. 127). Humana Press, Totowa, NJ..

"The Zebrafish Book," Edition 3. (1995). Eds. Monte Westerfield. University of Oregon Press, Eugene, OR.

"Normal Table of *Xenopus laevis* (Daudin): A Systematical and Chronological Survey of the Development from the Fertilized Egg till the End of Metamorphosis (1994). Eds. P. D. Nieuwkoop and J. Farber. Garland Publishing, New York, NY.

"*Xenopus laevis:* Practical Uses in Cell and Molecular Biology" (1991). Eds. Brian K. Kay and H. Benjamin Peng (Methods in Cell Biology, Vol. 36). Academic Press, San Diego, CA.

References

Allende, M. L., Amsterdam, A., Becker, T., Kawakami, K., Gaiano, N., and Hopkins, N. (1996). Insertional mutagenesis in zebrafish identifies two novel genes, pescadillo and dead eye, essential for embryonic development. *Genes Dev* **10**, 3141–3155.

Almouzni, G., and Wolffe, A. P. (1995) Constraints on transcriptional activator function contribute to transcriptional quiescence during early *Xenopus* embryogenesis. *EMBO J.* **14**, 1752–1765.

Amaya, E., and Kroll, K. L. (1999). A method for generating transgenic frog embryos. *Methods Mol. Biol.* **97**, 393–414.

Amaya, E., Musci, T. J., and Kirschner, M. W. (1991). Expression of a dominant negative mutant of the FGF receptor disrupts mesoderm formation in *Xenopus* embryos. *Cell* **66**, 257–270.

Amsterdam, A., Burgess, S., Golling, G., Chen, W., Sun, Z., Townsend, K., Farrington, S., Haldi, M., and Hopkins, N. (1999). A large-scale insertional mutagenesis screen in zebrafish. *Genes Dev.* **13**, 2713–2724.

Ando, H., and Mishina, M. (1998). Efficient mutagenesis of zebrafish by a DNA cross-linking agent. *Neurosci Lett* **244**, 81–84.

Austin, H. B. (1995). DiI analysis of cell migration during mullerian duct regression. *Dev. Biol.* **169**, 29–36.

Bass, B. L. (2000). Double-stranded RNA as a template for gene silencing. *Cell* **101**, 235–238.

Bayer, T. A., and Campos-Ortega, J. A. (1992). A transgene containing lacZ is expressed in primary sensory neurons in zebrafish. *Development* **115**, 421–426.

Beattie, C. E., Raible, D. W., Henion, P. D., and Eisen, J. S. (1999). Early pressure screens. *Methods Cell Biol* **60**, 71–86.

Brand, A. H., and Perrimon, N. (1993). Targeted gene expression as a means of altering cell fates and generating dominant phenotypes. *Development* **118**, 401–415.

Brand, M., Heisenberg, C. P., Jiang, Y. J., Beuchle, D., Lun, K., Furutani-Seiki, M., Granato, M., Haffter, P., Hammerschmidt, M., Kane, D. A., Kelsh, R. N., Mullins, M. C., Odenthal, J., van Eeden, F. J., and Nusslein-Volhard, C. (1996). Mutations in zebrafish genes affecting the formation of the boundary between midbrain and hindbrain. *Development* **123**, 179–190.

Brennan, H. C., Nijjar, S., and Jones, E. A. (1998). The specification of the pronephric tubules and duct in *Xenopus laevis*. *Mech. Dev.* **75**, 127–137.

Brennan, H. C., Nijjar, S., and Jones, E. A. (1999). The specification and growth factor inducibility of the pronephric glomus in *Xenopus laevis*. *Development* **126**, 5847–5856.

Brockerhoff, S. E., Hurley, J. B., Janssen-Bienhold, U., Neuhauss, S. C., Driever, W., and Dowling, J. E. (1995). A behavioral screen for isolating zebrafish mutants with visual system defects. *Proc. Natl. Acad. Sci. USA* **92**, 10545–10549.

Carroll, T., Wallingford, J., Seufert, D., and Vize, P. D. (1999a). Molecular regulation of pronephric development. *Curr. Top. Dev. Biol.* **44**, 67–100.

Carroll, T. J., and Vize, P. D. (1996). Wilms' tumor suppressor gene is involved in the development of disparate kidney forms: Evidence from expression in the *Xenopus* pronephros. *Dev. Dyn.* **206**, 131–138.

Carroll, T. J., and Vize, P. D. (1999). Synergism between Pax-8 and lim-1 in embryonic kidney development. *Dev. Biol.* **214**, 46–59.

Carroll, T. J., Wallingford, J. B., and Vize, P. D. (1999b). Dynamic patterns of gene expression in the developing pronephros of *Xenopus laevis*. *Dev. Genet.* **24**, 199–207.

Chakrabarti, S., Streisinger, G., Singer, F., and Walker, C. (1983). Frequency of gamma-ray induced specific locus and recessive lethal mutations in mature germ cells of the zebrafish, *Brachydanio rerio*. *Genetics* **103**, 109–124.

Chalmers, A. D., and Slack, J. M. (2000). The *Xenopus* tadpole gut: Fate maps and morphogenetic movements. *Development* **127**, 381–392.

Cheng, K. C., and Moore, J. L. (1997). Genetic dissection of vertebrate processes in the zebrafish: A comparison of uniparental and two-generation screens. *Biochem. Cell. Biol.* **75**, 525–533.

Chitnis, A., Henrique, D., Lewis, J., Ish-Horowicz, D., and Kintner, C. (1995). Primary neurogenesis in *Xenopus* embryos regulated by a homologue of the *Drosophila* neurogenic gene Delta. *Nature* **375**, 761–6.

Cho, K. W., Goetz, J., Wright, C. V., Fritz, A., Hardwicke, J., and De Robertis, E. M. (1988). Differential utilization of the same reading frame in a *Xenopus* homeobox gene encodes two related proteins sharing the same DNA-binding specificity. *EMBO J* **7**, 2139–2149.

Conlon, F. L., Sedgwick, S. G., Weston, K. M., and Smith, J. C. (1996). Inhibition of Xbra transcription activation causes defects in mesodermal patterning and reveals autoregulation of Xbra in dorsal mesoderm. *Development* **122**, 2427–2435.

Cooke, J., Smith, J. C., Smith, E. J., and Yaqoob, M. (1987). The organization of mesodermal pattern in *Xenopus laevis:* Experiments using a *Xenopus* mesoderm-inducing factor. *Development* **101**, 893–908.

Cornell, R. A., and Kimelman, D. (1994). Activin-mediated mesoderm induction requires FGF. *Development* **120**, 453–462.

Culp, P., Nusslein-Volhard, C., and Hopkins, N. (1991). High-frequency germ-line transmission of plasmid DNA sequences injected into fertilized zebrafish eggs. *Proc. Natl. Acad. Sci. USA* **88**, 7953–7957.

Dagle, J. M., Littig, J. L., Sutherland, L. B., and Weeks, D. L. (2000). Targeted elimination of zygotic messages in *Xenopus laevis* embryos by modified oligonucleotides possessing terminal cationic linkages. *Nucleic Acids Res.* **28**, 2153–2157.

Dale, L., and Slack, J. M. (1987). Fate map for the 32-cell stage of *Xenopus laevis*. *Development* **99**, 527–551.

Denecke, B., Bartkowski, S., Senkel, S., Klein-Hitpass, L., and Ryffel, G. U. (1993). Chimeric liver transcription factors LFB1 (HNF1) containing the acidic activation domain of VP16 act as positive dominant interfering mutants. *J. Biol. Chem.* **268**, 18076–18082.

Driever, W., Solnica-Krezel, L., Schier, A. F., Neuhauss, S. C., Malicki, J., Stemple, D. L., Stainier, D. Y., Zwartkruis, F., Abdelilah, S., Rangini, Z., Belak, J., and Boggs, C. (1996). A genetic screen for mutations affecting embryogenesis in zebrafish. *Development* **123**, 37–46.

Driever, W., Stemple, D., Schier, A., and Solnica-Krezel, L. (1994). Zebrafish: Genetic tools for studying vertebrate development. *Trends Genet.* **10**, 152–159.

Drummond, I. A. (2000). The zebrafish pronephros: A genetic system for studies of kidney development. *Pediatr. Nephrol.* **14**, 428–435.

Drummond, I. A., Majumdar, A., Hentschel, H., Elger, M., Solnica-Krezel, L., Schier, A. F., Neuhauss, S. C., Stemple, D. L., Zwartkruis, F., Rangini, Z., Driever, W., and Fishman, M. C. (1998). Early development of the zebrafish pronephros and analysis of mutations affecting pronephric function. *Development* **125**, 4655–4667.

Etkin, L. D. (1985). Regulation of the mid-blastula transition in amphibians. *Dev. Biol.* **5**, 209–225.

Fire, A., Xu, S., Montgomery, M. K., Kostas, S. A., Driver, S. E., and Mello, C. C. (1998). Potent and specific genetic interference by double-stranded RNA in *Caenorhabditis elegans*. *Nature* **391**, 806–811.

Fornzler, D., Her, H., Knapik, E. W., Clark, M., Lehrach, H., Postlethwait, J. H., Zon, L. I., and Beier, D. R. (1998). Gene mapping in zebrafish using single-strand conformation polymorphism analysis. *Genomics* **51**, 216–222.

Fritz, A., Rozowski, M., Walker, C., and Westerfield, M. (1996). Identification of selected gamma-ray induced deficiencies in zebrafish using multiplex polymerase chain reaction. *Genetics* **144**, 1735–1745.

Gammill, L. S., and Sive, H. (1997). Identification of otx2 target genes and restrictions in ectodermal competence during *Xenopus* cement gland formation. *Development* **124**, 471–481.

Gardner, R. L., and Davies, T. J. (2000). Mouse chimeras and the analysis of development. *Methods Mol. Biol.* **135**, 397–424.

Gates, M. A., Kim, L., Egan, E. S., Cardozo, T., Sirotkin, H. I., Dougan, S. T., Lashkari, D., Abagyan, R., Schier, A. F., and Talbot, W. S. (1999). A genetic linkage map for zebrafish: Comparative analysis and localization of genes and expressed sequences. *Genome Res.* **9**, 334–347.

Geisler, R., Rauch, G. J., Baier, H., van Bebber, F., Brobeta, L., Dekens, M. P., Finger, K., Fricke, C., Gates, M. A., Geiger, H., Geiger-Rudolph, S., Gilmour, D., Glaser, S., Gnugge, L., Habeck, H., Hingst, K., Holley, S., Keenan, J., Kirn, A., Knaut, H., Lashkari, D., Maderspacher, F., Martyn, U., Neuhauss, S., Haffter, P., *et al.* (1999). A radiation hybrid map of the zebrafish genome. *Nature Genet.* **23**, 86–89.

Gibbs, P. D., Gray, A., and Thorgaard, G. (1994). Inheritance of P element and reporter gene sequences in zebrafish. *Mol. Mar. Biol. Biotechnol.* **3**, 317–326.

Graff, J. M., Thies, R. S., Song, J. J., Celeste, A. J., and Melton, D. A. (1994). Studies with a *Xenopus* BMP receptor suggest that ventral mesoderm-inducing signals override dorsal signals *in vivo*. *Cell* **79**, 169–179.

Green, J. B., Howes, G., Symes, K., Cooke, J., and Smith, J. C. (1990). The biological effects of XTC-MIF: Quantitative comparison with *Xenopus* bFGF. *Development* **108**, 173–183.

Green, J. B., New, H. V., and Smith, J. C. (1992). Responses of embryonic *Xenopus* cells to activin and FGF are separated by multiple dose thresholds and correspond to distinct axes of the mesoderm. *Cell* **71**, 731–739.

Grinblat, Y., Gamse, J., Patel, M., and Sive, H. (1998). Determination of the zebrafish forebrain: Induction and patterning. *Development* **125**, 4403–4416.

Grinblat, Y., Lane, M. E., Sagerstrom, C., and Sive, H. (1999). Analysis of zebrafish development using explant culture assays. *Methods. Cell. Biol.* **59**, 127–156.

Grunwald, D. J., and Streisinger, G. (1992). Induction of recessive lethal and specific locus mutations in the zebrafish with ethyl nitrosourea. *Genet. Res.* **59**, 103–116.

Haffter, P., Granato, M., Brand, M., Mullins, M. C., Hammerschmidt, M., Kane, D. A., Odenthal, J., van Eeden, F. J., Jiang, Y. J., Heisenberg, C. P., Kelsh, R. N., Furutani-Seiki, M., Vogelsang, E., Beuchle, D., Schach, U., Fabian, C., and Nusslein-Volhard, C. (1996). The identification of genes with unique and essential functions in the development of the zebrafish, *Danio rerio*. *Development* **123**, 1–36.

Haffter, P., and Nusslein-Volhard, C. (1996). Large scale genetics in a small vertebrate, the zebrafish. *Int. J. Dev. Biol.* **40**, 221–227.

Halfon, M. S., Kose, H., Chiba, A., and Keshishian, H. (1997). Targeted gene expression without a tissue-specific promoter: Creating mosaic embryos using laser-induced single-cell heat shock. *Proc. Natl. Acad. Sci. USA* **94**, 6255–6260.

Halloran, M. C., Sato-Maeda, M., Warren, J. T., Su, F., Lele, Z., Krone, P. H., Kuwada, J. Y., and Shoji, W. (2000). Laser-induced gene expression in specific cells of transgenic zebrafish. *Development* **127**, 1953–1960.

Halpern, M. E., Thisse, C., Ho, R. K., Thisse, B., Riggleman, B., Trevarrow, B., Weinberg, E. S., Postlethwait, J. H., and Kimmel, C. B. (1995). Cell-autonomous shift from axial to paraxial mesodermal development in zebrafish floating head mutants. *Development* **121**, 4257–4264.

Harris, J., Honigberg, L., Robinson, N., and Kenyon, C. (1996). Neuronal cell migration in *C. elegans:* Regulation of Hox gene expression and cell position. *Development* **122**, 3117–3131.

He, X., Saint-Jeannet, J. P., Woodgett, J. R., Varmus, H. E., and Dawid, I. B. (1995). Glycogen synthase kinase-3 and dorsoventral patterning in *Xenopus* embryos [published erratum appears in *Nature* **375**(6528), 253(1995)]. *Nature* **374**, 617–622.

Heasman, J., Crawford, A., Goldstone, K., Garner-Hamrick, P., Gumbiner, B., McCrea, P., Kintner, C., Noro, C. Y., and Wylie, C. (1994).

Overexpression of cadherins and underexpression of beta-catenin inhibit dorsal mesoderm induction in early *Xenopus* embryos. *Cell* **79**, 791–803.

Heasman, J., Kofron, M., and Wylie, C. (2000). Beta-catenin signaling activity dissected in the early *Xenopus* embryo: A novel antisense approach. *Dev. Biol.* **222**, 124–134.

Henikoff, S. (1998). Conspiracy of silence among repeated transgenes. *Bioessays* **20**, 532–535.

Higashijima, S., Okamoto, H., Ueno, N., Hotta, Y., and Eguchi, G. (1997). High-frequency generation of transgenic zebrafish which reliably express GFP in whole muscles or the whole body by using promoters of zebrafish origin. *Dev. Biol.* **192**, 289–299.

Honig, M. G., and Hume, R. I. (1986). Fluorescent carbocyanine dyes allow living neurons of identified origin to be studied in long-term cultures. *J. Cell Biol.* **103**, 171–187.

Huang, H., and Brown, D. D. (2000a). Overexpression of *Xenopus laevis* growth hormone stimulates growth of tadpoles and frogs. *Proc. Natl. Acad. Sci. USA* **97**, 190–194.

Huang, H., and Brown, D. D. (2000b). Prolactin is not a juvenile hormone in *Xenopus laevis* metamorphosis. *Proc. Natl. Acad. Sci. USA* **97**, 195–199.

Huang, H., Marsh-Armstrong, N., and Brown, D. D. (1999). Metamorphosis is inhibited in transgenic *Xenopus laevis* tadpoles that overexpress type III deiodinase. *Proc. Natl. Acad. Sci. USA* **96**, 962–967.

Hudziak, R. M., Barofsky, E., Barofsky, D. F., Weller, D. L., Huang, S. B., and Weller, D. D. (1996). Resistance of morpholino phosphorodiamidate oligomers to enzymatic degradation. *Antisense Nucleic Acid Drug Dev.* **6**, 267–272.

Hukriede, N. A., Joly, L., Tsang, M., Miles, J., Tellis, P., Epstein, J. A., Barbazuk, W. B., Li, F. N., Paw, B., Postlethwait, J. H., Hudson, T. J., Zon, L. I., McPherson, J. D., Chevrette, M., Dawid, I. B., Johnson, S. L., and Ekker, M. (1999). Radiation hybrid mapping of the zebrafish genome. *Proc. Natl. Acad. Sci. USA* **96**, 9745–9750.

Ivics, Z., Hackett, P. B., Plasterk, R. H., and Izsvak, Z. (1997). Molecular reconstruction of Sleeping Beauty, a Tc1-like transposon from fish, and its transposition in human cells. *Cell* **91**, 501–510.

Ivics, Z., Izsvak, Z., and Hackett, P. B. (1993). Enhanced incorporation of transgenic DNA into zebrafish chromosomes by a retroviral integration protein. *Mol. Mar. Biol. Biotechnol.* **2**, 162–173.

Ivics, Z., Izsvak, Z., and Hackett, P. B. (1999). Genetic applications of transposons and other repetitive elements in zebrafish. *Methods. Cell. Biol.* **60**, 99–131.

Jaynes, J. B., and O'Farrell, P. H. (1991). Active repression of transcription by the engrailed homeodomain protein. *EMBO J.* **10**, 1427–1433.

Johnson, S. L., Africa, D., Horne, S., and Postlethwait, J. H. (1995). Half-tetrad analysis in zebrafish: Mapping the ros mutation and the centromere of linkage group I. *Genetics* **139**, 1727–1735.

Karlstrom, R. O., Trowe, T., Klostermann, S., Baier, H., Brand, M., Crawford, A. D., Grunewald, B., Haffter, P., Hoffmann, H., Meyer, S. U., Muller, B. K., Richter, S., van Eeden, F. J., Nusslein-Volhard, C., and Bonhoeffer, F. (1996). Zebrafish mutations affecting retinotectal axon pathfinding. *Development* **123**, 427–438.

Keller, R., Shih, J., and Sater, A. (1992). The cellular basis of the convergence and extension of the *Xenopus* neural plate. *Dev. Dyn.* **193**, 199–217.

Kelly, P. D., Chu, F., Woods, I. G., Ngo-Hazelett, P., Cardozo, T., Huang, H., Kimm, F., Liao, L., Yan, Y. L., Zhou, Y., Johnson, S. L., Abagyan, R., Schier, A. F., Postlethwait, J. H., and Talbot, W. S. (2000). Genetic linkage mapping of zebrafish genes and ESTs. *Genome Res.* **10**, 558–567.

Kennerdell, J. R., and Carthew, R. W. (1998). Use of dsRNA-mediated genetic interference to demonstrate that frizzled and frizzled 2 act in the wingless pathway. *Cell* **95**, 1017–1026.

Kessler, D. S. (1997). Siamois is required for formation of Spemann's organizer. *Proc. Natl. Acad. Sci. USA* **94**, 13017–13022.

Kimmel, C. B., Warga, R. M., and Schilling, T. F. (1990). Origin and organization of the zebrafish fate map. *Development* **108**, 581–594.

Knapik, E. W., Goodman, A., Atkinson, O. S., Roberts, C. T., Shiozawa, M., Sim, C. U., Weksler-Zangen, S., Trolliet, M. R., Futrell, C., Innes, B. A., Koike, G., McLaughlin, M. G., Pierre, L., Simon, J. S., Vilallonga, E., Roy, M., Chiang, P. W., Fishman, M. C., Driever, W., and Jacob, H. J. (1996). A reference cross DNA panel for zebrafish (*Danio rerio*) anchored with simple sequence length polymorphisms. *Development* **123**, 451–460.

Knox, B. E., Schlueter, C., Sanger, B. M., Green, C. B., and Besharse, J. C. (1998). Transgene expression in *Xenopus* rods. *FEBS Lett.* **423**, 117–121.

Kodjabachian, L., Karavanov, A., Hikasa, H., Hukriede, N. A., Aoki, T., Taira, M., and Dawid, I. B. (2000). A study of Xlim1 function in Spemann's organizer. *Int. J. Dev. Biol.*

Kolm, P. J., and Sive, H. L. (1995). Efficient hormone-inducible protein function in *Xenopus laevis*. *Dev Biol.* **171**, 267–272.

Kozlowski, D. J., Murakami, T., Ho, R. K., and Weinberg, E. S. (1997). Regional cell movement and tissue patterning in the zebrafish embryo revealed by fate mapping with caged fluorescein. *Biochem. Cell Biol.* **75**, 551–562.

Krieg, P. A., and Melton, D. A. (1987). *In vitro* RNA synthesis with SP6 RNA polymerase. *Methods Enzymol.* **155**, 397–415.

Kroll, K. L., and Amaya, E. (1996). Transgenic *Xenopus* embryos from sperm nuclear transplantations reveal FGF signaling requirements during gastrulation. *Development* **122**, 3173–3183.

Krone, P. H., Lele, Z., and Sass, J. B. (1997). Heat shock genes and the heat shock response in zebrafish embryos. *Biochem. Cell Biol.* **75**, 487–497.

LeDouarin, N. M., Renaud, D., Teillet, M.-A., and LeDouarin, G.-H. (1975). Cholinergic differentiation of presumptive adrenergic neuroblasts in interspecific chimeras after heterotopic transplantation. *Proc. Natl. Acad. Sci. USA* **72**, 728–732.

Lele, Z., Engel, S., and Krone, P. H. (1997). hsp47 and hsp70 gene expression is differentially regulated in a stress- and tissue-specific manner in zebrafish embryos. *Dev. Genet.* **21**, 123–133.

Lemaire, P., Darras, S., Caillol, D., and Kodjabachian, L. (1998). A role for the vegetally expressed Xenopus gene Mix.1 in endoderm formation and in the restriction of mesoderm to the marginal zone. *Development* **125**, 2371–2380.

Li, Y. X., Farrell, M. J., Liu, R., Mohanty, N., and Kirby, M. L. (2000). Double-stranded RNA injection produces null phenotypes in zebrafish [published erratum appears in Dev Biol. 220(2),432(2000)]. *Dev. Biol.* **217**, 394–405.

Lin, S., Yang, S., and Hopkins, N. (1994). lacZ expression in germline transgenic zebrafish can be detected in living embryos. *Dev. Biol.* **161**, 77–83.

Long, Q., Meng, A., Wang, H., Jessen, J. R., Farrell, M. J., and Lin, S. (1997). GATA-1 expression pattern can be recapitulated in living transgenic zebrafish using GFP reporter gene. *Development* **124**, 4105–4111.

Lun, K., and Brand, M. (1998). A series of no isthmus (noi) alleles of the zebrafish pax2.1 gene reveals multiple signaling events in development of the midbrain-hindbrain boundary. *Development* **125**, 3049–3062.

Macdonald, R., Scholes, J., Strahle, U., Brennan, C., Holder, N., Brand, M., and Wilson, S. W. (1997). The Pax protein Noi is required for commissural axon pathway formation in the rostral forebrain. *Development* **124**, 2397–2408.

Maeno, M., Ong, R. C., Suzuki, A., Ueno, N., and Kung, H. F. (1994). A truncated bone morphogenetic protein 4 receptor alters the fate of ventral mesoderm to dorsal mesoderm: roles of animal pole tissue in the development of ventral mesoderm. *Proc. Natl. Acad. Sci. USA* **91**, 10260–10264.

Majumdar, A., and Drummond, I. A. (1999). Podocyte differentiation in the absence of endothelial cells as revealed in the zebrafish avascular mutant, cloche. *Dev. Genet.* **24**, 220–229.

Majumdar, A., and Drummond, I. A. (2000). The zebrafish floating head mutant demonstrates podocytes play an important role in directing glomerular differentiation. *Dev. Biol.* **222,** 147–157.

Majumdar, A., Lun, K., Brand, M., and Drummond, I. A. (2000). Zebrafish no isthmus reveals a role for pax2.1 in tubule differentiation and patterning events in the pronephric primordia. *Development* **127,** 2089–2098.

Marsh-Armstrong, N., Huang, H., Berry, D. L., and Brown, D. D. (1999). Germ-line transmission of transgenes in *Xenopus laevis*. *Proc. Natl. Acad. Sci. USA* **96,** 14389–14393.

Mattioni, T., Louvion, J. F., and Picard, D. (1994). Regulation of protein activities by fusion to steroid binding domains. *Methods. Cell Biol.* **43,** 335–352.

Melby, A. E., Warga, R. M., and Kimmel, C. B. (1996). Specification of cell fates at the dorsal margin of the zebrafish gastrula. *Development* **122,** 2225–2237.

Meng, A., Tang, H., Yuan, B., Ong, B. A., Long, Q., and Lin, S. (1999). Positive and negative cis-acting elements are required for hematopoietic expression of zebrafish GATA-1. *Blood* **93,** 500–508.

Mills, K. R., Kruep, D., and Saha, M. S. (1999). Elucidating the origins of the vascular system: A fate map of the vascular endothelial and red blood cell lineages in *Xenopus laevis*. *Dev. Biol.* **209,** 352–368.

Moens, C. B., and Fritz, A. (1999). Techniques in neural development. *Methods Cell Biol.* **59,** 253–272.

Moens, C. B., Yan, Y. L., Appel, B., Force, A. G., and Kimmel, C. B. (1996). valentino: A zebrafish gene required for normal hindbrain segmentation. *Development* **122,** 3981–3990.

Moody, S. A. (1987a). Fates of the blastomeres of the 16-cell stage *Xenopus* embryo. *Dev. Biol.* **119,** 560–578.

Moody, S. A. (1987b). Fates of the blastomeres of the 32-cell-stage *Xenopus* embryo. *Dev. Biol.* **122,** 300–319.

Moriya, N., Uchiyama, H., and Asashima, M. (1993). Induction of pronephric tubules by activin and retinoic acid in presumptive ectoderm of *Xenopus laevis*. *Dev. Growth Differ.* **35,** 123–128.

Muller, M., Affolter, M., Leupin, W., Otting, G., Wuthrich, K., and Gehring, W. J. (1988). Isolation and sequence-specific DNA binding of the Antennapedia homeodomain. *EMBO J.* **7,** 4299–4304.

Nakano, H., Amemiya, S., Shiokawa, K., and Taira, M. (2000). RNA interference for the organizer-specific gene Xlim-1 in *Xenopus* embryos. *Biochem. Biophys. Res. Commun.* **274,** 434–439.

Ngo, H., Tschudi, C., Gull, K., and Ullu, E. (1998). Double-stranded RNA induces mRNA degradation in *Trypanosoma brucei*. *Proc. Natl. Acad. Sci. USA* **95,** 14687–14692.

Oates, A. C., Bruce, A. E., and Ho, R. K. (2000). Too much interference: Injection of double-stranded RNA has nonspecific effects in the zebrafish embryo. *Dev. Biol.* **224,** 20–28.

Offield, M. F., Hirsch, N., and Grainger, R. M. (2000). The development of *Xenopus tropicalis* transgenic lines and their use in studying lens developmental timing in living embryos. *Development* **127,** 1789–1797.

Paddock, S. W. (1994). To boldly glow ... applications of laser scanning confocal microscopy in developmental biology. *Bioessays* **16,** 357–365.

Phelps, C. B., and Brand, A. H. (1998). Ectopic gene expression in *Drosophila* using GAL4 system. *Methods* **14,** 367–379.

Pierce, S. B., and Kimelman, D. (1995). Regulation of Spemann organizer formation by the intracellular kinase Xgsk-3. *Development* **121,** 755–765.

Piette, J., and Hearst, J. (1985). Sites of termination of *in vitro* DNA synthesis on psoralen phototreated single-stranded templates. *Int. J. Radiat. Biol. Relat. Stud. Phys. Chem. Med.* **48,** 381–388.

Postlethwait, J. H., Johnson, S. L., Midson, C. N., Talbot, W. S., Gates, M., Ballinger, E. W., Africa, D., Andrews, R., Carl, T., Eisen, J. S., *et al.* (1994). A genetic linkage map for the zebrafish. *Science* **264,** 699–703.

Raz, E., van Luenen, H. G., Schaerringer, B., Plasterk, R. H. A., and Driever, W. (1998). Transposition of the nematode *Caenorhabditis elegans* Tc3 element in the zebrafish Danio rerio. *Curr. Biol.* **8,** 82–88.

Rubin, G. M., and Spradling, A. C. (1982). Genetic transformation of *Drosophila* with transposable element vectors. *Science* **218,** 348–353.

Sagerstrom, C. G., Grinbalt, Y., and Sive, H. (1996). Anteroposterior patterning in the zebrafish, *Danio rerio:* An explant assay reveals inductive and suppressive cell interactions. *Development* **122,** 1873–1883.

Sanchez Alvarado, A., and Newmark, P. A. (1999). Double-stranded RNA specifically disrupts gene expression during planarian regeneration. *Proc. Natl. Acad. Sci. USA* **96,** 5049–5054.

Sargent, T. D., and Mathers, P. H. (1991). Analysis of class II gene regulation. *Methods. Cell. Biol.* **36,** 347–365.

Sato, A., Asashima, M., Yokota, T., and Nishinakamura, R. (2000). Cloning and expression pattern of a Xenopus pronephros-specific gene, XSMP-30. *Mech. Dev.* **92,** 273–275.

Scheer, N., and Camnos-Ortega, J. A. (1999). Use of the Gal4-UAS technique for targeted gene expression in the zebrafish. *Mech. Dev.* **80,** 153–158.

Scherrer, L. C., Picard, D., Massa, E., Harmon, J. M., Simons, S. S., Jr., Yamamoto, K. R., and Pratt, W. B. (1993). Evidence that the hormone binding domain of steroid receptors confers hormonal control on chimeric proteins by determining their hormone-regulated binding to heat-shock protein 90. *Biochemistry* **32,** 5381–5386.

Seufert, D. W., Brennan, H. C., DeGuire, J., Jones, E. A., and Vize, P. D. (1999). Developmental basis of pronephric defects in *Xenopus* body plan phenotypes. *Dev. Biol.* **215,** 233–242.

Shawlot, W., and Behringer, R. R. (1995). Requirement for Lim1 in head-organizer function. *Nature* **374,** 425–430.

Shimoda, N., Knapik, E. W., Ziniti, J., Sim, C., Yamada, E., Kaplan, S., Jackson, D., de Sauvage, F., Jacob, H., and Fishman, M. C. (1999). Zebrafish genetic map with 2000 microsatellite markers. *Genomics* **58,** 219–232.

Shiokawa, K., Kurashima, R., and Shinga, J. (1994). Temporal control of gene expression from endogenous and exogenously-introduced DNAs in early embryogenesis of *Xenopus laevis*. *Int. J. Dev. Biol.* **38,** 249–255.

Sladek, F. M., Melian, A., and Howard-Flanders, P. (1989). Incision by UvrABC excinuclease is a step in the path to mutagenesis by psoralen crosslinks in *Escherichia coli*. *Proc Natl Acad Sci USA* **86,** 3982–3986.

Spemann, H. (1938). "Embryonic Development and Induction." Yale Univ. Press, New Haven.

Spradling, A. C., and Rubin, G. M. (1982). Transposition of cloned P elements into *Drosophila* germ line chromosomes. *Science* **218,** 341–347.

Streisinger, G., Singer, F., Walker, C., Knauber, D., and Dower, N. (1986). Segregation analyses and gene-centromere distances in zebrafish. *Genetics* **112,** 311–319.

Streisinger, G., Walker, C., Dower, N., Knauber, D., and Singer, F. (1981). Production of clones of homozygous diploid zebra fish *(Brachydanio rerio)*. *Nature* **291,** 293–296.

Stringham, E. G., and Candido, E. P. (1993). Targeted single-cell induction of gene products in *Caenorhabditis elegans:* A new tool for developmental studies. *J. Exp. Zool.* **266,** 227–233.

Stuart, G. W., McMurray, J. V., and Westerfield, M. (1988). Replication, integration and stable germ-line transmission of foreign sequences injected into early zebrafish embryos. *Development* **103,** 403–412.

Stuart, G. W., Vielkind, J. R., McMurray, J. V., and Westerfield, M. (1990). Stable lines of transgenic zebrafish exhibit reproducible patterns of transgene expression. *Development* **109,** 577–584.

Summerton, J., and Weller, D. (1997). Morpholino antisense oligomers: Design, preparation, and properties. *Antisense Nucleic Acid Drug Dev.* **7,** 187–195.

Suzuki, A., Thies, R. S., Yamaji, N., Song, J. J., Wozney, J. M., Murakami, K., and Ueno, N. (1994). A truncated bone morphogenetic protein receptor affects dorsal-ventral patterning in the early *Xenopus* embryo. *Proc. Natl. Acad. Sci. USA* **91,** 10255–10259.

Tada, M., O'Reilly, M. A., and Smith, J. C. (1997). Analysis of competence and of Brachyury autoinduction by use of hormone-inducible Xbra. *Development* **124**, 2225–2234.

Tadrous, P. J. (2000). Methods for imaging the structure and function of living tissues and cells. 3. Confocal microscopy and micro-radiology. *J. Pathol.* **191**, 345–354.

Talbot, W. S., and Schier, A. F. (1999). Positional cloning of mutated zebrafish genes. *Methods Cell Biol.* **60**, 259–286.

Talbot, W. S., Trevarrow, B., Halpern, M. E., Melby, A. E., Farr, G., Postlethwait, J. H., Jowett, T., Kimmel, C. B., and Kimelman, D. (1995). A homeobox gene essential for zebrafish notochord development. *Nature* **378**, 150–157.

Trowe, T., Klostermann, S., Baier, H., Granato, M., Crawford, A. D., Grunewald, B., Hoffmann, H., Karlstrom, R. O., Meyer, S. U., Muller, B., Richter, S., Nusslein-Volhard, C., and Bonhoeffer, F. (1996). Mutations disrupting the ordering and topographic mapping of axons in the retinotectal projection of the zebrafish, *Danio rerio. Development* **123**, 439–450.

Tsai, M. J., and O'Malley, B. W. (1994). Molecular mechanisms of action of steroid/thyroid receptor superfamily members. *Annu. Rev. Biochem.* **63**, 451–486.

Uochi, T., and Asashima, M. (1996). Sequential gene expression during pronephric tubule formation *in vitro* in *Xenopus* ectoderm. *Dev. Growth Differ.* **38**, 625–634.

van Eeden, F. J., Granato, M., Odenthal, J., and Haffter, P. (1999). Developmental mutant screens in the zebrafish. *Methods Cell Biol.* **60**, 21–41.

Vize, P. D., Melton, D. A., Hemmati-Brivanlou, A., and Harland, R. M. (1991). Assays for gene function in developing Xenopus embryos. *Methods Cell Biol.* **36**, 367–387.

Vogt, W. (1929). Gestaltungsanalyse am Amphibienkeim mit ortlicher Vitalfarbung. II. Teil Gastrulation und Mesodermbildung bei Urodelen und Anuren. *Wilhem Roux Arch. Entwicklungsmech. Org.* **120**, 384–706.

Walker, C. (1999). Haploid screens and gamma-ray mutagenesis. *Methods Cell Biol.* **60**, 44–68.

Walker, C., and Streisinger, G. (1983). Induction of mutations by gamma-rays in pregonial germ cells of zebrafish embryos. *Genetics* **103**, 126–136.

Wallingford, J. B., Carroll, T. J., and Vize, P. D. (1998). Precocious

expression of the Wilms' tumor gene xWT1 inhibits embryonic kidney development in *Xenopus laevis. Dev Biol.* **202**, 103–112.

Warga, R., and Kimmel, C. B. (1990). Cell movements during epiboly and gastrulation in zebrafish. *Development* **108**, 569–580.

Wargelius, A., Ellingsen, S., and Fjose, A. (1999). Double-stranded RNA induces specific developmental defects in zebrafish embryos. *Biochem. Biophys. Res. Commun.* **263**, 156–161.

Waterhouse, P. M., Graham, M. W., and Wang, M. B. (1998). Virus resistance and gene silencing in plants can be induced by simultaneous expression of sense and antisense RNA. *Proc. Natl. Acad. Sci. USA* **95**, 13959–13964.

Weinstein, B. M., Stemple, D. L., Driever, W. D., and Fishman, M. C. (1995). *gridlock,* a localized heritable vascular patterning defect in the zebrafish. *Nature Med.* **11**, 1143–1147.

Wetts, R., and Fraser, S. E. (1989). Slow intermixing of cells during *Xenopus* embryogenesis contributes to the consistency of the blastomere fate map. *Development* **105**, 9–15.

Wheeler, G. N., Hamilton, F. S., and Hoppler, S. (2000). Inducible gene expression in transgenic xenopus embryos. *Curr. Biol.* **10**, 849–852.

Wianny, F., and Zernicka-Goetz, M. (2000). Specific interference with gene function by double-stranded RNA in early mouse development. *Nature Cell Biol.* **2**, 70–75.

Wilson, E. T., Cretekos, C. J., and Helde, K. A. (1995). Cell mixing during early epiboly in the zebrafish embryo. *Dev. Genet.* **17**, 6–15.

Wright, C. V., Cho, K. W., Hardwicke, J., Collins, R. H., and De Robertis, E. M. (1989). Interference with function of a homeobox gene in *Xenopus* embryos produces malformations of the anterior spinal cord. *Cell* **59**, 81–93.

Xu, L., Schaffner, W., and Rungger, D. (1993). Transcriptional activation by recombinant GAL4-VP16 in the *Xenopus* oocyte. *Nucleic Acids Res.* **21**, 2775.

Xu, T., and Rubin, G. M. (1993). Analysis of genetic mosaics in developing and adult *Drosophila* tissues. *Development* **117**, 1223–1237.

Yandell, M. D., Edgar, L. G., and Wood, W. B. (1994). Trimethylpsoralen induces small deletion mutations in *Caenorhabditis elegans. Proc. Natl. Acad. Sci. USA* **91**, 1381–1385.

Yelon, D., and Stainier, D. Y. (1999). Patterning during organogenesis: Genetic analysis of cardiac chamber formation. *Semin. Cell Dev. Biol.* **10**, 93–98.

Submitted September 20, 2000

10

The Metanephros

Jonathan Bard

I. Introduction

This section of the book considers the development of the mammalian metanephric kidney, the third and final member of a triad of renal tissues whose development within the intermediate mesoderm begins soon after the formation of the anetrioposterior axis. While the pro- and mesonephroi are transitory in birds and mammals (Chapters 1, 3, 6), the meta-nephros is stable throughout life; indeed, the human kidney can carry out its roles reliably for over a century.

There is, however, a range of congenital and later anatomical abnormalities of the human kidney (reviewed in Chapter 21) whose origin and provenance are difficult to understand except in a developmental context (for review, see Davies and Bard, 1998). One important role for this section of the book is therefore to provide this context, but it is not the only one! As will become apparent, the development of the kidney, perhaps more than any other tissue, requires a wide variety of developmental mechanisms, and the metanephros thus provides a model system for studying them. The following chapters provide detailed reviews summarizing how the kidney develops, our current understanding of the underlying molecular mechanisms, and how

this knowledge has been discovered. As the availability of human congenital abnormalities provides natural probes for studying normal kidney development, there is thus an easy reciprocal relationship between investigating this developmental model and trying to understand human kidney disease.

The main purpose of this chapter on normal development is thus to give sufficient background about kidney formation and growth to enable the reader to appreciate the rest of this section. A second purpose is to provide necessary information for readers who wish to focus on the final section of the book to understand the normal basis of kidney development and so appreciate how congenital and other developmental dysmorphologies can arise.

II. Development of the Metanephros

Metanephros development in all higher vertebrates starts with the demarcation of the metanephric blastema in the caudal part of each intermediate mesoderm. This blastema is a domain of perhaps a few thousand cells, usually known as metanephric mesenchyme (MM), that expresses two key transcription factors, WT1 and Pax-2 (Kreidberg *et al.*, 1993; Torres *et al.*, 1995) and is 1–200 μm from the medial mesonephric (or Wolffian) duct that is coursing toward the cloaca and future bladder (Fig. 10.1). Demarcation takes place at about E10 in the mouse (all subsequent days are for the mouse; rat timings are about a day later than those for mice) and E30 in the human. Cells of this blastema give rise to nephron, stromal, neuronal, endothelial, and smooth muscle cells, as well as the juxtaglomerular complex and perhaps capillaries, although we still do not know how many independent lineages are represented in the blastema (See Chapter 13).

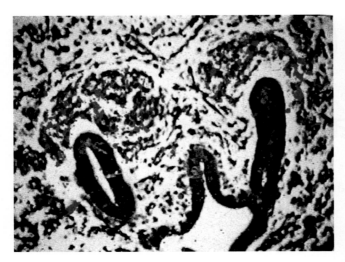

Figure 10.1 *In situ* hybridization of WT1 in a section through an E11 embryo showing the two ureteric buds and the metanephric mesenchyme. In this figure, which has been digitally enhanced, WT1 expression is shown in red.

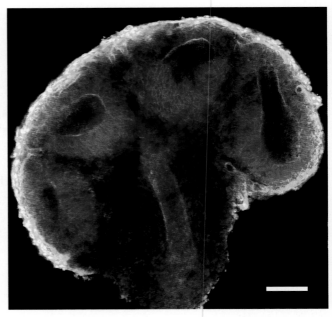

Figure 10.2 An E12 mouse kidney showing a ureteric bud that has branched twice from Bard *et al.* (2001), with permission.

Each metanephric blastema soon secretes glial-derived neural growth factor, (GDNF), and this signal interacts with the adjacent mesonephric (or Wolffian) duct that expresses c-ret, the GDNF receptor, and its co-receptor GFR-1a (Chapter 12). As a result of this interaction, the bud is induced to sprout (E10.5; Sainio *et al.*, 1997; Batourina *et al.*, 2001) and so form the ureteric bud that extends toward and then invades its nearby blastema (E11, Fig. 10.1). The blastema then secretes further signals that cause the bud to undergo successive bifurcations and eventually form the collecting duct system with their principal and intercalated cells, the calyces and the ureter. While the details of this story are still unknown, a recent major achievement has been the demonstration (Sakurai *et al.*, 2001) that an isolated ureteric bud will, when cultured with the pleiotrophin, a heparin sulfate-binding protein, undergo branching. This observation suggests that the essentials of epithelial branching in the developing kidney are autonomous to the ureteric bud (Chapter 12).

As the ureteric bud grows and bifurcates, the growing tips of the collecting ducts reciprocally induce the metanephric mesenchyme by secreting signals that have a range of inductive effects on the MM. These inductive effects include diverting the MM from the apoptotic fate shown by isolated blastemas (see later) and inducing it to undergo the complex series of interactions that lead to the formation of functional nephrons (Chapter 14). While the molecular details responsible for inducing these effects in the mouse metanephros remain unclear, Plisov *et al.* (2001) showed that tubule induction in isolated rat MM can be achieved by the addition of LIF, TGFβ2, and FGF2, with some involvement of Wnt receptors.

Nephrogenesis *in vivo* starts at E11.5–E12 with groups of MM cells that express NCAM, WT1, Pax2, and other proteins aggregating to form dense caps four to five cells deep around the tips of the buds of the ureteric tree (e.g., Klein *et al.*, 1988). These caps can be identified easily in standard sections stained with hematoxylin and eosin, but, as they express a range of antigens that include the adhesion molecule NCAM, they are visualized more dramatically when stained immunofluorescently and examined in a confocal microscope (Fig. 10.2; Bard *et al.*, 2001).

The first sign of pretubular aggregation occurs a few hours later (~E12.75, Fig. 10.3) when small but distinct condensations of six to eight cells can be identified at the proximal end of each cap, abutting the duct epithelium. These small NCAM-expressing mesenchymal aggregates enlarge and undergo a mesenchyme-to-epithelium transition, and the new nephron epithelium fuses rapidly to the duct epithelium, the process being mediated by the presence of Pax-6 (Mah *et al.*, 2000), to generate the distal link between the nephron and the duct. The duct basal lamina is lost in the region where fusion takes place while a new basal lamina is secreted by the primitive nephron; as a result, a continuous basal lamina soon bounds the nephron and the duct (Fig. 10.4).

As the ducts extend, they maintain a dense cap of MM at their tips, although it is not clear whether cap cells get carried forward with the duct or whether new cells are continually being recruited. The tips leave behind the primitive nephrons that still express NCAM (Klein *et al.*, 1988), which become surrounded by loose stroma that has lost NCAM expression. Details of the fates of cap cells that do not form nephrogenic aggregates are also unclear: some may extend toward the

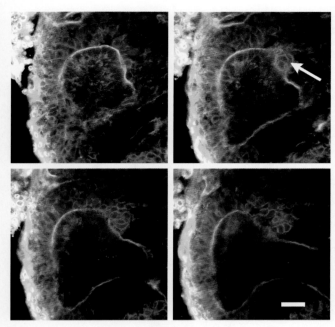

Figure 10.3 An early nephrogenic aggregate in an E12.5 mouse kidney. The specimen has been stained with NCAM (red), laminin (green), and the nuclear stain ToPro-3-iodide (blue) and viewed in a confocal microscope. The images have a vertical separation of 8 μm. From Bard *et al.* (2001), with permission.

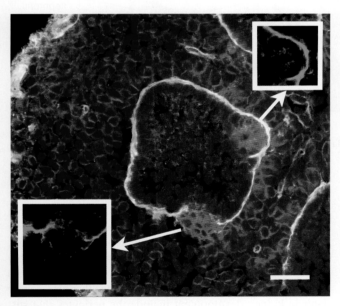

Figure 10.4 A confocal micrograph of a cross section of an E12.75 collecting duct proximal to the tip with two adjacent early nephrogenic aggregates about 10 cells apart. The main picture is stained with NCAM (red), laminin (green), and TO-PRO-3 (nuclei-blue). Cells in the aggregate to the right show clear signs of epithelialization (see insert) as the laminin of the basal lamina between the duct and the aggregate has almost gone, whereas that at the periphery of the aggregate is like that of the duct. In contrast, the other aggregate (left insert) has only a fine basal lamina on its outer surface, whereas the basal lamina between the aggregate and the duct is clearly being degraded. Bar: 25 μm. From Bard *et al.* (2001), with permission.

periphery of the kidney, forming the rind of blastemal cells that is the source of future nephrons, whereas others may become stromal cells that remain in the medullary region (for a dis-cussion of cell lineage, see Chapter 13).

While ducts are still extending toward the cortex of the kidney, the initial nephrogenic aggregates that are now near what will become the future corticomedullary border begin to differentiate. The epithelialized aggregates that have fused to adjacent ducts extend rapidly, forming a comma-shaped morphology. Capillaries invade the cleft of this proximal part of the nephron (which is distal to the connection with the duct) under the influence of angiotrophic factors such as VEGF (Tufro, 2000) and start to form the glomerulus (for a summary, see Fig. 10.5). As the cleft cells close around the capillaries, they form Bowman's capsule with the external part becoming the parietal epithelium and the internalized part (the visceral region) differentiating to form the podocytes (see Chapter 16). These are unusual cells whose phenotype has characteristics of both epithelial and mesenchymal cells; they, together with the endothelial cells of the glomerular tuft, soon lay down the thick glomerular basal membrane. This tuft includes a small vascular system and its associated mesangial cells: the former comprises efferent and afferent arterioles that provide the external link to the renal vascular system and that are connected within the glomerulus by anastomizing capillaries; the latter include pericyte-like mesenchymal cells embedded in stroma (see Chapter 16). The space on the podocyte region of the glomerular membrane is known as the Bowman's region and links directly to the lumen of the future nephron.

As the primitive nephron extends and acquires its S-shaped morphology, other features appear rapidly. Adjacent to the glomerulus and continuous with it is the proximal convoluted tubule, characterized by its high columnar epithelial cells; this leads to the distal tubule, the connecting duct, and so the collecting duct. Lineage studies have shown that in mature kidneys, the small length of tubule linking the distal tubule and the collecting duct itself derives from the connecting duct. The boundary between duct- and nephron-derived cells at the distal end of the nephron is not sharp (Qiao *et al.*, 1995), but whether this reflects cell mingling or an ability of blastema-derived cells to form connecting duct tissue remains unclear. The final major element of the nephron, the loop of Henle, forms rather later from the region connecting the proximal and distal parts of the tubule (see Chapter 11). By E15 or so, however, the essential structure of the early nephrons is in place, and there is the first evidence of renal function (see Chapter 18); this is manifested by the appearance of urine in the amniotic fluid, which is swallowed continually and then excreted by the embryo (Smith *et al.*, 1966). Nevertheless, nephron maturation is not a rapid process in the mouse and it seems to take well over a week for the fully developed morphology of the nephron tubule to form and longer for the full functional repertoire to develop (Chapter 14). There is thus a gradient of nephrons within the developing kidney

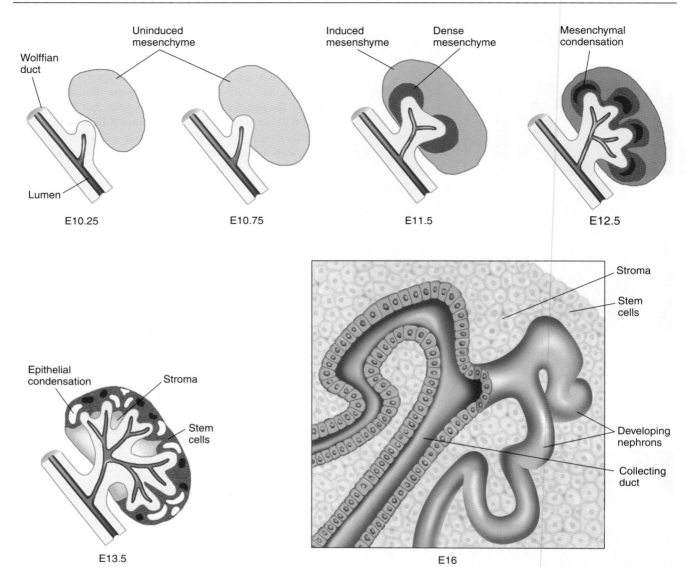

Figure 10.5 Key events in kidney morphogenesis (after Davies and Bard, 1998). Growth and apoptosis within the developing kidney are shown.

cortex from mature nephrons near the corticomedullary border to early aggregates at the periphery.

The collecting ducts and nephrons are initially embedded in a matrix of stromal cells, one of the descendants of the original metanephric blastema. Early stromal cells are characterized by the expression of *FoxB* (known previously as *BF2*), a transcription factor responsible for at least one of the factors that regulate nephrogenesis (Hatini *et al.,* 1996). There is also a second population of *FoxB*-expressing stromal cells near the periphery of the cortex (Hatini *et al.,* 1996), and these may become the renin-producing cells of the juxtaglomerular complex, a stromal derivative (see Chapter 13). Later, some of the stromal cells in the medulla start to express neuronal markers and will form the nerve supply to nephrons (Sainio

et al., 1994). Although these cells probably derive from stromal cells that do not undergo apoptosis (see later), their lineage remains unclear. As birth approaches in the mouse, most stromal cells in the medullary region are lost, probably through apoptosis, and the space that they occupied becomes filled with loops of Henle. Other cell types that also appear later in kidney development include the pericytes that surround capillaries and other smooth muscle cells and the mesangial cells within the glomerulus, most of which develop from stroma, although a subpopulation may derive from blood (Ito *et al.,* 2001). It is not until some weeks after birth that the mature mouse kidney, with its medulla packed with collecting ducts and nephric tubules, is observed; by then, virtually all stromal and blastemal cells have been lost (see later).

One of the more interesting features of kidney differentiation, and one of the reasons why it is such a good model system, is that so much of its development will take place *in vitro* where it can be manipulated in a variety of ways. If the early metanephros is cultured on a substratum at a medium–air interface, nephrogenesis will proceed to a quite considerable extent: a collecting duct arborization will develop with up to 50 nephrons fused to it, some of which will have capillary-free glomeruli and well-developed proximal convoluted tubules. It is not easy to estimate just how much development will take place in culture, but after 4 or 5 days, an initially E11.5 rudiment has many of the features of a flattened E14 kidney. Moreover, the ureteric bud and metanephric mesenchyme can be separated and recombined, or cultured with other tissues, and subjected to the whole gamut of molecular technology. To make such work easier and to standardize assays, Chapter 19 details the key experimental techniques for culturing kidney rudiments and for analyzing their molecular phenotype.

III. Growth

The development of the kidney requires rather more than just reciprocal inductive interactions between duct tips and MM. This is partly because of the sophisticated integration of the separate components of the kidney that takes place as the organ develops and partly because the structure has to allow nephrons to become functional as soon as possible *in utero* while, at the same time, enabling new ones continue to form and develop until, in the case of the mouse, well after birth. The key to understanding these relationships is in appreciating the importance of growth and apoptosis in kidney development, an importance emphasized by the fact that the most common renal phenotype in transgenic mice is growth impairment, which of course inhibits most of normal renal differentiation (for a review of transgenic phenotypes, see Davies and Brandli, 2001).

The way that growth is integrated with development is both simple and elegant, and a second look at kidney development, this time from a growth viewpoint, explains what happens (apoptosis is considered later). Once the ureteric bud has invaded the blastema and adjacent MM cells have condensed around the tip, the bud grows and bifurcates. As each tip extends, it remains surrounded by a cap of dense cells, but leaves behind small condensations, the future nephrons, that continue to express NCAM. Once the tip reaches the periphery of the kidney, the condensed cells spread out (Fig. 10.6) and form a population of blastemal cells that, after about E14, is the source of all future nephrons. These blastemal cells can be recognized both by their peripheral location, and by their large nucleus and small cytoplasm; their rapid growth rate ensures that most kidney enlargement is at the periphery. Whether these blastemal cells should be consi-

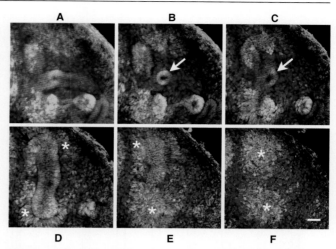

Figure 10.6 A stack of confocal images (separation about 10 μm) showing how the cap cells that express Pax2+ cells (green; the counterstain is propidium iodide, which stains nuclei) reach the kidney periphery as duct tips extend toward its surface. (A) This section (the deepest) shows a duct extending toward the medial surface of the kidney; the basal part of a small tubule is apparent. (B and C) The small tubule (white arrow) connecting the duct (shown in A) to a more superficial T-shaped duct. (D–E) The two domains of Pax2+ MM cells surround the tips of this duct, while an early Pax2+ epithelializing aggregate is visible in D (black arrowhead). (F) The surface of the kidney and two patches (white asterisks) of Pax2+ cells overlying the caps surrounding the duct tips. Bar: 50 μm. From Bard *et al.* (2001), with permission.

dered as *stem cells* or simply as *nephron-precursor cells* depends on whether they can give rise to more than one type of differentiated cell or just nephrons, a controversy that is discussed later (see Chapter 13).

As it enlarges, the position of the kidney changes: the small initial blastema is located within the intermediate mesoderm at the level of the hindlimb buds, but, by about E13 or so, it is cranial to the hindlimbs and protrudes into the coelomic cavity where it is big enough to be plucked from its surrounding tissues with sharpened forceps. By this stage, it is surrounded by coelomic mesothelium, which carries the protein phosphatase inhibitor 1 marker (McLaren *et al.*, 2000) and is the source of the kidney capsule. Differential growth of the embryo, together with other mechanisms that are still unclear, causes the relative position of the kidneys to become more cranial, although the right kidney may not rise as far as the left, perhaps because its ascent is blocked by the liver. This is other evidence that development of the two kidneys can be different: there can be notable differences in the blood supplies to the left and right kidneys. The most intriguing difference, however, comes from the behavior of the the *inv* mutant that reverses left–right asymmetry in the mouse: one aspect of the phenotype is cyst formation in the kidneys (Mochizuki *et al.*, 1998).

Growth of the mouse kidney is rapid: measurements show that it more than quadruples in volume every day over the period E12–E16 with growth only slowing a little between E16 and birth (Fig. 10.7). The way that each compartment of

Log growth of the developing mouse kidney

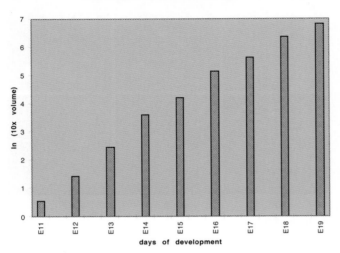

Figure 10.7 Growth of the kidney: a log plot showing that growth is exponential between E11 and about E16, then slows a little. From Davies and Bard (1998).

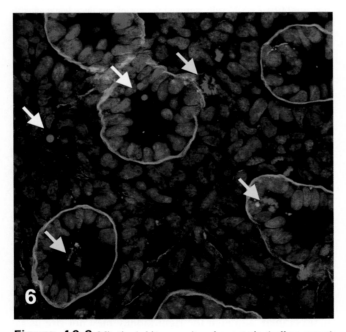

Figure 10.8 Mitotic (white arrow) and apoptotic (yellow arrow) nuclei (red: propidium iodide) in the E16.5 mouse kidney. The plane of focus of this micrograph is just below the blast cells and shows early nephron and collecting ducts bounded by lamin (green) surrounded by stromal cells. This micrograph is atypical in that it has abnormally large numbers of mitotic and apoptotic nuclei. Courtesy of J. Foleym P Sekaran and A. West. From Bard (2001), with permission.

the kidney contributes to this rapid growth is not yet known, but mitosis studies show that the peripheral and duct cells both grow rapidly (Fig. 10.8). The final size of the adult kidney is variable (the number of nephrons in a normal human kidney ranges from 600,000 to 1,000,000, whereas the mouse has

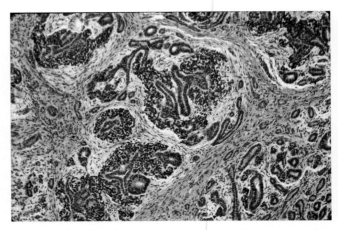

Figure 10.9 A section of a human Wilms' tumor, kindly provided by Dr. B. Beckwith. This example shows blastemal, stromal, and tubular elements mimicking embryogenesis.

about 1600 glomeruli at birth and about 11,000 when mature; Yuan *et al.,* 2001). In addition, the general nutritional condition of the embryo is known to have a major impact in the final renal mass (Merlet-Benichou, 1999).

The kidneys continue to grow until adulthood, but the interactions of growth and development that are marked by the appearance of new nephrons cease when the blastemal cells are lost. This happens about 2 weeks after birth in mice but about 6 weeks before birth in humans, although small clusters of blastemal cells known as *rests* are sometimes observed in postnatal kidneys. It is thought that mutations in the cells of these rests are responsible for Wilms' nephroblastomas (Pritchard-Jones, 1999; Chapter 22), which are characterized by uncontrolled proliferation and the presence of stem cells, immature nephrons, and stroma (Fig. 10.9). It is probably due to loss of blastemal cells that kidneys cannot regenerate new nephrons as a response to damage, and if, for example, one kidney is removed, functional compensation is achieved through the hypertrophy and hyperplasia of existing nephrons.

As to the molecular control of kidney growth, surprisingly little is known, even though a considerable number of signals that can act as growth factors are expressed in the developing kidney (Fig. 10.10). Mutations in BMP-7, FGF-7, the IGFs, and PDGF, in particular, have now all been shown to affect the size of the mouse kidney, suggesting that these are key mediators of kidney growth (for review, see Bard, 2001). While human congenital disorders are really the provenance of the next section, it is worth pointing out here that an inactivating mutation in p57 and activating mutations in insulin-like growth factor 2 (IGF2) are found in Beckwith–Wiedemann syndrome (Eggenschwiler *et al.,* 1997; Grandjean *et al.,* 2000), with both genes affecting the pathway that regulates G1 cell cycle progression. p57 does not directly regulate embryo size, but is involved in the cell cycle arrest that precedesthe terminal differentiation of tissues (Caspary *et al.*, 1999). IGF2, however, is a direct regulator of fetal growth (DeChiara *et al.*, 1990). It is not possible here to review all the signaling systems involved in the growth of the kidney, but the interested reader will find more detail in Bard (2001) and in the kidney development database.

Apoptosis modulates the rapid growth of the kidney (Koseki *et al.*, 1992; Coles *et al.*, 1993) with four separate events being distinguishable. First, the metanephric blastema dies if it fails to be induced by the ureteric bud, and targeted mutagenesis experiments demonstrate that expression of the transcription factor, WT1, in the MM is needed both for its growth and for its escape from apoptosis (Kreidberg *et al.*, 1993). Second, the periphery of the early blastema is delimited by an envelope of apoptosis that gives integrity to the early kidney (H. Sariola, personal communication). Third, there is cell death in the induced MM (Fig. 10.8), particularly around the developing nephrons (Koseki *et al.*, 1992), and apoptosis here may help sculpt these complex epithelial tubules (Coles *et al.*, 1993). Finally, there is apoptosis in the stromal cells toward the end of development (Koseki *et al.*, 1992), a process that may facilitate extension of the loops of Henle into the medullary area.

The molecular mechanisms by which kidney cells either activate or prevent apoptosis are poorly known. Bcl-2 is obviously a major inhibitor of apoptosis in pretubular condensates and in kidney tubules, and Bcl-2-deficient mice develop dysplastic kidneys with massive apoptosis (Veis *et al.*, 1993; Sorenson *et al.*, 1995). In addition, whole animal and organ culture experiments have shown that externally applied EGF and BMP-7 protect renal cells from apoptosis (Coles *et al.*, 1993; Dudley *et al.*, 1999).

IV. Investigating Regulatory Networks

Two names associated with the key experimental work that established the nature of the basic inductive interactions underpinning kidney development are Clifford Grobstein and Lauri Saxén: in California, Grobstein (1955) showed that the early mouse kidney would not only develop in culture, but that its two constituent tissues, the metanephric mesenchyme and the ureteric bud, could be separated and their inductive interactions separately investigated. In Helsinki, Saxén and his group investigated the way in which the interactions between these tissues are mediated, and this classic work is reviewed in his monograph (Saxén, 1987); this book marks the beginning of the era of modern molecular approaches to the study of kidney development. During the last decade, as the techniques of molecular genetics have been applied to this organ, the study of the kidney has broadened markedly, as the geographic spread of authors in this book demonstrates.

One key reason for this expanded interest in kidney development is that we now realize that its ontogeny requires a far wider variety of developmental events and regulatory

Table 10.1 Databases and Tools for Analyzing Kidney Development

Website	URL
Kidney Development Database	http://www.ana.ed.ac.uk/anatomy/database/ kidbase/kidhome.html
PubMed	http://www.ncbi.nlm.nih.gov/PubMed/
Mouse gene expression database, GXD	http://www.informatics.jax.org/menus/ expression_menu.shtml
Gene Homology Search Tool—GHOST	http://www.hgmp.mrc.ac.uk/Registered/ Webapp/ghost/
For databases and search tools	http://www.ebi.ac.uk/ (among many other sites)

networks than was anticipated. Investigating these networks in any depth has not, however, turned out to easy. The first stage, discovering those genes expressed as the kidney develops, is relatively straightforward: we now have the raw details of the expression patterns of several hundreds of ligands, receptors, transcription factors and other regulatory and structural genes. Analysis is helped by the ready availability of these data in the *Kidney Development Database* (Davies and Brandli, 1997; see Table 10.1), which also contains data on gene targeting. One measure of the sheer complexity of genetic data comes from the realization that there seems, on the basis of gene expression data, to be some 80 ligand–receptor pairings in the developing kidney (Fig. 10.10). While much is known about some of them (e.g., GDNF from the blastema binds to c-ret/GFR-1α on the nephric duct, stimulating the outgrowth of the ureteric bud), it is still proving difficult to elucidate the significance of most of these pairings.

The second stage of integrating genetic data into a set of coherent and interlocking molecular networks is thus extremely difficult, and we are still at the stage of recognizing the individual pieces of the jigsaw puzzle rather than fitting them together. This is particularly so in the identification of downstream targets of transcription factors whose activation leads to a change in the developmental phenotype. Chapter 20 summarizes and integrates this information and also points to ways in which our knowledge and understanding of these networks may increase in the future.

V. The Unsolved Problems of Kidney Development

The brief description of kidney development and growth given earlier reveals the complexity of metanephros formation and indicates how many distinct developmental events are required to produce a working organ. These include the regulation of lineage descent, the events taking place within

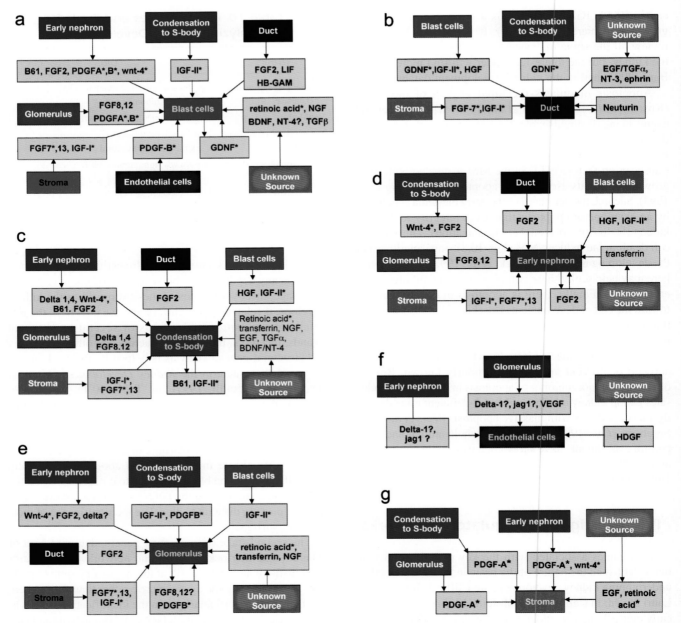

Figure 10.10 Signal–receptor interactions for the various compartments of the kidney identified on the basis of their expression patterns (a) blast cells; b) ducts; c) early stages of nephrogenesis; d) early nephrons; e) glomeruli; f) endothelial cells; g) stroma. An asterisk indicates that a mutation has a renal phenotype (see text). A question mark means that the link is likely but not proven. From Bard (2001), with permission.

the nephrogenic condensation, and the means by which the sophisticated pattern of differentiation and morphogenesis along the nephron is set up, the control of epithelial branching, and the regulation of growth—virtually the complete gamut of developmental phenomena.

It is one thing to identify these problems, but quite another to solve them and to identify the aspects that go awry in congenital kidney disease. There is now a wide range of experimental techniques that can be used to investigate these processes (Chapter 19), but it is worth mentioning here the

impressive bioinformatics resources that are accessible over the internet to expand our knowledge of kidney development (Table 10.1). These resources are not only unusually comprehensive, but also provide the means for keeping abreast of progress and for identifying candidate genes that might be involved in kidney development. The *Kidney Development Database* has already been mentioned, while literature sources such as *PubMe*d can always be searched, although the cataloguing of publications by tissue is not always adequate.

It is most unlikely that we yet know all the important genes whose expression is required for normal kidney development and several further resources (see Table 10.1 for websites) can provide data and tools for analyzing kidney development. One source of candidate genes may be from tissues whose development requires mechanisms similar to that of the kidney (e.g., mesenchyme-to-epithelial transitions and ductal branching) for which information is currently available in the *Mouse Gene Expression Database* (GXD). In addition, mouse or human homologues of genes known to be expressed in the nonhomologous kidneys (e.g., *Drosophila* Malpighian tubules or the zebrafish pronephros) can be identified easily using the *GHOST* interface. Alternatively, the *Sequence Retrieval System (SRS)* and homology recognition algorithms such as *BLAST* or *FASTA* can be used, although they are less user-friendly. In due course, the availability of the *Mouse Atlas Database* will provide graphic tools for linking signals and receptors.

It would, however, be naïve to think that knowledge of gene expression and gene function alone will be enough to provide an adequate understanding of the molecular basis of kidney development. We also need a clear understanding of the tissue and cellular events that define organogenesis, but some of the basic phenomenology still eludes us: we do not know, for example, how many lineages are represented in the metanephric mesenchyme, nor can we be sure that we have identified all of the tissue interactions needed for development to proceed normally. Without such information, it may be difficult to understand the significance of a particular expression pattern. A further likely complication is that there is likely to be far more intertissue molecular cross talk than we initially thought.

The net result is that the state of our understanding of kidney development is, as it has been for almost half a century, in a state of flux. Nevertheless, given our current knowledge and the effort currently being dedicated to this topic, it is hoped that many of the developmental problems raised here and discussed in the following chapters will be solved in the not too distant future. If so, the information gleaned in the academic context of trying to understand the ontogeny of this fascinating tissue will help cure congenital renal abnormalities in humans.

References

Bard, J. B. L. (2001). Growth and death in the developing mammalian kidney: Signal receptors and conversations. *BioEssays.* **24**, 72–82.

Bard, J. B. L., Gordon, G., Sharp, L. and Sellers, W. I. (2001). Early nephron formation in the developing mouse kidney. *J. Anat.* **199**, 385–392.

Batourina, E., Gim, S., Bello, N., Shy, M., Clagett-Dame, M., Srinivas, S., Costantini, F., and Mendelsohn, C. (2001). Vitamin A controls epithelial/mesenchymal interactions through *Ret* expression. *Nature Med.* **27**, 74–78.

Caspary, T., Michele, A., Cleary, M. A., Perlman, E. J., Zhang, P., Elledge, S. J., and Tilghman, S. M. (1999). Oppositely imprinted genes p57Kip2 and Igf2 interact in a mouse model for Beckwith-Wiedemann syndrome. *Genes Dev.* **13**, 3115–3124.

Coles, H. S. R., Burne, J. F., and Raff, M. C. (1993). Large scale normal cell death in the developing rat kidney and its reduction by epidermal growth factor. *Development* **118**, 777–784.

Davies, J. A., and Bard, J. B. L. (1998). The development of the kidney. *Curr. Top. Dev. Biol.* **39**, 245–301.

Davies, J. A., and Brandli, A. (1997). The Kidney Development Database. World Wide Web URL: *http://mbisg2.sbc.man.ac.uk/kidbase/kidhome.html*

Davies, J. A., Lyon, M., Gallagher, J., and Garrod, D. R. (1995). Sulphated proteoglycan is required for collecting duct growth and branching but not nephron formation during kidney development. *Development* **121**, 1507–1517.

DeChiara, T. M., Efstratiadis, A., and Robertson, E. J. (1990). A growth-deficiency phenotype in heterozygous mice carrying an insulin-like growth factor II gene disrupted by targeting. *Nature* **345**, 78–80.

Dudley, A. T., Godin, R. E., and Robertson, E. J. (1999). Interaction between FGF and BMP signaling pathways regulates development of metanephric mesenchyme. *Genes Dev.* **13**, 1601–1613.

Eggenschwiler, J., Ludwig, T., Fisher, P., Leighton, P. A., Tilghman, S. M., and Efstratiadis, A. (1997). Mouse mutant embryos overexpressing IGF-II exhibit phenotypic features of the Beckwith-Wiedemann and Simpson-Golabi-Behmel syndromes. *Genes Dev.* **11**, 3128–3142.

Grandjean, V., Smith, J., Schofield, P. N., and Ferguson-Smith, A. C. (2000). Increased IGF-II protein affects p57kip2 expression *in vivo* and *in vitro*: Implications for Beckwith-Wiedemann syndrome. *Proc. Natl. Acad. Sci. USA* **97**, 5279–5284.

Grobstein, C. (1955). Inductive interactions in the development of the mouse metanephros. *J. Exp. Zool.* **130**, 319–340.

Hatini, V., Huh, S. O., Herzlinger, D., Soares, V. C., and Lai, E. (1996). Essential role of stromal mesenchyme in kidney morphogenesis revealed by targeted disruption of Winged Helix transcription factor BF-2. *Genes Dev.* **10**, 1467–1478.

Ito, T., Akira Suzuki, Enyu Imai, E., Okabe, M., and Hori, M. (2001). Bone marrow is a reservoir of repopulating mesangial cells over the glomerular remodelling. *J. Am. Soc. Nephrol.* **12**, 2625–2635.

Klein, G., Langegger, M., Goridis, C., and Ekblom, P. (1988). Neural cell adhesion molecules during induction and development of the kidney. *Development* **102**, 749–761.

Koseki, C., Herzlinger, D., and Al-Awqati, Q. (1992). Apoptosis in metanephric development. *J. Cell Biol.* **119**, 1327–1333.

Kreidberg, J. A., Sariola, H., Loring, J. M., Maeda, M., Pelletier, J., Housman, D., and Jaenisch, R. (1993). WT-1 is required for early kidney development. *Cell* **74**, 679–691.

Mah, S. P., Saueressig, H., Goulding, M., Kintner, C., and Dressler, G. R. (2000). Kidney development in cadherin-6 mutants: Delayed mesenchyme-to-epithelial conversion and loss of nephrons. *Dev. Biol.* **223**, 38–53.

McLaren, L., Boyle, S., Mason, J. O., and Bard, J. B. L. (2000). Mouse protein phosphatase inhibitor-1: Genetic characterization and significance in mesothelial development. *Mech. Dev.* **96**, 237–241.

Merlet-Benichou, C. (1999). Influence of fetal environment on kidney development. *Int. J. Dev. Biol.* **43**, 453–456.

Mochizuki, T., Saijoh, Y., Tsuchiya, K., Shirayoshi, Y., Takai, S., Taya, C., Yonekawa, H., Yamada, K., Nihei, H., Nakatsuji, N., Overbeek, P. A., Hamada, H., and Yokoyama, T. (1998). Cloning of inv, a gene that controls left/right asymmetry and kidney development. *Nature* **395**, 177–181.

Plisov, S. Y., Yoshino, K., Dove L. F., Higinbotham, K. G., Rubin, J. S., and Perantoni, A. O. (2001). TGFβ2, LIF and FGF2 cooperate to induce nephrogenesis. *Development* **128**, 1045–1057.

Pritchard-Jones, K. (1999). The Wilms tumour gene, WT1, in normal and abnormal nephrogenesis. *Pediatr. Nephrol.* **13**, 620–625.

Qiao, J., Cohen, D., and Herzlinger, D. (1995). The metanephric blastema differentiates into collecting system and nephron epithelia *in vitro*. *Development* **121**, 3207–3214.

Sakurai, H., Bush, K. T., and Nigam, S. K. (2001). Identification of pleiotrophin as a mesenchymal factor involved in ureteric bud branching morphogenesis. *Development* **128**, 3283–3293.

Sainio, K., Nonclercq, D., Saarma, M., Palgi, J., Saxen, L., and Sariola, H. (1994). Neuronal characteristics in embryonic renal stroma. *Int. J. Dev Biol.* **38**, 77–84.

Sainio, K., Suvanto, P., Saarma, M., Arumäe, U., Lindahl, M., Davies, J. A., and Sariola, H. (1997). Glial cell-line derived neurotrophic factor is a morphogen for the ureteric bud epithelium. Submitted for publication.

Saxén, L. (1987). *"Organogenesis of the Kidney."* Cambridge Univ. Press, Cambridge.

Smith, F. G., Adams, F. H., Borden, M., and Hilburn, J. (1966). Studies of renal function in the intact fetal lamb. *Am. J. Obstet. Gynecol.* **96**, 240–246.

Sorenson, C. M., Rogers, S. A., Korsmeyer, S. J., and Hammerman, M. R. (1995). Fulminant metanephric apoptosis and abnormal kidney development in bcl-2-deficient mice. *Am. J. Physiol.* **268**, F73–F81.

Torres, M., Gomez-Pardo, E., Dressler, G. R., and Gruss, P. (1995). Pax-2 controls multiple steps of urogenital development. *Development* **121**, 4057–4065.

Tufro, A. (2000). VEGF spatially directs angiogenesis during metanephric development *in vitro*. *Dev. Biol.* **227**, 558–566.

Veis, D. J., Sorenson, C. M., Shutter, J. R., and Korsmeyer, S. J. (1993). Bcl-2-deficient mice demonstrate fulminant lymphoid apoptosis, polycystic kidneys, and hypopigmented hair. *Cell* **75**, 229–240.

Yuan, H. T., Chitra, S., Landon, D. N., Yancopoulos, G. D., and Woolf, A. S. (2000). Angiopoietin-2 is a site-specific factor in differentiation of mouse renal vasculature. *J. Am. Soc. Nephrol.* **11**, 1055–1056.

11

Anatomy and Histology of the Human Urinary System

Benedikt Hallgrímsson, Hallgrímur Benediktsson, and Peter D. Vize

I. Gross Anatomy of the Urinary System

A. Overview and Relations

The human urinary system consists of paired kidneys and ureters and a midline bladder and urethra (Fig. 11.1). The kidneys are bean shaped, and the adult kidney is approximately 115 to 170 g in weight, although this varies among individuals and between males and females (Tisher and Madsen, 2000). Each kidney has a superior and inferior pole. The parenchyma of the kidney is divided into a cortex and medulla. The human kidney is a multilobar structure, with approximately 14 lobes, each having a conical shape, with the base facing the renal capsule and the tip oriented to the corresponding renal calyx (Fig. 11.2). Lobules are more obvious in the kidneys of embryos and infants. The cortical areas of adjacent pyramids are not clearly demarcated, but the medullary portions are separated by "septa," composed of cortical tissue, also known as the columns of Bertin. The corticomedullary junction is usually distinct due to the difference in appearance of the cortex versus the medulla. Each pyramid connects with a minor calyx, and urine flows from the medulla into each calyx via a porous region, the cribriform plate. Minor calices open into major calices that coalesce to form the renal pelvis. The renal pelvis and calices reside within a cavity referred to as the renal sinus. The renal sinus also contains the branches and tributaries of the renal artery and vein, lymphatic vessels, autonomic nerves, and fat (Fig. 11.2). The renal sinus narrows into an ovoid opening referred to as the hilum of the kidney. Renal vessels, nerves, and the ureter enter the kidney through the hilum.

The kidneys reside in the posterior abdominal wall. Like other derivatives of intermediate mesoderm, they are primary retroperitoneal structures. They lie between the levels of the 12th thoracic and the 3rd lumbar vertebrae with the right kidney having a slightly lower position than the left due to its relation to the right lobe of the liver. Superiorly, the right kidney is overlain by the right lobe of the liver and by the spleen. The adrenal glands lie superomedially to the superior pole of each kidney (Figs. 11.1 and 11.2). The posterior relations of the kidney are the diaphragm superiorly and the *quadratus lumborum* muscle inferiorly. Anteriorly, the right kidney is bordered on the right by the liver and on the left by the stomach. The duodenum and tail of the pancreas are located immediately anterior to the hilum of the right and left kidneys, respectively. Medially, the kidneys are bordered by the *psoas* major muscle as well as the inferior *vena cava* and abdominal aorta.

The kidney is covered in a fibrous capsule (Fig. 11.2). Immediately surrounding the kidney there is a layer of fat called the perirenal fat. It is, in turn, surrounded by the renal or Gerota's fascia (Fig. 11.3). On its posterior, medial, and lateral sides, the renal fascia is surrounded by the pararenal fat, which lies between the kidney and the body wall proper.

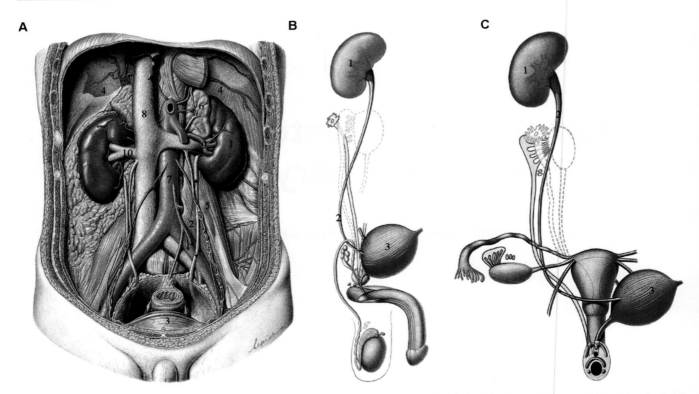

Figure 11.1 Overview and relations of the human urinary system: (1) kidney, (2) ureter, (3) bladder, (4) diaphragm, (5) psoas major muscle, (6) adrenal gland, (7) abdominal aorta, (8) inferior vena cava, (9) renal artery, and (10) renal vein. *With permission from Sobotta Atlas (Putz and Pabst, 1997).*

The pararenal fat is a region of the endoabdominal (or preperitoneal) fat and fascia that lies between the peritoneum and the *transversalis* fascia (Rosse and Gaddum-Rosse, 1997). This relationship is best appreciated in the cross-sectional view seen in Fig. 11.3.

The ureters are continuous with the renal pelvis. They transmit urine from the kidney to the bladder. Like the kidney, the ureters are retroperitoneal and course through the endo-abdominal fascia. As they course toward the pelvic brim, the ureters lie directly anterior to the *psoas* major muscle. They then pass over the common iliac artery and vein and then cross the pelvic brim to enter the true pelvis (Fig. 11.1). The ureters then pass through the endopelvic fascia along the posterior wall of the pelvis to enter the bladder at the posterolateral angles of the trigone.

The bladder receives and stores urine from the ureters. Because bladder development is not covered by other chapters, it is touched on here. Development of the bladder represents a structural specialization of the cloaca—the common receptacle of both urogenital and alimentary systems in early embryos. Bladder development in both males and females is depicted in Fig. 11.4. Early in development the mesonephric (= pronephric) duct is linked to the cloaca (Chapter 4). The ureteric bud branches from the mesonephric duct and induces formation of the metanephros. As development proceeds, the lower end of the mesonephric duct is absorbed into the cloaca, resulting in two distinct linkages to

the cloaca: the ureteric bud (Chapter 12)-derived metanephric duct and the original mesonephric duct. Urine in the mesonephric duct passes from the cloaca to the allantois, an extraembryonic structure. As the metanephric kidney ascends, the membrane separating the cloaca from the external enviroment breaks down and the cloaca divides into two structures:- the urogenital sinus and the rectum. In turn, the urogenital sinus forms another two structures—the future bladder (vesicourethral canal) and the urethra (definitive urogenital sinus). A small portion of the upper urethra forms from the vesicourethral canal. As the bladder begins to form, the position of the meso- and metanephric ducts shifts so that the metanephric duct, originally caudal to the mesonephric duct, is considerably more cranial (Fig. 11.4). This shift results in the metanephric duct (now the ureter) being linked directly to the anterior bladder. The mesonephric duct, while it remains, is now linked to the forming urethra. Bladder development from this point onward is very different in males and females. In males the urethra gives rise to the prostate, whereas the mesonephric duct forms the *ductus deferens.* In females the urethra contributes to the formation of the vestibule and a small portion of the vagina. The female mesonephric duct degenerates.

The adult bladder has an outer fibrous coat, which is continuous with that of the ureter. The majority of the wall of the bladder consists of an interlaced network of smooth muscle fibers referred to as the detrusor muscle. Contraction

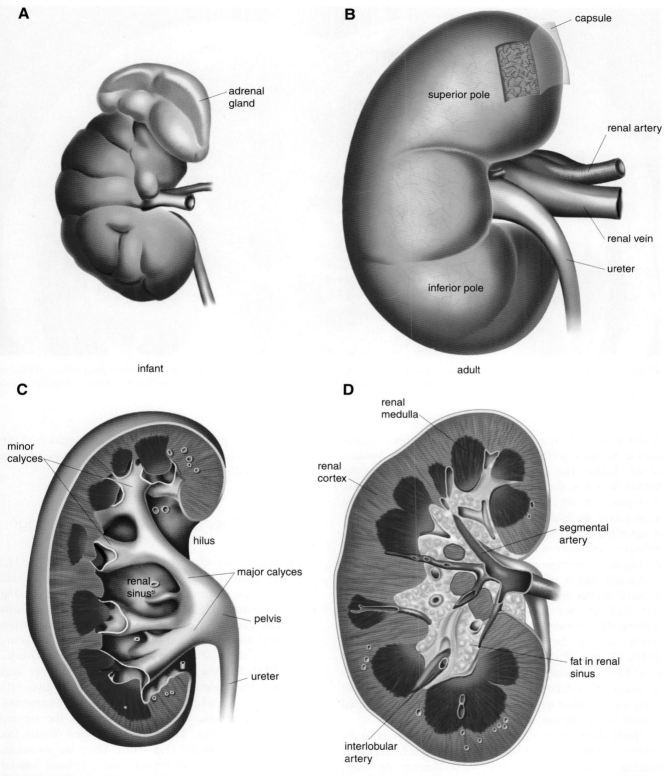

Figure 11.2 Gross structure of the human kidney: (A) fetal or neonatal kidney showing fetal lobulation; (B) external morphology of the adult kidney; (C) coronal section through the kidney showing the relations of the hilus, sinus, and renal pelvis; and (D) coronal section showing the relations of the contents of the renal sinus to the segmental and interlobular arteries.

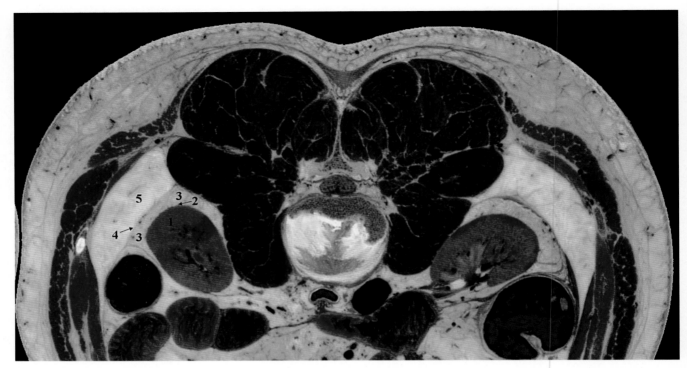

Figure 11.3   Transverse section of a human at level L1 showing the fascias surrounding the kidney. The labeled structures are as follows: (1) kidney, (2) renal capsule (not visible), (3) perirenal fat, (4) renal or Gerota's fascia, and (5) pararenal fat or endoabdominal fascia. *With permission from the National Library of Medicine, Visible Human Project.*

of this muscle, along with relaxation of the pubococcygeus muscle and the external urethral sphincter, produces voiding of the bladder. The ureters enter a triangular region of the bladder known as the trigone. Unlike the rest of the bladder, the trigone does not expand as the bladder fills with urine. The urethra exits the bladder in the center of the trigone a short distance (ca. 2 cm) from where the ureters enter. The longitudinal muscle layer of the ureter (see later) continues within the trigone as the trigone muscle. The function of this muscle is debated, but it may contribute to opening the internal *urethral meatus* and closing the *ostia* of the ureters (Rosse and Gaddum- Rosse, 1997). As the bladder fills, the horizontally oriented slit-like *ostia* of the ureters are compressed by the overlying muscular wall (Beckwith, 1997). Closing of the urethral *ostia* prevents backflow of urine from the bladder to the ureters during voiding. There is no anatomically defined urethral sphincter in the trigone, but a physiologic sphincter is thought to exist around the internal urethral *meatus* (Rosse and Gaddum-Rosse, 1997).

The bladder resides within the true pelvis and rests on the pubic symphysis. It is surrounded by endopelvic fascia and is overlain superiorly and posteriorly by peritoneum. As it fills, it expands superiorly into the peritoneal cavity. In males, the prostate gland is continuous with the neck of the bladder and surrounds the urethra. The *pubococcygeus musclea* portion of the pelvic diaphragm supports the neck of the bladder in

females and the prostate gland in males. Relaxation of this muscle is believed to be necessary to allow flow of urine through the urethra during micturition.

B. Blood Supply and Lymphatic Drainage

1. Kidneys

Blood enters the kidneys via the renal arteries (Figs. 11.1 and 11.2). There is usually a single renal artery on each side, but considerable variation in vascularization is observed (Fig. 11.5; Eustachius, 1552). Multiple renal arteries can arise through the persistence of transient renal arteries that form as the kidneys ascend from the pelvis. The kidneys in a young adult receive about 20% of the cardiac output at rest (Dvorkin *et al.*, 2000) and the renal arteries are consequently large branches of the abdominal aorta. Each renal artery gives off a suprarenal artery, which supplies the adrenal (suprarenal) gland, and a ureteric artery, which supplies the superior portion of the urethra. The renal artery then divides into several segmental arteries (Fig. 11.6). There are no anastomoses (cross connections) between the segmental arteries. Hence, if a segmental artery is occluded (blocked), that vascular segment will die, or infarct (Venkatachalam and Kriz, 1998). Within the renal sinus, the segmental arteries divide to form interlobar arteries that enter the renal parenchyma.

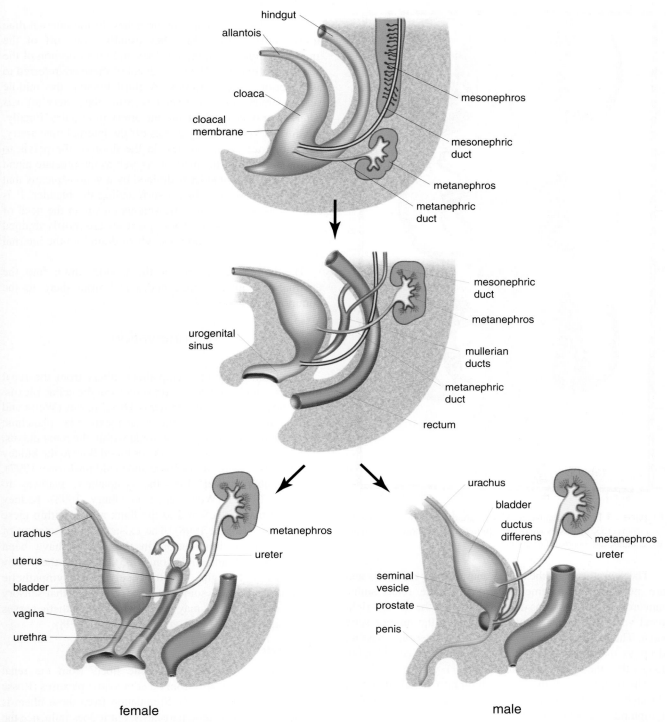

Figure 11.4 Development of the human bladder (redrawn after Netter, 1973). (Top) Early relationship of the nephric duct and cloaca. (Center) Urogenital system at around 4 weeks of development. (Bottom) Male and female urogenital systems at around 4 months of development.

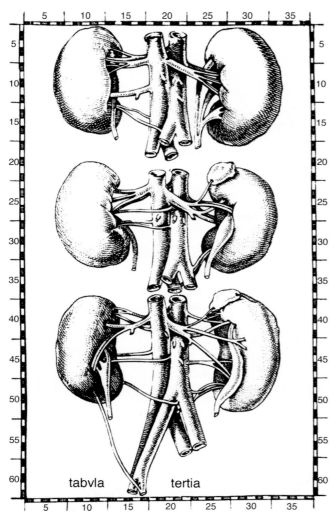

Figure 11.5 Variation in kidney arterial supply. Drawing by Eustachius (1552). Arteries are derived from the aorta, aortic bifurcations, and common and internal iliac arteries.

The veins that drain the kidney anastomose extensively and are not arranged segmentally. Within the renal sinus, numerous interlobar veins drain directly into the left and right renal vein. Both veins empty directly into the inferior *vena cava*. The left renal vein crosses the aorta anteriorly, passing deep to the superior mesenteric artery. It is considerably longer than the right renal vein (Fig. 11.1).

The lymphatics of the kidney drain into vessels that pass through the renal sinus and hilum to drain into the lumbar lymph nodes (Netter, 1973).

2. Ureters

The ureters receive small branches from the renal artery, the aorta, the testicular or ovarian arteries, and the common iliac arteries. These small branches form a longitudinal anastomotic network along the urethra. Venous drainage mirrors the arterial supply.

3. Bladder

The bladder is supplied by branches off the internal iliac artery. Two or three branches usually come off of the umbilical artery to supply the apex and upper portion of the body of the bladder. The more distal of these are referred to as superior vesical arteries. A third branch, the middle vesical artery, usually branches from the artery to *vas deferens* in males and the uterine artery in females. Finally, the inferior vesical artery comes off the internal iliac artery, or vaginal artery in females, in the floor of the pelvis to supply the neck of the bladder, as well as the prostate gland in the male. The bladder is drained by a venous plexus that lies in the endopelvic fascia surrounding the bladder. It is continuous with the prostatic venous plexus in the neck of the bladder in the male. These plexuses are mostly drained by the inferior vesical arteries, which drain into the internal iliac vein.

The lymphatic vessels of the bladder drain into the external and internal iliac nodes and from there to the common iliac nodes.

C. Innervation

1. Kidney

The kidney receives sympathetic fibers from the renal plexuses, which are lateral extension from the celiac plexus. These fibers originate in segments T8-L2 in rats (Weiss and Chowdhury, 1998), reach the renal plexus via splanchnic nerves, and synapse in renal ganglia within the renal plexus. Stimulation of these nerves reduces blood flow to the kidney through vasoconstriction (Rosse and Gaddum-Rosse, 1997). Sensory fibers travel along the sympathetic pathway to segments T10-11 (Weiss and Chowdhury, 1998). Kidney pain in humans is referred to the flank region within these dermatomes (Rosen, 1998). The existence and function of parasympathetic innervation in the kidney have been debated (Norvell and Anderson, 1983), although the kidney is currently thought not to receive parasympathetic innervation. There is some evidence, however, that sensory fibers from the kidney may travel in parasympathetic nerves to the vagus nerve (Weiss and Chowdhury, 1998).

2. Ureters

The ureter receives sympathetic fibers from the renal plexus and possibly also from other preaortic plexuses (Rosse and Gaddum-Rosse, 1997). Stimulation from these fibers is not necessary for ureteric peristalsis, but it does influence the rate and strength of contractile activity (Weiss *et al.*, 1978). Visceral afferents travel with the sympathetic nerves, and ureteric pain is referred to dermatomes T11-L2.

3. Bladder

Parasympathetic fibers reach the bladder from the pelvic splanchnic nerves (S2,S3,S4) via the inferior hypogastric plexus. These fibers innervate the detrusor muscle and are involved in reflex contraction of the bladder during micturition (Bradley *et al.*, 1974; De Groat,1975). Sympathetic innervation of the bladder is involved in urinary retention by inhibiting stimulation of the detrusor muscle and increasing urethral resistance (deGroat and Booth, 1980). Relaxation of the external urethral sphincter and pubococcygeus muscle is involved in initiating micturition. Visceral afferents travel along the pelvic splanchnic nerves (parasympathetic) (Rosse and Gaddum-Rosse, 1997).

II. Microanatomy of the Urinary System

A. Components of the Renal Parenchyma

1. Renal Microvasculature

To understand the function of the nephron, it is necessary to understand how the nephron relates to the renal microvasculature. Blood enters the kidney via the interlobar arteries (see earlier discussion), which are tertiary branches from the renal arteries. The interlobar arteries travel between the renal pyramids. They give off the arcuate arteries that travel along the corticomedullary junction (Fig. 11.6). The arcuate arteries in turn give off a series of small arteries, the interlobular arteries, which ascend the cortex. The afferent glomerular arterioles branch from the interlobular arteries, although some may branch directly from arcuate arteries. Afferent glomerular arterioles enter the Bowman's capsule and branch into the glomerular capillaries. The glomerular capillaries then drain into a portal vessel, the efferent glomerular arteriole, which transports the blood to a second capillary network, the peritubular capillaries. Capillary pressure of the glomerulus, and hence the glomerular filtration rate, is largely regulated by the independent vasoconstriction and dilation of the afferent and efferent arterioles. The anatomy of the afferent and efferent arteriole and the glomerulus are discussed later in the sections on the renal corpuscle and the juxtaglomerular apparatus.

The efferent glomerular arteriole gives rise to the peritubular capillary bed (Fig. 11.7). These capillaries surround the nephron tubules throughout the cortex and medulla. They are formed by a fenestrated (perforated) endothelium supported by a thin basement membrane and are specialized

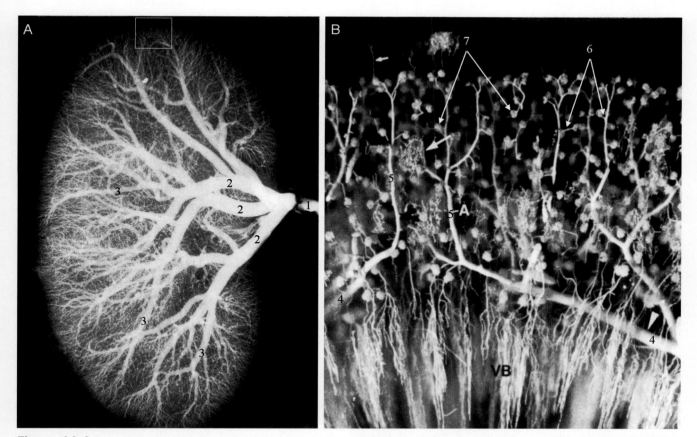

Figure 11.6 The arterial system of the kidney. (A) A radiograph of a contrast injected kidney showing the renal artery (1), segmental arteries (2), and interlobar arteries (3). (B) A scanning EM of an arterial corrosion cast showing the arcuate arteries (4), interlobular arteries (5), afferent arterioles (6), and glomerular capillaries (7). The box in A shows an area that might contain the structures shown in B. *With permission from Venkatachalam and Kriz (1998).*

for reabsorption of interstitial fluid. Within the cortex, the efferent arterioles give off direct branches to the cortical peritubular capillary beds. The arteriole then forms a straight vessel that descends into the medulla. These terminal branches of the efferent arteriole are the descending *arteriolae rectae*. These vessels enter the peritubular capillary network at various levels of the medulla. Peritubular capillaries are then drained by venules that ascend the medulla toward the cortex. These ascending *venae rectae* begin at various levels in medulla and thus mirror the arterial side. Collectively, these ascending and descending vessels are called the *vasae rectae* due to their vertical arrangement in the renal medulla. They are arranged in bundles such that ascending and descending vessels are intermingled. Adjacent arterial and venous limbs of the *vasae rectae* act as a countercurrent exchange systems and thus help maintain the osmotic gradient in the renal medulla (Figs. 11.7 and 11.8).

The nephron

The nephron is the basic structural and functional unit of the kidney. Each human kidney contains approximately 800,000 to 1,200,000 nephrons. The nephron comprises the renal corpuscle (Bowman's capsule and glomerulus) and the renal tubule. The renal tubule in turn consists of the proximal tubule, including the convoluted and straight parts, the thin limbs of the loop of Henle, the distal tubule, including a convoluted and straight portion, and the connecting segment, which empties into the collecting duct (Fig. 11.7). The tubules exhibit complex spatial arrangements between nephrons and the vascular tree. This structural organization is the basis of the countercurrent system, which allows concentration of the ultrafiltrate produced in the renal corpuscle by glomerular filtration.

Nephrons can be classified according to the length of the loop of Henle, and accordingly, there are two main populations of nephrons recognized, those with either a short or a long loop of Henle (Fig. 11.8). Nephrons that have glomeruli located in the outer or midcortical zone tend to have shorter loops of Henle, whereas deeper glomeruli, located in the juxtamedullary zone, will usually have longer loops of Henle, extending into the inner medulla.

There is considerable interspecies variation in the proportion of long versus short nephrons. In the human kidney, approximately 15% of the nephrons have long loops, whereas in the rat, which has a unilobar kidney, the corresponding figure is about 28%.

The cortex contains glomeruli, proximal tubules, and distal tubules, as well as portions of the loops of Henle, connecting segments, and cortical collecting ducts. The medulla can be divided into an outer and inner zone, based on the predominance of the specific segments of the renal tubule present. The outer medulla can in turn be divided into an outer stripe and inner stripe. The outer stripe contains the straight part (*pars recta*) of the proximal and distal tubules, as well as

the thick ascending limb of the loop of Henle and the collecting ducts. The inner stripe contains the thin descending limbs, the collecting ducts, and the thick ascending limbs. The inner medulla contains thin descending and ascending limbs, as well as large collecting ducts that drain into the ducts of Bellini, which in turn empty into the renal calyx. Closely related to the medullary portions of the nephron are the *vasae rectae*, which are capillaries that originate in the efferent arterioles of deep-seated (juxtaglomerular) nephrons.

3. Renal Corpuscle

The Bowman's capsule is the expanded proximal end of the nephron. It is invaded by a vascular tuft known as the glomerulus. The renal corpuscle consists of the Bowman's capsule and the glomerulus along with the connective tissue framework that supports them. The mean diameter of human renal corpuscles is about 200 μm (Venkatachalam and Kriz, 1998). Arterioles enter and exit the Bowman's capsule at the vascular pole. The glomerular capillaries are lined by a highly specialized visceral epithelium that is continuous with the simple squamous parietal epithelium, which lines the internal surface of the capsule. The space that separates the visceral and parietal linings of the Bowman's capsule, and thus surrounds the glomerulus, is referred to as Bowman's space. Ultrafiltrate is forced into Bowman's space and then drains out into the proximal convoluted tubule via the vascular pole.

The function of the glomerulus is the production of ultra-filtrate by forcing selective components of blood through the filtration barrier. The filtration barrier consists of endothelial cells of the glomerular capillaries, the glomerular basement membrane (Fig. 11.9), and the specialized visceral epithelium of Bowman's capsule (Fig. 11.10). Endothelial cells of the glomerular capillaries are specialized in that they exhibit numerous fenestrations that are 70–100 nm in diameter and do not have diaphragms. These fenestrations allow the endothelial cells to serve as a barrier to the cellular components of blood and to large macromolecules.

The glomerular basement membrane (GBM) consists of the fused basal lamina of the capillary endothelial cells and epithelial cells of visceral layer of Bowman's capsule. It is a meshwork of type IV collagen, heparin sulfate proteoglycans, laminin, and fibronectin (Abrahamson, 1987) and is divided into three layers: the *lamina rara interna,* the *lamina densa,* and the *lamina rara externa* (Fig. 11.9). The *lamina densa* consists mainly of type IV collagen. Laminin and fibronectin in the *rara interna* and *externa* attach the endothelial of the glomerulus and the epithelial cells of the Bowman's capsule to the *lamina densa.* Despite a considerable difference in structure, the GBM is continuous with the basement membrane of the parietal layer of the Bowman's capsule at the vascular pole. The GBM is quite dense and contributes significantly to the structural support of the glomerulus. It is also the key structure maintaining the structural integrity of the filtration barrier against the

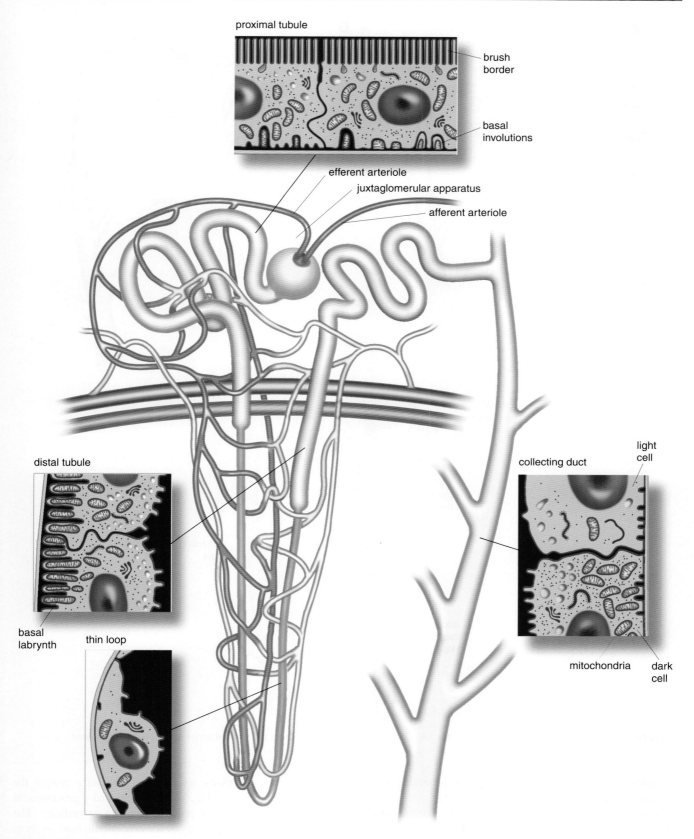

proximal tubule

brush border

basal involutions

efferent arteriole

juxtaglomerular apparatus

afferent arteriole

distal tubule

basal labrynth

thin loop

collecting duct

light cell

mitochondria

dark cell

Figure 11.7 Nephron structure. Callouts illustrate the cell types found in different parts of the nephron. The central tubule is surrounded by a peritubular capillary bed.

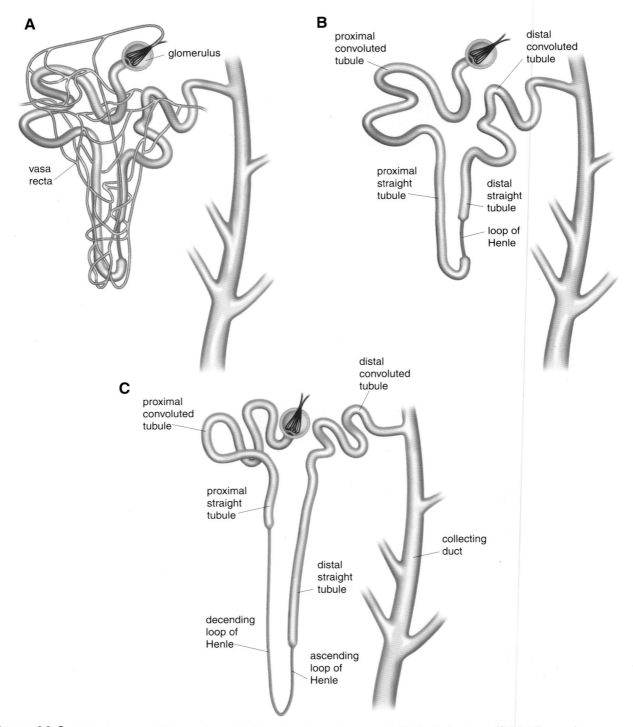

Figure 11.8 Cortical versus medullary nephrons. (A) Cortical nephron with *vasa recta.* (B) Cortical nephron. (C) Medullary nephron.

considerable hydrostatic pressure that is generated within the glomerular capillaries (Kriz *et al.,* 1995).

Finally, the outermost layer of the filtration barrier consists of the highly specialized epithelial cells of the visceral layer of Bowman's capsule. These cells, known as podocytes, send

long primary processes, which wrap obliquely around the glomerular capillaries (Fig. 11.10). The primary processes, in turn, send out secondary processes known as pedicles. The pedicles of adjacent primary processes interdigitate so as to form a network that covers the surface of the glomerual

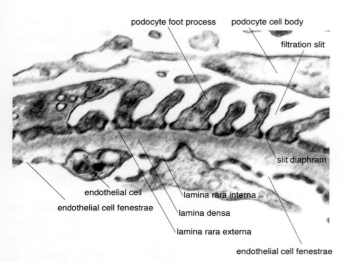

Figure 11.9 Electromicrograph showing the filtration barrier in the glomerulus.

capillaries that face the Bowman's space. The interdigitating pedicles are separated by narrow (30–40 nm) gaps referred to as filtration slits (Venkatachalam and Kriz, 1998). A thin filamentous sheet, known as the slit diaphragm, bridges the space between adjacent pedicles. The slit diaphragm consists of a central filament that runs the length of the filtration slit and cross-bridges that attach it to the pedicles on either side. Tiny (4 × 14 nm) pores in this diaphragm (Rodewald and Karnovsky, 1974) server as a barrier to macromolecules with an effective radius that is larger than about 4.0 nm (Venkatachalam and Kriz, 1998).

The filtration barrier works in two ways. First, it form a mesh-like mechanical filter that excludes all molecules larger than albumin (69 kDa, radius 3.6 nm). Second, many components of the filtration barrier, such as the glycocalyx of the pedicles, the GBM, and the internal membrane of the capillary endothelial cells, are charged due to the presence of negatively charged glycoproteins and heparin sulfate

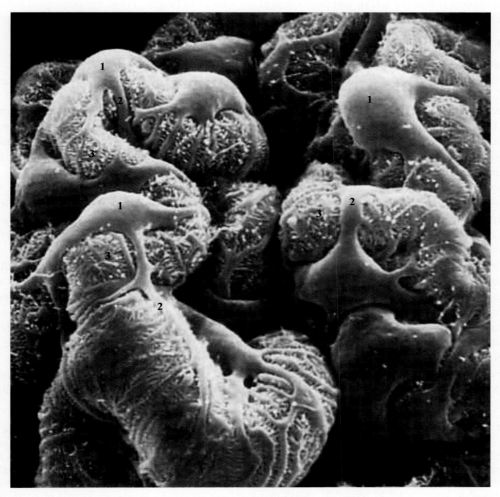

Figure 11.10 Scanning electromicrograph of a glomerular capillary tuft showing the arrangements of podocyte cell bodies (1), central processes (2), and interdigitating foot processes (3). *Reprinted with permission from Orth and Ritz (1998).*

proteoglycans. This repels negatively charged molecules such as plasma proteins and prevents them from crossing the filtration barrier.

Tucked in among the glomerular capillaries are mesangial cells. These are probably modified smooth muscle cells. In some mammalian species, including humans, they are known to contract *in vitro* in response to platelet-activating factor (Iglesias-De La Cruz *et al.*, 2000). Mesangial cells are irregularly shaped and surrounded by an extracellular matrix known as the mesangial matrix. The mesangial matrix lies between the mesangial cells and the capillary endothelium or the GBM. Figure 11.11 illustrates this relationship. The mesangial matrix contains a dense network of elastic fiber proteins and glycoproteins that serve to anchor the mesangial cells to the GBM and capillary endothelial cells (Sterzel *et al.*, 2000). Together, the mesangial cells and their matrix are important to maintaining the structural integrity of the glomerulus against the hydrostatic forces generated within the capillaries. Mesangial cells also perform a number of other functions, such as maintaining the GBM, phagocytosis of macromolecules trapped within the GBM, and production of prostaglandins and cytokines (Schlondorff, 1987).

4. Proximal Tubule

The proximal tubule is the first portion of the nephron to modify the ultrafiltrate. It accounts normally for all resorption of proteins, amino acids, glucose, and creatine and most resorption of water, as well as Cl^- and Na^+ ions. At the urinary pole of the Bowman's capsule, the epithelium lining the nephron makes an abrupt transition from a simple squamous to a simple cuboidal epithelium and this transition marks the beginning of the proximal tubule (Fig. 11.11). Epithelial cells of the proximal tubule are tall and have a dense brush border of microvilli that greatly expand the area available for absorption. At the bases of the microvilli, the cell membrane is pitted extensively and displays numerous apical caniluculi that lead into the cytoplasm. This network of pits and tubules is underlain by vesicles representing various stages of endocytosis and is related to resorption of proteins from the ultrafiltrate (Maunsbach, 1976). Epithelial cells of the proximal tubule are densely packed with mitochondria and exhibit well-developed Golgi complexes, as well as both rough and smooth endoplasmic reticula.

The brush border of the proximal tubule epithelium has abundant water channels, as attested to by the expression of aquaporin-1 (Maunsbach *et al.*, 1997). Expression of aquaporin-1 increases along the proximal tubule and is greatest in the straight descending portion (Schnermann *et al.*, 1998). Aquaporin-7 has also been localized in the proximal tubule in rats and mice and is thought to contribute to water reabsorption (Nejsum *et al.*, 2000). The brush border also exhibits a large number of membrane-bound enzymes and proteins related to ion and solute transport.

Lateral cell membranes of the proximal tubule epithelial cells are deeply infolded and elaborated and show complex interdigitating with neighboring cells (Fig. 11.7). This morphologic feature is consistent with cells with high levels of ion transport activity.

There is significant morphological variation along the length of the proximal tubule. In rats, it is divided into three zones with S1 and S2 dividing equally the convoluted portion and S3 consisting of the straight descending portion. In humans, this segmentation has not been defined, but there is a gradual morphological transition from proximal to distal. Along this transition, the brush borders get shorter, mitochondria become fewer in number, and the cells become shorter in height. These morphological differences correlate with functional specialization along the length of the proximal tubule.

5. The Loop of Henle

The loop of Henle (Fig. 11.8) is the anatomical basis for the countercurrent multiplier system that establishes and maintains the osmotic gradient in the renal medulla. It consists of three parts: the thin descending, thin ascending, and thick ascending limbs. Cortical nephrons tend to lack the thin ascending portion, whereas juxtamedullary nephrons have long thin ascending and descending limbs. The thin descending limb is freely permeable to water and ions, whereas the thin ascending limp is much less permeable to water but permeable to salts. The thick ascending limb is completely impermeable to water and actively transports salts from the filtrate.

In humans and rodents, the thin descending limb consists of a simple squamous epithelium that lacks a brush border of microvilli (Fig. 11.7). Lateral cell membranes are characterized by deep tight junctions and areas of membrane fusion (Venkatachalam and Kriz, 1998). The thin descending limb in longer (juxtamedullary) nephrons is more complex. Here, there is a transition between epithelial cells characterized by the presence of microvilli and a highly interdigitating lateral cell membrane to the simpler type described earlier. This

Figure 11.11 Glomerular and extraglomerular cells. Diagram (A) and lowpower EM (C) showing a cross-sectional view of the glomerular tuft. Internal mesangial cells are embedded in mesangial matrix and in turn are surrounded by glomerular capillaries formed by fenestrated endothelial cells. Capillaries and mesangial cells are surrounded by the glomerular basement membrane (shown in green), on which sit the podocyte foot processes (red). Podocyte cell bodies can be seen slightly elevated from the glomerular capillary. Diagram (B) and electron micrograph (D) of the juxtaglomerular apparatus. Extraglomerular mesangial cells (juxtaglomerular cells) are tucked in between the macula densa of the distal tubule and the vascular pole of the glomerulus. The afferent arteriole (AA) is surrounded by specialized smooth muscle cells called granular myoepithelioid cells or juxtaglomerular cells. Other components of the glomerulus are depicted but not labeled. *EM reprinted with permission from Venkatachalam and Kriz (1998).*

For figure see opposite page

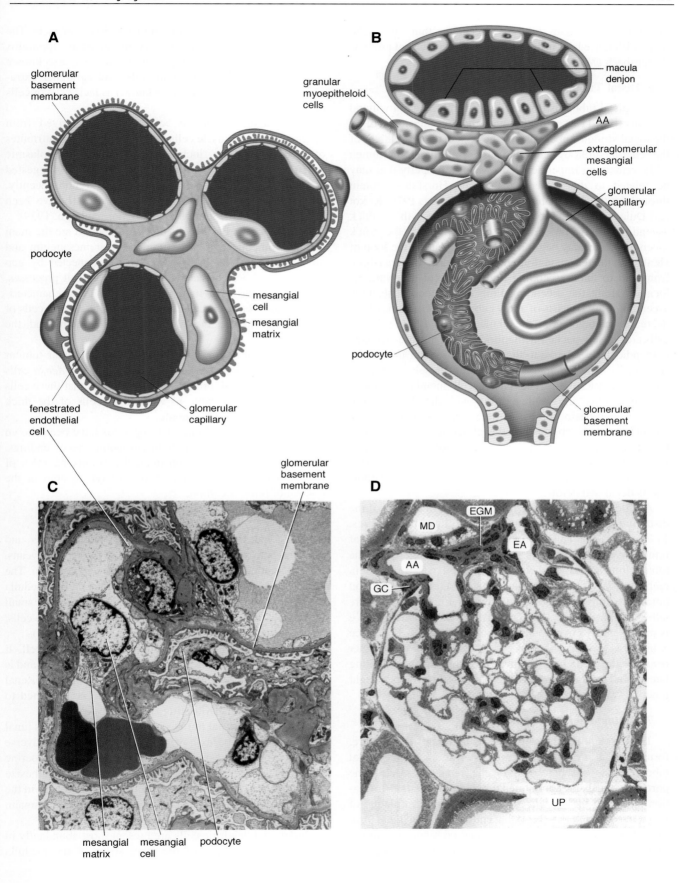

A

glomerular basement membrane

podocyte

fenestrated endothelial cell

glomerular capillary

mesangial cell

mesangial matrix

B

granular myoepitheloid cells

macula denjon

AA

extraglomerular mesangial cells

glomerular capillary

podocyte

glomerular basement membrane

C

glomerular basement membrane

mesangial matrix   mesangial cell   podocyte

D

MD   EGM

EA

AA

GC

UP

morphology continues into the ascending thin limb. The major difference here is the absence of water channel proteins (Maunsbach *et al.,* 1997).

6. Distal Tubule

The distal tubule is divided into straight and convoluted portions (Fig. 11.8). Together with the thin ascending limb, the distal straight tubule establishes the osmotic gradient of the renal medulla by removing salts from the tubule lumen while remaining impermeable to water, a property that may be related to the presence of Tamm–Horsfall's protein throughout the distal tubule (Hoyer and Seiler, 1979; Kokot and Dulawa, 2000). The distal tubule begins with the thick ascending limb. In short cortical nephrons, the thick ascending limb may begin near or at the bend in the loop of Henle. In longer juxtaglomerular nephrons, the transition from thin to thick ascending limb will occur at a variable location along the ascending limb of the loop. The thick ascending limb consists of a low cuboidal epithelium, which decreases in height along the tubule (Kone *et al.,* 1984). The cells are densely packed with mitochondria that are arranged perpendicularly relative to the tubular basement membrane (Fig. 11.7). The luminal surfaces are sparsely populated with short microvilli. The lateral cell membranes interdigitate extensively and exhibit well-developed tight junctions and desmosomes. At the point at which the thick ascending tubule approaches the glomerulus (Fig. 11.11), the tubular epithelium adjacent to the glomerulus forms a specialized "plaque-like" structure, the *macula densa.* This structure is a component of the juxtaglomerular apparatus and is discussed below.

A little past the *macula densa,* the straight portion of the distal tubule grades into the distal convoluted tubule (Figs. 11.7 and 11.11). This portion of the nephron is capable of removing the remaining sodium and chloride from the tubule lumen and it plays an important role in calcium reabsorption. It can also actively secrete hydrogen and potassium ions into the tubule lumen. Epithelial cells are taller than in the distal straight tubule. The luminal surface is more densely covered in microvilli and the apical surface exhibits a greater abundance of vesicles. In most other respects, the epithelium is similar, exhibiting interdigitating lateral cell membranes, as well as well-developed tight junctions and desmosomes.

7. Juxtaglomerular Apparatus

The juxtaglomerular apparatus is a composite structure, formed by afferent and efferent arterioles, and the distal tubule at the vascular pole of the renal corpuscle. The portion of the distal-convoluted tubule involved in its formation is termed the *macula densa,* consisting of a plaque of specialized tubular cells. Adjacent to the *macula densa,* lodged between the two arterioles and continuous with the glomerular mesangium, is the extraglomerular mesangium,

also known as the polar cushion or polkissen, or lacis. The vascular component of the juxtaglomerular apparatus contains two distinct cell types: granular cells, also known as epithelioid or myoepithelial cells, and agranular extraglomerular mesangial cells, also known as lacis cells or cells of Goormaghtigh.

The granular cells are believed to be derived from vascular smooth muscle cells, but have developed attributes of secretory activity, including a prominent endoplasmic reticulum and Golgi apparatus. Goormaghtigh suggested that these cells were the source of renin, and more recently, the presence of both renin and angiotensin II have been demonstrated in the granular cells (Goormaghtigh, 1939).

The lacis cells or the cells of Goormaghtigh are the main cellular component of the extraglomerular mesangium and are continuous with it. These cells are connected by gap junctions and are characterized by elaborate cell processes, embedded in basement membrane material. These characteristics suggest that lacis cells function as a link among cells of the *macula densa,* the glomerular mesangium, and the arterioles.

The *macula densa* is a plaque of low columnar tubular cells with an apically oriented nucleus. *Macula densa* cells lack Tamm Horsfall protein, which suggests that these cells may be more permeable to water than cells of the thick ascending limb or distal tubule.

Cells of the juxtaglomerular apparatus have been shown to have synaptic junctions with autonomic nerve endings, principally adrenergic, confirming the connected roles of this structure and the sympathetic nervous system in the control of renin secretion.

8. Interstitium

The renal interstitium is normally inconspicuous and contains both sulfated and nonsulfated glycosaminoglycans, as well as small amounts of type I and type III collagen. The relative volume of the interstitium is greater in the medulla than in the cortex. The interstitium also contains a small number of cells, including fibroblasts, lymphocyte-like cells, and interstitial dendritic cells (Grupp and Muller, 1999).

The fibroblast is the most abundant interstitial cell. It plays a role in the maintenance of extracellular matrix and is believed to play a central role in interstitial fibrosis. Renal fibroblasts are connected by junctions and are attached to tubules and vessels (Kaissling *et al.,* 1996).

Cortical interstitial fibrosis is a feature of numerous renal diseases and has been shown to be an independent adverse indicator of prognosis. Renal fibroblasts have endocrine functions, cortical fibroblasts are believed to synthesize erythropoietin, and medullary fibroblasts are involved in the regulation of water and electrolyte homeostasis (Bachmann *et al.,* 1993; Grupp and Muller, 1999).

Interstitial dendritic cells are present most abundantly in the peritubular spaces of the outer stripe of the medulla

(Kaissling *et al.,* 1996). Dendritic cells express abundant MCH class II protein, whereas fibroblast lack this marker and express ecto-5′-nucleotidase.

Macrophages are found in the connective tissue of the renal capsule and the pelvic wall. Lymphocytes are rare in healthy kidneys, but are frequently present in many renal disorders.

9. Collecting Duct System

The collecting system of the kidney differs in embryological origin from the rest of the kidney in that it is derived from the ureteric bud (Chapter 12) as opposed to the metanephric mesenchyme (Chapters 13 and 14). Despite the difference in origin, there is a gradual transition in the human kidney between distal convoluted tubules and the cortical collecting ducts that represent the beginning of the collecting system. This intermediate zone is referred to as the connecting segment and it contains cells characteristic of the distal tubule, as well as principal cells, which are characteristic of the collecting ducts. It also contains cells referred to as connecting tubule cells that are intermediate in morphology between these two types (Madsen and Brenner, 1994).

Past the connecting segment, the collecting ducts are divided into cortical, outer medullary, and inner medullary or papillary segments. The ducts increase in size toward the papilla as tributary branches come together. At the papilla, the collecting ducts in humans are 200–300 μm in diameter.

The collecting ducts contain two main cell types, principal (light) and intercalated (dark) cells, which vary in relative number throughout the collecting duct system (Figs. 11.7 and 11.12). Principal cells are the more abundant type. These cells are characterized by abundant cytoplasm with relatively few organelles. There are extensive infoldings on the basal surface, but the lateral cell walls are simple, showing no interdigitations between cells. Principal cells are connected to adjacent cells by deep tight junctions and numerous desmosomes. The luminal surface shows a single cilium and only a few short microvilli. Principal cells are specialized for water transport and respond to vasopressin by activating aquaporin 2 water channels on the luminal surface (Nielsen and Agre, 1995).

Intercalated cells are less abundant than principal cells. They exhibit a more electron-dense cytoplasm, which appears darker on electron micrographs. They are also characterized by more numerous mitochondria and better developed Golgi complexes and smooth endoplasmic reticulum. Like principal cells, lateral cell membranes are not interdigitated and the basal cell membrane may exhibit moderate infolding. In contrast to principal cells, the apical cytoplasm is highly vesicular and exhibits extensive tubovesicular structures. Intercalated cells show more morphological and functional heterogeneity than principal cells and are often divided into two subtypes. The A subtype

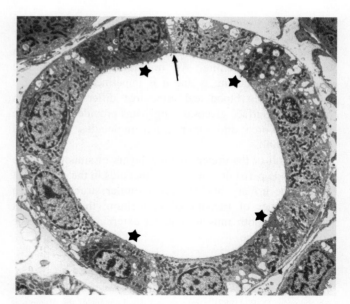

Figure 11.12 Low-power electron micrograph of the rat cortical collecting duct. Intercalated cells are marked by stars. The remainder are principal cells. *Reprinted with permission from Kaissling (1982).*

has a larger apical surface and is specialized for secreting hydrogen ions into the collecting duct lumen (Alper *et al.,* 1989; Madsen and Tisher, 1986). The B subtype possesses a smaller luminal surface that is vesiculated more extensively. These cells are specialized for bicarbonate secretion into the collecting duct lumen (Alper *et al.,* 1989; Madsen and Tisher, 1986). Intercalated cells thus play an important role in the regulation of acid–base balance.

The proportion and morphology of the collecting duct cell types vary along the collecting duct. In the cortical collecting duct, principal cells compose about two-thirds of the cell population. The remaining one-third is about equally divided between type A and B intercalated cells (Madsen and Brenner, 1994). Along the collecting duct system, the number of intercalated cells decreases from the cortex to the papilla and the proportionate representation of the A subtype intercalated cell increases. In the inner medullary collecting ducts, intercalated cells represent only 10% of the cell population and all of these cells are of the A subtype (Madsen and Brenner, 1994).

10. Renal Calyces, Pelvis, Ureter, and Urinary Bladder

The renal calyces, pelvis, ureters, and bladder are lined by transitional epithelium, a specialized type of epithelium that is adapted to the extreme chemical environment of the urine, as well as being able to undergo a remarkable degree of physical distention. This epithelium is two to three cells thick in the minor calyces and is up to six or more cells in the empty, collapsed bladder. The superficial layer of the epithelium consists of large, rounded cells, umbrella cells, that are able to stretch upon distention of the organ. This is aided by the presence of numerous V- shaped indentations and stacks of vesicles arranged at the luminal surface of the cell that can unfold with stretching. An important property of the transitional epithelium is that it is impermeable, forming a barrier between blood and urine that differs radically in osmolarity. Surface glycosaminoglycans provide an effective barrier to urea and other small molecules (Lilly and Parsons,1990).

The wall of the ureter contains layers of smooth muscle that provide peristaltic movement that aids in the passage of urine. The urinary bladder has a similar structure with a luminal layer of transitional epithelium, three indistinct layers of smooth muscle, and an external layer of loose connective tissue.

References

Abrahamson, D. R. (1987). Structure and development of the glomerular capillary wall and basement membrane. *Am. J. Physiol.* **253**, F783–F794.

Alper, S. L., Natale, J., Gluck, S., Lodish, H. F., and Brown, D. (1989). Subtypes of intercalated cells in rat kidney collecting duct defined by antibodies against erythroid band 3 and renal vacuolar H+-ATPase. *Proc. Natl. Acad. Sci. USA* **86**, 5429–5433.

Bachmann, S., Le Hir, M., and Eckardt, K. U. (1993). Co-localization of erythropoietin mRNA and ecto-5′-nucleotidase immunoreactivity in peritubular cells of rat renal cortex indicates that fibroblasts produce erythropoietin. *J. Histochem. Cytochem.* **41**, 335–341.

Beckwith, J. B. (1997). Urinary bladder, ureter, and renal pelvis. *In* "Histology for Pathologists" (S. S. Sternberg, ed.). Lippincott-Raven, Philadelphia.

Bradley, W. E., Timm, G. W., and Scott, F. B. (1974). Innervation of the detrusor muscle and urethra. *Urol. Clin. North. Am.* **1**, 3–27.

De Groat, W. C. (1975). Nervous control of the urinary bladder of the cat. *Brain Res* **87**, 201–211.

deGroat, W. C., and Booth, A. M. (1980). Physiology of the urinary bladder and urethra. *Ann. Intern. Med.* **92**, 312–315.

Dvorkin, L. D., Sun, A. M., and Brenner, B. M. (2000). The renal circulations. *In* "Brenner and Rector's the Kidney" (B. M. Brenner, ed.), pp. 277–318. Saunders, Philadelphia.

Eustachius, B. (1552). See Lancisi, G. M. (1714). *In* "Tabulae Anatomicae". Rome.

Goormaghtigh, N. (1939). Existence of an endocrine gland in the media of renal arterioles. *Proc. Soc. Exp. Biol. Med.* **42**, 688.

Grupp, C., and Muller, G. A. (1999). Renal fibroblast culture. *Exp. Nephrol.* **7**, 377–385.

Hoyer, J. R., and Seiler, M. W. (1979). Pathophysiology of Tamm-Horsfall protein. *Kidney Int.* **16**, 279–289.

Iglesias-De La Cruz, M. C., Ruiz-Torres, M. P., De Lucio-Cazana, F. J., Rodriguez-Puyol, M., and Rodriguez-Puyol, D. (2000). Phenotypic modifications of human mesangial cells by extracellular matrix: The importance of matrix in the contractile response to reactive oxygen species. *Exp. Nephrol.* **8**, 97–103.

Kaissling, B., Hegyi, I., Loffing, J., and Le Hir, M. (1996). Morphology of interstitial cells in the healthy kidney. *Anat. Embryol. (Berl)* **193**, 303–318.

Kaissling, B., and Kriz, W. (1982). Variability of intercellular spaces between macula densa cells: A transmission electron microscopic study in rabbits and rats. *Kidney Int. Suppl.* **12**, S9–S17.

Kokot, F., and Dulawa, J. (2000). Tamm-Horsfall protein updated. *Nephron* **85**, 97–102.

Kone, B. C., Madsen, K. M., and Tisher, C. C. (1984). Ultrastructure of the thick ascending limb of Henle in the rat kidney. *Am. J. Anat.* **171**, 217–226.

Kriz, W., Elger, M., Mundel, P., and Lemley, K. V. (1995). Structure-stabilizing forces in the glomerular tuft. *J. Am. Soc. Nephrol.* **5**, 1731–1739.

Lilly, J. D., and Parsons, C. L. (1990). Bladder surface glycosaminoglycans is a human epithelial permeability barrier. *Surg. Gynecol. Obstet* **171**, 493–496.

Madsen, K. M., and Brenner, B. M. (1994). Structure and function of the renal tubule and interstitium. *In* "Renal Pathology" (C. C. Tisher and B. M. Brenner, eds.), pp. 651–698. J. Lippincott, Philadelphia.

Madsen, K. M., and Tisher, C. C. (1986). Structural-functional relationship along the distal nephron. *Am. J. Physiol.* **250**, F1–F15.

Maunsbach, A. B. (1976). Cellular mechanisms of tubular protein transport. *Int. Rev. Physiol.* **11**, 145–167.

Maunsbach, A. B., Marples, D., Chin, E., Ning, G., Bondy, C., Agre, P., and Nielsen, S. (1997). Aquaporin-1 water channel expression in human kidney. *J. Am. Soc. Nephrol.* **8**, 1–14.

Nejsum, L. N., Elkjaer, M., Hager, H., Frokiaer, J., Kwon, T. H., and Nielsen, S. (2000). Localization of aquaporin-7 in rat and mouse kidney using RT-PCR, immunoblotting, and immunocytochemistry. *Biochem. Biophys. Res. Commun.* **277**, 164–170.

Netter, F. H. (1973). "Kidneys, Ureters, and Urinary Bladder." Volume 6 of the Ciba Collection of Medical Illustrations. CIBA, New Jersey.

Nielsen, S., and Agre, P. (1995). The aquaporin family of water channels in kidney. *Kidney Int.* **48**, 1057–1068.

Norvell, J. E., and Anderson, L. (1983). Assessment of possible para-sympathetic innervation of the kidney. *J. Auton. Nerv. Syst.* **8**, 291–294.

Orth, S. R., and Ritz, E. (1998). The nephrotic syndrome. *N. Engl. J. Med.* **338**, 1202–1211.

Putz, R., and Pabst, R. (eds.) "Sobotta Atlas of Human Anatomy," 12th English Ed. Williams & Wilkins, Baltimore, MD

Rodewald, R., and Karnovsky, M. J. (1974). Porous substructure of the glomerular slit diaphragm in the rat and mouse. *J. Cell. Biol.* **60**, 423–433.

Rosen, P. (1998). Emergency Medicine : Concepts and Clinical Practice. Mosby, St. Louis, MO.

Rosse, C., and Gaddum-Rosse, P. (1997). "Hollinshead's Textbook of Anatomy," 5th Ed. Lippincott-Raven, Philadelphia.

Schlondorff, D. (1987). The glomerular mesangial cell: An expanding role for a specialized pericyte. *FASEB. J.* **1**, 272–281.

Schnermann, J., Chou, C. L., Ma, T., Traynor, T., Knepper, M. A., and Verkman, A. S. (1998). Defective proximal tubular fluid reabsorption in transgenic aquaporin-1 null mice. *Proc. Natl. Acad. Sci. USA* **95**, 9660–9664.

Sterzel, R. B., Hartner, A., Schlotzer-Schrehardt, U., Voit, S., Hausknecht, B., Doliana, R., Colombatti, A., Gibson, M. A., Braghetta, P., and Bressan, G. M. (2000). Elastic fiber proteins in the glomerular mesangium *in vivo* and in cell culture. *Kidney Int.* **58**, 1588–1602.

Tisher, C. C., and Madsen, K. M. (2000). Anatomy of the kidney. *In* "Brenner and Rector's the Kidney" (B. M. Brenner ed.), pp. 3–67. Saunders, Philadelphia.

Venkatachalam, M. A., and Kriz, W. (1998) Anatomy. *In* "Hetinstall's Pathology of the Kidney," (J. C. Jennette, J. L. Olson, M. M. Schwartz, and F. G. Silva, eds.), 5th Ed., pp. 3–66. Lippincott-Raven, Philadelphia.

Weiss, M. L., and Chowdhury, S. I. (1998). The renal afferent pathways in the rat: A pseudorabies virus study. *Brain. Res.* **812**, 227–241.

Weiss, R. M., Bassett, A. L., and Hoffman, B. F. (1978). Adrenergic innervation of the ureter. *Invest Urol.* **16**, 123–127.

12

Development of the Ureteric Bud

Jamie Davies

I. Introduction

The ureter and collecting duct system of the metanephric kidney develop from the ureteric bud, an unbranched tubular outgrowth that arises from the Wolffian duct. The murine ureteric bud forms at E10 (around E28 in human) and grows directly into the metanephrogenic mesenchyme. Once in the mesenchyme, the ureteric bud does three things: (1) it grows and branches (arborizes) to form the collecting duct system, (2) it differentiates into distinct cell types, and (3) it organizes the rest of the kidney by inducing the formation of nephrons.

The first two actions are considered in this chapter, whereas the last one is considered in Chapters 13 and 14.

Presently, ureteric bud development is studied in three main ways:

(1) *in vivo*; (2) in organ culture, in which the kidney develops in isolation from other tissues, in relatively controlled conditions (Fig. 12.1); (3) in cell line models for branching morphogenesis, in which a pure population of cells is grown in a controlled artificial matrix.

In vivo approaches have the advantage of realism, but kidneys growing in mammalian embryos are inaccessible to most experimental manipulation. Organ culture allows experimenters to treat developing kidneys with growth factors, enzymes, antibodies, antisense oligonucleotides, and peptide and pharmacological reagents, but there is always debate, sometimes lively, about the extent to which organ culture results may be extrapolated back to the natural situation. There are, after all, several molecules that seem to be absolutely required for organ culture (e.g., HGF; Woolf *et al.,* 1995), but whose absence causes no renal phenotype in a transgenic knockout mouse (Schmidt *et al.,* 1995; Uehara *et al.,* 1995).

Cell line abstractions are generally based on a few cell types that can be induced to develop branching tubules in three-dimensional collagen gels (e.g., Montesano *et al.,* 1991a,b; Cantley *et al.,* 1994; Fig. 12.2). These systems are accessible to a full range of manipulations, including transfection with mutants and reporter genes, and there is only one cell type present, but the degree to which they reflect real ureteric bud branching remains a matter of faith. Qiao *et al.* (1999) reported a culture system, using an artificial matrix, some characterized medium additives, and some uncharacterized components from conditioned medium, that supports branching morphogenesis of isolated ureteric buds in primary culture. While this system will not

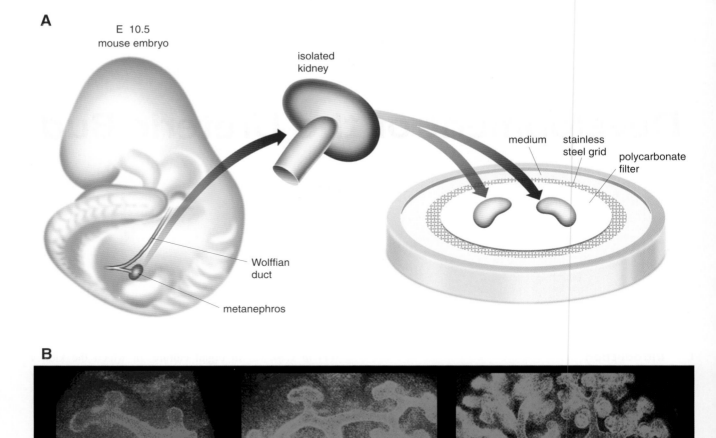

Figure 12.1   The method of organ culture outlined in Fig 1a works well for kidney rudiments, which develop a highly-branched collecting duct system and also excretory nephrons, over the course of about four days in culture.

Table 12.1 Principal Molecules Involved in Ureteric Bud Development

Molecules	Biochemical function	Function in ureteric bud development
Activin A	Signaling	Negative regulator of branching
BMP-2, BMP-7	Signaling	Concentration-dependent inhibition of ureteric bud branching, presumably part of a regulative balance of positive and negative factors *in vivo*
GDNF, c-Ret, GFR-α1	Signaling; c-Ret and GFR-α1 form a receptor complex for GDNF-signaling	1. Induction of ureteric bud outgrowth from Wolffian duct 2. Maintenance of ureteric bud branching
HGF, c-Met	Signaling; c-Met is the receptor for HGF	Supports ureteric bud elongation (but transgenic knockout is OK)
Heparan sulfate proteoglycans	Presentation of growth factors?	Required for elongation and branching of ureteric bud
Integrins	Binding of cells to matrix	Required for elongation and branching of ureteric bud
MAP-kinase	Intracellular signal transduction	Positive regulator of signal transduction; downstream of GDNF
Metalloproteinases MMP2, MMP9	Digestion of interstitial matrix	Required for elongation of ureteric buds—needed to clear a path?
Protein kinase A	Intracellular signal transduction	Negative regulator of branching; downstream of BMPs
Protein kinase C	Intracellular signal transduction	Positive regulator of branching
TGF-β	Signaling	Delays branching while permitting elongation

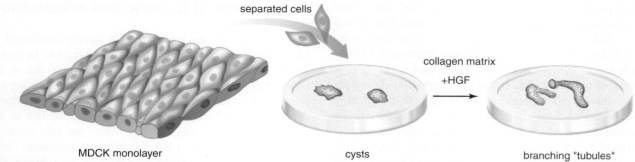

Figure 12.2   The MDCK model system.

allow the easy manipulation of cells that is afforded by the MDCK system and requires difficult microdissection, it might provide a valuable bridge between cell line models and organ culture.

II. Induction of Ureteric Bud Formation

Formation of the ureteric bud appears to be induced by the metanephrogenic mesenchyme; in embryos that lack metanephrogenic mesenchyme, due to surgical or genetic lesions, the ureteric bud does not form. The basis of this induction is now reasonably well understood thanks to a combination of experiments using transgenic knockouts, culture manipulations, and biochemical studies. The ureteric bud bears receptors (the c-Ret tyrosine kinase and the GFR1-a coreceptor) for glial cell-line derived neurotrophic factor (GDNF), whereas the mesenchyme produces GDNF. GDNF-soaked beads elicit the formation of supernumerary ureteric buds that project toward the beads (Sainio *et al.*, 1996), and GDNF–/– and c-Ret–/– transgenic mice lack the ureteric bud and its derivatives. The phenotypes of these mice are not, however, 100% penetrant, which suggests that some other molecules must also have bud-inducing activity. One other molecule, neurturin, has been found to induce supernumerary buds when applied on beads (Davies *et al.*, 1999), but because neurturin is normally expressed by the bud itself rather than by the mesenchyme, it could not be a natural paracrine inductive factor. It also signals via c-Ret so neurturin activity cannot explain the incomplete penetrance of the

c-Ret–/– phenotype. Presumably, other molecules still to be identified are also capable of inducing the bud, albeit at low efficiency.

The fact that only one ureteric bud normally forms from each Wolffian duct suggests some self-organizing property of the developing protuberance, perhaps based on lateral inhibition. The system for ensuring this is clearly not perfect because it can be overridden by high concentrations of GDNF or neurturin applied on beads. Double ureteric buds also form naturally with low frequency and can give rise to clinical complications when one of the ureters ends up emptying below, rather than into, the bladder.

Classically, all cells of the ureteric bud have been thought to originate in the Wolffian duct (which itself forms, much earlier, by the mesenchyme-to-epithelium transition from the intermediate mesoderm). Recently, however, the work of Herzlinger (1993) and Qiao (1995) has cast doubt on the separateness of the lineages and fates of ureteric bud and mesenchymal cells. Using Dil labeling of tissues and retroviral marking, this group found that labeled ureteric bud cells subsequently turn up in nephrons and that labeled mesenchyme cells become incorporated into the growing ureteric bud, implying that bud growth takes place partly by the recruitment of mesenchymal cells. If this model is an accurate reflection of reality, cell exchange may conceivably play a role in subsequent differentiation of collecting duct cells (see later).

Epithelial branching is a common feature of mammalian organogenesis and is very extensive in some epithelia, such as those of the lungs, kidneys, and breast. In all cases,

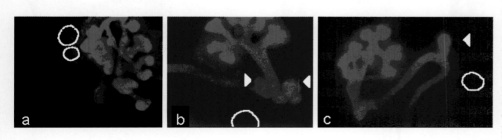

Figure 12.3 Morphogenetic effects of local sources of growth factors, in this case neurturin. Agarose beads soaked in neurturin cause local hyperplasia of the arborizing ureteric bud when placed close to it (a), cause production of supernumerary buds when placed near to the site of normal bud emergence (b; arrow), and cause ectopic ureteric buds to emerge when placed adjacent to anterior portions of the Wolffian duct (c; arrow). From Davies et al. (1999)

morphogenesis begins when an epithelial "bud" forms from a preexisting epithelium, either by evaginating from a tube (lungs, kidneys, salivary glands) or by invaginating from the ectoderm (sweat glands and also mammary glands, which are modified sweat glands). In this context, it is reasonable to ask to what extent the ureteric bud is unique and to what extent it shares properties with other branching epithelia. The idea of uniqueness is supported by experiments in which metanephrogenic mesenchyme is combined with branching epithelia other than ureteric bud; in no case will another epithelium substitute for the bud and induce nephron formation (for a review, see Saxén, 1987). However, ureteric bud arborization can take place in lung mesenchyme, and the morphology of the resulting epithelium is more reminiscent of lung than of kidney (Kispert et al., 1996), a result suggesting that ureteric buds and lung buds are similar enough to respond appropriately to the same signals.

How different is the ureteric bud from its parent tissue, the Wolffian duct? Both are simple, single-layered epithelia that express a range of developmentally relevant molecules, including the transcription factors Pax2, PFB3, LIM-1, GATA-3, the signaling molecules BMP-7 and Wnt 11, the signal receptors MDK1, c-met, c-ret, c-ros, and retinoic acid receptors, and the adhesion and matrix components L1, laminin a1, and a6 integrin (the foregoing list includes pooled data from a number of species because complete data sets are not yet available for any one species). They therefore have much in common. The ureteric bud is, however, different from the Wolffian duct in at least two important respects: only the ureteric bud can be induced to undergo branching morphogenesis by the metanephrogenic mesenchyme and only the ureteric bud will induce nephrons in the metanephrogenic mesenchyme.

III. Anatomy of Ureteric Bud Arborization

Arborization of the ureteric bud is usually considered to take place by simple iterative terminal bifurcation. In such a simple system, n branching events would give rise to 2n branching tips. Branching by iterative terminal bifurcation has been observed in time-lapse studies of early murine kidneys developing in culture, "early" meaning from the first to the third or fourth branching event, as either the only mode of branching (Saxén et al., 1965; Saxén and Wartiovaara, 1966) or as one of the modes of branching (Srinivas et al., 1999). The first bifurcation is at an angle of about 180°, to make a "T" shape, and subsequent branchings are at more acute angles (Saxén et al., 1965) (Fig. 12.1). Studies of ureteric bud branching based on post hoc reasoning from microdissected human kidneys have suggested a quite different "closed divided system of branching" (Oliver, 1968). This is based wholly on an ordered mixture of lateral branching and bifurcation in which a lateral branch arises from somewhere behind a terminus and bifurcates to form two new termini, neither of which branch, but their "stalks" give rise to the next generation of lateral branches (Fig. 12.4). In this system, n bifurcations give rise to 2 × 2n-2 tips in all. The creation of the ureteric bud itself, as a lateral outgrowth from the Wolffian duct that bifurcates, would be the first of approximately 15 rounds of lateral branching required in human renal organogenesis. Oliver's "closed divided" model for branching is closely associated with the hypothesis that branch termini are themselves blocked from branching by the nephrons they induce.

Discriminating between these possibilities experimentally remains difficult because it is difficult to film developing kidneys with time-lapse techniques at sufficient resolution to resolve bifurcation of a terminus from creation of a lateral

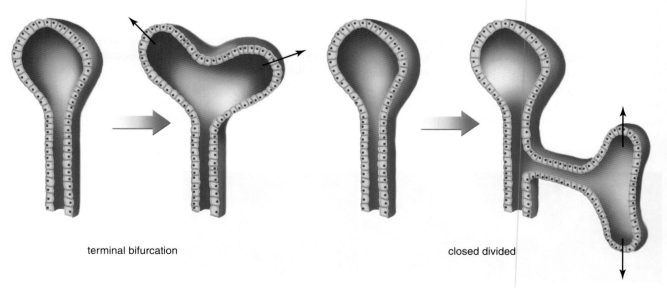

terminal bifurcation   closed divided

Figure 12.4   Two models of branching.

branch only a very short distance from a terminus (for a discussion, see Al-Awqati and Goldberg, 1998). It is, however, clear that lateral branching can happen in culture. This has been suggested by several authors (e.g., Davies and Bard 1998), based either on direct observation of developing kidneys or on experiments in which the treatment of cultured kidneys with branch-promoting substances results in apparently lateral branch initiation. The clearest evidence is probably that of Srinivas et al. (1999), who addressed the problem by making transgenic mice that express green fluorescent protein under the control of the Hox b7 promoter, the renal expression of which is restricted to the bud and its derivatives. They filmed the development of cultured kidneys from these mice by time-lapse fluorescent micrography and presented data that show, beyond a reasonable doubt, that internodal (lateral) branching can take place in the kidney and that it is at least a plausible mechanism for "filling in" volumes of space left by the terminal branching mechanism.

Later in development, after about 15 branch events in the human (which will have created a few tens of thousands of tips), there is a burst of nephron formation so that several nephrons form at the same level of the collecting duct tree and connect together in "arcades." In this phase, the collecting ducts branch much less often as they continue to grow cortically so that they connect to several arcades before making another branch. The extent to which the arcades are part of the collecting duct architecture rather than the nephrons' architecture is not yet certain — there is good circumstantial evidence, based on the expression of marker proteins such as the Tamm Horsfall glycoprotein and the PKK2 antigen (Howie et al., 1993), that connecting tubules are derived from the collecting duct and not the nephrons proper, but solid evidence of lineage has not yet been obtained.

IV. Mechanisms of Collecting Duct Arborization

When the ureteric bud has formed and penetrated the mesenchyme, its morphogenesis continues to depend on a large number of intrinsic and extrinsic molecules.

A. Transcription Factors

Developing ureteric buds express a large number of transcription factors, including Hox a3, a4, a5, a6, a11, b3, b5, b7, c6,c8,c9, d11, brg1, lfb3, Lim-1, Pax-2, kdn-1, evi-1, GATA-3, L-myc, and Emx-2 (for references, see Davies and Brandli, 2001). The functions of most of these remain quite unclear, at least in the context of renal development, but the importance of some (e.g., Lim-1, Emx-2, Pax-2) has been highlighted by the phenotypes of transgenic mice, whereas careful study of the expression patterns of some

transcription factors in kidneys deficient for others is uncovering apparent "cascades" of gene action.

Lim-1 is expressed by the Wolffian duct, the ureteric bud, and developing collecting duct system; it is also expressed transiently in developing nephrons, although not in uninduced mesenchyme (Barnes et al., 1994; Fuji et al., 1994; Shawbrot et al., 1995; Karavanov et al., 1998). Transgenic Lim-1–/– mice show renal agenesis (Shawbrot and Behringer, 1995). The homeobox-containing transcription factor Emx2 is expressed in the Wolffian duct and the ureteric bud and also in maturing nephrons. In Emx-2–/– mice, the ureteric bud forms as usual and even succeeds in invading the metanephric blastema, but then fails both to branch and to induce nephrogenesis (Miyamoto et al., 1997). Interestingly, Emx-2–/– ureteric buds show aberrant expression of other transcription factors discussed in this section; they show abnormally low expression of Pax-2, no Lim-1 at their tips, and the c-Ret receptor tyrosine kinase is absent. Emx-2 therefore acts "upstream" of these genes.

Pax-2 is a paired box transcription factor expressed by the ureteric bud and also by the renal mesenchyme (greatly increased mesenchymal Pax-2 expression is an early consequence of induction). The gene is expressed biallelically, and heterozygous mutations of it in both mice and humans result in renal hypoplasia, including that of the ureteric bud-derived compartment (Porteous et al., 2000). Pax-2–/– mice fail to form even mesonephric tubules, and Wolffian duct growth fails before it reaches the cloaca and gives rise to a ureteric bud (Torres et al., 1995); the mice are therefore not useful in analyzing specifically metanephric issues. The antisense experiments of Rothenpieler and Dressler (1993), which yield much information on the role of Pax-2 in differentiation of the renal mesenchyme, are relatively unhelpful about its role in the bud, as the antisense oligonucleotides are not taken up well enough by the epithelial cells of this tissue.

The homeobox proteins Hox a-11 and Hox d-11 are both expressed in developing ureteric bud, as are many other Hox genes. While the single knockouts of Hox a-11 and Hox d-11 are without apparent effect in the kidney, the double knockout has renal defects varying from hypoplasia to complete aplasia. All parts of the kidney are affected, however, and it is not yet clear whether the effects on the bud are direct or whether they arise from an effect on the mesenchyme (Davis et al., 1995).

Some transcription factors whose absence results in ureteric bud malformation are expressed only by tissues other than the bud itself and presumably act on the bud via paracrine intermediates such as growth factors. One example is the winged helix factor BF-2, a member of the HNF family of transcription factors. BF-2 is normally expressed in developing renal stroma, not in ureteric bud derivatives, but the absence of BF-2 causes hypoplasia of ureteric bud derivatives, together with a serious disturbance of the mesenchymal lineage (Hatini *et al.,* 1996). Another example is WT-1, a zinc finger transcription factor, which, in some alternate splice forms, also appears to function in the control of RNA splicing (for a review, see Hirose, 1999).

Germline mutations in the WT-1 gene are associated with congenital malformations of the urogenital system, and WT-1–/– mice fail to form kidneys in the first place (Kreidberg *et al.,* 1993). WT-1 is not expressed by the ureteric bud itself or its derivatives (Armstrong *et al.,* 1993) so the effect on bud morphogenesis is presumably indirect. Eya-1 and Six-2 are also transcription factors expressed by the mesenchyme. The Eya-1+/– heterozygotic mouse shows renal defects such as hypoplasia with low penetrance, whereas Eya-1–/– homogygotes show a complete absence of kidneys and ureteric buds, although the expression of WT-1 and Pax-2 is normal. Again, the effect seems to be indirect in that, in the absence of Eya-1, Six-2 is not transcribed and GDNF, a growth factor critical in inducing ureteric bud formation (see later), is absent from the mesenchyme.

B. Matrix Requirements

Development of the ureteric bud/collecting duct system requires the presence of particular constituents of the extracellular matrix, particularly large glycoconjugates, and also the expression of functioning matrix receptors.

Sulfated glycosaminoglycans, such as heparan and chondroitin sulfates, are expressed strongly in the basemement membranes and mesenchyme of developing kidneys, supported on various proteoglycans cores such as syndecan I, syndencan II, and dystroglycan (Vainio *et al.,* 1989; David 1993, Davies *et al.,* 1995; Durbeej *et al.,* 1995). Culture experiments in which the expression of sulfated glycosaminoglycans is reduced, using chlorate ions (which inhibit glycan sulphation by competing with sulfate) or using chondroitinases and heparanases (which hydrolyze chondroitin and heparan sulfates, respectively), block growth and branching of the ureteric bud (Fig. 12.5; Davies *et al.,* 1995). In these culture experiments, nephrons continue to form, but even this is abolished in a transgenic mouse that is missing the enzyme heparan sulfate-2-sulfotransferase and which is therefore completely devoid of a specific type of heparan sulfate chain (Bullock *et al.,* 1998). So far, we do not have much detailed information on the precise functions of glycosaminoglycans in this system, but we do know that they are at least required (however indirectly) to maintain expression of Wnt11 at the growing tips (Kispert *et al.,* 1996) and suspect that they may play a role in presenting growth factors to their receptors (see later).

Cells of the ureteric bud bear specific receptors for matrix components. Most of these receptors are of the integrin family — a2, a6, b1, and b4 integrin chains are all borne by the cells, at least in humans (Korhonen *et al.,* 1990, 1992). Knockout data on these ureteric bud integrins are not available, although

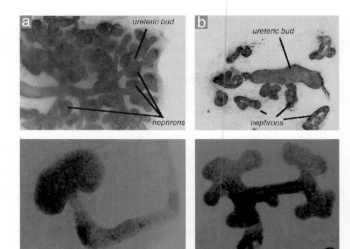

Figure 12.5   Matrix requirements of ureteric bud. Ureteric buds of kidney rudiments grown in normal medium arborize organotypically (a), but in medium in which the synthesis of matrix sulfated glycosaminoglycans has been inhibited, the ureteric bud fails to arborize and only produces a few fine processes (b); nephrons still form normally. Matrix receptors based on the b-1 integrin are also required for arborization, and antibodies to b-1 integrin block arborization of the ureteric bud (c), whereas control antibodies do not (d) Plates a and b are stained for laminin, which shows nephrons as well as the bud, whereas c and d are stained for calbindin-D-28K, which does not.

function-blocking antibodies to integrin b1 do inhibit ureteric bud branching in culture (Fig. 11.5). Knockout data are available for other integrins expressed by the tissues immediately surrounding the ureteric bud; the a8 integrin is expressed by condensing mesenchyme and developing nephrons and is required for bud arborization, as well as for nephron development (Muller *et al.,* 1997). Similarly, absence of the a3 integrin, normally expressed by developing glomeruli, results in a collecting duct system of reduced size (Kreidberg *et al.,* 1996). The effect of these mutations is clearly indirect, and the route between a direct effect on nephrogenesis and an indirect one on the bud may be very circuitous. In the MDCK cell model system, in which collecting duct-derived MDCK cells undergo branching "tubulogenesis" in three-dimensional collagen matrices, the a2b1 integrin is essential for morphogenesis to take place (Saelman *et al.,* 1995). In addition to integrins, ureteric bud cells express a 67/32-kDa laminin receptor (Laurier, 1989) and galectin-3 (Hughes, 1994). In the MDCK cell system, galectin-3 also plays a role in branching "tubulogenesis"; it is absent from the tips of new branches, and adding exogenous galectin-3 inhibits branching, whereas adding antigalectin-3 promotes it (Bao and Hughes, 1995).

As well as supporting ureteric bud development, the extracellular matrix stands physically in its path. Simple zymography studies have demonstrated a correlation between morphogenetic activity and the activity of certain matrix

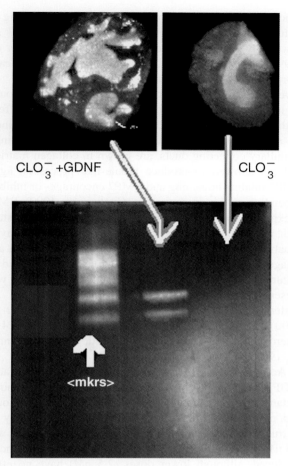

CLO$_3^-$ +GDNF   CLO$_3^-$

Figure 12.6 Matrix metalloproteinase expression in collecting duct morphogenesis. Kidneys in which ureteric bud arborization has been blocked using sodium chlorate show levels of activated MMP-2/9 that are undetectable by this zymography assay, whereas those in which arborization has been 'rescued' using GDNF show strong activity of MMP-2 and MMP-9.

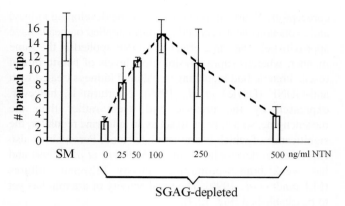

Figure 12.7 Supraphysiological concentrations of morphogenetic growth factors such as GDNF, neurturin, and persephin can partially "rescue" the branching of ureteric buds whose normal branching has been blocked by treatment with agents that remove sulfated glycosaminoglycans. In this example, natural branching has been prevented by treatment with chlorate ions, and branch initiation has been restored in a concentration-dependent manner by an application of exogenous neurturin. (see Milbrandt *et al.*, 1999).

metalloproteinases, such as MMP2 and MMP9, capable of digesting stromal matrix components (Fig. 12.6). Furthermore, culture studies using function-blocking antibodies and natural tissue inhibitors of metalloproteinases have shown that MMP9 activity is absolutely required for ureteric bud arborization to take place (LeLongt *et al.*, 1997). Similarly, matrix metalloproteinase activity is required for HGF-induced "tubulogenesis" by MDCK cells in three-dimensional collagen gel culture, and the cells upregulate the membrane-bound MMP, MT1-MMP, as tubulogenesis begins (Kadono *et al.*, 1998). Mere existence of the correct matrix is therefore not sufficient to support normal development — the matrix must also be in a dynamic state of synthesis coupled with destruction.

C. Growth Factors and Receptors

As well as requiring the correct extracellular matrix components, the ureteric bud/developing collecting duct

system requires a set of growth factors, without which it will not arborize normally. Many have been identified as being required in some assay systems (e.g., organ culture) but not in others (e.g., transgenic animals), and no growth factor or receptor has yet been reported to be absolutely essential for every kidney in every assay system (even transgenic nulls show variable phenotypes).

GDNF, which induces ureteric bud formation, continues to be produced by the undifferentiated metanephrogenic mesenchyme into which the growing tips will extend, and its c-Ret receptor tyrosine kinase continues to be expressed by the branching tip (Pachnis *et al.*, 1993; Schuchardt *et al.*, 1994; Sainio *et al.*, 1997). At least in culture, the branching of the ureteric bud requires GDNF, a fact that can be demonstrated by function-blocking anti-GDNF antibodies (Vega *et al.*, 1996) and soluble ret-immunoglobulin fusion proteins (which compete with native Ret; Ehrenfels *et al.*, 1999). Branching may also be to some extent guided by GDNF in that application of excess GDNF on a bead causes local, directed overgrowth of the arborizing bud system toward the bead (Sainio *et al.*, 1997). Also, supraphysiological levels of GDNF can initiate branching from ureteric buds whose branching has been blocked by depriving them of sulfated glycosaminoglycans. This suggests that one of the roles of these GAGs is presentation of GDNF to its high-affinity receptor [with presentation by GAGs, physiological concentrations of GDNF are sufficient; without this presentation, supraphysiological concentrations are required (Sainio *et al.*, 1997)].

The c-Ret receptor tyrosine kinase can transduce signals from a number of related growth factors (neurturin, persephin, and artemin), as well as from GDNF itself, with the overall specificity of the system being modulated by GFRa type coreceptors. Neurturin is expressed by the developing kidney, and exogenous neurturin and neurturin applied on beads have approximately the same effect as GDNF applied in the same manner, whereas supra-physiological levels of neurturin can rescue ureteric bud branching in cultured kidneys treated with anti-GDNF (Davies *et al.,* 1999). Neurturin is, however, expressed by the ureteric bud itself rather than the mesenchyme, so it appears to act as an autocrine regulator, the significance of which is not yet understood. Persephin is also expressed in developing kidneys (Milbrant *et al.,* 1999) and has weak branch-promoting activity in renal cultures (Milbrandt *et al.,* 1999). The renal activity of artemin has yet to be established (Fig. 12.7).

Hepatocyte growth factor (HGF) has strong effects on culture models of renal development, but the precise nature of its effect varies from model to model. When MDCK cells are suspended in collagen gels, aggregates of these cells form small cysts, and HGF causes these cysts to produce long branching processes (Montesano *et al.,* 1991a,b). HGF is produced by metanephrogenic mesenchyme, and its c-Met receptor tyrosine kinase is borne by cells of the ureteric bud/developing collecting duct system (Sonnenberg *et al.,* 1993; Woolf *et al.,* 1995). The addition of purified HGF to mouse kidney rudiments increases the extent of their branching, whereas treating cultured kidneys with anti-HGF inhibits development of the collecting duct tree (Woolf *et al.,* 1995). From this evidence, HGF therefore appears to be a powerful and necessary paracrine regulator of ureteric bud arborization. When supraphysiological concentration of HGF are examined for their ability to rescue the branching of ureteric buds in kidneys deprived of sulfated glycos-aminoglcyans (using the same logic described earlier for GDNF), HGF is found to induce buds to elongate without branching. This may indicate that HGF is simply an elongation factor and that in the other assays, there were enough ramogenic (branch-promoting) molecules that elongation was the rate-limiting factor in tree formation or that the ramogenic effect of HGF has a higher requirement for GAGs than the growth-promoting one. c-Met receptors for HGF are known to be assisted by specific glycosaminoglycan coreceptors in other systems (Deakin and Lyon, 1999). For all of its powerful effects in culture, HGF does not seem to be required by kidneys developing *in vivo,* at least not in the strains of transgenic mice examined so far (Schmidt *et al.,* 1995, Uehara *et al.,* 1995).

The bone morphogenetic proteins BMP2 and BMP7 are both produced in the kidney: BMP2 largely in the mesenchyme-derived compartment and BMP7 in both ureteric bud and mesenchymal compartments (Lyons *et al.,* 1995; Vukecivic *et al.,* 1996; Godin *et al.,* 1998.). BMP2 inhibits ureteric bud branching, partly by increasing apoptosis (Gupta *et al.,* 1999). BMP7 is more complicated in that inhibition of BMP7 synthesis using antisense oligonucleotides inhibits branching and that addition of low concentrations of the molecule stimulates branching in cultured kidneys, but high concentrations of BMP7 are powerfully inhibitory (Piscione *et al.,* 1997; Gupta *et al.,* 1999). BMP7 knockout mice seem to show normal early morphogenesis but, after approximately E15, the development of several parts of the kidney, most obviously the mesenchyme-derived compartment, fails. Similar effects to those elicited by BMP2 and BMP7 on cultured kidneys are seen when mIMCD3 cells, derived from mouse inner medullary collecting ducts, are grown in collagen matrices and are allowed to produce tubule-like processes; again BMP2 inhibits branching and BMP7 encourages or inhibits, depending on the concentration (Gupta *et al.,* 1999). BMP4 is expressed by stromal cells around the ureteric bud stalk, and its receptor is expressed in renal mesenchyme (Raatikainen-Ahokas *et al.,* 2000; Ikeda *et al.,* 1996). It also antagonizes ureteric bud arborization when applied in culture, but apparently does so indirectly by interfering with mesenchymal differentiation, particularly toward the posterior pole of the kidney (Raatikainen-Ahokas *et al.,* 2000). Underexpression of BMP4, in BMP4+/– heterozygotic mice, results in polycystic kidneys (Dunn *et al.,* 1997; see Chapters 13, 17, and 22).

Activin A, a member of the transforming growth factor (TGF)-β family of growth factors, is expressed by mouse metanephric mesenchyme (Ritvos *et al.,* 1995), although data of Roberts *et al.* (1994) suggest that the same may not be true for rat. Treatment of cultured kidney rudiments with 7.5 n*M* activin reduces the number of ureteric bud branches produced, although lower doses (2.5 n*M*) have no effect (Ritvos *et al.,* 1995). Elongation continues so that the arbor produced has too few branches, each of which is long and spindly. Activin therefore seems to act as a negative regulator of branching. TGF-β is expressed in the kidney from approximately E12.5 in mouse (about 2 days after entry of the ureteric bud), mainly in developing stroma (Lehnert and Ackhurst, 1988; Boivin *et al.,* 1995). Treatment of cultured kidneys with TGF-β causes an unusual anomaly of branching, in which the ureteric bud elongates without branching until it reaches the periphery of the kidney, at which time it makes one 180° bifurcation, and the two branches formed give off a number of apparently lateral branches (Ritvos *et al.* 1995), although this interpretation of lateral branching is from the final anatomy and not from time-lapse frames and is therefore uncertain. If true, however, it may imply that TGF-β can induce the switch to the "arcade" type of branching characteristic of late kidney development. TGF-β also acts on the MDCK and mIMCD-3 models, in which cell lines derived from collecting ducts undergo "branching tubulogenesis" in a three-dimensional collagen matrix. TGF-β inhibits branching and extension of MDCK cells strongly, whereas it inhibits branching but not extension in the mIMCD-3 system (Sakurai and Nigam 1997). In this system, the effect of TGF-β includes a reduction in matrix-degrading protease activity (see earlier discussion).

The developing kidney also expresses several genes of the Wnt family, members of which participate in a large variety of developmental processes during embryogenesis. Wnt4 is expressed only in the condensing mesenchyme, but Wnt7b and Wnt11 are both expressed in the developing collecting duct system. Wnt7b expression begins only during the maturation of collecting duct "trunks," but Wnt11 is expressed from the beginning of ureteric bud arborization and is located only in the branch tips (Kispert *et al.*, 1996). The function of Wnt11 in branching remains unknown, although removal of sulfated glycosaminoglycans also removes expression of Wnt11, a fact that strengthens a correlation between Wnt11 and branching activity.

While not strictly a growth factor, the vitamin A derivative all-*trans*-retinoic acid acts as a powerful promoter of kidney growth (ureretic bud and mesenchyme compartments) in culture. This observation is supported by the severe congenital renal abnormalities that result from dietary vitamin A deficiencies in rats (Wilson and Warkany, 1948) and from transgenic mice deficient for two nuclear retinoic acid receptors (Mendelsohn *et al.*, 1994). Work by Moreau *et al.* (1998) has shown that adding retinoic acid to culture medium increases the expression of c-Ret by kidney rudiments developing in that medium at both message and protein levels, whereas an absence of retinoic acid results in a precipitate decline in c-Ret. Furthermore, enforced expression of c-Ret in mice deficient for retinoic acid receptors restores renal development (Batourina, 2001). This observation therefore provides at least one plausible link between the retinoic acid story and the growth factors discussed earlier.

V. Integration of Influences

Cells of the ureteric bud are thus subject to control by a large number of growth factors and matrix molecules, and they have to integrate these signals to produce a coordinated morphogenetic response. Integration is likely to be achieved mainly by the network of interacting signal transduction systems within the cytoplasm of each cell. This important aspect of understanding the ureteric bud has only recently begun to receive much attention, and while several relevant signal transduction systems have now been identified, the

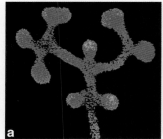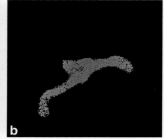

Figure 12.8 Arborization of the ureteric bud requires protein kinase C activity. Ureteric buds arborise normally in control media (a), but not in the presence of the protein kinase C inhibitor, bisindolylmaleimide (b).

precise ways in which they interact with each other and with external influences remain to be determined.

One of the first signal transduction mechanisms shown to have a possible role in ureteric bud branching was protein kinase C (PKC). Stimulation of PKC stimulates production of new branch tips (Davies *et al.*, 1995), whereas inhibition of PKC enzymes with bisindolylmaleimide blocks the development of ureteric buds in culture (Fig. 12.8). Activation of protein kinase A, using 8-bromo-cAMP, inhibits tubule formation and branching by mIMCD-3 collecting duct cells in collagen gel culture. Conversely, inhibition of PKA using H-89 stimulates branching of ureteric buds and mIMCD-3 collecting duct cells in culture and protects ureteric buds from the effects of exogenous BMP2 (Gupta *et al.*, 1999). Direct measurement of PKA activity in various culture conditions reveals that BMP2 activates PKA and hence at least one of its effectors, CREB, in mIMCD-3 cells, thus establishing at least part of one morphogenetically relevant signal transduction pathway (Gupta *et al.*, 1999).

The MAP kinase pathway also appears to be involved in collecting duct morphogenesis; inhibition of Erk1/2 MAP kinases using the drug PD98059 blocks branching of ureteric buds in culture while allowing elongation to continue (Fig. 12.9). Erk1 and Erk2 are normally active in developing ureteric bud, but their activity drops markedly when bud morphogenesis is blocked by depriving kidneys of sulfated GAGs in culture. MAP kinase activity also plays a role in the behavior of collecting duct-derived MDCK cells. When grown in two-dimensional culture, MDCK cells can be induced to "scatter" by treating them with hepatocyte growth factor, the very same growth factor that causes these cells to

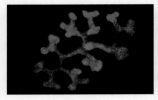

Std Medium   Std medium + PD98059   Std medium + PD98059

Figure 12.9 Arborization of the ureteric bud requires MAP kinase activity. Ureteric buds of kidney rudiments grown in standard medium arborize normally, but those grown in the presence of the drug PD98059, which inhibits activation of MAP kinase by the Erk MAP kinase kinase, show elongation with little or no branching.

undergo "tubulogenesis" in collagen gel culture. Inhibition of the (Mek) MAP kinase kinase blocks the ability of HGF to elicit scattering (Tanimura *et al.*, 1998). Furthermore, transfection of MDCK cells with constitutively active MAP kinase kinase (Mek-1) forces them to adopt a highly invasive phenotype (Montesano *et al.*, 1999).

Work using the MDCK model for branching tubulogenesis (MDCK cells in a collagen gel) has also shown that the tubulogenic growth factor, HGF, signals through its c-Met receptor tyrosine kinase to the signal transduction protein Grb2; mutant forms of c-Met unable to signal through Grb2, although still able to signal through other pathways, fail to support HGF-mediated tubulogenesis (Royal *et al.*, 1997). The response of MDCK monolayers, grown on plastic rather than in collagen gels, to HGF is scattering rather than tubulogensesis, which requires c-Met activation of PI-3-kinase rather than Grb-2. The fact that HGF-induced branching tubulogenesis of MDCK cells proceeds via Grb2 is particularly interesting given that the main inducer of branching morphogenesis of the ureteric bud, GDNF rather than HGF, is known to signal via Grb2, at least in other systems (Bornello *et al.*, 1994; van Weesing *et al.*, 1995). If GDNF also activated Grb2 (via c-Ret) in the kidney, it may be that the Grb2 pathway itself is what is critical to branching in a number of epithelial cell types; the external growth factors that can activate it (HGF, GDNF, etc.) may vary according to the specific type of cell.

The MDCK cell system also requires the activity of β-catenin, a molecule that has activity in both adherens junctions and signal transduction pathways (downstream of, for example, Wnt signals). In this case, the important feature of β-catenin seems not to be its interaction with adherens junctions, but rather its ability to organize the adenomatous polyposis coli (APC) protein at the tips of developing "tubules," where it possibly functions to organize the microtubule network (Pollack *et al.*, 1997).

As well as this culture evidence for involvement of the PKC, PKA, and MAP kinase pathways, there is also genetic evidence for the involvement of formins, which are also signal transduction proteins. The ld mutant, which lacks formins, shows a variable failure of ureteric bud outgrowth (Maas *et al.*, 1994). In other systems, formins are known to link signals to the machinery that organizes the actin cytoskeleton and thus directs cell movement (for a review, see Tanaka, 2000) and they may play a similar role in the kidney.

VI. Engines of Morphological Change

Having integrated morphogenetic signals, ureteric bud the cells have to produce the appropriate morphological change. This aspect of ureteric bud development — the mechanism of actual morphogenesis — remains very poorly understood.

One simple engine of morphogenetic change is cell multiplication, which is responsible for elongation of the bud and its derivatives. BrdU labeling shows that normally there is much cell multiplication in the developing ureteric bud, although this is largely shut off when ureteric bud development is halted by depriving it of sulfated GAGs (Davies *et al.*, 1995). Prevention of multiplication by methotrexate blocks the elongation of branches, although some initiation of new branches still takes place (Davies *et al.*, 1995). Loss of cells, through apoptosis, is another regulator of tissue size, and one effect of BMP-2 is to promote apoptosis in the ureteric bud, at least in culture (Gupta *et al.*, 1999). It is not yet clear whether BMP2 inhibits ureteric arborization by increasing apoptosis or whether the ureteric bud cells increase their apoptosis because their arborization is frustrated at another level by BMP-2. Transgenic mice that lack Bcl2, a protein that, when present as a homodimer, normally protects cells against apoptosis, show a peculiar defect in collecting duct development; the kidneys are small due to excess apoptosis, as might be expected, but the collecting ducts also become cystic (Veis *et al.*, 1993; Nagata *et al.*, 1996). BMP4+/– heterozygotes show a similar phenotype (Dunn *et al.*, 1997), although the extent to which these observations are connected is not yet clear.

Control of cell population size clearly cannot be the sole determinant of morphogenesis in the ureteric bud; simple growth would presumably result in a simple balloon-like expansion of the bud tip instead of controlled arborization. Growth has instead to be directed, and certain cells need to change their shapes in order to create a new branch point. Branching involves sharp curvature of the ureteric bud epithelium in a plane in which it was not previously curved; could this be driven by actin/myosin contraction as it is in amphibian neurulation (Jacobson, 1994)? In branching ureteric buds, as in neurulating amphibia, locations of tight cell curvature are characterized by the intense expression of actin microfilaments (Fig. 12.10), and normal branching morphogenesis will not take place in the presence of cytochalasin D (Fig. 12.10). Because actin is required for so many aspects of the life of cells, it would, however, be premature to conclude from this that the same mechanism has to be involved in both cases. Myosin is the usual motor protein that generates tensile forces on actin, and inhibition of myosin light chain kinase activation using butanedione monoxime also blocks branching of the ureteric bud (Michael and Davies, unpublished results), but again there remains the worry that the primary effect of the drug may lie in an aspect of the cells other than actin contraction, or even that the primary effect is on the mesenchyme and not the epithelium. As well as actin, microtubules may also be important in ureteric bud morphogenesis, although precisely how is not at all clear. Microtubules are required for tubulogenesis by MDCK cells growing in collagen gels, and in the presence of either taxol or colcemid, both of which interfere with microtubule

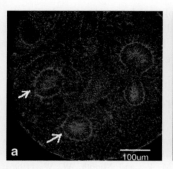
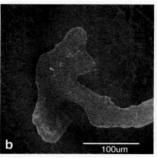

Figure 12.10 a) Filamentous actin, here stained with labelled phalloidin, is expressed very strongly in the apical domains of branching ureteric bud tips (arrows). It is also expressed near the basement membrane, though we do not yet know whether this is in the epithelial or mesenchymal cells. b) Disruption of actin microfilaments with 0.1mg/ml cytochalasin D prevents normal branching morphogenesis; the ureteric bud loses its natural shape and gains a swollen appearance instead of branching (image credit, Lydia Michael).

architecture, the cells are incapable of forming branched tubules in response to HGF (Dugina *et al.*, 1995).

Cytoskeletal changes can only directly control the morphology of a single cell — an integrated deformation of the sheet requires that forces be transmitted from cell to cell by adhesion molecules and junctions. The ureteric bud expresses a large number of cell–cell adhesion molecules, including A-CAM (chicken), desmocollins, desmogleins, gicerin (chicken), L1, and E-cadherin (Duband *et al.*, 1988; Garrod and Fleming, 1990; Takaha, 1995; Sainio, 1994). Because these molecules are important in early embryogenesis, transgenic knockout mice tend to suffer severe abnormalities for a long time before ureteric bud development begins, so what little relevant functional data are available come from antibody inhibition experiments in organ culture. Antibodies to L1 cause severe disruption of ureteric bud branching (Debiec *et al.*, 1998), although it is always possible that the effect on the bud is indirect because L1 is also expressed by neurons in the developing kidney. Antibodies to E-cadherin have apparently no effect on bud morphogenesis (Vestweber *et al.*, 1985), whereas transgenic knockout mice die much sooner before renal formation.

VII. Differentiation within the Maturing Collecting Duct

Although cells of the early ureteric bud appear to be of one single type, the tissue undergoes subsequent differentiation into different types of specialized cell. Arguably the first example of this differentiation is division of the branching ureteric bud system into tip cells, which express high levels of c-Ret, c-Ros, and Wnt11, and non-tip cells, which do not.

Later, the mature regions of the collecting duct give rise to a dominant population of "principal cells" and smaller populations of "intercalated cells." Principal cells possess a single cilium, a few very short microvilli, characteristically round mitochondria, and ion pumps to allow them to recover Na^+ and water and secrete K^+. Intercalated cells, of which there are three types (α, $\beta 1$, and $\beta 2$), have copious microvilli, large and prominent mitochondria, and are concerned with acid/base balance and some aspects of resorption. The zone of differentiation into these specific cell types lies a short distance behind the ampullary tip of the developing collecting duct, the section of duct between the tip and the zone of differentiation being composed of immature cells that divide rapidly (Aigner *et al.*, 1995).

Differentiation of immature collecting duct cells into both principal and intercalated cells will take place in culture, although the degree of differentiation depends on the type of culture medium used, particularly concentrations of NaCl (Aigner *et al.*, 1994, Schumacher *et al.*, 1999). In a culture system in which both sides of the epithelium are presented with normal salt concentrations, fewer than 5% of immature cells differentiate into either principal or intercalated cells, but when the apical side of the collecting duct epithelium is presented with a higher concentration of NaCl than the basal side, more than 80% of the immature cells differentiate (Minuth *et al.*, 1999). The ratio of intercalated to principal cells appears to be extraordinarily plastic for an essentially mature tissue. It can be controlled, at least in culture, by aldosterone, the addition of which shifts differentiation in the direction of intercalated cells (Minuth *et al.*, 1993). Understanding the patterns of differentiation is, however, complicated, and combining the results from different studies suggests that most cell types can give rise to most others. In primary cultures of rabbit cortical collecting ducts, the ratio of intercalated cells to principal cells changes in a manner that cannot be explained by different cell division rates, which suggests instead that principal cells can differentiate into intercalated cells (Jamous *et al.*, 1995). However, when pure populations of principal cells and b-intercalated cells that had been separated by fluorescence-activated cell sorting were cultured, the initially pure population of β-intercalated cells slowly gave rise to a culture consisting of more than 50% principal cells. Cell lines generated from β-intercalated cells can generate both principal and α-intercalated cells, suggesting that the b-intercalated cell might be the "stem cell" for both types (Fejes-Toth and Naray-Fejes-Toth, 1992). Thus, principal cells can apparently differentiate into intercalated cells and vice versa.

VIII. Some Outstanding Problems

This chapter has two purposes: the primary one of reviewing what is known about the development of the ureteric bud and a secondary one of pointing out what we still do not know but have a reasonable chance of discovering before too long.

One task, which is at least conceptually simple, is to complete the list of growth factors and matrix molecules that support ureteric bud development. The most promising method of finding a complete "minimal list of requirements" probably lies in analysis of the conditioned media that are currently necessary to support bud development in characterized matrices such as those described by Qiao *et al.* (1999). Results from transgenic knockouts may reveal additional factors that, while not absolutely required for any branching to take place, are nevertheless necessary for normal branching and correct integration of branching with the development of the rest of the kidney.

More difficult will be identification of the key morphogenetic mechanisms that ureteric bud cells use to change their shape and to create new branches. Strongly suggestive evidence implicates the cytoskeleton, some adhesion molecules and matrix receptors, and proteolytic enzymes, but understanding how they act together may be difficult in intact kidney rudiments. Culture models are more accessible, but there is always uncertainty about how safely one may extrapolate from the model to reality.

Once morphogenetic mechanisms have been identified, we will have a set of "end points" on which signal transduction pathways triggered by growth factor and matrix receptors must converge. This information will make it much easier to trace signal transduction pathways both "upward" and "downward" between receptors and ultimate effectors. It will then be possible to identify how signaling pathways triggered by the many regulators of morphogenesis converge and interact. Identification of the signaling pathways of collecting ducts and also of other systems, such as lungs, will allow us to assess the extent to which mechanisms of branching morphogenesis are "universal modules" that happen to be evoked by different extracellular factors in the different organs and to what extent the kidney really is unique.

Problems at a scale larger than single cells include the process by which branches are spaced out appropriately. In other systems, such as the tracheae of *Drosophila* or mammalian lungs, the spacing of branch tips appears to be regulated by a lateral inhibition process in which established tips produce a protein called "sprouty," which antagonizes action of the FGF-like molecule branchless, which would otherwise promote the formation of more new branches (Hacohen *et al.*, 1998). There is an analogous problem in ensuring that only one ureteric bud is produced by the Wolffian duct. Kidneys do not seem to produce a homologue of sprouty, but a similar system, based on another molecule, may act in the same way. Other large-scale events that are little understood include the "global" switch to arcade production (presumably mediated by a diffusible factor, possibly TGF-β) and of course the eventual cessation of collecting duct growth.

There is also a large set of problems concerned with the relationship between normal collecting duct development and the cystic diseases of kidneys, but this material is covered more fully elsewhere.

References

Aigner, J., Kloth, S., Kubitza, M., Kashgarian, M., Dermietzel, R., and Minuth, W. W. (1994). Maturation of renal collecting duct cells *in vivo* and under perifusion culture. *Epithelial Cell Biol.* **3**, 70–78.

Aigner, J., Kloth, S., Jennings, M. L., and Minuth, W. W. (1995). Transitional differentiation patterns of principal and intercalated cells during renal collecting duct development. *Epithelial Cell Biol.* **4**, 121–130.

Al-Awqati, Q., and Goldberg, M. R. (1998). Architectural patterns in branching morphogenesis in the kidney. *Kidney Int.* **54**, 1832–1842.

Armstrong, J. F., Pritchard-Jones, K., Bickmore, W. A., Hastie, N. D., and Bard, J. B. (1993). The expression of the Wilms' tumour gene, WT1, in the developing mammalian embryo. *Mech. Dev.* **40**, 85–97.

Bao, Q., and Hughes, R. C. (1995). Galectin-3 expression and effects on cyst enlargement and tubulogenesis in kidney epithelial MDCK cells cultured in three-dimensional matrices *in vitro*. *J. Cell Sci.* **108**, 2791–2800.

Barnes, J. D., Crosby, J. L., Jones, C. M., Wright, C. V., and Hogan, B. L. (1994). Embryonic expression of Lim-1, the mouse homolog of *Xenopus* Xlim-1, suggests a role in lateral mesoderm differentiation and neurogenesis. *Dev. Biol.* **161**, 168–178.

Batourina, E., Gim, S., Bello, N., Shy, M., Clagett-Dame, M., Srinivas, S., Costantini, F., and Mendelsohn, C. (2001). Vitamin A controls epithelial/mesenchymal interactions through Ret expression. *Nature Med.* **27**, 74–78.

Boivin, G. P., Otoole, B. A., Orisby, I. E., Diebold, R. J., Eis, M. J., Doetschman, T., and Kier, A. B. (1995). Onset and progression of pathological lesions in transforming growth factor beta 1 deficient mice. *Am. J??? Pathol.* **146**, 276–288.

Borrello, M. G., Pelicci, G., Arighi, E., De Filippis, L., Greco, A., Bongarzone, I., Rizzetti, M., Pelicci, P. G., and Pierotti, M. A. (1994). The oncogenic versions of the Ret and Trk tyrosine kinases bind Shc and Grb2 adaptor proteins. *Oncogene* **9**, 1661–1668.

Bullock, S. L., Fletcher, J. M., Beddington, R. S., and Wilson, V. A. (1998). Renal agenesis in mice homozygous for a gene trap mutation in the gene encoding heparan sulfate 2-sulfotransferase. *Genes Dev.* **12**, 1894–1906.

Cantley, L. G., Barros, E. J., Gandhi, M., Rauchman, M., and Nigam, S. K. (1994). Regulation of mitogenesis, motogenesis, and tubulogenesis by hepatocyte growth factor in renal collecting duct cells. *Am. J. Physiol.* **267**, F271–F280.

David, G., Bai, X. M., Van der Schueren, B., Marynen, P., Cassiman, J. J., and Van den Berghe, H. (1993). Spatial and temporal changes in the expression of fibroglycan (syndecan-2) during mouse embryonic development. *Development* **119**, 841–854.

Davies, J. A., Lyon, M., Gallagher, J., and Garrod, D. R. (1995). Sulphated proteoglycan is required for collecting duct growth and branching but not nephron formation during kidney development. *Development* **121**, 1507–1517.

Davies, J. A., and Brandli, A. W. (1996). A computer database for kidney development. *Trends Genet.* **12**, 322.

Davies, J. A., and Bard, J. B. L. (1998). Development of the metanephric kidney. *Curr. Top. Dev. Biol.* **39**, 245–302.

Davies, J. A., Millar, C. B., Johnson, E., and Milbrandt, J. (1999a). Regulation of renal collecting duct development by neurturin. *Dev. Genet.* **24**, 284–292.

Davies, J. A., Perera, A., and Walker, C. (1999b) Mechanisms of epithelial development and neoplasia in the developing kidney. *Int. Rev. Dev. Biol.* **43**, 473–478.

Davies, J. A. (1999). The Kidney Development Database. *Dev. Genet.* **24**, 194–198.

Davies, J. A., and Brändli, A. W. (2000). The Kidney Development Database http://www.ana.ed.ac.uk/anatomy/database/kidbase/ kidhome. html???

Davies, J. A., and Davey, M. D. (1999). Collecting duct morphogenesis. *Pediatr. Nephrol.* **13**, 535–541.

Davis, A. P., Witte, D. P., Hseih-Li, H. M., Potter, S. S., and Capecchi M. R. (1995). Absence of radius and ulna in mice lacking hox-a11 and hox-d11. *Nature* **375**, 791–795.

Deakin, J. A., and Lyon, M. (1999). Differential regulation of hepatocyte growth factor/scatter factor by cell surface proteoglycans and free glycosaminoglycan chains. *J. Cell Sci.* **112**, 1999–2009.

Debiec, H., Christensen, E. I., ???and Ronco, P. M. (1998). The cell adhesion molecule L1 is developmentally regulated in the renal epithelium and is involved in kidney branching morphogenesis. *J. Cell Biol.* **143**, 2067–2079.

Deuchar, E. M. (1975). "Cellular Interactions in Animal Development." Chapman and Hall, London.

Duband, J. L., Volberg, T., Sabanay, I., Thiery, J.-P., and Geiger, B. (1988). Spatial and temporal distribution of the adherens-junction-associated adhesion molecule A-CAM during avian embryogenesis. *Development* **103**, 325–344.

Dugina, V. B., Alexandrova, A. Y., Lane, K., Bulanova, E., and Vasiliev, J. M. (1995). The role of the microtubular system in the cell response to HGF/SF. *J. Cell Sci.* **108**, 1659–1667.

Dunn, N. R., Winnier, G. E., Hargett, L. K., Schrick, J. J., Fogo, A. B., and Hogan, B. L. (1997). Haploinsufficient phenotypes in Bmp4 heterozygous null mice and modification by mutations in Gli3 and Alx4. *Dev. Biol.* **188**, 235–247.

Durbeej, M., Larsson, E., Ibraghimov-Beskrovnaya, O., Roberds, S. L., Campbell, K. P., and Ekblom, P. (1995). Non-muscle alpha dystroglycan is involved in epithelial development. *J. Cell Biol.* **130**, 79–91.

Ehrenfels, C. W., Carmillo, P. J., Orozce, O., Cate, R. L., and Sanicola, M. (1999). Perturbation of RET signalling in the embryonic kidney. *Dev. Genet.* **24**, 263–272.

Fejes-Toth, G., and Naray-Fejes-Toth, A. (1992). Differentiation of renal beta-intercalated cells to alpha-intercalated and principal cells in culture. *Proc. Natl. Acad. Sci. USA* **89**, 5487–5491.

Fuji, T., Pichel, J. G., Taira, M., Toyama, R., Dawid, I. B., and Westphal, H. (1994). Expression patterns of the murine LIM class homeobox gene lim1 in the developing brain and excretory system. *Dev Dyn.* **199**, 73–83.

Garrod, D. R., and Fleming, S. (1990). Early expression of desmosomal components during kidney tubule morphogenesis in human and murine embryos. *Development* **108**, 313–321.

Godin, R. E., Takaesu, N. T., Robertson, E. J., and Dudley, A. T. (1998). Regulation of BMP7 expression during kidney development. *Development* **125**, 3473–3482.

Gupta, I. R., Piscione, T. D., Grisaru, S., Phan, T., Macias-Silva, M., Zhou, X., Whiteside, C., Wrana, J., and Rosenblum, N. D. (1999). Protein kinase A is a negative regulator of renal branching morphogenesis and modulates inhibitory and stimulatory bone morphogenetic proteins. *J. Biol. Chem.* **274**, 26305–26314.

Hacohen, N., Kramer, S., Sutherland, D., Hiromi, Y., and Krasnow, M. A. (1998). sprouty encodes a novel antagonist of FGF signaling that patterns apical branching of the *Drosophila* airways. *Cell* **92**, 253–263.

Hatini, V., Huh, S. O., Herzlinger, D., Soares, V. C., and Lai, E. (1996). Essential role of stromal mesenchyme in kidney morphogenesis revealed by targeted disruption of Winged Helix transcription factor BF-2. *Genes Dev.* **10**, 1467–1478.

Herzlinger, D., Abramson, R., and Cohen, D. (1993). Phenotypic conversions in renal development. *J. Cell Sci. Suppl.* **17**, 61–64.

Hirose, M. (1999). The role of Wilms' tumor genes. *J. Med. Invest.* **46**, 130–140.

Howie, A. J., Smithson, N., and Rollason, T. P. (1993). Reconsideration of the development of the distal tubule of the human kidney. *J. Anat.* **183**, 141–147.

Hughes, R. C. (1994). Mac-2: A versatile galactose-binding protein of mammalian tissues. *Glycobiology* **4**, 5–12.

Ikeda, T., Takahashi, H., Suzuki, A., Ueno, N., Yokose, S., Yamaguchi, A., and Yoshiki, S. (1996). Cloning of rat type I receptor cDNA for bone morphogenetic protein-2 and bone morphogenetic protein-4, and the localization compared with that of the ligands. *Dev. Dyn.* **206**, 318–329.

Jacobson, A. G. (1994). Normal neurulation in amphibians. *Ciba Found. Symp.* **181**, 6–21.

Jamous, M., Bidet, M., Tauc, M., Koechlin, N., Gastineau, M., Wanstok, F., and Poujeol, P. (1995). In young primary cultures of rabbit kidney cortical collecting ducts intercalated cells originate from principal or undifferentiated cells. *Eur. J. Cell. Biol.* **66**, 192–199.

Kadono, Y., Shibahara, K., Namiki, M., Watanabe, Y., Seiki, M., and Sato, H. (1998). Membrane type 1-matrix metalloproteinase is involved in the formation of hepatocyte growth factor/scatter factor-induced branching tubules in madin-darby canine kidney epithelial cells. *Biochem. Biophys. Res. Commun.* **251**, 681–687.

Karavanov, A. A., Karavanova, I., Perantoni, A., and Dawid, I. B. (1998) Expression pattern of the rat Lim-1 homeobox gene suggests a dual role during kidney development. *Int. J. Dev. Biol.* **42**, 61–66.

Kispert, A., Vainio, S., Shen, L., Rowitch, D. H., and McMahon, A. P. (1996). Proteoglycans are required for maintenance of Wnt-11 expression in the ureter tips. *Development* **122**, 3627–3637.

Korhonen, M., Laitenen, L., Ylanne, J., Gould, V. S., and Virtanen, I. (1992). Integrins in developing, normal and pathological human kidney. *Kidney Int.* **41**, 641–644.

Korhonen, M., Ylanne, J., Laitinen, L., and Virtanen, I. (1990). The a1-a6 subunits of integrins are characteristically expressed in distinct segments of developing and adult human nephron. *J. Cell Biol.* **111**, 1245–1254.

Kreidberg, J. A., Sariola, H., Loring, J. M., Maeda, M., Pelletier, J., Housman, D., and Jaenisch, R. (1993). WT-1 is required for early kidney development. *Cell* **74**, 679–691.

Kreidberg, J. A., Donovan, M. J., Goldstein, S. L., Rennke, H., Shepherd, K., Jones, R. C., and Jaenisch, R. (1996). Alpha 3 beta 1 integrin has a crucial role in kidney and lung organogenesis. *Development* **122**, 3537–3547.

Lako, M., Strachan, T., Bullen, P., Wilson, D. I., Robson, S. C., and Linsay, S. (1998). Isolation, characterisation and emrbyonic expression of Wnt11, a gene which maps to 11q13.5 and has possible roles in the development of skeleton, kidney and lung. *Gene* **219**, 101–110.

Laurie, G. W., Horikoshi, S., Killen, P. D., Segui-Real, B., and Yamada, Y. (1989). In situ hybridization reveals temporal and spatial changes in cellular expression of mRNA for a laminin receptor, laminin, and basement membrane (type IV) collagen in the developing kidney. *J. Cell Biol.* **109** , 1351–1362.

Lehnert, S. A., and Akhurst, R. J. (1988). Embryonic expression pattern of TGF beta type-1 RNA suggests both paracrine and autocrine mechanisms of action. *Development* **104**, 263–273.

Lelongt, B., Trugnan, G., Murphy, G., and Ronco, P. M. (1997). Matrix metalloproteinases MMP2 and MMP9 are produced in early stages of kidney morphogenesis but only MMP9 is required for renal organogenesis *in vitro*. *J. Cell Biol.* **136**, 1363–1373.

Lyons, K. M., Hogan, B. L. M., and Robertson, E. J. (1995). Colocalization of BMP7 and BMP2 mRNAs suggests that these factors cooperatively mediate tissue interactions during murine development. *Mech. Dev.* **50**, 71–83.

Maas, R., Elfering, S., Glaser, T., and Jepeal, L. (1994). Deficient outgrowth of the ureteric bud underlies the renal agenesis phenotype in mice manifesting the limb deformity (ld) mutation. *Dev. Dyn.* **199**, 214–228.

Mendelsohn, C., Lohnes, D., Decimo, D., Lufkin, T., LeMeur, M., Chambon, P., and Mark, M. (1994). Function of the retinoic acid receptors (RARs) during development (II): Multiple abnormalities at various stages of organogenesis in RAR double mutants. *Development* **120**, 2749–2771.

Milbrandt, J., de Sauvage, F., Fahrner, T. L., Baloh, R. H., Leitner, M. L., Tansey, M. L., Lampe, P. A., Heuckeroth, R. O., Kotzbauer, P. T., Simburger, K. S., Golden, J. P., Davies, J. A., Vejsads, R., Kato, A. C., Hynes, M., Sherman, D., Nishimura, M., Wang, L.-C., Vandlen, R., Moffat, B., Klein, R. D., Poulsen, K., Gray, C., Garces, A., Henderson, C. E., Phillips, H. S., and Johnson, E. (1998). Persephin, a neurotrophic factor related to GDNF and Neurturin. *Neuron* **20**, 1–20.

Minuth, W. W., Fietzek, W., Kloth, S., Aigner, J., Herter, P., Rockl, W., Kubitza, M., Stockl, G., and Dermietzel, R. (1993). Aldosterone modulates PNA binding cell isoforms within renal collecting duct epithelium. *Kidney Int.* **44**, 537–544.

Minuth, W. W., Steiner, P., Strehl, R., Schumacher, K., de Vries, U., and Kloth, S. (1999). Modulation of cell differentiation in perfusion culture. *Exp. Nephrol.* **7**, 394–406.

Miyamoto, N., Yoshida, M., Kuratani, S., Matsuo, I., and Aizawa, S. (1997). Defects of urogenital development in mice lacking Emx2. *Development* **124**, 1653–1664.

Montesano, R., Matsumoto, K., Nakamura, T., and Orci, L. (1991a). Identification of a fibroblast-derived epithelial morphogen as hepatocyte growth factor. *Cell* **67**, 901–908.

Montesano, R., Schaller, G., and Orci, L. (1991b). Induction of epithelial tubular morphogenesis *in vitro* by fibroblast-derived soluble factors. *Cell* **66**, 697–711.

Montesano, R., Soriano, J. V., Hosseini, G., Pepper, M. S., and Schramek, H. (1999). Constitutively active mitogen-activated protein kinase kinase MEK1 disrupts morphogenesis and induces an invasive phenotype in Madin-Darby canine kidney epithelial cells. *Cell Growth Differ.* **10**, 317–332.

Moreau, E., Vilar, J., Lelievre-Pegorier, M., Merlet-Benichou, C., and Gilbert, T. (1998). Regulation of c-ret expression by retinoic acid in rat metanephros: Implication in nephron mass control. *Am. J. Physiol.* **275**, F938–F945.

Muller, U., Wang, D., Denda, S., Meneses, J., Pedersen, R. A., and Reichardt, L. F. (1997). Integrin alpha8beta1 is critically important for epithelial-mesenchymal interactions during kidney morphogenesis. *Cell* **88**, 603–613.

Nagata, M., Nakauchi, H., Nakayama, K., Nakayama, K., Loh, D., and Watanabe, T. (1996). Apoptosis during an early stage of nephrogenesis induces renal hypoplasia in bcl-2-deficient mice. *Am. J. Pathol.* **148**, 1601–1611.

Oliver, J. (1968). "Nephrons and Kidneys." Hoeber Medical Division, Harper and Row, New York.

Pachnis, V., Mankoo, B., and Constantini, F. (1993). Expression of the c-ret proto-oncogene during mouse embryogenesis. *Development* **119**, 1005–1017.

Piscione, T. D., Yager, T. D., Gupta, I. R., Grinfeld, B., Pei, Y., Attisano, L., Wrana, J. L., and Rosenblum, N. D. (1997). BMP-2 and OP-1 exert direct and opposite effects on renal branching morphogenesis. *Am. J. Physiol.* **273**, F961–F975.

Pollack, A. L., Barth, A. I. M., Altschuler, Y., Nelson, W. J., and Mostov, K. E. (1997). Dynamics of beta-catenin interactions with APC protein regulate epithelial tubulogenesis. *J. Cell Biol.* **137**, 651–662.

Porteous, S., Torban, E., Cho, N. P., Cunliffe, H., Chua, L., McNoe, L., Ward, T., Souza, C., Gus, P., Giugliani, R., Sato, T., Yun, K., Favor, J., Sicotte, M., Goodyer, P., and Eccles, M. (2000). Primary renal hypoplasia in humans and mice with PAX2 mutations, evidence of increased apoptosis in fetal kidneys of Pax2(1Neu) +/– mutant mice. *Hum. Mol. Genet.* **9**, 1–11.

Qiao, J., Cohen, D., and Herzlinger, D. (1995). The metanephric blastema differentiates into collecting system and nephron epithelia *in vitro*. *Development* **121**, 3207–3214.

Qiao, J., Sakurai, H., and Nigam, S. K. (1999). Branching morphogenesis independent of mesenchymal-epithelial contact in the developing kidney. *Proc. Natl. Acad. Sci. USA* **96**, 7330–7335.

Raatikainen-Ahokas, A., Hytonen, M., Tenhunen, A., Sainio, K., and Sariola, H. (2000). BMP-4 affects the differentiation of metanephric mesenchyme and reveals an early anterior-posterior axis of the embryonic kidney. *Dev. Dyn.* **217**, 146–158.

Ritvos, O., Tuuri, T., Eramaa, M., Sainio, K., Hilden, K., Saxen, L., and Gilbert, S. F. (1995). Activin disrupts epithelial branching morphogenesis in developing glandular organs of the mouse. *Mech. Dev.* **50**, 229–245.

Roberts, V. J., and Barth, S. L. (1994). Expression of messenger ribonucleic acids encoding the inhibin/activin system during mid- and last-gestation rat embryogenesis. *Endocrinology* **134**, 914–923.

Rothenpieler, U. W., and Dressler, G. R. (1993). Pax-2 is required for mesenchyme-to-epithelium conversion during kidney development. *Development* **119**, 711–720.

Royal, I., Fournier, T. M., and Park, M. (1997). Differential requirement of Grb2 and PI3-kinase in HGF/SF-induced cell motility and tubulogenesis. *J. Cell Physiol.* **173**, 196–201.

Saelman, E. U., Keely, P. J., and Santoro, S. A. (1995). Loss of MDCK cell alpha 2 beta 1 integrin expression results in reduced cyst formation, failure of hepatocyte growth factor/scatter factor-induced branching morphogenesis, and increased apoptosis. *J. Cell Sci.* **108**, 3531–3540.

Sainio, K., Nonclercq, D., Saarma, M., Palgi, J., Saxen, L., and Sariola, H. (1994). Neuronal characteristics in embryonic renal stroma. *Int. J. Dev. Biol.* **38**, 77–84.

Sainio, K., Suvanto, P., Davies, J. A., Wartiovaara, J., Wartiovaara, K., Saarma, M., Arumäe, U., Meng, X., Lindahl, M., Pachnis, V., and Sariola, H. (1997). Glial cell-line derived neurotrophic factor is required for bud initiation from ureteric epithelium. *Development* **124**, 4077–4087.

Sakurai, H., and Nigam, S. K. (1997). Transforming growth factor-beta selectively inhibits branching morphogenesis but not tubulogenesis. *Am. J. Physiol.* **272**, F139–F146.

Sauer, B. (1998). Inducible gene targeting in mice using the Cre/lox system. *Methods* **14**, 381–392.

Saxén, L. (1986). "Organogenesis of the Kidney." Cambridge Univ. Press, Cambridge.

Schmidt, C., Bladt, F., Goedecke, S., Brinkmann, V., Zschiesche, W., Sharpe, M., Gherardi, E., and Birchmeier, C. (1995). Scatter factor/hepatocyte growth factor is essential for liver development. *Nature* **373**, 699–702.

Schuchardt, A., D'Agati, V., Larsson-Blomberg, L., Costantini, F., and Pachnis, V. (1994). Defects in the kidney and enteric nervous system of mice lacking the tyrosine kinase receptor Ret. *Nature* **367**, 380–383.

Schumacher, K., Strehl, R., Kloth, S., Tauc, M., and Minuth, W. W. (1999). The influence of culture media on embryonic renal collecting duct cell differentiation. *In Vitro Cell Dev. Biol. Anim.* **35**, 465–471.

Sonnenberg, E., Meyer, D., Weidner, K. M., and Birchmeier, C. (1993). Scatter factor/hepatocyte growth factor and its receptor, the c-met tyrosine kinase, can mediate a signal exchange between mesenchyme and epithelia during mouse development. *J. Cell Biol.* **123**, 223–235.

Srinivas, S., Goldberg, M. R., Watanabe, T., D'Agati, V., Al-Awqati, Q., and Costantini, F. (1999). Expression of green fluorescent protein in the ureteric bud of transgenic mice: A new tool for the analysis of ureteric bud morphogenesis. *Dev. Genet.* **24**, 241–251.

Takaha, N., Taira, E., Taniura, H., Nagino, T., Tsukamoto, Y., Makumoto, T., Kotani, T., Sakamura, S., and Miki, N. (1995). Expression of gicerin in development, oncogenesis and regeneration of the chick kidney.

Differentiation **58,** 313–320.

Tanaka, K. (2000). Formin family proteins in cytoskeletal control. *Biochem. Biophys. Res. Commun.* **267,** 479–481.

Tanimura, S., Chatani, Y., Hoshino, R., Sato, M., Watanabe, S., Kataoka, T., Nakamura, T., and Kohno, M. (1998). Activation of the 41/43 kDa mitogen-activated protein kinase signaling pathway is required for hepatocyte growth factor-induced cell scattering. *Oncogene* **17,** 57–65.

Torres, M., Gomez-Pardo, E., Dressler, G. R., and Gruss, P. (1995). Pax-2 controls multiple steps of urogenital development. *Development* **121,** 4057–4065.

Uehara, Y., Minowa, O., Mori, C., Shiota, K., Kuno, J., Noda, T., and Kitamura, N. (1995). Placental defect and embryonic lethality in mice lacking hepatocyte growth factor/scatter factor. *Nature* **373,** 702–705.

Vainio, S., Jalkanen, M., Bernfield, M., and Saxén, L. (1992). Transient expression of syndecan in mesenchymal cell aggregates of the embryonic kidney. *Dev. Biol.* **152,** 221–232.

Vainio, S., Lehtonen, E., Jalkanen, M., Bernfield, M., and Saxen, L. (1989). Epithelial-mesenchymal interactions regulate the stage-specific expression of a cell surface proteoglycan, syndecan, in the developing kidney. *Dev. Biol.* **134,** 382–391.

van Weering, D. H., Medema, J. P., van Puijenbroek, A., Burgering, B. M., Baas, P. D., and Bos, J. L. (1995). Ret receptor tyrosine kinase activates extracellular signal-regulated kinase 2 in SK-N-MC cells. *Oncogene* **7,** 2207–2214.

Vega, Q. C., Worby, C. A., Lechner, M. S., Dixon, J. E., and Dressler, G. R. (1996). Glial cell line-derived neurotrophic factor activates the receptor tyrosine kinase RET and promotes kidney morphogenesis. *Proc. Natl. Acad. Sci. USA.* **93,** 10657–10661.

Veis, D. J., Sorenson, C. M., Shutter, J. R., and Korsmeyer, S. J. (1993). Bcl-2-deficient mice demonstrate fulminant lymphoid apoptosis, polycystic kidneys, and hypopigmented hair. *Cell* **75,** 229–240.

Vestweber, D., Kemler, R., and Ekblom, P. (1985). Cell-adhesion molecule uvomorulin during kidney development. *Dev. Biol.* **112,** 213–221.

Vukicevic, S., Kopp, J. B., Luyten, F. P., and Sampath, T. K. (1996). Induction of nephrogenic mesenchyme by osteogenic proetin 1 (BMP-7). *Proc. Natl. Acad. Sci. USA.* **93,** 9021–9026.

Wilson, J. G., and Warkany, J. (1948). Malformations in the genitourinary tract induced by maternal vitamin A deficiency in the rat. *Am. J. Anat.* **83,** 357–407.

Woolf, A. S., Kotalsi-Joannou, M., Hardman, P., Andermacher, E., Moorby, C., Fine, L. G., Jat, P. S., Noble, M. D., and Gherardi, E. (1995). Roles of hepatocyte growth factor/scatter factor and the Met receptor in the early development of the metanephros. *J. Cell. Biol.* **128,** 171–184.

13

Fates of the Metanephric Mesenchyme

Hannu Sariola, Kirsi Sainio, and Jonathan Bard

I. Summary

The metanephric mesenchyme (MM) or blastema is composed of a few thousands of morphologically similar mesenchymal cells that give rise to virtually all of the kidney except the collecting ducts. While there is still no unequivocal lineage data determining the number of precursor (single fate) or stem (multifate) cell populations represented in the blastema, it is probable that the future neuronal and endothelial cells derive from separate cell lineages, with the former likely to be neural crest derived. Although it seems likely, the absence of early lineage markers means that it is still not possible to prove that a single cell within the MM can give rise to both loose stromal cells and dense cap cells that surround ureteric bud tips, a subset of which form the small pretubular aggregates that epithelialize and form nephrons. It is also possible that two populations of stromal cells differentiate from single FoxB2-expressing founders: the one in the medulla forms the clear cell type stroma that is later invaded by the loops of Henle and facilitates nephrogenesis, whereas the cortical stromal cell population may form part of the juxtaglomerular complex. These views should be considered provisional until it is possible to trace, isolate, and manipulate different cell types in the MM.

II. Introduction

The blastema is first seen as a condensation within the caudal part of the intermediate mesoderm at the posterior end of each urogenital ridge. The intermediate mesoderm, which originally extends from the cardiac region to the lower limb buds, is the common origin of the transient embryonic (pro- and mesonephroi) and permanent (metanephroi) renal organs, the epididymis, hematopoietic cells, and Müllerian duct derivatives in female genitals, as well as the somatic structures of testis and ovary (Sainio and Raatikainen-Ahokas, 1999).

The MM induces the formation of the ureteric bud, a diverticulum from the Wolffian or nephric duct that, in turn, invades the MM and triggers its differentiation to nephrons and other renal structures. In response, the MM induces branching of the bud that then becomes the collecting duct network that drains urine to the pelvis (see Chapter 12). The interplay between the ureteric bud and the MM is a classic example of a reciprocal inductive interaction, which leads to the former giving rise to the collecting duct system while the latter eventually differentiates to a range of cell types that

181

includes those of nephrons, stroma, juxtaglomerular complex, neurons, and capillaries (Chapter 11).

The purpose of this chapter is to review the early stage of these events at both cellular and molecular levels to obtain some understanding as to how a blastema of so few cells can give rise to so many cell types. This introduction is followed by a brief summary on the determination, commitment, and induction of the MM and then by a section describing the main transitory and permanent cell types to which the MM gives rise and the fundamental and still unresolved problem of their origin. It is not yet known whether MM represents a homogeneous group of multipotent stem cells with several developmental fates, a mixture of precursor cell lineages, or a mixture of the two. The following section discusses in some detail the origins of nephrons and stromal cells in the context of organ culture and other experimental work on induction (a topic considered in more detail but from a slightly different point of view in Chapter 14). The final section attempts to integrate our knowledge of the renal lineages and to consider how nephron polarity and segmental specificities are established on the basis of signaling from the duct epithelium and the adjacent dense metanephric mesenchyme that forms a cap around duct tips.

III. Early Stages of Kidney Formation

A. Determination

In vertebrate embryos, the MM develops near the posterior end of the Wolffian ducts within the urogenital ridge (Fig. 13.1). The series of developmental events that programs the MM for nephrogenesis is referred to as determination, but little is known of their molecular details. Microsurgical deletion and organ transplantation experiments in chick and amphibian embryos have, however, shown that a pro- and mesonephric bias is already apparent in the intermediate mesoderm region at the early neural plate stage (for review, see Chapter 3). The MM appears in mammalian embryos at the beginning of the organogenetic period and, although the few thousand cells of the metanephric blastema superficially appear indistinguishable from any other mesenchymal cells in the embryo, tissue recombination experiments have shown that the MM is the only embryonic cell population that can be induced to undergo nephrogenesis (Saxén, 1970).

B. Commitment

After determination, the MM is committed to nephrogenesis and is thus genetically programmed to respond to inductive cues from the ureteric bud and so ready to give rise to the range of cell types that characterize the kidneys. While relatively little is known of the determinants

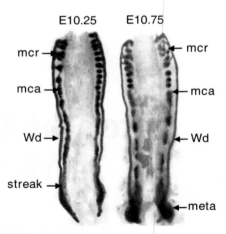

Figure 13.1 Structure of the murine urogenital ridge as shown by whole-mount *in situ* hybridization for Pax-2 that is expressed by the nephrogenic mesenchyme, ureteric bud, and Wolffian duct, but not by stromal cells. Note that at a very early stage, the nephrogenic mesenchyme in meso- and metanephros forms a continous nephrogenic streak. At E10.75, metanephric mesenchyme (meta) has already separated from the mesonephric mesenchyme. mcr, cranial mesonephric tubules; mca, caudal mesonephric tubules; Wd, Wolffian duct (a kind gift of Madis Jakobsson, University of Tarto).

responsible for this commitment, there can be little doubt that key early genes here are Wilms' tumor gene-1 (Kreidberg *et al.*, 1993) and Pax-2 (Dressler *et al.*, 1993; Torres *et al.*, 1995) (see also Chapter 14). Not only is WT1 an early marker of MM (Call *et al.*, 1990; Gessler *et al.*, 1990), but gene targeting has shown that the MM requires its expression to be able to initiate nephron differentiation (Kreidberg *et al.*, 1993) and also to induce formation of the ureteric bud. If WT1 fails to be expressed, no ureteric bud is induced and the cells of the MM soon undergo apoptosis and disappear.

WT1 was originally identified as a gene involved in the pathogenesis of Wilms' tumor or nephroblastoma, a pediatric renal malignancy (Call *et al.*, 1990; Gessler *et al.*, 1990; Chapter 22; Pritchard-Jones, 1999; Lee and Haber, 2001) where the gene is inactivated in about 15% of cases. The WT1 protein includes four zinc fingers and exists as four separate splice forms, two of which seem to act as transcription factors, whereas the other two are found in spliceosomes and seem to be involved in RNA processing (Larsson *et al.*, 1995). The role of WT1 in nephrogenesis has been studied experimentally through the use of cell lines, organ culture, and gene targeting. Such work shows that WT1 regulates many growth factors involved in kidney morphogenesis. These include fibroblast growth factor-2 (FGF-2), interleukin-11 (IL-11), and amphiregulin, which was identified as a WT1 target in a microarray screening (Lee *et al.*, 1999). Amphiregulin is a ligand for epidermal growth factor receptor and is unique in exhibiting bifunctional properties, enhancing proliferation of some epithelial cells while inhibiting that of many cancer cell lines (Shoyab *et al.*, 1988). Fibroblastic cells transfected with the WT1 gene partially epithelialize and show some features of renal epithelial differentiation (Hosono *et al.*, 1999).

When early, uninduced MM (E11 in mouse) is micro-dissected, isolated experimentally from inducing tissue, and cultured *in vitro,* it upregulates apoptosis and disappears (Grobstein, 1955, 1967). It is therefore clear that signaling from the ureteric bud plays a key role in switching MM from an apoptotic fate to producing the range of cell types that it generates for the mature kidney. Before considering such signaling, we first review these cell types, leaving aside any lineage relationships among them until the latter part of the chapter.

IV. Cell Types Derived from Metanephric Mesenchyme

A. Transitory Cells

A wide range of cell types eventually develop from the MM, including the epithelial cell subtypes of secretory nephrons, juxtaglomerular, neuronal, stromal, smooth muscle, and endothelial cells (Table 13.1). The various views about the ontogenesis of different renal cell types may, in some cases, derive from species differences, but they mainly reflect difficulties in identifying individual cell types and in tracing their origins within the blastema. It is notable that although MM cells are morphologically similar and behave synchronously, lectin-binding studies have revealed heterogeneity within the very early MM cell population (Laitinen *et al.,* 1987; 1988). While we still do not know how many cell lineages, or their precursors, are initially present within MM, it is clear that the developing kidney maintains two transitional embryonic cell populations that are lost at or soon after birth: blastemal cells, which give rise to secretory nephrons, and stromal cells, which are located mainly within the renal medulla.

1. Nephrogenic Blastemal Cells

There is still no unambiguous terminology for the names of the cells originally in the metanephric blastema and that have yet to differentiate. Before the initial inductive interaction that diverts them from apoptosis to growth and differentiation, they are known as blastemal cells or metanephric mesenchyme. Once induced, these cells partition into those that condense around duct tips (we will call these *cap* cells) and those that are looser (these are often called stromal cells, a convention we will follow).

Cap cells cover the surface of the tips of the ureteric bud to a depth of about four cells and may then be called either *stem* or *precursor* cells. The distinction is based on the standard definition of a stem cell. If a self-renewing cell can

Table 13.1 Metanephric Mesenchyme (MM)-Derived Cell Types

Tissue type	Possible embryonic origins
Secretory nephron	MM (with a possible minor contribution from the collecting duct)
Endothelial cell	Invading capillary network or MM
Juxtaglomerular complex	MM?
Vascular smooth muscle cells and pericytes	MM or invading blood vessels
Neuronal cell	Neural crest
Embryonic renal stroma	Neural crest or MM

give rise to more than one cell lineage *in vivo,* it can be considered a stem cell. If a putative stem cell is only able to give rise to itself and a single differentiated cell lineage, then it is viewed as a precursor cell. In the context of the kidney, the MM includes *stem* cells if a single cell can give rise to more than one phenotype, such as epithelium and stroma, but *precursor* cells if, for instance, an individual cell is fated to become only nephrogenic epithelium. Because we do not know the differentiation options of individual metanephric cells, it is preferable to use the term *blastemal,* a neutral word that describes the cells in terms of their origin. We will return to this problem at the end of the chapter.

Blastemal cells are easy to recognize in embryonic kidneys. They are unpolarized small cells, their cytoplasm is scant, and, once duct tips have extended to the kidney periphery (E13 in the mouse), they are situated as a thin rind in the outermost cortex (see Chapter 10). The process of nephron formation from the blastemal cell population continues at the kidney periphery for a considerable amount of time, and the process only ceases when the last of the blastemal cells have been induced. This occurs during late embryogenesis in human and soon after the birth in rat and mouse, and cessation may be due to an inability of the cells to divide sufficiently fast to maintain themselves so that almost all of them eventually form nephrons. In humans, occasional small groups of nephrogenic blastemal cells are, however, maintained well beyond birth and are known as *nephrogenic rests.* Oncogenic mutations in this postnatally persisting cell population are thought to lead to the abnormal growth and malignant transformation that is manifested as Wilms' tumor (Pritchard-Jones, 1999; Lee and Haber, 2001; see Chapter 22).

While previous studies have suggested that the transition from blastema to induced pretubular condensates would essentially be a switch event in molecular terms, analyses of microdissected MMs and kidney rudiments from WT1-deficient mice show that several molecular markers characteristic of the induced nephrogenic mesenchyme are already expressed by the blastema before ureteric bud invasion (Donovan *et al.,* 1999; Godin *et al.,* 1999). The appearance of "induction-associated molecules" in the MMs of WT1-deficient mice is particularly interesting, as it shows that while WT1 is essential for nephrogenesis, it is not the unique determinant of metanephric determination.

2. Clear Cell Type Stroma

A major cell type in the differentiating embryonic kidney is the clear cell type renal stroma, which is characterized by large spindle-shaped cells with an extensive, "clear" cytoplasm. This transitional cell type becomes morphologically recognizable soon after the first nephrons have been induced. By midgestation, clear cell type renal stroma is abundant, but it disappears almost completely during late nephrogenesis as a result of large-scale apoptosis. Although the clear cell type stroma is morphologically similar to other types of embryonic mesenchyme, it does not express such typical mesenchymal markers as vimentin; instead, most, if not all of these cells, express neuronal molecules such as neurofilaments, p75 neurotrophin receptor, and FoxB2 [known previously as brain factor-2 (BF-2); Sariola *et al.*, 1988; Sainio *et al.*, 1994; Hatini *et al.*, 1996; Sainio, 1996]. It is still, however, not clear how many distinct cell populations are present in the renal stroma.

Because of its neuronal characteristics and the fact that neuronal cells develop within this stroma, it has been suggested that clear cell type stromal cells are neural crest derivatives (Sainio *et al.*, 1994), much like those in the cranial region where the neural crest cells differentiate to a spectrum of mesenchymal cell types that populate the skull and cardiac outflow tract (LeDouarin, 1982; Fishman and Kirby, 1998). This hypothesis gains some support from neural crest transplantation experiments that take advantage of the nuclear differences between chick and quail. When quail crest is transplanted to chick embryos, some quail cells are found in the stroma of mesonephros (LeDouarin and Teillet, 1974). Chick neural crest-derived cells have also been traced to the metanephros by cell-labeling experiments (Bronner-Fraser and Fraser, 1988) where some probably give rise to the neuronal cell precursors that can be identified in cultured kidney rudiments (Sariola *et al.*, 1988; Sainio *et al.*, 1994). Direct experimental evidence for the neural crest origin of renal stromal cells in mammals is, however, still lacking.

An alternative view is that stromal cells represent a developmental end point for those blastemal cells that have not been induced to undergo tubulogenesis. The traditional role assigned to the clear cell type stroma was that before being lost toward birth by apoptosis or differentiation into other cell types such as the neurons, they make the loose extracellular matrix that is later occupied by the loops of Henle as they extend toward the renal pelvis. It is, however, now clear that stromal cells have another important, if still slightly unclear, role in nephron formation (Hatini *et al.*, 1996; Bard, 1996). The renal stroma (and not the MM,

secretory nephrons, or ureteric bud cells) specifically expresses GD3 ganglioside and a winged-helix transcription factor FoxB2. Both GD3 antibodies tested in organ culture (Sariola *et al.*, 1988a) and gene targeting of the FoxB2 gene (Hatini *et al.*, 1996) perturb nephron differentiation. Of particular interest is the fact that FoxB2-deficient mice, instead of forming normal small epithelializing condensates, produce a few, very large mesenchymal aggregates that, for reasons that are still unknown, fail to epithelialize (Hatini *et al.*, 1996).

It is now becoming clear that the stroma can also regulate branching morphogenesis in the collecting duct system. It has been known for some time that vitamin A deficiency can lead to renal abnormalities, and Batourina *et al.* (2001) demonstrated that this retinoic acid response is mediated via RAR receptors expressed in the stromal cells. They can initiate expression of an unknown signal that regulates expression of the Ret receptor tyrosine kinase at the duct tips. In particular, they have demonstrated that lesions associated with the deletion of RAR receptors can be rescued by forced expression of Ret and have put forward possible models to explain how the stroma can control duct morphogenesis (Fig. 13.2).

B. Mature Cell Types

1. Secretory Nephrons

Morphological and genetic evidence have shown (Gossens and Unsworth, 1972; Kreidberg *et al.*, 1993; Stark *et al.*, 1994) that, under the inductive influence of the ureteric bud, nephron differentiation is a multistep process (Fig. 13.3). The whole MM is first programmed to escape the apoptotic default pathway. Thereafter, MM cells condense around the ureteric bud tip. This cap-like structure surrounding the tip should not be confused with the small pretubular condensates that form a little later at the proximal edge of the cap and that will subsequently undergo epithelial transition on the way to becoming nephrons. Some authors have named this initial cap condensation a "primary condensate" (Bard *et al.*, 2001; for review, see Saxén, 1987). However, because most cap cells do not contribute to the pretubular condensates and because the cap is seen not only in the early kidney rudiment, but at each duct tip through much of development, we propose that the term primary condensate should be avoided. It is less misleading to refer to these cell aggregates simply as caps.

Within 36 h of the mouse blastema and ureteric bud coming into contact and when the bud has bifurcated twice to give four tips, small MM aggregates, a few cells in diameter, appear at the proximal ends of the caps and soon convert to nephron epithelium. *In vitro* studies have shown that after 24 h of contact with an inducer, the epithelialization process becomes autonomous and inducer independent (Grobstein,

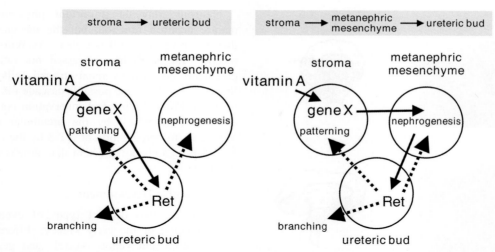

Figure 13.2 Two alternative hypotheses for the interplay between stroma and ureteric branching. In both alternatives, the first step is regulated by vitamin A that upregulates an unknown molecule (factor X) in the stroma, which then affects either directly or via nephrogenic mesenchyme the Ret expression by the ureteric bud and thereby ureteric branching.

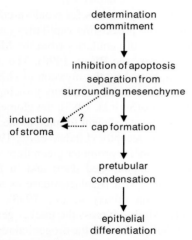

Figure 13.3 Multistep model for the differentiation of MM. The dotted arrows and question mark indicate the unknown origin of stromal cells (either as a separate cell lineage or as a derivative of those cap cells that do not undergo epithelial differentiation).

1967; Saxén *et al.,* 1983; see Chapter 10). The inducer-dependent period of the mesenchyme-to-epithelium conversion has been called the *induction* period and the inducer-independent period the *morphogenetic* period (Saxén, 1987). During the inducer-dependent period, the mitotic cycle of cells in pretubular condensates speeds up (Saxén *et al.,* 1983) and there are widespread changes in gene expression levels and the molecular profiles of the induced cells (Donovan *et al.,* 1999). During the morphogenetic period, the main events are their transformation from a mesenchymal to an epithelial phenotype, formation of a lumen, and generation of the primitive nephron tubule.

As a result of inductive and growth interactions between the MM and collecting duct arborization, an adult kidney can have 600,000 to 1,000,000 nephrons in human and about 2000 to 5000 in mouse. Each tip of the branching collecting ducts induces one or two pretubular condensates from the MM, which then become situated on the proximal aspect of the tip (Fig. 13.4; Saxén, 1987). The pretubular condensates first form a comma-shaped and then an S-shaped epithelial structure that soon (approximately in 3 days in the mouse) partitions itself into the different segments of the secretory nephrons: the glomerulus, the proximal tubule, the loop of Henle, and the distal tubule, which fuses with and opens up to the collecting duct. Although the majority of secretory nephron cells are clearly derivatives of the MM, a virus-targeted marker gene expression study has suggested that a short segment of the distal tubule of the nephron might originate from the collecting ducts (Qiao *et al.,* 1995). It is not yet clear whether this arises by growth, transdifferentiation, or the mixing of duct and nephron epithelial cells.

The great majority of the length of the nephron is associated with secretion, whereas the so-called proximal region becomes the glomerulus, the link between the nephron and the capillary supply. Perhaps the most interesting component here is the podocyte population, as the molecular phenotype of these unusual epithelial cells has distinct mesenchymal attributes. Instead of keratins, desmoplakins, and other epithelial cell characteristics, podocytes express such stromal markers as FoxB2, Pod-1, p75 low-affinity nerve growth factor receptor (NGFR), and neuronal cell adhesion protein (NCAM), as well as the mesenchymal sytosceletal marker vimentin (Davies and Brändli, 1997). This phenotype is compatible with these cells being an intermediate state between mesenchyme and epithelial cells, and we return to this point at the end of the chapter.

A

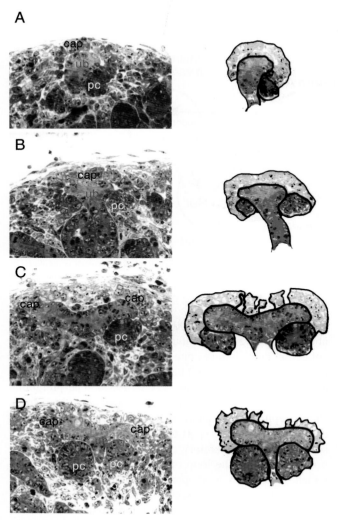

B

C

D

Figure 13.4 Early stages of nephron differentiation. Semithin sections showing different cell types at consecutive stages of nephrogenesis (A through D) [caps: yellow, stromal cells (S), pretubular condensates (pt, blue), ureteric buds (ub, red)].

2. Neuronal Cells

The metanephric kidney has both intrinsic (intrarenal) neuronal cells and extrinsic (extrarenal) nerve fibers. The kidney is innervated very early during its development, and the first neuronal precursor cells are already present in the undifferentiated metanephric kidney rudiment (Sariola *et al.*, 1988b). Organ culture experiments have shown that neurotrophic factor neurotrophin-3 (NT-3) maintains intrinsic neuronal precursor cells in the kidney and promotes their neuronal differentiation and neurite outgrowth (Karavanov *et al.*, 1995). NT-3 is expressed in many sites where there are epithelial–mesenchymal tissue interactions and its expression is regulated by Wnt signaling molecules (Patapoutian *et al.*, 1999) that are also important for nephron differentiation (Kuure *et al.*, 2000).

Neurons play an important physiological role in mature nephron function, but the mechanism by which they reach them is still not clear. As Wnt-4 and the trkC receptor for NT-3 are expressed not only by neuronal precursors but also by renal stromal cells (Sainio, 1996; Stark *et al.*, 1994), one possible cascade for the interplay of nephron differentiation and innervation could be transient expression of Wnt-4 by the pretubular condensates of MM leads to upregulation of NT-3 by the condensates and NT3 acts as a chemoattractant that attracts neurites toward the nephrons

3. Vascular Development

There are two distinct types of events in vascular differentiation of the early kidney: differentiation of the main arterial and venous vessels and generation of the capillary plexus. The former has not been studied closely in the embryonic kidneys, but the latter may well be regulated by chemotactic signals secreted by cleft cells at the comma stage of nephrogenesis (Chapter 16).

There are two main theories for renal vascular development: sprouting from preexisting capillaries (angiogenesis) and *in situ* formation of capillaries from the MM (vasculogenesis; for a review, see Risau, 1998). The angiogenetic mechanism is supported by the invasion of chick and quail capillaries into mouse kidney explants growing on chorioallantoic membranes (Sariola, 1985), the glomerular phenotype of transgenic mice deficient for platelet-derived growth factor (PDGF) or its receptor (Lindahl *et al.*, 1998), and the vascular patterning in Tie promoter green fluorescence mice (K. Sainio *et al.*, unpublished data) and in Flk1 heterozygous mutant mice with β-galactosidase as a marker for transgene expression (Barry *et al.*, 1998). These two transgenic mouse strains express the marker genes specifically in the endothelia, which in the urogenital region seem to originate directly from aorta (K. Sainio *et al.*, unpublished data) and contribute to glomerular vascular differentiation (Barry *et al.*, 1998). However, the vasculogenetic mechanism gains support from the phenotype of these transgenic mice with marker genes in blood vessel precursors (angioblasts). Such cells are detectable in early kidney rudiments and when such kidneys are cultured in the anterior chamber of eye where they develop blood vessels (Hyink and Abrahamson, 1995; Abrahamson, 1999, Woolf and Loughna, 1998). The present view is that both angiogenesis and vasculogenesis contribute to vessel formation in embryonic kidneys, and it is likely that that angiogenesis is particularly important in glomerular capillary formation (for further details, see Chapters 16).

As in many other organs, blood vessel development of the metanephric kidney is regulated by FGF-2, EphrinB-1, and vascular endothelial growth factor (Risau and Ekblom, 1986; Simon *et al.*, 1995; Takahashi *et al.*, 1998; Yancopoulos *et al.*, 1998). Some molecules that regulate vessel growth

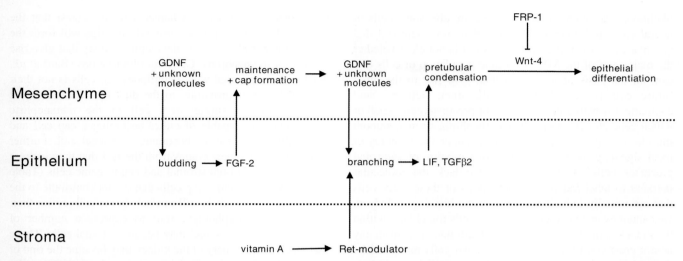

Figure 13.5 Sequence of inductive events during early nephrogenesis. Alternative views from the stromal contribution in ureteric branching are presented in Fig. 13.2. FRP-1: frizzled-related protein 1.

we would like to consider as separate are actually interrelated, we cannot exclude the possibility that addition of a single signal may induce an unexpectedly complicated result.

Nevertheless, organ culture does allow experimental investigation of signal function under controlled, if artificial, conditions and the results to date have given useful information and testable hypotheses about the regulation of MM differentiation. For example, organ culture experiments have shown that uninduced MM is rescued from apoptosis by FGF-2 (Barasch *et al.*, 1997), which may be the first step in the multistage cascade leading to nephron differentiation. Here, FGFs and bone morphogenetic proteins can act synergistically to maintain the competence of MM (Dudley *et al.*, 1999). The next step, epithelialization of nephrons, can be induced in primed MMs if FGF-2 is combined with conditioned medium from ureteric bud cell cultures (Karavanova *et al.*, 1998) or if the FGF-2 pulse is followed by leukemia inhibitory factor (LIF) treatment (Barasch *et al.*, 1999), and this epithelialization is obviously dependent on an autocrine action of Wnt-4 (Stark *et al.*, 1994). The difficulty of interpreting these results is highlighted by the fact that, under culture conditions, isolated rat but not mouse MM undergoes full epithelial differentiation in the presence of TGF-β2, LIF, and FGF2 (Plisov *et al.*, 2001), whereas the LIF receptor knockout mouse has functional kidneys, although they are smaller than in wild-type mice (Yoshida *et al.*, 1996). One possibility is that, as this sequential action of inductive signals promotes nephron differentiation efficiently in rat, other members of the interleukin-6 family, such as oncostatin M, could act as tubule inducers in mouse.

A further advantage of organ culture here is that it can throw up unexpected results whose significance may not be immediately clear but that provide clues for future thinking and experimentation. In this context, it is noteworthy that

cell lines transfected with various members of the Wnt-family can induce full epithelial differentiation of isolated MM (Herzlinger *et al.*, 1994; Kispert *et al.*, 1998), raising the possibility that Wnt proteins not only represent autocrine epithelializing factors, but may also promote full epithelial differentiation of the MM. At the moment, this seems unlikely as the Wnt proteins expressed by the ureteric bud, Wnt-6, Wnt-7B, and Wnt-11, do not induce MM in cell culture assays, whereas the Wnts that do induce tubulogenesis are either expressed by the pretubular condensates or not expressed by the embryonic kidney (Kuure *et al.*, 2000). A plausible explanation for *in vitro* data is that Wnt-transfected cells express other signals that prime the MM. The question will remain open until Wnt proteins have been purified and tested as purified recombinant proteins on the MM. In addition, these experiments may be interesting in studying the differentiation of other cell types, as recent data suggest that Wnts may also have roles in renal innervation and angiogenesis (S. Vainio, personal communication; Patapoutian *et al.*, 1999).

VI. How Many Cell Types Are Present in the Metanephric Blastema?

This is of course the key question that this chapter addresses and, as should now be clear, it is not one that the present state of knowledge allows us to answer properly. This section considers how far we are from providing an answer and what further information will be needed to come to a clear, unambiguous description of the original lineages and the hierarchy of cell types to which they give rise.

There are two major cell types in the metanephric rudiment that are clearly distinct, the bud and the MM, with both deriving from the intermediate mesoderm, with the likelihood that, within the MM, there are also some cells of neural crest and endothelial origin that form neuronal and vascular cells. Even if the latter is so, it is not clear whether the major part of the MM is a single population of cells that can, on the basis of local signaling, give rise to the other future cell types (apart from the duct, neurons, and capillaries); whether it has a range of precursor cells, each of which generates its own cell type (nephron, stroma, smooth muscle, juxtaglomerular, etc.), again under the stimulus of local signaling; or whether there is a mixture of stem and precursor cells. At the moment, we lack the molecular markers to label individual populations of these early cells that are initially indistinguishable to the histologist. Indeed, we cannot even be certain that all the cells found later within the kidney actually derive from the blastema: apart from any neural crest contribution, the endothelial cells may well be derived from the capillary network surrounding the kidney rudiment rather than the nephrogenic mesenchyme. If so, then the original question reduces to asking about the number of precursor/stem cell populations that give rise to the cells of nephrons, stroma, smooth muscle, and the juxtaglomerular complex. Little is known of the ontogeny of the last two of these cell populations, but there are some relevant data on the origins of nephrons and stroma; the two cell types that differentiate well in culture and that clearly derive from the MM.

The view that the field currently seems to hold is that a stem cell is defined as a cell that is capable of giving rise to at least two distinct differentiation pathways, as well as reproducing itself. If a cell can only reproduce itself and give rise to a single phenotype (e.g., skin ectoderm), then it should be considered a precursor cell. These definitions are hard to apply to kidney morphogenesis partly for two reasons. First, it has proven difficult to grow MM cells *in vitro* and it is therefore impossible to see if the cells can be directed down to two distinct pathways, a problem made all the harder as we still do not know what signals might be used for this purpose (see earlier discussion). Second, even if we could manipulate mesenchyme cells in culture by the addition of signals, this would not prove that we had accurately mimicked the events taking place *in vivo*. Nevertheless, observations of kidney development *in vivo* (see Chapter 10) do suggest that the dense cells of the cap that surround duct tips will form the cells at the periphery of the developing kidney that give rise to most of the nephrons. Confocal observations (Bard *et al.*, 2001) also imply that what determines cap cells is not their lineage, but their proximity to the duct tip. If this is so, it indicates that a single stem cell in the metanephric mesenchyme is capable of either becoming a cap cell and eventually a nephron or becoming a stromal cell. Further evidence for this view comes from the fact that they express early markers for both stromal and nephrogenic cells (Table 13.3). The fate of those cap cells that do not contribute to the pretubular aggregates is still unclear. Some cells retain their mesenchymal morphology, start to express a number of stromal cells markers, and may become stromal cells. Those that reach the periphery of the kidney may become the rind of blastemal cells that later become nephrons. Once the pretubular aggregate has started to form, it is likely that it gives rise to all of the cells and cell types of the nephron, apart from the endothelial and mesangial cells of the glomerulus. Less clear, however, is how the nephron becomes patterned to form the distal tubule, the loop of Henle, the proximal tubule to the glomerulus, and the podocytes, each of which has its own functional capacities (see Chapter 10). The segment-specific patterns of Notch receptors and their ligands may be critical in the segregation of different nephron segments (K. Sainio, unpublished data). Notch-2-deficient mice develop podocytes but never produce mature glomeruli (McCright *et al.*, 2001).

Two sorts of interactions can initiate patterning within the early nephron. The first is the interaction that takes place between the pretubular condensation and the surface of the duct epithelium, an interaction that certainly determines the location of the future nephron and distal tubule. There is, however, a second plausible pattern-forming interaction and that is via the abutting cap, which is also formed through an inductive interaction with the duct. As this is several cells deep, it may contain a gradient of some signal extending from the duct surface to the outer edge of the cap. If so, this gradient could help pattern the condensation, with condensate cells adjacent to the duct becoming distal

Table 13.3 Some Useful Markers for Different Cell Types or Stages of Nephrogenesis

Domain	Markers[a]	Comments
Metanephric mesenchyme	FoxC1, WT-1, Pax-2, Lim-1, GDNF	These genes are expressed by *uninduced* metanephric mesenchyme but also by different cell types in later nephrogenesis
Cap	NCAM, p75, neurotrophin receptor	These genes are expressed by cap cells but also by pretubular condensates and stromal cells
Stroma	RALDH2, FoxB2, ganglioside GD3,	These genes are specifically expressed by the stromal cells
Pretubular condensates	Eya-1, Pax-2, WT-1, Lim-1, Wnt-4, GDNF	These genes mark pretubular condensates but also caps and undifferentiated nephrogenic mesenchyme (where the expression levels are clearly low)

[a] For references, see Davies and Brändli (1997). Note that, except those for stromal cells, markers show overlapping patterns.

tubules, whereas those furthest away become podocytes. It is noteworthy here, that the phenotype of these cells is intermediate between epithelial and stromal cells (see earlier discussion and Fig. 13.4).

This dual signaling model is compatible with gene expression, organ culture, and gene-targeting studies. For instance, when the MM is induced in transfilter culture, where the MM is placed on the top of a porous filter and a piece of spinal cord serves as an inducer, the podocytes form a network instead of single podocyte-like bodies. In FoxB2-deficient mice, stromal differentiation proceeds normally, but epithelial differentiation is rudimentary, and mesenchymal structures, which have been referred to as large pretubular condensates, do not undergo epithelial differentiation (Hatini et al., 1996). These condensates may thus represent cap cell populations rather than pretubular condensates.

In conclusion, it should be emphasized that the lineage relationships within the developing metanephric mesenchyme, particularly as they reflect the formation of nephrons and stromal cells, are still not understood, but the problems can, in the light of burgeoning molecular data, be framed more accurately than has hitherto been possible. Things will, however, remain complicated until we can be sure that neuronal cells of the kidney derive from the neural crest and that capillary cells of the kidney derive from the surrounding vascular system, and thus have their own lineages. Once this is established, the focus will shift to deciding whether nephrons and stroma derive from a single blastemal cell. The key to solving these problems will be through the use of proper lineage markers. Once we have these, we will be in a far better position to investigate the wide range of signals that mediate kidney development and the way in which the different cell lineages interact with each other.

References

Abrahamson, D. R. (1999). Glomerular endothelial cell development. *Kidney Int.* **56**, 1597–1598.

Bard, J. B. L. (2001). Growth and death in the developing mammalian kidney: Signals receptors and conversations. *BioEssays*.

Bard, J. B. L., Gordon, G., Sharp. L., and Sellers, W. I. (2001). Early nephron formation in the developing mouse kidney. *J. Anat.* **199**, 385–392.

Barasch, J., Qiao, J., McWilliams, G., Chen, D., Oliver, J., and Hezlinger, D. (1997). Ureteric bud cells secrete multiple factors, incuding bFGF, which rescue renal progenitors from apoptosis. *Am. J. Physiol* **273**, F757–7676.

Barasch, J., Yang, J., Ware, C. B., Taga, T., Yoshida, K., Erdjument-Bromage, H., Tempst, P., Parravicini, E., Malach, S., Aranoff, T., and Oliver, J. A. (1999). Mesenchymal to epithelial conversion in rat metanephros is induced by LIF. *Cell* **99**, 377–386.

Bard, J. (1996). A new role for the stromal cells in kidney development. *Bioessays* **18**, 705–707.

Bard, J. B., McConnell, J. E., and Davies, J. A. (1994) Towards a genetic basis for kidney development. *Mech. Dev.* **48**, 3–11.

Batourina, E., Gim, S., Bello, N., Shy, M., Clagett-Dame, M., Srinivas S., Costantini, F., and Mendelsohn, C. (2001). Vitamin A controls epithelial/

mesenchymal interactions through Ret expression. *Nature Genet.* **27**, 74–78.

Blantz, R. C. (1993). Glomerular blood flow. *Semin. Nephrol.* **13**, 436–446.

Bronner-Fraser, M., and Fraser, S. E. (1988). Cell lineage analysis reveals multipotency of some avian neural crest cells. *Nature* **335**, 161–164.

Burrow, C. R. (2000). Regulatory molecules in kidney development. *Pediatr. Nephrol.* **14**, 240–253.

Call, K. M., Glaser, T., Ito, C. Y., Buckler, A. J., Pelletier, J., Haber D. A., Rose, E. A., Kral, A., Yeger, H., Lewis, W. H., Jones, C., and Housman, D. E. (1996). Isolation and characterisation of a zinc finger olypeptide gene at the human chromosome 11 Wilms' tumor locus. *Cell* **60**, 509–520.

Cancilla, B., Ford-Perriss, M. D., and Bertram, J. F. (1999). Expression and localization of fibroblast growth factors and fibroblast growth factor receptors in the developing rat kidney. *Kidney Int.* **56**, 2025–2039.

Caspary, T., Michele, A., Cleary, M. A., Perlman, E. J., Zhang, P., Elledge, S. J., and Tilghman, S. M. (1999). Oppositely imprinted genes p57Kip2 and Igf2 interact in a mouse model for Beckwith-Wiedemann syndrome. *Genes Dev.* **13**, 3115–3124.

Davies, J. A., and Bard, J. B. L. (1998). The development of the kidney. *Curr. Top. Dev. Biol.* **39**, 245–301.

Davies, J. A., and Brändli, A. W. (1997). "The kidney development database" http://mbisg2.sbc.man.ac.uk/kidbase/kidhome.html and http://www.ana. ed.ac. uk/anatomy/kidbase/kidhome. html.

Dressler, G. R. (1997). Genetic control of kidney development. *Adv. Nephrol. Necker Hosp.* **26**, 1–17.

Dressler, G. R., Wilkinson, J. E., Rothenpieler, U. W., Patterson, L. T., Williams-Simons, L., and Westphal, H. (1993). Deregulation of Pax-2 expression in transgenic mice generates severe kidney abnormalities. *Nature* **362**, 65–67.

Dudley, A. T., Godin, R. E., and Robertson, E. J., (1999). Interaction between FGF and BMP signaling pathways regulates development of metanephric mesenchyme. *Genes Dev.* **13**, 1601–1613.

Dudley, A. T., Lyons, K. M., and Robertson, E. J. (1995). A requirement for bone morphogenetic protein 7 during development of the mammalian kidney and eye. *Genes Dev.* **9**, 2795–2807.

Dudley, A. T., and Robertson, E. J. (1997). Overlapping expression domains of bone morphogenetic protein family members potentially account for limited tissue defects in BMP7 deficient embryos. *Dev. Dyn.* **208**, 349–362.

Fishman, M. C., and Kirby, M. L. (1998). Fallen arches, or how the vertebrate got its head. *J. Clin. Invest.* **102**, 1–3.

Gessler, M., Poustka, A., Cavenee, W., Neve, R. L., Orkin, S. H., and Bruns, G. A. (1990). Homozygous deletion in Wilms tumours of a zinc-finger gene identified by chromosome jumping. *Nature* **343**, 774–778.

Godin, R. E., Robertson, E. J., and Dudley, A. T. (1999). Role of BMP family members during kidney development. *Int. J. Dev. Biol.* **43**, 405–411.

Gossens, C., and Unsworth, B. (1972). Evidence for two-step mechanism operating during *in vitro* mouse kidney tubulogenesis. *J Embryol. Exp. Med.* **28**, 615–663.

Grobstein, C. (1953a). Inductive epithelio-mesenchymal interaction in cultured organ rudiments of the mouse. *Science* **118**, 52–55.

Grobstein, C. (1953b). Morphogenetic interaction between embryonic mouse tisssues separated by a membrane filter. *Nature* **172**, 869–871.

Grobstein, C. (1955). Inductive interaction in the development of the mouse metanephros. *J. Exp. Zool.* **130**, 319–340.

Grobstein, C. (1956). Trans-filter induction of tubules in mouse metanephrogenic mesenchyme. *Exp. Cell Res.* **10**, 424–440.

Grobstein, C. (1967). Mechanisms of organogenetic tissue interaction. *Natl. Cancer Inst. Monogr.* **26**, 107–119.

Hatini, V., Huh, S. O., Herzlinger, D., Soares, V. C., and Lai, E. (1996). Essential role of stromal mesenchyme in kidney morphogenesis revealed by targeted disruption of Winged Helix transcription factor BF-2. *Genes Dev.* **10**, 1467–1478.

Herzlinger, D., Qiao, J., and Cohen, D., Ramakrishna, N., and Brown, A. M. (1994). Induction of kidney epithelial morphogenesis by cells expressing Wnt-1. *Dev. Biol.* **166,** 815–818.

Hosono, S., Luo, X., Hyink, D. P., Schapp, L. M., Wilson, P. D., Burrow, C. R., Reddy, J. C., Atweh, G. F., and Licht, J. D. (1999). WT1 expression induces features of renal epithelial differentiation in mesenchymal fibroblasts. *Oncogene* **18,** 417–427.

Hyink, D. P., Abrahamson, D. R. (1995). Origin of the glomerular vasculature in the developing kidney. *Semin. Nephrol.* **15,** 300–314.

Ito, T., Suzuki, A., Imai, E., Okabe, M., and Hori, M. (2002). Bone marrow is a reservoir of repopulating cells over the glomerular remodelling. *J. Am. Soc. Nephrol.* **12,** 2625–2635.

Jena, N., Martin-Seisdedos, C., McCue, P., and Croce, C. M. (1997). BMP7 null mutation in mice: Developmental defects in skeleton, kidney, and eye. *Exp. Cell Res.* **230,** 28–37.

Karavanov, A., Sainio, K., Palgi, J., Saarma, M., Saxén, L., and Sariola, H. (1995). Neurotrophin-3 rescues neuronal precursors from apoptosis and promotes neuronal differentiation in the embryonic metanephric kidney. *Proc. Natl. Acad. Sci. USA* **92,** 11279–11283

Karavanova, I. D., Dove, L. F., Resau, J. H., and Perantoni, A. O. (1996). Conditioned medium from a rat ureteric bud cell line in combination with bFGF induces complete differentiation of isolated metanephric mesenchyme. *Development* **122,** 4159–4167.

Kispert, A., Vainio, S., and McMahon, A. P. (1998). Wnt-4 is a mesenchymal signal for epithelial transformation of metanephric mesenchyme in the developing kidney. *Development* **125,** 4225–4234.

Kreidberg, J. A., Sariola, H., Loring, J. M., Maeda, M., Pelletier, J., Housman, D., Jaenisch, R. (1993). WT1 is required for early kidney development. *Cell* **74,** 679–691.

Kuure, S., Vuolteenaho, R., and Vainio, S. (2000). Kidney morphogenesis: Cellular and molecular regulation. *Mech Dev.* **92,** 19–30.

Laitinen, L., Virtanen, I., Saxén, L., and Lehtonen, E. (1998). Expression of cellular glycoconjugates in transfilter-induced metanephric mesenchyme. *Anat. Rec.* **220,** 190–197.

Laitinen, L., Virtanen, I., and Saxén L. (1987). Changes in the glycosylation pattern during embryonic development of mouse kidney as revealed with lectin conjugates. *J Histochem Cytochem* **35,** 55–65.

Larsson, S. H., Charlieu, J. P., Miyagawa, K., Engelkamp, D., Rassoulzadegan, M., Ross, A., Cuzin, F., van Heyningen, V., and Hastie N. D. (1995). Subnuclear localization of WT1 in splicing or transcription factor domains is regulated by alternative splicing. *Cell* **81,** 391–401.

LeDouarin, N. (1982). "The Neural Crest." Cambridge Univ. Press, Cambridge.

LeDouarin, N., and Teillet, M.-A. (1974). Experimental analysis of the migration and differentiation of neuroblasts of the autonomic nervous system and of neuroectodermal derivatives, using a biological cell marking technique. *Dev. Biol.* **41,** 162–184.

Lee, S. B., Huang, K., Palmer, R., Truong, V. B., Herzlinger, D., Kolquist, K. A., Wong, J., Paulding, C., Yoon, S. K., Gerald, W., Oliner, J. D., and Haber, D. A. (1999). The Wilms tumor suppressor WT1 encodes a transcriptional activator of amphiregulin. *Cell* **98,** 663–673.

Lee, S. B., and Haber, D. A. (2001). Wilms tumor and the WT1 gene. *Exp. Cell Res.* **10,** 264:74–99.

Lehtonen, E., Wartiovaara, J., Nordling, S., and Saxén, L. (1975). Demonstration of cytoplasmic processes in Millipore filters permitting kidney tubule induction. *J. Embryol. Exp. Morphol.* **33,** 187–203.

Lindahl, P., Hellstrom, M., Kalen, M., Karlsson, L., Pekny, M., Pekna, M., Soriano, P., and Betsholtz, C. (1998). Paracrine PDGF-B/PDGF-Rbeta signaling controls mesangial cell development in kidney glomeruli. *Development* **125,** 3313–3322.

McCright, B., Gao, X., Shen, L., Lozier, J., Lan, Y., Maguire, M., Herzlinger, D., Weinmaster, G., Jiang R., and Gridley, T. (2002). Defects in development of the kidney, heart and eye vasculature in mice homozygous for a hypomorphic Notch2 mutation. *Development* **128,** 491–502.

Minowada, G., Jarvis, L. A., Chi, C. L., Neubuser, A., Sun, X., Hacohen, N., Krasnow, M. A., Martin, G. R. (1999). Vertebrate Sprouty genes are induced by FGF signaling and can cause chondrodysplasia when overexpressed. *Development* **126,** 4465–4475.

Nguyen, H. Q., Danilenko, D. M., Bucay, N., DeRose, M. L., Van G. Y., Thomason, A., and Simonet, W. S. (1996). Expression of keratinocyte growth factor in embryonic liver of transgenic mice causes changes in epithelial growth and differentiation resulting in polycystic kidneys and other organ malformations. *Oncogene* **10,** 2109–2119.

Oliver, J. A., and Al-Awqati, Q. (1998). An endothelial growth factor involved in rat renal development. *J. Clin. Invest.* **102,** 1208–1219.

Patapoutian, A., Backus, C., Kispert, A., and Reichardt, L. F. (1999). Regulation of neurotrophin-3 expression by epithelial-mesenchymal interactions: the role of Wnt factors. *Science* **283,** 1180–1183.

Pepper, M. S., Ferrara, N., Orci, L., and Montesano, R. (1995). Leukemia inhibitory factor (LIF) inhibits angiogenesis *in vitro. J. Cell Sci.* **108,** 73–83.

Plisov, S. Y., Yoshino, K., Dove, L. F., Higinbotham, K. G., Rubin, J. S., and Perantoni, A. O. (2001). TGF-β2, LIF, and FGF2 cooperate to induce nephrogenesis. *Development* **128,** 1045–1057.

Pritchard-Jones, K. (1999). The Wilms tumour gene, WT1, in normal and abnormal nephrogenesis. *Pediatr. Nephrol.* **13,** 620–625.

Qiao, J., Cohen, D., and Herzlinger, D. (1995). The metanephric blastema differentiates into collecting system and nephron epithelia *in vitro. Development* **121,** 3207–3214.

Qiao, J., Uzzo, R., Obara-Ishihara, T., Degenstein, L., Fuchs, E., and Herzlinger, D. (1999). FGF-7 modulates ureteric bud growth and nephron number in the developing kidney. *Development* **126,** 547–554.

Reich, A., Sapir, A., and Shilo, B. (1999). Sprouty is a general inhibitor of receptor tyrosine kinase signaling. *Development* **126,** 4139–4147.

Risau, W. (1998). Development and differentiation of endothelium. *Kidney Int.* **67,** S3–S6.

Risau, W., and Ekblom, P. (1986). Production of a heparin-binding angiogenesis factor by the embryonic kidney. *J. Cell Biol.* **103,** 1101–1107.

Saarma, M., and Sariola, H. (1999). Other neurotrophic factors: Glial cell line-derived neurotrophic factor (GDNF). *Microsc. Res. Tech.* **45,** 292–302.

Sainio, K. (1996). "Neuronal Characteristics in the Early Metanephric Kidney". Academic Dissertation, University of Helsinki.

Sainio, K., Nonclercq, D., Saarma, M., Palgi, J., Saxén, L., and Sariola, H. (1994). Neuronal characteristics in embryonic renal stroma. *Int. J. Dev. Biol.* **38,** 77–84.

Sainio, K., and Raatikainen-Ahokas, A. (1999). Mesonephric kidney: A stem cell factory? *Int. J. Dev. Biol.* **43,** 435–439.

Sainio, K., Suvanto, P., Davies, J., Wartiovaara, K., Saarma, M., Arumäe, U., Meng, X., Lindahl, M., Pachnis, V., and Sariola, H. (1997). Glial-cell-line-derived neurotrophic factor is required for bud initiation from ureteric epithelium. *Development* **124,** 4077–4087.

Sariola, H. (1985). Interspecies chimeras: An experimental approach for studies on embryonic angiogenesis. *Med. Biol.* **63,** 43–65.

Sariola, H., Aufderheide, E., Bernhard, H., Henke-Fahle, S., Dippold, W., Ekblom, P. (1998a). Antibodies to cell surface ganglioside GD3 perturb inductive epithelial -mesenchymal interactions. *Cell* **54,** 235–245.

Sariola, H., Holm, K., and Henke-Fahle, S. (1988b). Early innervation of the metanephric kidney. *Development* **104,** 589–599.

Sariola, H., and Sainio, K. (1997). Tip-top branching ureteric bud. *Curr. Opin. Cell Biol.* **9,** 877–884.

Saxén, L. (1987). "Organogenesis of the Kidney". Cambridge Univ. Press, Cambridge.

Saxén, L., Koskimies, O., Lahti, A., Miettinen, H., Rapola, J., and Wartiovaara, J. (1968). Differentiation of kidney mesenchyme in an experimental model system. *Adv. Morphol.* **7,** 251–293.

Saxén, L., and Lehtonen, E. (1978). Transfilter induction of kidney tubules as a function of the extent and duration of intercellular contacts. *J. Embryol. Exp. Morphol.* **47,** 97–109.

Saxén, L., Lehtonen, E., Karkinen-Jaaskelainen, M., Nordling, S., and Wartiovaara, J. (1976). Are morphogenetic tissue interactions mediated by transmissible signal substances or through cell contacts? *Nature* **259**, 662–663.

Saxén, L. (1970). Failure to show tubule induction in a heterologous mesenchyme. *Dev. Biol.* **23**, 511–523.

Saxén, L., Salonen, J., Ekblom, P., and Nordling, S. (1983). DNA synthesis and cell generation cycle during determination and differentiation of the metanephric mesenchyme. *Dev. Biol.* **98**, 130–138.

Shoyab, M., McDonald, V. L., Bradley, J. G., and Todaro, G. J. (1998). Amphiregulin: A bifunctional growth-modulating glycoprotein produced by the phorbol 12-myristate 13-acetate-treated human breast adenocarcinoma cell line MCF-7. *Proc. Natl. Acad. Sci. USA* **85**, 6528–6532.

Simon, M., Grone, H. J., Johren, O., Kullmer, J., Plate, K. H., Risau, W., and Fuchs, E. (1995). Expression of vascular endothelial growth factor and its receptor in human renal ontogenesis and in adult kidney. *Am. J. Physiol.* **268**, F240–F250.

Stark, K., Vainio, S., Vassileva, G., and McMahon, A. P. (1994). Epithelial transformation of metanephric mesenchyme in the developing kidney regulated by Wnt-4. *Nature* **372**, 679–683.

Takahashi, T., Huynh-Do, U., and Daniel, T. O. (1998). Renal microvascular assembly and repair: Power and promise of molecular definition. *Kidney Int.* **53**, 826–835.

Torres, M., Gomez-Pardo, E., Dressler, G. R., and Gruss, P. (1995). Pax-2 controls multiple steps of urogenital development. *Development* **121**, 4057–4065.

Towers, P. R., Woolf, A. S., and Hardman, P. (1998). Glial cell line-derived neurotrophic factor stimulates ureteric bud outgrowth and enhances survival of ureteric bud cells *in vitro*. *Exp. Nephrol.* **6**, 337–351.

Wartiovaara, J., Lehtonen, E., Nordling, S., and Saxén, L. (1972). Do membrane filters prevent cell contacts? *Nature* **238**, 407–408.

Woolf, A. S., and Loughna, S. (1998). Origin of glomerular capillaries: Is the verdict in? *Exp. Nephrol.* **6**, 17–21.

Yancopoulos, G. D., Klagsbrun, M., and Folkman, J. (1998). Vasculogenesis, angiogenesis, and growth factors: Ephrins enter the fray at the border. *Cell* **93**, 661–664.

Yoshida, K., Taga, T., Saito, M., Suematsu, S., Kumanogoh, A., Tanaka, T., Fujiwara, H., Hirata, M., Yamagami, T., and Nakahata, T. K. (1996). Targeted disruption of gp130, a common signal transducer for the interleukin-6 family of cytokines, leads to myocardial and hematological disorders. *Proc. Natl. Acad. Sci. USA* **93**, 407–411.

14

Formation and Development of Nephrons

Eun Ah Cho and Gregory R. Dressler

I. Introduction

The development of renal tubules in the adult kidney, or metanephros, begins with the reciprical, inductive inter-actions between the ureteric bud epithelium and the meta-nephric mesenchyme. The classical model predicts that signals emanating from the ureteric bud epithelial promote conversion of the metanephric mesenchyme to tubular epithelium, where-as signals from the mesenchyme promote growth and branch-ing morphogenesis of the ureteric bud. As growth and invasion of the ureteric bud have been discussed in previous chapters, this chapter focuses on the metanephric mesenchyme, its responses to inductive signals, and the factors required for the conversion of the mesenchyme to tubular epithelium.

Prior to induction, the metanephric mesenchyme lineage is already fixed within the environment of the intermediate mesoderm. Several key factors are now known to mark this as the renal anlagen and are required for epithelial specifi-

cation. The signals that induce the mesenchyme to become epithelium are also being elucidated, although many ques-tions still remain. Finally, the chapter discusses factors that promote growth, remodeling, and cell type specification of the nephron segments.

The conversion of metanephric mesenchyme to tubular epithelium has been studied intensively since the 1960s. The morphology of the developing tubules, from uninduced mesenchyme to fully differentiated epithelium, is described in detail in a classic monograph by Saxén (1987). With respect to the parts of the nephron, the prevailing view held that the glomerular, the proximal tubular, and the distal tubular epithelia were derived from the metanephric mesenchyme, whereas the branching ureteric bud generated the collecting ducts and tubules. However, precise cell lineage mapping has revealed more plasticity between the two renal primordial cell types than previously appreciated (Herzlinger *et al.*, 1992; Qiao *et al.*, 1995). Epithelial cells from the ureteric bud are able to delaminate and contribute to more proximal parts of the nephron. Thus, if we consider the proximal distal axis along the tubular epithelium of the nephron, from glomerulus to collecting duct, it becomes more difficult to draw a clear boundary between ureteric bud-derived and metanephric mesenchyme-derived tubular epithelium. Nevertheless, it is clear that much of the tubular epithelium is generated from the induced mesenchyme through a well-characterized sequence of events.

This chapter introduces genetic and biochemical mech-anisms that are implicated in the development of the renal epithelia. As with any scientific endeavor, there is some con-troversy and conflicting data regarding potential mechanisms that regulate early nephron formation. This chapter attempts to relate recent insights, gained primarily from the phenotypic analysis of mutant mice, with more classical *in vitro* and biochemical data regarding induction and differentiation. To put these factors into a morphological context, a brief description of the process of mesenchyme-to-epithelial conversion follows.

II. Morphogenesis

In the mouse and rat, the mammalian species studied most extensively, the ureteric bud makes contact with the metanephric mesenchyme at approximately embryonic days 11 (E11) and E13, respectively. As outlined in Fig. 14.1, the initial mesenchymal response to ureteric-bud-derived inductive signals is condensation of mesenchymal cells, three to four cell layers thick, around the tip of the ureteric bud. As the bud undergoes dichotomous branching, this condensed mesenchyme forms an aggregate that remains associated, at one side, with the ureteric bud epithelium. While the ureteric bud proliferates outward, along the radial axis of the kidney, the aggregate cells begin to show signs of epithelial polarity. Expression of new cadherin genes is induced and a laminin-containing basement membrane is laid down. The polarized aggregate is known as the renal vesicle. This vesicle develops two clefts, resulting in a characteristic S-shaped structure, whereas the part of the renal vesicle abutting the ureteric bud epithelium fuses to form a continuous epithelial lumen. At this stage, the S-shaped body is already compartmentalized into distinct groups of cells, based on the expression of a variety of marker genes. The cleft furthest from the ureteric bud becomes the glomerular epithelium, consisting of the visceral epithelium or podocyte layer and the parietal epithelium of Bowmans capsule. Endothelial cells invade this distal cleft–distal with respect to the ureteric bud, although it generates the proximal part of the nephron–and begin to form the glomerular tuft. The junction between the mesenchymal-derived epithelium and the ureteric bud epithelium is not entirely clear, as cell lineage tracing reveals a degree of cell mixing between these two precursor cell populations. Presumably, that part of the S-shaped body that connects to the ureteric bud will generate the distal and collecting tubules. Proximal tubule precursors are located between the two ends of the S-shaped structure.

As this rudimentary nephron continues to proliferate, the proximal tubules form a highly convoluted mass of epithelium that then leads straight into the medullary zone to generate the descending and ascending limbs of Henle's loop. The ascending limb of Henle's loop returns to establish contact with the juxtaglomerular cells of the same nephron at the macula densa, a system designed to provide feedback and control the release of intrarenal renin. Epithelial cells along the nephron are highly specialized and express many unique markers, depending on their position. Little is known about the terminal differentiation of epithelial subtypes in the nephron. However, the proper localization of transmembrane water and ion channels and ion pumps is essential for maintaining electrolyte and water homeostasis and regulating blood pressure.

Much of the early morphogenetic events can be mimicked *in vitro* using organ culture methods. The kidney rudiment is dissected out of an E11 or E13 mouse or rat embryo, just after induction when the ureteric bud has invaded the mesenchyme and formed an ampullae-shaped structure. After 2 days, the cultured kidneys exhibit extensive ureteric bud branching and mesenchymal condensations and epithelial derivatives (Fig. 14.2). Expression of early marker genes is very similar to the *in vivo* pattern of gene expression, with some notable differences. *In vitro*-cultured kidneys undergo only an initial burst of cell proliferation, do not develop the renal vasculature, and do not exhibit the characteristic ascending and descending limbs of Henle's loop. Despite these limitations, the early events of mesenchyme-to-epithelial conversion have been modeled quite successfully in organ culture up to about the S-shaped body stage.

III. Induction

A. Nature of Inductive Signals

In the developing kidney, epithelial and mesenchymal cell–cell interactions are necessary for the survival and differentiation of both cell types. The fundamental tenets governing these interactions were first proposed by Grobstein (1956), who utilized primarily *in vitro* organ culture methods to formulate his conclusions (see Fig. 14.3). He was able to microdissect the ureteric bud from the metanephric mesenchyme at a time when the mesenchyme had not shown any measurable response to the bud (E11.0) and culture the two tissues separately and in combination. In these experiments, neither the bud epithelium nor the metanephric mesenchyme were able to proliferate or differentiate *in vitro* because of the lack of inductive signals emanating from the other tissue. Thus, in the model of reciprocal inductive signaling, the mesenchyme provided essential signals to promote growth and branching of the ureteric bud, whereas the bud epithelium provided signals to promote aggregation and differentiation of mesenchyme into the epithelium of the developing nephron. Furthermore, the metanephric mesenchyme could be induced by tissues other than the ureteric bud. Among the heterologous tissues tried, the dorsal embryonic neural tube proved a powerful inducer of kidney mesenchyme that was easy to isolate and grow (Lombard and Grobstein, 1969; Unsworth and Grobstein, 1970). However, the metanephric mesenchyme would always form characteristic renal epithelium, no matter what inducing tissue was used, suggesting

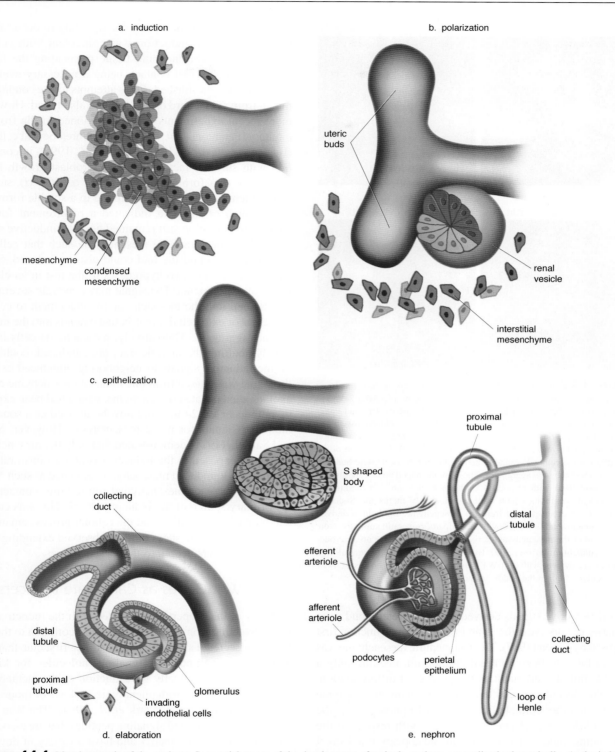

a. induction

mesenchyme

condensed mesenchyme

b. polarization

uteric buds

renal vesicle

interstitial mesenchyme

c. epithelization

S shaped body

collecting duct

distal tubule

proximal tubule

invading endothelial cells

glomerulus

d. elaboration

proximal tubule

efferent arteriole

afferent arteriole

podocytes

perietal epithelium

distal tubule

collecting duct

loop of Henle

e. nephron

Figure 14.1 Morphogenesis of the nephron. Sequential stages of the development of a single nephron are outlined schematically. At the induction phase, mesenchymal cells aggregate at the tips of the ureteric bud. Not all mesenchyme becomes aggregates, as some cells remain along the periphery as potential mesenchymal stem cells and others become interstitial mesenchyme. The aggregate becomes polarized into a renal vesicle, which remains closely associated with the ureteric bud. Meanwhile, a new aggregate forms at the growing tip of the bud as the next generation of nephron is induced. At the S-shaped body stage, the polarized vesicle elongates and forms two visible clefts, the most distal of which becomes infiltrated with endothelial precursor cells. The glomerulus is beginning to take shape and much of the epithelia now express specific markers for proximal or distal tubules. Proliferation of tubular epithelium generates a highly convoluted proximal and distal part, whereas the tubules of Henle's loop now extend toward the medullary zone along the radial axis of the kidney. The glomerular tuft takes shape as endothelial cells, and the visceral epithelia, or podocytes, come into contact to form the filtration unit.

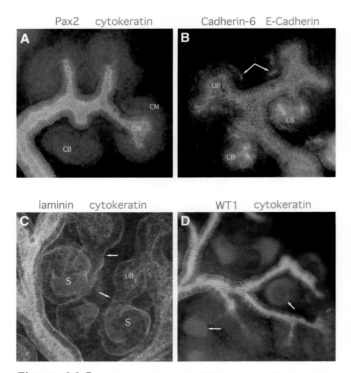

Figure 14.2 *in vitro* tubulogenesis. Whole-mount antibody staining against proteins expressed in the early kidney rudiment. (A) After 2 days in culture, multiple branches of the ureteric bud (UB) are evident and Pax2-positive cells of the condensed mesenchyme (CM) are closely associated with the bud tips. Some Pax2-positive cells have undergone polarization into renal vesicles and comma-shaped bodies (CB). (B) Expression of the cell adhesion molecules E-cadherin and cadherin-6 mark ureteric bud epithelia and mesenchymal aggregates. The comma-shaped bodies are already compartmentalized with respect to cadherin expression. (C) Staining of the basement membranes with antilaminin clearly marks the S-shaped structures (S) that are derived from the mesenchyme. Epithelium derived from the ureteric bud is stained with an antipan cytokeratin antibody. Note the junction of the mesenchymal-derived epithelium and the ureteric bud-derived epithelium (arrow). (D) By 4 days in culture, glomeruli-like structures are evident with anti-WT1 antibodies (arrows). Levels of WT1 are particularly high in the podocytes and their precursors.

that the inductive signals derived from the ureteric bud were primarily permissive rather than instructive, a distinction first proposed by Saxén (1987). The metanephric mesenchyme was already fated to become renal epithelium, needing only a signal to initiate and maintain the process of differentiation. Heterologous inducers could not reprogram the mesenchyme to generate other epithelial cell types. Morphologically, the metanephric mesenchyme is patterned with respect to the surrounding intermediate mesoderm and appears as a distinct aggregate of cells adjacent to the nephric duct. A number of specific genes, such as *WT1, GDNF,* and *Pax2,* are already expressed prior to induction and mark the metanephric anlagen.

Properties of the inductive signals emanating from the ureteric bud and spinal cord were characterized in detail using the transfilter organ culture system (Saxén, 1987). The inducers were not small, freely diffusible molecules, as a mini-

mum pore size was necessary for signaling to occur across the filter. Indeed, induction was coincident with cellular processes from the inducing tissue permeating the filters. Thus, the idea of cell contact, being a necessary event for induction, was established. Early attempts to use conditioned media from spinal cord or ureteric bud also failed. However, more recent experiments using conditioned media from rat pituitary (Perantoni *et al.,* 1991) or from ureteric bud cell lines (Barasch *et al.,* 1999; Karavanova *et al.,* 1996), in conjunction with survival factors such as fibroblast growth factor (FGF) and transforming growth factor α (TGF-α), suggest that inducing activity could be found in a soluble form.

As much of the data indicated a requirement for cell contact, the question still remained how the inductive signal was able to penetrate the mesenchyme such that cells not immediately around the bud could also be induced. Saxén and colleagues put two hypotheses to the test in an elegant series of experiments. The signal may penetrate several cell layers deep into the mesenchyme by attachment to cellular processes or extracellular matrix that extends into the mesenchymal aggregate. Alternatively, mesenchyme cells immediately abutting the ureteric bud, once induced, could pass on the inductive signals to neighboring uninduced cells. If the signal was passed from one induced mesenchyme cell to another, then isolated mesenchyme, which had been exposed to an inducer for 24 h, could now be utilized as a source of inductive signals for naive mesenchyme. However, Saxén and Saksela (1971) demonstrated that only the mesenchyme originally exposed to the inducer was able to form tubules, whereas the uninduced mesenchyme remained as such. Thus, it appears that each mesenchymal cell requires contact with the primary source of the inducing signal. This was consistent with the observation that the cellular process emanating from the inducing tissues could be observed extending three to four cell layers deep into the mesenchyme.

B. Wnt Genes as Candidate Inducers

Many of the properties associated with the inductive signals emanating from the ureteric bud are common to the wnt family of secreted signaling peptides. The first clue that Wnt genes may be important signaling molecules for kidney development, specifically for induction of the metanephric mesenchyme, came from phenotypic analysis of mouse mutants for the *wnt4* gene (Stark *et al.,* 1994). The Wnt gene family encodes secreted signaling peptides that are associated with the extracellular matrix and have a variety of developmental functions (Nusse and Varmus, 1992; Wodarz and Nusse, 1998). The prototype Wnt protein is *Drosophila* wingless, which controls segment polarity, patterning of the wing disc, and epithelial planer cell polarity. Wnt4 is expressed in the early mesenchymal aggregates (Fig. 14.4), just after induction by the ureteric bud and prior to formation of the polarized renal vesicle. Homozygous mice for a *wnt4* null allele exhibit severe renal agenesis due to a developmental arrest

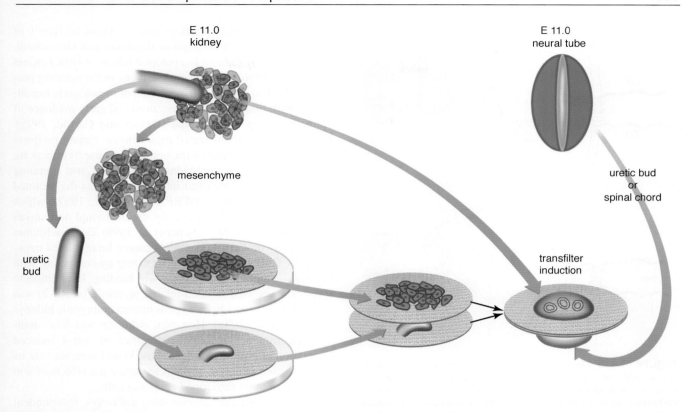

Figure 14.3 The transfilter induction assay. E11 or E13 mouse or rat kidneys are microdissected from the embryo, generally with fine tungsten needles or 30-gauge syringe needles. The tissues are cultured on a polycarbonate filter, with a defined pore size, and are suspended above the media on a solid support. For the mesenchyme induction assay, the inducer is fixed to the bottom of the filter and the mesenchyme is placed on top. The inducing signal permeates the pores of the filter and stimulates the conversion of mesenchyme to epithelium over a period of 24–48 h.

shortly after the ureteric bud invades the mesenchyme. The bud is able to branch a few times, and there is evidence of early aggregation of the mesenchyme. However, *wnt4* mutants

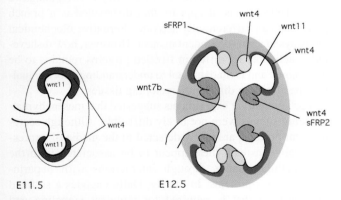

Figure 14.4 Expression of *Wnt* genes in the metanephric kidney. The secreted components of the Wnt signaling pathway are expressed in a temporal and spatial manner during induction and mesenchymal aggregation. At E11.5, Wnt4 is in the aggregates, whereas Wnt11 is in the ureteric bud, especially the tips. By E12, Wnt11 becomes restricted to the ureteric buds and Wnt4 continues to be in the aggregates and renal vesicles. The secreted frizzled-related proteins, potential inhibitors of Wnt signaling, are now observed in the comma-shaped structures (sFRP2) and in the uninduced mesenchyme (sFRP1).

show no evidence of a mesenchyme-derived epithelium. The *Pax2* gene is detectable, albeit at low levels, in these *wnt4* mutant aggregates, but there is no expression of the *Pax8* gene, which normally is activated in more developed aggregates. While it appears that *wnt4* is not the primary ureteric bud-derived inductive signal, based on its expression pattern and on the early aggregation of the *wnt4* mutant mesenchyme, *wnt4* is a secondary signal for the induced mesenchyme to undergo polarization. Alternatively, given the relationship between the stromal lineage and the epithelial lineage in the mesenchyme (see Section V), *wnt4* may act as a paracrine signal from the mesenchymal aggregates that stimulates stromal cells to secrete survival factors for both induced mesenchyme and ureteric bud epithelium. In the absence of this paracrine signaling, developmental arrest occurs.

Evidence that Wnt signals can function as the primary inducers of mesenchyme-to-epithelial conversion is also compelling. By expressing wnt1 in NIHT3 cells, Herzlinger *et al.* (1994) could confer the ability of 3T3 cells to induce metanephric mesenchyme in the transfilter assay, indicating that either Wnt1 can itself induce the mesenchyme or that wnt1 activates a factor in 3T3 cells capable of inducing mesenchyme. Similar experiments further demonstrated that many Wnt proteins are capable of inducing tubulogenesis in the metanephric mesenchyme. Indeed, the different Wnt proteins

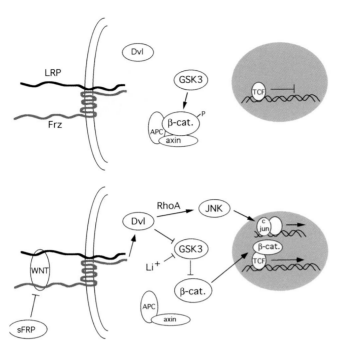

Figure 14.5 The simple model of Wnt signaling. Wnt proteins interact with the transmembrane receptor frizzled (Frz) and the LDL receptor-related protein (LRP) to activate the disheveled (Dvl) protein. Disheveled can inhibit the kinase GSK3, which is also inhibited by lithium. Disheveled can also activate the c-jun N-terminal kinase (JNK) pathway. GSK3 phosphorylates β-catenin such that β-catenin binds the proteins axin and APC and is targeted for degradation. Upon inhibition of GSK3, β-catenin levels rise and can trasnlocate to the nucleus where it interacts with the TCF/Lef family of DNA-binding proteins. The TCF/Lef family binds DNA but suppresses transcription in the absence of β-catenin, also through interactions with the groucho family of repressors. Nuclear β-catenin provides an activation domain *in trans* such that the complex can now activate target gene expression.

examined by Kispert *et al.* (1998) fall into two basic categories: those able to induce mesenchyme include wnt 1, 3a, 4, 7a, and 7b and those unable to induce the mesenchyme, including wnt5a and wnt11. Thus, the ability of the spinal chord to induce the mesenchyme can be explained by the prevalence and diversity of wnt genes expressed there. Unfortunately, the only wnt gene expressed in the ureteric bud tips, wnt11 (Kispert *et al.*, 1996), proved incapable of inducing tubules in the transfilter assay.

The canonical wnt signaling pathway is activated when the wnt protein binds to the frizzled family of transmembrane receptors (Figure 14.5; Wodarz and Nusse, 1998). This activates the cytoplasmic protein disheveled, which can inhibit the kinase, GSK3. The β-catenin protein is a target of GSK3 phosphorylation and, once phosphorylated, is shuttled to the ubiquitin degradation pathway via the β-catenin-binding proteins axin and APC. Inhibition of GSK3 results in the accumulation and nuclear translocation of β-catenin where it can interact with the TCF/LEF family of DNA-binding proteins and activate transcription. In the absence of nuclear β-catenin,

the TCF/LEF factors cooperate with the Groucho family of proteins to repress transcription (Eastman and Grosschedl, 1999). Strikingly, lithium is a potent inhibitor of GSK3 (Klein and Melton, 1996) and can simulate many of the inducing properties of wnt signals. Upon incubation of metanephric mesenchyme with lithium ions, aggregation and early evidence of epithelialization are observed (Davies and Garrod, 1995). However, the lithium-treated mesenchyme cannot progress to a fully polarized epithelium, perhaps because its effects are too broad. During normal kidney devlopment, wnt signaling may be modulated by local inhibitors, such as the secreted frizzled-related proteins (sFRP)(Leyns *et al.*, 1997), which are highly expressed in the early mesenchymal derivatives (Leimeister *et al.*, 1998; Lescher *et al.*, 1998). Because lithium acts directly on GSK3, its effects cannot be regulated negatively by sFRPs, which inhibit signaling by competing with the transmembrane frizzleds for wnt binding. In addition to Pax-8, the secreted frizzled-related protein-2 (sFRP-2) was not detected in wnt4 homozygous mutant embryonic kidneys (Lescher *et al.*, 1998). However, culturing wnt-4 –/– metanephric mesenchyme in the presence of wnt-4 induced expression of both molecules. Thus, Wnt-4 may activate its own potential inhibitor, perhaps to negate the effects of wnt signaling in more developed epithelial cells.

Wnt signaling can activate other pathways, independent of β-catenin. The specification of epithelial planar cell polarity in the fly is mediated partly through wnt activation of the c-jun N-terminal kinase (JNK; Mlodzik, 1999). Activation of JNK requires disheveled and appears to work through the GTPase RhoA. The disheveled protein has at least three different protein–protein interaction domains: the amino-terminal DIX domain stabilizes β-catenin, the carboxy-terminal DEP domain activates JNK, and between these terminal domains lies a PDZ domain whose function is less clear (Axelrod *et al.*, 1998; Boutros *et al.*, 1998; Yanagawa *et al.*, 1995). Thus, it appears that disheveled is a branch point that can differentially activate alternative biochemical pathways in a wnt-dependent manner. However, how disheveled is activated by different frizzled proteins remains to be determined and is fundamental to understanding WNT signaling specificity in different developing tissues.

Early transfilter experiments suggested that mesenchymal-inducing signals were not freely diffusible. Although secreted, wnt proteins are rarely detected in the media of expressing cells. Instead, they appear to be associated with the extra-cellular matrix through interactions with heparin-sulfated glycoproteins. In the fly, Dally encodes a sulfated proteoglycan that is required for wingless signaling and whose overexpression can mimic a wingless gain-of-function phenotype in the absence of extra wingless protein (Lin and Perrimon, 1999; Tsuda *et al.*, 1999). In the kidney, chlorate treatment inhibits Wnt-inducing activity and reduces the accumulation of wnt11 mRNA at the ureteric bud tips (Kispert *et al.*, 1996). In addition, mouse mutants carrying a null allele for the gene encoding heparan sulfate 2-sulfotransferase exhibit renal agenesis (Bullock *et al.,* 1998), reminiscent of a wnt4 phenotype. Thus, wnt signaling requires association with glycoproteins found in the extracellular matrix, consistent with the observation of contact and process extension into the mesenchyme as a prerequisite for induction.

Clearly there is a definitive role for wnt signaling in the early mesenchyme-to-epithelial transition. Although wnt signals can mimic the inductive signals emanating from the ureteric bud and the spinal chord, it has not been shown conclusively that a specific wnt signal is the primary inducer of the metanephric mesenchyme. Furthermore, the receptors for individual wnt proteins, of which at least six are expressed in the kidney, have not been described in sufficient detail so that individual ligands can be assigned for each frizzled family member.

C. Induction of Tubules by Soluble Factors

Since the initial observations by Grobstein, many growth factors have been tested for inducing activity (Weller *et al.,* 1991). Although early transfilter experiments were not consistent with the possibility of soluble, diffusible factors, several reports have demonstrated mesenchyme induction using a cocktail of growth factors and extracts. Rat pituitary extracts in combination with epidermal growth factor and extracellular matrix could stimulate epithelial conversion of rat metanephric mesenchyme (Perantoni *et al.,* 1991). More recently, basic fibroblast growth factor and transforming growth factor-α, together with conditioned media from a ureteric bud cell line, were used as an inducer of rat metanephric mesenchyme (Barasch *et al.,* 1996). The active ingredient in the conditioned medium of ureteric bud cell lines was determined to be leukocyte inhibitory factor (LIF), which, together with a defined concentration of bFGF and TGF-α, had inducing activity (Barasch *et al.,* 1999). While these data are thought provoking and cause one to reexamine some of the early transfilter experiments, a number of caveats make the interpretation difficult. The experiments done with rat tissue, and at least in one case, were not reproducible with the corresponding mouse mesenchyme (Perantoni *et al.,* 1991). Given the conservation of genetic pathways between organisms as disparate as humans and flies, this is rather unexpected and points to an inherent problem. Furthermore, mouse mutants do not support the role of any of these factors in kidney induction. There is no appreciable renal phenotype in bFGF mutant mice (Dono *et al.,* 1998). Knockout mice carrying a null allele of the LIF receptor gp130 show only a slight reduction in kidney size, with no real effect on induction or epithelial differentiation (Yoshida *et al.,* 1996). In the classic *in vitro* induction experiment, mesenchyme is microdissected away from the ureteric bud, yet it has already been in contact and may have received a primary signal. In culture, 24 h of contact was required before a full response was seen

in the mesenchyme (Saxén and Lehtonen, 1978), as judged by the formation of tubules. If the appropriate mixture of growth and survival factors is present, perhaps this need for extended contact is reduced. Thus, at the time of dissection and explant, cells that have been exposed to the primary inducer can now survive and generate tubules in sufficient number. These issues need to be examined before the role of LIF and FGFs during primary induction can be established definitively. Regardless of their potential as primary inducers, FGF and LIF may provide important survival or secondary signals for the mesenchyme and its derivatives.

IV. Intrinsic Factors That Control the Induction Response

Prior to induction by the ureteric bud, the metanephric mesenchyme is an aggregate of distinct cells, expressing a unique combination of genes. Some of these early genes are necessary for the survival of the mesenchyme and for its ability to respond to the inductive signals emanating from the bud. However, little is known about the positional cues that pattern the posterior intermediate mesoderm. In frog and chick embryos, paraxial mesoderm is required for formation of the pro- and mesonephric fields, as marked by the expression of *Pax2* and *lim1*, whereas lateral plate mesoderm or notochord is not required (Seufert *et al.,* 1999; Mauch *et al.,* 2000). What signals may preprogram the mesenchyme such that its developmental fate is restricted remain to be identified. However, it can be said with some assurance that the process of tubule development begins before induction of the mesenchyme, as genetic mechanisms have fixed, at least in part, the renal epithelial lineages.

A. *Pax2* Gene

An essential gene for early urogenital development is transcription factor *Pax2*. Members of the *Pax* gene family were first identified in *Drosophila* as genes regulating segmentation and neural cell identity. In the mouse, *Pax* genes are required for the development of the eye, the vertebral column, certain derivatives of the neural crest, B lymphocytes, the thyroid, and various neural structures (Dahl *et al.,* 1997; Noll, 1993; Stuart *et al.,* 1993). In both mice and humans, *Pax* genes are haploinsufficient, such that heterozygous individuals already exhibit phenotypes in organ systems regulated by the respective gene. The phenotype of heterozygous individuals is variable and less severe than that found among homozygotes. In at least one case, additional copies of a wild-type *Pax6* gene also disturb normal eye development (Schedl *et al.,* 1996). Thus, data indicate a strict requirement for Pax gene dosage during normal development.

In the developing kidney, *Pax2* expression is first detected in the intermediate mesoderm, prior to formation of the nephric duct, beginning at approximately E8.5. Mouse mutants lacking *Pax2* fail to develop mesonephric tubules, the metanephros, and the sex-specific epithelial components derived from this region (Torres *et al.*, 1995). Despite its early expression pattern, *Pax2* is not required for initiation and extension of the nephric duct epithelium. Within the area of the mesonephros, *Pax2* mutants show no evidence of tubule formation, indicating that the periductal mesenchyme is unable to generate epithelium or respond to signals derived from the nephric duct. Metanephric mesenchyme isolated from *Pax2* mutants is also unable to respond to inductive signals from heterologous-inducing tissues in the organ culture assay. Thus it appears that *Pax2* is necessary for specifying the region of intermediate mesoderm destined to undergo mesenchyme-to-epithelium conversion. However, the *Pax2* mutant nephric duct does not respond to GDNF and shows no evidence of ureteric bud outgrowth, despite expression of at least some markers of normal ductal epithelium, suggesting also that *Pax2* in the duct epithelium is required for maintenance or responsiveness (Brophy *et al.*, 2001).

Pax2-expressing cells in the metanephric mesenchyme are closely associated with the ureteric bud. This observation led to the proposition that *Pax2* expression was activated by inductive signals emanating from the bud (Dressler *et al.*, 1990; Dressler and Douglass, 1992). Further support for the dependence of *Pax2* expression on inductive signals came from studies with the *Danforth's short tail* (*Sd*) mouse, in which *Pax2* expression was detected in the nephric duct and ureteric bud but not in the mesenchyme, presumably due to lack of ureteric bud invasion (Phelps and Dressler, 1993). However, the primary defect in the *Sd* mouse is posterior degeneration of the notocord. Thus, gene expression in the intermediate mesoderm could be affected by loss of notocord-derived patterning signals. Until recently, it has proved difficult to separate *Pax2* expression in the mesenchyme from ureteric bud invasion, as it is not easy to define the mesenchyme morphologically prior to E11. However, we have observed *Pax2* expression in the metanephric mesenchyme of E11.5 RET homozygous mutants, which have no ureteric bud but essentially wild-type mesenchyme. Thus, *Pax2* expression in the metanephric mesenchyme predates ureteric bud invasion and marks the posterior intermediate mesoderm as the renal anlagen (Brophy *et al.*, 2001).

B. *WT1* Gene

One potential target of *Pax2* activation is the Wilms' tumor suppressor gene, *WT1*. Wilms' tumor is an embryonic kidney neoplasia that consists of undifferentiated mesenchymal cells, poorly organized epithelium, and surrounding stromal cells. Because the neoplastic cells of the tumor are able to differentiate into a wide variety of cell types, the genes responsible for Wilms' tumor were thought to be important regulators of early kidney development (van Heyningen and

Hastie, 1992; Chapter 22). Expression of *WT1* is regulated spatially and temporally in a variety of tissues and is further complicated by the presence of at least four isoforms, generated by alter-native splicing. In the developing kidney, *WT1* can be found in the uninduced metanephric mesenchyme and in differen-tiating epithelium after induction (Armstrong *et al.*, 1992; Pritchard-Jones *et al.*, 1990). Initial expression levels are low in the mesenchyme. Strikingly, *WT1* becomes upregulated at the S-shaped body stage in the precursor cells of the glome-rular epithelium, the podocytes. High *WT1* levels persists in the podocytes of the glomerulus into adulthood.

How *WT1* impacts kidney development is still under intense investigation. However, some conclusions can be drawn from the phenotypes of both mouse and human mutations. The homozygous null *WT1* mouse has complete renal age-nesis (Kreidberg *et al.*, 1993) because the ureteric bud fails to grow out of the nephric duct and the metanephric mesen-chyme undergoes apoptosis. The arrest of ureteric bud growth appears to be noncell autonomous and is most probably due to lack of signaling by the *WT1* mutant mesenchyme. How-ever, the mesenchyme is unable to respond to inductive signals even if a wild-type inducer is utilized in the organ culture assay. Taken together, it appears that *WT1* is required early in the mesenchyme to promote cell survival, such that cells can respond to inductive signals and express ureteric bud growth promoting factors.

What might the *WT1* gene be doing at later stages of kidney development? Increased expression levels in podocyte precursor cells correlate with repression of the *Pax2* gene (Ryan *et al.*, 1995). *WT1* binds to the promoters of a variety of genes, including *Pax2*, *IGF2*, and the IGF2 receptor, by direct contact of the zinc finger DNA-binding domain to a conserved GC-rich nanomer. Much data indicate that the amino-terminal region of *WT1* acts as a transcription repres-sion domain (Madden *et al.*, 1991), consistent with the increased levels of growth factor expression observed in Wilms' tumors. However, in other cases, *WT1* is able to func-tion as an activator. Furthermore, specific isoforms of WT1 proteins have been colocalized to the mRNA splicing machi-nery, suggesting a posttranscriptional role (Davies *et al.*, 1998). Thus, many questions still remain to be addressed. The expression levels of *WT1* are regulated precisely during kidney development and may indicate different functions depending on the threshold level of protein present. How-ever, it is not even clear if all WT1 isoforms are expressed equally in different cell types. This raises the possibility that a shift in the ratio of WT1 isoforms may specify repression, activation, or posttranscriptional regulatory functions.

C. *Eya-1*

The mammalian homologues of the *Drosophila eyes absent* gene constitute a small gene family, of which at least one member is an essential regulator of early kidney development.

In humans, mutations in the *Eya1* gene are associated with Branchio-Oto-Renal syndrome, a complex multifaceted phenotype (Abdelhak *et al.*, 1997). In mice homozygous for an *Eya1* mutation, kidney development is arrested at E11 because ureteric bud growth is inhibited and the mesenchyme remains uninduced (Xu *et al.*, 1999).

Eya1 expression during kidney development is first detected in both the mesonephric and the metanephric mesenchyme and appears to regulate the expression of *GDNF*. In *Eya1* mutants, *Pax2* and *WT1* expression appears normal, but *Six2* and *GDNF* expression are lost in the mesenchyme. The loss of *GDNF* expression most probably underlies the failure of ureteric bud growth. However, it is not clear if the mesenchyme is competent to respond to inductive signals if a wild-type inducer were to be used *in vitro*. Eya proteins share a conserved domain but do not have DNA-binding activity. Eya proteins appear to function through direct interactions with the Six family of homeodomain proteins. Mouse *Six* genes are homologues of the *Drosophila sina oculis* homeobox gene and encode DNA-binding proteins that also interact directly with the Eya family of nuclear factors. Eya proteins, in turn, can also bind to the Dachshund family of proteins. This cooperative interaction between Six and Eya proteins is necessary for nuclear translocation and transcriptional activation of Six target genes (Ohto *et al.*, 1999). In *Drosophila* eye development (Pignoni *et al.*, 1997) and during muscle cell specification in the chick embryo (Heanue *et al.*, 1999), the regulatory network involving *Pax*, *Eya*, *Six*, and *dachshund* genes is well conserved. Based on conservation of expression patterns and mutant phenotypes, a similar regulatory network has been proposed in the developing mouse kidney (Xu *et al.*, 1999), with *Pax2* being upstream of *Six2* and *Eya1*. A dachshund family homologue has not yet been described in the kidney, but there is no reason to believe it will not be found.

V. Factors That Drive Mesenchyme-to-Epithelial Conversion

The response of the metanephric mesenchyme to inductive signals has been under intense scrutiny for many years (Ekblom, 1989). The aggregation and polarization of the mesenchyme have been utilized as a readout for the presence of inductive signals and for the characterization of the cellular changes leading to a fully polarized epithelium. Presently, the ability to follow the cellular changes in response to induction has been advanced greatly by the identification of many new gene products that are activated in the induced cell and that may drive the process of epithelialization.

The mesenchyme undergoes extensive remodeling after induction. Proliferation and differentiation are tightly linked within the newly formed tubules. The activation of genes specific for epithelia and the repression of mesenchymal specific genes must be regulated by transcription factors that respond, at least initially, to the inductive interactions. The driving forces underlying cell aggregation and polarization would include the activation of new cellular adhesion molecules and extracellular matrix components.

A. Aggregation and Cell Adhesion

Among the most intensely studied cell adhesion molecules, cadherins are transmembrane proteins that promote homophillic adhesion between cells. Adhesion is mediated by an extracellular cadherin repeat whose structural conformation changes with increasing calcium concentration (Gumbiner, 1996; Koch *et al.*, 1999). Cadherins are classified into type I and type II subfamilies based on the sequence of the extracellular domain and functional properties of the intracellular domain. Type I cadherins, which include E-, N-, P-, and R-cadherin, contain five extracellular cadherin repeats, the first of which mediates specific homophilic interactions. This repeat promotes interactions between cadherins of the same type on opposing cells and quite possibly within the same cell (Shapiro *et al.*, 1995). The intracellular domains of type I cadherins bind to the catenins, a family of cytoplasmic proteins linking cadherins to the actin cytoskeleton (Kemler and Ozawa, 1989; Ozawa *et al.*, 1989). Within the same cell, association with the catenins is also important for clustering cadherins at the cell surface (Katz *et al.*, 1998; Yap *et al.*, 1997, 1998). The adherents junction, a specialized site of cell–cell contact that binds cells into solid tissue, forms as cadherins cluster at points of contact between two cells expressing like cadherins (Gumbiner, 1996). In epithelia, disruption of the adherents junction has a profound affect on the maintenance of other contacts and ultimately to a loss of epithelial polarity and integrity. Thus, adhesive forces can remodel developing tissues and compartmentalize the tubules into distinct units. The type II subfamily of cadherins, including cadherin-6, cadherin-11, and F-cadherin, are closely related to the classical cadherins in both structure and function (Cho *et al.*, 1998; Hoffmann and Balling, 1995; Kimura *et al.*, 1995). Type II cadherins can also mediate cell adhesion and bind to the actin-based cytoskeleton via catenins (Inoue *et al.*, 1997; Nakagawa and Takeichi, 1995).

During the development of renal tubules, cadherin expression is dynamic and complex (Fig. 14.6). Prior to induction, the metanephric mesenchyme expresses cadherin-11, whereas the ureteric bud expresses E-cadherin (Cho *et al.*, 1998). Upon aggregation of the metanephric mesenchyme, in response to inducing signals from the ureteric bud, the expression of cadherin-11 is downregulated and the expression of R-cadherin and cadherin-6 is upregulated (Cho *et al.*, 1998; Rosenberg *et al.*, 1997). The onset of R-cadherin and cadherin-6 expression is coincident with formation of the renal vesicle and the deposition of a basement membrane. Thus, formation of the first polarized epithelium derived

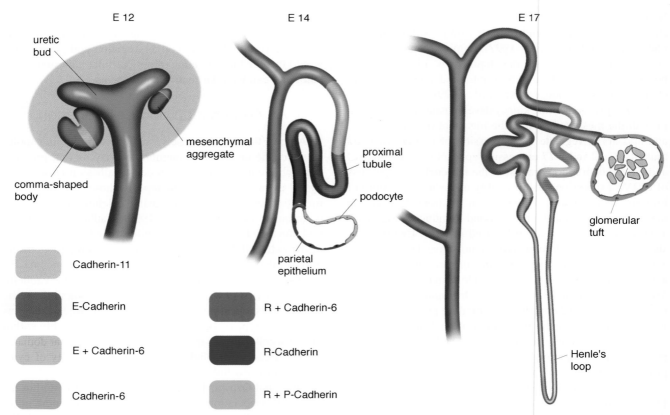

Figure 14.6 A model of cadherin gene switching during nephrogenesis. This schematic model presents a composite of published expression patterns (Cho *et al.,* 1998; Goto *et al.,* 1998; Mah *et al.,* 2000; Rosenberg *et al.,* 1997; Tassin *et al.,* 1994). The model tries to reconcile any discrepancies in the literature as fairly as possible. The uninduced mesenchyme expresses cadherin-11, which is widely distributed in other embryonic mesenchymal tissue. Upon aggregation of the induced mesenchyme, R-cadherin and cadherin-6 are activated. E-Cadherin is prevalent in the ureteric bud epithelium and can also be seen in part of the aggregates close to the bud tips. By the S-shaped body stage, R- and P-cadherin can be detected in the podocyte precursor cells, R-cadherin in much of the proximal part of the developing nephron, cadherin-6 in the proximal part of the nephron, and E-cadherin in the more distal parts of the developing nephron. In the maturing nephron, P- and R-cadherin are mostly in the glomerular tuft, the podocytes, whereas cadherin-6 becomes more restricted to the proliferating epithelia, such as the loop of Henle. E-cadherin is widely distributed both in more mature proximal and distal tubules and all collecting ducts.

from the mesenchyme correlates with profound changes in cadherin expression. As the renal vesicle proliferates to form the S-shaped body, R-cadherin is primarily restricted to the epithelial structures furthest from the ureteric bud, E-cadherin is expressed in the bud and most distal tubules, and cadherin-6 is located between E-cadherin and R-cadherin expression domains (Cho *et al.,* 1998; Rosenberg *et al.,* 1997). The expression of cadherin-6 marks the proximal tubule precursor cells, including the ascending and descending limbs of Henle's loop at later developmental stages. Within the glomerulus, R-cadherin becomes restricted to the parietal epithelium of Bowman's capsule and the most proximal tubule cells, whereas P-cadherin becomes evident in the podocytes of the visceral epithelium in the glomerular tuft. As proximal tubules mature, E-cadherin becomes more prominent and cadherin-6 expression is suppressed. These dynamic changes in the expression of type II and type I cadherins correlate with proliferation, cell polarity, and tissue remodeling, strongly indicating that

cadherins mediate aspects of the differential cell adhesion required for nephrogenesis.

Within the kidney, cadherin function has been addressed *in vitro* and *in vivo.* Cell adhesion can be neutralized *in vitro* using Fab fragments of antibodies against the cadherin extracellular domains. In kidney organ cultures, anti-cadherin-6 Fabs inhibit mesenchymal cell adhesion and the subsequent formation of comma and S-shaped structures, without affecting ureteric bud branching (Cho *et al.,* 1998). Similar experiments using neutralizing antibodies to E-cadherin have no apparent effect on the conversion of mesenchyme to epithelium (Vestweber *et al.,* 1985). Cadherin-6 mutant mice have also been described, yet their phenotype is not nearly as striking as the *in vitro* experiments might suggest (Mah *et al.,* 2000). The mesenchyme of cadherin-6 mutants appears delayed in its ability to form a fully polarized epithelium. This delay may compromise the ability of the mesenchymal-derived epithelium to fuse with the ureteric bud-derived epithelium and results in a high frequency of dead-ending nephrons. Despite some similarites between *in vitro* antibody-mediated disruption experiments and knockout cadherin-6 mutants, it is clear that the organ culture assay is much more sensitive to perturbations. *in vitro* development proceeds at a slower rate and lacks certain cell types, such as the vasculature. Alternatively, the antibody inhibition experiments might inhibit adhesion mediated not only by cadherin-6, but also by potential heterodimeric cadherins. The recent discovery that at least some cadherins can form heterodimers within the same cell and function as heterodimeric adhesion molecules between cells raises the possibility that antibodies may inhibit the function of more than just one type of cadherin protein (Shan *et al.*, 2000). In any event, *in vitro* inhibition studies must be evaluated with caution before any assignment of protein function can be defined.

B. Cell–Matrix Interactions

Changes in cell adhesion occur not only at the junctions between cells, but also at the interface between cells and the extracellular matrix (ECM). The integrin family of trans-membrane proteins regulates specific interactions among cells and various extracellular substrates. Integrins are hetero-dimeric receptors that consist of an α and β subunit, such that a single β subunit is able to heterodimerize with multiple α subunits (Giancotti and Ruoslahti, 1999). The specific combination of the α and the β subunits determines specificity of binding to the ECM, whereas the β subunit can interact with cytoplasmic signal transduction molecules. In the developing renal tubules, at least four different α subunits, in combination with the β1 subunit, are found in different cell types at different times. The α8β1 integrin is expressed in the undinduced mesenchyme and is upregulated to high levels in the induced mesenchyme surrounding the ureteric bud tips. Homozygous mice carrying a mutant allele for the α8 subunit show severe renal agenesis, with a varying degree of penetrance (Muller *et al.*, 1997). Most newborn α8 homozygous mutants had bilateral renal agenesis and no urethras present. However, some homozygous mutant animals had a single kidney rudiment and about 25% had a single, apparently normal, kidney. At E11 in 100% of the α8 mutants, the ureter had not invaded the mesenchyme and not undergone branching morphogenesis. Thus it would appear that the α8β1 integrin in the mesenchyme provides a permissive environment for optimal ureteric bud invasion. This permissive environment may be due to interactions of the α8β1 integrin with the ECM of the growing ureteric bud. Alternatively, the α8β1 integrin may affect the expression of bud guidance or survival factors, such as GDNF, through an integrin-linked signaling role in the mesenchyme.

Epithelial cells of the renal tubules are polarized (Chapter 16). The apical side faces the lumen of the tubules, whereas the basal side contacts the basement membrane that com-pletely sur-rounds the outside of individual tubules. The basement membrane is made up of collagens, laminins, entactin, and sulfated proteoglycans. During renal tubule development, composition of the basement membranes shifts both spatially and temporally such that specific isoforms of both collagen and laminin are found in different parts of the developing nephron (Miner, 1999). These isoforms of the matrix compo-nents appear to have unique functions, as genetic mutations of specific isoforms can have dramatic effects on both early and late development.

The assembly of the epithelial basement membrane can be detected as early as the renal vesicle stage. The hetero-trimeric protein laminin consists of a single α, β, and γ chain and is a major structural determinant of the basement membrane. The initial isoform detected in the renal vesicle is laminin-1, which encompasses an α1, β1, and γ1 chain. In the induced mesenchyme, expression of the laminin α1 chain may be a rate-limiting step in formation of the early basement membrane, as expression of the laminin β1 and γ1 chains can already be detected in the uninduced mesenchyme in a filamentous pattern. In kidney organ cultures, antibodies against the α1 chain of laminin-1 inhibit the conversion of mesenchyme to epithelium, presumably by disturbing the formation of the epithelial basement membrane or integrin–matrix interactions (Klein *et al.*, 1988).

As tubules develop to the S-shaped body stage, the glomerular epithelial precursors begin to express laminin-10 (α5β1γ1). In the mature glomerular basement membrane, an essential element of the filtration barrier, laminin-10 is replaced by laminin-11 (α5β2γ1) (Miner *et al.*, 1997). The function of these seemingly homologous isoforms has been examined in mouse knockout models by mutating individual laminin chains. Homozygous mice carrying a laminin β2 null allele develop severe proteinuria (Noakes *et al.*, 1995). In the β2 null mouse, podocyte cells of the glomerulus are abnormal and fused, yet the basement membrane appears normal at the ultratructural level. The laminin-β1 is found in the mature glomerular basement membrane, but cannot fully compensate for the lack of the laminin-β2 chain. Thus, the shift in laminin isoform expression has a profound effect on the visceral epithelium of the glomerulus. Perhaps the maturing podocytes receive signals, via integrins, from the glomerular-specific laminin chains. Alternatively, as integrin expression in the podocytes shifts, a compensatory shift in laminin chain expression is necessary to maintain tight cell–matrix adhesion.

Gene targeting of the laminin α5 chain also reveals an essential function during development of the glomerular epithelium. Although some mice homozygous for the lama5 allele show bilateral or unilateral renal agenesis, most of the affected mutants show some degree of nephrogenesis with a gross defect in the glomerulus (Miner and Li, 2000). The laminin α5 chain is expressed in the basement membrane of the S-shaped body when endothelial cells are first observed invading the prospective glomerular cleft. This is coincident with a shift from laminin-1 to laminin-10. At this time, upregulation of the *WT1* gene marks the podocyte precursor cells, which, form a crescent-shaped layer along the cleft and subsequently outline the developing Bowmans capsule. Rather than this single layer of podocyte precursors, lama5 mutants exhibit multiple layers that ultimately displace the invading endothelial cells from the glomerular cleft.

Another major structural component of the renal basement membrane is collagen IV. There are six distinct chains of collagen IV, $\alpha 1$–$\alpha 6$, which form a triple helical rod-like structure. Collagen IV isoforms are expressed differentially depending on the type of basement membrane and the developmental stage (Miner and Sanes, 1994). $\alpha 1$ and $\alpha 2$ chains are expressed in the basement membrane of the renal vesicle, the comma-shaped bodies, and the S-shaped bodies, whereas $\alpha 1$ and $\alpha 5$ are detected in the developing glomerulus at the capillary loop stage. However, the mature glomerular basement membrane contains mostly $\alpha 3$, $\alpha 4$, and $\alpha 5$. Although little is known about the function of distinct collagen IV chains in developing tubules, the glomerular basement membrane has been the focus of much attention due to the effect of collagen IV mutations on the filtration barrier.

In humans, mutations in any of the $\alpha 3$, $\alpha 4$, or $\alpha 5$ chains can cause Alport syndrome (Barker *et al.,* 1990; Hudson *et al.,* 1993), which is characterized by decreased glomerular filtration, increased mesangial matrix deposition, glomerular sclerosis, and tubule atrophy. The persistence of collagen IV $\alpha 1$ and $\alpha 2$ chains in the Alport syndrome kidneys appears to increase the susceptibility of the glomerular basement membrane to damage.

C. Epithelium–Stroma Interactions

The classic view of tubulogenesis focused on the role of inductive signaling from the ureteric bud and epithelial conversion of the mesenchyme. However, not all of the metanephric mesenchyme is fated to become epithelium. Certainly by E11.5 in the mouse there are epithelial and stromal precursor cells and there is limited evidence of endothelial precursors. The effects of nonepithelial cell types on the process of tubulogenesis are profound. Both survival and proliferation signals are now known to emanate from the stromal population within the developing kidney that act in a paracrine fashion on the developing tubules (for summary, see Fig. 14.7). The genetic factors that designate the stromal and

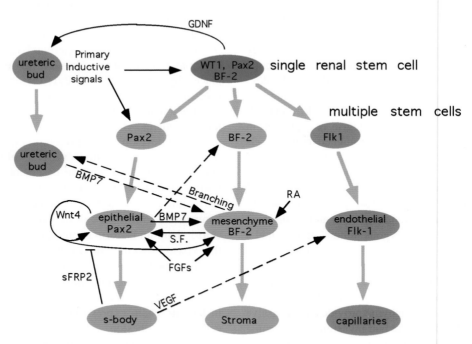

Figure 14.7 A model of cellular interactions among different lineage precursors. Inductive signals from the ureter bud act on either a single renal stem cell or on multiple progenitor cells whose fate may be restricted prior to induction. The epithelial lineage is marked by Pax2 expression, whereas the stromal lineage expresses BF-2. Endothelial cell precursors, which express the VEGF receptor Flk1, have been observed as early as E11.5 and may represent a third lineage. BMP7 is expressed in both the ureteric bud and the condensed mesenchyme and promotes survival of BF-2 positive stroma, and potentially also uninduced mesenchyme. Stromal cells feedback to the epithelial lineage by providing as yet unidentified survival factors. Retinoids also act on the stromal lineage and, indirectly, on the ureteric bud cells by activating a branching stimulant, which increases the expression of RET. FGFs, whose source is not clear, promote survival of the stromal lineage at the expense of the epithelial lineage. Wnt4 is expressed in the mesenchymal aggregates and may provide an autocrine signal for epithelial polarization or a paracrine signal to stromal cells. More differentiated structures, such as the S-shaped body, express potential inhibitors of Wnt4, perhaps to localize the effects of Wnt4 signaling to less differentiated cell types. VEGF becomes expressed in the presumptive glomerular epithelium and functions to attract endothelial cells into the glomerular tuft.

epithelial cell populations are now being addressed using a combination of genetic and tissue culture models.

Within the induced mesenchyme, the transcription factor BF-2 is restricted to those cells not undergoing epithelial conversion after induction (Hatini *et al.*, 1996). *BF-2* expres-sion is localized along the periphery of the kidney and in the interstitial mesenchyme, or stroma. Indeed, there is little over-lap between the expression domains of *BF-2* and *Pax2*, which marks the mesenchymal aggregates and early epithelial derivatives. Whether mesenchyme cells may already be partitioned into a *BF-2*-positive stromal precursor and a *Pax2*-positive epithelial precursor prior to induction remains to be deter-mined. However, the mouse knockout for BF-2 reveals an essential role for the stroma in pro-moting tubulogenesis (Hatini *et al.*, 1996). In homozygous BF-2 null mice, ureteric bud growth and initial branching are unaffected, as is the for-mation of the first mesenchymal aggregates. However, within 2 days after induction, mesen-chymal aggregates fail to dif-ferentiate into comma and S-shaped bodies at a rate similar to wild type. This results in reduced branching of the ureteric bud and coincidentally fewer new mesenchymal aggregates forming. Early mesenchymal aggregates are able to form epi-thelium and express the appropriate markers, such as *Pax2, wnt4,* and *WT1*. How-ever, in the absence of BF-2-expressing stromal cells, growth of both ureteric bud epithelium and mesenchymal aggre-gates is slowed or inhibited. BF-2 knockout studies reveal an essential role for stromal cells in supporting the survival and proliferation of epithelial precursor cells. As epithelial cells become induced to form aggregates, a self-renewing population of epithelial precursor cells must be maintained along the periphery of the developing kidney such that new aggregates can form around the tips of the next ureteric bud end point. Factors secreted from the stroma may provide the cues for this self-renewing population of epi-thelial precursors, in the absence of which the mesenchyme is quickly exhausted of all cells fated to become epithelia.

What might some of these stromal-derived factors be? Evidence points to both fibroblast growth factors and bone morphogenetic proteins (BMPs) as influential players in the decision to go stromal or epithelial. Among the early candi-dates for the inductive signal, *BMP7* is expressed in the ureteric bud epithelium and in the induced mesenchyme (Vukicevic *et al.*, 1996). However, BMP7 mutants do not show a fundamental defect in early mesenchyme induction. Rather, they are developmentally arrested after the induction phase, do exhibit ureteric bud branching morphogenesis, and express early markers of induced mesenchyme (Dudley *et al.*, 1995; Luo *et al.*, 1995). The ability of BMP7 to induce mesenchyme was not confirmed in more recent studies (Dudley *et al.*, 1999). Using the *in vitro* organ culture model, Dudley *et al.*, (1999) demonstrated the effect of BMP7 on mesenchymal cell survival. Thus, uninduced mesenchyme, which dies very quickly in culture, can survive for several days and still respond to inductive signals. In combination

with FGF2, the survival of the mesenchyme and the ability to respond to induction are enhanced greatly by BMP7. In fact, both factors together appear to increase the proportion of cells expressing BF-2, the stromal population, at the expense of the Pax2 expressing epithelial cells. These *in vitro* experiments point to a delicate balance between stromal and epithelial progenitor cells, the proportion of which must be well regu-lated by both autocrine and paracrine factors, such that neither cell population is exhausted prematurely.

Other factors within the stroma that may affect epithelial cell proliferation and tubulogenesis include retinoids and their receptors. That vitamin A deficiency leads to congenital renal defects, including a reduction in the number of nephrons, is well documented (Lelievre-Pegorier *et al.*, 1998). The stromal cell population expresses both retinoic acid receptors (RAR) α and β2. A striking renal phenotype has been reported in mice that are homozygous null for both receptors (Mendelsohn *et al.*, 1999). A general developmental arrest is evident that can be explained in large part due to the inhi-bition of ureteric bud branching. Consistent with this view, down-regulation of the RET receptor is evident in the ureteric bud epithelial of RAR α/β2 null animals. Reduced expression of BF-2 is also observed, indicating that proli-feration of the stromal population may be limited. Using exo-genous retinoids in the organ culture model, activation of RET and a stimulation of ureteric bud branching morpho-genesis is observed (Vilar *et al.*, 1996). Thus the effects of retinoids appear to be mediated by the stromal population. The stromal cells must provide signals that maintain RET expression and provide an environment for branching of the ureteric bud. Failure to do so results in a general develop-mental arrest and an inhibition of tubulogenesis.

VI. Summary

This chapter outlined the current state of knowledge with respect to the genetic and biochemical basis of tubule develop-ment in the nephron. Intrinsic factors, such as Pax2, WT1, and Eya1, specify the metanephric mesenchyme from sur-rounding intermediate mesoderm. Expression of these genes most likely depends on positional cues from somitic meso-derm and perhaps surface ectoderm. While Pax2 and WT1 are required to enable the mesenchyme to respond to induc-tive signals, they may also be necessary for the proliferation and differentiation of tubular epithelium after induction occurs. The expression of both genes is highly dynamic during the course of nephron development. Subsequent chapters discuss the relevance of Pax2 and WT1 to human renal disease, further underscoring the importance of activation and repression of Pax2 and WT1.

The inductive signals emanating from the ureteric bud have not been defined conclusively. However, evidence that the secreted Wnt proteins can induce the metanephric mesen-chyme is compelling. In contrast, the trio of secreted signaling molecules bFGF, TGF-α, and LIF has also been implicated as a mesenchymal inducer, although genetic evidence supporting their *in vivo* role is lacking. These issues need to be clarified. In theory, a mouse carrying a mutation in the ureteric bud-derived inducer should exhibit growth and branching of the ureteric bud without any evidence of primary induction in the mesenchyme. Unfortunately, such a genetic mutant has not been identified to date, leaving only the *in vitro* assays as biochemical evidence of inductive capacity.

The response of the mesenchyme to inductive signals has been characterized over many years. However, there are many critical elements missing. While activation of cell adhesion molecules must underlie the aggregation and perhaps the polarization of the early epithelia, how the inductive signals are translated into morphological changes remains obscure. As the nephron develops, compartmentalization occurs along the axis of the tubular epithelia. This is evident by segment-specific gene expression of cadherins, integrins, laminins, and other markers. How are changes in integrins and cadherins regulated so precisely along the developing nephron? Why is there a need for a shift in extracellular matrix collagen and laminin isoforms? What are the factors that determine specialized epithelial cell types along the axis of the nephron? These questions remain unanswered and will surely be addressed in the coming decade as a more complete understanding of nephrogenesis is at hand.

References

Abdelhak, S., Kalatzis, V., Heilig, R., Compain, S., Samson, D., Vincent, C., Weil, D., Cruaud, C., Sahly, I., Leibovici, M., Bitner-Glindzicz, M., Francis, M., Lacombe, D., Vigneron, J., Charachon, R., Boven, K., Bedbeder, P., Van Regemorter, N., Weissenbach, J., and Petit, C. (1997). A human homologue of the Drosophila eyes absent gene underlies branchio- oto-renal (BOR) syndrome and identifies a novel gene family. *Nature Genet.* **15**, 157–164.

Armstrong, J. F., Pritchard-Jones, K., Bickmore, W. A., Hastie, N. D., and Bard, J. B. L. (1992). The expression of the Wilms' tumor gene, *WT1*, in the developing mammalian embryo. *Mech. Dev.* **40**, 85–97.

Axelrod, J. D., Miller, J. R., Shulman, J. M., Moon, R. T., and Perrimon, N. (1998). Differential recruitment of Dishevelled provides signaling specificity in the planar cell polarity and Wingless signaling pathways. *Genes Dev.* **12**, 2610–2622.

Barasch, J., Pressler, L., Connor, J., and Malik, A. (1996). A ureteric bud cell line induces nephrogenesis in two steps by two distinct signals. *Am. J. Physiol.* **271**, F50–F61.

Barasch, J., Yang, J., Ware, C. B., Taga, T., Yoshida, K., Erdjument-Bromage, H., Tempst, P., Parravicini, E., Malach, S., Aranoff, T., and Oliver, J. A. (1999). Mesenchymal to epithelial conversion in rat metanephros is induced by LIF. *Cell* **99**, 377–386.

Barker, D. F., Hostikka, S. L., Zhou, J., Chow, L. T., Oliphant, A. R., Gerken, S. C., Gregory, M. C., Skolnick, M. H., Atkin, C. L., and Tryggvason, K. (1990). Identification of mutations in the COL4A5 collagen gene in Alport syndrome. *Science* **248**, 1224–1227.

Boutros, M., Paricio, N., Strutt, D. I., and Mlodzik, M. (1998). Dishevelled activates JNK and discriminates between JNK pathways in planar polarity and wingless signaling. *Cell* **94**, 109–118.

Brophy, P. D., Ostrom, L., Lang, K. M., and Dressler, G. R. (2001). Regulation of uteric bud outgrowth by PAX2-dependent activation of

the glial derived neurotrophic factor gene. *Development* **128**, 4747–4756.

Bullock, S. L., Fletcher, J. M., Beddington, R. S., and Wilson, V. A. (1998). Renal agenesis in mice homozygous for a gene trap mutation in the gene encoding heparan sulfate 2-sulfotransferase. *Genes Dev.* **12**, 1894–1906.

Cho, E. A., Patterson, L. T., Brookhiser, W. T., Mah, S., Kintner, C., and Dressler, G. R. (1998). Differential expression and function of cadherin-6 during renal epithelium development. *Development* **125**, 4806–4815.

Dahl, E., Koseki, H., and Balling, R. (1997). Pax genes and organogenesis. *Bioessays* **19**, 755–765.

Davies, J. A., and Garrod, D. R. (1995). Induction of early stages of kidney tubule differentiation by lithium ions. *Dev. Biol.* **167**, 50–60.

Davies, R. C., Calvio, C., Bratt, E., Larsson, S. H., Lamond, A. I., and Hastie, N. D. (1998). WT1 interacts with the splicing factor U2AF65 in an isoform-dependent manner and can be incorporated into spliceosomes. *Genes Dev.* **12**, 3217–3225.

Dono, R., Texido, G., Dussel, R., Ehmke, H., and Zeller, R. (1998). Impaired cerebral cortex development and blood pressure regulation in FGF-2-deficient mice. *EMBO J.* **17**, 4213–4225.

Dressler, G. R., Deutsch, U., Chowdhury, K., Nornes, H. O., and Gruss, P. (1990). *Pax2*, a new murine paired-box containing gene and its expression in the developing excretory system. *Development* **109**, 787–795.

Dressler, G. R., and Douglass, E. C. (1992). *Pax-2* is a DNA-binding protein expressed in embryonic kidney and Wilms tumor. *Proc. Natl. Acad. Sci. USA* **89**, 1179–1183.

Dudley, A. T., Godin, R. E., and Robertson, E. J. (1999). Interaction between FGF and BMP signaling pathways regulates development of metanephric mesenchyme. *Genes Dev.* **13**, 1601–1613.

Dudley, A. T., Lyons, K. M., and Robertson, E. J. (1995). A requirement for bone morphogenetic protein-7 during development of the mammalian kidney and eye. *Genes Dev.* **9**, 2795–2807.

Eastman, Q., and Grosschedl, R. (1999). Regulation of LEF-1/TCF transcription factors by Wnt and other signals. *Curr. Opin. Cell Biol.* **11**, 233–240.

Ekblom, P. (1989). Developmentally regulated conversion of mesenchyme to epithelium. *FASEB J.* **3**, 2141–2150.

Giancotti, F. G., and Ruoslahti, E. (1999). Integrin signaling. *Science* **285**, 1028–1032.

Goto, S., Yaoita, E., Matsunami, H., Kondo, D., Yamamoto, T., Kawasaki, K., Arakawa, M., and Kihara, I. (1998). Involvement of R-cadherin in the early stage of glomerulogenesis. *J. Am. Soc. Nephrol.* **9**, 1234–1241.

Grobstein, C. (1956). Trans-filter induction of tubules in mouse metanephric mesenchyme. *Exp. Cell Res.* **10**, 424–440.

Gumbiner, B. M. (1996). Cell adhesion: The molecular basis of tissue architecture and morphogenesis. *Cell* **84**, 345–357.

Hatini, V., Huh, S. O., Herzlinger, D., Soares, V. C., and Lai, E. (1996). Essential role of stromal mesenchyme in kidney morphogenesis revealed by targeted disruption of Winged Helix transcription factor BF-2. *Genes Dev.* **10**, 1467–1478.

Heanue, T. A., Reshef, R., Davis, R. J., Mardon, G., Oliver, G., Tomarev, S., Lassar, A. B., and Tabin, C. J. (1999). Synergistic regulation of vertebrate muscle development by Dach2, Eya2, and Six1, homologs of genes required for Drosophila eye formation. *Genes Dev.* **13**, 3231–3243.

Herzlinger, D., Koseki, C., Mikawa, T., and al-Awqati, Q. (1992). Metanephric mesenchyme contains multipotent stem cells whose fate is restricted after induction. *Development* **114**, 565–572.

Herzlinger, D., Qiao, J., Cohen, D., Ramakrishna, N., and Brown, A. M. (1994). Induction of kidney epithelial morphogenesis by cells expressing Wnt-1. *Dev. Biol.* **166**, 815–818.

Hoffmann, I., and Balling, R. (1995). Cloning and expression analysis of a novel mesodermally expressed cadherin. *Dev. Biol.* **169**, 337–346.

Hudson, B. G., Reeders, S. T., and Tryggvason, K. (1993). Type IV collagen: Structure, gene organization, and role in human diseases. Molecular basis of Goodpasture and Alport syndromes and diffuse leiomyomatosis. *J. Biol. Chem.* **268**, 26033–26036.

Inoue, T., Chisaka, O., Matsunami, H., and Takeichi, M. (1997). Cadherin-6 expression transiently delineates specific rhombomeres, other neural tube subdivisions, and neural crest subpopulations in mouse embryos. *Dev. Biol.* **183**, 183–194.

Karavanova, I. D., Dove, L. F., Resau, J. H., and Perantoni, A. O. (1996). Conditioned medium from a rat ureteric bud cell line in combination with bFGF induces complete differentiation of isolated metanephric mesenchyme. *Development* **122**, 4159–4167.

Katz, B. Z., Levenberg, S., Yamada, K. M., and Geiger, B. (1998). Modulation of cell-cell adherens junctions by surface clustering of the N-cadherin cytoplasmic tail. *Exp. Cell Res.* **243**, 415–424.

Kemler, R., and Ozawa, M. (1989). Uvomorulin-catenin complex: Cytoplasmic anchorage of a Ca^{2+}-dependent cell adhesion molecule. *Bioessays* **11**, 88–91.

Kimura, Y., Matsunami, H., Inoue, T., Shimamura, K., Uchida, N., Ueno, T., Miyazaki, T., and Takeichi, M. (1995). Cadherin-11 expressed in association with mesenchymal morphogenesis in the head, somite, and limb bud of early mouse embryos. *Dev. Biol.* **169**, 347–358.

Kispert, A., Vainio, S., and McMahon, A. P. (1998). Wnt-4 is a mesenchymal signal for epithelial transformation of metanephric mesenchyme in the developing kidney. *Development* **125**, 4225–4234.

Kispert, A., Vainio, S., Shen, L., Rowitch, D. H., and McMahon, A. P. (1996). Proteoglycans are required for maintenance of Wnt-11 expression in the ureter tips. *Development* **122**, 3627–3637.

Klein, G., Langegger, M., Timpl, R., and Ekblom, P. (1988). Role of laminin A chain in the development of epithelial cell polarity. *Cell* **55**, 331–341.

Klein, P. S., and Melton, D. A. (1996). A molecular mechanism for the effect of lithium on development. *Proc. Natl. Acad. Sci. USA* **93**, 8455–8459.

Koch, A. W., Bozic, D., Pertz, O., and Engel, J. (1999). Homophilic adhesion by cadherins. *Curr. Opin. Struct. Biol.* **9**, 275–281.

Kreidberg, J. A., Sariola, H., Loring, J. M., Maeda, M., Pelletier, J., Housman, D., and Jaenisch, R. (1993). WT1 is required for early kidney development. *Cell* **74**, 679–691.

Leimeister, C., Bach, A., and Gessler, M. (1998). Developmental expression patterns of mouse sFRP genes encoding members of the secreted frizzled related protein family. *Mech. Dev.* **75**, 29–42.

Lelievre-Pegorier, M., Vilar, J., Ferrier, M. L., Moreau, E., Freund, N., Gilbert, T., and Merlet-Benichou, C. (1998). Mild vitamin A deficiency leads to inborn nephron deficit in the rat. *Kidney Int.* **54**, 1455–1462.

Lescher, B., Haenig, B., and Kispert, A. (1998). sFRP-2 is a target of the Wnt-4 signaling pathway in the developing metanephric kidney. *Dev. Dyn.* **213**, 440–451.

Leyns, L., Bouwmeester, T., Kim, S. H., Piccolo, S., and De Robertis, E. M. (1997). Frzb-1 is a secreted antagonist of Wnt signaling expressed in the Spemann organizer. *Cell* **88**, 747–756.

Lin, X., and Perrimon, N. (1999). Dally cooperates with Drosophila Frizzled 2 to transduce Wingless signalling. *Nature* **400**, 281–284.

Lombard, M. N., and Grobstein, C. (1969). Activity in various embryonic and postembryonic sources for induction of kidney tubules. *Dev. Biol.* **19**, 41–51.

Luo, G., Hofmann, C., Bronckers, A. L. J. J., Sohocki, M., Bradley, A., and Karsenty, G. (1995). BMP-7 is an inducer of nephrogenesis, and is also required for eye development and skeletal patterning. *Genes Dev.* **9**, 2808–2820.

Madden, S. L., Cook, D. M., Morris, J. F., Gashier, A., Sukhatme, V. P., and Rauscher, F. J., III.??? (1991). Transcriptional repression mediated by the WT1 Wilms' tumor gene product. *Science* **253**, 1550–1553.

Mah, S. P., Saueressig, H., Goulding, M., Kintner, C., and Dressler, G. R. (2000). Kidney development in cadherin-6 mutants: Delayed mesenchyme-to-epithelial conversion and loss of nephrons. *Dev. Biol.* **223**, 38–53.

Mauch, T. J., Yang, G., Wright, M., Smith, D., and Schoenwolf, G. C. (2000). Signals from trunk paraxial mesoderm induce pronephros formation in chick intermediate mesoderm. *Dev. Biol.* **220**, 62–75.

Mendelsohn, C., Batourina, E., Fung, S., Gilbert, T., and Dodd, J. (1999).

Stromal cells mediate retinoid-dependent functions essential for renal development. *Development* **126**, 1139–1148.

Miner, J. H. (1999). Renal basement membrane components. *Kidney Int.* **56**, 2016–2024.

Miner, J. H., and Li, C. (2000). Defective glomerulogenesis in the absence of laminin alpha5 demonstrates a developmental role for the kidney glomerular basement membrane. *Dev. Biol.* **217**, 278–289.

Miner, J. H., Patton, B. L., Lentz, S. I., Gilbert, D. J., Snider, W. D., Jenkins, N. A., Copeland, N. G., and Sanes, J. R. (1997). The laminin alpha chains: Expression, developmental transitions, and chromosomal locations of alpha-5, identification of heterotrimeric laminins 8–11, and cloning of a novel alpha3 isoform. *J. Cell Biol.* **137**, 685–701.

Miner, J. H., and Sanes, J. R. (1994). Collagen IV alpha 3, alpha 4, and alpha 5 chains in rodent basal laminae: sequence, distribution, association with laminins, and developmental switches. *J. Cell Biol.* **127**, 879–891.

Mlodzik, M. (1999). Planar polarity in the Drosophila eye: A multifaceted view of signaling specificity and cross-talk. *EMBO J.* **18**, 6873–6879.

Muller, U., Wang, D., Denda, S., Meneses, J. J., Pedersen, R. A., and Reichardt, L. F. (1997). Integrin alpha8beta1 is critically important for epithelial-mesenchymal interactions during kidney morphogenesis. *Cell* **88**, 603–613.

Nakagawa, S., and Takeichi, M. (1995). Neural crest cell-cell adhesion controlled by sequential and subpopulation-specific expression of novel cadherins. *Development* **121**, 1321–1332.

Noakes, P. G., Miner, J. H., Gautam, M., Cunningham, J. M., Sanes, J. R., and Merlie, J. P. (1995). The renal glomerulus of mice lacking s-laminin/laminin beta 2: Nephrosis despite molecular compensation by laminin beta 1. *Nature Genet.* **10**, 400–406.

Noll, M. (1993). Evolution and role of Pax genes. *Curr. Opin. Genet. Dev.* **3**, 595–605.

Nusse, R., and Varmus, H. E. (1992). Wnt genes. *Cell* **69**, 1073–1087.

Ohto, H., Kamada, S., Tago, K., Tominaga, S. I., Ozaki, H., Sato, S., and Kawakami, K. (1999). Cooperation of six and eya in activation of their target genes through nuclear translocation of Eya. *Mol. Cell Biol.* **19**, 6815–6824.

Ozawa, M., Baribault, H., and Kemler, R. (1989). The cytoplasmic domain of the cell adhesion molecule uvomorulin associates with three independent proteins structurally related in different species. *EMBO J.* **8**, 1711–1717.

Perantoni, A. O., Dove, L. F., and Williams, C. L. (1991). Induction of tubules in rat metanephric mesenchyme in the absence of an inductive tissue. *Differentiation* **48**, 25–31.

Phelps, D. E., and Dressler, G. R. (1993). Aberrant expression of Pax-2 in Danforth's short tail (Sd) mice. *Dev. Biol.* **157**, 251–258.

Pignoni, F., Hu, B., Zavitz, K. H., Xiao, J., Garrity, P. A., and Zipursky, S. L. (1997). The eye-specification proteins So and Eya form a complex and regulate multiple steps in Drosophila eye development [published erratum appears in *Cell* 92(4):following 585(1998)]. *Cell* **91**, 881–891.

Pritchard-Jones, K., Fleming, S., Davidson, D., Bickmore, W., Porteous, D., Gosden, C., Bard, J., Buckler, A., Pelletier, J., Housman, D., van Heyningen, V., and Hastie, N. (1990). The candidate Wilms' tumor gene is involved in genitourinary development. *Nature* **346**, 194–197.

Qiao, J., Cohen, D., and Herzlinger, D. (1995). The metanephric blastema differentiates into collecting system and nephron epithelia *in vitro*. *Development* **121**, 3207–3214.

Rosenberg, P., Esni, F., Sjodin, A., Larue, L., Carlsson, L., Gullberg, D., Takeichi, M., Kemler, R., and Semb, H. (1997). A potential role of R-cadherin in striated muscle formation. *Dev. Biol.* **187**, 55–70.

Ryan, G., Steele-Perkins, V., Morris, J., Rauscher, F. J., III, and Dressler, G. R. (1995). Repression of Pax-2 by WT1 during normal kidney development. *Development* **121**, 867–875.

Saxén, L. (1987). Organogenesis of the kidney. *In* "Developmental and Cell Biology Series 19" (P. W. Barlow, P. B. Green, and C. C. White, eds.). Cambridge Univ. Press, Cambridge.

Saxén, L., and Lehtonen, E. (1978). Transfilter induction of kidney tubules as a function of the extent and duration of intercellular contacts. *J. Embryol. Exp. Morphol.* **47**, 97–109.

Saxén, L., and Saksela, E. (1971). Transmision and spread of embryonic induction. II. Exclusion of an assimilatory transmission mechanism in kidney tubule induction. *Exp. Cell Res.* **66**, 369–377.

Schedl, A., Ross, A., Lee, M., Engelkamp, D., Rashbass, P., van Heyningen, V., and Hastie, N. D. (1996). Influence of PAX6 gene dosage on development: Overexpression causes severe eye abnormalities. *Cell* **86**, 71–82.

Seufert, D. W., Brennan, H. C., DeGuire, J., Jones, E. A., and Vize, P. D. (1999). Developmental basis of pronephric defects in xenopus body plan phenotypes. *Dev. Biol.* **215**, 233–242.

Shan, W. S., Tanaka, H., Phillips, G. R., Arndt, K., Yoshida, M., Colman, D. R., and Shapiro, L. (2000). Functional cis-heterodimers of N- and R-cadherins. *J. Cell Biol.* **148**, 579–590.

Shapiro, L., Fannon, A. M., Kwong, P. D., Thompson, A., Lehmann, M. S., Grubel, G., Legrand, J. F., Als-Nielsen, J., Colman, D. R., and Hendrickson, W. A. (1995). Structural basis of cell-cell adhesion by cadherins. *Nature* **374**, 327–337.

Stark, K., Vainio, S., Vassileva, G., and McMahon, A. P. (1994). Epithelial transformation of metanephric mesenchyme in the developing kidney regulated by Wnt-4. *Nature* **372**, 679–683.

Stuart, E. T., Kioussi, C., and Gruss, P. (1993). Mammalian pax genes. *Annu. Rev. Genet.* **27**, 219–236.

Tassin, M.-T., Beziau, A., Gubler, M.-C., and Boyer, B. (1994). Spatiotemporal expression of molecules associated with junctional complexes during the *in vivo* maturation of renal podocytes. *Int. J. Dev. Biol.* **38**, 45–54.

Torres, M., Gomez-Pardo, E., Dressler, G. R., and Gruss, P. (1995). Pax-2 controls multiple steps of urogenital development. *Development* **121**, 4057–4065.

Tsuda, M., Kamimura, K., Nakato, H., Archer, M., Staatz, W., Fox, B., Humphrey, M., Olson, S., Futch, T., Kaluza, V., Siegfried, E., Stam, L., and Selleck, S. B. (1999). The cell-surface proteoglycan Dally regulates Wingless signalling in Drosophila. *Nature* **400**, 276–280.

Unsworth, B., and Grobstein, C. (1970). Induction of kidney tubules in mouse metanephrogenic mesenchyme by various embryonic mesenchymal tissues. *Dev. Biol.* **21**, 547–556.

van Heyningen, V., and Hastie, N. (1992). Wilms' tumour: Reconciling genetics and biology. *Trends Genet.* **8**, 16–21.

Vestweber, D., Kemler, R., and Ekblom, P. (1985). Cell-adhesion molecule uvomorulin during kidney development. *Dev. Biol.* **112**, 213–221.

Vilar, J., Gilbert, T., Moreau, E., and Merlet-Benichou, C. (1996). Metanephros organogenesis is highly stimulated by vitamin A derivatives in organ culture. *Kidney Int.* **49**, 1478–1487.

Vukicevic, S., Kopp, J. B., Luyten, F. P., and Sampath, T. K. (1996). Induction of nephrogenic mesenchyme by osteogenic protein 1 (bone morphogenetic protein 7). *Proc. Natl. Acad. Sci. USA* **93**, 9021–9026.

Weller, A., Sorokin, L., Illgen, E. M., and Ekblom, P. (1991). Development and growth of mouse embryonic kidney in organ culture and modulation of development by soluble growth factor. *Dev. Biol.* **144**, 248–261.

Wodarz, A., and Nusse, R. (1998). Mechanisms of Wnt signaling in development. *Annu. Rev. Cell Dev. Biol.* **14**, 59–88.

Xu, P. X., Adams, J., Peters, H., Brown, M. C., Heaney, S., and Maas, R. (1999). Eya1-deficient mice lack ears and kidneys and show abnormal apoptosis of organ primordia. *Nature Genet.* **23**, 113–117.

Yanagawa, S., van Leeuwen, F., Wodarz, A., Klingensmith, J., and Nusse, R. (1995). The dishevelled protein is modified by wingless signaling in Drosophila. *Genes Dev.* **9**, 1087–1097.

Yap, A. S., Brieher, W. M., and Gumbiner, B. M. (1997). Molecular and functional analysis of cadherin-based adherens junctions. *Annu. Rev. Cell Dev. Biol.* **13**, 119–146.

Yap, A. S., Niessen, C. M., and Gumbiner, B. M. (1998). The juxtamembrane region of the cadherin cytoplasmic tail supports lateral clustering, adhesive strengthening, and interaction with p120ctn. *J. Cell Biol.* **141**, 779–789.

Yoshida, K., Taga, T., Saito, M., Suematsu, S., Kumanogoh, A., Tanaka, T., Fujiwara, H., Hirata, M., Yamagami, T., Nakahata, T., Hirabayashi, T., Yoneda, Y., Tanaka, K., Wang, W. Z., Mori, C., Shiota, K., Yoshida, N., and Kishimoto, T. (1996). Targeted disruption of gp130, a common signal transducer for the interleukin 6 family of cytokines, leads to myocardial and hematological disorders. *Proc. Natl. Acad. Sci. USA* **93**, 407–411.

15

Establishment of Polarity in Epithelial Cells of the Developing Nephron

Sharon L. Karp and Bruce A. Molitoris

I. Summary

Research from multiple laboratories has documented the importance of establishing and maintaining cell polarity for normal cell function. Cell biological studies have defined many of the necessary proteins and processes that facilitate this specialized organization. In the developing kidney, little of this information has yet been utilized to further our understanding of how epithelial cells develop an apical basal axis of polarity. However, with the availability of new developmental techniques and the tremendous growth in the understanding of the establishment and maintenance of epithelial polarity elsewhere, substantial progress should soon be made in our understanding of this aspect of kidney development and function. Of particular importance here are new data regarding protein scaffolds and their potential role in mediating the establishment of polarity in developing epithelial cells of the nephron.

II. Introduction

Once the kidney anlagen has formed, nephrons develop from small condensations of metanephric mesenchyme under inductive signals from the tips of the collecting duct system. This process then continues through much of the rest of renal development. As each condensation starts to epithelialize, the polarization process begins afresh and the forming tubular epithelium starts to develop specific transport functions, soon establishing and maintaining cell surface membrane polarity (reviewed by Saxén, 1987).

Surface membrane polarity is the key to nephron function: apical and basolateral membrane (BLM) domains that are structurally, biochemically, and physiologically unique enable cells to separate the glomerular filtrate in the lumen from the interstitial compartment on the basal side of the epithelium. This separation has been found to require the proper placement and organization of the junctional complex at the interface between apical and basolateral membrane surfaces, so separating them. This junctional complex not only acts as a seal between adjacent cells, but also regulates the transport of molecules moving between the lumen and blood compartments across the paracellular pathway. Additional directional reabsorption, secretion, and exchange between the lumen and blood compartments are accomplished by domain-specific membrane components that include ion channels, transport proteins, enzymes, cytoskeletal associations, and

211

lipids. The intact epithelium required for normal renal function thus has apical and basolateral surface membrane domains capable of selective vectorial reabsorption and secretion unique to each segment of the nephron (reviewed in Wagner and Molitoris, 1999).

Establishment of these surface membrane apical and basolateral domains is a multistage process involving cell–cell and cell–extracellular matrix (ECM) interactions, together with multiple intracellular targeting steps. These dynamic events require the proper organization of the cytoskeleton in general and the actin cytoskeleton in particular. For instance, surface membrane amplification and the increased reabsorptive area of microvilli depend on the actin cytoskeleton; cell–cell and cell–substratum junctions require the participation of actin and intermediate filaments; whereas intracellular vesicle transport depends on both microtubules and the actin cytoskeleton. Any alteration that disrupts the polarity of the surface membrane domains will prevent normal cell function and may result in organ dysfunction and potentially a disease state. This chapter describes the components of a polarized renal epithelial cell, the mechanisms by which they are established and maintained, and what is known of the developmental aspects of this process.

III. Acquisition of Epithelial Polarity

During nephrogenesis, polarization of epithelial cells begins early in the process of cell differentiation from a mesenchymal to epithelial phenotype. Little is known of how this takes place *in vivo,* but the process has been studied in cell culture models, particularly in Madin-Darby canine kidney (MDCK) cells, an epithelial cell line that forms polarized cysts and tubules when grown with hepatocyte growth factor in a collagen gel (Montesano et al., 1991). Pollack et al. (1998) examined the development of polarity in MDCK cultures. MDCK cells form polarized cysts and tubules when grown with hepatocyte growth factor in a collagen gel (Montesano et al., 1991). Pollack et al., (1998) examined changes in cell-surface marker proteins in apical, basolateral, and junctional domains as tubules developed from MDCK cysts in this system. In the cyst stage, the apical marker gp135 and the basolateral protein E-cadherin were normally polarized. During the period of tubulogenesis, when cells grow asymmetrically off of the cyst, the apical and basolateral membranes initially remain distinct, but in later stages, E-cadherin lost its BLM specificity and gp135 disappeared in cells without a lumen, evidence for the loss of polarity. Later still, after a new lumen formed, gp135 localized to the luminal membrane, and later E-cadherin again became restricted to the basolateral membrane. Cells remain in contact with one another during this entire process. In this model, cell polarity was lost transiently during the process of tubulogenesis, whereas

plasma membrane subdomains underwent reorganization (Pollack et al., 1998).

The differential expression of specific surface membrane transporters during embryonic development has been reviewed (Horster, 2000). In several different species, expression was dependent on the specific transporter class and varied for specific subunits for the amiloride-sensitive epithelial sodium channel (EnaC), aquaporins, and chloride channels, including the cystic fibrosis transmembrane conductance regulator. Expression for EnaC and aquaporin mRNAs was seen at about embryonic day 15 in the rat and continued to increase postnatally. Unfortunately, little is known about differential apical and BLM expression of these proteins. Data do, however, indicate that a specific isoform of the B subunit of $Na^+K^+ATPase$ may determine the acquisition of basolateral sideness (Burrow et al., 1999). During embryonic development, the B_2 isoform is expressed and $Na^+K^+ATPase$ is found in both apical and BLM domains. Postnatally, the B2 subunit is downregulated, while the B1-subunit increases, and $Na^+K^+ATPase$ becomes distributed in a polar fashion. These events depend on molecular regulatory processes initiated by various inductive signals and thus reflect the downstream effects of several key transcription factors (Pax-2, WT-1, HNF-kA, SP1, and SP3), but their role in initiating the temporal and spatial expression of surface membrane transporters in kidney epithelium remains to be determined.

IV. Structural Organization

All epithelial cells share common characteristics. There is an apical domain that faces the external milieu and contains cell type-specific proteins mediating specific and selective processes of absorption and secretion. The basolateral domain is less cell specific, showing many functions that are common to all epithelial cells, and it can be subdivided into lateral and basal regions. The lateral region is involved in cell–cell interactions and is the site of the junctional complex that mediates cell–cell attachment. Communication between cells occurs both at the junctional complex and via gap junctions that are also present in the lateral region. The basal region contains hemidesmosomes and adhesion molecules that contact the basal lamina and serve to attach cells to the extracellular matrix. Together, these surface membrane domains allow the epithelium to dictate what is passed from the lumen to the interstitial compartment and vice versa (Mays et al., 1994, Caplan, 1997).

Renal tubular epithelial cells possess a highly specialized apical region, and a diagram of proximal tubular epithelial cells is shown in Fig. 15.1. The apical membrane domain consists of microvilli and terminal web regions (Bretscher, 1991). Each microvillus consists of 20 to 30 longitudinally oriented, polarized (barbed or "plus" end at the microvillus

Contact Naïve Cell

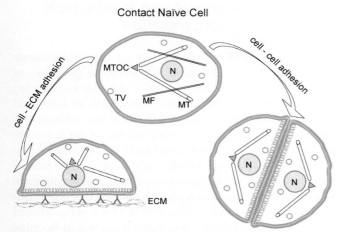

cell - ECM adhesion

cell - cell adhesion

MTOC

TV   MF   MT

N

N

N

N

ECM

Figure 15.1 Structural and functional aspects of epithelial cells.

tip) actin microfilaments that extend into the terminal web area and are cross-linked by specific actin-bundling proteins such as fimbrin and villin (Otto, 1994). Attachment of this core of microfilaments to the lateral portions of the microvillus membrane involves both ezrin and myosin I, and each of these proteins has actin and membrane-binding domains (Tsukita *et al.*, 1997; Colucio, 1997). Attachment of the microvillar actin core at the tip of the microvillus involves a poorly characterized protein complex that is likely to have similarities to the Z-line in muscles with which the plus end of actin also interacts.

Microvillar actin in proximal tubular cells is stabilized in the terminal web region by its interactions with villin, tropomyosin, spectrin, and nonmuscle myosin II. The terminal web contains a dense meshwork of actin and actin-associated proteins and intermediate filaments that are primarily oriented perpendicular to the microvilli. Although the proteins in the terminal web region are heavily cross-linked, they also need to be dynamic enough to permit the passage of molecules and vesicles through this active endocytic transport area. Understanding and better defining the transport mechanisms responsible for bidirectional movement to and from the apical surface are areas of intense investigation. Because this region lacks microtubules, a general hypothesis proposed by many laboratories is that the movement of vesicles is an actin-dependent process requiring the participation of an actin motor, e.g., myosin (Goodson *et al.*, 1997). Filaments of the terminal web make connections to the lateral junctions: actin arrives from the tight junction and microvilli, and the intermediate filaments originate from desmosomes. The apical arrangement of the cytoskeleton in the distal tubule epithelium is not as elaborate, but it also contains an actin cytoskeleton interacting with the plasma membrane through cross-linking proteins (Brown and Stow, 1996).

The lateral membrane is structurally separated from the apical domain by the junctional complex that consists of

four types of junctions. The first are the tight junctions (*zonula occludens*), which encircle the apex of the cell, forming a barrier to intradomain movement of intrinsic membrane proteins and outer leaflet phospholipids (the "fence" function) and to paracellular movement of solutes between biologic compartments (the "gate" function) (Lutz and Siahaan, 1997). Three cytoplasmic tight junction-associated proteins have been identified, ZO-1, ZO-2, and cingulin, as well as one transmembrane protein, occludin, which may comprise part of the tight junction strands observed by freeze-fracture analysis. An additional class of transmembrane tight junction proteins, the claudins, has been implicated in mediating the selective transport of ions across the tight junction (Simon *et al.*, 2000). Paracellin-1 has been shown to mediate the paracellular transport of Mg^{2+} across the tight junction of the thick ascending limb of Henle. Additional members of this family of proteins may be responsible for mediating the paracellular transport of other ions in a segment-specific pattern within the nephron. The "leakiest" tight junctions are found in proximal tubules, whereas the tightest and most complex are present in the collecting duct epithelium (Brown and Stow, 1996). Interestingly, tight junctions are present between the two different epithelial cell types, the intercalated and principal cells, in the collecting duct.

The second junction underneath the tight junction is the *zonula adherens*. It consists of a circumferential band of actin microfilaments of mixed polarity that encircles the cell. Myosin II, α-actinin, and tropomyosin have all been identified here and are likely to cross-link and provide contractile properties for this F-actin. The site of lateral membrane attachment for this F-actin is the adhesion plaque that contains α-actinin, zyxin, vinculin, and radixin and is linked to the catenin/cadherin complex that mediates cell–cell adhesion (Barth *et al.*, 1997). Cadherins are a superfamily of transmembrane glycoproteins that bind to each other in a homophilic manner to mediate Ca^{2+}-dependent cell–cell adhesion. Third, and below the *zonula adherens*, are the desmosomes (*macula adherens*): these are distributed along the lateral membrane forming cell-cell adhesions mediated by cadherins. These three junctional complexes are critical for establishing and maintaining structural and functional cell polarity. The final junction found on the lateral membrane is the gap junction (Nicholson and Bruzzone, 1997). This junction results in the formation of a protein channel between adjacent cells that permits the passage of low molecular weight molecules with an approximate molecular weight of 1000.

The basal portions of the cells are attached to the substratum via hemidesmosomes mediated by integrins, a family of heterodimeric transmembrane glycoproteins that mediate Ca^{2+}-dependent cell–substratum and heterophilic cell–cell binding (Clark and Brugge 1995). There are over 20 integrin members formed by the pairing of different α and β subunit isoforms. Binding to the extracellular matrix (fibronectin, laminin, collagen) usually involves the Arg-Gly-Asp (R-G-D) tripeptide sequence and is regulated by both intracellular and extracellular signals that ultimately result in interactions with the actin cytoskeleton. Focal adhesions are also present on the basal surface and help anchor cells to the substratum. Integrin heterodimers interact with a protein complex containing talin, vinculin, and actin capping proteins that in turn interact with actin filaments cross-linked by α-actinin.

While the actin cytoskeleton and intermediate filaments play a large role in maintaining the overall three-dimensional structure of the epithelial cells, microtubules are involved in providing a track or road for membrane vesicles to move along (Thaler and Haimo, 1996). Consequently, the microtubule cytoskeleton participates in the regulation and delivery of membrane components to their correct membrane domains. Microtubules are polar structures with a fast-growing, or "plus," end distal to the centriole region and a slow-growing, or "minus," end closer to the centriole. This polarity is thought to be important, as only unidirectional microtubule motors have been identified. It appears that, in epithelial cells, microtubules run parallel to the long axis of the cell; their minus ends are located in the apical aspect of the cell and extend, with their plus ends pointed distally, to basal regions of the cell. Shorter microtubules, whose origin and polarity are often mixed, reside closer to the plasma membrane. Importantly, there is no evidence that microtubules have direct contact with the epithelial plasma membrane. This reinforces the concept that a microtubule motor may "hand off" cargo to an actin-based motor for the final steps of plasma membrane-directed transport, although information indicates that E-cadherin may regulate the dynamic stability of microtubules (Chausovsky et al., 2000).

Vesicle movement along microtubules utilizes ATP-driven microtubule motors. Kinesin is a plus end-directed motor for basal-directed movement. The minus end-directed motor, dynein, transports vesicles to the apical surface (Vallee and Sheetz, 1996). Cytoplasmic dynein is regulated by dynactin through its interactions with vesicles and microtubules. In addition, Lis1 has been shown to interact with cytoplasmic dynein/dynactin and microtubules in mammalian brain (Smith, 2000). Lis1 is prominent at microtubule organizing centers and its overexpression results in the accumulation of microtubules at the cell periphery, with reductions at the microtubule organizing centers. It has therefore been proposed that Lis1 upregulates the dynein-driven translocation of microtubule segments to the surface membrane. Finally, Lis1 overexpression altered Golgi distribution, which has been shown to be regulated by dynein activity. Details of the motor interactions with their respective receptors, and regulation of the motor activity, are areas of active and important research.

V. Physiological and Biochemical Organization

As luminal fluid originating from the glomerular filtrate traverses the renal tubule, it encounters a gradation of epithelia whose different properties are due to differing surface membrane components. This results in individual tubule segments showing differing selective reabsorption and secretion. Table 15.1 compares some of the key components in the apical and basolateral membranes of epithelial cells. Proximal tubule cells have a high rate of endocytic activity at the apical membrane, thus requiring an equally active pathway to replace membrane that is internalized during endocytosis (Linas, 1997). This is accomplished via a rapid endosome recycling pathway, as well as insertion of newly synthesized membrane components arriving from the Golgi (Fig. 15.2).

A key function of the proximal tubule is sodium reabsorption. This is dependent on the polarized delivery of specific carrier proteins, such as the Na^+/H^+ antiporter and the Na^+-dependent glucose, amino acid, and phosphate cotransporters to the apical membrane, as well as localization of Na^+,K^+-ATPase to the basolateral membrane. There are also several polypeptide hormone receptors targeted to the apical membrane of the proximal tubule where they probably participate in the recovery of peptide hormones from the glomerular filtrate. Other receptors, such as Gp330, a glycoprotein and member of the LDL receptor family with multiple ligands, including B-LDL and plasminogen, are also found on the apical membrane. Following receptor internalization, dissociation occurs with the ligand being targeted to lysosomes, and the receptor recycled back to the apical surface (Mukherjee et al., 1997). Many of the recycling pathways depend on the actin and microtubule cytoskeleton for efficient and accurate targeting and delivery.

While much is known about the proximal tubule, the epithelium of the loop of Henle is the most poorly characterized tubule segment. One membrane protein that has been found in abundance on both apical and basolateral membranes in the thin descending limb is the aquaporin 1 (AQP1) water channel (Brown and Stow, 1996). This region has high water permeability in contrast to the thin ascending limb that is water impermeant and does not contain detectable AQP1 water channels.

The epithelium of the collecting duct contains two distinct cell types: intercalated cells and principal cells. Water reabsorption is a critical function in the collecting duct and is facilitated by principal cells that have AQP2 water channels in their apical membrane and AQP3 channels in their basolateral membrane. The intercalated cells are specialized for the transport of protons that depend on the polar distribution of both a H^+-ATPase pump and an anion exchanger. One of the mechanisms by which the epithelium

Table 15.1 Surface Membrane Asymmetry of Polarized Epithelial Cells

	Apical membrane	Basolateral membrane
Lipids		
Cholesterol	High	Low
Sphingomyelin	High	Low
Phosphatidylcholine	Low	High
Phosphatidylinositol	Low	High
Proteins		
ATPases	**H⁺-ATPase** (vacuolar type), PTC and thick acscending limb	**H⁺-ATPase**, β-intercalated cells in connecting and cortical collecting tubules
	Mg²⁺-ATPase	**Ca²⁺-ATPase**
	H⁺/K⁺-ATPase, several cell types in collecting ducts	
	Thiazide-sensitive Na⁺-Cl⁻ cotransport (TSC) present in distal convoluted tubule cells	**Na⁺,K⁺-ATPase**
Ion carriers	**Amiloride-sensitive Na⁺ channel**	**Na⁺-(HCO³⁻)₃** cotransport, proximal tubule and cortical thick ascending limb
	Cl⁻/HCO³⁻₃ exchanger, proximal tubule and β-intercalated cells	**Cl⁻/HCO³⁻₃ exchanger**
	Na⁺-dependent cotransporters	**Na⁺-independent glucose carrier**
	Na⁺/K⁺/2Cl⁻ cotransporter (BSC1 or NKCC2), thick ascending limb and macula densa cells	**Na⁺-K⁺-2Cl⁻ cotransporter** (BSC2 or NKCC1), outer and inner medullary collecting ducts
	Na⁺/H⁺ exchangers (NHE3 and NHE2 isoforms)	**Na⁺/H⁺ exchangers** (NHE1 and NHE4 isoforms) present in basolateral membranes in various nephron segments
Water channels	**Aquaporin 1 (AQP1)**, proximal tubule and thin descending limb of Henle	**AQP1**, basolateral membrane of proximal tubule
	AQP2, predominantly in principal cells of collecting duct	**AQP3**, principal cells of collecting duct
		AQP4, principal cells of collecting duct
Enzymes	**Leucine aminopeptidase**	**Adenylate cyclase**
	Maltase	
	GPI-linked proteins	
	Alkaline phosphatase	
Receptors	**Megalin**	**Insulin**
		Parathyroid hormone
		Epidermal growth factor
		Laminin

adjusts to physiological changes is by altering both the number and the location of their membrane transporter proteins. In addition, there are differences in lipid composition between apical and basolateral membranes that result in a more fluid basolateral membrane and lead to alterations in the function of several membrane proteins (Molitoris and Hoilien, 1987).

VI. Establishment and Maintenance of Epithelial Cell Polarity

In vitro studies examining cell polarity indicate that both cell–cell recognition, mediated by cadherins and catenins, and cell–ECM adhesion, mediated by integrins, are important initial events in the establishment of cell polarity (Fig. 15.2) as they each help initiate structural and molecular asymmetry at the cell surface. These signals induce the localized assembly of an actin cytoskeleton, which then leads to the development of a targeting patch for transport vesicles that have been sorted in the Golgi. These steps are followed by the reorganization of microtubules and actin filaments (for review, see Yeaman, 1999).

Cell–cell interactions initiate a series of events that result in the establishment of specific protein domains that serve as foci for the assembly of junctional and nonjunctional cytoskeletal structures. Extracellular contact and adhesion between two nonpolarized epithelial cells, even without cell–ECM interactions, initiate the segregation of apical and basolateral membrane domains. One result of this type of

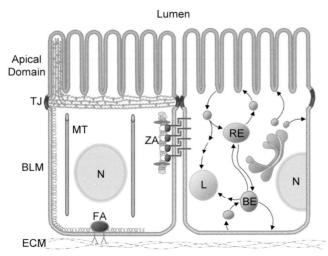

Figure 15.2 Development of asymmetric aspects of epithelial cells following attachment. Either cell–cell or cell–ECM adhesion establishes an axes of polarity with a contacting and a noncontacting surface. Cellular structural development proceeds with reorganization of the actin and microtubule cytoskeletons. Complete epithelial polarity requires both cell–cell and cell–ECM adhesion with formation of junctional complexes and protein scaffold complexes with actin networks and the polarized targeting of domain-specific transport vesicles from the TGN.

interaction is the establishment of a specific apical and basolateral membrane E-cadherin distribution that plays a central role in specifying membrane domain asymmetry (McNeill *et al.*, 1990). In addition, attachment of even single epithelial cells to the ECM initiates differences in protein localization in the surface membrane that distinguish areas of contact and contact-free surfaces (Ohno *et al.*, 1996). Therefore, cell adhesion to the ECM is also of key importance in establishing the apicobasal axis of polarity (Wang *et al.*, 1990). As integrins mediate these changes in the axis of polarity, both cell–cell and cell–ECM interactions are required for the establishment of complete epithelial polarity. This ultimately results in the formation of apical and basolateral domains separated by the junctional complex that also prevents the free passage of membrane components between domains (the "fence" function).

Following integrin and cadherin-mediated adhesion, specialized cytoskeletal and signaling protein networks accumulate at the sites of membrane contacts. The resulting protein–protein interactions then direct both local and global organization of the cell surface membrane. Such protein scaffolds then convert surface membrane spatial cues into local and regional information by facilitating the assembly of higher ordered structures anchored to the actin cytoskeleton. They also recruit and integrate components of signal transduction pathways to these same regions (Yeaman *et al.*, 1999). Finally, these scaffold sites function to organize docking sites for transport vesicles, thus providing a targeting patch specific for vesicles with

domain-specific surface markers. At sites of cell–cell junctions, protein scaffolds thus mediate cell–cell adhesion and apical–basal polarity, organize signaling pathways, and act as domain-specific docking sites.

The number of proteins known to participate in these protein scaffolds is growing rapidly. Of particular importance are LAP (leucine-rich repeats and PDZ domains) proteins that contain 16 leucine-rich repeats at their amino terminal and one to four copies of the PDZ (named after its presence in PSD-95, Discs-Large, and ZO-1 proteins) domain at the carboxy terminal. Their function is to bind other proteins critical for cell structure and for the organization and regulation of signal transduction pathways controlling growth and differentiation (Bryant and Huwe 2000). They localize to the junctional complex in both arthropods, studied primarily in *Drosophila*, and vertebrate epithelial cells. Scribble (Scrib) is one such protein that has been described as maintaining apical–basal polarity by localizing to the septate junction in *Drosophila* (the homologue of the tight junction in vertebrate epithelial cells). There it colocalizes with Dlg, another PDZ protein, to ensure the correct localization of Lgl (Bilder and Perrimon, 2000; Greaves, 2000). All three of these proteins are required for correct apical–basal polarity in *Drosophila*.

ERBIN is another basolateral PDZ protein that functions as an adapter protein for the basolateral localization of receptor ERBB2/HER2 in epithelia (Borg *et al.*, 2000). LET-413 is a basolateral PDZ protein required for the assembly of adherens junctions in *C. elegans*. Mutations to this protein result in alterations in adherens junction placement, epithelial polarity, and actin cytoskeleton disorganization (Legouis *et al.*, 2000). Finally, there is an interaction between PAR (partitioning-defective) proteins, PAR6 and PAR3, Cdc42/Rac1, and an atypical protein kinase C in mediating cell–polarity in C. elegans (Joberty *et al.*, 2000; Lin *et al.*, 2000). Interactions of these proteins have important implications for establishing and maintaining polarity, as Cdc42 plays a role in the fidelity of vesicle transport, and Rac1 is necessary for proper tight junction formation (for review, see Kim, 2000).

Apical and basolateral membrane proteins are synthesized in the endoplasmic reticulum, transported through the Golgi complex, and delivered to a *trans*-Golgi network (TGN) within which they are sorted separately into specific subdomains and packaged into transport vesicles (Rothman and Wieland, 1996; Mostov *et al.*, 2000, Brown and Stow, 1996; Yeaman *et al.*, 1999). Basolateral sorting signals frequently contain either a critical tyrosine residue or an amino acid sequence that forms a tight β turn in the cytoplasmic domain of the protein. Basolateral sorting itself requires P200 or myosin II. Sorting of protein into apical vesicles is dependent on motifs present in the luminal domain or membrane anchor and also involves VIP21/ caveolin, a cholesterol-binding protein that leads to the formation of glycosphingolipid rafts implicated in apical sorting (Mayor *et al.*, 1994).

A second method that the cell uses to sort is through the delivery of protein to both apical and basolateral domains, but with selective retention in only one of them. This type of sorting occurs for the Na$^+$,K$^+$-ATPase where interactions with the basolateral actin cortical cytoskeleton (Fig. 15.2) result in its retention there. There is a third method, used extensively by hepatocytes, that involves delivery of all membrane proteins to the basolateral surface and selective transcytosis of apical membrane proteins to the apical surface. The sorting and targeting of lipids are less well

understood than that of proteins, which is due to a lack of specific probes. There is, however, evidence that sphingolipids are transferred preferentially from the TGN to the apical domain, via vesicles containing co clustered GPI-anchored proteins. Proteins and lipids delivered to either the apical or the basolateral surface are restrained by the tight junction "fence" function and by interactions with the actin cytoskeleton (Fig. 15.3). Retention at a specific membrane region may also be achieved by exclusion from the endocytic pathway.

Transport of vesicles to either apical or basolateral membranes utilizes ATP and the cytoskeleton. Microtubules

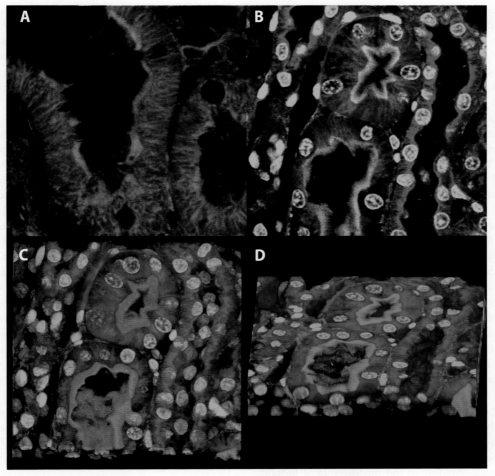

Figure 15.3 Filamentous actin and microtubles localized in perfusion fixed rat kidney sections and imaged by single-and two-photon confocal microscopy. Cortical rat kidney sections were stained with fluorescein phalloidin to localize filamentous actin (green), and microtubules were immunolabled using a monoclonal antibody against tubulin and a secondary antibody conjugated to Texas Red (red). The nuclear dye DAPI, which is detected only during the dual-photon acquisition mode, was used as a counterstain. A single plane image taken using the single-photon mode (A) and dual-photon mode (B) shows alignment and organization of the microtubules (red) to be apical–basal, as well as the large accumulation of filamentous actin in the apical microvillar region of proximal tubules, while showing little F-actin staining in the adjacent collecting duct. A three-dimensional (3D) frontal (C) and side view (D) reconstruction of a through focus series taken using two-photon microscopy shows the complex architecture of the tubular systems that comprise the kidney. In the 3D projection, the undulating topography of the filamentous actin rich microvilli and terminal web of the proximal tubule is seen readily in C, lower center; which stands in sharp contrast to the lumen of a collecting duct seen in D, right. The 3D projections were created using VOXX, a proprietary rendering software developed at the Indiana Center for Biological Microscopy (http://nephrology.iupui.edu/imaging/index.html). Figure courtesy of Ruben Sandavol.

and their associated motors, dynein and kinesin, are responsible for the long distance movement of vesicles throughout cells. The actin and associated myosin motors play a less well-defined role in short distance movement near the plasma membrane. Membrane receptors for kinesin and dynein, kinectin and dynactin, are likely to regulate the motor and vesicle interaction and may even affect the motor filament interaction. In addition, dynactin is a multisubunit complex, one of which is an actin-related protein, ARP, which may link microtubule transport to actin transport (Allan, 1996). Present research is directed toward understanding the events such as phosphorylation, calcium changes, and GTPase activity that have been shown to affect motor activity.

Maintenance of cell polarity also requires sorting and recycling of endocytosed plasma membrane domains. This enables cells to perform their specific absorptive functions and maintain surface membrane polarity. Receptor-mediated endocytosis utilizes clathrin-coated pits and dynamin for internalization of the receptor–ligand complex (Schekman and Orci, 1996; Warnock and Schmid, 1996). These early endosomes, defined functionally as having both cargo to be degraded and components to be recycled, can be handled in several ways upon arrival at the early endosome compartment. They can release their material, i.e., iron from transferrin, and then be recycled rapidly back to the cell surface, or the endosomes may be targeted to the lysosomal system. Some endosomes are also transcytosed to an alternate surface membrane. For example, the IgA receptor is inserted into the basolateral membrane, binds ligand, and then is endocytosed and targeted to the apical membrane where it releases the IgA (Apodaca et al., 1994). There is evidence for a rapid recycling or sorting endosome compartment in proximity to both apical and basolateral membranes. In proximal tubule cells, β_2-microglobulin internalized into apical and basolateral early endosomes has been shown to converge in a common, apically oriented, "late" endosomal compartment (multivesicular body) before delivery to lysosomes (Cohen et al., 1995). The movement, over long distances, from one membrane surface to another, requires microtubule-dependent transport, whereas short distance movement, such as rapid recycling, appears to be micro-tubule independent and actin dependent. This is not surprising given the location of these two cytoskeletal elements. Our understanding of the specific role of individual cytoskeletal motors in endosome movement is still limited and is under active investigation.

VII. Protein Trafficking in Embryonic Kidney

Little is known about the specific role of the many vesicle proteins in the kidney. Northern analysis has shown that several proteins are present (Lehtonen et al., 1999a), but cellular localization studies, essential to begin to determine function, have not been published. SNARE [soluble N-ethylmaleimide-sensitive (NSF) fusion protein receptor] proteins are thought to be involved in vesicular transport. The current paradigm is that vesicle membrane-associated SNAREs (v-SNAREs) specify the target of transport vesicles. Target membrane-associated SNAREs (t-SNAREs) are present on target membranes and interact specifically with v-SNARE molecules. In renal epithelial cells, syntaxin 4, a t-SNARE, is thought to regulate the fusion of aquaporin-2-containing vesicles in the apical membrane of principal cells in the collecting duct (Mandon et al., 1996). Another t-SNARE protein, SNAP 23, is present in the apical plasma membrane of principal cells, as well as in the proximal tubule and thick ascending limb cells (Inoue et al., 1998). SNARE proteins are likely to be important in the trafficking of aquaporin-containing vesicles and also in the trafficking of proton pumps that secrete hydrogen ions in intercalated cells of the inner medullary collecting duct (Lehtonen et al., 1999a).

Munc-18-2, a Sec1-related protein, and Syntaxin 3 have been studied in the developing mouse kidney by Lehtonen et al., (1999b). Munc–18-1 is expressed primarily in epithelial cells and interacts with syntaxin-3, a t-SNARE present in many cell types, including MDCK and other epithelial cells. By embryonic day 13 (E13) in the mouse kidney, in situ hybridization shows that both syntaxin–3 and Munc-18-2 mRNAs are present in epithelial cells. They were found in the proximal tubule and collecting duct with a greater abundance at E17 as compared to E13. Other t-SNARES present in embryonic kidney include syntaxin-2, which is more prominent in mesenchyme than epithelium, and syntaxins 1A, 4, and 5, which are present ubiquitously. Munc-18-2 was shown to be complexed with syntaxin 3 in adult mouse kidney by coimmunoprecipitation, with the pair localizing in the ureteric buds at E11, in collecting ducts at E13, and in S-shaped bodies at E15 and in the apical region of proximal tubules in the adult kidney. These proteins are likely to have a role in vectorial transport in developing renal epithelial cells, as well as in fully differentiated epithelia (Lehtonen et al., 1999b). Munc-18-2 and syntaxin-3 are probably involved in apical exocytosis or in recycling of endocytosed apical membrane constituents (Lehtonen et al., 1999a).

Rab17 is expressed in mouse kidney, liver, and intestine. In kidney, Rab17 mRNA is expressed in epithelial cells of developing kidney, starting first in the branching ureter and then the collecting duct, late S-shaped bodies, and more mature tubules with more mRNA present in proximal than distal tubules. In proximal tubules, Rab17 localized to the basolateral membrane and also stained tubular structures below the apical brush border (Lehtonen et al., 1999a; Lutcke et al., 1994). In studies in cultured epithelial cells,

Rab17 was postulated to function as a regulator of apical recycling and transcytosis (Lehtonen *et al.*, 1999a).

VIII. A Final Comment

It seems barely necessary to say that the establishment and maintenance of cell polarity are essential for normal cell function. As this chapter has shown, many of the proteins and processes necessary to facilitate this specialized organization have been identified and involve cell–cell interactions, cell–ECM interactions, and the actin cytoskeleton and microtubules, as well as multiple proteins involved in intracellular transport. While the molecular infrastructure of renal epithelia is now beginning to be characterized, there has been less investigation of the functional aspects of cell polarity in the developing kidney. With new developmental techniques and the tremendous growth in our understanding of epithelial polarity elsewhere, substantial progress in our understanding of these processes in the kidney is likely in the near future.

References

Allan, V. (1996). Motor proteins: A dynamic duo. *Curr. Biol.* **6**, 630–633.

Apodaca, G., Katz, L. A., and Mostov, K. E. (1994). Receptor-mediated transcytosis of IgA in MDCK cells is via apical recycling endosomes. *J. Cell Biol.* **125**, 67–86.

Barth, A. I., Nathke, I. S., and Nelson, W. J. (1997). Cadherins, catenins and APC protein: Interplay between cytoskeletal complexes and signaling pathways. *Curr. Opin. Cell Biol.* **9**, 683–690.

Bilder, D., Li, M., and Perrimon, N. (2000). Cooperative regulation of cell polarity and growth by Drosophila tumor suppressors. *Science* **289**, 113–115.

Borg, J. P., Marchetto, S., Le Bivic, A., Ollendorff, V., Jaulin-Bastard, F., and Hiroko, S. (2000). ERBIN: A basolateral PDZ protein that interacts with the mammalian ERBB2/HER2 receptor. *Nature Cell Biol.* **2**, 407–414.

Bretscher, A. (1991). Microfilament structure and function in the cortical cytoskeleton. *Annu. Rev. Cell Biol.* **7**, 337–374.

Brown, D., and Stow, J. L. (1996). Protein trafficking and polarity in kidney epithelium: From cell biology to physiology. *Physio. Rev.* **76**, 245–297.

Bryant P. J., and Huwe, A. (2000). LAP proteins: What's up with epithelia? *Nature Cell Biol.* **2**, E141–E145.

Burrow, C. R., Devuyst, O., Li, X., Gatti, L., and Wilson, P. D. (1999). Expression of the beta2-subunit and apical localization of Na+-K+-ATPase in metanephric kidney. *Am. J. Physiol.* **277**, F391–F403.

Caplan, M. J. (1997). Membrane polarity in epithelial cells: Protein sorting and establishment of polarized domains. *Am. J. Physiol.* **272**, F425–F429.

Chausovsky, A., Bershadsky, A. D., and Borisy, G. G. (2000). Cadherin-mediated regulation of microtuble dynamics. *Nature Cell Biol.* **2**, 797–804.

Clark, E. A., and Brugge, J. S. (1995). Integrins and signal transduction pathways: The road taken. *Science* **268**, 233–239.

Cohen, M., Sundin, D. P., Dahl, R., and Molitoris, B. A. (1995). Convergence of apical and basolateral endocytic pathways for beta-2-microglobulin in LLC-PK1 cells. *Am. J. Physiol.* **268**, F829–F838.

Colucio, L.M. (1997). Myosin I. *Am. J. Physiol.* **273**, C347–C359.

Goodson, H. V., Valetti, C, and Kreis, T. E. (1997). Motors and membrane traffic. *Curr. Opin. Cell Biol.* **9**, 18–28.

Greaves, S. (2000). Growth and polarity: The case for Scribble. *Nature Cell Biol.* **2**, E140.

Horster, M. (2000). Embryonic epithelial membrane transporters. *Am. J Physiol.* **279**, F982–F996.

Inoue, T., Nielsen, S., Mandon, B., Terris, J., Kishore, B. K. and Knepper, M. A. (1998). SNAP-23 in rat kidney: Colocalization with aquaporin-2 in collecting duct vesicles. *Am J. Physiol.* **275**, F752–F760.

Joberty, G., Petersen, C., Gao, L., and Macara, I. G. (2000). The cell-polarity protein Par6 links Par 3 and atypical protein kinase C to Cdc42. *Nature Cell Biol.* **2**, 531–538.

Kim, S. K. (2000). Cell polarity: New PARtners for Cdc42 and Rac. *Nature Cell Biol.* **2**, 143–145.

Legouis, R., Gansmuller, A., Sookhareea, S., Bosher, J. M., Baillie, D. M., and Laboues, M. (2000). LET-413 is a basolateral protein required for the assembly of adherens junctions in *Caenorhabditis elegans*. *Nature Cell Biol.* **2**, 415–422.

Lehtonen, S., Lehtonen, E., and Olkkonen, V. M. (1999a). Vesicular transport and kidney development. *Int. J. Dev. Biol.* **43**, 425–433.

Lehtonen, S., Riento, K., Olkkonen, V. M, and Lehtonen, E. (1999b). Syntaxin 3 and Munc- 18-2 in epithelial cells during kidney development. *Kidney Int.* **56**, 815–826.

Lin, D., Edwards, A. S., Fawcett, J. P., Mbamalu, G., Scott J. D., and Pawson, T. (2000). A mammalian PAR-3-PAR-6 complex implicated in Cdc42/Rac1 and aPKc signaling and cell polarity. *Nature Cell Biol.* **2**, 540–547.

Linas, S.L. (1997). Role of receptor mediated endocytosis in proximal tubule epithelial function. *Kidney Int.* **61**, S18–S21.

Lutz, K. L, and Siahaan, T. J. (1997). Molecular structure of the apical junction comnplex and its contribution to the paracellular barrier. *J. Pharm. Sci.* **86**, 977–984.

Mandon, B., Chou, C. L., Nielsen, S., and Knepper, M. A. (1996). Syntaxin-4 is localized to the apical plasma membrane of rat renal collecting duct cells: Possible role in aquaporin-2 trafficking. *J. Clin. Invest.* **98**, 906–913.

Mayor, S. K., Rothberg, K. G., and Maxfield, F. R. (1994). Sequestration of GPI- anchored proteins and caveolae triggered by cross-linking. *Science* **264**, 1948–1951.

Mays, R. W., Beck, K. A., and Nelson, W. J. (1994). Organization and function of the cytoskeleton in polarized epithelial cells: A component of the protein sorting machinery. *Curr. Opin. Cell Biol.* **6**, 16–24.

McNeill, H., Ozawa, M., Kemler, R., and Nelson W. J. (1990). Novel function of the cell adhesion molecule uvomorulin as an inducer of cell surface polarity. *Cell* **62**, 309–316.

Molitoris, B. A., and Hoilien, C. (1987). Static and dynamic components of renal cortical brush border and basolateral fluidity: Role of cholesterol. *J. Membr. Biol.* **99**, 165–172.

Montesano, R., Matsumoto, K., Nakamura, T., and Orci, L. (1991). Identification of a fibroblast derived epithelial morphogen as hepatocyte growth factor. *Cell* **67**, 901–908.

Mostov, K. E., Verges, M., and Altschuler, Y.(2000). Membrane traffic in polarized epithelial cells. *Curr. Opin. Cell Biol.* **12**, 483–490.

Mukherjee, S., Ghosh, R. N., and Maxfield, F. R. (1997). Endocytosis. *Physiol. Rev.* **77**, 759–803.

Nicholson, S. M., and Bruzzone, R. (1997). Gap junctions: Getting the message through. *Curr. Biol.* **7**, R340–344.

Ohno, H., Fournier, M. C., Poy, G., and Bonifacino, J. S. (1996). Structural determinants of interaction of tyrosine-based sorting signals with the adaptor medium chains. *J. Biol. Chem.* **271**, 29009–29015.

Otto, J. J. (1994). Actin-bundling proteins. *Curr. Opin. Cell Biol.* **6**, 105–109.

Pollack, A. L., Runyan, R. B., and Mostov K. E. (1998). Morphogenetic mechanisms of epithelial tubulogenesis: MDCK cell polarity is transiently rearranged without loss of cell-cell contact during scatter factor/hepatocyte growth factor-induced tubulogenesis. *Dev. Biol.* **204**, 64–79.

Rothman, J. E., and Wieland, F. T. (1996). Protein sorting by transport vesicles. *Science* **272**, 227–234.

Saxen, L. (1987), Organogensis of the kidney. In "Developmental and Cell Biology Series 19 P. W. Barlow, P. B. Green, and C. C. White, (eds)." Cambridge Univ. Press, Cambridge.

Schekman, R., and Orci, L. (1996). Coat proteins and vesicle budding. *Science* **271**,1526–1532.

Simon, D. B., Lu, Y Choate, K. A., Velazquez, H., Al-Sabban, E., Praga, M., Casari, G. and Bettinelli, A. (2000). Paracellin-1, a renal tight junction protein required for paracellular Mg (2+) resorption. *Science* **285**, 103–106.

Smith, D. S., Niethammer, M., Ayala, R., Zhou, Y., Gambello, M. J., Boris-Wynshaw, A., and Tsai, L. (2000). Regulation of cytoplasmic dynein behaviour and microtubule organization by mammalian Lis1. *Nature Cell Biol.* **2**, 767–775.

Thaler, C. D., and Haimo, L. T. (1996). Microtubules and microtubule motors: Mechanisms of regulation. *Int. Rev. Cytol.* **164**, 269–327.

Tsukita, S., Yonemura, S., and Tsukita, S. (1997). ERM proteins: Head-to-tail regulation of actin–plasma membrane interaction. *TIBS* **22**, 53–58.

Wagner, M. C. and Molitoris, B. A. (1999). Renal epithelial polarity in health and disease. *Pediatr. Nephro.* 13,163–170.

Wang, A. Z., Ojakian, G. K., and Nelson, W. J. (1990). Steps in the morphogenesis of a polarized epithelium. I. Uncoupling the roles of cell–cell and cell–substratum contact in establishing plasma membrane polarity in multicellular epithelial (MDCK) cysts. *J. Cell Sci.* **95**, 137–151.

Warnock, D. E., and Schmid, S. L. (1996). Dynamin GTPase, a force-generating molecular switch. *BioEssays* **18**, 885–893.

Vallee, R. B., and Sheetz, M. P. (1996) Targeting of motor proteins. *Science* **271**, 1539–1544.

Yeaman, C., Grindstaff, K. K., Hansen, M. D. H., and Nelson W. J. (1999). Cell polarity: Versatile scaffolds keep things in place. *Curr. Biol.* **9**, R515–R517.

16

Development of the Glomerular Capillary and Its Basement Membrane

Dale R. Abrahamson and Ruixue Wang

I. Introduction

Renal organogenesis proceeds through an integrated series of complex interactions among epithelial cells, mesenchymal cells, extracellular matrix, growth factors, cytokines, and other signaling molecules. The process of reciprocal induction between ureteric epithelium and the metanepheric mesenchyme is a key first step in generating the overall architecture of the kidney. Many of the molecular and cellular events governing early kidney development and progress in renal gene targeting studies have been reviewed (Kreidberg, 1996; Kanwar *et al.*, 1997; Miner, 1998; Lipschutz, 1998; Smoyer and Mundel, 1998; Orellana and Ellis, 1998) and are also covered in depth in this book. This chapter focuses specifically on the development of the renal glomerulus and summarizes much of the current data from gene targeting studies that affect glomerulogenesis

(summarized in Table 16.1). First, we discuss the principal cellular and extracellular matrix components of the glomerulus. Then, we describe how these elements originate in the developing kidney and interact with one another during glomerular formation. Finally, we briefly mention some of the more common glomerular diseases, especially those that have developmental significance. For a more detailed coverage of glomerular disease, see Chapter 27.

II. Glomerular Structure

The renal corpuscle consists of the Bowman's capsule, which is punctured at the so-called vascular pole by two arterioles. The afferent arteriole delivers blood into, and the efferent arteriole carries blood away from, a capillary tuft (the glomerulus) that is entirely contained within the Bowman's capsule. The primary function of the glomerulus is to form a high volume, acellular, and almost protein-free filtrate of the blood plasma. This filtrate crosses the glomerular capillary walls, enters the urinary space within the Bowman's capsule, and exits the capsule by way of the urinary pole, which leads into the proximal convoluted tubule of the nephron. The peripheral glomerular capillary walls therefore constitute the plasma filtration barrier. This structure is composed of three primary elements: a vascular endothelial cell layer, a middle layer of basement membrane, termed the glomerular basement membrane (GBM), and an outer epithelial cell layer (Fig. 16.1). Between the individual capillaries of the glomerular tuft are intercapillary mesangial cells. These contractile

Table 16.1 Selected Gene Targeting Studies Resulting in Glomerular Development Phenotypes

Deletion	Protein function	Phenotype	Reference
Nephrin	Slit diaphragm component	Foot process effacement, proteinuria, early postnatal death	Putaala et al. (2001)
CD2AP	Links cytoplasmic domain of nephrin to cytoskeleton?	Foot process effacement, proteinuria, death at 6 weeks	Shih et al. (1999)
Integrin α3 chain	Cell–matrix adhesion	Foot process effacement, GBM disorganization, perinatal death	Kreidberg et al. (1996)
GLEPP1(Ptpro)	Podocyte receptor tyrosine phosphatase	Minor foot process effacement, reduced GFR	Wharram et al. (2000)
PDGFB	Mesangial recruitment?	Fatal perinatal hemorrhaging, absent mesangium	Leveen et al. (1994)
PDGFβ	PDGF receptor	Fatal perinatal hemorrhaging, absent mesangium	Soriano (1994)
Collagen IV α3 chain	Forms heterotrimer with α4(IV) and α5(IV) chains in mature GBM	Absence of collagen a4(IV) and α5(IV) in GBM, Alport-like features, renal failure by 6–9 weeks	Cosgrove et al. (1996); Miner and Sanes (1996); Lu et al. (1999)
Laminin α5 chain	Forms heterotrimer with laminin β2 and γ1 chains	E17 lethality (neural tube and placental defects), avascular glomeruli	Miner ad Li. 2000
Laminin β2 chain	Forms heterotrimer with laminin α5 and γ1 chains	Proteinuria, foot process effacement, death by 1 month	Noakes et al. (1995)

cells resemble both vascular smooth muscle cells and pericytes and probably regulate diameters of the separate glomerular capillaries and hence help govern glomerular filtration rates.

The glomerular endothelium is unusual in that it is extensively fenestrated, but, unlike the fenestrated endothelial cells found in most capillaries, glomerular endothelial fenestrae are open and do not contain diaphragms (Fig. 16.1). Endothelial cell nuclei generally are located in the cell body, which lies near the mesangial waist region of each capillary.

The GBM is approximately twice the thickness of most basement membranes found elsewhere in the body and has classically been described as having three layers: an inner lamina rara interna, a central lamina densa, and an outer lamina rara externa (Fig. 16.1). As discussed presently, the dual thickness of the GBM is due, at least in part, to the fusion of endothelial and epithelial basement membranes during glomerular development. When viewed in the electron microscope, all layers of the GBM have a fine granular, meshwork texture, and thin fibrils extend from the lamina densa across the lamina rara interna and externa and insert into the plasma membrane of endothelial cells and podocytes, respectively (Figs. 11.19 and 16.1).

The outermost layer of the glomerular capillary is lined by the epithelial podocytes. These cells are composed of three structurally and functionally different segments: the cell body, major processes, and foot processes (Smoyer and Mundel, 1998). The podocyte cell body and major process are not directly attached to the GBM and therefore are completely bathed by the Bowman's space. The cell body cytoplasm contains numerous free ribosomes, extensive smooth and rough endoplasmic reticula, and Golgi complexes. The foot processes themselves, which interdigitate closely with those of other processes, are directly applied to the GBM

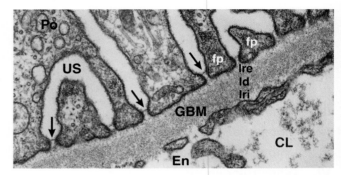

Figure16.1 Electron micrograph of glomerular capillary wall from mature kidney. The wall consists of an inner lining of fenestrated endothelium (En) and an outer, epithelial podocyte (Po) layer, which display interdigitating foot processes (fp). Slit diaphragms (arrows) span the intercellular space (sometimes referred to the filtration slits) between separate foot processes. Between endothelial and podocyte cell layers is the GBM, which contains an inner layer [lamina rara interna (lri)], central lamina densa (ld), and outer lamina rara externa (lre). CL, capillary lumen; US, urinary space. Modified from Abrahamson (1986), with permission.

(Fig. 16.1). They contain bundles of microfilaments that give them a contractile capacity and are linked to microtubules formed in the cell body (Mundel and Kriz, 1995). The space between individual foot processes (the filtration slit or slit pore) measures ~25 nm and this space is spanned by an extracellular membrane termed the slit diaphragm (Fig. 16.1; Smoyer and Mundel, 1998).

Podocytes, like epithelial cells in general, are polarized with apical and basolateral membrane domains (Mundel and Kriz, 1995). During podocyte differentiation, the tight junction and developing slit diaphragms help maintain podocyte cell polarity and separate the basal foot process membrane from the apical membrane domain (Schnabel et al., 1990). The basal plasma membrane of the foot process is completely embedded in the underlying GBM. The apical domain, which is located above the slit diaphragm, contains a well-developed glycocalyx layer of negatively charged glycoproteins (Smoyer and Mundel, 1998; Mundel and Kriz, 1995). The glycocalyx is composed of several sialoglycoproteins, including podocalyxin (Kerjaschki *et al.*, 1984) and GLEPP1 (Wiggins *et al.*, 1995), and is believed to be of critical importance for the formation and preservation of the characteristic cellular architecture of the podocyte.

As mentioned earlier, mesangial cells resemble vascular smooth muscle cells and probably function as pericytes specialized to enable the glomerulus to regulate its vascular tone. Mesangial cells are embedded in an extracellular matrix (mesangial matrix) that contains many of the same classes of molecules as the peripheral loop GBM. The inventory of basement membrane proteins between the GBM and the mesangium differs somewhat, however, and the mesangial matrix lacks defined laminae densae and rarae.

III. Glomerular Filtration Barrier

Normally, the glomerular filter very efficiently restricts the passage of macromolecules from the blood into the Bowman's space and discriminates on the basis of size (size selectivity) and molecular charge (charge selectivity). The flow of glomerular filtrate from the capillary lumen to the urinary space follows a strictly extracellular route, passing in sequence through the fenestrated endothelium, then the GBM, and finally across the slit diaphragms spanning the filtration slits whereupon the filtrate enters into the urinary (Bowman's) space. Because the glomerular endothelium contains open fenestrae, this cell layer is permeated easily and does not restrict the passage of plasma proteins (Daniels, 1993). Although the net negative charge of the endothelial plasma membrane may establish some barrier properties, the main structural barrier of the kidney filter is largely reliant on the GBM and glomerular podocytes. As discussed later, however, the GBM is the product of both endothelial cells and podocytes. In view of this evidence, all three elements of the glomerular capillary wall—the endothelium, GBM, and podocytes—are each essential for integrity of the filtration barrier. Indeed, the endothelium, GBM, and podocytes are directly linked to one another and damage to any one component in this series can result in filtration barrier defects.

The protein networks within the GBM act as a substrate for the adherent endothelial cells and podocytes. The GBM also provides a molecular sieve capable of restricting the penetration of molecules that have a molecular weight of more than ~70,000 or a high negative charge. A new ultrastructural model for glomerular filtration has shown that the size selectivity of glomeruli is dependent on hindrance coefficients of the GBM and on structural features of the slit diaphragm as well (Edwards *et al.*, 1999). Nevertheless, the charge

selectivity of the glomerulus is clearly associated with the presence of the polyanionic charged molecular components. In the GBM, heparan sulfate proteoglycans (HSPGs) provide a high negative charge due to the presence of numerous glycosaminoglycan side chains. The importance of the negative charge of HSPGs in the GBM has been evaluated by removing or blocking the heparan sulfates with heparatinase or antiheparan sulfate antibodies. In both cases, these maneuvers result in increased glomerular permeability to anionic ferritin and/or selective albuminuria (Kanwar *et al.*, 1980; van Den Born *et al.*, 1992). Other molecules, such as hyaluronic acid, chondroitin sulfate, hydroxyl groups on carbohydrates, and carboxyl groups of glutamic and aspartic acid, also contribute to the anionic charge of the GBM (Kanwar *et al.*, 1991). Further evidence for the existence of an anionic charge barrier has also been derived from experimental studies in which GBM anionic sites are neutralized by polycations such as protamine sulfate (Kelley and Cavallo, 1978), hexadimethrine (Londono and Bendayan, 1988), or polyethyleneimine (Barnes *et al.*, 1984). Infusion of these polycations results in increased permeability to albumin and IgG or augmented permeation of the GBM by anionic ferritin particles.

The podocyte cell layer acts as an additional filtration system superimposed on the GBM (Kriz, 1997). Podocytes are adherent to the GBM by numerous foot processes, which stabilize small patches of the underlying GBM and counteract local elastic distension (Kriz, 1997). Their location as the terminal element in the filtration barrier and their elaborate internal cytoskeleton help the podocyte withstand the large hydraulic properties of the glomerulus (Mundel and Kriz, 1995). Approximately half of the resistance to water flow has been calculated as being due to the GBM and half to the slit diaphragm, with endothelial resistance normally being negligible (Drumond and Deen, 1995; Lafayette *et al.*, 1998). Using dimensions derived from various electron microscopic examination of the filtration slit and slit diaphragms, and hydrodynamic analysis of the transport of spherical macromolecules, theoretical studies have led to the conclusion by some investigators that the slit diaphragm provides the dominant size selective element in glomerular filtration (Drumond and Deen, 1995). These results are also consistent with an ultrastructural model in which the slit diaphragm is proposed to be the most restrictive part of the filtration barrier (Edwards *et al.*, 1999). As discussed later, however, mutations that disrupt GBM proteins and cause proteinuria apparently do not affect slit diaphragm structure.

To establish the charge selectivity of podocytes, the negative charge of the glycocalyx has been demonstrated to play a critical role in maintaining slit-pore integrity (Mundel and Kriz, 1995; Daniels, 1993; Kriz, 1997), and there is good evidence that loss of this glycocalyx causes proteinuria. Removal of sialic acids from the glomerular filtration barrier by intraperitoneal injection of sialidase into mice results in a transient loss of charge from the endothelium and epithelium.

This loss of plasma membrane polyanion is accompanied by the effacement of foot processes and the formation of tight junctions between adjacent podocytes, and mice develop proteinuria and renal failure in a dose-dependent manner (Gelberg *et al.*, 1996). In humans, foot process effacement and other histologic alterations of glomerular epithelial cells are prominent features in almost all protein-uric renal diseases. In response to immunologic or toxic injury, podocytes withdraw their foot processes and often undergo cytoplasmic fusion to form binucleated cells (Smoyer and Mundel, 1998). This results in a marked reduction in slit pore surface area, which probably increases hydraulic fluxes across the remnant slit pores, exacerbating the filtration defect.

Maintenance of the normal relationship between the podocyte and the GBM is also critical for the normal barrier function as evidenced by the proteinuria that accompanies disruption of this association. For example, antibody targeted at β1-integrins results in epithelial cell detachment *in vitro* and marked proteinuria *in vivo* (Adler and Chen, 1992; Adler *et al.*, 1996). Charge neutraliza-tion of proteoglycans and other anionic GBM molecules with protamine sulfate also causes podocyte detachment and simplification of foot processes and proteinuria. Indeed, detachment of epithelial cells from the GBM results in increased macromolecular permeability at specifically those detachment sites and accounts for much of the proteinura in a variety of renal disease (Daniels, 1993).

IV. Glomerular Basement Membrane Proteins

Basement membranes divide tissues into compartments and act as supportive substrates for communities of cells, providing differentiation and migration information that are transmitted through cell–matrix receptors. Macromolecules of the basement membrane include both collagenous and noncollagenous glycoproteins. The collagenous component is predominantly type IV collagen and the major non-collagenous glycoprotein is laminin, both of which exist as various isoforms (Fig. 16.2; Colognato and Yurchenco, 2000; Timpl and Brown, 1996; Kuhn, 1994). Other ubiquitous basement membrane components are proteo-glycans and nidogen/entactin, and specific binding inter-actions between all of these components establish the basement membrane superstructure.

A. Laminins

Laminins represent a family of multifunctional macro-molecules that are the most abundant noncollagenous structural glycoproteins of basement membranes and play

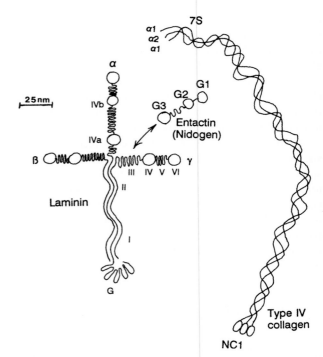

Figure 16.2 Scale diagram showing the multidomain structures of laminin-1, entactin-1, and type IV collagen. The G3 domain of entactin binds to domain III on the laminin γ1 chain as shown. Modified from Abrahamson *et al.*, (1997), with permission.

three essential roles. First, they are major structural elements of basement membrane, forming one of two self-assembling networks (the other is composed of collagen type IV) to which other glycoproteins and proteoglycans of the basement membrane attach (Colognato and Yurchenco, 2000). Second, they interact directly with cell surface receptors such as dystroglycan and integrins to attach cells to the extracellular matrix. Third, they are signaling molecules that, by interacting with integrins, convey morphogenetically important informa-tion to the interior cytoskeleton of the cell (Henry and Campbell, 1996).

Laminins are large disulfide-bonded heterotrimers composed of three distinct chains, α, β and γ (Fig. 16.2), whose genes are located on different chromosomes. The best charac-terized laminin molecule is laminin-1. Initially isolated from the mouse Engelbreth–Holm–Swarm (EHS) tumor, laminin-1 contains one ~400-kDa α1 chain and two ~200-kDa chains, β1 and γ1, which associate to form a heterotrimeric, asym-metric, cross-shaped molecule (Fig. 16.2; Burgeson *et al.*, 1994; Colognato and Yurchenco, 2000).

In addition to the mouse tumor laminin-1 trimer, five laminin α (α1–α5), four laminin β (β1–β4), and three γ(γ1–γ3) chains have been identified to date (Colognato and Yurchenco, 2000). In the current laminin nomenclature, laminin chains and laminin heterotrimers are named with arabic numerals in essentially their order of discovery

(Burgeson *et al.*, 1994). All native laminins isolated and characterized biochemically are composed of one α, one β, and one γ chain, and 12 distinct heterotrimers have been identified at this writing (Colognato and Yurchenco, 2000; Timpl, 1996).

The amino acid sequences of all laminin chains, mainly deduced from sequencing cDNA clones, predict several characteristic functional domains. The classic α chains have a large carboxyl-terminal globular domain, followed by domains I and II that contribute to the coiled coil of the laminin long arm. Alternating cysteine-rich repeats (domain IIIa, IIIb, and V) and globules (domains IVa, IVb, and VI) form the short arm. The classic β and γ chains also contain domains I+II, which participate in the long arm coiled coil structure, and domains III, IV, V, and VI, which correspond to those in the α chains. Analysis of intrachain ionic interactions within the coiled-coil domain have demonstrated that three different chains are indeed present in the native molecule and that homodimer and homotrimer formation in this region is energetically unfavorable (Beck *et al.*, 1993).

The α/β/γ trimers of laminin are formed intracellularly through several steps, including chain selection, assembly, and stabilization (Aumailley and Smyth, 1998). Following formation of the disulfide-linked β/γ dimer in the rough endoplasmic reticulum (ER), the α chain is incorporated to form the heterotrimer (Aumailley and Smyth, 1998; Yurchenco *et al.*, 1997). Laminin exits the ER only after assembly into a disulfide-bonded trimer (Niimi and Kitagawa, 1997). However, some studies have shown that the α chain of laminin-1 can be secreted independently, whereas the β and γ chains cannot (Yurchenco *et al.*, 1997).

Laminin binds to cells through a variety of cell surface receptors. A large number of integrins, including α1 β1, α2 β1, α3 β1, α6 β1, α7 β1, αv β3, and $α_{IIb}$ β3, have been shown to act as laminin receptors (Sonnenberg *et al.*, 1990; Aumailley and Krieg, 1996). Among them, the integrin family containing the β1 chain has been shown to be the most relevant counterpart in laminin-mediated specific mechanical functions and signaling. Integrins α6 β1 and α7 β1 recognize the long arm E8 (elastase resistant) domain of laminin-1 (Sorokin *et al.*, 1990). In addition to integrins, other proteins, including dystroglycan, a cell surface laminin receptor that was first identified as a component of the dystrophin–glycoprotein complex, heparin, and several galectins, which are members of the family of β-galactoside-specific lectins, can also bind laminins to form nonintegrin-mediated cellular interactions. Dystroglycan specifically binds the E3 fragment of the laminin α1 chain and to the E3-like fragment of the laminin α2 chain (Durbeej *et al.*, 1998). Galectin-3 binds laminin through its numerous poly-*N*-acetyllactosamine chains (van den Brule *et al.*, 1995), and the major heparin-binding site has been known to reside in the long arm globular domain (G domain) of the laminin α1 chain (Yoshida *et al.*, 1999).

In addition to binding to cells, laminin also binds specifically to other basement membrane components, including type IV collagen, HSPG, entactin, and fibulins (Colognato and Yurchenco, 2000; Utani *et al.*, 1997). As discussed later, entactin/nidogen can bind to both the laminin γ1 chain and the triple helix of type IV collagen. Domain IV of the laminin γ2 chain can interact with fibulin 2 (Utani *et al.*, 1997), and a sequence containing the IKLLI peptide from the long arm of the laminin α1 chain has been shown to represent a binding site for both integrin and HSPG (Tashiro *et al.*, 1999). Laminin also contains self-binding sites, which are crucial for the formation of laminin networks (Colognato and Yurchenco, 2000; Yurchenco and Cheng, 1993). Specifically, the laminin-binding sites are located within the short-arm N-terminal globules at each of the three laminin polypeptides. The most stable laminin polymers occur in a calcium-dependent way in which each short arm of laminin engages short arms from two other laminin molecules to form trimers of N-terminal domains (Colognato and Yurchenco, 2000; Yurchenco and Cheng, 1993).

Laminin is expressed in temporally and spatially regulated patterns in the kidney (Miner, 1998; Hansen and Abrass, 1999). As discussed later, the major laminin occurring in the mature kidney GBM is the laminin-11 isoform, which is composed of the α5, β2, and γ1 chains (Miner *et al.*, 1997). This laminin isoform also occurs in the basement membrane of the neuromuscular junction (Sanes, 1995). Laminin-1(α1 β1 γ1), laminin-8 (α4β1γ1), laminin-9 (α4β2γ1), and laminin-10 (α5β1γ1) all transiently appear in immature GBM (Miner *et al.*, 1997; Sorokin *et al.*, 1997). Laminin-2 and -4 have been also found as minor components in tubular basement membrane and mesangial matrix, but other laminins have not been observed in the kidney (Sanes *et al.*, 1990; Miner and Sanes, 1994; Durbeej *et al.*, 1996; Miner *et al.*, 1997; Miner, 1998).

B. Collagen Type IV

Collagen IV is the principal collagenous constituent of basement membranes. The monomer is a ~400-nm-long rod-shaped triple helix and consists of heterotrimers of three α chains with a combined molecular mass of approximately 550 to 600 kDa (Fig. 16.2). Each of the six genetically different α chains is about 1700 amino acids long and contains (a) a short collagenous domain at the N terminus (called the 7S domain), (b) a central relatively long collagenous domain, and (c) a C-terminal noncollagenous domain (NC1; Hudson *et al.*, 1993). Analysis of complementary cDNA-derived amino acid sequences from many vertebrate and invertebrate species has shown that the 7S and NC1 domains are the most highly conserved regions. Three α chains self-assemble to form monomers consisting of triple helical regions in both central and N-terminal 7S collagenous domains (Fig. 16.3; Kuhn *et al.*, 1994). Collagen type IV

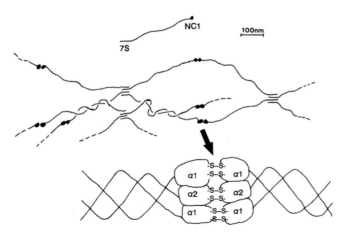

Figure 16.3 Diagram showing interactions between C-terminal NC1 and N-terminal 7S domains of separate monomers of type IV collagen to form an extended meshwork. Lateral interations along the lengths of the molecules can add additional complexities. Arrow points to a magnified diagram of C-terminal globules, which consist of a hexamer stabilized by disulfide cross-links. Modified from Abrahamson et al., (1997), with permission.

monomers then assemble themselves into a large polymerized network, which is stabilized by covalent interactions between NC1 domains, covalent associations of 7S domains, and lateral interactions between adjacent central collagenous segments (Fig. 16.3).

Unlike fibril-forming collagens (type I, II, and III), type IV chains contain interruptions of the typical Gly-X-Y collagenous repeats in the major collagenous domain. These interruptions are more or less evenly distributed, and, upon alignment of separate α chains, most are in the same position. These disturbances might provide the flexible sites observed by electron microscopy and hence may promote the formation of supertwisted collagen IV helices. These interruptions in the collagenous sequences are also sites for proteolytic attack and for the attachment of collagen IV to other basement membrane molecules (Kuhn et al., 1994; Sado et al., 1998).

To date, six genetically distinct chains, α1, α2, α3, α4, α5, and α6, have been completely sequenced and their chromosomal locations identified. Genes encoding the human α1(COL4A1) and α2(COL4A2) collagen type IV chains are found on human chromosome 13 and are orientated in a head-to-head antiparallel fashion (Poschl et al., 1988). A similar array has been found on human chromosome X for the genes COL4A5 and COL4A6 (Zhou and Reeders, 1996). COL4A3 and COL4A4 map to chromosome 2q35-37 and are also organized in a similar antiparallel arrangement (Mariyama et al., 1992; Kuhn, 1994).

Collagen type IV has been shown to comprise a gene family that includes tissue-restricted isoforms. The best studied and most widely distributed collagen IV is the trimer of α1 and α2 chains, which assemble in a 2:1 ratio,

respectively. Molecules containing the α3, α4, α5, and α6 chains are important components of specialized basement membranes (Kuhn, 1994; Sado et al., 1998). Exactly how the α3–α6(IV) chains assemble stoichiometrically is still not entirely clear, but evidence indicates that networks in basement membranes contain (α3₂ α4) and (α5/α6) heterotrimeric molecules (Kahsai et al., 1997; Sugimoto et al., 1994; Ninomiya et al., 1995). A heterotrimeric molecule of collagen type IV, α3(IV), α4(IV), and α5(IV), has also been found in a novel supramolecular GBM network that is cross-linked by disulfide bonds between triple helices (Gunwar et al., 1998).

C. Nidogen / Entactin

Nidogen (also called entactin) is a 150-kDa single-chain glycoprotein isolated originally from the EHS tumor matrix (Paulsson et al., 1986). It is a ubiquitous component in most basement membrane and has a strong affinity for laminin, and also binds to collagen type IV. The dumbbell-shaped nidogen molecule is composed of three globular domains (G1, G2, and G3) separated by two threads (Fig. 16.2; Olsen et al., 1989). G1 comprises the N-terminal end of the molecule and contains negatively charged residues possessing a calcium-binding EF hand motif. A second putative calcium-binding domain is located in a short rod-like segment that joins G1 to G2, and the binding sites for collagen IV, perlecan, and BM-90/fibulin are present in G2 (Aumailley et al., 1993). The carboxy-terminal globule (G3) is a cysteine-rich domain that bears resemblance to EGF and to the low-density lipoprotein (LDL) receptor. This G3 region on entactin binds to the fourth EGF-like repeat in domain III of the laminin γ1 chain (Fig. 16.2; Mayer et al., 1993). The human entactin gene exists as a single locus on human chromosome 1q43 (Olsen et al., 1989). In human kidney, entactin/nidogen is present in all renal basement membranes and is distributed through the full width of the GBM, as well as the peripheral regions of mesangium (Katz et al., 1991, Dziadek, 1995).

A homologue of nidogen/entactin, named nidogen-2, has been identified in both mouse and human osteoblasts. A structural comparison in each domain between entactin-2 and entactin-1 shows a similar domain arrangement and high homology between their individual G2 and G3 domains (31.2 and 38.4%, respectively) and rod-like structures (33.7%; Kimura et al., 1998). The tissue distribution of both entactin-1 and entactin-2 is also highly similar; immunofluorescence shows a close colocalization of these two proteins in kidney, testis, heart, and skeletal muscle (Kohfeldt et al., 1998). Entactin-2 can also bind laminin, collagen type IV, and perlecan but fails to bind to fibulins. The affinity of binding between the laminin γ1 chain to entactin-2 is 100 to 1000-fold lower than to entactin-1 (Kohfeldt et al., 1998).

D. Proteoglycans

Proteoglycans are heterogeneous macromolecules made up of glycosaminoglycan (GAGs) chains that are bound by O-glycosidic linkage to a core protein (Iozzo, 1998; Miner, 1998). Proteoglycans carrying either heparan sulfate and/or chondroitin sulfate are typical constituents of basement membranes. The presence of proteoglycans in the GBM confers negative charge permselectivity to the filtration barrier. Three distinct proteoglycans, perlecan, agrin, and bamacan, have been documented as integral GBM components, but others may occur as well (Groffen et al., 1997, 1998, Miner, 1998; Couchman et al., 1996).

1. Perlecan

Like laminin-1 and entactin-1, perlecan was originally purified from the mouse EHS tumor. Perlecan contains a 400 to 450-kDa core protein, with three heparan sulfate chains attached to its N-terminal domain (Noonan et al., 1991). The derived amino acid sequence from the full-length perlecan cDNA shows five different structural domains with the N-terminal domain I being the only region of perlecan that has no significant homology with any other protein. Within a 14 amino acid stretch of domain I are three repeats of the sequence SGD and each of these represents the attachment site for a heparan sulfate glycosaminoglycan side chain. Domain II contains four repeating subunits that are homologous to members of the LDL receptor family. Domain III has homology with the short arms of laminin and contains both cysteine-rich, rod-like regions and globular regions. Domain IV is the largest and contains a series of 14 and 21 repeats in mouse and human, respectively, that are similar to the immunoglobulin superfamily and to the neural cell adhesion molecule (N-CAM). Domain V consists of three globular repeats and also contains several SGXG sequences (Noonan et al., 1991). The perlecan gene has been localized to chromosome 1q36 (Kallunki et al., 1991).

Perlecan was initially considered the major HSPG in GBM and was thought to be crucially important for establishing the glomerular permselective filtration barrier. Growing evidence has, however, shown a relative paucity of perlecan in the adult human peripheral loop GBM (Groffen et al., 1997) and most data now point to agrin as the principal GBM HSPG (Tsen et al., 1995).

2. Agrin

Agrin is present throughout the full width of the mature GBM, whereas perlecan is restricted to the subendothelial aspect (Groffen et al., 1998). Agrin was identified and named on the basis of its involvement in the aggregation of acetylcholine receptors (AChRs) during synaptogenesis at the neuromuscular junction. The human agrin gene, named AGRN, has been assigned to the locus 1q36. 1-1pter (Rupp et al., 1992) and studies have demonstrated that it is a large extracellular matrix protein HSPG with a molecular mass in excess of 500 kDa and a protein core of 220 kDa (Tsen et al., 1995). The agrin core protein is glycosylated with heparan sulfate and contains three GAG attachment sites. The primary structure shows a multimodular composition containing a globular laminin-binding domain, nine follistatin-like protease inhibitor domains, two laminin-like EGF repeats, two serine/threonine-rich domains, a SEA module (a new extracellular domain associated with O-glycosylation first found in sperm protein, enterokinase and agrin), four EGF repeats, and three domains sharing homology with globules of laminin chains (Groffen et al., 1998; Rupp et al., 1991).

Agrin interacts with the basement membrane by binding to a central segment of the laminin long arm, and dystroglycan, a distinct type of laminin receptor, has also been identified as a high-affinity ligand of agrin. Emerging evidence indicates that the function of agrin is not limited to the formation of the neuromuscular junction. In kidney, agrin is present primarily along the glomerular capillary loops and throughout the width of the mature GBM. By binding dystroglycan and laminin, agrin may be one of the components involved in the strong adhesion of the foot processes of the podocyte to the GBM (Groffen et al., 1998). Agrin may also make an important contribution to the polyanionic charge of the GBM and could be essential for maintenance of the permselective charge barrier (Groffen et al., 1998). Furthermore, in the developing blood–brain barrier, agrin has been shown to play an important role in the formation and maintenance of cerebral microvascular impermeability (Barber and Lieth, 1997).

3. Bamacan

Bamacan is a chondroitin sulfate proteoglycan originally isolated from organ cultures of embryonic parietal yolk sac (rat Reichert's membrane) and has been identified as a component of the basement membrane in the EHS tumor matrix (Couchman et al., 1996). Using backcrossing experiments in mouse, the bamacan gene has been mapped to distal chromosome 19, a locus syntenic to human chromosome 10q25 (Ghiselli et al., 1999).

Amino acid sequence analysis of bamacan predicts five structural domains. Domain I, which lacks cysteine and is largely hydrophilic, is predicted to have a globular structure. Domains II and IV are each predicted to form coiled-coil structures that comprise >50% of the protein sequence. Domain III lies between the two coiled-coil regions and contains four cysteine and six proline residues that probably contribute flexibility to this protein. Domain V contains a 35 amino acid region that apparently contributes to a helix-loop-helix motif (Wu and Couchman, 1997). Sequence analysis of bamacan has also shown that domains I, III, and IV share a high degree of sequence homology with members of the SMC (structural maintenance of chromosome) protein family (Wu and Couchman, 1997; Ghiselli *et al.*, 1999). In contrast, domains II and IV are homologous to the myosin heavy chain, and secondary structure analysis of the protein reveals that the N–and C-terminal ends carry a P loop and a DA box motif, which are believed to act cooperatively to bind ATP. When this occurs, it triggers structural alteration of the molecule, leading to its contraction and subsequent condensation of the bound DNA in chromatin (Ghiselli *et al.*, 1999).

Based on the strong sequence homology with the SMC family of proteins and, in particular, the similarity to the SMC3 subclass, bamacan proteoglycan can be also assigned to the SMC3 subfamily (Ghiselli *et al.*, 1999). Members of the SMC protein family are involved in chromosome condensation, sister chromatid cohesion, and gene dosage compensation, an essential chromosome-wide regulatory process that modulates the expression of numerous genes related solely by their linkage to the same chromosome (Stursberg *et al.*, 1999). Whether the strong sequence homology between bamacan and the SMC family of proteins indicates similar functions among these proteins requires further study. However, this is not the only case in which proteoglycan molecules can display chromosomal binding. For example, glypican, a cell surface proteoglycan, has been detected immunologically in nuclei of neuronal cells and has been implicated in playing a specific function during the cell cycle (Liang *et al.*, 1997).

Antibodies against bamacan generally do not react with nuclei but instead exhibit a wide distribution in basement membranes. In kidney, bamacan antibodies label basement membranes of the S-shaped body and precapillary loop stages, and the labeling is eliminated gradually from the maturing GBM, but remains in the mesangial matrix of adult. Antibamacan binds all other basement membranes of the nephron (McCarthy *et al.*, 1993).

E. Other Basement Membrane Proteins

Fibulin-1 (BM-90) and fibulin-2 have been identified as a family of microfibrillar components of extracellular matrix proteins that interact with the laminin-1–nidogen complex, type IV collagen, and fibronectin (Utani *et al.*, 1997). Double-labeling immunofluorescence reveals that fibrillin-1 and fibrillin-2 are colocalized in skin, elastic intima of blood vessels, and the glomerulus (Reinhardt *et al.*, 1996). Fibronectin is a large dimeric molecule that adheres to many extracellular matrix components, including collagens, heparan sulfate, and fibrin, and, in the glomerulus, is mainly expressed in the mesangial matrix (Kanwar *et al.*, 1997). Tenascin, like fibronectin, is another mesenchymal protein that interacts with fibronectin and proteoglycans and it appears transiently around the condensates, S-shaped bodies, and tubules in the developing kidney. Previous studies have documented a strong expression of tenascin in embryonic

kidney and in both normal and abnormal mature glomeruli, implying a role for tenascin in nephrogenesis and glomerular scarring (Truong *et al.*, 1996). Nevertheless, no renal defects have been reported in tenascin-deficient mice (Saga *et al.*, 1992). Collagen XVIII is a recently discovered basal lamina heparan sulfate proteoglycan and is the first collagen/proteoglycan with heparan sulfate side chains (Halfter *et al.*, 1998). This nonfibrillar collagen contains frequent interruptions in the triple helical structure, and these may confer flexibility. Type XVIII collagen is present in highly vascularized organs such as the liver, lung, kidney, and placenta (Oh *et al.*, 1994) and *in situ* hybridization has shown that it is widely deposited in basement membranes, including those found associated with renal tubules and glomeruli. The function of collagen XVIII in basement membrane is unknown (Halfter *et al.*, 1998), but it is worth noting that a proteolytic product of collagen XVIII, named angiostatin, has potent antiangiogenic properties (Cao, 2001).

V. Unique Features of Podocytes

A. Surface Proteins

The coated pits and coated vesicles frequently observed in the basal plasma membrane and basal cytoplasm of foot processes indicate a high rate of endocytosis (Mundel and Kriz, 1995). The apical membrane surface of podocytes contains a well-developed glycocalyx layer of negatively charged glycoproteins (Smoyer and Mundel, 1998) that help maintain the registration of foot processes and prevent an adherence of podocytes to the parietal epithelium of the Bowman's capsule (Schnabel *et al.*, 1989; Kerjaschki, 1978). Charge neutralization with polycations, desialylation with neuraminidase, or treatment with puromycin amino-nucleoside all induce loss of interdigitating foot process structure, podocyte detachment, and leakiness of the glomerular filtration barrier (Andrews, 1988; Whiteside *et al.*, 1989). Podocalyxin is one of the major sialoproteins of the glycocalyx lining the foot processes, where it helps maintain foot process structure and function in part by virtue of its negative charge. Podocalyxin is synthesized at a high rate in the differentiating podocyte and its distribution is restricted to the apical surface of foot processes, as well as on the luminal membrane domain of endothelial cells (Schnabel *et al.*, 1989; Horvat *et al.*, 1986). A cDNA clone from the rabbit glomerulus has also revealed another core transmembrane siaglycoprotein named podocalyxin-like protein-1, which has many characteristics of podocalyxin, and is expressed in a similar pattern as podoclyaxin (Kershaw *et al.*, 1995).

Among the specific proteins of the apical surface membrane domain of podocytes, glomerular epithelial protein 1 (GLEPP1) has been identified as a receptor tyrosine phosphatase containing a large extracellular domain with several fibronectin type III-like repeats, a hydrophobic transmembrane segment, and a single PTPase domain. Northern blots show that GLEPP1 is expressed in both brain and kidney where its expression can be detected as early as the comma-shaped body stage. With glomerular maturation, GLEPP1 is seen predominantly on the apical surface of foot processes (Wiggins *et al.*, 1995; Sharif *et al.*, 1998), but, like many other receptor tyrosine phosphatases in this family, neither the ligand nor the substrate has been identified for GLEPP1.

B. Nephrin, a Slit Diaphragm Protein

Until relatively recently, only limited information has been available on the molecular components of the epithelial slit diaphragm, a structure that is unique to podocytes. A flurry of evidence published since the late 1990s has, however, shown that a unique protein named nephrin is a major component of this structure (reviewed by Tryggvason, 1999). Nephrin is the protein encoded by the *NPHS1* (congenital nephrotic syndrome of the Finnish type) gene and, when mutated, heavy proteinuria occurs before birth, progresses postnatally, and results in early death. Structural analysis shows that nephrin is a member of the immunoglobulin superfamily of cell adhesion molecules that mediate either cell–cell or cell–extracellular matrix interactions (Kestila *et al.*, 1998). Immunofluorescence and immunogold electron microscopy have shown that nephrin is targeted specifically to the slit pore and localizes to the lateral aspect of the foot process where it is concentrated at a basal site consistent with the position of the slit diaphragm (Holzman *et al.*, 1999). Using antibodies against the two terminal extracellular Ig domains of recombinant human nephrin, immunogold labeling has further confirmed that nephrin is located specifically at the glomerular podocyte slit diaphragm (Ruotsalainen *et al.*, 1999).

The nephrin intracellular domain is 155 residues in length but contains no definable motifs that would aid in identifying function. However, a mutation in the gene for this protein was found to be the underlying abnormality in the congenital nephrotic syndrome of Finnish type, where the glomerular podocyte fails to make normal foot processes or slit diaphragms (Kestila *et al.*, 1998). Additional evidence for a role of nephrin in establishing the slit diaphragm has come from knockout studies in mice in which the nephrin gene was deleted selectively (Putaala *et al.*, 2001). Homozygous null mutants are massively proteinuric at birth, and podocytes exhibit foot process effacement and an absence of slit diaphragms (Putaala *et al.*, 2001). Similarly, examination of mice that lack the CD2-associated protein (CD2AP), which in T lymphocytes tethers the CD2 receptor to the actin cytoskeleton, showed that these animals also underwent renal failure at birth and exhibited abnormal podocyte foot processes (Shih *et al.*, 1999). Together with data that the nephrin protein coimmunoprecipitates with CD2AP, a picture is emerging that the slit diaphragm protein, nephrin, is linked indirectly to the podocyte skeleton by CD2AP (Li *et al.*, 2000).

Proteins other than nephrin are also likely to be part of the slit diaphragm. Immunolocalization studies have shown that P-cadherin, another cell–cell adhesion molecule, can be found in the slit diaphragm region between foot processes (Somlo and Mundel, 2000), but whether this molecule is an integral element of the slit diaphragm structure itself remains to be proven. Another candidate is podocin, a product of the *NPHS2* gene, which, when mutated, results in an autosomal recessive steroid resistant nephrotic syndrome that affects young children and generally progresses to focal segmental glomerulosclerosis (Boute *et al.*, 2000). This protein resembles members of the band-7 stomatin family, which form hairpin-like structures through the plasma membrane with both N and C termini projecting into the cytoplasm (Boute *et al.*, 2000).

C. Other Podocyte Differentiation Markers

Several other proteins have been identified as potentially important podocyte-specific differentiation markers. The best documented include WT-1, vimentin, synaptopodin, and, podoplanin. WT-1, a gene transcription factor with zinc finger-binding domains, causes Wilms tumors when mutated. The pattern of WT-1 gene expression in renal development and tumor pathology, as well as its encoded protein structure, suggests that it is an early regulatory gene that arrests the proliferation of undifferentiated mesenchymal cells and promotes their development into specialized epithelium (Chapter 22; Pelletier *et al.*, 1991; Baird *et al.*, 1992). This has been indicated by the total failure of metanephric development in WT–1-deficient mice (Kreidberg *et al.*, 1993). WT-1 mRNA expression is first noted in the undifferentiated metanephric blastema and in renal vesicles during early fetal development and declines thereafter, although WT-1 protein can be detected readily in glomerular podocytes throughout adulthood, a persistence suggesting that WT-1 has a role in podocyte homeostasis (Mundlos *et al.*, 1993). One of the physiological targets of WT-1 is *amphiregulin,* a member of the EGF family, which can stimulate epithelial branching in organ cultures of embryonic mouse kidney (Lee *et al.*, 1999). WT-1 can bind directly to the *amphiregulin* promoter and results in potent transcriptional activation of this factor (Lee *et al.*, 1999).

Vimentin is an intermediate filament protein characteristic of mesenchymal cells (Oosterwijk *et al.*, 1990). Each vimentin filament is an elongated fibrous molecule composed of three domains: an amino-terminal head, a carboxyl-terminal tail, and a central rod domain. During nephrogenesis, vimentin is first expressed in undifferentiated mesenchymal cells but disappears shortly thereafter.

Vimentin reappears in developing podocytes at the early capillary loop stage and remains expressed during later stages of glomerulogenesis and in adult kidneys (Smoyer *et al.*, 1998; Mundel and Kriz, 1995; Oosterwijk *et al.*, 1990). The expression of vimentin is enhanced at sites where podocytes are proliferating, e.g., in diseases such as membranoproliferative or crescentic glomerulonephritis (Moll *et al.*, 1991). Vimentin is confined to the podocyte cell body and the major processes but is virtually absent from the foot processes themselves. This suggests that vimentin and its polymers may not be important for the structure and function of the foot processes themselves. In fact, animals homozygous for vimentin mutations develop normally and reproduce without an obvious phenotype (Colucci-Guyon, 1994), suggesting that other intermediate filament proteins can compensate for the loss of vimentin.

Synaptopodin is an actin-associated protein of differeniated podocytes that is also expressed in postsynaptic densities and in dendritic spines (Mundel *et al.*, 1991, 1997a). It constitutes a novel class of proline-rich proteins and appears to be a linear polypeptide. Due to the periodic distribution of a high amount of proline (20%), the formation of globular domains is not possible in this molecule. Synaptopodin expression in the kidney is first detectable at the capillary loop stage and, at the ultrastructural level, is associated with actin microfilaments of podocyte foot processes. This association is also seen *in vitro* in cultured podocytes where synaptopodin colocalizes with the microfilaments in a punctate pattern and is also detected in focal contacts (Mundel *et al.*, 1991, 1997a).

Podoplanin is a novel 43-kDa membrane protein that has been isolated from a rat glomeruli expression library. The expression of podoplanin was found exclusively in a linear pattern on cell membranes of podocytes and the parietal epithelial cells of the Bowman's capsule (Breiteneder-Geleff *et al.*, 1997). The function of podoplanin, as well as of its homologues in other organs, remains to be determined, but binding of antipodoplanin antibodies *in vivo* causes a flattening of foot processes and generates a situation similar to puromycin nephrosis in which podoplanin is down-regulated transcriptionally (Matsui *et al.*, 1999).

The cell bodies of podocytes contain small amounts of microtubule-associated proteins, such as MAP3, MAP4, and Tau, which have binding sites for both microtubules and microfilaments. In foot processes, a complete microfilament-based contractile apparatus is present, which contains actin, myosin-II, α-actinin, talin, vinculin, and synaptopodin (Mundel *et al.*, 1997b). This apparatus is anchored by cytoskeleton-associated proteins to the underlying GBM via cell membrane-associated integrins, thereby firmly attaching-foot processes to the substrate (Kriz, 1997).

Attachment of the podocyte to the GBM occurs via binding of cell-matrix-receptor molecules. Integrin and dystroglycan are two well-studied matrix protein receptors of podocytes

and, in human kidneys, integrin $\alpha3\beta1$ is apparently the only integrin present on podocytes. The extracellular domain of the integrin interacts with laminins in the GBM, and the cytoplasmic domain of the $\alpha3\beta1$ subunit interacts with the cytoskeletal protein actin through a heteromeric complex of actin-associated proteins, including talin, vinculin, α-actinin, and paxillin (Simon and McDonald, 1990; Otey *et al.*, 1993). Dystroglycan is present on kidney epithelial cell membranes from the moment of mesenchymal-to-epithelial transition and is also expressed prominently in adult glomeruli (Durbeej *et al.*, 1998).

VI. Glomerulogenesis

The mammalian kidney originates through the successive appearance of three embryonic stages: pronephros, mesonephros, and metanephros. The pronephros and mesonephros are transitory structures, whereas the metanephros matures to form the permanent, metanephric kidney (Saxén, 1987). Metanephrogenesis begins at embryonic day 11 (E11) in mice, E12 in rats, and 4th–5th week gestation in humans (Saxén, 1987). As the ureteric bud grows out from the mesonephric or Wolffian duct and enters undifferentiated metanephric mesenchyme (MM), mesenchymal cells are induced to aggregate around the advancing edge of the bud (Fig.16.4). The condensed mesenchyme then converts to an epithelial phenotype and proceeds through an orderly sequence of developmental structures, which are termed vesicle, comma shaped, S shaped, developing capillary loop, and maturing glomerulus stages (Fig.16.4; Abrahamson, 1991; Chapters 10 and 14). The process of nephron induction and differentiation continues repeatedly until the full complement of nephrons is formed, which, in rodents, concludes ~1 week after birth (Saxén, 1987).

Beginning with the vesicle stage, the induced mesenchyme cells switch to an epithelial phenotype and begin synthesizing basement membrane matrix glycoproteins, such as collagen type IV, laminin, and basement membrane proteoglycans (Abrahamson *et al.*, 1989; Ekblom, 1981; Lelongt *et al.*, 1987). As development progresses through the S-shaped stage, a primitive bilayered glomerular structure now becomes recognized: the outer (parietal) cell layer develops into Bowman's capsular epithelium, whereas the internal (visceral) layer of cells ultimately differentiates into podocytes (Fig.16.4 and 16.5). At the same time, invading mesenchymal endothelial precursor cells (angioblasts) enter the lower crevice (vascular cleft) of the developing glomerulus (Hyink and Abrahamson, 1995). Figure 16.6 shows this early vascularization process in newborn mouse kidney, in which developing endothelial cells expressing a bacterial transgene, *lacZ*, can be distinguished readily. [These mice were created by replacing the gene encoding *Flk1*, a VEGF receptor, with *lacZ*, which encodes the enzyme β-galactosidase and therefore serves as a

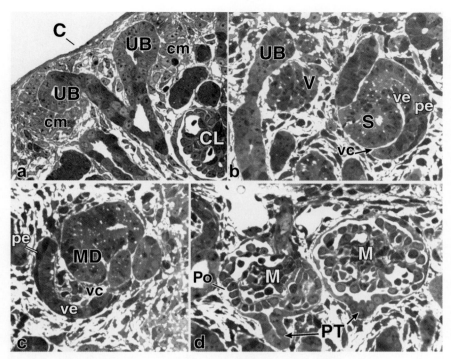

Figure 16.4 Light micrographs of newborn mouse kidney cortex showing early stages of glomerular development. (a) Immediately beneath the kidney capsule (C), the first stage of nephron formation can be seen and consists of condensing mesenchymal cells (cm), which aggregate at the distal tips of each branch of the ureteric bud (UB). Deeper within the cortex is a portion of a capillary loop stage (CL) glomerulus. (b) The condensing mesenchyme subsequently forms a ball or vesicle (V) of cells. After several rounds of mitosis, a comma- and then S-shaped nephric figure develops (S), which contains a vascular cleft (vc) at its lower aspect. Beneath the cleft are visceral epithelial cells (ve) (from which podocytes are derived) and a parietal epithelial cell layer (pe) (which develops into the Bowman's capsule). UB, ureteric bud. (c) A slightly more advanced S-shaped figure is shown. The glomerular vasculature can be seen emerging from the vascular cleft. ve, visceral epithelial cell layer; pe, parietal epithelial cells. (d) Two capillary loop stage glomeruli are visible. Central mesangial areas (M) and distinct podocytes (Po) can now be discriminated. Proximal convoluted tubules (PT) exiting the Bowman's capsule can also be seen.

histochemical reporter expressed by endothelial cells exclusively. Homozygous *Flk1* null animals die at embryonic day 8. 5 because of vascular insufficiencies, but heterozygous animals appear entirely normal (Shalaby *et al.*, 1995)]. With the progressive invasion and differentiation of endothelial cells, the developing capillary endothelium begins to form its own extracellular matrix scaffold (Fig.16.7). This newly synthesized basement membrane then fuses with the basement membrane layer beneath the podocytes to form the GBM, which therefore is derived from both epithelial cells and endothelial cells (Abrahamson, 1985).

During capillary loop formation, the glomerular epithelium adhering to the developing GBM continues to differentiate and begin foot process formation and endothelial cells gradually flatten and become fenestrated. At this time, both the endothelium and the epithelium are actively involved in synthesizing components for the dual GBM. This fact has been established by metabolic studies, by histochemical and immunohistochemical techniques, and by interspecies grafting experiments (Abrahamson 1985; Sariola, 1984; Kanwar *et al.*, 1984; Reeves *et al.*, 1980; Lelongt *et al.*, 1987) and has been reviewed in detail previously (Abrahamson,

1987). Dual unfused basement membranes are rarely seen in glomeruli at maturing glomerular stages. Instead, extensive loops and irregular segments of basement membrane are found beneath podocytes (Fig.16.8). These subepithelial outpockets of GBM are found in the highest concentration in areas where foot process interdigitation is still underway. *In vivo* injection studies with antilaminin antibodies have shown that these outpockets represent newly synthesized GBM, which is then spliced or inserted into the existing fused GBM as glomeruli mature (Abrahamson and Perry, 1986; Desjardins and Bendayan, 1991). Well after the initial fusion and splicing events are completed, continued modification and remodeling of the GBM occur with the appearance of new basement membrane protein isoforms. As discussed later, however, little is known about how these processes are regulated at either the gene or the protein level.

A. Origin and Differentiation of the Glomerular Endothelium

Until relatively recently, glomerular endothelial cells were thought to be derived from the ingress of externally derived

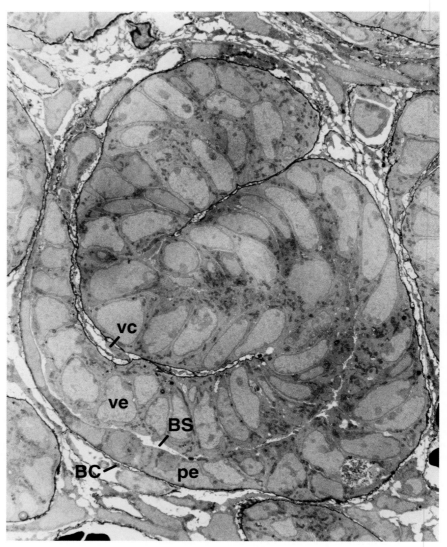

Figure 16.5 Electron micrograph of newborn mouse kidney S-shaped nephric figure labeled with antilaminin IgG-horseradish peroxidase (which renders basement membranes black). The vascular cleft (vc), visceral epithelial (ve), and parietal epithelial (pe) cells can all be seen. BS, Bowman's space; BC, Bowman's capsule. Cells that give rise to the distal segment of the nephron lie above the vascular cleft. Modified from Clapp and Abrahamson (1994), with permission.

vessels growing into the kidney during metanephrogenesis (Hyink and Abrahamson, 1995). This belief came principally from compelling studies in which embryonic mouse or quail kidney was grown *in ovo* upon avian chorioallantoic membranes (Sariola *et al.*, 1983, Sariola *et al.*, 1984). These studies showed convincingly that glomeruli developing within grafts were chimeras containing host-derived endothelial and mesangial cells, and graft-derived podocytes. Because the vascular elements of these glomeruli stemmed from the host, results from this particular graft system therefore signified an extrarenal origin of glomerular endothelial and mesangial cells. During the past few years, however, new evidence has shown that the metanephric mesenchyme (MM) contains progenitors not only for nephric epithelial cells, but for

endothelial cells as well (angioblasts). Proof of this has come from organ culture and grafting experiments making use of kidneys obtained from *Flk1-lacZ* heterozygous mice (Robert *et al.*, 1998; Chapter 17). When *Flk1-lacZ* kidneys from embryonic day 12 mice are grown under routine organ culture conditions, extensive tubulogenesis occurs but vascular development does not take place (Fig.16.9; Robert *et al.*, 1998). These metanephric organ cultures do not express *Flk1-lacZ* and the "glomeruli" that form *in vitro* are avascular (Fig.16.9). However, when avascular kidney rudiments that have been maintained in organ culture for 6 days are then grafted into anterior eye chambers of normal (nontransgenic) hosts, *Flk1-lacZ* is upregulated in the grafts, and microvessels and glomeruli lined by transgenic (graft-derived) endothelial

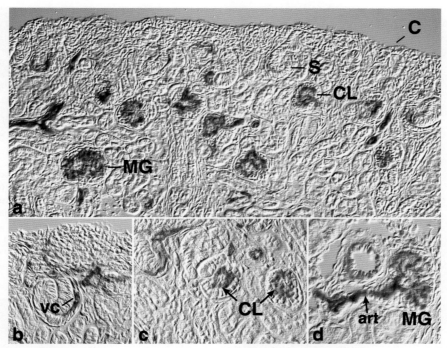

Figure 16.6 Micrographs of a newborn Flk1-lacZ kidney after processing β-galactosidase histochemistry to reveal sites of Flk1 transcription and thereby mark endothelial cells. Cells expressing the Flk1 line the developing vasculature and are blue. **(a)** View of cortex showing outer capsule (C), S-shaped nephric figures (S), capillary loop stage (CL), and maturing stage glomeruli (MG). **(b)** S-shaped figure containing endothelial cells within the vascular cleft (vc). **(c)** Capillary loop stage glomeruli (CL) contain endothelial cells expressing Flk1. The developing mesangial regions contained in central areas of these glomeruli are not blue, however. **(d)** Maturing stage glomerulus (MG) connected to arteriole (art). (Modified from Robert and Abrahamson, 2001, with permission.)

cells are observed (Fig.16.10; Robert *et al.*, 1998). These findings thus show that dissected metanephroi contain cells capable of establishing the complete glomerular and microvascular endothelium *in vivo*.

To address the apparent paradox between the findings with the organ culture anterior chamber grafts of *Flk1-lacZ* kidneys with those referred to earlier using the avian chorio-allantoic graft system, other studies have made use of the ROSA26 transgenic mouse in which *lacZ* is expressed ubiquitously by *all* cells. When embryonic day 12 kidneys from normal donors were grafted under the kidney capsule of adult ROSA26 hosts, a site that contains few or no angioblasts, all of the microvascular and glomerular endothelial cells within the graft were again derived from the engrafted kidney (Robert *et al.*, 1996). In contrast, however, when embryonic kidneys were grafted into the kidney cortex of newborn ROSA26 hosts, a site that contains abundant angioblasts, many glomeruli and other capillaries within the graft contained host-derived endothelial cells (Robert *et al.*, 1996). These results, in which embryonic kidneys are grafted into a quasi-embryonic setting (developing kidney cortex), are therefore exactly the same as when embryonic kidneys are grafted onto the avian chorioallantoic membrane (an extra-embryonic structure), and both the graft and the neighboring host tissue are probably rich sources of angioblasts.

Many different transcription factors, membrane receptor–ligand-signaling systems, and other proteins are known to be essential requirements for normal embryonic vascularization processes (Daniel and Abrahamson, 2000), but only a few of these have been examined in detail in the glomerulus. Nevertheless, the formation of the glomerular vasculature can be considered to progress through four interrelated stages: (1) angioblast proliferation/survival/ differentiation, (2) endothelial cell recruitment and maturation, (3) initial assembly of the glomerular capillary and mesangium, and (4) maturation of the glomerular capillaries (Daniel and Abrahamson, 2000, Robert and Abrahamson, 2000).

Developing podocytes are a key source of vascular endothelial growth factor (VEGF), which may be one of the most critical initiators for the development of glomerular capillaries. Release of VEGF from immature podocytes may attract angioblasts bearing VEGF receptors into vascular clefts of early nephric figures and subsequently lead to the organotypic formation of glomeruli. Indeed, injection of anti-VEGF antibodies into the developing mouse kidney cortex *in vivo* results in the formation of avascular glomerular structures (Kitamoto *et al.*, 1997). Podocytes do, however, continue to synthesize VEGF after glomeruli are fully mature. Because Flk1 expression is also maintained by glomerular endothelial cells of mature kidneys, VEGF-Flk1 signaling probably exerts

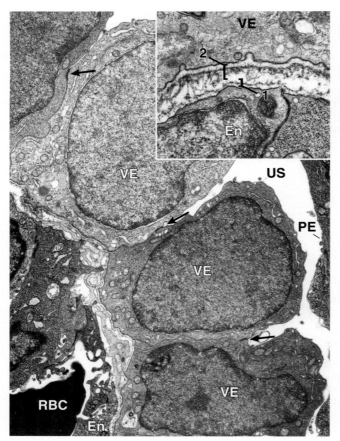

Figure 16.7 Electron micrograph of vascular cleft region showing endothelial (En) and visceral epithelial (VE) cell layers. Tight junctional complexes (arrows) exist between visceral epithelial cells in their apicolateral domains. RBC, red blood cell; US, evolving urinary space; PE, parietal epithelium. **(Inset)** High magnification view of the vascular cleft showing cytoplasm of nonfenestrated endothelium, subendothelial basement membrane (1), subendothelial basement membrane (2), and cytoplasm of visceral epithelium. The double basement membrane between the endothelium and the epithelium subsequently fuses. From Abrahamson (1987), with permission.

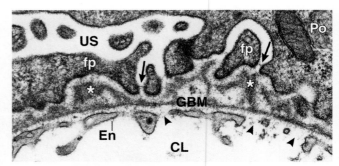

Figure 16.8 Developing capillary wall seen in capillary loop and maturing stage glomeruli. Open fenestrations (arrowheads) form in the endothelial layer (En). Foot process (fp) interdigitation occurs between podocytes (Po), and slit diaphragms become clearly distinguishable (arrows). Beneath developing foot processes are extensive scrolls and outpockets of basement membrane material (*). These segments become incorporated into the underlying GBM by unknown mechanisms. Modified from Abrahamson (1987), with permission.

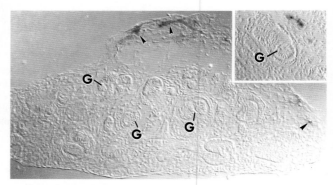

Figure 16.9 Micrograph of an embryonic day 12 Flk1-lacZ kidney that had been grown for 6 days in organ culture and then developed for β-galactosidase. A few cells expressing Flk1 are seen on the periphery of the organ culture (arrowheads), but no defined vascular structures can be seen inside. Note avascular glomerular epithelial tufts (G) that do not express Flk1. From Robert *et al.* (1998), with permission.

functions beyond those needed for initial establishment of the glomerular capillary. These may include maintenance of the differentiated state by endothelial cells, which, as described earlier, display fenestrations lacking diaphragms. Neuropilin-1, which in certain vessels is a coreceptor with Flk1 for VEGF (especially for the VEGF164 isoform), is also expressed in developing and mature glomerular endothelium (Robert *et al.*, 2000).

Many other receptor–ligand–signaling systems are probably crucial for glomerular capillary formation, including members of the Eph/ephrin, Tie/angiopoietin, PDGF/PDGF receptor families (Daniel and Abrahamson, 2000, Robert and Abrahamson, 2001). Specifically, the receptor tyrosine kinase EphB1 and its ligand ephrin-B1, which itself is a transmembrane protein receptor, are both expressed in similar distribution patterns on developing kidney microvascular endothelial cells (Daniel *et al.*, 1996). Although the precise

roles for Eph/eprhin signaling in the glomerulus are still uncertain, knockout mice display lethal vascular phenotypes, including defects in vessel patterning, sprouting, and remodeling (reviewed in Daniel and Abrahamson, 2000). Because reciprocal gradients of Eph receptor and ephrin counterreceptor concentrations have been documented as critical regulators for neuronal patterning in the developing brain, a similar function for this intercellular signaling system may exist in the developing vasculature. Hence Eph/ephrin signaling appears to convey spatial signals between differentiating endothelial cells, helping to promote cell targeting to appropriate anatomical domains (Daniel and Abrahamson, 2000).

The Tie receptor, angiopoietin ligand system is also vital for normal vascular formation (Woolf and Yuan, 2001). Tie-2 is expressed by developing glomerular endothelial cells, and one of its ligands, angiopoietin-1, has been shown to be

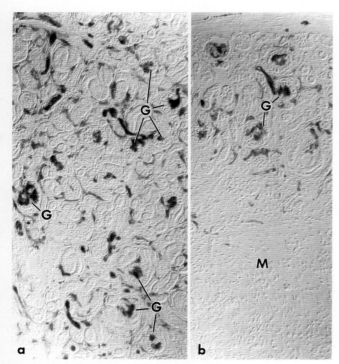

Figure 16.10 Section of E12 Flk1-lacZ that had first been maintained for 6 days in organ culture and then grafted into the anterior eye chamber of a wild-type host. Graft was removed 7 days after transplantation and developed for β-galactosidase histochemistry. **(a)** Unlike what was seen after organ culture (Fig.16.9), cultured and then grafted kidneys develop organotypic vascular patterns, and numerous glomeruli (G) contain Flk1-positive endothelial cells. **(b)** Areas within some grafts fail to undergo extensive tubulogenesis and glomerulogenesis. These regions appear rich in mesenchyme (M) and are largely avascular. From Robert *et al.* (1998), with permission.

important for vascular organization and remodeling. Another ligand, angiopoietin-2, appears to regulate vascular integrity and permeability, and the coordinated expression of these two related proteins may be essential for final maturation and stabilization phases of the glomerular capillary (reviewed in Woolf and Yuan, 2001; Chapter 17).

B. Platelet-Derived Growth Factor and Development of the Mesangium

Because of the relative ease with which glomerular mesangial cells can be purified and grown in culture and because of their central importance in a variety of glomerular diseases, a large number of insightful cell biological, pathological, and immunological studies have been carried out on these cells. Nevertheless, and in part because of the paucity of surface markers specific for mesangial cells, the differentiation of these cells remains mysterious, and very little is known about how the mesangial areas are formed initially during glomerulogenesis and maintained thereafter. Regardless, mesangial cells are derived from MM precursors that do not

express Flk1 and therefore are distinct from the endothelial cell developmental pathway. A critically important finding made in the early 1990s showed that a member of the PDGF family, PDGFB, is expressed strongly by mesangial cells and that its receptor, PDGFRβ, is also expressed by these cells (Alpers *et al.*, 1992). Further, examination of the expression of this ligand–receptor pair during glomerular development showed that immature podocytes produce PDGFB, which may help attract mesangial cell precursors expressing PDGFRβ into glomeruli. Later, both PDGF and PDGFRβ expression becomes confined to the mesangium, which may promote mesangial proliferation and/or maturation (Alpers *et al.*, 1992). Compelling studies implicating PDGFB/PDGFRβ signaling in the establishment of the mesangium have come from gene-targeting studies in mice. Null mutants for either protein produce similar, early postnatal lethality due to hemorrhaging (Soriano 1994, Leveen *et al.*, 1994). Glomeruli in these animals consist of one or few large, dilated capillaries that completely lack mesangial cells. Thus both the PDGFB ligand and its receptor, PDGFRβ, are required for normal development of the mesangium (Soriano, 1994; Leveen *et al.*, 1994; Table 16.1).

Interestingly, once the glomerulus has fully matured, there appears to be a small population of extraglomerular mesangial cells capable of completely repopulating the glomerulus if the intraglomerular mesangium becomes severely injured. These mesangial reserve cells reside in the juxtaglomerular apparatus and are distinct from renin-secreting cells, macrophages, vascular smooth muscle cells, and endothelial cells (Hugo *et al.*, 1997). In an anti-Thy 1 model of proliferative glomerulonephritis in rats, these extra-glomerular mesangial reserve cells migrate into the glomerulus and entirely restore the depleted intraglomerular mesangium (Hugo *et al.*, 1997). Perhaps additional studies based on these fascinating observations can shed more light on the origin and development of mesangial cells during glomerulogenesis. Similarly, much more work needs to be done on the assembly and maintenance of the mesangial matrix. Although this matrix undergoes morphologic and compositional changes throughout glomerulogenesis, it has not yet been subject to thorough study. However, considerable progress has been made in understanding assembly of the GBM.

C. Developmental Transitions in GBM Proteins

1. Changes in Laminin Composition

During glomerulogenesis, the expression of laminin chains in the GBM is under strict temporal control. Three different laminin α chains have been detected so far in the GBM. Laminin α1 chain is the least widely expressed α chain and is specific for epithelial cells. Antibodies reacting with the elastase-derived E8 or E3 fragment of laminin-1 (which bind near the carboxy terminus of the "1 chain) have been applied to organ cultures of embryonic kidney and shown to block the mesenchyme-to-epithelium conversion *in vitro* (Durbeej *et al.*, 1996; Ekblom *et al.*, 1998; Klein *et al.*, 1988; Sorokin *et al.*, 1992). During nephrogenesis, laminin α1 and α4 chains can be detected in the single continuous basement membrane surrounding the newly induced vesicle, and the expression of these chains remains in the vascular cleft of the comma- and S-shaped stages (Miner *et al.*, 1997). The laminin α1 chain then disappears abruptly from the early GBM before defined glomerular capillary loops form (Fig.16.11; St. John and Abrahamson, 2001). During the capillary loop stage, the laminin α5 chain becomes strongly expressed in the GBM and no laminin α1 chain is seen (Figs 16.11 and 16.12). With further maturation of the kidney, laminin β1 chains persist and laminin α5 becomes the only α chain observed in the GBM, which now presumably contains the laminin-11 isoform (consisting of laminin α5, β2, and γ1 chains; Fig.16.13; Miner, 1998; Miner *et al.*, 1997; St. John and Abrahamson, 2001). At this time, laminin α4 is absent from the renal cortex and the laminin α1 chain is found only in kidney mesangium.

As described earlier, the GBM originates in comma- and S-shaped figures through fusion of a subendothelial basement membrane with a basement membrane layer beneath developing podocytes. Two recent lines of evidence indicate that glomerular endothelial cells remain active in basement membrane biosynthesis into glomerular maturation stages. First, immunohistochemical studies have shown that the normal laminin–isoform switching patterns do not occur within the avascular glomeruli that form when embryonic kidneys are maintained in organ culture. When these cultured kidneys are, however, grafted, endothelial cells (of graft origin) vascularize glomeruli and normal laminin–isoform switching patterns are restored (St. John *et al.*, 2001). Second, direct immunoelectron microscopy of native kidneys has shown that glomerular endothelial cells and podocytes jointly synthesize laminin-1 and laminin-11 chains (St. John and Abrahamson, 2001). The general sequence of glomerular capillary wall development and the cellular sources for the laminin components are diagrammed in Fig.16.13. Although these results demonstrate an important and sustained role for the glomerular endothelium in GBM biosynthesis, the relative amounts of laminin secreted by endothelial cells and podocytes are still unknown.

The fundamental importance of laminin α5 chain in kidney development as well as other tissues, is evident from the α5 knockout mouse. In null mutants, mice die by E17 with defects in anterior neural tube closure and placental

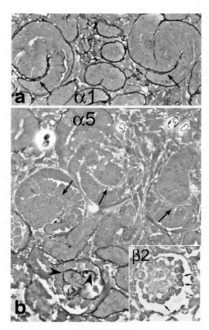

Figure 16.11 Micrographs showing immunoperoxidase labeling for the laminin α1 (top) and α5 chains (bottom), as marked. (**a**) The laminin α1 chain, representing laminin-1, is present within vascular clefts of S-shaped figures. (**b**) The laminin α5 chain, representing laminin-11, is absent in vascular clefts (arrows) but is expressed abruptly in GBMs of capillary loop stage glomeruli (arrowheads). (**Inset**) Immunolabeling for laminin-β2 chain, also representing laminin-11, is detected intracellularly within developing podocytes (small arrows). From St. John and Abrahamson (2001), with permission.

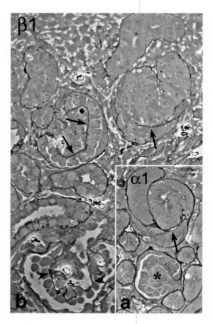

Figure 16.12 Immunolabeling for laminin α1 chain (**a**, lower right) and β1 chain (**b**). (**a**) The laminin α1 chain is present within the vascular cleft of the S-shaped nephric figure (arrow) but disappears abruptly from GBMs of capillary loop stage glomeruli (*). (**b**) The laminin β1 chain, in contrast, is expressed in vascular clefts (arrows) and GBMs of capillary loop stage glomeruli (arrowheads). From St.John and Abrahamson (2001), with permission.

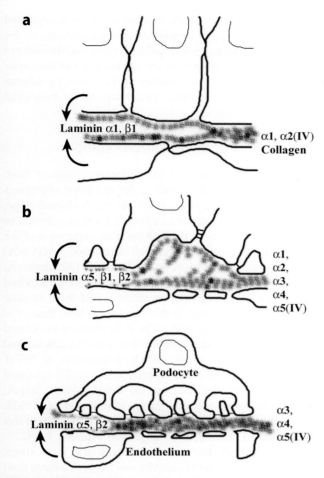

Figure 16.13 Diagram summarizing the morphologic features of glomerular capillary wall development and GBM protein isoform transitions. **(a)** Within vascular clefts of comma- and S-shaped nephric figures, laminin α1 and β1 chains are present within subendothelial and subepithelial basement membranes. Both chains are synthesized by developing endothelial cells and podocytes (curved arrows). **(b)** In capillary loop stage glomeruli, the laminin α1 chain is no longer detectable. Laminin α5 and β2 chains appear for the first time, while the laminin β1 chain persists, and all three chains are derived from both endothelial cells and podocytes. **(c)** As glomeruli mature, laminin α1 chain disappears, whereas laminin α5 and β2 chains persist. The approximate sequence for the collagen type IV transitions appears on the right. Although the cellular origins for these molecules is uncertain, the endothelial cells and podocytes may jointly participate jointly in collagen isoform synthesis and substitution. From St. John and Abrahamson (2001), with permission.

dysmorphogenesis. Among other problems, the kidney is also small or absent (Miner *et al.*, 1998). The few glomeruli that form are unusual in that they lack endothelial and mesangial cells, suggesting that the laminin α5 chain is important for endothelial recruitment and/or adherence to the GBM (Miner and Li, 2000).

The expression of laminin β and γ chains during kidney development also occurs in dynamic and complex patterns. The β1 and γ1 chains are first detected within embryos at the two to four cell stage. During nephron development, the laminin γ1 chain is found throughout all developmental stages of GBM and mesangial matrix assembly (Miner *et al.*, 1997; Miner and Sanes, 1994; Sorokin *et al.*, 1997) and its critical role in early development has been confirmed by deletion of the laminin γ1 gene (LAMC1), which causes early lethality, presumably shortly after implantation (Timpl, 1996). With the maturation of GBM, the laminin β2 chain gradually replaces the fetal isoform of laminin β1 and becomes the sole laminin β chain in the fully mature GBM (Miner and Sanes, 1994; Noakes *et al.*, 1995; Virtanen *et al.*, 1995). At this time, however, both laminin β1 and β2 chains can still be detected in the mesangial matrix. Elimination of the laminin β2 chain in mice causes massive proteinuria and abnormal motor nerve terminals in muscle (Virtanen *et al.*, 1995; Noakes *et al.*, 1995). The GBM of β2 mutant mice appears normal ultrastructurally but is rich in the laminin β1 chain, indicating persistence of the immature phenotype (Noakes *et al.*, 1995). A particularly noteworthy finding is that massive proteinuria is also associated with the effacement of foot processes, which closely resembles what is seen in humans with minimal change nephrosis (Noakes *et al.*, 1995). This suggests that the laminin β2 chain might be necessary for the proper differentiation and/or maintenance of podocyte foot processes.

2. Transitions in Type IV Collagen

During kidney development in rodents, expression of the collagen α1(IV) and α2(IV) chains is first detected at the vesicle stage and persists into the early stages of glomerular development. In contrast, α3(IV)–α5 (IV) chains first appear at the capillary loop stage where they colocalize with the α1(IV) and α2(IV) chains (Fig. 16.13; Miner, 1998; Miner and Sanes, 1994). With the maturation of glomeruli, the collagen type IV α1 and α2 chains are eliminated progressively from the GBM, and the α3(IV)–α5(IV) chains become the only collagen type IV chains present (Fig. 16.13; Miner, 1998; Miner and Sanes, 1994). In humans, the expression of α3(IV)–α5(IV) chains is first seen after the capillary loop stage, or later than in mice, whereas the α1(IV) and α2(IV) chains remain at very low levels in fully mature glomeruli (Miner, 1998; Zhou and Reeders, 1996; Lohi *et al.*, 1997; Kalluri *et al.*, 1997). The collagen type IV (α6) chain is only found in the Bowman's capsular basement membrane in kidney, and α3(IV)–α5 (IV) chains are also present in proximal TBMs (Miner, 1998). In adult human, mouse, and rat kidneys, the α1, α2(IV) are still strongly expressed in TBMs, basement membranes of blood vessels, and glomerular mesangial matrix, whereas α3(IV)–α5(IV) chains are absent from the mesangium (Lohi *et al.*, 1997). Cellular sources for the different collagen chains within the glomerulus are not yet defined, but, like laminins, both the endothelium and the podocyte cell layers may synthesize the full spectrum of collagen type IV.

The importance of the α3(IV)–α5(IV) chains to the proper function of the GBM is evident from the effects of mutations in the genes that encode these chains. The α3(IV) chain contains the major autoantigen in autoimmune glomerulonephritis known as Goodpasture syndrome (Sado *et al.* 1998), whereas the α5(IV) chain gene is mutated in the hereditary glomerulonephritis, Alport syndrome (Tryggvason *et al.*, 1993; Chapter 27). Homozygote mutant mice with defects in the α3 of type IV collagen die of renal failure 3–4 months after birth (Cosgrove *et al.*, 1996; Miner and Sanes, 1996; Lu *et al.*, 1999). GBMs in type IV collagen α3 mutants at the time of death are multilaminated and closely resemble those of humans with Alport syndrome. There are also defects in podocyte foot process architecture, with widespread effacement occurring during the late stages of the disease. Interestingly, numerous changes in the molecular composition of GBM are also accompanied with the onset of renal dysfunction in this mutant, which include loss of collagens α4 and α5(IV), retention of collagen α1(IV) and α2(IV) chains, appearance of fibronectin and collagen VI, and increased levels of perlecan (Miner and Sanes, 1996). When double knockout mice containing null mutations for both the α3 chain of type IV collagen and integrin α1 are generated, much less severe defects are seen (Cosgrove *et al.*, 2000).

3. Developmental Changes in Other GBM Proteins

Although entactin appears to play an important stabilizing role in basement membrane networks, only a few studies have examined its synthetic patterns in the kidney. Immunofluorescence microscopy has shown that the expression of entactin-1 and -2 in the kidney is mainly localized in the basement membrane zone of proximal and distal tubules, glomeruli, and Bowman's capsule (Kohfeldt *et al.*, 1998). *In situ* hybridization studies indicate that entactin actually comes from MM and not epithelial cells (Ekblom *et al.*, 1994). This finding therefore suggests that mesenchymally derived entactin influences renal epithelial development. Because of the widespread expression of entactins and their putative roles of cross-linking laminin and collagen IV networks, the absence of an obvious phenotype after the removal of entactin-1 by gene targeting was unexpected (Murshed *et al.*, 2000). While it is possible that entactin-2 could substitute for entactin-1, it is worth noting that no compensatory upregulation of entactin-2 was observed in the entactin-1 knockouts (Murshed *et al.*, 2000).

As discussed earlier, the α1, α4, and β1 laminin chains of the immature GBM are gradually substituted by α5 and β2, and the collagen IV composition shifts from α1–α2 to α3–α4–α5. Similarly, the proteoglycan composition also changes between fetal and adult stages. Perlecan transcripts and protein expression are observed from the onset of nephrogenesis (Handler *et al.*, 1997) and bamacan is expressed transiently during the S-shaped body and capillary stage, but gradually disappears in the maturing GBM (McCarthy *et al.*, 1993). Although the time of onset of

agrin expression is not yet clear, agrin mRNA is upregulated as kidney development progresses and accumulates in the mature GBM. Mechanisms regulating these matrix-isoform switches during development are not known, but the variation in the molecular composition of basement membrane undoutedly leads to dramatic differences in the biological properties of the GBM.

D. Podocyte Differentiation

During kidney development, podocytes originate from the nephrogenic vesicle, which is derived from aggregates of induced MM cells (Fig. 16.4). At the S-shaped stage, primitive podocytes appear as a simple layer of cuboidal or columnar epithelial cells that exhibit high expression of the proliferation marker proliferating cell nuclear antigen (PCNA) and possess apical tight junctions (Fig. 16.7) (Abrahamson, 1987; Nagata *et al.*, 1993). The bases of the cells are attached to a subepithelial basement membrane, and there is little basolateral interdigitation at this stage (Fig. 16.7; Reeves *et al.*, 1978; Abrahamson, 1991). The tight junction-associated protein ZO-1 is found in conjunction with the apical tight junctions (Schnabel *et al.*, 1990). Expression of WT-1 mRNA, which can be noted as early as the undifferentiated metanephric blastema and in the renal vesicle stages, continues in developing podocytes (Mundlos *et al.*, 1993). Podocalyxin is also seen in these immature podocytes at this time (Schnabel *et al.*, 1989; Smoyer and Mundel, 1998).

As the glomerulus enters the capillary loop stage, podocytes lose their mitotic activity and begin to assume their complex architecture (Nagata *et al.*, 1993). As these cells flatten, basal cytoplasmic extensions begin to interdigitate more extensively with those from adjacent podocytes to form the foot processes. The intercellular junctional complexes at apical margins between cells migrate basolaterally and either (a) are gradually eliminated and replaced by the slit diaphragms or (b) convert themselves into slit diaphragms, which bridge the filtration slits at the bases of the forming foot process (Fig. 16. 8; Schnabel *et al.*, 1990; Reeves *et al.*, 1978; Smoyer and Mundel, 1998, Ruotsalainen *et al.*, 2000). The role of the slit diaphragm complex in formation and/or maintenance of registration of the foot processes is uncertain, but studies in which nephrin has been deleted in mice result in massive proteinuria, early postnatal death, widespread effacement of foot processes, and absence of the slit diaphragm (Table 16.1) (Putaala *et al.*, 2001). Mice with deficiencies in the nephrin-associated protein CD2AP exhibited similar, although much less severe, defects and die 6–7 weeks after birth (Shih *et al.*, 1999).

As the junctional complexes migrate basally and ultimately form epithelial slit diaphragms, ZO-1 descends from its apical location to the cytoplasmic surface immediately adjacent to the basal slit diaphragms (Schnabel *et al.*, 1990). At about this time, GLEPP1 becomes distributed to the apical surface of the forming foot processes and is present in lesser amounts on the body of the maturing podocyte (Sharif *et al.*, 1998; Wiggins *et al.*, 1995). When GLEPP1 is knocked out, mice exhibit widening of foot processes and reduced filtration rates, but no proteinuria (Wharram *et al.*, 2000). Hypertension occurred in uninephrectomized mice, however, indicating that GLEPP1 might play a role in glomerular filtration and podocyte structure (Wharram *et al.*, 2000). During podocyte maturation, podocalyxin extends along the lateral plasmalemma above the migrating junctional complexes (Schnabel *et al.*, 1989). As the foot processes form, other proteins start to be expressed, including synaptopodin, a novel proline-rich actin-associated protein (previously termed pp44), and integrin β1 (Mundel *et al.*, 1991, 1997a; Korhonen *et al.*, 1990a,). The intermediate filament protein, vimentin, which had disappeared from epithelial cells shortly after the vesicle stage, now reappears at capillary loop stages in the cell bodies of podocytes (Holthofer *et al.*, 1984; Oosterwijk *et al.*, 1990).

During the maturing glomerulus stage, the podocyte no longer undergoes cytokinesis *in vivo* (Nagata *et al.*, 1993; Kriz *et al.*, 1994, 1996), although the cell can be induced by FGF-2 to reenter the cell cycle and undergo nuclear division (Kriz *et al.*, 1995). Podocalyxin is now seen on lateral surfaces of developing foot processes, but is always distributed above the filtration slits (Schnabel *et al.*, 1989). Expression of the WT-1 gene appears to decline, but podocytes continue to express considerable levels of the WT-1 protein (Mundlos *et al.*, 1993). The unique interdigitating foot process architecture of the podocyte is now fully developed.

E. Assembly of the GBM

The GBM develops by processes unlike those occurring in basement membranes found in most locations elsewhere in the body. As described earlier, the GBM is formed first by fusion of the subendothelial and subepithelial basement membrane layers found in the vascular cleft region of early nephric figures (Fig.16.7). With glomerular maturation, new basement membrane segments present beneath podocyte foot processes are spliced or inserted into the existing fused basement membrane (Fig.16.8; Abrahamson, 1985; 1986). Mechanisms of GBM fusion and splicing are not understood, but may involve intermolecular binding interactions between proteins in the subendothelial and subepithelial basement membrane layers. To test this hypophysis experimentally, purified mouse laminin-1 was injected intravenously into newborn rats and was found to become incorporated into the developing GBM (Wang *et al.*, 1998). No such incorporation was seen when laminin was injected into adult rats, which have fully assembled GBMs. Although these studies indicate that laminin-binding sites are indeed present within immature GBMs, the domains on

laminin recognized by these sites, and the protein(s) to which laminin binds, are undefined (Wang *et al.*, 1998).

I. Laminin and Type IV Collagen Associations

The supramolecular architecture and the assembly of the various proteins to form the sheet-like basement membrane arrays have been studied by two separate but convergent approaches. First, highly purified preparations of the separate proteins have been incubated together, rotary shadowed with platinum, and then examined by electron microscopy to visualize possible intermolecular relationships (Yurcheno and O'Rear, 1994; Colognato and Yurchenco, 2000). These techniques have indicated that two independent but interwoven polymers probably exist within native basement membranes *in vivo*. One polymer consists of a three-dimensional network of collagen type IV and the other polymer consists of laminin. These two networks are thought to be connected by entactin/nidogen. The second approach has been to use antibodies against defined epitopes or domains on the various intrinsic proteins and immunomicroscopic techniques to map the distributions of these epitopes in tissue sections. In general, the results from this approach have shown that most of the known basement membrane molecules codistribute throughout the full width of tubular basement membranes, as well as the GBM (Abrahamson *et al.*, 1989; Desjardins *et al.*, 1989), and these results are therefore in good agreement with the rotary shadowing findings.

In vitro studies have shown that laminin-1 aggregates into a larger polymer in a temperature-, time-, concentration-, and calcium-dependent manner. Polymerization in solution shows a critical laminin concentration of about 60 nM, and about a 10 μM concentration of calcium (or 100 μM of magnesium) is required for half-maximal aggregation (Timpl, 1996). The polymerization of laminin-1 occurs through a three-arm interaction model (Yurchenco and Cheng, 1993; Colognato and Yurchenco, 2000). This suggests that bonds are made between β1 chain N-terminal domain VI and domain VI of the α1 and γ1 chains. Theoretically, six different laminin isoforms are present during GBM development: laminins-1,-3,-8,-9,-10, and -11. The individual laminin chains in these isoforms include α1,4,5, β1,2, and γ1. Each of these contains the N-terminal globular domain VI, except the laminin α4 chain, which only has an internal globular domain IV. It is therefore reasonable to predict that laminin assembly in GBM follows the three arm interaction model.

Like laminin self-assembly, the triple helical monomers of type IV collagen also self-associate, forming a three-dimensional basement membrane network, and this structure is stabilized by three types of interactions. First, the amino-terminal segments from four collagen IV monomers (the 7S domain) produce a four-armed structure. Second, end-to-end joining and disulfide cross-linking occur between the carboxyl-terminal globules (the NC1 domain) of two collagen IV monomers. Finally, irregular lateral associations (noncovalent, side-by-side interactions) of the triple helical regions, apparently arranged as supercoils, also occur (Kuhn, 1994; Yurchenco and Schittny, 1990).

When collagen IV molecules assemble via N termini, the initial step is an antiparallel 25-nm overlap controlled by hydrophobic interactions between the 7S segments. This arrangement is stabilized by intermolecular disulfide bridges to produce tetramers (Siebold *et al.*, 1987). For assembly via C termini, a hexameric complex is formed and stabilized by the interchain disulfide bond between NC1 domains of separate chains. At the same time, conformational changes occur that stabilize the hydrophobic interactions between the NC1 domains.

2. Networks of GBM

The laminin and type IV collagen networks are cross-linked by entactin/nidogen. Specifically, the nidogen G2 domain binds to collagen IV triple helices, whereas the G3 domain interacts with a single EGF-like motif, γ1III4, in the laminin γ1 chain, thus leading to the formation of tertiary complexes among collagen IV, nidogen, and laminin (Mayer *et al.*, 1993; Yurchenco and O'Rear, 1994). This three-dimensional architecture is then modified further through the binding of proteoglycan and other basement membrane components as described earlier (Noonan *et al.*, 1991; Wu and Couchman, 1997; Battaglia *et al.*, 1992).

In the proteoglycan family, perlecan may bind to the laminin–nidogen complex via the nidogen G2 domain, as well as by direct binding of heparan sulfate chains to the laminin-1 fragment E3 (Hopf *et al.*, 1999). Agrin displays different binding capabilities for laminin β chains in the neuromuscular junction, and it binds more strongly to the laminin β2 chain than to the laminin β1 chain through its NH^2-terminal region (Denzer *et al.*, 1997). As described earlier, the binding affinity of entactin-2 for γ1III4 of laminin-1 is lower than that for entactin-1. This may indicate that entactin-2 binds to laminin-1 through an epitope unrelated to γ1III4. If so, entactin-2 may provide a second bridging system between collagen IV and laminin. Indeed, the coexpression pattern of entactin-2 and entactin-1 in the same tissues suggests a role for entactin-2 in basement membrane assembly (Kimura *et al.*, 1998; Kohfeldt *et al.*, 1998).

GBM assembly not only involves interactions between various matrix proteins, but is also reliant on the balance of basement membrane protein synthesis and degradation (Johnson *et al.*, 1992). As described earlier, expression of GBM components and their various isoforms takes place in temporally and spatially regulated patterns, and these gradients of matrix proteins may be partly due to the expression of matrix-degrading enzymes. Indeed, several endogenous GBM-degrading proteinases have been identified in the metanephros, including MMP (matrix metallo-proteinase)-2 and -9 (Lelongt *et al.*, 1997) and MT-1

(membrane type 1) -MMP (Ota *et al.*, 1998). Additionally, tPA(tissue-type plasminogen activator) has been detected in S-shaped bodies and glomeruli, uPA (urokinase-type) in renal tubular epithelia (Sappino *et al.*, 1991), and a 72-kDa neutral type IV collagenase in mesangial cells (Reponen *et al.*, 1992). Although the precise functions of these enzymes are not clear, they may participate in normal GBM turnover and may help regulate the GBM fusion and splicing steps that occur during initial GBM assembly.

F. Cellular Interactions during GBM assembly

1. Integrins

Assembly of the GBM matrix is also reliant on at least two types of cell surface receptors, integrins and dystro-glycan, which help coordinate the formation of laminin networks. The specific laminin-binding integrin, $\alpha6\beta1$, appears during nephrogenesis at the onset of basement membrane formation and is expressed transiently on developing podocytes (Ekblom *et al.*, 1998; Korhonen *et al.*, 1990b). This integrin serves as the receptor for the E8 fragment of laminin-1, and anti-integrin $\alpha6$ antibodies can perturb the formation of polarized nephric epithelium and branching morphogenesis of the metanephros *in vitro* (Falk *et al.*, 1996).

The $\alpha3\beta1$ integrin has also been shown to function specifically by binding laminins, type collagen IV, fibro-nectin, and entactin/nidogen (Kreidberg *et al.*, 1996). Because $\alpha3\beta1$ is apparently the only integrin present on basolateral domains of podocytes, it probably operates as a critical receptor for adherence of the podocyte foot process to the GBM and may also help organize the GBM. Indeed, homo-zygous mice deficient in $\alpha3$ integrin die during the first day after birth with severe abnormalities in the kidneys, including foot process effacement and glomerular basement membrane disorganization, as well as fewer, and wider glomerular capillary loops (Kreidberg *et al.*, 1996).

2. Dystroglycan

Dystroglycan (a cell surface protein, which, in muscle, links laminin-2 to the intracellular cytoskeleton) is present on the membrane of kidney epithelial cells from the moment of mesenchymal-to epithelial transition. Additionally, antibodies against dystroglycan can perturb epithelial development in kidney organ culture (Durbeej *et al.*, 1998), whereas disruption of the dystroglycan gene in the mouse causes early embryonic lethality. The dystroglycan null embryos fail to progress beyond the early egg cylinder stage of development and are characterized by structural and functional perturbations of Reichert's membrane, which is one of the earliest basement membranes that form in the rodent embryo. In this case the predominant basement membrane components–laminin, type collagen IV, and proteoglycans–are mislocalized (Henry and Campbell, 1998).

Although dystroglycan is produced by podocytes in adult kidney, neither laminin α1 or α2 chains are present in fully mature GBM and α5 is the only laminin α chain at this stage. Whether the laminin α5 chain also binds dystroglycan therefore needs to be determined. However, another study has reported that a laminin-10/11 mixture provides a strong adhesive component for epithelial cells, although high concentrations of antibodies (50 μg/ml) against integrin β1 and α6, or α3 or dystroglycan did not inhibit the attachment of cells to laminin-10/11 (Ferletta et al., 1999). Because laminin-11 (α5β2γ1) is the only laminin isoform in the mature GBM, other adhesive mechanisms may therefore be involved.

G. Regulation of GBM Protein Genes

In order for basement membranes to assemble and function properly, the synthesis of various polypeptide chains of the several basement membrane proteins must somehow be coordinated. Although it is not clear how the transcription machinery operates to produce collagen IV monomers, many studies have shown that genes coding for the six α(IV) chains are arranged in pairs of transcription units not found for other collagen genes. The transcription start sites for COL4A1 and COL4A2 are separated by only 130 bp on chromosome 13 (Poschl et al., 1988). Starting from the short intergenic space that harbors common bidirectional promoters, the genes are transcribed from the opposite DNA strands. These promoters in α1(IV) and α2(IV) do not, however, possess transcriptional activity, and each gene contains an activator region in the area around the first exon/intron border, which activates the promoter in either the α1(IV) or the α2(IV) direction (Poschl et al., 1988). Instead of containing a TATA box, which is a sequence present in most eukaryotic promoters transcribed by RNA polymerase II, the promoter in α1(IV) and α2(IV) contains a GC box, a CAAT motif, a CTC box, and Sp1 and retinoic acid binding sites (Kuhn, 1994). The COL4A5 and COL4A6 genes are also aligned head to head, but on chromosome X. Interestingly, the COL4A6 gene is transcribed from two alternate promoters in a tissue-specific manner (Sugimoto et al., 1994). The structural relationship between the COL4A3 and COL4A4 genes is less well characterized, but they are arranged head-to-head on chromosome 2 (Kuhn, 1994).

Promoter regions for laminin β and γ chains lack a TATA or CAAT box. Cloning of the 5′ region of the laminin γ1 chain gene revealed that the human and rodent promoters of these genes contain several GC boxes and a number of potential Sp1-binding sites, an arrangement seen commonly in TATA-less promoters (O'Neill et al., 1997). A bcn-1 element identified in the laminin γ1 promoter may potentially cooperate with the intron enhancer elements to yield the high phorbol 12-myristate 13-acetate-induced laminin γ1 mRNA levels seen in glomerular cells (O'Neill et al., 1997).

The laminin β1 and β2 chain genes must have acquired different regulatory elements in order to produce their distinctive tissue distribution patterns. The human and mouse β2 chain genes contain an AT-rich sequence upstream of the predicted transcription start site that could serve as a TATA box. In contrast, the promoter regions of human and mouse laminin β1 chain genes lack TATA box-like elements (Vuolteenaho et al., 1990). Both the human β1 and β2 chain genes possess potential binding sites for Sp1 and AP-2 (Durkin et al., 1996).

Unlike COL4A1 and COL4A2, as well as at least some of the laminin genes, TATA boxes are found in human and mouse entactin/nidogen (Fazio, 1991). Multiple positive and negative cis-acting elements in the nidogen promoter region have been found to control expression of this gene. Experiments have demonstrated that Sp1-like transcription factors in positive regulatory domains are essential for high expression of the human nidogen gene, and a novel silencer element located at −1333 and −1322 in the negative regulatory domains is bound by a distinct nuclear factor (Zedlacher et al., 1999). Regulatory regions of the human perlecan gene and mouse bamacan gene, in contrast, lack TATA motifs, but several GC sites for binding Sp1 transcription factors are found (Cohen et al., 1993; Ghiselli et al., 1999).

Based on what is known about the regulation of collagen type IV and some of the laminin genes, they probably are under completely different transcriptional controls. Experiments to quantify laminin message expression have shown that much greater amounts of β1 chain mRNA are detected than those for α or γ chains, although posttranscriptional regulation events cannot be ruled out (Vanden Heuvel et al., 1993). Furthermore, analysis of mRNA from newborn and adult kidneys showed severalfold decreases in adult RNA encoding the three laminin chains (α1, β1, γ1) and the α1 chain type IV collagen. However, the HSPG core protein mRNA persisted at relatively high levels in adult rat kidneys (Vanden Heuvel et al., 1993). In conjunction with a previous study on the rapid loss of labeled HSPG from adult GBM (Beavan et al. 1989), these results therefore document that although there is a marked decline of certain laminin and collagen IV mRNAs in the adult kidney, HSPG core protein mRNA is synthesized at much higher rates in vivo.

VII. Glomerular Defects

Decreased water filtration and increased permeation of albumin are the hallmarks of abnormal glomerular function and are reflected in nephrotic syndrome and other renal diseases by the dramatic changes in the structure and function of the GBM and podocytes. Many of these glomerular pathologies appear to reflect disturbances in development and/or maintenance of the differentiated state.

A. Goodpasture (Anti-GBM) Disease

Goodpasture, or anti-GBM, disease is an uncommon and usually severe disorder caused by autoimmunity to a component of certain basement membranes, and patients with this disease usually present with a crescentic glomerulonephritis and lung hemoptysis (Turner,1996). Affinity chromatography has shown that 1% of the Goodpasture patients' total IgG are antitype IV collagen antibodies and that 90% of these antibodies are specific for the $\alpha 3$(IV) chain. The identity of the $\alpha 3$ chain as the Goodpasture autoantigen was established through serial biochemical experiments, including the digestion of GBM preparations by bacterial collagenase and the identification of a 28-kDa protein, which was subsequently identified as the NC1 domain of $\alpha 3$ (IV) (Butkowski et al., 1987). These earlier biochemical experiments were later confirmed by molecular cloning of the antigen and generation of the human recombinant $\alpha 3$(IV) NC1 domain in various in vitro expression systems (Neilson et al., 1993). By synthesis of multiple $\alpha 3$(IV) NC1 peptides, site-directed mutagenesis, and measurement of their antibody binding capacity, three hydrophobic residues in the carboxyl-terminal region of $\alpha 3$(IV) chain have been shown to be critical for the immunodominant epitope of the Goodpasture autoantibody (David et al., 2001).

B. Alport Syndrome

Alport syndrome is a progressive hereditary kidney disease characterized by hematuria, sensorineural hearing loss, and ocular lesions with structural defects in the GBM (Kashtan, 1998, 1999). Depending on the age at which end stage renal disease occurs, patients are classified as either juvenile- or adult-onset Alport syndrome. More rarely, patients develop diffuse esophageal leiomyomatosis (a rare condition characterized by proliferation of smooth muscle in the upper gastrointestinal tract) or macrothrombocytopenia (a disorder characterized by giant platelets). Alport syndrome is a clinically and genetically heterogeneous disease and affects 1 in 5000 individuals. The predominant form (85%) of Alport syndrome is X linked, the autosomal-recessive form is responsible for about 15% of cases, and the autosomal dominant form is rare (<1%). Approximately 1%–5% of transplanted Alport patients have been reported to develop anti-GBM nephritis leading to loss of the renal allograft.

Patients with X-linked Alport syndrome, which is not an autoimmune disease but leads to progressive renal failure and is often associated with deafness, have been shown to carry mutations and/or deletions in the type IV collagen $\alpha 5$ and/or $\alpha 6$ chains (Tryggvason, 1993). Autosomally inherited forms of Alport syndrome seem to be caused by mutations in the $\alpha 3$(IV) and $\alpha 4$(IV) chains. In patients with Alport syndrome and leiomyomatosis, large deletions covering the 5′ region of both the COL4A5 and the COL4A6 genes are found (Heidet et al., 1995). Associated with the juvenile form of Alport syndrome are mutations that cause absent or truncated regions of the terminal NC1 domain of COL4A5 chains (Flinter et al., 1988). In most cases of anti-GBM nephritis, antibodies are directed against the COL4A3 gene product, whereas the majority of the mutations leading to Alport disease have been identified in the COL4A5 gene. A similar situation exists in dogs, with a mutation in the collagen $\alpha 5$(IV) gene (Thorner et al., 1996), and in mice, with targeted mutations in the collagen $\alpha 3$(IV) gene (Cosgrove et al., 1996).

C. Nephrotic Syndrome

Nephrotic syndrome is a common kidney disease seen in both children and adults and is characterized clinically by massive proteinura, hypoalbuminemia, and edema. These clinical findings correlate with glomerular morphological changes, including retraction and effacement of podocyte foot processes, narrowing of the filtration slits, apical displacement and duplication of the slit diaphragms, and, in more severe cases, detachment of podocytes from the underlying GBM (Smoyer and Mundel, 1998). Features of several different kinds of nephrotic syndrome are summarized as follows.

1. Minimal Change Disease

Minimal change disease is an important cause of nephrotic syndrome and accounts for 85% of nephrotic cases in children. The clinical entity of this disorder is heavy proteinuria (over 5 g and often as great as 10 g/day). No evidence of glomerular damage is observed on routine light microscopy, but electron micrographs of biopsy specimens show widespread effacement of epithelial foot processes (Andrews, 1988). The majority of patients who fail to respond to corticosteroid therapy almost invariably progress to end stage renal disease. At present, the pathogenetic mechanisms that underly minimal change disease are unknown.

2. Congenital Nephrotic Syndrome of the Finnish Type (CNF)

CNF disorder is often associated with a consanguinous familial background. It is caused by mutations of nephrin, which is expressed specifically by glomerular podocytes and is an important element of the slit diaphragm (Ruotsalainen et al., 1999). Typical clinical features include a premature delivery with an enlarged placenta and proteinuria starting in fetal stages. Ultrastructural analysis shows diffuse absence of foot processes similar to that seen in minimal change disease (Sharif *et al.*, 1998).

3. Focal Segmental Glomerulosclerosis (FSGS)

FSGS has been distinguished from minimal change disease by hematuria, hypertension, and renal insufficiency at presentation. When analyzed by electron microscopy, GBM thickening can be observed, whereas podocytes, which are often adherent to the Bowman's capsule, are sometimes detached from the GBM. One form of FSGS, HIV-associated nephropathy, displays rapidly progressive collapse and sclerosis of the entire glomerular tuft (Humphreys *et al.*, 1995). The pathogenesis of FSGS is unknown but may be the result of circulating "permeability" factors, possibly a lymphokine or cytokine, which leads to glomerular podocyte injury (Savin *et al.*, 1996). Autosomal-dominant or -recessive patterns of inheritance have been described for the familial forms of FSGS, but the genetics of these hereditary forms of FSGS are unknown, although it has been reported that the autosomal dominant form of FSGS is linked to a region of chromosome 19q (Winn *et al.*, 1999).

4. Podocyte Alterations

Molecular mechanisms accounting for most of the nephrotic syndromes are almost entirely unknown, but the loss of multiple and specific structural features of podocytes has been described (D'Agati, 1994; Sharif *et al.*, 1998; Barisoni *et al.*, 1999). The phosphorylation of tight junction and other proteins contributes to rapid changes in epithelial cell shape. After perfusion with protamine sulfate, immuno-blotting studies showed that the phosphotyrosine signal was increased dramatically in the glomerular epithelium and that ZO-1 is one of the phosphorylation targets (Kurihara, 1995). In addition to the changing of negative charges and phosphorylation status, the expression pattern of some other podocyte proteins is also changed in renal disease. Podoplanin is downregulated in puromycin nephrosis (Breiteneder-Geleff *et al.*, 1997), and in puromycin aminonucleoside or protamine sulfate treated rats, expression of ZO-1 can be found concentrated along both the newly formed occludens-type junctions and the remaining slit diaphragms (Breiteneder-Geleff *et al.*, 1997). The increased immunofluorescence staining of α-actinin in podocytes has also been reported during foot process effacement in nephrotoxic serum nephritis (Shirato *et al.*, 1996).

VIII. Closing Remarks

Glomerulogenesis is an active morphogenic process of cell proliferation and functional differentiation that is tightly controlled by a series of genes expressed in temporally and spatially regulated patterns. These processes are heavily involved with various extracellular matrix proteins, receptors, cell adhesion molecules, intracellular cytoskeletal proteins, enzymes and growth factors (Kanwar *et al.*, 1997; Lipschutz, 1998; Miner, 1998, Orellana and Ellis, 1998; Smoyer and Mundel, 1998).

Laminin is the major noncollagenous component of the GBM. A growing body of evidence in mice and other species has provided important insights into the complexity of GBM laminins, and changes in laminin expression that occur with development and these intricacies are likely to play a role in and/or influence kidney disease. During kidney development, the switches between different isoforms occur not only for laminin, but also for type IV collagen. These changes are clearly important for the morphological and physiological maturation of the glomerulus, but the mechanisms by which the appearance and disappearance of different GBM proteins are regulated during development are still not clear. Furthermore, how the newly formed GBM is added and the old GBM removed during turnover also remain as mysteries.

Various factors contribute to the process of glomerular endothelial cell, mesangial cell, and podocyte differentiation during nephron development, but little is known about the details of the cellular and molecular mechanisms that occur within these cells in both normal and nephrotic situations. The interactions between these cells and their associated extracellular matrices undoubtedly contribute to cell differentiation, but whether the presence of an organized basement membrane is permissive or instructive for endothelial cell, mesangial cell, and podocyte differentiation needs to be determined. Because both glomerular endothelial cells and podocytes contribute to the GBM, these cells may produce their own unique cell-specific matrix, and perhaps the accumulation of certain matrix proteins governs cell proliferation and maturation within the glomerulus. This may be one reason for the isoform switching of laminin and collagen type IV in developing GBM, and why failures in these transitions cause such prominent glomerular defects.

Acknowledgments

We thank Eileen Roach for help with preparation of the figures. Funds came from NIH Grants DK-34792 and DK-52483.

References

Abrahamson, D. R. (1985). Origin of the glomerular basement membrane visualized after *in vivo* labeling of laminin in newborn rat kidneys. *J. Cell Biol.* **100**, 1988–2000.

Abrahamson, D. R. (1986). Recent studies on the structure and pathology of basement membranes. *J. Pathol.* **149**, 257–278.

Abrahamson, D. R. (1987). Structure and development of the glomerular capillary wall and basement membrane. *Am. J. Physiol.* **253**, F783–F794.

Abrahamson, D. R. (1991). Glomerulogenesis in the developing kidney. *Semin. Nephrol.* **11**, 375–389.

Abrahamson, D. R., Irwin, M. H., St. John, P. L., Perry, E. W., Accavitti, M. A., Heck, L. W., and Couchman, J. R. (1989). Selective immuno-reactivities of kidney basement membranes to monoclonal antibodies against laminin: Localization of the end of the long arm and the short arms to discrete microdomains. *J. Cell Biol.* **109**, 3477–3491.

Abrahamson, D. R., and Perry, E. W. (1986). Evidence for splicing new basement membrane into old during glomerular development in newborn rat kidneys. *J. Cell Biol.* **103**, 2489–2498.

Abrahamson, D. R., Vanden Heuvel, G. B., and Clapp, W. L. (1997). Nephritogenic antigens in the glomerular basement membrane. *In* "Immunologic Renal Diseases" (E. G. Neilson and W. G. Couser, eds.), pp. 217–234. Lippincott-Raven, Philadelphia.

Adler, S., Sharma, R., Savin, V. J., Abbi, R., and Eng, B. (1996). Alteration of glomerular permeability to macromolecules induced by cross-linking of beta 1 integrin receptors. *Am. J. Pathol.* **149**, 987–996.

Adler, S., and Chen, X. (1992). Anti-Fx1A antibody recognizes a beta 1-integrin on glomerular epithelial cells and inhibits adhesion and growth. *Am. J. Physiol.* **262**, F770–F776.

Alpers, C. E., Seifert, R. A., Hedkins, K. L., Johnson, R. J., and Bowen-Pope, D. R. (1992) Developmental patterns of PDGF B-chain, PDGF-receptor, and "-actin expression in human glomerulogenesis. Kidney Int. 42, 390–399.

Andrews, P. (1988). Morphological alterations of the glomerular (visceral) epithelium in response to pathological and experimental situations. *J. Electron Microsc. Tech.* **9**, 115–144.

Aumailley, M., and Smyth, N. (1998). The role of laminins in basement membrane function. *J. Anat.* **193**, 1–21.

Aumailley, M., Battaglia, C., Mayer, U., Reinhardt, D., Nischt, R., Timpl, R., and Fox, J. W. (1993). Nidogen mediates the formation of ternary complexes of basement membrane components. *Kidney Int.* **43**, 7–12.

Aumailley, M., and Krieg, T. (1996). Laminins: A family of diverse multi-functional molecules of basement membranes. *J. Invest. Dermatol.* **106**, 209–214.

Baird, P. N., Groves, N., Haber, D. A., Housman, D. E., and Cowell, J. K. (1992). Identification of mutations in the WT1 gene in tumours from patients with the WAGR syndrome. *Oncogene* **7**, 2141–2149.

Barber, A. J., and Lieth, E. (1997). Agrin accumulates in the brain micro-vascular basal lamina during development of the blood-brain barrier. *Dev. Dyn.* **208**, 62–74.

Barisoni, L., Kriz, W., Mundel, P., and D'Agati, V. (1999). The dys-regulated podocyte phenotype: A novel concept in the pathogenesis of collapsing idiopathic focal segmental glomerulosclerosis and HIV-associated nephropathy. *J. Am. Soc. Nephrol.* **10**, 51–61.

Barnes, J. L., Radnik, R. A., Gilchrist, E. P., and Venkatachalam, M. A. (1984). Size and charge selective permeability defects induced in glomerular basement membrane by a polycation. *Kidney Int.* **25**, 11–9.

Battaglia, C., Mayer, U., Aumailley, M., and Timpl, R. (1992). Basement-membrane heparan sulfate proteoglycan binds to laminin by its heparan sulfate chains and to nidogen by sites in the protein core. *Eur. J. Biochem.* **208**,359–66.

Beavan, L. A., Davies, M., Couchman, J. R., Williams, M. A., and Mason, R. M. (1989). In vivo turnover of the basement membrane and other heparan sulfate proteoglycans of rat glomerulus. *Arch. Biochem. Biophys.* **269**, 576–585.

Beck, K., Dixon, T. W., Engel, J., and Parry D (1993). Ionic interactions in the coiled-coil domain of laminin determine the specificity of chain assembly. *J. Mol. Biol.* **231**, 311–323.

Boute, N., Gribouval, O., Roselli, S., Benessy, F., Lee, H., Fuchshuber, A., Dahan, K., Gubler, M. C., Niaudet, P., and Antignac, C. (2000). NPHS2, encoding the glomerular protein podocin, is mutated in autosomal

recessive steroid-resistant nephrotic syndrome. *Nature. Genet.* **24**, 349–354.

Breiteneder-Geleff, S., Matsui, K., Soleiman, A., Meraner, P., Poczewski, H., Kalt, R., Schaffner, G., and Kerjaschki, D. (1997). Podoplanin: Novel 43-kd membrane protein of glomerular epithelial cells, is down-regulated in puromycin nephrosis. *Am J. Pathol.* **151**, 1141–1152.

Burgeson, R. E., Chiquet, M., Deutzmann, R., Ekblom, P., Engel, J., Kleinman, H., Martin, G. R., Meneguzzi, G., Paulsson, M., Sanes, J., et al. (1994). A new nomenclature for the laminins. *Matrix Biol.* **14**,209–211.

Butkowski, R. J., Langeveld, J. P., Wieslander, J., Hamilton, J., and Hudson, B. G. (1987). Localization of the Goodpasture epitope to a novel chain of basement membrane collagen. *J. Biol. Chem.* **262**, 7874–7877.

Cao, Y. (2001). Endogenous angiogenesis inhibitors and their therapeutic implications. *Int. J. Biochem. Cell Biol.* **33**, 357–369.

Clapp, W. L., and Abrahamson, D. R. (1994). Development and gross anatomy of the kidney. *In* "Renal Pathology" (C. C. Tisher and B. M. Brenner, eds.), Vol. I, pp. 3–59. Lippincott, Philadelphia.

Cohen, I. R., Grassel, S., Murdoch, A. D., and Iozzo, R. V. (1993). Structural characterization of the complete human perlecan gene and its promoter. *Proc. Natl. Acad. Sci. USA* **90**, 10404–10408.

Colognato, H., and Yurchenco. (2000). Form and function: The laminin family of heterotrimers. *Dev. Dyn.* **218**, 213–234.

Colucci-Guyon, E., Portier, M. M., Dunia, I., Paulin, D., Pournin, S., and Babinet, C. (1994). Mice lacking vimentin develop and reproduce without an obvious phenotype. *Cell* **79**, 679–694.

Cosgrove, D., Meehan, D. T., Grunkemeyer, J. A., Kornak, J. M., Sayers, R., Hunter, W. J., and Samuelson, G. C. (1996). Collagen COL4A3 knockout: A mouse model for autosomal Alport syndrome. *Genes Dev.* **10**, 2981–2992.

Cosgrove, D., Rodgers, K., Meehan, D., Miller, C., Bovard, K., Gilroy, A., Gardner, H., Kotelianski, V., Gotwals, P., Amatucci, A., and Kalluri, R. (2000). Integrin alpha1beta1 and transforming growth factor-beta1 play distinct roles in alport glomerular pathogenesis and serve as dual targets for metabolic therapy. *Am. J. Pathol.* **157**, 1649–1659.

Couchman, J. R., Kapoor, R., Sthanam, M., and Wu, R. R. (1996). Perlecan and basement membrane-chondroitin sulfate proteoglycan (bamacan) are two basement membrane chondroitin/dermatan sulfate proteo-glycans in the Engelbreth-Holm-Swarm tumor matrix. *J. Biol. Chem.* **27**, 9595–9602.

D'Agati, V (1994). The many masks of focal segmental glomerulo-sclerosis. *Kidney Int.* **46**, 1223–1241.

Daniel, T. O., and Abrahamson, D. R. (2000). Endothelial signal integra-tionin vascular assembly. *Annu. Rev. Physiol.* **62**, 649–671.

Daniel, T. O., Stein, E., Cerretti, D. P., St. John, P. L., Robert, B., and Abrahamson, D. R. (1996). ELK and LERK-2 in developing kidney and microvascular endothelial assembly. *Kidney Int.* **50**, 73–81.

Daniels, B. S. (1993). The role of the glomerular epithelial cell in the maintenance of the glomerular filtration barrier. *Am. J. Nephrol.* **13**, 318–323.

David, M., Borza, D. B., Leinonen, A., Belmont, J. M., and Hudson, B. G. (2001). Hydrophobic amino acid residues are critical for the immuno-dominant epitope of the Goodpasture autoantigen. A molecular basis for the cryptic nature of the epitope. *J. Biol. Chem.* **276**, 6370–6377.

Denzer, A. J., Brandenberger, R., Gesemann, M., Chiquet, M., and Ruegg, M. A. (1997). Agrin binds to the nerve-muscle basal lamina via laminin. *J. Cell Biol.* **137**, 671–683.

Desjardins, M., and Bendayan, M. (1991). Ontogenesis of glomerular basement membrane: Structural and functional properties. *J. Cell Biol.* **113**, 689–700.

Desjardins ,M., and Bendayan, M. (1989). Heterogenous distribution of type IV collagen, entactin, heparan sulfate proteoglycan, and laminin among renal basement membranes as demonstrated by quantitative immunocytochemistry. *J. Histochem. Cytochem.* **37**, 885–897.

Drumond, M. C., and Deen, W. M. (1995). Hindered transport of macromolecules through a single row of cylinders: application to glomerular filtration. *J. Biomechan. Eng.* 117, 414–422.

Durbeej, M., Fecker, L., Hjalt, T., Zhang, H.-Y., Salmivirta, K., Klein, G., Timpl, R., Sorokin, L., Ebendal, T., Ekblom, P., and Ekblom, M. (1996). Expression of laminin α1, α5 and β2 chains during embryogenesis of the kidney and vasculature. *Matrix Biol.* **15**, 397–413.

Durbeej, M., Henry, M. D., Ferletta, M., Campbell, K. P., and Ekblom, P. (1998). Distribution of dystroglycan in normal adult mouse tissues. *J. Histochem. Cytochem.* **46**, 449–457.

Durkin, M. E., Gautam, M., Loechel, F,. Sanes, J. R., Merlie, J. P., Albrechtsen, R., and Wewer, U. M. (1996). Structural organization of the human and mouse laminin beta2 chain genes, and alternative splicing at the 5' end of the human transcript. *J. Biol. Chem.* **271**, 13407–13416.

Dziadek, M. (1995). Role of laminin-nidogen complexes in basement membrane formation during embryonic development. *Experientia* **51**, 901–13

Edwards, A., Daniels, B. S., and Deen, W. M. (1999). Ultrastructural model for size selectivity in glomerular filtration. *Am. J. Physiol.* **276**, F892–902.

Ekblom, P. (1981). Formation of basement membranes in the embryonic kidney: An immunohistological study. *J. Cell Biol.* **91**, 1–10.

Ekblom, P., Ekblom, M., Fecker, L., Klein, G., Zhang, H. Y., Kadoya, Y., Chu, M. L., Mayer, U., and Timpl, R. (1994). Role of mesenchymal nidogen for epithelial morphogenesis *in vitro*. *Development* **120**, 2003–2014.

Ekblom, M., Falk, M., Salmivirta, K., Durbeej, M., and Ekblom, P. (1998). Laminin isoforms and epithelial development. *Ann. N. Y. Acad. Sci.* **857**,194–211

Falk, M., Salmivirta, K., Durbeej, M., Larsson, E., Ekblom, M., Vestweber, D., and Ekblom, P. (1996). Integrin alpha 6B beta 1 is involved in kidney tubulogenesis *in vitro*. *J. Cell Sci.* **109**, 2801–2810.

Fazio, M. J., O'Leary, J., Kahari, V. M., Chen, Y. Q., Saitta, B., and Uitto, J. (1991). Human nidogen gene: Structural and functional characterization of the 5'-flanking region. *J. Invest. Dermatol.* 97, 281–285.

Ferletta, M., and Ekblom, P. (1999). Identification of laminin-10/11 as a strong cell adhesive complex for a normal and a malignan human epithelial cell line. *J. Cell Sci.* **112**, 1–10.

Flinter, F. A., Cameron, J. S., Chantler, C., Houston, I., and Bobrow, M. (1988). Genetics of classic Alport's syndrome. *Lancet* **8618**, 1005–1007.

Gelberg, H., Healy, L., Whiteley, H., Miller, L. A., and Vimr, E. (1996). In vivo enzymatic removal of alpha 2-6-linked sialic acid from the glomerular filtration barrier results in podocyte charge alteration and glomerular injury. *Lab. Invest.* **74**, 907–20.

Ghiselli, G., Siracusa, L. D., and Iozzo, R. V. (1999). Complete cDNA cloning, genomic organization, chromosomal assignment, functional characterization of the promoter, and expression of the murine Bamacan gene. *J. Biol. Chem.* **274**, 17384–17393.

Groffen, A. J., Buskens, C. A., van Kuppevelt, T. H,, Veerkamp, J. H., Monnens, L. A., and van den Heuvel, L. P. (1998). Primary structure and high expression of human agrin in basement membranes of adult lung and kidney. *Eur. J. Biochem.* **254**, 123–128.

Groffen, A. J., Hop, F. W., Tryggvason, K., Dijkman, H., Assmann, K. J., Veerkamp, J. H., Monnens, L. A., and van den Heuvel, L. P. (1997). Evidence for the existence of multiple heparan sulfate proteoglycans in the human glomerular basement membrane and mesangial matrix. *Eur. J. Biochem.* **247,** 175–182.

Gunwar, S., Ballester, F., Noelken, M. E., Sado, Y., Ninomiya, Y., and Hudson, B. G. (1998). Glomerular basement membrane. Identification of a novel disulfide-cross-linked network of alpha3, alpha4, and alpha5 chains of type IV collagen and its implications for the pathogenesis of Alport syndrome. *J. Biol. Chem.* **273**, 8767–8775.

Halfter, W., Dong, S., Schurer, B., and Cole, G. J. (1998). Collagen XVIII Is a basement membrane heparan sulfate proteoglycan. *J. Biol. Chem.* **273**, 25404–25412

Handler, M., Yurchenco, P. D., and Iozzo, R. V. (1997). Developmental expression of perlecan during murine embryogenesis. *Dev. Dyn.* **210**, 130–145.

Hansen, K., and Abrass, C. K. (1999). Role of laminin isoform in glomerular structure *Pathobiology* **67**, 84–91.

Heidet, L., Dahan, K., Zhoum J., Xu, Z., Cochat, P., Gould, J. D., Leppig, K. A., Proesmans, W., Guyot, C., and Guillot, M. (1995). Deletions of both alpha 5(IV) and alpha 6(IV) collagen genes in Alport syndrome and in Alport syndrome associated with smooth muscle tumours. *Hum. Mol Genet.* **4**, 99–108.

Henry, M. D., and Campbell, K. P. (1996). Dystroglycan: An extracellular matrix receptor linked to the cytoskeleton. *Curr. Opin. Cell Biol.* **8**, 625–631.

Henry, M. D., and Campbell, K. P. (1998). A role for dystroglycan in basement membrane assembly. *Cell* **95**, 859–70.

Holthofer, H., Miettinen, A., Lehto, V. P., Lehtonen, E., and Virtanen, I. (1984). Expression of vimentin and cytokeratin types of intermediate filament proteins in developing and adult human kidneys. *Lab. Invest.* **50**, 552–559.

Hopf, M., Gohring, W., Kohfeldt, E., Yamada, Y., and Timpl, R. (1999). Recombinant domain IV of perlecan binds to nidogens, Laminin-nidogen complex, fibronectin, fibulin-2 and heparin. *Eur. J. Biochem.* **259**, 917–925.

Holzman, L. B., St. John, P. L., Kovarila, A., Verma, R., Holthofer, H., and Abrahamson, D. R. (1999). Nephrin localizes to the slit pore of the glomerular epithelial cell. *Kidney Int.* **56**, 1481–1491.

Horvat, R., Hovorka, A., Dekan, G., Poczewski, H., and Kerjaschki D (1986). Endothelial cell membranes contain podocalyxin: The major sialoprotein of visceral glomerular epithelial cells. *J. Cell Biol.* **102**, 484–491.

Hudson, B. G., Reeders, S. T., and Tryggvason, K. (1993). Type IV collagen: Structure, gene organization, and role in human diseases. Molecular basis of Goodpasture and Alport syndromes and diffuse leiomyomatosis. *J. Biol. Chem.* **268**, 26033–26036.

Hugo, C., Shankland, S. J., Bowen-Pope, D. F., Couser, W. G., and Johnson, R. J. (1997). Extraglomerular origin of the mesangial cell after injury: A new role of the juxtaglomerular apparatus. *J. Clin. Invest.* **100**, 786–794.

Humphreys, M. H. (1995). Human immunodeficiency virus-associated glomerulosclerosis. *Kidney Int.* **48**, 311–320.

Hyink, D. P., and Abrahamson, D. R. (1995). Origin of the glomerular vasculature in the developing kidney. *Semin. Nephrol.* **15**, 300–314.

Iivanainen, A., Vuolteenaho, R., Sainio, K., Eddy, R., Shows, T. B., Sariola, H., and Tryggvason, K. (1995). The human laminin beta 2 chain (S-laminin): Structure, expression in fetal tissues and chromosomal assignment of the LAMB2 gene. *Matrix Biol.* **14**, 489–497.

Iozzo, R. V. (1998). Matrix proteoglycans: From molecular design to cellular function. *Annu. Rev. Biochem.* **67**, 609–652.

Johnson, R., Yamabe, H., Chen, Y. P., Campbell, C., Gordon, K., Baker, P., Lovett, D., and Couser, W. G. (1992). Glomerular epithelial cells secrete a glomerular basement membrane-degrading metalloproteinase. *J. Am. Soc. Nephrol.* **2**, 1388–1397.

Kahsai, T. Z., Enders, G. C., Gunwar, S., Brunmark, C., Wieslander, J., Kalluri, R., Zhou, J., Noelken, M. E., and Hudson, B. G. (1997). Seminiferous tubule basement membrane. Composition and organization of type IV collagen chains, and the linkage of alpha3(IV) and alpha5(IV) chains. *J. Biol. Chem.* **272**, 17023–17032.

Kallunki, P., Eddy, R. L., Byers, M. G., Kestila, M., Shows, T. B., and Trygvasson, K. (1991). Cloning of human heparan sulfate proteoglycan core protein, assignment of the gene (HSPG2) to 1p36. 1-p35 and identification of BamHI restriction fragment polymorphism. *Genomics* **2**, 389–396.

Kalluri, R., Shield, C. F., Todd, P., Hudson, B. G., and Neilson, E. G.

(1997). Isoform switching of type IV collagen is developmentally arrested in X-linked Alport syndrome leading to increased susceptibility of renal basement membranes to endoproteolysis. *J. Clin. Invest.* **99**, 2470–2478.

Kanwar, Y. S., Carone, F. A., Kumar, A., Wada, J., Ota, K., and Wallner, E. I. (1997). Role of extracellular matrix, growth factors and proto-oncogenes in metanephric development. *Kidney Int.* **52**, 589–606.

Kanwar, Y. S., Liu, Z. Z., Kashihara, N., and Wallner, E. I. (1991). Current status of the structure and functional basis of glomerular filtration and proteinuria. *Semin. Nephrol.* **11**, 390–413.

Kanwar, Y. S., Jakubowski, M. L., Rosenzweig, L. J., and Gibbons, J. T. (1984). De novo cellular synthesis of sulfated proteoglycans of the developing renal glomerulus in vivo. *Proc. Natl. Acad. Sci. USA.* **81**, 7108–7111.

Kanwar, Y. S., Linker, A., and Farquhar, M. G. (1980). Increased permeability of the glomerular basement membrane to ferritin after removal of glycosaminoglycans (heparan sulfate) by enzyme digestion. *J. Cell Biol.* **86**, 688–693.

Kashtan, C. E. (1998). Alport syndrome and thin glomerular basement membrane disease. *J. Am. Soc. Nephrol.* **9**, 1736–1750.

Kashtan, C. E. (1999). Alport syndrome: An inherited disorder of renal, ocular, and cochlear basement membranes. *Medicine* **78**, 338–360.

Katz, A., Fish, A. J., Kleppel, M. M., Hagen, S. G., Michael, A. F., and Butkowski, R. J. (1991). Renal entactin (nidogen): Isolation, characterization and tissue distribution. *Kidney Int.* **40**, 643–52.

Kawachi, H., Kurihara, H., Topham, P. S., Brown, D., Shia, M. A., Orikasa, M., Shimizu, F., and Salant, D. J. (1997). Slit diaphragm-reactive nephritogenic MAb 5-1-6 alters expression of ZO-1 in rat podocytes. *Am. J. Physiol.* **273**, F984–F993.

Kelley, V. E., and Cavallo, T. (1978). Glomerular permeability: Transfer of native ferritin in glomeruli with decreased anionic sites. *Lab. Invest.* **39**, 547–553.

Kerjaschki, D., Sharkey, D. J., and Farquhar, M. G. (1984). Identification and characterization of podocalyxin: The major sialoprotein of the renal glomerular epithelial cell. *J. Cell Biol.* **98**, 1591–1596.

Kerjaschki, D. (1978). Polycation-induced dislocation of slit diaphragms and formation of cell junctions in rat kidney glomeruli: The effects of low temperature, divalent cations, colchicine, and cytochalasin B. *Lab. Invest.* **39**, 430–440.

Kershaw, D. B., Thomas, P. E., Wharram, B. L., Goyal, M., Wiggins, J. E., Whiteside, C. I., and Wiggins, R. C. (1995). Molecular cloning, expression, and characterization of podocalyxin-like protein 1 from rabbit as a transmembrane protein of glomerular podocytes and vascular endothelium. *J. Biol. Chem.* **270**, 29439–29446.

Kestila, M., Lenkkeri, U., Mannikko, M., Lamerdin, J., McCready, P., Putaala, H., Ruotsalainen, V., Morita, T., Nissinen, M., Herva, R., Kashtan, C. E., Peltonen, L., Holmberg, C., Olsen, A., and Tryggvason, K. (1998). Positionally cloned gene for a novel glomerular protein—nephrin—is mutated in congenital nephrotic syndrome. *Mol. Cell* **1**, 575–82.

Kimura, N., Toyoshima, T., Kojima, T., and Shimane, M. (1998). Entactin-2: A new member of basement membrane protein with high homology to entactin/nidogen. *Exp. Cell Res.* **241**, 36–45.

Kitamoto, Y., Tokunaga, H., and Tomita, K. (1997). Vascular endothelial growth factor is an essential molecule for mouse kidney development: Glomerulogenesis and nephrogenesis. *J. Clin. Invest.* **99**, 2351–2357.

Klein, G., Langegger, M., Timpl, R., and Ekblom, P/ (1988). Role of laminin A chain in the development of epithelial cell polarity. *Cell* **55**, 331–341.

Kohfeldt, E., Sasaki, T., Gohring, W., and Timpl, R. (1998). Nidogen-2: A new basement membrane protein with diverse binding properties. *J. Mol. Biol.* **282**, 99–109.

Korhonen, M., Ylanne, J., Laitinen, L., and Virtanen, I. (1990a). Distribution of beta 1 and beta 3 integrins in human fetal and adult kidney. *Lab. Invest.* **62**, 616–625.

Korhonen, M., Ylanne, J., Laitinen, L., and Virtanen, I. (1990b). The alpha 1-alpha 6 subunits of integrins are characteristically expressed in distinct segments of developing and adult human nephron. *J. Cell Biol.* **111**, 1245–1254.

Kreidberg, J.A., Donovan, M.J., Goldstein, S.L., Rennke, H., Shepherd, K., Jones, R.C., Jaenisch, R. (1996). Alpha 3 beta 1 integrin has a crucial role in kidney and lung organogenesis. *Development* 122:3537–47

Kreidberg, J. A., Sariola, H., Loring, J. M., Maeda, M., Pelletier, J., Housman, D., and Jaenisch, R. (1993). WT-1 is required for early kidney development. *Cell* **74**, 679–691.

Kriz, W. (1997). The role of the podocyte in the degeneration of a renal glomerulus. *Adv. Nephrol.* **27**, 1–13.

Kriz, W., Elger, M., Nagata, M., Kretzler, M., Uiker, S., Koeppen-Hageman, I., Tenschert, S., and Lemley, K. V. (1994). The role of podocytes in the development of glomerular sclerosis. *Kidney Int.* **45**, S64–S72

Kriz, W., Hahnel, B., Rosener, S., and Elger, M. (1995). Long term treatment of rats with FGF-2 results in focal segmental glomerulosclerosis. *Kidney Int.* **48**, 1435–1450.

Kriz, W., Kretzler, M., Nagata, M., Provoost, A. P., Shirato, I., Uiker, S., Sakai, T., and Lemley, K. V. (1996). A frequent pathway to glomerulosclerosis: Deterioration of tuft architecture-podocyte damage-segmental sclerosis. *Kidney Blood Press. Res.* **19**, 245–253.

Kühn, K. (1994). Basement membrane (type IV) collagen. *Matrix Biol.* **14**, 439–445.

Kurihara, H., Anderson, J. M., and Farquhar, M. G. (1995). Increased Tyr phosphorylation of ZO-1 during modification of tight junctions between glomerular foot processes. *Am. J. Physiol.* **268**, F514–24

Lafayette, R. A., Druzin, M., Sibley, R., Derby, G., Malik, T., Huie, P., Polhemus, C., Deenm, W. M., and Myers, B. D. (1998). Nature of glomerular dysfunction in pre-eclampsia. *Kidney Int.* **54**, 1240–1249.

Lee, S. B., Huang, K., Palmer, R., Truong, V. B., Herzlinger, D., Kolquist, K. A., Wong, J., Paulding, C., Yoon, S. K., Gerald, W., Oliner, J. D., and Haber, D. A. (1999). The Wilms tumor suppressor WT1 encodes a transcriptional activator of amphiregulin. *Cell* **98**, 663–673.

Lelongt, B., Trugnan, G., Murphy, G., and Ronco, P. M. (1997). Matrix metalloproteinases MMP2 and MMP9 are produced in early stages of kidney morphogenesis but only MMP9 is required for renal organogenesis in vitro. *J. Cell Biol.* **136**, 1363–1373.

Lelongt, B., Makino, H., and Kanwar, Y. S. (1987). Maturation of the developing renal glomerulus with respect to basement membrane proteoglycans. *Kidney Int.* **32**, 498–506.

Leveen, P., Pekny, M., Gebre-Medhin, S., Swolin, B., Larsson, E., and Betsholtz, C., (1994). Mice deficient for PDGF B show renal, cardiovascular, and hematological abnormalities. *Genes Dev.* **8**, 1875–1887.

Li, C., Ruotsalainen, V., Tryggvason, K., Shaw, A. S., and Miner, J. H. (2000). CD2AP is expressed with nephrin in developing podocytes and is found widely in mature kidney and elsewhere. *Am. J. Physiol. Renal Physiol.* **279**, F785–F792.

Liang, Y., Haring, M., Roughley, P. J., Margolis, R. K., and Margolis, R. U. (1997). Glypican and biglycan in the nuclei of neurons and glioma cells: Presence of functional nuclear localization signals and dynamic changes in glypican during the cell cycle. *J. Cell Biol.* **139**, 851–864.

Lipschutz, H. J. (1998). Molecular development of the kidney: A review of the results of gene disruption studies. *Am. J. Kidney Dis.* **31**, 383–397.

Lohi, J., Korhonen, M., Leivo, I., Kangas, L., Tani, T., Kalluri, R., Miner, J. H., Lehto, V. P., and Virtanen, I. (1997). Expression of type IV collagen alpha1(IV)-alpha6(IV) polypeptides in normal and developing human kidney and in renal cell carcinomas and oncocytomas. *Int. J. Cancer* **72**, 43–49.

Londono, I., and Bendayan, M. (1988). High-resolution cytochemistry of neuraminic and hexuronic acid-containing macromolecules applying the enzyme-gold approach. *J. Histochem. Cytochem.* **36**, 1005–1014.

Lu, W., Phillips, C. L., Killen, P. D., Hlaing, W. R., Elder, F. F., Miner, J. H., Overbeek, P. A., and Meisler, M. H. (1999). Insertional mutation of the collagen genes Col4a3 and Col4a4 in a mouse model for Alport syndrome. *Genomics* 15, 113–124.

Mallick, N. P., Brenchley, P. E., and Webb, N. J. (1997). Minimal change nephropathy and focal segmental glomerulosclerosis. *Kidney Int.* 58, S80–S82.

Mariyama, M., Zheng, K., Yang-Feng, T. L., and Reeders, S. T. (1992). Colocalization of the genes for the alpha 3(IV) and alpha 4(IV) chains of type IV collagen to chromosome 2 bands q35-q37. *Genomics* 13, 809–813.

Matsui, K., Breitender-Geleff, S., Soleiman, A., Kowalski, H., and Kerjaschki, D. (1999). Podoplanin: A novel 43–kDa membrane protein, controls the shape of podocytes. *Nephrol. Dial. Transplant.* 14(Suppl.1), 9 –11.

Mayer, U., Nischt, R., Poschl, E., Mann, K., Fukuda, K., Gerl, M., Yamada, Y., and Timpl, R. (1993). A single EGF-like motif of laminin is responsible for high affinity nidogen binding. *EMBO J.* 12, 1879–85.

McCarthy, K. J., Bynum, K., St. John, P. L., Abrahamson, D. R., and Couchman, J. R. (1993). Basement membrane proteoglycans in glomerular morphogenesis: Chondroitin sulfate proteoglycan is temporally and spatially restricted during development. *J. Histochem. Cytochem.* 41, 401–414.

Miner, J. H. (1998). Developmental biology of glomerular basement membrane components. *Curr. Opin. Nephrol. Hypertens.* 7, 13–19.

Miner, J. H., Cunningham, J., and Sanes, J. R. (1998). Roles for laminin in embryogenesis: Exencephaly, syndactyly, and placentopathy in mice lacking the laminin alpha5 chain. *J. Cell Biol.* 143, 1713–1723.

Miner, J. H., and Li, C. (2000). Defective glomerulogenesis in the absence of laminin alpha 5 demonstrates a developmental role for the kidney glomerular basement membrane. *Dev. Biol.* 217, 278–289.

Miner, J. H., Patton, B. L., Lentz, S. I., Gilbert, D. J., Snider, W. D., Jenkins, N. A., Copeland, N. G., and Sanes, J. R. (1997). The laminin alpha chains: Expression, developmental transitions, and chromosomal locations of alpha1-5, identification of heterotrimeric laminins 8–11, and cloning of a novel alpha3 isoform. *J. Cell Biol.* 137, 685–701.

Miner J. H., and Sanes, J. R. (1994). Collagen IV "3, "4, and "5 chains in rodent basal laminae: Sequence, distribution, association with laminins, and developmental switches. *J. Cell Biol.* 127, 879–891.

Miner, J. H., and Sanes, J. R. (1996). Molecular and functional defects in kidneys of mice lacking collagen "3(IV): Implications for Alport syndrome. *J. Cell Biol.* 135, 1403–1413.

Moll, R., Hage, C., and Thoenes, W. (1991). Expression of intermediate filament proteins in fetal and adult human kidney: Modulations of intermediate filament patterns during development and in damaged tissue. *Lab. Invest.* 65, 74–86.

Mundel, P., Gilbert, P., and Kriz, W. (1991). Podocytes in glomerulus of rat kidney express a characteristic 44 KD protein. *J. Histochem. Cytochem.* 39, 1047–1056.

Mundel, P., Heid, H. W., Mundel, T. M., Kruger, M., Reiser, J., and Kriz, W. (1997a). Synaptopodin: An actin-associated protein in telencephalic dendrites and renal podocytes. *J. Cell Biol.* 139, 193–204.

Mundel, P., and Kriz, W. (1995). Structure and function of podocytes: An update. *Anat. Embryol.* 192, 385–97.

Mundel, P., Reiser, J., and Kriz, W. (1997b). Induction of differentiation in cultured rat and human podocytes. *J. Am. Soc. Nephrol.* 8, 697–705.

Mundlos, S., Pelletier, J., Darveau, A., Bachmann, M., Winterpacht, A., and Zabel, B. (1993). Nuclear localization of the protein encoded by the Wilms' tumor gene WT1 in embryonic and adult tissues. *Development* 119, 1329–1341.

Murshed, M., Smyth, N., Miosge, N., Karolat, J., Krieg, T., Paulsson, M., and Nischt, R. (2000). The absence of nidogen 1 does not affect murine basement membrane formation. *Mol. Cell Biol.* 18, 7007–7012.

Nagata, M., Yamaguchi, Y., and Ito, K. (1993). Loss of mitotic activity and the expression of vimentin in glomerular epithelial cells of developing human kidneys. *Anat. Embryol.* 187, 275–279.

Neilson, E. G., Kalluri, R., Sun, M. J., Gunwar, S., Danoff, T., Mariyama, M., Myers, J. C., Reeders, S. T., and Hudson, B. G. (1993). Specificity of Goodpasture autoantibodies for the recombinant noncollagenous domains of human type IV collagen. *J. Biol. Chem.* 268, 8402–8405.

Niimi, T., and Kitagawa, Y. (1997). Distinct roles of mouse laminin beta1 long arm domains for alpha1beta1gamma1 trimer formation. *FEBS Lett.* 400, 71–74.

Ninomiya, Y., Kagawa, M., Iyama, K., Naito, I., Kishiro, Y., Seyer, J. M., Sugimoto, M., Oohashi, T., and Sado, Y. (1995). Differential expression of two basement membrane collagen genes, COL4A6 and COL4A5, demonstrated by immunofluorescence staining using peptide-specific monoclonal antibodies. *J. Cell. Biol.* 130, 1219–1229.

Noakes, P. G., Miner, J. H., Gautam, M., Cunningham J. M., Sanes, J. R., and Merlie, J. P. (1995). The renal glomerulus of mice lacking s-laminin/laminin β2: nephrosis despite molecular compensation by laminin β1. *Nat. Genet.* 10, 400–406.

Noonan, D. M., Fulle, A., Valente, P., Cai, S., Horigan, E., Sasaki, M., Yamada, Y., and Hassell, J. R. (1991). The complete sequence of perlecan, a basement membrane heparan sulfate proteoglycan, reveals extensive similarity with laminin A chain, LDL receptor and N-CAM. *J. Biol. Chem.* 266:22939–22947.

O'Neill, B. C., Suzuki, H., Loomis, W. P., Denisenko, O., and Bomsztyk, K. (1997). Cloning of rat laminin gamma 1-chain gene promoter reveals motifs for recognition of multiple transcription factors. *Am. J. Physiol.* 273, F411–F420.

Oosterwijk, E., Van Muijen, G. N., Oosterwijk-Wakka, J. C., and Warnaar, S. O. (1990). Expression of intermediate-sized filaments in developing and adult human kidney and in renal cell carcinoma. *J. Histochem. Cytochem.* 38, 385–392.

Olsen, D. R., Nagayoshi, T., Fazio, M., Mattei, M. G., Passage, E., Weil, D., Timpl, R., Chu, M. L., and Uitto, J. (1989). Human nidogen: cDNA cloning, cellular expression, and mapping of the gene to chromosome Iq43. *Am. J. Hum. Genet.* 44, 876–885.

Orellana, A. S., and Ellis, A. D. (1998). Cell and molecular biology of kidney Development. *Semin. Nephrol.* 18, 233–243.

Ota, K., Stetler-Stevenson, W. G., Yang, Q., Kumar, A., Wada, J., Kashihara, N., Wallner, E. I., and Kanwar, Y. S. (1998). Cloning of murine membrane-type-1-matrix metalloproteinase (MT-1-MMP) and its metanephric developmental regulation with respect to MMP-2 and its inhibitor. *Kidney Int.* 54, 131–142.

Otey, C. A., Vasquez, G. B., Burridge, K., and Erickson, B. W. (1993). Mapping of the alpha-actinin binding site within the beta 1 integrin cytoplasmic domain *J. Biol. Chem.* 268, 21193–21197.

Paulsson, M., Deutzmann, R., Dziadek, M., Nowack, H., Timpl, R., Weber, S., and Engel, J. (1986). Purification and structural characterization of intact and fragmented nidogen obtained from a tumor basement membrane. *Eur. J. Biochem.* 156, 467–478.

Pelletier, J., Bruening, W., Li, F. P., Haber, D., Glaser, T., and Housman, D. E. (1991). WT1 mutations contribute to abnormal genital system development and hereditary Wilms' tumour. Nature 353,431–434

Poschl, E., Pollner, R, and Kuhn, K. (1988). The genes for the alpha 1(IV) and alpha 2(IV) chains of human basement membrane collagen type IV are arranged head-to-head and separated by a bidirectional promoter of unique structure. *EMBO.*7, 2687–2695

Putaala, H., Soininen, R., Kilpelainen, P., Wartiovaara, J., and Tryggvason, K. (2001). The murine nephrin gene is specifically expressed in kidney, brain and pancreas: Inactivation of the gene leads to massive proteinuria and neonatal death. *Hum. Mol. Genet.* 10, 1–8.

Reeves, W., Caulfield, J. P., and Farquhar, M. G. (1978). Differentiation of epithelial foot processes and filtration slits. Sequential appearance of occluding junctions, epithelial polyanion, and slit membranes in developing glomeruli. *Lab. Invest.* 39, 90–100.

Reeves, W. H., Kanwar, Y. S., and Farquhar, M. G. (1980). Assembly of the glomerular filtration surface. Differentiation of anionic sites in glomerular capillaries of newborn rat kidney. *J. Cell Biol.* **85**, 735–753.

Reinhardt, D. P., Sasaki, T., Dzamba, B. J., Keene, D. R., Chu, M. L., Gohring, W., Timpl, R., and Sakai, L. Y. (1996). Fibrillin-1 and fibulin-2 interact and are colocalized in some tissues. *J. Biol. Chem.* **271**, 19489–19496.

Reponen, P., Sahlberg, C., Huhtala, P., Hurskainen, T., Thesleff, I., and Tryggvason, K. (1992). Molecular cloning of murine 72-kDa type IV collagenase and its expression during mouse development. *J. Biol. Chem.* **267**, 7856–7862.

Robert, B. and Abrahamson, D.R. (2001). Control of glomerular capillary development by growth factor/receptor kinases. *Pediatr. Nephrol.* **16**, 294–301.

Robert, B., St. John, P.L., and Abrahamson, D.R. (1998). Direct visualization of renal vascular morphogenesis in Flk1 heterozygous mutant mice. *Am. J. Physiol.* **275**, F164–F172.

Robert, B., St. John, P.L., Hyink, D.P., and Abrahamson, D.R. (1996). Evidence that embryonic kidney cells expressing flk-1 are intrinsic, vasculogenic angioblasts. *Am. J. Physiol.* **271**, F744–F753.

Robert, B., Zhao, X. and Abrahamson, D.R. (2000). Coexpression of neuropilin-1, Flk1, and VEGF(164) in developing and mature mouse kidney glomeruli. *Am. J. Physiol. Renal. Physiol.* 279, F275–F282.

Ruotsalainen, V., Ljungberg, P., Wartiovaara, J., Lenkkeri, U., Kestila, M., Jalanko, H., Holmberg, C., and Tryggvason, K. (1999). Nephrin is specifically located at the slit diaphragm of glomerular podocytes. *Proc. Natl. Acad. Sci. USA* **96**, 7962–7967.

Ruotsalainen, V., Patrakka, J., Tissari, P., Reponen, P., Hess, M., Kestila, M., Holmberg, C., Salonen, R., Heikinheimo, M., Wartiovaara, J., Tryggvason, K., and Jalanko, H. (2000). Role of nephrin in junction formation in human nephrogenesis. *Am. J. Pathol.* **157**, 1905–1916.

Rupp, F., Ozcelik, T., Linial, M., Peterson, K., Francke, U., and Scheller, R. (1992). Structure and chromosomal localization of the mammalian agrin gene. *J. Neurosci.* **12**, 3535–3544.

Sado, Y., Kagawa, M., Naito, I., Ueki, Y., Seki, T., Momota, R., Oohashi, T., and Ninomiya Y (1998). Organization and expression of basement membrane collagen IV genes and their roles in human disorders. *J. Biochem.* **123**, 767–776.

Saga, Y., Yagi, T., Ikawa, Y., Sakakura, T., and Aizawa, S. (1992). Mice develop normally without tenascin. *Genes Dev.* **6**, 1821–1831.

Sanes, J. R. (1995). The synaptic cleft of the neuromuscular junction. *Semin. Dev. Biol.* **6**, 163–173.

Sanes, J. R., Engvall, E., Butkowski, R., and Hunter, D. D. (1990). Molecular heterogeneity of basal laminae: isoforms of laminin and collagen IV at the neuromuscular junction and elsewhere. *J. Cell Biol.* **111**, 1685–1699.

Sappino, A. P., Huarte, J., Vassalli, J. D., and Belin D (1991). Sites of synthesis of urokinase and tissue-type plasminogen activators in the murine kidney. *J. Clin. Invest.* **87**, 962–970.

Sariola, H., Ekblom, P., Lehtonen, E., and Saxen, L. (1983). Differentiation and vascularization of the metanephric kidney grafted on the chorioallantoic membrane. *Dev. Biol.* **96**, 427–435.

Sariola, H., Timpl, R., von der Mark, K., Mayne, R., Fitch, J. M., Linsenmayer, T. F., and Ekblom, P. (1984). Dual origin of the glomerular basement membrane. *Dev. Biol.* **101**, 86–96.

Savin, V. J., Sharma, R., Sharma, M., McCarthy, E. T., Swan, S. K., Ellis, E., Lovell, H., Warady, B., Gunwar, S., Chonko, A. M., Artero, M., and Vincenti, F. (1996). Circulating factor associated with increased glomerular permeability to albumin in recurrent focal segmental glomerulosclerosis. *N. Eng. J. Med.* **334**, 878–883.

Saxen, L. (1987). "Organogenesis of Kidney," pp. 1–173. Cambridge Univ. Press, Cambridge.

Schnabel, E., Anderson, J. M., and Farquhar, M. G. (1990). The tight junction protein ZO-1 is concentrated along slit diaphragms of the glomerular epithelium. *J. Cell Biol.* **111**, 1255–1263.

Schnabel, E., Dekan, G., Miettnen, A., and Farquhar, M. G. (1989). Biogenesis of podocalyxin: The major glomerular sialoglycoprotein in rat kidney. *Eur. J. Cell Biol.* **48**, 313–326.

Shalaby, F., Rossant, J., Yamaguchi, T. P., Gertsenestein, M., Wu, X. F., Breitman, M. L., and Schuh, A. C. (1995). Failure of blod-island formation and vasculogenesis in Flk-1-deficient mice. *Nature* **376**, 62–66.

Sharif, K., Goyal, M., Kershaw, D., Kunkel, R., and Wiggins, R. (1998). Podocyte phenotypes as defined by expression and distribution of Glepp1 in the developing glomeruli from MCD, CNF, and FSGS. *Exp. Nephrol.* **6**, 324–344.

Shih, N. Y., Li, J., Karpitskii, V., Nguyen, A., Dustin, M. L., Kanagawa, O., Miner, J. H., and Shaw, A. S. (1999). Congenital nephrotic syndrome in mice lacking CD2-associated protein. *Science* **286**, 312–315.

Shirato, I., Sakai, T., Kimura, K., Tomino, Y., and Kriz, W. (1996). Cytoskeletal changes in podocytes associated with foot process effacement in Masugi nephritis. *Am. J. Pathol.* **148**, 1283–1296.

Siebold, B., Qian, R. A., Glanville, R. W., Hofmann, H., Deutzmann, R., and Kuhn, K. (1987). Construction of a model for the aggregation and cross-linking region (7S domain) of type IV collagen based upon an evaluation of the primary structure of the alpha 1 and alpha 2 chains in this region. *Eur. J. Biochem.* **168**, 569–575.

Simon, E. E., and McDonald, J. A. (1990). Extracellular matrix receptors in the kidney cortex. *Am. J. Physiol.* **259**, F783–F792.

Somlo, S., and Mundel, P. (2000). Getting a foothold in nephrotic syndrome. *Nature Genet.* **24**, 333–335.

Smoyer, W. E., and Mundel, P. (1998). Regulation of podocyte structure the development of nephrotic syndrom. *J. Mol. Med.* **76**, 172–183.

Sonnenberg, A., Linders, C. J., Modderman, P. W., Damsky, C. H., Aumailley, M., and Timpl, R. (1990). Integrin recognition of different cell-binding fragments of laminin (P1, E3, E8) and evidence that alpha 6 beta 1 but not alpha 6 beta 4 functions as a major receptor for fragment E8. *J. Cell Biol.* **110**, 2145–2155.

Soriano, P. (1994). Abnormal kidney development and hematological disorders in platelet-derived growth factor receptor knock out mice. *Genes Dev.* **8**, 1888–1896.

Sorokin, L., Sonnenberg, A., Aumailley, M., Timpl, R., and Ekblom, P. (1990). Recognition of the laminin E8 cell-binding site by an integrin possessing the alpha 6 subunit is essential for epithelial polarization in developing kidney tubules. *J. Cell Biol.* **111**, 1265–1273.

Sorokin, L. M., Conzelmann, S., Ekblom, P., Battaglia, C., Aumailley, M., and Timpl, R. (1992). Monoclonal antibodies against laminin A chain fragment E3 and their effects on binding to cells and proteoglycan and on kidney development. *Exp. Cell Res.* **201**, 137–144.

Sorokin, L. M., Pausch, F., Durbeej, M., and Ekblom, P. (1997). Differential expression of five laminin alpha (1–5) chains in developing and adult mouse kidney. *Dev. Dyn.* **210**, 446–462.

St. John, P. L., Wang, R., Yin, Y., Miner, J. H., Robert, B., and Abrahamson, D. R. (2001). Glomerular laminin isoform transitions: Errors in metanephric culture are corrected by grafting. *Am. J. Physiol. Renal Physiol.* **280**, F695–F705.

St. John, P. L., and Abrahamson, D. R. (2001). Glomerular endothelial cells and podocytes jointly synthesize laminin-1 and -11 chains. *Kidney Int.* **60**, 1037–1046.

Stursberg, S., Riwar, B., and Jessberger, R. (1999). Cloning and characterization of mammalian SMC1 and SMC3 genes and proteins, components of the DNA recombination complexes RC-1. *Gene* **228**, 1–12.

Sugimoto, M., Oohashi, T., and Ninomiya, Y. (1994). The genes COL4A5 and COL4A6, coding for basement membrane collagen chains alpha 5(IV) and alpha 6(IV), are located head-to-head in close proximity on human chromosome Xq22 and COL4A6 is transcribed from two alternative promoters. *Proc. Natl. Acad. Sci. USA* **91**, 11679–11683.

Tashiro, K., Monji, A., Yoshida, I., Hayashi, Y., Matsuda, K., Tashiro, N., and Mitsuyama, Y. (1999). An IKLLI-containing peptide derived from the laminin alpha1 chain mediating heparin–binding, cell adhesion, neurite outgrowth and proliferation, represents a binding site for integrin alpha3beta1 and heparan sulphate proteoglycan. *Biochem. J.* **340**, 119–126.

Thorner, P. S., Zheng, K., Kalluri, R., Jacobs, R., and Hudson, B. G. (1996). Coordinate gene expression of the alpha3, alpha4, and alpha5 chains of collagen type IV: Evidence from a canine model of X-linked nephritis with a COL4A5 gene mutation. *J. Biol. Chem.* **271**, 13821–13828.

Timpl, R. (1996) Macromolecular organization of basement membranes. *Curr. Opin. Cell Biol.* **8**, 618–624.

Timpl, R., and Brown, J. C. (1996). Supramolecular assembly of basement membranes. *Bioessays* **18**, 123–132.

Truong, L. D., Pindur, J., Foster, S., Majesky, M., and Suki, W. N. (1996). Tenascin expression in nephrogenesis and in normal or pathologic glomerulus morphologic features and functional implications. *Nephron* **72**, 499–506.

Tryggvason, K. (1999). Unraveling the mechanisms of glomerular ultrafiltration: Nephrin, a key component of the slit diaphragm. *J. Am. Soc. Nephrol.* **10**, 2440–2445.

Tryggvason, K., Zhou, J., Hostikka, S. L., and Shows, T. B. (1993). Molecular genetics of Alport syndrome. *Kidney Int.* **43**, 38–44.

Tsen, G., Halfter, W., Kroger, S., and Cole, G. J. (1995). Agrin is a heparan sulfate proteoglycan. *J. Biol. Chem.* **270**, 3392–3399.

Turner, A. N., and Rees, A. J. (1996). Goodpasture's disease and Alport's syndromes. *Annu. Rev. Med.* **47**, 377–386.

Utani, A., Nomizu, M., and Yamada, Y. (1997). Fibulin-2 binds to the short arms of laminin–5 and laminin–1 via conserved amino acid sequences. *J. Biol. Chem.* **272**, 2814–2820.

van den Born, J., van den Heuvel, L. P., Bakker, M. A., Veerkamp, J. H., Assmann, K. J., and Berden, J. H. (1992). A monoclonal antibody against GBM heparan sulfate induces an acute selective proteinuria in rats. *Kidney Int.* **41**, 115–123.

van den Brule, F. A., Buicu, C., Sobel, M. E., Liu, F. T., and Castronovo, V. (1995). Galectin-3, a laminin binding protein, fails to modulate adhesion of human melanoma cells to laminin. *Neoplasma* **42**, 215–219.

Vanden Heuvel, G. B., and Abrahamson, D. R. (1993). Quantitation and localization of laminin A, B1 and B2 chain RNA transcripts in developing kidney. *Am. J. Physiol.* **265**, F293–F299.

Virtanen, I., Laitinen, L., and Korhonen, M. (1995). Differential expression of laminin polypeptides in developing and adult human kidney. *J. Histochem. Cytochem.* **43**, 621–628.

Vuolteenaho, R., Chow, L. T., and Tryggvason, K. (1990). Structure of the human laminin B1 chain gene. *J. Biol. Chem.* 265, 15611–15616.

Wang, R., Moorer-Hickman, D., St. John, P. L., and Abrahamson, D. R. (1998). Binding of injected laminin to developing kidney glomerular mesangial matrices and basement membranes *in vivo*. *J. Histochem. Cytochem.* **46**, 291–300.

Wharram, B. L., Goyal, M., Gillespie, P. J., Wiggins, J. E., Kershaw, D. B., Holzman, L. B., Dysko, R. C., Saunders, T. L., Samuelson, L. C., and Wiggins, R. C. (2000). Altered podocyte structure in GLEPP1 (Ptpro)-deficient mice associated with hypertension and low glomerular filtration rate. *J. Clin. Invest.* **106**, 1281–1290.

Whiteside, C., Prutis, K., Cameron, R., and Thompson, J. (1989). Glomerular epithelial detachment, not reduced charge density, correlates with proteinuria in adriamycin and puromycin nephrosis. *Lab. Invest.* **61**, 650–660.

Wiggins, R. C., Wiggins, J. E., Goyal, M., Wharram, B. L., and Thomas, P. E. (1995). Molecular cloning of cDNAs encoding human GLEPP1, a membrane protein tyrosine phosphatase: Characterization of the GLEPP1 protein distribution in human kidney and assignment of the GLEPP1 gene to human chromosome 12p12-p13. *Genomics.* **27**, 174–181.

Winn, M. P., Conlon, P. J., Lynn, K. L., Howell, D. N., Gross, D. A., Rogala, A. R., Smith, A. H., Graham, F. L., Bembe, M., Quarles, L. D., Pericak-Vance, M. A., and Vance, J. M. (1999). Clinical and genetic heterogeneity in familial focal segmental glomerulosclerosis. International Collaborative Group for the Study of Familial Focal Segmental Glomerulosclerosis. *Kidney Int.* **55**, 1241–1246.

Woolf, A. S., and Yuan, H. T. (2001). Angiopoietin growth factors and Tie receptor tyrosine kinases in renal vascular development. *Pediatr. Nephrol.* **16**, 177–184.

Wu, R.-R., and Couchman, J. R. (1997). cDNA cloning of the basement membrane chondroitin sulfate proteoglycan core protein, bamacan: A five domain structure including coiled-coil motifs. *J. Cell Biol.* **136**, 433–444.

Yurchenco, P. D., and Cheng, Y. S. (1993). Self-assembly and calcium-binding sites in laminin. A three-arm interaction model. *J. Biol. Chem.* **268**, 17286–17299.

Yurchenco, P. D., and O'Rear, J. J. (1994). Basal lamina assembly. *Curr. Opin. Cell Biol.* **6**, 674–681.

Yurchenco, P. D., Quan, Y., Colognato, H., Mathus, T., Harrison, D., Yamada, Y., and O'Rear, J. J. (1997). The alpha chain of laminin-1 is independently secreted and drives secretion of its beta- and gamma-chain partners. *Proc. Natl. Acad. Sci. USA* **94**, 10189–10194.

Yurchenco, P. D., and Schittny, J. C. (1990). Molecular architecture of basement membrane. *FASEB J.* 4, 1577–15790.

Zedlacher, M., Schmoll, M., Zimmermann, K., Horstkorte, O., and Nischt, R. (1999). Differential regulation of the human nidogen gene promoter region by a novel cell-type-specific silencer element. *Biochem. J.* 338, 343–350.

Zhou, J., and Reeders, S. T. (1996). The alpha chains of type IV collagen. *Contr. Nephrol.* **117**, 80–104.

17

Development of Kidney Blood Vessels

Adrian S. Woolf, and Hai T. Yuan

The adult mammalian kidney is a highly vascular organ, receiving 20% of the cardiac output. This chapter focuses on the anatomy of developing kidney vessels, including the genesis of renal arteries, glomerular capillaries, and the vasa recta microcirculation, in the most often used experimental model, the mouse metanephros. Studies have been performed to address the origin of metanephric vessels, and some experimental evidence supports the existence of both angiogenesis, the ingrowth of capillaries into the embryonic organ, and vasculogenesis, the *in situ* differentiation of endothelia. Diverse vascular growth factors are expressed in the developing kidney, and these molecules may direct the

growth of renal blood vessels: they include vascular endothelial growth factor (VEGF) and the angiopoietins, which signal through receptor tyrosine kinases expressed by endothelial precursors. Less is known about cell adhesion molecules and transcription factors in the context of metanephric blood vessel development, although these classes of molecule are certainly important in vessel formation elsewhere in the embryo.

I. Introduction

Several informative reviews have been published regarding the general aspects of construction of embryonic blood vessels (Hungerford and Little, 1999; Gale and Yancopoulos, 1999; Daniel and Abrahamson, 2000; Carmeliet, 2000). In this chapter, after a brief summary of general aspects of embryonic vessel development, our specific focus is on the morphogenesis of the renal vasculature. Specifically, we first review the anatomy of developing kidney vessels in the mouse metanephros, and then address two related questions: (1) what is the origin of renal vessels and (2) which molecules control the differentiation and proliferation of these structures? To date, several studies have investigated the expression and roles of vascular growth factors in the developing kidney, and hence the content of the current chapter reflects this information. Much less is known about cell adhesion molecules and transcription factors in this context, although these classes of molecule are certainly important in vessel formation elsewhere in the embryo.

II. Blood Vessel Formation in the Embryo

A. The Two Modes of Embryonic Blood Vessel Formation Vasculogenesis and Angiogenesis

During development, there are two ways by which endothelial cells are known to be formed (Carmeliet, 2000; Pardanaud et al., 1996). In vasculogenesis, or in situ vessel formation, "naïve" mesenchymal precursor cells differentiate to acquire an endothelial phenotype, characterized by the expression of proteins such as platelet endothelial cell adhesion molecule (PECAM; also known as CD31), CD34, and factor VIII-related antigen (von Willebrand factor), as well as the ability to take up acetylated low-density lipoprotein. During this maturation process, a characteristic series of receptor tyrosine kinases are expressed on the cell surface. As discussed in detail later, these bind to, and transduce signals from, locally produced vascular growth factors, hence mediating the survival, proliferation, and differentiation of the endothelial lineage.

During the process of angiogenesis, in contrast, endothelial cells sprout or delaminate from existing vessels after their systematic destabilization and then migrate to form new vessels elsewhere. Although vasculogenesis had classically been considered to be limited to the embryonic period, it has become apparent that bone marrow-derived cells, which have the phenotype of endothelial cell precursors, circulate in the bloodstream postnatally and can contribute to sites of new vessel formation, e.g., during recovery from vascular injury (Murohara et al., 2000).

B. Experimental Models Used to Study Embryonic Vessel Formation

The first embryonic endothelial cells, including those of the yolk sac and heart primordium, must, by necessity, form by vasculogenesis. Later in gestation, although the results of descriptive studies may be interpreted as being consistent with either vasculogenesis or angiogenesis, such studies cannot alone definitively assign vascular growth to either mechanism. If, however, avascular tissue can be transplanted into a site where it will continue to differentiate and if the donor and host tissue can be distinguished, then the origin of endothelial cells can be established.

Avian vessel formation has been studied in transplants between quail and chick. These cell lineage experiments are feasible because avian embryos can be manipulated in ovo and because quail cells have a characteristic nucleolar phenotype and there is a quail-specific endothelial cell marker, QH1 (Pardanaud et al., 1996). Such studies have demonstrated that vessels in tissues derived from splanchnopleure (mesoderm and endoderm), such as the liver, arise by vasculogenesis. In contrast, vessels in the limb and brain, structures that derive

from the somatopleure (mesoderm and ectoderm), form by sprouting angiogenesis in which migratory endothelia invade lateral plate and head mesoderm. In the context of this chapter, it is interesting to note that avian mesonephric vessels originate by angiogenesis.

Unfortunately, the lack of markers for mammalian endothelial cell precursors, together with difficulties in embryo manipulation as compared to avian species, until recently precluded comparable analyses in mice or rats.

C. Further Maturation of Embryonic Vessels into Capillary Networks, Arteries and Veins

In both vasculogenesis and angiogenesis, individual cells interact to form primitive capillary tubes, which are later subject to remodeling by a balance of further formation of new vessels, coalescence of small vessels into larger channels, and apoptosis-mediated regression of parts of the network. In some of these vessels, further differentiation leads to the formation of specialized arteries or veins. It is now thought that demarcation of precursors into arterial and venous domains is determined, at least in part, by two-way signaling through ephrin ligands and Eph receptors (Adams et al., 1999).

During the formation of arteries, smooth muscle cells are integrated into the vessel wall (Hungerford and Little, 1999; Carmeliet, 2000). Hungerford and Little (1999) have reviewed the ontogeny of vascular smooth muscle, which differentiates as "fibroblast-like" cells, which lack myofilaments and basement membranes and appear in aggregates near embryonic endothelium. In avian embryos, where lineage studies have been performed, aortic arch vascular smooth muscle cells derive from the neural crest (i.e., ectoderm), whereas vascular smooth muscles in the rest of the embryo have been considered to originate from the mesoderm. An additional alternative mechanism was proposed by De Ruiter et al. (1999), who showed that embryonic avian endothelia in the dorsal aorta could transdifferentiate into mesenchymal cells that expressed myofilaments. Once recruited into the lineage, vessel wall precursors express marker proteins for smooth muscle maturation. One such structural protein is heavy caldesmon (Frid et al., 1992), whereas another is a smooth muscle actin (αSMA), which constitutes up to 40% of total protein in mature cells (Hungerford and Little, 1999).

III. Anatomy of Kidney Blood Vessels

A. Overview of Structure of the Vasculature of the Mature Kidney

In adult mammals, approximately 20% of the cardiac output flows through the kidneys: in adult humans, for instance, a liter of blood is delivered to the kidneys every minute. The great majority of this flow is targeted via a branching arterial system to approximately one million glomeruli located in the renal cortex.

The mature vasculature of the kidney consists of four main types of vessel (Tisher and Madsen, 1996; Risdon and Woolf, 1998).

1. Arteries, which act as conduits for incoming blood and consist of an endothelial cell lining surrounded by multiple layers of vascular smooth muscle cells.
2. Veins, which are low-resistance vessels through which blood flows out of the organ.
3. Capillaries, which comprise microscopic channels. There are three specialized renal capillary microcirculations: (a) the glomerular capillary loops, (b) the cortical capillary labyrinth surrounding proximal tubules, and (c) the vasa rectae located in the medulla of the organ.
4. Lymphatic channels: these have been little studied.

In each human kidney, the renal artery, a branch of the abdominal aorta, enters at the hilum and branches into inter-lobar arteries deep within the organ (Chapter 11). At the junction of the renal medulla and cortex, interlobar arteries branch into arcuate arteries, which are arranged at right angles to the interlobar branches. The arcuate arteries themselves branch into small cortical interlobular arteries, which terminate in afferent glomerular arterioles. The larger arteries, to the level of interlobular vessels, are accompanied by renal veins.

In the mature kidney, the wall of the arteriole immediately proximal to the glomerulus is modified to form a structure called the juxtaglomerular apparatus: this secretes renin, a hormone important in blood pressure homeostasis (Reddi et al., 1998). Within the glomeruli, blood passes at high pressure through the glomerular capillaries, where a plasma ultrafiltrate is generated, traversing the slit diaphragms of glomerular podocytes (also called visceral epithelial cells) and entering Bowman's space. This is the first step in the formation of urine. Endothelial cells in glomerular capillaries are "fenestrated" by microscopic pores, which allow unimpeded transfer of water and solutes, and they are supported by mesangial cells, a specialized form of pericyte or smooth muscle cell (Dubey et al., 1997). During development, both the glomerular endothelia and the adjacent podocytes contribute to the formation of the intervening glomerular basement membrane, described in detail in Chapter 16.

Blood leaving the glomerulus drains into the efferent glomerular arteriole. The diameter of this vessel is variable, with angiotensin II an important vasocontrictor, and this variation can modulate hydrostatic pressure within the glomerular capillary bed and hence the rate of plasma ultrafiltration. Blood draining from more superficial glomeruli enters the cortical capillary microcirculation, a fenestrated network around the proximal tubules. These vessels receive water and solutes, which have been reclaimed from the glomerular filtrate by the proximal tubules.

However, blood draining from deeper, juxtamedullary, glomeruli enter the descending vasa rectae, vessels that travel into the medulla. These "capillaries" have some arteriole-like features, as their proximal sections are surrounded by pericytes, which are modified smooth muscle cells (Park et al., 1997). At their deepest points in the medulla, the descending vasa rectae vessels feed into ascending vasa rectae; these are true capillaries with fenestrated walls. The vasa rectae are intimately associated with descending and ascending loops of Henle, as well as collecting ducts; together, these structures comprise a functional unit involved in urine concentration, with reabsorbed water entering the ascending vasa rectae.

B. Vascular Anatomy of the Mouse Metanephric Kidney

Although the differentiation of embryonic kidney vessels has begun to be studied in relatively "simple" animals such as fish (Majumdar and Drummond, 1999; Majumdar et al., 2000), amphibians (Tufro-McReddie et al., 1995), and birds (Pardanaud et al., 1996), most work in this area has focused on the mammalian kidney. While studies on the development of kidney vessels in humans have only been descriptive to date (Risdon and Woolf, 1998; Kaipainen et al., 1993; Simon et al., 1995), the mouse provides a precisely staged model of renal vascular development, which can be manipulated in vivo and in vitro, that is most likely little different from that of the human.

At mouse embryonic day 11 (E11), approximately equivalent to 5 weeks after fertilization in humans and E12–13 in rats, the metanephros is formed when nephrogenic mesenchyme, a distal blastema of intermediate mesoderm, condenses around ureteric bud epithelium, a branch of the mesonephric, or Wolffian, duct. At this stage, the metanephric mesenchyme is avascular as assessed by light and electron microscopy, although capillaries are detected in the loose connective tissue between the early rudiment and the mesonephric duct and on the outer perimeter of the renal mesenchyme (Loughna et al., 1996, 1997). A day later, at E12, when the mouse ureteric bud has branched a few times, capillary-sized vessels are detected on the perimeter of the organ and around the stalk of the ureteric bud in the metanephric hilum (Figs. 17.1A and 17.1B; Figs. 17.2A and 17.2B). Apoptosis is prominent around hilar capillaries in the early mouse metanephros, and it is likely that there is much remodeling of this primitive vasculature (Fig. 17.2A).

Between E13 and E15, approximately equivalent to weeks 8–10 of human gestation, the metanephric vasculature undergoes a considerable increase in complexity, as depicted

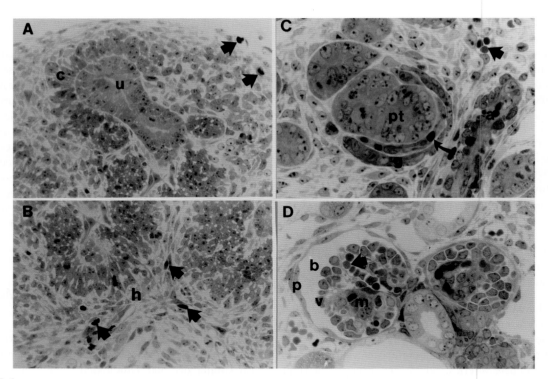

Figure 17.1 The anatomy of mouse metanephric vessel formation as assessed by light microscopy. (A) The cortex of an E12 metanephros contains condensing mesenchyme (c) adjacent to a ureteric bud branch tip (u). There are a few patent capillaries (arrows) located in the periphery of the organ; these are seen easily when their lumens contain red blood cells. (B) The hilum (h) of the same E12 organ depicted in A. Note the loose network of capillaries (arrows) in this region. (C) One day later, at E13, the first S-shaped bodies, or nephron precursors, have formed. These comprise a primitive proximal tubule (pt) and a crescent of epithelial cells, which will differentiate into glomerular epithelia (g); between these structures are located endothelial precursors (curved arrow), which differentiate into glomerular capillaries. Note a nearby arterial structure (a) and capillaries (straight arrow). (D) By E14–15, the first glomeruli with primitive capillary loops (arrow) have formed. Note the Bowman's capsule of parietal epithelium (p), the Bowman's space (b), the podocytes, or visceral epithelium (v), and the core of mesangial cells (m).

by comparison of Figs. 17.3A with 17.3B. By E13, patent capillaries surround nephron precursors (Fig. 17.3A), the most mature of which have formed S-shaped bodies, each with a few endothelial cells in the cleft between maturing podocytes and the primitive proximal tubule (Fig. 17.1C; Fig. 17.4A). Up to this stage, metanephric vessels form loose networks that express markers such as CD31 and VEGF receptors, but show no clear morphological differentiation into arterial or venous structures.

Around E13–14 a single renal artery can be seen, running from the abdominal aorta to the hilum of the metanephros (Yuan *et al.*, 2000a). After entering the kidney, this vessel branches into smaller arteries (Fig. 17.1C; Figs. 17.2C and 17.D; Fig. 17.5A), which run to the corticomedullary junction where they form arcades. From these structures, small cortical arteries branch and terminate in afferent glomerular arterioles (Figs. 17.5B and 17.5C). From this stage onward, αSMA is widely expressed in the arterial vessel walls of the developing kidney (Figs. 17.5B and 17.5C) (Yuan *et al.*, 2000a; Carey *et al.*, 1992), and prominent intracellular fibrils are evident by electron microscopy (Fig. 17.2D).

From mouse E15, the most mature glomerular structures, those located toward the center of the metanephros, begin to acquire capillaries and, as their loops mature, are supported by mesangial cells (Fig. 17.4B). At birth, about E20–21, the mouse kidney contains around 1.5×10^3 glomeruli, but new layers of glomeruli are generated in the first postnatal week, resulting in approximately 10×10^3 glomeruli in the mature organ (Yuan *et al.*, 2000a). In contrast to the development of main renal arterial branches and the formation of glomeruli, the mouse vasa recta microcirculation does not become well differentiated until after birth. During the first three postnatal weeks, the deep medulla grows considerably with the maturation of loops of the descending and ascending loops of Henle that closely flank ascending and descending vasa rectae (Figs. 17.6A and 17.6B) (Yuan *et al.*, 2000a).

In the normal adult kidney, there is thought to be a very low endothelial cell turnover, although both proliferation and apoptosis are upregulated in certain models of acute and chronic nephropathies (Kitamura *et al.*, 1998; Ostendorf *et al.*, 1999; Pillebout *et al.*, 2001; Long *et al.*, 2001; Yuan *et al.*, 2002).

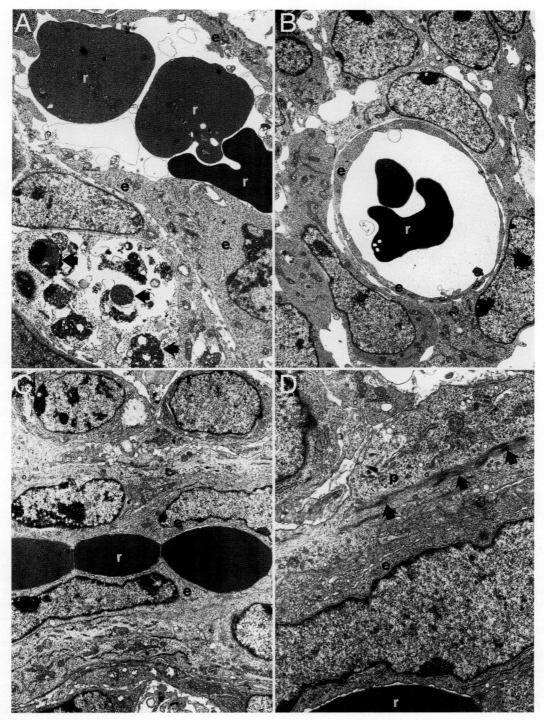

Figure 17.2 Anatomy of mouse metanephric vessel formation as assessed by electron microscopy. (A) In the hilum of the E12 metanephros, endothelial cells (e) form capillaries, which contain red blood cells. This area is subject to remodeling, with prominent apoptosis (arrows) of adjacent cells. Red blood cells are labeled 'r'. (B) The peripheral cortex of the organ depicted in A contains thin-walled capillaries surrounded by undifferentiated mesenchyme. (C) By E14, the metanephros contains arterial structures in which endothelial cells appear to become enveloped by surrounding cells, which are pericyte/smooth muscle precursors. (D) A higher power view of C in which the lumen of the arterial vessel is lowermost. An endothelial cell (e) is adjacent to a pericyte (p), which contains dense filaments (arrows) representing cytoskeletal elements characteristic of the smooth muscle lineage.

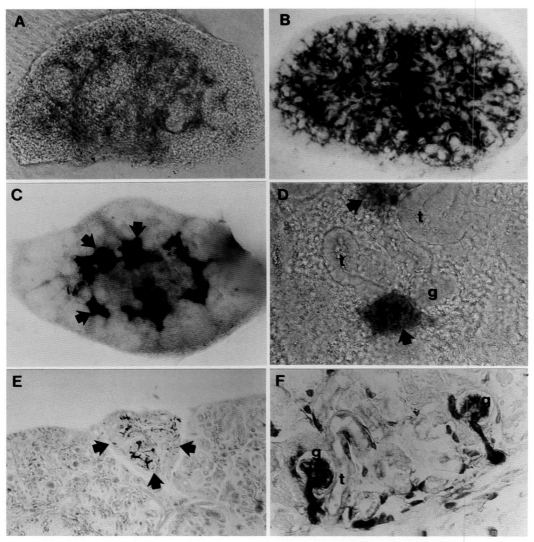

Figure 17.3 Mouse metanephric endothelia as assessed by expression of a Tie-1/LacZ transgene. The transgene is expressed by all metanephric endothelia cells and is detected in tissues by the X-gal reaction, which produces a blue color. (A) Promoter activity (blue) was detected *in vivo* at E13 *in vivo* in loose networks of vessels around nascent nephrons; glomeruli have yet to form at this stage. (B) At E15 *in vivo*, Tie-1 expression is markedly upregulated in the forming renal arterial system, as well as in glomerular capillaries. (C) When the E13 organ depicted in A was grown in hypoxic organ culture for a few days, the normal pattern of vessel branching was lost, yet masses of Tie-1-expressing endothelial cells (arrows) formed between tubules. (D) Histology of the organ depicted in C. Unstructured endothelial cell masses (arrows) were located near tubules (t). Glomeruli (g) remained avascular. (E) Implantation of an avascular E11 Tie-1/LacZ metanephros into the nephrogenic cortex of a wild-type mouse; 1 week after surgery, the transplant (arrowed) has differentiated alongside the host kidney. (F) Histology of the transplant depicted in E. Note that endothelial cell precursors derived from the donor have differentiated *in situ* and undergo normal patterning to form glomerular (g) arterioles and capillaries, as well as interstitial vessels located between tubules (t). For original experiments, see Loughna *et al.* (1997, 1998).

IV. Experiments that Address the Origins of Metanephric Blood Vessels

A. Studies that Support Metanephric Angiogenesis

With regard to murine metanephric vessel development, some experimental data available to date are compatible with both angiogenesis and vasculogenesis (Woolf and Loughna, 1998). Results from two classical sets of experiments have been used to support the case for metanephric angiogenesis.

First, although the explanted murine metanephros undergoes considerable epithelial morphogenesis and differentiation in organ culture, the glomeruli that form in this context appear to be avascular (Bernstein *et al.*, 1981). This argues against the presence of endothelial cell precursors in the rudiment.

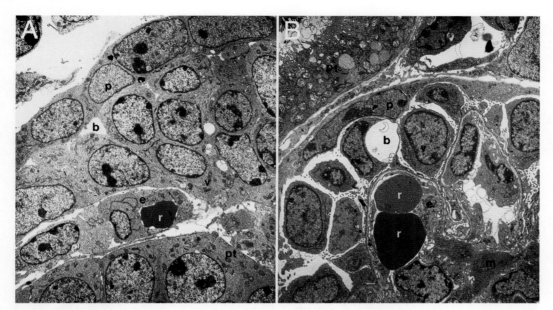

Figure 17.4 Mouse metanephric glomerular vessels as assessed by electron microscopy. (A) An S-shaped body in an E13 metanephros is depicted. Note the primitive proximal tubule (pt) and a crescent of epithelial cells, which will differentiate into glomerular parietal (p) and visceral (v) epithelia (or åpodocytes), separated by the Bowman's space (b). Between these structures are located endothelia (e), which will differentiate into glomerular capillaries. Note the red cell (r) in the forming capillary. (B) By E14, the first glomeruli with endothelia (e) forming primitive capillary loops are detected. Note the Bowman's capsule of parietal epithelium (p), the Bowman's space (b), the visceral epithelium (v), and the core of mesangial cells (m). A proximal tubule (pt) with apical microvilli lies adjacent to the glomerulus.

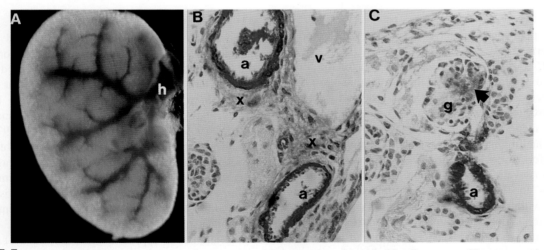

Figure 17.5 Arterial system of the mouse metanephros as assessed by expression of an Ang-2/LacZ transgene. The transgene is expressed by metanephric precursor cells, which form vascular pericytes and smooth muscle, and is detected in tissues by the X-gal reaction, which produces a blue color. (A) Whole mount of a neonatal kidney showing the main renal artery at the hilum (h) of the organ; this divides serially into major renal arteries and arterioles. (B) Histology of the organ depicted in A, immunostained with an antibody against αSMA (red); note that the walls of arteries (a) deep in the cortex have an inner layer of αSMA-rich smooth muscle cells surrounded by Ang-2-expressing cells (x); the latter cells may represent pericyte/smooth muscle cell precursors. (C) A glomerulus in the same organ depicted in A. Note the Ang-2-expressing mesangial cell core of the glomerulus; these cells are related to the pericyte lineage. A small arteriole (a) runs nearby the glomerulus. For original data, see Yuan *et al.*, 2000a.

Second, when the avascular mouse metanephros is transplanted onto the quail chorioallantoic membrane, the glomeruli that develop are invaded by the avian host endothelia, which can be identified by their characteristic nucleolar marker (Sariola *et al.*, 1983). In similar experiments in which mouse metanephric kidneys were transplanted into chick hosts, staining with a species-specific antiserum to collagen IV, a component of basement membranes, led to a similar conclusion (Sariola *et al.*, 1984). An inference from this experiment is that mesenchymal or ureteric bud-derived cells produce growth factors, which summon endothelial cells from distant sites,

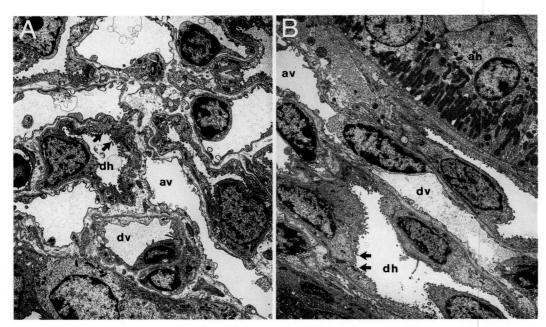

Figure 17.6 Maturing mouse vasa rectae as assessed by electron microscopy. A 3 week postnatal organ is shown. A is a transverse section and B is a longitudinal section through the vasa recta area in the outer medulla. Descending vasa rectae capillaries (dv) are surrounded by pericytes (p) whereas ascending versa recta (av), which receive reabsorbed water, are characterized by fenestrated endothelia without pericytes. These vessels are adjacent to descending limbs of loops of Henle (dh), which comprise epithelia with thin cytoplasm and sparse microvilli (arrows), and to ascending limbs of loops of Henle (ah), which comprise thicker epithelial cells rich in mitochondria.

and hence that angiogenesis is the mode of glomerular capillary formation.

It could, however, be argued that neither of the aforementioned experiments provides the correct milieu for renal vasculogenesis. The metanephros is conventionally cultured in a 20% oxygen atmosphere (i.e., air) and fed by serum-containing media and it is conceivable that oxygen-rich conditions and inhibitory factors in serum might prevent *in situ* vessel formation. Furthermore, even if some endothelial cells were to develop in organ culture, they could be difficult to identify in the absence of definitive markers for precursors within the lineage. Finally, in the case of avian work, chorioallantoic membrane vessels may be intrinsically highly invasive, and clearly the culture of mammalian kidney in this location represents a highly contrived scenario.

B. Studies That Support Metanephric Vasculogenesis

More recent studies have used the expression of endothelial cell-specific receptor tyrosine kinases as metanephric vascular markers, which are discussed in detail here. These experiments have confirmed that endothelial cells do not thrive when metanephric explants are grown in a standard, normoxic (i.e., 20% O_2) atmosphere in defined media (Robert *et al.*, 1998; Tufro-McReddie *et al.*, 1997;

Tufro *et al.*, 1999). However, when cultured in hypoxia, metanephric explants upregulate the expression of VEGF, and endothelial cell proliferation is enhanced within the explant (Tufro-McReddie *et al.*, 1997), whereas the addition of VEGF to rat explants, and angiopoietin 1 to mouse explants, grown in air also enhances vessel formation (Tufro *et al.*, 1999; Kolatsi-Joannou *et al.*, 1997). Similar conclusions, regarding the endogenous origin of glomerular capillaries, were reached by Hyink and colleagues (1996) after transplantation and differentiation of murine metanephroi into rat anterior eye chambers and subsequent analysis with species-specific antisera to basement membrane components.

An alternative approach was followed by Loughna and colleagues (1997, 1998), who exploited a Tie-1/LacZ transgenic mouse to follow kidney endothelial cell development. This anatomically normal strain expresses LacZ (bacterial β-galactosidase) driven by the Tie-1 receptor tyrosine kinase promoter in endothelial cell precursors and their mature derivatives, thus allowing gene activity to be visualized *in situ* using a simple enzymatic (X-gal) color reaction (Fig. 17.3). Although networks of Tie-1/LacZ-expressing capillaries are detected around developing nephrons at E13 (Fig. 17.3A), these delicate vessels regress when organs are cultured in air for a few days. In contrast, glomeruli from E15 kidneys that have developed *in vivo* have prominent Tie-1 expressing capillary loops (Fig. 17.3B). When kidney rudiments are explanted in hypoxic conditions, Tie-1 expression is maintained in large, but relatively unstructured, masses of endothelial cells located between tubules (Figs. 17.3C and 17.3D) (Loughna et al., 1998). Strikingly, however, when avascular E11 Tie-1/LacZ metanephric kidneys are implanted into the nephrogenic cortex of wild-type neonatal kidneys (Loughna et al., 1997), a site allowing the differentiation of nephrogenic precursors into filtering glomeruli (Woolf et al., 1990), transgene-expressing glomerular and stromal capillaries develop within transplants (Figs. 17.3E and 17.3F). Conversely, in experiments in which wild-type metanephroi were implanted into Tie-1/LacZ neonates, transgene expression in implants was minimal.

This work demonstrates the possibility of endothelial cell differentiation from Tie-1-positive precursors present at the inception of the metanephros. The results, using Tie-1 as a marker for endothelial cell development, generally support the conclusions about a possible vasculogenic origin of kidney capillaries based on organ culture and in oculo transplantation studies using VEGF receptor tyrosine kinases as markers (Robert et al., 1998; Tufro-McReddie et al., 1997; Tufro et al., 1999).

Collectively, these experiments suggest that endothelial cell precursors are indeed present in the early metanephros and that their growth is enhanced by VEGF. However, it remains to be demonstrated whether these precursors represent "naïve" renal mesenchymal cells, which have begun to enter the endothelial lineage, or whether they are scattered angioblasts that have invaded the organ at the same time that the intermediate mesoderm is condensing to form renal mesenchyme. It should also be noted that as neither the medullary (vasa rectae) microcirculation nor large vessels develop in organ culture or within transplanted rudiments, the origins of these structures thus remain unknown.

V. Growth Factors and Embryonic Kidney Vessel Development

A. Endothelial Growth Factors

In the past decade, evidence has emerged that diverse growth factors, generally signaling through cell surface receptors, including receptor tyrosine kinases, are critical for epithelial nephron formation and collecting duct maturation (Woolf and Cale, 1997). In fact, some of the same "epithelial" growth factors expressed in the metanephros could also directly affect the growth of endothelial cells. These include fibroblast growth factor 2 (Kloth et al., 1998), hepatocyte growth factor (Grant et al., 1993; Woolf et al., 1995), and transforming growth factor β1 (Dickson et al., 1995; Liu et al., 1999), as well as other secreted molecules (Oliver and Al-Awqati, 1998). Hence, such secreted factors could have effects on multiple metanephric cell lineages.

Indirect effects probably explain the observation made by Kloth and Suter-Crazzolara (2000) that capillary growth in cultures of rabbit nephrogenic cortex could be enhanced by the addition of glial cell line-derived neurotrophic factor, a prototypic morphogen with a direct action on the ureteric bud (Towers et al., 1998).

Elsewhere in the embryo, there are two main growth factor signaling systems that show specificity for the endothelial cell lineage. These are mediated by VEGF and angiopoietin (Ang) growth factors, and evidence now implicates these molecules in renal blood vessel growth (Gale and Yancopoulos, 1999; Carmeliet, 2000). The complex ephrin/Eph signaling system is also likely to be important in metanephric vessel development (Daniel et al., 1996; Adams et al., 1999), but is not further described further here.

B. The VEGF Axis in Metanephric Development

VEGF is a major enhancer of angiogenesis and vasculogenesis in health and disease (Carmeliet, 2000). There are four secreted isoforms, $VEGF_{206}$, $VEGF_{189}$, $VEGF_{165}$ and $VEGF_{121}$, produced by differential mRNA splicing. $VEGF_{165}$ and $VEGF_{121}$ are soluble, whereas $VEGF_{206}$ and $VEGF_{189}$ are associated with cell surface heparan sulfate proteoglycans. The mature factors exist as dimeric glycoproteins of 34–43 kDa. VEGF induces the proliferation and sprouting of adult endothelial cells and also increases capillary permeability, at least partly by forming fenestrae (Roberts and Palade, 1995; Esser et al., 1998), hence its alternative name, vascular permeability factor. During embryogenesis, VEGF is expressed by a wide variety of nonendothelial cells in locations where endothelia are proliferating. VEGF is also expressed by macrophages and various tumor cells, indicating a role in the angiogenesis of wound healing and tumor growth.

VEGF receptor 2 (VEGFR2, also know as flk-1, the murine homologue of human KDR) and VEGFR1 (also known as flt-1) belong to class III of the receptor tyrosine kinase superfamily. They are expressed predominantly by the endothelial lineage, although hematopoietic precursors also express VEGFR2. VEGFR2 undergoes autophosphorylation on binding to VEGF and transduces survival and differentiation signals into endothelial cell precursors (Millauer et al., 1993), whereas VEGFR1 may function mainly to "mop up" extra VEGF and hence fine-tune VEGF-induced signaling (Hiratsuka et al., 1998). The expression of VEGF and its receptors is enhanced by hypoxia by a variety of mechanisms (Levy et al., 1995; Brogi et al., 1996; Tuder et al., 1995).

The expression patterns of VEGF and its receptors during embryogenesis are consistent with a paracrine role for the factor in vasculogenesis and angiogenesis (Miquerol et al., 1999), an impression confirmed by the phenotypes of homozygous null mutant mice. VEGFR2-deficient mice die around E9 due to defective vasculogenesis (Shalaby *et al.*, 1995). Interestingly, despite the lack of embryonic endothelial cells, a LacZ reporter gene that had been "knocked into" the VEGFR2 locus was expressed in subsets of mesenchymal cells around E9; these may represent precursors that have been "earmarked" to enter the endothelial lineage. An impaired generation of hematopoietic progenitors was also reported here, supporting the hypothesis that endothelial and red blood cells have a common precursor, the "hemangioblast". A different phenotype, however, was detected for mice homozygous for a VEGFR1 null mutation (Fong *et al.*, 1995; Fong *et al.*, 1999). In this model, vessels were present but they were large and malformed. Both VEGFR2 and VEGFR1 mouse null mutants die *in utero* before the inception of the metanephros.

Mice expressing LacZ driven by a VEGFR2 promoter have also been used to study the normal development of mouse kidney vasculature. On late E10, as metanephric mesenchyme condenses from intermediate mesoderm, a VEGFR2-expressing network of cells surrounds the blastema. At E11, a transgene expressing vessel is located on the ventral aspect alongside the ureteric bud, at a stage when aortic branches have yet to connect with the metanephros itself (Robert *et al.*, 1998). VEGF receptor transcripts are expressed by E11 mouse renal mesenchymal tissues and subsequently by all metanephric endothelial cells (Loughna *et al.*, 1997, 1998; Tufro-McReddie *et al.*, 1997; Tufro *et al.*, 1999). Furthermore, at the onset of mouse nephrogenesis, VEGF transcripts could be detected in both renal mesenchyme and ureteric bud (Loughna *et al.*, 1997). Clonal mesenchymal cell lines isolated from the avascular mouse E11 metanephros express either VEGF alone or both the ligand and its receptors (Loughna *et al.*, 1997), but these cells lack the characteristics of mature endothelial cells. The expression of VEGF and its receptors clearly demonstrates that the molecular machinery to support angio-and/or vasculogenesis is present in the renal mesenchyme from the onset of nephrogenesis. Of note, VEGF-soaked beads implanted into the cortex of neonatal mouse kidneys, a site where new nephrons are generated, elicit a dense network of capillaries (Fig. 17.7).

High levels of VEGF transcripts are expressed in nascent glomeruli (Breier *et al.*, 1992). Studies using genetic and immunological strategies to block VEGF *in vivo* have led to the conclusion that this factor is critical for glomerular capillary growth at later stages of nephrogenesis (Kitamoto *et al.*, 1998; Gerber *et al.*, 1999; Carmeliet *et al.*, 1999). Intriguingly, podocytes continue to express VEGF after nephrogenesis is complete (Brown *et al.*, 1992), even though endothelial cell proliferation is exceeding low in this location in adults (Kitamura *et al.*, 1998). There is emerging evidence that VEGF may have a role in

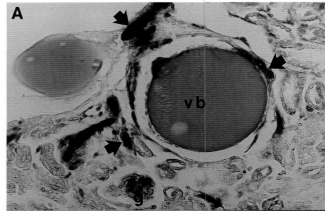

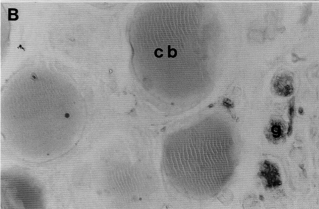

Figure 17.7 Effect of VEGF-soaked beads implanted into the cortex of neonatal Tie-1/LacZ transgenic mice. (A) A VEGF-soaked bead (vb) was implanted into the neonatal kidney cortex, and the effects were assessed after 1 week. Note that transgene-expressing capillaries are prominent around the implant. (B) A control bead (cb) soaked in vehicle failed to elicit a similar effect. In both conditions, adjacent glomeruli (g) of the host kidney contain Tie-1/LacZ-expressing capillary loops. Personal observations by A.S.Woolf and S. Loughna.

endothelial cell survival and proliferation in acquired glomerular diseases (Ostendorf *et al.*, 1999). In addition, because VEGF can induce fenestrae in endothelial cells *in vitro* (Roberts and Palade, 1995; Esser *et al.*, 1998), the factor has been postulated as having a physiological role in maintaining glomerular capillary fenestrae *in vivo*. Futhermore, because VEGF enhances capillary permeability (Roberts and Palade, 1995), the molecule has also been considered as a candidate for enhancing proteinuria (i.e., the pathological leakage of excess protein from the glomerular capillary lumen into the urinary space) in glomerular diseases. However, Gerber *et al*, (1999) were unable to elicit histological changes in adult glomeruli even after 4 weeks of systemic VEGF blockade, whereas Ostendorf *et al.* (1999) failed to modulate pathological glomerular proteinuria by inhibiting $VEGF_{165}$.

There is less information about the expression of VEGF and its receptors in human as compared to mouse nephrogenesis. Kaipainen *et al.* (1993) used *in situ* hybridization in human fetuses of 17–20 weeks gestation. They localized VEGF transcripts to glomerular epithelial cells, VEGFR2 to the endothelium of glomerular capillaries, and VEGFR1 to glomerular and peritubular capillaries. Simone and co-workers (1995) used *in situ* hybridization and immunohistochemistry to show that VEGF was expressed by epithelia in glomeruli and in collecting ducts in the human embrygenesis, whereas the two receptors were expressed by endothelial cells in glomeruli and around tubules.

The VEGF axis has appeared as an increasingly complex signaling system and an extending series of receptors have been defined, such as VEGFR3 and neurophilin-1 (Dumont *et al.*, 1998; Irrthum *et al.*, 2000; Soker *et al.*, 1998). Neurophilin-1, for example, appears to be expressed during nephrogenesis (Robert *et al.*, 2000), but the roles of these molecules in kidney development are currently unknown.

C. The Angiopoietin Axis in Metanephric Development

The second growth factor signalling system on which we focus is that mediated by the angiopoietin (Ang) ligands and the Tie (tyrosine kinase containing immunoglobulin-like loops and epidermal growth factor similar domains) receptors. As endothelia differentiate, the onset of Tie expression postdates VEGFR2 but precedes maturity. Hence, Tie receptors modulate turnover and morphogenesis of precursors that have already entered the endothelial lineage (Sato *et al.*, 1995; Puri *et al.*, 1999; Partanen *et al.*, 1996). Tie-1 is currently an "orphan receptor tyrosine kinase", meaning that its growth factor ligand has yet to be defined. Tie-2 is a Tie-1 homologue and its ligands are the angiopoietins, a family of secreted factors (Valenzuela *et al.*, 1999) that possess an amino-terminal coiled-coil domain that mediates the formation of dimers and higher order multimers between specific family members together with carboxy-terminal fibrinogen-like domains that mediate the differential effects of the factor on Tie-2 phosphorylation (Procopio *et al.*, 1999). Ang-1 binds to Tie-2 (Davis *et al.*, 1996) and the subsequent tyrosine phosphorylation transduces signals for endothelial cell survival and, in synergy with VEGF, capillary sprouting (Koblizek *et al.*, 1998; Papapetropoulos *et al.*, 1999), Both Ang-1 and Tie-2 null mutant mouse embryos have abnormal vascular networks with growth-retarded vascular smooth muscle precursors (Sato *et al.*, 1995; Dumont *et al.*, 1994; Suri *et al.*, 1996).

In addition to the direct effects on endothelia, Tie-2 activation is thought to cause reciprocal, maturational effects on adjacent smooth muscle precursors elicited by endothelial cell-derived factors such as platelet-derived growth factor-B (PDGF-B; Hellstrom *et al.*, 1999). Ang-1

also inhibits capillary permeability (Thurston *et al.*, 1999; Thurston *et al.*, 2000), preventing plasma leakage in response to VEGF and mustard oil, and the factor also has a role in blood formation (Takakura *et al.*, 1998). Upregulated signaling though a constitutively active mutant Tie-2 receptor causes cutaneous vascular malformations in humans (Vikkula *et al.*, 1996) whereas transgenic Ang-1 overexpression in mouse skin elicits larger, more numerous, and highly branched vessels versus normals (Suri *et al.*, 1998). Tie-2 expression is upregulated by hypoxia by endothelial Per-ARNT/AhR-Sim (PAS) domain protein 1, an endothelial cell-specific transcription factor (Tian *et al.*, 1999).

Ang-2, the second member of the ligand family, binds Tie-2 without causing tyrosine phosphorylation (Maisonpierre *et al.*, 1997); instead, it antagonizes Ang-1-induced Tie-2 phosphorylation. Ang-2 overexpression *in vivo* causes defects resembling Tie-2 and Ang-1 null mutants (Maisonpierre *et al.*, 1997). In the presence of abundant VEGF, Ang-2 is thought to destabilize vascular networks and facilitate sprouting, e.g., during tumor growth (Tanaka et al., 1999; Stratmann *et al.*, 1998). Conversely, with low ambient VEGF levels, Ang-2 may cause vessel regression, e.g., in corpus luteum involution (Maisonpierre *et al.*, 1997). Hypoxia has been demonstrated to upregulate the transcription of Ang-2 (Mandriota and Pepper, 1998). A third member of ligand family, Ang-3, has been cloned in mice: it is postulated to have a similar action to Ang-2 (Valenzuela *et al.*, 1999).

Ang-1/Tie-2 signaling acts as a stabilizing influence during the later stages of capillary formation, whereas Ang-2 acts as a natural inhibitor of Ang-1. Expression studies in nonrenal tissues are consistent with the hypothesis that Ang-1 and Ang-2 are expressed by smooth muscle cells and their precursors and exert paracrine effects on Tie-2-expressing endothelia (Davis *et al.*, 1996; Maisonpierre *et al.*, 1997). Colen *et al.*, (1999) reported that Ang-1 and-2 were expressed during lung and pancreas vessel growth, and Akeson *et al.* (2000) reported that subsets of lung precursor cells expressed these ligands as well as Tie genes. Furthermore, after injection of these embryonic lung mesenchymal cells into blastocysts, they contributed to endothelial cell development within the developing lung and heart (Akeson *et al.*, 2000).

Tie-1 is expressed in differentiating endothelia in mouse metanephroi from E11 with transcript levels peaking in the first few weeks after birth (Figs. 17.3A and 17.3B) (Loughna *et al.*, 1997, 1998; Yuan *et al.*, 1999). In mice genetically engineered so that both Tie-1 alleles are ablated, the null mutant embryos die in mid- to late gestation with impaired vessel integrity (Sato *et al.*, 1995; Puri *et al.*, 1999). Furthermore, mutant cells in animals derived from chimeras between -/- and normal cells (Tie-1^{lcz}/Tie-1^{lczn-} chimeric mice) fail to contribute to renal vasculature, suggesting a nephrogenic role for this gene (Partanen *et al.*, 1996). Ang-1, Ang-2, and Tie-2 are expressed during mouse meta-nephrogenesis with peak levels in the first three postnatal weeks; subsequently, they are downregulated (Yuan *et al.*, 1999a). Ang-1 transcripts are found in condensing mesenchyme, maturing glomeruli, proximal tubules, and outer medullary tubules. Tie-2 is expressed by capillaries in nephrogenic cortex, glomerular tufts, and vasa rectae. Preliminary data (M. Kolatsi-Joannou *et al.*, 2001) demonstrated that addition of Ang-1 to cultured embryonic mouse kidneys enhances the formation of glomerular vessels.

Ang-2 is expressed in the walls of differentiating renal arteries (Fig. 17.5) (Yuan *et al.*, 2000a, b); some of these cells probably represent smooth muscle cell precursors condensing from surrounding mesenchyme. Of note, walls of developing kidney vessels also show widespread expression of another secreted protein, renin (Reddi *et al.*, 1998). However, whereas Ang-2 expression is continuous through the fetal kidney arterial system (Yuan *et al.*, 2000a), renin-expressing cells are discontinuous, often localized to branch points (Reddi *et al.*, 1998). Ang-2 is also expressed in cores of maturing glomeruli where mesangial cells reside (Yuan *et al.*, 2000a, b). Because mesangial cells share biosynthetic and structural properties with vascular smooth muscle, Ang-2 expression is consistent with this relationship. Cultured mesangial cells from juvenile mice expressed Ang-2 and Ang-3, but not Ang-1 or Tie-2; these cells upregulate Ang-2 in response to hypoxia, concomitantly with an increase in VEGF (Yuan *et al.*, 2000b). The correct differentiation of mesangial cells is critically dependent on PDGFB signaling, as described in detail by Soriano (1994) and Lindhahl *et al.* (1998).

Ang-2 transcripts are also detected in a subset of thin descending limbs of loops of Henle (Yuan *et al.*, 1999a, 2000a). In this location in the outer medulla, Ang-2-expressing tubules show the striking configuration of a "fence" that surrounds ascending and descending vasa rectae capillaries. Furthermore, levels of this ligand increase postnatally at the same time as medullary expression of Ang-1 is declining (Yuan *et al.*, 1999, 2000a). We speculate that perpetual hypoxia, which is known to occur in the medulla, may maintain high levels of Ang-2 in this locality, as demonstrated *in vitro* for other cells that express Ang-2 (Yuan *et al.*, 2000b; Mandriota and Pepper, 1998; Oh *et al.*, 1999).

There is currently little information about the expression of the Ang/Tie axis in renal diseases. Preliminary evidence, however, shows upregulation of Tie-2 expression in capillaries of Wilms' tumors, the commonest type of childhood kidney cancer (Yuan *et al.*, 1999b; Fig. 17.8). In the same study, epithelial elements of Wilms' tumors expressed Ang-1 transcripts. In addition, we have found that Ang-1 protein is upregulated markedly after folic acid-induced acute renal failure in mice (Long *et al.*, 2000), which is accompanied by an increased expression of kidney

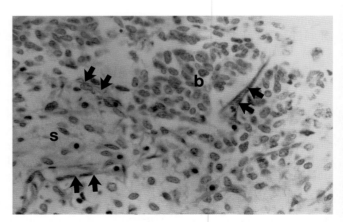

Figure 17.8 Tie-2-expressing capillaries in a childhood kidney Wilms' tumor. The histology section has been immunostained brown with an antibody against Tie-2. Note the rich capillary network (arrows), which expresses this receptor tyrosine kinase (brown color) between stromal (s) and blastemal (b) cells of the tumor. Personal observations by A.S. Woolf and H.T. Yuan.

VEGF. Conversely, it is down regulated in a form of glomerular injury (Yuan *et al.*, 2002). In these contexts, Ang signaling may enhance endothelial cell survival and morphogenesis.

VI. Other Molecules Involved in Vascular Growth

A. Transcription Factors

Apart from growth factors and their receptors, other classes of molecule are implicated in embryonic vascular growth (Carmeliet, 2000). For example, transcription factors are key players in vascular morphogenesis. Ema *et al.* (1997) reported a novel basic helix-loop-helix PAS factor with close sequence similarity to hypoxia-inducible factor 1α: it was shown to interact with the aryl hydrocarbon receptor nuclear translocator (ARNT) and high levels of expression in developing embryos correlated with VEGF expression. Mice that lack ARNT itself demonstrate defective angiogenesis in the yolk sac and abnormal responses to glucose and oxygen deprivation (Maltepe *et al.*, 1997). Another transcription factor is COUP-TF11 (Pereira *et al.*, 1999) which is expressed by the metanephric mesenchyme and which has been implicated in the control of Ang-1 expression. Similarly, null mutants for MEF2C, a member of the MADS box family, have reduced fetal myocardial expression of VEGF and Ang-1 (Bi *et al.*, 1999). Finally, an intriguing report of a zebrafish mutant with defective PAX2 expression in differentiating kidney provided descriptive data consistent with the hypothesis that this transcription factor has a role in down-regulating VEGF expression in renal epithelial cells (Majumdar *et al.*, 2000; Chapter 5).

B. Cell Adhesion Molecules

Changes in cell adhesion play a critical part of embryonic vessel formation. For example, in mice defective for vascular endothelial cadherin, blood islands of the yolk sac and clusters of angioblasts in the allantois failed to establish a capillary plexus and remained isolated (Gory-Faure *et al.*, 1999), whereas diverse experiments *in vivo* and *in vitro* have implicated $\alpha_v\beta_3$ integrin in angiogenesis (Brooks *et al.*, 1994; Drake *et al.*, 1995). However, the expression and roles of these and other adhesion molecules have yet to be investigated in renal vascular growth and differentiation.

VII. Conclusions and Perspectives

From the aforementioned discussion, it is apparent that the molecular mechanisms of kidney blood vessel development are only just beginning to unravel. Indeed, this field of study has been relatively slow to gain momentum compared with other aspects of nephrogenesis, such as mesenchymal induction, epithelial polarization, and morphogenesis of the ureteric tree. We suggest that the following may be important and interesting areas of investigation.

i. Mice that are mutant for molecules active in embryonic vascular growth have often been uninformative concerning renal vascular development, as affected embryos die before the metanephros is established. Therefore, there will need to be an increased use of strategies that involve the *in vivo* disruption of specific signaling systems later in renal development, such as those used by Gerber and colleagues (1999).

ii. The relative failure of glomerular capillary growth in metanephric organ culture versus capillary differentiation after metanephric transplantation (e.g., into the neonatal renal cortex) is striking. If the optimal conditions to enhance renal capillary growth *ex vivo* could be defined, we would surely learn something fundamental about normal renal vessel development. The observation that hypoxic metanephric organ culture can enhance vessel growth leads us to suggest that, soon after its inception, metanephros may be hypoxic *in vivo*; in the future, techniques will need to be devised to measure tissue oxygen tension through metanephrogenesis, perhaps as have been applied by tumor biologists (Raleigh *et al.*, 1996).

iii. Just as insights have been made regarding the mechanisms of serial branching of the ureteric bud tree, there must be specific molecules and conditions that control the complex branching patterns of the renal arterial tree. Descriptive evidence shows that the renin is expressed at such branch points (Reddi *et al.*, 1998),

although there is, as yet, no firm data to functionally implicate this molecule in the branching process. In addition, nothing is understood about the patterning of renal arteries and veins, especially in the vasa rectae where descending arterial-like vessels and ascending venous-like vessels are arranged in a checkerboard-like pattern. Perhaps the Eph/ephrin signaling system will be found to be important in generating this type of pattern (Adams *et al.*, 1999).

iv. It is theoretically feasible that molecules exist that specifically control renal vessel growth, rather than vascular growth in general. In this respect, there are paradigms in other organ systems, such as cloche, a gene reported to affect endothelium differentiation in the heart more than in extracardiac tissues (Stainier *et al.*, 1995). In this regard, a human developmental disorder called renal artery stenosis leads to narrowing of the main kidney arteries and their initial branches, and this is associated with the development of severe high blood pressure in early childhood (Guignard *et al.*, 1989). In some cases, this disorder exists as part of a genetic multiorgan malformation syndrome (e.g., Williams syndrome caused by mutations of elastin); in other cases, however, the disease appears restricted to the kidney, and we speculate that these children could have mutations of renal vessel-specific genes.

v. In the adult kidney, endothelial cell turnover is exceedingly low, but both apoptosis and proliferation have been documented to occur in certain animals models of acquired disease. For example, glomerular capillary regeneration occurs during recoverry from the experimental hemolytic uremic syndrome (Kim *et al.*, 2000), whereas cortical peritubular proliferation and remodeling occur in a toxic model of acute renal failure (Long *et al.*, 2001) and after subtotal nephrectomy, i.e., surgical removal of renal tissue (Pillebout *et al.*, 2002). In the future, the administration of vascular growth factors may be used to enhance renal recovery in these settings (Kim *et al.*, 2000).

Acknowledgment

This work was supported by Wellcome Trust Project Grant 058005, the Kidney Research Aid Fund, and the National Kidney Research Fund.

References

Adams, R. H., Wilkinson, G. A., Weiss, C., Diella, F., Gale, N. W., Deutsch, U., Risau, W., and Klein, R. (1999). Roles of ephrinB ligands and EphB receptors in cardiovascular development: Demarcation of arterial/venous domains, vascular morphogenesis, and sprouting angiogenesis. *Genes. Dev.* **13**, 295–306.

Akeson, A. L., Wetzel, B., Thompson, F. Y., Brooks, S. K., Paradis, H., Gendron, R. L., and Greenberg, J. M. (2000). Embryonic vasculogenesis by endothelial precursor cells derived from lung mesenchyme. *Dev. Dyn.* **217**, 11–23.

Bernstein, J., Cheng, F., and Roszka, J. (1981). Glomerular differentiation in metanephric culture. *Lab. Invest.* **45**, 183–190.

Bi, W., Drake, C. J., and Schwarz, J. J. (1999). The transcription factor MEF2C-null mouse exhibits complex vascular malformations and reduced cardiac expression of angiopoietin 1 and VEGF. *Dev. Biol.* **211**, 255–267.

Breier, G., Albrecht, U., Sterrer, S., and Risau, W. (1992). Expression of vascular endothelial growth factor during embryonic angiogenesis and endothelial cell differentiation. *Development* **114**, 521–532.

Brogi, E., Schatteman, G., Wu T., Kim, E.A., Varticovski, L., Keyt B., and Isner, J.M. (1996). Hypoxia-induced paracrine regulation of vascular endothelial growth factor receptor expression. *J. Clin. Invest.* **97**, 469–476.

Brooks, P. C., Clark, R. A. F., and Cheresh, D. A. (1994). Requirement of vascular integrin $\alpha_v\beta_3$ for angiogenesis. *Science* **264**, 569–571.

Brown, L. F., Berse, B., Tognazzi, K., Manseau, E. J., Van de Water, L., Senger, D. R., Dvorak, H. F., and Rosen, S. (1992). Vascular permeability factor mRNA and protein expression in human kidney. *Kidney Int* **42**, 1457–1461.

Carmeliet, P., Ng, Y. S., Nuyens, D., Theilmeier, G., Brusselmans, K., Cornelissen, I., Ehler, E., Kakkar, V. V., Stalmans, I., Mattot, V., Perriard, J. C., Dewerchin, M., Flameng, W., Nagy, A., Lupu, F., Moons, L., Collen, D., D'Amore, P. A., and Shima, D.T. (1999). Impaired myocardial angiogenesis and ischemic cardiomyopathy in mice lacking vascular endothelial growth factor isoforms VEGF$_{164}$ and VEGF$_{188}$. *Nature Med.* **5**, 495–502.

Carmeliet, P. (2000). Mechanisms of angiogenesis and arteriogenesis. *Nature Med.* **6**, 389–395.

Carey, A. V., Carey, R. M, and Gomez, R. A. (1992). Expression of α-smooth muscle actin in the developing kidney vasculature. *Hypertension* **19**(2 Suppl), II168–II175.

Colen, K. L., Crisera, C. A., Rose, M. I., Connelly, P. R., Longaker, M. T., and Gittes, G. K. (1999). Vascular development in the mouse embryonic pancreas and lung. *J. Pediatr. Surg.* **34**,781–785.

Daniel, T. O., and Abrahamson, D. (2000). Endothelial signal integration in vascular assembly. *Annu. Rev. Physiol.* **62**, 649–671.

Daniel, T. O., Stein, E., Cerretti, D. P., St. John, P. L., Robert, B, and Abrahamson, D. R. (1996). ELK and LERK-2 in developing kidney and microvascular endothelial assembly. *Kidney Int. Suppl.* **57**, S73–S81.

Davis, S., Aldrich, T. H., Jones, P. F., Acheson, A., Compton, D. L., Jain, V., Ryan, T. E., Bruno, J., Radziejewski, C., Maisonpierre, P. C., and Yancopoulos, G. D. (1996). Isolation of angiopoietin-1, a ligand for the TIE2 receptor, by secretion-trap expression cloning. *Cell* **87**, 1161–1170.

De Ruiter, M. C., Poelmann, R. E., Van Munsteren, J. C., Mironov, V., Markwald, R. R., and Gitternberger-de Groot, A. C. (1997). Embryonic endothelial cells transdifferentiate into mesenchymal cells expressing smooth muscle actins in vivo and in vitro. *Circ. Res.* **80**, 444–451.

Dickson M. C., Martin J. S., Cousins F. M., Kulkarni A. B., Karsson S., and Akhurst R. J. (1995) Defective haematopoiesis and vasculogenesis in transforming growth factor-β1 knock out mice. *Development* **121**, 1845–1854.

Drake, C. J., Cheresh, D. A., and Little, C. D. (1995). An antagonist of integrin $\alpha_v\beta_3$ prevents maturation of blood vessels during embryonic neovascularisation. *J. Cell Sci.* **108**, 2655–2661.

Dubey, R. K., Jackson, E. K., Rupprecht, H. D., and Sterzel, R. B. (1997). Factors controlling growth and matrix production in vascular smooth muscle and glomerular mesangial cell. *Curr. Opin. Nephrol. Hypertens.* **6**, 88–105.

Dumont, D. J., Gradwohl, G., Fong, G. H., Puri, M. C., Gerstenstein, M., Auerbach, A., and Breitman, M. L. (1994). Dominant-negative and targeted null-mutations in the endothelial receptor tyrosine kinase, tek, reveal a critical role in vasculogenesis of the embryo. *Genes Dev.* **8**, 1897–1909.

Dumont, D. J., Jussila, L., Taipale, J., Lymboussaki, A., Mustonen, T.,

Pajusola, K., Breitman, M., and Alitalo, K. (1998). Cardiovascular failure in mouse embryos deficient in VEGF receptor-3. *Science* **282**, 946–949.

Esser, S., Wolburg, K., Wolburg, H., Breier, G., Kurzchalia T., and Risau, W. (1998). Vascular endothelial growth factor induces endothelial fenestrations in vitro. *J. Cell Biol.* **140**,947–959.

Ema, M., Taya, S., Yokotani, N., Sogawa, K., Matsuda, Y., and Fujii-Kuriyama, Y. (1997). A novel βHLH-PAS factor with close sequence similarity to hypoxia-inducible factor 1a regulates the VEGF expression and is potentially involved in lung and vascular development. *Proc. Natl. Acad. Sci. USA* **94**, 4273–4278.

Fong, G. H., Rossant, J., Gertsenstein, M., and Breitman, M. L. (1995). Role of the Flt-1 receptor tyrosine kinase in regulating the assembly of vascular endothelium. *Nature* **376**, 66–70.

Fong, G. H., Zhang, L., Bryce, D. M., and Peng, J. (1999). Increased hemangioblast commitment, not vascular disorganisation, is the primary defect in flt-1 knock out mice. *Development* **126**, 3015–3025.

Frid, M. G., Shekhonin, B. V., Koteliansky, V. E., and Glukhova, M. A. (1992). Phenotypic changes of human smooth muscle cells during development: Late expression of heavy caldesmon and calponin. *Dev. Biol.* **153**, 185–193.

Gale, N. W., and Yancopoulos, G. D. (1999). Growth factors acting via endothelial cell-specific receptor tyrosine kinases: VEGFs, angiopoietins, and ephrins in vascular development. *Genes Dev.* **13**, 1055–1066.

Gerber, H. P., Hillan, K. J., Ryan, A. M., Kowalski, J., Keller, G.-A., Rangell, L., Wright, B. D., Radtke, F., Aguet, M., Ferrara, N. (1999). VEGF is required for growth and survival in neonatal mice. *Development* **126**, 1149–1159.

Gory-Faure, S., Prandini, M. H., Pointu, H., Roullot, V., Pignot-Paintrand, I., Vernet, M., and Huber, P. (1999). Role of vascular endothelial-cadherin in vascular morphogenesis. *Development* **126**, 2093–2102.

Grant, D. S., Kleinman, H. K., Goldberg, I. D., Bhargava, M. M., Nickoloff, B. J., Kinsella, J. L., Polverini, P., and Rosen, E. M. (1993). Scatter factor induces blood vessel formation in vivo. *Proc. Natl. Acad. Sci. USA* **90**, 1937–1941.

Guignard, J. P., Gouyon, J. B., and Adelman, R. D. (1989). Arterial hypertension in the newborn infant. *Biol. Neonate* **55**, 77–83.

Hellstrom, M., Kaln M., Lindahl, P., Abramsson, A., Betsholtz, C. (1999). Role of PDGF-B and PDGFR-β in recruitment of vascular smooth muscle cells and pericytes during embryonic blood vessel formation in the mouse. *Development* **126**, 3047–3055.

Hiratsuka, S., Minowa, O., Kuno, J., Noda, T., and Shibuya, M. (1998). Flt-1 lacking the tyrosine kinase domain is sufficient for normal development and angiogenesis in mice. *Proc. Natl. Acad. Sci. USA* **95**, 9349–9354.

Hungerford, J. E., and Little, C. D. (1999). Developmental biology of the vascular smooth muscle cell: Building a multilayered vessel wall. *J. Vasc. Res.* **36**, 2–27.

Hyink, D. P., Tucker, D. C., St.John, P. L., Leardkamolkarn, V., Accavitti, M. A., Abrass, C. K., and Abrahamson, D. R. (1996). Endogenous origin of glomerular endothelial and mesangial cells in grafts of embryonic kidneys. *Am. J. Physiol.* **270**, F886–F899.

Irrthum, A., Karkkainen, A., Karkkainen, M. J., Devriendt, K., Alitalo, K., and Vikkula, M. (2000). Congenital hereditary lymphedema caused by a mutation that inactivates VEGFR3 tyrosine kinase. *Am. J. Hum. Genet.* **67**, 295–301.

Kaipainen, A., Korhonen, Pajusola K., Aprelikova, O., Persico, M. G., Terman, B. I., and Alitalo, K. (1993) The related FLT4, FLT1, and KDR receptor tyrosine kinases show distinct expression patterns in human fetal endothelial cells. *J. Exp. Med.* **178**, 2077–2088.

Kim, Y. G., Suga, S. I., Kang, D. H., Jefferson, J. A., Mazzali, M., Gordon, K. L., Matsui, K., Breiteneder-Geleff, S., Shankland, S. J., Hughes, J., Kerjaschki, D., Schreiner, G. F., and Johnson, R. J. (2000). Vascular endothelial growth factor accelerates renal recovery in experimental thrombotic microangiopathy. *Kidney Int.* **58**, 2390–2399.

Kitamoto, Y., Tokunaga, H., and Tomita, K. (1997). Vascular endothelial growth factor is an essential molecule for mouse kidney development: Glomerulogenesis and nephrogenesis. *J. Clin. Invest.* **99**, 2351–2357.

Kitamura, H., Shimizu, A., Masuda, Y., Ishizaki, M., Sugisaki, Y., and Yamanaka, N. (1998). Apoptosis in glomerular endothelial cells during the development of glomerulosclerosis in the remnant-kidney model. *Exp. Nephrol.* **6**, 328–336.

Kloth, S., Gerdes, J., Wanke, C., and Minuth, W. W. (1998). Basic fibroblast growth factor is a morphogenic modulator in kidney vessel development. *Kidney Int.* **53**, 970–978.

Kloth, S., and Suter-Crazzolara, C. (2000). Modulation of renal blood vessel formation by glial cell line-derived neurotrophic factor. *Microvasc Res* **59**, 190–194.

Koblizek, T. I, Weiss, C., Yancopoulos, G. D., Deutsch U., and Risau, W. (1998). Angiopoietin-1 induces sprouting angiogenesis in vitro. *Curr. Biol.* **8**, 529–532.

Kolatsi-Joannou M., Yuan H. T., Li X., Suda T., Woolf A. S. Expression and potential role of angiopoietins and Tie-2 in early development of the mouse metanephros. *Dev. Dyn.* **222**, 120–126, 2001.

Liu, A., Dardik, A., and Ballermann, B. J. (1999). Neutralizing TGFb1 antibody infusion in the neonatal rat delays in vivo glomerular capillary formation. *Kidney Int.* **56**, 1334–1348.

Levy, A. P., Levy, N. S., Wegner, S., and Goldberg, M. A. (1995). Transcriptional regulation of the rat vascular endothelial growth factor gene by hypoxia. *J. Biol. Chem.* **270**, 13333–13340.

Lindhahl, P., Hellstrom, M., Kalen, M., Karlsson, L., Pekny, M., Pekna, M., Soriano, P., and Betsholtz, C. (1998). Paracrine PDGF-B/PDGF-Rβ signaling controls mesangial cell development in kidney glomeruli. *Development* **125**, 3313–3322, 1998.

Long, D. A., Woolf, A. S., Suda, T., and Yuan, H. T. (2001) Deregulated renal angiopoietin-1 expression in folic acid-induced nephrotoxicity. Sumitted for publication. *J. Am. Soc. Nephrol.* **12**, 2721–2731.

Loughna, S., Landels, E. C., and Woolf, A. S. (1996) Growth factor control of developing kidney endothelial cells. *Exp. Nephrol.* **4**,112–118.

Loughna, S., Hardman, P., Landels, E., Jussila, L., Alitalo, K., and Woolf, A. S. (1997). A molecular and genetic analysis of renal glomerular capillary development. *Angiogenesis* **1**,84–101.

Loughna, S., Yuan H. T., and Woolf, A. S. (1998). Effects of oxygen on vascular patterning in *Tie1/LacZ* metanephric kidneys in vitro. *Biochem. Biophys. Res. Commun.* **247**, 361–366.

Maisonpierre, P. C., Suri, C., Jones, P. F., Bartunkova, S., Wiegand, S. J., Radziejewski, C., Compton, D., McClain, J., Aldrich, T. H., Papadopoulos, N., Daly, T. J., Davis, S., Sato, T. N., and Yancopoulos, G. D. (1997). Angiopoietin-2, a natural antagonist for Tie2 that disrupts in vivo angiogenesis. *Science* **277**, 55–60.

Majumdar, A., and Drummond, I. A. (1999). Podocyte differentiation in the absence of endothelial cells as revealed in the zebrafish avascular mutant, *cloche. Dev. Genet.* **24**, 220–229.

Majumdar, A., Lun, K., Brand, M., Drummond, I. A. (2000). Zebrafish *no isthmus* reveals a role for *pax2.1* in tubule differentiation and patterning events in the pronephric primordia. *Development* **127**, 2089–2098.

Maltepe, E., Schmidt, J. V., Baunoch, D., Bradfield, C. A., and Simon, M. C. (1997). Abnormal angiogenesis and responses to glucose and oxygen deprivation in mice lacking the protein ARNT. *Nature* **386**, 403–407.

Mandriota, S. J., and Pepper, M. S. (1998). Regulation of angiopoietin-2 mRNA levels in bovine microvascular endothelial cells by cytokines and hypoxia. *Circ. Res.* **19**, 852–859.

Millauer, B., Wizigmann-Voos, S., Schnurch, H., Martinez, R., Moller, N. P. H., Risau, W., and Ullrich, A. (1993) High affinity VEGF binding and developmental expression suggest Flk-1 as a major regulator of vasculogenesis and angiogenesis. *Cell* **72**, 835–846.

Miquerol, L., Gertsenstein, M., Harpal, K., Rossant, J., and Nagy, A. (1999). Multiple developmental roles of VEGF suggested by a LacZ-tagged allele. *Dev. Biol.* **212**, 307–322.

Murohara, T., Ikeda, H., Duan, J., Shintani, S., Sasaki, K. I., Eguchi, H., Initsuka, I., Matsui, K., and Imaizumi, T. (2000). Transplanted cord blood-derived endothelial precursor cells augment postnatal neovascularization. *J. Clin. Invest.* **105**,1527–1536.

Oh, H., Takagi, H., Suzuma, K., Otani, A., Matsumura, M., and Honda, Y. (1999). Hypoxia and vascular endothelial growth factor selectively upregulate angiopoietin-2 in bovine microvascular endothelial cells. *J. Biol. Chem.* **274**, 15732–15739.

Oliver, J. A., and Al-Awqati, Q. (1998). An endothelial growth factor involved in rat renal development. *J. Clin. Invest.* **102**, 1208–1219.

Ostendorf, T., Kunter, U., Eitner, F., Loos, A., Regele, H., Kerjaschki, D., Henninger, D. D., Janjic, N., and Floege, J. (1999). VEGF$_{165}$ mediates glomerular endothelial repair. *J. Clin. Invest.* **104**, 913–923.

Papapetropoulos, A., Garcia-Cardena, G., Dengler, T. J., Maisonpierre, P. C., Yancopoulos, G. D., Sessa, W. C. (1999). Direct actions of angiopoietin-1 on human endothelium: Evidence for network stabilisation, cell survival, and interaction with other angiogenic growth factors. *Lab. Invest.* **79**, 213–223.

Pardanaud, L., Luton, D., Prigent, M., Bourcheix, L. M., Catala, M., and Dieterlen-Lievre, F. (1996). Two distinct endothelial lineages in ontogeny, one of them related to hemopoiesis. *Development* **122**, 1363–1371.

Park, F., Mattson, D. L., Roberts, L. A., Cowley, A. W., Jr. (1997). Evidence for the presence of smooth muscle alpha-actin within pericytes of the renal medulla. *Am. J. Physiol.* **273**, R1742–1748.

Partanen, J., Puri, M. C., Schwartz, L., Fischer, K. D., Bernstein, A., and Rossant, J. (1996). Cell autonomous functions of the receptor tyrosine kinase TIE in late phases of angiogenic capillary growth and endothelial cell survival during murine development. *Development* **122**, 3013–3021.

Pereira, F. A., Qiu, Y., Zhou, G., Tsai, M. J., and Tsai, S. Y. (1999). The orphan nuclear receptor COUP-TFII is required for angiogenesis and heart development. *Genes Dev.* **13**, 1037–1049.

Pillebout, E., Burtin, M., Yuan, H. T., Briand, P., Woolf, A. S., Friedlander, G., and Terzi, F. (2001). Proliferation and remodeling of the peritubular microcirculation after nephron reduction: Association with the progression of renal lesions. *Am. J. Pathol.* **159**, 547–560.

Procopio, W. N., Pelavin, P. I., Lee, W. M., and Yeilding, N. M. (1999). Angiopoietin-1 and -2 coiled coil domains mediate distinct homo-oligomerisation patterns, but fibrinogen-like domains mediate ligand activity. *J. Biol. Chem.* **274**, 30196–30201.

Puri, M. C., Partanen, J., Rossant, J., and Bernstein, A. (1999). Interaction of the TEK and TIE receptor tyrosine kinases during cardiovascular development. *Development* **126**, 4569–4580.

Raleigh, J. A., Dewhirst, M. W., and Thrall, D. E. (1996). Measuring tumor hypoxia. *Semin. Radiat. Oncol.* **6**, 37–45.

Reddi, V., Zaglul, A., Pentz, E. S., and Gomez, R. A. (1998). Renin-expressing cells are associated with branching of the developing kidney vasculature. *J. Am. Soc. Nephrol.* **9**, 63–71.

Risdon, R. A., and Woolf, A. S. (1998). Development of the kidney. In *Heptinstall's Pathology of the Kidney.* J. C. Jennette, J. L, Olson, M. M, Schwartz, and F.6. Silva, eds., 5th ed., pp. 67–84.Lippincott-Raven, Philadelphia.

Robert, B., St. John, P. L., and Abrahamson, D. R. (1998). Direct visualisation of renal vascular morphogenesis in Flk1 heterozygous mutant mice. *Am. J. Physiol.* **275**, F164–F172.

Robert, B., Zhao, X., and Abrahamson, D. R. (2000). Coexpression of neuropilin-1, Flk1 and VEGF(164) in developing and mature mouse kidney glomeruli. *Am. J. Physiol. Renal Physiol.* **279**, F275–F282.

Roberts, W. G., and Palade, G. E. (1995). Increased microvascular permeability and endothelial fenestration induced by vascular endothelial growth factor. *J. Cell. Sci.* **108**, 2369–2379.

Sariola, H., Ekblom, P., Lehtonen, E. and Saxen, L. (1983). Differentiation and vascularization of the metanephric kidney grafted on the chorio-allantoic membrane. *Dev. Biol.* **96**, 427–435.

Sariola, H., Timpl, R., von der Mark, K., Mayne, R., Fitch, J. M,. Lisenmayer, T. F., and Ekblom, P. (1984). Dual origin of glomerular basement membrane. *Dev. Biol.* **101**, 86–96.

Sato, T. N., Tozawa, Y., Deutsch, U., Wolburg-Buchholz, K., Fujiwara, Y., Gendron-Maguire, M., Gridley, T., Wolburg, H., Risau, W., and Qin, Y. (1995). Distinct roles for the receptor tyrosine kinases Tie-1 and Tie-2 in blood vessel formation. *Nature* **376**, 70–74.

Shalaby, F., Rossant, J., Yamaguchi, T. P., Gerstenstein, M., Wu, X.-F., Breitman, M. L., and Schuh, A. C. (1995). Failure of blood island formation and vasculogenesis in flk-1 deficient mice. *Nature* **376**, 62–66.

Simon, M., Grone, H.-J., Johren, O., Kullmer, J., Plate, H. K., Risau, W., and Fuchs, E. (1995). Expression of vascular endothelial growth factor and its receptors in human renal ontogenesis and in adult kidney. *Am. J. Physiol.* **268**, F240–F250.

Soker, S., Takashima, S., Miao, H. Q., Neufeld, G., and Klagsbrun, M. (1998). Neurophilin-1 is expressed by endothelial and tumor cells as an isoform-specific receptor for vascular endothelial growth factor. *Cell* **92**, 735–745.

Soriano, P. (1994). Abnormal kidney development and hematological disorders in platelet derived growth factor β-receptor mutant mice. *Genes Dev.* **8**, 1888–1896.

Stainier, D. Y., Weinstein, B. M., Detrich, H. W., III, Zou, L.I, and Fishman, M. C. (1995). Cloche, an early acting zebrafish gene, is required by both the endothelial and hemopoietic lineages. *Development* **121**, 3141–3150.

Stratmann, A., Risau, W., and Plate, K. H. (1998). Cell type-specific expression of angiopoietin-1 and angiopoietin-2 suggests a role in glioblastoma angiogenesis. *Am. J. Pathol.* **153**, 1459–1466.

Suri, C., Jones, P. F., Patan, S., Bartunkova, S., Maisonpierre, P. C., Davis, S., Sato, T. N,. and Yancopoulos, G. D. (1996). Requisite role of angiopoietin-1, a ligand for the TIE2 receptor, during embryonic angiogenesis. *Cell* **87**, 1171–1180.

Suri, C., McClain, J., Thurston, G., McDonald, D. M., Zhou, H., Oldmixon, E. H., Sato, T. N., and Yancopoulos, G. D. (1998). Increased vascularization in mice overexpressing angiopoietin-1. *Science* **282**, 468–471.

Takakura, N., Huang, X. L., Naruse, T., Hamguchi, I., Dumont, D. J., Yancopoulos, G. D., and Suda, T. (1998) Critical role of the TIE2 endothelial cell receptor in the development of definitive hematopoiesis. *Immunity* **9**, 677–686.

Tanaka, S., Mori, M., Sakamoto, Y., Makuuchi, M., Sugimachi, K., and Wands, J. R. (1999). Biologic significance of angiopoietin-2 expression in human hepatocellular carcinoma. *J. Clin. Invest.* **103**, 341–345.

Tian, H., McKnight, S. L, and Russell, D. W. (1999). Endothelial PAS domain protein 1 (EPAS1), a transcription factor selectively expressed in endothelial cells. *Genes Dev.* **11**, 72– 82.

Tisher, C. C., and Madsen, K. M. (1996). Anatomy of the kidney. In *"The Kidney"*, B. M. Brenner, ed. 5th ed., pp. 3–71. Saunders, Philadelphia.

Thurston, G., Suri, C., Smith, K., McClain, J., Sato, T. N., Yancopoulos, G. D., and McDonald, D. M. (1999). Leakage-resistant blood vessels in mice transgenically overexpressing angiopoietin-1. *Science* **286**, 2511–2514.

Thurston, G., Rudge, J. S. M., Ioffe, E., Zhou, H., Ross, L., Croll, S. D., Glazer, N., Holash, J., McDonald, D. M., and Yancopoulos, G. D. (2000). Angiopoietin-1 protects the adult vasculature against plasma leakage. *Nature Med.* **6**, 460–463.

Towers, P. R., Woolf, A. S., and Hardman, P. (1998). Glial cell line-derived neurotrophic factor stimulates ureteric bud outgrowth and enhances survival of ureteric bud cells in vitro. *Exp. Nephrol.* **6**, 337–351.

Tuder, R. M., Flook, B. E., and Voelkel, N. F. (1995). Increased gene expression for VEGF and the VEGF receptors KDR/Flk and Flt in lungs exposed to acute or chronic hypoxia: Modulation of gene expression by nitric oxide. *J. Clin. Invest.* **95**, 1798–1807.

Tufro, A., Norwood, V. F., Carey, R. M., and Gomex, R. A. (1999). Vascular endothelial growth factor induces nephrogenesis and vasculogenesis. *J. Am. Soc. Nephrol.* **10**, 2125–2134.

Tufro-McReddie, A. Norwood, V. F., Carey, R. M., and Gomez, R. A. (1997). Oxygen regulates vascular endothelial growth factor-mediated vasculogenesis and tubulogenesis. *Dev. Biol.* **183**, 139–149.

Tufro-McReddie, A., Romano, L. M., Harris, J. M., Ferder, L., and Gomez, R. A. (1995). Angiotensin II regulates nephrogenesis and renal vascular development. *Am. J. Physiol.* **269**, F110–115.

Valenzuela, D. M., Griffiths, J. A., Rojas, J., Aldrich, T. H., Jones, P. F., Zhou, H., McClain, J., Copeland, N. G., Gilbert, D. J., Jenkins, N. A., Huang, T., Papadopoulos, N., Maisonpierre, P. C., Davis, S., and Yancopoulos, G. D. (1999). Angiopoietins 3 and 4: Diverging gene counterparts in mice and humans. *Proc. Natl. Acad. Sci. USA* **96**, 1904–1909.

Vikkula, M., Boon, L. M., Carraway, K. L., III, Calvert, J. T., Diamonti, A. J., Goumnerov, B., Pasyk, K. A., Marchuk, D. A., Warman, M. L., Cantley, L. C., Mulliken, J. B., and Olsen, B. R. (1996). Vascular dysmorphogenesis caused by activating mutation in the receptor tyrsoine kinase TIE2. *Cell* **87**, 1181–1190.

Woolf, A. S. (1999). Embryology of the kidney. In *"Pediatric Nephrology"*, T. M. Barratt, A. Avner, and W. Harmon, eds., 4th Ed., pp. 1–19. Williams and Wilkins, Baltimore, MD.

Woolf, A. S., and Cale, C. M. (1997). Roles of growth factors in renal development. *Curr. Opin. Nephrol. Hypertens.* **6**, 10–14.

Woolf, A. S., Kolatsi-Joannou, M., Hardman, P., Andermarcher, E., Moorby, C., Fine, L. G., Jat, P. S., Noble, M. D., and Gherardi, E. (1995). Roles of hepatocyte growth factor/scatter factor and the met receptor in the early development of the metanephros. *J. Cell. Biol.* **128**, 171–184.

Woolf, A. S., and Loughna, S. (1998). Origin of glomerular capillaries: Is the verdict in? *Exp. Nephrol.* **6**,17–21.

Woolf, A. S., Palmer, S. J., Snow, M. L., and Fine, L. G. (1990). Creation of a functioning chimeric mammalian kidney. *Kidney Int.* **38**, 991–997.

Yuan, H. T., Suri, C., Landon, D. N., Yancopoulos, G. D., and Woolf, A. S. (2000a). Angiopoietin-2 is a site specific factor in differentiation of mouse renal vasculature. *J. Am. Soc. Nephrol.* **11**, 1055–1066.

Yuan, H. T., Yang, S. P., and Woolf, A. S. (2000b). Hypoxia upregulates angiopoietin-2, a Tie-2 ligand, in mouse mesangial cells. *Kidney Int.* **58**, 1912–1919.

Yuan, H. T., Suri, C., Yancopoulos, G. D., and Woolf, A. S. (1999a). Expression of angiopoietin-1, angiopoietin-2 and the Tie-2 receptor tyrosine kinase during mouse kidney maturation. *J. Am. Soc. Nephrol.* **10**, 1722–1736.

Yuan, H. T., Suri, C., Yancopoulos, G. D., and Woolf, A. S. (1999b). Tie2 and its angiopoietin-1 ligand in normal nephrogenesis and Wilms' tumors. *J. Am. Soc. Nephrol.* **10**, 413A. [Abstract]

Yuan, H. T., Tipping, P. G., Li, X. Z., Long, D. A., Woolf, A. S. (2002) Angiopoietin correlates with glomerular capillary loss in anti-glomerular basement membrane glomerulonephritis. *Kidney Int.* **61**, 2078–2089.

18

Development of Function in the Metanephric Kidney

Lisa M. Satlin, Craig B. Woda, and George J. Schwartz

I. Introduction

Embryologic development of the definitive mammalian kidney, the metanephros, is preceded by the transient and sequential appearance of two primitive kidneys: the rudimentary pronephros (Chapter 1), which appears at 3 weeks and regresses by 5 weeks of human fetal life, and the mesonephros, which appears at 4 weeks, is maximally developed by 8 weeks, and degenerates by 12 weeks (Chapters 6 and 7; de Martino and Zamboni, 1966; Saxén, 1987). The metanephros starts to form at 5 weeks (Chapter 10; Kazimierczak, 1970; Saxén, 1987) when the metanephric blastema, which derives from caudal intermediate mesoderm, induces the ureteric bud to form a spur off the mesonephric (or Wolffian) duct. Reciprocal inductive interactions between the two primordia induce the dichotomous branching of the ureteric bud, initiating morphogenesis of the collecting duct system

(Grobstein, 1955; Horster *et al.*, 1999; Osathanondh and Potter, 1966; Saxén, 1987; see Fig. 14.1). At the tips of the branching ureteric buds, induced mesenchyme condenses and aggregates to form a renal vesicle, which then develops into comma and S-shaped bodies (Chapter 14; Kazimierczak, 1970; Saxén, 1987). Endothelial cells invade the lower cleft of the S-shaped body, eventually giving rise to the glomerular capillary loops (Osathanondh and Potter, 1966; Saxén, 1987). Each S-shaped body fuses with the ureteric bud-derived collecting duct. The S-shaped body ultimately differentiates into the glomerulus, proximal convoluted tubule, loop of Henle, distal convoluted tubule, and probably the connecting segment, whereas derivatives of the ureteric bud ultimately form the ureter, pelvis, calyces, and collecting ducts of the metanephros (Chapter 11).

The processes of induction, morphogenesis, and differentiation of the metanephros occurs in a centrifugal pattern, proceeding from the center to the periphery (Evan *et al.*, 1983, 1991; Potter, 1943). Within the nephrogenic zone, the most outer cortical portion of the developing kidney, are located the youngest, most immature tubules; the oldest, most mature nephrons reside in deeper regions of the kidney. Nephrons in the outer cortex of the neonate may become midcortical nephrons of the adult as centrifugal maturation proceeds. The full complement of approximately 1 million nephrons per kidney in the human is achieved at approximately 35 weeks gestational age (GA) (MacDonald, 1959; Potter, 1943). A full term infant of 40 weeks conceptional age (gestational plus postnatal age) thus possesses all the nephrons he/she will ever have. In contrast, nephronogenesis, or the formation of new nephrons, continues after birth in premature infants of less than 35 weeks conceptional age.

The metanephric kidney first begins to produce urine by 10 weeks of gestation (Rabinowitz *et al.*, 1989). Fetal urine contributes to the formation of amniotic fluid (Potter, 1943; Rabinowitz *et al.*, 1989), necessary for symmetrical fetal growth and lung development (Moritz and Wintour, 1999; Perlman and Levin, 1974). Although human renal function is not necessary for regulation of fetal water and electrolyte homeostasis, a process assumed by the placenta, adaptation to the extrauterine environment requires that the kidneys assume responsibility for these functions immediately after birth.

In animal models used to study ontogeny of renal function, nephronogenesis is complete before term birth in sheep (Robillard *et al.*, 1981) and guinea pig (Spitzer and Brandis, 1974). The fetal sheep, in particular, provides an excellent model of human nephronogenesis as its program of renal development closely parallels that of the human. Nephronogenesis continues for ~1 week after term birth in rat (Larsson *et al.*, 1980; Solomon, 1977) and mouse (Davies and Bard, 1996) and ~2 weeks in rabbit (Evan *et al.*, 1983, 1991) and dog (Kleinman and Reuter, 1973). Thus, functional analyses of the differentiating kidney in these latter species are complicated by the concurrent presence of nephrons in diverse stages of differentiation (Fig. 18.1). Furthermore, physiologic, biochemical, and enzymatic maturation of newly formed nephrons in all species may lag

behind anatomic maturation by weeks or months. For example, functional maturity of the rat proximal tubule is not achieved until after the fifth week of postnatal life (Aperia and Larsson, 1979). A comparison of the developmental changes in renal function among various species is best considered not as pre- or postnatal events, but in terms of their relationship to the completion of nephrogenesis.

The focus of this chapter is on characterizing the normal functional development of the metanephric kidney. Although the bulk of our fundamental knowledge about these processes derives from investigations in animal models, available data from studies in the human fetus and neonate are also discussed.

II. Methods to Study Developmental Renal Physiology

The first studies of function in the developing kidney were performed in the 1930s. In these early investigations, vital dyes were injected into fetal mesonephric and metanephric kidneys of a variety of animals. Thereafter, histologic specimens were prepared and analyzed to demonstrate tubular function (Chambers and Kempton, 1933; Gersh, 1937). With the advent of clearance methodology at about this same time, investigators compared the amount of substance excreted in the urine with that filtered by the kidneys. Barnett (1940), among the first to apply this technique to the study of the developing kidney, demonstrated that the glomerular filtration rate (GFR), assessed by measuring the inulin clearance, was low in infants. Elegant clearance and metabolic studies performed in the chronically catheterized fetal lamb (Robillard *et al.*, 1975,1993; Robillard and Nakamura, 1988a,b, Robillard *et al.*, 1988a,b,c, 1987, 1992a,b, 1977a, 1990, Robillard and Weiner, 1988) have provided the foundation for much of our understanding of the developmental regulation of fetal renal hemodynamics.

Introduction of the micropuncture technique in the 1950s made it possible to localize and analyze the transport characteristics of single nephron segments in superficial nephrons accessible at the surface of the kidney. While much of our fundamental knowledge of the characteristics of tubular transport in the mature subject derives from micropuncture studies in the rat (Aperia and Herin, 1975; Aperia and Larsson, 1979; Baker and Solomon, 1976; Chevalier, 1982; Dlouha, 1976; Elinder, 1981; Zink and Horster, 1977), dog (Horster and Valtin, 1971; Wong *et al.*, 1983), and guinea pig (Kaskel *et al.*, 1988; Spitzer and Brandis, 1974; Spitzer and Edelmann, 1971), the widespread application of this technique to the study of the developing kidney has been limited for several reasons. First, application of this method to the immature kidneys of many laboratory animals is impossible due to the absence

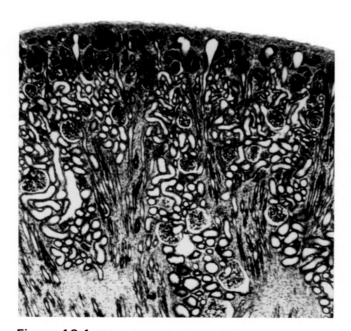

Figure 18.1 Light micrograph of a saggital section of a neonatal rabbit kidney. The outer cortex possesses a nephrogenic zone. The glomeruli and tubules become progressively larger from the outer to the inner cortex. After Evan *et al.*, (1983).

of functioning nephrons in the relatively undifferentiated superficial nephrogenic zone that is accessible to micropuncture. Second, postnatal changes in factors extrinsic to the renal tubule (e.g., GFR, hormonal milieu) can complicate the interpretation of these *in vivo* studies.

The advent of *in vitro* microperfusion of isolated tubules in the 1970s (Burg *et al.*, 1966) provided a means of examining the transport characteristics of tubular segments of known identity and location, including those not accessible to micropuncture (e.g., thick ascending limb of Henle and collecting duct). Using this technique, tubular transport has been measured under standardized conditions throughout development in the absence of potentially confounding luminal (e.g., filtered load) and peritubular (e.g., circulating hormones) factors (Baum, 1990, 1992; Baum *et al.*, 1997; Mehrgut *et al.*, 1990; Satlin, 1994; Schwartz and Evan, 1983; Schwartz *et al.*, 1978). Manipulation of the extracellular environment can be easily accomplished, thus providing a method to identify factors that control and regulate transport. The widespread use of *in vitro* microperfusion in developmental studies has been hampered, however, by its technical difficulty, the short length of segments in the immature kidney, and long collection times necessary to generate detectable concentration gradients across the epithelium.

Electrophysiologic methods, including short circuit current, ion sensitive microelectrode studies, and the patch clamp technique, have rarely been applied to the developing kidney. The latter technique, which allows one to define the characteristics of single ion channels in the native epithelium (Palmer, 1986), has been used to examine ontogenic changes in distal nephron ion channel activity in the developing collecting duct (Huber *et al.*, 1999; Huber and Horster, 1996; Satlin and Palmer, 1996, 1997). Such an analysis of the apical membrane of the differentiating principal cell in the rabbit cortical collecting duct (CCD) has identified a temporal dissociation between the early appearance of amiloride-sensitive sodium channels and later detection of low-conductance potassium channels (Satlin and Palmer, 1996, 1997).

Morphologic and ultrastructural analyses of the maturing kidney have provided a solid basis for the understanding of the functional development of the nephron. The observation of large, irregular sinusoidal capillaries in the cortex of the puppy kidney instead of the peritubular capillary networks characteristic of the adult suggested the presence of a functional postglomerular shunt that would direct blood flow away from the proximal tubule (Evan *et al.*, 1979). This vascular arrangement was considered to account for the low extraction ratio and transport rates for secreted solutes characteristic of the neonatal kidney. Morphologic studies of developing proximal tubules have shown that the complexity of cell shape and surface area of basolateral cell membranes are, in general, directly related to the rates of net

fluid absorption (Horster and Larsson, 1976; Larsson and Horster, 1976).

Biochemical assays using preparations such as isolated nephron segments and cells have provided insight into function, hormonal response and associated intracellular events. Such studies have identified that Na-K-ATPase activity in proximal and distal tubular segments from neonatal animals is far lower than that in their adult counterparts (Constantinescu *et al.*, 2000; Schmidt and Horster, 1977). Transport studies in plasma membrane vesicles isolated from apical and basolateral membranes have confirmed developmental changes in the stoichiometry, mechanism, and kinetics of several transport pathways.

Finally, the advent and rapid progress of the techniques of molecular biology have provided powerful new tools and strategies with which to explore normal renal development, physiology, and the pathophysiological basis of renal disease. The recent molecular cloning of renal-specific proteins has, in particular, enhanced our understanding of the mechanism of action and regulation of these proteins in the healthy and diseased kidney. Methodologies such as immunohisto- and cyto-chemistry, *in situ* hybridization, and reverse transcriptase–polymerase chain reaction (RT-PCR) have provided information about the presence and localization of specific genes and their protein products at unique developmental stages. Molecular studies have allowed for the analysis of transcriptional and translational control of the developmental expression of various channels, transporters, enzymes, autocoids, and receptors. The identification of mutations in the genes encoding these proteins has provided unequivocal evidence for the physiologic relevance of these genes (and their encoded proteins). The two most commonly used methods for analyzing the function of a gene *in vivo*, that of targeted gene disruption and overexpression, have proved to be extremely useful tools (Babinet, 2000; Jaisser, 2000) in establishing the functional relevance of specific genes in development, maintenance of homeostasis, and genetic disorders, such as hypertension. Advances in DNA array technology, coupled with the completion of several genome-sequencing projects, have made it possible to define broad patterns of gene expression during kidney organogenesis, including those related to renal function (Stuart *et al.*, 2001).

III. Development and Regulation of Renal Blood Flow

Renal blood flow (RBF) is low in fetus and neonate, whether normalized to body weight, surface area, or kidney weight. In the human, RBF increases approximately 3-fold between 25 and 40 weeks gestation (Rudolph and Heymann, 1967; Veille *et al.*, 1993), reaching adult levels by 2 yrs of age (Rubin, 1949). The perinatal increase in RBF is gradual

(Aperia *et al.*, 1977; Aperia and Herin, 1975; Calcagno and Rubin, 1963; Robillard *et al.*, 1981), consistent with the premise that clamping of the umbilical cord and the immediate imposition of functional demand at birth do not in themselves account for the increase in RBF. Nor can the maturational increase in RBF be completely explained by increases in renal mass (Jose *et al.*, 1971) or by glomerulogenesis, which is complete well before maximal levels of RBF are achieved. The overall maturational increase in RBF (~6-fold between birth and maturity in a variety of species) is related to both an increase in mean arterial blood pressure (MAP; ~2-fold) associated with an increase in cardiac output, and a decrease in renal vascular resistance (RVR; ~3-fold) (Fig. 18.2) (Gruskin *et al.*, 1970; Jose *et al.*, 1971; Spitzer and Edelmann, 1971). Although developmental changes in all hemodynamic variables contribute to the

age-related increase in RBF, the decrease in RVR is the most important determinant (Gruskin *et al.*, 1970; Jose *et al.*, 1974).

The two kidneys of the adult, which comprise 0.5% of the total body mass, receive 25% of the total cardiac output. In contrast, the kidneys of the 10- to 20-week previable human fetus receive only about 5% of the cardiac output and the 1-week-old term infant, about 9% (Rudolph *et al.*, 1971).

The sum of afferent and efferent arteriolar resistances ultimately determines the RVR and RBF, while the relative resistances of the afferent and efferent arterioles determine glomerular capillary pressure and GFR. RVR is much higher in the newborn than in the adult (Gruskin *et al.*, 1970; Ichikawa *et al.*, 1979). The rapid maturational decrease occurs at a time when the systemic vascular resistance

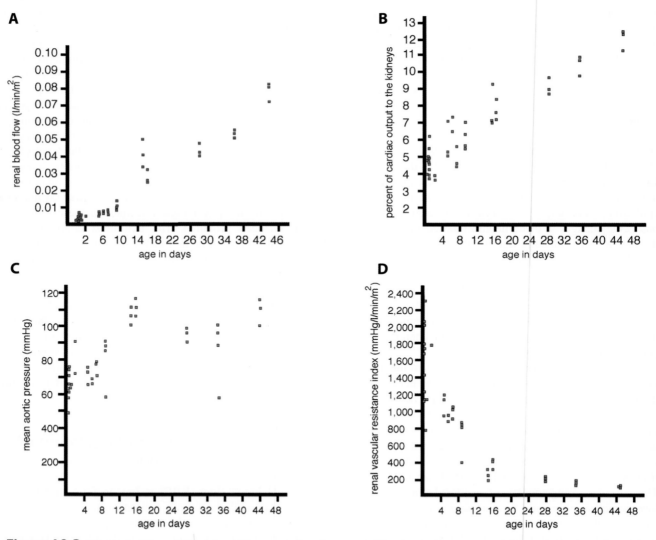

Figure 18.2 Changes in (A) renal blood flow, (B) percent of cardiac output, (C) mean aortic pressure, and (D) renal vascular resistance in maturing postnatal piglets. The overall maturational increase in renal blood flow is related to both an increase in mean arterial blood pressure associated with an increase in cardiac output and a decrease in renal vascular resistance. After Gruskin *et al.* (1975)

increases approximately 6-fold (Gruskin *et al.*, 1970). Cumulative evidence suggests that changes in the balance of the vasoconstrictor renin–angiotensin system (RAS) and renal sympathetic nervous system, both of which are highly active during early development, and vasodilatory humoral factors such as nitric oxide (NO) contribute to the developmental reduction in RVR and the increase in RBF.

The intrarenal distribution of RBF changes during maturation, reflecting the centrifugal pattern of increase in relative size, number, and distribution of glomeruli present in the different regions of the kidney at each maturational stage. Whereas the perfusion rate for nephrons in all regions of the cortex remains constant during the last trimester of gestation (Robillard *et al.*, 1981), RBF in the newborn is distributed primarily to the inner cortex and medulla (Fig. 18.3). Thereafter, intrarenal blood flow to glomeruli residing in the outer cortex increases significantly whether expressed in absolute rates or as a percentage of blood flow to juxtamedullary nephrons (Aperia *et al.*, 1977; Aschinberg *et al.*, 1975; Jose *et al.*, 1971; Kleinman and Reuter, 1973; Olbing *et al.*, 1973; Robillard *et al.*, 1981).

In addition to these hemodynamic changes, anatomic factors contribute to the developmental increases and redistribution of RBF. The intrarenal vascular system distal to the afferent arteriole in the neonate differs from that in the adult. Variability in the complexity of the glomerular capillary network exists early in postnatal life (see Chapter 16). Efferent arterioles in the outer cortex of the neonatal kidney join the venous system by way of venous channels or sinusoids (Evan *et al.*, 1979). Few of the efferent arterioles

descend into the medulla and divide to form the vasa rectae and peritubular capillaries. Thus, the renal vasculature of the neonatal kidney is characterized not only by fewer vessels than present in the adult, but also by efferent arterioles that connect directly to the venous system (arteriovenous shunting), thereby bypassing the proximal tubules. The observation that RBF can be increased acutely during the neonatal period by volume expansion (Drukker *et al.*, 1980; Elinder *et al.*, 1980) indicates that anatomic immaturity is not solely responsible for the low renal perfusion rates of the newborn kidney.

In sum, the adaptation of the kidney to extrauterine life is characterized by the evolution of a low-flow, high-resistance fetal organ, with most of the blood supplying the inner cortex, into a high-flow, low-resistance organ, with most of the blood supplying the outer cortex. A variety of vasoactive substances (Table 18.1) have been suggested to play a role in the developmental regulation of RBF and GFR. The most important factors are discussed below. The effects of these factors on the renal vasculature are often species- and developmental age-specific due to ontogenic changes in anatomic factors (e.g., innervation), receptor density and subtypes, postreceptor events (e.g., second messengers), and/or interaction with other vasoactive systems. Among these, interactions among the RAS, catecholamines, and NO appear to play the most important role in the perinatal changes in renal hemodynamics. Also relevant to the focus of this book is the general theme that vasoconstrictors stimulate the growth of renal cells (mitogenesis and hypertrophy), whereas vasodilators inhibit the growth response, perhaps by the activation of common second messenger pathways.

A. Renin–Angiotensin System

All components of the RAS are present in the fetal metanephric kidney (Butkus *et al.*, 1997a; Gomez *et al.*, 1990; Lumbers, 1995; Schutz *et al.*, 1996; Wintour *et al.*, 1996). Renin, angiotensinogen, and angiotensin-converting enzyme (ACE) are first detected in fetal metanephroi by ~6 weeks gestation in the human and sheep (Schutz *et al.*, 1996; Wintour *et al.*, 1996). The localization of renin within the kidney is developmental stage specific. Whereas most renin-containing cells are located in the juxtaglomerular apparatus in the newborn and adult, renin message and protein are also present in the arcuate and interlobular arteries and glomeruli in the fetus (Gomez *et al.*, 1988, 1989; Graham *et al.*, 1992). Stimulation of renin release in fetal lambs leads to generation and secretion of the potent vasoconstrictor angiotensin II (ANG II)(Broughton Pipkin *et al.*, 1974; Broughton-Pipkin, 1974; Smith *et al.*, 1974).

The activity of the RAS is inversely related to gestational age in the fetus and postnatal age in the newborn (Smith *et al.*, 1974). The observations that administration of ACE

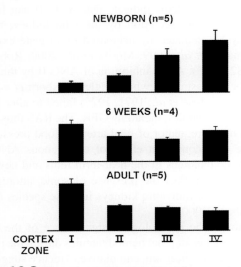

Figure 18.3 Maturational changes in blood flow per glomerulus in quartiles (subcapsular zone I to juxtamedullary zone IV) of the renal cortex of dogs. Renal blood flow in the newborn is distributed primarily to the inner cortex. Thereafter, intrarenal blood flow to glomeruli residing in the outer cortex increases significantly. Reproduced with permission from Olbing *et al.* (1973).

inhibitors during late gestation can result in anuria and oligohydramnios, renal tubular abnormalities, pulmonary hypoplasia, growth retardation, and increased fetal loss (Broughton Pipkin *et al.*, 1982; Guron *et al.*, 1997; Pryde *et al.*, 1993) support a critical role for the RAS in normal fetal growth and development. Genetic disruption experiments in mice have underscored the importance of this axis in the normal growth and development of the kidney (Coffman, 1998). Mice that lack the gene for angiotensinogen exhibit a delay in glomerular maturation, hypoplastic papillae, and develop lesions in the renal cortex similar to those of hypertensive nephrosclerosis (Niimura *et al.*, 1995). Disruption of the gene encoding ACE leads to distorted architecture of the renal vasculature and increased levels of renin mRNA expression, tubular obstruction, dilatation, and atrophy (Hilgers *et al.*, 1997). Mice lacking the ANG II type 1 (AT$_1$) receptor exhibit grossly abnormal kidney morphology similar to that described in the angiotensinogen mutant mouse (Tsuchida *et al.*, 1998). In contrast, mice with disruption of the AT$_2$ receptor gene have intact litters of normal newborns with no renal histologic abnormalities. A role for the AT$_2$ receptor in blood pressure control is suggested by the observation that deletion of this gene leads to a rise in MAP and an enhanced pressor response to ANG II *in vivo* (Hein *et al.*, 1995; Ichiki *et al.*, 1995).

ANG II mediates its biologic effects by interacting with two major subtypes of receptors, each of which has unique patterns of expression, signal transduction mechanisms, and effects (Table 18.1). Autoradiographic studies revealed that the fetal kidney predominantly expresses the AT$_2$ receptor, whereas the AT$_1$ receptor is most abundant in the adult (Grady *et al.*, 1991). ANG II receptors appear early in the metanephric kidney of the human and lamb (Butkus *et al.*, 1997a; Schutz *et al.*, 1996). At this early stage, the AT$_1$ receptor is detected at sites destined to contribute to the regulation of vascular tone and salt balance, including presumptive mesangial cells of developing glomeruli, medulla, and medullary rays (Grady *et al.*, 1991). Expression of AT$_2$ receptor mRNA, localized to the macula densa (Butkus *et al.*, 1997a; Kakuchi *et al.*, 1995; Robillard *et al.*, 1995), and protein (Grady *et al.*, 1991) decreases as nephrogenesis nears completion, such that AT$_2$ receptors are virtually undetectable in most adult mammalian kidneys. AT$_1$ mRNA and receptor increase during the third trimester (Aguilera *et al.*, 1994; Grady *et al.*, 1991; Grone *et al.*, 1992; Robillard *et al.*, 1995; Tufro-McReddie *et al.*, 1993).

Reflecting the high levels of expression of the renal renin gene at birth, plasma renin activity (PRA) is three to five times higher in human infants (Broughton Pipkin *et al.*, 1981; Dillon, 1980; Godard *et al.*, 1979; Kotchen *et al.*, 1972; Richer *et al.*, 1977; Sulyok *et al.*, 1979a; Van Acker *et al.*, 1979) and newborn animals (Broughton-Pipkin, 1974; Drukker *et al.*, 1980; Siegel and Fisher, 1977; Wallace *et al.*, 1980) than in their adult counterparts. PRA increases, as

expected, in response to volume depletion (e.g., hemorrhage, administration of furosemide) and hypoxia (John *et al.*, 1981b; Siegel and Fisher, 1977; Sulyok *et al.*, 1980b). Activity levels decrease after volume expansion with isotonic saline or administration of substances such as the β-adrenergic antagonist propranolol and the prostaglandin synthetase inhibitor indomethacin (Drukker *et al.*, 1980; Siegel and Fisher, 1977). The high levels of PRA in the newborn human and animal are generally associated with circulating levels of ANG II and aldosterone that significantly exceed those in the adult (Catt *et al.*, 1970; Pipkin *et al.*, 1971a,b, 1974; Siegel *et al.*, 1974; Sulyok *et al.*, 1979b). These high levels may reflect high rates of secretion, low metabolic clearance rates relative to body size, and/or a relative end organ unresponsiveness to aldosterone (Beitins *et al.*, 1972; Kowarski *et al.*, 1974). Despite this, in contrast to the adult, systemic blood pressure is low and systemic vascular resistance is very low.

Administration of ANG II to the near-term fetus increases MAP and RVR (renal vasoconstriction) and reduces RBF, effects similar to those observed in the adult (Robillard *et al.*, 1982); the absence of change in GFR (discussed later) in these subjects suggests that ANG II preferentially increases efferent arteriolar resistance, as it does in the adult. However, inhibition of endogenous RAS activity by the administration of ACE inhibitors or AT$_1$ receptor antagonists to fetal (Lumbers *et al.*, 1993; Robillard *et al.*, 1983; Stevenson *et al.*, 1996) and newborn (Robillard *et al.*, 1983) sheep decreases MAP and RVR, but leads to variable effects on RBF and GFR. The reduction in plasma renin concentration and decline in renin and AT$_1$ and AT$_2$ receptor gene expression observed following ANG II infusion into the fetus indicate that ANG II can feedback to decrease renin secretion by the fetal kidney, and that this may be mediated by decreased renin gene expression (Giammattei *et al.*, 1999; Moritz *et al.*, 2000; Robillard *et al.*, 1982, 1995). Acute inhibition of ANG II by the peptide inhibitor saralasin in newborn piglets (Osborn *et al.*, 1980) and puppies (Jose *et al.*, 1971, 1975) failed to alter MAP or renal hemodynamics, including GFR. The RAS thus appears to be a major regulator of fetal arterial blood pressure. The absence of a consistent effect of endogenous ANG II on renal hemodynamics in the near-term fetus and newborn is considered to reflect the presence of some autoregulatory capacity of the immature kidneys in these species (Nilsson and Friberg, 2000).

Although the physiological significance of the RAS in the maintenance of renal hemodynamics during the perinatal period remains uncertain, cumulative evidence suggests that this axis may be important under conditions of stress in the near–term fetus. Fetal hemorrhage and hypovolemia result in significant increases in renin and circulating levels of ANG II (Broughton Pipkin *et al.*, 1974; Iwamoto and Rudolph, 1981; Robillard *et al.*, 1979b; Rosnes *et al.*,

1998; Scroop et al., 1992; Siegel and Fisher, 1977). ACE inhibition blocks the increase in RVR and reduction in RBF produced by hemorrhage in near term fetuses (Gomez et al., 1984; Gomez and Robillard, 1984).

B. Renal Nerves and the Adrenergic System

Renal sympathetic nerve activity regulates multiple aspects of renal function, including RBF, GFR, solute transport, and renin secretion. Neurotransmitters released at the sympathetic nerve terminal–neuroeffector junctions interact with specific postjunctional receptors coupled to defined intracellular signaling and effector systems (Table 18.1).

The renal vascular bed of the fetus is less reactive to renal nerve stimulation than that of the newborn and adult. Renal nerve stimulation in the fetal sheep leads to a fall in RBF and a rise in RVR of lesser magnitude than observed in newborn animals (Robillard et al., 1987). In turn, the renal vasculature of newborns is also less responsive to renal nerve stimulation than it is in its adult counterpart (Buckley et al., 1979). However, renal nerve stimulation during α-adrenoceptor blockade produces renal vasodilation and an increase in RBF in fetal and newborn sheep but not in adults (Robillard et al., 1986, 1987). This response is blocked by selective β$_2$-adrenergic receptor antagonists but not those directed against cholinergic or dopaminergic receptors (Robillard et al., 1987). In sum, these results demonstrate that, in contrast to observations made in adult animals, renal nerve stimulation in the fetal and newborn kidney activates noradrenergic pathways, that norepinephrine stimulates β$_2$-adrenoceptors to produce renal vasodilation, and that maturation of the renal adrenergic system is associated with downregulation of the renal β-adrenoceptor response (Nakamura et al., 1987b).

Surgical or pharmacologic renal denervation does not alter fetal renal hemodynamics and function, suggesting that tonic renal nerve activity is low before birth (Robillard et al., 1986; Smith et al., 1990). However, in the fetus, renal nerves contribute to the response to stress associated with hypoxemia and hemorrhage (Gomez et al., 1984; McMullen et al., 1998; Robillard et al., 1986). In the newborn piglet, renal denervation is associated with an increase in RBF and a decrease in RVR, consistent with a role of sympathetic innervation exerting tonic vasoconstriction (Buckley et al., 1979), presumably via the α-adrenergic system.

Fetal and neonatal kidneys, demonstrates enhanced sensitivity to catecholamines compared to that in the adult. Thus, intrarenal infusion of epinephrine or α$_1$-adrenergic agonists lead to a greater renal vasoconstrictor response, manifest as a reduction in RBF (and GFR), in the near-term fetus than in the newborn or adult sheep or dog (Gitler et al., 1991; Guillery et al., 1994b; Jose et al., 1974; McKenna and

Angelakos, 1970). However, α-adrenergic blockade increases RBF and GFR and decreases RVR to a greater extent in puppies than adult dogs (Fildes et al., 1985; Jose et al., 1974). The observations that during α-adrenergic blockade, epinephrine or norepinephrine leads to a greater renal vasodilatory response in the fetus than in newborn or adult sheep and that this vasodilatory response is inhibited by selective β$_2$ blockade (Nakamura et al., 1988) underscore the contribution of the β$_2$-adrenergic vasodilatory mechanism to the maintenance of renal hemodynamics in fetal life.

Circulating catecholamine levels are elevated just before and immediately after birth (Kotchen et al., 1972; Lagercrantz and Bistoletti, 1977; Nakamura et al., 1987c), falling to adult values within a few days of life. The high circulating plasma levels of catecholamines, especially that of norepinephrine, acting directly to increase afferent arteriolar tone and indirectly, via stimulation of renin and ANG II release, to increase efferent resistance have been proposed to contribute to the maintenance of the high RVR characteristic of the neonatal kidney (Jose et al., 1974). The sensitivity of the immature renal vasculature to circulating catecholamines may be related in part to developmental differences in adrenergic receptor density. Innervation of the developing kidney is incomplete at birth (McKenna and Angelakos, 1968; McKenna and Angelakos, 1970), and the identity and density of adrenergic receptors is regulated ontogenically (Table 18.1).

C. Nitric Oxide

Nitric oxide is a major vasodilator in the developing kidney, contributing to the maintenance of intrarenal vascular tone and serving to buffer the highly activated endogenous vasoconstrictor state (Simeoni et al., 1997; Solhaug et al., 1996). The sensitivity of renal hemodynamics in the fetal and neonatal kidney to inhibition of NO synthesis is significantly greater than that of the adult kidney. Intrarenal infusion of NO synthesis inhibitors at doses that do not alter systemic blood pressure produces a ~30% reduction in RBF and GFR and a ~45% increase in RVR in fetal and newborn animals (Ballevre et al., 1996; Bogaert et al., 1993; Simeoni et al., 1997; Solhaug et al., 1993) compared to 10% changes in both variables in the adult. The differential effect of ANG II on afferent and efferent arteriolar tone may be due to NO modulation of afferent arteriolar tone (Ito et al., 1993).

D. Dopamine

Fetal and neonatal kidneys are less sensitive than their adult counterparts to the renal vasodilatory and, as discussed later, the natriuretic effects of dopamine (Felder et al., 1989; Nakamura et al., 1987a; Pelayo et al., 1984). Agonist binding to dopamine D$_1$-like receptors results in an

Table 18.1 Effects of various vasoactive substances on renal hemodynamics and GFR.

mediators	site of production	receptor	late trimester fetus	newborn	adult	general effects	references
angiotensin II	lung, vascular endothelial cells, glomerulus, proximal tubule	AT₁: • predominates in newborn and adult • increases phospholipase C to produce IP3 and DAG; decreases adenylyl cyclase activity AT₂: • predominates in fetus • increases guanylyl cyclase activity to produce cGMP	• inconsistent effect on RBF	• inconsistent effect on RBF • no change in GFR	• ↓ RBF via preferential ↑ in efferent arteriolar resistance • no change in GFR	AT₂: • vasodilation • growth suppression • apoptosis AT₁: • vasoconstriction • stimulation of growth • aldosterone release leading to sodium absorption	Jose et al., 1971; Jose et al., 1975; Lumbers et al., 1993; Osborn et al., 1980; Robillard, 1983; Robillard, 1982; Robillard et al., 1988; Stevenson et al. 1996
catecholamines (norepinephrine and epinephrine)	sympathetic nerves and adrenal medulla	β-adrenergic: • fetal kidney expresses only β2 • adult expresses both β1 and β2 • increases adenylyl cyclase activity to produce cAMP α-adrenergic: • predominate in newborn and adult • increases phospholipase C to produce IP3 and DAG, and decreases adenylyl cyclase activity	• enhanced sensitivity to β-adrenergic stimulation compared to adult • ↓ RBF and GFR and ↑ RVR of greater magnitude than observed in newborns	• enhanced sensitivity to β-adrenergic stimulation compared to adult • ↓ RBF and GFR and ↑ RVR of greater magnitude than observed in adults	• ↓ RBF and GFR and ↑ RVR	β-adrenergic: • vasodilation α-adrenergic: • vasoconstriction	Buckley et al., 1979; Felder et al., 1990; Felder et al., 1983; Fildes et al., 1985; Gitler et al., 1991; Guillery, 1993; Guillery et al., 1994; Jose et al., 1974; McKenna et al., 1970; Nakamura et al., 1988; Robillard et al., 1986
nitric oxide	endothelial and macula densa cells and nephron segments specific to each isoform	stimulates guanylyl cyclase activity to produce cGMP	maintains renal vasodilation, RBF and GFR	maintains renal vasodilation, RBF and GFR	tonic release maintains renal vasodilation, RBF and GFR	• vasodilation • synthesized from L-arginine by nitric oxide synthase (NOS) • 3 isoforms of NOS: neuronal (nNOS), inducible (iNOS) and endothelial (eNOS)	Ballevre et al., 1996; Bogaert et al., 1993; Baylis et al., 1990; Granger et al., 1992; Lahera et al., 1990; Solhaug et al., 1993; Solhaug et al., 1996; Simeoni et al., 1997
dopamine (low dose)	arterioles, proximal tubule, thick ascending limb of Henle, and cortical collecting duct	D₁-like: • increases adenylyl cyclase activity to produce cAMP D₂-like: • inhibits adenylyl cyclase	• blunted ↑ RBF and GFR compared to adult • high density of renal D₂-like receptors	• blunted ↑ RBF and GFR compared to adult • low density of renal D₁-like receptors compared to adult	↑ RBF and GFR	D₁ • vasodilation D₂ • vasoconstriction/vasodilation	Cadnapaphornchai et al., 1977; Felder et al., 1989; Meyer et al., 1967; Nakamura et al., 1987; Nguyen et al., 1999; Pelayo et al., 1984; Segar et al., 1992

Table 18.1 Effects of various vasoactive substances on renal hemodynamics and GFR—cont'd

mediators	site of production	receptor	late trimester fetus	newborn	adult	general effects	references
prostaglandins (PGs)	endothelial cells	EP (PGE2) FP (PGF2α) IP (PGI2 or prostacyclin) TP (thromboxane A2) DP (PGD2)	maintain RBF and GFR	maintain RBF and GFR	maintain RBF and GFR by buffering ANG II- and norepinephrine-induced vasoconstriction	• vasodilation mediated by PGE2, PGD2 and PGI2 which ↓ RVR and ↑ RBF • vasoconstriction mediated by PGF2α and thromboxane A2	Chamaa et al., 2000; Hendricks et al., 1990; Kaplan et al., 1994; Kirshon et al., 1988; Matson et al., 1981; Seyberth et al., 1983; van den Anker et al., 1994
endothelin	endothelial cells	ET$_A$: • predominate in newborn and adult • increases phospholipase C to produce IP3 and DAG ET$_B$: • predominate in fetus	↓ RVR and ↑ RBF, probably due to release of NO	• high urinary excretion and high receptor density • ↑ RVR, ↓ RBF and GFR	↑ RVR, ↓ RBF and GFR	ET$_A$: • vasoconstriction • mitogenesis ET$_B$: • NO release • vasodilation	Abadie et al., 1996; Bhat et al., 1995; Bogaert et al., 1996; Edwards et al., 1990; Mattyus et al., 1997; Semama et al., 1993a; Semama et al., 1993b
adenosine		A$_1$: • decreases adenylyl cyclase activity and cAMP production A$_2$: • increases adenylyl cyclase activity to produce cAMP	unknown	prevents hypoxemia-induced ↑ RVR, ↓ RBF and GFR	prevents hypoxemia-induced ↑ RVR, ↓ RBF and GFR	A$_1$: • vasoconstriction A$_2$ • vasodilation	Gouyon et al., 1988; Jenik et al., 2000
atrial natriuretic peptide (ANP)	atrial cardiocytes	ANP$_A$ ANP$_B$ • increases guanylyl cyclase activity to produce cGMP NPR$_c$ • inhibits adenylyl cyclase • clearance receptor	• circulating levels in the fetus>newborn and adult • intrarenal infusion: ↑ RBF • systemic infusion: ↓ MAP and RBF, ↑ RVR, ↔ GFR	• high circulating levels fall in immediate postnatal period • intrarenal infusion: ↑ RBF • systemic infusion: ↓ MAP and RBF, ↑ RVR, ↔ GFR	↑ RBF and GFR • systemic infusion: ↔ RBF, RVR, GFR	• vasodilation	Castro et al., 1988; Chevalier et al., 1988; Chevalier, 1993; Fujino et al., 1992; Robillard et al., 1988; Smith et al., 1989; Tulassay et al., 1986; Varille et al., 1989; Wei et al., 1987; Yamaji et al., 1986

All of these agents are linked to G proteins except for nitric oxide and the atrial natriuretic peptide receptors (NPR$_A$ and NPR$_B$); NPR$_C$ may be linked to inhibitor G proteins. GFR, glomerular filtration rate; DAG, diacylglycerol; IP3, inositol 1,4,5-triphosphate; MAP, mean arterial pressure; RBF, renal blood flow; RVR, renal vascular resistance.

increase in RBF and may increase GFR in the adult but is without effect early in life (Nguyen *et al.*, 1999; Segar *et al.*, 1992). This blunted response is considered to reflect limited generation of the vasodilatory second messenger cAMP (Felder *et al.*, 1989; Feltes *et al.*, 1987) and, as demonstrated by radioligand binding studies in rat, a low density of renal D_1-like receptors in the neonatal kidney (Tenore *et al.*, 1997). The density of renal D_2-like receptors, high in fetal life, decreases after birth (Felder *et al.*, 1989).

The pharmacological effects of dopamine are dose dependent. At low concentrations, the renal response to dopamine generally reflects dopaminergic receptor-mediated effects. At increasing concentrations, actions secondary to the stimulation of α-adrenergic receptors also appear. Thus, the renal vasoconstriction observed following intrarenal dopamine infusion in fetal and neonatal animals (Buckley *et al.*, 1983a; Nakamura *et al.*, 1987a) probably reflects dopamine binding to functional α-adrenoceptors, which are abundant by term (Felder *et al.*, 1983; Fildes *et al.*, 1985).

E. Prostaglandins

Prostaglandins are important in maintaining RBF and GFR, especially during conditions of enhanced vasoconstrictor activity. Complex interactions between prostaglandins and the RAS and kinin system have been described, occasionally complicating the assignation of specific effects of the former on the regulation of renal hemodynamics and transport. Prostaglandin synthesis from arachidonic acid requires cyclooxygenase (COX), the inhibitory target of aspirin and various nonsteroidal anti-inflammatory drugs (NSAIDs) (Vane *et al.*, 1998). Two isoforms of COX have been identified, each representing a different gene product and subject to differential regulation. COX-1 is expressed constitutively in renal collecting duct and vascular smooth muscle cells of pre- and postglomerular vessels in the mature nephron. In fetal kidney, expression of COX-1, localized primarily to podocytes and collecting duct cells, increases markedly between the subcapsular and the juxtamedullary cortical regions (Komhoff *et al.*, 1997). The expression of COX-1 in podocytes of the fetal kidney and its absence in adult glomeruli suggests that this isoform might be involved in glomerulogenesis (Komhoff *et al.*, 1997). COX-2, induced rapidly by a number of environmental, mitogenic, and hormonal stimuli, is expressed in the macula densa, a specialized group of ~15 cells located within the region of the thick ascending limb of the loop of Henle (TAL) that is closely apposed to its own glomerulus, and adjacent cortical TAL as well as medullary interstitial cells (Breyer and Harris, 2001; Komhoff *et al.*, 1997). The presence of COX-2 in the macula densa in the developing rat nephron (Zhang *et al.*, 1997) is consistent with a role of this enzyme in the regulation of renal perfusion and

glomerular hemodynamics. The patterns of COX-1 and -2 localization differ between adult and fetal human kidneys, an observation that may account for differences in renal responses to prostaglandin inhibition between the two groups (Komhoff *et al.*, 1997).

The urinary excretion of prostaglandins is high in the fetus (Walker and Mitchell, 1978; Walker and Mitchell, 1979), presumably reflecting a high rate of renal synthesis. Although rates of excretion decrease after birth, urinary prostaglandin excretion remains significantly higher in neonates, especially premature infants, than in children or adults (Benzoni *et al.*, 1981; Joppich *et al.*, 1979).

Prostaglandins produced by the fetal and neonatal kidney may contribute to the regulation of renal perfusion and glomerular hemodynamics. Maternal administration of NSAIDs such as indomethacin, a nonselective inhibitor of COX-1 and -2, reduces RBF in the fetus and is associated with an increase in RVR and MAP (Matson *et al.*, 1981). Although indomethacin does not induce a reduction of GFR in most animal models, it is well established that intrauterine exposure of the human fetus to COX inhibitors can lead to a decrease in GFR with a consequent reduction in fetal output and oligohydramnios (Hendricks *et al.*, 1990; Kaplan *et al.*, 1994; Kirshon *et al.*, 1988; van den Anker *et al.*, 1994). Administration of indomethacin postnatally to human infants (Seyberth *et al.*, 1983), to induce closure of patent ductus arteriosis, or to newborn animals (Chamaa *et al.*, 2000) may also compromise renal function, leading to a significant reduction in RBF, GFR, and urine volume. The renal dysfunction observed postnatally is generally dose dependent and is affected by the cardiovascular and renal status of the infant prior to treatment. Controversy exists as to the importance of the vasodilatory prostaglandins in regulating renal hemodynamics in early life, as some studies show no effect of prostaglandin synthesis blockade on MAP, RVR, or RBF in either normoxic or hypoxic animals (Arnold-Aldea *et al.*, 1991; Osborn *et al.*, 1980).

IV. Development and Regulation of Glomerular Filteration

Glomerular filtration begins between 9 and 12 weeks gestation in the human fetus and thereby contributes to the accumulation of amniotic fluid (Gersh, 1937). The glomerular filtration rate (GFR), assessed by inulin or creatinine clearance, generally correlates well with conceptional age in the human or animal, whether the fetus remains *in utero* or is born prematurely (Aperia *et al.*, 1981a; Brion *et al.*, 1986; Coulthard, 1985; Fawer *et al.*, 1979a, Guignard *et al.*, 1980; Leake and Trygstad, 1977). GFR is low during fetal life, even when corrected for body weight (Robillard *et al.*, 1975). The inulin clearance in fetal

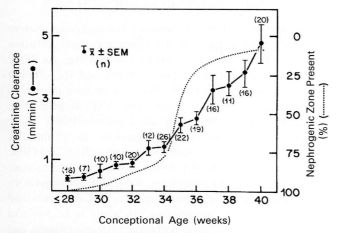

Figure 18.4 Developmental increases in glomerular filtration rate, estimated by creatinine clearance, and nephrogenic activity in the renal cortex. After ~34 weeks conceptional age, regardless of postnatal age, creatinine clearance increases rapidly as nephrogenesis is completed. Reproduced with permission from Arant. Neonatal adjustments to extrauterine life. In: (C.M. Edelmann Jr., ed.), p. 1021, Pediatric Kidney Disease. Little, Brown and Co., Boston.

sheep increases 10-fold in the last trimester of gestation (Robillard *et al.*, 1975). In the human infant, GFR increases only slightly before 34 weeks conceptional age (Fig. 18.4), although body size and kidney weight increase appreciably during this time. After about 34 weeks conceptional age, regardless of postnatal age, GFR increases rapidly, often by 3- to 5-fold within 1 week (Arant, 1978) as nephrogenesis is completed. Thus, an infant born prematurely at 28 weeks GA shows little increase in GFR until it is about 6 weeks old (i.e., until a conceptional age of 34 weeks is attained). After birth, GFR continues to increase rapidly relative to kidney weight, body size, and surface area until GFR, corrected for body surface area, reaches adult levels by about 2 years of age. A similar nonlinear pattern of increase in GFR, with an accelerated rate of increase in GFR around the time that nephrogenesis is complete, has been noted in a variety of other mammalian species (Guignard *et al.*, 1975; Kleinman and Lubbe, 1972; Nakamura *et al.*, 1987c; Smith and Lumbers, 1989). Overall, the developmental increase in absolute GFR (in ml/min) from term birth in a number of species is about 25-fold.

The developmental maturation of single nephron GFR follows a centrifugal pattern, similar to that described for RBF. At birth, the more mature glomeruli in the juxtamedullary cortex have higher filtration rates than the most recently formed glomeruli in the superficial cortex, some of which may not begin filtration for some time (Aperia *et al.*, 1977). With both anatomic growth of the nephrons and maturation of the physical forces that regulate filtration, GFR increases, with most of the surge due to enhanced perfusion (Aperia *et al.*, 1974a; Aperia and Herin,

1975; de Rouffignac and Monnens, 1976; Robillard *et al.*, 1981; Spitzer and Brandis, 1974) related temporally to the increase in total RBF and its centrifugal redistribution within the renal cortex.

The number of functional nephrons and the filtration rate of each nephron determine total kidney GFR (TKGFR). The addition of newly functioning nephrons is, however, insufficient to account for the large increase in GFR observed during maturation. A substantial rise in the rate of filtration per nephron must also occur. Measurements of single nephron GFR (SNGFR) performed using micropuncture techniques in superficial nephrons of the guinea pig show little increase in SNGFR in the first 2 weeks of postnatal life, followed by a surge that continues until ~5 weeks when the 25-fold greater SNGFR of the adult is reached (Spitzer and Brandis, 1974; Fig. 18.5). Similar developmental increases in SNGFR have been reported in other species (Aperia and Herin, 1975; Horster and Valtin, 1971; Tucker and Blantz, 1977), during which time TKGFR increases at a constant rate. These data indicate that the increase in TKGFR arises initially from an increase in SNGFR of deep nephrons; superficial nephrons appear to contribute to TKGFR later.

The process of urine formation begins with the ultrafiltration of plasma through the glomerular capillary membrane. The determinants of the rate of filtration include the filtration characteristics of the membrane, the net ultrafiltration pressure or P_{uf}, represented by the difference between the mean glomerular transcapillary hydraulic (glomerular capillary minus the Bowman's space hydraulic pressures) and the mean colloid osmotic pressures over the length of the glomerular capillary, and the glomerular plasma flow. The capillary hydrostatic pressure promotes filtration, whereas both the colloid osmotic pressure within the capillary and the hydrostatic pressure in Bowman's space

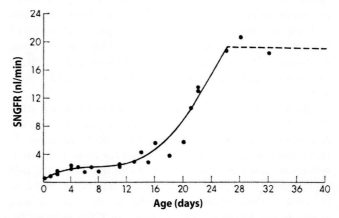

Figure 18.5 Postnatal maturation of superficial nephron glomerular filtration rate (SNGFR) in guinea pig. SNGFR increases rapidly between ~2 and 4 weeks of postnatal life in this species. Reproduced with permission from Spitzer and Brandis (1974).

oppose it. The variables that must be considered in describing maturational changes in GFR are described by the following formula: SNGFR = $K_f \times P_{uf}$, where K_f is the glomerular ultrafiltration coefficient and P_{uf} is the net ultrafiltration pressure, defined earlier. K_f, the product of the total surface area available for filtration and the hydraulic conductivity of the capillary, provides an index of the net permeability of the glomerular capillary.

The systemic arterial pressure in the newborn is generally lower than that in the adult and may be accompanied by a low glomerular capillary hydrostatic pressure (Allison *et al.*, 1972; Spitzer and Edelmann, 1971). The plasma oncotic pressure and proximal tubular pressure, an index of the pressure in the Bowman's space, are also low early in life (Allison *et al.*, 1972; Spitzer and Edelmann, 1971). Although the postnatal increases in the latter two variables would tend to oppose glomerular filtration, the net balance of changes results in an overall modest 2.5-fold increase in P_{uf} between birth and 50 days of life in the guinea pig (Spitzer and Edelmann, 1971). This increase in P_{uf}, however, accounts for only about 10% of the 25-fold increase in GFR observed during maturation (Spitzer and Edelmann, 1971). Glomerular plasma flow is low in neonatal animals (Aperia and Herin, 1975; Ichikawa *et al.*, 1979; Tucker and Blantz, 1977) due to the high afferent and efferent arteriolar resistances prevailing in the immature kidney. Whereas P_{uf} changes little after ~40 days in a variety of animal models, glomerular plasma flow increases 3-fold in rats between 40 days and maturity (Ichikawa *et al.*, 1979).

The 10- to-15-fold increase in K_f during development (Ichikawa *et al.*, 1979; Spitzer and Brandis, 1974; Tucker and Blantz, 1977; Tufro *et al.*, 1999) can be accounted for predominantly by an increase in glomerular surface area (8-fold) and thus surface area available for filtration. The glomerular capillary surface area increases by ~4-fold in the juxtamedullary glomeruli and 10-fold in the superficial glomeruli during postnatal life (John *et al.*, 1981a; Johnson and Spitzer, 1986). Analysis of the clearance of dextrans of various sizes in dogs (Goldsmith *et al.*, 1986) suggests that the permeability characteristics of the glomerular capillary membrane increase only modestly (~1.3-fold) with advancing age.

Many of the vasoactive substances known to affect RBF (Table 18.1) also influence GFR. ANG II, catecholamines, and vasopressin (AVP) modulate SNGFR by reducing K_f (Kon and Ichikawa, 1985). The target for these vasoactive substances and hormones appears to be the mesangial cell (Schlondorff, 1987). Contraction of mesangial cells in response to these agents reduces the capillary surface area available for filtration. The increase in GFR induced by glucocorticoids is secondary to vasodilatation of both preglomerular and efferent resistances, resulting in an increase in glomerular plasma flow (Baylis *et al.*, 1990a).

A. Autoregulation of RBF and GFR

Autoregulation is the mechanism that maintains relative constancy of RBF and GFR in the human as renal perfusion pressure varies from about 80 to 150 mm Hg. As MAP falls, the renal afferent arteriole dilates and the efferent arteriole constricts, the latter effect due, at least in part, to the stimulation of renin release and ANG II generation (Arendshorst *et al.*, 1999; Badr and Ichikawa, 1988). These adjustments prevent large fluctuations in glomerular capillary hydrostatic pressure and preserves RBF and GFR. Although systemic blood pressure in the fetus and neonate is generally below the lower limit of the adult autoregulatory range, experimental evidence suggests that the fetus and newborn are able to autoregulate efficiently in response to modest increases in renal perfusion pressure (Aperia and Herin, 1976; Buckley *et al.*, 1983b; Chevalier and Kaiser, 1983; Jose *et al.*, 1975). For example, a 16% increase in fetal arterial blood pressure elicited by AVP infusion does not alter RBF (Robillard and Weitzman, 1980). Both a myogenic response of the afferent arteriole (Gilmore *et al.*, 1980) and the tubuloglomerular feedback control mechanism (Moore, 1982) are responsive to changes in the rate and composition of fluid delivered to the macula densa and may mediate autoregulation of RBF and GFR (Schnermann *et al.*, 1998). However, the autoregulatory range in the newborn is set at a lower perfusion pressure than in the adult.

B. Tubuloglomerular Feedback

The tubuloglomerular feedback mechanism couples distal nephron flow and SNGFR. A stimulus (e.g., increase in tubular flow rate and/or sodium chloride concentration) at the macula densa is transmitted to the vascular structures of the nephron that control GFR (Navar and Mitchell, 1990; Schnermann *et al.*, 1998). This feedback system contributes to maintaining a constant rate of water and salt delivery to distal segments involved in the final regulation of fluid and electrolyte balance. As GFR increases with maturation, the maximal response and flow range of maximum sensitivity also increase so that the relative sensitivity of the tubuloglomerular feedback mechanism is unaltered during growth (Briggs *et al.*, 1984; Muller-Suur *et al.*, 1983). An intact RAS appears to be critical for the tubuloglomerular feedback signaling pathway (Schnermann *et al.*, 1998); NO may play a modulatory role (Welch *et al.*, 1999; Wilcox *et al.*, 1992).

V. Ontogeny of Tubular Funtion

The distribution of exchangers, cotransporters, and channels along specific segments of the nephron allows the kidney to reabsorb the bulk of glomerular filtrate proximally and then, in more distal segments, adjust the solute and water content of the urine to maintain homeostasis. The fully differentiated kidney is generally a reabsorptive organ. However, transport of some ions and solutes by individual nephron segments is bidirectional. Thus, sodium, bicarbonate, phosphate, amino acids, and glucose are reabsorbed, hydrogen ions are secreted, and potassium and organic acids are both reabsorbed and secreted.

A thorough understanding of the ontogeny of renal ion and solute transport is required to best understand the contribution of the differentiating kidney to growth and homeostasis, as well as the role of dysregulated transport in the genesis of disease. However, studies addressing these issues are complicated by a number of the factors. For example, the anatomic and functional definition of specific nephron segments in the maturing kidney is not as straightforward as in the adult because segments in the process of differentiation frequently exhibit intermediate features (Table. 18.2). Likewise, the molecular detection of a message or protein specific for an exchanger or channel may not be accompanied by the coincident expression of a functionally mature protein. Nonetheless, with the exponential growth in the number of transporting proteins cloned, a clearer understanding of the developmental regulation of tubular function is beginning to emerge.

A. Sodium

The fractional excretion of sodium (FENa), or clearance of sodium normalized to the creatinine clearance, may exceed 20% during early fetal life but falls towards the end of gestation to a value of ~0.2% (Nakamura *et al.*, 1987c; Robillard *et al.*, 1977b; Siegel and Oh, 1976). The efficient renal retention of sodium characteristic of full-term neonates is associated with a state of positive sodium balance (Aperia *et al.*, 1972, 1979; Spitzer, 1982), a condition essential for somatic growth. Premature infants of less than 30 weeks GA continue to show elevated values of FENa similar to those observed in the fetus (Aperia *et al.*, 1974b; Engelke *et al.*, 1978; Siegel and Oh, 1976; Fig. 18.6). The inability of these subjects to conserve sodium frequently leads to a state of negative sodium balance and hyponatremia during the first few weeks of postnatal life (Al-Dahhan *et al.*, 1983; Sulyok *et al.*, 1980a, 1981). Studies in laboratory animals show similar decreases in the absolute and fractional excretion of sodium during the transition from fetal to newborn life (Merlet-Benichou and de Rouffignac, 1977; Nakamura *et al.*, 1987c).

The tendency of the kidney of the full-term infant to retain sodium is also manifest as a limited capacity to excrete a sodium load. Expansion of the extracellular fluid space that follows exogenous administration of a sodium load to an adult signals the kidney to decrease tubular sodium reabsorption, resulting in an increased excretion of sodium and a relatively rapid return of the extracellular fluid

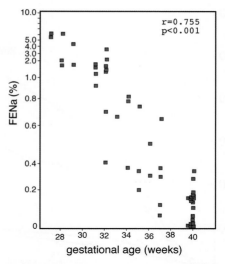

Figure 18.6 Relationship between fractional excretion of sodium (FENa) and gestational age in the human infant. The FENa, or clearance of sodium normalized to creatinine clearance, generally exceeds 5% during late fetal life, falling to a value of ~0.2% in the full–term infant. After Siegel and Oh (1976).

Table 18.2

Nephron segment	Specialized function
proximal tubule	• bulk reabsorption of >70% of the filtered sodium, chloride, water, bicarbonate, glucose, amino acids, potassium, phosphate, calcium • secretion of organic anions (e.g., uric acid) and cations • site of ammonia production
loop of Henle	• reabsorption of up to 25% of the filtered sodium and chloride • site of urinary dilution • regulatory site for magnesium excretion
distal tubule	• reabsorption of sodium chloride • regulatory site for calcium excretion
connecting tubule and cortical collecting duct	• aldosterone-responsive sodium absorption and potassium secretion by principal cells • vasopressin-sensitive water absorption by principal cells • acid-base transport and potassium absorption by intercalated cells
medullary collecting duct	• final site of sodium and chloride and vasopressin-sensitive water reabsorption • proton secretion • regulatory site for potassium excretion

space to baseline. In contrast, full-term newborns given a sodium load experience a rise in serum sodium levels, increase in weight, and generalized edema (Aperia *et al.*, 1972; Dean and McCance, 1949; Kleinman, 1975). A similar blunted diuretic and natriuretic response to acute saline expansion has also been shown to occur in young rats

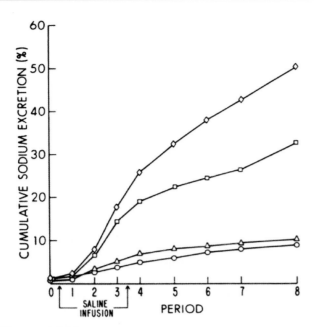

Figure 18.7 Cumulative excretion of sodium (expressed as percentage of total amount administered) over a 2-h period in the dog. ○, 1 week; □, 2 weeks; △, 3 weeks; ◇, adult. Two hours after administration of an isotonic saline load equal to 10% of body weight, puppies excreted only 10% of the load whereas adult dogs excreted 50%. Reproduced with permission from Goldsmith *et al.* (1979).

(Aperia and Elinder, 1981; Misanko *et al.*, 1979), guinea pigs (Merlet-Benichou and de Rouffignac, 1977), and dogs (Goldsmith *et al.*, 1979). Two hours after administration of an isotonic saline load equal to 10% of body weight, puppies excreted only 10% of the load, whereas adult dogs excreted 50% (Fig. 18.7). The difference in cumulative sodium excretion persisted even after correction for GFR, suggesting that it is likely mediated by the maturation of tubular, rather than glomerular, function.

Paradoxically, preterm infants of less than 36 weeks of GA excrete a sodium load more efficiently than full-term newborns, but not as efficiently as adults (Aperia *et al.*, 1974b, 1979, 1981a). Premature subjects exhibit a lower capacity for proximal reabsorption under basal conditions than adults. The further depression in proximal reabsorption that accompanies saline loading increases distal delivery of sodium to a degree that presumably exceeds the sodium reabsorptive capacity of distal tubular segments.

1. Overview of Sodium Handling by the Maturing Kidney

Sodium is filtered freely at the glomerulus. The initial two-thirds of the proximal tubule of the suckling rat absorbs approximately 50% of the filtered load of sodium and water (Aperia and Elinder, 1981; Celsi *et al.*, 1986; Lelievre-Pegorier *et al.*, 1983; Solomon, 1974a), a value only slightly less than the 60–70% reported in the adult (Corman and Roinel, 1991). Within this segment, one-half of the sodium reabsorption is passive (Baum and Berry, 1984). Sodium reabsorption from the luminal fluid drives the reabsorption of water and other filtered solutes through the paracellular spaces.

Glomerulotubular balance describes the functional correlate of the morphologic relationship between the surface area available for filtration and that available for reabsorption, i.e., the balance between capacity of the glomerulus to filter and the tubule to reabsorb the filtrate. The observations that the ratio of glomerular surface area to proximal tubule volume is 28 in the full-term newborn, 13 in the 3-month-old infant, and 3 in the adult (Fetterman *et al.*, 1965) point to a morphologic preponderance of glomeruli over tubules early in life. However, micropuncture studies in a variety of laboratory animals ingesting a normal salt diet demonstrate parallel and proportionate increases in GFR and the reabsorptive capacity of the proximal tubule after birth (Elinder, 1981; Hill and Lumbers, 1988; Horster and Valtin, 1971; Lelievre-Pegorier *et al.*, 1983; Spitzer and Brandis, 1974). These data, consistent with the maintenance of glomerulotubular balance during postnatal development, suggest either that the tubular capacity for reabsorption controls glomerular filtration or that glomerular size is an unreliable indicator of the capacity for filtration.

In the adult, 5–10% of the filtered load of sodium reaches the superficial early distal tubule, reflecting further net reabsorption of sodium in the loop of Henle (Malnic *et al.*, 1966b). In contrast, up to 35% of the filtered load of sodium reaches the superficial distal tubule of the young (2-week-old) rat (Lelievre-Pegorier *et al.*, 1983), exceeding the distal delivery measured in the adult. The fractional reabsorption of sodium along the loop of Henle, expressed as a percentage of delivered load, increases from 60% in 2-week-old rats to 80% by 6 weeks of age (Lelievre-Pegorier *et al.*, 1983), consistent with functional maturation of the loop of Henle during postnatal life.

The sodium reabsorptive capacity of the neonatal distal nephron is more efficient than that of more proximal segments (Aperia and Elinder, 1981; Kleinman, 1975; Rodriguez-Soriano *et al.*, 1983; Schoeneman and Spitzer, 1980). Micropuncture studies in developing rats have shown that although the fraction of filtered sodium delivered to the early distal tubule is significantly higher in younger than older animals, the fraction of filtered sodium remaining at the late distal tubule is equivalent in both age groups (Aperia and Elinder, 1981). These data indicate that the fractional reabsorption of sodium along the distal tubule is relatively greater in younger than in older animals and provide support for the assertion that the distal tubule contributes to the sodium retention and blunted response to sodium loading characteristic of the young animal.

2. Sites of Sodium Transport along the Nephron

a. Proximal Tubule The absolute amount of sodium and water reabsorbed along the proximal tubule increases significantly during renal development (Schwartz and Evan, 1983; Shah *et al.*, 1998). The determinants of proximal tubular reabsorption include the hydrostatic pressure gradient across the proximal tubule wall (higher in young vs adult) (Allison *et al.*, 1972; Horster and Valtin, 1971), the balance between hydrostatic and oncotic pressures across the peritubular capillary membranes, the surface area available for reabsorption, the number of apical and basolateral membrane transporters, and the activity of the basolateral Na-K-ATPase. In the newborn period, a time when the sum of forces favoring reabsorption is low, a high hydraulic conductance and permeability across the paracellular pathway may favor transepithelial solute and water movement (Horster and Larsson, 1976; Kaskel *et al.*, 1987; Larsson, 1975; Larsson and Horster, 1976). As the size of the proximal tubule, the capacity for active transport, and the hydrooncotic pressure favoring reabsorption increase, the paracellular permeability decreases. The net effect is to maintain the proportionality between the rate of glomerular filtration and proximal reabsorption.

Transcellular sodium absorption throughout the nephron is ultimately driven by the basolateral Na-K-ATPase. The Na-K pump is a heterodimeric membrane protein composed of one of at least four catalytic α subunits and one of three glycosylated β subunits (Blanco and Mercer, 1998). The composition of the pump expressed at the basolateral membrane of fully differentiated absorptive epithelial cells is generally composed of the α1 and β1 isoforms (Farman, 1996). This enzyme is the molecular target for a number of hormones that regulate sodium metabolism during health, disease, and development. In guinea pig, the 5-fold increase in pump activity during the transition from fetal to postnatal life is associated with a 4- to 7-fold increase in α and β subunit abundance in basolateral membranes (Guillery *et al.*, 1997), consistent with an increase in pump production or decrease in pump degradation. In a number of species, postnatal maturation of proximal tubular function is accompanied by a doubling of the basolateral surface area per millimeter tubular length (Aperia and Larsson, 1979, 1984; Evan *et al.*, 1983; Larsson and Horster, 1976; Schwartz and Evan, 1983) and a 3- to 4-fold increase in Na-K-ATPase activity (Aperia *et al.*, 1981b; Celsi *et al.*, 1986; Schmidt and Horster, 1977; Schwartz and Evan, 1984). The gradual increase in abundance of mRNA encoding the Na-K-ATPase subunits in the rat kidney during the first month of postnatal life (Orlowski and Lingrel, 1988) implies a transcriptional upregulation that accompanies the completion of nephronogenesis. The rate of ATP production does not appear to be limiting for pump activity, as there are only small differences between neonatal and adult animals in the rate of

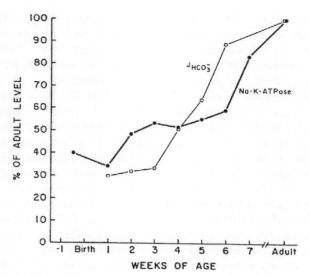

Figure 18.8 Development of Na-K-ATPase activity (●) and bicarbonate absorption (○) in isolated microperfused rabbit juxtamedullary proximal convoluted tubules. A temporal lag exists between the postnatal surge in rate of fluid and solute reabsorption and the maturation of Na-K-ATPase activity. Reproduced with permission from Schwartz and Evan (1984).

oxygen consumption (Caldwell and Solomon, 1974; Dicker and Shirley, 1971; Elinder and Aperia, 1982).

The temporal lag between the surge in the rate of fluid and solute reabsorption and the maturation of Na-K-ATPase activity (Fig. 18.8; Larsson *et al.*, 1990; Schwartz and Evan, 1983, 1984) suggests that activation of the basolateral pump is a consequence, rather than a cause, of the postnatal increase in proximal tubular transport. This interpretation would predict that net reabsorption of sodium and water in the neonatal proximal tubule is not limited by the activity of the Na-K-ATPase, but by the influx of sodium at the luminal membrane. In support of this hypothesis is the observation that an increase in cell sodium concentration in proximal tubule cells in culture, generated by the stimulation of Na/H exchange activity, results in the stimulation of Na-K-ATPase activity (Harris *et al.*, 1986). Similarly, chronic increases in Na/H antiporter activity in proximal tubules isolated from weanling, but not adult rats, stimulate Na-K-ATPase activity, also by increasing cell sodium activity (Fukuda and Aperia, 1988).

Net sodium reabsorption across the apical membrane of the fully differentiated proximal tubule is mediated by the parallel operation of Na/H and chloride/base exchangers (Aronson and Giebisch, 1997; Baum, 1987; Lorenz *et al.*, 1999; Schultheis *et al.*, 1998). Targeted disruption of the apical Na/H exchanger NHE3 indicates that this isoform accounts for ~50% of sodium-dependent proton secretion (Choi *et al.*, 2000). Several studies have revealed a maturational increase in proximal tubule Na/H exchange activity

(Baum, 1990, 1992; Beck *et al.*, 1991; Guillery *et al.*, 1994a). The activity of this antiporter in the fetal and newborn rabbit is only ~25% of that measured in the adult (Baum, 1990), increasing after the second week of life due to an increase in V_{max} associated with an increase in abundance of NHE3 message and protein (Baum *et al.*, 1995). The K_m does not change with development (Beck *et al.*, 1991). Immuno-cytochemical studies localize NHE3 protein to the apical membrane of the proximal anlagen of the S-shaped body and, in later stages of proximal tubule development, along both microvilli and apical cytoplasmic vesicles (Biemesderfer *et al.*, 1997). Those species in which nephrogenesis is complete at birth and the FENa declines immediately after birth (Merlet-Benichou and de Rouffignac, 1977; Nakamura *et al.*, 1987c), including guinea pig and sheep, demonstrate a rapid upregulation of proximal tubule apical Na/H exchange within days of birth (Guillery and Huss, 1995) due to an increase in antiporter V_{max} (Guillery *et al.*, 1994a).

A maturational increase in the activity of additional sodium-coupled cotransporters, including the sodium phosphate cotransporter (NaP$_i$-2), in the proximal tubule may also contribute to the developmental increase in proximal tubular sodium absorption. The functional and molecular basis for these developmental processes is addressed later.

b. Thick Ascending Limb of Loop of Henle   The TAL of the adult reabsorbs ~25% of the filtered load of sodium and chloride via the apical bumetanide (and furosemide)-sensitive electroneutral Na-K-2Cl cotransporter (NKCC2) and Na/H exchanger (NHE). Sodium reabsorption is driven by the favorable electrochemical gradient generated by the basolateral Na-K-ATPase (Kikeri *et al.*, 1990; Shirley *et al.*, 1998). Water is not reabsorbed in this segment. Reabsorp-tion of sodium chloride in the absence of water allows for the excretion of dilute urine, thus accounting for the identity of the TAL as the "diluting" segment. Net movement of sodium across the tubule also requires recycling of potassium across the apical membrane of the tubular cell through the ROMK potassium channel and the transport of chloride across the basolateral voltage-gated chloride channel ClC-Kb. Mutations in the genes encoding NKCC2, ROMK, or ClC-Kb have been identified in patients with Bartter's syndrome (Simon *et al.*, 1996a, b, c, 1997; Simon and Lifton, 1996; Simon *et al.*, 1996c), an autosomal-recessive disorder characterized by hypokalemic metabolic alkalosis, salt wasting, hyperreninism and hyperaldoster-onism, but normal blood pressure, hypomagnesemia, and hypercalciuria (Rodriguez-Soriano, 1999).

Whereas the mature TAL expresses the NHE isoforms NHE2 and NHE3 at the apical and NHE1 and NHE4 at the basolateral membranes, the specialized macula densa cells express only apical NHE2 and basolateral NHE4 (Biemesderfer *et al.*, 1992, 1997; Peti-Peterdi *et al.*, 2000).

Apical NKCC2 is present in both TAL and macula densa cells. Within the developing nephron, NKCC2 mRNA is detected in the TAL in the post-S shape stage once nephro-genesis is complete (on ~postnatal day 8 in rat) (Igarashi *et al.*, 1995; Schmitt *et al.*, 1999); expression increases significantly between postnatal days 10 and 40 (Yasui *et al.*, 1996). The earliest signal for NKCC2 mRNA is identified in the future macula densa, the site responsible for renin secretion and sensing tubular sodium chloride concentration for tubuloglomerular feedback (Schmitt *et al.*, 1999), with NKCC2 expression extending proximally thereafter. The pattern of immunolabeling for NHE3 in medullary regions of the 1-day old rat kidney is similar to that in the adult with label present along the apical membrane of thin and thick loops of Henle (Biemesderfer *et al.*, 1997).

The well-described natriuretic and diuretic response of newborn infants and animals (Melendez *et al.*, 1991; Ross *et al.*, 1978; Smith and Abraham, 1995; Sulyok *et al.*, 1980b; Yeh *et al.*, 1985) to furosemide is consistent with the early expression of TAL cotransporter activity. However, the capacity of the TAL to establish a transepithelial sodium gradient, in the absence of tubular flow or water reabsorp-tion, is limited in the newborn rabbit (Horster, 1978). The maximal reduction in osmolality achieved by the neonatal TAL perfused with an isosmotic solution was less than 25% of that observed by segments isolated from 1-month-old animals (Horster, 1978); net fluid movement in the TAL was not detected in either age group. The steady-state sodium concentration gradient generated in the absence of flow averaged 45 mmol/liter at 1 week and 75 mmol/liter at 1 month (Horster, 1978). To the extent that the Na-K-2Cl cotransporter generates the transepithelial sodium gradient, these data suggest that activity of this cotransporter is limited early in life. The developmental increase in diluting capacity of this segment has been further demonstrated in free flow micropuncture studies that revealed a fall in tubular fluid osmolarity in the early distal tubule from 285 mosmol/liter in 2-week-old rats to 180 mosmol/liter by 1 month of age (Zink and Horster, 1977).

Na-K-ATPase activity (V_{max}) in the neonatal TAL is only 20% of that reported in the mature nephron when expressed per unit of dry weight (Schmidt and Horster, 1977). The postnatal increase in medullary pump activity is accompa-nied by an increase in the message encoding the α1 subunit of the Na-K-ATPase (Yasui *et al.*, 1996).

c. Distal Convoluted Tubule (DCT)   The DCT of the adult reabsorbs ~5% of the filtered load of sodium chloride. The observation that the DCT of ~20-day-old saline-loaded rats reabsorb more sodium than that of 40-day-old animals is consistent with avid sodium reabsorption in this segment during early life.

Sodium reabsorption in the early distal tubule of the adult is mediated by a thiazide-sensitive NaCl cotransporter (TSC or NCC) (Ellison *et al.*, 1987), driven by the basolateral Na-K-ATPase. The distal part of this segment also expresses a sodium-calcium (NaCa) exchanger, the amiloride-sensitive epithelial sodium channel ENaC (described in further detail later), and 11β-hydroxysteroid dehydrogenase (11β-HSD), (Schmitt *et al.*, 1999), an enzyme that confers mineralocorticoid specificity to aldosterone target tissues (discussed in detail later). In the developing nephron, apical TSC expression first appears in the post-S shape stage within the distal portion of the DCT, extending back to the post-macula densa segment during the stage of tubule elongation (Schmitt *et al.*, 1999). Immunodetectable apical TSC and ENaC, basolateral NaCa (see later), and cytoplasmic 11β-HSD can be first detected in the developing nephron at the stage of tubule elongation (Schmitt *et al.*, 1999).

Mutations in the gene encoding TSC are associated with Gitelman's syndrome (Simon *et al.*, 1996c), an autosomal–recessive disorder that typically presents with hypokalemic alkalosis, hypomagnesemia, and hypocalciuria. In the absence of functional TSC, apical sodium uptake is limited, leading to enhanced basolateral sodium entry mediated by the NaCa exchanger.

d. Distal Nephron Within the fully differentiated distal nephron, the connecting tubule (CNT) and cortical collecting duct are responsible for the final regulation of sodium balance. The CNT expresses proteins characteristic of segments derived from the nephrogenic blastema as well as the ureteric bud, and has thus been suggested to represent a hybrid epithelium arising from mutual induction of these adjoining structures (Schmitt *et al.*, 1999). Nascent CNTs forming arcades are composed of a more mature "stem" portion, which includes a major proximal part of the arcade and a distally adjoining "neck" portion composed of immature cells (Neiss, 1982). Whereas immunoreactive NaCa is observed solely in the stem portion, ENaC and

11β-HSD are expressed very early and exclusively in the neck region (Schmitt *et al.*, 1999). These observations are consistent with the derivation of the CNT from both the ureteric bud and the nephrogenic blastema. Although the function of the CNT has not been examined in the fetal or neonatal kidney, the abundant expression of mineralocorticoid receptor (MR) mRNA in the neonate, threefold higher than that detected in the adult (Kalinyak *et al.*, 1992), suggests that components of mineralocorticoid-sensitive sodium absorption in this segment may be active in early development.

CCDs isolated from 1-week-old rabbits show no significant net sodium transport (Satlin, 1994; Fig. 18.9). These results are consistent with the clinical observation of the obligatory natriuresis (increase in sodium and water excretion) that occurs after birth and the excessive urinary sodium losses that prevail in premature human infants in whom, like in the neonatal rabbit, nephrogenesis is incomplete at birth. Both the net reabsorptive flux of sodium and that inhibited by ouabain, reflecting the active component of reabsorption driven by the basolateral Na-K-ATPase, are low at birth, but increase significantly after 2 weeks of age in this species (Satlin, 1994; Vehaskari, 1994).

The CCD is a heterogeneous segment composed of both principal and the less numerous intercalated cells (Giebisch, 1998; Satlin, 1999). Active sodium reabsorption in the CCD requires apical sodium entry into the principal cell through the amiloride-sensitive sodium channel (ENaC) and its extrusion at the basolateral membrane by the Na-K-ATPase. The rate-limiting step in this transport process appears to be located at the apical channel. Patch clamp analysis has shown that the postnatal increase in net sodium transport in the CCD is associated with an increase in the number and open probability of apical sodium channels (ENaCs) in the differentiating principal cell (Fig. 18.10) (Huber *et al.*, 1999; Satlin and Palmer, 1996).

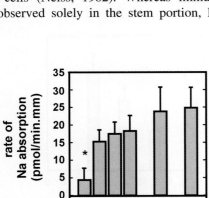

Figure 18.9 Relationship between postnatal age and net sodium absorption in isolated microperfused rabbit cortical collecting ducts. Most of the increase in net sodium absorption occurs between the first and the second weeks of postnatal life. Reproduced with permission from Satlin (1999).

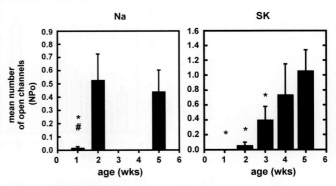

Figure 18.10 Postnatal changes in number of conducting Na (ENaC) and secretory potassium (SK) channels per cell-attached patch of apical membrane of rabbit principal cells. The number of conducting Na channels increased ~30–fold between the first and second weeks of life, whereas the number of functional SK channels increased gradually after the second postnatal week. Reproduced with permission from Satlin (1999).

The fully functional ENaC is composed of three homologous subunits, which are encoded by three different genes (Canessa *et al.*, 1994; Hansson *et al.*, 1995; Shimkets *et al.*, 1994). Expression of message for the α, β, and γ ENaC subunits increases in the near-term fetal rat kidney (Huber *et al.*, 1999; Vehaskari *et al.*, 1998; Watanabe *et al.*, 1999), due, at least in part, to increases in the abundance of mRNA in the ureteric bud and CCD (Huber *et al.*, 1999). Of interest is that the α-ENaC message and functional sodium channel protein are expressed in the ureteric bud well before filtrate is delivered to this site (Huber *et al.*, 1999). Mature levels of expression are achieved within the first few days after birth (Vehaskari *et al.*, 1998; Watanabe *et al.*, 1999). Immunodetectable α-ENaC is identified at the apical membrane of a majority of cells in CCDs in rats older than 1 week of age, whereas β- (Schmitt *et al.*, 1999; Zolotnitskaya and Satlin, 1999) and γ-ENaC (Zolotnitskaya and Satlin, 1999) subunit expression is detected routinely, albeit in a mostly cytoplasmic distribution, at birth.

Ultrastructural and morphometric analyses of principal cells in the neonatal CCD provide evidence for an incompletely differentiated epithelium (Evan *et al.*, 1991; Satlin *et al.*, 1988). On transmission electron microscopy, mid-cortical principal cells possess few organelles, simple apical and basolateral cell surfaces, and occasional areas of glycogen deposition. In contrast, principal cells in the mature CCD possess more organelles, mitochondria, and basolateral infoldings than observed earlier in life. The ultrastructural finding in neonatal principal cells of simple basolateral membrane surfaces is compatible with the observation that the rate of basolateral ouabain-inhibitable rubidium uptake, an index of basolateral Na-K pump activity, in neonatal CCDs is only 50% of that measured in segments from mature animals (Constantinescu *et al.*, 2000; Fig. 18.11).

The ~2.5-fold maturational increase in Na-K-ATPase hydrolytic activity in the CCD (Constantinescu *et al.*, 2000; Schmidt and Horster, 1977) is consistent with an increase in the abundance of active enzyme. Comparisons between the rates of basolateral Rb uptake and Na-K-ATPase activity suggests that the basolateral pump operates far below V_{max} at all ages, probably reflecting the low intracellular sodium concentration (rate-limiting condition) in intact tubules.

Developmental differences in the polarity and subunit composition of the Na-K-ATPase also exist. The apical localization of immunodetectable α subunit in the distal nephron of human (Burrow *et al.*, 1999) and rodent (Horster *et al.*, 1997) during early gestation is associated with expression of the β2 isoform at the same membrane (Burrow *et al.*, 1999). At birth, the α subunit of the Na-K-ATPase is expressed along the basolateral membranes of cortico-medullary collecting ducts, as well as both apical and basolateral membranes of CCDs adjacent to the ampullae, and is absent in the ampullae (Minuth *et al.*, 1987). This axial heterogeneity of pump distribution early in life may represent one stage in the differentiation of embryonic ampullary cells to functionally mature collecting duct principal cells.

Mutations of ENaC are associated with two distinct diseases. Gain-of-function mutations in the β and γ subunits of ENaC lead to Liddle's syndrome, a rare form of salt-sensitive hypertension (Lifton, 1996; Schild *et al.*, 1995; Shimkets *et al.*, 1994). The hypertension is due to constitutive activation of ENaC at the apical membrane, leading to unregulated sodium reabsorption, plasma volume expansion, and inhibition of renin and aldosterone secretion. Loss-of-function mutations in the α or β subunits of ENaC result in autosomal-recessive pseudohypoaldosteronism type 1, a rare genetic disorder characterized by renal salt wasting

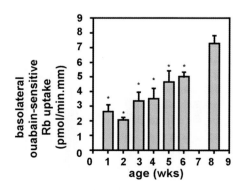
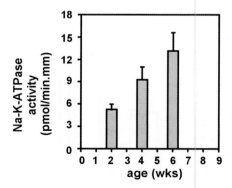

Figure 18.11 Postnatal maturation of basolateral Na-K-ATPase-mediated rubidium (Rb, marker for K) uptake (left) and total hydrolytic activity (right) in single rabbit cortical collecting ducts. The rate of basolateral ouabain-inhibitable Rb uptake, an index of basolateral Na-K pump activity, in neonatal CCDs is only 50% of that measured in segments from mature animals. The ~2.5-fold maturational increase in Na-K-ATPase hydrolytic activity in the CCD is consistent with an increase in abundance of active enzyme. Comparison of the rates of basolateral Rb uptake and Na-K-ATPase activity suggests that the basolateral pump operates far below V_{max} at all ages. Adapted with permission from Constantinescu *et al.* (2000).

with hyperkalemia, variable degrees of metabolic acidosis, and an end organ resistance to the high circulating levels of aldosterone (Chang *et al.*, 1996; Geller *et al.*, 1998; Strautnieks *et al.*, 1996).

e. Medullary Collecting Duct (MCD) Active sodium transport gradually disappears as the collecting duct descends into the medulla of the adult kidney (Sonnenberg *et al.*, 1990; Stokes, 1982). The sodium transport capacity of the fetal and neonatal MCD is unknown. Studies identify a progressive shift in the intensity of expression of ENaC mRNA (β and γ subunits) and immunoreactive protein from the inner medulla to cortex during the transition from fetal to neonatal life in the rat (Duc *et al.*, 1994; Schmitt *et al.*, 1999; Watanabe *et al.*, 1999). The physiologic significance of the latter observation remains to be defined.

3. Regulation of Renal Sodium Transport

a. Sympathetic Nervous System Renal sympathetic nerve activity plays a part not only in the regulation of renal hemodynamics, as described earlier, but also in the regulation of renal sodium excretion. Direct renal nerve stimulation in fetal and newborn sheep leads to sodium retention (Robillard and Nakamura, 1988b), a response qualitatively similar to that observed in adult animals and attributed to norepinephrine acting on α-adrenergic receptors (Hayashi *et al.*, 1999). Surgical or pharmacological sympathectomy of fetal lambs (Smith *et al.*, 1991) or newborns rats (Appenroth and Braunlich, 1981; Slotkin *et al.*, 1988), respectively, leads to a larger natriuresis than that observed in age-matched controls. Stimulation of proximal tubule sodium transport in response to norepinephrine is mediated by an increase in basolateral Na-K-ATPase activity (Aperia *et al.*, 1992; Beach *et al.*, 1987) and apical Na/H exchange (Gesek *et al.*, 1989; Nord *et al.*, 1987)

Although the predominance of α-adrenergic influence in the newborn kidney (Felder *et al.*, 1983) predicts that α-adrenergic blockade should result in a larger natriuresis in younger compared to adult subject, the response is species specific. Intrarenal infusion of an α_1-adrenoceptor agonist into sheep leads to a greater reduction in urinary volume and sodium excretion in newborn than fetal or adult animals (Guillery *et al.*, 1994b). A different effect is observed in dogs (Fildes *et al.*, 1985); the natriuresis observed in adults, which is not associated with changes in RVR, RBF, and GFR, exceeds that in pups. This finding may reflect the high distal avidity for sodium reabsorption or the absence of an endogenous β-adrenergic and dopamine-coupled natriuretic response early in postnatal life in this species (Fildes *et al.*, 1985).

b. Dopamine Dopamine inhibits sodium reabsorption in the nephron by activating D_1-like dopamine receptors. Receptor activation results in the inhibition of Na-K-ATPase activity via a phospholipase C-linked protein kinase C (PKC) cascade within the proximal tubule, medullary thick ascending limb, and CCD and triggers a cAMP-linked protein kinase A (PKA) pathway that inhibits NHE3 Na/H exchange activity in the proximal tubule (Aperia, 2000; Baum and Quigley, 1998; Bello-Reuss *et al.*, 1982; Felder *et al.*, 1990a; Missale *et al.*, 1998; Ominato *et al.*, 1996; Wiederkehr *et al.*, 2001). The natriuretic effect of dopamine is less pronounced in the fetus and neonate kidney of many species, including human, compared to the adult due to blunted renal hemodynamic and tubular effects (Felder *et al.*, 1989; Jaton *et al.*, 1992; Kaneko *et al.*, 1992; Pelayo *et al.*, 1984; Segar *et al.*, 1992; Seri, 1990). The inhibitory effect of dopamine on NHE3 and Na-K-ATPase activity in the neonatal proximal tubule and medullary TAL is attenuated compared to that observed in adult rats (Aperia, 2000; Fryckstedt *et al.*, 1993; Fukuda *et al.*, 1991; Kaneko *et al.*, 1992; Li *et al.*, 2000). The limited natriuretic effect early in life may be due to an inefficiency in the production of second messengers (Fryckstedt *et al.*, 1993; Kinoshita and Felder, 1990) or the prevalence of alternate pathways that regulate the function of sodium-coupled transporters (Li *et al.*, 2000). The density of D_1-like receptors, low in the newborn proximal tubule (Kaneko *et al.*, 1992; Kinoshita and Felder, 1990; Tenore *et al.*, 1997), does not change in the brush border during maturation (Li *et al.*, 2000).

c. Atrial Natriuretic Factor (ANF) In the mature kidney, ANF increases sodium excretion and GFR, antagonizes renal vasoconstriction, and inhibits renin secretion (Maack, 1996). ANF release from atrial cardiocytes is stimulated in the fetus by an increase in intracardiac pressure and atrial distention (Panos *et al.*, 1989; Robillard and Weiner, 1988; Ross *et al.*, 1987, 1988a) and levels fall in response to a decrease in central venous pressure, such as hemorrhage (Cheung and Brace, 1991). The attenuated natriuretic and diuretic response to systemically infused ANF is lower in newborns than in adult counterparts (Chevalier *et al.*, 1988; Hargrave *et al.*, 1989; Robillard *et al.*, 1988b; Semmekrot *et al.*, 1990; Tulassay *et al.*, 1986). Whereas specific ANF receptors have been identified on near-term fetal glomerular membranes of rabbit, the ANF-binding capacity is age dependent, increasing seven-fold between fetal and adult life (Castro *et al.*, 1991). The blunted response of the immature kidney to ANF has been attributed to ineffective production of the second messenger cGMP (Chevalier *et al.*, 1988; Muchant *et al.*, 1995), a rapid systemic clearance of ANF (Chevalier *et al.*, 1988; Robillard *et al.*, 1988b), and a greater ANF-induced reduction in RBF in fetuses and newborns than in adults (Robillard *et al.*, 1988b). In support of the latter mechanism is the observation that renal perfusion pressure is a major determinant of the natriuretic response to ANF in adult animals (Paul *et al.*, 1987; Seymour *et al.*, 1987).

d. Glucocorticoids Circulating levels of glucocorticoids, including cortisol and corticosterone, surge in many species during or just before the period of weaning (Henning, 1978, 1981; Malinowska *et al.*, 1972; Malinowska and Nathanielsz, 1974). Glucocorticoids activate cytoplasmic receptors; the hormone–receptor complex is then transported into the nucleus where it binds to specific glucocorticoid response elements (GREs) on the 5′ flanking region of target genes, and thereby induces transcription (Beato *et al.*, 1989, 1996). Glucocorticoid receptors appear early in the fetal metanephric kidney and increase to adult levels shortly after birth (Ellis *et al.*, 1986; Kalinyak *et al.*, 1989). The receptors are ubiquitous with message and protein detected along the entire nephron (Bonvalet, 1987; Farman *et al.*, 1991; Lee *et al.*, 1983; Todd-Turla *et al.*, 1993).

Both endogenous gluco- and mineralocorticoids bind to the mineralocorticoid receptor (MR) with equal affinity (Farman, 1999). The observation that blood glucocorticoid concentrations are ~100-fold higher than aldosterone concentrations predicts that occupancy of the MR by circulating glucocorticoids could markedly enhance sodium retention. However, in aldosterone target tissues, such as the principal cells of the CCD and epithelial cells of the distal colon, the MR is protected from glucocorticoids by 11β-hydroxysteroid dehydrogenase type 2 (11β-HSD2), an enzyme that metabolizes cortisol (corticosterone in rodents) into inactive derivatives that have only a low affinity for the MR (Farman, 1999). Thus, 11β-HSD2 plays a crucial role in protecting aldosterone responsive cells from the relative excess of circulating glucocorticoids, and allows aldosterone to bind to its receptor and activate downstream target genes.

The importance of 11β-HSD2 and the MR in the regulation of salt and blood pressure homeostasis is evident from the examination of human subjects and mouse models with mutations in the respective genes. Homozygous mutations in the gene encoding 11β-HSD2 lead to a syndrome of apparent mineralocorticoid excess that mimics hyperaldosteronism with hypertension and hypokalemic alkalosis (Kotelevtsev *et al.*, 1999; Mune *et al.*, 1995; White *et al.*, 2000). Patients with mutations in the MR gene present with pseudohypoaldosteronism type I (PHA1), albeit a milder disease than that associated with a mutation in ENaC (Berger *et al.*, 1998; Geller *et al.*, 1998). MR mutant mice generally die by the second week of life due to renal salt wasting, leading to hyponatremia, hyperkalemia, and activation of the renin–angiotensin–aldosterone system (RAAS) (Berger *et al.*, 1998, 2000; McDonald *et al.*, 1999).

The level of expression of MR mRNA in the neonatal kidney exceeds that detected in the adult (Kalinyak *et al.*, 1992); receptors are present in fetal life (Pasqualini *et al.*, 1972). 11β-HSD2 activity is lower in the newborn than in the adult kidney, due in part to low levels of expression in the proximal tubule (Brem *et al.*, 1994). Expression of 11β-HSD2 weak in the CCD and strong in the MCD during early development, becomes stronger in the CCD than MCD with maturation (Bachmann *et al.*, 1999; Schmitt *et al.*, 1999). The presence of ENaC, MR, and low levels of 11β-HSD2 in the CCD suggests that glucocorticoids may act as sodium-retaining steroids during early postnatal life (Bostanjoglo *et al.*, 1998; Farman, 1992).

Chronic exposure to synthetic or endogenous glucocorticoids stimulates transcription and activity of a variety of sodium-transporting proteins in the fetal and neonatal kidney. Administration of glucocorticoids to pregnant rabbits enhances volume absorption (equivalent to sodium absorption) and bicarbonate transport in the proximal tubules of the fetal and neonatal kidneys by stimulating activity of the apical NHE3, basolateral $Na(HCO_3)_3$ symporter, and basolateral Na-K-ATPase (Aperia *et al.*, 1981b; Baum *et al.*, 1995; Baum and Quigley, 1991; Beck *et al.*, 1991; Guillery *et al.*, 1995; Gupta *et al.*, 2001; Igarashi *et al.*, 1983; Petershack *et al.*, 1999). These functional changes are associated with increases in the expression of message and protein for NHE3 (Baum *et al.*, 1995) and both the α1 and β1 subunits of the Na-K-ATPase (Celsi *et al.*, 1991; Petershack *et al.*, 1999; Wang and Celsi, 1993; Wang *et al.*, 1994). Postnatal systemic glucocorticoid treatment also stimulates proximal tubular Na-K-ATPase activity and volume absorption in the neonate (Aperia and Larsson, 1984; Aperia *et al.*, 1981b; Igarashi *et al.*, 1983; Schwartz and Evan, 1984). Transcriptional regulation of Na-K-ATPase subunits by glucocorticoids is tissue- and age-dependent, occurring only during a limited period of fetal and neonatal life, being maximal at 10 days postnatal age in the rat (Celsi *et al.*, 1993; Wang *et al.*, 1994; Yasui *et al.*, 1996). The time-limited window of responsiveness to the Na-K-ATPase glucocorticoid effect is mediated by the appearance of an ontogenically regulated auxiliary factor that modulates the interaction between the glucocorticoid receptor and the Na-K-ATPase GRE (Wang *et al.*, 1994).

In addition to proximal tubular effects, glucocorticoids induce transepithelial sodium transport in the CCD (Laplace *et al.*, 1992; Naray-Fejes-Toth and Fejes-Toth, 1990). This biological response apparently reflects glucocorticoid crossover binding to MRs (Naray-Fejes-Toth and Fejes-Toth, 1990). When 11β-HSD2 is absent or suppressed, as observed following ingestion of licorice derivatives or other competitive inhibitors, endogenous glucocorticoids can occupy both glucocorticoid receptors and MRs in the distal nephron, leading to enhanced sodium reabsorption, expansion of the extracellular fluid volume, and hypertension (Brem *et al.*, 1992; Gaeggeler *et al.*, 1989; Morris and Souness, 1992; Ulmann *et al.*, 1975; Whorwood *et al.*, 1993).

e. Renin–Angiotensin–Aldosterone System Angiotensin II (ANG II) and aldosterone play major roles in the regulation of renal sodium reabsorption, and of the extracellular fluid volume and blood pressure, independent of its effect on GFR. ANG II regulates sodium transport via binding to AT1 receptors in the proximal tubule, TAL, and collecting duct (Mujais *et al.*, 1986; Schlatter *et al.*, 1995; Wang and Giebisch, 1996; Weiner *et al.*, 1995). The binding capacity for ANG II is higher in the proximal convoluted tubule than in all other segments (Mujais *et al.*, 1986). Within this segment, its effects are mediated by binding to basolateral receptors (Baum *et al.*, 1997; Harrison-Bernard *et al.*, 1997; Miyata *et al.*, 1999). Proximal tubular synthesis and secretion of ANG II into the lumen and binding to luminal receptors also regulate transport in this segment in an autocrine or paracrine fashion, independent of circulating ANG II levels (Braam *et al.*, 1993; Quan and Baum, 1996). Low concentrations of ANG II stimulate, whereas supraphysiologic high concentrations of ANG II inhibit sodium and water transport (Feraille and Doucet, 2001; Harris and Young, 1977). Luminal ANG II receptor blockade or luminal inhibition of ANG II synthesis reduces the rate of proximal tubule reabsorption by ~40% (Quan and Baum, 1996). The latter effect can be reversed by the administration of luminal ANG II. The sensitivity of the neonatal kidney to ANG II is demonstrated by the observation that chronic treatment of rat pups with an AT1 receptor blocker leads to increases in basal urine flow and sodium excretion compared to control littermates (Chevalier *et al.*, 1996).

ANG II regulates proximal tubular sodium and fluid reabsorption in the adult via the coordinated modulation of apical Na/H exchange and basolateral $Na(HCO_3)_3$ cotransport (Baum *et al.*, 1997; Bloch *et al.*, 1992; Geibel *et al.*, 1990; Houillier *et al.*, 1996; Quan and Baum, 1997). These effects probably result from complex interactions between multiple signaling pathways, including G-protein-mediated depression of adenosine 3'5'-cyclic monophosphate (cAMP) production (Liu and Cogan, 1989). In contrast to the transporter-specific effects observed in the adult, exogenous ANG II does not alter the V_{max} of Na/H antiport activity in cortical brush border membrane vesicles or NHE3 mRNA abundance in the renal cortex of near-term fetal sheep (Guillery *et al.*, 1996). These data suggest it unlikely that ANG II plays an important role in the regulation of NHE3 activity in the neonate.

Aldosterone action requires its initial binding to the MR, followed by translocation of the hormone–receptor complex to the nucleus where specific genes are stimulated to code for physiologically active proteins (e.g., Na-K-ATPase). Although the collecting duct is considered to be the primary target for mineralocorticoid action, the DCT also responds to chronic aldosterone treatment with an increase in expression of the NaCl cotransporter TSC (Kim *et al.*, 1998). Cellular effects within the fully differentiated CCD include increases in the density of conducting sodium channels (Bonvalet, 1998; Pacha *et al.*, 1993) due to *de novo* synthesis of the α-subunit of ENaC (Masilamani *et al.*, 1999; Mick *et al.*, 2001), activation of preexisting channels

and enhancement of basolateral Na-K-ATPase activity (Doucet, 1992). The net effect of these actions is the stimulation of net sodium absorption (Fig. 18.12) (Schwartz and Burg, 1978). The increase in sodium absorption additionally leads to an increase in lumen-negative transepithelial voltage, thereby facilitating net potassium secretion. The effects of aldosterone on ENaC and the Na-K pump appear to be indirect, mediated by aldosterone-induced proteins (Garty and Palmer, 1997; Verrey, 1995).

Plasma aldosterone concentrations in the newborn are high when compared to the adult (Van Acker *et al.*, 1979) and have been proposed to contribute to the avid sodium absorptive state characteristic of the full term infant. However, clearance studies in fetal (Robillard *et al.*, 1985) and newborn (Stephenson *et al.*, 1984) animals and premature infants (Aperia *et al.*, 1979; Sulyok *et al.*, 1979b) identify a blunted responsiveness of the immature kidney to aldosterone. Mineralocorticoid administration to neonatal rabbits fails to stimulate sodium absorption in the CCD (Vehaskari, 1994). Given that the density of aldosterone binding sites, receptor affinity, and degree of nuclear binding of hormone-receptor are believed to be similar in mature and immature rats (Stephenson *et al.*, 1984), the early hyposensitivity to aldosterone is considered to represent a postreceptor phenomenon. Overall, these data suggest that endogenous mineralocorticoids play a limited role in the regulation of cation transport in the CCD early in postnatal life.

Urinary aldosterone excretion is lower in premature than term infants, despite the high PRA in the former. This suggesting that premature infants can augment their PRA in response to salt wasting, but that their adrenals initially fail to respond adequately with the stimulation of aldosterone synthesis (Robillard *et al.*, 1982; Siegel *et al.*, 1981; Spitzer, 1982). The relative hypoaldosteronism results in an inability to conserve sodium, manifested clinically by weight loss and hyponatremia.

f. Other The developmental increases in both active and passive sodium absorption and apical Na/H exchange activity in the proximal tubule require thyroid hormone (Baum *et al.*, 1998a; Shah *et al.*, 2000). Serum thyroid hormone levels, low in the immediate postnatal period, increase at about the time of weaning (Walker *et al.*, 1980). Administration of thyroid hormone to hypothyroid adult rodents restores proximal tubule Na-K-ATPase activity and volume absorption (Barlet and Doucet, 1986; Capasso *et al.*, 1985; De Santo *et al.*, 1980; Kinsella and Sacktor, 1985).

Nitric oxide induces a natriuresis secondary to inhibition of proximal tubular Na/H exchange and Na-K-ATPase activity and, in the CCD, via an effect on ENaC (Stoos *et al.*, 1994, 1995; Stoos and Garvin, 1997). Inhibition of basal NO production in the fetus and neonate blocks the natriuresis that accompanies volume loading (Bogaert *et al.*, 1993).

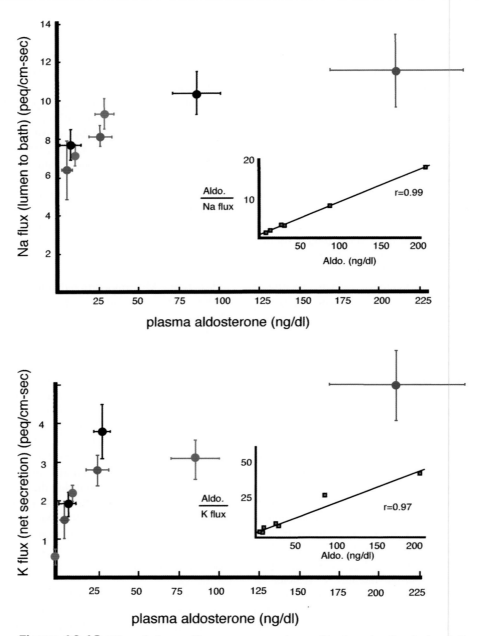

Figure 18.12 Effect of plasma aldosterone concentration on Na absorptive flux (top) and K secretory flux (bottom) across isolated perfused rabbit cortical collecting ducts. An increase in the circulating mineralocorticoid level is associated with an increase in sodium absorption and potassium secretion. After Schwartz and Burg (1978).

Although prostaglandin E_2 (PGE_2), PGD_2, and PGI_2 reduce sodium chloride and water absorption in the adult, promoting a natriuresis and diuresis, inhibition of prostaglandin production in the neonate by the administration of indomethacin does not affect salt and water excretion (Osborn *et al.*, 1980; Winther *et al.*, 1980). The detection in fetal sheep treated with indomethacin of increases in urinary sodium excretion despite a decrease in RBF (Matson *et al.*, 1981) suggests that renal prostaglandins in the fetal kidney may have tubular effects on sodium and chloride absorption that are opposite to those generally ascribed to adult kidneys.

Vasopressin (AVP), which is secreted in response to either an increase in plasma osmolality or a decrease in plasma volume, not only increases the water permeability of the collecting duct, but also stimulates sodium reabsorption in the TAL (Knepper *et al.*, 1999; Wittner *et al.*, 1988) and collecting duct (Schafer and Hawk, 1992). Binding of AVP to basolateral V_2 receptors activates adenylyl cyclase and cAMP generation. Acute effects of receptor–ligand binding include an increase in the number of conducting ENaCs at the apical membrane (Marunaka and Eaton, 1991). Chronic exposure to AVP upregulates the expression of the Na-K-2Cl cotransporter (NKCC2) in the TAL (Kim et al., 1999) and increases the abundance of ENaC protein in the collecting duct (Ecelbarger et al., 2000). The ANF-induced diuresis and natriuresis discussed earlier is due, at least in part, to inhibition of the V_2 receptor-mediated action of AVP in the collecting ducts.

B. Potassium

Premature infants and neonatal subjects maintain a state of positive potassium balance for somatic growth, unlike adults who are in zero balance. Cumulative evidence indicates that the neonatal kidney contributes to this potassium retention. The renal potassium clearance is low in newborns, even when corrected for their low GFR (Satlin, 1999; Sulyok et al., 1979b). In response to potassium loading, infants, like adults, can excrete potassium at a rate that exceeds its filtration (Tuvdad et al., 1954), indicating a capacity for net tubular secretion. However, the rate of potassium excretion expressed either per unit body or kidney weight in infants and young animals subject to exo-genous potassium loading is less than that observed in older animals (Lorenz et al., 1986; McCance and Widdowson, 1958). In general, the limited potassium secretory capacity of the immature kidney becomes clinically relevant only under conditions of potassium excess.

1. Overview of Potassium Handling by the Maturing Kidney

Filtered potassium is reabsorbed almost entirely in proximal segments of the nephron, with urinary potassium being derived predominantly from distal potassium secretion. Therefore, potassium balance, at least in the adult, is maintained by renal secretion rather than reabsorption.

Potassium is freely filtered at the glomerulus. Approx-imately 50% of the filtered load of potassium is reabsorbed passively along the initial two-thirds of the proximal tubule of both suckling (Lelievre-Pegorier et al., 1983; Solomon, 1974a; Solomon, 1974b) and adult (Malnic et al., 1966a,b, 1971) rats. In the adult rat, only 5–15% of the filtered load of potassium reaches the superficial early distal tubule, reflecting significant further net reabsorption of this cation in the intervening nephron segments (Malnic et al., 1966b). In contrast, up to 35% of the filtered load of potassium reaches the superficial distal tubule of the 2-week-old rat compared to 10% in 5-week-old animals (Lelievre-Pegorier et al., 1983). The fractional reabsorption of potassium along the loop of Henle, expressed as a percentage of delivered load, increases from 60% at 13 days to 80% by 40 days of

postnatal age (Lelievre-Pegorier et al., 1983), suggesting that this segment is functionally immature early in life. Although the collecting duct is inaccessible to micro-puncture, clearance studies provide evidence, albeit indirect, for a low rate of potassium secretion and an enhanced fractional reabsorption of potassium in young dogs when compared to the adult (Kleinman and Banks, 1983; Lorenz et al., 1986).

2. Sites of Potassium Transport Along the Nephron

a. Proximal Tubule   The major driving forces for potassium reabsorption in the proximal tubule in the adult are solvent drag and diffusion (Giebisch, 1998). Potassium ions within the tubular fluid are reabsorbed predominantly along the paracellular pathways, driven in part by the positive transepithelial voltage that prevails along part of the proximal tubule. Given the similarities in fractional water, sodium, and potassium absorption along the proximal tubule of young and adult rats (Lelievre-Pegorier et al., 1983), it is presumed, but not proven, that similar pathways mediate net potassium movement in both age groups.

b. Thick Ascending Limb of Loop of Henle   In the fully differentiated TAL, potassium reabsorption from the tubular fluid is mediated by the apical electroneutral Na-K-2Cl cotransporter (NKCC2) in series with basolateral potassium exit pathways, including potassium-selective channels or cotransport with chloride or bicarbonate (Giebisch, 1985, 1998). Potassium is also reabsorbed across the paracellular pathway, driven by the lumen-positive transepithelial voltage (Schwab and Oberleithner, 1988). An apical small conductance potassium (SK) channel, encoded by ROMK (also present in the CCD; see later), recycles potassium across the luminal membrane to ensure an abundant supply of substrate for the cotransporter (Greger et al., 1991; Greger and Schlatter, 1981; Wang, 1994). The observations that both diluting capacity (reflecting Na-K-2Cl cotransport) and Na-K-ATPase activity increase postnatally, although NKCC2 mRNA (Igarashi et al., 1995; Schmitt et al., 1999) and immunodetectable apical ROMK protein (Zolotnitskaya and Satlin, 1999) are already present in the neonatal rat TAL, are consistent with developmental maturation of potassium absorptive pathways in this segment. Note, however, that direct functional analysis of the potassium transport capacity of the TAL in the developing nephron has not been performed.

c. Distal Nephron and Collecting Duct   Urinary potas-sium excretion is derived almost entirely from potassium secretion in distal segments of the nephron, including the CNT and CCD (Giebisch, 1998). In contrast to the high rate of potassium secretion observed in CCDs isolated from adults, segments isolated from neonatal animals and perfused at physiological flow rates show no significant net

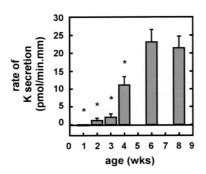

Figure 18.13 Relationship between postnatal age and net potassium secretion in isolated microperfused rabbit cortical collecting ducts. Net potassium secretion, absent in the immediate postnatal period, increased significantly after the third week of postnatal life. Reproduced with permission from Satlin (1999).

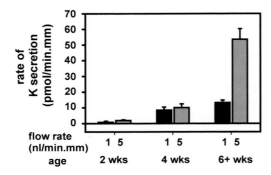

Figure 18.14 Effect of luminal flow rate on net potassium secretion in microperfused cortical collecting ducts isolated from 2-, 4-, and ≥6-week-old rabbits. Whereas potassium secretion in the mature CCD is strongly stimulated by an increase in the luminal flow rate, a similar relationship is not obtained in early life. Reproduced with permission from Satlin (1999).

potassium transport until after the third week of postnatal life (Fig. 18.13) (Satlin, 1994). By 6 weeks of age, the rate of net potassium secretion in the CCD is comparable to that observed in the adult (Satlin, 1994). Whereas potassium secretion in the mature CCD is strongly stimulated by an increase in luminal flow rate, a similar relationship does not obtain in early life (Fig. 18.14). The developmental appearance of flow-stimulated potassium secretion appears ~2 weeks after baseline potassium secretion is first detected (Satlin, 1994). These results indicate that the low rates of potassium excretion characteristic of the newborn kidney are due, at least in part, to a low secretory capacity of the CCD.

As discussed earlier, the fully differentiated principal cell, the majority cell type in the CCD, absorbs sodium (and water in the presence of AVP) and secretes potassium. Acid–base transporting intercalated cells may play a role in potassium reabsorption under certain conditions (Satlin, 1999). The direction and magnitude of net potassium transport in this segment reflect the balance of potassium

secretion and absorption, opposing processes mediated by principal and intercalated cells, respectively.

Principal cell. Potassium secretion in the CCD requires the active uptake of potassium into the principal cell in exchange for sodium by the basolateral Na-K pump. The high cell potassium concentration and lumen-negative voltage within the CCD, generated by apical sodium entry and its electrogenic basolateral extrusion, favor the passive diffusion of cell potassium into the luminal fluid. The magnitude of potassium secretion in the CCD is determined by its electrochemical gradient and the permeability of the apical membrane to potassium. Two apical potassium-selective channels have been identified in the CCD by patch clamp analysis. The small conductance potassium (SK) channel, encoded by ROMK (Ho *et al.*, 1993; Zhou *et al.*, 1994), has a high open probability at the resting membrane potential and is considered to mediate baseline potassium secretion (Frindt and Palmer, 1989; Satlin and Palmer, 1997; Wang *et al.*, 1990). Although the high-conductance, stretch- and calcium-activated maxi-K channel is closed at the physiologic membrane potential and intracellular calcium concentration (Pacha *et al.*, 1991; Satlin and Palmer, 1996), two observations suggest that this channel contributes to urinary potassium secretion. First, patients with Bartter's syndrome due to loss-of-function mutations in ROMK have modest hypokalemia (Rodriguez-Soriano, 1999; Simon and Lifton, 1996), and not hyperkalemia, as might be expected in the absence of functional secretory potassium channels. Second, the developmental appearance of flow–dependent potassium lags behind that of baseline secretion, suggesting that baseline and flow–dependent potassium secretion are mediated by distinct channels with unique developmental patterns of expression (Satlin, 1994). Indeed, evidence suggests that the maxi-K channel mediates flow-stimulated potassium secretion in a calcium-dependent manner (Woda *et al.*, 2001b).

The limited capacity of the neonatal CCD for baseline net potassium secretion appears not to be due to an unfavorable electrochemical gradient (Satlin, 1999) but, at least in part, to a paucity of apical conducting SK channels (Fig. 18.10) (Huber and Horster, 1996; Satlin and Palmer, 1997). Patch clamp analysis of maturing rabbit principal cells identified an increase in the number of conducting SK channels per patch after the second week of life, ~1 week after an increase in ENaC activity is detected (Satlin and Palmer, 1996, 1997). The developmental increase in number of SK channels appears to derive primarily from an increase in transcription and translation of functional ROMK channel proteins (Benchimol *et al.*, 2000; Satlin, 1994; Zolotnitskaya and Satlin, 1999); the open probability of the SK channel did not change between 3 and 5 weeks of age, a time interval during which the number of conducting channels increased almost three-fold (Satlin and Palmer, 1997). Immuno-detectable (Zolotnitskaya and Satlin, 1999) and functional

(Satlin and Palmer, 1997) apical SK channels, absent in neonatal principal cells, are detected routinely by patch clamp analysis and immunofluorescence microscopy, respectively, in CCDs isolated from weanling animals.

Embryonic renal expression of message encoding ROMK, the inwardly rectifying potassium channel Kir6.1, and its associated sulfonylurea receptor have been proposed to reflect their role in cell cycle (Horster, 2000), mitosis, and mesenchyme-to-epithelium transition.

Intercalated cell. Functional (pharmacologic sensitivity) and immunohistochemical studies indicate that intercalated cells possess an apical H-K-ATPase, an enzyme that couples potassium reabsorption to proton secretion (Constantinescu *et al.*, 1997; Silver *et al.*, 1996; Wingo and Smolka, 1995; Zhou *et al.*, 2000). Potassium deficiency is associated with selective hypertrophy of the apical surfaces of medullary intercalated cells (Fig. 18.15). Associated with this structural adaptation is the stimulation of H-K-ATPase activity and potassium absorption in the distal nephron.

Indirect evidence suggests that the neonatal distal nephron absorbs potassium. Saline-expanded newborn dogs were found to absorb 25% more of the distal potassium load than adult animals (Lorenz *et al.*, 1986). Functional analysis of the rabbit CCD has shown that the activity of apical H-K-

ATPase in neonatal intercalated cells is equivalent to that in mature cells (Constantinescu *et al.*, 1997). These data alone do not predict transepithelial potassium absorption under physiologic conditions. However, stimulation of apical H-K-ATPase in series with a basolateral potassium conductance may enhance potassium absorption in the CCD (Wingo and Smolka, 1995; Zhou *et al.*, 2000). Also, an important determinant of the rate of H-K-ATPase-mediated potassium absorption in the collecting duct is the potassium concentration of the tubular fluid delivered to that site. *In vivo* measurements of distal tubular fluid potassium concentration in the young rat, obtained from micropuncture experiments, are approximately twice those reported in the adult (Lelievre-Pegorier *et al.*, 1983). A high tubular fluid potassium concentration may facilitate lumen-to-cell potassium absorption mediated by the H-K-ATPase.

3. Effects of Aldosterone on Potassium Secretion

The role of aldosterone in stimulating potassium secretion in the fully differentiated CCD is well established (Fig. 18.12) (Schwartz and Burg, 1978). The mineralocorticoid-induced stimulation of renal potassium secretion in the adult is due primarily to an increase in the electrochemical driving force favoring potassium exit across the apical membranes

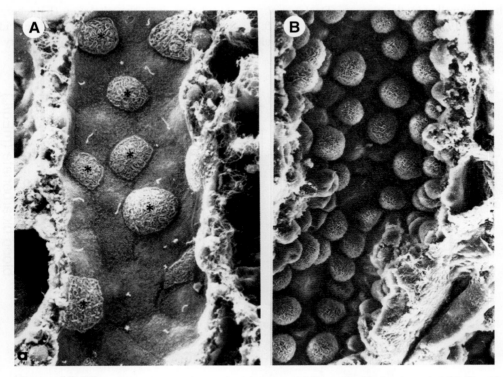

Figure 18.15 Scanning electron micrograph of the outer medullary collecting duct in a control (left) and potassium depleted (right) rat. Intercalated cells are characterized by an elaborate display of microplicae and occasional microvilli (* in left panel); the principal cell possesses a central cilium. Potassium deficiency is associated with selective hypertrophy of the apical surfaces of intercalated cells. Reproduced with permission from Hansen, G.P., Tisher, C.C., and Robinson, R.R. (1980). Response of the collecting duct to disturbances of acid-base and potassium balance. *Kidney Int.* **17**:326–337.

generated by activation of the sodium conductance and mineralocorticoid-stimulated sodium entry (Palmer *et al.*, 1994). In some species, aldosterone may directly increase activity of the SK channel (Ling *et al.*, 1991).

Chronic administration of exogenous aldosterone to fetal sheep increases sodium absorption and decreases plasma renin activity, yet fails to increase urinary potassium excretion (Robillard *et al.*, 1985). However, administration of the aldosterone receptor antagonist spironolactone or sodium channel blocker amiloride, both without a consistent effect on sodium excretion, inhibits renal potassium excretion in fetal sheep, suggesting that endogenous mineralocorticoids influence renal potassium excretion *in utero* (Kairaitis and Lumbers, 1990).

C. Acid–base

Infants maintain lower values of blood pH and bicarbonate concentration than older children and adults (Fig. 18.16) (Weisberg, 1982; Weisbrot *et al.*, 1958). In addition, the infant is prone to develop acidosis during periods of sickness and poor dietary intake. Although the neonatal kidney can maintain acid–base homeostasis, it is limited in its response to eliminate exogenous acid loads.

At birth the acid–base state of the neonate reflects that of the fetus, which is controlled by the placenta (Daniel *et al.*, 1972; Moore *et al.*, 1972a; Smith and Schwartz, 1970; Vaughn *et al.*, 1968). Fetal blood pH is less than maternal

pH by 0.1 to 0.2 units, and pCO_2 is 10 to 15 mm Hg higher, whereas no differences are apparent in bicarbonate concentrations (Daniel *et al.*, 1972; Kesby and Lumbers, 1986; Vaughn *et al.*, 1968; Weston *et al.*, 1970; Yamada, 1970). Shortly after birth, the term neonate is in a state of mild metabolic acidosis (Fig. 18.16) (Graham *et al.*, 1951; Weisberg, 1982; Weisbrot *et al.*, 1958). By 24 h of age, the term infant has a blood pH that is comparable to that of the adult, but lower pCO_2 (~34 mm Hg) and bicarbonate concentrations (~20 mEq/liter), due in large part to a centrally driven hyperventilation. As the kidney matures, the respiratory drive diminishes (Schwartz, 1992). The plasma bicarbonate concentration, lower in preterm than term neonates during the first few weeks of postnatal life (Arant, 1987; Schwartz *et al.*, 1979; Sulyok, 1971), increases progressively after birth (Fig. 18.16) (Albert and Winters, 1966; Edelmann *et al.*, 1967). Consequently, infants have ~80% of the mature amount of bicarbonate buffer per unit of plasma.

The lower levels of buffer base concentration in the blood of infants can be accounted for in part by the inability to completely excrete the by products of growth and metabolism (Edelmann and Spitzer, 1969; Schwartz, 1992). The infant has a larger load than the adult of endogenous acid generated by the metabolism of protein and deposition of calcium into the skeleton (Kildeberg *et al.*, 1969; Svenningsen and Lindquist, 1973). The acid load generated by endogenous acid production, urinary organic anion

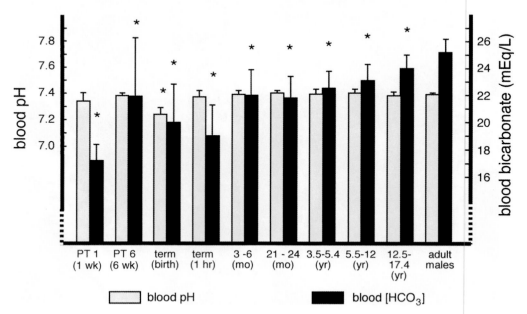

Figure 18.16 Blood acid–base measurements as a function of age (open bars, pH; filled bars, bicarbonate concentration; *, significantly different from adult males). Blood pH does not change with maturation except for being significantly lower in term infants at birth. Blood bicarbonate concentrations are significantly lower than those of adult males throughout maturation. Data taken from Schwartz, G.J. (1999) Potassium and acid-base. *In*: "Pediatric Nephrology" 4th Ed., p.174. (T.M. Barratt, E.D. Avner, and W.E. Harmon), Lippincott Williams & Wilkins, Baltimore.

excretion, and production of sulfuric acid in healthy premature infants amounts to ~2 mEq/kg/day (Chan, 1980; Kildeberg *et al.*, 1969; Kildeberg and Winters, 1978). Balance studies performed in healthy premature infants indicate that the incorporation of 1 g of calcium into the skeleton is associated with the release of 20 mEq of protons that must be excreted by the kidney or be neutralized by gastrointestinal absorption of the base (Kildeberg *et al.*, 1969; Kildeberg and Winters, 1978; Wamberg *et al.*, 1976). In total, endogenous acid production per kilogram body weight in premature infants and growing children is 50 to 100% higher than that of adult subjects.

1. Overview of Acid–Base Handling by the Neonatal Kidney

It is helpful to analyze the maturation of acid–base transport in terms of two major processes: *bicarbonate reabsorption*, primarily a function of the proximal tubule, and *net acid excretion*, a function of proton secretion by the distal nephron resulting in the titration of acids and ammonia.

i. Bicarbonate reabsorption: The concentration of bicarbonate in plasma is determined predominantly by the renal bicarbonate threshold, a set point of the proximal tubule.

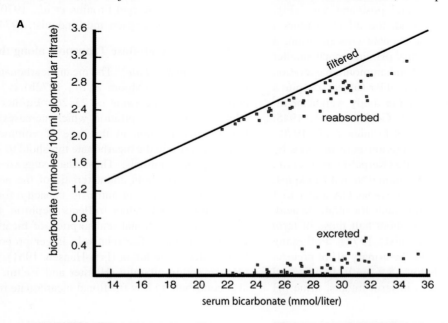

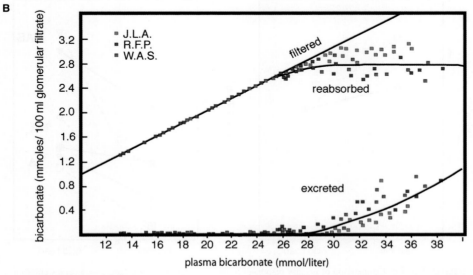

Figure 18.17 Reabsorption and excretion of filtered bicarbonate in (A) infants and (B) adults undergoing bicarbonate titration. Infants have a lower threshold for bicarbonate reabsorption, despite comparable rates of maximal bicarbonate reabsorption per milliliter GFR. After Edelmann *et al.* (1967) and Pitts *et al.* (1949).

Both term and preterm infants have a lower bicarbonate threshold than adults, despite having maximal rates of bicarbonate reabsorption that are in the range reported for adults (Fig. 18.17) (Edelmann *et al.*, 1967; Pitts *et al.*, 1949; Schwartz *et al.*, 1979; Svenningsen, 1974; Tuvdad *et al.*, 1954). These results indicate that the intrinsic capacity of the neonatal proximal tubule to reabsorb bicarbonate exceeds the rate of bicarbonate reabsorption observed at threshold and, relative to the prevailing GFR, appears to be more than adequate to handle the filtered load of bicarbonate.

ii. Net acid excretion: The renal response to acid loading increases with both gestational and postnatal ages (Fig. 18.18). In response to an acid load, the infant exhibits a larger fall in blood pH and bicarbonate concentration, a smaller and less rapid fall in urinary pH, and much smaller increases in urinary titratable acid and ammonium excretion per m^2 (body surface area) than the older subject, despite a comparable dose of acid per kilogram body weight or m^2 surface area (Fomon *et al.*, 1959; Gordon *et al.*, 1948; Hatemi and McCance, 1961; Kerpel-Fronius *et al.*, 1970). Titratable acid and ammonium excretion both increase by ~50% during the first 3 weeks of life (Kerpel-Fronius *et al.*, 1970; Sulyok and Heim, 1971; Svenningsen and Lindquist, 1974). Premature infants of 34 to 36 weeks GA and 1 to 3 weeks postnatal age have excretion rates for titratable acid, ammonium, and net acid that are about half those of term babies (Svenningsen and Lindquist, 1974). Following loading with ammonium chloride, the urinary pH of preterm infants seldom decreases below 5.9 until the second postnatal month of life, whereas in term infants, urine pH

decreases to below 5 within the first few weeks of life (Schwartz *et al.*, 1979; Sulyok and Heim, 1971; Sulyok *et al.*, 1972; Svenningsen, 1974). The limitation in ammonium excretion by premature infants is significantly greater than that in titratable acid excretion.

In summary, infants operate at close to the maximum rate of net urinary acid excretion, due in part to the acid-generating process of growth. The renal response to acid loading increases with both gestational and postnatal ages. Net acid excretion, including both titratable acid and ammonium excretion, increases by ~50% during the first 3 weeks of life (Kerpel-Fronius *et al.*, 1970; Sulyok and Heim, 1971; Svenningsen and Lindquist, 1974).

2. Sites of Acid–Base Transport along the Nephron

a. Proximal Tubule The mean bicarbonate threshold in newborn puppies (Moore *et al.*, 1972b) is ~18 mEq/liter compared with the value of 23–26 mEq/liter measured in adult dogs. Gastric aspiration, which increases plasma bicarbonate concentration in the face of volume contraction, results in a rise in the bicarbonate threshold to ~25 mEq/liter in the young animals. These data suggest that the low bicarbonate threshold characteristic of the newborn is not due to a limitation in intrinsic capacity for bicarbonate reabsorption, but rather reflects nephron heterogeneity and/or a low fractional reabsorption of bicarbonate in the immature kidney. The relatively larger proportion of total body water in the infant (Fris-Hansen, 1961), in addition to a lower serum albumin (Horster and Valtin, 1971), could also contribute to low fractional bicarbonate reabsorption.

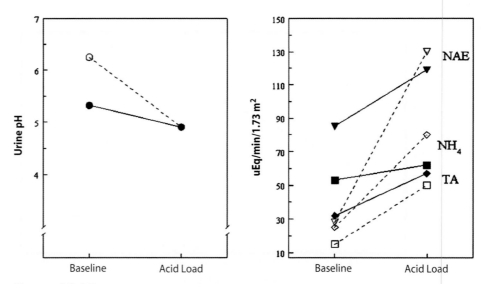

Figure 18.18 Comparison of urinary responses to an acute acid load in infants (——) versus children (– – – –): circles, pH; squares, titratable acid (TA) excretion; diamonds, ammonium (NH_4) excretion; triangles, net acid excretion (NAE). Infants show a much smaller rise in titratable acid excretion, ammonium excretion, and net acid excretion than observed in children. Data given as mean±SD for baseline and after acid loading and are taken from Edelmann *et al.* (1967).

Comparison of the capacity of the neonatal and mature proximal tubule for bicarbonate absorption is complicated by the heterogeneity of nephron development. This heterogeneity has been suggested to be the cause of excessive splay in bicarbonate titration studies performed in infants, meaning that there are populations of nephrons with varying capacities for bicarbonate reabsorption (Edelmann et al., 1967). Therefore, to compare rates of bicarbonate reabsorption as a function of age, it is necessary to examine the same nephron segment at various stages of maturation. Analysis of microperfused rabbit juxtamedullary proximal convoluted tubules has identified a maturational increase in the rate of sodium-dependent bicarbonate absorption, as well as net fluid absorption, during the first 2 months of postnatal life (Fig. 18.8; Schwartz and Evan, 1983).

The bicarbonate permeability of the neonatal proximal tubule, measured using imposed bicarbonate gradients and carbonic anhydrase inhibition, is severalfold lower than observed in mature segments (Quigley and Baum, 1990). This surprising observation contrasts with the findings of a high passive reabsorption of inert substances in neonatal proximal tubules and a maturational increase in length of the intercellular channels (Kaskel et al., 1987). Thus, the lower rate of bicarbonate absorption and the smaller bicarbonate gradient observed in the neonatal compared to the mature juxtamedullary proximal convoluted tubule cannot be explained by high rates of bicarbonate backleak, but rather by low rates of active bicarbonate transport.

Proton secretion. Apical proton secretion in the adult proximal tubule is mediated primarily by the apical Na/H exchanger (Baum, 1989; Kinsella and Aronson, 2001; Murer et al., 1976), discussed in detail earlier, and to a lesser extent by sodium-independent proton secretion mediated by an apical H-ATPase (Baum, 1992; Burg and Green, 1977). In neonates, most apical proton secretion is sodium dependent and is thus ascribed to apical Na/H exchange (Baum, 1990, 1992; Schwartz and Evan, 1983).

Bicarbonate exits the proximal tubule cell primarily via the $Na(HCO_3)_3$ symporter (Alpern, 1985; Baum, 1989; Krapf et al., 1987; Soleimani et al., 1987). In contrast to the relatively modest Na/H exchange activity prevailing in the neonatal juxtamedullary proximal convoluted tubule, the rate of basolateral $Na(HCO_3)_3$ symporter activity in this segment is already ~60% of the adult level during the first month of life (Baum, 1990). These data suggest that the postnatal maturation of the proximal tubular capacity for bicarbonate absorption (proton secretion) is probably due to postnatal increases in Na/H exchange, H-ATPase, and/or the ability to generate cellular ATP, whereas the basolateral $Na(HCO_3)_3$ symporter is not rate limiting for bicarbonate absorption early in life.

Carbonic anhydrase. Carbonic anhydrase facilitates renal acidification by catalyzing the interconversion of carbon dioxide and water to carbonic acid. Approximately 95% of renal carbonic anhydrase is cytosolic (isoform II)(Brion et al., 1994; Dobyan and Bulger, 1982), which is present in proximal tubule cells and serves to catalyze the hydration of carbon dioxide. Deficiency of carbonic anhydrase II results in proximal renal tubular acidosis, and inhibition of enzyme activity markedly diminishes proximal tubular bicarbonate reabsorption (Burg and Green, 1977; Cogan et al., 1979; DuBose and Lucci, 1983; Lucci et al., 1980). An additional isoform, carbonic anhydrase IV, comprising ~5% of renal carbonic anhydrase, is membrane bound and has been found on both apical and basolateral membranes of proximal tubule cells (Brown et al., 1990; Schwartz et al., 1999, 2000) where it catalyzes the dehydration of carbonic acid in the luminal fluid. Inhibition of carbonic anhydrase IV substantially reduces the rate of proximal tubular bicarbonate reabsorption (Lucci et al., 1983; Tsuruoka et al., 2001). The role of other membrane-bound carbonic anhydrases has not been examined in neonatal proximal tubules. Thus, the limitation in neonatal bicarbonate reabsorption could be due to a paucity of carbonic anhydrase II or IV or another membrane-bound isoform in neonatal proximal tubules. Low activity of cytosolic carbonic anhydrase II could reduce the buffering of intracellular base, whereas low activity of the membrane-bound enzyme may lead to accumulation of acid in the lumen or alkali in the intercellular spaces, thereby generating a disequilibrium pH.

Carbonic anhydrase activity is present in the early human fetal kidney (Day and Franklin, 1951; Larsson and Lonnerholm, 1985; Lonnerholm and Wistrand, 1983). Whereas some studies indicate that total renal cortical carbonic anhydrase activity does not change during human development (Day and Franklin, 1951; Lonnerholm and Wistrand, 1983), histochemical examination of fetal kidneys reveals low enzyme activity, especially in nephrogenic areas (Lonnerholm and Wistrand, 1983) and in differentiating proximal tubules (Larsson and Lonnerholm, 1985; Lonnerholm and Wistrand, 1983). Interestingly, the immature proximal tubules show primarily cytoplasmic activity and, with maturation, show more extensive brush border and basolateral membrane activity (Larsson and Lonnerholm, 1985). The newborn kidneys of several experimental animals (Brion et al., 1991; Fisher, 1961; Maren, 1967) exhibit less carbonic anhydrase activity than mature kidneys, and there is a substantial postnatal maturation. Total cortical carbonic anhydrase II activity and IV hydratase activity increase by 100 and 230%, respectively, in rabbit kidney during maturation (Brion et al., 1991; Schwartz et al., 1999), with about half the increase occurring by 3 weeks of age (Schwartz et al., 1999).

Carbonic anhydrase II protein abundance at 1 week of age in the rat, measured by an enzyme-linked immunosorbent assay and normalized to 1 mm tubular length, is 11% of the corresponding level at 12 weeks of age (Karashima et al.,

1998). Of this nine-fold maturational increment, the most rapid increase, to 41% of mature levels, occurs between the second and third weeks of life. The concentration of carbonic anhydrase II at 7 weeks of age is higher in S1 than in S2 and S3 proximal tubules (Karashima *et al.*, 1998).

Carbonic anhydrase IV protein abundance in the neonatal rabbit kidney cortex (presumably representing proximal tubular enzyme) is 15–30% of adult levels during the first 2 weeks of life before increasing to adult levels by 5 weeks of age (Schwartz *et al.*, 1999). Immunohistochemical analyses identify a large increase in carbonic anhydrase IV labeling of apical and basolateral membranes with maturation (Schwartz *et al.*, 1999; Winkler *et al.*, 2001), a developmental change associated with an increase in mRNA (Winkler *et al.*, 2001). There is little staining of medullary rays of rabbit kidneys before the age of 2 weeks, whereas mature kidneys show heaviest staining in these medullary rays in the S2 proximal straight tubules (Fig. 18.19). There is only faint staining in the labyrinths where early proximal tubules (S1 segments) are found, and even juxtamedullary tubules from neonatal kidneys are poorly labeled during the first month of life (Schwartz *et al.*, 1999; Winkler *et al.*, 2001). The major increases in carbonic anhydrases II and IV probably contribute to the three-fold increase in bicarbonate absorption observed in the maturing proximal tubule.

b. Thick Ascending Limb of Loop of Henle   The capacity of the mature TAL for bicarbonate absorption is species specific. Thus, although there is no net absorption of bicarbonate in mature TAL of rabbit (Iino and Burg, 1981), the adult rat TAL absorbs bicarbonate at rates approaching 10–15% of the filtered load (Good, 1985). Bicarbonate absorption in this segment is mediated primarily by apical Na/H exchange (NHE3) (Biemesderfer *et al.*, 1997; Good,

1985). Presently, there are no published data concerning the maturation of bicarbonate transport in the loop of Henle in any species.

Cytosolic carbonic anhydrase activity and staining are demonstrable in the thin descending limb and TAL of rat kidney (Brown *et al.*, 1983; Lonnerholm and Wistrand, 1983; Ridderstrale *et al.*, 1992). Carbonic anhydrase IV has been detected on both apical and basolateral membranes of TAL cells of rat (Brown *et al.*, 1990). Expression of the protein increases with maturation in these segments (Brown *et al.*, 1983; Lonnerholm and Wistrand, 1983).

c. Distal Convoluted Tubule and Connecting Segment   The roles of these segments in acid–base homeostasis are not clear. The distal tubule is unlikely to reabsorb much bicarbonate under normal circumstances (Levine and Jacobson, 1986). Microperfusion studies indicate that the connecting segment of the mature rabbit secretes net bicarbonate (Tsuruoka and Schwartz, 1999). Data in immature segments are lacking.

d. Cortical Collecting Duct   The collecting duct is the primary distal segment involved in acid–base transport. The mature CCD can secrete protons or bicarbonate depending on the acid–base status of the animal (Lombard *et al.*, 1983; McKinney and Burg, 1977; Schwartz *et al.*, 1985). Proton and bicarbonate secretion are mediated by α- and β-intercalated cells, respectively. The α-intercalated cell is characterized by apical H-ATPase and basolateral band 3 chloride-bicarbonate anion exchangers (Bastani *et al.*, 1991; Brown *et al.*, 1988; Madsen *et al.*, 1992; Schuster *et al.*, 1986, 1991), cytosolic carbonic anhydrase II (Kim *et al.*, 1990, 1994; Matsumoto and Schwartz, 1992), and possibly luminal carbonic anhydrase IV (Winkler *et al.*, 2001). The

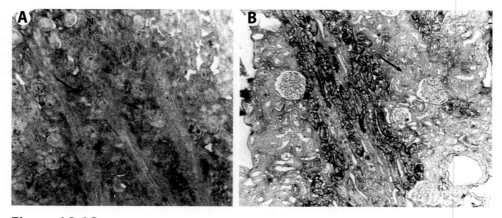

Figure 18.19 Low–power views of 10-day (A) and adult (B) rabbit kidney cortices stained immunohistochemically for CA IV. Immature kidney stains faintly over the cortical labyrinth (A, arrow) and not over medullary rays (*) or the nephrogenic zone (N). Mature kidney cortex (B) shows heavy CA IV staining along medullary rays in apical and basolateral membranes of proximal straight tubules and less so in proximal convoluted tubules of the cortical labyrinth (arrow). Glomeruli and thick limbs did not stain at either age. Reprinted with permission from Schwartz *et al.* (1999).

β-intercalated cell possesses basolateral or nonpolarized H-ATPase, but no band 3 labeling (Bastani *et al.*, 1991; Brown *et al.*, 1988; Schuster *et al.*, 1986, 1991); these cells in the rabbit express cytoplasmic carbonic anhydrase II (Kim *et al.*, 1990; Ridderstrale *et al.*, 1988, 1992). Intercalated cells comprise 20–35% of total cells in the CCD, with the remainder accounted for by principal cells. Although the number of intercalated cells per millimeter tubular length doubles during maturation (Satlin *et al.*, 1992; Satlin and Schwartz, 1987), the percentage of intercalated cells that are β type (90%) is similar in the midcortical collecting duct of newborn, weanling and mature rabbits (Satlin *et al.*, 1992; Schwartz *et al.*, 1985). Neonatal CCDs in a variety of species show no intercalated cells in the outer cortex (Clark, 1957; Evan *et al.*, 1991; Kim *et al.*, 1994; Matsumoto *et al.*, 1996; Satlin and Schwartz, 1987), consistent with the pattern of centrifugal nephron development.

Immunostaining of 18-day fetal rat kidneys shows a simultaneous appearance of carbonic anhydrase II and the vacuolar H-ATPase in a subpopulation of cells in the CNT and medullary collecting duct (Kim *et al.*, 1994). These data suggest that intercalated cells differentiate from two separate foci, one in the nephron and one in the collecting duct. Cells with apical (H-secreting or α) and basolateral (bicarbonate-secreting or β) H-ATPase are seen at both foci, indicating that both types of intercalated cells differentiate simultaneously during development. Negligible band 3 chloride-bicarbonate exchanger is present in fetal kidney, although immunolabeling is evident at 3 days of age, presumably reflecting the activation of α-type intercalated cells.

Ultrastructural studies of 19-day fetal and neonatal rat kidney demonstrate α-intercalated cells in the inner medulla and renal pelvis (Narbaitz *et al.*, 1991). Within 2 weeks of birth, α-intercalated cells were no longer present in these regions but had become numerous in outer medullary collecting ducts (OMCDs) and, to a lesser extent, in CCDs. β-intercalated cells first appeared 3 weeks after birth in the CCD (Narbaitz *et al.*, 1991). Maternal alkalosis resulted in an earlier appearance and in an increase in number of β-intercalated cells in neonatal rats (Narbaitz *et al.*, 1993). Because the rabbit diet provides a larger alkaline load than that of rats, it is reasonable to expect a higher percentage of β-intercalated cells in the maturing rabbit kidney compared to the rat.

Renal development is accompanied not only by an increase in the number of intercalated cells, but also changes in the morphology and function of these specialized cells. The pH of neonatal CCD intercalated cells is ~0.15 pH units less alkaline than that in mature cells (Satlin and Schwartz, 1987). The number of acidic cytoplasmic vesicles is less in immature intercalated cells, consistent with fewer functioning proton pumps, and there is a smaller mitochondrial potential (Satlin and Schwartz, 1987). Immunocytochemical

studies reveal little labeling for H-ATPase and the basolateral band 3 anion exchanger, as well as carbonic anhydrase and β-intercalated cell surface markers (Holthofer, 1987; Kim *et al.*, 1994; Matsumoto *et al.*, 1996; Satlin *et al.*, 1992). Carbonic anhydrase II expression (ng/mm tubule length) in rat CCDs increases nine-fold during maturation, with nearly half the increase occurring during the first 3 weeks of life (Karashima *et al.*, 1998). Ultrastructurally, intercalated cells from newborn rabbit CCDs show smaller apical perimeters and markedly reduced vesicular profiles and mitochondrial volume percentage (Evan *et al.*, 1991). Differentiation proceeds such that a nearly mature ultrastructural appearance is achieved by 3 weeks of age; however, at this age the number of intercalated cells per mid or outer cortical zone is still only about half of what is observed at maturity (Evan *et al.*, 1991).

Functional studies of β-intercalated cells, which possess apical chloride-bicarbonate exchangers, reveal a lower cell pH during the first month of life (Satlin *et al.*, 1992; Satlin and Schwartz, 1987) and a markedly reduced capacity for chloride-dependent bicarbonate extrusion (Fig. 18.20; Satlin *et al.*, 1992). These observations are consistent with a reduced number and/or activity of apical anion exchangers in β-intercalated cells during the first month of life. In addition, the low cell pH of immature β-intercalated cells, coupled with the evidence of low expression of immuno-detectable H-ATPase in cortical intercalated cells (noted earlier), suggests that a paucity of basolateral proton pumps exists to generate intracellular bicarbonate for secretion. Accordingly, CCDs during the first month of life fail to show spontaneous bicarbonate secretion and may contribute to a mild metabolic alkalosis of the neonatal rabbit (Mehrgut *et al.*, 1990).

The maturational changes in intercalated cell structure and function are reflected in significant differences in

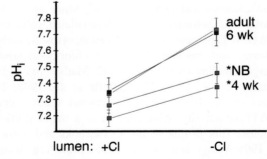

Figure 18.20 Effect of luminal chloride (Cl⁻) removal on β-intercalated cell pH (pH_i) in maturing CCDs. Cell pH was measured in each cell in the presence of luminal Cl⁻ (+Cl) and then after luminal Cl⁻ was replaced with gluconate (-Cl). Changes in pH_i in response to luminal Cl⁻ removal in CCDs from newborn and 4-week animals were significantly less than those measured in CCDs from older rabbits (*, p<0.05). Cell pH in the absence of luminal Cl⁻ was significantly higher than that measured under baseline conditions (+Cl) in all age groups. After Satlin *et al.* (1992).

net bicarbonate transport during the postnatal period. Net bicarbonate transport is not observed in CCDs isolated from newborn and 4-week-old rabbits, whereas CCDs from 6-week-old and adult rabbits show net bicarbonate secretion (Mehrgut *et al.*, 1990). Removal of bath chloride, which inhibits the basolateral chloride-bicarbonate anion exchanger of α-intercalated cells and depletes β-type intercalated cells of chloride, markedly stimulates chloride-dependent bicarbonate secretion in CCDs isolated from animals greater than 6 weeks of age (Mehrgut *et al.*, 1990; Yasoshima *et al.*, 1992). In contrast, this same experimental maneuver leads to no significant stimulation of bicarbonate transport in CCDs from animals in the first month of life. These data indicate that β-type intercalated cells are not fully differentiated and are not capable of secreting significant amounts of bicarbonate in the postnatal period.

Mutations in the collecting duct intercalated cell anion exchanger AE1 (Bruce *et al.*, 1997; Jarolim *et al.*, 1998; Tanphaichitr *et al.*, 1998) or B1 subunit of the vacuolar H-ATPase (Karet *et al.*, 1999a,b) result in distal renal tubular acidosis. A recessively inherited form of renal tubular acidosis (RTA) which appears in the first year of life, presenting with failure to thrive, osteopetrosis and cerebral calcification, is associated with mutations in cytosolic carbonic anhydrase II (Schwartz *et al.*, 1991; Sly *et al.*, 1985). There is also a mutation in the gene encoding the Na(HCO$_3$)$_3$ cotransporter, NBC1, which has been identified in proximal RTA with corneal calcification (Shayakul and Alper, 2000).

e. Outer Medullary Collecting Duct

The mature OMCD is composed of principal cells and α-type inter-calated cells; the latter comprise 10–15% of total cells in the outer stripe and a third to half in the inner stripe of the outer medulla (Evan *et al.*, 1991; Holthofer *et al.*, 1987; Kaissling and Kriz, 1979; Satlin and Schwartz, 1987; Schuster *et al.*, 1986, 1991). This segment regularly secretes protons at high rates (McKinney and Davidson, 1987; Stone *et al.*, 1983; Tsuruoka and Schwartz, 1997). Intercalated cells in the mature OMCD possess apical H-ATPase, cytosolic carbonic anhydrase, and basolateral band 3 chloride-bicarbonate exchanger labeling (Kim *et al.*, 1992; Madsen *et al.*, 1992; Matsumoto *et al.*, 1996; Ridderstrale *et al.*, 1988, 1992; Schuster *et al.*, 1986, 1991). Functional studies identify H-K-ATPase activity, which mediates a major portion of bicarbonate flux in this segment (Tsuruoka and Schwartz, 1997, 1998; Wingo and Smolka, 1995). Some regions of the OMCD express functional luminal carbonic anhydrase activity (Star *et al.*, 1987; Tsuruoka and Schwartz, 1998), which is likely to be attributable to carbonic anhydrase IV (Schwartz *et al.*, 2000). OMCDs from newborn rats and rabbits show fainter and less apical polarization of H-ATPase and less basolateral polarization of band 3 compared to animals greater than 3 weeks of age (Kim

et al., 1994; Matsumoto *et al.*, 1996). Carbonic anhydrase II expression (ng/mm tubule length) in rat OMCDs increases 7-fold during maturation, with nearly half the increase occurring during the first 3 weeks of life (Karashima *et al.*, 1998). Carbonic anhydrase II activity in outer medullary homogenates doubles during maturation, with most of the increase occurring after 4 weeks of age (Brion *et al.*, 1991). Carbonic anhydrase IV protein, expressed primarily by medullary collecting ducts, increases up to 10-fold, with more than half the increase occurring within the first month of life (Schwartz *et al.*, 1999).

Immunostaining of 18 day fetal rat kidneys shows carbonic anhydrase II and H-ATPase in cells throughout the medullary collecting duct and papillary surface (Kim *et al.*, 1994). After birth the immunostaining disappears from the terminal IMCD collecting duct and papillary surface. Intercalated cells with apical H-ATPase or carbonic anhydrase IV labeling (α-intercalated cells) are extruded from the epithelium into the lumen (Kim *et al.*, 1996; Schwartz *et al.*, 1999). Cells with basolateral H-ATPase labeling (β-intercalated cells) disappear gradually from the OMCD and initial IMCD. These intercalated cells are deleted by apoptosis and subsequent phagocytosis by neighboring principal cells or IMCD cells (Kim *et al.*, 1996).

The expression of differentiated proteins by the immature OMCD appears to be more advanced than in the CCD (Evan *et al.*, 1991; Kim *et al.*, 1994; Matsumoto *et al.*, 1996). This is confirmed by cell pH studies, which show that neonatal medullary intercalated cells have a cell pH similar to that of mature cells (Satlin and Schwartz, 1987). In addition, mature numbers of intercalated cells are observed in the neonatal OMCD (Satlin and Schwartz, 1987). Neonatal OMCDs are capable of luminal acidification (Satlin and Schwartz, 1987). Ultrastructural studies of intercalated cells in the outer stripe confirm that they are present in mature numbers, but there are slightly smaller vesicular and mitochondrial volumes and much smaller apical perimeters compared to mature segments (Evan *et al.*, 1991).

Microperfusion studies indicate that the neonatal OMCD absorbs bicarbonate (secretes protons) at 70% of the mature rate (Mehrgut *et al.*, 1990). Removal of bath chloride, which inhibits basolateral chloride-bicarbonate exchanger activity, completely inhibits net bicarbonate absorption in both newborn and adult OMCDs. Each of these studies indicates the relative maturity of α-type intercalated cells in the neonatal OMCD.

f. Inner Medullary Collecting Duct (IMCD)

The IMCD acidifies the urine (Graber *et al.*, 1981; Ullrich and Papavassiliou, 1981), despite a paucity of intercalated cells (Clapp *et al.*, 1987; Kaissling and Kriz, 1979; Wall *et al.*, 1990). In this segment, proton secretion is accomplished primarily via the H-K-ATPase (Wall *et al.*, 1996). Cytoplasmic carbonic anhydrase activity has been demonstrated not only in intercalated cells in the initial IMCD, but also in IMCD cells comprising the entire terminal IMCD (Kleinman *et al.*, 1992). This has been confirmed at the mRNA level (Tsuruoka *et al.*, 1998), as well as by activity assays of inner medullary homogenates (Brion *et al.*, 1991). In addition, there is functional evidence for luminal carbonic anhydrase activity in the initial, but not the terminal, IMCD (Wall *et al.*, 1991), and this luminal enzyme is likely to be carbonic anhydrase IV (Schwartz *et al.*, 2000). There are no published maturational acid–base studies in the IMCD. In rabbit kidneys, inner medullary homogenate carbonic anhydrase activity doubles (but not statistically significantly) after the fourth week of postnatal life (Brion *et al.*, 1991).

D. Phosphate

Inorganic phosphate (P_i) homeostasis in the adult reflects the balance of efficient intestinal P_i absorption and the capacity of the kidneys to excrete P_i into the urine. Clearance studies reveal that the maximum capacity of the kidney to reabsorb P_i is high in the neonatal rat (Mulroney and Haramati, 1990; Pastoriza-Munoz *et al.*, 1983; Trohler *et al.*, 1976) and declines progressively with advancing age (Fig. 18.21) (Caverzasio *et al.*, 1982; Haramati *et al.*, 1988). This high renal reabsorptive capacity for P_i early in life allows the infant to retain a large portion (~40%) of the P_i absorbed from the gut and to sustain a state of positive P_i balance (Hohenauer *et al.*, 1970). This facilitates growth and development, due to the role of P_i as a constituent of bone, muscle, membrane phospholipids, and cellular processes involving ATP.

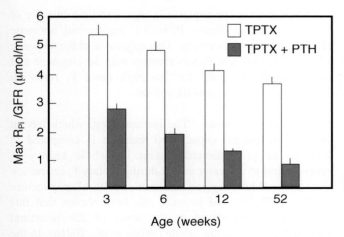

Figure 18.21 Effect of age on the maximum capacity of the kidney to reabsorb phosphate (TmPi expressed as Max R_{Pi}/GFR) in thyroparathyroidectomized (TPTX) rats given either parathyroid hormone (PTH) or vehicle. PTH infusion decreased TmPi significantly in all age groups, although the TmPi was highest early in life and decreased progressively with age. After Haramati *et al.* (1988).

1. Overview of Renal P_i Handling

The low urinary P_i excretion characteristic of the developing animal was originally attributed to the low GFR prevailing early in life (Brodehl *et al.*, 1982). However, a doubling of GFR in immature rats, effected by infusing arginine into the kidney, does not raise renal P_i excretion (Haramati and Mulroney, 1987) nor does a reduction in the GFR of adult rats to levels seen in the immature animal, achieved by constricting the abdominal aorta above the renal arteries, raise the capacity of the kidney to retain P_i (Haramati and Mulroney, 1987). These observations suggest that neonatal P_i retention is due to unique characteristics in the tubular handling of P_i reabsorption, the presence of factors that promote tubular P_i uptake, and/or the resistance of the immature kidney to phosphaturic stimuli (Haramati, 1989).

Kidneys of the normal adult animal reabsorb approximately 80% of the filtered P_i (Kaskel *et al.*, 1988), predominantly in the proximal convoluted tubule (~60–70%) (Agus *et al.*, 1971; Amiel *et al.*, 1970; Dennis *et al.*, 1976, 1977; Greger *et al.*, 1977; Strickler *et al.*, 1964; Suzuki *et al.*, 1988). Micropuncture studies in the young rat (Woda *et al.*, 2001a) and guinea pig (Kaskel *et al.*, 1988) demonstrate avid P_i uptake in the proximal convoluted tubule compared to adults. While the loop of Henle does not reabsorb P_i due to its low P_i permeability (Lang *et al.*, 1977; Rocha *et al.*, 1977), the DCT, CCD, and IMCD can reabsorb another 10–20% of the filtered load under certain states of P_i conservation in the adult (Bengele *et al.*, 1979; Dennis *et al.*, 1977; Haas *et al.*, 1978; Harris *et al.*, 1977; Pastoriza-Munoz *et al.*, 1978; Shareghi and Agus, 1982b).

2. Sites of P_i Transport along the Nephron

a. Proximal Tubule P_i reabsorption in the proximal tubule is mediated by a sodium-dependent P_i transport (NaPi) system localized to the apical brush border membrane (BBM) (Murer, 1992). This transport process is driven by the sodium gradient established by the basolateral Na-K-ATPase. The rates of P_i reabsorption are highest in the early proximal tubule and decline gradually along the proximal straight segment (Brunette *et al.*, 1984). The high brush border membrane vesicle (BBMV) sodium-dependent P_i uptake during growth (Karlen, 1989; Ladas *et al.*, 1993; Neiberger *et al.*, 1989; Woda *et al.*, 2001a) occurs in the absence of any maturational differences in the affinity of the transporter for either sodium or P_i (Ladas *et al.*, 1993). P_i transport across the basolateral membrane is mediated by a sodium-independent anion exchanger and a NaPi cotransporter that can be distinguished from the apical protein by its unique stoichiometry (Murer *et al.*, 1991).

There are several intrinsic, developmental stage-specific characteristics of the proximal tubule that may account for the high rates of P_i reabsorption observed early in life.

First, intracellular P_i concentrations are significantly lower in growing animals compared to adults (Barac-Nieto *et al.*, 1991, 1993) and are associated with a higher V_{max} of BBM sodium-dependent P_i cotransport (Barac-Nieto *et al.*, 1991). Presumably, the low cytosolic P_i promotes the expression and insertion of BBMV NaPi transporters into the apical membrane (Barac-Nieto *et al.*, 1990). In addition, the content of cholesterol, sphingomyelin, and phosphatidyl-inositol in BBMV harvested from young animals is lower than that present in the adult and may account for a high membrane fluidity (Levi *et al.*, 1989; Pratz and Corman, 1985; Pratz *et al.*, 1987). Dietary P_i restriction in the adult, which places the animal in a state of P_i conservation, leads to a reduction in BBM cholesterol which in turn, is associated with an increase in the V_{max} for NaPi uptake (Molitoris *et al.*, 1985). Thus, a decrease in membrane lipid content and a subsequent increase in membrane fluidity may directly regulate NaPi cotransporter activity in the proximal tubule (De Smedt and Kinne, 1981; Friedlander *et al.*, 1988; Levi *et al.*, 1990; Molitoris *et al.*, 1985). Finally, nephron heterogeneity may also explain, in part, the limited urinary P_i excretion seen in the rapidly growing animal. Because deep nephrons reabsorb more P_i than cortical nephrons (Haas *et al.*, 1978; Haramati *et al.*, 1984) and nephrogenesis begins in the juxtamedullary region, the kidney of the immature animal may contain a relatively high number of functioning nephrons with a high capacity for P_i reabsorption.

The NaPi cotransporter has been cloned and classified into two distinct types: type I (called NaPi-1) and type II (referred to as NaPi-2 in rat, NaPi-3 in human, and NaPi-6 in rabbit) (Biber *et al.*, 1993a; Chong *et al.*, 1993; Collins and Ghishan, 1994; Hartmann *et al.*, 1995; Magagnin *et al.*, 1993; Miyamoto *et al.*, 1995; Murer and Biber, 1994; Sorribas *et al.*, 1994; Verri *et al.*, 1995; Werner *et al.*, 1994b). The NaPi-1 protein is uniformly expressed along the proximal tubule of both superficial and deep nephrons (Biber *et al.*, 1993b; Custer *et al.*, 1993, 1994). In contrast, the abundance of NaPi-2 is highest in the proximal convoluted tubule and decreases gradually along the axial length of the proximal straight tubule (Biber *et al.*, 1993b; Custer *et al.*, 1993, 1994). While the role of the type I NaPi cotransporter is still not well understood, the type II transporter is proposed to represent a target for the physiological regulation of proximal tubular P_i reabsorption (Busch *et al.*, 1996). Thus, the expression of type II protein in the BBM is modulated by changes in dietary P_i intake and PTH infusions (Biber *et al.*, 1993a; Hansch *et al.*, 1993; Hoag *et al.*, 1999; Levi *et al.*, 1994; Lotscher *et al.*, 1997; Ritthaler *et al.*, 1999; Verri *et al.*, 1995; Werner *et al.*, 1994a; Woda *et al.*, 2001a). Targeted disruption of the type II NaPi cotransporter reveals that 80–85% of renal P_i transport is mediated by this transporter (Hoag *et al.*, 1999).

During nephrogenesis, NaPi-2 mRNA is first detected in proximal convolutions in the post-S-shape stage with signal ending at the transition to the descending limb of the primitive loop of Henle (Schmitt *et al.*, 1999). In newborn rats, the expression of NaPi-2 mRNA is highest in proximal convoluted tubules of juxtamedullary and intermediate (glomeruli present but tubular system undifferentiated) nephrons and virtually disappears along the straight segments (Traebert *et al.*, 1999). NaPi-2 protein is expressed only in those proximal tubular cells that have differentiated to the extent that an apical brush border is detectable (Traebert *et al.*, 1999). The abundance of NaPi-2 mRNA in whole kidney and protein in apical BBM is higher in juvenile compared to adult animals (Woda *et al.*, 2001a). The ontogeny of NaPi-1 expression has not been rigorously investigated.

b. Distal Nephron Micropuncture and microperfusion experiments directed at specific nephron sites in the adult animal have shown that segments beyond the proximal convoluted tubule, such as the DCT (Haramati *et al.*, 1983; Pastoriza-Munoz *et al.*, 1978, 1983), CCD (Peraino and Suki, 1980), and IMCD (Magaldi *et al.*, 1992), have the capacity to reclaim filtered P_i. Analysis of differences in the fractional delivery of P_i to the beginning of the distal convoluted tubule compared to the final urine suggests that significant P_i reabsorption occurs in the distal nephron of the developing rat (Woda *et al.*, 2001a). However, nephron heterogeneity, as described earlier, may account for most of this discrepancy. While little is known about the mechanism of P_i uptake in the distal nephron, P_i reabsorption in the IMCD is sodium independent (Magaldi *et al.*, 1992).

3. Regulation of Renal P_i Transport

P_i transport in the nephron is regulated by a number of factors, including dietary P_i intake, parathyroid hormone (PTH), and growth hormone. The magnitude of the response of the kidney to these factors changes with development and may account, in part, for the high renal P_i retention characteristic of the growing animal.

a. Dietary P_i Intake The immature rat, which has an intrinsically greater capacity to reabsorb P_i compared to adults, can proportionally adapt its whole kidney P_i reabsorption to a greater extent during dietary P_i restriction than adults (Mulroney and Haramati, 1990). Micropuncture studies in P_i-deprived juvenile rats have shown that this adaptation is mediated at the level of the proximal convoluted and straight tubule (Woda *et al.*, 2001a). In the juvenile rat, the V_{max} for Na-dependent P_i transport in the proximal convoluted tubule increased ~80% during dietary P_i restriction compared to controls (Ladas *et al.*, 1993). This adaptation was associated with an increase in immuno-detectable proximal tubular protein (Woda *et al.*, 2001a).

The observation that NaPi-2 mRNA levels did not change is consistent with an increase in the stability of message or other posttranscriptional modification (Woda *et al.*, 2001a). In contrast to these results in rat, BBMV prepared from proximal tubules of P_i-deprived neonatal guinea pig do not exhibit an adaptive increase in P_i uptake (Neiberger *et al.*, 1989), suggesting that in this species, the reabsorptive capacity for P_i in this segment is already maximal in the neonate.

Chronic administration of a high P_i diet caused the NaPi-2 transporter to be removed from the BBM in juvenile rats (Woda *et al.*, 2001a). Whereas NaPi-2 transporters in proximal tubular cells are internalized and transported to lysosomes for degradation in P_i-loaded adults (Keusch *et al.*, 1998), the internalized NaPi-2 transporters in the P_i-loaded juvenile animals are not degraded. The mechanism responsible for this age-related difference remains unknown.

b. Parathyroid Hormone PTH acts on the proximal convoluted tubule to increase urinary P_i excretion by inhibiting BBM sodium-dependent P_i transport (Murer, 1992; Murer and Biber, 1994). PTH binds to a basolateral receptor, which in turns leads to an increase in intracellular cAMP (Agus *et al.*, 1971; Chase and Aurbach, 1967), phospholipase C (Hruska *et al.*, 1987), and phospholipase A_2 (Zhang *et al.*, 1999). Evidence suggests that the PTH-induced reduction in proximal tubular P_i transport is due to the endocytotic removal of the NaPi-2 cotransporters in the rat (Dousa *et al.*, 1976; Hoppe *et al.*, 1991; Kempson *et al.*, 1995; Pfister *et al.*, 1997, 1998). Following the internalization of these cotransporters, these protein are routed to lysosomes where they are subsequently degraded (Keusch *et al.*, 1998).

Growing or dietary P_i-restricted animals show a minimal increase in urinary P_i excretion following PTH infusions, despite intact PTH second messenger systems (Corn *et al.*, 1989; Johnson and Spitzer, 1986; Linarelli, 1972; Webster and Haramati, 1985). Micropuncture studies in the juvenile rat show that the effects of PTH are attenuated in both proximal convoluted and straight tubules (Woda *et al.*, 2001a). The mechanisms underlying the relative insensitivity of the neonatal proximal tubule to PTH remain to be clarified.

c. Growth Hormone (GH) GH influences the renal uptake of P_i and modulates the expression of BBM NaPi-2 cotransporters. Administration of a GH-releasing factor antagonist (GRF-AN), which suppresses the pulsatile release of GH from the anterior pituitary, to immature animals results in a two-fold increase in the urinary excretion of P_i (Mulroney *et al.*, 1989). The maximum capacity for P_i reabsorption in these animals is similar to that observed in the adult (Fig. 18.22) (Mulroney *et al.*, 1989) due to a reduction in sodium-dependent P_i transport in

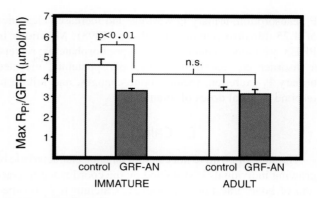

Figure 18.22 Effect of a growth hormone releasing factor antagonist (GRF-AN; shaded bars) on the transport maximum to reabsorb phosphate (TmPi) in immature and adult rats. Inhibition of the pulsatile release of GH in the immature rat significantly reduced the TmPi to levels observed in the adult. In contrast, administration of GRF-AN to adults had no effect on TmPi. After Haramati *et al.* (1990).

proximal tubular BBMV (Ladas *et al.*, 1993). Micropuncture studies confirm that suppression of the pulsatile release of GH in the juvenile rat reduces P_i reabsorption in the proximal-convoluted tubule and, furthermore, abolishes P_i uptake along the proximal straight segment (Woda *et al.*, 1999). The abundance of NaPi-2 protein in proximal tubular BBM harvested from these animals is reduced (Woda *et al.*, 1999). Of note, there is no change in P_i reabsorption in the GH-suppressed adult animal (Haramati *et al.*, 1990).

It is uncertain whether the effect of GH on the proximal tubule represents a direct effect of GH or an indirect effect of GH-stimulated insulin-like growth factor-1 (IGF-1) release. While GH receptors are present on the basolateral surface of the proximal tubule and their activation leads to an increase in phospholipase C (Rogers and Hammerman, 1989), IGF-1 receptors have been localized to both apical and basolateral surfaces of this segment (Hammerman and Gavin, 1986; Hammerman and Rogers, 1987). IGF-1, and not GH, directly stimulates P_i uptake in isolated proximal convoluted tubules (Quigley and Baum, 1991) and increases the V_{max} for sodium-dependent P_i transport in proximal tubular BBMV prepared from hypophysectomized rats (Caverzasio *et al.*, 1990). However, the intracellular mechanism by which GH/IGF-1 stimulates P_i uptake in these tubular cells remains unknown.

Proximal tubular P_i reabsorption may also be regulated by the putative hormone phosphatonin. This hormone stimulates NaPi-2 and downregulates renal 24-hydroxylase activity, thereby decreasing the metabolism of 1,25-dihydroxyvitamin D (Roy *et al.*, 1994). PHEX (phosphate-regulating gene with homologies to endopeptidases on the X chromosome) has been proposed to suppress the activity of phosphatonin (Rowe, 1998). Mutations in PHEX presumably lead to the release of incorrectly processed phosphatonin into the circulation, resulting in inhibition of

P_i reabsorption, urinary P_i wasting, and increased clearance of 1,25-dihydroxyvitamin D (Rowe, 1998). Mutations in PHEX are found in patients with hypophosphatemic rickets, a disorder characterized by hypophosphatemia, elevated urinary P_i excretion, rachitic bone changes, and failure to respond to usual doses of vitamin D.

E. Calcium

The state of positive net calcium balance characteristic of growing individuals is sustained by the coordinated interaction of bone, intestine, and kidney. Calcium is reabsorbed throughout the nephron such that less than 2% of the filtered load is excreted in the urine of the adult. Within the adult kidney, approximately 70% of the filtered calcium is reabsorbed in the proximal tubule, 20% in the TAL, 5–10% in the distal tubule, and less than 5% in the collecting duct (Friedman, 1998). Absorption in both the proximal tubule (Bourdeau and Burg, 1979; Ng *et al.*, 1984; Shareghi and Agus, 1982a) and TAL is coupled predominantly to sodium absorption and is a passive process through the paracellular pathway. Within the TAL, sodium, potassium and chloride reabsorption by the Na-K-2Cl cotransporter (NKCC2; see earlier discussion) generates a lumen positive potential difference. This, together with the high paracellular permeability to calcium, drives the paracellular reabsorption of this ion. Thus, in disease states such as Bartter's syndrome (see earlier discussion), salt wasting is accompanied by hypercalciuria, consistent with the parallel nature of sodium and calcium absorption in the TAL.

Calcium reabsorption within the distal nephron, including the DCT, CNT, and initial portion of the CCD, is active, transcellular, and is regulated independently of sodium (Friedman, 1998). Luminal calcium entry in these segments appears to be mediated by apical calcium channels (Bacskai and Friedman, 1990; Bourdeau and Lau, 1989; Gesek and Friedman, 1992; Matsunaga *et al.*, 1994). Among the molecular candidates for these channels are the recently cloned epithelial calcium channel (ECaC) (Hoenderop *et al.*, 2000) and CaT (Peng *et al.*, 2000). Calcium diffuses passively into the tubular cells down its concentration gradient, as well as by diffusion facilitated by calcium-binding protein carriers, such as calbindin (Koster *et al.*, 1995). Calcium exit across the basolateral membrane into the extracellular fluid, which occurs against an electrochemical gradient and is energy dependent, occurs primarily through the sodium-calcium (NaCa) exchanger in the DCT and CNT and the Ca-ATPase in the DCT (White *et al.*, 1998).

The fractional reabsorption of calcium in the cortical TAL of the neonatal and weanling rat and mouse is low, increasing significantly with advancing postnatal age (Lelievre-Pegorier *et al.*, 1983; Wittner *et al.*, 1997). The observation that the sodium reabsorptive capacity of the TAL reaches maturity before that for calcium (Wittner

et al., 1997) provides evidence for distinct programs of differentiation for transcellular compared to paracellular transport pathways in the TAL.

The principal hormones that regulate renal calcium excretion in the adult are PTH, 1,25-dihydroxy vitamin D3, and calcitonin (reviewed in Friedman, 2000). Due to the paucity of information available about the hormonal regulation of calcium transport in the differentiating kidney, the actions of these hormones on the fully differentiated nephron are presented only briefly.

PTH reduces renal calcium excretion by stimulating calcium reabsorption across the cortical TAL, DCT, and CNT (Bourdeau and Burg, 1980; Gesek and Friedman, 1992; Greger *et al.*, 1978; Shareghi and Stoner, 1978; Shimizu *et al.*, 1990; Suki and Rouse, 1981) by both cAMP–protein kinase A and phospholipase A–protein kinase C coupled signaling pathways (Friedman, 2000). Additionally, the PTH-induced reduction in K_f leads to a decrease in GFR (Ichikawa *et al.*, 1978), which, in turn, reduces the filtered load of calcium. Vitamin D is required for normal calcium reabsorption in the distal nephron, the site in which vitamin D-dependent calbindin-D has been localized, although this effect is species specific (Kurokawa, 1987). Calcitonin is another renal calcium-conserving hormone (Carney, 1995, 1997). In the rabbit, calcitonin increases calcium transport in the distal tubule by opening apical calcium channels and stimulating the basolateral NaCa exchanger; both actions depend on the activation of adenylate cyclase (Zuo *et al.*, 1997).

Physiologic changes in the extracellular calcium concentration that regulates PTH secretion are detected by the G-protein-coupled calcium-sensing receptor, CaSR. CaSR, expressed in the parathyroid gland and in kidney, is responsive to both calcium and magnesium, although the sensitivity to extracellular calcium is greater than that for magnesium (Brown *et al.*, 1993). Within the parathyroid gland, CaSR mediates the inhibitory effects of extracellular calcium on PTH secretion and expression of the PTH gene (Brown *et al.*, 1998). Within the kidney, CaSR is most abundant on the basolateral membrane of the TAL (Riccardi *et al.*, 1998) where it tonically inhibits the activity of the apical ROMK channel (Brown *et al.*, 1993), thereby limiting sodium and chloride reabsorption and minimizing the driving force for divalent cation reabsorption. Apical CaSR has been identified in the rat IMCD where it has been proposed to reduce AVP-elicited osmotic water permeability when luminal calcium rises (Sands *et al.*, 1997).

PTH-responsive adenylate cyclase is present in renal cortical homogenates from preterm rabbits (Linarelli *et al.*, 1973) and TALs of newborn rats (Imbert-Teboul *et al.*, 1984). Although neonatal hypocalcemia has been attributed to end organ unresponsiveness to PTH, administration of exogenous PTH to premature and full-term newborns increases urinary excretion of cAMP (Linarelli, 1972; Mallet *et al.*, 1978) and results in a calcemic response, but affects calciuria and phosphaturia only minimally (Tsang *et al.*, 1973). Evidence suggests that the developmentally regulated changes in renal handling of divalent cations and water described in this chapter may be due to a perinatal increase in the expression of renal CaSR, leading to inhibitory effects on the actions of PTH and AVP (Chattopadhyay *et al.*, 1996). There is little expression of CaSR in the fetal kidney (Chattopadhyay *et al.*, 1996). Steady-state abundance of CaSR mRNA and protein increases significantly during the first week of life, related principally to an increase in the expression of the receptor in the TAL and, to a lesser extent, in the collecting duct (Chattopadhyay *et al.*, 1996).

F. Magnesium

Little is known about the ontogeny of renal magnesium handling. Approximately 80% of the total plasma magnesium is filtered across the glomerulus, of which 97% is reabsorbed throughout the mature nephron (de Rouffignac and Quamme, 1994; Quamme and Dirks, 1980, 1986). Micropuncture studies show that the proximal tubule of the adult animal reabsorbs only ~10% of the filtered magnesium, whereas that of the young (13- to 15-day) rat reabsorbs ~60% of the filtered load (Lelievre-Pegorier *et al.*, 1983). The cortical TAL and the distal tubule are avid sites of magnesium transport, responsible for reabsorption of 60 and 10%, respectively, of the filtered load of magnesium (Dai *et al.*, 2001; Mandon *et al.*, 1993). Magnesium transport in these segments is efficient early in life. The fractional reabsorption of magnesium in the TAL of 2-week-old rats is similar to that measured in 6-week-old animals (Lelievre-Pegorier *et al.*, 1983). Studies in mouse cortical TALs microperfused *in vitro* suggest that the capacity for transepithelial magnesium absorption may increase further after weaning until mature levels are reached by the eighth week of postnatal life (Wittner *et al.*, 1997).

Magnesium reabsorption in the TAL is passive and driven through the paracellular pathway by the lumen-positive transepithelial voltage (Di Stefano *et al.*, 1993; Mandon *et al.*, 1993; Quamme and de Rouffignac, 2000; Shareghi and Agus, 1982a). Transport in the superficial distal tubule is transcellular and active (Dai *et al.*, 2001; Quamme and de Rouffignac, 2000). Magnesium reabsorption in these segments is regulated by a number of hormones, including PTH, calcitonin, glucagon, and AVP (Bailly *et al.*, 1985, 1990; Dai *et al.*, 2001; Di Stefano *et al.*, 1990; Kang *et al.*, 2000; Wittner *et al.*, 1988). In addition, dietary magnesium restriction or loading stimulates (Shafik and Quamme, 1989) or inhibits magnesium reabsorption, as appropriate, a response mediated by the CaSR in the cortical TAL and distal tubule (Hebert *et al.*, 1997). Loop diuretics such as furosemide inhibit magnesium absorption and increase magnesium excretion due to their inhibition of sodium chloride transport and modification of the transepithelial

voltage in the TAL (Di Stefano *et al.*, 1993). Inactivating mutations of the recently cloned gene paracellin (Simon *et al.*, 1999), which encodes a tight junctional protein of the paracellular pathway, is associated with the hypomagnesemia hypercalciuria syndrome (Weber *et al.*, 2000). The latter disorder is characterized by hypomagnesemia, hypercalciuria, advanced nephrocalcinosis, hyposthenuria, and progressive renal failure (Weber *et al.*, 2000). Loss-of-function mutations in the distal tubule Na-Cl cotransporter gene NCC (see Gitelman's syndrome; see earlier discussion), or CaSR, as has been reported in autosomal dominant hypoparathyroidism (Cole and Quamme, 2000), are also associated with hypomagnesemia.

G. Urinary Concentration and Dilution

The fetal metanephric kidney of a variety of species, including human, produces large volumes of hypotonic urine that contribute significantly to the volume and composition of amniotic fluid (McCance and Stainer, 1960; McCance and Widdowson, 1953; Merlet-Benichou and de Rouffignac, 1977; Nakamura *et al.*, 1987c; Robillard *et al.*, 1979a). Although the urine is generally hypotonic during early development, the fetal nephron is able to concentrate urine under conditions of stress, such as that induced by maternal water deprivation (Ross *et al.*, 1988b), hemorrhage (Gomez *et al.*, 1984), or infusion of AVP (Horne *et al.*, 1993; Woods *et al.*, 1986). However, the maximum urine osmolality that can be achieved is only about 20% of that in the adult (Lingwood *et al.*, 1978; Robillard and Weitzman, 1980).

Urine voided at or shortly after birth continues to be hypotonic with respect to plasma (McCance and Widdowson, 1953; Strauss *et al.*, 1981). Despite the ability of preterm and term neonates to produce a hypotonic urine, their ability to excrete a water load is limited, presumably because of their low GFR. The capacity to maximally concentrate the urine (~1000–1200 mOsmol/kg) is generally not attained before 6 months of age in humans (Edelmann *et al.*, 1960; Polacek *et al.*, 1965) and a postnatal age of 3 weeks in rat (Yasui *et al.*, 1996). After fluid deprivation for 12 to 24 h, the maximal urine osmolality achieved in both premature infants and full-term newborns is 600 to 800 mOsm/kg, respectively (Calcagno *et al.*, 1954; Edelmann *et al.*, 1960; Hansen and Smith, 1953), a value roughly 60% that observed in older children and adults. Administration of AVP or 1-desamino-8-D-AVP (DDAVP) to healthy 1- to 3-week-old newborns leads to a response, albeit of shorter duration and reduced magnitude than that observed at 4 to 6 weeks (Svenningsen and Aronson, 1974).

1. Urinary Concentration

The collecting duct is responsible for the reabsorption of approximately 10 to 15% of the filtered water, a process that is regulated by AVP, synthesized in the hypothalamus. In the absence of AVP, the collecting duct is impermeable to water. When present, AVP binds to basolateral vasopressin V_2 receptors that, via stimulation of adenylyl cyclase activity, cAMP production, and phosphorylation of PKA (Lolait et al., 1992), induce the translocation of vesicles containing aquaporin-2 (AQP2) water channels into the apical membrane. An increase in osmotic water permeability ensues. Water that enters the cell can then move into the hypertonic renal interstitium through the basolateral water channels AQP3 and AQP4 (Knepper, 1997). Osmotic equilibration between the collecting duct tubular fluid and the hypertonic interstitium in the inner medulla is essential to maximally concentrate the urine. Thus, urinary concentration requires a hypertonic medullary interstitium, expression of functional V_2 receptors in the collecting duct, efficient coupling between receptor binding and cAMP generation, and the presence of appropriate collecting duct aquaporin water channels.

Cumulative evidence, summarized later, suggests that the blunted sensitivity of the fetal and neonatal kidney to AVP and limited concentrating ability of the neonatal animal is not due to a paucity of V_2 receptors, AQP channels or efficiency of coupling to second messengers after the first week of postnatal life, but is limited primarily by the inability to develop a high corticopapillary osmotic gradient.

a. Medullary Gradient Anatomic maturation of the metanephric kidney is characterized by an elongation of the loops of Henle and their penetration into the inner medulla (Bankir and de Rouffignac, 1985; Horster et al., 1984; Speller and Moffat, 1977; Trimble, 1970). In the rat, the 1.6-fold postnatal increase in length of the renal medulla correlates well with the 1.5-fold increase in urine osmolality observed between 10 and 20 days of age (Trimble, 1970). Concomitant with loop elongation is the generation of an axial corticomedullary osmotic gradient (Horster et al., 1984), as is necessary for urinary concentration. The postnatal increases in interstitial sodium and urea concentrations in the rat renal medulla (Rane et al., 1985) presumably reflect maturational increases in active sodium reabsorption by the TAL (see earlier discussion), countercurrent flow in the vasa rectae and loops of Henle, and recycling of urea reabsorbed in the distal nephron. The developmental activation of aldose reductase, an enzyme necessary for the generation of intracellular osmolytes, is also important for the maintenance of cell function in the concentrated milieu (Edelmann et al., 1966; Horster et al., 1984; Schwartz et al., 1992).

Within the renal inner medulla, urea is transported by both facilitated and active mechanisms. The AVP-regulated facilitated urea transporter UT-A in the IMCD permits high rates of transepithelial urea transport. This results in the delivery of large quantities of urea into the deepest regions of the inner medulla where it is essential to maintain a high interstitial osmolality (Sands, 1999). Four cDNA isoforms of the UT-A urea transporter family have been cloned. In addition, there are three secondary active, sodium-dependent urea transport mechanisms in the IMCD (Sands, 1999). The functional development of these transporters has not been studied.

b. Vasopressin (AVP) The limited ability of the immature kidney to concentrate urine is not due to an inability of the fetus or neonate to synthesize and secrete AVP. Circulating levels of AVP are elevated in preterm and term infants and decrease rapidly in term infants within 24 h of birth (Hadeed et al., 1979; Rees et al., 1980). Both fetal and newborn animals (Leake et al., 1979; Robillard et al., 1979b; Weitzman et al., 1978), as well as human infants (DeVane and Porter, 1980; Rees et al., 1980), exhibit a qualitatively appropriate response to osmolar or volume stimuli known to affect AVP release. The gene encoding the V_2 receptor to which AVP binds in the collecting duct is detected early in gestation (Ostrowski et al., 1993); the ontogeny of protein expression has not been characterized.

Although there appears to be inefficient coupling between AVP receptor binding and cAMP generation in the first week of postnatal life (Imbert-Teboul et al., 1984; Rajerison et al., 1976; Schlondorff et al., 1978), the signal transduction pathway appears to mature rapidly thereafter. Osmotic water permeability in microperfused IMCDs isolated from 1- to 2-week-old rats increases two-fold in response to basolateral exposure to AVP, a response that is significantly less robust than the six-fold increase observed in tubules from adult animals (Siga and Horster, 1991). However, forskolin, which directly activates adenylyl cyclase, independent of the state of the AVP receptors and early signaling events, increases the water permeability in the immature IMCD to a comparable degree to that elicited by AVP (Siga and Horster, 1991). These data are consistent with an intact cAMP-mediated signal transduction pathway by the second week of postnatal life and suggest that the low osmotic water permeability of the neonatal IMCD is not due to immaturity of the receptor–ligand interaction or early postreceptor signal transduction cascade (Siga and Horster, 1991).

In vivo, AVP-stimulated cAMP production in the immature CCD may be inhibited by prostaglandins via activation of inhibitory G proteins or direct inhibition of the catalytic subunit of adenylyl cyclase (Bonilla-Felix and John-Phillip, 1994). The high levels of renal PGE that prevail in the immature kidney may additionally maintain high rates of medullary blood flow, inhibit sodium chloride transport in the TAL (see earlier discussion), and decrease urea absorption in the collecting duct, the net effect would be to reduce the medullary solute content. However, inhibition of prostaglandin synthesis in fetal lambs has no effect on the urinary concentrating ability (Matson *et al.*, 1981) and, at the level of the single collecting duct isolated from 1- to 2-week old rabbits, does not substantially increase AVP-stimulated water permeability (Bonilla-Felix *et al.*, 1999). Data suggest that prostaglandins contribute little to the postnatal maturation of renal concentrating ability.

Aquaporin-1, located in proximal tubules and descending thin limbs of Henle, and collecting duct AQP2 are the channels involved in water transport across renal tubule epithelia (Harris and Zeidel, 1993). The genes encoding AQP1 and AQP2 can be detected in the metanephric kidney of the ovine fetus by ~6 weeks gestation (Butkus *et al.*, 1997b; Moritz and Wintour, 1999). In this species, gene expression for AQP1 increases seven-fold by term (140 days of gestation), reaching adult levels 6 weeks after birth (Wintour *et al.*, 1998). Gene expression for AQP2 reaches 40% of the adult level by term (Wintour *et al.*, 1998). In the human fetus, AQP1 protein is detected in the proximal convoluted tubule by 15 weeks of gestation, reaching levels approximating 50% of those in the adult renal cortex at birth. AQP2 is present in fetal ureteric bud and collecting duct by week 12 of fetal life (Devuyst *et al.*, 1996). While total kidney AQP2 and AQP3 proteins, particularly their glycosylated forms, continue to increase after birth (Baum *et al.*, 1998b), the detection of apical AQP2 and basolateral AQP3 protein in the ureteric bud and collecting duct of fetal rat (by embryonic day 20) (Baum *et al.*, 1998b; Bonilla-Felix and Jiang, 1997; Devuyst *et al.*, 1996; Yamamoto *et al.*, 1997) suggests that the low AVP responsiveness and concentrating capacity of the fetal and neonatal kidney are not limited by water channel expression, but may be due to the absence of an adequate corticomedullary concentration gradient.

Mutations in the genes encoding either the V_2 receptor (Cheong *et al.*, 1997) and AQP2 (Deen *et al.*, 1994; van Lieburg *et al.*, 1994) have been detected in patients with nephrogenic diabetes insipidus, a disorder in which the collecting duct is resistant to the action of AVP.

c. Regulation AQP2 expression is regulated by several factors. In a variety of mature native and cultured collecting duct cells, either prolonged dehydration or chronic vasopressin exposure increases AQP2 gene expression, presumably secondary to an effect on a cAMP response element (CRE) and an activator protein 1 (AP1) site (Knepper, 1997; Matsumura *et al.*, 1997; Terris *et al.*, 1996; Yasui *et al.*, 1997); water loading reduces AQP2 mRNA expression rapidly (Saito *et al.*, 1997). Similarly, 10-day-old rats treated with a V_2 receptor agonist exhibit an increase in urine osmolality and renal medullary AQP-2 mRNA levels within 6 h (Yasui *et al.*, 1996).

Prenatal administration of dexamethasone to pregnant ewes increases both AQP1 and AQP2 gene expression in their offspring (Wintour *et al.*, 1998). Glucocorticoid treat-

ment of neonatal, but not adult, rats is also associated with an increase in AQP2 mRNA and protein expression (Yasui *et al.*, 1996).

The RAS is an important regulator of water homeostasis. ANG II stimulates the synthesis and release of AVP by the hypothalamus (Phillips, 1987; Qadri *et al.*, 1993) and also modulates the renal concentrating mechanisms through its effects on hemodynamics and perhaps through direct effects on renal epithelia. ANG II not only causes vasoconstriction of efferent (and afferent) arterioles, which may modulate physical forces in the peritubular capillaries of the proximal tubule to favor solute and water reabsorption, but also decreases medullary blood flow and thus may affect osmolar gradients (Cupples *et al.*, 1988; Faubert *et al.*, 1987). The role of the RAS in regulating renal water excretion is underscored by the observation that ACE-deficient mice generate high urine volumes of low osmolality (Okubo *et al.*, 1998). Although the atrophy of the renal papilla detected in these animals would be expected to limit the IMCD surface area available for water and urea reabsorption, mice lacking AT_{1A} receptors, who have no structural abnormality of the kidney, also have a defect in urinary concentration (Oliverio *et al.*, 2000). The reduction in medullary AQP2 expression observed in some of these rodent models may contribute to the concentrating defect (Guron *et al.*, 1999).

2. Urinary Dilution

Premature infants of less than 35 weeks GA studied under conditions of maximal water diuresis can decrease their urine osmolality to 70 mOsm/kg, whereas infants over 35 weeks of GA are able to reduce their urine osmolality to 50 mOsm/kg (Rodriguez-Soriano *et al.*, 1983). Thus, premature infants are unable to dilute their urine as efficiently as term infants or adults. Although there is greater proximal sodium rejection in preterm than term infants, the high avidity of the distal nephron for sodium reabsorption allows the neonate, especially the preterm infant, to generate a free water clearance greater than that in adults (Aperia and Elinder, 1981; Kleinman, 1975; Kleinman and Banks, 1983). Despite the greater capacity for free water clearance, the ability of the neonate to excrete a hypotonic load is limited, presumably due to their low GFR.

VI. Summary

Functional maturation of the metanephric kidney is a gradual process that lags behind anatomic maturation and is not completed *in utero*. During fetal life, urine produced by the metanephric kidney contributes to the volume and composition of the amniotic fluid. Cumulative evidence indicates that the fetal kidney is functional and participates in the regulation of blood pressure, fluid and electrolyte homeostasis, acid–base transport, and autocrine/paracrine signaling. However, the placenta is primarily responsible for maintaining homeostasis during fetal life. The transition from the intrauterine to the extrauterine environment necessitates that the kidney rapidly assumes responsibility for these functions.

As summarized in this chapter, substantial progress has been made over the past few decades in expanding our knowledge base regarding the maturation of RBF, GFR, and tubular transport in the differentiating kidney. We now understand the physiologic basis underlying the evolution of the low-flow, high-resistance fetal organ, with most of the blood supplying the inner cortex, into a high-flow, low-resistance organ, with most of the blood supplying the outer cortex. Interactions among the RAAS, catecholamines, and NO appear to be most important in the perinatal changes in renal hemodynamics. The developmental maturation of GFR follows a centrifugal pattern, similar to that described for RBF. At birth, the more mature glomeruli in the juxtamedullary cortex have higher filtration rates than the most recently formed glomeruli in the superficial cortex, some of which may not begin filtration for some time. The developmental increase in absolute GFR from term birth to maturity in a number of species can be accounted for predominantly by an increase in glomerular surface available for filtration.

Growth requires a positive balance for a variety of ions and minerals. Indeed, tubular function in the kidney of the full-term newborn is uniquely suited to meet these physiologic needs. Thus, the kidney of the newborn reabsorbs sodium and phosphate avidly and secretes little potassium. Because of the rapid deposition of calcium into the skeleton, the newborn is required to excrete an additional load of protons such that neonatal acid excretion rates are near maximal, even in the absence of exogenous acid loads. Analyses of newborn animals, including rabbit, mouse, and rat, which serve as excellent models for premature human infants in that nephronogenesis continues postnatally, reveal that the "preterm" kidney lacks the tubular maturity necessary to maintain homeostasis. Preterm infants, unlike their more mature counterparts, exhibit substantial renal salt wasting and a markedly diminished ability to excrete potassium, acid, and water loads. While some hormones, such as renin, aldosterone, and AVP, are at high levels in the newborn period, the transport processes that they effect generally function less potently than those observed in the adult. This phenomenon may reflect hyporesponsiveness of the transducers or effectors.

The application of a variety of physiological techniques to the differentiating metanephric kidney has provided us with a fundamental understanding of renal functional maturation. However, there remain many unanswered questions in the discipline of developmental renal physiology. The exploding field of molecular biology, especially as it relates to the identification of maturation-specific and disease-related genes, promises to yield further insight into the mechanisms underlying normal renal development, physiology, and the pathophysiological basis of renal disease. A truly integrated picture of renal development demands collaboration among developmental physiologists, biologists, and pathologists, as well as cell and molecular biologists, a perspective reflected in the scope of this book.

References

Abadie, L., Blazy, I., Roubert, P., Plas, P., Charbit, M., Chabrier, P. E., and Dechaux, M. (1996). Decrease in endothelin-1 renal receptors during the 1st month of life in the rat. *Pediatr. Nephrol.* **10**, 185–189.

Aguilera, G., Kapur, S., Feuillan, P., Sunar-Akbasak, B., and Bathia, A. J. (1994). Developmental changes in angiotensin II receptor subtypes and AT1 receptor mRNA in rat kidney. *Kidney Int.* **46**, 973–979.

Agus, Z. S., Puschett, J. B., Senesky, D., and Goldberg, M. (1971). Mode of action of parathyroid hormone and cyclic adenosine $3',5'$-monophosphate on renal tubular phosphate reabsorption in the dog. *J. Clin. Invest.* **50**, 617–626.

Albert, M. S., and Winters, R. W. (1966). Acid-base equilibrium of blood in normal infants. *Pediatrics* **37**, 728–732.

Al-Dahhan, J., Haycock, G. B., Chantler, C., and Stimmler, L. (1983). Sodium homeostasis in term and preterm neonates. I. Renal aspects. *Arch. Dis. Child* **58**, 335–342.

Allison, M., Lipham, E., and Gottschalk, C. (1972). Hydrostatic pressure in the rat kidney. *Am. J. Physiol.* **223**, 975–983.

Alpern, R. J. (1985). Mechanism of basolateral membrane H+/OH-/HCO-3 transport in the rat proximal convoluted tubule: A sodium-coupled electrogenic process. *J. Gen. Physiol.* **86**, 613–636.

Amiel, C., Kuntziger, H., and Richet, G. (1970). Micropuncture study of handling of phosphate by proximal and distal nephron in normal and parathyroidectomized rat. Evidence for distal reabsorption. *Pflug. Arch.* **317**, 93–109.

Aperia, A., Broberger, O., Elinder, G., Herin, P., and Zetterstrom, R. (1981a). Postnatal development of renal function in pre-term and full-term infants. *Acta. Paediatr. Scand.* **70**, 183–187.

Aperia, A., Broberger, O., and Herin, P. (1974a). Maturational changes in glomerular perfusion rate and glomerular filtration rate in lambs. *Pediatr. Res.* **8**, 758-765.

Aperia, A., Broberger, O., Herin, P., and Joelsson, I. (1977). Renal hemodynamics in the perinatal period. A study in lambs. *Acta. Physiol. Scand.* **99**, 261-269.

Aperia, A., Broberger, O., Herin, P., and Zetterstrom, R. (1979). Sodium excretion in relation to sodium intake and aldosterone excretion in newborn pre-term and full-term infants. *Acta. Paediatr. Scand.* **68**, 813–817.

Aperia, A., Broberger, O., Thodenius, K., and Zetterstrom, R. (1972). Renal response to an oral sodium load in newborn full term infants. *Acta. Paediatr. Scand.* **61**, 670–676.

Aperia, A., Broberger, O., Thodenius, K., and Zetterstrom, R. (1974b). Developmental study of the renal response to an oral salt load in preterm infants. *Acta. Paediatr. Scand.* **63**, 517–524.

Aperia, A., and Elinder, G. (1981). Distal tubular sodium reabsorption in the developing rat kidney. *Am. J. Physiol.* **240**, F487–F91.

Aperia, A., and Herin, P. (1975). Development of glomerular perfusion rate and nephron filtration rate in rats 17-60 days old. *Am. J. Physiol.* **228**, 1319–1325.

Aperia, A., and Herin, P. (1976). Effect of arterial blood pressure reduction on renal hemodynamics in the developing lamb. *Acta. Physiol. Scand.* **98**, 387–394.

Aperia, A., Ibarra, F., Svensson, L. B., Klee, C., and Greengard, P. (1992). Calcineurin mediates alpha-adrenergic stimulation of Na$^+$,K$^+$-ATPase activity in renal tubule cells. *Proc. Natl. Acad. Sci. USA* **89**, 7394–7397.

Aperia, A., and Larsson, L. (1979). Correlation between fluid reabsorption and proximal tubule ultrastructure during development of the rat kidney. *Acta. Physiol. Scand.* **105**, 11–22.

Aperia, A., and Larsson, L. (1984). Induced development of proximal tubular Na-K-ATPase, basolateral cell membranes and fluid reabsorption. *Acta. Physiol. Scand.* **121**, 133–141.

Aperia, A., Larsson, L., and Zetterstrom, R. (1981b). Hormonal induction of Na,K-ATPase in developing proximal tubular cells. *Am. J. Physiol.* **241**, F356–F360.

Aperia, A. C. (2000). Intrarenal dopamine: A key signal in the interactive regulation of sodium metabolism. *Annu. Rev. Physiol.* **62**, 621–647.

Appenroth, D., and Braunlich, H. (1981). Effect of sympathectomy with 6-hydroxydopamine on the renal excretion of water and electrolytes in developing rats. *Acta. Biol. Med. Ger.* **40**, 1715–1721.

Arant, B. S., Jr. (1978). Developmental patterns of renal functional maturation compared in the human neonate. *J. Pediatr.* **92**, 705–712.

Arant, B. S., Jr. (1987). Postnatal development of renal function during the first year of life. *Pediatr. Nephrol.* **1**, 308–313.

Arendshorst, W. J., Brannstrom, K., and Ruan, X. (1999). Actions of angiotensin II on the renal microvasculature. *J. Am. Soc. Nephrol.* **10 (Suppl. 11)**, S149–S161.

Arnold-Aldea, S. A., Auslender, R. A., and Parer, J. T. (1991). The effect of the inhibition of prostaglandin synthesis on renal blood flow in fetal sheep. *Am. J. Obstet. Gynecol.* **165**, 185–190.

Aronson, P. S., and Giebisch, G. (1997). Mechanisms of chloride transport in the proximal tubule. *Am. J. Physiol.* **273**, F179-F192.

Aschinberg, L. C., Goldsmith, D. I., Olbing, H., Spitzer, A., Edelmann, C. M., Jr., and Blaufox, M. D. (1975). Neonatal changes in renal blood flow distribution in puppies. *Am. J. Physiol.* **228**, 1453–1461.

Babinet, C. (2000). Transgenic mice: an irreplaceable tool for the study of mammalian development and biology. *J. Am. Soc. Nephrol.* **11 (Suppl 16)**, S88–S94.

Bachmann, S., Bostanjoglo, M., Schmitt, R., and Ellison, D. H. (1999). Sodium transport-related proteins in the mammalian distal nephron:-Distribution, ontogeny and functional aspects. *Anat. Embryol. (Berl)* **200**, 447–468.

Bacskai, B. J., and Friedman, P. A. (1990). Activation of latent Ca^{2+} channels in renal epithelial cells by parathyroid hormone. *Nature* **347**, 388–391.

Badr, K. F., and Ichikawa, I. (1988). Prerenal failure: A deleterious shift from renal compensation to decompensation. *N. Engl. J. Med.* **319**, 623–629.

Bailly, C., Imbert-Teboul, M., Roinel, N., and Amiel, C. (1990). Isoproterenol increases Ca, Mg, and NaCl reabsorption in mouse thick ascending limb. *Am. J. Physiol.* **258**, F1224-F1231.

Bailly, C., Roinel, N., and Amiel, C. (1985). Stimulation by glucagon and PTH of Ca and Mg reabsorption in the superficial distal tubule of the rat kidney. *Pflug. Arch.* **403**, 28–34.

Baker, J. T., and Solomon, S. (1976). Maturation of the renal response to hypertonic sodium chloride loading in rats: Micropuncture and clearance studies. *J. Physiol. (Lond)* **258**, 83–98.

Ballevre, L., Thonney, M., and Guignard, J. P. (1996). Nitric oxide modulates glomerular filtration and renal blood flow of the newborn rabbit. *Biol. Neonate* **69**, 389-398.

Bankir, L., and de Rouffignac, C. (1985). Urinary concentrating ability: Insights from comparative anatomy. *Am. J. Physiol.* **249**, R643–666.

Barac-Nieto, M., Corey, H., Liu, S. M., and Spitzer, A. (1993). Role of intracellular phosphate in the regulation of renal phosphate transport during development. *Pediatr. Nephrol.* **7**, 819–822.

Barac-Nieto, M., Dowd, T. L., Gupta, R. K., and Spitzer, A. (1991). Changes in NMR-visible kidney cell phosphate with age and diet: Relationship to phosphate transport. *Am. J. Physiol.* **261**, F153–F162.

Barac-Nieto, M., Gupta, R. K., and Spitzer, A. (1990). NMR studies of phosphate metabolism in the isolated perfused kidney of developing rats. *Pediatr. Nephrol.* **4**, 392–398.

Barlet, C., and Doucet, A. (1986). Kinetics of triiodothyronine action on Na-K-ATPase in single segments of rabbit nephron. *Pflug. Arch.* **407**, 27–32.

Barnett, H. (1940). Renal physiology in infants and children. I. Method for estimation of glomerular filtration rate. *Proc. Soc. Exp. Biol. Med.* **44**, 654–658.

Bastani, B., Purcell, H., Hemken, P., Trigg, D., and Gluck, S. (1991). Expression and distribution of renal vacuolar proton-translocating adenosine triphosphatase in response to chronic acid and alkali loads in the rat. *J. Clin. Invest.* **88**, 126–136.

Baum, M. (1987). Evidence that parallel Na$^+$-H$^+$ and Cl$^-$-HCO$_3^-$(OH$^-$) antiporters transport NaCl in the proximal tubule. *Am. J. Physiol.* **252**, F338–F345.

Baum, M. (1989). Axial heterogeneity of rabbit proximal tubule luminal H$^+$ and basolateral HCO$_3^-$ transport. *Am. J. Physiol.* **256**, F335–F341.

Baum, M. (1990). Neonatal rabbit juxtamedullary proximal convoluted tubule acidification. *J. Clin. Invest.* **85**, 499–506.

Baum, M. (1992). Developmental changes in rabbit juxtamedullary proximal convoluted tubule acidification. *Pediatr. Res.* **31**, 411–414.

Baum, M., and Berry, C. A. (1984). Evidence for neutral transcellular NaCl transport and neutral basolateral chloride exit in the rabbit proximal convoluted tubule. *J. Clin. Invest.* **74**, 205–211.

Baum, M., Biemesderfer, D., Gentry, D., and Aronson, P. S. (1995). Ontogeny of rabbit renal cortical NHE3 and NHE1: Effect of glucocorticoids. *Am. J. Physiol.* **268**, F815–F820.

Baum, M., Dwarakanath, V., Alpern, R. J., and Moe, O. W. (1998a). Effects of thyroid hormone on the neonatal renal cortical Na$^+$/H$^+$ antiporter. *Kidney Int.* **53**, 1254–1258.

Baum, M., and Quigley, R. (1991). Prenatal glucocorticoids stimulate neonatal juxtamedullary proximal convoluted tubule acidification. *Am. J. Physiol.* **261**, F746–F752.

Baum, M., and Quigley, R. (1998). Inhibition of proximal convoluted tubule transport by dopamine. *Kidney Int.* **54**, 1593–1600.

Baum, M., Quigley, R., and Quan, A. (1997). Effect of luminal angiotensin II on rabbit proximal convoluted tubule bicarbonate absorption. *Am. J. Physiol.* **273**, F595–F600.

Baum, M. A., Ruddy, M. K., Hosselet, C. A., and Harris, H. W. (1998b). The perinatal expression of aquaporin-2 and aquaporin-3 in developing kidney. *Pediatr. Res.* **43**, 783–790.

Baylis, C., Handa, R. K., and Sorkin, M. (1990a). Glucocorticoids and control of glomerular filtration rate. *Semin. Nephrol.* **10**, 320–329.

Baylis, C., Harton, P., and Engels, K. (1990b). Endothelial derived relaxing factor controls renal hemodynamics in the normal rat kidney. *J. Am. Soc. Nephrol.* **1**, 875–881.

Beach, R. E., Schwab, S. J., Brazy, P. C., and Dennis, V. W. (1987). Norepinephrine increases Na$^+$-K$^+$-ATPase and solute transport in rabbit proximal tubules. *Am. J. Physiol.* **252**, F215–F220.

Beato, M., Chalepakis, G., Schauer, M., and Slater, E. P. (1989). DNA regulatory elements for steroid hormones. *J Steroid Biochem* **32**, 737–747.

Beato, M., Chavez, S., and Truss, M. (1996). Transcriptional regulation by steroid hormones. *Steroids* **61**, 240–251.

Beck, J. C., Lipkowitz, M. S., and Abramson, R. G. (1991). Ontogeny of Na/H antiporter activity in rabbit renal brush border membrane vesicles. *J. Clin. Invest.* **87**, 2067–2076.

Beitins, I. Z., Bayard, F., Levitsky, L., Ances, I. G., Kowarski, A., and Migeon, C. J. (1972). Plasma aldosterone concentration at delivery and during the newborn period. *J. Clin. Invest.* **51**, 386–394.

Bello-Reuss, E., Higashi, Y., and Kaneda, Y. (1982). Dopamine decreases fluid reabsorption in straight portions of rabbit proximal tubule. *Am. J. Physiol.* **242**, F634-F640.

Benchimol, C., Zavilowitz, B., and Satlin, L. M. (2000). Developmental expression of ROMK mRNA in rabbit cortical collecting duct. *Pediatr. Res.* **47**, 46–52.

Bengele, H. H., Lechene, C. P., and Alexander, E. A. (1979). Phosphate transport along the inner medullary collecting duct of the rat. *Am. J. Physiol.* **237**, F48–F54.

Benzoni, D., Vincent, M., Betend, B., and Sassard, J. (1981). Urinary excretion of prostaglandins and electrolytes in developing children. *Kidney Int.* **20**, 386–388.

Berger, S., Bleich, M., Schmid, W., Cole, T. J., Peters, J., Watanabe, H., Kriz, W., Warth, R., Greger, R., and Schutz, G. (1998). Mineralocorticoid receptor knockout mice: Pathophysiology of Na$^+$ metabolism. *Proc. Natl. Acad. Sci. USA* **95**, 9424–9429.

Berger, S., Bleich, M., Schmid, W., Greger, R., and Schutz, G. (2000). Mineralocorticoid receptor knockout mice: lessons on Na$^+$ metabolism. *Kidney Int.* **57**, 1295–1298.

Bhat, R., John, E., Chari, G., Shankararao, R., Fornell, L., Gulati, A., and Vidyasagar, D. (1995). Renal actions of endothelin-1 in newborn piglets: Dose-effect relation and the effects of receptor antagonist (BQ-123) and cyclooxygenase inhibitor (indomethacin). *J. Lab. Clin. Med.* **126**, 458–469.

Biber, J., Caderas, G., Stange, G., Werner, A., and Murer, H. (1993a). Effect of low-phosphate diet on sodium/phosphate cotransport mRNA and protein content and on oocyte expression of phosphate transport. *Pediatr. Nephrol.* **7**, 823–826.

Biber, J., Custer, M., Werner, A., Kaissling, B., and Murer, H. (1993b). Localization of NaPi-1, a Na/Pi cotransporter, in rabbit kidney proximal tubules. II. Localization by immunohistochemistry. *Pflug. Arch.* **424**, 210–215.

Biemesderfer, D., Reilly, R. F., Exner, M., Igarashi, P., and Aronson, P. S. (1992). Immunocytochemical characterization of Na$^+$-H$^+$ exchanger isoform NHE-1 in rabbit kidney. *Am. J. Physiol.* **263**, F833–F840.

Biemesderfer, D., Rutherford, P. A., Nagy, T., Pizzonia, J. H., Abu-Alfa, A. K., and Aronson, P. S. (1997). Monoclonal antibodies for high-resolution localization of NHE3 in adult and neonatal rat kidney. *Am. J. Physiol.* **273**, F289–F299.

Blanco, G., and Mercer, R. W. (1998). Isozymes of the Na-K-ATPase: Heterogeneity in structure, diversity in function. *Am. J. Physiol.* **275**, F633–F450.

Bloch, R. D., Zikos, D., Fisher, K. A., Schleicher, L., Oyama, M., Cheng, J. C., Skopicki, H. A., Sukowski, E. J., Cragoe, E. J., Jr., and Peterson, D. R. (1992). Activation of proximal tubular Na$^+$-H$^+$ exchange by angiotensin II. *Am. J. Physiol.* **263**, F135–F143.

Bogaert, G. A., Kogan, B. A., and Mevorach, R. A. (1993). Effects of endothelium-derived nitric oxide on renal hemodynamics and function in the sheep fetus. *Pediatr. Res.* **34**, 755–761.

Bogaert, G. A., Kogan, B. A., Mevorach, R. A., Wong, J., Gluckman, G. R., Fineman, J. R., and Heymann, M. A. (1996). Exogenous endothelin-1 causes renal vasodilation in the fetal lamb. *J. Urol.* **156**, 847–853.

Bonilla-Felix, M., and Jiang, W. (1997). Aquaporin-2 in the immature rat: Expression, regulation, and trafficking. *J. Am. Soc. Nephrol.* **8**, 1502–1509.

Bonilla-Felix, M., and John-Phillip, C. (1994). Prostaglandins mediate the defect in AVP-stimulated cAMP generation in immature collecting duct. *Am. J. Physiol.* **267**, F44–F48.

Bonilla-Felix, M., Vehaskari, V. M., and Hamm, L. L. (1999). Water transport in the immature rabbit collecting duct. *Pediatr. Nephrol.* **13**, 103–107.

Bonvalet, J. P. (1987). Binding and action of aldosterone, dexamethasone, 1-25(OH)$_2$D$_3$, and estradiol along the nephron. *J. Steroid Biochem.* **27**, 953–961.

Bonvalet, J. P. (1998). Regulation of sodium transport by steroid hormones. *Kidney Int. Suppl.* **65**, S49–S56.

Bostanjoglo, M., Reeves, W. B., Reilly, R. F., Velazquez, H., Robertson, N., Litwack, G., Morsing, P., Dorup, J., Bachmann, S., Ellison, D. H., and Bostonjoglo, M. (1998). 11Beta-hydroxysteroid dehydrogenase,

mineralocorticoid receptor, and thiazide-sensitive Na-Cl cotransporter expression by distal tubules. *J. Am. Soc. Nephrol.* **9**, 1347–1358.

Bourdeau, J. E., and Burg, M. B. (1979). Voltage dependence of calcium transport in the thick ascending limb of Henle's loop. *Am. J. Physiol.* **236**, F357–F364.

Bourdeau, J. E., and Burg, M. B. (1980). Effect of PTH on calcium transport across the cortical thick ascending limb of Henle's loop. *Am. J. Physiol.* **239**, F121–F126.

Bourdeau, J. E., and Lau, K. (1989). Effects of parathyroid hormone on cytosolic free calcium concentration in individual rabbit connecting tubules. *J. Clin. Invest.* **83**, 373–379.

Braam, B., Mitchell, K. D., Fox, J., and Navar, L. G. (1993). Proximal tubular secretion of angiotensin II in rats. *Am. J. Physiol.* **264**, F891–F898.

Brem, A. S., Bina, B., Matheson, K. L., Barnes, J. L., and Morris, D. J. (1994). Developmental changes in rat renal 11 beta-hydroxysteroid dehydrogenase. *Kidney Int.* **45**, 679–683.

Brem, A. S., Matheson, K. L., and Morris, D. J. (1992). Effect of carbenoxolone sodium on steroid-induced sodium transport in the toad bladder: further studies. *J. Steroid Biochem. Mol. Biol.* **42**, 911–914.

Breyer, M. D., and Harris, R. C. (2001). Cyclooxygenase 2 and the kidney. *Curr. Opin. Nephrol. Hypertens.* **10**, 89–98.

Briggs, J. P., Schubert, G., and Schnermann, J. (1984). Quantitative characterization of the tubuloglomerular feedback response: effect of growth. *Am. J. Physiol.* **247**, F808–F815.

Brion, L. P., Fleischman, A. R., McCarton, C., and Schwartz, G. J. (1986). A simple estimate of glomerular filtration rate in low birth weight infants during the first year of life: Noninvasive assessment of body composition and growth. *J. Pediatr.* **109**, 698–707.

Brion, L. P., Zavilowitz, B. J., Rosen, O., and Schwartz, G. J. (1991). Changes in soluble carbonic anhydrase activity in response to maturation and NH$_4$Cl loading in the rabbit. *Am. J. Physiol.* **261**, R1204–1213.

Brion, L. P., Zavilowitz, B. J., Suarez, C., and Schwartz, G. J. (1994). Metabolic acidosis stimulates carbonic anhydrase activity in rabbit proximal tubule and medullary collecting duct. *Am. J. Physiol.* **266**, F185–F195.

Brodehl, J., Gellissen, K., and Weber, H. P. (1982). Postnatal development of tubular phosphate reabsorption. *Clin. Nephrol.* **17**, 163–171.

Broughton Pipkin, F., Lumbers, E. R., and Mott, J. C. (1974). Factors influencing plasma renin and angiotensin II in the conscious pregnant ewe and its foetus. *J. Physiol. (Lond)* **243**, 619–636.

Broughton Pipkin, F., Smales, O. R., and O'Callaghan, M. (1981). Renin and angiotensin levels in children. *Arch. Dis. Child.* **56**, 298–302.

Broughton Pipkin, F., Symonds, E. M., and Turner, S. R. (1982). The effect of captopril (SQ14 225) upon mother and fetus in the chronically cannulated ewe and in the pregnant rabbit. *J. Physiol. (Lond)* **323**, 415–422.

Broughton-Pipkin, F., Kirkpatrick, S.M., *et al.* (1974). Renin and angiotensin-like levels in foetal, newborn, and adult sheep. *J. Physiol.* **243**, 619–636.

Brown, D., Hirsch, S., and Gluck, S. (1988). Localization of a proton-pumping ATPase in rat kidney. *J. Clin. Invest.* **82**, 2114–2126.

Brown, D., Kumpulainen, T., Roth, J., and Orci, L. (1983). Immunohisto-chemical localization of carbonic anhydrase in postnatal and adult rat kidney. *Am. J. Physiol.* **245**, F110–F118.

Brown, D., Zhu, X. L., and Sly, W. S. (1990). Localization of membrane-associated carbonic anhydrase type IV in kidney epithelial cells. *Proc. Natl. Acad. Sci. USA* **87**, 7457–7461.

Brown, E. M., Gamba, G., Riccardi, D., Lombardi, M., Butters, R., Kifor, O., Sun, A., Hediger, M. A., Lytton, J., and Hebert, S. C. (1993). Cloning and characterization of an extracellular Ca^{2+}-sensing receptor from bovine parathyroid. *Nature* **366**, 575–580.

Brown, E. M., Pollak, M., and Hebert, S. C. (1998). The extracellular calcium-sensing receptor: Its role in health and disease. *Annu. Rev. Med.* **49**, 15–29.

Bruce, L. J., Cope, D. L., Jones, G. K., Schofield, A. E., Burley, M., Povey, S., Unwin, R. J., Wrong, O., and Tanner, M. J. (1997). Familial distal renal tubular acidosis is associated with mutations in the red cell anion exchanger (Band 3, AE1) gene. *J. Clin. Invest.* **100**, 1693–1707.

Brunette, M. G., Chan, M., Maag, U., and Beliveau, R. (1984). Phosphate uptake by superficial and deep nephron brush border membranes. Effect of the dietary phosphate and parathyroid hormone. *Pflug. Arch.* **400**, 356–362.

Buckley, N. M., Brazeau, P., and Frasier, I. D. (1983a). Cardiovascular effects of dopamine in developing swine. *Biol. Neonate* **43**, 50–60.

Buckley, N. M., Brazeau, P., and Frasier, I. D. (1983b). Renal blood flow autoregulation in developing swine. *Am. J. Physiol.* **245**, H1–H6.

Buckley, N. M., Brazeau, P., Gootman, P. M., and Frasier, I. D. (1979). Renal circulatory effects of adrenergic stimuli in anesthetized piglets and mature swine. *Am. J. Physiol.* **237**, H690–H695.

Burg, M., Grantham, J., Abramow, M., and Orloff, J. (1966). Preparation and study of fragments of single rabbit nephrons. *Am. J. Physiol.* **210**, 1293–1298.

Burg, M., and Green, N. (1977). Bicarbonate transport by isolated perfused rabbit proximal convoluted tubules. *Am. J. Physiol.* **233**, F307–F314.

Burrow, C. R., Devuyst, O., Li, X., Gatti, L., and Wilson, P. D. (1999). Expression of the beta2-subunit and apical localization of Na^+-K^+-ATPase in metanephric kidney. *Am. J. Physiol.* **277**, F391–F403.

Busch, A., Biber, J., Murer, H., and Lang, F. (1996). Electrophysiological insights of type I and II Na/Pi transporters. *Kidney Int.* **49**, 986–987.

Butkus, A., Albiston, A., Alcorn, D., Giles, M., McCausland, J., Moritz, K., Zhuo, J., and Wintour, E. M. (1997a). Ontogeny of angiotensin II receptors, types 1 and 2, in ovine mesonephros and metanephros. *Kidney Int.* **52**, 628–636.

Butkus, A., Alcorn, D., Earnest, L., Moritz, K., Giles, M., and Wintour, E. M. (1997b). Expression of aquaporin-1 (AQP1) in the adult and developing sheep kidney. *Biol. Cell* **89**, 313–320.

Cadnapaphornchai, P., Taher, S. M., and McDonald, F. D. (1977). Mechanism of dopamine-induced diuresis in the dog. *Am. J. Physiol.* **232**, F524–F528.

Calcagno, P., and Rubin, M. (1963). Renal extraction of para-amino-hippurate in infants and children. *J. Clin. Invest.* **42**, 1632.

Calcagno, P. L., Rubin, M. I., and Weintraub, D. H. (1954). Studies on the renal concentrating and diluting mechanisms in the premature infant. *J. Clin. Invest.* **33**, 91–96.

Caldwell, T., and Solomon, S. (1974). Changes in oxygen consumption of kidney during maturation. *Biol. Neonate* **25**, 1–9.

Canessa, C. M., Schild, L., Buell, G., Thorens, B., Gautschi, I., Horisberger, J. D., and Rossier, B. C. (1994). Amiloride-sensitive epithelial Na^+ channel is made of three homologous subunits. *Nature* **367**, 463–467.

Capasso, G., Lin, J. T., De Santo, N. G., and Kinne, R. (1985). Short term effect of low doses of tri-iodothyronine on proximal tubular membrane Na-K-ATPase and potassium permeability in thyroidectomized rats. *Pflug. Arch.* **403**, 90–96.

Carney, S. L. (1995). Acute effect of endogenous calcitonin on rat renal function. *Miner. Electrolyte Metab.* **21**, 411–416.

Carney, S. L. (1997). Calcitonin and human renal calcium and electrolyte transport. *Miner. Electrolyte Metab.* **23**, 43–47.

Castro, L. C., Lam, R. W., Ross, M. G., Ervin, M. G., Leake, R. D., Hobel, C. J., and Fisher, D. A. (1988). Atrial natriuretic peptide in the sheep. *J. Dev. Physiol.* **10**, 235–246.

Castro, R., Leake, R. D., Ervin, M. G., Ross, M. G., and Fisher, D. A. (1991). Ontogeny of atrial natriuretic factor receptors and cyclic GMP response in rabbit renal glomeruli. *Pediatr. Res.* **30**, 45–49.

Catt, K. J., Cain, M. D., Coghlan, J. P., Zimmet, P. Z., Cran, E., and Best, J. B. (1970). Metabolism and blood levels of angiotensin II in normal subjects, renal disease, and essential hypertension. *Circ. Res.* **27**, (Suppl. 2), 177+.

Caverzasio, J., Bonjour, J. P., and Fleisch, H. (1982). Tubular handling of Pi in young growing and adult rats. *Am. J. Physiol.* **242**, F705–F710.

Caverzasio, J., Montessuit, C., and Bonjour, J. P. (1990). Stimulatory effect of insulin-like growth factor-1 on renal Pi transport and plasma 1,25-dihydroxyvitamin D3. *Endocrinology* **127**, 453–459.

Celsi, G., Larsson, L., and Aperia, A. (1986). Proximal tubular reabsorption and Na-K-ATPase activity in remnant kidney of young rats. *Am. J. Physiol.* **251**, F588–F593.

Celsi, G., Nishi, A., Akusjarvi, G., and Aperia, A. (1991). Abundance of Na^+-K^+-ATPase mRNA is regulated by glucocorticoid hormones in infant rat kidneys. *Am. J. Physiol.* **260**, F192–F197.

Celsi, G., Wang, Z. M., Akusjarvi, G., and Aperia, A. (1993). Sensitive periods for glucocorticoids' regulation of Na^+-K^+-ATPase mRNA in the developing lung and kidney. *Pediatr. Res.* **33**, 5–9.

Chamaa, N. S., Mosig, D., Drukker, A., and Guignard, J. P. (2000). The renal hemodynamic effects of ibuprofen in the newborn rabbit. *Pediatr. Res.* **48**, 600–605.

Chambers, R., and Kempton, R. T. (1933). Indications of function of the chick mesonephros in tissue culture with phenol red. *J. Cell Comp. Physiol.* **3**, 131–167.

Chan, J. C. (1980). Acid-base and mineral disorders in children: A review. *Int. J. Pediatr. Nephrol.* **1**, 54–63.

Chang, S. S., Grunder, S., Hanukoglu, A., Rosler, A., Mathew, P. M., Hanukoglu, I., Schild, L., Lu, Y., Shimkets, R. A., Nelson-Williams, C., Rossier, B. C., and Lifton, R. P. (1996). Mutations in subunits of the epithelial sodium channel cause salt wasting with hyperkalaemic acidosis, pseudohypoaldosteronism type 1. *Nature Genet.* **12**, 248–253.

Chase, L. R., and Aurbach, G. D. (1967). Parathyroid function and the renal excretion of 3'5'-adenylic acid. *Proc. Natl. Acad. Sci. USA* **58**, 518–525.

Chattopadhyay, N., Baum, M., Bai, M., Riccardi, D., Hebert, S. C., Harris, H. W., and Brown, E. M. (1996). Ontogeny of the extracellular calcium-sensing receptor in rat kidney. *Am. J. Physiol.* **271**, F736–F743.

Cheong, H. I., Park, H. W., Ha, I. S., Moon, H. N., Choi, Y., Ko, K. W., and Jun, J. K. (1997). Six novel mutations in the vasopressin V2 receptor gene causing nephrogenic diabetes insipidus. *Nephron.* **75**, 431–437.

Cheung, C. Y., and Brace, R. A. (1991). Hemorrhage-induced reductions in plasma atrial natriuretic factor in the ovine fetus. *Am. J. Obstet. Gynecol.* **165**, 474–481.

Chevalier, R. L. (1982). Functional adaptation to reduced renal mass in early development. *Am. J. Physiol.* **242**, F190–F196.

Chevalier, R. L. (1993). Atrial natriuretic peptide in renal development. *Pediatr. Nephrol.* **7**, 652–656.

Chevalier, R. L., Gomez, R. A., Carey, R. M., Peach, M. J., and Linden, J. M. (1988). Renal effects of atrial natriuretic peptide infusion in young and adult rats. *Pediatr. Res.* **24**, 333–337.

Chevalier, R. L., and Kaiser, D. L. (1983). Autoregulation of renal blood flow in the rat: Effects of growth and uninephrectomy. *Am. J. Physiol.* **244**, F483–F487.

Chevalier, R. L., Thornhill, B. A., Belmonte, D. C., and Baertschi, A. J. (1996). Endogenous angiotensin II inhibits natriuresis after acute volume expansion in the neonatal rat. *Am. J. Physiol.* **270**, R393–R397.

Choi, J. Y., Shah, M., Lee, M. G., Schultheis, P. J., Shull, G. E., Muallem, S., and Baum, M. (2000). Novel amiloride-sensitive sodium-dependent proton secretion in the mouse proximal convoluted tubule. *J. Clin. Invest.* **105**, 1141–1146.

Chong, S. S., Kristjansson, K., Zoghbi, H. Y., and Hughes, M. R. (1993). Molecular cloning of the cDNA encoding a human renal sodium phosphate transport protein and its assignment to chromosome 6p21.3-p23. *Genomics* **18**, 355–359.

Clapp, W. L., Madsen, K. M., Verlander, J. W., and Tisher, C. C. (1987). Intercalated cells of the rat inner medullary collecting duct. *Kidney Int.* **31**, 1080–1087.

Clark, S. L., Jr. (1957). Cellular differentiation in the kidneys of newborn mice studied with the electron microscope. *J. Biophys. Biochem. Cytol.* **3**, 349–362.

Coffman, T. M. (1998). Gene targeting in physiological investigations: Studies of the renin-angiotensin system. *Am. J. Physiol.* **274**, F999–F1005.

Cogan, M. G., Maddox, D. A., Warnock, D. G., Lin, E. T., and Rector, F. C., Jr. (1979). Effect of acetazolamide on bicarbonate reabsorption in the proximal tubule of the rat. *Am. J. Physiol.* **237**, F447–F454.

Cole, D. E., and Quamme, G. A. (2000). Inherited disorders of renal magnesium handling. *J. Am. Soc. Nephrol.* **11**, 1937–1947.

Collins, J. F., and Ghishan, F. K. (1994). Molecular cloning, functional expression, tissue distribution, and in situ hybridization of the renal sodium phosphate (Na$^+$/P$_i$) transporter in the control and hypophosphatemic mouse. *FASEB J.* **8**, 862–868.

Constantinescu, A., Silver, R. B., and Satlin, L. M. (1997). H-K-ATPase activity in PNA-binding intercalated cells of newborn rabbit cortical collecting duct. *Am. J. Physiol.* **272**, F167–F177.

Constantinescu, A. R., Lane, J. C., Mak, J., Zavilowitz, B., and Satlin, L. M. (2000). Na$^+$-K$^+$-ATPase-mediated basolateral rubidium uptake in the maturing rabbit cortical collecting duct. *Am. J. Physiol. Renal. Physiol.* **279**, F1161–F1168.

Corman, B., and Roinel, N. (1991). Single-nephron filtration rate and proximal reabsorption in aging rats. *Am. J. Physiol.* **260**, F75–F80.

Corn, P. G., Mulroney, S. E., and Haramati, A. (1989). Restoration of a phosphaturic response to parathyroid hormone in the immature rat. *Pediatr. Res.* **26**, 54–57.

Coulthard, M. G. (1985). Maturation of glomerular filtration in preterm and mature babies. *Early Hum. Dev.* **11**, 281–292.

Cupples, W. A., Sakai, T., and Marsh, D. J. (1988). Angiotensin II and prostaglandins in control of vasa recta blood flow. *Am. J. Physiol.* **254**, F417–F424.

Custer, M., Lotscher, M., Biber, J., Murer, H., and Kaissling, B. (1994). Expression of Na-P(i) cotransport in rat kidney: Localization by RT-PCR and immunohistochemistry. *Am. J. Physiol.* **266**, F767–F774.

Custer, M., Meier, F., Schlatter, E., Greger, R., Garcia-Perez, A., Biber, J., and Murer, H. (1993). Localization of NaPi-1, a Na-Pi cotransporter, in rabbit kidney proximal tubules. I. mRNA localization by reverse transcription/polymerase chain reaction. *Pflug. Arch.* **424**, 203–209.

Dai, L. J., Ritchie, G., Kerstan, D., Kang, H. S., Cole, D. E., and Quamme, G. A. (2001). Magnesium transport in the renal distal convoluted tubule. *Physiol. Rev.* **81**, 51–84.

Daniel, S. S., Baratz, R. A., Bowe, E. T., Hyman, A. I., Morishima, H. O., Sarcia, S. R., and James, L. S. (1972). Elimination of hydrogen ion by the lamb fetus and newborn. *Pediatr. Res.* **6**, 584–592.

Davies, J., and Bard, J. (1996). Inductive interactions between the mesenchyme and the ureteric bud. *Exp. Nephrol.* **4**, 77–85.

Day, R., and Franklin, J. (1951). Renal carbonic anhydrase in premature and mature infants. *Pediatrics* **7**, 182–185.

de Martino, C., and Zamboni, L. (1966). A morphologic study of the mesonephros of the human embryo. *J. Ultrastruct. Res.* **16**, 399–427.

de Rouffignac, C., and Monnens, L. (1976). Functional and morphologic maturation of superficial and juxtamedullary nephrons in the rat. *J. Physiol.* **262**, 119–129.

de Rouffignac, C., and Quamme, G. (1994). Renal magnesium handling and its hormonal control. *Physiol. Rev.* **74**, 305–322.

De Santo, N. G., Capasso, G., Paduano, C., Carella, C., and Giordano, C. (1980). Tubular transport processes in proximal tubules of hypothyroid rats. Micropuncture studies on isotonic fluid, amino acid and buffer reabsorption. *Pflug. Arch.* **384**, 117–122.

De Smedt, H., and Kinne, R. (1981). Temperature dependence of solute transport and enzyme activities in hog renal brush border membrane vesicles. *Biochim. Biophys. Acta.* **648**, 247–253.

Dean, R. F. A., and McCance, R. A. (1949). The renal response of infants and adults to the administration of hypertonic solutions of sodium chloride and urea. *J. Physiol. (Lond)* **109**, 81–87.

Deen, P. M., Verdijk, M. A., Knoers, N. V., Wieringa, B., Monnens, L. A.,

van Os, C. H., and van Oost, B. A. (1994). Requirement of human renal water channel aquaporin-2 for vasopressin- dependent concentration of urine. *Science* **264**, 92–95.

Dennis, V. W., Bello-Reuss, E., and Robinson, R. R. (1977). Response of phosphate transport to parathyroid hormone in segments of rabbit nephron. *Am. J. Physiol.* **233**, F29–F38.

Dennis, V. W., Woodhall, P. B., and Robinson, R. R. (1976). Characteristics of phosphate transport in isolated proximal tubule. *Am. J. Physiol.* **231**, 979–985.

DeVane, G. W., and Porter, J. C. (1980). An apparent stress-induced release or arginine vasopressin by human neonates. *J. Clin. Endocrinol. Metab.* **51**, 1412–1416.

Devuyst, O., Burrow, C. R., Smith, B. L., Agre, P., Knepper, M. A., and Wilson, P. D. (1996). Expression of aquaporins-1 and -2 during nephrogenesis and in autosomal dominant polycystic kidney disease. *Am. J. Physiol.* **271**, F169–F183.

Di Stefano, A., Roinel, N., de Rouffignac, C., and Wittner, M. (1993). Transepithelial Ca^{2+} and Mg^{2+} transport in the cortical thick ascending limb of Henle's loop of the mouse is a voltage-dependent process. *Ren. Physiol. Biochem.* **16**, 157–166.

Di Stefano, A., Wittner, M., Nitschke, R., Braitsch, R., Greger, R., Bailly, C., Amiel, C., Roinel, N., and de Rouffignac, C. (1990). Effects of parathyroid hormone and calcitonin on Na$^+$, Cl$^-$, K$^+$, Mg^{2+} and Ca^{2+} transport in cortical and medullary thick ascending limbs of mouse kidney. *Pflug. Arch.* **417**, 161–167.

Dicker, S. E., and Shirley, D. G. (1971). Rates of oxygen consumption and of anaerobic glycolysis in renal cortex and medulla of adult and newborn rats and guinea-pigs. *J. Physiol.* **212**, 235–243.

Dillon, M. J. (1980). Renin-angiotensin-aldosterone system. *Eur. J. Clin. Pharmacol.* **18**, 105–108.

Dlouha, H. (1976). A micropuncture study of the development of renal function in the young rat. *Biol. Neonate* **29**, 117–128.

Dobyan, D. C., and Bulger, R. E. (1982). Renal carbonic anhydrase. *Am. J. Physiol.* **243**, F311–F324.

Doucet, A. (1992). Na-K-ATPase in the kidney tubule in relation to natriuresis. *Kidney. Int. Suppl.* **37**, S118–S124.

Dousa, T. P., Duarte, C. G., and Knox, F. G. (1976). Effect of colchicine on urinary phosphate and regulation by parathyroid hormone. *Am. J. Physiol.* **231**, 61–65.

Drukker, A., Goldsmith, D. I., Spitzer, A., Edelmann, C. M., Jr., and Blaufox, M. D. (1980). The renin angiotensin system in newborn dogs: developmental patterns and response to acute saline loading. *Pediatr. Res.* **14**, 304–307.

DuBose, T. D., Jr., and Lucci, M. S. (1983). Effect of carbonic anhydrase inhibition on superficial and deep nephron bicarbonate reabsorption in the rat. *J. Clin. Invest.* **71**, 55–65.

Duc, C., Farman, N., Canessa, C. M., Bonvalet, J. P., and Rossier, B. C. (1994). Cell-specific expression of epithelial sodium channel alpha, beta, and gamma subunits in aldosterone-responsive epithelia from the rat: localization by in situ hybridization and immunocytochemistry. *J. Cell Biol.* **127**, 1907–1921.

Ecelbarger, C. A., Kim, G. H., Terris, J., Masilamani, S., Mitchell, C., Reyes, I., Verbalis, J. G., and Knepper, M. A. (2000). Vasopressin-mediated regulation of epithelial sodium channel abundance in rat kidney. *Am. J. Physiol. Renal Physiol.* **279**, F46–F53.

Edelmann, C. M., Jr Barnett , H. L., and Troupkou, V. (1960). Renal concentrating mechanisms in newborn infants: effect of dietary protein, and water content, role of urea and responsiveness to antidiuretic hormone. *J. Clin. Invest.* **39**, 1062.

Edelmann, C. M., Jr, Rodriguez-Soriano, J., Boichis, H., Gruskin, A. B., and Acosta, M. (1967). Renal bicarbonate reabsorption and hydrogen ion excretion in infants. *J. Clin. Invest.* **46**, 1309–1317.

Edelmann, C. M., Jr., Barnett, H. L., and Stark, H. (1966). Effect of urea on concentration of urinary nonurea solute in premature infants. *J. Appl. Physiol.* **21**, 1021–1025.

Edelmann, C. M., Jr., and Spitzer, A. (1969). The maturing kidney. A modern view of well-balanced infants with imbalanced nephrons. *J. Pediatr.* **75**, 509–519.

Edwards, R. M., Trizna, W., and Ohlstein, E. H. (1990). Renal microvascular effects of endothelin. *Am. J. Physiol.* **259**, F217–221.

Elinder, G. (1981). Effect of isotonic volume expansion on proximal tubular reabsorption of Na and fluid in the developing rat kidney. *Acta Physiol. Scand.* **112**, 83–88.

Elinder, G., and Aperia, A. (1982). Renal oxygen consumption and sodium reabsorption during isotonic volume expansion in the developing rat. *Pediatr. Res.* **16**, 351–353.

Elinder, G., Aperia, A., Herin, P., and Kallskog, O. (1980). Effect of isotonic volume expansion on glomerular filtration rate and renal hemodynamics in the developing rat kidney. *Acta Physiol. Scand.* **108**, 411–417.

Ellis, D., Turocy, J. F., Sweeney, W. E., Jr., and Avner, E. D. (1986). Partial characterization and ontogeny of renal cytosolic glucocorticoid receptors in mouse kidney. *J. Steroid Biochem.* **24**, 997–1003.

Ellison, D. H., Velazquez, H., and Wright, F. S. (1987). Thiazide-sensitive sodium chloride cotransport in early distal tubule. *Am. J. Physiol.* **253**, F546–F554.

Engelke, S. C., Shah, B. L., Vasan, U., and Raye, J. R. (1978). Sodium balance in very low-birth-weight infants. *J. Pediatr.* **93**, 837–841.

Evan, A. P., Gattone, V. H. 2nd., and Schwartz, G. J. (1983). Development of solute transport in rabbit proximal tubule. II. Morphologic segmentation. *Am. J. Physiol.* **245**, F391–407.

Evan, A. P., Jr., Stoeckel, J. A., Loemker, V., and Baker, J. T. (1979). Development of the intrarenal vascular system of the puppy kidney. *Anat. Rec.* **194**, 187–199.

Evan, A. P., Satlin, L. M., Gattone, V. H. 2nd., Connors, B., and Schwartz, G. J. (1991). Postnatal maturation of rabbit renal collecting duct. II. Morphological observations. *Am. J. Physiol.* **261**, F91–F107.

Farman, N. (1992). Steroid receptors: Distribution along the nephron. *Semin. Nephrol.* **12**, 12–17.

Farman, N. (1996). Na,K-pump expression and distribution in the nephron. *Miner. Electrolyte Metab.* **22**, 272–278.

Farman, N. (1999). Molecular and cellular determinants of mineralocorticoid selectivity. *Curr. Opin. Nephrol. Hypertens.* **8**, 45–51.

Farman, N., Oblin, M. E., Lombes, M., Delahaye, F., Westphal, H. M., Bonvalet, J. P., and Gasc, J. M. (1991). Immunolocalization of gluco- and mineralocorticoid receptors in rabbit kidney. *Am. J. Physiol.* **260**, C226–C233.

Faubert, P. F., Chou, S. Y., and Porush, J. G. (1987). Regulation of papillary plasma flow by angiotensin II. *Kidney Int.* **32**, 472–478.

Fawer, C. L., Torrado, A., and Guignard, J. P. (1979a). Maturation of renal function in full-term and premature neonates. *Helv. Paediatr. Acta.* **34**, 11–21.

Fawer, C. L., Torrado, A., and Guignard, J. P. (1979b). Single injection clearance in the neonate. *Biol. Neonate* **35**, 321–324.

Felder, C. C., Campbell, T., Albrecht, F., and Jose, P. A. (1990a). Dopamine inhibits Na$^+$-H$^+$ exchanger activity in renal BBMV by stimulation of adenylate cyclase. *Am. J. Physiol.* **259**, F297–F303.

Felder, C. C., Piccio, M. M., McKelvey, A. M., Nakamura, K. T., Robillard, J. E., and Jose, P. A. (1990b). Ontogeny of renal beta adrenoceptors in the sheep. *Pediatr. Nephrol.* **4**, 635–639.

Felder, R. A., Felder, C. C., Eisner, G. M., and Jose, P. A. (1989). The dopamine receptor in adult and maturing kidney. *Am. J. Physiol.* **257**, F315–F327.

Felder, R. A., Pelayo, J. C., Calcagno, P. L., Eisner, G. M., and Jose, P. A. (1983). Alpha-adrenoceptors in the developing kidney. *Pediatr. Res.* **17**, 177–180.

Feltes, T. F., Hansen, T. N., Martin, C. G., Leblanc, A. L., Smith, S., and Giesler, M. E. (1987). The effects of dopamine infusion on regional blood flow in newborn lambs. *Pediatr. Res.* **21**, 131–136.

Feraille, E., and Doucet, A. (2001). Sodium-potassium-adenosinetri-phosphatase-dependent sodium transport in the kidney: Hormonal control. *Physiol. Rev.* **81**, 345–418.

Fetterman, G. H., Shuplock, N. A., Phillip, F. J., and Gregg, H. S. (1965). The growth and maturation of human glomeruli and proximal convolutions from term to adulthood. *Pediatrics* **35**, 601–619.

Fildes, R. D., Eisner, G. M., Calcagno, P. L., and Jose, P. A. (1985). Renal alpha-adrenoceptors and sodium excretion in the dog. *Am. J. Physiol.* **248**, F128–F133.

Fisher, D. A. (1961). Carbonic anhydrase activity in fetal and young rhesus monkeys. *Proc. Soc. Exp. Biol. Med.* **107**, 359–363.

Fomon, S., Harris, D., and Jensen, R. (1959). Acidification of the urine by infants fed human milk and whole cow's milk. *Pediatrics* **23**, 113–120.

Friedlander, G., Shahedi, M., Le Grimellec, C., and Amiel, C. (1988). Increase in membrane fluidity and opening of tight junctions have similar effects on sodium-coupled uptakes in renal epithelial cells. *J. Biol. Chem.* **263**, 11183–11188.

Friedman, P. A. (1998). Codependence of renal calcium and sodium transport. *Annu. Rev. Physiol.* **60**, 179–197.

Friedman, P. A. (2000). Mechanisms of renal calcium transport. *Exp. Nephrol.* **8**, 343–350.

Frindt, G., and Palmer, L. G. (1989). Low-conductance K channels in apical membrane of rat cortical collecting tubule. *Am. J. Physiol.* **256**, F143–F151.

Fris-Hansen, B. (1961). Body water compartment in children: Changes during growth and related changes in body composition. *Pediatrics* **28**, 169–181.

Fryckstedt, J., Svensson, L. B., Linden, M., and Aperia, A. (1993). The effect of dopamine on adenylate cyclase and Na$^+$-K$^+$-ATPase activity in the developing rat renal cortical and medullary tubule cells. *Pediatr. Res.* **34**, 308–311.

Fujino, Y., Ross, M. G., Ervin, M. G., Castro, R., Leake, R. D., and Fisher, D. A. (1992). Ovine maternal and fetal glomerular atrial natriuretic factor receptors: response to dehydration. *Biol. Neonate* **62**, 120–126.

Fukuda, Y., and Aperia, A. (1988). Differentiation of Na$^+$-K$^+$ pump in rat proximal tubule is modulated by Na$^+$-H$^+$ exchanger. *Am. J. Physiol.* **255**, F552–F557.

Fukuda, Y., Bertorello, A., and Aperia, A. (1991). Ontogeny of the regulation of Na$^+$,K$^+$-ATPase activity in the renal proximal tubule cell. *Pediatr. Res.* **30**, 131–134.

Gaeggeler, H. P., Edwards, C. R., and Rossier, B. C. (1989). Steroid metabolism determines mineralocorticoid specificity in the toad bladder. *Am. J. Physiol.* **257**, F690–F695.

Garty, H., and Palmer, L. G. (1997). Epithelial sodium channels: Function, structure, and regulation. *Physiol. Rev.* **77**, 359–396.

Geibel, J., Giebisch, G., and Boron, W. F. (1990). Angiotensin II stimulates both Na$^+$-H$^+$ exchange and Na$^+$/HCO$_3^-$ cotransport in the rabbit proximal tubule. *Proc. Natl. Acad. Sci. USA* **87**, 7917–7920.

Geller, D. S., Rodriguez-Soriano, J., Vallo Boado, A., Schifter, S., Bayer, M., Chang, S. S., and Lifton, R. P. (1998). Mutations in the mineralo-corticoid receptor gene cause autosomal dominant pseudohypoaldo-steronism type I. *Nature Genet.* **19**, 279–281.

Gersh, I. (1937). The correlation of structure and function in the developing mesonephros and metanephros. *Contrib. Embryol.* **153**, 35–58.

Gesek, F. A., Cragoe, E. J., Jr., and Strandhoy, J. W. (1989). Synergistic alpha-1 and alpha-2 adrenergic stimulation of rat proximal nephron Na$^+$/H$^+$ exchange. *J. Pharmacol. Exp. Ther.* **249**, 694–700.

Gesek, F. A., and Friedman, P. A. (1992). On the mechanism of parathyroid hormone stimulation of calcium uptake by mouse distal convoluted tubule cells. *J. Clin. Invest.* **90**, 749–758.

Giammattei, C. E., Strandhoy, J. W., and Rose, J. C. (1999). Regulation of in vitro renin secretion by ANG II feedback manipulation in vivo in the ovine fetus. *Am. J. Physiol.* **277**, R1230–R1238.

Giebisch, G. (1998). Renal potassium transport: Mechanisms and regulation. *Am. J. Physiol.* **274**, F817–F833.

Gilmore, J. P., Cornish, K. G., Rogers, S. D., and Joyner, W. L. (1980). Direct evidence for myogenic autoregulation of the renal microcirculation in the hamster. *Circ. Res.* **47**, 226–230.

Gitler, M. S., Piccio, M. M., Robillard, J. E., and Jose, P. A. (1991). Characterization of renal alpha-adrenoceptor subtypes in sheep during development. *Am. J. Physiol.* **260**, R407–R412.

Godard, C., Geering, J. M., Geering, K., and Vallotton, M. B. (1979). Plasma renin activity related to sodium balance, renal function and urinary vasopressin in the newborn infant. *Pediatr. Res.* **13**, 742–745.

Goldsmith, D. I., Drukker, A., Blaufox, M. D., Edelmann, C. M., Jr., and Spitzer, A. (1979). Hemodynamic and excretory response of the neonatal canine kidney to acute volume expansion. *Am. J. Physiol.* **237**, F392–397.

Goldsmith, D. I., Jodorkovsky, R. A., Sherwinter, J., Kleeman, S. R., and Spitzer, A. (1986). Glomerular capillary permeability in developing canines. *Am. J. Physiol.* **251**, F528–F531.

Gomez, R. A., Chevalier, R. L., Carey, R. M., and Peach, M. J. (1990). Molecular biology of the renal renin-angiotensin system. *Kidney Int. Suppl.* **30**, S18–S23.

Gomez, R. A., Lynch, K. R., Chevalier, R. L., Wilfong, N., Everett, A., Carey, R. M., and Peach, M. J. (1988). Renin and angiotensinogen gene expression in maturing rat kidney. *Am. J. Physiol.* **254**, F582–F587.

Gomez, R. A., Lynch, K. R., Sturgill, B. C., Elwood, J. P., Chevalier, R. L., Carey, R. M., and Peach, M. J. (1989). Distribution of renin mRNA and its protein in the developing kidney. *Am. J. Physiol.* **257**, F850–F858.

Gomez, R. A., Meernik, J. G., Kuehl, W. D., and Robillard, J. E. (1984). Developmental aspects of the renal response to hemorrhage during fetal life. *Pediatr. Res.* **18**, 40–46.

Gomez, R. A., and Robillard, J. E. (1984). Developmental aspects of the renal responses to hemorrhage during converting-enzyme inhibition in fetal lambs. *Circ. Res.* **54**, 301–312.

Good, D. W. (1985). Sodium-dependent bicarbonate absorption by cortical thick ascending limb of rat kidney. *Am. J. Physiol.* **248**, F821–F829.

Gordon, H., McNamara, H., and Benjamin, H. (1948). The response of young infants to ingestion of ammonium chloride. *Pediatrics* **2**, 290–302.

Gouyon, J. B., and Guignard, J. P. (1988). Theophylline prevents the hypoxemia-induced renal hemodynamic changes in rabbits. *Kidney Int.* **33**, 1078–1083.

Graber, M. L., Bengele, H. H., Schwartz, J. H., and Alexander, E. A. (1981). pH and pCO$_2$ profiles of the rat inner medullary collecting duct. *Am. J. Physiol.* **241**, F659–F668.

Grady, E. F., Sechi, L. A., Griffin, C. A., Schambelan, M., and Kalinyak, J. E. (1991). Expression of AT2 receptors in the developing rat fetus. *J. Clin. Invest.* **88**, 921–933.

Graham, B., Wilson, J., Tsao, M., Baumann, M., and Brown, S. (1951). Development of neonatal electrolyte homeostasis. *Pediatrics* **8**, 68–78.

Graham, P. C., Kingdom, J. C., Raweily, E. A., Gibson, A. A., and Lindop, G. B. (1992). Distribution of renin-containing cells in the developing human kidney: An immunocytochemical study. *Br. J. Obstet. Gynaecol.* **99**, 765–769.

Granger, J. P., Alberola, A. M., Salazar, F. J., and Nakamura, T. (1992). Control of renal hemodynamics during intrarenal and systemic blockade of nitric oxide synthesis in conscious dogs. *J. Cardiovasc. Pharmacol.* **20**, S160–S162.

Greger, R. (1985). Ion transport mechanisms in thick ascending limb of Henle's loop of mammalian nephron. *Physiol. Rev.* **65**, 760–797.

Greger, R., Bleich, M., and Schlatter, E. (1991). Ion channel regulation in the thick ascending limb of the loop of Henle. *Kidney Int. Suppl.* **33**, S119–S124.

Greger, R., Lang, F., Marchand, G., and Knox, F. G. (1977). Site of renal phosphate reabsorption. Micropuncture and microinfusion study. *Pflug. Arch.* **369**, 111–118.

Greger, R., Lang, F., and Oberleithner, H. (1978). Distal site of calcium reabsorption in the rat nephron. *Pflug. Arch.* **374**, 153–157.

Greger, R., and Schlatter, E. (1981). Presence of luminal K$^+$, a prerequisite for active NaCl transport in the cortical thick ascending limb of Henle's loop of rabbit kidney. *Pflug. Arch.* **392**, 92–94.

Grobstein, C. (1955). Inductive interactions in the development of of the mouse metanephros. *J. Exp. Zool.* **130**, 319–340.

Grone, H. J., Simon, M., and Fuchs, E. (1992). Autoradiographic characterization of angiotensin receptor subtypes in fetal and adult human kidney. *Am. J. Physiol.* **262**, F326–F331.

Gruskin, A. B., Edelmann, C. M., Jr., and Yuan, S. (1970). Maturational changes in renal blood flow in piglets. *Pediatr. Res.* **4**, 7–13.

Guignard, J. P., Torrado, A., Da Cunha, O., and Gautier, E. (1975). Glomerular filtration rate in the first three weeks of life. *J. Pediatr.* **87**, 268–272.

Guignard, J. P., Torrado, A., Feldman, H., and Gautier, E. (1980). Assessment of glomerular filtration rate in children. *Helv. Paediatr. Acta.* **35**, 437–447.

Guillery, E. N., and Huss, D. J. (1995). Developmental regulation of chloride/formate exchange in guinea pig proximal tubules. *Am. J. Physiol.* **269**, F686–F695.

Guillery, E. N., Huss, D. J., McDonough, A. A., and Klein, L. C. (1997). Posttranscriptional upregulation of Na$^+$-K$^+$-ATPase activity in newborn guinea pig renal cortex. *Am. J. Physiol.* **273**, F254–F263.

Guillery, E. N., Karniski, L. P., Mathews, M. S., Page, W. V., Orlowski, J., Jose, P. A., and Robillard, J. E. (1995). Role of glucocorticoids in the maturation of renal cortical Na$^+$/H$^+$ exchanger activity during fetal life in sheep. *Am. J. Physiol.* **268**, F710–F717.

Guillery, E. N., Karniski, L. P., Mathews, M. S., and Robillard, J. E. (1994a). Maturation of proximal tubule Na$^+$/H$^+$ antiporter activity in sheep during transition from fetus to newborn. *Am. J. Physiol.* **267**, F537–F545.

Guillery, E. N., Mathews, M. S., Orlowski, J., and Robillard, J. E. (1996). Angiotensin II and the maturation of renal cortical Na$^+$/H$^+$ exchanger activity during fetal life in sheep. *Am. J. Physiol.* **271**, R1507–R1513.

Guillery, E. N., Porter, C. C., Page, W. V., Jose, P. A., Felder, R., and Robillard, J. E. (1993). Developmental regulation of the alpha 1B-adrenoceptor in the sheep kidney. *Pediatr. Res.* **34**, 124–128.

Guillery, E. N., Segar, J. L., Merrill, D. C., Nakamura, K. T., Jose, P. A., and Robillard, J. E. (1994b). Ontogenic changes in renal response to alpha 1-adrenoceptor stimulation in sheep. *Am. J. Physiol.* **267**, R990–R998.

Gupta, N., Tarif, S. R., Seikaly, M., and Baum, M. (2001). Role of glucocorticoids in the maturation of the rat renal Na$^+$/H$^+$ antiporter (NHE3). *Kidney Int.* **60**, 173–181.

Guron, G., Adams, M. A., Sundelin, B., and Friberg, P. (1997). Neonatal angiotensin-converting enzyme inhibition in the rat induces persistent abnormalities in renal function and histology. *Hypertension* **29**, 91–97.

Guron, G., Nilsson, A., Nitescu, N., Nielsen, S., Sundelin, B., Frokiaer, J., and Friberg, P. (1999). Mechanisms of impaired urinary concentrating ability in adult rats treated neonatally with enalapril. *Acta Physiol. Scand.* **165**, 103–112.

Haas, J. A., Berndt, T., and Knox, F. G. (1978). Nephron heterogeneity of phosphate reabsorption. *Am. J. Physiol.* **234**, F287–F290.

Hadeed, A. J., Leake, R. D., Weitzman, R. E., and Fisher, D. A. (1979). Possible mechanisms of high blood levels of vasopressin during the neonatal period. *J. Pediatr.* **94**, 805–808.

Hammerman, M. R., and Gavin, J. R., 3rd (1986). Binding of IGF I and IGF I-stimulated phosphorylation in canine renal basolateral membranes. *Am. J. Physiol.* **251**, E32–E31.

Hammerman, M. R., and Rogers, S. (1987). Distribution of IGF receptors in the plasma membrane of proximal tubular cells. *Am. J. Physiol.* **253**, F841–F847.

Hansch, E., Forgo, J., Murer, H., and Biber, J. (1993). Role of microtubules in the adaptive response to low phosphate of Na/Pi cotransport in opossum kidney cells. *Pflug. Arch.* **422**, 516–522.

Hansen, J. D. L., and Smith, C. A. (1953). Effects of withholding fluid in the immediate postnatal period. *Pediatrics* **12**, 99.

Hansson, J. H., Nelson-Williams, C., Suzuki, H., Schild, L., Shimkets, R., Lu, Y., Canessa, C., Iwasaki, T., Rossier, B., and Lifton, R. P. (1995). Hypertension caused by a truncated epithelial sodium channel gamma subunit: genetic heterogeneity of Liddle syndrome. *Nature Genet.* **11**, 76–82.

Haramati, A. (1989). Phosphate handling by the kidney during development: Immaturity or unique adaptations for growth. *News Physiol. Sci.* **4**, 234–238.

Haramati, A., Haas, J. A., and Knox, F. G. (1983). Adaptation of deep and superficial nephrons to changes in dietary phosphate intake. *Am. J. Physiol.* **244**, F265–F269.

Haramati, A., Haas, J. A., and Knox, F. G. (1984). Nephron heterogeneity of phosphate reabsorption: Effect of parathyroid hormone. *Am. J. Physiol.* **246**, F155–F158.

Haramati, A., and Mulroney, S. E. (1987). Enhanced tubular capacity for phosphate reabsorption in immature rats: Role of glomerular filtration rate. *Fed. Proc.* **48**, 1288.

Haramati, A., Mulroney, S. E., and Lumpkin, M. D. (1990). Regulation of renal phosphate reabsorption during development: Implications from a new model of growth hormone deficiency. *Pediatr. Nephrol.* **4**, 387–391.

Haramati, A., Mulroney, S. E., and Webster, S. K. (1988). Developmental changes in the tubular capacity for phosphate reabsorption in the rat. *Am. J. Physiol.* **255**, F287–F291.

Hargrave, B. Y., Iwamoto, H. S., and Rudol, A. M. (1989). Renal and cardiovascular effects of atrial natriuretic peptide in fetal sheep. *Pediatr. Res.* **26**, 1–5.

Harris, C. A., Sutton, R. A., and Dirks, J. H. (1977). Effects of hypercalcemia on calcium and phosphate ultrafilterability and tubular reabsorption in the rat. *Am. J. Physiol.* **233**, F201–F206.

Harris, H. W., Jr., and Zeidel, M. L. (1993). Water channels. *Curr. Opin. Nephrol. Hypertens.* **2**, 699–707.

Harris, P. J., and Young, J. A. (1977). Dose-dependent stimulation and inhibition of proximal tubular sodium reabsorption by angiotensin II in the rat kidney. *Pflug. Arch.* **367**, 295–297.

Harris, R. C., Seifter, J. L., and Lechene, C. (1986). Coupling of Na-H exchange and Na-K pump activity in cultured rat proximal tubule cells. *Am. J. Physiol.* **251**, C815–C824.

Harrison-Bernard, L. M., Navar, L. G., Ho, M. M., Vinson, G. P., and el-Dahr, S. S. (1997). Immunohistochemical localization of ANG II AT1 receptor in adult rat kidney using a monoclonal antibody. *Am. J. Physiol.* **273**, F170–F177.

Hartmann, C. M., Wagner, C. A., Busch, A. E., Markovich, D., Biber, J., Lang, F., and Murer, H. (1995). Transport characteristics of a murine renal Na/Pi-cotransporter. *Pflug. Arch.* **430**, 830–836.

Hatemi, N., and McCance, R. (1961). Renal aspects of acid-base control in the newly born. III. Response to acidifying drugs. *Acta Paediatr. Scand.* **50**, 603–616.

Hayashi, Y., Chiba, K., Matsuoka, T., Suzuki-Kusaba, M., Yoshida, M., Hisa, H., and Satoh, S. (1999). Renal nerve stimulation induces alpha2-adrenoceptor-mediated antinatriuresis under inhibition of prostaglandin synthesis in anesthetized dogs. *Tohoku J. Exp. Med.* **188**, 335–346.

Hebert, S. C., Brown, E. M., and Harris, H. W. (1997). Role of the Ca^{2+}-sensing receptor in divalent mineral ion homeostasis. *J. Exp. Biol.* **200**, 295–302.

Hein, L., Barsh, G. S., Pratt, R. E., Dzau, V. J., and Kobilka, B. K. (1995). Behavioural and cardiovascular effects of disrupting the angiotensin II type-2 receptor in mice [published erratum appears in *Nature* **380**(6572),366 (1996)]. *Nature* **377**, 744–747.

Hendricks, S. K., Smith, J. R., Moore, D. E., and Brown, Z. A. (1990). Oligohydramnios associated with prostaglandin synthetase inhibitors in preterm labour. *Br. J. Obstet. Gynaecol.* **97**, 312–316.

Henning, S. J. (1978). Plasma concentrations of total and free corticosterone during development in the rat. *Am. J. Physiol.* **235**, E451–E456.

Henning, S. J. (1981). Postnatal development: coordination of feeding, digestion, and metabolism. *Am. J. Physiol.* **241**, G199–G214.

Hilgers, K. F., Reddi, V., Krege, J. H., Smithies, O., and Gomez, R. A. (1997). Aberrant renal vascular morphology and renin expression in mutant mice lacking angiotensin-converting enzyme. *Hypertension* **29**, 216–221.

Hill, K. J., and Lumbers, E. R. (1988). Renal function in adult and fetal sheep. *J. Dev. Physiol.* **10**, 149–159.

Ho, K., Nichols, C. G., Lederer, W. J., Lytton, J., Vassilev, P. M., Kanazirska, M. V., and Hebert, S. C. (1993). Cloning and expression of an inwardly rectifying ATP-regulated potassium channel. *Nature* **362**, 31–38.

Hoag, H. M., Martel, J., Gauthier, C., and Tenenhouse, H. S. (1999). Effects of Npt2 gene ablation and low-phosphate diet on renal Na^+/phosphate cotransport and cotransporter gene expression. *J. Clin. Invest.* **104**, 679–686.

Hoenderop, J. G., Hartog, A., Stuiver, M., Doucet, A., Willems, P. H., and Bindels, R. J. (2000). Localization of the epithelial Ca^{2+} channel in rabbit kidney and intestine. *J. Am. Soc. Nephrol.* **11**, 1171–1178.

Hohenauer, L., Rosenberg, T. F., and Oh, W. (1970). Calcium and phosphorus homeostasis on the first day of life. *Biol Neonate.* **15**, 49–56.

Holthofer, H. (1987). Ontogeny of cell type-specific enzyme reactivities in kidney collecting ducts. *Pediatr. Res.* **22**, 504–508.

Holthofer, H., Schulte, B. A., Pasternack, G., Siegel, G. J., and Spicer, S. S. (1987). Three distinct cell populations in rat kidney collecting duct. *Am. J. Physiol.* **253**, C323–328.

Hoppe, A., Lin, J. T., Onsgard, M., Knox, F. G., and Dousa, T. P. (1991). Quantitation of the Na^+-Pi cotransporter in renal cortical brush border membranes. [^{14}C]phosphonoformic acid as a useful probe to determine the density and its change in response to parathyroid hormone. *J. Biol. Chem.* **266**, 11528–11536.

Horne, R. S., MacIsaac, R. J., Moritz, K. M., Tangalakis, K., and Wintour, E. M. (1993). Effect of arginine vasopressin and parathyroid hormone-related protein on renal function in the ovine foetus. *Clin. Exp. Pharmacol. Physiol.* **20**, 569–577.

Horster, M. (1978). Loop of Henle functional differentiation: In vitro perfusion of the isolated thick ascending segment. *Pflug. Arch.* **378**, 15–24.

Horster, M. (2000). Embryonic epithelial membrane transporters. *Am. J. Physiol. Renal Physiol.* **279**, F982–F996.

Horster, M., Huber, S., Tschop, J., Dittrich, G., and Braun, G. (1997). Epithelial nephrogenesis. *Pflug. Arch.* **434**, 647–660.

Horster, M., and Larsson, L. (1976). Mechanisms of fluid absorption during proximal tubule development. *Kidney Int.* **10**, 348–363.

Horster, M., and Valtin, H. (1971). Postnatal development of renal function: Micropuncture and clearance studies in the dog. *J. Clin. Invest.* **50**, 779–795.

Horster, M. F., Braun, G. S., and Huber, S. M. (1999). Embryonic renal epithelia: Induction, nephrogenesis, and cell differentiation. *Physiol. Rev.* **79**, 1157–1191.

Horster, M. F., Gilg, A., and Lory, P. (1984). Determinants of axial osmotic gradients in the differentiating countercurrent system. *Am. J. Physiol.* **246**, F124–F132.

Houillier, P., Chambrey, R., Achard, J. M., Froissart, M., Poggioli, J., and Paillard, M. (1996). Signaling pathways in the biphasic effect of angiotensin II on apical Na/H antiport activity in proximal tubule. *Kidney Int.* **50**, 1496–1505.

Hruska, K. A., Moskowitz, D., Esbrit, P., Civitelli, R., Westbrook, S., and Huskey, M. (1987). Stimulation of inositol trisphosphate and diacylglycerol production in renal tubular cells by parathyroid hormone. *J. Clin. Invest.* **79**, 230–239.

Huber, S. M., Braun, G. S., and Horster, M. F. (1999). Expression of the epithelial sodium channel (ENaC) during ontogenic differentiation of the renal cortical collecting duct epithelium. *Pflug. Arch.* **437**, 491–497.

Huber, S. M., and Horster, M. F. (1996). Ontogeny of apical membrane ion conductances and channel expression in cortical collecting duct cells. *Am. J. Physiol.* **271**, F698–F708.

Ichikawa, I., Humes, H. D., Dousa, T. P., and Brenner, B. M. (1978). Influence of parathyroid hormone on glomerular ultrafiltration in the rat. *Am. J. Physiol.* **234**, F393–F401.

Ichikawa, I., Maddox, D. A., and Brenner, B. M. (1979). Maturational development of glomerular ultrafiltration in the rat. *Am. J. Physiol.* **236**, F465–F471.

Ichiki, T., Labosky, P. A., Shiota, C., Okuyama, S., Imagawa, Y., Fogo, A., Niimura, F., Ichikawa, I., Hogan, B. L., and Inagami, T. (1995). Effects on blood pressure and exploratory behaviour of mice lacking angiotensin II type-2 receptor. *Nature* **377**, 748–750.

Igarashi, P., Vanden Heuvel, G. B., Payne, J. A., and Forbush, B., 3rd (1995). Cloning, embryonic expression, and alternative splicing of a murine kidney-specific Na-K-Cl cotransporter. *Am. J. Physiol.* **269**, F405–F418.

Igarashi, Y., Aperia, A., Larsson, L., and Zetterstrom, R. (1983). Effect of betamethasone on Na-K-ATPase activity and basal and lateral cell membranes in proximal tubular cells during early development. *Am. J. Physiol.* **245**, F232–F237.

Iino, Y., and Burg, M. B. (1981). Effect of acid-base status in vivo on bicarbonate transport by rabbit renal tubules in vitro. *Jpn. J. Physiol.* **31**, 99–107.

Imbert-Teboul, M., Chabardes, D., Clique, A., Montegut, M., and Morel, F. (1984). Ontogenesis of hormone-dependent adenylate cyclase in isolated rat nephron segments. *Am. J. Physiol.* **247**, F316–F325.

Ito, S., Arima, S., Ren, Y. L., Juncos, L. A., and Carretero, O. A. (1993). Endothelium-derived relaxing factor/nitric oxide modulates angiotensin II action in the isolated microperfused rabbit afferent but not efferent arteriole. *J. Clin. Invest.* **91**, 2012–2019.

Iwamoto, H. S., and Rudolph, A. M. (1981). Effects of angiotensin II on the blood flow and its distribution in fetal lambs. *Circ. Res.* **48**, 183–189.

Jaisser, F. (2000). Inducible gene expression and gene modification in transgenic mice. *J. Am. Soc. Nephrol.* **11 (Suppl 16)**, S95–S100.

Jarolim, P., Shayakul, C., Prabakaran, D., Jiang, L., Stuart-Tilley, A., Rubin, H. L., Simova, S., Zavadil, J., Herrin, J. T., Brouillette, J., Somers, M. J., Seemanova, E., Brugnara, C., Guay-Woodford, L. M., and Alper, S. L. (1998). Autosomal dominant distal renal tubular acidosis is associated in three families with heterozygosity for the R589H mutation in the AE1 (band 3) Cl-/HCO3- exchanger. *J. Biol. Chem.* **273**, 6380–6388.

Jaton, T., Thonney, M., Gouyon, J. B., and Guignard, J. P. (1992). Renal effects of dopamine and dopexamine in the newborn anesthetized rabbit. *Life Sci.* **50**, 195–202.

Jenik, A. G., Ceriani Cernadas, J. M., Gorenstein, A., Ramirez, J. A., Vain, N., Armadans, M., and Ferraris, J. R. (2000). A randomized, double-blind, placebo-controlled trial of the effects of prophylactic theophylline on renal function in term neonates with perinatal asphyxia. *Pediatrics* **105**, E45.

John, E., Goldsmith, D. I., and Spitzer, A. (1981a). Quantitative changes in the canine glomerular vasculature during development: physiologic implications. *Kidney Int.* **20**, 223–229.

John, E. G., Zeis, P. M., Samayoa, C., Chan, L., Wojdula, L., and Aschinberg, L. C. (1981b). Renin-aldosterone system response to chronic salt loading and volume contraction in puppies. *Int. J. Pediatr. Nephrol.* **2**, 9–15.

Johnson, V., and Spitzer, A. (1986). Renal reabsorption of phosphate during development: whole-kidney events. *Am. J. Physiol.* **251**, F251–F256.

Joppich, R., Scherer, B., and Weber, P. C. (1979). Renal prostaglandins: relationship to the development of blood pressure and concentrating capacity in pre-term and full term healthy infants. *Eur. J. Pediatr.* **132**, 253–259.

Jose, P. A., Logan, A. G., Slotkoff, L. M., Lilienfield, L. S., Calcagno, P. L., and Eisner, G. M. (1971). Intrarenal blood flow distribution in canine puppies. *Pediatr. Res.* **5**, 335–344.

Jose, P. A., Slotkoff, L. M., Lilienfield, L. S., Calcagno, P. L., and Eisner, G. M. (1974). Sensitivity of neonatal renal vasculature to epinephrine. *Am. J. Physiol.* **226**, 796–799.

Jose, P. A., Slotkoff, L. M., Montgomery, S., Calcagno, P. L., and Eisner, G. (1975). Autoregulation of renal blood flow in the puppy. *Am. J. Physiol.* **229**, 983–988.

Kairaitis, K., and Lumbers, E. R. (1990). The influence of endogenous mineralocorticoids on the composition of fetal urine. *J. Dev. Physiol.* **13**, 347–351.

Kaissling, B., and Kriz, W. (1979). Structural analysis of the rabbit kidney. *Adv. Anat. Embryol. Cell Biol.* **56**, 1–123.

Kakuchi, J., Ichiki, T., Kiyama, S., Hogan, B. L., Fogo, A., Inagami, T., and Ichikawa, I. (1995). Developmental expression of renal angiotensin II receptor genes in the mouse. *Kidney Int.* **47**, 140–147.

Kalinyak, J. E., Bradshaw, J. G., and Perlman, A. J. (1992). The role of development and adrenal steroids in the regulation of the mineralocorticoid receptor messenger RNA. *Horm. Metab. Res.* **24**, 106–109.

Kalinyak, J. E., Griffin, C. A., Hamilton, R. W., Bradshaw, J. G., Perlman, A. J., and Hoffman, A. R. (1989). Developmental and hormonal regulation of glucocorticoid receptor messenger RNA in the rat. *J. Clin. Invest.* **84**, 1843–1848.

Kaneko, S., Albrecht, F., Asico, L. D., Eisner, G. M., Robillard, J. E., and Jose, P. A. (1992). Ontogeny of DA1 receptor-mediated natriuresis in the rat: in vivo and in vitro correlations. *Am. J. Physiol.* **263**, R631–R638.

Kang, H. S., Kerstan, D., Dai, L. J., Ritchie, G., and Quamme, G. A. (2000). beta-adrenergic agonists stimulate Mg^{2+} uptake in mouse distal convoluted tubule cells. *Am. J. Physiol. Renal Physiol.* **279**, F1116–F1123.

Kaplan, B. S., Restaino, I., Raval, D. S., Gottlieb, R. P., and Bernstein, J. (1994). Renal failure in the neonate associated with in utero exposure to non-steroidal anti-inflammatory agents. *Pediatr. Nephrol.* **8**, 700–704.

Karashima, S., Hattori, S., Ushijima, T., Furuse, A., Nakazato, H., and Matsuda, I. (1998). Developmental changes in carbonic anhydrase II in the rat kidney. *Pediatr. Nephrol.* **12**, 263–248.

Karet, F. E., Finberg, K. E., Nayir, A., Bakkaloglu, A., Ozen, S., Hulton, S. A., Sanjad, S. A., Al-Sabban, E. A., Medina, J. F., and Lifton, R. P. (1999a). Localization of a gene for autosomal recessive distal renal tubular acidosis with normal hearing (rdRTA2) to 7q33–34. *Am. J. Hum. Genet.* **65**, 1656–1665.

Karet, F. E., Finberg, K. E., Nelson, R. D., Nayir, A., Mocan, H., Sanjad, S. A., Rodriguez-Soriano, J., Santos, F., Cremers, C. W., Di Pietro, A., Hoffbrand, B. I., Winiarski, J., Bakkaloglu, A., Ozen, S., Dusunsel, R., Goodyer, P., Hulton, S. A., Wu, D. K., Skvorak, A. B., Morton, C. C., Cunningham, M. J., Jha, V., and Lifton, R. P. (1999b). Mutations in the gene encoding B1 subunit of H⁺-ATPase cause renal tubular acidosis with sensorineural deafness. *Nature Genet.* **21**, 84–90.

Karlen, J. (1989). Renal response to low and high phosphate intake in weanling, adolescent and adult rats. *Acta Physiol. Scand.* **135**, 317–322.

Kaskel, F. J., Kumar, A. M., Feld, L. G., and Spitzer, A. (1988). Renal reabsorption of phosphate during development: tubular events. *Pediatr. Nephrol.* **2**, 129–134.

Kaskel, F. J., Kumar, A. M., Lockhart, E. A., Evan, A., and Spitzer, A. (1987). Factors affecting proximal tubular reabsorption during development. *Am. J. Physiol.* **252**, F188–F197.

Kazimierczak, J. (1970). Histochemical observations of the developing glomerulus and juxtaglomerular apparatus. *Acta Pathol. Microbiol. Scand. A* **78**, 401–413.

Kempson, S. A., Lotscher, M., Kaissling, B., Biber, J., Murer, H., and Levi, M. (1995). Parathyroid hormone action on phosphate transporter mRNA and protein in rat renal proximal tubules. *Am. J. Physiol.* **268**, F784–F791.

Kerpel-Fronius, E., Heim, T., and Sulyok, E. (1970). The development of the renal acidifying processes and their relation to acidosis in low-birth-weight infants. *Biol. Neonate* **15**, 156–168.

Kesby, G. J., and Lumbers, E. R. (1986). Factors affecting renal handling of sodium, hydrogen ions, and bicarbonate by the fetus. *Am. J. Physiol.* **251**, F226–F231.

Keusch, I., Traebert, M., Lotscher, M., Kaissling, B., Murer, H., and Biber, J. (1998). Parathyroid hormone and dietary phosphate provoke a lysosomal routing of the proximal tubular Na/Pi-cotransporter type II. *Kidney Int.* **54**, 1224–1232.

Kikeri, D., Azar, S., Sun, A., Zeidel, M. L., and Hebert, S. C. (1990). Na⁺-H⁺ antiporter and Na⁺-(HCO₃⁻) symporter regulate intracellular pH in mouse medullary thick limbs of Henle. *Am. J. Physiol.* **258**, F445–456.

Kildeberg, P., Engel, K., and Winters, R. W. (1969). Balance of net acid in growing infants. Endogenous and transintestinal aspects. *Acta. Paediatr. Scand.* **58**, 321–329.

Kildeberg, P., and Winters, R. W. (1978). Balance of net acid: concept, measurement and applications. *Adv. Pediatr.* **25**, 349–381.

Kim, G. H., Ecelbarger, C. A., Mitchell, C., Packer, R. K., Wade, J. B., and Knepper, M. A. (1999). Vasopressin increases Na-K-2Cl cotransporter expression in thick ascending limb of Henle's loop. *Am. J. Physiol.* **276**, F96–F103.

Kim, G. H., Masilamani, S., Turner, R., Mitchell, C., Wade, J. B., and Knepper, M. A. (1998). The thiazide-sensitive Na-Cl cotransporter is an aldosterone-induced protein. *Proc. Natl. Acad. Sci. USA* **95**, 14552–14557.

Kim, J., Cha, J. H., Tisher, C. C., and Madsen, K. M. (1996). Role of apoptotic and nonapoptotic cell death in removal of intercalated cells from developing rat kidney. *Am. J. Physiol.* **270**, F575–F592.

Kim, J., Tisher, C. C., Linser, P. J., and Madsen, K. M. (1990). Ultrastructural localization of carbonic anhydrase II in subpopulations of intercalated cells of the rat kidney. *J. Am. Soc. Nephrol.* **1**, 245–256.

Kim, J., Tisher, C. C., and Madsen, K. M. (1994). Differentiation of intercalated cells in developing rat kidney: An immunohistochemical study. *Am. J. Physiol.* **266**, F977–F990.

Kim, J., Welch, W. J., Cannon, J. K., Tisher, C. C., and Madsen, K. M. (1992). Immunocytochemical response of type A and type B intercalated cells to increased sodium chloride delivery. *Am. J. Physiol.* **262**, F288–F302.

Kinoshita, S., and Felder, R. A. (1990). Ontogeny of DA1 receptor-adenylate cyclase coupling in proximal convoluted tubules. *Am. J. Physiol.* **259**, F971–F976.

Kinsella, J., and Sacktor, B. (1985). Thyroid hormones increase Na⁺-H⁺ exchange activity in renal brush border membranes. *Proc. Natl. Acad. Sci. USA* **82**, 3606–3610.

Kinsella, J. L., and Aronson, P. S. (2001). Properties of the Na⁺-H⁺ exchanger in renal microvillus membrane vesicles, 1980. *J. Am. Soc. Nephrol.* **12**, 1085–1095.

Kirshon, B., Moise, K. J., Jr., Wasserstrum, N., Ou, C. N., and Huhta, J. C. (1988). Influence of short-term indomethacin therapy on fetal urine output. *Obstet. Gynecol.* **72**, 51–53.

Kleinman, J. G., Bain, J. L., Fritsche, C., and Riley, D. A. (1992). Histochemical carbonic anhydrase in rat inner medullary collecting duct. *J. Histochem. Cytochem.* **40**, 1535–1545.

Kleinman, L. I. (1975). Renal sodium reabsorption during saline loading and distal blockade in newborn dogs. *Am. J. Physiol.* **228**, 1403–1408.

Kleinman, L. I., and Banks, R. O. (1983). Segmental nephron sodium and potassium reabsorption in newborn and adult dogs during saline expansion. *Proc. Soc. Exp. Biol. Med.* **173**, 231–237.

Kleinman, L. I., and Lubbe, R. J. (1972). Factors affecting the maturation of glomerular filtration rate and renal plasma flow in the new-born dog. *J. Physiol. (Lond)* **223**, 395–409.

Kleinman, L. I., and Reuter, J. H. (1973). Maturation of glomerular blood flow distribution in the new-born dog. *J. Physiol. (Lond)* **228**, 91–103.

Knepper, M. A. (1997). Molecular physiology of urinary concentrating mechanism: regulation of aquaporin water channels by vasopressin. *Am. J. Physiol.* **272**, F3–F12.

Knepper, M. A., Kim, G. H., Fernandez-Llama, P., and Ecelbarger, C. A. (1999). Regulation of thick ascending limb transport by vasopressin. *J. Am. Soc. Nephrol.* **10**, 628–634.

Komhoff, M., Grone, H. J., Klein, T., Seyberth, H. W., and Nusing, R. M. (1997). Localization of cyclooxygenase-1 and -2 in adult and fetal human kidney: Implication for renal function. *Am. J. Physiol.* **272**, F460–F468.

Kon, V., and Ichikawa, I. (1985). Hormonal regulation of glomerular filtration. *Annu. Rev. Med.* **36**, 515–531.

Koster, H. P., Hartog, A., Van Os, C. H., and Bindels, R. J. (1995). Calbindin-D28K facilitates cytosolic calcium diffusion without interfering with calcium signaling. *Cell Calcium* **18**, 187–196.

Kotchen, T. A., Strickland, A. L., Rice, T. W., and Walters, D. R. (1972). A study of the renin-angiotensin system in newborn infants. *J. Pediatr.* **80**, 938–46.

Kotelevtsev, Y., Brown, R. W., Fleming, S., Kenyon, C., Edwards, C. R., Seckl, J. R., and Mullins, J. J. (1999). Hypertension in mice lacking 11beta-hydroxysteroid dehydrogenase type 2. *J. Clin. Invest.* **103**, 683–689.

Kowarski, A., Katz, H., and Migeon, C. J. (1974). Plasma aldosterone concentration in normal subjects from infancy to adulthood. *J. Clin. Endocrinol. Metab.* **38**, 489–491.

Krapf, R., Alpern, R. J., Rector, F. C., Jr., and Berry, C. A. (1987). Basolateral membrane Na/base cotransport is dependent on CO₂/HCO₃ in the proximal convoluted tubule. *J. Gen. Physiol.* **90**, 833–853.

Kurokawa, K. (1987). Calcium-regulating hormones and the kidney. *Kidney Int.* **32**, 760–771.

Ladas, J., Mulroney, S., Jiang, G., Winaver, J., and Haramati, A. (1993). Adaptation of renal Pi transport to dietary Pi deprivation in weanling rats: Independence from growth hormone (GH) release. *J. Am. Soc. Nephrol.* **4**, 710 abstract.

Lagercrantz, H., and Bistoletti, P. (1977). Catecholamine release in the newborn infant at birth. *Pediatr. Res.* **11**, 889–893.

Lahera, V., Salom, M. G., Fiksen-Olsen, M. J., Raij, L., and Romero, J. C. (1990). Effects of NG-monomethyl-L-arginine and L-arginine on acetylcholine renal response. *Hypertension* **15**, 659–663.

Lang, F., Greger, R., Marchand, G. R., and Knox, F. G. (1977). Stationary microperfusion study of phosphate reabsorption in proximal and distal nephron segments. *Pflug. Arch.* **368**, 45–48.

Laplace, J. R., Husted, R. F., and Stokes, J. B. (1992). Cellular responses to steroids in the enhancement of Na⁺ transport by rat collecting duct cells in culture. Differences between glucocorticoid and mineralocorticoid hormones. *J. Clin. Invest.* **90**, 1370–1378.

Larsson, L. (1975). Ultrastructure and permeability of intercellular contacts of developing proximal tubule in rat kidney. *J. Ultrastruct. Res.* **52**, 100–113.

Larsson, L., Aperia, A., and Wilton, P. (1980). Effect of normal development on compensatory renal growth. *Kidney Int.* **18**, 29–35.

Larsson, L., and Horster, M. (1976). Ultrastructure and net fluid transport in isolated perfused developing proximal tubules. *J. Ultrastruct. Res.* **54**, 276–285.

Larsson, L., and Lonnerholm, G. (1985). Carbonic anhydrase in the metanephrogenic zone of the human fetal kidney. *Biol. Neonate* **48**, 168–171.

Larsson, S. H., Rane, S., Fukuda, Y., Aperia, A., and Lechene, C. (1990). Changes in Na influx precede postnatal increase in Na, K-ATPase activity in rat renal proximal tubular cells. *Acta. Physiol. Scand.* **138**, 99–100.

Leake, R. D., and Trygstad, C. W. (1977). Glomerular filtration rate during the period of adaptation to extrauterine life. *Pediatr. Res.* **11**, 959–962.

Leake, R. D., Weitzman, R. E., Weinberg, J. A., and Fisher, D. A. (1979). Control of vasopressin secretion in the newborn lamb. *Pediatr. Res.* **13**, 257–260.

Lee, S. M., Chekal, M. A., and Katz, A. I. (1983). Corticosterone binding sites along the rat nephron. *Am. J. Physiol.* **244**, F504–F509.

Lelievre-Pegorier, M., Merlet-Benichou, C., Roinel, N., and de Rouffignac, C. (1983). Developmental pattern of water and electrolyte transport in rat superficial nephrons. *Am. J. Physiol.* **245**, F15–F21.

Levi, M., Baird, B. M., and Wilson, P. V. (1990). Cholesterol modulates rat renal brush border membrane phosphate transport. *J. Clin. Invest.* **85**, 231–237.

Levi, M., Jameson, D. M., and van der Meer, B. W. (1989). Role of BBM lipid composition and fluidity in impaired renal Pi transport in aged rat. *Am. J. Physiol.* **256**, F85–F94.

Levi, M., Lotscher, M., Sorribas, V., Custer, M., Arar, M., Kaissling, B., Murer, H., and Biber, J. (1994). Cellular mechanisms of acute and chronic adaptation of rat renal P_i transporter to alterations in dietary P_i. *Am. J. Physiol.* **267**, F900–F908.

Levine, D. Z., and Jacobson, H. R. (1986). The regulation of renal acid secretion: new observations from studies of distal nephron segments. *Kidney Int.* **29**, 1099–1109.

Li, X. X., Albrecht, F. E., Robillard, J. E., Eisner, G. M., and Jose, P. A. (2000). Gbeta regulation of Na/H exchanger-3 activity in rat renal proximal tubules during development. *Am. J. Physiol. Regul. Integr. Comp. Physiol.* **278**, R931–R936.

Lifton, R. P. (1996). Molecular genetics of human blood pressure variation. *Science* **272**, 676–680.

Linarelli, L. G. (1972). Newborn urinary cyclic AMP and developmental renal responsiveness to parathyroid hormone. *Pediatrics* **50**, 14–23.

Linarelli, L. G., Bobik, J., and Bobik, C. (1973). The effect of parathyroid hormone on rabbit renal cortex adenyl cyclase during development. *Pediatr. Res.* **7**, 878–882.

Ling, B. N., Hinton, C. F., and Eaton, D. C. (1991). Potassium permeable channels in primary cultures of rabbit cortical collecting tubule. *Kidney Int.* **40**, 441–452.

Lingwood, B., Hardy, K. J., Horacek, I., McPhee, M. L., Scoggins, B. A., and Wintour, E. M. (1978). The effects of antidiuretic hormone on urine flow and composition in the chronically-cannulated ovine fetus. *Q. J. Exp. Physiol. Cogn. Med. Sci.* **63**, 315–330.

Liu, F. Y., and Cogan, M. G. (1989). Angiotensin II stimulates early proximal bicarbonate absorption in the rat by decreasing cyclic adenosine monophosphate. *J. Clin. Invest,* **84**, 83–91.

Lolait, S. J., O'Carroll, A. M., McBride, O. W., Konig, M., Morel, A., and Brownstein, M. J. (1992). Cloning and characterization of a vasopressin V2 receptor and possible link to nephrogenic diabetes insipidus. *Nature* **357**, 336–339.

Lombard, W. E., Kokko, J. P., and Jacobson, H. R. (1983). Bicarbonate transport in cortical and outer medullary collecting tubules. *Am. J. Physiol.* **244**, F289–F296.

Lonnerholm, G., and Wistrand, P. J. (1983). Carbonic anhydrase in the human fetal kidney. *Pediatr. Res.* **17**, 390–397.

Lorenz, J. M., Kleinman, L. I., and Disney, T. A. (1986). Renal response of newborn dog to potassium loading. *Am. J. Physiol.* **251**, F513–F519.

Lorenz, J. N., Schultheis, P. J., Traynor, T., Shull, G. E., and Schnermann, J. (1999). Micropuncture analysis of single-nephron function in NHE3-deficient mice. *Am. J. Physiol.* **277**, F447–F453.

Lotscher, M., Kaissling, B., Biber, J., Murer, H., and Levi, M. (1997). Role of microtubules in the rapid regulation of renal phosphate transport in response to acute alterations in dietary phosphate content. *J. Clin. Invest.* **99**, 1302–1312.

Lucci, M. S., Pucacco, L. R., DuBose, T. D., Jr., Kokko, J. P., and Carter, N. W. (1980). Direct evaluation of acidification by rat proximal tubule: role of carbonic anhydrase. *Am. J. Physiol.* **238**, F372–F379.

Lucci, M. S., Tinker, J. P., Weiner, I. M., and DuBose, T. D., Jr. (1983). Function of proximal tubule carbonic anhydrase defined by selective inhibition. *Am. J. Physiol.* **245**, F443–F449.

Lumbers, E. R. (1995). Functions of the renin-angiotensin system during development. *Clin. Exp. Pharmacol. Physiol.* **22**, 499–505.

Lumbers, E. R., Burrell, J. H., Menzies, R. I., and Stevens, A. D. (1993). The effects of a converting enzyme inhibitor (captopril) and angiotensin II on fetal renal function. *Br. J. Pharmacol.* **110**, 821–827.

Maack, T. (1996). Role of atrial natriuretic factor in volume control. *Kidney Int.* **49**, 1732–1737.

MacDonald, M. S. A. E., J.L. (1959). The late intrauterine and postnatal development of human renal glomeruli. *J. Anatomy* **93**, 331–340.

Madsen, K. M., Kim, J., and Tisher, C. C. (1992). Intracellular band 3 immunostaining in type A intercalated cells of rabbit kidney. *Am. J. Physiol.* **262**, F1015–F1022.

Magagnin, S., Werner, A., Markovich, D., Sorribas, V., Stange, G., Biber, J., and Murer, H. (1993). Expression cloning of human and rat renal cortex Na/Pi cotransport. *Proc. Natl. Acad. Sci. USA* **90**, 5979–5983.

Magaldi, A. J., Oyamaguchi, M. N., Kudo, L. H., and Rocha, A. S. (1992). Phosphate transport in isolated rat inner medullary collecting duct. *Pflug. Arch.* **420**, 544–550.

Malinowska, K. W., Hardy, R. N., and Nathanielsz, P. W. (1972). Plasma adrenocorticosteroid concentrations immediately after birth in the rat, rabbit and guinea-pig. *Experientia* **28**, 1366–1367.

Malinowska, K. W., and Nathanielsz, P. W. (1974). Plasma aldosterone, cortisol and corticosterone concentrations in the new-born guinea-pig. *J. Physiol.* **236**, 83–93.

Mallet, E., Basuyau, J. P., Brunelle, P., Devaux, A. M., and Fessard, C. (1978). Neonatal parathyroid secretion and renal receptor maturation in premature infants. *Biol. Neonate* **33**, 304–308.

Malnic, G., De Mello Aires, M., and Giebisch, G. (1971). Potassium transport across renal distal tubules during acid-base disturbances. *Am. J. Physiol.* **221**, 1192–1208.

Malnic, G., Klose, R. M., and Giebisch, G. (1966a). Microperfusion study of distal tubular potassium and sodium transfer in rat kidney. *Am. J. Physiol.* **211**, 548–559.

Malnic, G., Klose, R. M., and Giebisch, G. (1966b). Micropuncture study of distal tubular potassium and sodium transport in rat nephron. *Am. J. Physiol.* **211**, 529–547.

Mandon, B., Siga, E., Roinel, N., and de Rouffignac, C. (1993). Ca^{2+}, Mg^{2+} and K^+ transport in the cortical and medullary thick ascending limb of the rat nephron: influence of transepithelial voltage. *Pflug. Arch.* **424**, 558–560.

Maren, T. H. (1967). Carbonic anhydrase: chemistry, physiology, and inhibition. *Physiol. Rev.* **47**, 595–781.

Marunaka, Y., and Eaton, D. C. (1991). Effects of vasopressin and cAMP on single amiloride-blockable Na channels. *Am. J. Physiol.* **260**, C1071–C1084.

Masilamani, S., Kim, G. H., Mitchell, C., Wade, J. B., and Knepper, M. A. (1999). Aldosterone-mediated regulation of ENaC alpha, beta, and gamma subunit proteins in rat kidney. *J. Clin. Invest.* **104**, R19–R23.

Matson, J. R., Stokes, J. B., and Robillard, J. E. (1981). Effects of inhibition of prostaglandin synthesis on fetal renal function. *Kidney Int.* **20**, 621–627.

Matsumoto, T., Fejes-Toth, G., and Schwartz, G. J. (1996). Postnatal differentiation of rabbit collecting duct intercalated cells. *Pediatr. Res.* **39**, 1–12.

Matsumoto, T., and Schwartz, G. J. (1992). Novel method for performing carbonic anhydrase histochemistry and immunocytochemistry on cryosections. *J. Histochem. Cytochem.* **40**, 1223–1227.

Matsumura, Y., Uchida, S., Rai, T., Sasaki, S., and Marumo, F. (1997). Transcriptional regulation of aquaporin-2 water channel gene by cAMP. *J. Am. Soc. Nephrol.* **8**, 861–867.

Matsunaga, H., Stanton, B. A., Gesek, F. A., and Friedman, P. A. (1994). Epithelial Ca^{2+} channels sensitive to dihydropyridines and activated by hyperpolarizing voltages. *Am. J. Physiol.* **267**, C157–C165.

Mattyus, I., Zimmerhackl, L. B., Schwarz, A., Brandis, M., Miltenyi, M., and Tulassay, T. (1997). Renal excretion of endothelin in children. *Pediatr. Nephrol.* **11**, 513–521.

McCance, R. A., and Stainer, M. W. (1960). The function of the metanephros of foetal rabbits and pigs. *J. Physiol.* **151**, 479.

McCance, R. A., and Widdowson, E. M. (1953). Renal function before birth. *Proc. R. Soc. (Lond)* **141**, 488.

McCance, R. A., and Widdowson, E. M. (1958). The response of the newborn piglet to an excess of potassium. *J. Physiol.* **141**, 88–96.

McDonald, F. J., Yang, B., Hrstka, R. F., Drummond, H. A., Tarr, D. E., McCray, P. B., Stokes, J. B., Welsh, M. J., and Williamson, R. A. (1999). Disruption of the beta subunit of the epithelial Na$^+$ channel in mice: hyperkalemia and neonatal death associated with a pseudohypoaldosteronism phenotype. Proc. Natl. Acad. Sci. USA 96, 1727–1731.

McKenna, O. C., and Angelakos, E. T. (1968). Adrenergic innervation of the canine kidney. Circ. Res. 22, 345–354.

McKenna, O. C., and Angelakos, E. T. (1970). Development of adrenergic innervation in the puppy kidney. Anat. Rec. 167, 115–125.

McKinney, T. D., and Burg, M. B. (1977). Bicarbonate transport by rabbit cortical collecting tubules. Effect of acid and alkali loads in vivo on transport in vitro. J. Clin. Invest 60, 766–768.

McKinney, T. D., and Davidson, K. K. (1987). Bicarbonate transport in collecting tubules from outer stripe of outer medulla of rabbit kidneys. Am. J. Physiol. 253, F816–F822.

McMullen, J. R., Gibson, K. J., and Lumbers, E. R. (1998). Effects of intravenous infusions of noradrenaline on renal function in chronically catheterised fetal sheep. Biol. Neonate 73, 254–253.

Mehrgut, F. M., Satlin, L. M., and Schwartz, G. J. (1990). Maturation of HCO$_3^-$ transport in rabbit collecting duct. Am. J. Physiol. 259, F801–F808.

Melendez, E., Reyes, J. L., and Melendez, M. A. (1991). Effects of furosemide on the renal functions of the unanesthetized newborn rat. Dev. Pharmacol. Ther. 17, 210–219.

Merlet-Benichou, C., and de Rouffignac, C. (1977). Renal clearance studies in fetal and young guinea pigs: effect of salt loading. Am. J. Physiol. 232, F178–F185.

Meyer, M. B., McNay, J. L., and Goldberg, L. I. (1967). Effects of dopamine on renal function and hemodynamics in the dog. J. Pharmacol. Exp. Ther. 156, 186–192.

Mick, V. E., Itani, O. A., Loftus, R. W., Husted, R. F., Schmidt, T. J., and Thomas, C. P. (2001). The alpha-subunit of the epithelial sodium channel is an aldosterone- induced transcript in mammalian collecting ducts, and this transcriptional response is mediated via distinct cis-elements in the 5′-flanking region of the gene. Mol. Endocrinol. 15, 575–588.

Minuth, W. W., Gross, P., Gilbert, P., and Kashgarian, M. (1987). Expression of the alpha-subunit of Na/K-ATPase in renal collecting duct epithelium during development. Kidney Int. 31, 1104–1112.

Misanko, B. S., Evan, A. P., Bengele, H. H., and Solomon, S. (1979). Renal response to Ringer expansion in developing rats. Biol. Neonate 35, 52–59.

Missale, C., Nash, S. R., Robinson, S. W., Jaber, M., and Caron, M. G. (1998). Dopamine receptors: From structure to function. Physiol. Rev. 78, 189–225.

Miyamoto, K., Tatsumi, S., Sonoda, T., Yamamoto, H., Minami, H., Taketani, Y., and Takeda, E. (1995). Cloning and functional expression of a Na$^+$-dependent phosphate co-transporter from human kidney: cDNA cloning and functional expression. Biochem. J. 305, 81–85.

Miyata, N., Park, F., Li, X. F., and Cowley, A. W., Jr. (1999) Distribution of angiotensin AT1 and AT2 receptor subtypes in the rat kidney. AJP-Renal Physiol. 277, 437–446.

Molitoris, B. A., Alfrey, A. C., Harris, R. A., and Simon, F. R. (1985). Renal apical membrane cholesterol and fluidity in regulation of phosphate transport. Am. J. Physiol. 249, F12–F19.

Moore, E. S., De Lannoy, C. W., Paton, J. B., and Ocampo, M. (1972a). Effect of Na$_2$SO$_4$ on urinary acidification in the fetal lamb. Am. J. Physiol. 223, 167–171.

Moore, E. S., Fine, B. P., Satrasook, S. S., Vergel, Z. M., and Edelmann, C. M., Jr. (1972b). Renal reabsorption of bicarbonate in puppies: Effect of extracellular volume contraction on the renal threshold for bicarbonate. Pediatr. Res. 6, 859–867.

Moore, L. C. (1982). Interaction of tubuloglomerular feedback and proximal nephron reabsorption in autoregulation. Kidney Int. Suppl. 12, S173–S178.

Moritz, K., Koukoulas, I., Albiston, A., and Wintour, E. M. (2000).

Angiotensin II infusion to the midgestation ovine fetus: Effects on the fetal kidney. Am. J. Physiol. Regul. Integr. Comp. Physiol. 279, R1290–R1297.

Moritz, K. M., and Wintour, E. M. (1999). Functional development of the meso- and metanephros. Pediatr. Nephrol. 13, 171–178.

Morris, D. J., and Souness, G. W. (1992). Protective and specificity-conferring mechanisms of mineralocorticoid action. Am. J. Physiol. 263, F759–F768.

Muchant, D. G., Thornhill, B. A., Belmonte, D. C., Felder, R. A., Baertschi, A., and Chevalier, R. L. (1995). Chronic sodium loading augments natriuretic response to acute volume expansion in the preweaned rat. Am. J. Physiol. 269, R15–R22.

Mujais, S. K., Kauffman, S., and Katz, A. I. (1986). Angiotensin II binding sites in individual segments of the rat nephron. J. Clin. Invest. 77, 315–318.

Muller-Suur, R., Ulfendahl, H. R., and Persson, A. E. (1983). Evidence for tubuloglomerular feedback in juxtamedullary nephrons of young rats. Am. J. Physiol. 244, F425–F431.

Mulroney, S. E., and Haramati, A. (1990). Renal adaptation to changes in dietary phosphate during development. Am. J. Physiol. 258, F1650–F1656.

Mulroney, S. E., Lumpkin, M. D., and Haramati, A. (1989). Antagonist to GH-releasing factor inhibits growth and renal Pi reabsorption in immature rats. Am. J. Physiol. 257, F29–F34.

Mune, T., Rogerson, F. M., Nikkila, H., Agarwal, A. K., and White, P. C. (1995). Human hypertension caused by mutations in the kidney isozyme of 11 beta-hydroxysteroid dehydrogenase. Nature Genet. 10, 394–399.

Murer, H. (1992). Homer Smith Award. Cellular mechanisms in proximal tubular Pi reabsorption: some answers and more questions. J. Am. Soc. Nephrol. 2, 1649–1665.

Murer, H., and Biber, J. (1994). Renal sodium-phosphate cotransport. Curr. Opin. Nephrol. Hypertens. 3, 504–510.

Murer, H., Hopfer, U., and Kinne, R. (1976). Sodium/proton antiport in brush-border-membrane vesicles isolated from rat small intestine and kidney. Biochem. J. 154, 597–604.

Murer, H., Werner, A., Reshkin, S., Wuarin, F., and Biber, J. (1991). Cellular mechanisms in proximal tubular reabsorption of inorganic phosphate. Am. J. Physiol. 260, C885–C899.

Nakamura, K. T., Felder, R. A., Jose, P. A., and Robillard, J. E. (1987a). Effects of dopamine in the renal vascular bed of fetal, newborn, and adult sheep. Am. J. Physiol. 252, R490–R497.

Nakamura, K. T., Matherne, G. P., Jose, P. A., Alden, B. M., and Robillard, J. E. (1987b). Ontogeny of renal beta-adrenoceptor-mediated vasodilation in sheep: comparison between endogenous catecholamines. Pediatr. Res. 22, 465–470.

Nakamura, K. T., Matherne, G. P., Jose, P. A., Alden, B. M., and Robillard, J. E. (1988). Effects of epinephrine on the renal vascular bed of fetal, newborn, and adult sheep. Pediatr. Res. 23, 181–186.

Nakamura, K. T., Matherne, G. P., McWeeny, O. J., Smith, B. A., and Robillard, J. E. (1987c). Renal hemodynamics and functional changes during the transition from fetal to newborn life in sheep. Pediatr. Res. 21, 229–234.

Naray-Fejes-Toth, A., and Fejes-Toth, G. (1990). Glucocorticoid receptors mediate mineralocorticoid-like effects in cultured collecting duct cells. Am. J. Physiol. 259, F672–F678.

Narbaitz, R., Kapal, V. K., and Levine, D. Z. (1993). Induction of intercalated cell changes in rat pups from acid- and alkali-loaded mothers. Am. J. Physiol. 264, F415–F420.

Narbaitz, R., Vandorpe, D., and Levine, D. Z. (1991). Differentiation of renal intercalated cells in fetal and postnatal rats. Anat. Embryol. 183, 353–361.

Navar, L. G., and Mitchell, K. D. (1990). Contribution of the tubuloglomerular feedback mechanism to sodium homeostasis and interaction with the renin-angiotensin system. Acta Physiol. Scand. Suppl. 591, 66–73.

Neiberger, R. E., Barac-Nieto, M., and Spitzer, A. (1989). Renal reabsorption of phosphate during development: Transport kinetics in BBMV. *Am. J. Physiol.* **257**, F268–F274.

Neiss, W. F. (1982). Morphogenesis and histogenesis of the connecting tubule in the rat kidney. *Anat. Embryol. (Berl.)* **165**, 81–95.

Ng, R. C., Rouse, D., and Suki, W. N. (1984). Calcium transport in the rabbit superficial proximal convoluted tubule. *J. Clin. Invest.* **74**, 834–842.

Nguyen, L. B., Lievano, G., Radhakrishnan, J., Fornell, L. C., Jacobson, G., and John, E. G. (1999). Renal effects of low to moderate doses of dopamine in newborn piglets. *J. Pediatr. Surg.* **34**, 996–999.

Niimura, F., Labosky, P. A., Kakuchi, J., Okubo, S., Yoshida, H., Oikawa, T., Ichiki, T., Naftilan, A. J., Fogo, A., Inagami, T., *et al.* (1995). Gene targeting in mice reveals a requirement for angiotensin in the development and maintenance of kidney morphology and growth factor regulation. *J. Clin. Invest.* **96**, 2947–2954.

Nilsson, A. B., and Friberg, P. (2000). Acute renal responses to angiotensin-converting enzyme inhibition in the neonatal pig. *Pediatr. Nephrol.* **14**, 1071–1076.

Nord, E. P., Howard, M. J., Hafezi, A., Moradeshagi, P., Vaystub, S., and Insel, P. A. (1987). Alpha 2 adrenergic agonists stimulate Na$^+$-H$^+$ antiport activity in the rabbit renal proximal tubule. *J. Clin. Invest.* **80**, 1755–1762.

Okubo, S., Niimura, F., Matsusaka, T., Fogo, A., Hogan, B. L., and Ichikawa, I. (1998). Angiotensinogen gene null-mutant mice lack homeostatic regulation of glomerular filtration and tubular reabsorption. *Kidney Int.* **53**, 617–625.

Olbing, H., Blaufox, M. D., Aschinberg, L. C., Silkalns, G. I., Bernstein, J., Spitzer, A., and Edelmann, C. M., Jr. (1973). Postnatal changes in renal glomerular blood flow distribution in puppies. *J. Clin. Invest.* **52**, 2885–2895.

Oliverio, M. I., Delnomdedieu, M., Best, C. F., Li, P., Morris, M., Callahan, M. F., Johnson, G. A., Smithies, O., and Coffman, T. M. (2000). Abnormal water metabolism in mice lacking the type 1A receptor for ANG II. *Am. J. Physiol. Renal. Physiol.* **278**, F75–F82.

Ominato, M., Satoh, T., and Katz, A. I. (1996). Regulation of Na-K-ATPase activity in the proximal tubule: Role of the protein kinase C pathway and of eicosanoids. *J. Membr. Biol.* **152**, 235–243.

Orlowski, J., and Lingrel, J. B. (1988). Tissue-specific and developmental regulation of rat Na,K-ATPase catalytic alpha isoform and beta subunit mRNAs. *J. Biol. Chem.* **263**, 10436–10442.

Osathanondh, V., and Potter, E. L. (1966). Development of human kidney as shown by microdissection. IV. Development of tubular portions of nephrons. *Arch. Pathol.* **82**, 391–402.

Osborn, J. L., Hook, J. B., and Bailie, M. D. (1980). Effect of saralasin and indomethacin on renal function in developing piglets. *Am. J. Physiol.* **238**, R438–R442.

Ostrowski, N. L., Young, W. S., 3rd, Knepper, M. A., and Lolait, S. J. (1993). Expression of vasopressin V1a and V2 receptor messenger ribonucleic acid in the liver and kidney of embryonic, developing, and adult rats. *Endocrinology* **133**, 1849–1859.

Pacha, J., Frindt, G., Antonian, L., Silver, R. B., and Palmer, L. G. (1993). Regulation of Na channels of the rat cortical collecting tubule by aldosterone. *J. Gen. Physiol.* **102**, 25–42.

Pacha, J., Frindt, G., Sackin, H., and Palmer, L. G. (1991). Apical maxi K channels in intercalated cells of CCT. *Am. J. Physiol.* **261**, F696–F705.

Palmer, L. G. (1986). Patch-clamp technique in renal physiology. *Am. J. Physiol.* **250**, F379–F385.

Palmer, L. G., Antonian, L., and Frindt, G. (1994). Regulation of apical K and Na channels and Na/K pumps in rat cortical collecting tubule by dietary K. *J. Gen. Physiol.* **104**, 693–710.

Panos, M. Z., Nicolaides, K. H., Anderson, J. V., Economides, D. L., Rees, L., and Williams, R. (1989). Plasma atrial natriuretic peptide in human fetus: Response to intravascular blood transfusion. *Am. J. Obstet. Gynecol.* **161**, 357–361.

Pasqualini, J. R., Sumida, C., and Gelly, C. (1972). Mineralocorticosteroid receptors in the foetal compartment. *J. Steroid. Biochem.* **3**, 543–556.

Pastoriza-Munoz, E., Colindres, R. E., Lassiter, W. E., and Lechene, C. (1978). Effect of parathyroid hormone on phosphate reabsorption in rat distal convolution. *Am. J. Physiol.* **235**, F321–F330.

Pastoriza-Munoz, E., Mishler, D. R., and Lechene, C. (1983). Effect of phosphate deprivation on phosphate reabsorption in rat nephron: Role of PTH. *Am. J. Physiol.* **244**, F140–F149.

Paul, R. V., Kirk, K. A., and Navar, L. G. (1987). Renal autoregulation and pressure natriuresis during ANF-induced diuresis. *Am. J. Physiol.* **253**, F424–F431.

Pelayo, J. C., Fildes, R. D., and Jose, P. A. (1984). Age-dependent renal effects of intrarenal dopamine infusion. *Am. J. Physiol.* **247**, R212–R216.

Peng, J. B., Chen, X. Z., Berger, U. V., Vassilev, P. M., Brown, E. M., and Hediger, M. A. (2000). A rat kidney-specific calcium transporter in the distal nephron. *J. Biol. Chem.* **275**, 28186–28194.

Peraino, R. A., and Suki, W. N. (1980). Phosphate transport by isolated rabbit cortical collecting tubule. *Am. J. Physiol.* **238**, 358–362

Perlman, M., and Levin, M. (1974). Fetal pulmonary hypoplasia, anuria, and oligohydramnios: Clinicopathologic observations and review of the literature. *Am. J. Obstet. Gynecol.* **118**, 1119–1123.

Petershack, J. A., Nagaraja, S. C., and Guillery, E. N. (1999). Role of glucocorticoids in the maturation of renal cortical Na$^+$-K$^+$-ATPase during fetal life in sheep. *Am. J. Physiol.* **276**, R1825–R1832.

Peti-Peterdi, J., Chambrey, R., Bebok, Z., Biemesderfer, D., St John, P. L., Abrahamson, D. R., Warnock, D. G., and Bell, P. D. (2000). Macula densa Na(+)/H(+) exchange activities mediated by apical NHE2 and basolateral NHE4 isoforms. *Am. J. Physiol. Renal. Physiol.* **278**, F452–F463.

Pfister, M. F., Lederer, E., Forgo, J., Ziegler, U., Lotscher, M., Quabius, E. S., Biber, J., and Murer, H. (1997). Parathyroid hormone-dependent degradation of type II Na$^+$/P$_i$ cotransporters. *J. Biol. Chem.* **272**, 20125–20130.

Pfister, M. F., Ruf, I., Stange, G., Ziegler, U., Lederer, E., Biber, J., and Murer, H. (1998). Parathyroid hormone leads to the lysosomal degradation of the renal type II Na/Pi cotransporter. *Proc. Natl. Acad. Sci. USA* **95**, 1909–1914.

Phillips, M. I. (1987). Functions of angiotensin in the central nervous system. *Annu. Rev. Physiol.* **49**, 413–435.

Pipkin, F. B., Kirkpatrick, S. M., Lumbers, E. R., and Mott, J. C. (1974). Renin and angiotensin-like levels in foetal, newborn and adult sheep. *J. Physiol. (Lond).* **241**, 575–588.

Pipkin, F. B., Kirkpatrick, S. M., and Mott, J. C. (1971a). Angiotensin II-like activity in arterial blood in new-born lambs. *J. Physiol. (Lond).* **218**, 61P–62P.

Pipkin, F. B., Mott, J. C., and Roberton, N. R. (1971b). Resting concentration of angiotensin II-like activity in the arterial blood of rabbits of different ages. *J. Physiol. (Lond).* **214**, 21P–22P.

Pitts, R., Ayer, J., and Schiess, W. (1949). The renal regulation of acid-base balance in man. III. The reabsorption and excretion of bicarbonate. *J. Clin. Invest.* **28**, 35–44.

Polacek, E., Vocel, J., and Neugebauerova, L. (1965). The osmotic concentrating ability in healthy infants and children. *Arch. Dis. Child.* **40**, 291.

Potter, E. L. a. T. S. T. (1943). Glomerular development in the kidney as an index of foetal maturity. *J. Pediatr.* **22**, 695.

Pratz, J., and Corman, B. (1985). Age-related changes in enzyme activities, protein content and lipid composition of rat kidney brush-border membrane. *Biochim. Biophys. Acta.* **814**, 265–273.

Pratz, J., Ripoche, P., and Corman, B. (1987). Cholesterol content and water and solute permeabilities of kidney membranes from aging rats. *Am. J. Physiol.* **253**, R8–R14.

Pryde, P. G., Sedman, A. B., Nugent, C. E., and Barr, M., Jr. (1993). Angiotensin-converting enzyme inhibitor fetopathy. *J. Am. Soc. Nephrol.* **3**, 1575–1582.

Qadri, F., Culman, J., Veltmar, A., Maas, K., Rascher, W., and Unger, T. (1993). Angiotensin II-induced vasopressin release is mediated through alpha-1 adrenoceptors and angiotensin II AT1 receptors in the supraoptic nucleus. *J. Pharmacol. Exp. Ther.* **267**, 567–574.

Quamme, G. A., and de Rouffignac, C. (2000). Epithelial magnesium transport and regulation by the kidney. *Front Biosci.* **5**, D694–D711.

Quamme, G. A., and Dirks, J. H. (1980). Intraluminal and contraluminal magnesium and calcium transfer in the rat nephron. *Am. J. Physiol.* **238**, F187–F198.

Quamme, G. A., and Dirks, J. H. (1986). The physiology of renal magnesium handling. *Ren. Physiol.* **9**, 257–269.

Quan, A., and Baum, M. (1996). Endogenous production of angiotensin II modulates rat proximal tubule transport. *J. Clin. Invest.* **97**, 2878–2882.

Quan, A., and Baum, M. (1997). Regulation of proximal tubule transport by angiotensin II. *Semin. Nephrol.* **17**, 423–430.

Quigley, R., and Baum, M. (1990). Developmental changes in rabbit juxtamedullary proximal convoluted tubule bicarbonate permeability. *Pediatr. Res.* **28**, 663–666.

Quigley, R., and Baum, M. (1991). Effects of growth hormone and insulin-like growth factor I on rabbit proximal convoluted tubule transport. *J. Clin. Invest.* **88**, 368–374.

Rabinowitz, R., Peters, M. T., Vyas, S., Campbell, S., and Nicolaides, K. H. (1989). Measurement of fetal urine production in normal pregnancy by real-time ultrasonography. *Am. J. Obstet. Gynecol.* **161**, 1264–1266.

Rajerison, R. M., Butlen, D., and Jard, S. (1976). Ontogenic development of antidiuretic hormone receptors in rat kidney: Comparison of hormonal binding and adenylate cyclase activation. *Mol. Cell. Endocrinol.* **4**, 271–285.

Rane, S., Aperia, A., Eneroth, P., and Lundin, S. (1985). Development of urinary concentrating capacity in weaning rats. *Pediatr. Res.* **19**, 472–475.

Rees, L., Forsling, M. L., and Brook, C. G. (1980). Vasopressin concentrations in the neonatal period. *Clin. Endocrinol.* **12**, 357–362.

Riccardi, D., Hall, A. E., Chattopadhyay, N., Xu, J. Z., Brown, E. M., and Hebert, S. C. (1998). Localization of the extracellular Ca^{2+}/polyvalent cation-sensing protein in rat kidney. *Am. J. Physiol.* **274**, F611–F622.

Richer, C., Hornych, H., Amiel-Tison, C., Relier, J. P., and Giudicelli, J. F. (1977). Plasma renin activity and its postnatal development in preterm infants. Preliminary report. *Biol. Neonate* **31**, 301–304.

Ridderstrale, Y., Kashgarian, M., Koeppen, B., Giebisch, G., Stetson, D., Ardito, T., and Stanton, B. (1988). Morphological heterogeneity of the rabbit collecting duct. *Kidney Int.* **34**, 655–670.

Ridderstrale, Y., Wistrand, P. J., and Tashian, R. E. (1992). Membrane-associated carbonic anhydrase activity in the kidney of CA II- deficient mice. *J. Histochem. Cytochem.* **40**, 1665–1673.

Ritthaler, T., Traebert, M., Lotscher, M., Biber, J., Murer, H., and Kaissling, B. (1999). Effects of phosphate intake on distribution of type II Na/Pi cotransporter mRNA in rat kidney. *Kidney Int.* **55**, 976–983.

Robillard, J. E., Gomez, R. A., VanOrden, D., and Smith, F. G., Jr. (1982). Comparison of the adrenal and renal responses to angiotensin II fetal lambs and adult sheep. *Circ. Res.* **50**, 140–147.

Robillard, J. E., Guillery, E. N., Segar, J. L., Merrill, D. C., and Jose, P. A. (1993). Influence of renal nerves on renal function during development. *Pediatr. Nephrol.* **7**, 667–671.

Robillard, J. E., Kulvinskas, C., Sessions, C., Burmeister, L., and Smith, F. G., Jr. (1975). Maturational changes in the fetal glomerular filtration rate. *Am. J. Obstet. Gynecol.* **122**, 601–606.

Robillard, J. E., Matson, J. R., Sessions, C., and Smith, F. G., Jr. (1979a). Developmental aspects of renal tubular reabsorption of water in the lamb fetus. *Pediatr. Res.* **13**, 1172–1176.

Robillard, J. E., and Nakamura, K. T. (1988a). Hormonal regulation of renal function during development. *Biol. Neonate* **53**, 201–211.

Robillard, J. E., and Nakamura, K. T. (1988b). Neurohormonal regulation of renal function during development. *Am. J. Physiol.* **254**, F771–F779.

Robillard, J. E., Nakamura, K. T., and DiBona, G. F. (1986). Effects of renal denervation on renal responses to hypoxemia in fetal lambs. *Am. J. Physiol.* **250**, F294–F301.

Robillard, J. E., Nakamura, K. T., and Lawton, W. J. (1985). Effects of aldosterone on urinary kallikrein and sodium excretion during fetal life. *Pediatr. Res.* **19**, 1048–1052.

Robillard, J. E., Nakamura, K. T., Matherne, G. P., and Jose, P. A. (1988a). Renal hemodynamics and functional adjustments to postnatal life. *Semin. Perinatol.* **12**, 143–150.

Robillard, J. E., Nakamura, K. T., Varille, V. A., Andresen, A. A., Matherne, G. P., and VanOrden, D. E. (1988b). Ontogeny of the renal response to natriuretic peptide in sheep. *Am. J. Physiol.* **254**, F634–F641.

Robillard, J. E., Nakamura, K. T., Varille, V. A., Matherne, G. P., and McWeeny, O. J. (1988c). Plasma and urinary clearance rates of atrial natriuretic factor during ontogeny in sheep. *J. Dev. Physiol.* **10**, 335–346.

Robillard, J. E., Nakamura, K. T., Wilkin, M. K., McWeeny, O. J., and DiBona, G. F. (1987). Ontogeny of renal hemodynamic response to renal nerve stimulation in sheep. *Am. J. Physiol.* **252**, F605–F612.

Robillard, J. E., Page, W. V., Mathews, M. S., Schutte, B. C., Nuyt, A. M., and Segar, J. L. (1995). Differential gene expression and regulation of renal angiotensin II receptor subtypes (AT1 and AT2) during fetal life in sheep. *Pediatr. Res.* **38**, 896–904.

Robillard, J. E., Segar, J. L., Smith, F. G., and Jose, P. A. (1992a). Regulation of sodium metabolism and extracellular fluid volume during development. *Clin. Perinatol.* **19**, 15–31.

Robillard, J. E., Sessions, C., Burmeister, L., and Smith, F. G., Jr. (1977a). Influence of fetal extracellular volume contraction on renal reabsorption of bicarbonate in fetal lambs. *Pediatr. Res.* **11**, 649–655.

Robillard, J. E., Sessions, C., Kennedey, R. L., Hamel-Robillard, L., and Smith, F. G., Jr. (1977b). Interrelationship between glomerular filtration rate and renal transport of sodium and chloride during fetal life. *Am. J. Obstet. Gynecol.* **128**, 727–734.

Robillard, J. E., Smith, F. G., Nakamura, K. T., Sato, T., Segar, J., and Jose, P. A. (1990). Neural control of renal hemodynamics and function during development. *Pediatr. Nephrol.* **4**, 436–441.

Robillard, J. E., Smith, F. G., Segar, J. L., Guillery, E. N., and Jose, P. A. (1992b). Mechanisms regulating renal sodium excretion during development. *Pediatr. Nephrol.* **6**, 205–213.

Robillard, J. E., and Weiner, C. (1988). Atrial natriuretic factor in the human fetus: Effect of volume expansion. *J. Pediatr.* **113**, 552–555.

Robillard, J. E., Weismann, D. N., Gomez, R. A., Ayres, N. A., Lawton, W. J., and VanOrden, D. E. (1983). Renal and adrenal responses to converting-enzyme inhibition in fetal and newborn life. *Am. J. Physiol.* **244**, R249–R256.

Robillard, J. E., Weismann, D. N., and Herin, P. (1981). Ontogeny of single glomerular perfusion rate in fetal and newborn lambs. *Pediatr. Res.* **15**, 1248–1255.

Robillard, J. E., and Weitzman, R. E. (1980). Developmental aspects of the fetal renal response to exogenous arginine vasopressin. *Am. J. Physiol.* **238**, F407–F414.

Robillard, J. E., Weitzman, R. E., Fisher, D. A., and Smith, F. G., Jr. (1979b). The dynamics of vasopressin release and blood volume regulation during fetal hemorrhage in the lamb fetus. *Pediatr. Res.* **13**, 606–610.

Rocha, A. S., Magaldi, J. B., and Kokko, J. P. (1977). Calcium and phosphate transport in isolated segments of rabbit Henle's loop. *J. Clin. Invest.* **59**, 975–983.

Rodriguez-Soriano, J. (1999). Bartter's syndrome comes of age. *Pediatrics* **103**, 663–664.

Rodriguez-Soriano, J., Vallo, A., Oliveros, R., and Castillo, G. (1983). Renal handling of sodium in premature and full-term neonates: A study using clearance methods during water diuresis. *Pediatr. Res.* **17**, 1013–1016.

Rogers, S. A., and Hammerman, M. R. (1989). Growth hormone activates phospholipase C in proximal tubular basolateral membranes from canine kidney. *Proc. Natl. Acad. Sci. USA* **86**, 6363–6366.

Rosnes, J. S., Valego, N., Wang, J., Zehnder, T., and Rose, J. C. (1998). Active renin, prorenin, and renin gene expression after reduced renal perfusion pressure in term ovine fetuses. *Am. J. Physiol.* **275**, R141–R147.

Ross, B. S., Pollak, A., and Oh, W. (1978). The pharmacologic effects of furosemide therapy in the low-birth-weight infant. *J. Pediatr.* **92**, 149–152.

Ross, M. G., Ervin, M. G., Lam, R. W., Castro, L., Leake, R. D., and Fisher, D. A. (1987). Plasma atrial natriuretic peptide response to volume expansion in the ovine fetus. *Am J. Obstet. Gynecol.* **157**, 1292–1297.

Ross, M. G., Ervin, M. G., Lam, R. W., Leake, R. D., and Fisher, D. A. (1988a). Fetal atrial natriuretic factor and arginine vasopressin responses to hyperosmolality and hypervolemia. *Pediatr. Res.* **24**, 318–321.

Ross, M. G., Sherman, D. J., Ervin, M. G., Castro, R., and Humme, J. (1988b). Maternal dehydration-rehydration: Fetal plasma and urinary responses. *Am. J. Physiol.* **255**, E674–E679.

Rowe, P. S. (1998). The role of the PHEX gene (PEX) in families with X-linked hypophosphataemic rickets. *Curr. Opin. Nephrol. Hypertens.* **7**, 367–376.

Roy, S., Martel, J., Ma, S., and Tenenhouse, H. S. (1994). Increased renal 25-hydroxyvitamin D3-24-hydroxylase messenger ribonucleic acid and immunoreactive protein in phosphate-deprived Hyp mice: A mechanism for accelerated 1,25-dihydroxyvitamin D3 catabolism in X-linked hypophosphatemic rickets. *Endocrinology* **134**, 1761–1767.

Rubin, M., Bruck, E., Rapoport, M.J. (1949). Maturation of renal function in childhood: Clearance studies. *J. Clin. Invest.* **28**, 1144–1162.

Rudolph, A., Heymann, M., and Teramo, K. (1971). Studies on the circulation of the previable human fetus. *Pediatr. Res.* **5**, 452.

Rudolph, A. M., and Heymann, M. A. (1967). The circulation of the fetus in utero. Methods for studying distribution of blood flow, cardiac output and organ blood flow. *Circ. Res.* **21**, 163–184.

Saito, T., Ishikawa, S. E., Sasaki, S., Fujita, N., Fushimi, K., Okada, K., Takeuchi, K., Sakamoto, A., Ookawara, S., Kaneko, T., and Marumo, F. (1997). Alteration in water channel AQP-2 by removal of AVP stimulation in collecting duct cells of dehydrated rats. *Am. J. Physiol.* **272**, F183–F191.

Sands, J. M. (1999). Regulation of renal urea transporters. *J. Am. Soc. Nephrol.* **10**, 635–646.

Sands, J. M., Naruse, M., Baum, M., Jo, I., Hebert, S. C., Brown, E. M., and Harris, H. W. (1997). Apical extracellular calcium/polyvalent cation-sensing receptor regulates vasopressin-elicited water permeability in rat kidney inner medullary collecting duct. *J. Clin. Invest.* **99**, 1399–1405.

Satlin, L. M. (1994). Postnatal maturation of potassium transport in rabbit cortical collecting duct. *Am. J. Physiol.* **266**, F57–F65.

Satlin, L. M. (1999). Regulation of potassium transport in the maturing kidney. *Semin. Nephrol.* **19**, 155–165.

Satlin, L. M., Evan, A. P., Gattone, V. H. D., and Schwartz, G. J. (1988). Postnatal maturation of the rabbit cortical collecting duct. *Pediatr. Nephrol.* **2**, 135–145.

Satlin, L. M., Matsumoto, T., and Schwartz, G. J. (1992). Postnatal maturation of rabbit renal collecting duct. III. Peanut lectin-binding intercalated cells. *Am. J. Physiol.* **262**, F199–F208.

Satlin, L. M., and Palmer, L. G. (1996). Apical Na$^+$ conductance in maturing rabbit principal cell. *Am. J. Physiol.* **270**, F391–F397.

Satlin, L. M., and Palmer, L. G. (1997). Apical K$^+$ conductance in maturing rabbit principal cell. *Am. J. Physiol.* **272**, F397–F404.

Satlin, L. M., and Schwartz, G. J. (1987). Postnatal maturation of rabbit renal collecting duct: Intercalated cell function. *Am. J. Physiol.* **253**, F622–F635.

Saxén, L. (1987). "Organogenesis of the Kidney." Harvard Univ. Press, Cambridge, MA.

Schafer, J. A., and Hawk, C. T. (1992). Regulation of Na$^+$ channels in the cortical collecting duct by AVP and mineralocorticoids. *Kidney Int.* **41**, 255–268.

Schild, L., Canessa, C. M., Shimkets, R. A., Gautschi, I., Lifton, R. P., and Rossier, B. C. (1995). A mutation in the epithelial sodium channel causing Liddle disease increases channel activity in the *Xenopus laevis* oocyte expression system. *Proc. Natl. Acad. Sci. USA* **92**, 5699–5703.

Schlatter, E., Haxelmans, S., Ankorina, I., and Kleta, R. (1995). Regulation of Na$^+$/H$^+$ exchange by diadenosine polyphosphates, angiotensin II, and vasopressin in rat cortical collecting duct. *J. Am. Soc. Nephrol.* **6**, 1223–1229.

Schlondorff, D. (1987). The glomerular mesangial cell: An expanding role for a specialized pericyte. *FASEB J.* **1**, 272–281.

Schlondorff, D., Weber, H., Trizna, W., and Fine, L. G. (1978). Vasopressin responsiveness of renal adenylate cyclase in newborn rats and rabbits. *Am. J. Physiol.* **234**, F16–F21.

Schmidt, U., and Horster, M. (1977). Na-K-activated ATPase: Activity maturation in rabbit nephron segments dissected in vitro. *Am. J. Physiol.* **233**, F55–F60.

Schmitt, R., Ellison, D. H., Farman, N., Rossier, B. C., Reilly, R. F., Reeves, W. B., Oberbaumer, I., Tapp, R., and Bachmann, S. (1999). Developmental expression of sodium entry pathways in rat nephron. *Am. J. Physiol.* **276**, F367–F381.

Schnermann, J., Traynor, T., Yang, T., Arend, L., Huang, Y. G., Smart, A., and Briggs, J. P. (1998). Tubuloglomerular feedback: new concepts and developments. *Kidney Int. Suppl.* **67**, S40–S45.

Schoeneman, M. J., and Spitzer, A. (1980). The effect of intravascular volume expansion of proximal tubular reabsorption during development. *Proc Soc Exp Biol Med* **165**, 319–322.

Schultheis, P. J., Clarke, L. L., Meneton, P., Miller, M. L., Soleimani, M., Gawenis, L. R., Riddle, T. M., Duffy, J. J., Doetschman, T., Wang, T., Giebisch, G., Aronson, P. S., Lorenz, J. N., and Shull, G. E. (1998). Renal and intestinal absorptive defects in mice lacking the NHE3 Na$^+$/H$^+$ exchanger. *Nature Genet.* **19**, 282–285.

Schuster, V. L., Bonsib, S. M., and Jennings, M. L. (1986). Two types of collecting duct mitochondria-rich (intercalated) cells: Lectin and band 3 cytochemistry. *Am. J. Physiol.* **251**, C347–C355.

Schuster, V. L., Fejes-Toth, G., Naray-Fejes-Toth, A., and Gluck, S. (1991). Colocalization of H$^+$-ATPase and band 3 anion exchanger in rabbit collecting duct intercalated cells. *Am. J. Physiol.* **260**, F506–F517.

Schutz, S., Le Moullec, J. M., Corvol, P., and Gasc, J. M. (1996). Early expression of all the components of the renin-angiotensin-system in human development. *Am. J. Pathol.* **149**, 2067–2079.

Schwab, A., and Oberleithner, H. (1988). Trans- and paracellular K$^+$ transport in diluting segment of frog kidney. *Pflug. Arch* **411**, 268–272.

Schwartz, G. (1992). Acid-base homeostasis. *In* "Pediatric Kidney Disease" (J. CM Edelmann, ed.), pp. 201–230. Little, Brown and Company, Boston.

Schwartz, G. J., Barasch, J., and Al-Awqati, Q. (1985). Plasticity of functional epithelial polarity. *Nature* **318**, 368–371.

Schwartz, G. J., Brion, L. P., Corey, H. E., and Dorfman, H. D. (1991). Case report 668. Carbonic anhydrase II deficiency syndrome (osteopetrosis associated with renal tubular acidosis and cerebral calcification). *Skel. Radiol* **20**, 447–52.

Schwartz, G. J., and Burg, M. B. (1978). Mineralocorticoid effects on cation transport by cortical collecting tubules in vitro. *Am. J. Physiol.* **235**, F576–F585.

Schwartz, G. J., and Evan, A. P. (1983). Development of solute transport in rabbit proximal tubule. I. HCO$_3^-$ and glucose absorption. *Am. J. Physiol.* **245**, F382–F390.

Schwartz, G. J., and Evan, A. P. (1984). Development of solute transport in rabbit proximal tubule. III. Na-K-ATPase activity. *Am. J. Physiol.* **246**, F845–F852.

Schwartz, G. J., Goldsmith, D. I., and Fine, L. G. (1978). p-aminohippurate transport in the proximal straight tubule: Development and substrate stimulation. *Pediatr. Res.* **12**, 793–796.

Schwartz, G. J., Haycock, G. B., Edelmann, C. M., Jr., and Spitzer, A. (1979). Late metabolic acidosis: A reassessment of the definition. *J. Pediatr.* **95**, 102–107.

Schwartz, G. J., Kittelberger, A. M., Barnhart, D. A., and Vijayakumar, S. (2000). Carbonic anhydrase IV is expressed in H⁺-secreting cells of rabbit kidney. *Am. J. Physiol. Renal. Physiol.* **278**, F894–F904.

Schwartz, G. J., Olson, J., Kittelberger, A. M., Matsumoto, T., Waheed, A., and Sly, W. S. (1999). Postnatal development of carbonic anhydrase IV expression in rabbit kidney. *Am. J. Physiol.* **276**, F510–F520.

Schwartz, G. J., Zavilowitz, B. J., Radice, A. D., Garcia-Perez, A., and Sands, J. M. (1992). Maturation of aldose reductase expression in the neonatal rat inner medulla. *J. Clin. Invest.* **90**, 1275–1283.

Scroop, G. C., Stankewytsch-Janusch, B., and Marker, J. D. (1992). Renin-angiotensin and autonomic mechanisms in cardiovascular homeostasis during haemorrhage in fetal and neonatal sheep. *J. Dev. Physiol.* **18**, 25–33.

Segar, J. L., Smith, F. G., Guillery, E. N., Jose, P. A., and Robillard, J. E. (1992). Ontogeny of renal response to specific dopamine DA1-receptor stimulation in sheep. *Am. J. Physiol.* **263**, R868–R873.

Semama, D. S., Thonney, M., and Guignard, J. P. (1993a). Role of endogenous endothelin in renal haemodynamics of newborn rabbits. *Pediatr. Nephrol.* **7**, 886–890.

Semama, D. S., Thonney, M., Guignard, J. P., and Gouyon, J. B. (1993b). Effects of endothelin on renal function in newborn rabbits. *Pediatr. Res.* **34**, 120–123.

Semmekrot, B. A., Wiesel, P. H., Monnens, L. A., and Guignard, J. P. (1990). Age differences in renal response to atrial natriuretic peptide in rabbits. *Life Sci.* **46**, 849–856.

Seri, I. (1990). Dopamine and natriuresis. Mechanism of action and developmental aspects. *Am. J. Hypertens.* **3**, 82S–86S.

Seyberth, H. W., Rascher, W., Hackenthal, R., and Wille, L. (1983). Effect of prolonged indomethacin therapy on renal function and selected vasoactive hormones in very-low-birth-weight infants with symptomatic patent ductus arteriosus. *J. Pediatr.* **103**, 979–984.

Seymour, A. A., Smith, S. G., 3rd, and Mazack, E. K. (1987). Effects of renal perfusion pressure on the natriuresis induced by atrial natriuretic factor. *Am. J. Physiol.* **253**, F234–F238.

Shafik, I. M., and Quamme, G. A. (1989). Early adaptation of renal magnesium reabsorption in response to magnesium restriction. *Am. J. Physiol.* **257**, F974–F977.

Shah, M., Quigley, R., and Baum, M. (1998). Maturation of rabbit proximal straight tubule chloride/base exchange. *Am. J. Physiol.* **274**, F883–F888.

Shah, M., Quigley, R., and Baum, M. (2000). Maturation of proximal straight tubule NaCl transport: role of thyroid hormone. *Am. J. Physiol. Renal. Physiol.* **278**, F596–F602.

Shareghi, G. R., and Agus, Z. S. (1982a). Magnesium transport in the cortical thick ascending limb of Henle's loop of the rabbit. *J. Clin. Invest.* **69**, 759–769.

Shareghi, G. R., and Agus, Z. S. (1982b). Phosphate transport in the light segment of the rabbit cortical collecting tubule. *Am. J. Physiol.* **242**, F379–F384.

Shareghi, G. R., and Stoner, L. C. (1978). Calcium transport across segments of the rabbit distal nephron in vitro. *Am. J. Physiol.* **235**, F367–F375.

Shayakul, C., and Alper, S. L. (2000). Inherited renal tubular acidosis. *Curr. Opin. Nephrol. Hypertens.* **9**, 541–546.

Shimizu, T., Yoshitomi, K., Nakamura, M., and Imai, M. (1990). Effects of PTH, calcitonin, and cAMP on calcium transport in rabbit distal nephron segments. *Am. J. Physiol.* **259**, F408–F414.

Shimkets, R. A., Warnock, D. G., Bositis, C. M., Nelson-Williams, C., Hansson, J. H., Schambelan, M., Gill, J. R., Jr., Ulick, S., Milora, R. V., Findling, J. W., et al. (1994). Liddle's syndrome: Heritable human hypertension caused by mutations in the beta subunit of the epithelial sodium channel. *Cell* **79**, 407–414.

Shirley, D. G., Walter, S. J., Unwin, R. J., and Giebisch, G. (1998). Contribution of Na⁺-H⁺ exchange to sodium reabsorption in the loop of henle: A microperfusion study in rats. *J. Physiol.* **513**, 243–249.

Siegel, S. R., and Fisher, D. A. (1977). The renin-angiotensin-aldosterone system in the newborn lamb: Response to furosemide. *Pediatr. Res.* **11**, 837–839.

Siegel, S. R., Fisher, D. A., and Oh, W. (1974). Serum aldosterone concentrations related to sodium balance in the newborn infant. *Pediatrics* **53**, 410–413.

Siegel, S. R., Oaks, G., and Palmer, S. (1981). Effects of angiotensin II on blood pressure, plasma renin activity, and aldosterone in the fetal lamb. *Dev. Pharmacol. Ther.* **3**, 144–149.

Siegel, S. R., and Oh, W. (1976). Renal function as a marker of human fetal maturation. *Acta Paediatr. Scand.* **65**, 481–485.

Siga, E., and Horster, M. F. (1991). Regulation of osmotic water permeability during differentiation of inner medullary collecting duct. *Am. J. Physiol.* **260**, F710–F716.

Silver, R. B., Mennitt, P. A., and Satlin, L. M. (1996). Stimulation of apical H-K-ATPase in intercalated cells of cortical collecting duct with chronic metabolic acidosis. *Am. J. Physiol.* **270**, F539–F547.

Simeoni, U., Zhu, B., Muller, C., Judes, C., Massfelder, T., Geisert, J., and Helwig, J. J. (1997). Postnatal development of vascular resistance of the rabbit isolated perfused kidney: Modulation by nitric oxide and angiotensin II. *Pediatr. Res.* **42**, 550–555.

Simon, D. B., Bindra, R. S., Mansfield, T. A., Nelson-Williams, C., Mendonca, E., Stone, R., Schurman, S., Nayir, A., Alpay, H., Bakkaloglu, A., Rodriguez-Soriano, J., Morales, J. M., Sanjad, S. A., Taylor, C. M., Pilz, D., Brem, A., Trachtman, H., Griswold, W., Richard, G. A., John, E., and Lifton, R. P. (1997). Mutations in the chloride channel gene, CLCNKB, cause Bartter's syndrome type III. *Nature Genet.* **17**, 171–178.

Simon, D. B., Karet, F. E., Hamdan, J. M., DiPietro, A., Sanjad, S. A., and Lifton, R. P. (1996a). Bartter's syndrome, hypokalaemic alkalosis with hypercalciuria, is caused by mutations in the Na-K-2Cl cotransporter NKCC2. *Nature Genet.* **13**, 183–188.

Simon, D. B., Karet, F. E., Rodriguez-Soriano, J., Hamdan, J. H., DiPietro, A., Trachtman, H., Sanjad, S. A., and Lifton, R. P. (1996b). Genetic heterogeneity of Bartter's syndrome revealed by mutations in the K+ channel, ROMK. *Nature Genet.* **14**, 152–156.

Simon, D. B., and Lifton, R. P. (1996). The molecular basis of inherited hypokalemic alkalosis: Bartter's and Gitelman's syndromes. *Am. J. Physiol.* **271**, F961–F966.

Simon, D. B., Lu, Y., Choate, K. A., Velazquez, H., Al-Sabban, E., Praga, M., Casari, G., Bettinelli, A., Colussi, G., Rodriguez-Soriano, J., McCredie, D., Milford, D., Sanjad, S., and Lifton, R. P. (1999). Paracellin-1, a renal tight junction protein required for paracellular Mg²⁺ resorption. *Science* **285**, 103–106.

Simon, D. B., Nelson-Williams, C., Bia, M. J., Ellison, D., Karet, F. E., Molina, A. M., Vaara, I., Iwata, F., Cushner, H. M., Koolen, M., Gainza, F. J., Gitleman, H. J., and Lifton, R. P. (1996c). Gitelman's variant of Bartter's syndrome, inherited hypokalaemic alkalosis, is caused by mutations in the thiazide-sensitive Na-Cl cotransporter. *Nature Genet.* **12**, 24–30.

Slotkin, T. A., Lau, C., Kavlock, R. J., Gray, J. A., Orband-Miller, L., Queen, K. L., Baker, F. E., Cameron, A. M., Antolick, L., Haim, K., et al. (1988). Role of sympathetic neurons in biochemical and functional development of the kidney: Neonatal sympathectomy with 6-hydroxydopamine. *J. Pharmacol. Exp. Ther.* **246**, 427–433.

Sly, W. S., Whyte, M. P., Sundaram, V., Tashian, R. E., Hewett-Emmett, D., Guibaud, P., Vainsel, M., Baluarte, H. J., Gruskin, A., Al-Mosawi, M., and et al. (1985). Carbonic anhydrase II deficiency in 12 families with the autosomal recessive syndrome of osteopetrosis with renal tubular acidosis and cerebral calcification. *N. Engl. J. Med.* **313**, 139–145.

Smith, F. G., and Abraham, J. (1995). Renal and renin responses to furosemide in conscious lambs during postnatal maturation. *Can. J. Physiol. Pharmacol.* **73**, 107–112.

Smith, F. G., Jr., Lupu, A. N., Barajas, L., Bauer, R., and Bashore, R. A. (1974). The renin-angiotensin system in the fetal lamb. *Pediatr. Res.* **8**, 611–620.

Smith, F. G., Jr., and Schwartz, A. (1970). Response of the intact lamb fetus to acidosis. *Am. J. Obstet. Gynecol.* **106**, 52–58.

Smith, F. G., and Lumbers, E. R. (1989). Comparison of renal function in term fetal sheep and newborn lambs. *Biol. Neonate* **55**, 309–316.

Smith, F. G., Sato, T., McWeeny, O. J., Klinkefus, J. M., and Robillard, J. E. (1990). Role of renal sympathetic nerves in response of the ovine fetus to volume expansion. *Am. J. Physiol.* **259**, R1050–R1055.

Smith, F. G., Sato, T., Varille, V. A., and Robillard, J. E. (1989). Atrial natriuretic factor during fetal and postnatal life: A review. *J. Dev. Physiol.* **12**, 55–62.

Smith, F. G., Smith, B. A., Guillery, E. N., and Robillard, J. E. (1991). Role of renal sympathetic nerves in lambs during the transition from fetal to newborn life. *J. Clin. Invest.* **88**, 1988–1994.

Soleimani, M., Grassi, S. M., and Aronson, P. S. (1987). Stoichiometry of Na+-HCO-3 cotransport in basolateral membrane vesicles isolated from rabbit renal cortex. *J. Clin. Invest.* **79**, 1276–1280.

Solhaug, M. J., Ballevre, L. D., Guignard, J. P., Granger, J. P., and Adelman, R. D. (1996). Nitric oxide in the developing kidney. *Pediatr. Nephrol.* **10**, 529–539.

Solhaug, M. J., Wallace, M. R., and Granger, J. P. (1993). Endothelium-derived nitric oxide modulates renal hemodynamics in the developing piglet. *Pediatr. Res.* **34**, 750–754.

Solomon, S. (1974a). Absolute rates of sodium and potassium reabsorption by proximal tubule of immature rats. *Biol. Neonate.* **25**, 340–351.

Solomon, S. (1974b). Maximal gradients of na and k across proximal tubules of kidneys of immature rats. *Biol. Neonate.* **25**, 327–339.

Solomon, S. (1977). Developmental changes in nephron number, proximal tubular length and superficial nephron glomerular filtration rate of rats. *J. Physiol. (Lond).* **272**, 573–589.

Sonnenberg, H., Honrath, U., and Wilson, D. R. (1990). In vivo microperfusion of inner medullary collecting duct in rats: Effect of amiloride and ANF. *Am. J. Physiol.* **259**, F222–F226.

Sorribas, V., Markovich, D., Hayes, G., Stange, G., Forgo, J., Biber, J., and Murer, H. (1994). Cloning of a Na/Pi cotransporter from opossum kidney cells. *J. Biol. Chem.* **269**, 6615–6621.

Speller, A. M., and Moffat, D. B. (1977). Tubulo-vascular relationships in the developing kidney. *J. Anat.* **123**, 487–500.

Spitzer, A. (1982). The role of the kidney in sodium homeostasis during maturation. *Kidney Int.* **21**, 539–545.

Spitzer, A., and Brandis, M. (1974). Functional and morphologic maturation of the superficial nephrons. Relationship to total kidney function. *J. Clin. Invest.* **53**, 279–287.

Spitzer, A., and Edelmann, C. M., Jr. (1971). Maturational changes in pressure gradients for glomerular filtration. *Am. J. Physiol.* **221**, 1431–1435.

Star, R. A., Burg, M. B., and Knepper, M. A. (1987). Luminal disequilibrium pH and ammonia transport in outer medullary collecting duct. *Am. J. Physiol.* **252**, F1148–F1157.

Stephenson, G., Hammet, M., Hadaway, G., and Funder, J. W. (1984). Ontogeny of renal mineralocorticoid receptors and urinary electrolyte responses in the rat. *Am. J. Physiol.* **247**, F665–F671.

Stevenson, K. M., Gibson, K. J., and Lumbers, E. R. (1996). Effects of losartan on the cardiovascular system, renal haemodynamics and function and lung liquid flow in fetal sheep. *Clin. Exp. Pharmacol. Physiol.* **23**, 125–133.

Stokes, J. B. (1982). Ion transport by the cortical and outer medullary collecting tubule. *Kidney Int.* **22**, 473–484.

Stone, D. K., Seldin, D. W., Kokko, J. P., and Jacobson, H. R. (1983). Anion dependence of rabbit medullary collecting duct acidification. *J. Clin. Invest.* **71**, 1505–1508.

Stoos, B. A., Carretero, O. A., and Garvin, J. L. (1994). Endothelial-derived nitric oxide inhibits sodium transport by affecting apical membrane channels in cultured collecting duct cells. *J. Am. Soc. Nephrol.* **4**, 1855–1860.

Stoos, B. A., Garcia, N. H., and Garvin, J. L. (1995). Nitric oxide inhibits sodium reabsorption in the isolated perfused cortical collecting duct. *J. Am. Soc. Nephrol.* **6**, 89–94.

Stoos, B. A., and Garvin, J. L. (1997). Actions of nitric oxide on renal epithelial transport. *Clin. Exp. Pharmacol. Physiol.* **24**, 591–594.

Strauss, J., Daniel, S. S., and James, L. S. (1981). Postnatal adjustment in renal function. *Pediatrics* **68**, 802–808.

Strautnieks, S. S., Thompson, R. J., Gardiner, R. M., and Chung, E. (1996). A novel splice-site mutation in the gamma subunit of the epithelial sodium channel gene in three pseudohypoaldosteronism type 1 families. *Nature Genet.* **13**, 248–250.

Strickler, J., Thompson, D., Klose, R., and Giebisch, G. (1964). Micropuncture study of inorganic phosphate excretion in the rat. *JCI* **43**, 1596–1607.

Stuart, R. O., Bush, K. T., and Nigam, S. K. (2001). Changes in global gene expression patterns during development and maturation of the rat kidney. *Proc. Natl. Acad. Sci. USA* **98**, 5649–5654.

Suki, W. N., and Rouse, D. (1981). Hormonal regulation of calcium transport in thick ascending limb renal tubules. *Am. J. Physiol.* **241**, F171–F174.

Sulyok, E. (1971). The relationship between electrolyte and acid-base balance in the premature infant during early postnatal life. *Biol. Neonate.* **17**, 227–237.

Sulyok, E., and Heim, T. (1971). Assessment of maximal urinary acidification in premature infants. *Biol. Neonate.* **19**, 200–210.

Sulyok, E., Heim, T., Soltesz, G., and Jaszai, V. (1972). The influence of maturity on renal control of acidosis in newborn infants. *Biol. Neonate.* **21**, 418–435.

Sulyok, E., Nemeth, M., Tenyi, I., Csaba, I., Gyory, E., Ertl, T., and Varga, F. (1979a). Postnatal development of renin-angiotensin-aldosterone system, RAAS, in relation to electrolyte balance in premature infants. *Pediatr. Res.* **13**, 817–820.

Sulyok, E., Nemeth, M., Tenyi, I., Csaba, I. F., Varga, F., Gyory, E., and Thurzo, V. (1979b). Relationship between maturity, electrolyte balance and the function of the renin-angiotensin-aldosterone system in newborn infants. *Biol. Neonate.* **35**, 60–65.

Sulyok, E., Nemeth, M., Tenyi, I., Csaba, I. F., Varga, L., and Varga, F. (1981). Relationship between the postnatal development of the renin-angiotensin-aldosterone system and the electrolyte and acid-base status in the sodium chloride supplemented premature infant. *Acta Paediatr. Acad. Sci. Hung.* **22**, 109–121.

Sulyok, E., Varga, F., Csaba, I. F., Nemeth, M., Tenyi, I., Gyory, E., and Ertl, T. (1980a). Function of the renin-angiotensin-aldosterone system in relation to electrolyte balance in the small-for-date neonate. *Acta Paediatr. Acad. Sci. Hung.* **21**, 153–157.

Sulyok, E., Varga, F., Nemeth, M., Tenyi, I., Csaba, I. F., Ertl, T., and Gyory, E. (1980b). Furosemide-induced alterations in the electrolyte status, the function of renin-angiotensin-aldosterone system, and the urinary excretion of prostaglandins in newborn infants. *Pediatr. Res.* **14**, 765–768.

Suzuki, M., Capparelli, A., Jo, O. D., Kawaguchi, Y., Ogura, Y., Miyahara, T., and Yanagawa, N. (1988). Phosphate transport in the in vitro cultured rabbit proximal convoluted and straight tubules. *Kidney Int.* **34**, 268–272.

Svenningsen, N. W. (1974). Renal acid-base titration studies in infants with and without metabolic acidosis in the postneonatal period. *Pediatr. Res.* **8**, 659–672.

Svenningsen, N. W., and Aronson, A. S. (1974). Postnatal development of renal concentration capacity as estimated by DDAVP test in normal and asphyxiated neonates. *Biol. Neonate* **25**, 230–241.

Svenningsen, N. W., and Lindquist, B. (1973). Incidence of metabolic acidosis in term, preterm and small-for-gestational age infants in relation to dietary protein intake. *Acta Paediatr. Scand.* **62**, 1–10.

Svenningsen, N. W., and Lindquist, B. (1974). Postnatal development of renal hydrogen ion excretion capacity in relation to age and protein intake. *Acta Paediatr. Scand.* **63**, 721–731.

Tanphaichitr, V. S., Sumboonnanonda, A., Ideguchi, H., Shayakul, C., Brugnara, C., Takao, M., Veerakul, G., and Alper, S. L. (1998). Novel AE1 mutations in recessive distal renal tubular acidosis. Loss-of-function is rescued by glycophorin A. *J. Clin. Invest.* **102**, 2173–2179.

Tenore, G., Barili, P., Sabbatini, M., Tayebati, S. K., and Amenta, F. (1997). Postnatal development of dopamine D1-like and D2-like receptors in the rat kidney: A radioligand binding study. *Mech. Ageing. Dev.* **95**, 1–11.

Terris, J., Ecelbarger, C. A., Nielsen, S., and Knepper, M. A. (1996). Long-term regulation of four renal aquaporins in rats. *Am. J. Physiol.* **271**, F414–F422.

Todd-Turla, K. M., Schnermann, J., Fejes-Toth, G., Naray-Fejes-Toth, A., Smart, A., Killen, P. D., and Briggs, J. P. (1993). Distribution of mineralocorticoid and glucocorticoid receptor mRNA along the nephron. *Am. J. Physiol.* **264**, F781–F791.

Traebert, M., Lotscher, M., Aschwanden, R., Ritthaler, T., Biber, J., Murer, H., and Kaissling, B. (1999). Distribution of the sodium/phosphate transporter during postnatal ontogeny of the rat kidney. *J. Am. Soc. Nephrol.* **10**, 1407–1415.

Trimble, M. E. (1970). Renal response to solute loading in infant rats: Relation to anatomical development. *Am. J. Physiol.* **219**, 1089–1097.

Trohler, U., Bonjour, J. P., and Fleisch, H. (1976). Inorganic phosphate homeostasis. Renal adaptation to the dietary intake in intact and thyroparathyroidectomized rats. *J. Clin. Invest.* **57**, 264–273.

Tsang, R. C., Light, I. J., Sutherland, J. M., and Kleinman, L. I. (1973). Possible pathogenetic factors in neonatal hypocalcemia of prematurity. The role of gestation, hyperphosphatemia, hypomagnesemia, urinary calcium loss, and parathormone responsiveness. *J. Pediatr.* **82**, 423–429.

Tsuchida, S., Matsusaka, T., Chen, X., Okubo, S., Niimura, F., Nishimura, H., Fogo, A., Utsunomiya, H., Inagami, T., and Ichikawa, I. (1998). Murine double nullizygotes of the angiotensin type 1A and 1B receptor genes duplicate severe abnormal phenotypes of angiotensinogen nullizygotes. *J. Clin. Invest.* **101**, 755–760.

Tsuruoka, S., Kittelberger, A. M., and Schwartz, G. J. (1998). Carbonic anhydrase II and IV mRNA in rabbit nephron segments: stimulation during metabolic acidosis. *Am. J. Physiol.* **274**, F259–F267.

Tsuruoka, S., and Schwartz, G. J. (1997). Metabolic acidosis stimulates H$^+$ secretion in the rabbit outer medullary collecting duct (inner stripe) of the kidney. *J. Clin. Invest.* **99**, 1420–1431.

Tsuruoka, S., and Schwartz, G. J. (1998). HCO$_3^-$ absorption in rabbit outer medullary collecting duct: Role of luminal carbonic anhydrase. *Am. J. Physiol.* **274**, F139–F147.

Tsuruoka, S., and Schwartz, G. J. (1999). Mechanisms of HCO$_3^-$ secretion in the rabbit connecting segment. *Am. J. Physiol.* **277**, F567–F574.

Tsuruoka, S., Swenson, E. R., Petrovic, S., Fujimura, A., and Schwartz, G. J. (2001). Role of basolateral carbonic anhydrase in proximal tubular fluid and bicarbonate absorption. *Am. J. Physiol. Renal. Physiol.* **280**, F146–F154.

Tucker, B. J., and Blantz, R. C. (1977). Factors determining superficial nephron filtration in the mature, growing rat. *Am. J. Physiol.* **232**, F97–F104.

Tufro, A., Norwood, V. F., Carey, R. M., and Gomez, R. A. (1999). Vascular endothelial growth factor induces nephrogenesis and vasculogenesis. *J. Am. Soc. Nephrol.* **10**, 2125–2134.

Tufro-McReddie, A., Harrison, J. K., Everett, A. D., and Gomez, R. A. (1993). Ontogeny of type 1 angiotensin II receptor gene expression in the rat. *J. Clin. Invest.* **91**, 530–537.

Tulassay, T., Rascher, W., Seyberth, H. W., Lang, R. E., Toth, M., and Sulyok, E. (1986). Role of atrial natriuretic peptide in sodium homeostasis in premature infants. *J. Pediatr.* **109**, 1023–1027.

Tuvdad, F., McNamara, H., and Barnett, H. (1954). Renal response of premature infants to administration of bicarbonate and potassium. *Pediatrics* **13**, 4–16.

Ullrich, K. J., and Papavassiliou, F. (1981). Bicarbonate reabsorption in the papillary collecting duct of rats. *Pflug. Arch.* **389**, 271–275.

Ulmann, A., Menard, J., and Corvol, P. (1975). Binding of glycyrrhetinic acid to kidney mineralocorticoid and glucocorticoid receptors. *Endocrinology* **97**, 46–51.

Van Acker, K. J., Scharpe, S. L., Deprettere, A. J., and Neels, H. M. (1979). Renin-angiotensin-aldosterone system in the healthy infant and child. *Kidney Int.* **16**, 196–203.

van den Anker, J. N., Hop, W. C., de Groot, R., van der Heijden, B. J., Broerse, H. M., Lindemans, J., and Sauer, P. J. (1994). Effects of prenatal exposure to betamethasone and indomethacin on the glomerular filtration rate in the preterm infant. *Pediatr. Res.* **36**, 578–581.

van Lieburg, A. F., Verdijk, M. A., Knoers, V. V., van Essen, A. J., Proesmans, W., Mallmann, R., Monnens, L. A., van Oost, B. A., van Os, C. H., and Deen, P. M. (1994). Patients with autosomal nephrogenic diabetes insipidus homozygous for mutations in the aquaporin 2 water-channel gene. *Am. J. Hum. Genet.* **55**, 648–652.

Vane, J. R., Bakhle, Y. S., and Botting, R. M. (1998). Cyclooxygenases 1 and 2. *Annu. Rev. Pharmacol. Toxicol.* **38**, 97–120.

Varille, V. A., Nakamura, K. T., McWeeny, O. J., Matherne, G. P., Smith, F. G., and Robillard, J. E. (1989). Renal hemodynamic response to atrial natriuretic factor in fetal and newborn sheep. *Pediatr. Res.* **25**, 291–294.

Vaughn, D., Kirschbaum, T. H., Bersentes, T., Dilts, P. V., Jr., and Assali, N. S. (1968). Fetal and neonatal response to acid loading in the sheep. *J. Appl. Physiol.* **24**, 135–141.

Vehaskari, V. M. (1994). Ontogeny of cortical collecting duct sodium transport. *Am. J. Physiol.* **267**, F49–F54.

Vehaskari, V. M., Hempe, J. M., Manning, J., Aviles, D. H., and Carmichael, M. C. (1998). Developmental regulation of ENaC subunit mRNA levels in rat kidney. *Am. J. Physiol.* **274**, C1661–C1666.

Veille, J. C., Hanson, R. A., Tatum, K., and Kelley, K. (1993). Quantitative assessment of human fetal renal blood flow. *Am. J. Obstet. Gynecol.* **169**, 1399–1402.

Verrey, F. (1995). Transcriptional control of sodium transport in tight epithelial by adrenal steroids. *J. Membr. Biol.* **144**, 93–110.

Verri, T., Markovich, D., Perego, C., Norbis, F., Stange, G., Sorribas, V., Biber, J., and Murer, H. (1995). Cloning of a rabbit renal Na-Pi cotransporter, which is regulated by dietary phosphate. *Am. J. Physiol.* **268**, F626–F633.

Walker, D. W., and Mitchell, M. D. (1978). Prostaglandins in urine of foetal lambs. *Nature* **271**, 161–162.

Walker, D. W., and Mitchell, M. D. (1979). Presence of thromboxane B2 and 6-keto-prostaglandin F1 alpha in the urine of fetal sheep. *Prostaglandins Med.* **3**, 249–250.

Walker, P., Dubois, J. D., and Dussault, J. H. (1980). Free thyroid hormone concentrations during postnatal development in the rat. *Pediatr. Res.* **14**, 247–249.

Wall, S. M., Flessner, M. F., and Knepper, M. A. (1991). Distribution of luminal carbonic anhydrase activity along rat inner medullary collecting duct. *Am. J. Physiol.* **260**, F738–F748.

Wall, S. M., Sands, J. M., Flessner, M. F., Nonoguchi, H., Spring, K. R., and Knepper, M. A. (1990). Net acid transport by isolated perfused inner medullary collecting ducts. *Am. J. Physiol.* **258**, F75–F84.

Wall, S. M., Truong, A. V., and DuBose, T. D., Jr. (1996). H$^+$-K$^+$-ATPase mediates net acid secretion in rat terminal inner medullary collecting duct. *Am. J. Physiol.* **271**, F1037–F1044.

Wallace, K. B., Hook, J. B., and Bailie, M. D. (1980). Postnatal development of the renin-angiotensin system in rats. *Am. J. Physiol.* **238**, R432–R437.

Wamberg, S., Kildeberg, P., and Engel, K. (1976). Balance of net base in the rat. II. Reference values in relation to growth rate. *Biol. Neonate* **28**, 171–190.

Wang, T., and Giebisch, G. (1996). Effects of angiotensin II on electrolyte transport in the early and late distal tubule in rat kidney. *Am. J. Physiol.* **271**, F143–F149.

Wang, W. H. (1994). Two types of K[+] channel in thick ascending limb of rat kidney. *Am. J. Physiol.* **267**, F599–F605.

Wang, W. H., Schwab, A., and Giebisch, G. (1990). Regulation of small-conductance K[+] channel in apical membrane of rat cortical collecting tubule. *Am. J. Physiol.* **259**, F494–F502.

Wang, Z. M., and Celsi, G. (1993). Glucocorticoids differentially regulate the mRNA for Na[+]-K[+]-ATPase isoforms in infant rat heart. *Pediatr. Res.* **33**, 1–4.

Wang, Z. M., Yasui, M., and Celsi, G. (1994). Glucocorticoids regulate the transcription of Na[+]-K[+]-ATPase genes in the infant rat kidney. *Am. J. Physiol.* **267**, C450-C455.

Watanabe, S., Matsushita, K., McCray, P. B., and Stokes, J. B. (1999). Developmental expression of the epithelial Na[+] channel in kidney and uroepithelia. *Am. J. Physiol.* **276**, F304–F314.

Weber, S., Hoffmann, K., Jeck, N., Saar, K., Boeswald, M., Kuwertz-Broeking, E., Meij, II, Knoers, N. V., Cochat, P., Sulakova, T., Bonzel, K. E., Soergel, M., Manz, F., Schaerer, K., Seyberth, H. W., Reis, A., and Konrad, M. (2000). Familial hypomagnesaemia with hypercalciuria and nephrocalcinosis maps to chromosome 3q27 and is associated with mutations in the PCLN-1 gene. *Eur. J. Hum. Genet.* **8**, 414–422.

Webster, S. K., and Haramati, A. (1985). Developmental changes in the phosphaturic response to parathyroid hormone in the rat. *Am. J. Physiol.* **249**, F251–F255.

Wei, Y. F., Rodi, C. P., Day, M. L., Wiegand, R. C., Needleman, L. D., Cole, B. R., and Needleman, P. (1987). Developmental changes in the rat atriopeptin hormonal system. *J. Clin. Invest.* **79**, 1325–1329.

Weiner, I. D., New, A. R., Milton, A. E., and Tisher, C. C. (1995). Regulation of luminal alkalinization and acidification in the cortical collecting duct by angiotensin II. *Am. J. Physiol.* **269**, F730–F738.

Weisberg, H. F. (1982). Acid-base pathophysiology in the neonate and infant. *Ann. Clin. Lab. Sci.* **12**, 245–253.

Weisbrot, I., LS, J., Prince, C., Holaday, D., and Apgar, V. (1958). Acid-base homeostasis in the newborn infant during the first 24 hr of life. *J. Pediatr.* **53**, 395–403.

Weitzman, R. E., Fisher, D. A., Robillard, J., Erenberg, A., Kennedy, R., and Smith, F. (1978). Arginine vasopressin response to an osmotic stimulus in the fetal sheep. *Pediatr. Res.* **12**, 35–38.

Welch, W. J., Wilcox, C. S., and Thomson, S. C. (1999). Nitric oxide and tubuloglomerular feedback. *Semin. Nephrol.* **19**, 251–262.

Werner, A., Kempson, S. A., Biber, J., and Murer, H. (1994a). Increase of Na/Pi-cotransport encoding mRNA in response to low Pi diet in rat kidney cortex. *J. Biol. Chem.* **269**, 6637–6639.

Werner, A., Moore, M. L., Mantei, N., Biber, J., Semenza, G., and Murer, H. Cloning and expression of cDNA for a Na/Pi cotransport system of kidney cortex.

Werner, A., Murer, H., and Kinne, R. K. (1994b). Cloning and expression of a renal Na-Pi cotransport system from flounder. *Am. J. Physiol.* **267**, F311–F317.

Weston, P., Brinkmann, C. R., 3rd, Ladner, C., Kirschbaum, T. H., and Assali, N. S. (1970). Fetal and neonatal response to base loading. *Biol. Neonate.* **16**, 261–277.

White, K. E., Gesek, F. A., Reilly, R. F., and Friedman, P. A. (1998). NCX1 Na/Ca exchanger inhibition by antisense oligonucleotides in mouse distal convoluted tubule cells. *Kidney Int.* **54**, 897–906.

White, P. C., Agarwal, A. K., Nunez, B. S., Giacchetti, G., Mantero, F., and Stewart, P. M. (2000). Genotype-phenotype correlations of mutations and polymorphisms in HSD11B2, the gene encoding the kidney isozyme of 11beta-hydroxysteroid dehydrogenase. *Endocr. Res.* **26**, 771–780.

Whorwood, C. B., Sheppard, M. C., and Stewart, P. M. (1993). Licorice inhibits 11 beta-hydroxysteroid dehydrogenase messenger ribonucleic acid levels and potentiates glucocorticoid hormone action. *Endocrinology* **132**, 2287–2292.

Wiederkehr, M. R., Di Sole, F., Collazo, R., Quinones, H., Fan, L., Murer, H., Helmle-Kolb, C., and Moe, O. W. (2001). Characterization of acute inhibition of Na/H exchanger NHE-3 by dopamine in opossum kidney cells. *Kidney Int.* **59**, 197–209.

Wilcox, C. S., Welch, W. J., Murad, F., Gross, S. S., Taylor, G., Levi, R., and Schmidt, H. H. (1992). Nitric oxide synthase in macula densa regulates glomerular capillary pressure. *Proc. Natl. Acad. Sci. USA* **89**, 11993–11997.

Wingo, C. S., and Smolka, A. J. (1995). Function and structure of H-K-ATPase in the kidney. *Am. J. Physiol.* **269**, F1–F16.

Winkler, C. A., Kittelberger, A. M., Watkins, R. H., Maniscalco, W. M., and Schwartz, G. J. (2001). Maturation of carbonic anhydrase IV expression in rabbit kidney. *Am. J. Physiol. Renal. Physiol.* **280**, F895–F903.

Winther, J. B., Hoskins, E., Printz, M. P., Mendoza, S. A., Kirkpatrick, S. E., and Friedman, W. F. (1980). Influence of indomethacin on renal function in conscious newborn lambs. *Biol. Neonate* **38**, 76–84.

Wintour, E. M., Alcorn, D., Butkus, A., Congiu, M., Earnest, L., Pompolo, S., and Potocnik, S. J. (1996). Ontogeny of hormonal and excretory function of the meso- and metanephros in the ovine fetus. *Kidney Int.* **50**, 1624–1633.

Wintour, E. M., Earnest, L., Alcorn, D., Butkus, A., Shandley, L., and Jeyaseelan, K. (1998). Ovine AQP1: cDNA cloning, ontogeny, and control of renal gene expression. *Pediatr. Nephrol.* **12**, 545–553.

Wittner, M., Desfleurs, E., Pajaud, S., Moine, G., Simeone, S., de Rouffignac, C., and Di Stefano, A. (1997). Calcium and magnesium transport in the cortical thick ascending limb of Henle's loop: influence of age and gender. *Pflug. Arch.* **434**, 451–456.

Wittner, M., di Stefano, A., Wangemann, P., Nitschke, R., Greger, R., Bailly, C., Amiel, C., Roinel, N., and de Rouffignac, C. (1988). Differential effects of ADH on sodium, chloride, potassium, calcium and magnesium transport in cortical and medullary thick ascending limbs of mouse nephron. *Pflug. Arch.* **412**, 516–523.

Woda, C., Mulroney, S. E., Halaihel, N., Sun, L., Wilson, P. V., Levi, M., and Haramati, A. (2001a). Renal tubular sites of increased phosphate transport and NaPi-2 expression in the juvenile rat. *Am. J. Physiol. Regul. Integr. Comp. Physiol.* **280**, R1524–R1533.

Woda, C. B., Bragin, A., Kleyman, T. R., and Satlin, L. M. (2001b). Flow-dependent K[+] secretion in the cortical collecting duct is mediated by a maxi-K channel. *Am. J. Physiol. Renal. Physiol.* **280**, F786–F793.

Woda, C. W., Mulroney, S. E., Levi, M., Halaihel, N., and Haramati, A. (1999). Nephron sites of phosphate reabsorption in the juvenile rat. *J. Am. Soc. Nephr.* **10**, 614A.

Wong, N. L., Dirks, J. H., and Quamme, G. A. (1983). Tubular reabsorptive capacity for magnesium in the dog kidney. *Am. J. Physiol.* **244**, F78-83.

Woods, L. L., Cheung, C. Y., Power, G. G., and Brace, R. A. (1986). Role of arginine vasopressin in fetal renal response to hypertonicity. *Am. J. Physiol.* **251**, F156–F163.

Yamada, N. (1970). Respiratory environment and acid-base balance in the developing fetus.

Yamaji, T., Hirai, N., Ishibashi, M., Takaku, F., Yanaihara, T., and Nakayama, T. (1986). Atrial natriuretic peptide in umbilical cord blood: Evidence for a circulating hormone in human fetus. *J. Clin. Endocrinol. Metab.* **63**, 1414–1417.

Yamamoto, T., Sasaki, S., Fushimi, K., Ishibashi, K., Yaoita, E., Kawasaki, K., Fujinaka, H., Marumo, F., and Kihara, I. (1997). Expression of AQP family in rat kidneys during development and maturation. *Am. J. Physiol.* **272**, F198–F204.

Yasoshima, K., Satlin, L. M., and Schwartz, G. J. (1992). Adaptation of rabbit cortical collecting duct to in vitro acid incubation. *Am. J. Physiol.* **263**, F749–F756.

Yasui, M., Marples, D., Belusa, R., Eklof, A. C., Celsi, G., Nielsen, S., and Aperia, A. (1996). Development of urinary concentrating capacity: Role of aquaporin-2. *Am. J. Physiol.* **271**, F461–F468.

Yasui, M., Zelenin, S. M., Celsi, G., and Aperia, A. (1997). Adenylate cyclase-coupled vasopressin receptor activates AQP2 promoter via a dual effect on CRE and AP1 elements. *Am. J. Physiol.* **272**, F443–F450.

Yeh, T. F., Raval, D., John, E., and Pildes, R. S. (1985). Renal response to frusemide in preterm infants with respiratory distress syndrome during the first three postnatal days. *Arch. Dis. Child.* **60**, 621–626.

Zhang, M. Z., Wang, J. L., Cheng, H. F., Harris, R. C., and McKanna, J. A. (1997). Cyclooxygenase-2 in rat nephron development. *Am. J. Physiol.* **273**, F994–F1002.

Zhang, Y., Norian, J. M., Magyar, C. E., Holstein-Rathlou, N. H., Mircheff, A. K., and McDonough, A. A. (1999). In vivo PTH provokes apical NHE3 and NaPi2 redistribution and Na-K- ATPase inhibition. *Am. J. Physiol.* **276**, F711–F719.

Zhou, H., Tate, S. S., and Palmer, L. G. (1994). Primary structure and functional properties of an epithelial K channel. *Am. J. Physiol.* **266**, C809–C824.

Zhou, X., Lynch, I. J., Xia, S. L., and Wingo, C. S. (2000). Activation of H^+-K^+-ATPase by CO^2 requires a basolateral Ba^{2+}-sensitive pathway during K restriction. *Am. J. Physiol. Renal. Physiol.* **279**, F153–F160.

Zink, H., and Horster, M. (1977). Maturation of diluting capacity in loop of Henle of rat superficial nephrons. *Am. J. Physiol.* **233**, F519–F524.

Zolotnitskaya, A., and Satlin, L. M. (1999). Developmental expression of ROMK in rat kidney. *Am. J. Physiol.* **276**, F825–F836.

Zuo, Q., Claveau, D., Hilal, G., Leclerc, M., and Brunette, M. G. (1997). Effect of calcitonin on calcium transport by the luminal and basolateral membranes of the rabbit nephron. *Kidney Int.* **51**, 1991–1999.

19

Experimental Methods for Studying Urogenital Development

Kirsi Sainio

I. Introduction

The purpose of the chapter is to give detailed technical instructions on how to undertake experimental work with urogenital material, with the intention of giving those with limited experience the basic tools to initiate experimental work in the area. The instructions are thus much more detailed than those usually given in research publications and include methods, reagents and additional information for working with embryonic urogenital tissues. The main references describing these protocols are given at the end of the chapter.

The chapter has three main sections. First, instruction are given on how, where and at what time point of embryonic development the urogenital organs can be dissected out. The second section gives details of how to culture and manipulate urogenital tissues, whereas the third show's how to assay this material, and protocols are given for immunohistochemistry and *in situ* hybridization, as well as for analyzing apoptosis and cell proliferation.

All of these techniques have been used extensively since the mid–1980s in the kidney development laboratories at the Haartman Institute and the Institute of Biotechnology at the University of Helsinki. Many of the culture techniques have already been published (Lehtonen *et al.*, 1995), whereas the whole–mount *in situ* hybridization has been adopted from that of Dr. Illar Pata (University of Tartu, Estonia). The *in situ* hybridization protocol has been written by Dr. Rainer Krebs and incorporates modifications made at the Institute of Biotechnology to the original protocols (Wilkinson and Green, 1990; Wilkinson, 1993).

II. Tissue Dissection and Separation

Most students and researchers seem to find that the most difficult aspect of obtaining early rudiments is not dissection, but actually identifying the tissue of interest. At earlier stages of the mouse development (E9–E10), the urogenital cord covers most of the dorsoventral area of the embryo from the level of forelimb buds to the tail. The shape, location, and pattern of the cord change dramatically as organogenesis proceeds, with some areas of the urogenital region becoming more apparent, such as gonads and kidney, whereas the others totally disappear, such as most of the mesonephros and Wolffian duct. The situation is further complicated by the region shortening and compacting as it develops. This means that the process of tissue separation depends on the stage of development as something that is apparent at E14, such as gonads or kidneys, are not visible at E9, or look totally different at E11.

A. The Mesonephric Kidney

In the mouse, this starts to develop at late E9.5 (20–25 somites). The first tubular structures are seen at the level of the 8–10th somite where the rudimentary tubules of the pronephros and the most cranial vesicles of the mesonephros are visible. At this point, the Wolffian duct is extending caudally toward the presumptive hindlimb and cloaca. The urogenital block (including Wolffian ducts, genital ridges, mesonephric kidneys, and presumptive metanephric area) can be removed first by cutting the embryo at the level of heart and removing most of the ventral abdomen. By turning the embryo dorsal side down, bilateral rows of small mesonephric tubules can be seen subjacent to the neural tube and somites. To isolate them, dissect away the lateral walls of the abdomen and then turn the embryo and remove the neural tube and as much of the somites as possible. The remaining area covers the mesonephros and the descending nephric duct. If a single mesonephros is needed, the caudal part of the embryo can be bisected longtitudinally with needles before removing the neural tube.

At E11 the urogenital area has moved downward. To remove the mesonephros, cut the embryo below the forelimbs. Remove the heart, lungs, liver, and intestine, and the urogenital area is then visible at the lateral side of the midline, ventral to the neural tube and somites. The mesonephros consists of a cranial part, where tubules are connected to the Wolffian duct and of caudal tubules unconnected to the duct (Chapter 6). The genital ridge is clearly visible in the lateral side of the mesonephros and the metanephric kidneys can be seen in front of the cloaca at the level of hindlimbs. Turn the embryo sideways and remove the neural tube and somites. Keep the embryo still by holding the other hindlimb with your needle. Turn the embryo dorsal side down and remove the abdominal walls and the hindlimbs.

By E13, the urogenital area has further compacted and become wider and can be seen easily, once the liver and intestinal organs have been removed. The genitals have developed so that the seminiferous tubules of the testis are clearly visible; the ovary is amorphic and not as compact as the testis. Metanephric kidneys are moving upward and the presumptive adrenal glands are visible close to the anterior part of the kidney. Caudal mesonephric tubules have started to regress, whereas cranial mesonephric tubules are close to the anterior part of the testis, where they will form the epididymal ducts of the male. Müllerian ducts are visible on the lateral side of the Wolffian duct in both sexes, but will soon disappear from the male embryos.

B. The Early Metanephric Kidney

The metanephric or definitive kidney consists of the ureteric bud and its surrounding metanephric mesenchyme (MM). The metanephros can be first dissected out from the embryo at E11, when it is still uninduced (this can be recognized by the fact that the ureteric bud has not expanded at its tip—this is the bud stage). By E11.5, the ureteric bud has started to branch and has formed a T-bud. At this point the MM has also been induced. If the initial stages of either the ureter or MM are required, bud-stage rudiments should be used.

At these early stages, metanephric kidneys are located in front of the cloaca at the level of the hindlimbs. The cloaca, hindgut, and the dorsal aorta in the midline of the embryo are good landmarks. To obtain the kidneys, the whole urogenital blocks are first removed (see earlier discussion). Once the neural tube, somites, tail, and lateral body walls have been removed, the metanephric kidneys are visible close to the proximal hindlimb. The Wolffian ducts and the kidneys are removed from the block and separated from the midline. Again, single metanephroi can also be removed by bisecting longitudinally the caudal part of the embryo and dissecting out the kidney.

The metanephric kidney at this stage comprises more than just the ureter and the condensed mesenchyme (transparent area), it is also surrounded by uncondensed mesenchyme. The whole mesenchyme is a faintly visible, shoe-like structure within which the ureter is located at the posterior part. If the whole MM is not dissected, the kidney will remain relatively undifferentiated and small in culture. Dissect out more rather than less tissue!

From E12 onward the metanephric kidneys are clearly visible in the posterior part of the embryo, but as development proceeds, they move rostrally. By E13.5, it is usually possible to use fine forceps to pluck out the metanephroi from the ventral surface of caudal embryos, once the gut and gonadal tissues have been removed.

After dissection, remove the tissues from the dish using a Pasteur pipette connected by tubing to a mouth piece and store in fresh Dulbecco's phosphate–buffered saline (PBS) or in culture media (see later) at room temperature.

C. Separating Ureteric Buds from Metanephric Mesenchyme

Dissect out E11–E12 kidney rudiments and keep them in Dulbecco's PBS. Avoid using serum in any step prior to enzymatic treatment. Treat the kidneys in pancreatin–trypsin solution as follows. Thaw 1 ml of ready-made enzyme solution and place the Eppendorf tube on ice (see later). Place kidney rudiments (10–20 may be treated simultaneously) in a tube or small-diameter glass dish. Replace Dulbecco's PBS with pancreatin–trypsin solution. Keep on ice for 80 s and then transfer the kidney rudiments to standard culture medium supplemented with 10 to 20% fetal bovine serum (FBS) for 15 min at room temperature. Gently separate each mesenchyme from its epithelium with needles under a stereomicroscope (Fig. 19.1).

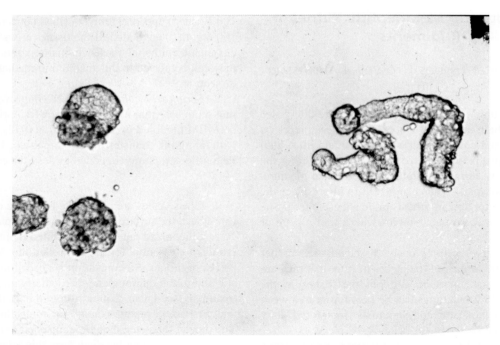

Figure 19.1 E11 mouse kidneys that have been cultured, treated with pancreatin–trypsin, and then separated to give metanephric mesenchyme (left) and ureteric buds (right). From Saxén and Wartiovaara (1966).

Alternatively, MM can simply be dissected free of the ureteric bud with fine needles. This technique yields clean MM, but the buds may be contaminated with small amounts of mesenchyme.

Reagents and Materials Needed for Dissection

1. Disposable, stainless-steel sterile injection needles and a 1-ml syringe are used for dissection. The syringes can be reused.

2. Glass dishes, sterile scissors and forceps, siliconized Pasteur pipettes, microburner, mouthpiece (a plastic micropipette tip with a cut and adequate cotton filling), and tubing for micropipette. Thin micropipettes pulled from the siliconized Pasteur pipettes after heating with the microburner are used for handling smaller tissue rudiments.

3. Dulbecco's PBS, pH 7.2 –7.4 (with Ca^{2+} and Mg^{2+}). Tissues are dissected in this buffer at room temperature under a stereomicroscope.

4. 2.25% pancreatin, 0.75% trypsin in Ca-Mg-free Tyrode's solution, pH 7.4, supplemented with penicillin and streptomycin.

   Trypsin 1:250 (Sigma, cell culture grade, T-4799, store dessicated at 4°C)
   Pancreatin 4xNF [GibcoBRL (Lifetech.) 15725-013, lyophilized, store at 4°C]
   Dissolve to 20 ml sterileH_2O and divide to 1 ml/tube, store then at -20°C

Tyrode's solution (Ca-Mg free) pH 7.4:
8.0 g/liter   NaCl,
0.2 g/liter   KCl,
0.05 g/liter (MW 137,99) $NaH_2PO_4 + H_2O$,
1.0 g/liter (D(+)-glucose, MW 180.16, H_2O free) and glucose,
1.0 g/liter $NaHCO_3$.
Sterile filter with 0.22–μm Millipore, store at 4°C
To make 10 ml of pancreatin-trypsin, pH 7.4, use
0.225 g   trypsin
1.0 ml   pancreatin
(penicillin–streptomycin may be added, if required)
Keep the enzymes on ice and dilute to 10 ml with Tyrode's solution.

Sterilize the solution by passing it through a 0.22–μm Millipore filter. Store in 1-ml aliquots in Eppendorf tubes at -20°C for a week; longer storage should be at -70°C. Enzyme quality can be determined by monitoring the tissues after treatment. If the tissues look smeary, they have either been incubated for too long or (and more likely) the enzymes are no longer working adequately.

Note. The pH may vary from 7.2 to 7.8, but it is important that the solution is roughly neutral. If the pH is higher than 7.8, discard the solution and make a fresh batch, having first checked the pH of the Tyrode's solution (this is affecting the final pH).

III. Culturing Metanephric Kidney Rudiments

A. Urogenital Tissues (Trowell–Type Tissue Culture)

To prepare culture dishes, wash the Nuclepore filters (see later) and cut them into small pieces. Place these pieces on a metal grid in a 3-cm-diameter culture dish and fill it with the medium to just below the level of the grid. This holds the filters in place by surface tension (with too much medium, the filters start to float). When using 1.0-μm filters, place the mat side of the filter upward (this is more adhesive for tissue fragments). Remove any air bubbles with a thin pipette or needle.

Transfer the tissues (previously dissected—see earlier discussion) to the surface of the pieces of filter that they are at the medium–gas interface and put the dishes in the incubator (Fig. 19.2). Cultures can be kept for up to a week in a humidified incubator and should be monitored daily under a stereomicroscope.

Note: Because E9-E10 urogenital tissue is fragile, use a minimum of medium when picking them up and placing them on the grid. At E11 the urogenital tissues are still rather long and fragile. Try to place them dorsal side toward the surface of the filter (gut primordium upward).

Reagents and Materials Needed for Culture

1. *Standard culture Medium.* Eagle's minimum essential medium (MEM) with Earle's salts or, alternatively, Dulbecco's modified minimum essential medium (D-MEM), supplemented with 10% FBS and Glutamax-1 (GibcoBRL). Penicillin (final concentration

100 IU ml^{-1}) and streptomycin (final concentration 100 mg ml^{-1}) are added. If the tissues are not used in enzymatic epithelial–mesenchymal separations (see later), they can be placed to this medium immediately after dissection.

2. *Serum-free cultures.* I-MEM (improved Eagle's minimum essential medium, GibcoBRL) originally based on MEM modified by Richter *et al.*, (1972), supplemented with 50 μg ml^{-1} transferrin (Ekblom *et al.*, 1983). Penicillin–streptomycin added as describe earlier.

Notes

1. Glass dishes are preferable to plastic ones as needles can scratch the surface and cause pieces of plastic to be shed.

2. Siliconized pipettes, used with a mouth piece and tubing, are essential for the controlled moving of small pieces of tissues. Siliconization is needed to avoid tissue sticking to the micropipette, particularly after enzymatic treatment (see later). Cotton filling in both the mouthpiece and the Pasteur pipette is good for controlling the flow. The pipette diameter should be matched to the size of the piece of tissue being handled; very thin edges can easily damage small fragments of tissue. Test Pasteur pipettes carefully: the quality of glass used in the pipettes provided by different manufacturers can be variable and may not make good, thin micropipettes.

3. Prewashed Nuclepore (Whatman) filters should be stored in 70% ethanol and rinsed three times for 5 min in sterile PBS or distilled water before use. Pore sizes vary from 0.1 to 1.0 μm, and while filters with smaller pores are transparent, tissues attach better to filters with larger pores. Larger pore size filters can be used for transfilter cultures (see Fig 19.3).

4. Filters are placed on stainless–steel grids, matched to the size of culture dish used (a good standard size is 30 mm). Small holes can be made in the grids, appropriate to the size of the tissue and filter, so facilitating the monitoring of the tissue during culture. The surface of the grids should be kept as smooth as possible (Fig. 19.2).

5. A humidified 37°C incubator supplied with a 5% CO_2/air mix. A high–quality stereo microscope with transmitted light.

Pretreatment of Nuclepore Filters

Place the filters with forceps to large decanter and wash in 1% Alconox (Alconox Inc., New York) or corresponding cleaner designed for cell and tissue culture purpose. Rinse in *slowly* running tap water for at least 6 h and cover the top of the decanter with gauze. Rinse extensively (at least 10 times in sterile distilled or MilliQ water — this can be poured in and out through the gauze), rinse once in 70 % ethanol, replace the gauze with a proper seal, and store in 70 % ethanol until required.

Note. Filters are fragile, handle them with care!

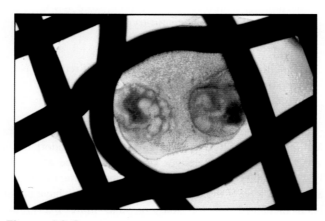

Figure 19.2 Trowell-type tissue culture. For this, pieces of metal grid, bent to give ~3-mm "legs," are placed in a 3-cm-diameter culture dish and medium is added to the top of the grid. Tissue are cultured on fragments of Millipore or Nuclepore filters at the gas–medium interface. Here, two whole E11 mouse kidneys have been recombined with a piece of dorsal spinal cord and cultured on a Nuclepore 0.1-μm filter (phase contrast).

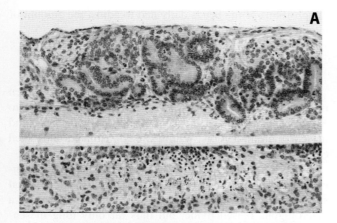

Figure 19.3 Transfilter cultures. (A) Section of a 3-day transfilter (1.0-mm Nuclepore) culture of mouse E11 metanephric mesenchyme (top) and a piece of dorsal E11 spinal cord (bottom). The nephric tubules that have been induced are well differentiated. Hematoxylin-eosin staining, Photograph courtesy of H. Sariola. (B) SEM micrograph of the top surface of a Nuclepore filter overlying a mouse spinal cord cultured for 3 days. Material growing from the filter is mostly neurons. Such neurons are also present in the filter between the spinal cord and the metanephric tubules (Sariola *et al.*, 1988).

B. Transfilter and Tissue Recombination Cultures

Dissect kidney rudiments from E11 bud stage embryos, separate each mesenchyme from its ureteric epithelium with pancreatin–trypsin (see earlier discussion), and keep them in culture medium at 37°C. Dissect a small piece of dorsal spinal cord (Grobstein, 1956) from the same embryo and place it on the matted side of a Nuclepore 1.0 μm filter. Melt 2% agarose solution (with care!) and pipette a small amount of the agarose on the spinal cord with micropipette. This is needed to glue the spinal cord to the filter.

Turn the filter so that the spinal cord is underneath and place it on the grid (ideally, over an enlarged hole in the grid) and then place a piece of mesenchyme (or several) above the spinal cord. After 24 h, the mesenchyme should be fully induced and the spinal cord can be removed carefully by needle. Culture and monitor the mesenchyme for up to 7 days (Fig. 19.3).

The dorsal spinal cord is the only reliable heterologous inducer for transfilter cultures, although other heterologous inducers can be used in tissue recombination experiments. Here the tissues are placed next to one other on the Nuclepore filter and are cultured for up to 7 days.

Additional reagents needed are 2% agarose (Sigma A-4018) in MEM (or D-MEM). Dissolve by boiling, autoclave in glass tubes in 1–ml aliquots, and store at 4°C.

C. Hanging Drop Cultures

Once induced, metanephric mesenchymes can be trypsinized to give a suspension of single cells that will reaggregate in hanging-drop culture and subsequently form tubular structures.

The procedure is as follows: dissect out E11 bud stage kidney rudiments, incubate them with pancreatin–trypsin solution, and separate the mesenchymes (as described earlier), and then transfilter-induce them for 24 h. Remove the spinal cord and make a single cell suspension of the mesenchyme by trypsinizing. Invert the lid from a sterile, tissue culture-grade plastic dish and place drops culture medium on it (maximum 30 μl). Pipette cells into each drop with a fine micropipette using a mouthpiece, minimizing the amount of medium added. Turn the lid carefully so that the drops start to hang (do not move the lid sideways while turning it) and place it on the dish that has been filled to about 30% with either Dulbecco's PBS or culture medium. Culture the drops for up to 4 days, changing the medium in the drop daily if required.

Note. The yield from one mesenchyme is few thousand cells; the epithelium of the ureteric bud contains only 200–300 cells.

D. Bead experiments

Dissect whole urogenital blocks, kidney rudiments, or metanephric mesenchymes and place them on filters on Trowell grids in the standard way. Pipette protein-treated agarose beads (Vainio *et al*, 1991; Sainio *et al.*, 1997) or heparin-coated beads (Kettunen and Thesleff, 1998) onto the appropriate region of the tissue using a micropipette and mouthpiece. Culture for several days and monitor development under a stereomicroscope daily. Beads can be replaced if fresh ones are needed, but this is not normally necessary, as they contain a high dose of protein.

Affi-Gel Blue Agarose (100–200 mesh, diameter 75–100 µm, Bio–Rad Laboratories) or heparin-coated acrylic beads (Sigma H-5263) should be washed three times in sterile PBS in Eppendorf tubes, spun, and placed in a glass petri dish in PBS. Pick up ~100 beads to the Eppendorf tube with a siliconized *thin* micropipette and mouthpiece under stereomicroscopic control. Remove PBS and incubate in 5 µl of concentrated protein solution (100 ng/µl) for 1–2 h at room temperature or, alternatively, for 30 min in 37°C. Pick up the beads from the tube by a thin micropipette, avoiding protein solution contamination to the tissue culture dish, and place in the appropriate place on the rudiment (Figs. 19.4a and 19.4b).

The same beads can be used for up to several days if kept at 4°C in concentrated protein solution (the time depends on the degradation rate of the protein in question). Avoid freezing and thawing.

Excessive amounts of culture medium in the dish can cause beads to move and it is therefore important to monitor the positions of the beads daily and to adjust the amount of medium in the dish if necessary.

E. Culturing Ureteric Buds

Separated ureteric buds cultured on filters soon flatten into single cell layer and these cells lose their ability to express the genes needed for branching morphogenesis and mesenchyme induction. Within 2 days, isolated buds shed epithelial cells and the tissue virtually disappears.

Culturing ureteric buds without metanephric mesenchyme so that they survive and keep their three-dimensional structure thus needs some care. One approach is to use a heterologous mesenchyme, but thus far only lung mesenchyme has been shown to support ureteric branching morphogenesis and gene expression (Kispert *et al.*, 1996; Sainio *et al.*, 1997). The other approach is to modify the culture conditions: buds retain their three-dimensional structure relatively well in hanging-drop cultures (Sainio

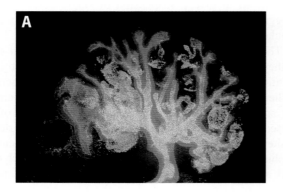
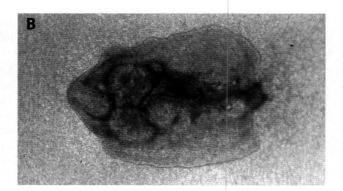
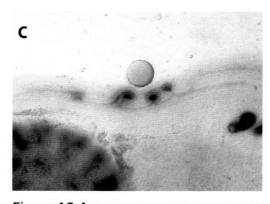
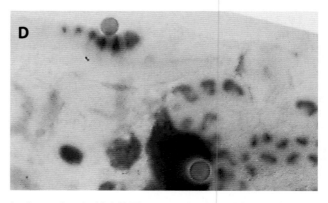

Figure 19.4 Bead experiments. (**A**) A mouse E11 kidney that has been cultured with Affi-Blue agarose bead containing GDNF. Ureter bud growth in the region of the bead (on left) has been stimulated and the branching pattern altered. [Whole-mount fluorescent immunohistochemistry with antibodies to proximal tubule–specific brush border antigens (green) and to cytokeratin 8/18 (Red).] (**B**) Affi-Blue agarose beads have been soaked in GDNF. Multiple ectopic "ureters" are growing from the Wolffian duct (See more in detail (**D**)). (**C**) Whole-mount *in situ* hybridization for glial cell line-derived neurotrophic factor family receptor GFRα1. A mouse E11 day urogenital block has been cultured for 3 days with a heparin-coated bead that has been soaked in GDNF and placed next to the Wolffian duct. Ectopic "ureteric buds" express GFRα1 (mRNA visualized by digoxigenin coupled to UTP with an anti-DIG antibody coupled to alkaline phosphatase). Photograph courtesy of M. Hytönen. (**D**) Whole-mount *in situ* of (**B**) with *ret* (receptor tyrosine kinase) cRNA (digoxigenin visualization). In the metanephric area (below), the bead has induced strong ureteric epithelial growth, it also induces neuronal growth within the mesenchyme. Photograph courtesy of M. Hytönen.

et al., 1997), in collagen gels (Montenano *et al.*, 1991) and Matrigel. These three-dimensional cultures do not, however, seem to support branching morphogenesis of the ureter even with the addition of growth factors such as GDNF (Sainio *et al.*, 1997).

The technique for coculturing ureteric buds with lung mesenchyme is as follows: dissect E11 (bud) or E11.5 (T-bud) kidney rudiments and separate each mesenchyme from its ureteric epithelium with pancreatin–trypsin treatment. Similarly, remove the lung bud from the same embryo and separate out its mesenchyme with pancreatin–trypsin. Place the ureter on a Nuclepore filter and cover the whole bud with the lung tissue to mimic the metanephric mesenchyme–ureteric epithelium structure. Culture with (bud stage) or without (T-bud stage) such growth factors as HGF or GDNF for up to 7 days. The typical concentration for HGF is 25–50 ng/ml and for GDNF is 25–100 ng/ml of culture medium. The protein concentration depends on the protein modification (native protein/recombinant *Escherichia coli* produced protein, baculovirus recombinant protein, etc.) and should be tested separately for each protein.

F. Angiogenesis and Chorioallantoic Cultures

By E11 of mouse development, kidney rudiments already contain endothelial cell precursors, whereas angiogenesis can be triggered relatively easily. Avascular mouse E11 bud stage kidneys can be grafted onto the chorioallantoic membrane of a 7-day-old Leghorn chicken egg where it is vascularized by the chicken allantoic vessels. Angiogenesis can be monitored directly and by endothelial cell markers (Fig. 19.5).

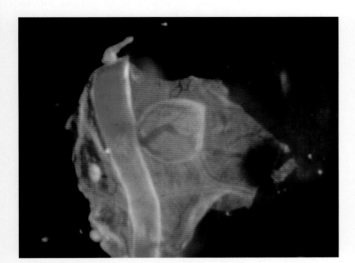

Figure 19.5 Angiogenesis of an initially avascular E11 metanephric kidney after *10* days of culture on the chorioallantoic membrane of a 7-day preincubated Leghorn chicken egg. Photograph courtesy of H. Sariola.

Incubate fertilized eggs on their side in a humidified (~80%) incubator at 37°C, turning them daily. At least 3 h before use, set the eggs with their blunt end up (this end has the air space). Take the eggs out of the incubator and clean them with 70% alcohol. Push a small hole in the top of the shell (with a needle) and, with scissors, cut, starting from the hole, a small window out of the shell. Carefully remove the white shell membrane with forceps to reveal the underlying chorioallantoic membrane. Pipette the dissected kidneys onto the membrane with a small pipette and mouthpiece close to a large vessel and seal the window with a piece of Scotch or Sello-tape (any nontoxic tape will do). Place the eggs window side up and incubate at 37°C for up to 10 days. Then open the seal and transfer the chorionallantoic membrane with the rudiments to a glass dish containing Dulbecco's PBS in which the kidneys can be dissected away from the membrane. For further details, see Ekblom *et al.*, (1982); Sariola *et al.*, (1983).

Note. Small pieces of egg shell that may drop to the chorioallantoic membrane are not harmful, but try to avoid this happening. Remove large pieces with forceps. Do not use eggs whose chorioallantoic membrane is badly bleeding.

G. Inhibition of Protein and Gene Expression *in vitro*

Over the last decade, the development of gene–targeting techniques has enabled gene activity and hence protein production to be manipulated in living animals. It is also possible to disrupt both protein and gene expression *in vitro*, and embryonic kidneys have been one target in such experiments. Because these techniques have provided inconsistent data, a few references rather than reliable protocols are provided.

In case of protein inhibition, specific antibodies can be used *in vitro* to manipulate the protein expression in the embryonic kidney (Sariola *et al.*, 1988). There is, however, only a small set of antibodies that efficiently inhibit their specific epitopes *in vitro* without also causing non-specific effects. Care should therefore be taken to use nonspecific immunoglobulins or control serum to rule out the possible undefined results.

Gene expression can be manipulated *in vitro* by the addition of specific antisense oligonucleotides (Sariola *et al.*, 1991; Sainio *et al.*, 1994), although the possible toxic effects of the chemical modifications that are needed to stabilize such nucleotides may themselves cause problems. The development of new, less toxic modifications or the use of the RNAi technique for mammalian tissues could, in the future, make this approach more relevant. However, the antisense oligonucleotide inhibition technique is reliable in controlled conditions.

IV. Tissue Analysis

A. Whole-Mount Immunohistochemistry

The simplest method of whole tissue staining is based on permeabilizing fixation and prolonged incubation times (Sariola *et al.*, 1988).

Whole tissue explants, attached on Nuclepore filters, are fixed in ice-cold methanol for 5 min in Eppendorf tubes and washed in PBS for 20 min. For antibodies that require aldehyde fixation, samples should be fixed for 1–3 h in 4% paraformaldehyde (PFA; this should be freshly made up, but freezes well). If permeabilization is required (e.g., for antibodies against transcription factors), samples are then washed in PBS with 1% Triton-X for 20 min (with care being taken to avoid the formation of bubbles).

Reagents

Paraformaldehyde (PFA)**:** Dissolve paraformaldehyde in PBS at 65°C in 4% solution or prepare a 20% stock of PFA (convenient, as it requires less freezer space). Use immediately or freeze in aliquots (45 ml) and melt only once.

Note. PFA should be freshly prepared or stored at -20°C, as it deteriorates rapidly in solution. PFA solution should not be boiled and should be heated in a water bath in a fume hood to minimize the damage in case of breakage or spillage. Aldehyde fixatives are very irritating and should be kept in a hood and handled with gloves.

Antibody incubations are done in Eppendorf tubes at 4°C for 16 h (overnight) and antibody volumes should not be less than about 150 µl. The actual time of incubation depends on the location of the antigen: extracellular matrix and cell surface antigens only require a few hours for small tissues (no more than about 2–300 µm thick), whereas cytoplasmic and nuclear antigens may need up to 48 h of incubation at 4°C or room temperature, as may larger samples. Wash extensively with PBS (at least three times for 2 h each; for large tissues, the total washing time may need to be 24 h). Incubate in secondary antibody coupled with a suitable fluorochrome overnight at 4°C. Again wash extensively and mount with Immumount (Shandon; Fig. 19.6). This mounting medium hardens and preserves the fluorescence extremely well (for several months, protected from light). Avoid bubbles during mounting.

Note. Lectins conjugated with various fluorochoromes can be used instead of antibodies to detect specific tubular structures: *Dolichos biflorus*-agglutinin can be used for ureteric epithelium and peanut agglutinin for proximal tubules (see also Laitinen *et al*, 1987; Holthöfer and Virtanen, 1987). It should, however, be noted that glycosylation patterns differ between species and that the pattern of lectin binding may thus be species specific. Use lectins as secondary antibodies so that fluorescence can be detected

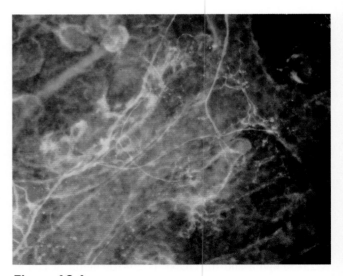

Figure 19.6 Whole-mount immunofluorescence. An E11 mouse kidney cultured for 4 days and stained with L1 neural cell adhesion molecule (green FITC) and laminin (orange TRITC) that respectively highlight neurofilaments and ureter, and epithelial cell basement membrane.

after only a single stain. Wash extensively (see earlier discussion) and mount with Immumount (Shandon). Whole-mount staining can be monitored under conventional fluorescence microscope with epifluorescence or with confocal microscopy.

B. Cell Proliferation and Apoptosis Assays

The nucleotide analog–bromodeoxyuridine (BrdU) is the best way of monitoring the rate of cell proliferation during tissue culture experiments, as it can be used to stain all cells that were in S phase while the BrdU was present.

Tissue fragments such as dissected urogenital blocks, kidney rudiments, and transfilter cultures are cultured in Trowell-type tissue culture in the presence of exogenous growth factors, beads, etc. 5-Bromo-2-deoxyuridine (Amersham; stock solution 13 µg/ml) is added to the culture medium in 1:1000 (1.3 ng/ml) and incorporation is allowed to take place for 30 min to 1 h in the incubator. The samples are then fixed in 95% ethanol/5% acetic acid and stained with the mouse monoclonal antibody against BrdU (Amersham) in whole-mount immunofluorescence. The number of cells and their nuclear morphology can be determined independently by a Hoechst 33342 fluorochrome (2 µg/ml) visible with UV light. This stains the nuclei when the fluorochrome is added to the last PBS wash. If a three-channel confocal microscope with long-wave (far red) detection is being used, a useful nuclear marker is TOPO-3-Iodide (Molecular Probes) used at 1:200 dilution.

Alternatively, pregnant mice at almost any day of gestation can be injected intraperitoneally with BrdU (1 ml/100 g of 3 mg/ml BrdU). The label should be incorporated for

2–3 h and the embryos or their tissues fixed with Bouin's fixative (see later) and processed for paraffin sectioning and immunohistochemistry in the standard way.

Fixation Fix whole tissue or material cultured on a Nuclepore filter with 95% EthOH/5% acetic acid (ice cold) for 5–10 min on ice (depending on the size of the tissues). Rinse at least three times for 5 min each in PBS and use for whole-mount staining or frozen/paraffin histology. Alternatively, use Bouin's fixative as it gives excellent histological detail:

750 ml saturated picric acid, aqueous solution (1.2%), 250 ml formaldehyde (~ 37% – 40%), and 50 ml acetic acid, glacial.

Fix cultured tissue samples for 1–2 h and larger tissues or whole embryos for 12 h (overnight is the maximum time). Picric acid is then washed out in 50–70% ethanol until the ethanol remains colorless. The tissues are processed for paraffin histology.

Picric acid can also be dissolved in absolute ethanol to give ethanol-Bouin, and this fixative works well for BrdU-labeling using the standard fixation times and washing procedure for normal Bouin's.

Note. Bouin-fixed tissues can only be used for bright-field immunohistochemistry and histology, as it is not suitable for immunofluorescence or *in situ* hybridization. However, while the detection of some antigens is excellent in this fixative, other antigens/epitopes do not tolerate picric acid treatment. As ever, the optimal fix for each antigen has to be determined experimentally.

C. TUNEL-Staining Technique for Apoptosis

To visualize apoptotic cells in cultured kidneys, the commercial ApopTag labeling kit (Intergen) based on the TUNEL technique (*in situ* terminal transferase end labeling of fragmented DNA; Gacrieli *et al.*, 1992) is very useful. Samples are fixed as whole mounts in 10% formalin (or 4% PFA) and postfixed in 95% ethanol/5% acetic acid for 30 min. Thereafter, the manufacturer's step-by-step protocol can be followed, but with increased washing times (see whole-mount immunohistochemistry). Fluorescent marking is often appropriate here (Fig. 19.7).

V. *In Situ* Hybridization of mRNA

A. Whole-Mount *in situ* hybridization for Embryos and Cultured Embryonic Tissues

Whole-mount *in situ* hybridization requires the use of nonradioactive probes. Several types of probes and labels are available that include single-stranded DNA probes, oligonucleotides, and RNA probes generated by *in vitro*

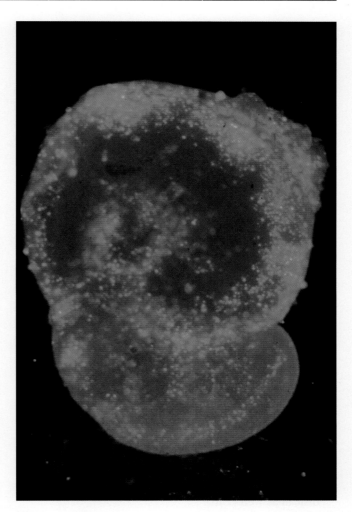

Figure 19.7 Apoptosis detected by TUNEL staining in a mouse E11 metanephric kidney cultured for 2 days. Apoptotic cells (green) are mainly located to the perifery of the kidney (whole-mount immunofluoresence) Photograph courtesy of A. Raatikainen-Ahokas.

transcription. This protocol focuses on the use of RNA probes, as these can be synthesized easily in large amounts and offer the best sensitivity and specificity, as the unbound probe can be digested by the RNase A treatment. The hapten label of choice is digoxigenin coupled to UTP (DIG-11-UTP, from Boehringer-Mannheim). The bound probe is visualized by incubating the embryos with anti-DIG-antibody coupled to alkaline phosphatase, followed by incubation in substrate for alkaline phosphatase that yields an insoluble colour product (Figs. 19.4c and 19.4d). Other hapten labels (e.g., biotin and fluorescein) have been used in this system with lower efficacy, but different labels offer the possibility of colocalization studies in double-labeling experiments. The protocol described here is based on that of David Wilkinson (1993) with some modifications made by Dr. Illar Pata (University of Tartu, Estonia) and our own laboratory. This protocol has been used successfully for *in situ* hybridization work on urogenital tissues, both freshly removed and after culture on Nuclepore filters.

Synthesis of the RNA Probe (Done in Advance)

Mix the reagents in the following order at room temperature: Nuclease-free H_2O to final volume 20 μl, 4 μl 5× transcription buffer (Promega), 2 μl nucleotide mix, 1–3 μl linearized plasmid (1 μg/ml), 0.5 μl RNase inhibitor (Promega), 20 units T7 or T3 RNA polymerase

1. Incubate at 37°C for 2 h.
2. Analyse a 1-μl aliquot of the reaction on 1% agarose/TBE gel. Run the gel for 15 min at 5–10 V/cm. A RNA band at least 10-fold more intense than the plasmid should be seen, indicating that 10–15 μg of probe has been synthesized.
3. Add 1 μl of RNase free-DNase I and incubate at 37°C for 15 min.
4. Add 100 μl H_2O, 10 μl 4 M LiCl, and 300 μl ethanol. Keep at -20°C (or -70°C) for 30 min.
5. Spin in microfuge at 4°C for 10 min, wash the pellet once with 70% ethanol, and air dry.
6. Redissolve the pellet in 50% formamide at about 0.1 μg/μl (100 μl) and store at 20°C.

Reagents

1. 5× transcription buffer is provided ready-made by Promega. Boehringer Mannheim provides 10× buffer that contains 400 mM Tris–HCl, pH 8.0, 60 mM MgCl2, 100 mM dithioerythritol (DTE), 20 mM spermidine, 100 mM NaCl, 1 unit/ml RNase inhibitor. This can be used instead as well (2 μl).
2. Nucleotide mix: 10 mM ATP, 10 mM CTP, 10 mM GTP, 6.5 mM UTP, and 3.5 mM DIG-11-UTP (Boehringer-Mannheim).

Notes. The optimal size of the probe is 1 kb but can be up to 4 kb and this does not need size reduction by hydrolysis. During plasmid DNA isolation procedures for template preparation, use low concentrations of RNase A (0.5–1 μg/ml). After the restriction digest, linearized plasmid DNA should be purified with two phenol extractions, followed by chloroform treatment and ethanol precipitation. Avoid using SP6 RNA polymerase (this gives low yield in transcription). The probe is stable in 50% formamide for months.

Prehybridization Treatments of Samples

1. Whole embryos should be dissected out in Dulbecco under the dissecting microscope and their extraembryonic membranes removed (they can be used for genotyping).
2. Fix material in 4% PFA overnight at 4°C.

Notes. It is beneficial to keep the embryos in culture medium provided with 10% FCS in the cell culture

incubator for 15–30 min so that the heart starts to beat. This restores the mRNA.

Prior to fixation of whole embryos, the cavities should be opened (to avoid the trapping of reagents) by rupturing the membrane of the fourth ventricle and puncturing the heart and forebrain vesicle of the embryo.

If embryos have to be processed separately, this is done most conveniently in a 24-well cell culture plate using 1 ml of reagents. It is possible to pool up to five E9.5 or two E10.5 embryos in a single well. Similarly, 10–15 embryos can be processed together in a 15-ml centrifuge tube.

3. Urogenital tissues on Nuclepore filters: Fix the samples on filters briefly in ice cold methanol, rinse in PBS, and postfix in 4% PFA for 1–3 h.
4. Wash the samples with PBT.
5. Dehydrate the embryos in 25, 50, and 75% methanol in PBT, and then twice with 100% methanol (5 min each).

Unless otherwise stated, washes are for 5 min at room temperature with gentle rocking.

Embryos older than 11 days should be dissected thoroughly, and washing and incubation steps should be extended to 10 min.

Samples can be stored in methanol at -20°C for a few days but for prolonged storage process the embryos to the prehybridization mix without SDS (step 13). Omitting the methanol step generates air bubbles within the embryos during subsequent procedures.

Prehybridization (Day 1)

1. Rehydrate the embryos by the methanol/PBT series in reverse and then wash twice in PBT.
2. Bleach with 6% hydrogen peroxide in PBT for 1 h.
3. Wash three times with PBT.
4. Treat with 10 mg/ml proteinase K in PBT (for E8.5, 5 min; E9.5, 8 min; E 10.5, 12 min).
5. Wash in freshly prepared 2 mg/ml glycine in PBT and then twice in PBT.
6. Refix in 4% PFA/0.2% glutaraldehyde for 20 min.
7. Wash twice with PBT.
8. Wash with prehybridization mix and transfer the samples to Eppendorf 2-ml safe-lock tubes.

Note. Samples can be stored for months at -20°C in prehybridization mix without SDS.

9. Prehybridize at 70°C for at least 1 h. For hybridizations and high-temperature washings, use a shaking water bath.
10. Heat the aliquot of probe at 80°C for 5 min and chill on ice. Mix the probe with hybridization mix at a concentration of 1 μg/ml. Replace the prehybridization mix with 0.4 ml hybridization mix per vial.
11. Incubate at 70°C overnight with gentle rocking.

Reagents

1. PBT: PBS containing 0.1% Tween-20.
2. DEPC treatment of water and solutions: DEPC (diethylpyrocarbonate) is used to avoid any RNase contamination in all liquids prior to posthybridization washes, except for Tris solutions. Tris based solutions have to be made to inactivated DEPC-H_2O. DEPC is added to the solutions (1:1000), mixed well, allowed to stand, and then inactivated by autoclaving.

Note. Do not use any DEPC solution that bubbles after the bottle is opened as it will already have started to hydrolyze to ethanol and CO_2 and is no longer usable. The manufacturer will replace the bottle by request. DEPC is carcinogenic, irritating, and should be handled with care and only in the hood.

3. Proteinase K. This digests proteins and thus increases the accessibility of the target RNA.

Proteinase K stock: 10 mg/ml, store in aliquots at –20°C
Proteinase K buffer stock solutions:
1 *M* Tris–HCl, pH 8.0; sterilized by autoclaving.
0.5 *M* EDTA, pH 8.0; DEPC treated.

Notes. Proteinase K may be kept at room temperature briefly without harm, but should be replaced in the freezer as quickly as possible.

Two racks may be treated with the same solution if they follow each other "immediately."

The amount of proteinase K to be used depends on the size of the tissue (maximum 30 µg/ml). The reaction may take place at 37°C or the reaction time may be prolonged. The optimal conditions have to be found by trial and error.

4. Glycine: A stock solution (100 mg/ml) can be prepared and stored at room temperature.
5. Prehybridization mix: 50% formamide, 5×SSC (pH 5), 1% SDS, 50 µg/ml yeast tRNA, and 50 µg/ml heparin. The hybridization solution containing the probe can be reused. Keep at -20°C.
6. 20×SSC pH 5: add 1 ml of 1 *M* citric acid to 40 ml 20×SSC pH 7.
7. Heparin: The stock solution (50 mg/ml) is kept at 4°C.

Posthybridization Washings and Antibody Incubation (Day 2)

1. Wash the samples 3×30 min with preheated solution 1 at 70°C with rocking. During the last 30 min wash, turn the temperature down to 65°C.
2. Wash with preheated solution 2 at 65°C 3×30 min
3. Wash with maleic acid buffer (MAB) + 0.1% Tween 3×5 min.
4. Preblock the samples by incubating with MAB + 2% Boehringer Mannheim blocking reagent (BBR) + 10% sheep serum. (Heat MAB + BBR briefly to dissolve, will remain cloudy. Add serum after heating.) Incubate for 1–3 h (or overnight) at room temperature.

5. Replace with 0.7 ml of MAB + 2% BBR + 1% sheep serum + anti DIG-antibody coupled to alkaline phosphatase (Boehringer), final dilution 1:2000. Rock gently overnight at 4°C.

Reagents

Solution 1: 50% formamide, 5× SSC, pH 4,5, 1% SDS
Solution 2: 50% formamide, 2× SSC, pH 4,5
MAB: 100 m*M* maleic acid, 150 m*M* NaCl, pH 7.5, 2 m*M* levamisole

Postantibody Washes and Histochemistry (Day 3)

1. Wash 3×5 min in MABT (MAB + 0.1% Tween).
2. Wash five times for 1 h with MABT and then leave to wash overnight on a rotator at 4°C.

Postantibody Washes and Histochemistry Continued (Day 4)

3. Wash three times for 10 min in NTMT at room temperature.
4. Transfer embryos to the 24-well culture dish. Add either 0.4 ml BM purple + 2 m*M* levamisoleor or 1 ml NTMT containing 4.5 µl NBT (Boehringer Mannheim) and 3.5 ml BCIP (Boehringer Mannheim) per each well.

Keep in the dark. Rock gently for the first 20 min. Monitor the color development after 30 min and then every hour. Replace the staining solution as it turns pink.

A strong signal usually appears in 30 min and staining for 2–4 h is optimal. Detecting weak expression may take 10–48 h staining, but embryos should not be left in the staining solution without regular monitoring. If the stain has not come up by the end of the day, wash out the substrate and restart the color reaction the next day.

5. After obtaining satisfactory results, wash in PBT overnight. For photography, samples are post fixed in 4% PFA and can be transferred to 80% glycerol in PBT through the row (30 and 50% glycerol, 1 hr, and 80% for several hours), which makes embryos more transparent. For storage, keep in PBT at 4°C.

Reagents

1. BM purple is manufactured by Roche
2. NTMT is 100 m*M* Tris–HCl, pH 9.5, 100 m*M* NaCl, 50 m*M* MgCl2, 0.1% Tween 20, 2 m*M* levamisole. Make from stock on the day of use.
3. NBT is 4-nitro blue tetrazolium chloride (Boehringer) dissolved at 75 mg/ml in 70% dimethylformamide
4. BCIP (X-phosphate) is 5-bromo-4-chloro-3indolyl-phosphate (Boehringer) at 50 mg/ml in 100% dimethylformamide.

Note. Boehringer non-radioactive ISH reagents are now provided by Roche.

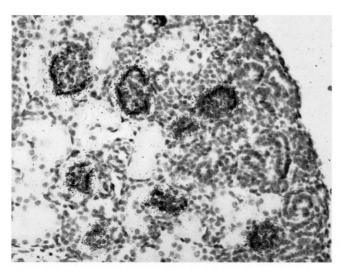

Figure 19.8 Radioactive mRNA in situ hybridization. The expression of nephrin mRNA in glomerular podocytes of the newborn mouse kidney. The micrograph is a superposition of bright-field and dark-field digital photographs; in the latter, autoradioraphy grains are shown as red. (This paraffin section was exposed to emulsion for 10 days and then counterstained with hematoxylin.)

B. Radioactive *in situ* hybridization for Paraffin Sections (Fig. 19.8)

The following protocol is basically that of Wilkinson and Green (1990) with some modifications introduced.

Preparation of the Probe (Done in Advance)

For Preparation of the template cDNA linearize the plasmid (completely!) with an appropriate restriction endonuclease extract (to remove any proteins): twice with phenol:chloroform:isoamylalcohol (25 : 24 : 1) twice with chloroform : isoamylalcohol (24 : 1).

Precipitate the DNA, wash it once with 80 % ethanol and once with 100 % ethanol, and measure the concentration.

Transcription

2 µl 5 x transcription buffer (Promega)
1 µl 0.1 M DTT
1 µl 2.5 mM ATP/CTP/GTP
x µl linearized plasmid (1 µg)
0.5 µl Rnasin (Promega)
3-5 µl ^{35}S-UTP(Amersham)
0.5 µl appropriate RNA polymerase (10 U/ml)

1. Mix, spin down, and incubate at 37°C for 30 min, add another 0.5 µl of RNA polymerase, and incubate another 30 min.
2. Add 0.5 µl of DNase, mix, spin down, and incubate at 37°C for 15 min.

Add an equal volume of 80 mM NaHCO$_3$, and 120 mM Na$_2$CO$_3$. Mix, spin down, and incubate at 60°C for t min, where $t = (L\text{-}0.1)/0.011\,L$ (L is the original probe length in kb). Theoretically, this gives an average length of 100 bp, but may cause some background effects. We have, however, successfully used cDNAs that are 3–4 kb long without any hydrolysation and the result has usually been better than after hydrolysis.

3. Gel filtration of the cRNA probe
Preparation of the gel filtration buffer (25 ml):

Final concentration	Stock solutions	Volumes taken
10 mM Tris–HCl	1 M Tris–HCl, pH 7.5	250 µl
1 mM EDTA	0.5 M EDTA	50 µl
0.1 % SDS	10 % SDS	250 µl
10 mM DTT	1 M DTT	250 µl
	DEPC-H$_2$O add	25 ml

4. Fractionation on a Sephadex G50 (Pharmacia) column: Pour off the storage buffer from the column. Equilibrate with 3 ml buffer and allow to elute. Add all the labeled probe (approximately 10 µl), and let it sink in. Rinse the probe tube with 10 µl DEPC-H$_2$O and add to the column. Add 400 µl buffer and collect fraction 1. Add 400 µl buffer and collect fraction 2 (contains labeled RNA). Add 400 µl buffer and collect fraction 3 (may also contain some labeled RNA, but mainly unincorporated nucleotides). Add 400 µl buffer and collect fraction 4.

Note. filtration buffer contains SDS and thus tends to produce bubbles; these should be avoided.

5. For measuring radioactivity, use 2 ml of scintillation liquid (OptiPhase "Highsafe"3, Wallac) and add 1–3 µl of liquid from fraction 2. Mix well by vortexing. Measure in β counter with an appropriate program for the ^{35}S label.

Note. Counts normally exceed 100,000 cmp/µl. If this is not the case, measure fraction number 3. If most of the counts are in this fraction, the transcription has failed.

6. Divide fraction 2 equally into two Eppendorf tubes (200 µl). To each add 20 µl 3 M NaOAc (filtered, pH 5.2) and 550 µl absolute ethanol (2.5×), mix, and take to -20°C (overnight).
7. Centrifuge the precipitates for 30 min/full speed/4°C and wash the pellets first with 80% ethanol and then with absolute ethanol. Centrifuge after each wash as described earlier and finally let the pellets air dry.
8. Resuspend one pellet at about 2–4 × 10^4 cpm/µl in hybridization mix (1:10 1 M DTT and hybridization buffer).

Dissolve the pellet stepwise. Add 200 µl of the hybridization mix, mix carefully by tapping, let stand on ice for about 15 min, and measure the activity. If necessary, add more hybridization mixture, mix, etc. until the desired cpm/ml is achieved. You may speed up the process by heating (37°C) for 10–15 min.

To the second tube just add a small amount (50 µl) of hybridization mix or 1 M DTT (20 µl) and store at -70°C for up to 4 weeks.

Notes. For cloning, a plasmid with promoter sites for T3 or/and T7 polymerase (e.g., Bluescript, Stratagene) is better, as SP6 polymerase has proved to be less active when used according to this protocol. However, Bluescript and its derivatives should ***not*** be used with embryonic rat tissues (see Raatikainen-Ahokas *et al.*, 2000)

It is important to avoid linearizing the template with restriction enzymes that generate a 3′ overhang. Always try to use an enzyme that leaves a 5′ protruding end. If there is no alternative, then you can use the 3′→5′ exonuclease activity of the Klenow DNA polymerase to convert the 3′ overhang to a blunt end.

A possible cause of trouble is a GC-rich transcript. In case of doubt, the procedure should be tried with a different transcript.

The plasmid has to be linearized with an appropriate restriction enzyme at a site downstream of the RNA polymerase promoter. The optimal size of the transcript is about 500 to 700 bp, but successful *in situ* has been done with templates of 4 kb. The digestion has to be complete. Each template/probe should be treated individually to optimise the *in situ* hybridization.

Reagents

1. The ATP/CTP/GTP mix is made by combining 1 µl of each stock (10 mM, e.g., Promega) and 1 µl of nuclease-free H_2O. Do not use DEPC-treated water in the transcription reaction, as even trace amounts of DEPC apparently inhibit RNA polymerase activity. In our protocol, however, no water is added to the transcription mixture. The transcription mixture is prepared just before use. Stock solutions are kept in -20°C and are used exclusively for preparing the mixture.

2. Commercial Pharmacia Biotech, nick Sephadex 50 column (17-0855-02)

3. Hybridization buffer

60% formamide	6 ml formamide
0.3 M NaCl	0.6 ml 5.0 M NaCl
20 mM Tris-HCl, pH 8.0	0.2 ml 1.0 M Tris–HCl
5 mM EDTA, pH 8.0	0.1 ml 0.5 M EDTA
10 % dextranesulfate (MW 500,000)	1.0 g dextransulfate
0.5 mg/ml yeast tRNA	5.0 mg yeast tRNA
1 × Denhardts solution:	0.2 ml 50xDenhardt's sol.
0.02% BSA	

0.02% Ficoll
0.02% polyvinylpyrrolidon
DEPC-H_2O add to total volume of 10 ml

The buffer is filtered, stored in aliquots at -70°C, and melted only once.

Dextransulfate does not always dissolve easily and it is best to dissolve all components except formamide first, pass the solution through a Millipore filter, measure the remaining volume, and add formamide up to 60% final concentration.

Deionization of formamide (needed only for the hybridization buffer) is as follows:

Put 1 liter of formamide into a big beaker. Add two tablespoonfuls of Duolite MB6113 resin and mix with a magnetic stirrer for several hours until the color of the blue resin turns yellow. Filter through thin Whatman and store at -20°C in brown bottles or 50-ml Falcon tubes. Make suitable portions. Melted deionized formamide is stored in the refrigerator and can be used for the humidifying solution (see hybridization procedure).

4. 1 M DTT: 3.09 g DTT in 20 ml of 0.01 M Na-acetate (pH 5.2), sterilize by filtration (no autoclaving!), and store in aliquots at -20°C.

Silinazing slides for *in situ* hybridization

1. Wash slides in hot soap water.
2. Rinse in hot water for 2 h.
3. Dip slides in 10% HCl/70% ethanol.
4. Rinse in distilled water.
5. Dip slides in 94% ethanol.
6. Dry slides in oven at 150°C for 5 min and allow to cool.
7. Dip slides in 2% TESPA in acetone for exactly 10 sec (3-amino-propyl-triethoxy-silane).
8. Wash slides twice in acetone for 10 sec.
9. Wash slides in DEPC–treated distilled water for 10 sec.
10. Dry slides at 37°C or 42°C
11. Store the silanized slides in clean boxes at room temperature.
12. Cut at 5–10 µm (preferably 7 µm) sections for *in situ* hybridization and mount on TESPAed slides. Keep the slides at 37°C overnight to ensure that the sections stick tightly to them.

Alternatively, commercially available, precleaned SuperFrost Plus (Menzel-Gläser) slides may be used. They do not require pretreatment, but are more expensive.

Pretreatment of Paraffin Sections (Day 1)

Prepare the solutions and buffers in advance.

1. Dewax and rehydrate the slides: xylene (2 × 10 min), absolute ethanol (2 × 5 min), 94 % ethanol (1 × 3 min) and 70 % ethanol (1 × 3 min).

2. Wash 5 min in DEPC-PBS.

3. Fix in fresh 4% paraformaldehyde in PBS for 20 min (do not discard). This protects the mRNA.

4. Wash 5 min in PBS. The wash is needed to remove all PFA, traces of which might inactivate proteinase K!

5. Drain the slides, and place in proteinase K (7–14 μg/ml) for 15 min at room temperature in 50 mM Tris–HCl, 5 mM EDTA, pH 8.0. This step is crucial for paraffin sections.

6. Wash in PBS for 5 min.

7. Repeat fixation in PFA, using the same solution (20 min), wash 5 min in PBS, and then rinse briefly in DEPC-treated water. This refixation protects the new mRNA sites that may have revealed after proteinase K treatment.

8. Treat the slides with 0.1 M triethanolamine HCl (pH 8.0)/acetic anhydride (final concentration 2.5 ml/liter). Place the slide rack to a jar with 250 ml of buffer, add acetic anhydride to the buffer, and dissolve with magnetic stirrer until it has fully dispersed. Turn off the stirrer. Let the slides stand in the liquid for 10 min.

This step neutralizes any charges from the sections, as acetic anhydride inactivates the proteinase and acetylates positive amino groups. The TEA-acetic anhydride mixture should be used for a maximum of two racks of slides.

9. Wash with DEPC-PBS and then with DEPC-H_2O (5 min each).

10. Dehydrate the slides by transferring through ethanol dilutions: 50 % (2 min), 70 % (5 min), 94 % (2 min), and 2 × 100 % (2 min).

11. Air dry at room temperature for 30 min. For drying, keep slides on a tray and cover it to protect the slides against dust. Do not touch the samples, not even with gloves!

12. Use for hybridization or store at -20°C.

Hybridization

The labeled cRNA probes have been mixed with the hybridization buffer/DTT mixture to the desired amounts of cpm/ml (see earlier discussion). When using a probe for the first time, however, you may wish to use a range of concentrations to find the optimum: high levels of labeled probe may actually lead to a lower signal and a higher background.

1. Heat the hybridization mixture at 80°C for 2 min in a heating block. Mix gently and spin briefly. Place on ice.
2. Apply the hybridization mix (50–100 μl) to the sections (the amount depends on the size and number of the sections).
3. Cover the sections with pieces of Parafilm of appropriate size (a little larger than the sample area), taking care to avoid bubbles.
4. Place the slides horizontally in a plastic slide box together with tissue soaked in 60 % formamide and 5× SSC. Close the box tightly and wrap it in foil.
5. Incubate overnight at 50°C (or up to 70°C).

Slides must not be allowed to dry, as this may produce a high background signal.

Posthybridization Washings (Day 2)

Before starting the posthybridization treatment, prepare all necessary solutions and warm them.

1. Take Parafilm off with forceps and collect the slides in a rack. Discard the Parafilm in radioactive waste.
2. *Low stringency wash*: place the slides in a solution of 5× SSC, 10 mM DTT (warmed to 50°C) for 30 min. Removes excess hybridization mixture.
3. *High stringency wash*: place the slides in a solution (warmed to 65°C) of 50 % formamide, 2 × SSC, 20 mM DTT for 30 min. The buffer will be reused so do not discard it, but leave it in the water bath to stay warm.
4. Wash three times in NTE buffer at 37°C for 10 min each.
5. Treat with 20 μg/ml ribonuclease A in 250 ml NTE buffer at 37°C for 30 min.
6. Wash with NTE buffer at 37°C for 15 min.
7. Repeat the high stringency wash using the same solution.
8. Wash in 2× SSC and then in 0.1× SSC for 15 min each at room temperature (37°C in case of need). *From now on, powdered gloves must not be used; indeed, it is better to use powderless gloves during the whole procedure.*
9. Dehydrate slides by transferring quickly them through 30% (2 min), 60% (2 min), 80% (5 min), 95% (2 min) and 2 × 100% (2 min each) EtOH.
10. Air dry on a foil with a lid as protection against dust. Then use for autoradiography. If necessary, the slides may be stored at 4°C overnight or at -20°C for up to 1–2 weeks before autoradiography.

Notes. All instruments used prior to hybridization (also the magnet for stirring, etc.) have to be wrapped in aluminium foil and heated in an oven (250°C overnight) before use to destroy RNases. The glassware (including measuring cylinders and bottles used for storing solutions) also has to be so treated (at 250°C overnight).

Buffer washes (PBS, etc.) may be extended, but the time for enzymatic treatments and high stringency washings is aquarete.

The alcohol dilutions may be used several times (solutions can be kept in marked bottles to avoid evaporation). However, the second absolute ethanol after xylene treatment should be changed after two to three racks (the second ethanol wash becomes first, and fresh ethanol is then used for the second wash).

The high stringency wash removes double-stranded RNA, which has not hybridized completely (nonspecific hybridization). Formamide enables the high stringency to be achieved at lower temperature than the T_m (= the melting temperature) of hybrids.

The high stringency washing step is the most critical one in controlling the background. The amount of formamide, DTT (up to 40 mM), and SSC (possible decrease to 0.5 × SSC), the temperature, and the time (up to 1 hour) are all adjustable, and optimal conditions have to be found by trial and error. Although longer washes, the lower the background, they may also lower the signal.

The ribonuclease degrades the single strand probe, leaving double strand hybrids intact.

Excessive amounts of background can be lowered using a mixture of RNase A and RNase T, but this is not recommended as it may also remove signal.

Reagents

1. The 1 M triethanolamine stock solution is sterilized by autoclaving and is stored in 4°C.
2. DTT for washings: 10 mM DTT = 386 mg/250 ml.
3. NTE buffer: 0.5 M NaCl, 10 mM Tris–Cl, pH 8.0, 5 mM EDTA, pH 8.0. For 1 liter use, 100 ml 5 M NaCl, 10 ml 1 M Tris–HCl, pH 8.0, and 10 ml 0.5 M EDTA. Add 1000 ml sterile H_2O. This wash is done three times, as DTT is an efficient RNase inhibitor and must be removed completely.

C. *In situ* Hybridization for Frozen Sections

Sections are cut, fixed with 4%PFA, dehydrated, and stored at -20ºC until needed.

1. Sections are rinsed in PBS for 5 min.
2. Proteinase K 0.5 µg/ml 5 min at room temperature.
3. Rinse in PBS 5 min.
4. 4% PFA fixation 5 min.
5. 50 % formamide + 2 x SSC 5 min.
6. 0.1 M TEA, pH 8.0, + acetic anhydride 10 min.
7. 50% formamide + 2× SSC until hybridization.
8. Prehybridization 30 min at 50°C (not necessarily needed).
9. Hybridization overnight.
10. Posthybridization washings as described earlier.

Note. As frozen sections should not be dried before hybridization, slides should be taken from the 50% FA/2×SSC solutions one at the time. Remove excess solution (hybridization mix if prehybridization has been done) by turning the slide sideways and draining the liquid with a soft tissue. Neither touch the sections nor let them dry!!

D. Dipping of Slides (Both Paraffin and Frozen Sections)

Autoradiography of the section *in situ* is not always easy. The author has seen hundreds of *in situ* slides destroyed by bad autoradiography processing. This step may be the last, but is by no means the least important!

We use Kodak NTB-2 autoradiography emulsion. This should be stored in the dark at 4°C, well away from any contaminating radiation.

The following steps are all performed in the dark (Kodak provides suitable red filters for light sources). The diluted emulsion is melted at 43°C and is poured into the pre-warmed dipping container standing in a water bath. The slides are dipped twice for 3 (count 3) or once for 7 (count 7), left to dry in the dark for at least 2 h, preferably in a light protected box overnight, and then packed in dark slide boxes with a tube of silica gel. The boxes are wrapped in foil and stored at 4°C. Slides are exposed for 8–30 days depending on the amount of expression in the sample. Exposure times of 8–12 weeks have been used in case of low expression levels of the transcript.

Note. Emulsion may be diluted in water 1:1 and, as this makes the emulsion layer thinner, it usually gives a sharper signal. The emulsion must be treated very gently, without any shaking or rocking, to avoid producing secondary grains that give background. Once the emulsion has completely dried on sections, it is stable. A small amount of glycerol can be used to avoid extensive bubbling of the emulsion during the dipping, 1–2% is enough. This can be dissolved into the water that is used to mixing of the emulsion.

E. Developing Slides

The slides are developed in jars. Pour developer into the first jar, distilled water into the second, and fix into the third jar.

Develop for 2 min at room temperature or 4 min at 15°C, rinse in water 30 sec to 1 min, and fix for 4–5 min. Check that the emulsion has turned transparent before putting the lights on, if not, shake slides and continue the fixation. Rinse in water for at least 10 min before counterstaining with hematoxylin.

Reagents

1. Developer: Kodak D-19 Working solution as in manufacturer's instructions: 800 g in 5 liters tap water, pH 9–12. Keep in dark bottles in a dark room. Can be used several times until the liquid turns yellow-brown. *Note.* Any "slow" developer can be used.
2. Fix: Kodak sodium fixative (so-called slow fixative). Powder diluted as in manufacturer's instructions. Can be used several times.

F. Counterstaining the Slides

Rinse the slides in distilled water for 10 min. Stain them in hematoxylin (e.g. Shandon's instant hematoxylin) for 30 sec–2 min. Filter the stain prior to use! Rinse in distilled water to check the staining, if the sections need more color, restain them accordingly. Rinse slides in running tap water for 10 min. Rinse in distilled water and dehydrate the slides, rinse in xylene, and mount them in DePex or Mountex ect. Use clean, dust-free coverslips.

Note. Toluidine blue is not recommended for counter-staining as it decreases the amount of grains.

References

Ekblom, P., Sariola, H., Karkinen-Jääskeläinen, M., and Saxén, L. (1982). The origin of the glomerular endothelium. *Cell Differ.* **10**, 281–288.

Gacrieli, Y., Sherman, Y., and Ben-Sasson, S.A. (1992). Identification of programmed cell death in situ via specific labeling of nuclear DNA fragmentation. *J. Cell Biol.* **119**, 493–501.

Grobstein, C. (1956). Transfilter induction of tubules in mouse metanephrogenic mesenchyme. *Exp. Cell Res.* **10**, 424–440.

Kettunen, P., and Thesleff, I. (1998). Expression and fuction of FGFs –4,-8 and –9 suggest functional redundancy and repetitive use as epithelial signals during tooth development. *Dev. Dyn.* **211**, 256–268.

Kispert, A., Vainio, S., Shen, L., Rowitch, D., and McMahon, A. (1996), Proteoglycans are required for maintenance of Wnt-11 expression in the ureter tips. *Development* **122**, 3627–3637.

Laitinen, L., Virtanen, I., and Saxén, L. (1987). Changes in the glyco-sylation pattern during embryonic development of mouse kidney as revealed with lectin conjugates. *J. Histochem. Cytochem.* **35**, 55–65.

Lehtonen, E., Saxén, L., and Tuomi, A. (1995). Embryonic kidney in organ culture. *In* "Cell and Tissue Culture: Laboratory Procedures." (J.B. Griffiths, A. Doyle, and D.G. Newell, eds.) pp. 14A: 1.1-1.7. Wiley, Chichester, England.

Montesano, R., Schaller, G., and Orci L. (1991). Induction of epithelial tubular morphogenesis in vitro by fibroblast-derived soluble factor. *Cell* **66**, 697–711.

Raatikainen-Ahokas, A., Immonen, T., Rossi, P., Sainio, K., and Sariola, H. (2000). An artifactual in sity hybridization signal in rat embryo. *J. Histochem. Cytochem.* **48**, 955–961.

Sainio, K., Saarma, M., Nonclercq, D., Paulin, L., and Sariola, H. (1994). Antisense inhibition of low-affinity nerve growth factor receptor in kidney cultures: Power and pitfalls. *Cell Mol. Neurobiol.* **14**, 439–457.

Sainio, K., Suvanto, P., Davies, J., Wartiovaara, J., Wartiovaara, K., Saarma, M., Arumäe, U., Meng, X., Lindahl, M., Pachnis, V., and Sariola H. (1997). Glial-cell-line-derived neurotrophic factor is required for bud initiation from ureteric epithelium. *Development* **124**, 4077–4087.

Sariola, H., Aufderheide, E., Bernhard, H., Henke-Fahle, S., Dippold, W., and Ekblom, P. (1988). Antibodies to cell surface ganglioside G$_{D3}$ perturd inductive epitehlial-mesenchymal interactions. *Cell* **54**, 235–245.

Sariola, H., Ekblom, P., Lehtonen, E., and Saxén, L. (1983). Differentiation and vascularization of the metanephric kidney grafted on chorioallantoic membrane. *Dev. Biol.* **96**, 427–435.

Sariola, H., Holm, K., and Henke-Fahle, S. (1988). Early innervation of the metanephric kidney. *Development* **104**, 589–599.

Sariola, H., Saarma, M., Sainio, K., Arumäe, U., Palgi, J., Vaahtokari, A,, Thesleff, I., and Karavanov, A. (1991). Dependence of kidney morpho-genesis on the expression of the nerve growth factor receptor. *Science* **254**, 571–573.

Saxén, L. (1987). "Organogenesis of the Kidney." Cambridge Uni. Press, Cambridge.

Vainio, S., Karavanova, I., Jowett, A., and Thesleff, I., (1993). Identi-fication of BMP-4 as a signal mediating secondary induction between epithelial and mesenchymal tissues during early tooth development. *Cell* **75**, 45–58.

Wilkinson, D. (1993). Whole-mount in situ hybridization of vertebrate embryos. *In situ* Hybridization: A Practical Approach (D. Wilkinson, eds., pp. 75–83 IRL Press, Oxford).

Wilkinson, D., and Green, P. (1990). *In situ* hybridization and the three-dimensional reconstruction of serial sections. In "Postimplantation Mammalian Embryos." (New York. A.J. Copp and D.L. Cockroft, eds.,). pp. 155–171, Oxford Univ. Press, New York.

20

Overview: The Molecular Basis of Kidney Development

Thomas J. Carroll and Andrew P. McMahon

understanding of this process, we need only to compare the number of molecules whose roles have been characterized to the number of molecules that are expressed in the developing kidney (see http://golgi.ana.ed.ac.uk/kidhome.html) to appreciate how little we fully understand.

The current models for the molecular regulation of metanephric development are largely based on gene ablation and cell culture work performed in mice and rats. Advances in the technology available to nephrologists promise an explosion of data that will advance the field rapidly in the very near future (necessarily making this chapter obsolete). That being said, this section reviews our current understanding of the molecules involved in the regulation of metanephrogenesis and, where possible, attempts to assign specific cellular roles to molecules based on their null or misexpression phenotypes. Finally, we attempt to address the future of this dynamic field; where are we going and how can we get there?

I. Introduction

As the previous chapters have illustrated, the formation of a fully differentiated, functional kidney requires the precise regulation and integration of a number of distinct developmental processes. Over the last two decades, developmental biologists have sought to identify and characterize molecules regulating processes that had been described previously through embryological manipulation and/or detailed observation. The myriad complexities of nephrogenesis have provided us with unexpected discoveries and unique challenges. Although we are slowly gaining a greater

II. Specification of Nephrogenic Mesenchyme

One of the topics that have long puzzled nephrologists is induction of the metanephric mesenchyme (MM). It is now quite evident that induction of the metanephros involves a permissive signal to a prespecified region of the intermediate mesoderm known as the metanephric anlage. Although investigators have identified a number of secreted molecules that may be involved in the inductive signal, we are just now beginning to understand the nature of the cell autonomous factors required to respond to this signal, and this section discusses some of molecules that seem to be involved in specification of the anlage.

A. Lim1

Lim1 encodes a LIM domain containing homeobox protein that is expressed in the early lateral plate and intermediate mesoderm of most vertebrates (for review, see Curtiss and Heilig, 1998; Bach, 2000; Hobert and Westphal, 2000). In 8.5 day (E8.5) mouse embryos, *Lim1* transcripts are detected readily in the lateral plate and intermediate mesoderm. Expression remains throughout the nephrogenic cord and mesonephric tubules and duct throughout the next few days but is lost in the lateral plate. At E11.5, *Lim1* is expressed in the ureteric bud as it invades MM, although no expression is detected within MM itself. As the bud branches, *Lim1* expression within the ureteric bud weakens while it is activated in the renal vesicles and S-shaped bodies. Although weak expression is maintained in the ureteric bud tips, the remainder of the collecting duct system lacks detectable *Lim1* transcripts (Fujii *et al.*, 1994).

The pattern of *Lim1* expression suggests several roles for this factor in development of the urogenital system. Unfortunately, the majority of mice lacking the *Lim1* gene product die prior to nephrogenesis, but those that do survive seem to completely lack all intermediate mesodermal derivatives, including the pro-, meso-, and metanephros (Shawlot and Behringer, 1995). It was initially unclear whether this was a direct effect due to ablation of *Lim1* in the intermediate mesoderm or a secondary effect caused by mesodermal patterning defects. Tsang and co-workers (2000) have reinvestigated the intermediate mesodermal phenotype. Expression analysis of *Pax-2*, an early intermediate mesodermal marker, showed that although it was expressed, levels were decreased and the expressing tissue was extremely disorganized (Fig. 20.1). Analysis of *Hoxb6* was similarly affected in the intermediate mesoderm, although its lateral plate expression was unaffected (Fig. 20.1; Tsang *et. al.*, 2000). This suggests that the *Lim1* phenotype is a direct effect due to defects in the specification and/or differentiation of the intermediate mesoderm. Because these mice completely lack a mesonephros, it is difficult to determine what later role this gene may play in metanephric induction or patterning. The expression pattern suggests that it may be involved in ureteric bud formation and tubulogenesis, but, at the moment, *Lim1* can only be implicated in differentiation of the intermediate mesoderm.

Over the years, several LIM-containing homeodomain proteins (the current nomenclature refers to these factors as Lmx and Lhx proteins depending on their class) have been identified in a wide variety of organisms (for review, see Bach, 2000). Mutational analysis suggests that these molecules are involved in a number of developmental

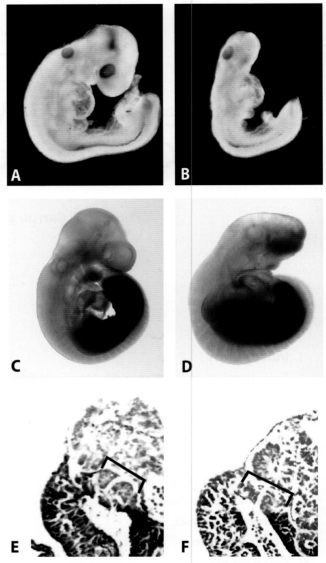

Figure 20.1 Phenotype of *Lim1* mutants. Wild-type(A,C,E) and *Lim1* mutant (B,C,F,) E9.5 embryos. Expression of *Pax-2*. (A) *Pax-2* is expressed throughout the entire length of the intermediate mesoderm. (B) *Pax-2* expression is reduced and nearly absent from the posterior portion of the mutant intermediate mesoderm. (C, E) *Hoxb6*-lac-z is expressed throughout the lateral plate and intermediate mesoderm of wild-type embryos. Sections show expression in the nephrogenic cord (bracket in E). (D, F) In Lim1 mutants, although expression in the lateral plate is maintained, it is lost in the intermediate mesoderm (bracket in F). Adapted from Tsang *et al.*, (2000). Photographs courtesy of Tania Tsang.

processes. In fact, mutation of another LIM homeoprotein, *Lmx1b*, also results in kidney defects, although they have not yet been well characterized (Chen *et al.*, 1998; Dreyer *et al.*, 1998). Assigning a cellular function to this class of proteins is quite difficult as little is known of the upstream mediators or downstream targets of *Lim1*, especially within the intermediate mesoderm. Recent studies have, however, allowed us to gain some insight into the roles of the Lim protein function (for review, see Bach, 2000), as the the Lim domain seems to mediate protein/protein interactions (Agulnick *et al.*, 1996; Schmeichel and Beckerle, 1994). Several factors, including the POU class homeodomain protein Pit-1, have now been shown to interact physically and functionally with LIM proteins (Bach, 2000; Bach *et al.*, 1997). Formation of such complexes may require or be facilitated by the Lim domain-binding protein Ldb1 (also known as CLIM), which seems to act as a scaffold for assembling these complexes (Agulnick *et al.*, 1996; Bach *et al.*, 1997). Although *Ldb1* is expressed in most embryonic tissues, *Pit-1* is not expressed in the metanephros suggesting that some other factor(s) cooperates with *Lim1* to determine the intermediate mesoderm. In the pronephros, this may be the paired box-containing protein Pax-8 (Carroll and Vize, 1999), although this interaction has not been demonstrated in the meso- or metanephros. Two hybrid studies with intermediate mesodermal or kidney libraries should identify further cofactors required for this important family of proteins in the intermediate mesoderm.

B. Eya1

Eya1 is a member of a growing family of genes that are homologous to the *Drosophila eyes absent (eya)* gene. The human ortholog of this gene has been implicated in the dominant-inherited disorder branchio-oto-renal (BOR) syndrome (Abdelhak *et al.*, 1997). Along with craniofacial disorders and hearing loss, mutation of this gene in humans leads to severe kidney defects.

In early stage mouse embryos, *Eya1* transcripts can be detected in the nephrogenic cord and the uninduced MM (Xu *et al.*, 1999). At later stages, *Eya1* is detected in the condensing MM but is absent from the differentiated tubules and ductal network (Kalatzis *et al.*, 1998). Ablation of this gene results in multiple organ defects, including a complete lack of kidneys (Xu *et al.*, 1999), with mutant mice failing to form a ureteric bud. Several markers of the induced MM, including *Pax-2*, *Six-2*, and *GDNF*, are not expressed, although *Pax-2* expression in the mesonephric duct and nephrogenic cord is maintained. The gene encoding the transcription factor *WT1* is expressed in the mutant MM before it undergoes apoptosis at E12.5 (Xu *et al.*, 1999). The phenotype and molecular profile of the mutant MM suggest that *Eya1* is required in MM downstream of *WT1*, but its precise role is unclear. One possibility is that its role in MM specification is to activate *GDNF* expression. Without the expression of *GDNF* in the anlage, the ureteric bud cannot invade (see section VI). If so, MM from *Eya1* mutants should be refractory to tubule-inducing signals in culture, but this remains to be determined. An alternative view is that *Eya1* plays an essential but limited role in the regulation of mesenchymal gene activity in response to MM determinants.

In the *Drosophila* eye, the expression of *eya* and *sine oculis* (a *Six2* homologue) is regulated by *eyeless* (*Pax-6*). In addition, *eya* and *sine oculis* form a regulatory complex that may autoregulate their own expression, as well as positively regulating *eyeless* expression (for review, see Wawersik and Maas, 2000). Molecular analysis of *Eya1* mutants suggests that this regulatory cascade may be conserved during metanephric development with *Pax-2* or *Pax-8* substituting for *eyeless* and *Six2* substituting for *sine oculis*. The possibility that these factors cooperate in the specification of the MM certainly merits further investigation.

C. Pax-2

Pax-2 encodes a paired box–containing transcription factor expressed in multiple embryonic tissues (for reviews, see Torban and Goodyer, 1998; Mansouri *et al.*, 1999; Underhill, 2000). Like *Lim1*, *Pax-2* is expressed in the nephrogenic cord at E8.5 and is maintained in the mesonephric duct and tubules. *Pax-2* is also expressed in the ureter as it invades the metanephric anlage. Whether *Pax-2* is expressed in the uninduced metanephric anlage is a topic of some debate (see later). Shortly before or immediately after invasion of the bud, *Pax-2* is strongly expressed in the condensed MM (Dressler *et al.*, 1990).

Targeted ablation of *Pax-2* reveals that both copies of the gene are required for normal nephrogenesis. Heterozygous mutant mice have hypoplastic kidneys with reduced calyces and cortical nephrons, suggesting defects in cell survival or proliferation and/or the epithelial-to-mesenchymal transition. *Pax-2* homozygous mutants completely lack a metanephros. However, in contrast to the situation with *Lim1*, *Pax-2* mutant mice do form a mesonephric duct, but this may be due to redundancy with the closely related gene *Pax-8*, which is also expressed in the early nephrogenic ridge (Asano and Gruss, 1992; Plachov *et al.*, 1990). In *Pax-2* mutants, the ureteric bud fails to form (Fig. 20.2; Torres *et al.*, 1995). If *Pax-2* is not expressed in the MM prior to induction, then failure of the ureteric bud to form would seem to be due to defects within the mesonephric duct, but it is also possible that *Pax-2* expression within the MM is required for bud invasion.

In addition to its role in early nephrogenesis, antisense and misexpression data suggest that *Pax-2* may play a more general role in epithelialization (Dressler *et al.*, 1993; Rothenpieler and Dressler, 1993). However, given the dynamic and broad expression of *Pax-2* and the closely related *Pax-8* in multiple cell lineages during normal nephrogenesis, caution must be taken in interpreting experiments that modify gene activity throughout the kidney. Given the justified skepticism regarding antisense approaches and the difficulty in sufficiently controlling these experiments, precise temporal and spatial regulation of *Pax-2* expression will be required in order to fully understand the role(s) of this factor in nephrogenesis.

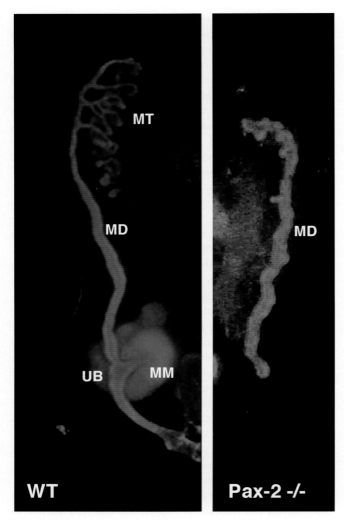

Figure 20.2 Phenotype of *Pax-2* mutant kidneys. Whole-mount E10.5 nephric regions from wild-type (WT, left) and *Pax-2* mutant (*Pax-2 -/-*, right) nephric regions (anterior at top). Tissues are stained with cytokeratin (green), which recognizes the ductal components (meso- and metanephric), and Pax-2 (red), which recognizes ductal and mesenchymal components (overlap of the two stains is orange). At this stage in normal development there are well-formed mesonephric tubules (MT) and ducts (MD), whereas the ureteric bud (UB) has invaded the mesenchyme (MM), causing it to condense. In *Pax-2* mutants, there are neither mesonephric tubules nor a ureteric bud. Photograph courtesy of Greg Dressler.

Pax-2 has been implicated in a number of cellular processes, including cell proliferation, apoptosis, and epithelial-to-mesenchymal transition (see later). As the transcriptional targets of this factor most likely depend on its cellular context, *Pax-2* may be involved in all of these processes. Identification of target genes should greatly facilitate our understanding of the role of this protein, and one candidate target is the *WT1* gene. *Pax-2* has been shown to be capable of activating expression of *WT1 in vitro* while it is required to repress *WT1* expression (either directly or

indirectly) *in vivo* (Dehbi *et al.*, 1996; Majumdar *et al.*, 2000). The mutually exclusive expression patterns of these genes during later nephrogenesis supports their coregulation (Ryan *et al.*, 1995).

In *Drosophila* larvae, the Pax gene *gooseberry* (*gsb*) is essential for establishing cuticular segmentation. *gsb* seems to be involved in a feedback loop with *Hh* and *Wg* (for review, see DiNardo *et al.*, 1994). The expression of homologues of all of these factors within the developing kidney suggests that such a loop may exist during nephrogenesis, although currently there are no data to support it.

D. WT1

Transcripts encoding the zinc finger domain containing protein *WT1* are first detected in the intermediate mesoderm lateral to the coelomic cavity at E9.0 (Armstrong *et al.*, 1993). Expression remains strong in the nephrogenic ridge and mesonephric mesenchyme prior to differentiation, but is excluded from the mesonephric duct. *WT1* is also expressed (although at low levels) in the uninduced MM. Transcript levels are increased greatly within the MM once it is induced and condenses (Armstrong *et al.*, 1993; Pellegrini *et al.*, 1997). In addition, WT1 transcripts are present at high levels in the differentiated podocytes of the glomerulus (see Section IX).

WT1 null mutant mice completely lack kidneys. Although a mesonephric duct forms, the ureteric bud never invades the MM. In the absence of invasion, the MM undergoes apoptosis quickly. Recombination experiments between mutant MM and the spinal cord suggest that *WT1* is required within the MM for ureteric bud invasion (Kreidberg *et al.*, 1993). However, in light of our current understanding of the nature of the spinal cord induction, caution must be taken in interpreting these data (Kispert *et al.*, 1998).

The seemingly normal expression of several genes, including *Pax-2*, *Six-2*, and *GDNF*, within *WT1* mutant MM suggests that the regulation of these factors is independent of WT1 (Donovan *et al.*, 1999). It is particularly interesting to note the expression of *Pax-2* in *WT1* mutants. As mentioned previously, *Pax-2* expression may be activated in response to induction by the ureteric bud. The expression of *Pax-2* in mutant MM suggests that its expression is not only independent of *WT1* but also of ureteric bud invasion. However, while it is possible that MM induction does not require invasion by the bud, the fact that the MM of *Ret* mutant mice does not condense suggests that bud invasion is required for this process (see Section VI). As mentioned previously, *Eya1* may form a complex with *Pax-2* in the kidney, while it is also believed to act downstream of *WT1* within the MM (Xu *et al.*, 1999). It would therefore be worth examining the pattern of *Eya1* expression in *WT1* mutants.

The role of *WT1* in nephrogenesis may be quite complicated, as it has been shown to be both a transcriptional repressor and an activator (for reviews, see Little *et al.*, 1999; Mrowka and Schedl, 2000; Lee and Haber, 2001). Potential targets of its repressive activity include *IGF-1*, *IGF-2*, *Pax-2*, and *Pax-8*, but demonstrating that these genes are actual targets *in vivo* has proven difficult. The transcription of *amphiregulin*, a ligand of the EGF receptor, has been shown to be activated by WT1 (Lee *et al.*, 1999). This result is rather satisfying due to the observation that EGF receptor ligands promote cell survival in MM, a process that is defective in *WT1* mutants (see Section III).

WT1 may activate the transcription of some factor (or factors, including *amphiregulin*) required for MM survival or competence. Alternatively, or in addition, WT1 may be required to repress the transcription of a factor that blocks invasion by the ureter. This issue is complicated by data suggesting that WT1 also functions as part of the RNA–splicing machinery and may regulate the export of certain mRNAs from the nucleus (Larsson *et al.*, 1995). The only certainty is that the initial expression of *WT1* within MM is absolutely required for nephrogenesis to continue. Partial loss-of-function analysis supports an additional role for this gene in later steps of nephrogenesis (see Section IX-A and Moore *et al.*, 1999).

E. Foxc1/Foxd1

Many of the factors that play a role in metanephric specification are initially expressed throughout the nephrogenic ridge and then localize to the metanephric anlage. *In vivo*, the MM only forms in the caudal-most portion of the intermediate mesoderm. Most hypotheses have suggested that some instructional cue is activated at a specific somite level that either permits or directs ureteric bud branching, but we still know little about such a signal. In fact, the regulation of positional specification of organs has proven to be a general enigma in developmental biology, but progress in defining the mechanisms by which regions of the lateral mesoderm are specified to form the fore- or hindlimb buds (Kawakami *et al.*, 2001; Rodriguez-Esteban *et al.*, 1999; Takeuchi *et al.*, 1999), together with the following data, offers encouragement that this puzzle can be solved.

Foxc1 and *Foxc2* (known previously as *Mf1* and *Mf2*, respectively) are winged-helix-type transcription factors that are expressed in largely overlapping patterns throughout the intermediate mesoderm at E8.5. At E9.5, they are expressed in the mesonephric anlage adjacent to but excluded from the mesonephric duct, whereas at E10.5, they are expressed at high levels in the uninduced MM. Levels are maintained in the condensed MM after induction (Kume *et al.*, 2000a,b).

Evidence that *Foxc1/2* play a role in specifying positional information within the intermediate mesoderm comes from mice lacking functional *Foxc1*. In certain genetic backgrounds, *Foxc1* mutant mice display an increased number of mesonephric tubules, extending caudally nearly to the position of the future ureteric bud (Kume *et al.*, 2000b) (Fig. 20.3). In addition, *GDNF* and *Eya1* expression in the MM extends anteriorly, suggesting an expansion of the metanephric anlage (Fig. 20.3). This seems to result in an expansion of the ureteric bud and several ectopic regions of *c-Ret* expression (Fig. 20.3). Subsequently, multiple mesonephric duct-derived ureteric buds invade, resulting in multilobed kidneys with additional ureters (Fig. 20.4). These ectopic ureters frequently display hydroureter, possibly due to incorrect positioning of the ectopic ureters within the bladder. Although mutation of *Foxc2* has only a mild kidney phenotype (Kume *et al.*, 2000a), compound heterozygous mutants for *Foxc1* and *Foxc2* display similar phenotypes to the *Foxc1* homozygotes.

There are several possible roles that *Foxc1* and *Foxc2* might play in the intermediate mesoderm and metanephros. One hypothesis is that *Foxc1* and *Foxc2* either directly or indirectly repress *GDNF* expression in the MM posterior to the mesonephros and anterior to the metanephros. Mice lacking these factors express *GDNF* ectopically, and so form ectopic ureteric buds (Section VI). Another possibility is that *Foxc1* and *Foxc2* play a role in the posterior migration of nephrogenic mesenchyme. In this scenario, the posterior migration of the mesenchyme is delayed, resulting in an anterior expansion of *GDNF*. While both mechanisms could lead to the observed phenotype, there is no clear evidence to support or refute either of them and further investigation will be needed to clarify the roles of these genes in nephrogenesis (Table 20.1).

III. Cell Survival

Experiments have indicated that MM induction is a multistep process (Perantoni *et al.*, 1991). One of the first required signals rescues the MM from a default apoptotic pathway (Coles *et al.*, 1993; Koseki, 1993; Koseki *et al.*, 1992). Failure of the MM to receive this signal, presumably from the ureteric bud, results in rapid programmed cell death. However, in the absence of the MM, the ureteric bud fails to survive. The mutual dependence of these two cell populations on one another has a pleasing logic, but its molecular basis is not well understood. This section characterizes a number of molecules that seem to be involved in regulating cell survival within the developing kidney.

A. FGFs

In attempts to isolate molecules capable of inducing MM to undergo tubulogenesis, various growth factors have been incubated with cultured MM. Ideally these experiments would have employed uninduced MM, which is placed into

WT   Foxc1 -/-

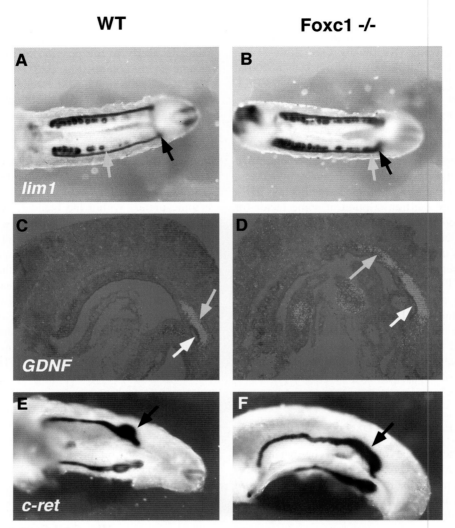

Figure 20.3 Expansion of mesonephric and metanephric tissue in *Foxc1* mutants (anterior is to the left). Wild-type (A,C,E) and *Foxc1* -/- (B,D,F) E10.5 embryos. (A, B, E, and F) Ventral views of whole-mount *in situ* – hybridized embryos. (C and D) ^{35}S *in situ* hybridizations on sagittal sections. (A, B) *Lim1* expression reveals a caudal expansion of mesonephric tubules in *Foxc1* mutants. Note a gap between the caudal-most mesonephric tubules (yellow arrow in A) and the ureteric bud (black arrow) in wild-type embryos. No such gap is detected in Foxc1 mutants (B). *GDNF* expression shows an anterior expansion of the metanephric mesenchyme in mutants vs wild type (compare C to D). The yellow arrow indicates *GDNF* expression in the mesenchyme, whereas the white arrow indicates the mesonephric duct. The anterior expansion of *GDNF* expression corresponds to a broadening of the ureteric bud as revealed by *c-ret* expression (black arrows in E and F). Adapted from Kume *et al.*, 2000. Photograph courtesy of Tsutomu Kume.

serum-free, defined tissue culture media in the presence of the factor to be tested. However, this level of rigor has been difficult to achieve. In practice, these experiments utilize MM that has been removed after the ureteric bud has already invaded and branched and thus presumably received some inductive signals. Even under these conditions, very few molecules have been found that are capable of inducing tubulogenesis. However, several molecules possess the ability to promote survival and/or condensation of the MM (Barasch *et al.*, 1996, 1997; Perantoni *et al.*, 1991, 1995).

One such molecule is FGF-2 (also known as basic FGF; Barasch *et al.*, 1997; Perantoni *et al.*, 1995). FGF-2 rescues mesenchymal precursors from apoptosis and may lead to some condensation and tubulogenesis (see Section IV). Molecular analysis indicates that MM treated with FGF-2 expresses low levels of *WT1*, *c-Met*, *BMP-7* and *Pax-2* (Perantoni et al., 1995; Godin, 1998; Dudley, 1999). Because *BMP-7* and perhaps *Pax-2* are immediate early markers of induction by the ureteric bud, it has been hypothesized that FGF may represent the signal for condensation.

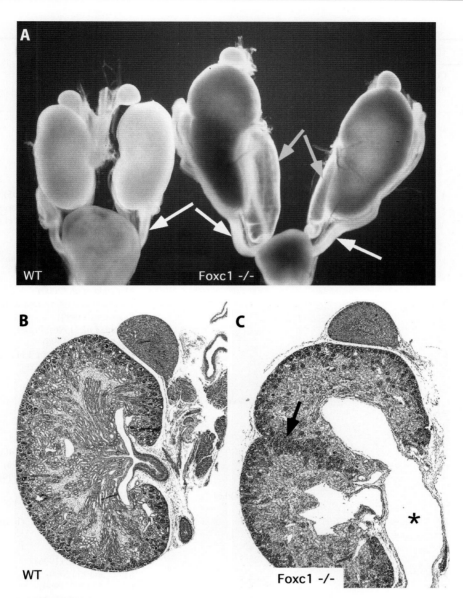

Figure 20.4 *Foxc1* newborns kidneys display double ureters and duplexed kidneys. (A) Formation of a secondary ureter is seen frequently in *Foxc1* mutant kidneys (white arrows indicate primary ureter, whereas yellow arrows indicate ectopic ureter). Ectopic ureters are always dilated due to a failure to connect properly to the bladder. Histology on *Foxc1* mutants reveals duplex kidneys. Hematoxylin/eosin staining on sections of newborn wild-type (B) and mutant (C) kidneys. The black arrow in C marks the area where the two kidneys have fused. The ureter of one kidney is clearly dilated (asterisk in C). Adapted from Kume *et al.* (2000) Photograph courtesy of Tsutomu Kume.

However, as these experiments used MM that was isolated after ureteric bud invasion, and hence likely to have been subjected to inductive signaling, data imply that *FGF-2* supports primarily the survival of both induced and uninduced MM but only promotes condensation indirectly.

In addition to *FGF-2*, several other FGFs are expressed in the developing kidney, including *FGF-1,-4,-5,- 7,- 8,-9,- 12* and *-13* (Barasch *et al.*, 1997; Dono and Zeller, 1994;

Hartung *et al.*, 1997; Mason *et al.*, 1994). However, only *FGF-2* and *-9* are known to be expressed in the ureteric bud at the time required of a survival or condensation factor (Barasch *et al.*, 1997). Three FGF receptors, *FGFR1*, *FGFR2*, and *FGFR4*, are expressed in the kidney, but only *FGFR1* is expressed in the MM (Dudley *et al.*, 1999; Stark *et al.*, 1991). *FGF-2* has been ablated by homologous recombination but does not display a mutant kidney

Table 20.1   Genes with a Mutant Kidney Phenotype

Gene name	Ur. bud tip	Coll. ducts	Unind. Mes.	Cond. Mes	Stroma	Tubule	Glomerulus	Reference
								Transcription Factors
AP2β	-	+	-	-	-	+	-	Moser et al. (1997)
BF2	-	-	-	-	+	-	-	Hatini et al. (1996)
Emx2	+	+	-	-/+[b]	-	+[c]	-	Miyamoto et al. (1997)
Eya 1	-	-	+	+	-	-	-	Xu et al. (1999)
Foxc1/2	-	-	+	+	-	-	-	Kume et al. (2000)
Lim1	+	+	-	-	-	+	-	Shawlot and Behringer (1995); Tsang et al.(2000)
Pax-2	+	+	+?	+	-	+	-	Sanyanusin et al. (1995); Torres et al. (1995); Favor et al. (1996); Ostrom et al. (2000); Porteous et al. (2000)
Pod1	-	-	+[d]	+	-	-	+	Quaggin et al. (1999)
WT1	-	-	+	+	-	+	+	Kreidberg et al. (1993); Donovan et al. (1999)
								Growth factors
BMP4			+[d]				+?	Miyazaki et al. (2000)
BMP7	+	+	-	+	-	+	-	Dudley et al. (1995); Luo et al. (1995)
FGF7	-	-	-	-	+	-	-	Qiao et al. (1999)
GDNF	-	-	+	+	-	-	-	Moore et al. (1996); Pichel et al. (1996); Sanchez et al. (1996)
IGF1/2	-	-	-	+	-	+	+	DeChiara et al. (1990); Liu et al. (1993)
PDGF-B	-	-	-	-	-	-	+	Lindahl et al. (1998)
Wnt4	-	-	-	+[b]	-	-	-	Stark et al. (1994)
								Receptors
GDNFR-α	+	+	+	+	-	-	-	Tomac et al. (2000)
IGF-1R	+	+	-	-	-	+	+	Liu et al. (1993)
Notch2	-	+	-	-	-	+	+	McCright et al. (2001)
PDGFR-β	-	-	-	-	-	-	+	Soriano, (1994); Lindahl et al. (1998)
Ret	+	-	-	-	-	-	-	Schuchardt et al. (1994, 1996)
RAR/RXR	-	+	-	-	+	+	-	Mendelsohn et al. (1994, 1999); Batourina et al. (2001)
								Other
Bcl-2	?	+	?	?	?	+	?	Sorenson et al. (1995)
Glypican 3	+	+	+	+	+	+	+	Cano-Gauci et al. (1999); Grisaru et al. (2001)
HS2ST	+/-	-	+	+	-	-	-	Bullock et al. (1998)
Integrin. α8	-	-	-	+	-	+[b]	-	Muller et al. (1997)
Integrin. α3β1	+	+	-	-	-	-	+	Kreidberg et al. (1996)
Laminin α5	+	+	-	-	-	+	+	Miner and Li (2000)
PKD1	-	+	-	-	-	+	-	Lu et al. (1997, 1999)
PKD2	-	+	-	-	-	+	-	Wu et al. (1998, 2000)

[a] +, normal; –, abnormal or missing.
[b] Pretubular aggregates.
[c] Commma and S-shaped bodies only.
[d] mesenchyme that lies adjacent to collecting ducts.

phenotype, possibly due to molecular redundancy with one of the other ligands. *FGFR1* has also been deleted but results in early embryonic lethality precluding analysis of its role in nephrogenesis (Yamaguchi *et al.*, 1994).

B. EGF Receptor and Ligands

Several molecules capable of preventing cell death in cultured MM have been identified as ligands of the epidermal growth factor receptor (EGFR), including EGF itself, transforming growth factor α (TGF-α), heparin binding EGF, betacellulin, and amphiregulin (Koseki *et al.*, 1992; Perantoni *et al.*, 1991; Sakurai *et al.*, 1997; Weller *et al.*, 1991). In contrast to the results obtained with FGFs, EGF-type ligands do not trigger condensation of the MM nor do they maintain expression of early response genes (Godin *et al.*, 1998). This seems to be due to an inability of the EGF ligands to promote survival of any mesenchymal cells other than stromal precursors (Godin *et al.*, 1998; Weller *et al.*, 1991). We do not yet know whether EGFR signaling mediates the specification of the stromal compartment or simply the survival of these cells. In fact, it may do both as, in the presence of a strong inducer (spinal cord), EGF promotes the stromal fate at the expense of tubules (Weller *et al.*, 1991).

Although the EGFR ligands are expressed broadly during nephrogenesis, *EGFR* is only detected in the collecting ducts and stroma (Bernardini *et al.*, 1996; Partanen and Thesleff, 1987). All three EGFR ligands expressed in the kidney (*amphiregulin*, *TGF-α*, and *EGF*) have been ablated, both individually and as a group, but do not yield a mutant kidney phenotype (Luetteke *et al.*, 1999). Ablation of *EGFR*, although early embryonic lethal in certain backgrounds, causes mild kidney defects in other backgrounds (Sibilia and Wagner, 1995; Threadgill *et al.*, 1995). It is therefore difficult to say what, if any role this pathway has in normal nephrogenesis. Findings that WT1 activates the transcription of *amphiregulin* give further support to some role for these factors in normal nephrogenesis (Lee *et al.*, 1999) but they are clearly not required. It is possible that redundancy exists with another unrelated pathway.

C. BMP-7

The transforming growth factor β (TGF-β) family of molecules, which includes TGF-β itself, activins, bone morphogenetic proteins (BMP), and growth/differentiation factors (GDF), are involved in regulating a wide variety of developmental processes. Several members of this family and its signal transduction pathway are expressed during normal nephrogenesis (Bosukonda *et al.*, 2000; Dick *et al.*, 1998; Dudley and Robertson, 1997). Genetic analysis in mice has demonstrated a clear requirement for one family member, *BMP7*, during nephrogenesis. *BMP-7* is expressed in the mesonephric duct, invading ureteric bud and induced MM (Dudley *et al.*, 1995; Godin *et al.*, 1998), and *BMP-7* remains expressed in the cortical collecting ducts, MM, and mesenchymal derivatives throughout nephrogenesis (Dudley *et al.*, 1995; Godin *et al.*, 1998). Its expression in the MM is regulated by ureteric bud signaling and requires proteoglycan synthesis. As expression in the ureter is maintained by heterotypic mesenchymes, including lung and limb, there is clearly no MM-specific factor needed here (Godin *et al.*, 1998). In fact, the MM may have quite a limited role in patterning of genes within the collecting ducts (Lin *et al.*, 2001).

Targeted ablation of *BMP-7* results in renal dysplasia and severe hydroureter (Dudley *et al.*, 1995; Luo *et al.*, 1995; Fig. 20.5). The initial stages of nephrogenesis occur normally in *BMP-7* mutants; the ureteric bud invades, branches, and induces nephron formation in the MM (Fig. 20.5). At E14.5, normal expression of *c-Ret*, *Pax-2*, *Pax-8*, *Wnt4*, *laminin A*, and *WT1* are observed, although at lower levels than wild type. By E16.5, this expression is gone in all but a few

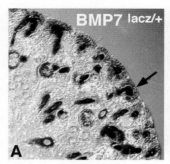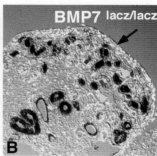

Figure 20.5 Phenotype of *BMP7* mutant kidneys. Sections of E13.5 kidneys stained for β-galactosidase activity from mice with the Lac-z gene knocked in to the *BMP7* locus. Lac-z expression occurs normally in the collecting ducts (yellow arrowhead) of both mutant and wild-type kidneys. Note that no *BMP7*-Lac-z is detected in the mutants condensates [compare black arrows in A to B and the decreased size of mutants (B)]. Photograph courtesy of Robert Godin.

clusters of cells. One probable cause of this phenotype is a drastic increase in the number of cells undergoing apoptosis, mainly in the cortical mesenchyme (Luo *et al.*, 1995).

The addition of exogenous *BMP-7* to cultured MM leads to increased survival, and these effects are enhanced greatly by the addition of *FGF-2* to the culture medium (Dudley *et al.*, 1999). MM treated with both factors can be induced to undergo tubulogenesis by the addition of spinal cord, whereas MM treated with either alone cannot. In addition to promoting survival, these two factors inhibit tubulogenesis and this seems to be due to an expansion of the MM precursor or stromal population (Dudley *et al.*, 1999). All this data taken together suggest that *BMP-7* and FGF cooperatively play a role in regulating growth of the MM by inhibiting tubulogenesis and promoting survival of the uncondensed MM. As mentioned earlier, the uncondensed MM may consist of at least two cell lineages, stroma and a possible stem cell population (see Chapter 13). It has been shown previously that the stromal lineage is required for normal nephrogenesis to occur, and a lack of stromal proliferation may partially explain the *BMP-7* phenotype. In addition, BMP signaling may be required to maintain the pool of stem cells competent to form nephrons by preventing tubulogenesis from spreading throughout the MM. A similar role has been suggested for BMPs in feather bud formation where expression of *BMP-2* in the feather bud precursors inhibits induction of further buds in the interbud regions, a process termed lateral inhibition (Noramly and Morgan, 1998).

As mentioned earlier, several members of the TGF-β family are expressed during embryonic development and seem to be involved in a wide variety of cellular processes. The effect that these factors have on a particular cell probably depends on the cellular environment in which the signal is received. TGF-β signaling is transduced by members of the Smad family of transcription factors (for review, see Miyazono, 2000; Miyazono *et al.*, 2001). Smads have been shown to interact functionally with a number of transcription factors, including Lef/Tcf, β-catenin, jun, Fos, Fast1 and Fast2, and others (Miyazono *et al.*, 2001). A full understanding of the roles this family plays in nephrogenesis will require a more detailed analysis of its signal transduction cascade and cooperating factors.

D. Bcl-2

Both positive and negative factors influence the balance between cell death and survival in the developing kidney (for review, see Ortiz *et al.*, 2000). *Bcl-2* encodes an inhibitor of apoptosis that is widely expressed in the developing kidney (Sorenson and Sheibani, 1999; Veis *et al.*, 1993). Mice lacking *Bcl-2* display abnormal levels of apoptosis throughout all segments of the kidney, especially the uninduced MM (Nagata *et al.*, 1996; Sorenson *et al.*, 1995). *Bcl-2* mutant kidneys are small with a reduced nephrogenic zone and decreased branching of the ureteric bud. The kidneys contain cysts in all epithelial lineages (Veis *et al.*, 1993). Cystic tissues seem to be hyperproliferative and display an abnormal distribution of cell adhesion molecules and improper nuclear localization of β-catenin (Sorenson, 1999). Although the result that excess apoptosis leads to hyperproliferation and cyst formation seems quite surprising, it is actually a common characteristic of cyst-causing mutations. It is possible that the excess cell death in *Bcl-2* mutants is in cells that would not normally contribute to the nephron (i.e., the stromal lineage) and that these cells are necessary for maintenance of a differentiated state. Alternatively, it is possible that increased cell death is merely representative of the mesenchymal cells maintaining an embryonic phenotype (see later). Either phenotype may be due to *Bcl-2* in some way regulating the nuclear accumulation of β-catenin, which has been shown to modify Wnt signaling (reviewed in Miller *et al.*, 1999).

Both *Pax-8* and *WT1* have been shown to directly activate the expression of *Bcl-2 in vitro* (Heckman *et al.*, 1997; Hewitt *et al.*, 1997; Mayo *et al.*, 1999). This agrees nicely with the fact that *WT1* mutant MM undergoes apoptosis (Kreidberg *et al.*, 1993) and suggests that Pax-8 as well as Pax-2 (see Section III) may act as an inhibitor of programmed cell death.

E. AP-2β

The AP-2 transcription factor is formed by hetero- or homodimers of three closely related genes: AP-2α, β, and γ (for review, see Hilger-Eversheim *et al.*, 2000). Targeted deletion has demonstrated that *AP-2α* and *AP-2β* are required for a number of developmental processes (reviewed in Hilger-Eversheim *et al.*, 2000). Both genes are expressed during normal nephrogenesis, with *AP-2α* confined to the proximal tubules and *AP-2β* in the distal tubules and collecting ducts (Moser *et al.*, 1997). Mice lacking *AP-2β* die shortly after birth. Up until E16.5, mutant and wild-type kidneys are virtually indistinguishable, but, at about this time, numerous small cysts are detected in the distal tubules and collecting ducts of mutants. At birth, a drastic increase in apoptosis is observed in the distal tubules and collecting ducts, resulting in a reduction in the size of the medullary region (Moser *et al.*, 1997). Although mutant kidneys normally express markers of patterning and differentiation, the expression of several factors involved in negatively regulating apoptosis, including *Bcl-2*, is reduced (Moser *et al.*, 1997).

AP-2 has been shown to positively regulate the transcription of a number of genes involved in cell proliferation/survival, including *p21, WAF/CIP, TGF-α*, and *c-KIT* (reviewed in Hilger-Eversheim *et al.*, 2000). In addition, it has been shown to negatively regulate the transcription of *C-Myc*, as well as inhibiting the ability of this factor to bind to its target sites. Transfection *of AP-2β* into cells overexpressing *C-Myc* (see Section V) blocks the ability of Myc to induce apoptosis (Moser *et al.*, 1997). It thus seems that *AP-2β* is required for the maintenance and survival of kidney epithelia and that it does this at least partially through the repression of *C-Myc* and the activation or maintenance of *Bcl-2*.

F. Pax-2

As mentioned previously, *Pax-2* is expressed in both the ureteric bud and the induced MM during nephrogenesis. Because mice containing a null mutation for *Pax-2* display complete agenesis of the kidneys, analysis of the role of this gene in later events has been hampered. However, embryos containing a hypomorphic allele of *Pax-2* (*Pax-2*[1neu]) undergo some aspects of nephrogenesis, revealing another role for this factor in kidney development (Favor *et al.*, 1996; Porteous *et al.*, 2000; Sanyanusin *et al.*, 1995). Mice heterozygous for this mutation contain hypoplastic kidneys with fewer nephrons and reduced ureter branching. The main defect seems to be a drastic increase in the number of cells undergoing apoptosis within the ureteric bud derivatives (Porteous *et al.*, 2000; Torban *et al.*, 2000). Similar results are obtained in mice that have had the closely related *Pax-5* gene "knocked in" to the *Pax-2* locus (Pfeffer *et al.*, 2000). This allele is only able to substitute for *Pax-2* in certain developmental contexts, including the earliest events in nephrogenesis. Complementary work has demonstrated that exogenous *Pax-2* can protect cultured cells from apoptosis mediated by caspase-2 (Torban *et al.*, 2000) and that removal of one functional allele of *Pax-2* in *cpk* mice partially rescues the cystic phenotype by increasing apoptosis (Fig. 20.6; Ostrom *et al.*, 2000). Thus, *Pax-2* may have multiple roles during nephrogenesis, including the suppression of apoptosis.

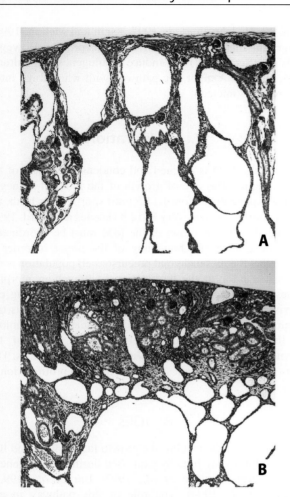

Figure 20.6 Pax-2 inhibits apoptosis. H/E staining of kidney sections from 10–day postpartum cpk -/- (A) or cpk -/-; Pax-2+/- (B) mice. Congenital polycystic kidney (Cpk) mutant mice (A) develop sever cysts due to increased proliferation of the epithelium. Removal of one copy of *Pax-2* significantly delays the progression of cystic disease due to an increase in apoptosis (B). Note the relative lack of cysts in the cortical region of *cpk -/-; Pax-2 +/-* kidneys (B) compared to A. Photograph courtesy of Greg Dressler.

G. p53

p53 encodes a nuclear phosphoprotein implicated in a wide variety of cellular activities. Mice lacking *p53* develop normally, suggesting that this gene is not required during embryogenesis (Donehower and Bradley, 1993, but also see Choi and Donehower, 1999). However, expression of a wild-type p53 transgene in the developing collecting duct system results in kidney hypoplasia (Godley *et al.*, 1996). Although proliferation of the ureteric bud derivatives was normal, they failed to undergo differentiation. In addition, *c-Ret* was expressed ectopically in the distal collecting ducts. The ectopic expression of *c-Ret* is a phenotype that has been observed in both *BF2* and *Pod-1* (see later) mutant mice. In all cases, the collecting ducts failed to undergo proper differentiation and so induced fewer nephrons. In

p53 transgenic mice, the high levels of apoptosis observed in the cortical MM suggested a defect in ureteric bud signaling.

The relevance of this phenotype to normal nephrogenesis is unclear. *Pax-2* and *Pax-8* may inhibit expression of *p53* and so inhibit cells from undergoing differentiation (possibly by keeping them in a proliferative state), and p53 expression may therefore need to be repressed in order for development to proceed. Further analysis is now needed to determine the mechanisms that underlie this phenotype and to test the possibility that this effect is purely artifactual, resulting from the binding of other transcriptional regulators needed for normal differentiation of the collecting ducts (squelching).

H. Neurotrophins

Neurotrophins have been implicated in the growth, migration, and survival of various neuronal cells. The role of neurotrophin *GDNF* and its receptor is discussed in Section (VI). However, several other neurotrophins and their receptors are expressed during nephrogenesis and may be involved in neuron survival and kidney innervation (Davies *et al.*, 1999; Durbeej *et al.*, 1993; Huber and Chao, 1995; Karavanov *et al.*, 1995; Patapoutian *et al.*, 1999). Specifically, the neurotrophin NT-3 may rescue cultured MM from undergoing apoptosis and so lead to the differentiation of nephrogenic neurons (Karavanov *et al.*, 1995). While this occurs in the absence of tubulogenesis, it is still not clear whether NT-3 acts directly on MM or on neural crest cells that are known to be NT-3 responsive and that might contaminate the kidney culture.

IV. Mesenchymal Condensation

In addition to regulating its survival, the ureteric bud induces a subset of the MM to condense around its tip (it is these cells that nephrologists usually think of as induced MM). There is some evidence that the signal(s) that promotes survival is distinct from that which induces condensation. In culture systems, at least, it is possible to add factors that support survival but do not initiate mesenchymal condensation. However, some factors are capable of inducing condensation but only in MM that has been rescued from apoptosis (Barasch *et al.*, 1996, 1997, 1999; Perantoni *et al.*, 1995; Perantoni *et al.*, 1991). This section, reviews what little we know of the molecular mechanisms regulating condensation.

A. FGF-2

The process of condensation is difficult to define at the molecular level. Upon receiving the inductive signal, mesenchymal cells adjacent to the ureteric bud tips condense around the bud. Although this process is visible morphologically, identifying a molecular marker of this event has proven difficult. Several factors, such as *Pax-2* and high levels of *WT1*, have been described as markers of the induced MM, but it is not clear whether these molecules regulate the condensation process itself, as assaying for condensation can be difficult.

One molecule that has been reported to be sufficient for condensation is FGF-2 (Barasch *et al.*, 1997; Karavanova *et al.*, 1996; Perantoni *et al.*, 1995). MM cultured with FGF-2 undergoes condensation as assayed morphologically and by the expression of a variety of markers, including *Pax-2*, *WT1* and *BMP-7* (Barasch *et al.*, 1999; Dudley *et al.*, 1999; Godin *et al.*, 1998; Perantoni *et al.*, 1995). Some reports suggest that FGF-2 can also induce some tubulogenesis (Karavanov *et al.*, 1998); however, in contrast to the case with the spinal cord where tubulogenesis occurs within 72 h, FGF-2 does not manifest its effects for nearly a week. These results suggest that while FGF-2 (or another FGF) may act as a condensation factor, additional molecules are required to support tubulogenesis *in vivo*.

FGF-2 has already been discussed in this chapter as a potential survival factor for the MM (Section III), but it is difficult to differentiate which (if either) process the molecule is involved in. As mentioned previously, the MM used in these experiments was taken from rudiments in which the ureteric bud has already invaded, and we do not know what other factors the MM has already been exposed to at this point and how they influence the response to FGF-2. As MM cannot undergo condensation without being rescued from the apoptotic pathway, it is possible that the role of FGF-2 is to support mesenchymal survival, which secondarily leads to condensation due to the presence of previously provided signals. Conversely, it is possible that FGF-2 directly induces condensation that secondarily leads to survival. Until further molecular analysis is performed, it is difficult to determine where FGF acts in this complicated series of events. There is currently no genetic data supporting a role for any FGF or member of its signal transduction pathway in condensation.

B. LiCl

Like FGF-2, LiCl has also been shown to trigger the condensation of cultured MM (Davies and Garrod, 1995). At the molecular level, FGF-2 and LiCl induce many of the same markers, although LiCl appears to act much more efficiently (Godin *et al.*, 1998). However, morphologically, the two molecules have different effects. FGF-2-treated MM sometimes forms tubules, but this never occurs with LiCl.

LiCl has been shown to mimic the downstream events of the canonical Wnt signal transduction pathway by inhibiting GSK-3β (Hedgepeth *et al.*, 1997; Klein and Melton, 1996), suggesting that Wnts are involved in this process. However, as Wnt4 (Section VI) can induce tubulogenesis in cultured MM, and LiCl cannot, the pathway with which LiCl interacts remains unclear.

V. Proliferation

Once the MM is rescued and condensed, growth of the organ begins. The rate of growth of the induced kidney is nearly exponential. From E11.5 until shortly after birth, the organ doubles in size every 8–12 h (Davies and Bard, 1998; Chapter 10). Proliferation of the MM must be coordinated with the growth and branching of the ureter in order to assure that the mesenchymal precursor cell population is not exhausted prior to the termination of branching morphogenesis. This may be regulated in part by controlling cell survival, as there is a significant amount of apoptosis that occurs in later stages of kidney development. Morphology of the kidney may ultimately be the result of the counteracting forces of cell proliferation and apoptosis. This section reviews molecules implicated in cell proliferation of the kidney.

A. IGFs

Members of the insulin-like growth factor (IGF) and IGF receptor family are broadly expressed during nephrogenesis (Lindenbergh-Kortleve *et al.*, 1997; Liu *et al.*, 1993b; Matsell *et al.*, 1994). The role of this pathway in cell proliferation is well established (for review, see Rother and Accili, 2000). In kidney explant cultures, the addition of IGF1 or insulin leads to an increase in the size of the organ due to increased cell proliferation (Liu *et al.*, 1997; Weller *et al.*, 1991). IGF signaling is transduced through the IGF and insulin receptors, which seem to play partially redundant roles during embryogenesis. Mice lacking *IGF1*, *IGF2*, or their common receptor, *IGF1r*, contain kidneys that although reduced in size are morphologically normal (Liu *et al.*, 1993a), suggesting that the IGF pathway is not involved in other aspects of nephrogenesis, as had been suggested previously. As all organs of the body are similarly affected, it is likely that IGF signaling plays a general role in organ growth. At this point it is unknown what factors lie upstream of IGF signaling in the kidney, although *WT1* has been shown to negatively regulate this pathway *in vitro*. As was mentioned at the beginning of this section, proliferation within the different regions of the kidney must be tightly regulated and coordinated. It will be interesting to determine what factors influence this pathway positively and how IGF signaling interacts with components that affect such processes as ductal growth and branching or proximal/distal growth and patterning of the nephron.

B. FGF7

Although FGF signaling has been implicated in both condensation and survival of the MM, the only genetic evidence for an FGF having any role in nephrogenesis comes from the ablation of *FGF7* (Qiao *et al.*, 1999). As discussed, the stromal cell layer plays a crucial role in the growth and differentiation of the metanephros. *FGF7* is expressed in the stroma surrounding the collecting ducts beginning at E14.5 (Ichimura *et al.*, 1996; Qiao *et al.*, 1999), whereas its high-affinity receptor, *FGFR2*, is expressed in the ureteric bud and the collecting ducts throughout development (Dudley *et al.*, 1999; Ichimura *et al.*, 1996). Ablation of *FGF7* results in kidneys that are reduced in size (Qiao *et al.*, 1999) and this derives from a decrease in cell proliferation within the collecting duct system that ultimately results in fewer branches of the collecting ducts and a corresponding decrease in the number of nephrons. As recombinant FGF7 supports survival but not branching of cultured collecting duct cells, the branching defect seems to be a secondary effect (Qiao *et al.*, 1999). A transient exposure of cultured kidneys to FGF7 results in increased proliferation within the collecting ducts and an increase in nephron number, whereas constant exposure results in increased proliferation but an inhibition of terminal differentiation (Qiao *et al.*, 1999). It thus seems that *FGF7* is involved in the proliferation and/or survival of the collecting duct system. The fact that *FGF7* is not expressed until E14.5 and the relatively mild phenotype of the mutant kidneys suggest that other FGFs may play compensatory roles during earlier stages of nephrogenesis. A full understanding of the role of FGFs in the early steps of nephrogenesis may require kidney-specific ablation of the receptors or their downstream signaling molecules.

C. N-Myc

The Myc proteins encode a family of basic helix-loop-helix transcription factors that have been implicated in a number of developmental processes, including cell proliferation, growth, and apoptosis (for review, see Henriksson and Luscher, 1996; Grandori *et al.*, 2000). All three mammalian members of the myc family, *C-myc*, *L-myc* and *N-myc*, are expressed, during nephrogenesis: *C-* and *N-myc* are widely expressed whereas *L-myc* is expressed solely in the collecting ducts (Hatton *et al.*, 1996; Mugrauer and Ekblom, 1991).

The *myc* genes were first identified due to their upregulation in human cancers. Since then, an enormous amount of work has been performed elucidating the cellular role of these factors. Myc proteins form heterodimers with the ubiquitously expressed protein Max (for review, see Grandori *et al.*, 2000). Several targets of this complex have been identified, including factors regulating the cell cycle and DNA synthesis (Grandori and Eisenman, 1997).

Although all three *Myc* genes have been targeted (Charron *et al.*, 1992; Davis *et al.*, 1993; Hatton *et al.*, 1996), only *N-myc* has been shown to play a role in nephrogenesis. Despite the fact that the null mutation of *N-myc* is early embryonic lethal, the embryos survive long enough to allow the intermediate mesoderm to be removed and cultured. In addition, embryos harboring a hypomorphic *N-myc* allele, along with the null allele (N/H), do not die until E12.5–14.5 (Moens *et al.*, 1993), allowing some characterization of the kidneys. In order to analyze the role of *N-myc* in kidney development, whole kidney explants of *N-myc^{N/H}* and *N-myc* homozygous null embryos were cultured (Bates *et al.*, 2000). Kidneys with no or decreased levels of *N-myc* are hypoplastic due to a significant reduction in cell proliferation (as assessed by BrdU incorporation) in the mesenchymal and ureteric bud tissues. Apoptosis is not affected in these mutants, suggesting that the primary role of N-myc (and perhaps C and L-myc) during nephrogenesis is to regulate cell proliferation. However, transgenic mice overexpressing *C-myc* in the kidney develop polycystic kidneys with increased cellular proliferation and apoptosis (Trudel *et al.*, 1998, 1991, 1997). In some contexts, Bcl-2 is capable of inhibiting C-Myc-induced apoptosis, but *Bcl-2* expression is unaffected here (Trudel *et al.*, 1998). This suggests that several factors are involved in mediating apoptosis in both an inductive and a repressive capacity and the extent to which apoptosis occurs may depend on their relative levels.

As mentioned earlier, the seemingly contradictory processes of increased proliferation and apoptosis are common results in polycystic kidney models. In the case of Myc, there are two possible explanations for this phenotype, each with some experimental support (Grandori *et al.*, 2000). The first model suggests that cells that receive conflicting signals (proliferate vs exit the cell cycle) recognize that there is a problem and undergo apoptosis. In this case, apoptosis is a secondary effect of incorrectly instructing a cell to proliferate. The second model suggests that Myc may be involved in activating genes involved in cell proliferation, as well as activating genes involved in apoptosis, and the pathway followed depends on the cellular context. In the event that Myc is activated incorrectly in a cell, both processes are initiated. These possibilities are not mutually exclusive and both may be involved in determining whether a cell proliferates or dies.

D. Glypican-3

Proteoglycans are sugar-modified proteins that have long been known to be components of the extracellular matrix and connective tissue, but there is mounting evidence for a role for these molecules in embryonic development. Investigations in a number of systems have shown that proteoglycans are involved in modifying the signaling of several extracellular molecules, including members of the Hedgehog, Wnt, FGF, and TGF-β families (for review, see Baeg and Perrimon, 2000; Selleck, 2000).

Proteoglycans are grouped according to their core protein, as well as their sugar moieties. Glypicans are members of the heparin sulfate proteoglycan (HSPG) subgroup. In humans, mutation of *glypican-3* has been shown to be the cause of the Simpson–Golabi–Behmel-dysmorphia (SGBS) syndrome, which is characterized by tissue overgrowth and tumor susceptibility (Selleck, 2000). Symptoms of this syndrome include an increased susceptibility to Wilms' tumors and polycystic kidneys, suggesting that glypicans may act as growth suppressors.

Mouse *glypican-3* is expressed throughout the MM and epithelia of developing kidneys. This gene has been targeted and results in an initial overgrowth of the kidneys and cyst formation due to enhanced cell proliferation, consistent with the view that this gene normally plays a role in growth suppression (Cano-Gauci *et al.*, 1999). Eventually, the medullary collecting ducts of mutants degenerate due to increased apoptosis (Grisaru *et al.*, 2001).

In *Drosophila*, the glypicans *Dally* (*Division abnormally delayed*) and *Dlp* (*Dally like protein*) have been shown to be involved in the reception of the Wnt molecule, *Wg*, and the BMP, *Decapentaplegic* (for review, see Baeg and Perrimon, 2000). Mutations in two genes involved in heparin sulfate synthesis, *sugarless* (*sgl*) and *sulfateless* (*sfl*), result in defective signaling through the Wnt, Hh, and FGF pathways (Haerry *et al.*, 1997; Lin *et al.*, 1999). As Wnt, BMP, Hh, and FGF family members are expressed in the developing collecting ducts and in the surrounding MM (Barasch *et al.*, 1997; Dudley *et al.*, 1995; Karavanova *et al.*, 1996; Kispert *et al.*, 1996), it is possible that the tubule overgrowth phenotype of *glypican-3* mutants arises at least partially from perturbation of the BMP and FGF pathways. The basis for the increase in apoptosis is less clear, but it may also involve these pathways and/or the Wnt and Hh pathways.

VI. Branching of the Ureteric Bud

The ureteric bud forms as a dorsal branch of a caudal portion of the mesonephric duct and soon invades the MM, causing the two tissues to undergo a series of reciprocal inductions and the bud to branch several thousand times (in humans). The derivatives of the ureteric bud eventually form the collecting ducts, renal papillae (including the minor and major calyces), pelvis, ureter, and trigone of the bladder. A great deal of information has been collected on the factors expressed within either the MM or the ureteric epithelium that are required for branching to occur (see Chapter 12).

A. GDNF/Ret

One molecule that is absolutely essential for normal branching of the ureter is glial cell line-derived neurotrophic factor (*GDNF*; for review, see Robertson and Mason, 1997). *GDNF* is first expressed at E8.5 in the nephrogenic cord (Sanchez *et al.*, 1996; Suvanto *et al.*, 1996), and its expression is later restricted to the mesonephric mesenchyme and at E10.5 to the metanephric anlage (Miyamoto *et al.*, 1997; Sainio *et al.*, 1997). After invasion of the ureteric bud, *GDNF* expression is maintained in the condensed and condensing cortical MM, but is downregulated in the differentiating vesicles and stroma (Hellmich *et al.*, 1996; Sainio *et al.*, 1997).

The *GDNF* signal is most likely transduced through a receptor complex that includes the *GDNF* receptor α (*GDNFRα*) and the receptor tyrosine kinase *c-Ret*. *GDNFRα* is expressed in both the ureteric bud epithelium and the condensed MM and vesicles (Sainio *et al.*, 1997; Srinivas *et al.*, 1999b). *Ret* transcripts are expressed initially throughout the mesonephric duct and the ureteric bud (Pachnis *et al.*, 1993; Tsuzuki *et al.*, 1995), but after several rounds of branching, transcripts are localized to the tips of the ureter and are maintained there throughout the remainder of nephrogenesis.

Mice mutant for *GDNF*, *GDNFRα*, or *c-Ret* lack kidneys due to the failure of the ureteric bud to form. The phenotypes are of varying penetrance (Fig. 20.7), suggesting there may be the possibility of redundant or modifier genes within the system (see Davies *et al.*, 1999). *GDNF* and *Ret* are involved in the interplay between the MM and the ureteric bud that promotes branching of the ureter, rather than induction or growth (Pepicelli *et al.*, 1997; Sainio *et al.*, 1997). The culturing of metanephric explants with recombinant *GDNF* or the closely related neurturin results in increased or ectopic branching by the ureteric bud (Davies *et al.*, 1999; Pepicelli *et al.*, 1997; Pichel *et al.*, 1996; Sainio *et al.*, 1997; Vega *et al.*, 1996). This phenotype is strikingly similar to that of *Foxc1* mutants in which the ectopic expression of *GDNF* within the mesonephric mesenchyme also results in supernumerary buds. Mice containing a transgene with the *Hoxb7* enhancer-driven expression of wild-type *c-Ret* throughout the collecting ducts display a phenotype similar to *GDNF* nulls, and this phenotype may be the result of the ectopic *Ret* exhausting the pool of mesenchymal *GDNF* available (Srinivas *et al.*, 1999b). In contrast, the same enhancer-driven expression of an activated form of *c-Ret* leads to multilobed kidneys (Fig. 20.7; Srinivas *et al.*, 1999b).

It is currently unknown at what point *GDNF/Ret* mutants fail. In mice mutant for *Ret* or *GDNF*, the metanephric blastema clearly forms but does not condense (Fig. 20.7). One would expect that since no ureteric bud forms, this MM remains uninduced. However, Pichel *et al* (1996) reported

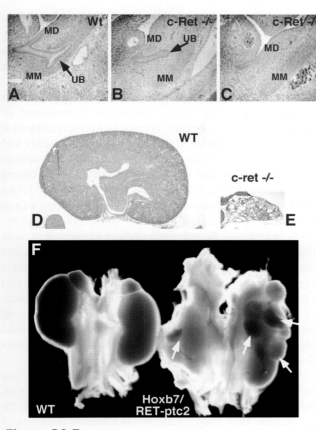

Figure 20.7 Defects in *Ret* mutant kidneys. Histological analysis of E11.5 metanephroi from wild-type (WT) and *Ret* mutant embryos (*c-Ret -/-*). (A) In the wild-type metanephros, the ureteric bud (UB) has branched from the mesonephric duct (MD), invaded the metanephric mesenchyme (MM), and branched once. Note that the metanephric mesenchyme has condensed around the ureteric bud. (B, C) *Ret* mutant mice show variable defects in branching of the ureteric bud from the mesonephric duct. In B, the ureteric bud has branched from the mesonephric duct but has not reached the metanephric mesenchyme. In C, no ureteric bud forms. The uncondensed metanephric mesenchyme is visible at E11.5 but eventually undergoes apoptosis. (D, E) Histology on newborn wild-type (D) and *Ret -/-* kidneys (E). *Ret* mutant kidneys are severely dysplastic and cystic (E). (F) Expression of a ligand-independent form of *Ret* (*RET-ptc2*) under control of the *Hoxb7* enhancer leads to the formation of multilobular kidneys. Lobes (white arrows) contain excess ureteric bud derived tissue surrounded by undifferentiated mesenchyme. Adapted from Srinivas *et al.*, (1999) and Schuchardt *et al.*, (1996), Photographs courtesy of Frank Constantini.

that MM from *GDNF* mutants expresses *Pax-2*, a gene considered to be a marker of induced MM. If correct, this suggests that induction can occur without invasion of the bud, a result consistent with the analysis of *WT1* mutant kidneys (Donovan *et al.*, 1999). However, this MM does not condense, suggesting that further signals from the bud are required (see Figs. 20.7B and 20.7C). As *GDNF* is expressed in the *WT1* mutant but not *Eya-1* mutants (see Section II), it is possible that Eya-1 itself activates the transcription of *GDNF*.

B. Emx2

Emx2, a mammalian paralog of the *Drosophila empty spiracles* gene (Simeone *et al.*, 1992a,b) is initially expressed throughout the nephrogenic cord at E8.5. By E9.5, transcripts are detected in the mesonephric duct and tubules and by E11.5 in the ureteric bud, but not the mesenchyme or pretubular aggregates. At E13.5, *Emx2* is not only expressed in the branching ureter, but also in the condensing tubules (comma and/or S-shaped bodies). At later stages, while expression in the collecting ducts is lost, expression in the cortical condensing tubules is maintained (Miyamoto *et al.*, 1997).

Targeted ablation of *Emx2* results in an interesting mutant phenotype: the urogenital system appears normal up to E11.5 when MM is invaded by the ureteric bud, but the tips of the ureteric bud fail to dilate and branch in *Emx2* mutants and the MM fails to condense or epithelialize. Molecular analysis reveals that by E11.5, ureteric bud expression of *Pax-2* and *Lim1* is lost, although the mesenchymal expression of *Pax-2* appears normal. *c-Ret* expression is reduced greatly and completely absent by E12.0. Within the MM, *WT1* and *GDNF* are initially expressed normally, but *GDNF* levels are reduced greatly by E11.5, whereas *Wnt4* expression (see later) is never detected (Miyamoto *et al.*, 1997).

The loss both of normal branching and of the expression of genes known to be involved in the branching process in the mutant suggests that *Emx2* plays a key role in branching morphogenesis. However, because tubulogenesis also fails, it is possible that EMX2 also has a role in nephron induction. Evidence to confirm this comes from recombination assays showing that the mutant ureteric bud cannot induce tubulogenesis in wild-type MM, whereas mutant MM can be induced to epithelialize and does support branching of the wild-type ureteric bud (Miyamoto *et al.*, 1997). *Emx2* may thus be involved in regulating the interaction between the MM and the branching bud, a role similar to that hypothesized for the mesenchymally expressed integrin α8. It is also possible that Emx2 regulates the expression of the integrin α8 ligand in the bud, but there is as yet no evidence to support this hypothesis.

C. Wnt-11

Wnt-11 encodes a member of the Wnt family of secreted glycoproteins (for review, see the Wnt homepage www.stanford.edu/~rnusse/wntwindow.html). The expression of *Wnt-11* in the metanephros closely resembles that of *c-Ret*. *Wnt-11* is initially expressed throughout the mesonephric duct and invading ureteric bud, but, after the bud has branched, *Wnt-11* expression is restricted to the tips of the bud. As the bud prepares to branch, *Wnt-11* expression is upregulated at the periphery of the ampullae while it is decreased in the center. At the T-shape stage, *Wnt-11* is maintained only at the tips and is lost in the fork of the branch (Fig. 20.10; Kispert *et al.*, 1996).

Although Wnts have been shown to be capable of inducing tubulogenesis in cultured MM (see later), *Wnt-11* does not seem to possess such activity (Kispert *et al.*, 1998). Instead, experiments studying the regulation of *Wnt-11* expression suggest that it may have a role in branching morphogenesis. Previous work has shown signals from the MM together with proteoglycans are required for normal bud branching and growth. However, culturing kidney explants in the presence of compounds that block the function of proteoglycans completely abolishes *Wnt-11* expression (and bud branching) (Kispert *et al.*, 1996). Interestingly, *Wnt-11* expression is downregulated more rapidly than that of *c-Ret*, suggesting that either its expression is directly downstream of proteoglycan function or it functions upstream of *c-Ret*. No genetic evidence yet supports a role for *Wnt-11* in metanephric development, and further work will be needed to fully understand its role here.

D. Retinoic Acid

A role for retinoids in metanephric development has been suspected since it was discovered that rats fed a vitamin A-deficient (VAD) diet gave birth to pups with multiple, severe birth defects that included urogenital abnormalities

(Wilson and Warkany, 1948). The derivative of vitamin A thought to play a role in embryogenesis is retinoic acid, as its addition to rat kidney cultures leads to excessive branching of the ureters and upregulation of *c-Ret* (Moreau *et al.*, 1998; Vilar *et al.*, 1996).

The retinoic acid signal is transduced through a nuclear receptor complex formed by hetero- or homodimers of members of the retinoic acid receptor (RAR) and retinoid X receptor (RXR) families (for review, see Mangelsdorf and Evans, 1995; Chambon, 1996). Each family has several different isoforms that are expressed differentially during development: *RARα, β*, and *γ* and *RXRα, β*, and *γ*. *RARα* and *β2* (not *γ*) are expressed during mouse nephrogenesis in the stromal cell layer (Dolle *et al.*, 1990; Mendelsohn *et al.*, 1999). In addition, the enzyme retinaldehyde dehydrogenase (*Raldh2*), which is involved in the synthesis of active retinoic acid, is also expressed in the stromal cell layer (Fig. 20.8), suggesting that this pathway is active in this cell layer. *RXRs* are also expressed during nephrogenesis, but, unlike *RARs*, *RXRs* are expressed in the epithelial components of the kidney, especially the proximal and distal tubules (*α* and *β*) and the inner medullary collecting ducts (*β* only; Sugawara *et al.*, 1997; Yang *et al.*, 1999).

All isoforms of the RAR and RXR families have been deleted through homologous recombination; however, none display urogenital phenotypes. This has led to the speculation that there may be functional redundancy

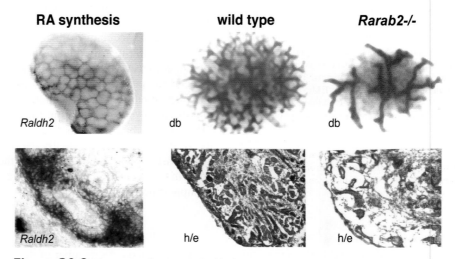

RA synthesis   **wild type**   *Rarab2-/-*

Raldh2   db   db

Raldh2   h/e   h/e

Figure 20.8 Phenotype of *RARα/β* double homozygous mutant mice. Retinoic acid (RA) synthesis in renal stroma revealed by retinaldehyde dehydrogenase-2 *(Raldh2)* expression. Wholemount *in situ* hybridization of an E14 wild-type embryonic kidney showing the honeycombed pattern of *Raldh2* expression in stromal cells. Section of an E14 embryonic kidney showing *Raldh2* expression in stromal mesenchyme. Branching morphogenesis in an E14 kidney from a wild-type embryo revealed by staining with the lectin *dolichos biflorus* (db). H/E staining of a wild-type E18 kidney. Impaired branching morphogenesis and nephron formation in *RARα-/-:RARβ2 -/-* double mutants *(Raraβ2-)*. An E14 *Rar αβ2-* embryonic kidney stained with db shows impaired ureteric bud branching. H/E staining of an E18 *Rar αβ2-* kidney showing lack of glomeruli, collecting ducts, and nephrogenic zone. Courtesy of Cathy Mendelsohn.

between the various isoforms. In support of this hypothesis, compound mutants for various isoforms do display defects that are consistent with those observed in VAD embryos. Although *RARα1/γ* mutant embryos display agenic or hypoplastic kidneys, no γ isoforms have been detected in the developing kidney. This result is therefore considered to be a secondary effect due to patterning defects in the embryo. However, as mentioned earlier, both *RARα* and *β2* are expressed during nephrogenesis, and *RARα/β2* double mutants display renal hypoplasia, hydronephrosis, ureter agenesis, and kidneys that fail to ascend (Mendelsohn *et al.*, 1994). These kidneys display decreased branching of the ureter, as well as a decreased nephrogenic zone (Fig. 20.8). Molecular analysis suggests that MM forms normally but the stroma is patterned abnormally. Although expression of *c-Ret* in the ureteric bud is initiated, it is reduced greatly and completely absent by E14 (Mendelsohn *et al.*, 1999). These results are intriguing because they implicate the stroma in mediating ureteric branching through the regulation of *c-Ret* expression (Mendelsohn *et al.*, 1999). In support of this hypothesis, *RARα/β2* mutant kidneys are rescued in mice driving *c-Ret* behind the *Hoxb7* enhancer element (Batourina *et al.*, 2001). It thus seems that retinoic acid mediates kidney development through the binding of RARα and β2, which subsequently regulate *c-Ret* expression. At this point it is unknown whether the regulation of *c-Ret* is direct or works through a mesenchymal intermediary (Batourina *et al.*, 2001).

E. BMPs

The mammalian lung represents another organ that undergoes extensive epithelial branching. Paralogs of many of the molecules known to be involved in lung branching morphogenesis and also expressed in the developing kidney include Hedgehog, BMP, and FGF family members (for review, see Warburton *et al.*, 2000). It is therefore tempting to speculate that many of the same mechanisms controlling branching morphogenesis will be conserved between these two organs. Although there are currently no data suggesting a role for either Hedgehog or FGF members in branching of the kidney, data do suggest that BMPs may play such a role. In the lung, BMP4 may inhibit the budding influences of FGF10, and expression analysis is consistent with it playing a similar role in the kidney collecting ducts (Dudley *et al.*, 1995). *BMP4* is expressed at high levels in the MM surrounding the distal portion of the duct epithelium (Dudley *et al.*, 1995). Although *BMP4* null mice die during early development, heterozygotes frequently contain small, cystic kidneys, while an ectopic ureteric bud sometimes forms, producing a second ureter (Miyazaki *et al.*, 2000). Addition of exogenous BMP4 to kidney cultures seems to inhibit branching of the ureteric bud and collecting ducts and possibly to promote growth (although this may be a secondary effect of branching inhibition; Miyazaki *et al.*,

2000; Raatikainen-Ahokas *et al.*, 2000). Thus, at least one component of the branching pathway may be conserved between lungs and kidneys. However, any inhibitory effect of BMP4 on branching may be indirect: BMP4-treated kidneys seem to have ectopic smooth muscle (Miyazaki *et al.*, 2000; Raatikainen-Ahokas *et al.*, 2000), and this muscle may either form from mesenchymal cells that would normally play a role in branching or block branching. It is worth noting here that the ureteric bud derivatives that are normally lined with smooth muscle (the ureter and pelvis) do not undergo extensive branching.

F. Pod-1

Another factor that may play a role in regulating branching in both the lung and the kidney is *Pod-1*, a bHLH encoding transcription factor. Its expression during nephrogenesis has been characterized through mRNA *in situ* hybridization and a lac-z knock-in allele (Quaggin *et al.*, 1998, 1999). Lac-z expression can be detected as early as E10.5 in the MM (Quaggin *et al.*, 1999). At E12.5, expression is seen in the condensing MM but not in the renal vesicles and uncondensed MM. Like *BMP4*, strong expression of *Pod-1* is seen in the MM immediately adjacent to the distal collecting ducts (Quaggin *et al.*, 1998), whereas expression in the nephron is in both precursor and differentiated podocytes (Quaggin *et al.*, 1999).

Embryos lacking *Pod-1* have phenotypically normal kidneys until E14, but, by E14.5, ureteric bud branching is reduced greatly. In addition, ureteric branches are frequently crowded together and ureteric bud tips are seen in the medullary region. Although *Pod-1* mutants display morphological defects in the MM, this does not seem to be the cause of the branching defect. Molecular analysis reveals that *c-Ret* expression extends frequently into (or is not repressed in) the distal collecting ducts of mutant kidneys. Both *BF2* mutant (see Section VIII) and *p53* transgenic mice (see Section III) have similar effects on *c-Ret* expression. In addition, transgenic mice driving *c-Ret* expression in the collecting ducts under the control of the *Hoxb7* enhancer have severely reduced branching. Results of this knockout suggest that signals from the adjacent MM may be required to pattern the distal collecting ducts by repressing *c-Ret* and allowing proper growth and/or differentiation to occur. It is thus possible that ectopic *c-Ret* is the cause of the *Pod-1* mutant kidney phenotype.

It is also interesting to note that, in *Pod-1* mutants, *BMP4* expression in the lungs is reduced greatly. As mentioned earlier, *BMP4* is also expressed in the MM adjacent to the distal collecting ducts and ureter and can repress branching of the ureteric bud *in vivo*. Although *BMP4* expression in the kidney has not been assessed in these mutants, it is possible that *Pod-1* regulates its expression or that the two genes function in the same or parallel pathways.

G. Heparin Sulfate 2-sulfotransferase

The growth and branching of the ureteric epithelium require constant remodeling of the extracellular matrix and changes in cell affinity. It is therefore sensible to conclude that molecules involved in the formation, maintenance, and degradation of the ECM and cell adhesion complexes will be involved in nephrogenesis. Indeed, the presence of several components of the ECM and its receptors in the developing kidney has been well documented (Ekblom, 1981; for reviews, see Rabb, 1994; Wallner et al., 1998; Muller and Brandli, 1999; Kreidberg and Symons, 2000). Not surprisingly, there is an enormous volume of in vitro work supporting a role for these molecules in nephrogenesis. For example, molecules that block formation of the basement membrane or interfere with integrin function have been shown to have profound effects in branching morphogenesis (Davies et al., 1995; Lelongt et al., 1997). However, to date, genetic data supporting roles for these molecules in nephrogenesis have been limited. As will be demonstrated earlier, this is probably due to redundancies within this system.

Proteoglycans make up part of the extracellular matrix, including a large part of the basement membranes, and are widely expressed during normal nephrogenesis (Davies and Garrod, 1995; Ekblom, 1981; Vainio et al., 1989). A role for proteoglycans in kidney development has been suggested on the basis of treating kidney rudiments with various compounds that block their assembly. These treatments result in a failure of the ureteric bud to branch and grow and, in some cases, interfere with normal nephron development (Davies et al., 1995; Lelongt et al., 1988).

Heparin sulfate 2-sulfotransferase (hs2st) is an enzyme necessary for the proper synthesis of heparin sulfate, a crucial component of heparin sulfate proteoglycans. A role for this molecule in kidney development was revealed in a gene-trapping experiment that inserted the lac-z gene into the Hs2st coding region (Bullock et al., 1998). Analysis of lac-z expression demonstrated that Hs2st is expressed during early stages of nephrogenesis: at E10.75, lac-z activity is detected within the MM, mesonephric duct, and forming ureteric bud. At E11.5, although expression remains intense in the MM, it is almost undetectable in the bud. Expression is maintained in the MM throughout later stages of development but is downregulated in the differentiating tubules (Bullock et al., 1998).

Disruption of this gene results in an interesting kidney phenotype. Although the ureteric bud invades the MM, it fails to branch, but continues to grow and seems to maintain a population of undifferentiated MM at its tip. Molecular analysis shows that mutant kidneys fail to maintain GDNF, c-Ret, and Wnt-11 expression, whereas Pax-2 expression in the duct persists. In addition, a small population of Pax-2-positive cells is maintained at the periphery of the MM (Bullock et al., 1998).

At this point, it is uncertain where Hs2st is required during kidney development. The expression data and knockout phenotype are most consistent with it having a role in the MM that regulates branching but not growth of the ureter (which also suggests that these two processes are genetically distinct) and in normal nephron formation. Proteoglycans have been shown previously to act as coreceptors for various growth factors, including Wnts and FGFs. It is tempting to speculate that Hs2st functions in receiving and/or concentrating a signal from the bud that is necessary for proper development of the MM and for ureter branching, possibly even GDNF.

H. Integrin α8

Further evidence supporting a role for the ECM in ureter branching comes from the ablation of integrin α8. Integrins comprise a family of transmembrane molecules that form heterodimers between α and β subunits. Integrins act as receptors for extracellular matrix proteins such as laminins, collagen, and fibronectin (for review, see Rabb, 1994; Kreidberg and Symons, 2000). In addition to providing a link to the cytoskeleton, integrins are also thought to transduce signals from the ECM (for review, see Richardson and Parsons, 1995). Several integrins are expressed during normal nephrogenesis (Kreidberg and Symons, 2000), including Integrin α8, which is expressed in the MM surrounding the ureteric bud from E11 pc onward (Muller et al., 1997). α8 is believed to dimerize exclusively with integrin β1.

The kidney phenotype in α8 null pups varies from wild type to complete agenesis of the kidneys (Muller et al., 1997). In animals with an intermediate phenotype, the ureteric bud forms one long, unbranched structure topped by a rudimentary mesenchyme, similar to the Hs2st phenotype. Analysis of E11.5 embryos shows that in most cases the ureteric bud does not branch from the mesonephric duct and, when it does, it frequently does not reach the MM (Muller et al., 1997). Unlike Hs2st, integrin α8 is only expressed in the mesenchymal compartment. It is currently postulated that it is required either for transmitting a signal from the mesenchyme back to the ureter or for receiving a signal from the ureter. As α8β1 mesenchyme is impaired in its ability to epithelialize, both situations may be correct.

Although α8β1 is known to act as a receptor for the ECM molecules fibronectin, vitronectin, and tenascin-C (Muller et al., 1997), none of these molecules is expressed in the appropriate manner to function as a ligand during kidney branching morphogenesis. An unidentified ligand was proposed to be expressed at the tip of the ureteric bud (Muller et al., 1997). This molecule appears to be the recently identified ECM molecule nephronectin (Brandenberger et al., 2001; Miner, 2001). Nephronectin binds to integrin α8 in vitro and is expressed in the ureteric bud from

E10.5 onward. This molecule is thus an excellent candidate for the *α8β1* ligand that mediates kidney branching morphogenesis.

I. Integrin α3β1

Further genetic evidence for integrins playing a role in branching morphogenesis comes from integrin *α3α1* double mutants. *Integrin α3β1* is expressed weakly in the ureteric bud and collecting ducts and in the presumptive podocytes (for review, see Kreidberg and Symons, 2000).

Although ablation of either integrin alone has no phenotype, double mutants display kidney defects (Kreidberg *et al.*, 1996). Although the kidneys are of nearly normal size and have a normal number of nephrons, they seem to have fewer collecting ducts than wild-type kidneys, possibly due to a decrease in ureter branching. It has also been suggested that due to the lack of branching, renal arcades may form in order to compensate for the lost induction of nephrons and so generate the normal number of nephrons present in these mutants (Kreidberg *et al.*, 1996). It is, however, important to note that the formation of arcades has not yet been documented in mice. Further support for a role for integrin α3β1 in branching comes from ablation of its probable ligand, Laminin α5 (*Lama5*) (Miner and Li, 2000). Approximately 20% of *Lama5* mutant mice demonstrated a loss of one or both kidneys and, as *Lama5* is expressed exclusively in the ureteric bud and derivatives, this may be due to a defect in branching (Miner and Li, 2000). It should, however, be pointed out that this phenotype has not been well characterized, which needs to be done if the role of these molecules in branching morphogenesis is to be clarified.

J. HGF

Another molecule hypothesized to be involved in branching of the ureter is hepatocyte growth/scatter factor (HGF). HGF is the founding member of the plasminogen-related growth factor family (for review, see Birchmeier and Gherardi, 1998) and its signal is received by the receptor tyrosine kinase c-Met. *HGF* is expressed in the condensing mesenchyme surrounding the ureteric bud while its receptor is expressed in both the ureteric tips and the mesenchyme (Sonnenberg *et al.*, 1993). In culture assays, HGF can induce the tubulogenesis, growth, and branching of Madine-Darby canine kidney cells grown in a collagen matrix (Montesano *et al.*, 1991; Santos *et al.*, 1994; Santos and Nigam, 1993). The effect of this molecule can be enhanced by some ECM components, whereas TGFβ and other ECM components repress it (Santos and Nigam, 1993). However, neither *Met* nor *HGF* mutants display gross defects in kidney development (Bladt *et al.*, 1995; Schmidt *et al.*, 1995; Uehara *et al.*, 1995). It has been suggested that this

may be due to redundancy with other molecules. Although no clear mouse *HGF* paralog has been identified, it has been shown that other unrelated molecules have similar effects to HGF in cell culture assays.

VII. Mesenchyme-to-Epithelial Transition

Nephrologists frequently talk about the process of induction. Induction is usually defined as all events that occur in the MM after invasion by the ureteric bud. However, as mentioned previously, this is most likely not a simple, one-signal process. There are most likely several, distinct signaling events that must occur in order for the MM to form an epithelium (for review, see Davies, 1996). Once the ureter invades, the MM undergoes a complicated series of morphological events that eventually lead to the formation of a polarized, segmented epithelium. The series of events that lead to the functioning nephron are: (1) condensation of the MM, (2) aggregation of the condensates (or a subset thereof) into an epithelial tubule (i.e., tubulogenesis), and (3) morphogenesis of this tubule (known at this point as the renal vesicle) along its proximal/distal axis from the glomerulus to the collecting duct. This section reviews our current understanding of these processes.

A. Leukemia Inhibitory Factor

MM cultured with FGF-2 condenses and, after an abnormally long time, may undergo some tubulogenesis (Karavanov *et al.*, 1998; Perantoni *et al.*, 1995). One explanation for why tubulogenesis is retarded in these assays is that other factors normally cooperate with FGF in this process. In support of this, supplementation of *FGF-2* with either an extract from pituitary gland (Perantoni *et al.*, 1991) or conditioned media from ureteric bud cell lines (Barasch *et al.*, 1997; Karavanova *et al.*, 1996) leads to complete and extensive tubulogenesis. Recent discoveries indicate that the active molecule from at least one of these cell lines is leukemia inhibitory factor (LIF; Barasch *et al.*, 1999; Plisov *et al.*, 2001).

LIF is expressed in both the ureteric bud and the adjacent MM while its receptor is expressed solely in the MM. Culturing rat mesenchymal rudiments with LIF in the presence of FGF-2 and TGF-α leads to complete and rapid tubulogenesis (Barasch *et al.*, 1999; Fig. 20.9). In these experiments, FGF-2 and TGF-α appeared to make mesenchymal cells competent to respond to LIF, whereas treatment of MM with LIF prior to incubation with FGF and TGF did not lead to tubule formation. FGF-2 and TGF-α both act as survival factors, maintaining a low expression of early markers of condensation, such as *Pax-2* and *Wnt4* (Barasch *et al.*, 1999). Interestingly, addition of TGF-β to

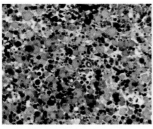
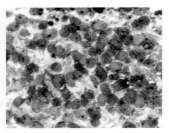

No treatment   FGF2

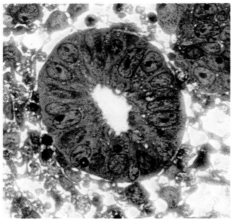

FGF2 + LIF

Figure 20.9 Induction of epithelia in rat metanephric mesenchyme cultured with FGF2 and LIF. Untreated metanephric mesenchyme. Without treatment, the mesenchyme undergoes apoptosis. Addition of fibroblast growth factor 2 (FGF2) to the culture medium rescues the mesenchyme from apoptosis, but does not induce the formation of epithelial structures. Mesenchyme cultured with both FGF2 and LIF forms epithelial structures. Courtesy of Jonathan Barasch.

this culture system was synergistic, leading to tubulogenesis within 72 h (Plisov *et al.*, 2001). As is the case *in vivo*, LIF–induced tubulogenesis is *Wnt4* dependent (Plisov *et al.*, 2001).

These results are consistent with the multisignal requirement for nephrogenesis that has been discussed previously. However, there are some inconsistencies between *in vivo* and *in vitro LIF* data. Although *LIF* and its receptor are expressed in the developing kidney, ablation of either gene has no affect on nephrogenesis (Li *et al.*, 1995; Stewart *et al.*, 1992; Ware *et al.*, 1995). Indeed, even mice lacking *gp130*, the common signal transducer for all IL-6 cytokines, show only a slight decrease in the size of the kidney (Barasch *et al.*, 1999; Yoshida *et al.*, 1996). Furthermore, LIF does not induce tubulogenesis in mouse MM. As TGF-β interacts with FGF-2 nearly as well as LIF to generate tubules, it is possible that there are redundant, parallel pathways operating in this process (Plisov *et al.*, 2001).

B. Wnt4

After the initial "induction" of the MM, further signals are needed to trigger complete epithelialization. At E11.5, *Wnt4* transcripts are expressed in a subset of the mesenchymal condensate on either side of the ureteric bud (Stark *et al.*, 1994) that will form the kidney tubules. *Wnt4* expression continues within the tubules throughout their morphogenesis into comma-shaped bodies and within the most distal portion of the S-shaped body but is lost on fusion of the tubule with the future collecting duct epithelium (Stark *et al.*, 1994; Fig. 20.10).

Wnt4 mutants display severe renal agenesis as the result of a failure to generate nephric tubules. Initial inductive events are not *Wnt4* dependent as assessed by condensation of the MM and normal expression of *Pax-2*, *WT1*, *N-myc*, and *c-Ret* (Stark *et al.*, 1994). However, markers of the pretubular aggregates, *Pax-8* and *Wnt4* itself, are not expressed, revealing that aggregation does not occur and that *Wnt4* signaling is needed to maintain its own transcription.

Placing LIF within this cascade is more difficult (Carroll and McMahon, 2000). As was mentioned previously, *Wnt4* is expressed in cultured MM prior to the addition of LIF, but such MM does not undergo tubulogenesis until LIF is added. In addition, *Wnt4* signaling seems to be necessary for LIF-induced tubulogenesis. It seems that these two genes may be involved in parallel pathways affecting tubulogenesis.

Interestingly, the ureteric bud undergoes significant branching in *Wnt4* mutant embryos (Figs. 20.10E and 20.10F; Stark *et al.*, 1994). Although signals from the MM are known to be necessary for branching, this indicates that tubulogenesis is not essential for at least the first several rounds of the branching process.

Several Wnts have been shown to induce tubulogenesis in cultured MM. Initially this was believed to indicate the presence of an inducing Wnt in the ureteric bud, but no Wnt has yet been identified that is expressed in the invading ureteric bud at the time of induction and capable of inducing nephrogenesis (Kispert *et al.*, 1996). In addition, a great deal of evidence suggests that tubulogenesis is a multistep process that most likely involves more than one molecule. How then are the Wnts able to induce tubulogenesis? It seems that the Wnts are probably mimicking an autoinductive property of the MM rather than the ureteric bud (Kispert *et al.*, 1996). This is supported by the fact that both spinal cord and cell lines expressing Wnts can rescue the *Wnt4* mutant phenotype (Kispert *et al.*, 1996). If Wnts were mimicking a ureteric bud inducer, this would not occur. The fact that the spinal cord can rescue a *Wnt4* mutant and that Wnts show many of the same properties as the well-characterized spinal cord suggests that it is the Wnts expressed in the spinal cord that account for its inductive activity (Kispert *et al.*, 1996). These results have led to a revised view of the classic experiments that utilized heterologous tissues as a ready

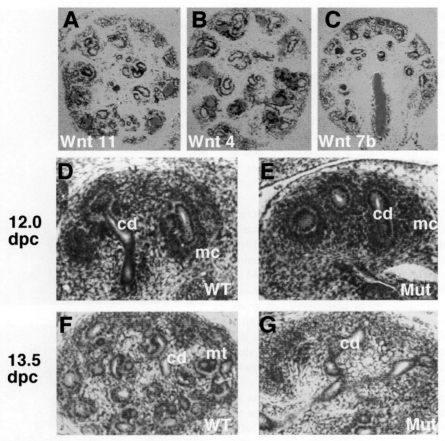

Figure 20.10 Phenotype of *Wnt4* mutant mice. E13.5 kidneys hybridized with ^{35}S-labeled probes for *Wnt11* (A), *Wnt4* (B), and *Wnt7b* (C). *Wnt11* is expressed at the tips of the ureter, *Wnt4* is expressed in mesenchymal aggregates and comma-shaped bodies, and *Wnt7b* is expressed in distal collecting ducts and the ureter. H/E staining on sections of E12 (D and E) and E13.5 wild-type (D and F) and *Wnt4* mutant (E and G) kidneys. Sections of E12 kidneys reveal that metanephric condensates (mc) form and that the first rounds of ureter branching occur normally in *Wnt4* mutants (compare D to E). At E13.5, wild-type kidneys have undergone extensive branching of the duct (cd) and formed mesenchymal aggregates and comma-shaped bodies (mt in F). In E13.5 *Wnt4* mutants, the mesenchyme does not aggregate or form tubules (G). Adapted from Kispert *et al.*, (1996).

source of inductive factors and suggest caution in interpreting experiments in which mutant MM is recombined with the spinal cord (Carroll and McMahon, 2000).

One central question that remains is how *Wnt4* activity leads to tubule formation. One possibility is that Wnt signals may regulate cell adhesion factors that are required for epithelialization, although there is currently no direct evidence to support such a model. A second question is how the activity of *Wnt4* is prevented from spreading throughout the entire MM, immediately exhausting the pool of mesenchymal cells that are available to form tubules. The answer here may lie in the expression of Wnt antagonists. These molecules encode proteins with homology to the extracellular binding domain of the Wnt receptor molecules, the Frizzleds. It has been demonstrated that the gene encoding one of these secreted Frizzled-related proteins, *sFRP-2*, is

expressed in the MM in a pattern similar to *Wnt4* (Lescher *et al.*, 1998). Additionally, *sFRP-2* has been shown to be a target of *Wnt4* signaling and is able to bind to the Wnt4 protein (Lescher *et al.*, 1998). This suggests that Wnt4 may activate its own antagonist in order to spatially modulate its signaling activity. In addition, as was mentioned previously, BMP and FGF molecules seem to antagonize tubulogenesis (Dudley *et al.*, 1999) and it is possible that these molecules have their effect through antagonism of the Wnt pathway.

C. BF2 (Foxd1)

One of the most significant recent contributions to our understanding of kidney development comes from the ablation of *BF2* (Hatini *et al.*, 1996), a member of the forkhead family of winged helix transcription factors. Its expression commences just after invasion of the ureteric bud in a rind of cells surrounding the condensed MM, the presumptive stromal precursors (Hatini *et al.*, 1996). At later stages of development, *BF-2* is expressed in both the cortical and the medullary stroma but expression is never detected in the induced MM or its derivatives. As mentioned in earlier chapters, cortical MM may consist of a bipotential stem cell population that can give rise to nephrogenic or stromal lineages. Expression of *BF2* reveals at least two cell populations within the cortex, a *BF2*-positive and a *BF2*-negative lineage, and the latter may represent a mesenchymal stem cell population, although there is no futher data to support this hypothesis yet.

The question of how and when the stromal lineage are specified has long puzzled nephrologists. The early expression of *BF2* suggests that the stromal fate is specified as an early response to induction, although it remains unclear whether the ureteric bud or the condensed MM is the inducer (Gossens and Unsworth, 1972; Sariola *et al.*, 1988). Furthermore, the question of how many cell types exist in the morphologically indistinct MM has also been a topic of debate. These data, along with the results from the EGF culture experiments discussed earlier, suggest that there are two or more cell populations within the newly induced MM (see later).

Prior to the knockout of *BF2*, the role of the stroma in nephrogenesis was largely overlooked as it was thought to mainly consist of fibroblasts that functioned as support cells for the epithelium. However, some early observations did suggest that uninduced MM, probably stromal precursors, was required to support normal nephrogenesis (Gossens and Unsworth, 1972; Sariola *et al.*, 1988; Saxén, 1970). *BF2* mutants support the hypothesis that these cells play a central role in nephrogenesis.

BF2 null kidneys display a number of defects in nephrogenesis. The kidneys are hypoplastic with approximately 7% the number of wild-type nephrons (Fig. 20.11) and a cortex consisting mainly of large aggregates of undifferentiated MM. This region of the mesenchyme normally expresses a number of markers, including Wnt4, suggesting that it has undergone normal survival, condensation, and perhaps aggregation, but that tubulogenesis is prevented. The expression of tenascin, a marker of stromal differentiation, in *BF2* mutants suggests that the defects in these mice are not simply due to a failure of the stroma to form.

As mentioned earlier, several embryological experiments have suggested that an uninduced mesenchymal population (i.e., stroma) is required for normal nephrogenesis to occur (Gossens and Unsworth, 1972; Sariola *et al.*, 1988). Furthermore, recombination experiments between spinal cord and MM have shown that induction will not take place if the MM is under a certain minimum size. It is possible that without a certain cell number, all induced cells are of nephrogenic character and a stromal population does not

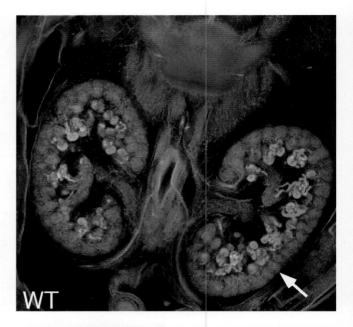

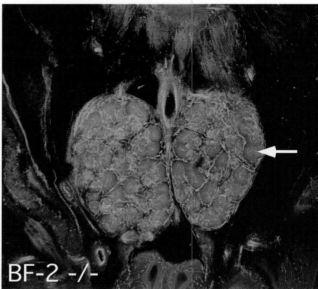

Figure 20.11 Phenotype of *BF-2* mutant kidneys. Confocal images of 100 μ*M* vibratome sections of E15.5 wild-type and *BF-2* mutant kidneys. Kidneys have been stained with the lectins *tetragonolobus purpureas* conjugated to FITC, which marks the proximal tubules (in green), and *arachis hypogaea* conjugated to TRITC, which marks glomeruli (in red). Note the lack of glomeruli and proximal tubules and the increase in size of condensed mesenchymal aggregates (white arrows) in *BF-2* mutants. Courtesy of Randy Levinson.

exist, so inhibiting nephrogenesis. These data suggest that *BF2* is required within the stroma to mediate this interaction between these cells and the MM.

In addition to mesenchymal defects, the ureteric bud of *BF2* null mice contains far fewer branches than the wild type and does not elongate. It shows expanded expression of *c-Ret* into the distal portions of the collecting ducts. This does not seem to be a secondary defect of perturbed tubulogenesis. Other mutants that have more severe defects in tubulogenesis, such as *Wnt4* (see Section VII), do not display similar defects in the collecting duct system. The expansion of *c-Ret* into the distal portions of the collecting ducts in both *BF2* and *Pod-1* (see Section VI) mutants implies a role for the stromal MM that lies adjacent to the duct in its normal patterning or differentiation. The *BF2* mutant reveals multiple roles for the stromal lineage in nephrogenesis.

D. Pod-1

In addition to its role in branching (see Section VI), *Pod-1* also seems to play a role in mesenchymal condensation. The first morphological sign of a defect in *Pod-1* mutants is seen within the MM at stage 14.5. In wild-type kidneys, the condensed MM surrounding the ureteric bud is 2 to 3 cell layers thick, but it is up to 15 cell layers thick in *Pod-1* mutants. In addition, the conversion of MM into epithelia is delayed and terminal differentiation of the proximal tubules and glomeruli does not occur (Quaggin *et al.*, 1999). *Pod-1* thus seems to play a role in regulating the number of cells that contribute to the condensate and is essential for their differentiation, but it is not yet clear how *Pod-1* regulates this process. As was mentioned previously, Pod-1 may be involved in the regulation of BMP expression and BMPs seem to be involved in the repression of tubulogenesis. It is therefore possible that in *Pod-1* mutants, BMP levels are decreased and the inductive signal is not repressed, allowing more cells to be recruited to the condensate. It is also possible that Pod-1 normally suppresses cell division and that the increase in cells surrounding the ureteric bud is simply due to increased proliferation. The mutant clearly needs further analysis.

VIII. Proximal/Distal Patterning

Once the tubules have formed, they undergo a process of elongation and segmentation eventually forming the functional nephron. The nephron is patterned from the proximal-most Bowman's capsule to the distal tubule, which is connected to the collecting duct. The ultimate function of the nephron relies on the proper organization of functional domains along this axis and the correct size and differentiation of each segment

Surprisingly little is known of the developmental regulation of proximal distal patterning of the nephron. Currently it is unknown whether the tubule is patterned intrinsically or whether it is patterned by extrinsic cues from the adjacent stroma or collecting ducts. One hypothesis suggests that patterning is controlled by signals from the bud (Gossens and Unsworth, 1972). In these experiments,

MM cultured for 24 h with spinal cord formed tubules but did not elongate. Lengthening the coincubation period by 6 h led to extensive elongation, but this may simply be due to a lack of a growth and/or survival signal secreted by the bud (Saxén, 1987).

Work has revealed when the proximal/distal pattern is first established. Expression analysis of a number of factors, including cadherins, *Wnt4* and *WT1*, shows a proximal distal bias to the nephron as early as formation of the renal vesicle (Armstrong *et al.*, 1993; Cho *et al.*, 1998; Stark *et al.*, 1994). The differential expression of cadherins is particularly intriguing in view of shape changes that underlie tubule morphogenesis (Cho *et al.*, 1998). However, at this point, with the exception of the *WT1* gene product (Moore *et al.*, 1999; Section IX,A), genetic studies have not uncovered a clear role for any of these genes in P/D patterning. One report suggests that the length of the proximal and/or distal tubules may be decreased in integrin $\alpha3\beta1$ mutants (Kreidberg *et al.*, 1996), but this phenotype has yet to be well characterized.

IX. Glomerulogenesis

As described earlier, the glomerulus is a complex structure made up of cells that are both epithelial and endothelial in origin (Chapter 16). Because the podocytes represent the most proximal derivative of the nephron epithelia, discussion of this structure could be included in the section on nephron patterning. However, because the glomerulus integrates cells from at least two distinct lineages, we will consider it separately. Targeted ablation studies have revealed a great deal about how the development of this structure is regulated.

A. WT1

The expression of *WT1* suggests that it plays multiple roles during nephrogenesis. One area where *WT1* is strongly expressed is in the podocytes of the glomerulus (Armstrong *et al.*, 1993), and genetic mutations in humans suggest that it may play a role in the differentiation of these cells. Further evidence for such a role came from attempts to rescue the *WT1* mutant phenotype in mice with a YAC that contained portions of the human *WT1* allele (Moore *et al.*, 1999). Although the YAC was able to rescue much of the early function of *WT1* (e.g., invasion of the bud and some nephrogenesis), differentiated glomeruli never formed as the podocytes failed to form (Moore *et al.*, 1999).

There has been a great deal published on the interactions between *WT1* and *Pax-2* (Dehbi *et al.*, 1996; McConnell *et al.*, 1997; Ryan *et al.*, 1995). Although the initial expression of *WT1* may depend on *Pax-2*, at later stages of nephrogenesis the patterns of the two genes become mutually exclusive (Ryan *et al.*, 1995), with both genes being able to negatively regulate the expression of the other (Majumdar *et al.*, 2000; Ryan *et al.*, 1995). It is attractive to postulate that the mutually exclusive expression domains of these two factors may play a role in proximal/distal patterning, but there is currently little evidence to support this hypothesis.

B. PDGF-B/PDGFR-β

Platelet-derived growth factor B (PDGF-B) is a member of a small family of connective tissue mitogens. *PDGF-B* and its receptor *PDGFR-β* are expressed, among other places, in the vasculature of the kidney (Leveen *et al.*, 1994; Lindahl *et al.*, 1998; Soriano, 1994). While *PDGF-B* is expressed in endothelial cells, its receptor is expressed in the smooth muscle and later the mesangial cells (Figs. 20.12A and 20.12B; Lindahl *et al.*, 1998). The ablation of either *PDGF* or its receptor results in a similar phenotype: improper formation of the glomerular capillary tufts, probably due to a lack of mesangial cells within the glomerulus (Figs. 20.12C and 20.12D) (Leveen *et al.*, 1994; Lindahl *et al.*, 1998; Soriano, 1994). Careful analysis of these mutants appears to have answered a long-standing question on the origin of the mesangial cells. It seems that signaling from the endothelial cells to the smooth muscle results in the recruitment of smooth muscle cells into the glomerulus along with the capillaries (Lindahl *et al.*, 1998). The mesangial cells thus arise from the smooth muscle cells of the vasculature.

PDGF has been shown to affect several cellular processes, including cell proliferation, survival, and migration, any of which could explain the mutant phenotype. In *PDGF-B* mutants, a small number of mesangial cell do form but are rarely found within the glomerulus, suggesting that determination of this lineage is not affected while perhaps proliferation or migration of the lineage is (Figs. 20.12E and 20.12F). As the number of cells undergoing apoptosis is not affected, impaired cell survival does not seem to be the basis of the phenotype. In addition, although the number of dividing cells within the glomerulus is decreased, this may be explained by a decrease in the migration of mesangial precursors as opposed to a decrease in proliferation (Lindahl *et al.*, 1998). Thus the cellular role of PDGF signaling in glomerular development remains unknown.

C. Notch2

The Notch family of transmembrane receptors is involved in a wide variety of cellular processes (for review, see Weinmaster, 2000), and four Notch paralogs have been identified in mammals. Although *Notch2* is expressed in the developing kidney, early embryonic lethality in null mutants has precluded any analysis of its role in the development of

this organ (Hamada *et al.*, 1999). However, examination of a hypomorphic allele has revealed a role for this factor in nephrogenesis (McCright *et al.*, 2001).

Notch2 is expressed in the developing comma and S-shaped bodies, the developing collecting ducts, and the podocytes of the glomerulus. Several potential ligands for *Notch2*, including *Delta1* and *Jagged2*, are also expressed in the kidney (McCright *et al.*, 2001). Kidneys of *Notch2* hypomorphic mice are hypoplastic with severe defects in the glomerulus and, like *PDGF-B/PDGFR-β* knockout mice, have few if any differentiated mesangial cells within the glomeruli, probably due to defects in proliferation. In addition, glomeruli of these mutants fail to form good cup-shaped epithelia due to a disorganization of the podocytes. *Notch2* also seems to be involved in proper migration of the endothelial and mesangial cells into the glomerulus (McCright *et al.*, 2001). How *Notch2* acts here is not clear, but perturbed expression of *PDGFR-β* may help explain the mesangial cell deficiency (McCright *et al.*, 2001). Because this characterization has been performed with a hypomorphic allele, it will be interesting to investigate whether *Notch2* plays additional roles in nephrogenesis.

D. Laminin α5

Laminin α5 is expressed in the basement membranes of the invading ureteric bud, tubular epithelium, and glomerular basement membrane (Miner, 1999; Miner *et al.*, 1995, 1997; Miner and Li, 2000). Ablation of this gene has multiple effects on nephrogenesis (Miner and Li, 2000), one of which is in formation of the glomerulus. Mutant glomeruli have disorganized basement membranes, which seem to result in a failure of the Bowman's space to form or be maintained. The podocytes stack up on one another instead of aligning in their normal, single cell layer thick organization and, as a result, capillaries and mesangial cells are extruded from the Bowman's capsule (Miner and Li, 2000). This mutant provides good evidence that the glomerular basement membrane (GBM) plays a role in maintaining the structure of the glomerulus.

E. Integrin α3/β1

Integrin α3β1 is expressed in multiple nephrogenic cell types (Ekblom *et al.*, 1991), including the glomerular podocytes (Adler, 1992). Ablation of this gene results in multiple defects during nephrogenesis (see Section VI, A) but most strikingly affects the development of the glomeruli. Like laminin α5 knockouts (a probable ligand for α3β1), α3β1 mutants show a disorganized GBM with the two layers of the GBM failing to fuse. In addition, podocytes fail to adhere to each other and do not form foot processes, whereas the capillary tuft fails to branch normally within these mutant glomeruli. Integrin α3β1 expression in the

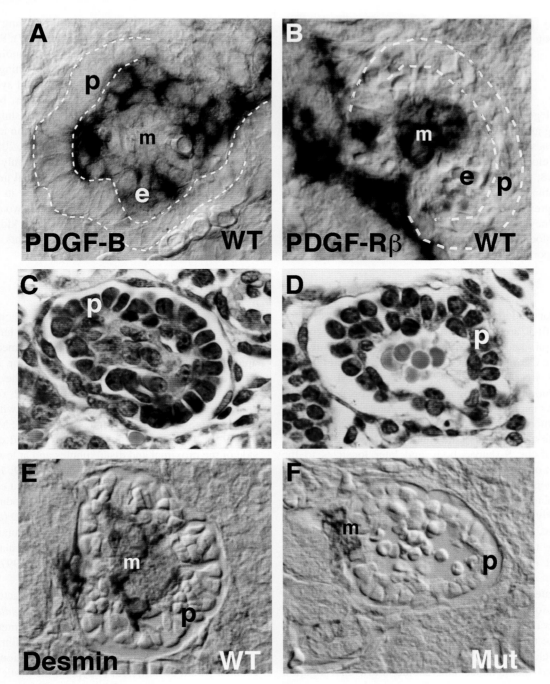

Figure 20.12 Phenotype of *PDGF-Rβ* mice. Sections through wild-type (A, B, C, and E) and mutant (D and F) glomeruli. (A) *PDGF-B* expression in a "cup"-staged glomerulus. Note that *PDGF-B* is expressed in the endothelial cells (**e**) but is absent from podocytes (**p**, outlined in white in A and B) and mesangial cells (**m**). (B) *PDGF-Rβ* is expressed in the vasculature and in mesangial cells, but not in the endothelial cell layer and podocytes. (C) Histology of a wild-type glomerulus at the "cup" stage. (D) A *PDGF-B* mutant glomerulus of the same developmental stage. Note the absence of a capillary tuft in the core of the glomerulus. (E) A wild-type glomerulus stained for the mesangial cell marker desmin. Note mesangial cells in the core of the glomerulus. (F) A *PDGF-B* mutant glomerulus. Note that although mesangial cells are present, they are absent from the glomerular core. **e**, endothelial cells, **m**, mesangial cells: **p**, podocytes. Adapted from Lindahl *et al.*, (1998). Courtesy of Christer Betsholtz.

GBM may play a role in adhesion, differentiation, and proliferation of the podocytes.

X. Vascularization

Several genes believed to be involved in angiogenesis (branching and migration of previously formed vessels) and vasculogenesis (*de novo* synthesis of vessels) are expressed in the developing kidney (Carmeliet and Collen, Takahashi *et al.*, 1998; Chapters 16 and 17). However, in most cases, targeted ablation of these genes results in early embryonic lethality, preventing analysis of their role in vascularization of the kidney. Experiments using conditional ablation and function blocking antibodies are extending our knowledge of this important process.

VEGF

Vascular endothelial growth factor and its receptors, flk-1 and flt-1, are believed to be involved in the migration, proliferation, and differentiation of the vasculature (for review, see Carmeliet and Collen, 1998). Flk-1 and Flt-1 are expressed in small pockets within the MM prior to vascularization, supporting a role for vasculogenesis in this process. As nephrogenesis proceeds, expression is observed in the glomerulus and endothelial cells (Tufro *et al.*, 1999). VEGF is first expressed in the condensed MM and proximal portions of the S-shaped body and later localizes to the epithelium of the glomerulus and the tubules (Brown *et al.*, 1992; Simon *et al.*, 1995).

As embryos that lack even one copy of VEGF die during early embryogenesis due to severe defects in their vasculature, standard knockout technology cannot be used to analyze the role of this protein in kidney developemt (Carmeliet *et al.*, 1996; Ferrara *et al.*, 1996). However, glomerular endothelial cells cultured in the presence of VEGF-coated beads migrated toward the beads. In addition, the migration of glomerular endothelial cells into kidney explants was blocked by an anti-VEGF antibody (Tufro, 2000). The recent generation of mice with a conditionally mutant allele of VEGF has demonstrated a role in vascular growth and maintenance and should allow investigation of the requirements for VEGF at various stages of nephrogenesis (Gerber *et al.*, 1999).

XI. Cell Polarity

Cell polarization is probably required for multiple processes involved in nephrogenesis, including induction, branching morphogenesis, and establishment of the epithelium. Failure to maintain a polarized epithelium results in severe malfunction of the kidney and may lie at the base of a number of human diseases, including polycystic kidney disease (for review, see Wilson, 1997). Surprisingly, although we have gained some understanding of how polarity is maintained, we know almost nothing about how it is established. Discoveries in other model systems, such as *Drosophila* and *C. elegans*, are, however, beginning to identify molecules involved in the establishment of cell polarity, which may also apply to the situation in the kidney (for review, see Yeaman *et al.*, 1999; Knust, 2000).

Establishing cell polarity requires not only the proper establishment and maintenance of adhesional junctions, but also proper vesicular targeting. Advances in our understanding of the function of PKD1 and 2, two genes associated with autosomal-dominant polycystic kidney disease (ADPKD), have provided unique insights into these processes.

PKD1 and 2

The cloning and characterization of two genes responsible for the vast majority of cases of ADPKD have provided us with a fascinating paradigm on the mechanism of apical/basal cell polarity in the kidney (for review, see Koptides and Deltas, 2000; Chapter 24). The products of *PKD1* and *-2*, polycystin-1 and polycystin-2, respectively, were found to encode large, novel proteins with predicted transmembrane domains (Hughes *et al.*, 1995; Mochizuki *et al.*, 1996). Polycystin-1 and *-2* are expressed on the basal side of the proximal and distal tubules and at lower levels in the collecting ducts (Guillaume *et al.*, 1999; Ibraghimov-Beskrovnaya *et al.*, 1997; Lu *et al.*, 1999; Wu *et al.*, 1998). Mutations in the *PKD1* and *PKD2* gene products account for approximately 95% of all cases of ADPKD (a third locus has yet to be identified). Not surprisingly, mutation of both of these genes in mice results in cyst formation in the kidneys as well as other organs (Lu *et al.*, 1997; Wu *et al.*, 1998, 2000; Chapter 24).

Studies aided by the cloning and characterization of homologous genes in *C. elegans*, *Drosophila*, and sea urchins suggest that these two proteins may interact to form an ion channel (for review, see Somlo and Ehrlich, 2001). One might expect that the improper function of an ion channel would be sufficient to explain the cystic phenotype, but, while the exact mechanism remains unknown, cysts frequently show improper localization of polarized membrane proteins, including several growth factors and their receptors (for review, see Wilson, 1997). This improper localization may explain the diseased phenotype, which includes increased cell proliferation rates and apoptosis within the cysts. One potential explanation for the improper localization is the observation that ADPKD cells display defective vesicular trafficking (Charron *et al.*, 2000). As was mentioned previously, proper membrane trafficking is necessary for the accurate sorting of membrane proteins, including channels, transporters, and growth factors. It seems quite reasonable to hypothesize that the PKD gene products are involved in establishing apical basal polarity through vesicular trafficking and that the lack of this polarity results in tubular cysts. Nevertheless, the way in which a potential ion channel regulates trafficking and establishes cell polarity still remains unknown.

XII. Future of the Field

The kidney is a complex organ and, not surprisingly, understanding the developmental mechanisms responsible for its construction is a challenging problem. As the preceding pages indicate, progress has been made, but the field remains in a relatively immature state. Although the kidney has several advantages as a developmental system, there are also several problems intrinsic to the system. Here, we discuss briefly some important issues with regard to the future of the field.

One very basic problem is how the mesenchyme of the metanephric kidney is specified; an event that underpins all subsequent steps in kidney development. Several genes have been implicated in this role from loss-of-function studies, but their exact roles are complicated by the frequent death of the MM in their absence. A helpful alternative approach, one that mirrors the extremely effective analysis of cell fate specification in the mesoderm of amphibians, is to find some other part of the body plan, presumably another area of intermediate mesoderm, which can adopt a kidney fate when supplied with the appropriate gene products. Perhaps the best option here is in the simpler model systems such as the fish and frog where widespread ectopic expression of factors of interest is relatively straightforward. However, it is still not clear whether the simple pronephric kidney of these embryos is really an appropriate model for the metanephric kidney of the mammal.

The general approach of ectopic expression or ectopic application of factors to MM in culture has led to some fascinating results, but these approaches have their own problems. For example, while a number of factors have been implicated *in vitro* in the induction of renal tubules, their involvement has not been supported by mutant studies in the mouse itself. While it is easy to invoke redundancy within a family of related signals or the overlapping roles of parallel pathways, the disconnection between several *in vitro* and *in vivo* analyses does lead to a certain level of discomfort. That a cell can respond in a certain way when factor "x" is added in culture does not mean that it normally responds in the same way in a developing embryo. Furthermore, the complexity of cell interactions within the system can complicate interpretations. For example, several studies have identified non kidney tissues that induce tubulogenesis. The activity of these heterologous inducers led to the reasonable

suggestion that signaling by this tissue mimics the inductive action of the ureteric bud. However, a reinterpretation of these studies in light of experiments on *Wnt4* mutants suggests that one of the most widely used heterologous inducers, the spinal cord, most likely mimics the later action of *Wnt4* within the MM rather than a primary inductive signal from the ureteric bud.

Culture models have played an important role in addressing problems such as tubule induction and ureter branching, early events within the continuum of kidney development. However, they have made little headway into later steps of pattern formation and differentiation within the epithelia of the renal vesicle and ureter, cellular events that are central to the establishment of a physiologically active nephron.

Here it is likely that three approaches will be important. First, a high-resolution, dynamic, analysis of nephric development should be undertaken. Progress has been made in visualizing ureteric branching using green fluorescent protein reporter strains (Srinivas *et al.*, 1999a). When similar analyses within the renal vesicles are integrated with this information into a four-dimensional picture of kidney development, a more thorough consideration of cellular events within the developing organ will be possible. It is no accident that the developmental systems we understand best are those that are underpinned by excellent descriptive (not a popular word these days) studies. However, in order for these analyses to occur, more kidney-specific enhancer elements must be defined. The complete sequencing of several vertebrate genomes should allow a relatively simple comparative approach, identifying conserved regulatory elements within orthologous genes expressed in the developing kidney. However, until more sequence data from different organisms are generated, we may have to rely on the more labor intensive but, at this point, more feasible approach of identifying enhancers from large fragments of genomic DNA. The investment will be worthwhile as the enhancers will certainly find many uses in the field.

A second approach is the use of sophisticated genetic tools. The fact that the kidney is an essential organ is of course one of the reasons why studying its development is important. However, the lethality that so often results from genetic manipulations within the developing organ is a considerable challenge to its investigation (and to the well-being of the investigator). Nevertheless, genetics will probably play the central role in future functional analyses. To be most effective, genetic strategies must allow the investigator to regulate gene activity in the kidney with precise spatial and temporal control. This applies to both loss-of-function and gain-of-function approaches. In so doing it will be possible to address the activity of a gene that has complex expression in several distinct cell populations within the kidney (e.g., *Pax-2*) or within a given population of cells at different stages of its development (e.g., *Wnt4*).

In addition, this will allow us to investigate the role of genes that may play essential roles elsewhere in the body plan prior to kidney development (e.g., *BMP4*). Clearly, engineering new strains of mice that express inducible regulatory factors (e.g., Gal4, Tet) or DNA recombinases (e.g., Cre and Flp) that allow temporal and cell type specific modification of gene activity within the kidney will be an important priority.

Finally, there is the general area of genomics. Across biology, we are in an information-gathering mode and the kidney is no exception. Applying genomic tools that allow wide-scale (in principle whole-genome) profiling of transcription will provide valuable new leads and may in themselves facilitate the process of hypotheses building and model testing that are needed if we are to understand how genes interact and function in generating the kidney.

Acknowledgments

The authors thank the following people for generously providing figures: Jonathan Barasch, Christer Betsholtz, Frank Constantini, Gregory Dressler, Robert Godin, Brigid Hogan, Tsutomu Kume, Eseng Lai, Randy Levinson, Cathy Mendelsohn, Elizabeth Robertson, Patrick Tam, and Tania Tsang. We also thank Ondine Cleaver, Arindam Majumdar, Todd Valerius, and Jing Yu for critical reading and scintillating discussions. Finally, thanks to Renate Hellmiss and David Smith for assistance with the figures. T.C. is supported by a postdoctoral fellowship from the American Cancer Society. Work in the McMahon laboratory is supported by grants from the National Institute of Health.

References

Abdelhak, S., Kalatzis, V., Heilig, R., Compain, S., Samson, D., Vincent, C., Weil, D., Cruaud, C., Sahly, I., Leibovici, M., Bitner-Glindzicz, M., Francis, M., Lacombe, D., Vigneron, J., Charachow, R., Boven, K., Bedbeder, P., Van Regemorter, N., Weissenbach, J., Petit, C. (1997). A human homologue of the drosophila eyes absent gene underlies branchio-oto-renal (BOR) syndrome and identifies a novel gene family. *Nat. Genet.* **15**(2), 157–164.

Adler, S. (1992). Characterization of glomerular epithelial cell matrix receptors. *Am. J. Pathol.* **141**, 571–578.

Agulnick, A. D., Taira, M., Breen, J. J., Tanaka, T., Dawid, I. B., and Westphal, H. (1996). Interactions of the LIM-domain-binding factor Ldb1 with LIM homeodomain proteins. *Nature* **384**, 270–272.

Armstrong, J. F., Pritchard-Jones, K., Bickmore, W. A., Hastie, N. D., and Bard, J. B. (1993). The expression of the Wilms' tumour gene, WT1, in the developing mammalian embryo. *Mech. Dev.* **40**, 85–97.

Asano, M., and Gruss, P. (1992). Pax-5 is expressed at the midbrain-hindbrain boundary during mouse development. *Mech. Dev.* **39**, 29–39.

Bach, I. (2000). The LIM domain: Regulation by association. *Mech. Dev.* **91**, 5–17.

Bach, I., Carriere, C., Ostendorff, H. P., Andersen, B., and Rosenfeld, M. G. (1997). A family of LIM domain-associated cofactors confer transcriptional synergism between LIM and Otx homeodomain proteins. *Genes Dev.* **11**, 1370–1380.

Baeg, G. H., and Perrimon, N. (2000). Functional binding of secreted molecules to heparan sulfate proteoglycans in Drosophila. *Curr. Opin. Cell Biol.* **12**, 575–580.

Barasch, J., Pressler, L., Connor, J., and Malik, A. (1996). A ureteric bud cell line induces nephrogenesis in two steps by two distinct signals. *Am. J. Physiol.* **271**, F50–F61.

Barasch, J., Qiao, J., McWilliams, G., Chen, D., Oliver, J. A., and Herzlinger, D. (1997). Ureteric bud cells secrete multiple factors, including βFGF, which rescue renal progenitors from apoptosis. *Am. J. Physiol.* **273**, F757–F767.

Barasch, J., Yang, J., Ware, C. B., Taga, T., Yoshida, K., Erdjument-Bromage, H., Tempst, P., Parravicini, E., Malach, S., Aranoff, T., and Oliver, J. A. (1999). Mesenchymal to epithelial conversion in rat metanephros is induced by LIF. *Cell* **99**, 377–386.

Bates, C. M., Kharzai, S., Erwin, T., Rossant, J., and Parada, L. F. (2000). Role of N-myc in the developing mouse kidney. *Dev. Biol.* **222**, 317–325.

Batourina, E., Gim, S., Bello, N., Shy, M., Clagett-Dame, M., Srinivas, S., Costantini, F., and Mendelsohn, C. (2001). Vitamin A controls epithelial/mesenchymal interactions through Ret expression. *Nature Genet.* **27**, 74–78.

Bernardini, N., Bianchi, F., Lupetti, M., and Dolfi, A. (1996). Immuno-histochemical localization of the epidermal growth factor, transforming growth factor alpha, and their receptor in the human mesonephros and metanephros. *Dev. Dyn.* **206**, 231–238.

Birchmeier, C., and Gherardi, E. (1998). Developmental roles of HGF/SF and its receptor, the c-Met tyrosine kinase. *Trends Cell Biol.* **8**, 404–410.

Bladt, F., Riethmacher, D., Isenmann, S., Aguzzi, A., and Birchmeier, C. (1995). Essential role for the c-met receptor in the migration of myogenic precursor cells into the limb bud. *Nature* **376**, 768–771.

Bosukonda, D., Shih, M. S., Sampath, K. T., and Vukicevic, S. (2000). Characterization of receptors for osteogenic protein-1/bone morphogenetic protein-7 (OP-1/BMP-7) in rat kidneys. *Kidney Int.* **58**, 1902–1911.

Brandenberger, R., Schmidt, A., Linton, J., Wang, D., Backus, C., Denda, S., Muller, U., and Reichardt, L. F. (2001). Identification and characterization of a novel extracellular matrix protein nephronectin that is associated with integrin alpha8beta1 in the embryonic kidney. *J. Cell Biol.* **154**, 447–458.

Brown, L. F., Berse, B., Tognazzi, K., Manseau, E. J., Van de Water, L., Senger, D. R., Dvorak, H. F., and Rosen, S. (1992). Vascular permeability factor mRNA and protein expression in human kidney. *Kidney Int.* **42**, 1457–1461.

Bullock, S. L., Fletcher, J. M., Beddington, R. S., and Wilson, V. A. (1998). Renal agenesis in mice homozygous for a gene trap mutation in the gene encoding heparan sulfate 2-sulfotransferase. *Genes Dev.* **12**, 1894–1906.

Cano-Gauci, D. F., Song, H. H., Yang, H., McKerlie, C., Choo, B., Shi, W., Pullano, R., Piscione, T. D., Grisaru, S., Soon, S., Sedlackova, L., Tanswell, A. K., Mak, T. W., Yeger, H., Lockwood, G. A., Rosenblum, N. D., and Filmus, J. (1999). Glypican-3-deficient mice exhibit developmental overgrowth and some of the abnormalities typical of Simpson-Golabi-Behmel syndrome. *J. Cell Biol.* **146**, 255–264.

Carmeliet, P., and Collen, D. (1998). Vascular development and disorders: Molecular analysis and pathogenic insights. *Kidney Int.* **53**, 1519–1549.

Carmeliet, P., Ferreira, V., Breier, G., Pollefeyt, S., Kieckens, L., Gertsenstein, M., Fahrig, M., Vandenhoeck, A., Harpal, K., Eberhardt, C., Declercq, C., Pawling, J., Moons, L., Collen, D., Risau, W., and Nagy, A. (1996). Abnormal blood vessel development and lethality in embryos lacking a single VEGF allele. *Nature* **380**, 435–439.

Carroll, T. J., and McMahon, A. P. (2000). Secreted molecules in metanephric induction. *J. Am. Soc. Nephrol.* (**11 Suppl 16**), S116–S119.

Carroll, T. J., and Vize, P. D. (1999). Synergism between Pax-8 and lim-1 in embryonic kidney development. *Dev. Biol.* **214**, 46–59.

Chambon, P. (1996). A decade of molecular biology of retinoic acid receptors. *FASEB J.* **10**, 940–954.

Charron, A. J., Nakamura, S., Bacallao, R., and Wandinger-Ness, A. (2000). Compromised cytoarchitecture and polarized trafficking in autosomal dominant polycystic kidney disease cells. *J. Cell Biol.* **149**, 111–124.

Charron, J., Malynn, B. A., Fisher, P., Stewart, V., Jeannotte, L., Goff, S. P., Robertson, E. J., and Alt, F. W. (1992). Embryonic lethality in mice homozygous for a targeted disruption of the N-myc gene. *Genes Dev.* **6**, 2248–2257.

Chen, H., Lun, Y., Ovchinnikov, D., Kokubo, H., Oberg, K. C., Pepicelli, C. V., Gan, L., Lee, B., and Johnson, R. L. (1998). Limb and kidney defects in Lmx1b mutant mice suggest an involvement of LMX1B in human nail patella syndrome. *Nature Genet.* **19**, 51–55.

Cho, E. A., Patterson, L. T., Brookhiser, W. T., Mah, S., Kintner, C., and Dressler, G. R. (1998). Differential expression and function of cadherin-6 during renal epithelium development. *Development* **125**, 803–812.

Choi, J., and Donehower, L. A. (1999). p53 in embryonic development: Maintaining a fine balance. *Cell Mol. Life Sci.* **55**, 38–47.

Coles, H. S., Burne, J. F., and Raff, M. C. (1993). Large-scale normal cell death in the developing rat kidney and its reduction by epidermal growth factor. *Development* **118**, 777–784.

Curtiss, J., and Heilig, J. S. (1998). DeLIMiting development. *Bioessays* **20**, 58-69.

Davies, J., Lyon, M., Gallagher, J., and Garrod, D. (1995). Sulphated proteoglycan is required for collecting duct growth and branching but not nephron formation during kidney development. *Development* **121**, 1507–1517.

Davies, J. A., and Bard, J. B. (1998). The development of the kidney. *Curr. Top. Dev. Biol.* **39**, 245–301.

Davies, J. A., and Garrod, D. R. (1995). Induction of early stages of kidney tubule differentiation by lithium ions. *Dev. Biol.* **167**, 50-60.

Davies, J. A., Millar, C. B., Johnson, E. M., Jr., and Milbrandt, J. (1999). Neurturin: An autocrine regulator of renal collecting duct development. *Dev. Genet.* **24**, 284–292.

Davis, A. C., Wims, M., Spotts, G. D., Hann, S. R., and Bradley, A. (1993). A null c-myc mutation causes lethality before 10.5 days of gestation in homozygotes and reduced fertility in heterozygous female mice. *Genes Dev.* **7**, 671–682.

Dehbi, M., Ghahremani, M., Lechner, M., Dressler, G., and Pelletier, J. (1996). The paired-box transcription factor, PAX2, positively modulates expression of the Wilms' tumor suppressor gene (WT1). *Oncogene* **13**, 447–453.

Dick, A., Risau, W., and Drexler, H. (1998). Expression of Smad1 and Smad2 during embryogenesis suggests a role in organ development. *Dev. Dyn.* **211**, 293–305.

DiNardo, S., Heemskerk, J., Dougan, S., and O'Farrell, P. H. (1994). The making of a maggot: Patterning the Drosophila embryonic epidermis. *Curr. Opin. Genet. Dev.* **4**, 529–534.

Dolle, P., Ruberte, E., Leroy, P., Morriss-Kay, G., and Chambon, P. (1990). Retinoic acid receptors and cellular retinoid binding proteins. I. A systematic study of their differential pattern of transcription during mouse organogenesis. *Development* **110**, 1133–1151.

Dono, R., and Zeller, R. (1994). Cell-type-specific nuclear translocation of fibroblast growth factor-2 isoforms during chicken kidney and limb morphogenesis. *Dev. Biol.* **163**, 316–330.

Donovan, M. J., Natoli, T. A., Sainio, K., Amstutz, A., Jaenisch, R., Sariola, H., and Kreidberg, J. A. (1999). Initial differentiation of the metanephric mesenchyme is independent of WT1 and the ureteric bud. *Dev. Genet.* **24**, 252–262.

Dressler, G. R., Deutsch, U., Chowdhury, K., Nornes, H. O., and Gruss, P. (1990). Pax2, a new murine paired-box-containing gene and its expression in the developing excretory system. *Development* **109**, 787–795.

Dressler, G. R., Wilkinson, J. E., Rothenpieler, U. W., Patterson, L. T., Williams-Simons, L., and Westphal, H. (1993). Deregulation of Pax-2 expression in transgenic mice generates severe kidney abnormalities. *Nature* **362**, 65–67.

Dreyer, S. D., Zhou, G., Baldini, A., Winterpacht, A., Zabel, B., Cole, W., Johnson, R. L., and Lee, B. (1998). Mutations in LMX1B cause abnormal skeletal patterning and renal dysplasia in nail patella syndrome. *Nature Genet.* **19**, 47–50.

Dudley, A. T., Godin, R. E., and Robertson, E. J. (1999). Interaction between FGF and BMP signaling pathways regulates development of metanephric mesenchyme. *Genes Dev.* **13**, 1601–1613.

Dudley, A. T., Lyons, K. M., and Robertson, E. J. (1995). A requirement for bone morphogenetic protein-7 during development of the mammalian kidney and eye. *Genes Dev.* **9**, 2795–807.

Dudley, A. T., and Robertson, E. J. (1997). Overlapping expression domains of bone morphogenetic protein family members potentially account for limited tissue defects in BMP7 deficient embryos. *Dev. Dyn.* **208**, 349–362.

Durbeej, M., Soderstrom, S., Ebendal, T., Birchmeier, C., and Ekblom, P. (1993). Differential expression of neurotrophin receptors during renal development. *Development* **119**, 977–989.

Ekblom, P. (1981). Formation of basement membranes in the embryonic kidney: An immunohistological study. *J. Cell Biol.* **91**, 1–10.

Ekblom, P., Klein, G., Ekblom, M., and Sorokin, L. (1991). Laminin isoforms and their receptors in the developing kidney. *Am. J. Kidney Dis.* **17**, 603–605.

Favor, J., Sandulache, R., Neuhauser-Klaus, A., Pretsch, W., Chatterjee, B., Senft, E., Wurst, W., Blanquet, V., Grimes, P., Sporle, R., and Schughart, K. (1996). The mouse Pax2(1Neu) mutation is identical to a human PAX2 mutation in a family with renal-coloboma syndrome and results in developmental defects of the brain, ear, eye, and kidney. *Proc. Natl. Acad. Sci. USA* **93**, 13870–13875.

Ferrara, N., Carver-Moore, K., Chen, H., Dowd, M., Lu, L., O'Shea, K. S., Powell-Braxton, L., Hillan, K. J., and Moore, M. W. (1996). Heterozygous embryonic lethality induced by targeted inactivation of the VEGF gene. *Nature* **380**, 439–442.

Fujii, T., Pichel, J. G., Taira, M., Toyama, R., Dawid, I. B., and Westphal, H. (1994). Expression patterns of the murine LIM class homeobox gene lim1 in the developing brain and excretory system. *Dev. Dyn.* **199**, 73–83.

Gerber, H. P., Hillan, K. J., Ryan, A. M., Kowalski, J., Keller, G. A., Rangell, L., Wright, B. D., Radtke, F., Aguet, M., and Ferrara, N. (1999). VEGF is required for growth and survival in neonatal mice. *Development* **126**, 1149–1159.

Godin, R. E., Takaesu, N. T., Robertson, E. J., and Dudley, A. T. (1998). Regulation of BMP7 expression during kidney development. *Development* **125**, 3473–3482.

Godley, L. A., Kopp, J. B., Eckhaus, M., Paglino, J. J., Owens, J., and Varmus, H. E. (1996). Wild-type p53 transgenic mice exhibit altered differentiation of the ureteric bud and possess small kidneys. *Genes Dev.* **10**, 836–850.

Gossens, C. L., and Unsworth, B. R. (1972). Evidence for a two-step mechanism operating during in vitro mouse kidney tubulogenesis. *J. Embryol. Exp. Morphol.* **28**, 615–631.

Grandori, C., Cowley, S. M., James, L. P., and Eisenman, R. N. (2000). The Myc/Max/Mad network and the transcriptional control of cell behavior. *Annu. Rev. Cell Dev. Biol.* **16**, 653–699.

Grandori, C., and Eisenman, R. N. (1997). Myc target genes. *Trends Biochem. Sci.* **22**, 177–181.

Grisaru, S., Cano-Gauci, D., Tee, J., Filmus, J., and Rosenblum, N. D. (2001). Glypican-3 modulates BMP- and FGF-mediated effects during renal branching morphogenesis. *Dev. Biol.* **231**, 31–46.

Guillaume, R., D'Agati, V., Daoust, M., and Trudel, M. (1999). Murine Pkd1 is a developmentally regulated gene from morula to adulthood: Role in tissue condensation and patterning. *Dev. Dyn.* **214**, 337–348.

Haerry, T. E., Heslip, T. R., Marsh, J. L., and O'Connor, M. B. (1997). Defects in glucuronate biosynthesis disrupt Wingless signaling in Drosophila. *Development* **124**, 3055–3064.

Hamada, Y., Kadokawa, Y., Okabe, M., Ikawa, M., Coleman, J. R., and Tsujimoto, Y. (1999). Mutation in ankyrin repeats of the mouse Notch2 gene induces early embryonic lethality. *Development* **126**, 3415–3424.

Hartung, H., Feldman, B., Lovec, H., Coulier, F., Birnbaum, D., and Goldfarb, M. (1997). Murine FGF-12 and FGF-13: Expression in embryonic nervous system, connective tissue and heart. *Mech. Dev.* **64**, 31–39.

Hatini, V., Huh, S. O., Herzlinger, D., Soares, V. C., and Lai, E. (1996). Essential role of stromal mesenchyme in kidney morphogenesis revealed by targeted disruption of Winged Helix transcription factor BF-2. *Genes Dev.* **10**, 1467–1478.

Hatton, K. S., Mahon, K., Chin, L., Chiu, F. C., Lee, H. W., Peng, D., Morgenbesser, S. D., Horner, J., and DePinho, R. A. (1996). Expression and activity of L-Myc in normal mouse development. *Mol. Cell. Biol.* **16**, 1794–1804.

Heckman, C., Mochon, E., Arcinas, M., and Boxer, L. M. (1997). The WT1 protein is a negative regulator of the normal bcl-2 allele in t(14;18) lymphomas. *J. Biol. Chem.* **272**, 19609–19614.

Hedgepeth, C. M., Conrad, L. J., Zhang, J., Huang, H. C., Lee, V. M., and Klein, P. S. (1997). Activation of the Wnt signaling pathway: A molecular mechanism for lithium action. *Dev. Biol.* **185**, 82–91.

Hellmich, H. L., Kos, L., Cho, E. S., Mahon, K. A., and Zimmer, A. (1996). Embryonic expression of glial cell-line derived neurotrophic factor (GDNF) suggests multiple developmental roles in neural differentiation and epithelial-mesenchymal interactions. *Mech. Dev.* **54**, 95–105.

Henriksson, M., and Luscher, B. (1996). Proteins of the Myc network: Essential regulators of cell growth and differentiation. *Adv. Cancer Res.* **68**, 109–182.

Hewitt, S. M., Hamada, S., Monarres, A., Kottical, L. V., Saunders, G. F., and McDonnell, T. J. (1997). Transcriptional activation of the bcl-2 apoptosis suppressor gene by the paired box transcription factor PAX8. *Anticancer Res.* **17**, 3211–3215.

Hilger-Eversheim, K., Moser, M., Schorle, H., and Buettner, R. (2000). Regulatory roles of AP-2 transcription factors in vertebrate development, apoptosis and cell-cycle control. *Gene* **260**, 1–12.

Hobert, O., and Westphal, H. (2000). Functions of LIM-homeobox genes. *Trends Genet.* **16**, 75–83.

Huber, L. J., and Chao, M. V. (1995). Mesenchymal and neuronal cell expression of the p75 neurotrophin receptor gene occur by different mechanisms. *Dev. Biol.* **167**, 227–238.

Hughes, J., Ward, C. J., Peral, B., Aspinwall, R., Clark, K., San Millan, J. L., Gamble, V., and Harris, P. C. (1995). The polycystic kidney disease 1 (PKD1) gene encodes a novel protein with multiple cell recognition domains. *Nature Genet.* **10**, 151–160.

Ibraghimov-Beskrovnaya, O., Dackowski, W. R., Foggensteiner, L., Coleman, N., Thiru, S., Petry, L. R., Burn, T. C., Connors, T. D., Van Raay, T., Bradley, J., Qian, F., Onuchic, L. F., Watnick, T. J., Piontek, K., Hakim, R. M., Landes, G. M., Germino, G. G., Sandford, R., and Klinger, K. W. (1997). Polycystin: In vitro synthesis, in vivo tissue expression, and subcellular localization identifies a large membrane-associated protein. *Proc. Natl. Acad. Sci. USA* **94**, 6397–6402.

Ichimura, T., Finch, P. W., Zhang, G., Kan, M., and Stevens, J. L. (1996). Induction of FGF-7 after kidney damage: A possible paracrine mechanism for tubule repair. *Am. J. Physiol.* **271**, F967–F976.

Kalatzis, V., Sahly, I., El-Amraoui, A., and Petit, C. (1998). Eya1 expression in the developing ear and kidney: towards the understanding of the pathogenesis of Branchio-Oto-Renal (BOR) syndrome. *Dev. Dyn.* **213**, 486–499.

Karavanov, A., Sainio, K., Palgi, J., Saarma, M., Saxen, L., and Sariola, H. (1995). Neurotrophin 3 rescues neuronal precursors from apoptosis and promotes neuronal differentiation in the embryonic metanephric kidney. *Proc. Natl. Acad. Sci. USA* **92**, 11279–11283.

Karavanov, A. A., Karavanova, I., Perantoni, A., and Dawid, I. B. (1998). Expression pattern of the rat Lim-1 homeobox gene suggests a dual role during kidney development. *Int. J. Dev. Biol.* **42**, 61–66.

Karavanova, I. D., Dove, L. F., Resau, J. H., and Perantoni, A. O. (1996). Conditioned medium from a rat ureteric bud cell line in combination with bFGF induces complete differentiation of isolated metanephric mesenchyme. *Development* **122**, 4159–4167.

Kawakami, Y., Capdevila, J., Buscher, D., Itoh, T., Rodriguez Esteban, C., and Izpisua Belmonte, J. C. (2001). WNT signals control FGF-dependent limb initiation and AER induction in the chick embryo. *Cell* **104**, 891–900.

Kispert, A., Vainio, S., and McMahon, A. P. (1998). Wnt-4 is a mesenchymal signal for epithelial transformation of metanephric mesenchyme in the developing kidney. *Development* **125**, 4225–4234.

Kispert, A., Vainio, S., Shen, L., Rowitch, D. H., and McMahon, A. P. (1996). Proteoglycans are required for maintenance of Wnt-11 expression in the ureter tips. *Development* **122**, 3627–3637.

Klein, P. S., and Melton, D. A. (1996). A molecular mechanism for the effect of lithium on development. *Proc. Natl. Acad. Sci. USA* **93**, 8455–8459.

Knust, E. (2000). Control of epithelial cell shape and polarity. *Curr. Opin. Genet. Dev.* **10**, 471–475.

Koptides, M., and Deltas, C. C. (2000). Autosomal dominant polycystic kidney disease: Molecular genetics and molecular pathogenesis. *Hum. Genet.* **107**, 115–126.

Koseki, C. (1993). Cell death programmed in uninduced metanephric mesenchymal cells. *Pediatr. Nephrol.* **7**, 609–611.

Koseki, C., Herzlinger, D., and al-Awqati, Q. (1992). Apoptosis in metanephric development. *J. Cell Biol.* **119**, 1327–1333.

Kreidberg, J. A., Donovan, M. J., Goldstein, S. L., Rennke, H., Shepherd, K., Jones, R. C., and Jaenisch, R. (1996). Alpha 3 beta 1 integrin has a crucial role in kidney and lung organogenesis. *Development* **122**, 3537–3547.

Kreidberg, J. A., Sariola, H., Loring, J. M., Maeda, M., Pelletier, J., Housman, D., and Jaenisch, R. (1993). WT-1 is required for early kidney development. *Cell* **74**, 679–691.

Kreidberg, J. A., and Symons, J. M. (2000). Integrins in kidney development, function, and disease. *Am. J. Physiol. Renal Physiol.* **279**, F233–F242.

Kume, T., Deng, K., and Hogan, B. L. (2000a). Minimal phenotype of mice homozygous for a null mutation in the forkhead/winged helix gene, Mf2. *Mol. Cell. Biol.* **20**, 1419–1425.

Kume, T., Deng, K., and Hogan, B. L. (2000b). Murine forkhead/winged helix genes Foxc1 (Mf1) and Foxc2 (Mfh1) are required for the early organogenesis of the kidney and urinary tract. *Development* **127**, 1387–1395.

Larsson, S. H., Charlieu, J. P., Miyagawa, K., Engelkamp, D., Rassoulzadegan, M., Ross, A., Cuzin, F., van Heyningen, V., and Hastie, N. D. (1995). Subnuclear localization of WT1 in splicing or transcription factor domains is regulated by alternative splicing. *Cell* **81**, 391–401.

Lee, S. B., and Haber, D. A. (2001). Wilms tumor and the WT1 gene. *Exp. Cell Res.* **264**, 74–99.

Lee, S. B., Huang, K., Palmer, R., Truong, V. B., Herzlinger, D., Kolquist, K. A., Wong, J., Paulding, C., Yoon, S. K., Gerald, W., Oliner, J. D., and Haber, D. A. (1999). The Wilms tumor suppressor WT1 encodes a transcriptional activator of amphiregulin. *Cell* **98**, 663–673.

Lelongt, B., Makino, H., Dalecki, T. M., and Kanwar, Y. S. (1988). Role of proteoglycans in renal development. *Dev. Biol.* **128**, 256–276.

Lelongt, B., Trugnan, G., Murphy, G., and Ronco, P. M. (1997). Matrix metalloproteinases MMP2 and MMP9 are produced in early stages of kidney morphogenesis but only MMP9 is required for renal organogenesis in vitro. *J. Cell Biol.* **136**, 1363–1373.

Lescher, B., Haenig, B., and Kispert, A. (1998). sFRP-2 is a target of the Wnt-4 signaling pathway in the developing metanephric kidney. *Dev. Dyn.* **213**, 440–451.

Leveen, P., Pekny, M., Gebre-Medhin, S., Swolin, B., Larsson, E., and Betsholtz, C. (1994). Mice deficient for PDGF B show renal, cardiovascular, and hematological abnormalities. *Genes Dev.* **8**, 1875–1887.

Li, M., Sendtner, M., and Smith, A. (1995). Essential function of LIF receptor in motor neurons. *Nature* **378**, 724–727.

Lin, X., Buff, E. M., Perrimon, N., and Michelson, A. M. (1999). Heparan sulfate proteoglycans are essential for FGF receptor signaling during Drosophila embryonic development. *Development* 126, 3715–3723.

Lin, Y., Zhang, S., Rehn, M., Itaranta, P., Tuukkanen, J., Heljasvaara, R., Peltoketo, H., Pihlajaniemi, T., and Vainio, S. (2001). Induced repatterning of type XVIII collagen expression in ureter bud from kidney to lung type: association with sonic hedgehog and ectopic surfactant protein C. *Development* 128, 1573–1585.

Lindahl, P., Hellstrom, M., Kalen, M., Karlsson, L., Pekny, M., Pekna, M., Soriano, P., and Betsholtz, C. (1998). Paracrine PDGF-B/PDGF-Rbeta signaling controls mesangial cell development in kidney glomeruli. *Development* 125, 3313–3322.

Lindenbergh-Kortleve, D. J., Rosato, R. R., van Neck, J. W., Nauta, J., van Kleffens, M., Groffen, C., Zwarthoff, E. C., and Drop, S. L. (1997). Gene expression of the insulin-like growth factor system during mouse kidney development. *Mol. Cell. Endocrinol.* 132, 81–91.

Little, M., Holmes, G., and Walsh, P. (1999). WT1: What has the last decade told us? *Bioessays* 21, 191–202.

Liu, J. P., Baker, J., Perkins, A. S., Robertson, E. J., and Efstratiadis, A. (1993a). Mice carrying null mutations of the genes encoding insulin-like growth factor I (Igf-1) and type 1 IGF receptor (Igf1r). *Cell* 75, 59–72.

Liu, Z. Z., Kumar, A., Ota, K., Wallner, E. I., and Kanwar, Y. S. (1997). Developmental regulation and the role of insulin and insulin receptor in metanephrogenesis. *Proc. Natl. Acad. Sci. USA* 94, 6758–6763.

Liu, Z. Z., Wada, J., Alvares, K., Kumar, A., Wallner, E. I., and Kanwar, Y. S. (1993b). Distribution and relevance of insulin-like growth factor-I receptor in metanephric development. *Kidney Int.* 44, 1242–1250.

Lu, W., Fan, X., Basora, N., Babakhanlou, H., Law, T., Rifai, N., Harris, P. C., Perez-Atayde, A. R., Rennke, H. G., and Zhou, J. (1999). Late onset of renal and hepatic cysts in Pkd1-targeted heterozygotes. *Nature Genet.* 21, 160–161.

Lu, W., Peissel, B., Babakhanlou, H., Pavlova, A., Geng, L., Fan, X., Larson, C., Brent, G., and Zhou, J. (1997). Perinatal lethality with kidney and pancreas defects in mice with a targetted Pkd1 mutation. *Nature Genet.* 17, 179–181.

Luetteke, N. C., Qiu, T. H., Fenton, S. E., Troyer, K. L., Riedel, R. F., Chang, A., and Lee, D. C. (1999). Targeted inactivation of the EGF and amphiregulin genes reveals distinct roles for EGF receptor ligands in mouse mammary gland development. *Development* 126, 2739–2750.

Luo, G., Hofmann, C., Bronckers, A. L., Sohocki, M., Bradley, A., and Karsenty, G. (1995). BMP-7 is an inducer of nephrogenesis, and is also required for eye development and skeletal patterning. *Genes Dev.* 9, 2808–2820.

Majumdar, A., Lun, K., Brand, M., and Drummond, I. A. (2000). Zebrafish no isthmus reveals a role for pax2.1 in tubule differentiation and patterning events in the pronephric primordia. *Development* 127, 2089-2098.

Mangelsdorf, D. J., and Evans, R. M. (1995). The RXR heterodimers and orphan receptors. *Cell* 83, 841–850.

Mansouri, A., Goudreau, G., and Gruss, P. (1999). Pax genes and their role in organogenesis. *Cancer Res.* 59, 1707s–1709s; discussion 1709s–1710s.

Mason, I. J., Fuller-Pace, F., Smith, R., and Dickson, C. (1994). FGF-7 (keratinocyte growth factor) expression during mouse development suggests roles in myogenesis, forebrain regionalisation and epithelial-mesenchymal interactions. *Mech. Dev.* 45, 15–30.

Matsell, D. G., Delhanty, P. J., Stepaniuk, O., Goodyear, C., and Han, V. K. (1994). Expression of insulin-like growth factor and binding protein genes during nephrogenesis. *Kidney Int.* 46, 1031–1042.

Mayo, M. W., Wang, C. Y., Drouin, S. S., Madrid, L. V., Marshall, A. F., Reed, J. C., Weissman, B. E., and Baldwin, A. S. (1999). WT1 modulates apoptosis by transcriptionally upregulating the bcl-2 proto-oncogene. *EMBO J.* 18, 3990–4003.

McConnell, M. J., Cunliffe, H. E., Chua, L. J., Ward, T. A., and Eccles, M. R. (1997). Differential regulation of the human Wilms tumour

suppressor gene (WT1) promoter by two isoforms of PAX2. *Oncogene* 14, 2689–2700.

McCright, B., Gao, X., Shen, L., Lozier, J., Lan, Y., Maguire, M., Herzlinger, D., Weinmaster, G., Jiang, R., and Gridley, T. (2001). Defects in development of the kidney, heart and eye vasculature in mice homozygous for a hypomorphic Notch2 mutation. *Development* 128, 491–502.

Mendelsohn, C., Batourina, E., Fung, S., Gilbert, T., and Dodd, J. (1999). Stromal cells mediate retinoid-dependent functions essential for renal development. *Development* 126, 1139–1148.

Mendelsohn, C., Lohnes, D., Decimo, D., Lufkin, T., LeMeur, M., Chambon, P., and Mark, M. (1994). Function of the retinoic acid receptors (RARs) during development (II). Multiple abnormalities at various stages of organogenesis in RAR double mutants. *Development* 120, 2749–2771.

Miller, J. R., Hocking, A. M., Brown, J. D., and Moon, R. T. (1999). Mechanism and function of signal transduction by the Wnt/beta-catenin and Wnt/Ca^{2+} pathways. *Oncogene* 18, 7860–7872.

Miner, J. H. (1999). Renal basement membrane components. *Kidney Int.* 56, 2016–2024.

Miner, J. H. (2001). Mystery solved: Discovery of a novel integrin ligand in the developing kidney. *J. Cell Biol.* 154, 257–260.

Miner, J. H., Lewis, R. M., and Sanes, J. R. (1995). Molecular cloning of a novel laminin chain, alpha 5, and widespread expression in adult mouse tissues. *J. Biol. Chem.* 270, 28523–28526.

Miner, J. H., and Li, C. (2000). Defective glomerulogenesis in the absence of laminin alpha5 demonstrates a developmental role for the kidney glomerular basement membrane. *Dev. Biol.* 217, 278–289.

Miner, J. H., Patton, B. L., Lentz, S. I., Gilbert, D. J., Snider, W. D., Jenkins, N. A., Copeland, N. G., and Sanes, J. R. (1997). The laminin alpha chains: Expression, developmental transitions, and chromosomal locations of alpha1-5, identification of heterotrimeric laminins 8-11, and cloning of a novel alpha3 isoform. *J. Cell Biol.* 137, 685–701.

Miyamoto, N., Yoshida, M., Kuratani, S., Matsuo, I., and Aizawa, S. (1997). Defects of urogenital development in mice lacking Emx2. *Development* 124, 1653–1664.

Miyazaki, Y., Oshima, K., Fogo, A., Hogan, B. L., and Ichikawa, I. (2000). Bone morphogenetic protein 4 regulates the budding site and elongation of the mouse ureter. *J. Clin. Invest.* 105, 863–873.

Miyazono, K. (2000). Positive and negative regulation of TGF-beta signaling. *J. Cell Sci.* 113, 1101–1109.

Miyazono, K., Kusanagi, K., and Inoue, H. (2001). Divergence and convergence of TGF-beta/BMP signaling. *J. Cell Physiol.* 187, 265–276.

Mochizuki, T., Wu, G., Hayashi, T., Xenophontos, S. L., Veldhuisen, B., Saris, J. J., Reynolds, D. M., Cai, Y., Gabow, P. A., Pierides, A., Kimberling, W. J., Breuning, M. H., Deltas, C. C., Peters, D. J., and Somlo, S. (1996). PKD2, a gene for polycystic kidney disease that encodes an integral membrane protein. *Science* 272, 1339–1342.

Moens, C. B., Stanton, B. R., Parada, L. F., and Rossant, J. (1993). Defects in heart and lung development in compound heterozygotes for two different targeted mutations at the N-myc locus. *Development* 119, 485–499.

Montesano, R., Matsumoto, K., Nakamura, T., and Orci, L. (1991). Identification of a fibroblast-derived epithelial morphogen as hepatocyte growth factor. *Cell* 67, 901–908.

Moore, A. W., McInnes, L., Kreidberg, J., Hastie, N. D., and Schedl, A. (1999). YAC complementation shows a requirement for Wt1 in the development of epicardium, adrenal gland and throughout nephrogenesis. *Development* 126, 1845–1857.

Moreau, E., Vilar, J., Lelievre-Pegorier, M., Merlet-Benichou, C., and Gilbert, T. (1998). Regulation of c-ret expression by retinoic acid in rat metanephros: Implication in nephron mass control. *Am. J. Physiol.* 275, F938–F945.

Moser, M., Pscherer, A., Roth, C., Becker, J., Mucher, G., Zerres, K., Dixkens, C., Weis, J., Guay-Woodford, L., Buettner, R., and Fassler, R.

(1997). Enhanced apoptotic cell death of renal epithelial cells in mice lacking transcription factor AP-2beta. *Genes Dev.* **11**, 1938–1948.

Mrowka, C., and Schedl, A. (2000). Wilms' tumor suppressor gene WT1: From structure to renal pathophysiologic features. *J. Am. Soc. Nephrol.* **11 (Suppl 16)**, S106–S115.

Mugrauer, G., and Ekblom, P. (1991). Contrasting expression patterns of three members of the myc family of protooncogenes in the developing and adult mouse kidney. *J. Cell Biol.* **112**, 13–25.

Muller, U., and Brandli, A. W. (1999). Cell adhesion molecules and extra-cellular-matrix constituents in kidney development and disease. *J. Cell Sci.* **112**, 3855–3867.

Muller, U., Wang, D., Denda, S., Meneses, J. J., Pedersen, R. A., and Reichardt, L. F. (1997). Integrin alpha8beta1 is critically important for epithelial-mesenchymal interactions during kidney morphogenesis. *Cell* **88**, 603–613.

Nagata, M., Nakauchi, H., Nakayama, K., Loh, D., and Watanabe, T. (1996). Apoptosis during an early stage of nephrogenesis induces renal hypoplasia in bcl-2-deficient mice. *Am. J. Pathol.* **148**, 1601–1611.

Noramly, S., and Morgan, B. A. (1998). BMPs mediate lateral inhibition at successive stages in feather tract development. *Development* **125**, 3775–3787.

Ortiz, A., Lorz, C., Catalan, M. P., Justo, P., and Egido, J. (2000). Role and regulation of apoptotic cell death in the kidney. Y2K update. *Front. Biosci.* **5**, D735–D749.

Ostrom, L., Tang, M. J., Gruss, P., and Dressler, G. R. (2000). Reduced Pax2 gene dosage increases apoptosis and slows the progression of renal cystic disease. *Dev. Biol.* **219**, 250–258.

Pachnis, V., Mankoo, B., and Costantini, F. (1993). Expression of the c-ret proto-oncogene during mouse embryogenesis. *Development* **119**, 1005–1017.

Partanen, A. M., and Thesleff, I. (1987). Localization and quantitation of 125I-epidermal growth factor binding in mouse embryonic tooth and other embryonic tissues at different developmental stages. *Dev. Biol.* **120**, 186–197.

Patapoutian, A., Backus, C., Kispert, A., and Reichardt, L. F. (1999). Regulation of neurotrophin-3 expression by epithelial-mesenchymal interactions: The role of Wnt factors. *Science* **283**, 1180–1183.

Pellegrini, M., Pantano, S., Lucchini, F., Fumi, M., and Forabosco, A. (1997). Emx2 developmental expression in the primordia of the reproductive and excretory systems. *Anat. Embryol. (Berl).* **196**, 427–433.

Pepicelli, C. V., Kispert, A., Rowitch, D. H., and McMahon, A. P. (1997). GDNF induces branching and increased cell proliferation in the ureter of the mouse. *Dev. Biol.* **192**, 193–198.

Perantoni, A. O., Dove, L. F., and Karavanova, I. (1995). Basic fibroblast growth factor can mediate the early inductive events in renal development. *Proc. Natl. Acad. Sci. USA* **92**, 4696–4700.

Perantoni, A. O., Dove, L. F., and Williams, C. L. (1991). Induction of tubules in rat metanephrogenic mesenchyme in the absence of an inductive tissue. *Differentiation* **48**, 25-31.

Pfeffer, P. L., Bouchard, M., and Busslinger, M. (2000). Pax2 and homeodomain proteins cooperatively regulate a 435 bp enhancer of the mouse Pax5 gene at the midbrain-hindbrain boundary. *Development* **127**, 1017–1028.

Pichel, J. G., Shen, L., Sheng, H. Z., Granholm, A. C., Drago, J., Grinberg, A., Lee, E. J., Huang, S. P., Saarma, M., Hoffer, B. J., Sariola, H., and Westphal, H. (1996). Defects in enteric innervation and kidney development in mice lacking GDNF. *Nature* **382**, 73–76.

Plachov, D., Chowdhury, K., Walther, C., Simon, D., Guenet, J. L., and Gruss, P. (1990). Pax8, a murine paired box gene expressed in the developing excretory system and thyroid gland. *Development* **110**, 643–651.

Plisov, S. Y., Yoshino, K., Dove, L. F., Higinbotham, K. G., Rubin, J. S., and Perantoni, A. O. (2001). TGF beta 2, LIF and FGF2 cooperate to induce nephrogenesis. *Development* **128**, 1045–1057.

Porteous, S., Torban, E., Cho, N. P., Cunliffe, H., Chua, L., McNoe, L.,

Ward, T., Souza, C., Gus, P., Giugliani, R., Sato, T., Yun, K., Favor, J., Sicotte, M., Goodyer, P., and Eccles, M. (2000). Primary renal hypoplasia in humans and mice with PAX2 mutations: Evidence of increased apoptosis in fetal kidneys of Pax2(1Neu) +/- mutant mice. *Hum. Mol. Genet.* **9**, 1–11.

Qiao, J., Uzzo, R., Obara-Ishihara, T., Degenstein, L., Fuchs, E., and Herzlinger, D. (1999). FGF-7 modulates ureteric bud growth and nephron number in the developing kidney. *Development* **126**, 547–554.

Quaggin, S. E., Schwartz, L., Cui, S., Igarashi, P., Deimling, J., Post, M., and Rossant, J. (1999). The basic-helix-loop-helix protein pod1 is critically important for kidney and lung organogenesis. *Development* **126**, 5771–5783.

Quaggin, S. E., Vanden Heuvel, G. B., and Igarash, P. (1998). Pod-1, a mesoderm-specific basic-helix-loop-helix protein expressed in mesenchymal and glomerular epithelial cells in the developing kidney. *Mech. Dev.* **71**, 37–48.

Raatikainen-Ahokas, A., Hytonen, M., Tenhunen, A., Sainio, K., and Sariola, H. (2000). BMP-4 affects the differentiation of metanephric mesenchyme and reveals an early anterior-posterior axis of the embryonic kidney. *Dev. Dyn.* **217**, 146–158.

Rabb, H. A. (1994). Cell adhesion molecules and the kidney. *Am. J. Kidney Dis.* **23**, 155–166.

Richardson, A., and Parsons, J. T. (1995). Signal transduction through integrins: A central role for focal adhesion kinase? *Bioessays* **17**, 229–236.

Robertson, K., and Mason, I. (1997). The GDNF-RET signalling partnership. *Trends Genet.* **13**, 1–3.

Rodriguez-Esteban, C., Tsukui, T., Yonei, S., Magallon, J., Tamura, K., and Izpisua Belmonte, J. C. (1999). The T-box genes Tbx4 and Tbx5 regulate limb outgrowth and identity. *Nature* **398**, 814–818.

Rothenpieler, U. W., and Dressler, G. R. (1993). Pax-2 is required for mesenchyme-to-epithelium conversion during kidney development. *Development* **119**, 711–720.

Rother, K. I., and Accili, D. (2000). Role of insulin receptors and IGF receptors in growth and development. *Pediatr. Nephrol.* **14**, 558–561.

Ryan, G., Steele-Perkins, V., Morris, J. F., Rauscher, F. J., 3rd, and Dressler, G. R. (1995). Repression of Pax-2 by WT1 during normal kidney development. *Development* **121**, 867–875.

Sainio, K., Suvanto, P., Davies, J., Wartiovaara, J., Wartiovaara, K., Saarma, M., Arumae, U., Meng, X., Lindahl, M., Pachnis, V., and Sariola, H. (1997). Glial-cell-line-derived neurotrophic factor is required for bud initiation from ureteric epithelium. *Development* **124**, 4077–4087.

Sakurai, H., Barros, E. J., Tsukamoto, T., Barasch, J., and Nigam, S. K. (1997). An in vitro tubulogenesis system using cell lines derived from the embryonic kidney shows dependence on multiple soluble growth factors. *Proc. Natl. Acad. Sci. USA* **94**, 6279–6284.

Sanchez, M. P., Silos-Santiago, I., Frisen, J., He, B., Lira, S. A., and Barbacid, M. (1996). Renal agenesis and the absence of enteric neurons in mice lacking GDNF. *Nature* **382**, 70–73.

Santos, O. F., Barros, E. J., Yang, X. M., Matsumoto, K., Nakamura, T., Park, M., and Nigam, S. K. (1994). Involvement of hepatocyte growth factor in kidney development. *Dev. Biol.* **163**, 525–529.

Santos, O. F., and Nigam, S. K. (1993). HGF-induced tubulogenesis and branching of epithelial cells is modulated by extracellular matrix and TGF-beta. *Dev. Biol.* **160**, 293–302.

Sanyanusin, P., McNoe, L. A., Sullivan, M. J., Weaver, R. G., and Eccles, M. R. (1995). Mutation of PAX2 in two siblings with renal-coloboma syndrome. *Hum. Mol. Genet.* **4**, 2183–2184.

Sariola, H., Aufderheide, E., Bernhard, H., Henke-Fahle, S., Dippold, W., and Ekblom, P. (1988). Antibodies to cell surface ganglioside GD3 perturb inductive epithelial-mesenchymal interactions. *Cell* **54**, 235–245.

Saxén, L. (1970). Failure to demonstrate tubule induction in a heterologous mesenchyme. *Dev. Biol.* **23**, 511–523.

Saxén, L. (1987). "Organogenesis of the Kidney." Cambridge Univ. Press, New York.

Schmeichel, K. L., and Beckerle, M. C. (1994). The LIM domain is a modular protein-binding interface. *Cell* **79**, 211–219.

Schmidt, C., Bladt, F., Goedecke, S., Brinkmann, V., Zschiesche, W., Sharpe, M., Gherardi, E., and Birchmeier, C. (1995). Scatter factor/hepatocyte growth factor is essential for liver development. *Nature* **373**, 699–702.

Selleck, S. B. (2000). Proteoglycans and pattern formation: Sugar biochemistry meets developmental genetics. *Trends Genet.* **16**, 206–212.

Shawlot, W., and Behringer, R. R. (1995). Requirement for Lim1 in head-organizer function. *Nature* **374**, 425–430.

Sibilia, M., and Wagner, E. F. (1995). Strain-dependent epithelial defects in mice lacking the EGF receptor. *Science* **269**, 234–238.

Simeone, A., Acampora, D., Gulisano, M., Stornaiuolo, A., and Boncinelli, E. (1992a). Nested expression domains of four homeobox genes in developing rostral brain. *Nature* **358**, 687–690.

Simeone, A., Gulisano, M., Acampora, D., Stornaiuolo, A., Rambaldi, M., and Boncinelli, E. (1992b). Two vertebrate homeobox genes related to the Drosophila empty spiracles gene are expressed in the embryonic cerebral cortex. *EMBO J.* **11**, 2541–2550.

Simon, M., Grone, H. J., Johren, O., Kullmer, J., Plate, K. H., Risau, W., and Fuchs, E. (1995). Expression of vascular endothelial growth factor and its receptors in human renal ontogenesis and in adult kidney. *Am. J. Physiol.* **268**, F240–F250.

Somlo, S., and Ehrlich, B. (2001). Human disease: Calcium signaling in polycystic kidney disease. *Curr. Biol.* **11**, R356–R360.

Sonnenberg, E., Meyer, D., Weidner, K. M., and Birchmeier, C. (1993). Scatter factor/hepatocyte growth factor and its receptor, the c-met tyrosine kinase, can mediate a signal exchange between mesenchyme and epithelia during mouse development. *J. Cell Biol.* **123**, 223–235.

Sorenson, C. M. (1999). Nuclear localization of beta-catenin and loss of apical brush border actin in cystic tubules of bcl-2 -/- mice. *Am. J. Physiol.* **276**, F210–F217.

Sorenson, C. M., Rogers, S. A., Korsmeyer, S. J., and Hammerman, M. R. (1995). Fulminant metanephric apoptosis and abnormal kidney development in bcl-2-deficient mice. *Am. J. Physiol.* **268**, F73–F81.

Sorenson, C. M., and Sheibani, N. (1999). Focal adhesion kinase, paxillin, and bcl-2: analysis of expression, phosphorylation, and association during morphogenesis. *Dev. Dyn.* **215**, 371–382.

Soriano, P. (1994). Abnormal kidney development and hematological disorders in PDGF beta-receptor mutant mice. *Genes Dev.* **8**, 1888–1896.

Srinivas, S., Goldberg, M. R., Watanabe, T., D'Agati, V., al-Awqati, Q., and Costantini, F. (1999a). Expression of green fluorescent protein in the ureteric bud of transgenic mice: A new tool for the analysis of ureteric bud morphogenesis. *Dev. Genet.* **24**, 241–251.

Srinivas, S., Wu, Z., Chen, C. M., D'Agati, V., and Costantini, F. (1999b). Dominant effects of RET receptor misexpression and ligand-independent RET signaling on ureteric bud development. *Development* **126**, 1375–1386.

Stark, K., Vainio, S., Vassileva, G., and McMahon, A. P. (1994). Epithelial transformation of metanephric mesenchyme in the developing kidney regulated by Wnt-4. *Nature* **372**, 679–683.

Stark, K. L., McMahon, J. A., and McMahon, A. P. (1991). FGFR-4, a new member of the fibroblast growth factor receptor family, expressed in the definitive endoderm and skeletal muscle lineages of the mouse. *Development* **113**, 641–651.

Stewart, C. L., Kaspar, P., Brunet, L. J., Bhatt, H., Gadi, I., Kontgen, F., and Abbondanzo, S. J. (1992). Blastocyst implantation depends on maternal expression of leukaemia inhibitory factor. *Nature* **359**, 76–79.

Sugawara, A., Sanno, N., Takahashi, N., Osamura, R. Y., and Abe, K. (1997). Retinoid X receptors in the kidney: Their protein expression and functional significance. *Endocrinology* **138**, 3175–3180.

Suvanto, P., Hiltunen, J. O., Arumae, U., Moshnyakov, M., Sariola, H., Sainio, K., and Saarma, M. (1996). Localization of glial cell line-derived neurotrophic factor (GDNF) mRNA in embryonic rat by in situ hybridization. *Eur. J. Neurosci.* **8**, 816–822.

Takahashi, T., Huynh-Do, U., and Daniel, T. O. (1998). Renal microvascular assembly and repair: power and promise of molecular definition. *Kidney Int.* **53**, 826–835.

Takeuchi, J. K., Koshiba-Takeuchi, K., Matsumoto, K., Vogel-Hopker, A., Naitoh-Matsuo, M., Ogura, K., Takahashi, N., Yasuda, K., and Ogura, T. (1999). Tbx5 and Tbx4 genes determine the wing/leg identity of limb buds. *Nature* **398**, 810–814.

Threadgill, D. W., Dlugosz, A. A., Hansen, L. A., Tennenbaum, T., Lichti, U., Yee, D., LaMantia, C., Mourton, T., Herrup, K., Harris, R. C., *et al.* (1995). Targeted disruption of mouse EGF receptor: Effect of genetic background on mutant phenotype. *Science* **269**, 230–234.

Torban, E., Eccles, M. R., Favor, J., and Goodyer, P. R. (2000). PAX2 suppresses apoptosis in renal collecting duct cells. *Am. J. Pathol.* **157**, 833–842.

Torban, E., and Goodyer, P. (1998). What PAX genes do in the kidney. *Exp. Nephrol.* **6**, 7–11.

Torres, M., Gomez-Pardo, E., Dressler, G. R., and Gruss, P. (1995). Pax-2 controls multiple steps of urogenital development. *Development* **121**, 4057–4065.

Trudel, M., Barisoni, L., Lanoix, J., and D'Agati, V. (1998). Polycystic kidney disease in SBM transgenic mice: role of c-myc in disease induction and progression. *Am. J. Pathol.* **152**, 219–229.

Trudel, M., D'Agati, V., and Costantini, F. (1991). C-myc as an inducer of polycystic kidney disease in transgenic mice. *Kidney Int.* **39**, 665–671.

Trudel, M., Lanoix, J., Barisoni, L., Blouin, M. J., Desforges, M., L'Italien, C., and D'Agati, V. (1997). C-myc-induced apoptosis in polycystic kidney disease is Bcl-2 and p53 independent. *J. Exp. Med.* **186**, 1873–1884.

Tsang, T. E., Shawlot, W., Kinder, S. J., Kobayashi, A., Kwan, K. M., Schughart, K., Kania, A., Jessell, T. M., Behringer, R. R., and Tam, P. P. (2000). Lim1 activity is required for intermediate mesoderm differentiation in the mouse embryo. *Dev. Biol.* **223**, 77–90.

Tsuzuki, T., Takahashi, M., Asai, N., Iwashita, T., Matsuyama, M., and Asai, J. (1995). Spatial and temporal expression of the ret proto-oncogene product in embryonic, infant and adult rat tissues. *Oncogene* **10**, 191–198.

Tufro, A. (2000). VEGF spatially directs angiogenesis during metanephric development in vitro. *Dev. Biol.* **227**, 558–566.

Tufro, A., Norwood, V. F., Carey, R. M., and Gomez, R. A. (1999). Vascular endothelial growth factor induces nephrogenesis and vasculogenesis. *J. Am. Soc. Nephrol.* **10**, 2125–2134.

Uehara, Y., Minowa, O., Mori, C., Shiota, K., Kuno, J., Noda, T., and Kitamura, N. (1995). Placental defect and embryonic lethality in mice lacking hepatocyte growth factor/scatter factor. *Nature* **373**, 702–705.

Underhill, D. A. (2000). Genetic and biochemical diversity in the Pax gene family. *Biochem. Cell Biol.* **78**, 629–638.

Vainio, S., Lehtonen, E., Jalkanen, M., Bernfield, M., and Saxen, L. (1989). Epithelial-mesenchymal interactions regulate the stage-specific expression of a cell surface proteoglycan, syndecan, in the developing kidney. *Dev. Biol.* **134**, 382–391.

Vega, Q. C., Worby, C. A., Lechner, M. S., Dixon, J. E., and Dressler, G. R. (1996). Glial cell line-derived neurotrophic factor activates the receptor tyrosine kinase RET and promotes kidney morphogenesis. *Proc. Natl. Acad. Sci. USA* **93**, 10657–10661.

Veis, D. J., Sorenson, C. M., Shutter, J. R., and Korsmeyer, S. J. (1993). Bcl-2-deficient mice demonstrate fulminant lymphoid apoptosis, polycystic kidneys, and hypopigmented hair. *Cell* **75**, 229–240.

Vilar, J., Gilbert, T., Moreau, E., and Merlet-Benichou, C. (1996). Metanephros organogenesis is highly stimulated by vitamin A derivatives in organ culture. *Kidney Int.* **49**, 1478–1487.

Wallner, E. I., Yang, Q., Peterson, D. R., Wada, J., and Kanwar, Y. S. (1998). Relevance of extracellular matrix, its receptors, and cell adhesion molecules in mammalian nephrogenesis. *Am. J. Physiol.* **275**, F467–F477.

Warburton, D., Schwarz, M., Tefft, D., Flores-Delgado, G., Anderson, K. D., and Cardoso, W. V. (2000). The molecular basis of lung morphogenesis. *Mech. Dev.* **92**, 55–81.

Ware, C. B., Horowitz, M. C., Renshaw, B. R., Hunt, J. S., Liggitt, D., Koblar, S. A., Gliniak, B. C., McKenna, H. J., Papayannopoulou, T., Thoma, B., *et al.* (1995). Targeted disruption of the low-affinity leukemia inhibitory factor receptor gene causes placental, skeletal, neural and metabolic defects and results in perinatal death. *Development* **121**, 1283–1299.

Wawersik, S., and Maas, R. L. (2000). Vertebrate eye development as modeled in Drosophila. *Hum. Mol. Genet.* **9**, 917–925.

Weinmaster, G. (2000). Notch signal transduction: A real rip and more. *Curr. Opin. Genet. Dev.* **10**, 363–369.

Weller, A., Sorokin, L., Illgen, E. M., and Ekblom, P. (1991). Development and growth of mouse embryonic kidney in organ culture and modulation of development by soluble growth factor. *Dev. Biol.* **144**, 248–261.

Wilson, J.G., and Warkany, J. (1948) Malformations in the genito-urinary tract induced by maternal vitamin A deficiency in the rat. *Am. J. Anat.* **83**, 357–407.

Wilson, P. D. (1997). Epithelial cell polarity and disease. *Am. J. Physiol.* **272**, F434–F442.

Wu, G., D'Agati, V., Cai, Y., Markowitz, G., Park, J. H., Reynolds, D. M., Maeda, Y., Le, T. C., Hou, H., Jr., Kucherlapati, R., Edelmann, W., and Somlo, S. (1998). Somatic inactivation of Pkd2 results in polycystic kidney disease. *Cell* **93**, 177–188.

Wu, G., Markowitz, G. S., Li, L., D'Agati, V. D., Factor, S. M., Geng, L., Tibara, S., Tuchman, J., Cai, Y., Park, J. H., van Adelsberg, J., Hou, H., Jr., Kucherlapati, R., Edelmann, W., and Somlo, S. (2000). Cardiac defects and renal failure in mice with targeted mutations in Pkd2. *Nature Genet.* **24**, 75–78.

Xu, P. X., Adams, J., Peters, H., Brown, M. C., Heaney, S., and Maas, R. (1999). Eya1-deficient mice lack ears and kidneys and show abnormal apoptosis of organ primordia. *Nature Genet.* **23**, 113–117.

Yamaguchi, T. P., Harpal, K., Henkemeyer, M., and Rossant, J. (1994). fgfr-1 is required for embryonic growth and mesodermal patterning during mouse gastrulation. *Genes Dev.* **8**, 3032–3044.

Yang, T., Michele, D. E., Park, J., Smart, A. M., Lin, Z., Brosius, F. C., 3rd, Schnermann, J. B., and Briggs, J. P. (1999). Expression of peroxisomal proliferator-activated receptors and retinoid X receptors in the kidney. *Am. J. Physiol.* **277**, F966–F973.

Yeaman, C., Grindstaff, K. K., and Nelson, W. J. (1999). New perspectives on mechanisms involved in generating epithelial cell polarity. *Physiol. Rev.* **79**, 73–98.

Yoshida, K., Taga, T., Saito, M., Suematsu, S., Kumanogoh, A., Tanaka, T., Fujiwara, H., Hirata, M., Yamagami, T., Nakahata, T., Hirabayashi, T., Yoneda, Y., Tanaka, K., Wang, W. Z., Mori, C., Shiota, K., Yoshida, N., and Kishimoto, T. (1996). Targeted disruption of gp130, a common signal transducer for the interleukin 6 family of cytokines, leads to myocardial and hematological disorders. *Proc. Natl. Acad. Sci. USA* **93**, 407–411.

21

Maldevelopment of the Human Kidney and Lower Urinary Tract: An Overview

Adrian S. Woolf, Paul J. D. Winyard, Monika M. Hermanns, and Simon J. M. Welham

I. Introduction

Normal development of the human kidney and lower urinary tract is a highly complex process that not uncommonly goes wrong. Furthermore, congenital malformations of these structures, involving absent, immature, or poorly-grown organs, account for most young children with long-term kidney failure who require dialysis and transplantation. Evidence is emerging that some of these individuals have mutations of genes, such as transcription factors, expressed in prenatal development of the kidney and ureter. In other patients, physical obstruction of the lower tracts causes urinary flow impairment, an event that triggers diverse aberrations of kidney development, and teratogens can occasionally be identified. It is also possible that modifications of maternal diet may affect kidney development, although this hypothesis is currently more persuasive for animal models than for human disease. Understanding the mechanisms of pathogenesis of these diseases is important for genetic counseling, avoidance of possible teratogens, and planning possible therapeutic interventions, including surgical decompression of obstructed, fetal urinary tracts.

I. Normal Development of Human Kidney and Lower Urinary Tract

A. Introduction

For convenience, in this chapter we have divided the whole urinary tract into the kidney and the "lower urinary tract," which includes the ureter and urinary bladder. As documented exhaustively elsewhere in this volume, normal development of these structures in animals involves differentiation and morphogenesis accompanied by a fine balance between mitosis and programmed cell death, or apoptosis; these processes are driven by diverse molecules, such as growth factors and cell adhesion molecules, the expression of which is controlled by transcription factors. Accumulating evidence, in part based on our increasing understanding of the pathogenesis of congenital disease, leads to the suggestion that the same processes, and often the same genes, are implicated in development of the human kidney and lower urinary tract.

In humans, as in other mammals, three pairs of "kidneys" develop from the intermediate mesoderm on the dorsal body wall: the pronephros, mesonephros, and metanephros. The pronephros and mesonephros degenerate during fetal life while the metanephros develops into the adult kidney. The anatomy of these structures has been described by several authors (Potter, 1972; Larsen, 1993; Gilbert, 1997; Moore and Persaud, 1998; Risdon and Woolf, 1998a). A time table of important events during kidney development is outlined in Table 21.1 and morphological events are depicted in Figs. 21.1 and 21.2.

Table 21.1   Nephrogenic Time Table: Human verus Mouse

	Human	Mouse
Pronephros		
First appears at	22 days	9 days
Regresses by	25 days	10 days
Mesonephros		
First appears at	24 days	10 days
Regresses by	16 weeks	14 days
Metanephros appears	28–32 days	11 days
Renal pelvis	33 days	12 days
Collecting tubules and nephrons	44 days	13 days
Glomeruli	8–9 weeks	14 days
Nephrogenesis ceases	34–36 weeks	7–10 days after birth
Length of gestation	40 weeks	20–21 days

a Summary of timing of nephrogenesis. All times are embryonic days/ weeks (i.e., *in utero*) aside from completion of nephrogenesis in mice, which occurs after birth.

B. The Pronephros

The human pronephros appears at the 10 somite stage on day 22 of gestation (i.e., 22 days after fertilization), which is morphologically equivalent to embryonic day 9 in mice. Initially, it consists of a small group of nephrotomes with segmental condensations, grooves, and vesicles between the second and sixth somites. Nephrotomes are nonfunctional, probably representing vestiges of the pronephric kidney of lower vertebrates (Chapter 3). This stage is essential for subsequent kidney development, however, because the pronephric duct develops from the intermediate mesoderm lateral to the notochord, elongates caudally to reach the cloacal wall on day 26, and is then renamed the mesonephric, or Wolffian duct, as mesonephric tubules develop. The nephrotomes and pronephric part of the duct involute and cannot be identified by 24 or 25 days gestation.

C. The Mesonephros

In humans, the elongated sausage-like mesonephros appears around 24 days after fertilization and consists of the mesonephric duct and adjacent mesonephric tubules (Chapter 1). The duct is initially a solid rod of cells, which then canalizes in a caudocranial direction after fusion with the cloaca. Mesonephric tubules develop medial to the duct by a "mesenchymal-to-epithelial" transformation from the intermediate mesoderm. This phenotypic transformation is subsequently reiterated during nephron formation in the metanephros. In humans, a total of around 40 pairs of mesonephric tubules are produced (several per somite), but the cranial tubules regress as caudal ones are forming so that there are never more than 30 pairs at any time. Each tubule

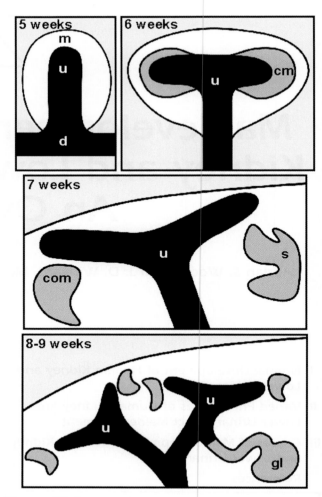

Figure 21.1 Early branching of the ureteric bud and nephron formation. Mesonephric duct (d) and ureteric bud (u) in black, uninduced metanephric mesenchyme (m) in white, and condensing mesenchyme (cm) and nephron precursors, including comma-shaped (com), S-shaped bodies (s), and immature glomeruli (gl) in gray. At 5 weeks gestation, the ureteric bud grows out from the mesonephric duct into the uninduced mesenchyme of the metanephric blastema. The bud starts branching by week 6 and mesenchyme condenses around the ampullae. Comma and S-shaped bodies are formed in week 7. The first glomeruli are formed by 8–9 weeks.

consists of a medial cup-shaped sac encasing a knot of capillaries, which appears analagous to the Bowman's capsule and glomerulus of the mature kidney, and a lateral portion in continuity with the mesonephric duct. Other segments of the tubule resemble mature proximal and distal tubules histologically but there is no loop of Henle. The human mesonephros is reported to produce small quantities of urine between 6 and 10 weeks of gestation, which drains via the mesonephric (Wolffian) duct; the murine organ is much more rudimentary and does not contain well-differentiated glomeruli. Mesonephric structures involute during the third month of human gestation, although caudal mesonephric tubules contribute to the efferent ducts of the

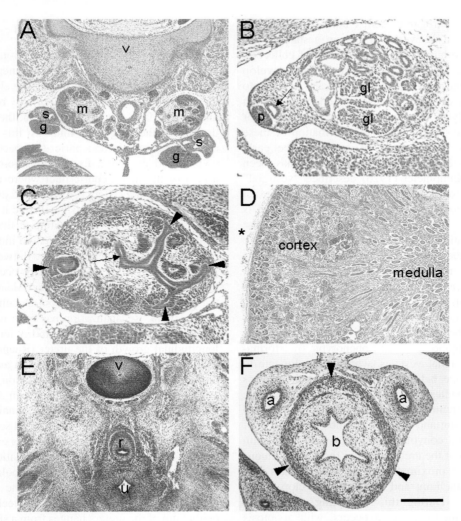

Figure 21.2 Histology of normal human kidney and bladder development. Hematoxylin and eosin-stained sections of human kidneys at 6 weeks (A–C) and 20 weeks of gestation (D) and the urinary bladder at 6 (E) and 10 (F) weeks. (A) Transverse section showing vertebral body (v), metanephros (m), mesonephros (s), and gonadal ridges (g). (B) Section of mesonephros showing large glomeruli (gl), tubules, and mesonephric duct (arrow); note also the paramesonephric duct (p). (C) Section of metanephros showing central ureteric stalk (arrow) and peripheral branching ampullae (arrowheads) surrounded by condensing mesenchyme. (D) Low power view of midgestation metanephros demonstrating the nephrogenic cortex at the edge of the kidney (*) with several layers of glomeruli deeper in the cortex and large collecting ducts in the medulla. (E) Urogenital sinus (u) anterior to rectum (r) and vertebral body (v). (F) Bladder (b) with developing detrusor smooth muscle layer (arrowheads) flanked by umbilical arteries (a). Bar corresponds to 250 μm in A, D, and E, and 100 μm in the remainder.

epididymis and the mesonephric duct contributes to the duct of the epididymis, the seminal vesicle, and ejaculatory duct.

D. The Metanephros

Human metanephric development begins at day 28 after fertilization when the ureteric bud sprouts from the distal part of the mesonephric duct. The tip of the bud then penetrates a specific area of sacral intermediate mesenchyme, termed the metanephric blastema around day 32. Next the mesenchyme condenses around the tip of the growing ureteric bud. The ureteric bud branches sequentially to form the ureter, renal pelvis, calyces, and collecting tubules,

while the mesenchyme undergoes an epithelial conversion to form the nephrons from the glomerulus to the distal tubule. The ureteric bud also gives rise to the urothelium of the renal pelvis and the ureter, and the junction of the bud and the mesonephric duct becomes incorporated into the cloaca, forming the urinary bladder trigone (see later).

As the epithelial ureteric bud grows into the metanephric blastema, it becomes invested with condensed mesenchyme, and interactions between the two cell types are associated with proliferation and branching of the bud and its derivatives. The branch tips are highly proliferative, as are the surrounding nephrogenic mesenchymal cells (Winyard *et al.*, 1996a and b). Branching continues at the tips of the bud in the outer (nephrogenic) cortex until completion of nephrogenesis around 34 weeks of human gestation. The end result is an arborialized (tree-like) system of collecting ducts, which is connected to nephrons that develop concurrently from the mesenchymal condensates adjacent to the ampullary tips (see later). Mature collecting ducts drain first into minor calyces, which then empty into the major calyces of the renal pelvis and ureter. Formation of these structures requires "remodeling" of many of the early branches of the ureteric bud, a process that has been recognized since Kampmeier (1926) described "vestigial" and "provisional" zones in the human medulla. The exact number of generations of branches that are remodeled is unknown, although Potter (1972) estimated that the first three to five generations form the pelvis and the next three to five give rise to the minor calyces and papillae. Nephrons that were initially attached to these early branches are thought to either transfer to a later branch or degenerate during development.

Each nephron develops from the mesenchyme that condenses around the ampullary tips of the ureteric bud. The condensed mesenchyme undergoes phenotypic transformation into epithelial renal vesicles, which then elongate to form a comma shape before folding back on themselves to form S-shaped bodies. The distal portion of the S shape elongates and differentiates into the proximal convoluted tubule, the descending and ascending limbs of the loop of Henle, and the distal convoluted tubule, which fuses with the adjacent branch of the ureteric bud to form a continuous functional unit. The proximal S shape differentiates into the glomerular epithelium, and capillaries develop in the glomerular crevice. During this "sculpting" process, a degree of programmed cell death occurs, with apoptosis prominent in the nephrogenic zone, proximate to primitive nephrons (Winyard *et al.*, 1996a). The first vascularized metanephric glomeruli are formed around 8 to 9 weeks of human gestation, and new nephron formation ceases around 34 weeks of gestation. The number of nephrons, as assessed by counting glomeruli, varies widely in different species, probably reflecting the number of branches of the ureteric bud required to both induce these nephrons and form the collecting ducts to drain urine from them. It has been calculated that 9 to 10 rounds of branching occur in mice, which generates 10–20,000 nephrons per kidney, and a further 10 branching generations occur in human to give rise to approximately a million nephrons in each kidney (Ekblom *et al.*, 1994). In fact, studies suggest that about two-thirds of nephrons are generated in the last third of human gestation and that the normal range of nephrons found in healthy human kidneys is rather large, around 0.5–1.0 million (Hinchliffe *et al.*, 1991, 1992, 1993). During the second half of human gestation, fetal urine is the main component of amniotic fluid and is thought to be important for fetal lung growth.

E. The Ureter

As reciprocal inductive interactions begin to occur between the ureteric bud and the metanephric mesenchyme, the urinary bladder starts to form. The cloaca separates into the urogenital sinus (primordial urinary bladder) and rectum and, by 28 days of human gestation, the mesonephric duct drains into the urogenital sinus. At this time, the epithelia of the sinus and mesonephric duct become fused and the ureteric bud arises as a diverticulum from the posteromedial aspect of the mesonephric duct where the duct enters the primordial bladder. Between 28 and 35 days of gestation the entire length of the ureter is patent and it has been assumed, because the cloaca is imperforate at that time, that mesonephric urine might lead to increased intraluminal pressure, thus maintaining ureteral patency. Between 37 and 47 days gestation, a membrane temporarily occludes the junction between the ureter and the bladder (Alcaraz *et al.*, 1991). In the same period, the ureter becomes occluded throughout its entire length. Recanalization has been reported to begin in the middle third of the ureter and is temporally related to longitudinal growth of the ureter (Ruano-Gil *et al.*, 1975). The cause of the apparent obstruction and recanalization is not understood. At 8 weeks of gestation the ureter is a non muscularized but patent tube. Between 32 and 56 days, the metanephros "ascends" relative to spinal landmarks from a position at the level of the upper sacral segments to the level of the upper lumbar vertebrae. This is explained by caudal growth of the spine and ureteric elongation. The developing kidney also rotates medially on its polar axis so that the renal pelvis faces the spine and the hilum is directed anteromedially. Between 10 and 14 weeks of gestation, the epithelium of the ureter changes from a simple monolayer to a more mature pseudo-multilayered arrangement, and by the end of this period the ureter begins its submucosal course as it enters the urinary bladder. Production of metanephric urine at 8–9 weeks of gestation, coinciding with the first layer of vascularized glomeruli, precedes muscularization of the upper ureter and may stimulate ureteric myogenesis, which starts at 12 weeks (Matsuno *et al.*, 1984; Escala *et al.*, 1989). The important anatomical landmark, the ureteropelvic junction, is apparent at 18 weeks.

F. The Urinary Bladder

On day 28 of gestation, the urorectal septum (Tourneux's fold) extends caudally through the cloaca toward the cloacal membranes. In addition, two tissue folds (Rathke's plicae) advance from the lateral aspect meeting by 44 days of gestation. This separation of the cloaca into the primitive urogenital sinus and rectum is the first step in the development of the urinary bladder. The portion of the urogenital sinus cranial to the mesonephric ducts is called the vesico-urethral canal, and the part caudal to the mesonephric ducts is called the urogenital sinus. By 33 days of gestation, the mesonephric duct below the ureteric bud dilates, and this "common excretory duct" is absorbed into the urogenital sinus as the precursor of the trigone. In this manner, the origin of the ureteric bud "enters" the bladder directly by day 37 to become the ureteric orifice; thereafter, the ureteric orifices migrate in a cranial and lateral direction. Continued growth of the epithelium of the absorbed common excretory ducts separates the ureteral orifices laterally and establishes the framework of the primitive trigone. The mesonephric duct above the ureteric bud becomes the vas deferens in the male, but this portion of the duct involutes in the female. The urogenital membrane ruptures on day 48 of gestation, hence providing a connection between the emerging bladder and the outside of the body. The human allantois, another potential outflow tract on the anterior of the developing bladder, appears at 21 days of gestation; the allantois will involute by 12 weeks of gestation, persisting as a remnant, the median umbilical ligament. After 10 weeks of gestation, the urinary bladder appears as a cylindrical tube lined by connective tissue. By 13 weeks of gestation the bladder mesenchyme has developed into circular and longitudinal smooth muscle fibers and, by 16 to 20 weeks, discrete inner and outer longitudinal layers and a middle circular muscle layer exist.

II. Varied Phenotypes of Human Kidney and Lower Urinary Tract Maldevelopment

As might be expected from such an anatomically and functionally complex entity as the kidney and lower urinary tract, there are several disease manifestations which arise when normal development goes wrong. The detailed histology of these malformations has been well described (Risdon and Woolf, 1998b). Some of the main types of structural disorders are discussed briefly.

A. Renal Agenesis

The most basic structural anomaly is "agenesis" when the organ is absent. In theory, in order to prove such a diagnosis, one would have to provide evidence that the metanephros had never formed. Clearly, this is impossible in humans unless a fetus is available for direct anatomical inspection just as the metanephros should be forming; this would constitute a very rare event. In the "imperfect world" of clinical medicine, renal agenesis is generally diagnosed radiologically, e.g., at the routine midgestation fetal ultrasonography or, postnatally, during investigation of renal tract disease. Less commonly, unilateral renal agenesis is noted incidentally at autopsy. Renal agenesis is usually accompanied by an absence of the ureter on the same side.

Unilateral renal agenesis is considered a rather common congenital anomaly with an incidence of 1:1000. Being born with a solitary functioning kidney is not generally associated with significant morbidity, although there is some evidence that the incidence of hypertension (high blood pressure) and renal functional impairment is greater than expected (Woolf, 1998; Duke *et al.*, 1998). Bilateral renal agenesis is an order of magnitude less common than unilateral disease and would inevitably lead to neonatal death without medical intervention. Because urine is the main constituent of amniotic fluid in the second half of gestation, a reduced amount of amniotic fluid (oligohydramnios) will accompany bilateral renal agenesis. It is notable that fetal renal excretory function is not required for prenatal life, as waste products are effectively dialyzed across the placenta into the maternal circulation; soon after birth, however, affected individuals would die from uremia unless treated by dialysis. Renal agenesis can occur as an isolated anomaly or be part of a multiorgan syndrome, some of which have defined genetic bases; a good example is Kallmann's syndrome, as discussed later.

B. Renal Dysplasia

This term refers to a kidney that has begun to form, but which fails to undergo normal cellular differentiation. Dysplastic kidneys can be considerably larger than normal, when they take the form of the multicystic dysplastic kidney; here, the organ is distended by cysts and generally attached to an "atretic" ureter that contains at least one segment with no lumen. With the increasing use of fetal and infant ultrasonography, multicystic kidneys have often been noted to involute prenatally or in the first year of life; in fact, some cases of human "renal agenesis" diagnosed by radiography may represent multicystic kidneys, which have "disappeared" beyond the lower limit of detection (Mesrobian *et al.*, 1993). Hence, multicystic dysplastic kidneys can be likened to "supernovae," which have spectacular phases of growth and regression. However, dysplastic organs can be much smaller than normal; these varieties are sometimes associated with abnormal lower urinary tracts, e.g., obstructive lesions including ureterocoeles (a pouch at the lower end of the ureter, which often has an ectopic opening into the urinary bladder or urethra) and posterior urethral valves (occluding leaflets in the male urethra). The definition of renal dysplasia is technically a histological one, with poorly branched tubules surrounded by stromal/mesenchymal-type cells; often, islands of cartilage can be seen in the latter compartment (Daikha-Dahmane *et al.*, 1997; Risdon and Woolf, 1998b). The appearances hold whatever the gross size of the malformed organ. As with renal agenesis, renal dysplasia can occur either as an isolated anomaly or be part of a multiorgan syndrome; an example is the renal cysts and diabetes syndrome, as discussed later.

C. Renal Hypoplasia

This is another histological term that refers to a kidney with significantly fewer nephrons than normal. Hypoplastic kidneys are small but, on biopsy, do not contain undifferentiated tissues, as does the dysplastic kidney. Although there are too few nephrons in hypoplastic kidneys, the glomeruli and tubules are sometimes enlarged greatly, a condition called oligomeganephronia (Salomon *et al.*, 2001). Renal hypoplasia can occur as part of a multiorgan congenital syndrome; e.g., it occurs in the renal coloboma syndrome, discussed in detail in Chapter 23.

D. Polycystic Kidney Disease

This term describes a large group of human diseases in which the major early steps of renal development are normal but an aberration of terminal epithelial differentiation follows so that the kidney becomes filled with fluid-filled cysts. One common variety is called autosomal-dominant polycystic kidney disease (ADPKD), which is discussed in detail in Chapter 24. Most cases are associated with mutations of *PKD1*, a gene coding for a large membrane-spanning protein called polycystin 1; in this condition, cysts can arise from all parts of the nephron. Another well-recognized variety is called autosomal-recessive polycystic kidney disease (ARPKD); here cystic dilatation of the ureteric bud-derived collecting ducts is the major feature. The gene for ARPKD has been defined; it is called polycystic kidney and hepatic disease 1 (*PKHD1*) and encodes a receptor-like protein (Ward *et al.*, 2002). Histopathologists have recognized a particular form of PKD called glomerulocystic kidney disease; here, cystic dilatation of the glomerular Bowman's space is the key feature. It has been discovered that some cases of glomerulocystic kidney disease are associated with mutations of a transcription factor called *hepatocyte nuclear factor 1β (HNF1β)*, as discussed later.

E. Other Disorders of Kidney Differentiation

Other kidney diseases can be considered disorders of renal differentiation, although they do not lead to gross structural congenital malformations. Among these diseases, which are addressed in detail later in this volume, are congenital nephrotic syndromes, diseases in which the differentiation of podocytes in the glomerular tuft are abnormal and lead to massive leakage of protein into the urinary space (Chapters 22 and 27); congenital diseases in which tubule physiology is abnormal (Chapter 26); and childhood and adult renal neoplasms, such as Wilms' tumor and clear cell carcinomas (Chapters 22 and 25) in which there is a major deregulation of growth, sometimes involving genes active in normal nephrogenesis.

F. Malformations of the Lower Urinary Tract

In many cases, major renal malformations are accompanied by aberrant development of the lower urinary tract. Because development of the ureteric bud and renal mesenchyme depends on mutual interactions (Grobstein, 1967), it is not surprising that a primary defect of either component will affect its partner's development *in vivo*. This provides one explanation for the clinical observation that an abnormal position of the insertion of the ureter into the bladder is associated with congenital renal parenchymal defects (Schwartz *et al.*, 1981; Vermillion and Heale, 1973). In such cases, the altered ureteric anatomy may reflect a delayed or premature branching of the ureteric bud from the mesonephric duct. During early organogenesis, this abnormal trajectory could result in the bud failing to fully engage the segment of intermediate mesoderm destined to become nephrogenic mesenchyme. The end result would be defective or absent nephron formation together with a failure of mesenchyme-driven ureteric bud growth and branching morphogenesis. In other cases, renal dysplasia can occur with physical obstruction of the lower urinary tract at the level of the ureter or urethra; hence, it can also be postulated that the renal anomaly occurs secondary to the impairment of urine flow, which is discussed in detail later in this chapter. Another, not mutually exclusive, hypothesis for the coexistence of upper and lower urinary tract anomalies is that the same genes are expressed in both tissues and that mutations would therefore perturb development of the kidney and the lower urinary tract.

A very common malformation of the lower urinary tract is called primary vesicoureteric reflux, which is thought to affect about 1% of all young children (Feather *et al.*, 1996). In this condition, there is a retrograde passage of urine from the bladder into the ureter, renal pelvis, and sometimes into the renal parenchyma itself. This is associated with a leaky valve mechanism at the junction of the ureter and urinary bladder. Another type of anomaly, also with an incidence of about 1%, are double or "duplex" ureters and kidneys (Whitten and Wilcox, 2001) in which the upper part of the kidney drains into a ureter with an ectopic, lower insertion into either the urinary bladder or even the urethra; the lower half of a duplex kidney drains into a separate ureter, which crosses the lower ureter, to drain into an ectopic, lateral insertion into the urinary bladder. Up to 40% of young children with renal failure requiring long-term dialysis and renal transplantation have dysplastic kidneys associated with lower urinary tract obstruction; in these cases, posterior urethral valves, a disorder unique to males, is the commonest specific diagnosis (Woolf and Thiruchelvam, 2001). Obstructive lesions can also occur at the junction of the renal pelvis and the top of the ureter (Josephson *et al.*, 1993), or at the junction of the ureter and the urinary bladder, the latter category including obstructive megaureters (Liu *et al.*, 1994) and ureterocoeles (Austin *et al.*, 1998).

G. Clinical Impact of Human Kidney and Lower Urinary Tract Malformations

Routine midgestation human fetal ultrasonographic scanning is used increasingly to detect organ malformations. Hence, more renal tract malformations are being diagnosed before birth (Anderson *et al.*, 1997; Hiraoka *et al.*, 1997; Yeung *et al.*, 1997; Jaswon *et al.*, 1999), accounting for up to 30% of all anomalies diagnosed prenatally (Noia *et al.*, 1996). Many have little clinical significance, but for conditions such as bilateral renal agenesis or dysplasia, early diagnosis allows consideration of termination or active therapeutic intervention such as surgical decompression of obstructed fetal urinary tracts (e.g., by forming an artificial conduit between the fetal bladder and the amniotic cavity) (Freedman *et al.*, 1999). It is well established that renal malformations are the major cause of chronic renal failure in children (Ehrich *et al.*, 1992; Drozdz *et al.*, 1998; Lewis, 1999; Woolf and Thiruchelvam, 2001). With advances in technology, babies with minimal renal function can be dialyzed from birth and toddlers can receive kidney transplants from the age of 1 year. These strategies, together with general improvements in the care of children with chronic renal failure, will ultimately mean an increase in long term survival into the adult period for these individuals. However, in the severest cases, accompanying lung hypoplasia can be life-threatening in the neonatal period even if dialysis is technically feasible.

III. Causes of Maldevelopment of Human Kidney and Lower Urinary Tract

In general terms, we can divide the possible causes of human kidney and lower urinary tract malformations into two categories: (1) mutations, and possibly polymorphisms, of genes expressed during developmental and (2) environmental influences on development, which can be subdivided into (a) changes that originate outside the fetus, such as alterations of maternal diet, and (b) changes within the fetus, which disrupt normal development, e.g., impairment of normal fetal urinary flow due to physical obstruction of the urinary tract. We will now explore these themes in more detail.

A. Mutations

1. Studies of Mutant Mice: Lessons for Human Disease

Kuure *et al.* (2000) (also see Chapter 20) listed known renal tract developmental defects associated with mutations of about 20 genes encoding diverse molecules, several of which are thought to modulate cell survival/apoptosis, proliferation, differentiation, and morphogenesis. It is certain that the eventual number of key genes involved in kidney and urinary tract is higher than this. In common with human disease, the resulting malformations include absent, poorly differentiated and small kidneys, respective phenotypes which represent defects in the regulation of formation of the metanephros, tubule differentiation, and nephron numbers. With regard to human disease, several general conclusions from mouse genetic experiments provide potentially important paradigms for understanding the genetics of human kidney and lower urinary tract malformations.

1. Many of the genes implicated in mouse kidney and lower urinary tract development are also expressed in other differentiating organs. Consequently, mouse mutants often have multiorgan malformation syndromes. For instance, the PAX2 transcription factor is expressed not only in the embryonic kidney, but also in the eye; hence, mice with *PAX2* mutations have both renal and ocular malformations (Favor *et al.*, 1996). The same disease spectrum applies to humans with the renal-coloboma syndrome; these individuals also have *PAX2* mutations (Sanyanusin *et al.*, 1995; Salomon *et al.*, 2001; Chapter 23). In fact, there are a considerable number of human malformation syndromes that involve both the kidney and lower urinary tract, as well as the central nervous system, the heart, the limbs, and other organs. Some of these conditions, several of which are inherited and hence must have a defined genetic basis, are listed in Table 21.2. For a complete listing of such syndromes, the reader is referred to the constantly updated McKusick's Online Mendelian Inheritance in Man (http:www4.ncbi.nlm.nih.gov/Omim/).

2. Mutant mice also demonstrate that the same kidney and urinary tract malformation can result from the mutation of different genes. For example, null mutations of either of the two transcription factor genes, *WTI* or *PAX2*, cause renal agenesis (Kreidberg *et al.*, 1993; Torres *et al.*, 1995), and the same phenotype is generated in mice with a genetically deleted glial cell line-derived neurotrophic factor signaling system (Schuchardt *et al.*, 1994; Sanchez *et al.*, 1996). Hence, one would predict in human disease that similar disease phenotypes might result from mutations of different genes. As an example of this phenomenon, unilateral renal agenesis has been described with mutations of both *KAL-1*, which codes for a cell–cell signaling molecule, and *HNF1β*, a transcription factor gene (see later for details).

3. Mouse experiments also demonstrate that kidney and lower urinary tract malformation phenotypes can vary between different mouse strains (Threadgill *et al.*, 1995), an observation that might be explained by polymorphisms, or variants, of genes that act during development. In many

Table 21.2   Genetics of Some Human Renal Tract Malformations and Inherited Cystic Diseases

Apert syndrome (*FGFR2*[a] mutation — growth factor receptor): hydronephrosis and duplicated renal pelvis with premature fusion of cranial sutures and digital anomalies (Cohen and Kreiborg, 1993; Wilkie *et al.*, 1996)

Autosomal-dominant polycystic kidney disease (*PKD1* and *PKD2* mutations — membrane proteins, the latter may be a cation channel): cysts arise from all nephron segments, with liver cysts and cerebral aneurysms (Hughes *et al.*, 1995; Gonzalez-Perret *et al.*, 2001; Chapter 24)

Autosomal-recessive polycystic kidney disease (*PKHD1* — possible cell surface receptor): kidney collecting duct cysts and hepatic fibrosis (Ward *et al.*, 2002)

Bardet–Biedl syndrome (several loci/genes involved — functions unknown): renal dysplasia and calyceal malformations with retinopathy, digit anomalies, obesity, diabetes mellitus, and male hypgonadism (Harnett *et al.*, 1988; Katsanis *et al.*, 2001)

Beckwith–Wiedemann syndrome (in a minority of patients, *p57KIP2* mutation — cell cycle gene): widespread somatic overgrowth with large kidneys, cysts, and dysplasia (Lam *et al.*, 1999; Choyke *et al.*, 1999)

Branchio-oto-renal syndrome (*EYA1*[a] mutation — possible transcription factor): renal agenesis and dysplasia with deafness and branchial arch defects such as neck fistulae (Konig *et al.*, 1994; Abdelhak *et al.*, 1997)

Campomelic dysplasia (*SOX9* mutation — transcription factor): diverse renal and skeletal malformations (Houston *et al.*, 1983; Wagner *et al.*, 1994)

Carnitine palmitoyltransferase II deficiency (gene for this enzyme is mutated): renal dysplasia (North *et al.*, 1995)

CHARGE association (genetic basis unknown): coloboma, heart malformation, choanal atresia, retardation, genital and ear anomalies; diverse urinary tract malforations can occur (Regan *et al.*, 1999)

Congenital anomalies of the kidney and urinary tract (CAKUT) syndrome (*AT2* polymorphism — growth factor receptor): diverse renal and lower urinary tract malformations (Nishimura *et al.*, 1999)

Denys Drash syndrome (*WT1*[a] mutation — transcription/splicing factor): mesangial cell sclerosis and calyceal defects (Jadresic *et al.*, 1990; Coppes *et al.*, 1993; Chapter 22)

Di George syndrome (microdeletion at 22q11 — probably several genes involved): renal agenesis, dysplasia, vesicoureteric reflux, with heart and branchial arch defects (Budarf *et al.*, 1995; Czarnecki *et al.*, 1998; Stewart *et al.*, 1999)

Glutaric aciduria type II (*glutaryl-CoA dehydrogenase* mutation): cystic and dysplastic disease (Wilson *et al.*, 1989)

Hypoparathroidism, sensorineural deafness, and renal anomalies (HDR) syndrome (*GATA3* mutation — transcription factor): renal agenesis, dysplasia, and vesicoureteric reflux (van Esch and Bilous, 2001)

Fanconi anaemia (six mutatant genes reported — involved DNA repair): renal agenesis, ectopic/horseshoe kidney, anemia and limb malformations (Lo Ten Foe *et al.*, 1996; Yamashita and Nakahata, 2001)

Kallmann's syndrome (*KAL1* mutation — cell signaling molecule): renal agenesis (Duke *et al.*, 1998; Hardelin, 2001; see this chapter)

Meckel syndrome (loci at 11q and 17q — genes unknown): cystic renal dysplasia, central nervous system, and digital malformations (Salonen, 1984; Paavola *et al.*, 1995; Roume *et al.*, 1998)

Nail-patella syndrome (*LMX1B*[a] mutation — transcription factor): malformation of the glomerulus and renal agenesis (Gubler and Levy, 1993; Haga *et al.*, 1997; Dreyer *et al.*, 1998)

Oral facial digital syndrome type 1 (*OFD1* mutation — function unknown): glomerular cysts with facial and digital anomalies (Feather *et el.*, 1997; Ferrante *et al.*, 2001)

Renal-coloboma syndrome (*PAX2*[a] mutation — transcription factor): renal hypoplasia and vesicoureteric reflux (Sanyanusin *et al.*, 1995; Salomon *et al.*, 2001; Chapter 23)

Renal cysts and diabetes syndrome (*HNF1β* mutation — transcription factor): renal dysplasia, cysts, and hypoplasia (Bingham *et al.*, 2000, 2001; Kolatsi-Joannou *et al.*, 2001; see this chapter)

Simpson–Golabi–Behmel syndrome (*GPC3*[a] mutation — proteoglycan): renal overgrowth, cysts, and dysplasia (Hughes-Benzie *et al.*, 1994; Pilia *et al.*, 1996; Gonzalez *et al.*, 1998)

Smith–Lemli–Opitz syndrome (*d(7)-dehydrocholesterol reductase* mutation — cholesterol biosynthesis): renal cysts and dysplasia (Akl *et al.*, 1977; Wallace *et al.*, 1994)

Townes–Brockes syndrome (*SALL1* mutation — transcription factor): renal dysplasia and lower urinary tract malformations (Salerno *et al.*, 2000)

Tuberous sclerosis (*TSC1* or *TSC2* mutations — tumor suppressor genes): kidney cysts with neuroectodermal features; a severe renal phenoype occurs with contiguous gene defect of *TSC2* and *PKD1* (Kleymenova *et al.*, 2001; Martigoni *et al.*, 2002)

Urofacial (Ochoa) syndrome (locus on 10q — gene unknown): congenital obstructive bladder and kidney malformation with abnormal facial expression (Wang *et al.*, 1997)

Urogenital adysplasia syndrome (some cases associated with *HNF1β* mutation): renal dysplasia and uterine anomalies (Bingham *et al.*, 2002)

VACTERL association (basis unknown apart from one report of mitochondial gene mutation): vertebral, cardiac, tracheoesophageal, renal, radial, and other limb anomalies (Damian *et al.*, 1996)

von Hippel Lindau disease (*VHL* mutation — tumor suppressor gene): renal and pancreatic cysts, renal tumors (Chatha *et al.*, 2001; Chapter 25)

WAGR syndrome (*WT1* and *PAX6* contiguous gene defect — transciption factors): Wilms' tumor, aniridia and genital and renal malformations (Pritchard-Jones, 1999; Chapter 22).

Zellweger syndrome (peroxisomal protein mutation): cystic dysplastic kidneys (Powers and Moser, 1998; Shimozawa *et al.*, 1992)

---

[a] Mutations of this gene also implicated in mouse kidney and/or lower urinary tract malformations.

human kidney and lower urinary tract malformations, simple inheritance patterns cannot be discerned, yet more than one member in an extended family can be affected; this suggests that modifying genetic influences could be in operation. One example was discovered by Nishimura *et* *al.* (1999), who reported that a common variant of the *angiotensin II type 2 receptor* (*AT2*) gene, which is located on the X chromosome and encodes a molecule that affects apoptosis, occurred more commonly in individuals with varied anomalies of the kidney and lower urinary tract than in normal controls. The variant polymorphism affects RNA splicing and mRNA transcript levels (Nishimura *et al.*, 1999). Furthermore, male mice with null mutation of AT2 display a low penetrance of diverse renal tract malformations (Nishimura *et al.*, 1999).

4. In mice, mutations of two homologous genes (e.g., *homeobox* or *retinoic acid receptor*) may summate to generate a major disruption of kidney differentiation (Mendelsohn *et al.*, 1994; Davis *et al.*, 1995; Patterson *et al.*, 2001). Furthermore, metanephric organ culture studies have shown that different growth factors can have complex, interacting effects on development (Plisov *et al.*, 2001). There are several human examples of kidney anomalies in which more than one gene is mutated. For example, in Bardet Biedl syndrome, a condition in which kidney malformations are associated with retinopathy, digit anomalies, obesity, diabetes mellitus, and male hypogonadism, more than one gene can be mutated in an affected individual (Katsanis *et al.*, 2001). In addition, clinically severe polycystic kidney disease has been noted in individuals who have a mutation of both *PKD1* and *PKD2* genes (Pei *et al.*, 2001) or who have contiguous gene defects that delete *PKD1* and *TSC2* (Kleymenova *et al.*, 2001; Martigoni *et al.*, 2002).

2. X-Linked Kallmann's Syndrome: A Genetically Defined Example of Human Renal Agenesis

The X-linked form of this disease is caused by mutations of *KAL-1* (Franco *et al.*, 1991; Legouis *et al.*, 1991; Hardelin, 2001). In affected males, anosmia (absent sense of smell) and hypogonadotrophic hypogonadism (testicular failure secondary to defective release of stimulating hormones from the hypothalamus) occur because of defective fetal elongation of olfactory neuron axons and migration of gonadotrophin-releasing hormone synthesizing neurons from the nasal placode into the forebrain; furthermore, the olfactory bulb fails to grow and is hypoplastic (Schwanzel-Fukuda *et al.*, 1989). About one-third of patients with the syndrome have a solitary functioning kidney, with presumed unilateral renal agenesis (Kirk *et al.*, 1994; Duke *et al.*, 1998). Other anomalies occur less commonly, including duplex systems, hydronephrosis, and vesicoureteric reflux. Patients with urinary tract agenesis can also lack the vas deferens, a structure derived from the mesonephric duct, which also gives rise to the ureteric bud and its derivatives. Deeb *et al.* (2001) reported two children with the syndrome and unilateral multicystic dysplastic kidney: they suggested that the apparent "agenesis" phenotype seen in older individuals with the syndrome might be the result of spontaneous involution of multicystic organs. *In vitro* studies demonstrate an adhesive role for the protein coded by *KAL-1* (Soussi-Yanicostas *et al.*, 1998). *In vivo* KAL-1 is expressed in the embryonic human central nervous and excretory systems

(Duke *et al.*, 1995; Hardelin *et al.*, 1999). The protein immunolocalizes to the epithelial interstitial matrix and basement membranes of the mesonephric collecting tubules and mesonephric duct and to the first generations of metanephric collecting duct branches of the ureteric bud. It is possible that failure of growth of either the ureteric bud or its first branches, perhaps due to alterations in cell adhesion, would lead to a failure of metanephric formation and hence renal agenesis. Certainly, other adhesion molecules have been considered critical for normal nephrogenesis in murine models (Klein *et al.*, 1988; Noakes *et al.*, 1995; Kreidberg *et al.*, 1996; Muller *et al.*, 1997; Bullock *et al.*, 2001).

3. Renal Cysts and Diabetes (RCAD) Syndrome: A Genetically Defined Cause of Renal Dysplasia

HNF1β mutations cause the recently described RCAD syndrome. Human mutations of this transcription factor gene were initially associated with MODY (maturity onset diabetes mellitus of the young): affected individuals have a failure of pancreatic insulin secretion. Diverse mutations, both inherited and occurring *de novo*, in the DNA-binding and transactivating domains of this transcription factor have been reported in patients with kidney malformations including absent kidney (Bingham *et al.*, 2002), renal dysplasia (Bingham *et al.*, 2000, 2002), renal hypoplasia with oligomeganephronia (Lindner *et al.*, 1999), and the glomerulocystic type of polycystic kidney disease (Bingham *et al.*, 2001), a form of polycystic kidney in which glomerular cysts predominate. Embryonic mouse null mutants die before the urinary tract is formed, but in *Xenopus*, the introduction of one of the human mutations perturbs kidney precursor development in a dominant-negative manner with the formation of cyst-like structures (Wild *et al.*, 2000). Furthermore, experimental disruption of the homologous zebrafish gene (Sun and Hopkins, 2001) apparently implicates it in the determination of the border between the pronephric glomerulus and the tubule; normal gene expression seems necessary for the maintenance of PAX2 and the downregulation of WT1 in the tubule. HNF1β is expressed in murine embryonic kidney and lower urinary tract (Coffinier *et al.*, 1999; Barbacci *et al.*, 1999). Kolatsi-Joannou *et al* (2001) reported that the gene is widely expressed in human embryos; e.g., in normal human metanephroi the highest levels of transcripts localized to fetal medullary and cortical collecting ducts, and low levels of expression in nephrogenic cortex mesenchyme, primitive nephron tubules, and immature glomeruli. It has been postulated that a primary defect in development of the ureteric bud/collecting duct could lead to either a very severe disruption of development (dysplasia) or to a more mild phenotype, namely the generation of glomerular cysts, perhaps secondary to the impairment of fetal urine flow from the proximal to distal part of the developing tubule (Woolf *et al.*, 2002). The gene is also expressed in the human embryonic pancreas, liver, gut, and lung (Kolatsi-Joannou *et al.*, 2001). Most likely, this is an example of a gene involved in epithelial differentiation and branching in several organ systems.

4. Genetics of Human Kidney and Lower Urinary Tract Malformations Not Associated with Multiorgan Syndromes

Kindreds have been reported with multiple individuals affected by kidney and/or lower tract malformations that are not associated with multiorgan syndromes. These disease phenotypes include families with renal agenesis and/or renal dysplasia (Cain *et al.*, 1974; Roodhooft *et al.*, 1984; McPherson *et al.*, 1987; Murugasu *et al.*, 1991; Arfeen *et al.*, 1993), multicystic dysplastic kidney (Moazin *et al.*, 1997; Murakami *et al.*, 1992), oligomeganephronic renal hypoplasia (Moerman *et al.*, 1984), pelviureteric junction obstruction (Izquierdo *et al.*, 1992), duplex kidneys and ureters (Atwell, 1985), and posterior urethral valves (Farrar and Pryor, 1976). A good example is primary vesicoureteric reflux — it is called "primary" because there is no outflow tract obstruction. It is associated with a "nephropathy" that represents a mixture of congenital dysplasia and/or hypoplasia (Risdon *et al.*, 1993) and the renal disease accounts for about 10% of both children and adults who require treatment with renal dialysis and kidney transplantation (Lewis, 1999; Smellie *et al.*, 2001). In some families, primary vesicoureteric reflux is clearly inherited in a dominant manner, with evidence in some families for a locus on chromosome 1p13 (Feather *et al.*, 1996; 2000); at present the causative gene is undefined.

B. Effects of Maternal Diet and Diverse Teratogens on Urinary Tract Development

On occasion, human kidney or lower urinary tract malformation have been reported in association with teratogens. These include angiotensin converting enzyme inhibitors, drugs used to treat high blood pressure (Pryde *et al.*, 1993; Barr *et al.*, 1994; Sedman *et al.*, 1995), cocaine (Battin *et al.*, 1995), corticosteroids (Hulton and Kaplan, 1995), ethanol (Moore *et al.*, 1997; Taylor *et al.*, 1994), gentamycin, an antibiotic (Hulton and Kaplan, 1995), glucose (Novak and Robinson, 1994; Lynch *et al.*, 1997; Woolf, 2000), nonsteroidal anti-inflammatory drugs (Voyer *et al.*, 1994), and vitamin A and its derivatives (Von Lennep *et al.*, 1985; Rothman *et al.*, 1995). In several cases, however, evidence for association is anecdotal (e.g. for gentamycin and nonsteriodal anti-inflammatory drugs) and may not be based on hard epidemiological data. Clearly, the effect of individual agents can be very complicated. For example, animal experiments indicate that there may be an optimal intake of vitamin A and its retinoid derivatives for the development of several organs with both too much and too little inhibiting normal differentiation; retinoids

may affect organogenesis by altering patterning (e.g., by perturbing homeobox gene expression) or by altering cell–cell signaling by growth factors (e.g., the GDNF/RET axis) (Padmanabhan, 1998; Moreau *et al.*, 1998; Mendelsohn *et al.*, 1999; Pitera *et al.*, 2001).

More intriguingly, relatively subtle changes in maternal diet, in terms of protein intake, may have an effect on metanephric development. Imposition of a severe dietary protein restriction (5–6% versus 18% control diet) during pregnancy reduced the final numbers of glomeruli generated in rat kidneys (Levy and Jackson, 1993; Merlet-Benichou *et al.*, 1994). Studies have also shown that milder maternal protein restriction is also associated with a final nephron deficit and high systemic blood pressure in young adult animals (Langley-Evans *et al.*, 1996, 1999; Welham *et al.*, 2002). How such an effect might be mediated, and when it might occur during nephrogenesis, has, until recently, been unknown. Kwong *et al.* (2000) demonstrated that maternal low-protein diets reduced cell numbers in the inner cell mass of preimplantation embryos. Using the study of Kwong *et al.* (2000) as a paradigm, Welham *et al.* (2002) postulated that low-protein maternal diets might affect cell turnover in early metanephrogenesis (Fig. 21.3). Rats were supplied with one of three isocaloric diets from day 0 of pregnancy: control (18% protein) or low protein (9 or 6%) diets. At 2 weeks postnatally, when nephrogenesis has finished in this species, controls had a mean of 17×10^3 glomeruli/kidney, whereas offspring exposed to either 9 or 6% protein diets had significantly fewer nephrons. At embryonic day 13, when the rat metanephros has just formed, metanephroi in all dietary groups contained the same number of cells (about 2×10^4). In all diets, apoptosis was noted in condensing mesenchyme (nephron/glomeruli precursors) and loose mesenchyme (interstitial precursors), yet it was increased significantly in the mesenchyme in developing kidneys exposed to low-protein diets. At embryonic day 15, when rat mesenchyme begins forming primitive nephrons but glomeruli are still absent, the average control metanephros had 2×10^6 cells, whereas time-matched embryos exposed to either 6 or 9% protein had significantly fewer. Other studies emphasize that the deregulation of apoptosis affects metanephic growth. Caspase inhibitors reduce renal mesenchymal cell death and perturb epithelial morphogenesis in mouse metanephric organ culture (Araki *et al.*, 1999), and null mutation of *AT2*, which decreased cell death around the lower urinary tract, causes a variety of kidney and ureter malformations (Nishimura *et al.*, 1999). In addition, an excess of metanephric cell death can be triggered genetically, e.g., in *PAX2+/-* and *BCL2 -/-* mutant mice (Sorenson *et al.*, 1995; Porteus *et al.*, 2000), or by exposure to specific cytokines, such as tumor necrosis factor α (Cale *et al.*, 1998); all these maneuvers lead to hypoplastic organs with too few nephrons.

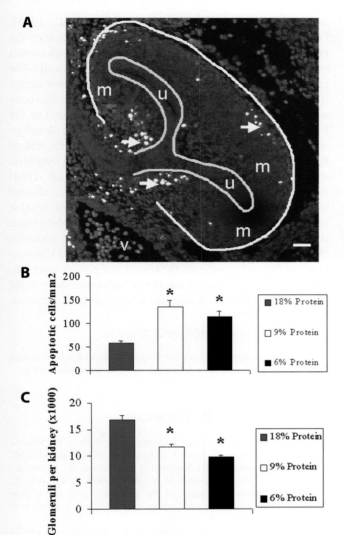

Figure 21.3 Effect of maternal diet on kidney development. (A) Confocal image of E13 metanephric rat kidney (white outline) probed for apoptosic nuclei using the *in situ* end-labeling technique and counterstained with propidium iodide for examination of nuclear morphology. Apoptotic cells (arrows) are visible within the mesenchyme (m) but not the ureteric bud (u). A blood vessel is also shown (v). Bar: 30 μm. (B) Metanephric apoptosis at E13. Bars represent the mean number of apoptotic cells per mm^2 of metanephric kidneys from rats exposed to a maternal 18% protein diet (control), a 9% protein diet (low protein) or a 6% protein diet (very low protein) throughout gestation. Significant differences are represented by an asterisk* ($P<0.05$). (C) Glomerular complement of offspring at 2 weeks of age after exposure to either a maternal 18% protein, a 9% protein, or a 6% protein diet throughout gestation. Significant differences are represented by an asterisk* ($P<0.05$).

Hence, in rats, maternal low protein diets reduce the final numbers of glomeruli in association with the enhanced deletion of mesenchymal cells at the start of kidney development. In humans, the equivalent developmental time frame would be 5 to 7 weeks gestation (Risdon and Woolf, 1998a), which might represent a critical window when kidney morphogenesis might be affected by dietary influences. Of note, Hinchliffe *et al.* (1993) reported nephron deficits, as assessed by counting glomeruli, in infants with intrauterine growth retardation, and epidemiological studies have found that individuals born to mothers with poor diets are prone to hypertension (Barker 1998); furthermore, others have speculated that congenital "nephron deficits" predispose individuals to hypertension later in life (Brenner *et al.*, 1988).

C. Fetal Urinary Flow Impairment

The presence of cartilage and smooth muscle cells, as well as epithelial/mesenchymal intermediate forms, in human dysplastic kidneys indicates major aberrations of cell differentiation. In addition, the abnormal growth and involution seen in some of these organs are likely to be related to the balance between proliferation and death at different stages of evolution of the disease. Studies have been made regarding gene expression and cell turnover in human dysplastic kidneys, some of which are attached to obstructed lower urinary tracts (Bussieres *et al.*, 1995; Winyard *et al.*, 1996a, b; Granata *et al.*, 1997; Matsell *et al.*, 1997; Kolatsi-Joannou *et al.*, 1997; Winyard *et al.*, 1997; Cale *et al.*, 2000; Poucell-Hatton *et al.*, 2000; Yang *et al.*, 2001; MacRae *et al.*, 2000; Woolf and Winyard, 2000) (Fig. 21.4). These studies show that cystic epithelia express PAX2 and BCL2, both of which enhance cell survival, and these cells tend to be hyperproliferative, as assessed by expression of the proliferating cell nuclear antigen. Indeed, genetically engineered mice with forced overexpression of PAX2 (Dressler *et al.*, 1993) acquire renal cysts, suggesting a role for this gene in epithelial overgrowth. In contrast, stromal cells show a high level of apoptosis, as assessed by *in situ* end-labeling and the presence of pyknotic nuclei; although some of these stromal cells express WT1, suggesting they might have been induced to form nephrons, BCL2 expression is low. Some stromal cells proximate to dysplastic tubules show a shift to a smooth muscle lineage, as assessed by expression of a smooth muscle actin (αSMA) (Yang *et al.*, 2000). Yang *et al.* (2000) postulated that an upregulation of transforming growth factor-β1 (TGF-β1), found in human dysplastic organs, might have a biological role in the disease; they found that this cytokine induced an epithelial to smooth muscle shift in cultured dysplastic tubule cells and also demonstrated *in vivo* that rare dysplastic epithelial cells expressed αSMA, whereas rare stromal cells expressed cytokeratin, an epithelial marker.

Experimental urinary tract obstruction in fetal sheep generates a histological appearance that resembles human renal dysplasia, with malformed tubules, a loss of the normal nephrogenic zone, transformation of precursor cells towards a smooth muscle phenotype, and the generation of cysts (Peters *et al.*, 1992; Attar *et al.*, 1998; Yang *et al.*,

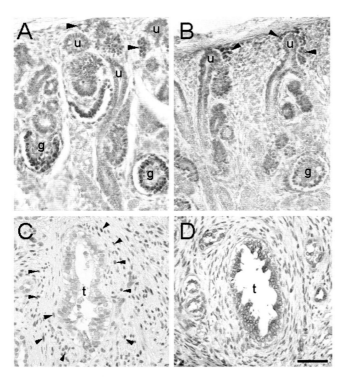

Figure 21.4 Comparison of WT1 and BCL2 expression in normal human kidney development and in dysplastic kidneys. Sections of the midgestation metanephric outer (nephrogenic) cortex (A and B) and postnatal dysplastic kidneys (C and D) immunostained for WT1 (A and C) or BCL2 (B and D) counterstained with methyl green (A–C) or hematoxylin (D). (A) Very low levels of WT1 protein are detected in condensing mesenchyme (arrowheads), but expression increases in nephron precursors, reaching maximal intensity in glomerular (g) podocytes. (B) Conversely, BCL2 levels are highest in the mesenchymal condensates and decrease as the nephron matures. Note that neither WT1 nor BCL2 is expressed in the ureteric bud branch tips (u). (C) WT1 is expressed in the majority of cells (arrowheads indicate some of these) around aberrant tubules (t) in dysplastic kidneys, but not in the dysplastic epithelium. (D) In contrast, BCL2 expression is predominantly in the dysplastic epithelium. Bar: 30 μm.

2001). It also induces a complex deregulation of cell turnover in the metanephros, with enhanced apoptosis in some structures, as well as increased proliferation in epithelia lining cysts; PAX2 expression is lost in the nephrogenic mesenchyme but is prominent in cystic epithelia. Furthermore, the expression of growth factors is altered, including an upregulation of TGF-β1 and its receptor, TGF-βRI (Medjebeur *et al.*, 1997; Attar *et al.*, 1998; Yang *et al.*, 2001). When urine outflow from the developing ovine urinary bladder is prevented, bladder muscle structure and innervation are perturbed, in addition to the generation of a cystic renal dysplasia (Nyirady *et al.*, 2002), emphasizing how a simple physical insult can be transduced in a complex aberration of morphology, cell biology, and physiology.

In a marsupial model, the administration of insulin-like growth factor reduces renal fibrosis, tubular cystic changes, and calyceal dilatation, which follow obstruction of the

developing opossum kidney (Steinhardt *et al.*, 1995). Furthermore, after experimental ureteric obstruction of the neonatal rat kidney (in rats, only about 10% of glomeruli have formed at birth), administration of either insulin-like growth factor or epidermal growth factor reduces damage (Chevalier *et al.*, 1998, 2000). Animal models have also been used to answer the question whether decompression of the fetal obstructed urinary tract can rescue normal development. *In utero* decompression performed before the end of the nephrogenic period can partially prevent renal dysplasia and the loss of glomeruli in fetal sheep (Glick *et al.*, 1984; Edouga *et al.*, 2001). In the neonatal rat model alluded to earlier, decompression attenuated but did not reverse renal injury resulting from 5 days of ureteric obstruction (Chevalier *et al.*, 1999). Probably, in the human scenario, effective therapeutic intervention, e.g., by surgical shunting to allow flow of urine from the obstructed urinary tract into the amniotic cavity, would need to be performed early in the genesis of this disease, before severe changes have occurred.

Acknowledgments

S. Welham is supported by a grant from the National Kidney Research Fund and M. Hermanns by a fellowship from the European Urological Scholarship Programme.

Refereances

Abdelhak, S., Kalatzis, V., Heilig, R., Compain, S., Samson. D., Vincent, C., Weil, D., Cruaud, C., Sahly, I., Leibovici, M., Bitner-Glindzicz, M., Francis, M., Lacombe, D., Vigneron, J., Charachon, R., Boven, K., Bedbeder, P., Van Regemorter, N. V., Weissenbach, J., and Petit, C. (1997). A human homologue of the drosophila *eyes absent* gene underlies branchio-oto-renal syndrome and identifies a novel gene family. *Nature Genet.* **15**, 157–164.

Akl, K. F., Khudr, G, S., Der Kaloustian, V. M., and Najjar, S. S. (1977). The Smith-Lemli-Opitz syndrome. Report of a consanguineous Arab infant with bilateral focal renal dysplasia. *Clin. Pediatr.* **16**, 665–668.

Alcaraz, A., Vinaixa, F., Tejedo-Mateu, A., Fore´s, Gotzens, V., Mestres C. A., Oliveira, J., and Caretero, P. (1991). Obstruction and recanalization of the ureter during embryonic development. *J. Urol.* **145**, 410–461.

Anderson, N. G., Abbott, G. D., Mogridge, N., Allan, R. B., Maling, T. M., and Wells, J. E. (1997) Vesicoureteric reflux in the newborn: Relationship to fetal renal pelvic diameter. *Pediatr. Nephrol.* **11**, 610–616.

Araki, T., Saruta, T., Okano, H., and Miura, M. (1999). Capsase activity is required for nephrogenesis in the developing mouse metanephros. *Exp. Cell Res.* **248**, 423–429.

Arfeen, S., Rosborough, D., Luger, A. M., and Nolph, K. D. (1993) Familial unilateral renal agenesis and focal and segmental glomerulosclerosis. *Am. J. Kidney Dis.* **21**, 663–668.

Attar, R., Quinn, F., Winyard, P. J, Mouriquand, P. D., Foxall, P., Hanson, M. A., and Woolf, A. S. (1998). Short-term urinary flow impairment deregulates PAX2 and PCNA expression and cell survival in fetal sheep kidneys. *Am. J. Pathol.* **152**, 1225–1235.

Atwell, J. D.(1985). Familial pelviureteric junction hydronephrosis and its association with a duplex pelvicalyceal system and vesicoureteric reflux. A family study. *Br. J. Urol.* **57**, 365–369.

Austin, P. F., Cain, M. P., Casale, A. J., Hiett, A. K., and Rink, R. C. (1998) Prenatal bladder outlet obstruction secondary to ureterocele. *Urology* **52**, 1132–1135.

Barbacci, E., Reber, M., Ott, M., Breillat, C., Huetz, F., and Cereghini, S. (1999) Variant hepatocyte nuclear factor 1 is required for visceral endoderm specification. *Development* **126**, 4795–4805.

Barker, D. J. P. (1998). "Mothers, Babies, and Health in Later Life." Churchill Livingston, Edinburgh.

Barr, M., Jr. (1994). Teratogen update: Angiotensin-converting enzyme inhibitors. *Teratology* **50**, 399–409.

Battin, M., Albersheim, S., and Newman, D. (1995) Congenital genito-urinary tract abnormalities following cocaine exposure in utero. *Am. J. Perinatol.* **12**, 425–428.

Bingham, C., Bulman, M. P., Ellard, S., Allen, L. I. S., Lipkin, G. W., van't Hoff, W. G., Woolf, A. S., Rizzoni, G., Novelli, G., Nicholls, A. J., Hattersley, A.T. (2001) Mutations in the hepatocyte nuclear factor-1β gene are associated with familial hypoplastic glomerulocystic kidney disease. *Am. J. Hum. Genet.* **68**, 219–224.

Bingham, C., Ellard, S., Allen, L., Bulman, M., Shepherd, M., Frayling, T., Berry, P. J., Clark, P. M., Lindner, T., Bell, G. I., Ryffel, G. U., Nicholls, A. J., and Hattersley, A. T. (2000). Abnormal nephron development associated with a frameshift mutation in the transcription factor hepatocyte nuclear factor-1β. *Kidney Int.* **57**, 898–907.

Bingham, C., Ellard, S., Cole, T. R. P., Jones, K. E., Allen, L. I. S., Goodship, J. A., Goodship, T. H. J., Bakalinova-Pugh, D., Russell, G. I., Woolf, A. S., Nicholls, A. J., and Hattersely, A. T. (2000). Solitary functioning kidney and diverse genital tract malformations associated with hepatocyte nuclear factor-1β mutations. *Kidney Int.* **61**, 1243–1251.

Brenner, B. M., and MacKenzie, H. S. (1997). Nephron mass as a risk factor for progression of renal disease. *Kidney Int.* **63**, S-124–S-127.

Budarf, M., Collins., J., Gong, W., Roe, B., Wang, Z., Bailey, L. C., Sellinger, B., Michaud, D., Driscoll, D., and Emanuel, B. S. (1995) Cloning a balanced translocation associated with Di George syndrome and identification of a disrupted candidate gene. *Nature Genet.* **10**, 269–271.

Bullock, S. L., Johnson, T., Bao, Q., Winyard, P. J. D., Hughes, R. C. Woolf, A. S. (2001) Galectin-3 modulates ureteric bud branching in organ culture of the developing mouse kidney. *J. Am. Soc. Nephrol.* **12**, 515–523.

Bussieres, L., Laborde, K., Souberbielle, J. C., Muller, F., Dommergues, M., and Sachs, C. (1995). Fetal urinary insulin-like growth factor I and binding protein 3 in bilateral obstructive uropathies. *Prenat. Diagn.* **15**, 1047–1055.

Cain, D. R., Griggs, D., Lackey, D. A,. and Kagan, B. M. (1974). Familial renal agenesis and total dysplasia. *Am. J. Dis. Child.* **128**, 377–380.

Cale, C M., Klein, N. J., Morgan, G., and Woolf, A. S. (1998). Tumour necrosis factor-α inhibits epithelial differentiation and morphogenesis in the mouse metanephric kidney in vitro. *Int. J. Dev. Biol.* **42**, 663–674.

Cale, C. M., Klein, N. J., Winyard, P. J. D., and Woolf, A. S. (2000). Inflammatory mediators in human renal dysplasia. *Nephrol. Dial. Transplant.* **15**, 173–183.

Chatha, R. K., Johnson, A. M., Rothberg, P. G., Townsend, R. R., Neumann, H. P., and Gabow, P. A. (2001). Von Hippel-Lindau disease masquerading as autosomal dominant polycystic kidney disease. *Am. J. Kidney Dis.* **37**, 852–858.

Chevalier, R. L., Goyal, S., Kim, A., Chang, A. Y., Landau, D., and LeRoith, D. (2000). Renal tubulointersitial injury from ureteral obstruction in the neonatal rat is attenuated by IGF-1. *Kidney Int.* **57**, 882–890.

Chevalier, R. L., Goyal, S., Wolstenholme, J. T., and Thornhill, B. A. (1998) Obstructive nephropathy in the neonatal rat is attenuated by epidermal growth factor. *Kidney Int.* **54**, 38–47.

Chevalier, R. L., Kim, A., Thornhill, B. A., and Wolstenholme, J. T. (1999). Recovery following relief of unilateral obstruction in the neonatal rat. *Kidney Int.* **55**, 793–807.

Choyke, P. L., Siegel, M. J., Oz, O., Sotelo-Avila, C., and DeBaun, M. R.

(1999). Nonmalignant renal disease in pediatric patients with Beckwith-Wiedemann syndrome. *Am. J. Roentgenol.* **171**, 733–737.

Coffinier, C., Barra, J., Babinet, C., and Yaniv, M. (199). Expression of vHNF/HNF1β homeoprotein gene during mouse organogenesis. *Mech. Dev.* **89**, 211–213.

Cohen, M. M., and Kreiborg, S. (1993). Visceral abnormalities in the Apert syndrome. *Am. J. Med. Genet.* **45**, 758–760.

Coppes, M. J., Huff, V., and Pelletier, J. (1993). Denys-Drash syndrome: Relating a clinical disorder to alterations in the tumor suppressor gene WT1. *J. Paediatr.* **123**, 673–678.

Czarnecki, P. M., Van Dyke, D. L., Vats, S., and Feldman, G. L. (1998). A mother with VCFS and unilateral dysplastic kidney and her fetus with multicystic dysplastic kidneys: Additional evidence to support the association of renal malformations and VCFS. *J. Med. Genet.* **35**, 348.

Daikha-Dahmane, F., Dommergues, M., Muller, F., Narcy, F., Lacoste, M., Beziau, A., Dumez, Y., and Gubler, M. C. (1997). Development of human fetal kidney in obstructive uropathy: Correlations with ultra-sonography and urine biochemistry. *Kidney Int.* **52**, 21–32.

Damian, M. S., Seibel, P., Schachenmayr, W., Reichmann, H., and Dorndorf, W. (1996), VACTERL with the mitochondial np 3243 point mutation. *Am. J. Med. Genet.* **62**, 398–403.

Davis, A. P., Witte, D. P., Hsieh-Li, H. M., Potter, S. S., and Capecchi, M. R. (1995) Absence of radius and ulna in mice lacking *hoxa-11* and *hoxd-11*. *Nature* **375**, 791–795.

Deeb, A., Robertson, A., MacColl, G., Bouloux, P. M., Gibson, M., Winyard, P. J. D., Woolf, A. S., Moghal. N. E., and Cheetham, T. D. (2001). Multicystic kidney and X-linked Kallmann's syndrome: A new association? *Nephrol. Dial. Transplant.* **16**, 1170–1175.

Dressler, G. R., Wilkinson, J. E., Rothenpieler, U. W., Patterson, L. T., Williams-Simons, L., and Westphal, H.(1993). Deregulation of *Pax-2* expression in transgenic mice generates severe kidney abnormalities. *Nature* **362**, 65–67.

Dreyer, S. D., Zhou, G., Baldini, A., Winterpacht, A., Zabel, B., Cole, W., Johnson, R. L., and Lee, B. (1998). Mutations in LMX1B cause abnormal skeletal patterning and renal dysplasia in nail patella syndrome. *Nature Genet.* **19**, 47–50.

Drozdz, D., Drodz, M., Gretz, N., Mohring, K., Mehls, O., and Scharer, K. (1998). Progression to end-stage renal disease in children with posterior urethral valves. *Pediatr. Nephrol.* **12**, 630–636.

Duke, V., Quinton, R., Gordon, I., Bouloux, P. M. G., and Woolf, A. S. (1998). Proteinuria, hypertension and chronic renal failure in X-linked Kallmann's syndrome, a defined genetic cause of solitary functioning kidney. *Nephrol, Dial. Transplant.* **13**, 1998–2003.

Duke, V. M., Winyard, P. J. D., Thorogood. P., Soothill, P., Bouloux, P. M. G., and Woolf, A. S. (1995). KAL, a gene mutated in Kallmann's syndrome, is expressed in the first trimester of human development. *Mol. Cell. Endocrinol.* **110**, 73–79.

Edouga, D., Huygueny, B., Gasser, B., Bussieres L., and Laborde, K. (2001) Recovery after relief of fetal urinary obstruction: Morphological, functional and molecular apsects. *Am. J. Physiol.* F26–F37.

Ehrich, J. H. H., Rizzoni, G., Brunner, F. P., Fassbinder, W., Geerlings, W., Mallick, N. P., Raine A. E. G., Selwood, N. H., and Tufveson, G. (1992). Renal replacement therapy for end-stage renal failure before 2 years of age. *Nephrol. Dial. Transplant.* 7, 1171–1177.

Ekblom, P., Ekblom, M., Fecker, L., Klein, G., Zhang, H.-Y., Kadoya, Y., Chu, M.-L., Mayer, U., and Timpl, R. (1993), Role of mesenchymal nidogen for epithelial morphogenesis in vitro. *Development* **120**, 2003–2014.

Escala, J. M., Keating, M. A., Boyd, G., Pierce, A., Hutton, J. L., and Lister, J. (1989). Development of elastic fibres in the upper urinary tract. *J. Urol.* **141**, 969–973.

Farrar, D. J., and Pryor, J. S. (1976). Posterior urethral valves in siblings. *Br. J. Urol.* **48**, 76–77.

Favor, J., Sandulache, R., Neuhauser-Klaus, A., Pretsch, W., Chaterjee, B., Senft, E., Wurst, W., Blanquet, V., Grimes, P., Sporle, R., and

Schughart, K. (1996). The mouse *Pax2^{1Neu}* mutation is identical to a human *PAX2* mutation in a family with renal-coloboma syndrome and results in developmental defects of the brain, ear, eye and kidney. *Proc. Natl. Acad. Sci. USA* **93**, 13870–13875.

Feather, S., Woolf, A. S., Gordon, I., Risdon, R. A., and Verrier-Jones, K. (1996). Vesicoureteric reflux: All in the genes? *Lancet* **348**, 725–728.

Feather, S. A., Malcolm, S., Woolf, A. S., Wright, V., Blaydon, D., Reid, C. J. D., Flinter, F. A., Proesmans, W., Devriendt, K., Carter, J., Warwicker, P., Goodship, T. H. J., and Goodship, J. A. (2000). Primary, non-syndromic vesicoureteric reflux and its nephropathy is genetically heterogeneous with a locus on chromosome 1. *Am. J. Hum. Genet.* **66**, 1420–1425.

Feather, S. A., Winyard, P. J. D., Dodd, S., and Woolf, A.S . (1997). Oral-facial-digital syndrome type 1 is another dominant polycystic kidney disease: Clinical, radiological and histopathological features of a new kindred. *Nephrol. Dial. Transplant.* **12**, 1354–1361.

Ferrante, M. I., Giorgio, G., Feather, S. A., Bulfone, A., Wright, V., Ghiani, M., Selicorni, A., Gammaro, L., Scolari, F., Woolf, A.S ., Odent, S., Le Marec, B., Malcolm, S., Winter, R., Ballabio, A., and Franco, B. (2001). Identification of the gene for oral facial digital syndrome type 1 (OFD1). *Am. J. Hum. Genet.* **68**, 569–576.

Franco, B., Guioli, S., Pragliola, A., Incerti, B., Bardoni, B., Tonlorenzi, R., Carrozo, R., Maestrini, E., Pieretti, M., Tiallon-Miller, P., Brown, C. J., Willard, H. F., Lawrence, C., Persico, M. G., Camerino, G., and Ballabio, A. (1991). A gene deleted in Kallmann's syndrome shares homology with neural cell adhesion and axonal path-finding molecules. *Nature* **353**, 529–536.

Freedman, A. L., Johnson, M. P., Smith, C. A., Gonzalez, R., and Evans, M. I. (1999). Long-term outcome in children after antenatal intervention for obstructive uropathies. *Lancet 1999* **354**, 374–377.

Gilbert, S. F. "(1997). Developmental Biology." Sinauer Associates, Sunderland, MA.

Glick, P. L., Harrison, M. R., Adzick, N. S., Noall, R. A., and Villa, R. L. (1984). Correction of congenital hydronephrosis in utero. IV. In utero decompression prevents renal dysplasia. *J. Peditr. Surg.* **19**, 649–657.

Gonzalez, A. D., Kaya, M., Shi, W., Song, H., Testa, J. R., Penn, L. Z., and Filmus, J. (1998). OCI-5/GPC3, a glycipan encoded by a gene that is mutated in the Simpson-Golabi-Behmel overgrowth syndrome, induces apoptosis in a cell line-specific manner. *J. Cell Biol.* **141**, 1407–1414.

Gonzalez-Perret, S., Kim, K., Ibarra, C., Damiano, A. E., Zotta, E., Batelli, M., Harris, P. C., Reisin, I. L., Arnaout, M. A., and Cantiello, H.F. (2001). Polycystin-2, the protein mutated in autosomal dominant polycystic kidney disease (ADPKD), is a Ca^{2+}-permeable nonselective cation channel. *Proc. Natl. Acad. Sci. USA* **98**, 1182–1187.

Granata, C., Wang, Y., Puri, P., Tanaka, K., and O'Briain, D. S. (1997). Decreased *bcl-2* expression in segmental renal dysplasia suggests a role in its morphogenesis. *Br. J. Urol.* **80**, 140–144.

Grobstein, C. (1967). Mechanisms of organotypic tissue interaction. *Natl. Cancer. Inst. Monogr.* **26**, 279–299.

Gubler, M. C., and Levy, M. (1993). Prenatal diagnosis of nail-patella syndrome by intrauterine kidney biopsy. *Am. J. Med. Genet.* **47**, 122–124.

Haga, W., Lee, K., Nakamura, K., Okazaki, Y., Mamada, K., and Kurokawa, T. (1997). Congenital deficiency of the fibula with ipsilateral iliac horn and absence of the kidney. *Clin. Dysmorphol.* **6**, 177–180.

Hardelin, J. P. (2001). Kallmann syndrome: Towards molecular pathogenesis. *Mol. Cell. Endocrinol.* **179**, 75–81.

Hardelin, J. P., Juillard, A. K., Moniot, B., Soussi-Yanicostas, N., Verney, C., Schwanzel-Fukuda, M., Ayer-Le Lievre, C., and Petit, C. (1999). Anosmin-1 is a regionally restricted component of basement membranes and interstitial matrices during organogenesis: Implications for the developmental anomalies of X chromosome-linked Kallmann syndrome. *Dev, Dyn.* **215**, 26–44.

Harnett, J. D., Green, J. S., Cramer, B. C., Johnson, G., Chafe, L., McManman, R., Farid, N. R., Pryse-Phillips, W., and Parfrey, P. S.

(1988). The spectrum of renal disease in Laurence-Moon-Biedl syndrome. *N. Engl. J. Med.* **319**, 615–618.

Hinchliffe, S. A., Howard, C. V., Lynch, M. R., Sargent, P. H., Judd, B. A., and van Velzen, D. (1993). Renal developmental arrest in sudden infant death syndrome. *Pediatr. Pathol* **13**, 333–343.

Hinchliffe, S. A., Lynch, M. R. J., Sargent, P. H., Howard, C. V., and van Velzen, D. (1992). The effects of intrauterine growth retardation on the development of renal nephrons. *Br. J. Obstet. Gynaecol.* **99**, 296–301.

Hinchliffe, S. A., Sargent, P. H., Howard, C. V., Chan, Y. F., Hutton, J. L., Rushton, D. I., and van Velzen, D. (1991). Human intra-uterine renal growth expressed in absolute number of glomeruli assessed by 'Disector' method and Cavalieri principle. *Lab. Invest.* **64**, 777–784.

Hiraoka, M., Hori, C., Tsukahara, H., Kasuga, K., Ishihara, Y., and Sudo, M. (1997). Congenitally small kidneys with reflux as a common cause of nephropathy in boys. *Kidney Int.* **52**, 811–816.

Houston, C. S., Opitz, J. M., Spranger, J. W., Macpherson, R. I., Reed, M. H., Gilbert, E. F., Herrmann, J., and Schinzel, A (1983). The campomelic syndrome: Review, report of 17 cases, and follow-up on the currently 17-year old boy first reported by Maroteaux. *Am. J. Med. Genet.* **15**, 3–28.

Hughes, J., Ward, C. J., Peral, B., Aspinwall, R., Clark, K., San Millan, J. L., Ganble, V., and Harris, P. C. (1995). The polycystic kidney disease 1 (PKD1) gene encodes a novel protein with multiple cell recognition domains. *Nature Genet.* **10**, 151–156.

Hughes-Benzie, R. M., Hunter, A. G., Allanson, J. E., and MacKenzie, E. A. (1994). Simpson-Golabi-Behmel syndrome associated with renal dysplasia and embryonal tumors: Localisation of the gene to Xqcen-Xq21. *Am. J. Med. Genet.* **43**, 428–435.

Hulton, S. A. and Kaplan, B. A. (1995). Renal dysplasia associated with in utero exposure to gentamycin and corticosteroids. *Am. J. Med. Genet.* **58**, 91–93.

Izquierdo, L., Porteous, M., Paramo, P. G., and Connor, J. M. (1992). Evidence for genetic heterogeneity in hereditary hydronephrosis caused by pelviureteric junction obstruction, with one locus assigned to chromosome 6p. *Hum. Genet.* **89**, 557–560.

Jadresic, L., Leake, J., Gordon, I., Dillon, M. J., Grant, D. B., Pritchard, J., Risdon, R. A., and Barratt, T. M. (1990). Clinico-pathologic review of 12 children with nephropathy, Wilms' tumour and genital abnormalities (Drash syndrome) *J. Pediatr.* **117**, 717–725.

Jaswon, M. S., Dibble, L., Puri, S., Davis, J., Young, J., Dave, R., and Morgan, H. (1999). Prospective study of outcome in antenatally diagnosed renal pelvis dilatation. *Arch. Dis. Child. Neonatal. Ed.* **80**, F135–138.

Josephson, S., Dhillon, H. K., and Ransley, P. G. (1993), Postnatal management of antenatally detected, bilateral hydronephrosis. *Urol. Int.* **51**, 79–84.

Kampmeier, O. F. (1926). The metanephros or so-called permanent kidney in part provisional and vestigial. *Anat. Rec.* **33**, 115–120.

Katsanis, N., Ansley, S. J., Badano, J. L., Eichers, E. R., Lewis, R. A., Hoskins, B. E., Scambler, P. J., Davidson, W. S., Beales, P. L., and Lupski, J. R. (2001). Triallelic inheritance in Bardet-Biedl syndrome, a Mendelian recessive disorder. *Science* **293**, 2256–2259.

Katsanis, N., Beales, P. L., Woods, M. O., Lewis, R .A., Green, J. S., Parfrey, P. S., Ansley, S. J., Davidson, W. S., and Lupski, J. R. (2000). Mutations of MKKS cause obesity, retinal dystriphy and renal malformations associated with Bardet-Biedl syndrome. *Nature Genet.* **26**, 67–70.

Kirk, J. M. W., Grant, D. B., Besser, G. M., Shalet, S., Quinton, R., Smith, C. S., White, M., Edwards, O., and Bouloux, P. M. J. (1994). Unilateral renal aplasia in X-linked Kallmann's syndrome. *Clin. Genet.* **46**, 260–262.

Klein, G., Langegger, M., Timpl, R., and Ekblom, P. (1998). Role of laminin A chain in the development of epithelial cell polarity. *Cell* **55**, 331–341.

Kleymenova, E., Ibraghimov-Beskrovnaya, O., Kugoh, H., Everitt, J., Xu, H., Kiguchi, K., Landes, G., Harris, P., and Walker, C. (2001). Tuberin-dependent membrane localization of polycystin-1: A functional link between polycystic kidney disease and the TSC2 tumor suppressor gene. *Mol. Cell* **7**, 823–832.

Kolatsi-Joannou, M., Bingham, C., Ellard, S., Bulman, M. P., Allen, L. I. S., Hattersley, A. T., and Woolf, A .S. (2001). Hepatocyte nuclear factor 1β: A new kindred with renal cysts and diabetes and gene expression in normal human development. *J. Am. Soc. Nephrol.* **12**, 2175–2180.

Kolatsi-Joannou, M., Moore, R., Winyard, P. J. D., and Woolf, A. S. (1997). Expression of hepatocyte growth factor/scatter factor and its receptor, MET, suggests roles in human embryonic organogenesis. *Pediatr. Res.* **41**, 657–665.

Konig, R., Fuchs, S., and Dukiet, C. (1994). Branchio-oto-renal syndrome: Variable expressivity in a five generation pedigree. *Eur. J. Pediatr.* **153**, 446–450.

Kreidberg, J. A., Donovan, M. J., Goldstein, S. L., Rennke, H., Shepard, K., Jones, R. C., and Jaenisch, R. (1996). α3β1 integrin has a crucial role in kidney and lung organogenesis. *Development* **122**, 3537–3547.

Kreidberg, J. A., Sariola, H., Loring, J. M., Maeda, M., Pelletier, J., Housman, D., and Jaenisch, R. (1993). WT-1 is required for early kidney development. *Cell* **74**, 679–691.

Kuure, S., Vuolteenaho, R., and Vainio, S., (2000). Kidney morphogenesis: Cellular and molecular regulation. *Mech. Dev.* **92**, 19–30.

Kwong, W. Y., Wild, A. E., Roberts, P., Willis, A. C., and Fleming, T. P. (2000). Maternal undernutrition during the preimplantation period of rat development causes blastocyst abnormalities and programming of postnatal hypertension. *Development* **127**, 4195–4202.

Lam, W. W., Hatada, I., Ohishi, S., Mukai, T., Joyce, J. A., Cole, T. R., Donnai, D., Reik, W., Schofield, P. N., and Maher, E. R. (1999). Analysis of germline CDKN1C (p57KIP2) provides a novel genotype-phenotype correlation. *J. Med. Genet.* **36**, 518–523.

Langley-Evans, S. C., Welham, S. J. M., and Jackson, A. A. (1999). Fetal exposure to a maternal low protein diet impairs nephrogenesis and promotes hypertension in the rat. *Life Sci.* **64**, 965–974.

Langley-Evans, S. C., Welham, S. J. M., Sherman, R. C., and Jackson, A. A. (1996). Weanling rats exposed to maternal low protein diets during discrete periods of gestation exhibit differing severity of hypertension. *Clin. Sci.* **91**, 607–615.

Larsen, W. J. "(1993), Human embryology." Churchill Livingstone, New York.

Legouis, R., Hardelin, J. P., Levilliers, J., Claverie, J. M., Compain, S., Wunderle, V., Millasseau, P., Le-Paslier, D., Cohen, D., Caterina, D., Bougueleret, L., Delemarre-Van de Waal, H., Lutfalla. G., Weissenbach, J., and Petit, C. (1991). The candidate gene for the X-linked Kallmann syndrome encodes a protein related to adhesion molecules. *Cell* **67**, 423–435.

Levy, L., and Jackson, A. A. (1993). Modest restriction of dietary protein during pregnancy in the rat: Fetal and placental growth. *J. Dev. Physiol.* **19**, 113–118.

Lewis, M. (1999). Report of the pediatric renal registry. In "The UK Renal Registry. The Second Annual Report." D. Ansell T. Feest, eds., pp. 175–187, The Renal Association, Bristol, UK.

Lindner, T. H., Njolstad, P. R., Horikawa, Y., Bostad, L., Bell, G. I., and Sovik, O. (1999). A novel syndrome of diabetes mellitus, renal dysfunction and genital malformation associated with a partial deletion of the pseudo-POU domain of hepatocyte nuclear factor-1β. *Hum. Mol. Genet.* **8**, 2001–2008.

Liu H. Y., Dhillon H. K, Yeung C. K, Diamond D. A, Duffy P. G., Ransley P. G. (1994). Clinical outcome and management of prenatally diagnosed primary megaureters. *J. Urol.* **152**, 614–617.

Lo Ten Foe, J. R., Rooimans, M. A., Bosnoyan-Collins, L., Alon, N., Wijker, M., Parker, L., Lightfoot, J., Carreau, M., Callen, D. F., Savoia, A., Cheng, N. C., van Berkel C. G. M., Strunk, M. H. P., Gille, J. J. P., Pals, G., Kruyt, F. A. E., Pronk, J. C., Arwert, F., Buchwald, M., Joenje, H. (1996): Expression cloning for the major Fanconi anaemia gene, FAA. *Nature Genet.* **14**, 320–323.

Lynch, S. A., and Wright, C. (1997). Sirenomelia, limb reduction defects, cardiovascular malformation, renal agenesis in an infant born to a diabetic mother. *Clin. Dysmorphol.* **6**, 75–80.

MacRae Dell, K., Hoffman, B. B., Leonard, M. B., Ziyadeh, F. N., and Schulman, S. L. (2000). Increased urinary transforming growth factor-β1 excretion in children with posterior urethral valves. *Urology* **56**, 311–314.

Martignoni, G., Bonetti, F., Pea, M., Tardanico, R., Brunelli, M., and Eble, J. N. (2002). Renal disease in adults with TSC2/PKD1 contiguous gene syndrome. *Am. J. Surg. Pathol.* **26**, 198–205.

Matsell, D. G., Bennett, T., Armstrong, R. A., Goodyer, P., Goodyer, C., and Han, V. K. (1997). Insulin-like growth factor (IGF) and IGF binding protein gene expression in multicystic renal dysplasia. *J. Am. Soc. Nephrol.* **8**, 85–94.

Matsuno, T., Tokunaka, S., and Koyanagi, T. (1984), Muscular development in the urinary tract. *J. Urol.* **132**, 148–152.

McKusick, V. A. "Online Mendelian Inheritance in Man. National Center for Biotechnology Information (http://www4.ncbi.nlm.nih.gov/Omim/).

McPherson, E., Carey, J., Kramer, A., Hall, J. G., Pauli, R. M., Schimke, R. N., and Tasin, M. H. (1989), Dominantly inherited renal adysplasia. *Am. J. Med. Genet.* **26**, 863–872.

Medjebeur, A. A., Bussieres, L., Gasser, B., Gimonet, V., and Laborde, K. (1997). Experimental bilateral urinary obstruction in fetal sheep: Transforming growth factor-β1 expression. *Am. J. Physiol.* **1273**, F372–F379.

Mendelsohn, C., Lohnes, D., Decimo, D., Lufkin, T., LeMeur, M., Chambon, P., and Mark, M. (1994). Function of the retinoic acid receptors during development. *Development* **120**, 2749–2771.

Mendelsohn, C., Batourina E., Fung, S., Gilbert, T., and Dodd, J. (1999). Stromal cell mediate retinoid-dependent functions essential for renal development. *Development* **126**, 1139–1148.

Merlet-Bernichou, C., Gilbert, T., Muffat-Joly, M., Lelievre-Pergorier, M., and Leroy, B. (1994), Intrauterine growth retardation leads to a permanent nephron deficit in the rat. *Pediatr. Nephrol.* **8**, 175–180.

Mesrobian, H. G, Rushton, H. G., and Bulas, D. (1993). Unilateral renal agenesis may result from in utero regression of multicystic renal dysplasia. *J. Urol.* **150**, 793–794.

Moazin, M. S., Ahmed, S., and Fouda-Neel, K. (1997). Multicystic kidney in siblings. *J. Pediatr. Surg.* **32**, 119–120.

Moerman, P., Van Damme, B., Proesmans, W., Devlieger, H., Goddeeris, P., and Lauweryns, J. (1984). Oligomeganephronic renal hypoplasia in two siblings. *J. Pediatr.* **105**, 75–77.

Moore, C. A., Khoury, M. J., and Liu, Y. (1997). Does light-to-moderate alcohol consumption during pregnancy increase the risk for renal anomalies among offspring? *Pediatrics* **99**, E11.

Moore, K., L., and Persaud, T. V. N. (1998). The Developing Human. Clinically Oriented Embryology. Saunders, Philadelphia.

Moreau, E., Villar, J., Lelievre-Pegorier, M., Merlet-Benichou, C., and Gilbert, T. (1998). Regulation of c-ret- expression by retinoic acid in rat metanephros: Implication in nephron mass control. *Am. J. Physiol.* **275**, F938–F945.

Muller, U., Wang, D., Denda, S., Meneses, J. J., Pedersen, R. A., and Reichardt, L. F. (1997). Integrin α8β1 is critically important for epithelial-mesenchymal interactions during kidney development. *Cell* **88**, 603–613.

Murakami, T., Kawakami, H., Kimoto, J., and Sase, M. (1992). Multicystic renal dysplasia in two consecutive male infants. *Am. J. Kidney Dis.* **20**, 676.

Murugasu, B., Cole, B. R., Hawkins, E. P., Blanton, S. H., Conley, S. B., and Portman, R. J. (1991). Familial renal adysplasia. *Am. J. Kidney Dis.* **18**, 495–496.

Nishimura, H., Yerkes, E., Hohenfellner, K., Miyazaki, Y., Ma, J., Huntley, T. E., Yoshida, H., Ichiki T., Threadgill, D., Phillips, J. A., 3rd, Hogan, B. M., Fogo, A., Brock, J. W., 3rd, Inagami, T., and Ichikawa, I. (1999). Role of the angiotensin type 2 receptor gene in congenital anomalies of the kidney and urinary tract, CAKUT, of mice and men. *Mol. Cell.* **3**, 1–10.

Noakes, P. G., Miner, J. H., Gautam, M., Cunningham, J. M., Sanes, J., and Merlie, J. P. (1995). The renal glomerulus of mice lacking s-laminin/laminin β2: Nephrosis despite molecular compensation by laminin β1. *Nature Genet.* **10**, 400–406.

Noia, G., Masini, L., De Santis, M., and Caruso, A. (1996). The impact of invasive procedures on prognostic, diagnostic and therapeutic aspects of urinary tract anomalies. In "Neonatal Nephrology in Progress" L. Cataldi, V. and Fanos, U. Simeoni, eds., pp. 67–84. Agora, Lecce, Italy.

North, K. N., Hoppel, C. L., De Girolami, U., Kozalewich, H. P., and Korson, M. S. (1995). Lethal neonatal deficiency of carnitine palmitoyltransferase II associated with dysgenesis of the brain and kidneys. *J. Pediatr.* **127**, 414–420.

Novak, R. W., and Robinson, H. B. (1994). Coincident Di George anomaly and renal agenesis and its relation to maternal diabetes. *Am. J. Med. Genet.* **50**, 311–312.

Nyirady, P., Thiruchelvam, N., Fry, C. H., Godley, M. L., Winyard, P. J. D., Peebles, D. M., Woolf, A. S., and Cuckow, P. M. (2002). Fetal sheep bladder outflow obstruction perturbs detrusor contractility, compliance and innervation. *J. Urol.* **168**, 1615–1620.

Paavola, P., Salonen, R., Weissenbach, J., and Peltonen, L. (1995). The locus for Meckel syndrome with multiple congenital anomalies maps to chromosome 17q21–q24. *Nature Genet.* **11**, 213–215.

Padmanabhan, R. (1998). Retinoic acid-induced caudal regression syndrome in the mouse fetus. *Reprod. Toxicol.* **12**, 139–151.

Patterson, L. T., Pembaur, M., and Potter, S. S. (2001). Hoxa11 and Hoxd11 regulate branching morphogenesis of the ureteric bud in the developing kidney. *Development* **128**, 2153–2161.

Pei, Y., Paterson, A. D., Wang, K. R., He, N., Hefferton, D., Watnick, T., Germoni, G. G., Parfrey, P., Somlo, S., St. George-Hyslop, P. (2001). Bilineal disease and trans-heterozygotes in autosomal dominant polycystic kidney disease. *Am. J. Hum. Genet.* **68**, 355–363.

Peters, C. A., Carr, M. C., Lais, A., Retik, A. B., and Mandell, J. (1992). The response of the fetal kidney to obstruction. *J. Urol.* **148**, 503–509.

Pilia, G., Hughes-Benzie, R. M., MacKenzie, A., Baybayan, P., Chen, E. Y., Huber, R., Neri G., Cao, A., Forabosco, A., and Schlessinger, D. (1996). Mutations in *GPC3*, a glypican gene, cause the Simpson-Golabi-Behmel overgrowth syndrome. *Nature Genet.* **12**, 241–247.

Pitera, J. E., Smith, V. V., Woolf, A. S., and Milla, P. J. (2001), Embryonic gut anomalies in a mouse model of retinoic acid-induced caudal regression syndrome. Delayed gut looping, rudimentary cecum and anorectal anomalies. *Am. J. Pathol.* **159**, 2321–2329.

Plisov, S. Y., Yoshino, K., Dive, L. F., Higinbotham, K. G., Rubin, J. S., and Perantoni, A. O. (2001), TGFβ2, LIF and FGF2 cooperate to induce nephrogenesis. *Development* **128**, 1045–1057.

Porteus, S., Torban, E., Cho, N. P., Cunliffe, H., Chua, L., McNoe, L., Ward, T., Souza, C., Gus, P., Giugliani, R., Sato, T., Yun, K., Favor, J., Sicotte, M., Goodyer, P., and Eccles, M. (2000). Primary renal hypoplasia in humans and mice with PAX2 mutations: Evidence of increased apoptosis in fetal kidneys of Pax2(1Neu)+/− mutant mice. *Hum. Mol. Genet.* **9**, 1–11.

Potter, E. L. "(1972). Normal and Abnormal Development of the Kidney." Year Book Medical, Chicago.

Poucell-Hatton, S., Huang, M., Bannykh, S., Benirschke, K., and Masliah, E. (2000). Fetal obstructive uropathy: Patterns of renal pathology. *Pediatr. Dev. Pathol.* **3**, 223–231.

Powers, J. M., and Moser, H. W. (1998), Peroxisomal disorders: Genotype, phenotype, major neurological lesions, and pathogenesis. *Brain Pathol.* **8**, 101–120.

Pritchard-Jones, K. (1999). The Wilms' tumour gene, *WT1*, in normal and abnormal nephrogenesis. *Pediatr. Nephrol.* **13**, 620–625.

Pryde, P. G., Sedman. A. B., Nugent, C. E., and Barr, M. (1993). Angiotensin-converting enzyme inhibitor fetopathy. *J. Am. Soc. Nephrol.* **3**, 1575–1582.

Regan, D. C., Casale, A. J., Rink, R. C., Cain, M.P., and Weaver, D. (1999). Genitourinary anomalies in the CHARGE association. *J. Urol.* **161**, 622–625.

Risdon, R. A., and Woolf, A. S. (1998a). Development of the kidney. In "Heptinstall's Pathology of the Kidney." J.C.Jennette J.L. Olson, M.M. Schwartz, and F.G. Silva, eds., 5Th Ed., pp. 67–84. Lippincott-Raven, Philadelphia.

Risdon, R. A., and Woolf, A. S. (1998b). Developmental defects and cystic diseases of the kidney. In "Heptinstall's Pathology of the Kidney." J.C. Jennette, J. L. Olson, M. M. Schwartz, and F. G. Silva, eds., 5th Ed., pp. 1149–1206, Lippincott-Raven, Philadelphia

Risdon, R. A., Yeung, C. K., and Ransley, P. (1993). Reflux nephropathy in children submitted to nephrectomy: A clinicopathological study. *Clin. Nephrol.* **40**, 308–314.

Roodhooft, A. M., Birnholz, J. C., and Holmes, L. B. (1984). Familial nature of congenital absence and severe dysgenesis of both kidneys. *N. Engl. J. Med.* **24**, 1341–1345.

Rothman, K. J., Moore, L. L., Singer, M. R., Nguyen, U.-S. D. T., Mannino, S., and Milunsky, A. (1995). Teratogenicity of high vitamin A intake. *N. Engl. J. Med.* **333**, 1369–1373.

Roume, J., Genin, E., Cormier-Daire, V., Ma, H.W., Mehaye, B., Attie, T., Razavi-Encha, F., Fallet-Bianco, C., Buenerd, A., Clerget-Darpoux, F., Munnich, A., and Le Merrer, (1998). M. A gene for Meckel syndrome maps to chromosome 11q13. *Am. J. Hum. Genet.* **63**, 1095–1101.

Ruano-Gil, D., Coca-Payeras, A., and Tejedo-Mateu, A. (1975). Obstruction and normal recanalization of the ureter in the human embryo: Its relation to congenital ureteric obstruction. *Eur. Urol.* **1**, 287–293.

Salerno, A., Kohlhase, J., and Kaplan, B. S. (2000). Townes-Brockes syndrome and renal dysplasia: A novel mutation in the *SALL1* gene. *Pediatr. Nephrol.* **14**, 25–28.

Salomon, R., Tellier, A. L., Attie-Bitach, T., Amiel, J., Vekemans, M., Lyonnet, S., Dureau, P., Niaudet, P., Gubler, M. C., and Broyer, M. (2001). PAX2 mutations in oligomeganephronia. *Kidney Int.* **59**, 457–462.

Salonen, R. (1984), The Meckel syndrome: Clinicopathological findings in 67 patients. *Am. J. Med. Genet.* **18**, 671–689.

Sanchez, M. P., Silos-Santiago, I., Frisen, J., He, B., Lira, S. A., and Barbacid, M. (1996). Renal agenesis and the absence of enteric neurones in mice lacking GDNF. *Nature* **382**, 70–73.

Sanyanusin P, Schimmentl L. A, McNoe L. A, Ward T. A, Pierpoint M. E. M, Sullivan M. J, Dobyns W. B, and Eccles M. R. (1995) Mutations of the PAX2 gene in a family with optic nerve colobomas, renal anomalies and vesicoureteral reflux. *Nature Genet.* **9**, 358–364.

Schuchardt, A., D'Agati, V., Larsson-Blomberg, L., Constnatini F., and Pachnis, V. (1994). Defects in the kidney and nervous system of mice lacking the receptor tyrosine kinase receptor Ret. *Nature* **367**, 380–383.

Schwanzel-Fukuda, M., Bick, D., and Pfaff, D. W. (1989), Leutinizing hormone-releasing hormone-expressing cells do not migrate normally in an inherited hypogonadal (Kallmann) syndrome. *Mol. Brain. Res.* **6**, 311–326.

Schwartz, R. D., Stephens, F. D., and Cussen, L. J. (1981). The pathogenesis of renal dysplasia. II. The significance of lateral and medial ectopy of the ureteric orifice. *Invest. Urol.* **19**, 97–100.

Sedman, A. B., Kershaw, D. B., and Bunchman, T. E. (1995). Recognition and management of angiotensin converting enzyme fetopathy. *Pediatr. Nephrol.* **9**, 382–385.

Shimozawa, N., Tsukamoto, T., Suzuki, Y., Orii, T., Shirayoshi, Y., Mori, T., and Fujiki, Y. (1992). A human gene responsible for Zellweger syndrome that affects peroxisome assembly. *Science* **255**, 1132–1134.

Smellie, J., Barratt, T. M., Chantler, C., Gordon, I., Prescod, N. P., Ransley, P. G., and Woolf, A. S. (2001). Medical versus surgical treatment in children with severe bilateral vesicoureteric reflux and bilateral nephropathy: A randomised trial. *Lancet* **357**, 1329–1333.

Sorenson, C. M., Rogers, S. A., Korsmeyer, S. J., and Hammerman, M. R. (1995). Fulminant metanephric apoptosis and abnormal kidney development in bcl-2-deficient mice. *Am. J. Physiol.* **268**, F73–F81.

Soussi-Yanicostas, N., Faivre-Sarrailh, C., Hardelin, J. P., Levilliers, J., Rougon, G., and Petit, C. (1998). Anosmin-1 underlying the X chromosome-linked Kallmann syndrome is an adhesion molecule that can modulate neurite growth in a cell-type specific manner. *J. Cell. Sci.* **111**, 2953–2965.

Steinhardt, G. F., Lipais, H., Phillips, B., Vogler, G., Nag, M., and Yoon, K. W. (1995). Insulin-like growth factor improves renal architecture of fetal kidneys with complete ureteric obstruction. *J. Urol.* **154**, 690–693.

Stewart, T. L., Irons, M. B., Cowan, J. M., and Bianchi, D. W. (1999). Increased incidence of renal anomalies in patients with chromosome 22q11 microdeletion. *Teratology* **59**, 20–22.

Sun, Z., and Hopkins, N. (2001). *vhnf1*, the MODY5 and familial GCKD-associated gene, regulates specification of the zebrafish gut, pronephros, and hindbrain. *Genes Dev.* **15**, 3217–3229.

Taylor, C. L., Jones, K. L., Jones, M. C., and Kaplan, G. W. (1994). Incidence of renal anomalies in children prenatally exposed to ethanol. *Pediatrics* **94**, 209–212.

Threadgill, D. W., Dlugosz, A. A., Hansen, L. A., Tennenbaum, T., Lichti, U., Yee, D., La Mantia, C., Mourton, T., Herrup, K., Harris, R. C., Barnard, J. A., Coffey, S. H., and Magnuson, T. (1995). Targeted disruption of mouse EGF receptor. Effect of genetic background on mutant phenotype. *Science* **269**, 230–234, 1995.

Torres, M., Gomez-Pardo, E., Dressler, G. R., and Gruss, P. (1995). *Pax-2* controls multiple steps of urogenital development. *Development* **121**, 4057–4065.

van Esch, H., and Bilous, R. W. (2001). GATA3 and kidney development: Why case reports are still important. *Nephrol. Dial. Transplant.* **16**, 2130–2132.

von Lennep, E., El Khazen, N., De Pierreux, G., Amy, J. J., Rodesch, F., and Van Regemorter, N. (1985). A case of partial sirenomelia and possible vitamin A teratogenesis. *Prenat. Diagn.* **5**, 35–40.

Voyer, L. E., Drut, R., and Mendez, J. H. (1994). Fetal maldevelopment with oligohydramnios following maternal use of piroxicam. *Pediatr. Nephrol.* **8**, 592–594.

Vermillion, C. D., and Heale, W. F. (1973). Position and configuration of the ureteral orifice and its relationship to renal scarring in adults. *J. Urol.* **109**, 579–584.

Wagner, T., Wirth, J., Meyer, J., Zabel, B., Held, M., Zimmer, J., Pasantes, J., Bricarelli, F. D., Keutel, J., Hustert, E., Wolf, U., Tommerup, N., Schempp, W., and Scherer, G. (1994). Autosomal sex reversal and campomelic dysplasia are caused by mutations in and around the SRY-related gene SOX9. *Cell* **79**, 1111–1120.

Wallace, M., Zori, R. T., Alley, T., Whidden, E., Gray, B. A. *et al.* (1994), Smith-Lemli-Opitz syndrome in a female with a de novo, balanced translocation involving 7q32: Probable disruption of a SLOS gene. *Am. J. Med. Genet.* **50**, 368–374.

Wang, C. Y., Hawkins-Lee, B., Ochoa, B., Walker, R. D., and She, J. X. (1997). Homozygosity and linkage-disequilibrium mapping of the urofacial (Ochoa) syndrome gene to a 1-cM interval on chromosome 10q23–q24. *Am. J. Hum. Genet.* **60**, 1461–1467.

Ward, C. J., Hogan, M. C., Rossetti, S., Walker, D., Sneddon, T., Wang, X., Kubly, V., Cunningham, J. M., Bacallao, R., Ishibashi, M., Milliner, D. S., Tirres, V. E., and Harris, P. C. (2002). The gene mutated in autosomal recessive polycystic kidney disease encodes a large, receptor-like protein. *Nature Genet* **30**, 259–269.

Welham, S. J. M., Wade, A., and Woolf, A. S. (2002). Protein restriction in pregnancy is associated with increased apoptosis of mesenchymal cells at the start of rat metanephrogenesis. *Kidney Int*, **61**, 1232–1242.

Whitten, S. M., and Wilcox. D. T. (2001). Duplex systems. *Prenat. Diag.* **21**, 952–957.

Wild, W., von Strandmann, E. P., Nastos, A, Senkel, S., Lingott-Frieg A., Bulman, M., Bingham, C., Ellard, S., Hattersley, A. T., and Ryffel, G. U. (2000). The mutated human gene encoding hepatocyte nuclear factor 1β inhibits kidney formation in developing *Xenopus* embyos. *Proc. Natl. Acad. Sci. USA* **97**, 4695–4700.

Wilkie, A. O. M., Slaney, S. F., Oldridge, M., Poole, M. D., Ashworth, G. J., Hockley, A. D., Hayward, R. D., David, D. J., Pulleyn, L. J., Rutland, P., Malcolm, S., Winter, R. M., and Reardon, R. (1996). Apert syndrome results from localised mutations of *FGFR2* and is allelic with Crouzon syndrome. *Nature Genet.* **9**, 165–172.

Wilson, G. N., de Chadarevian, J. P., Kaplan, P., Loehr, J. P., Frerman, F. E., and Goodman, S. I. (1989). Glutaric aciduria type II: Review of the phenotype and report of an unusual glomerulopathy. *Am. J. Med. Genet.* **32**, 395–401.

Winyard, P. J. D., Bao, Q., Hughes, R. C., and Woolf, A. S. (1997), Epithelial galectin-3 during human nephrogenesis and childhood cystic diseases. *J. Am. Soc. Nephrol.* **8**, 1647–1657.

Winyard, P. J. D., Nauta, J., Lirenman, D. S., Hardman, P., Sams, V. R., Risdon, R. A., and Woolf, A. S. (1996a). Deregulation of cell survival in cystic and dysplastic renal development. *Kidney Int.* **49**, 135–146.

Winyard, P. J. D., Risdon, R. A., Sams, V. R., Dressler, G., and Woolf, A. S. (1996b). The PAX2 transcription factor is expressed in cystic and hyperproliferative dysplastic epithelia in human kidney malformations. *J. Clin. Invest.* **98**, 451–459.

Woolf, A. S. (1998). The single kidney. *In* "Pediatric Surgery and Urology: Long-Term Outcomes." M.D. Stringer, P.D.E. Mouriquand, K.T. Oldham, E.R. Howard, eds., pp. 625–631. Saunders, C. London.

Woolf, A. S. (2000). Diabetes, genes and kidney development. *Kidney Int.* **57**, 1202–1203.

Woolf, A. S., Bingham, C., and Feather, S. A. (2002). Recent insights into kidney diseases associated with glomerular cysts. *Pediatr. Nephrol.* **17**, 229–235.

Woolf, A. S., and Thiruchelvam, N. (2001). Congenital obstructive uropathy: Its origin and contribution to end-stage renal failure in children. *Adv. Ren. Replace. Ther.* **8**, 157–163.

Woolf, A. S. and Winyard, P. J. D. (2000). Gene expression and cell turnover in human renal dysplasia. *Histol. Histopathol.* **15**, 159–166.

Yamashita, T., and Nakahata, T. (2001) Current knowledge on the pathophysiology of Fanconi anemia: From genes to phenotypes. *Int. J. Hematol.* **74**, 33–41.

Yang, S. P., Woolf, A. S., Quinn, F., and Winyard, P. J .D. (2001). Deregulation of renal transforming growth factor-β1 after experimental short-term ureteric obstruction in fetal sheep. *Am. J. Pathol.* **159**, 109–117.

Yang, S. P., Woolf, A. S., Yuan, H. T., Scott, R. J., Risdon, R. A., O'Hare, M. J, and Winyard, P. J. D. (2000). Potential biological role of transforming growth factor-β1 in human congenital kidney malformations. *Am. J. Pathol.* **157**, 1633–1647.

Yeung, C. K., Godley, M. L., Dhillon, H. K., Gordon, I., Duffy, P. G., and Ransley, P. G. (1997). The characteristics of primary vesico-ureteric reflux in male and female infants with pre-natal hydronephrosis. *Br. J. Urol.* **80**, 319–327.

22

WT1-Associated Disorders

Marie-Claire Gubler and Cécile Jeanpierre

I. Introduction

The Wilms' tumor suppressor gene, *WT1*, encodes a zinc finger transcription factor regulating the expression of numerous genes. It plays a major role in renal and genital development. Contrary to the initial hypothesis, WT1 germline mutations are found in only 10% of patients with Wilms' tumors. Most of them are located within the first exons of the gene and lead to the synthesis of a truncated protein. Somatic WT1 mutations are also rare, but WT1 splicing alterations are observed in nearly 90% of sporadic isolated Wilms' tumors, suggesting that they play a role in the tumorigenic process. Other candidate genes for Wilms' tumor have been localized but not yet identified. More recently, WT1 has been shown to be involved in other cancers, including acute myeloid leukemia.

WT1 is implicated in several syndromes characterized by the association of kidney and genital defects. Complete deletion of the gene in the WAGR syndrome is responsible for genitourinary malformations and the occurrence of Wilms' tumor. Germline WT1 point mutations located in exons 8 or 9, coding for zinc fingers 2 or 3, are found in patients with Denys–Drash syndrome (DDS), a disorder characterized by the association of male pseudohermaphroditism, infantile nephrotic syndrome with diffuse mesangial sclerosis and a high risk of developing Wilms' tumor. These mutations change the DNA-binding affinity of the mutated protein. Mutations in the donor splice site of intron 9 leading to the loss of +KTS isoforms have been observed in all patients with Frasier syndrome, the association of male pseudohermaphroditism with complete sex reversal, proteinuria with focal and segmental glomerulosclerosis, and possible development of gonadoblastoma.

Most reported WT1 mutations are *de novo* mutations. They are dominant and the patients who now survive must receive genetic counseling.

WT1, one of the Wilms' tumor suppressor genes, was identified in 1990 by positional cloning through the study of children presenting the WAGR syndrome. This syndrome is characterized by the association of Wilms' tumor (W), aniridia (A), genitourinary malformations (G), and mental retardation (R). The finding of large cytogenetically visible deletions of chromosome band 11p13 in these patients suggested that a tumor suppressor gene was located at this locus (Riccardi *et al.*, 1978). Actually, one candidate gene, encoding a Krüppel-like zinc finger protein able to bind DNA, was identified at this locus (Call *et al.*, 1990; Gessler *et al.*, 1990). The finding of a germline intragenic deletion in patients with sporadic unilateral (Haber *et al.*, 1990) or bilateral nephroblastoma (Huff *et al.*, 1991) confirmed that loss of function of this gene was an important factor in tumor development. In addition, the presence of genitourinary malformations in children with Wilms' tumor due to constitutional intragenic mutations strongly suggested that they also resulted from WT1 mutations (Pelletier *et al.*, 1991).

Since 1991, a tremendous amount of information has accumulated on the structure, tissue expression, and function of this gene, and we now know that WT1 plays a major role not only in tumorigenesis, but also in kidney and gonad development, as demonstrated by its strong expression during embryofetal life, the results of targeted gene disruption in mice, and the occurrence of gonad and kidney disorders in humans with constitutional mutation of the gene.

II. The *WT1* Gene

WT1 contains 10 exons spanning approximately 50 kb of genomic DNA (Buckler *et al.*, 1991; Coppes *et al.*, 1993; Gessler *et al.*, 1992; Haber *et al.*, 1991). It encodes a protein of 52–54 kDa with the structure of a transcription factor (Rauscher, 1993) (Fig. 22.1). Exons 1 to 6 encode a proline/glutamine-rich region involved in the repression or activa-

tion of transcription (Wang *et al.* 1993, 1995) and a domain involved in homodimerization of the protein (Moffett *et al.*, 1995). Exons 7 to 10 encode the four contiguous zinc fingers of the DNA-binding domain, very similar to those found in the EGR1 (early growth response) family of transcription factors. Each zinc finger consists of a short sequence of 28–30 amino acids in which a pair of cysteines and histidines are linked to a zinc atom, resulting in the formation of a finger. One basic amino acid, usually arginine, is located at the top of the finger (Fig. 22.2).

Twenty-four potential WT1 protein isoforms may be synthesized as a result of alternative start sites (Bruening *et al.*, 1996; Scharnhorst *et al.*, 1999), alternative splicing (Haber *et al.*, 1991), and RNA editing (Sharma *et al.*, 1994). There are two alternative splicing sites, the first one leading to the presence/absence of the 17 amino acids encoded by exon 5, the second including/excluding a short sequence of three amino acids, lysine, threonine and serine (KTS), between zinc fingers 3 and 4 (Fig. 22.1). The KTS motif is completely conserved in vertebrates (Schedl *et al.*, 1998). The four resulting isoforms are expressed in stable and definite proportion, with the +KTS isoforms corresponding to 80% of the transcripts (Haber *et al.*, 1991; van Heyningen *et al.*, 1998). They have different DNA-binding and transactivation activities (Hewitt *et al.*, 1996) and distinct subnuclear localizations revealing different functions (Drummond *et al.*, 1994; Englert *et al.*, 1995; Larsson *et al.*, 1995). The increased flexibility of the linker between ZF3 and ZF4, provided by the KTS tripeptide insertion, suppresses the binding of ZF4 to its cognate DNA site (Laity *et al.*, 2000).

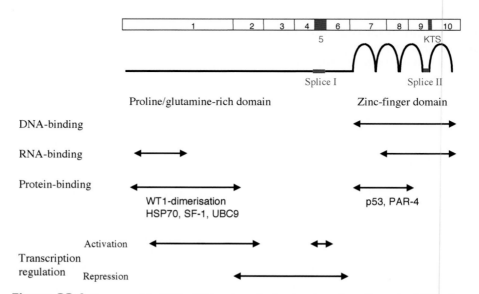

Figure 22.1 Structure of the WT1 mRNA and protein showing alternative splice region I (17 amino acids encoded by exon 5) and region II (amino acids KTS encoded by the 3′ end of exon 9) and the different functional domains of the protein.

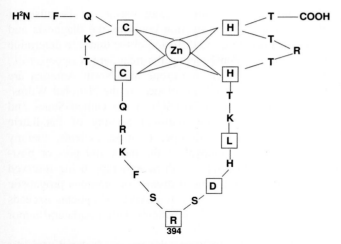

Figure 22.2 Schematic representation of WT1 zinc finger 3 encoded by exon 9. Boxed letters correspond to the most frequent missense mutations in Denys–Drash syndrome.

In vitro studies led to the identification of numerous potential target genes, regulated, most often negatively, by WT1 (Table 22.1) (reviews in Rauscher *et al.*, 1993; Reddy *et al.*, 1996). The presence of the 17 amino acids of exon 5 in the [WT1(+KTS)] isoforms adds to the suppressor function of the protein (Wang *et al.*, 1995). However, WT1 activation or repression of promoter reporters depends on experimental conditions, and the physiological targets remain uncertain, as the relevance of *in vitro* findings has not been established *in vivo* (Reddy *et al.*, 1996; Menke *et al.*, 1998; Little *et al.*, 1999). In addition, the regulatory activity of WT1 may be modulated by interaction with other cellular proteins such as par-4, Hsp70, p53 (Rauscher *et al.*, 1993), Ciao 1 (Johnstone *et al.*, 1998), and WTAP (Little *et al.*, 2000a,b; review in Little *et al.*, 1999). Using oligo-

nucleotide arrays to search for endogenous genes regulated by WT1(–KTS), none of the previously identified targets have been recovered, whereas the major target of WT1 was found to be *amphiregulin*, a member of the epidermal growth factor (EGF) family (Lee *et al.*, 1999). This gene is coexpressed with *WT1* in fetal kidney suggesting a role in kidney, development, but its function may be compensated by other EGF family members as the kidney develops normally in amphiregulin mutant mice (Luetteke *et al.*, 1999). In addition to a transcription regulation function, the +KTS isoforms of WT1 could play a role in posttranscriptional regulation as suggested by their nuclear pattern of distribution and their association with splicing factors (Larsson *et al.*, 1995, Caricasole *et al.*, 1996).

III. *WT1* and Development

WT1 is strongly expressed during embryofetal life (Pritchard-Jones *et al.*, 1990; Buckler *et al.*, 1991; Moore *et al.*, 1999). In the developing kidney, the expression, faint in mesenchymal blastema, increases in condensates, renal vesicles, and comma-shaped bodies simultaneously with the repression of other genes such as *PAX2*, *Lim1*, *N-Myc* or *IGF2* (Rauscher *et al.*, 1993). It is maximum in prepodocytes, whereas it disappears from tubular cells. In the mature kidney, *WT1* expression persists only in podocytes and epithelial cells of the Bowman's capsule (Fig. 22.3). The other main sites of expression are the spleen, mesothelium, epicardium, genital ridges, and gonads where it persists in Sertoli or granulosa cells during adult life. WT1 has also

Table 22.1 Candidate WT1 Target Genes

Growth factors/growth factor receptors
  IGF2, IGF1R
  PDGF-A
  TGF-β1
  CSF-1
  EGFR
  RARα
  Insulin R

Transcription factors
  WT1
  EGR1
  PAX2
  c-myc
  Bcl-2

Others
  novH
  Syndecan

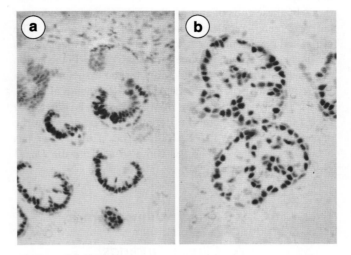

Figure 22.3 Immunoperoxidase labeling with anti-WT1 antibodies. (a) Normal fetal kidney: WT1 is faintly expressed in nuclei of mesenchymal blastema. The nuclear expression is more intense in condensates, renal vesicles, and prepodocytes of S-shaped bodies. (b) In the mature kidney, strong and uniform staining of podocyte nuclei is associated with less intense staining of Bowman's capsule epithelial cells.

been found in breast tissue (Silberstein *et al.*, 1997). As expected for a transcription factor, WT1 has a nuclear localization (Mundlos *et al.*, 1993).

The major role of *WT1* in renal and extrarenal development has been demonstrated by targeted gene disruption in mice (Kreidberg *et al.*, 1993). No anomaly has been reported to date in heterozygous mice, but homozygotes died between days 13 and 15 of gestation from severe heart, lung, and mesothelium anomalies. Failure of spleen development is observed in mice with a different genetic background, allowing delayed fetal lethality (Herzer *et al.*, 1999). In any case, homozygous null mice had complete renal and gonadal agenesis. Precise analysis from day 9 of gestation showed normal development of Wolffian ducts and induction of a reduced number of normal mesonephric structures, but there is no ureteric bud formation and the metanephric blastema, normally induced by the ureteric bud, undergoes apoptosis. At 12 days of gestational age, the Wolffian duct is the only remnant of the urogenital system. *In vitro* experiments showed that mesenchymal blastema from null mice was not induced by WT1+/+ spinal cord, a potent inducer of metanephric blastema differentiation. The human-derived *WT1* YAC construct has been shown to be able to completely rescue heart defects but only partially urogenital anomalies (Moore *et al.*, 1999). All of these data indicate that *WT1* plays a major role in the induction of the ureteric bud, survival, and epithelial differentiation of the mesenchymal blastema, the progression of nephrogenesis, and the maintenance of podocyte function.

Several transcription factors, including SRY and SOX8, are involved in the testicular determination of the bipotent embryonal gonads and the differentiation of Sertoli cells that express *WT1* (de Santa Barbara *et al.*, 2000). The synthesis by these cells of anti-Mullerian hormone, a product of the *MIS* gene, regulates male genital development. The initial and persistent expression of the *MIS* gene depends on the nuclear receptor SF-1 (steroidogenic factor-1) the action of which is increased considerably by the association with WT1, whereas it is inhibited by DAX-1 (Nachtigal *et al.*, 1998). The synergetic action of WT1 and SF-1 is therefore necessary for male genital differentiation and seems to be dependent on the WT1(–KTS) isoforms. However, it has been shown that WT1(–KTS) can also activate the *Dax-1* promotor (Lim *et al.*, 1999).

IV. *WT1* and Wilms' Tumor

A. Clinical Findings

Wilms' tumor (nephroblastoma), one of the most common solid tumors of childhood, affects approximately 1 child in 10,000, usually under the age of 6 years. It is revealed by an abdominal mass found by the parents or detected by the physician during routine examination or because of abdominal pain. Investigation confirms the diagnosis and evaluates the extent and diffusion of the tumor to determine the staging and consequently the treatment (Coppes *et al.*, 1999). Most children in Europe and North America are treated in clinical trials developed by the National Wilms' Tumor Study Group (NWTSG) in the United States and Canada or by the International Society of Paediatric Oncology (SIOP) in Europe. In most patients, therapy includes surgical removal of the mass and pre- or postoperative chemotherapy, with radiotherapy being reserved for children with advanced disease or ominous prognostic features. To date, the overall survival of patients exceeds 80% and is around 95% in patients with a unilateral tumor limited to the kidney.

In the majority of patients, the tumor is isolated, sporadic and unilateral, but it may be bilateral at diagnosis (5%) or secondarily (2%), familial (2%) or part of WAGR, Denys–Drash, or Beckwith Wiedemann syndromes. These last patients are at a higher risk of developing bilateral tumors.

B. Pathology

Diversity is the histological characteristic of this embryonic tumor, originating from pluripotential metanephric blastema or nephrogenic rests (Beckwith *et al.*, 1998). Classical "triphasic" nephroblastoma consists of the various combinations of blastemal cells, epithelial cells often mimicking normal nephrogenesis, and stromal cells. However, biphasic and monophasic patterns may be seen. In contrast, some tumors contain smooth or skeletal muscle, adipocytes, cartilage, or bone. Anaplasia, observed in about 10% of cases, is characterized by the presence of large polyploid hyperchromatic nuclei and abnormal multipolar mitotic figures. When diffuse, it has an unfavorable prognostic significance (Beckwith *et al.*, 1978).

C. Genetics

The study of malformative syndromes and of familial cases of Wilms' tumors led to the identification of genes or loci implicated in the occurrence of Wilms' tumors. As indicated previously, *WT1* was initially identified as a Wilms' tumor suppressor gene, the deletion of which was responsible for the occurrence of uni or bilateral nephroblastoma in WAGR patients. According to Knudson (1971) and the retinoblastoma model, somatic inactivation of the second allele is necessary for tumor initiation. Loss or intragenic mutation of the *WT1* wild-type allele has been found in some tumors of WAGR patients, supporting the "two-hit" hypothesis. However, the frequency of such a second event is not precisely known and mutations at other loci may account for the occurrence of nephroblastoma (review in Little *et al.*, 1997). In WAGR patients, aniridia is not related to *WT1* but to the deletion of the contiguous gene *PAX6* (Ton *et al.*, 1991), and genital abnormalities vary from cryptorchidism and/or hypospadias to male pseudo-hermaphroditism. Classically, the adjacent kidney is normal. However, an unexpected occurrence of renal failure has been observed in some WAGR patients. Analysis of nearly 6000 patients enrolled in clinical trials of the NWTSG showed that the cumulative risk of renal failure at 20 years was 38% in WAGR patients versus <1% in patients with isolated unilateral nephroblastoma (Breslow *et al.*, 2000). No differences in therapeutic regimen were detected. This observation suggests that an unidentified slowly progressive nephropathy could be associated with *WT1* heterozygous deletion.

WT1 germline mutations are observed in virtually all Denys–Drash patients, who are at high risk of developing Wilms' tumor (see later). In contrast, the estimated frequency of germline *WT1* mutation in sporadic nonsyndromic Wilms' tumors is about 10%, (Gessler *et al.*, 1994; Reddy *et al.*, 1995; Diller *et al.*, 1998; Schumacher *et al.*, 1997; Little *et al.*, 1997), strikingly lower than initially hypothesized. They are mostly large deletions or insertions, small deletions, or nonsense mutations located in the first exons of the gene and leading to the synthesis of truncated proteins (Fig. 22.4). They seem to be associated with tumors of stromal-predominant histology (Schumacher *et al.*, 1997; Baudry *et al.*, 2000). Interestingly, most patients with the

WT1 mutation also have genitourinary anomalies (Diller *et al.*, 1998; Schumacher *et al.*, 1997), whereas patients with Wilms' tumor and no genitourinary malformations are at low risk for carrying a *WT1* mutation.

Ten percent of patients with the overgrowth condition Beckwith–Wiedemann syndrome (gigantism, macroglossia, omphalocele, hyperinsulinism, and predisposition to several tumors), develop Wilms' tumor. These patients can show anomalies at the 11p15 locus, suggesting the presence of another gene, *WT2*, involved in the susceptibility to develop Wilms' tumors. At this locus are several imprinted candidate genes, H19, p57(Kip2), and the insulin-like growth factor IGF2, but their implication in tumorigenesis has not been demonstrated.

Germinal mutation of *WT1* has been observed in some cases of familial Wilms' tumors (Jeanpierre *et al.*, 1997; Kaplinsky *et al.*, 1996; Pelletier *et al.*, 1991; Pritchard-Jones *et al.*, 2000). In addition, two genes, *FWT1* located on chromosome 17q12-q21 and *FWT2* on 19q13 and not yet identified, are linked to familial Wilms' tumor predisposition, inherited as an autosomal-dominant trait. Data indicate that one or more additional familial Wilms' tumor genes remain to be found (Rapley *et al.*, 2000).

Multiple genetic events have been implicated in the biology of sporadic Wilms' tumors. Somatic mutations of *WT1* have been observed in only 10% of tumors, suggesting that other genes are more important in the pathogenesis of nephroblastomas (Gessler *et al.*, 1994). However, disruption

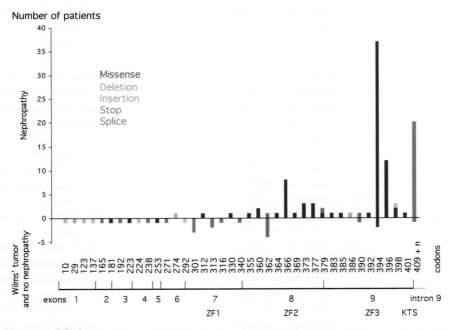

Figure 22.4 Comparison of the distribution of the WT1 mutations between patients with nephropathy (DDS and Frasier syndrome) and patients with Wilms' tumor and no nephropathy. The (409+n) mutations regroup mutations affecting the fourth, fifth, or sixth base in intron 9. These mutations affect alternative splicing of the KTS domain and are found in all Frasier syndromes.

of the alternative splicing of exon 5 or aberrant splicing of exon 2 has been described in some tumors (Haber *et al.*, 1993; Simms *et al.*, 1995). By studying a large series of 50 tumors, we have confirmed that *WT1* splicing alterations, affecting mostly exon 5, are present in about 90% of Wilms' tumors, suggesting that WT1 isoform imbalance, another form of somatic alteration, could play a role in the tumorigenic process (Baudry *et al.*, 2000). Several other chromosomic regions are concerned by loss (1p, 7p, 11p, 11q, 16q, 22q) or duplication (12) of alleles, but the genes have not been identified (Coppes *et al.*, 1999). The NTWTS5 trial in progress aims to determine if loss of heterozygosity (LOH) for chromosomes 16q and 1p predicts an adverse outcome (Grundy *et al.*, 1994). Analysis of these mutations in Wilms' tumor indicates that LOH at 11p13 and 11p15 and *WT1* mutations are early events, detected in intralobular rests as well as in the tumors, whereas LOH at 16q occurs late in tumor development, suggesting a multistep model of Wilms' tumor pathogenesis (Charles *et al.*, 1998).

V. *WT1* and Other Malignancies

WT1 mutations have been detected in some patients with other types of solid tumors, e.g. in juvenile granulosa cell tumor and in mesothelioma where they are far from being constant (review in Little *et al.*, 1997; Langerak *et al.*, 1995). Persistent expression of WT1 in mesotheliomas provides a specific histological marker for differenciation from other pleural tumors (Amin *et al.*, 1995). Decreased expression of WT1 or alteration in the relative proportions of the different isoforms has been observed in breast cancers (Silberstein *et al.*, 1997).

Interestingly, in desmoplastic small round cell tumor, a rare sarcoma of youth, a chromosomal translocation is observed that results in the synthesis of a chimeric protein containing the transactivation domain of the Ewing sarcoma protein (EWS) fused to zinc fingers 2–4 of WT1. This new protein induces the expression of a potent fibroblast growth factor, PDGFA (Lee *et al.*, 1997).

Two series of observations have led to the idea that *WT1* is involved in the pathogenesis of acute leukemias. There is an increased rate of leukemias in WAGR patients with *WT1* deletions, and *WT1* mutations have been found in leukemia, in acute myeloid (14%), or undifferentiated leukemia (17%) and rarely in acute lymphoid leukemia (King-Underwood and Pritchard-Jones, 1998). Most of the mutations are heterozygous, suggesting a dominant or dominant-negative mode of action of the mutated protein in haematopoietic cells. Their precise role in leukemogenesis, initiation or progression of the disease, remains to be evaluated, as well as their relationship with drug resistance. However, a high level of *WT1* expression is seen in leukemia, representing aberrant overexpression or blockage of the leukemic cells

at a very early stage of differentiation (Inoue *et al.*, 1997; King-Underwood and Pritchard-Jones, 1998). This finding is of clinical relevance, as expression of *WT1* transcripts may be a significant tumor marker for the evaluation of remission status and early relapse of the disease. In addition, because of this high expression, also observed in some other cancers, WT1 appears an interesting novel target for immunotherapy (Gao *et al.*, 2000).

VI. *WT1* and Denys–Drash Syndrome

A. Clinical Findings

Denys–Drash syndrome is a rare sporadic syndrome characterized by the triad of severe glomerulopathy progressing rapidly to end-stage renal disease (ESRD), male pseudohermaphroditism, and Wilms' tumor (Denys *et al.*, 1967; Drash *et al.*, 1970). The specific kidney lesion in DDS is diffuse mesangial sclerosis (DMS) (Habib *et al.*, 1985, 1993).

The nephropathy, nephrotic syndrome often preceded by proteinuria, is discovered within the first months of life, usually before the age of 2 years (Habib *et al.*, 1993). In rare cases, the onset of the nephropathy is antenatal (Maalouf *et al.*, 1998). It may be detected by prenatal ultrasonography showing oligohydramnios and large hyperechogenic kidneys, suggesting the diagnosis of polycystic disease (unpublished data). Later occurrence of the nephropathy is quite rare. In all cases the nephrotic syndrome is steroid resistant and is associated frequently with hypertension. Progression to ESRD before the age of 4 years is the rule. No recurrence of the original disease is observed after renal transplantation. Uni- or bilateral Wilms' tumor may be the first symptom of the disease or discovered after the nephrotic syndrome, supporting systematic echography in patients presenting with DMS. All 46,XY patients have either ambiguous genitalia or female phenotype with dysgenetic testis.

Clinical expression of DDS may be incomplete, with the common denominator being glomerular involvement. Actually, 46,XX patients have a normal female phenotype and may have normal genital development characterized by the occurrence of normal puberty (Jeanpierre *et al.*, 1998). Gonad anomalies have been described but seem to be rare. However, Wilms' tumor may not develop, especially if systematic bilateral nephrectomy before renal transplatation is performed early in life because of the rapid progression to ESRD. An additional symptom, diaphragmatic hernia, has been described in one DDS patient with the complete triad and *WT1* mutation (Devriendt *et al.*, 1995). This symptom is likely to result from *WT1* mutation, as *WT1* is expressed in the mesothelium and *WT1* null mice have severe mesothelial anomalies (Kreidberg *et al.*, 1993).

B. Pathology

In the early stages of the disease, diffuse mesangial sclerosis is characterized by a fibrillar increase in mesangial matrix with no mesangial cell proliferation and no apparent changes of the capillary walls that are lined by hypertrophied podocytes (Fig. 22.5a). However, segmental thickening of the glomerular basement membrane (GBM) due to the subepithelial apposition of thin layers of basal lamina may be seen by electron microscopy (Fig. 22.5d). In the fully developed lesion, there is a combination of thickening of the GBM, massive enlargement of mesangial areas, and retraction of the glomerular tuft, leading to reduction in the patency of the capillary lumens (Fig. 22.5b). By electron microscopy, some collagen fibrils may be seen in the enlarged mesangial matrix. In advanced stages, the glomerular tuft is contracted and sclerotic but is still surrounded by a layer of hypertrophied and vacuolized podocytes in a urinary space that often appears dilated (Fig. 22.5c). Capsular adhesions are rare. These various stages may coexist in the same specimen with a corticomedullary gradient of involvement, with the deepest glomeruli being the least severely affected. By immunofluorescence, deposits of IgM, C3 and C1q are present in the mesangium of the least affected glomeruli, whereas they outline the periphery of more sclerotic glomeruli (Fig. 22.5e). Tubular lesions are associated, mostly characterized by dilatations involving proximal tubules that are lined by a flat epithelium, and contain protein casts. An interesting feature, consistent with the role of WT1 in kidney development, is the presence, in the subcapsular zone, of small glomeruli with no more than two to four capillary loops, presenting various degrees of glomerulosclerosis (Fig. 22.5g). In the rare patients with antenatal expression of the nephropathy, tubular dilatations are prominent and diffuse, making difficult the identification of glomerular lesions (Fig. 22.6).

A striking morphological feature in DMS is the massive accumulation of glomerular extracellular matrix material. However, immunohistochemical studies did not disclose any early change in the distribution of the different proteinic constituents of the glomerular extracellular matrix. Progression to sclerosis is associated with unspecific mesangial accumulation of the normal components of the mesangial matrix and of proteins normally restricted to the GBM, agrin, nidogen/entactin, and laminin chains $\alpha1/5$, $\beta2$ and $\gamma1$, and with strong mesangial and subendothelial accumulation of chondroitin sulfate glycosaminoglycans (Yang *et al.*, 2001; Yang *et al.*, 1999). The only immunohistochemical change observed in early DMS lesions is the marked decrease in the expression of GBM heparan sulfate glycosaminoglycans.

C. Genetics

Eighty-three germline *de novo* mutations have now been reported in patients presenting with complete or incomplete DDS (http://www.umd.necker.fr) (Pelletier *et al.*, 1991a, b; Jeanpierre *et al.*, 1998a, b; Schumacher *et al.*, 1998; Kikuchi

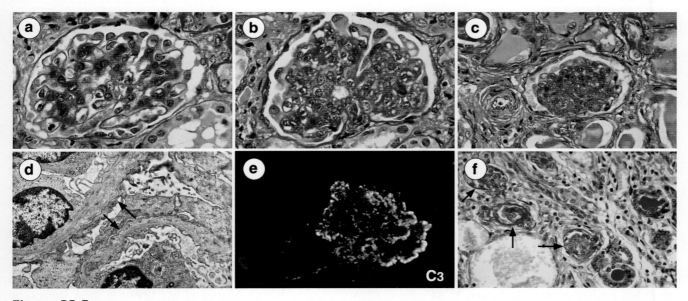

Figure 22.5 Diffuse mesangial sclerosis. Light microscopy. Periodic acid–Schiff. (a) Early stage showing the increase in mesangial matrix and hypertophied podocytes. (b) Typical lesion showing the contracted glomerular tuft surrounded by a crown of enlarged podocytes. (c) Complete sclerosis of the glomerular tuft and tubular dilatations. (d) Electron microscopy: Irregular thickening of the glomerular basement membrane due to apposition of the subepithelial strand of basement membrane material delineating small electron-lucent zones. (e) Massive subendothelial C3 deposits outlining the peripheral capillary loops. (f) Presence of small immature glomeruli in the subcapsular zone, most of which are sclerosed.

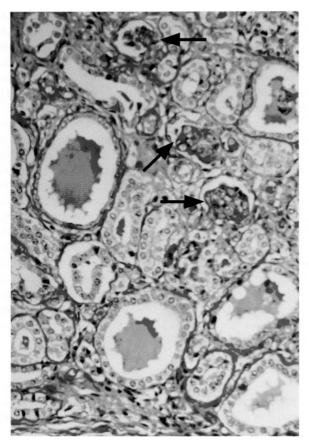

Figure 22.6 Diffuse mesangial sclerosis in a 35-week fetus with antenatal expression of the nephropathy. Tubular dilations are massive and diffuse. Glomeruli are small and sclerotic.

milder phenotype: genital anomalies less severe than in DDS and absent or, according to recent observations, delayed glomerular symptoms (Bresslow *et al.*, 2000). Regarding male genital development, WT1 mutations could additionally act at another level, as expression of the MIF gene involved in male genital differentiation depends on the synergic action of WT1 and SF-1, and this effect is reduced considerably by *WT1* mutation (Natchigal *et al.*, 1998).

In parallel with these *in vitro* studies, analysis of the podocyte distribution of WT1 in the kidney of DDS patients showed, in most patients, absence or reduction of the nuclear expression of WT1, suggesting that the binding capacities of the mutated protein (or the complex wild-type mutated protein) were also altered *in vivo* (Yang *et al.*, 1999). However this alteration was not constant, as expression was found to be normal in two DDS patients, one of them bearing the classical R394W *WT1* mutation. Abnormal expression of *WT1* was associated in some patients with the podocyte synthesis (RNA) and nuclear expression (protein) of *PAX2*. This finding suggests that the loss of downregulation by WT1 allows the persistent expression of *PAX2* that normally is no longer detected in the podocytes from the S-shaped body stage. The abnormal expression of *PAX2* may participate in the pathogenesis of glomerular lesions, as deregulation of *Pax2* expression in transgenic mice generates severe congenital nephrotic syndrome with prominent tubulointerstitial lesions (Dressler *et al.*, 1993).

One DDS mouse model has been generated by gene targeting, which truncated ZF3 at codon 396. Heterozygous and chimeric mice develop mesangial sclerosis, male genital defects, and Wilms' tumor in one example (Patek *et al.*, 1999). A surprising finding was the low level (5%) of the mutant WT1 protein, implying that a small amount of abnormal protein is sufficient to disturb normal genitourinary development, raising the question of its mode of action.

VII. *WT1* and Isolated Diffuse Mesangial Sclerosis

Infantile nephrotic syndrome with diffuse mesangial sclerosis also occurs without the other elements of the triad (Habib *et al.*, 1993). It could be an incomplete form of DDS in genetically female patients without Wilms' tumor. Actually, *WT1* mutations, similar to those observed in DDS, have been detected in one male patient who developed normal puberty (Jeanpierre *et al.*, 1997) and in some isolated diffuse mesangial sclerosis (IDMS) female patients (Jeanpierre *et al.*, 1998; Schumacher *et al.*, 1998), but this finding is rare (Koziell *et al.*, 1999). In these studies, no *WT1* mutation was identified in 6/10, 2/4, and 20/20 IDMS patients, respectively (Jeanpierre *et al.*, 1998; Schumacher *et al.*, 1998; Koziell *et al.*, 1999). At the clinical level, the possible occurrence of IDMS in XY males with normal

et al., 1998; Koziell *et al.*, 1999). They are mostly missense mutations located within exon 8 or 9 encoding zinc fingers 2 and 3 (ZF2, ZF3), respectively (Fig. 22.4). The most common *WT1* lesion is a missense 1180 C-to-T transition converting the arginine located at the top of ZF3 to tryptophan (R394W) (Fig. 22.2). Other less common *WT1* missense mutations alter conserved amino acids, arginine 366 located at the top of ZF2, histidines or cysteins located at the base of ZF2 or ZF3. Other mutations, deletions, insertions, and nonsenses result in the synthesis of a truncated protein. All of these *WT1* lesions change the structural organization of the respective zinc fingers and, consequently, result in loss or alteration of their DNA-binding ability as demonstrated by *in vitro* studies (Little *et al.*, 1995). In addition, dimerization of the wild type with the mutant protein, through the conserved N-terminal domain, abolishes binding of the ZF domain to its normal DNA target, suggesting that the mutated protein acts in a dominant negative fashion (Little *et al.*, 1993; Reddy *et al.*, 1995; Moffett *et al.*, 1995). Actually, the complete deletion of the gene observed in the WAGR syndrome results in a genitalia and the frequent familial incidence and/or parental consanguinity observed in IDMS patients suggest that it could be a specific entity, with autosomal recessive inheritance (Habib *et al.*, 1993; review in Gubler *et al.*, 1999). Because of the decreased podocyte expression of *WT1* associated with the strong expression of *PAX2* (Yang *et al.*, 1999), it may be hypothesized that one of the genes implicated in the cascade of WT1/PAX2 regulation could be involved in IDMS. This is also supported by findings in mice overexpressing Pax2: they develop early and severe glomerulopathy with nephrotic syndrome and glomerular and tubular changes reminiscent of those observed in DMS (Dressler *et al.*, 1993).

Familial data and new molecular information clearly indicate that the distinct glomerulopathy named diffuse mesangial sclerosis is genetically heterogenous, with one of the causative genes being *WT1* within the context of DDS. From a practical point of view, karyotype analysis in phenotypic females and ultrasonography in order to detect Wilms' tumor should be performed in any infant presenting with isolated DMS. Even if the results of these investigations are normal, analysis of *WT1* should be performed because of the potential risk of secondary development of Wilms' tumor in patients with the *WT1* mutation.

VIII. *WT1* and Frasier Syndrome

A. Clinical and Pathological Findings

Frasier syndrome (FS), described in 1964, is a rare disorder that is also characterized by the association of male pseudohermaphroditism and progressive glomerulopathy (Frasier *et al.*, 1964; Moorthy *et al.*, 1987). It is distinct from Denys–Drash syndrome by its slow progression to ESRD, the lack of specificity of glomerular lesions, and the usual absence of Wilms' tumor (Table 22.1). Patients have female external genitalia, despite a 46,XY karyotype, and streak gonads frequently at the origin of gonadoblastomas. It has been shown that the intronic mutation of *WT1* is responsible for the occurrence of Frasier syndrome (Barbaux *et al.*, 1997).

At present, about 30 cases of FS have been published (review in Gubler *et al.*, 1999; Barbaux *et al.*, 1997; Kikuchi *et al.*, 1998; Klamt *et al.*, 1998; Barbosa *et al.*, 1999; Okuhara *et al.*, 1999; Demmer *et al.*, 1999). Glomerular symptoms, less severe than in DDS, reveal the disease in most patients. The first sign is proteinuria detected before 7 years of age in the majority of patients or, subsequently, between 14 and 17 years in a few others. It increases progressively with age, resulting in nephrotic syndrome, and does not respond to treatment. The disease has a slowly progressive but unremitting course to ESRD, which usually occurs in the second or third decade of life. No recurrence

of the nephrotic syndrome has been observed after renal transplantation. Renal biopsy shows normal glomeruli or, most often, focal and segmental glomerular sclerosis/hyalinosis (FSGS). In our experience, the segmental lesion has a peculiar retracted appearance, (Fig. 22.7) and the GBM is thickened focally at the electron microscopic level. Wilms' tumor is classically absent even if renal survival exceeds 20 years. However, a unilateral Wilms' tumor had developed at 3 years of age in a patient recognized secondarily as affected with Frasier syndrome (Barbosa *et al.*, 1999).

Abnormal genital development is constant in Frasier patients, who all have normal female external genitalia. In these phenotypic females with steroid-resistant nephrotic syndrome, the evaluation of primary amenorrhea leads to the diagnosis of 46,XY gonadal dysgenesis. Follicle stimulating hormone (FSH) and luteinizing hormone (LH) plasma level are very high, the uterus is small and infantile, fallopian tubes, vagina, and external genitalia are normal, and there are streak gonads. Bilateral gonadectomy performed in 24 of 26 informative patients led to the detection of microscopic uni- or bilateral gonadoblastoma in 10 of them.

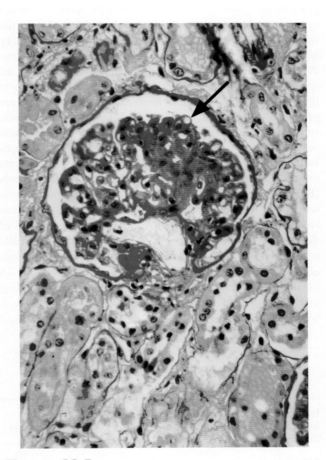

Figure 22.7 Frasier syndrome. Light microscopy. Periodic acid–Schiff base. Segmental lesion of the glomerular tuft with a retracted appearance.

In only 1 patient, a macroscopic gonadoblastoma revealed the syndrome at 6 years of age (Frasier *et al.*, 1964). No unfavorable tumoral course has been reported after gonadectomy.

On the whole, the diagnosis of FS is based on the association of steroid-resistant proteinuria/nephrotic syndrome with focal and segmental glomerular sclerosis and late-onset ESRD, together with 46,XY gonadal dysgenesis and complete sex reversal. Involvement of *WT1* was hypothesized because of the common clinical features with DDS.

B. Genetics

To date, point mutations in the donor splice site in intron 9 of *WT1* have been detected in all patients with the complete Frasier syndrome tested for the mutation (Table 22.2) (Barbaux *et al.*, 1997; Kikuchi *et al.*, 1998; Klamt *et al.*, 1998; Barbosa *et al.*, 1999; Demmer *et al.*, 1999). Similar mutations had been described previously in three unrelated patients initially classified as affected with DDS (Bruening *et al.*, 1992; König *et al.*, 1993; Barbeesy *et al.*, 1994) but whose clinical features (late-onset ESRD, no Wilms' tumor, and normal female phenotype with 46,XY gonadal dysgenesis) were consistent with those of Frasier syndrome. These results indicate that donor splice-site mutations in *WT1* intron 9 are constant in FS (Table 22.1). Most of them are located at position +4 (cytosine to thymidine transition) or +5 (guanine to adenine mutation) of the splicing donor site. They are heterozygote, and *de novo* in the patients whose parents have been studied. They are predicted to result in the loss of +KTS isoforms. Indeed, semiquantitative reverse transcription polymerase chain reaction (RT-PCR) analysis of +KTS/–KTS transcripts isolated from patient lymphocytes indicated a significant reduction in +KTS transcripts compared to normal controls (Barbaux *et al.*, 1997). Also, the +KTS/–KTS isoform ratio was found to be reduced in gonadal or tumor tissues from two Frasier patients with a G>A mutation at position +5 of the splicing site (Klamt *et al.*, 1998). In COS cells transfected with minigenes reproducing the point mutations observed in Frasier patients, at position +2, +4, or +5 of intron 9, only one amplified product corresponding to the transcript without nine nucleotides was obtained, whereas two products (with or without the nine nucleotides)

were obtained from the normal construct (Kikuchi *et al.*, 1998; Bruening *et al.*, 1992).

As indicated previously, the ratio of WT1 isoforms is constant and the isoforms that either contain or lack the KTS motif have different DNA-binding affinities and therefore, possibly, different regulatory functions. Absence or haploinsufficiency of the +KTS isoform caused by mutation in the splice donor site appears to be responsible for podocyte dysfunction and a major alteration in male genital development, demonstrating that a strict equilibrium between the different WT1 isoforms is required for normal renal and testicular development. In XY Frasier patients, it is possible that the increased amount of WT1(–KTS) secondary to the decrease in WT1(+KTS) isoforms results in the expression of abnormally high levels of DAX-1 and the inhibition of SRY activity. XY individuals with duplication (and high expression) of the DAX-1 gene show male-to-female sex reversal (Bardoni *et al.*, 1994).

IX. *WT1* Intronic Mutation (Frasier Mutation) in 46,XX Females and in Primary Steroid-Resistant Nephrotic Syndrome

The classical definition of Frasier syndrome, as delineated by Moorthy *et al.* (1987), includes only 46,XY patients with a female phenotype. The fate of 46,XX patients with the same *WT1* intronic mutation has been determined thanks to the precise analysis of three families. The first observation concerns the interesting family reported initially by Kinberg *et al.* (1987) and studied at the molecular level by Klamt *et al.* (1998). In this family, three sisters had a glomerular disease and developed ESRD at 2, 11, and 20 years, respectively. In one of them, with focal and segmental glomerular sclerosis, primary amenorrhea led to the diagnosis of 46,XY gonadal dysgenesis, whereas genital developement was normal in her 46,XX sister. In both patients, a G>A mutation was identified at position +5 in intron 9 of *WT1*. The second observation is very similar: two phenotypic females had persistent progressive proteinuria with focal segmental glomerulosclerosis. In one of them, primary amenorrhea revealed a 46,XY genotype and led to the diagnosis of Frasier syndrome and the identification of a *WT1* mutation at position +4 of the splice donor site of intron 9. The sister with a 46,XX genotype and the same mutation developed normal puberty (Demmer *et al.*, 1999). The third observation concerns a mother and her child (Denamur *et al.*, 1999). The 30-year-old mother had proteinuria from the age of 6 years, with focal and segmental glomerular sclerosis on the renal biopsy specimen obtained at 28 years. She has a normal female phenotype and normal genital development allowing a normal pregnancy. Her

Table 22.2   Splice Site Mutations in Frasier Syndrome[a]

Intron 9 (+2)	CAAgtgcgtaac	(t → c)	1 patient
Intron 9 (+4)	CAAgtgcgtaac	(c → t)	12 patients
Intron 9 (+5)	CAAgtgcgtaac	(g → t)	1 patient
	CAAgtgcgtaac	(g → a)	5 patients
Intron 9 (+6)	CAAgtgcgtaac	(t → a)	1 patient

[a] *From Barbaux et al.*

child, 46,XY, is affected with DDS characterized by the association of male pseudohermaphroditism and nephrotic syndrome detected at 9 months of age, with diffuse mesangial sclerosis on renal biopsy. Both of them have *WT1* mutation at location +5 of intron 9.

These observations clearly demonstrate that, in 46,XX females, *WT1* splice-site mutations (Frasier mutations) are responsible for the occurrence of isolated persistent glomerulopathy, with focal and segmental glomerular sclerosis. They do not impair female genital development. Consequently, it should be logical to search for *WT1* intronic mutation in every female presenting with "idiopathic" steroid-resistant proteinuria/nephrotic syndrome progressing to ESRD. From a practical point of view, this research is imperative when the karyotype is 46,XY or when the disease is familial. This systematic research has been done in a series of 37 children (17 boys and 20 girls) with non-familial primary steroid-resistant nephrotic syndrome. Only one patient (surprisingly a male with diaphramatic hernia, proximal hypospadias, and unilateral testicular ectopia) was found to carry a heterozygous splice-site mutation at position +4 in intron 9 (Denamur *et al.*, 2000). This study indicates that *WT1* mutations are not a common cause of isolated nephrotic syndrome but have to be suspected if genital or diaphragmatic anomalies are associated.

X. Conclusions

Since its identification in 1990 as a Wilms' tumor suppressor gene, major progress has been made in the understanding of *WT1* structure, function, and role in pathology. However, several questions remain regarding its mode of action, protein partners, and *in vivo* targets. For example, we do not know if and how WT1 regulates the expression of genes encoding proteins expressed specifically by the podocyte, such as nephrin (Kestilä *et al.*, 1998), α-actinin-4 (Kaplan *et al.*, 2000), or podocine (Boute *et al.*, 2000), whose mutations are responsible for nephrotic syndromes.

At the clinical level, a large spectrum of phenotypes has been found to be associated with *WT1* mutations. Overall, there is a good correlation between the type of mutation and the type of renal and extrarenal defects (Table 22.3). Mutations in the first exons of the gene resulting in the synthesis of truncated proteins are observed in Wilms' tumors isolated or associated with moderate genitourinary anomalies. Missense mutations in exons 8 or 9 result in DDS with severe nephropathy characterized by diffuse mesangial sclerosis and rapid progression to ESRD. However, some exceptions have been reported: focal and segmental glomerulosclerosis in patients with DDS mutation progressing rapidly to ESRD (Schumacher *et al.*, 1998); anunusually slow progression of glomerulopathy in three 46,XY Japanese patients (Ohta *et al.*, 2000; Kohsaka *et al.*, 1999), associated with incomplete DDS (absence of Wilms' tumor despite a long follow-up). More surprisingly, the typical R394Q mutation responsible for complete DDS in a boy is totally asymptomatic in his father (Coppes *et al.*, 1992) and has been found in a child with acute undifferentiated leukemia (King-Underwood *et al.*, 1998). Another striking example of the possible phenotypic heterogeneity is given by the different clinical expressions of the C-to-T transition generating a stop codon at codon 362. They vary from no symptoms, uni- or bilateral sporadic or familial Wilms' tumor with or without male genital anomalies, to complete DDS (Diller *et al.*, 1998, Kaplinsky *et al.*, 1996, Kohler *et al.*, 1999, Little *et al.*, 1993). Intronic mutations, affecting splicing of WT1 exon 9 and leading to the loss of the +KTS isoforms, have been found in all Frasier syndrome patients. However, the same mutation has also been

Table 22.3   Characteristics of WAGR, Denys–Drash, and Frasier Syndromes

	WAGR syndrome[a]	Denys–Drash syndrome	Frasier syndrome
Renal pathology	Absent or slow propression to ESRD	Early onset nephrotic ESRD < 4 years Diffuse mesagial sclerosis	Proteinuria/nephrotic syndrome ESRD 6–35 years MGC/FSGS[b]
Genital development			
In 46,XY patients	Hypospadias Cryptorchidism	Sexual ambiguity and gonadal dysgenesis Rare gonadoblastomas	Complete sex reversal "Streak gonads" Gonadoblastomas ++
In 46,XX patients		Normal female phenotype Gonadal dysgenesis or normal genital development	Normal female phenotype Normal genital development
Wilms' tumor	Present	High risk	Absent (one exception)
WT1 gene	Large deletion of the WT1 region	Exon 8 or 9 point mutation	WT1 intronic mutation

[a] In WAGR syndrome, aniridia due to PAX6 deletion and mental retardation are not linked to WT1 deletion.

[b] MGC, minimal glomerular changes; FSGS, focal and segmental glomerular sclerosis.

observed in a few patients with DDS or isolated diffuse mesangial sclerosis (Jeanpierre *et al.*, 1998) and also, as indicated previously, in a 46,XX mother with isolated persistent proteinuria (the female expression of Frasier syndrome) and her 46,XY child with DDS (Denamur *et al.*, 1999). Thus different phenotypes may result from the shift in the relative proportion of WT1 isoforms.

Until recently, DDS and Frasier syndrome were regarded as sporadic diseases. Obviously, males with severe genital anomalies, including dysgenetic testis or streak gonads leading to gonadectomy, are infertile. Conversely, female development appears normal in most females with DDS or Frasier syndrome. They now survive to ESRD because of hemodialysis and renal transplantation and are able to become pregnant. They have a 50% risk of transmitting the mutated gene and the disease to their children. Early prenatal diagnosis by direct genetic testing is now available and should be offered to patients with an identified *WT1* mutation.

References

Amin, K. M., Litzky, L. A., Smythe, W. R., Mooney, A. M., Morris, J. M., Mews, D. J., Pass, H. I., Kari, C., Rodeck, U., Rauscher, F. J. III, *et al.* (1995). Wilms' tumor 1 susceptibility (WT1) gene products are selectively expressed in malignant mesothelioma. *Am. J. Pathol.* **146**, 344–356.

Barbaux, S., Niaudet, P., Gubler, M. C., Grünfeld, J. P., Jaubert, F., Kuttenn, F., Nihoul-Fékété, C., Souleyreau-Therville, N., Thibaud, E., Fellous, M., and McElreavey, K. (1997). Donor splice site mutations in the *WT1* gene are responsible for Frasier syndrome. *Nature Genet.* **17**, 467–469.

Bardeesy, N., Zabel, B., Schmitt, K., and Pelletier, J. (1994). WT1 mutations associated with incomplete Denys–Drash syndrome define a domain predicted to behave in a dominant-negative fashion. *Genomics* **21**, 663–665.

Bardoni, B., Zanaria, E., Guioli, S., Floridia, G., Worley, K. C., Tonini, G., Ferrante, E., Chiumello, G., McCabe, E. R., Fraccaro, M., Zuffargdi, O., and Camerino, G. (1994). A dosage sensitive locus at chromosome Xp21 is involved in male to female sex reversal. *Nature Genet.* **7**, 497–501.

Baudry, D., Hamelin, M., Cabanis, M. O., Fournet, J. C., Tournade, M. F., Sarnacki, S., Junien, C., and Jeanpierre, C. (2000). WT1 splicing alterations in Wilms' tumors. *Clin. Cancer Res.* **6**, 3957–3965.

Beckwith, J. B. (1998). Nephrogenic rests and the pathogenesis of Wilms' tumor: Developmental and clinical considerations. *Am. J. Med. Genet.* **79**, 268–273.

Beckwith, J. B., and Palmer, N. (1978). Histopathology and prognosis of Wilms' tumor: Results of the National Wilms' Tumor Study. *Cancer* **41**, 1937–1948.

Blanchet, P., Daloze, P., Lesage, R., Papas, S., and Van Campenhout, J. (1977). XY gonadal dysgenesis with gonadoblastoma discovered after renal transplantation. *Am. J. Obstet. Gynecol.* **129**, 221–222.

Boute, N., Gribouval, O., Roselli, S., Benessy, F., Lee, H., Fuchshuber, A., Dahan, K., Gubler, M. C., Niaudet, P., and Antignac, C. (2000) NPHS2, encoding the glomerular protein podocin, is mutated in autosomal recessive steroid-resistant nephrotic syndrome. *Nature. Genet.* **24**, 349–354.

Breslow, N., Takashima, J. R., Ritchey, M. L., Strong, L. C., and Green, D. M. (2000). Renal failure in the Denys–Drash and Wilms' tumor-aniridia syndromes. *Cancer Res.* **60**, 430–432.

Bruening, W., Bardeesy, N., Silverman, B., Cohn, R. A., Machin, G. A., Aronson, A. J., Housman, D., and Pelletier, J. (1992). Germline intronic and exonic mutations in the Wilms' tumor gene (WT1) affecting urogenital development. *Nature. Genet.* **1**, 144–148.

Bruening, W., and Pelletier, J. (1994). Denys–Drash Syndrome: A role for the WT1 tumor suppressor gene in urogenital development. *Semini Dev. Biol.* **5**, 333–343.

Bruening, W., and Pelletier, J. (1996). A non-AUG translational initiation event generates novel WT1 protein isoforms. *J. Biol. Chem.* **15**, 8646–8654.

Bruening, W., Moffett, P., Chia, S., Heinrich, G., and Pelletier, J. (1996). Identification of nuclear localization signals within the zinc fingers of the WT1 tumor suppressor gene product. *FEBS Lett.* **393**, 41–47.

Buckler, A. J., Pelletier, J., Haber, D. A., Glaser, T., and Housman, D. E. (1991). Isolation, characterization, and expression of the murine Wilms' tumor gene (WT1) during kidney development. *Mol. Cell Biol.* **11**, 1707–1712.

Call, K. M., Glaser, T., Ito, C. Y., Buckler, A. J., Pelletier, J., Haber, D. A., Rose, E. A., Krai, A., Yeger, H., Lewis, W. H., Jones, C., and Housman, D. E. (1990). Isolation and characterization of a zinc finger polypeptide gene at the human chromosome 11 Wilms' tumor locus. *Cell* **60**, 509–520.

Caricasole, A., Duarte, A., Larsson, S. H., Hastie, N. D., Little, M., Holmes, G., Todorov, I., and Ward, A. (1996). RNA binding by the Wilms tumor suppressor zinc finger proteins. *Proc. Natl. Acad. Sci. USA* **93**, 7562–7566.

Caspary, T., Cleary, M. T., Perlman, E. J., Zhang, P., Elledge, S. J., and Tilghman, S. M. (1999). Oppositely imprinted genes p57(Kip2) and igf2 interact in a mouse model for Beckwith-Wiedemann syndrome. *Genes Dev.* **13**, 3115–3124.

Charles, A. K., Brown, K. W., and Berry, P. J. (1998). Microdissecting the genetic events in nephrogenic rests and Wilms' tumor development. *Am. J. Pathol.* **153**, 991–1000.

Coppes, M. J., Campbell, C. E., and Williams, B. R. G. (1993). The role of WT1 on Wilms' tumorigenesis. *FASEB J.* **7**, 886–893.

Coppes, M. J., and Egeler, R. M. (1999b). Genetics of Wilms' tumor. *Semin. Urol. Oncol.* **17**, 2–10.

Coppes, M. J., Liefers, G., Higuchi, M., Zinn, A., Balfe, J., and Williams, B. (1992). Inherited WT1 mutation in Denys–Drash syndrome. *Cancer Res.* **52**, 6125–6128.

Coppes, M. J., Wolff, E. A., and Ritchey, M. L. (1999a). Wilms tumor. Diagnosis and treatment. *Paediatr. Drugs* **1**, 251–262.

De Santa Barbara, P., Moniot, B., Poulat, F., and Berta, P. (2000). Expression and subcellular localization of SF-1, SOX9, WT1, and AMH proteins during early human testicular development. *Dev. Dyn.* **217**, 293–298.

Demmer, L., Primack, W., Loik, V., Brown, R., Therville, N., and McElreavey, K. (1999). Frasier syndrome: A cause of focal segmental glomerulosclerosis in a 46, XX female. *J. Am. Soc. Nephrol.* **10**, 2215–2218.

Denamur, E., Bocquet, N., Baudouin, V., Da Silva, F., Veitia, T., Peuchmaur, M., Elion, J., Gubler, M. C., Fellous, M., Niaudet, P., and Loirat, C. (2000). WT1 splice site mutations are rarely associated with primary steroid-resistant focal and segmental glomerulosclerosis. *Kidney Int.* **57**, 1868–1872.

Denamur, E., Bocquet, N., Mougenot, B., Da Silva, F., Martinat, L., Loirat, C., Elion, J., Bensman, A., and Ronco, P. M. (1999). Mother-to-child transmitted WT1 splice-site mutation is responsible for distinct glomerular diseases. *J. Am. Soc. Nephrol.* **10**, 2219–2223.

Denys, P., Malvaux, P., van den Berghe, H., Tanghe, W., and Proesmans, W. (1967). Association d'un syndrome anatomo-pathologique de pseudo-hermaphrodisme masculin, d'une tumeur de Wilms, d'une néphropathie parenchymateuse et d'un mosaicisme XX/XY. *Arch. Fr. Pediatr.* **24**, 729–739.

Devriendt, K., Deloof, E., Moerman, P., Legius, E., Vanhole, C., de Zegher, F., Proesmans, W., and Devlieger, H. (1995). Diaphragmatic hernia in Denys–Drash syndrome. *Am. J. Med. Genet.* **57**, 97–101.

Diller, L., Ghahremani, M., Morgan, J., Grundy, P., Reeves, C., Breslow, N., Green, D., Neuberg, D., Pelletier, J., and Li, F. P. (1998). Constitutional WT1 mutations in Wilms' tumor patients. *J. Clin. Oncol.* **16**, 3634–3640.

Drash, A., Sherman, F., Hartmann, W., and Blizzard, R. M. A. (1970). A syndrome of pseudohermaphroditism, Wilms' tumor, hypertension, and degenerative renal disease. *J. Pediatr.* **76**, 585–593.

Dressler, G. R., Wilkinson, J. E., Rothenpieler, U. W., Patterson, L. T., Williams-Simons, L., and Westphal, H. (1993). Deregulation of Pax-2 expression in transgenic mice generates severe kidney abnormalities. *Nature* **362**, 65–67.

Drummond, I. A., Rupprecht, H. D., Rower-Nutter, P., Lopez-Guisa, J. M., Madden, S. L., Rauscher, F., Jr., and Sikhatme, V. P. (1994). DNA recognition by splicing variants of the Wilms tumor suppressor WT1. *Mol. Cell Biol.* **14**, 3800–3809.

Frasier, S., Bashore, R. A., and Mosier, H. D. (1964). Gonadoblastoma associated with pure gonadal dysgenesis in monozygotic twins. *J. Pediatr.* **64**, 740–745.

Gao, L., Bellantuono, I., Elsasser, A., Marley, S. B., Gordon, M. Y., Goldman, J. M., and Stauss, H. J. (2000). Selective elimination of leukemic CD34(+) progenitor cells by cytotoxic T lymphocytes specific for WT1. *Blood* **95**, 2198–2203.

Gessler, M., Konig, A., Arden, K., Grundy, P., Orkin, S., Sallan, S., Peters, C., Ruyle, S., Mandell, J., Li F,Cavenee, W., and Bruns, G. (1994). Infrequent mutation of the WT1 gene in 77 Wilms' Tumors. *Hum. Mutat.* **3**, 212–322.

Gessler, M., König, A., and Bruns, G. A. P. (1992). The genomic organization and expression of the WT1 gene. *Genomics* **12**, 807–813.

Gessler, M., Poustka, A., Cavenee, W., Neve, R. L., Orkin, S. H., and Bruns, G. A. P. (1990). Homozygous deletion in Wilms' tumors of a zinc-finger gene identified by chromosome jumping. *Nature* **343**, 774–778.

Grundy, P. E., Telzerow, R. E., Dreslow, N., Mokness, J., Huff, V., and Paterson, M. C. (1994). Loss of heterozygosity for chromosomes 16q and 1p in Wilms' tumors predicts an adverse outcome. *Cancer Res.* **54**, 2331–2333.

Gubler, M. C., Yang, Y., Jeanpierre, C., Barbaux, S., and Niaudet, P. (1999). WT1, renal developpement, and glomerulopathies. *Adv. Nephrol. Necker Hosp.* **29**, 299–315.

Haber, D. A., Buckler, A. J., Glaser, T., Call, K. M., Pelletier, J., Sohn, R. L., Douglass, E., and Housman, D. E. (1990). An internal deletion within an 11p13 zinc finger gene contributes to the development of Wilms' tumor. *Cell* **61**, 1257–1269.

Haber, D. A., Park, S., Maheswaran, S., Englert, C., Re, G. G., Hazen-Martin, D. J., Sens, D. A., and Garvin, A. J. (1993). WT1-mediated growth suppression of Wilms' tumor cells expressing a WT1 splicing variant. *Science* **262**, 2057–2059.

Haber, D.A, Sohn, R. L., Buckler, A. J., Pelletier, J., Call, K. M., and Housman, D. E. (1991). Alternative splicing and genomic structure of the Wilms' tumor gene WT1. *Proc. Natl. Acad. Sci. USA* **88**, 9618–9622.

Habib, R., Loirat, C., Gubler, M. C., Niaudet, P., Bensman, A., Levy, M., and Broyer, M. (1985). The nephropathy associated with male pseudo-hermaphroditism and Wilms' tumor (Drash syndrome): A distinctive glomerular lesion, report of 10 cases. *Clin. Nephrol.* **24**, 269–278.

Habib, R., Gubler, M. C., Antignac, C., and Gagnadoux MF. (1993). Diffuse mesangial sclerosis: A congenital glomerulopathy with nephrotic syndrome. *Adv. Nephrol.* **22**, 43–56.

Haning, R. V., Chesney, R. W., Moorthy, A. V., and Gilbert, E. N. (1985). A syndrome of chronic renal failure and XY gonadal dysgenesis in young phenotypic females without genital ambiguity. *Am. J. Kidney Dis.* **6**, 40–48.

Harkins, P. G., Haning, R. V., and Shapiro, S. S. (1980). Renal failure with XY gonadal dysgenesis: Report of the second case. *Obstet. Gynecol.* **56**, 751–752.

Herzer, U., Crocoll, A., Barton, D., Howells, N., and Englert, C. (1999). The Wilms tumor suppressor gene wt1 is required for development of the spleen. *Curr. Biol.* **9**, 937–840.

Hewitt, S. M., Fraizer, G. C., Wu, Y. J., Rauscher, F. J., III, and Saunders, G. F. (1996). Differential function of Wilms' tumor gene WT1 splice isoforms in transcriptional regulation. *J. Biol. Chem.* **271**, 8588–8592.

Huff, V., Miwa, H., Haber, D. A., Call, K. M., Housman, D., Strong, L. C., and Saunders, G. F. (1991). Evidence for WT1 as a Wilms' tumor (WT) gene: Intragenic deletion in bilateral Wilms' tumor. *Am. J. Hum. Genet.* **48**, 997–1003.

Inoue, K., Ogawa, H., Sonoda, Y., Kimura, T., Sakabe, H., Oka, Y., Miyake, S., Tamaki, H., Oji, Y., Yamagami, T., Tatekawa, T., Soma, T., Kishimoto, T., and Sugiyama, H. (1997). Aberrant overexpression of the Wilms tumor gene (WT1) in human leukemia. *Blood* **89**, 1405–1412.

Jeanpierre, C., Béroud, C., Niaudet, P., and Junien, C. (1998a). Software and database for the analysis of mutations in the human WT1 gene. *Nucleic. Acids Res.* **26**, 273–277.

Jeanpierre, C., Denamur, E., Henry, I., Cabanis, M. O., Luce, S., Cécille, A., Elion, J., Peuchmaur, M., Loirat, C., Niaudet, P., Gubler, M. C., and Junien, C. (1998b). Identification of constitutional WT1 mutations in patients with isolated diffuse mesangial sclerosis (IDMS) and anlysis of genotype-phenotype correlations using a computerized mutation database. *Am. J. Hum. Genet.* **62**, 824–833.

Johnstone, R. W., Wang, J., Tommerup, N., Vissing, H., Roberts, T., and Shi, Y. (1998). Ciao 1 is a novel WD40 protein that interacts with the tumor suppressor protein WT1. *J. Biol. Chem.* **273**, 10880–10887.

Kaplan, J. M., Kim, S. H., North, K. N., Rennke, H., Correia, L. A., Tong, H. Q., Mathis, B. J., Rodriguez-Perez, J. C., Allen, P. G., Beggs, A. H., Ara Pollak, M. R. (2000). Mutations in ACTN4, encoding alpha-actinin-4, cause familial focal segmental glomerulosclerosis. *Nature. Genet.* **24**, 251–256.

Kaplinsky, C., Ghahremani, M., Frishberg, Y., Rechavi, G., and Pelletier, J. (1996). Familial Wilms' tumor associated with a WT1 zinc finger mutation. *Genomics* **38**, 451–453.

Kestila, M., Lenkkeri, U., Mannikko, M., Lamerdin, J., McCready, P., Putaala, H., Ruotsalainen, V., Morita, T., Nissinen, M., Herva, R., Kashtan, C. E., Peltonen, L., Holmberg, C., Olsen, A., and Tryggvason, K. (1998). Positionally cloned gene for a novel glomerular protein—nephrin—is mutated in congenital nephrotic syndrome. *Mol. Cell.* **1**, 575–582.

Kikuchi, H., Takata, A., Akasaka, Y., Fukuzawa, R., Yoneyama, H., Kurosawa, Y., Honda, M., Kamiyama, Y., and Hata, J. (1998). Do intronic mutations affecting splicing of WT1 exon 9 cause Frasier syndrome? *J. Med. Genet.* **35**, 54–48.

Kim, J., Prawitt, D., Bardeesy, N., Torban, E., Vicaner, C., Goodyer, P., Zabel, B., and Pelletier, J. (1999). The Wilms' tumor suppressor gene (wt1) product regulates Dax-1 gene expression during gonadal differentiation. *Mol. Cell Biol.* **19**, 2289–2299.

Kinberg, J. A., Angle, C. R., and Wilson, R. B. (1987). Nephropathy-gonadal dysgenesis, type 2: Renal failure in three siblings with XY dysgenesis in one. *Am. J. Kidney Dis.* **9**, 507–510.

King-Underwood, L., and Pritchard-Jones, K. (1998). Wilms' tumor (WT1) gene mutations occur mainly in acute myeloid leukemia and may confer drug resistance. *Blood* **91**, 2961–2968.

Klamt, B., Koziell, A., Poulat, F., Wieacker, P., Scambler, P., Berta, P., and Gessler, M. (1998). Frasier syndrome is caused by defective alternative splicing of WT1 leading to an altered ratio of WT1+/–KTS splice isoforms. *Hum. Mol. Genet.* **7**, 709–714.

Knudson, A. G. (1971). Mutation and cancer: Statistical study of retinoblastema. *Proc. Natl. Acad. Sci. USA* **4**, 820–823.

Kohler, B., Schumacher, V., Schulte-Overberg, U., Biewald, W., Lennert, T., l'Allemand, D., Royer-Pokora, B., and Gruters, A. (1999). Bilateral

Wilms tumor in a boy with severe hypospadias and cryptorchidism due to a heterozygous mutation in the WT1 gene. *Pediatr. Res.* **45**, 187–190.

Kohsaka, T., Tagawa, M., Takekoshi, Y., Yanagisawa, H., Tadokoro, K., and Yamada, M. (1999). Exon 9 mutations in the WT1 gene, without influencing KTS splice isoforms, are also responsible for Frasier syndrome. *Hum. Mutat.* **14**, 466–470.

König, A., Jakubiczka, S., Wieacker, P., Schlösser, H. W., and Gessler, M. (1993). Further evidence that imbalance of WT1 isoforms may be involved in Denys–Drash syndrome. *Hum. Mol. Genet.* **2**, 1967–1968.

Koziell, A. B., Grundy, R., Barratt, T. M., and Scambler, P. (1999). Evidence for genetic heterogeneity of nephropathic phenotypes associated with Denys–Drash and Frasier syndromes. *Am. J. Hum. Genet.* **64**, 1778–1781.

Kreidberg, J. A., Sariola, H., Loring, J. M., Maeda, M., Pelletier, J., Housman, D., and Jaenisch, R. (1993). WT1 is required for early kidney development. *Cell* **74**, 679–691.

Laity, J. H., Dyson, H. J., and Wright, P. E. (2000). Molecular basis for modulation of biological function by alternate splicing of the Wilms' tumor suppressor gene. *Proc. Natl. Acad. Sci. USA* **97**, 11932–11935.

Langerak, A. W., Williamson, K. A., Miyagawa, K., Hagemeijer, A., Versnel, M. A., and Hastie, N. D. (1995). Expression of the Wilms' tumor gene WT1 in human malignant mesothelioma cell lines and relationship to platelet-derived growth factor A and insulin-like growth factor 2 expression. *Genes Chromosomes Cancer* **12**, 87–96.

Larsson, S. H., Charlieu, J. P., Miyagawa, K., Engelkamp, D., Rassoulzadegan, M., Ross, A., Cuzin, F., van Heyningen, V., and Hastie, N. D. (1995). Subnuclear localization of WT1 in splicing or transcription factor domains is regulated by alternative splicing. *Cell* **81**, 391–401.

Lee, S. B., Huang, K., Palmer, R., Truong, V. B., Herzlinger, D., Kolquist, K. A., Wong, J., Paulding, C., Yoon, S. K., Gerald, W., Oliner, J. D., and Haber, D. A. (1999). The Wilms tumor suppressor gene *WT1* encodes a transcriptional activator of *amphiregulin*. *Cell* **98**, 663–673.

Lee, S. B., Kolquist, K. A., Nichols, K., Englert, C., Maheswaran, S., Ladanyi, M., Gerald, W. L., and Haber, D. A. (1997). The EWS-WT1 translocation product induces PDGFA in desmoplastic small round-cell tumor. *Nature. Genet.* **17**, 309–313.

Little, M., Carman, G., and Donaldson, E. (2000a). Novel WT1 exon 9 mutation (D396Y) in a patient with early onset Denys Drash syndrome. *Hum. Mutat.* **15**, 389.

Little, M., Holmes, G., Bickmore, W., Heyningen, V. V., Hastie, N., and Wainwright, B. (1995). DNA binding capacity of the WT1 protein is abolished by Denys–Drash syndrome WT1 point mutations. *Hum. Mol. Genet.* **4(3)**, 351–358.

Little, M., Holmes, G., and Walsh, P. (1999). WT1: What has the last decade told us? *Bioessays* **21**, 191–202.

Little, M., and Wells, C. (1997). A clinical overview of WT1 gene mutations. *Hum. Mutat.* **9**, 209–225.

Little, M. H., Williamson, K. A., Mannens, M., Kelsey, A., Gosden, C., Hastie, N. D., and van Heyningen, V. (1993). Evidence that WT1 mutations in Denys–Drash syndrome patients may act in a dominant-negative fashion. *Hum. Mol. Genet.* **2(3)**, 259–264.

Little, N. A., Hastie, N. D., and Davies, R. C. (2000b). Identification of WTAP, a novel Wilms' tumour-associated protein. *Hum. Mol. Genet.* **9**, 2231–2239.

Luetteke, N. C., Qiu, T. H., Fenton, S. E., Troyer, K. L., Riedel, R. F., Chang, A., and Lee, D. C. (1999). Targeted inactivation of the EGF and amphiregulin genes reveals distinct roles for EGF receptor ligands in mouse mammary gland development. *Development* **126**, 2739–2750.

Menke, A., McInnes, L., Hastie, N. D., and Schedl, A. (1998). The Wilms' tumor suppressor WT1: Approaches to gene function. *Kidney Int.* **53**, 1512–1518.

Moffett, P., Bruening, W., Nakagama, H., Bardeesy, N., Housman, D. E., and Pelletier, J. (1995). Antagonism of WT1 activity by protein self-association. *Proc. Natl. Acad. Sci. USA* **92**, 11105–11109.

Moore, A. W., McInnes, L., Kreiberg, J., Hastie, N. D., and Schedl, A. (1999). YAC complementation shows a requirement for WT1 in the development of epicardium, adrenal gland and throughout nephrogenesis. *Development* **126**, 1845–1857.

Moorthy, A. V., Chesney, R. W., and Lubinsky, M. (1987). Chronic renal failure and XY gonadal dysgenesis: "Frasier" syndrome—a commentary on reported cases. *Am. J. Med. Genet. Suppl* **3**, 297–302.

Mundlos, S., Pelletier, J., Darveau, A., Bachmann, M., Winterpacht, A., and Zabel, B. (1993). Nuclear localization of the protein encoded by the Wilms' tumor gene WT1 in embryonic and adult tissues. *Development* **119**, 1329–1341.

Murray, S. C. (1998). XY gonadal dysgenesis in an adolescent with chronic renal failure: A case of Frasier syndrome. *J. Pediatr. Adolesc. Gynecol.* **11**, 89–91.

Nachtigal, M. W., Hirokawa, Y., Enyeart-VanHouten, D. L., Flanagan, J. N., Hammer, G. D., and Ingraham, H. A. (1998). Wilms' tumor 1 and Dax-1 modulate the orphan nuclear receptor SF-1 in sex-specific gene expression. *Cell* **93**, 445–454.

Ohta, S., Ozawa, T., Izumino, K., Sakuragawa, N., and Fuse, H. (2000). A novel missense mutation of the WT1 gene causing Denys–Drash syndrome with exceptionally mild renal manifestations. *J. Urol.* **163**, 1857–1858.

Patek, C. E., Little, M. H., Fleming, S., Miles, C., Charlieu, J. P., Clarke, A. R., Miyagawa, K., Christie, S., Doig, J., Harrison, D. J., Porteous, D. J., Brookes, A. J., Hooper, M. L., and Hastie, N. D. (1999). A zinc finger truncation of murine WT1 results in the characteristic urogenital abnormalities of Denys–Drash syndrome. *Proc. Natl. Acad. Sci. USA.* **96**, 2931–2936.

Pelletier, J., Bruening, W., Kashtan, C. E., Mauer, S. M., Manivel, J. C., Striegel, J. E., Houghton, D. C., Junien, C., Habib, R., Fouser, L., Fine, R. N., Silverman, B. L., Haber, D. A., and Housman, D. (1991a). Germline mutations in the Wilms' tumor suppressor gene are associated with abnormal urogenital development in Denys–Drash syndrome. *Cell* **67**, 437–447.

Pelletier, J., Bruening, W., Li, F. P., Haber, D. A., Glaser, T., and Housman, D. (1991b). WT1 mutations contribute to abnormal genital system development and hereditary Wilms' tumor. *Nature* **353**, 431–434.

Pritchard-Jones, K., Fleming, S., Davidson, D., Bickmore, W., Porteous, D., Gosden, C., Bard, J., Buckler, A., Pelletier, J., Housman, D., van Heyningen, V., and Hastie, N. (1990). The candidate Wilms' tumor gene is involved in genitourinary development. *Nature* **346**, 194–197.

Pritchard-Jones, K., Rahman, N., Gerrard, M., Variends, D., and King-Underwood, L. (2000). Familial Wilms' tumor resulting from WT1 mutation: Intronic polymorphism causing artefactual constitutional homozygosity. *J. Med. Genet.* **37**, 377–379.

Rapley, E. A., Barfoot, R., Bonaiti-Pellie, C., Chompret, A., Foulkes, W., Perusinghe, N., Reeve, A., Royer-Pokora, B., Schumacher, V., Shelling, A., Skeen, J., de Tourreil, S., Weirich, A., Pritchard-Jones, K., Stratton, M. R., and Rahman, N. (2000). Evidence for susceptibility genes to familial Wilms tumor in addition to WT1, FWT1 and FWT2. *Br. J. Cancer* **83**, 177–183.

Rauscher, F. J., III, (1993). The WT1 Wilms tumor gene product: A developmentally regulated transcription factor in the kidney that functions as a tumor suppressor. *FASEB J.* **7**, 896–903.

Reddy, J. C., and Licht, J. D. (1996). The WT1 Wilms' tumor suppressor gene: How much do we really know? *BBA* **1287**, 1–28.

Reddy, J. C., Morris, J. C., Wang, J., English, M. A., Haber, D. A., Shi, Y., and Licht, J. D. (1995). WT1-mediated transcriptional activation is inhibited by dominant negative mutant proteins. *J. Biol. Chem.* **270**, 10878–10884.

Riccardi, V. M., Sujansky, Y., Smith, A. C., and Francke, U. (1978). Chromosome imbalance in the aniridia-Wilms' tumor association: 11p interstitial deletion. *Pediatrics* **61**, 604–610.

Scharnhorst, V., Dekker, P., van der Eb, A., and Jochemsen, A. G. (1999). Internal translation initiation generates novel WT1 isoforms with distinct biological properties. *J. Biol. Chem.* **274**, 23456–23462.

Schedl, A., and Hastie, N. (1998). Multiple roles for Wilms' tumor suppressor gene, WT1 in genitourinary development. *Mol. Cell Endocr.* **140**, 65–69.

Schumacher, V., Schärer, K., Wühl, E., Altrogge, H., Bonzel, K. E., Guschmann, M., Neuhaus, T. J., Pollastro, R. M., Kuwertz-Broking, E., Bulla, M., Tondera, A. M., Mundel, P., Helmchen, U., Waldherr, R., Weirich, A., and Royer-Pokora, B. (1998). Spectrum of early onset nephrotic syndrome associated with WT1 missense mutations. *Kidney Int.* **53**, 1594–1600.

Schumacher, V., Schneuder, S., Figge, A., Wildhart, G., Harms, D., Schmidt, D., Weirich, A., Ludwig, R., and Royer-Pokora, B. (1997). Correlation of germ-line mutations and two-hit inactivation of the WT1 gene with Wilms tumors of stromal-predominant histology. *Proc. Natl. Acad. Sci. USA* **94**, 392–3977.

Sharma, P. M., Bowman, M., Madden, S. L., Rauscher, F. J., and Sukumar, S. L. (1994). RNA editing in the Wilms' tumor susceptibility gene, WT1. *Genes Dev.* **8**, 720–731.

Silberstein, G. B., Van Horn, K., Strickland, P., Roberts, C. T., Jr., and Daniel, C. W. (1997). Altered expression of the WT1 Wilms' tumor suppressor gene in human breast cancer. *Proc. Natl. Acad. Sci. USA.* **94**, 8132–8137.

Simms, L. A., Algar, E. M., and Smith, P. J. (1995). Splicing of exon 5 in the WT1 gene is disrupted in Wilms' tumors. *Eur. J. Cancer* **31A**, 2270–2276.

Simpson, J. L., Chaganti, R. S., Mouradian, J., and German, J. (1982). Chronic renal disease, myotonic dystrophy, and gonadoblastoma in XY gonadal dysgenesis. *J. Med. Genet.* **19**, 73–76.

Thomalla, J. V. (1990). Chronic renal failure in an obese adolescent. *Hosp. Prac.* **25**, 128–133.

Ton, C. C. T., Hirvonen, H., Miwa, H., Weil, M. M., Monaghan, P., Jordan, P., van Heyningen, V., Hastie, M. D., Meijers-Heijboer, H., Drechsler, M., Royer-Pokara, B., Collins, F., S waroop, A. Strong, L. C., and Saunders, G. F. (1991). Positional cloning and characterization of a paired box- and homeobox-containing gene from the aniridia region. *Cell* **67**, 1059–1074.

Van Heyningen, V., and Hastie, N. D. (1992). Wilms' tumor: Reconciling genetics and biology. *Trends Genet.* **8**, 16–21.

Wang, Z. Y., Qiu, Q. Q., Huang, J., Gurrieri, M., and Deuel, T. F. (1995). Products of alternative spliced transcripts of the Wilms' tumor suppressor gene, WT1, have altered DNA binding specificity and regulate transcription in different ways. *Oncogene* **10**, 415–422.

Yang, Y., Jeanpierre, C., Dressler, G. R., Lacoste, M., Niaudet, P., and Gubler, M. C. (1999). WT1 and PAX-2 podocyte expression in Denys–Drash syndrome and isolated diffuse mesangial sclerosis. *Am. J. Pathol.* **54**, 181–192.

Yang, Y., Zhang, S. H., Sich, M., Beziau, A., Van den Heuvel, L., Gubler, M. C. (2001). Glomerular extracellular matrix, platelet-derived growth factor A, transforming growth factor betal in diffuse mesangial sclerosis. *Pediatr. Neprol.* **16**, 429–438.

PAX2 and Renal-Coloboma Syndrome

Michael Eccles, Nicholas Bockett, and Cherie Stayner

Mutations in the *PAX2* gene cause renal-coloboma syndrome (RCS), a rare syndrome characterized by optic nerve colobomas and renal abnormalities. *PAX2* mutations have also been associated with oligomeganephronia (primary renal hypoplasia). Approximately 63% of patients with RCS have renal hypoplasia in addition to optic nerve colobomas, with or without additional abnormalities, including vesico-ureteric reflux (VUR), high-frequency hearing loss, central nervous system (CNS) anomalies, or genital anomalies. Nine different *PAX2* mutations have been identified to date, which cause either RCS or oligomeganephronia. Several animal models of mutations in *Pax2* homologous sequences have been developed, including *Caenorhabditis elegans*, *Drosophila melanogaster*, zebrafish, and mice. Mice are the most robust in terms of recapitulating the human disease. One murine *Pax2* mutant in particular, *Pax2*[1Neu], harbors a mutation that is identical to the most commonly occurring

PAX2 mutation in humans. Furthermore, heterozygous *Pax2*[1Neu] mutant mice often have small kidneys, with markedly elevated levels of programmed cell death in the collecting duct epithelial cells. Much has been learned about kidney development and renal-coloboma syndrome from an analysis of *PAX2* mutations. Detailed analysis of animal models of RCS will provide further insight into the molecular basis of renal disease and development.

I. Introduction

Renal and eye abnormalities are relatively common birth defects, sometimes occurring together as a syndrome, either with or without other anomalies (Weaver *et al.*, 1988; Clarke *et al.*, 1992; Antignac *et al.*, 1993). The association of congenital eye and kidney defects together as a syndrome suggests that these two organ systems may share common developmental pathways. Indeed, identification of the genetic cause of renal-coloboma syndrome, a condition resulting from abnormal kidney and optic nerve morphogenesis (Schimmenti *et al.*, 1995), confirms that mutation of a single gene can disrupt development of both eyes and kidneys. Renal-coloboma syndrome (entry number 120330 in McKusick's Mendelian Inheritance in Man Online, OMIM) is a relatively rare disease resulting from mutations in the *PAX2* gene (Sanyanusin *et al.*, 1995). *PAX2* mutations have also been associated with oligomeganephronia (Salomon *et al.*, 1999). Patients with either RCS or bilateral renal hypoplasia have a high risk of end-stage renal disease, yet the underlying physiological mechanism for the onset of the renal disease in RCS remains obscure at the present time. This chapter describes the clinical presentations of RCS and oligomeganephronia and summarizes what is known about the role of *PAX2* in kidney development.

II. Pathologic Analysis of Renal-Coloboma Syndrome and Oligomeganephronia

A. Renal-Coloboma Syndrome; a Relatively New Syndrome

Renal-coloboma syndrome accounts for the majority of *PAX2* mutations identified to date. The term "renal-coloboma syndrome" has been adopted only since the late 1990s. In earlier reports, names such as morning glory syndrome (Karcher, 1979), papillorenal dysplasia (Bron *et al.*, 1988), coloboma-ureteral-renal syndrome (Schimmenti *et al.*, 1995), and optic nerve coloboma with renal disease (ONCR) (Narahara *et al.*, 1997) were used. In relation to this, it should be noted that *PAX2* mutations have been reported in association with morning glory syndrome but not papillorenal dysplasia (Bron *et al.*, 1988). While it has been assumed that each of these reports are of the same disease and simply reflect the variable phenotype associated with renal-coloboma syndrome, such an assumption may not necessarily be correct. That papillorenal dysplasia has not yet been shown to be caused by *PAX2* mutations, whereas patients described as having ONCR, and coloboma-ureteral-renal syndrome were shown to have *PAX2* mutations, may be of significance because there is presently debate as to whether dysplastic discs, rather than optic nerve colobomas, describe the eye abnormalities in patients with RCS more accurately and whether papillorenal syndrome is a more appropriate designation. This chapter uses the term renal-coloboma syndrome, because in the series of patients that have been analyzed at the University of Otago with *PAX2* mutations, optic nerve colobomas were the diagnosis based on ophthalmoscopic evaluation. We acknowledge, however, that there is wide variation in the clinical presentation of patients who have *PAX2* mutations.

Reiger (1997) was the first to identify the association of optic nerve and renal abnormalities. He described a father with optic nerve anomalies (Handmann anomaly) and chronic nephritis, a son with isolated Handmann anomaly, and a daughter with normal eyes and renal hypoplasia, who died of chronic renal failure. In 1979, Karcher described a father and son with "morning glory" disc anomaly and glomerulonephritis. In 1988, Weaver and colleagues described two brothers with optic nerve colobomas and renal disease. Following this, Bron and colleagues (1988), described a family with papillary abnormalities and glomerulonephritis, calling the disease "papillorenal dysplasia".

The exact incidence of RCS (or *PAX2* mutations) is unknown. Among the Belgian Flemish (who number some 6 million people), four different *PAX2* mutations have been identified in four separate families (Schimmenti *et al.*, 1997; Devriendt *et al.*, 1998). However, this figure is likely to represent an underestimate of the real incidence, as the disease is not well known. Furthermore, some patients may only present with renal abnormalities (e.g., oligomeganephronia), and the full significance of *PAX2* mutations in such patients has not yet been determined.

B. Oligomeganephronia

PAX2 mutations have been identified in association with isolated renal hypoplasia (oligomeganephronia) (Salomon *et al.*, 1999). Renal hypoplasia is usually sporadic, although on occasion it may be inherited dominantly, or it may be associated with other abnormalities. Renal hypoplasia refers to the failure of the kidneys to develop to a normal size. This anomaly may occur bilaterally, but more commonly it is observed unilaterally. True renal hypoplasia is the reduction in the number of renal lobules and calyces (five or fewer) in contrast with the normal development of 10 or more. In oligomeganephronia the kidney is small and the remaining nephrons are hypertrophied (Royer *et al.*, 1967).

Congenital renal hypoplasia occurs in approximately 1 in 500 liveborn infants, although the frequency of bilateral renal hypoplasia is much less than that. A significant proportion of infants with bilateral congenital renal hypoplasia develop end-stage renal failure within 5–10 years, whereas patients with severe cases of bilateral congenital renal hypoplasia die in infancy or early childhood. One school of thought is that the stress of another renal disease, such as stone formation or infection superimposed on renal hypoplasia, reduces renal function further. Another possibility is that the molecular mechanisms underlying renal hypoplasia contribute to the progressive renal failure. An important issue to consider in patients with renal hypoplasia and oligomeganephronia is the involvement of eye anomalies, which may suggest that they have RCS. All patients with renal hypoplasia should therefore be considered for an eye examination and *PAX2* mutation screening.

C. Clinical Features of Renal-Coloboma Syndrome

Renal-coloboma syndrome is extremely variable phenotypically, even in members of the same family who have an identical *PAX2* mutation (Schimmenti *et al.*, 1997). RCS is, nevertheless, distinct from other similar syndromes with overlapping characteristics, such as CHARGE syndrome (Tellier *et al.*, 2000) or a syndrome of optic nerve colobomas, renal abnormalities, and arthrogryposis multiplex (Al-Gazali *et al.*, 2000). The predominant abnormalities

Table 23.1   Distribution of Phenotypic Abnormalities in 32 Patients with *PAX2* Mutations

Abnormality	% of patients with *PAX2* mutation showing the abnormality
Ocular anomalies	100
Proteinuria or renal failure	90
Small kidneys	63
Vesicoureteral reflux	25
High frequency hearing loss	13
Joint laxity/soft skin	9
Mid/hindbrain anomalies	0.3

associated with RCS are bilateral optic nerve colobomas and renal hypoplasia, with or without renal failure (Table 23.1). Patients with RCS and *PAX2* mutations may also have additional congenital anomalies that occur with variable penetrance (Table 23.1) (Sanyanusin *et al.*, 1995a,b; Narahara *et al.*, 1997; Schimmenti *et al.*, 1997; Cunliffe *et al.*, 1998; Devriendt *et al.*, 1998; Porteous *et al.*, 2000).

1. Ocular Anomalies

The majority of patients with RCS have optic nerve colobomas and/or related developmental defects affecting the posterior globe. These defects range in severity from bilateral optic nerve colobomas (profound excavation of the optic nerve) to optic disc dysplasia involving the optic nerve head. Eye abnormalities may include anomalous retinal vessel patterns, retinal depigmentation, hyperpigmented macula, macula colobomas, or redundant fibroglial remnants (Sanyanusin *et al.*, 1995a,b; Narahara *et al.*, 1997; Schimmenti *et al.*, 1997; Cunliffe *et al.*, 1998; Devriendt *et al.*, 1998; Porteous *et al.*, 2000). The appearance of a typical optic nerve coloboma from a patient with RCS is shown in Fig. 23.1. Patients with RCS have also been observed with orbital cysts and microphthalmia (Fig. 23.1) (Schimmenti *et al.*, 1997). All of these defects are typical of the spectrum of developmental abnormalities associated with defective closure of the optic fissure during eye development (Pagon, 1981).

It has been well documented that optic nerve colobomas may occur without renal involvement in some families as an autosomal-dominant condition in isolation or in association with other unrelated abnormalities or syndromes. In this regard, it has been shown that patients with optic nerve colobomas, or other types of eye abnormalities with or without additional anomalies, do not have *PAX2* mutations and probably do not have RCS (Cunliffe *et al.*, 1998). *PAX2* mutations are found most commonly in association with "typical" RCS, involving optic nerve colobomas, renal failure, and/or renal hypoplasia with or without vesico-ureteric reflux.

Eye abnormalities in patients with RCS are frequently associated with a loss of visual acuity, although not all patients with optic nerve colobomas have visual defects. Visual-evoked potentials may be reduced in some patients with RCS, although remarkable variation in visual acuity may occur, even in patients from the same family (Devriendt *et al.*, 1998; Porteous *et al.*, 2000). In some patients, eye abnormalities may be very subtle and visual defects completely absent. For example, several patients with *PAX2* mutations and renal hypoplasia had no discernible visual abnormalities (Devriendt *et al.*, 1998). This reinforces the notion that all patients with primary renal hypoplasia should undergo a complete clinical assessment, including examination of both eyes.

2. Urogenital Anomalies

Urological abnormalities in patients with *PAX2* mutations characteristically involve hypoplastic kidneys, with decreased renal function and occasionally unilateral renal agenesis. The involvement of urological abnormalities in patients with RCS is a significant contributor to the renal disease. Approximately 63% of patients with RCS have small kidneys with a reduction in the size of both the renal pelvis and the renal cortex (Porteous *et al.*, 2000). The renal disease is usually progressive and frequently necessitates renal transplantion following chronic renal failure. Patients with *PAX2* mutations and mild or no renal disease can also occur within the same family as those with end-stage renal disease (Devriendt *et al.*, 1998). The kidneys may be variably affected, with one kidney being more severely affected than the other, and occasionally unilateral renal agenesis may be observed in patients with RCS (Sanyanusin *et al.*, 1995). One patient with RCS was found to have bilateral cryptorchidism, suggesting that genital anomalies may also sometimes be involved (Porteous *et al.*, 2000).

3. Vesicoureteric Reflux

VUR has been observed in 8/32 (25%) patients with *PAX2* mutations (Sanyanusin *et al.*, 1995; Devriendt *et al.*, 1998; Porteous *et al.*, 2000), although the frequency may be higher because VUR has a tendency to resolve with age (Eccles and Jacobs, 2000). In individuals who do not have RCS or *PAX2* mutations, VUR has been estimated to occur in approximately 1% of Caucasians. The presence of VUR in patients with RCS should be considered as a possible determinant of loss of renal function (Eccles *et al.*, 1996). VUR, regurgitation of urine from the bladder to the kidney as a result of an abnormal bladder/ureter junction (Fig. 23.2), is thought to result in physical damage to the kidney and an increased risk of pyelonephritis (Eccles and Jacobs, 2000). The defect leading to primary VUR arises during development of the ureter and consequently VUR may have its origins *in utero*. Primary VUR can cause both renal failure and scarring, and the resulting renal defects may look

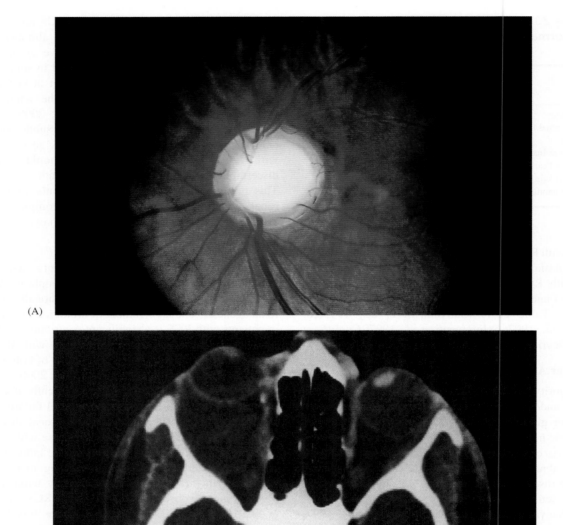

(A)

(B)

Figure 23.1 Eye abnormalities associated with *PAX2* mutations. (A) Fundus of a patient with RCS showing the typical appearance of an optic nerve coloboma [emanation of blood vessels from the periphery of a pale (excavated) optic nerve.] (B) CT scan of a patient with a PAX2 G619 insertion mutation showing a microphthalmic left eye and a retrobulbar cyst in the left optic nerve.

Continued

similar to those caused by *PAX2* mutations. As VUR is familial with a dominant inheritance pattern, it has been suggested that primary VUR might be caused by mutations in the *PAX2* gene (Eccles *et al.*, 1996). However, several studies have now shown that patients with primary familial VUR do not have mutations in *PAX2*, and VUR is not linked to the *PAX2* locus (Choi *et al.*, 1998; Feather *et al.*, 2000).

4. Hearing Loss

High–frequency hearing loss has been observed in 4/32 (13%) of patients with RCS (Sanyanusin *et al.*, 1995;

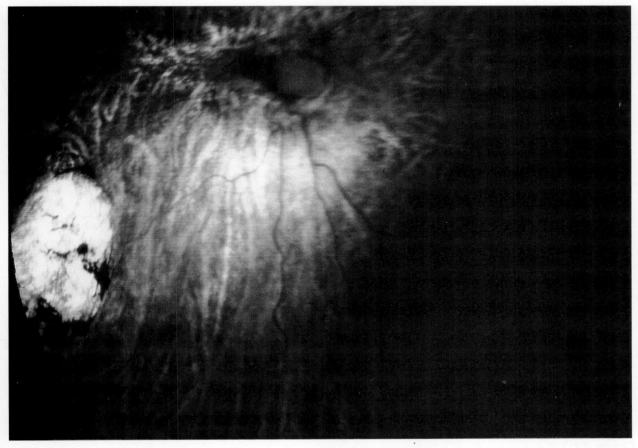

(C)

Figure 23.1 Eye abnormalities associated with *PAX2* mutations. (C) Fundus photograph of the right eye from the aforementioned patient with the retrobulbar cyst. Note the hypoplastic optic nerve and retinochoroidal coloboma. Reproduced with permission from Schimmenti *et al.* (1997).

Schimmenti *et al.*, 1997; Porteous *et al.*, 2000), although the anatomical and molecular basis for hearing loss in patients with *PAX2* mutations is as yet unknown.

5. Central Nervous System Anomalies

Generally, patients with RCS have been found to be of normal or above normal intelligence, although one patient with a *PAX2* mutation and RCS was mentally retarded (Schimmenti *et al.*, 1997). Two unrelated patients with normal intelligence and *PAX2* mutations presented with seizure disorders (Sanyanusin *et al.*, 1995; Schimmenti *et al.*, 1997). A patient with a G619 insertion *PAX2* mutation was found to have hydrocephalus, severe Chiari malformation type 1, and platybasia in addition to RCS (Schimmenti *et al.*, 1999). This phenotype correlates with the expression of *PAX2* at the mid/hindbrain boundary. *PAX2* mutation screening in a small cohort of four anencephalic infants did not reveal mutations in the coding region of *PAX2* by single strand conformational polymorphism analysis (SSCP) (Schimmenti *et al.*, 1999).

6. Skin and Joint Anomalies

Several patients with *PAX2* mutations have presented with "soft" skin and joint laxity (Sanyanusin *et al.*, 1995; Schimmenti *et al.*, 1997). Because *PAX2* expression has not been demonstrated in the skin or in mesodermal structures leading to limb formation, the reason for this is not known.

D. Histologic Features in Kidneys of Patients with PAX2 Mutations

Renal cortical biopsies from patients with RCS and chronic renal failure contained mesangial fibrosis and glomerulosclerosis, involuting glomeruli, hyalinization of glomeruli, atrophic tubules, and hyperplastic glomeruli with an overall decreased number of glomeruli (Fig. 23.3) (Weaver *et al.*, 1988; Schimmenti *et al.*, 1997; Devriendt *et al.*, 1998). Pathologic examination of kidneys revealed cortical thinning, hypoplastic papillae with low numbers of

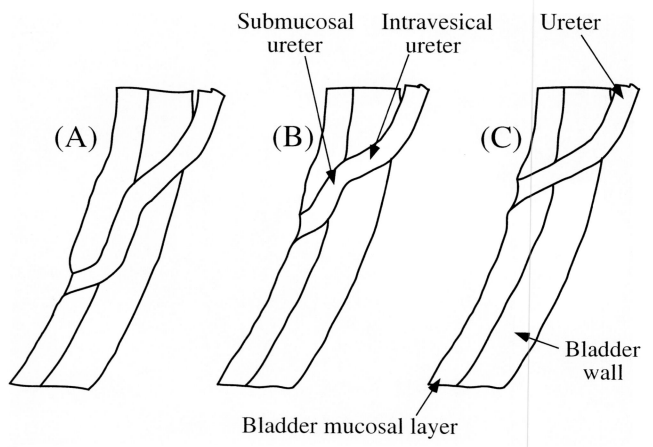

Figure 23.2 Features of nonrefluxing and refluxing ureters from patients with vesicoureteric reflux (VUR). The ureter traverses the bladder wall (intravesical portion, A, B, or C) and tunnels obliquely through the submucosa (submucosal portion, A, B, or C) before exiting at the ureteric orifice. The nonrefluxing ureteric orifice has a longer submucosal tunnel (shown in A) compared with the refluxing orifice in a patient with VUR (shown in B or C). In some patients, the ureteric orifice may be placed extremely laterally on the bladder wall (C). Reproduced with permission from Eccles and Jacobs (2000).

glomeruli and collecting ducts in the cortex and papilla, respectively, consistent with renal hypoplasia (Weaver *et al.*, 1988; Narahara *et al.*, 1997; Devriendt *et al.*, 1998). One patient with RCS presented with a cystic renal mass, which was unilateral (Cunliffe *et al.*, 1998).

Biopsies from patients with oligomeganephronia showed reduced numbers of normal glomeruli that appeared larger than usual (Salomon *et al.*, 1999).

Weaver *et al.* (1988) reported an immune complex disease in the kidneys of a patient with RCS, including interstitial fibrosis, tubular atrophy, and increased mesangial matrix and mesangial cell numbers associated with renal failure (Weaver *et al.*, 1988). A second renal biopsy taken from the same person 7 years later showed an increased amount of collagen-positive connective tissue, increased tubular atrophy, and a marked increase in mesangial matrix and mesangial cell number. There was also periglomerular fibrosis, glomerulosclerosis, electron-dense deposits within the mesangium and subendothelial spaces, and focal podocyte foot process fusion (Weaver *et al.*, 1988).

It is worth noting that several of the histological changes observed in renal biopsies from patients with *PAX2* mutations, which occur in a proportion of individuals with RCS (see Section II,C,3), have also been observed in patients with VUR.

An understanding of the pathologic basis of RCS has been enhanced greatly by a molecular analysis of the *PAX2* gene and other members in the *PAX* gene family. This is covered in the following sections.

III. Molecular Analysis of the *PAX2* Gene and Its Involvement in Renal-Coloboma Syndrome

A. The PAX2 Gene and the PAX Gene Family

The nine *PAX* genes (Fig. 23.4) encode a family of transcription factors with critical roles in embryogenesis (Dahl

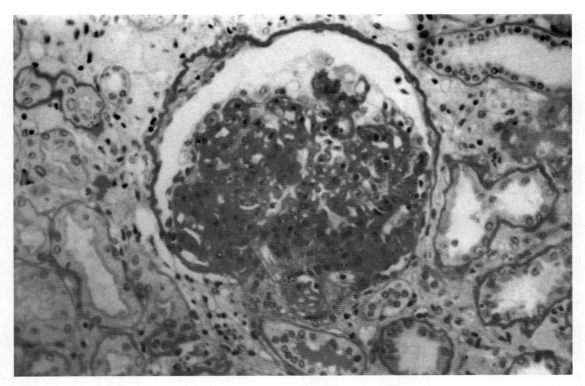

Figure 23.3 Section of a renal biopsy from a patient with renal-coloboma syndrome. The glomerulus exhibits mesangial fibrosis and glomerulosclerosis. PAS stain. ×200.

et al., 1997). *PAX* genes are grouped into four subclasses as defined by the combination of domains that they contain (Mansouri *et al.*, 1996). All *PAX* genes contain the characteristic paired box *(PAX)* DNA-binding domain, whereas some contain an octapeptide domain and a homeodomain (Noll, 1993). *PAX2* is the second member of the *PAX* gene family and is located on human chromosome band 10q24 (Eccles *et al.*, 1992; Narahara *et al.*, 1997). The *PAX2* gene is composed of 12 exons spanning approximately 70 kb of genomic DNA (Sanyanusin *et al.*, 1996). *PAX2* encodes a protein of 43 kDa in size that contains several domains, including the paired box, the octapeptide domain, a partial homeodomain, and a less conserved transactivation domain (Dressler and Douglass, 1992; Eccles *et al.*, 1992). Few direct targets for transcriptional activation by PAX2 are known, although a recently identified target is the *PAX5* gene, especially during mid/hindbrain development (Pfeffer *et al.*, 2000). The following paragraphs describe each of the subdomains of *PAX* genes.

1. Paired Domain

The paired box is a highly conserved domain of 128 amino acids, the function of which is DNA binding (Bopp *et al.*, 1986; Treisman *et al.*, 1991). The paired domain contains two sets of helix-turn-helix DNA-binding motifs, arranged in such a way as to be composed of two distinct subdomains; the C-terminal and the N-terminal (sometimes referred to as the PAI and RED domains, respectively). The crystal structure of the paired domain of *Drosophila* paired protein suggests that the main DNA contacts are via the N-terminal subdomain (Xu *et al.*, 1995). In contrast, studies of human or mouse *PAX* genes involving disruption of the N- or C-terminal subdomains suggest that both N- and C-terminal domains are important in DNA binding (Epstein *et al.*, 1994; Kozmik *et al.*, 1997).

A number of mutations in the *PAX2* gene have been identified within the N-terminal portion of the paired domain (see Section III, C, 1). As has been found for *PAX2*, several other *PAX* genes are mutated in different human diseases (Ton *et al.*, 1991; Baldwin *et al.*, 1992; Glaser *et al.*, 1992; Jordan *et al.*, 1992; Tassabehji *et al.*, 1992). A well-characterized example is mutation of the *PAX6* gene in patients with aniridia (Ton *et al.*, 1991; Glaser *et al.*, 1992; Jordan *et al.*, 1992). The crystal structure of the human PAX6 paired domain–DNA complex revealed that both the N- and the C-terminal subdomains were bound to DNA (Xu *et al.*, 1999). The N-terminal subdomain makes the majority of DNA contacts and is the dominant binding subdomain (Xu *et al.*, 1999). The linker region between the two subdomains of PAX6 is highly ordered and also makes extensive DNA contacts. In the human *PAX6* gene, it was found that mutations in residues involved in DNA binding

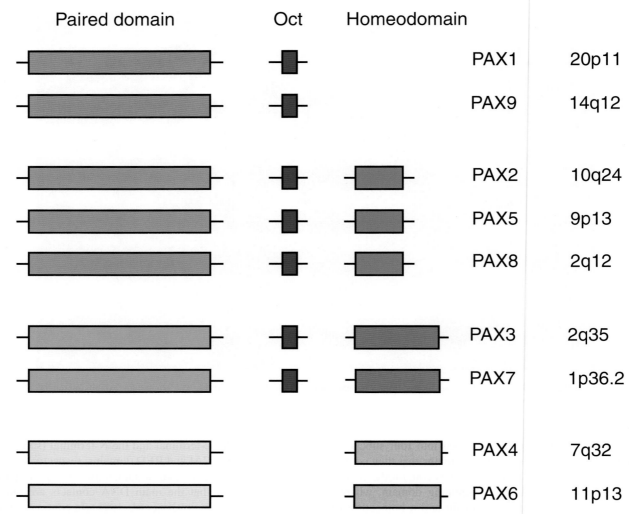

Figure 23.4 Members of the *PAX* gene family. *PAX* genes are grouped into four groups. Group I (red paired domain) consists of *PAX1* and *PAX9*, whereas group II (light green paired domain) consists of *PAX2*, *PAX5*, and *PAX8*. *PAX3* and *PAX7* are in group III (orange paired domain), whereas *PAX4* and *PAX6* are in group IV (yellow paired domain). Several *PAX* genes contain an octapeptide domain (Oct; purple boxes), while others contain a homeobox domain (depicted as green, red, or orange boxes), which is specific for each subclass. The right-hand side shows the chromosomal locations of each of the *PAX* genes in humans. Reproduced with permission from Eccles and Schimmenti (1999).

by the N subdomain occurred in patients with aniridia, whereas patients with missense mutations that allowed some DNA binding had less severe phenotypes (Xu *et al.*, 1999).

2. Homeodomain

Several PAX proteins contain a full homeodomain that functions in DNA binding in cooperation with the paired box (Fortin *et al.*, 1998), whereas the class III PAX proteins (PAX 2/5/8) contain only a partial homeodomain (Ward *et al.*, 1994). However, in class I *PAX* genes, the homeodomain is totally absent (Mansouri *et al.*, 1996). Cooperation among the three DNA-binding units (the N-

and C-terminal subdomains of the paired domain and the homeodomain) allows a variation of binding properties. The partial homeodomain in class III *PAX* genes was previously thought to be a nonfunctional remnant consisting of only the first α helix of the homeodomain, with no DNA-binding activity (Adams *et al.*, 1992). It has now been demonstrated in *PAX5* that the partial homeodomain functions to interact with retinoblastoma protein (RB) and the TATA-binding protein (TBP), providing a link to the cell cycle and transcriptional regulation (Eberhard and Busslinger, 1999). It is likely that other members of class III *PAX* genes *(PAX2/8)* also have the ability to bind RB and TBP.

3. Octapeptide

The octapeptide is a conserved domain present in seven of the nine *PAX* genes (Burri *et al.*, 1989; Mansouri *et al.*, 1996) and is involved in protein–protein interactions. In an interesting comparative study, Ziman and Kay (1998) showed the presence of a conserved nucleotide motif within the octapeptide domain, consisting of TN_8TCCT. It is postulated that this may be an element for recognition by another protein (Ziman and Kay, 1998). The octapeptide domain of *PAX2* is involved in repressing the transcription-promoting activity of the transactivation domain (Lechner and Dressler, 1996).

4. Transactivation Domain

The transactivation domain is located at the carboxy-terminal end of the PAX2 protein and is responsible for the ability of PAX2 to regulate the transcription of target genes (Dorfler and Busslinger, 1996). Transactivation domains in class III *PAX* genes contain a conserved 55 amino acid sequence rich in serine/threonine/proline (PST), composed of both inhibitory and activating regions (Dorfler and Busslinger, 1996). Transcription factors with this sort of transactivation domain are characteristic of those involved in intracellular signaling.

B. Pattern of *PAX2* Expression

The primary site of *Pax2* expression is in the embryo, during brain, eye, ear, spinal cord, urogenital tract, and kidney development (Dressler *et al.*, 1990; Nornes *et al.*, 1990; Eccles *et al.*, 1992; Tellier *et al.*, 2000). In addition, *Pax2* is expressed in a restricted fashion in adult tissues, particularly within the gonads and pancreas (Fickenscher *et al.*, 1993; Oefelein *et al.*, 1996; Ritz-Laser *et al.*, 2000). The following paragraphs describe the expression of *Pax2* during fetal mouse development or within the indicated species. The pattern of *PAX2* expression during human fetal development is less certain because of ethical constraints concerning tissue collection. Nevertheless, three studies of *PAX2* expression during human development have been reported, as mentioned later (Eccles *et al.*, 1992; Terzic *et al.*, 1998; Tellier *et al.*, 2000).

1. *Pax2* in the Developing Brain

Pax2 is first detected in the late primitive streak stage prior to formation of the somites, 7.5 days postcoitum (pc). Expression is distributed diffusely in the anterior region of the mouse embryo and encompasses most of the presumptive forebrain down to the prospective anterior hindbrain (Rowitch and McMahon, 1995). At 8.5 days pc, *Pax2* is detected in the mouse embryonic midbrain (Puschel *et al.*, 1992). At 10 days pc, expression of *Pax2* is observed in the neural tube in a nested expression pattern with *Pax5* and

Pax6, at the mid/hindbrain boundary (Nornes *et al.*, 1990). It has been shown that the expression of *Pax5* is in part dependent on the expression of *Pax2* (Pfeffer *et al.*, 2000), and the expression of both *Pax2* and *Pax5* is necessary to define the midbrain vesicle (Schwarz *et al.*, 1999). *Pax2* expression is also observed in two compartments of cells in the ventricular zone that run either side of the sulcus limitans (midway between the floor plate and the roof plate) and in the ventricular zone along the entire hindbrain and spinal cord (Nornes *et al.*, 1990). In the chick embryo, *Pax2* expression is observed at the mid/hindbrain boundary (isthmus) (Hidalgo-Sanchez *et al.*, 1999), and *Pax2* is required as an organizer for the development of the optic tectum (Okafuji *et al.*, 1999). *Pax2* is also expressed in differentiating interneurons in the spinal cord in chicks (Burrill *et al.*, 1997) and in a subset of GABAergic interneurons in the cerebellum of mice (Maricich and Herrup, 1999).

2. *Pax2* in the Developing Ear

Pax2 expression occurs very early in the anlage of the inner ear, the otic placode (Torres and Giraldez, 1998; Groves and Bronner-Fraser, 2000). The otic vesicle is then formed by invagination of an ectodermal placode in the region of the hindbrain. Following this, *Pax2* is expressed in distinct portions of the developing inner ear along with other transcription factors (Hidalgo-Sanchez *et al.*, 2000). At 9 days pc, *Pax2* is present in the otic vesicles and is restricted to the region flanking the neural tube. Soon after, *Pax2* is localized to the more ventral region of the otic vesicle, which later forms the saccular and cochlear (neurogenic) portions of the ear (Nornes *et al.*, 1990).

3. *Pax2* Expression in the Developing Eye

Pax2 is expressed in the ventral half of the optic cup and stalk and later in the optic disc and nerve (Nornes *et al.*, 1990). *Pax2* is initially observed in the most distal part of the optic vesicle, which on invagination forms the optic cup, closing ventrally at the optic fissure. At this point, expression of *Pax2* is located in cells of the optic fissure, in the retinal tissue surrounding the fissure, and in the ventral half of the optic stalk. *Pax2* expression disappears after the closure of the optic fissure, but is retained in the optic stalk, a transitory structure, providing the substratum for axons to extend from the eye to the diencephalon (forebrain). The optic stalk is gradually replaced by the optic nerve (Macdonald *et al.*, 1997). The expression of *Pax2* in the developing optic nerve is continuous with *Pax2* expressing neuroepithelium in the ventral forebrain.

4. *Pax2* Expression in the Developing Kidney

Pax2 is a primary regulator of the developing urogenital system, the importance of which is demonstrated by the fact that homozygous *Pax2* mutant mice lack ureters, kidneys, and the entire genital tract (Torres *et al.*, 1995). *Pax2* is expressed in two cell lineages during nephrogenesis, in epithelial cells of the Wolffian duct, and in induced nephrogenic mesenchyme (Dressler *et al.*, 1990). At 10 days pc in the mouse, the Pax2 protein is detected in the Wolffian duct, as well as the pronephric and mesonephric tubules (Dressler *et al.*, 1990; Dressler and Douglass, 1992). Metanephric kidney development starts at ~13.5 days pc, when the ureteric bud grows out from the caudal part of the Wolffian duct and contacts the nephrogenic mesenchyme.

Pax2 is expressed in the epithelium of the advancing and branching ureteric bud and in the condensing mesenchyme (Dressler *et al.*, 1990), which is induced through reciprocal interactions with the epithelial bud (Fig. 23.5). Following induction, the condensing mesenchyme differentiates to form renal vesicles, comma-shaped bodies, S-shaped bodies, and subsequently the glomeruli and tubules, which drain into the collecting ducts. *Pax2* expression continues in the differentiating renal vesicle and the early mesenchyme-derived epithelial structures, such as the comma- and S-shaped bodies (Dressler *et al.*, 1990), but expression declines in the proximal part of the S-shaped body from where the podocyte cells originate. Little or no *Pax2* expression is detected in the epithelium of the glomeruli or the proximal or distal tubules. *Pax2* expression continues in this manner around the edge of the developing kidney (the nephrogenic zone) and is downregulated in more differentiated structures. A significant level of *Pax2* expression is detectable in the collecting duct and renal pelvis (Fig. 23.5), but this expression diminishes following birth when the collecting duct is fully mature (Dressler *et al.*, 1990; Eccles *et al.*, 1992). Very little expression of *Pax2* is observed in adult kidney, except in the case of regenerating tubules of adult kidney following kidney damage induced by nephrotoxins, suggesting that adult kidney cells have the ability to reactivate *Pax2* expression during the regenerative process (Imgrund *et al.*, 1999).

As well as contributing to kidney and ureter tissue, the intermediate mesoderm also gives rise to the genital tract, including the oviducts in females and the epididymis, vas deferens, and seminal vesicles in males. *Pax2* is normally expressed in the Wolffian duct, and homozygous mutant mice were found to lack Wolffian duct-derived structures, including the epidydimis, vas deferens, and ductuli efferentes vesicles (Torres *et al.*, 1995). The uterus and vagina also do not form as these tissues are derived from the Müllerian duct, which is also observed to express *Pax2* (Torres *et al.*, 1995). *Pax2* continues to be expressed in the adult genital system (Fickenscher *et al.*, 1993; Oefelein *et al.*, 1996).

C. *PAX2* Mutations in Patients with Renal-Coloboma Syndrome

The identification of *PAX2* mutations in patients with RCS came from an unusual approach to disease gene discovery. Usually the identification of disease genes occurs by positional cloning, or a candidate gene approach. In contrast, mutations leading to RCS were identified by an inverse approach; i.e., the gene was identified first and then the mutant phenotype was identified later by analysis of patients for mutations in *PAX2* (Sanyanusin *et al.*, 1995a,b).

An important clue to the identification of the human *PAX2* mutation syndrome was the expression pattern of *PAX2* during development (see Section III,B) (Dressler *et al.*, 1990; Eccles *et al.*, 1992; Chapter 14). An additional clue was the identification of congenital abnormalities in *Krd* mice carrying a large interstitial transgene-induced deletion of chromosome 19, which included the mouse *Pax2* gene (Keller *et al.*, 1994). *Krd* mutant mice have kidney and retinal defects, including small under-developed kidneys and poorly developed retinas (Keller *et al.*, 1994). Soon after the *Krd* mouse was reported, researchers at the University of Minnesota reported a father and three sons with a potentially novel syndrome involving optic nerve colobomas, high-frequency hearing loss, vesicoureteric reflux, small dysplastic kidneys, and renal failure (Schimmenti *et al.*, 1995).

SSCP analysis of mutations in the Minnesota family led to the identification of a heterozygous frameshift mutation in exon 5 of *PAX2* in affected family members (Sanyanusin *et al.*, 1995b). The identification of this mutation firmly established the causal link between *PAX2* and the disease phenotype in this family. Additional *PAX2* mutations were later identified in families with typical RCS (Sanyanusin *et al.*, 1995a,b; Narahara *et al.*, 1997; Schimmenti *et al.*, 1997; Cunliffe *et al.*, 1998; Devriendt *et al.*, 1998; Porteous *et al.*, 2000).

1. *PAX2* Mutations Identified in Humans to Date

To date, 32 individuals from 13 unrelated families have been identified with a *PAX2* gene mutation (Sanyanusin *et al.*, 1995a,b; Narahara *et al.*, 1997; Schimmenti *et al.*, 1997; Cunliffe *et al.*, 1998; Devriendt *et al.*, 1998; Shim *et al.*, 1999; Porteous *et al.*, 2000). Seven of these patients presented with no family history of RCS. In 6 families, more than one generation was affected, whereas the 2 largest families showed two or three generations of inheritance (Devriendt *et al.*, 1998; Porteous *et al.*, 2000). Interestingly, mutations in the 2 largest families were associated with a less severe phenotype, on average, than in the other families in whom *PAX2* mutations were identified. These observations suggest that in the larger families the mutations show a greater degree of dissemination throughout the families than the other mutations, which may be due to the increased survival of the affected individuals in the larger families as a result of the less severe mutations. While RCS was no doubt a cause of significant childhood mortality in some families, this situation has now changed, presumably as more individuals with RCS survive into adulthood.

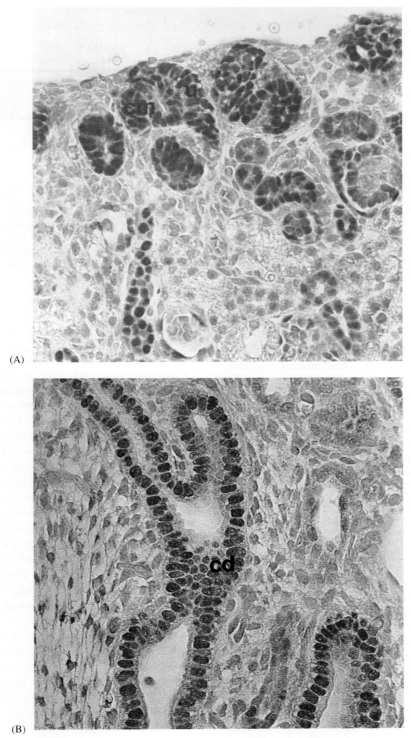

(A)

(B)

Figure 23.5 Pax2 expression in the kidneys of wild-type embryonic day 18 (E18) mice. The photomicrographs show immunohistochemical analysis of Pax2 expression in a fetal kidney section of wild-type mice using an anti-Pax2 antibody (Porteous *et al.*, 2000). (A) A portion from the renal cortex, Pax2-expressing cells, which stained brown in the sections, are in the ureteric bud (u) and the condensing mesenchyme cells (cm). (B) A portion of the renal medulla is shown, and Pax2-expressing cells are in the collecting ducts (cd). Sections were counterstained with methyl green (magnification 100X).

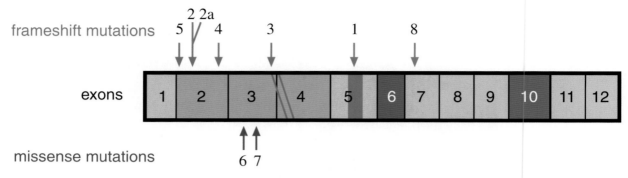

Figure 23.6 Mutations identified in the human *PAX2* gene to date. The *PAX2* gene is depicted as an orange, yellow, and blue bar, divided into 12 segments, corresponding to 12 exons identified in human *PAX2* (Sanyanusin *et al.*, 1996). The paired box (orange segments) and the octapeptide (red box within exon 5) are shown. Exons 6 and 10 (blue boxes) are alternatively spliced. Frameshift and protein-truncating mutations (mutations 1–5 and 8) are depicted above the bar, whereas missense mutations (mutations 6 and 7) are shown below the bar. Mutation 2 is an expansion mutation in a homonucleotide tract of 7 G's (Sanyanusin *et al.*, 1995 a,b; Schimmenti *et al.*, 1997, 1999; Porteous *et al.*, 2000), whereas mutation 2a is a contraction mutation in the same homonucleotide tract (Schimmenti *et al.*, 1999). The translocation mutation (mutation 3) affects intron 3 (Narahara *et al.*, 1997), but is represented by the double parallel line across exon 4. Further details on each mutation are presented in the legend of Fig 23.7. Reproduced with permission from Eccles and Schimmenti (1999).

In total, nine different *PAX2* gene mutations have been reported (summarized in Fig. 23.6). These mutations include five frameshift mutations, a missense mutation, a hexanucleotide duplication, a chromosomal translocation, and a nonsense mutation (Sanyanusin *et al.*, 1995a,b; Narahara *et al.*, 1997; Schimmenti *et al.*, 1997; Cunliffe *et al.*, 1998; Devriendt *et al.*, 1998; Shim *et al.*, 1999; Porteous *et al.*, 2000). Most of the mutations are clustered within exons 2 and 3, which encode the N-terminal portion of the paired box domain. Mutations have also been observed in the octapeptide domain (exon 5) (Sanyanusin *et al.*, 1995b) and in exon 7 in the transactivation domain (Porteous *et al.*, 2000).

2. A "Hot-Spot" Mutation in *PAX2*

One mutation in particular in *PAX2*, the G619 insertion mutation, has occurred at least five times independently in unrelated families (Sanyanusin *et al.*, 1995a; Schimmenti *et al.*, 1997; Shim *et al.*, 1999; Porteous *et al.*, 2000). This G-insertion mutation occurs within a poly-G nucleotide tract of 7 G's and expands the sequence to 8 G's. It is thought that DNA polymerase slippage during replication is the most likely cause of the mutation, as this phenomenon has also been observed with mutations in other genes. A contraction mutation to 6 G's has also been identified in the same poly-G tract in one patient (Shim *et al.*, 1999), and the *Pax2*[1Neu] mouse (which carries a spontaneous mutation in the mouse *Pax2* gene) also has a G-insertion mutation in the same tract (Favor *et al.*, 1996). Therefore, the poly-G tract in exon 2 appears to be a "hot spot" for new mutations in *PAX2*, arising from both contractions and expansions of the homonucleotide tract.

3. Effect of Mutations in *PAX2*

Frameshift mutations lead to stretches of an abnormal reading frame followed by a premature stop codon. The occurrence of frameshift mutations early in the paired domain coding region is predicted to result in an extremely truncated open reading frame, whereas missense mutations or hexanucleotide duplications within the paired box DNA-binding domain would presumably result in a protein with decreased DNA-binding potential. The reason for the relative preponderance of either frameshift or nonsense mutations resulting in the production of truncated PAX2 proteins is not known. It is possible that missense mutations produce either a much less severe phenotype or a different phenotype. An alternative hypothesis is that the protein truncation mutations have a dominant-negative effect, obscuring the function of the protein from the normal allele, whereas missense mutations associated with altered amino acids are less likely to produce this phenotype. The predicted mutant PAX2 proteins arising from *PAX2* gene mutations are shown in Fig. 23.7, although the nature of the mutant proteins has not yet been characterized. It is presumed, based on the dosage-sensitivity hypothesis (see discussion in Section III, E on haploinsufficiency), that truncated mutant PAX2 proteins exhibit little or no functional activity.

4. The *PAX2* Mutation Database

Most of the *PAX2* mutations identified to date are associated with a variable phenotype. Consequently, a human *PAX2* mutation database has been created as a reference source for information about *PAX2*, including all known poly-

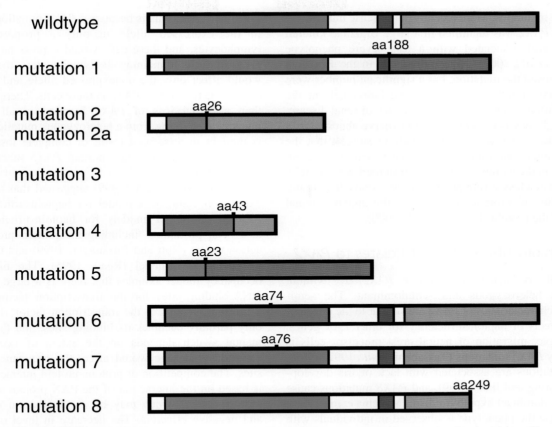

Figure 23.7 Predicted mutant PAX2 proteins associated with *PAX2* gene mutations reported to date in patients with renal-coloboma syndrome. Boxed segments represent different protein domains. The paired box domain (green portion), the octapeptide sequence (dark blue), and the transactivation domain (orange) are shown. Protein sequences that differ from the wild-type sequence result from a frameshift mutation in the *PAX2* gene and are depicted by red, purple, brown, and lilac-colored boxes. These mutant proteins (mutations 1, 2, 4, and 5) are truncated. Wild type is the normal wild-type PAX2 protein. Mutation 1 is a cytosine deletion at position 1104 in the octapeptide domain in exon 5, resulting in a frameshift (Sanyanusin *et al.*, 1995b); mutations 2 and 2a are a guanosine nucleotide insertion and deletion, respectively, at position 619 in exon 2 resulting in a frameshift (Sanyanusin *et al.*, 1995a; Schimmenti *et al.*, 1997, 1999; Porteous *et al.*, 2000); mutation 3 is a translocation (10;13)(q24;q12.3) with a breakpoint within intron 3 or 4 of the *PAX2* gene(Narahara *et al.*, 1997); mutation 4 is a 22 base deletion in exon 2 resulting in a frameshift (Schimmenti *et al.*, 1997); mutation 5 is a thymidine deletion at position 611, which results in a frameshift (Cunliffe *et al.*, 1998); mutation 6 is a missense guanosine-to-adenosine mutation at position 719 in exon 3 (Devriendt *et al.*, 1998); mutation 7 is a hexanucleotide duplication at positions 763–768 in exon 3 resulting in duplication of glutamic acid and threonine residues at amino acid positions 74 and 75 (Devriendt *et al.*, 1998); and mutation 8 is a nonsense mutation at position 1289, resulting in a stop-codon insertion at amino acid 249 (Porteous *et al.*, 2000). Reproduced with permission from Eccles and Schimmenti (1999).

morphisms and mutations. The URL for the human *PAX2* allelic variant database web site is http://www.hgu.mrc.ac. uk/Softdata/PAX2/. This database allows investigation of phenotypic features associated with *PAX2* mutations, as well as features of the *PAX2* gene sequence and related data.

D. Correlations between Genotype and Phenotype for Mutations in *PAX2*

Inspection of the mutations and their locations in *PAX2* shows that in most cases neither the severity nor the exact nature of the abnormalities correlates with the position of

the mutation within the *PAX2* gene. Also, the occurrence of anomalies such as CNS abnormalities, auditory abnormalities, or vesicoureteric reflux does not correlate with either the type of mutation or the position of the mutation. However, three mutations in *PAX2* have been clearly associated with a less severe phenotype. Patients with a missense mutation or a hexanucleotide duplication in exon 3 were either much older at the time of diagnosis or had symptoms that were less severe, on average, than patients with a frameshift mutation in *PAX2* (Devriendt *et al.*, 1998). Perhaps the mutant PAX2 protein in these patients was more functional than the truncated proteins associated with frameshift mutations. More recently, a nonsense mutation in exon 7 of *PAX2* was identified in a large Brazilian kindred that was also associated with a less severe phenotype (Porteous *et al.*, 2000). Nine individuals in the Brazilian kindred carried the mutation, but a significant number were affected only mildly. Moreover, some individuals in the kindred with the mutation had no evidence of renal disease, even though they had identifiable optic nerve abnormalities with normal visual acuity. These findings suggest that the nonsense mutation, which is predicted to cause a protein truncation in the carboxy-terminal half of the PAX2 protein, may result in a less severe phenotype on average than mutations leading to protein truncation in the amino-terminal portion of the protein (Porteous *et al.*, 2000).

E. Haploinsufficiency as It Relates to *PAX2*

Naturally occurring mutations in *PAX2* are, without exception, heterozygous and semidominant. The semidominant characteristics of *PAX2* are similar to the developmental effects of haploinsufficiency for other *PAX* genes. For example, migration of neural crest precursor cells is affected by *PAX3* mutations (Tassabehji *et al.*, 1992, 1993). *PAX6* mutations are associated with lack of iris development (Gehring and Ikeo, 1999), and *PAX8* mutations cause autosomal-dominant hypothyroidism (Macchia *et al.*, 1998). In each case the phenotype is observed in individuals with heterozygous mutations. In animal models (see discussion in Section IV,A,1) the extent of the mutant phenotype when both alleles of *Pax2* are mutated is much more severe than for the heterozygous phenotype and suggests that *Pax2* is partially functional when a single wild-type allele is present. This phenomenon is referred to as haploinsufficiency. It is believed that most mammalian genes are transcribed from both alleles, meaning that for recessive genes, if a mutation occurs in one allele, then the remaining allele will produce sufficient levels of the transcript and protein to maintain function. However, for dominant genes, a mutation in one allele disrupts the function of the gene, probably because there is a critical reliance on the full complement of gene product to maintain proper function. Haploinsufficient phenotypes are semidominant when one allele is mutated. The phenotype is fully expressed only when both alleles are mutated, which indicates a critical reliance on gene dosage. However, the notion that most genes are transcribed from both alleles may not necessarily be true.

Nutt and Busslinger (1999) and Nutt *et al.* (1999) make a strong argument for the cause of haploinsufficiency associated with *PAX5* mutations. They show that *PAX5* is expressed from one randomly selected allele during certain stages of B lymphocyte development (Nutt *et al.*, 1999). According to Nutt and Busslinger (1999), the combination of random monoallelic expression of *PAX5* and a mutation of one allele causes *PAX5* haploinsufficiency in

B lymphocytes. This is because a *PAX5* mutation would be in the expressed allele in only a proportion of B lymphocytes, and these cells would express no functional *PAX5*. In the remaining B lymphocytes, the mutation would affect only the nonexpressed allele and would not disrupt the function of *PAX5* in these cells. Therefore, rather than a partial loss of *PAX5* function in all cells, the phenotype resulting from a heterozygous mutation in *PAX5* is thought to arise as a result of complete loss of *PAX5* function in some cells and normal *PAX5* function in the remaining cells.

Nutt and Busslinger (1999) suggested that monoallelic expression could be a model for haploinsufficieny of all *PAX* genes. Other models for haploinsufficiency have also been postulated, including the protein–protein interaction model (Nutt and Busslinger, 1999) and the binding site occupancy model (Read, 1995). The binding site occupancy model assumes that there is a large number of DNA-binding sites for the transcription factor but only a limiting amount of the transcription factor. The protein only partially occupies its binding sites, and the developmental switch depends on the extent of occupation; a variation of dosage would affect the occupancy of some sites. The competition or protein–protein interaction model is based on the interaction of the PAX protein with one or more partners, which may be a combination of positive and negative effectors. The decrease in level of the PAX protein would affect many downstream genes due to the partners. Interestingly, there is already evidence that PAX proteins may interact with Ets-1 (Fitzsimmons *et al.*, 1996; Wheat *et al.*, 1999) and Rb-1 (Wiggan *et al.*, 1998; Eberhard and Busslinger, 1999). The monoallelic model suggested by Nutt and Busslinger (1999) allows for extreme variance in the penetration of mutant phenotypes observed in patients with *PAX* gene mutations because the ratio of inactivated mutant to wild-type alleles could be set randomly. This could explain the large variance in phenotypes observed in patients who have inherited the same *PAX2* mutation. Porteous *et al.* (2000) have, however, demonstrated that *Pax2* is expressed biallelically in *Pax2*[1Neu] mouse kidneys, suggesting that the monoallelic expression observed in *PAX5* does not occur with *Pax2*. Patients with RCS are therefore thought to have partial function of *PAX2* in all cells, rather than a total lack of expression in some cells as would occur in the monoallelic expression model.

IV. Animal Models to Investigate *PAX2* Function

A. Animal Models for Loss of *Pax2* Function

Several animal models have been developed in which the *Pax2* homologous gene has been disrupted. These include several strains of *Pax2* mutant mice, a zebrafish model with a mutation of the *no-isthmus (Pax2)* allele, an insect model, *Drosophila melanogaster*, with mutations of the *Sparkling* or *Shaven* alleles, and a worm model, *C. elegans*, with a mutation of the *Egl-38* gene. Investigation of the developmental abnormalities in these animal models may give clues as to the molecular mechanisms surrounding the *PAX2* mutation in humans with RCS.

1. Mouse Models of Renal-Coloboma Syndrome

Homozygous and heterozygous *Pax2* mutations in mice are associated with ocular and urogenital anomalies. Three different *Pax2* mutant mouse models have been described: *Krd* (Keller *et al.*, 1994), *Pax2*[1Neu] (Favor *et al.*, 1996), and a targeted disruption of the *Pax2* gene (Torres *et al.*, 1995, 1996). Mice heterozygous for a *Pax2* mutation have kidney and eye abnormalities similar to those in humans with a *PAX2* mutation, including a marked reduction in kidney size with reduced metanephric epithelial transformation and metanephric tubule formation, sometimes with renal agenesis or renal cysts, and dilated ureters (Chapters 14, 20).

The lack of kidney development in heterozygous *Pax2* mutant mice was similar to changes seen in mouse kidney organ cultures after the addition of *Pax2* antisense oligonucleotides (Rothenpieler and Dressler, 1993). In these cultures the degree of growth was inhibited, including an inhibition of the mesenchyme to epithelium transformation during early kidney development.

a. Krd Mutant Mice   *Krd* mice carry a chromosomal deletion induced by the insertion of a transgene on mouse chromosome 19, which involves loss of the entire *Pax2* gene, as well as approximately 7 cM of flanking DNA (Keller *et al.*, 1985). *Krd* mice exhibit a range of aplastic, hypoplastic, and cystic kidneys. Heterozygous mutant mice with small kidneys exhibit a decreased number of glomeruli and an attenuated nephrogenic zone. Eye defects in *Krd* mice involved abnormalities in electroretinograms, suggesting alterations in photoreceptor and bipolar cell function, and reduction of cell numbers in the inner cell and ganglion cell layers (Keller *et al.*, 1985). *Krd* mutant mice had altered morphogenetic movements of *Pax2*-expressing epithelial cells in the posterior optic cup and optic stalk. As a result, the optic fissure was not formed properly, and there was disorganization of the ganglion cell axons in the optic disc (Otteson *et al.*, 1998). Homozygous *Krd* mutants were preimplantation lethal (Keller *et al.*, 1985), which is not surprising considering that the 7 cM deletion may encompass some 400 loci.

b. Pax2[1Neu] *and targeted Pax2 Knockout Mice*   Both *Pax2*[1Neu] and targeted *Pax2* knockout mice have mutations that affect the amino-terminal portion of the Pax2 protein within the paired box region. *Pax2*[1Neu] mice carry a spontan-

eously occurring G-insertion mutation within a string of 7 G's in exon 2 (Favor *et al.*, 1996), identical to the G619 insertion mutation in humans. In targeted *Pax2* knockout mice the Pax2 gene contains a neomycin cassette, which has been inserted between exons 1 and 2, resulting in deletion of the intervening DNA (Torres *et al.*, 1995).

Kidney defects similar to those seen in *Krd/+* mice were noted in *Pax2*[1Neu/+] and targeted heterozygous mutant mice, whereas genital tracts appeared normal (Favor *et al.*, 1996). Kidney hypoplasia in targeted heterozygous *Pax2* mutant mice was mainly due to reduced calyces and poor development of the upper part of the ureter, implicating defects in ureteric branching and/or growth. The associated expression of Pax2 in *Pax2*[1Neu/+] mutant mice was spatially and temporally normal, although apparently reduced in level (Fig. 23.8). The optic discs showed colobomatous changes similar to optic nerve coloboma in humans (Torres *et al.*, 1995).

Homozygous mutant *Pax2*[1Neu] mice and *Pax2* knockout mice died at or soon after birth, with agenesis of the urogenital tract, including lack of kidneys and abnormalities of the midhindbrain, inner ear, and optic nerve tracts (Favor *et al.*, 1996; Torres *et al.*, 1996). The Wolffian duct in homozygous mutant animals appeared normal at 9–10 days pc, but failed to extend caudally, degenerating at 12.5 days pc. The homozygous mutants lacked mesonephric tubules, suggesting a requirement for *Pax2* in epithelial transformation of the nephrogenic cord. Failed development of the mesonephric tubules and the posterior parts of the Müllerian and Wolffian ducts in homozygous mutants meant that the ureteric bud could not branch from the posterior Wolffian duct.

The homozygous *Pax2* mutant mice lacked not only the ureters and kidneys, but also parts of the genital tract, including oviducts in the females and the epididymis, vas deferens, and ductuli efferentes in the males. Gonads were present, but were surrounded by a blunt-ended remnant of the genital ridge (Favor *et al.*, 1996; Torres *et al.*, 1996). The uterus and vagina also did not form, as these tissues are derived from the Müllerian duct, which expresses *Pax2* during development and degenerates prematurely in homozygous female *Pax2* mutant mice. Tissues affected by the knockout mutation of *Pax2* were derived from the intermediate mesoderm and included the Wolffian duct in males and the Müllerian duct in females, whereas endoderm-derived structures such as the urethra, bladder, and prostatic glands appeared normal. As well as contributing to kidney and ureter tissue, the intermediate mesoderm also gives rise to portions of the genital tract.

Homozygous *Pax2* mutant mice also exhibited exencephaly resulting from failure of the closure of the neural folds at the midbrain, as well as an enlarged chamber of the inner ear, abnormal organs of Corti, and an abnormal cochlear region that lacked innervation (Torres *et al.*, 1996). Homozygous *Pax2* mutant mice showed failure of the optic fissure to close, resulting in bilateral coloboma in *Pax2*

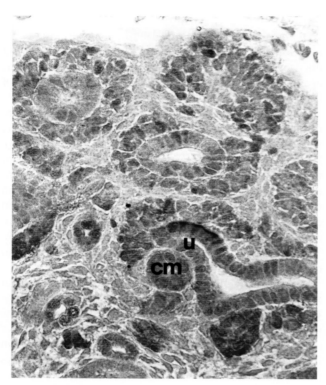

Figure 23.8 Pax2 expression in kidneys of Pax2 heterozygous mutant embryonic day 18 mice. The photomicrograph shows immunohistochemical analysis of Pax2 expression in a fetal kidney section of $Pax2^{1Neu/+}$ mutant mice using an anti-Pax2 antibody and diaminobenzidine staining (Porteous *et al.*, 2000). Pax2-expressing cells, which stained brown in the sections, are of cells in the collecting duct and ureteric bud (u), as well as the condensing mesenchymal cells (cm). Lighter staining with the antibody as compared with wild-type kidneys (see Fig. 23.5) was noted, as well as fewer mature epithelial structures. Sections were counterstained with methyl green (magnification 100X).

homozygous null mice. Axons of the optic nerve were reduced in number and were unable to form a chiasm and hence projected only ipsilaterally (Torres *et al.*, 1996). From the nature of the tissues affected in homozygous mutant mice, it seems that *Pax2* is required during multiple stages of embryonic development.

2. Chicken, Zebrafish, *Xenopus*, *Drosophila*, and *C. elegans* Models of *Pax2* Function

The chick is a useful model to investigate *Pax2* function because the developmental processes involving *Pax2* occur in the fertilized egg. It has been found that during neuronal differentiation in chicks, *Pax2* expression specifies neuronal differentiation to a particular fate (Baker and Bronner-Fraser, 2000). It has also been shown in the chick that *Pax2* plays an important role in development of the inner ear (Hutson *et al.*, 1999) and in formation of the mid/hindbrain junction (Zhang *et al.*, 2000).

Zebrafish have two *Pax2* genes, *Pax2.1* and *Pax2.2*, and also *Pax5* and *Pax8* genes (Pfeffer *et al.*, 1998). In zebrafish

the "*no isthmus*" (*noi*) mutation in the *Pax2.1* gene is associated with failed development of the mid/hindbrain boundary (isthmus) (Lun and Brand, 1998) and of the optic stalk (Pfeffer *et al.*, 1998). The *noi* mutation is also associated with disrupted development of the pronephros. Fish do not have the equivalent of a mammalian metanephros, and the phenotype does not exactly resemble the renal abnormalities in *Pax2* mutant mice. *Noi* mutants lack pronephric tubules, whereas glomerular cell differentiation is unaffected (Majumdar *et al.*, 2000). In addition, the *noi* mutation affects hair cell development in the ear, such that *noi* mutants have double the number of hair cells. Riley *et al.* (1999) suggested that in hair cell development, Pax2.1 is epistatic to both Delta and Notch signaling pathways.

Development of the visual system in *Xenopus* depends on the correct expression of *Pax2*. For example, the injection of *Shh* mRNA into *Xenopus* embryos results in expanded *Vax1* and *Pax2* domains at the expense of the *Pax6* expression domain and results in dysgenesis of the optic nerve (Hallonet *et al.*, 1999). The expression patterns of orthologues of the *Pax* genes can vary in a species-specific manner. For example, *Pax8* is expressed in *Xenopus* auditory and excretory system development, whereas *Pax2* is expressed in *Xenopus* thyroid, an organ that expresses *Pax8* in mammals (Heller and Brandli, 1999).

A homologue of the vertebrate *Pax2* gene in *Drosophila melanogaster*, called *D-Pax2*, has been identified in which a *Pax2* enhancer mutation called *Sparkling* results in eye abnormalities, indicating a role for *Pax2* in normal development of the *Drosophila* eye (Fu and Noll, 1997). Therefore, like *Pax6*, the involvement of *Pax2* in eye development may have been conserved through evolution. *D-Pax2* is also expressed in developing and mature bristles, which are mechanosensory organs covering the body surface of the fly. A loss-of-function mutation in a second enhancer of *D-Pax2*, called *Shaven* (sv), causes failure of development of the shaft structures of the bristle (Fu *et al.*, 1998), and misexpression of *D-Pax2* can induce the development of ectopic shaft structures (Kavaler *et al.*, 1999). It is thought that *D-Pax2* acts downstream of the Notch signaling pathway (Kavaler *et al.*, 1999).

The *Egl-38* phenotype in the worm *C. elegans* was shown to be due to a mutation in a *Pax2/5/8* homologous gene (Chamberlin *et al.*, 1997). This mutant phenotype involves defective development of the male tail and the egg-laying apparatus in hermaphrodites.

These animal studies show that *Pax2* has been involved in embryogenesis over a long evolutionary time scale and that *Pax2* was recruited in several different tissues, depending on the phylogenetic background. In vertebrates, the role of *Pax2* appears to be conserved in eye, genitourinary tract, and CNS development, suggesting that extrapolation from studies of *Pax2* in mice, for example to the role of *PAX2* in human development, is a reasonable assumption.

B. A Mouse Model of *Pax2* Gain of Function

Transgenic mice overexpressing *Pax2* during kidney development had cystic kidneys, demonstrating the need for *Pax2* repression during the differentiation of kidney cells (Dressler *et al.*, 1993). Transgenic *Pax2* offspring exhibited open eyes at birth, decreased renal function, proteinuria, glomerular atrophy, lack of podocyte foot processes, immature endothelial fenestrae, dilated proximal tubules, and multifocal microcystic tubular dilatation (Dressler *et al.*, 1993). The renal phenotype in the transgenic mice was similar to congenital nephrotic syndrome, although the mice died 3–4 days after birth, and *Pax2* was not linked to this condition in humans. These studies suggest that deregulated expression of *Pax2* is incompatible with kidney differentiation.

V. What Is the Function of *PAX2* in Kidney Development?

Although renal-coloboma syndrome is commonly associated with end-stage renal failure, little is known about how *PAX2* mutations could cause renal failure. The PAX2 protein binds DNA and regulates gene expression, but its exact function during fetal development is unknown. Porteous *et al.* (2000) analyzed patients with RCS in an attempt to identify the underlying abnormality and concluded that renal hypoplasia was the most common congenital renal abnormality in RCS patients. Furthermore, on investigating the cause of the renal hypoplasia, it was found that renal hypoplasia in *Pax2* mutant mice was associated with excessive amounts of apoptosis during development (Porteous *et al.*, 2000; Torban *et al.*, 2000). In contrast, there was no significant difference observed in the proliferation rate of fetal kidney cells between mutant and wild-type mice, suggesting that the increased level of apoptosis may be the main cause of the renal hypoplasia in *Pax2* mutant mice, (Fig. 23.9).

In *Pax2* mutant mice, the effect of the *Pax2* mutation was a decreased rate of new nephron induction, as the total number of early epithelial structures (at the tips of ureteric buds) and glomeruli (representing more advanced nephrons) was reduced strikingly in the mutant kidneys. In contrast, the nephrons that were formed, although reduced in number, appeared to have normal morphology (Porteous *et al.*, 2000; Torban *et al.*, 2000).

Apoptosis is known to occur during kidney development (Koseki *et al.*, 1992), but it is significant to note that the

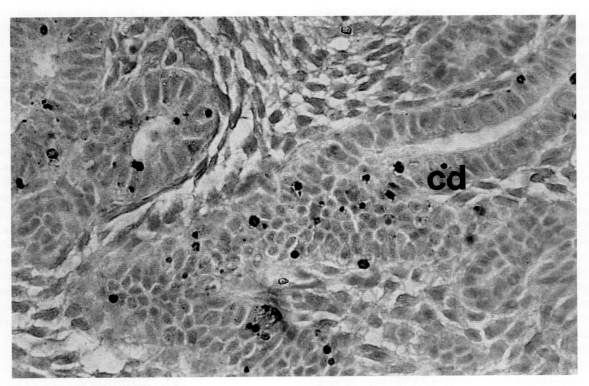

Figure 23.9 Excessive programmed cell death in developing kidneys of fetal mice with a *Pax2* mutation. Apoptotic cells (brown-black spots) can be seen in the collecting duct epithelia (cd) of kidneys from *Pax2*[1Neu] heterozygous mutant embryonic day 15 mice (Porteous *et al.*, 2000). Apoptotic cells were detected by the TUNEL (TdT-mediated dUTP nick-end labeling incorporating biotin dATP) reaction followed by detection with extravidin peroxidase, diaminobenzidine staining, and counterstaining with methyl green (magnification 100X).

collecting ducts of $Pax2^{1Neu}$ mutant mice had nine-fold higher levels of apoptosis than in wild-type fetal kidneys at the same age (Porteous *et al.*, 2000). It is possible that enhanced apoptosis is associated with the degeneration of and the lack of caudal extension of the Wolffian duct in homozygous *Pax2* mutant mice (Torres *et al.*, 1995). If enhanced apoptosis in the Wolffian duct was able to prevent its formation during early fetal development, then this may explain the lack of development of the epididymis, vas deferens, ureteric epithelium, and kidneys in *Pax2* homozygous mutants. These structures all require proper formation of the Wolffian duct.

In agreement with the observed increase in apoptosis in *Pax2* mutant mouse kidneys, Torban *et al.* (2000) showed that antisense *PAX2* expression constructs compromised the survival of cultured collecting duct cells, called mIMCD-3, which express high levels of *Pax2*. In contrast, COS-7 or HEK-293 cells, which do not express *PAX2*, were not affected by the antisense *PAX2* construct. If, however, HEK-293 cells containing an inducible *PAX2* expression construct were induced, and the cells transfected with a proapoptotic gene encoding caspase-2, it was found that the presence of *PAX2* promoted an increase in survival of these cells (Torban *et al.*, 2000).

It is interesting to compare the aforementioned data with mutant mice. In *Pax2* mutant mice, the amount and localization of the apoptosis correlated well with the known expression patterns of *Pax2* in the ureteric epithelium. However, apoptosis was not observed in mesenchyme-derived epithelial structures, even though *Pax2* was expressed in the differentiating mesenchyme and early mesenchyme-derived structures (Porteous *et al.*, 2000; Torban *et al.*, 2000). Lack of apoptosis in *Pax2*-expressing mesenchyme structures may be due to the fact that *Pax8* was co-expressed with *Pax2* in the differentiating mesenchyme (Eccles *et al.*, 1995). It is therefore possible that the expression of *Pax8* in mesenchymally derived structures of kidneys was able to compensate for the heterozygous *Pax2* mutation. This phenomenon has been described for the *Pax3* gene (Borycki *et al.*, 1999). The expression of *Pax* genes may therefore directly modulate the survival of cells.

Pax2 expression has been found to be highest in cells that are actively dividing during kidney development (Dressler and Douglass, 1992; Eccles *et al.*, 1992), and in some renal tumors and cystic kidney tissue, *Pax2* expression is strongly associated with proliferating cells (Gnarra and Dressler, 1995; Winyard *et al.*, 1996). In addition, following nephro-toxic kidney damage, *Pax2* expression was associated with regenerating kidney cells (Imgrund *et al.*, 1999). It is possible that the direct involvement of PAX2 in cell survival is permissive for cell proliferation through the trophic action of growth factors or branching morphogens.

The downstream gene targets of *Pax2* are largely unknown, and it is not clear whether Pax2 directly regulates genes in apoptotic pathways during kidney development. Interestingly, Pax2, Pax5, and Pax8 have all been reported to inhibit *p53* transcription (Stuart *et al.*, 1995), which can promote apoptosis (Levine, 1997). Furthermore, transgenic mice overexpressing wild-type p53 have altered differentiation of the ureteric bud and have small kidneys (Godley *et al.*, 1996). In addition, *Pax3* mutant mice have been shown to have enhanced apoptosis in the somitic mesoderm during embryonic development (Borycki *et al.*, 1999), and treatment of rhabdomyosarcoma cells lines with *PAX3* antisense oligonucleotides has been shown to result in apoptosis (Bernasconi *et al.*, 1996). *PAX3* has been shown to be able to transcriptionally activate the antiapoptotic protein BCL-XL (Margue *et al.*, 2000). *Pax5* has also been identified to have an antiapoptotic role, and in *Pax5* knockout mice, B cells fail to differentiate and undergo apoptotic cell death at an early stage (Urbanek *et al.*, 1994). Similarly, the *eyeless* mutation *(ey²)* (homologous to the mammalian *Pax6* gene) has been associated with large-scale apoptosis of photoreceptor precursors in the eye discs of *Drosophila* (Halder *et al.*, 1998).

Why is *Pax2* involved in regulating cell survival, considering that it has a strict dependence on gene (and presumably protein) dosage? Would this not cause a tenuous balance of cell fate in the kidney? One potential reason for the dosage sensitivity of *PAX2* is that in tissues such as kidney, brain, and optic nerve there are abundant pluripotent precursor cells, and that to form the final organ of the appropriate size with the right number of mature structures, a finely balanced amount of apoptosis must occur. This hypothesis would ascribe a function of *PAX2* associated with regulating the total cell number, or mass of a differentiated tissue. This may mean that there is a complex interplay among proliferation, apoptosis, and differentiation during the process of organogenesis. To form a structure, there is a need for commitment, proliferation, and organization of the cells. If there is not enough proliferation or too much apoptosis, then there may not be enough cells to form the structure, and the structure might then be aborted or form a blind vestige. Therefore, a balance between proliferation and apoptosis may be important for the developmental process. Cell number may be regulated by gene and protein dosage, and the same genes may have a concomitant role in cell differentiation. Aside from regulating cell survival in the kidney, *Pax2* may also function in other ways during the developmental process, although these functions have yet to be identified.

VI. Summary

PAX2 mutations underlie the pathological mechanism causing RCS, but exactly how the mutations cause the morphological changes associated with RCS is not yet clear.

Patients with RCS show significant phenotypic variation, even among those with the same mutation. Regarding *PAX* gene function, there is a need to better understand the mechanism of dosage sensitivity shown by *PAX* genes. It is likely that animal models of *Pax2*-loss-of-function will continue to prove useful in determining *Pax2* function and will also undoubtedly provide a fascinating glimpse into the mechanisms of organ development and disease. Of particular interest is the observation that *Pax2* seems to reside at the top of a hierarchy of genes controlling kidney development. It will therefore be of great interest to determine the molecular pathways that *Pax2* regulates and how these pathways impact on kidney development and disease.

References

Adams, B., Dorfler, P., Aguzzi, A., Kozmik, Z., Urbanek, P., Maurer-Fogy, I., and Busslinger, M. (1992). *Pax-5* encodes the transcription factor BSAP and is expressed in B lymphocytes, the developing CNS, and adult testis. *Genes Deve.* **6**, 1589–1607.

Al-Gazali, L. I., Bakir, M., Hamid, Z. M., Nair, D. K., Haas, D., Amirlak, I., and Rushdi, A. (2000). A new syndrome of optic nerve colobomas and renal abnormalities associated with arthrogryposis multiplex. *Clin. Dysmorphol.* **9**, 183–188.

Antignac, C., Arduy, C. H., Beckmann, J. S., Benessy, F., Gros, F., Medhioub, M., Hildebrandt, F., Dufier, J. L., Kleinknecht, C., Broyer, M., Weissenbach, J., Habib, R., and Cohen, D. (1993). A gene for familial juvenile nephronophthisis (recessive medullary cystic kidney disease) maps to chromosome 2p. *Nature Genet.* **3**, 342–345.

Baker, C. V., and Bronner-Fraser, M. (2000). Establishing neuronal identity in vertebrate neurogenic placodes. *Development* **127**, 3045–3056.

Baldwin, C. T., Hoth, C. F., Amos, J. A., Da-Silva, E. O., and Milunsky, A. (1992). An exonic mutations in the HuP2 paired domain gene causes Waardenburg's syndrome. *Nature* **355**, 637–638.

Bernasconi, M., Remppis, A., Fredericks, W. J., Iii, F. J. R., and Schafer, B. W. (1996). Induction of apoptosis in rhabdomyosarcoma cells through down-regulation of PAX proteins. *Proc. Natl. Acad. Sci. USA.* **93**, 13164–13169.

Bopp, D., Burri, M., Baumgartner, S., Frigerio, G., and Noll, M. (1986). Conservation of a large protein domain in the segmentation gene paired and in functionally related genes of Drosophila. *Cell* **47**, 1033–1040.

Borycki, A. G., Li, J., Jin, F., Emerson, C. P., and Epstein, J. A. (1999). Pax3 functions in cell survival and in pax7 regulation. *Development* **126**, 1665–1674.

Bron, A. J., Burgess, S. E. P., Awdry, P. N., Oliver, D., and Arden, G. (1988). An inherited association of optic disc dysplasia and renal disease: Report and review of the literature. *Ophthalm. Paediatr.* **10**, 185–198.

Burri, M., Tromvoukis, Y., Bopp, D., Frigerio, G., and Noll, M. (1989). Conservation of the paired domain in metazoan and its structure in three isolated human genes. *EMBO J.* **8**, 1183–1190.

Burrill, J. D., Moran, L., Goulding, M. D., and Saueressig, H. (1997). PAX2 is expressed in multiple spinal cord interneurons, including a population of EN1+ interneurons that require PAX6 for their development. *Development* **124**, 4493–4503.

Chamberlin, H. M., Palmer, R. E., Newman, A. P., Sternberg, P. W., Baillie, D. L., and Thomas, J. H. (1997). The PAX gene egl-38 mediates developmental patterning in Caenorhabditis elegans. *Development* **124**, 3919–3928.

Choi, K. L., Mcnoe, L. A., French, M. C., Guilford, P. J., and Eccles, M. R. (1998). Absence of PAX2 gene mutations in patients with primary familial vesicoureteric reflux. *J. Med. Genet.* **35**, 338–339.

Clarke, M. P., Sullivan, T. J., Francis, C., Baumal, R., Fenton, T., and Pearce, W. G. (1992). Senior-Loken syndrome. Case reports of two siblings and association with sensorineural deafness. *Br. J. Ophthalmo.* **76**, 171–172.

Cunliffe, H. E., Mcnoe, L. A., Ward, T. A., Devriendt, K., Brunner, H. G., and Eccles, M. R. (1998). The prevalence of PAX2 mutation in patients with isolated colobomas associated with urogenital anomalies. *J. Med. Genet.* **35**, 806–812.

Dahl, E., Koseki, H., and Balling, R. (1997). Pax genes and organogenesis. *Bioessays* **19**, 755–765.

Devriendt, K., Matthijs, G., Damme, B. V., Caesbroeck, D. V., Eccles, M., Vanrenterghem, Y., Fryns, J. P., and Leys, A. (1998). Missense mutation and hexanucleotide duplication in the PAX2 gene in two unrelated families with renal-coloboma syndrome (MIM 120330). *Hum. Genet.* **103**, 149–153.

Dorfler, P., and Busslinger, M. (1996). C-terminal activating and inhibitory domains determine the transactivation potential of BSAP (Pax-5), Pax-2 and Pax-8. *EMBO J.* **15**, 1971–1982.

Dressler, G. R., Deutsch, U., Chowdhury, K., Nornes, H. O., and Gruss, P. (1990). Pax2, a new murine paired-box containing gene and its expression in the developing excretory system. *Development* **109**, 787–795.

Dressler, G. R., and Douglass, E. C. (1992). Pax-2 is a DNA-binding protein expressed in embryonic kidney and Wilms tumor. *Proc. Natl. Acad. Sci. USA.* **89**, 1179–1183.

Dressler, G. R., Wilkinson, J. E., Rothenpieler, U. W., Patterson, L. T., Williams-Simons, L., and Westphal, H. (1993). Deregulation of *Pax-2* expression in transgenic mice generates severe kidney abnormalities. *Nature* **362**, 65–67.

Eberhard, D., and Busslinger, M. (1999). The partial homeodomain of the transcription factor Pax-5 (BSAP) is an interaction motif for the retinoblastoma and TATA-binding proteins. *Cancer Res.* **59**, 1716s–1724s.

Eccles, M. R., Bailey, R. R., Abbott, G. D., and Sullivan, M. J. (1996). Unravelling the genetics of vesicoureteric reflux: A common familial disorder. *Hum. Mol. Genet.* **5**, 1425–1429.

Eccles, M. R., and Jacobs, G. H. (2000). The genetics of primary vesicoureteric reflux. *Ann. Acad. Med. Singapore* **29**, 337–345.

Eccles, M. R., and Schimmenti, L. A. (1999). Renal-coloboma syndrome: A multisystem developmental disorder caused by PAX2 mutations. *Clin. Genet.* **56**, 1–9.

Eccles, M. R., Wallis, L. J., Fidler, A. E., Spurr, N. K., Goodfellow, P. J., and Reeve, A. E. (1992). Expression of the *PAX2* gene in human fetal kidney and Wilms tumour. *Cell Growth Differ.* **3**, 279–289.

Eccles, M. R., Yun, K., Reeve, A. E., and Fidler, A. E. (1995). Comparative in situ hybridization analysis of PAX2, PAX8, and WT1 gene transcription in human fetal kidney and Wilms tumours. *Am. J. Pathol.* **146**, 40–45.

Epstein, J. A., Glaser, T., Cai, J., Jepal, L., Walton, D. S., and Maas, R. L. (1994). Two independent and interactive DNA-binding subdomains of the Pax6 paired domain are regulated by alternative splicing. *Genes Dev.* **8**, 2022–2034.

Favor, J., Sandulache, R., Neuhauser-Klaus, A., Pretsch, W., Chatterjee, B., Senft, E., Wurst, W., Blanquet, V., Grimes, P., Sporle, R., and Schughart, K. (1996). The mouse Pax2 1Neu mutation is identical to a human PAX2 mutation in a family with renal-coloboma syndrome and results in developmental defects of the brain, ear, eye, and kidney. *Proc. Natl. Acad. Sci. USA* **93**, 13870–13875.

Feather, S. A., Malcolm, S., Woolf, A. S., Wright, V., Blaydon, D., Reid, C. J., Flinter, F. A., Proesmans, W., Devriendt, K., Carter, J., Warwicker, P., Goodship, T. H., and Goodship, J. A. (2000). Primary, nonsyndromic vesicoureteric reflux and its nephropathy is genetically heterogeneous, with a locus on chromosome 1. *Am. J. Hum. Genet.* **66**, 1420–1425.

Fickenscher, H. R., Chalepakis, G., and Gruss, P. (1993). Murine Pax-2 protein is a sequence-specific trans-activator with expression in the genital system. *DNA Cell Biol.* **12**, 381–391.

Fitzsimmons, D., Hodsdon, W., Wheat, W., Maira, S. M., Wasylyk, B., and Hagman, J. (1996). Pax-5 (BSAP) recruits Ets proto-oncogene family proteins to form functional ternary complexes on a B-cell-specific promoter. *Genes Dev.* **10**, 2198–2211.

Fortin, A. S., Underhill, D. A., and Gros, P. (1998). Helix 2 of the paired domain plays a key role in the regulation of DNA- binding by the Pax-3 homeodomain. *Nucleic Acids Res.* **26**, 4574–4581.

Fu, W., Duan, H., Frei, E., and Noll, M. (1998). shaven and sparkling are mutations in separate enhancers of the Drosophila Pax2 homolog. *Development* **125**, 2943–2950.

Fu, W., and Noll, M. (1997). The Pax2 homolog sparkling is required for development of cone and pigment cells in the Drosophila eye. *Genes Dev.* 2066–2077.

Gehring, W. J., and Ikeo, K. (1999). Pax 6: Mastering eye morphogenesis and eye evolution. *Trends Genet.* **15**, 371–377.

Glaser, T., Walton, D. S., and Maas, R. L. (1992). Genomic structure, evolutionary conservation and aniridia mutations in the human PAX6 gene. *Nature Genet.* **2**, 232–238.

Gnarra, J. R., and Dressler, G. R. (1995). Expression of Pax-2 in human renal cell carcinoma and growth inhibition by antisense oligonucleotides. *Cancer Res.* **55**, 4092–4098.

Godley, L. A., Kopp, J. B., Eckhaus, M., Paglino, J. J., Owens, J., and Varmus, H. E. (1996). Wild-type p53 transgenic mice exhibit altered differentiation of the ureteric bud and possess small kidneys. *Genes Dev.* **10**, 836–850.

Groves, A. K., and Bronner-Fraser, M. (2000). Competence, specification and commitment in otic placode induction. *Development* **127**, 3489–3499.

Halder, G., Callaerts, P., Flister, S., Walldorf, U., Kloter, U., and Gehring, W. J. (1998). Eyeless initiates the expression of both sine oculis and eyes during Drosophila compound eye development. *Development* **125**, 2181–2191.

Hallonet, M., Hollemann, T., Pieler, T., and Gruss, P. (1999). Vax1, a novel homeobox-containing gene, directs development of the basal forebrain and visual system. *Genes Dev.* **13**, 3106–3114.

Heller, N., and Brandli, A. W. (1999). Xenopus Pax-2/5/8 orthologues: Novel insights into Pax gene evolution and identification of Pax-8 as the earliest marker for otic and pronephric cell lineages. *Dev. Genet.* **24**, 208–219.

Hidalgo-Sanchez, M., Alvarado-Mallart, R., and Alvarez, I. S. (2000). Pax2, Otx2, Gbx2 and Fgf8 expression in early otic vesicle development. *Mech. Dev.* **95**, 225–229.

Hidalgo-Sanchez, M., Millet, S., Simeone, A., and Alvarado-Mallart, R. M. (1999). Comparative analysis of Otx2, Gbx2, Pax2, Fgf8 and Wnt1 gene expressions during the formation of the chick midbrain/hindbrain domain. *Mech. Dev.* **81**, 175–178.

Hutson, M. R., Lewis, J. E., Nguyen-Luu, D., Lindberg, K. H., and Barald, K. F. (1999). Expression of Pax2 and patterning of the chick inner ear. *J. Neurocytol.* **28**, 795–807.

Imgrund, M., Grone, E., Grone, H. J., Kretzler, M., Holzman, L., Schlondorff, D., and Rothenpieler, U. W. (1999). Re-expression of the developmental gene Pax-2 during experimental acute tubular necrosis in mice. *Kidney Int.* **56**, 1423–1431.

Jordan, T., Hanson, I., Zaletayev, D., Hodgson, S., Prosser, J., Seawright, A., Hastie, N., and Van Heyningen, V. (1992). The human PAX6 gene is mutated in two patients with aniridia. *Nature Gene.* **1**, 328–332.

Karcher, H. (1979). Zum morning glory syndrome. *Klin Mbl Augenheik.* **175**, 835–840.

Kavaler, J., Fu, W., Duan, H., Noll, M., and Posakony, J. W. (1999). An essential role for the Drosophila Pax2 homolog in the differentiation of adult sensory organs. *Development* **126**, 2261–2272.

Keller, G., Paige, C., Gilboa, E., and Wagner, E. F. (1985). Expression of a foreign gene in myeloid and lymphoid cells derived from multipotent haematopoietic precursors. *Nature* **319**, 149–154.

Keller, S. A., Jones, J. M., Boyle, A., Barrow, L. L., Kilen, P. D., Green, D. G., Kapousta, N. V., Hitchcock, P. F., Swank, R. T., and Meisler, M. H.

(1994). Kidney and retinal defects (Krd) a transgene induced mutation with a deletion of mouse chromosome 19 that includes the Pax-2 locus. *Genomics* **23**, 309–320.

Koseki, C., Herzlinger, D., and Al-Awqati, Q. (1992). Apoptosis in metanephric development. *J. Cell Biol.* **119**, 1327–1333.

Kozmik, Z., Czerny, T., and Busslinger, M. (1997). Alternatively spliced insertions in the paired domain restrict the DNA sequence specificity of Pax6 and Pax8. *EMBO J.* **16**, 6793–6803.

Lechner, M. S., and Dressler, G. R. (1996). Mapping of the Pax-2 transcription activation domains. *J. Biol. Chem.* **271**, 21088–21093.

Levine, A. J. (1997). p53, the cellular gatekeeper for growth and division. *Cell* **88**, 323–331.

Lun, K., and Brand, M. (1998). A series of *no isthmus* (*noi*) alleles of the zebrafish *pax2.1* gene reveals multiple signaling events in development of the midbrain-hindbrain boundary. *Development* **125**, 3049–3062.

Macchia, P. E., Lapi, P., Krude, H., Pirro, M. T., Missero, C., Chiovato, L., Souabni •, A., Baserga, M., Tassi, V., Phichera, A., Fenzi, G., Grüters, A., Busslinger, M., and Lauro, R. D. (1998). *PAX8* mutations associated with congenital hypothyroidism caused by thyroid dysgenesis. *Nature Gene.* **19**, 83–86.

Macdonald, R., Scholes, J., Strähle, U., Brennan, C., Holder, N., Brand, M., and Wilson, S. W. (1997). The Pax protein Noi is required for commissural axon pathway formation in the rostral forebrain. *Development* **124**, 2397–2408.

Majumdar, A., Lun, K., Brand, M., and Drummond, I. A. (2000). Zebrafish no isthmus reveals a role for pax2.1 in tubule differentiation and patterning events in the pronephric primordia. *Development* **127**, 2089–2098.

Mansouri, A., Hallonet, M., and Gruss, P. (1996). *Pax* genes and their roles in cell differentiation and development. *Curr. Opin. Cell Biol.* **8**, 851–857.

Margue, C. M., Bernasconi, M., Barr, F. G., and Schafer, B. W. (2000). Transcriptional modulation of the anti-apoptotic protein BCL-XL by the paired box transcription factors PAX3 and PAX3/FKHR. *Oncogene* **19**, 2921–2929.

Maricich, S. M., and Herrup, K. (1999). Pax-2 expression defines a subset of GABAergic interneurons and their precursors in the developing murine cerebellum. *J. Neurobiol.* **41**, 281–294.

Narahara, K., Baker, E., Ito, S., Yokoyama, Y., Yu, S., Hewitt, D., Sutherland, G. R., Eccles, M. R., and Richards, R. I. (1997). Localisation of a 10q breakpoint within the PAX2 gene in a patient with a de novo t(10:13) translocation and optic nerve coloboma-renal disease. *J. Med. Genet.* **34**, 213–216.

Noll, M. (1993). Evolution and role of Pax genes. *Curr. Opin. Gene. Dev.* **3**, 595–605.

Nornes, H. O., Dressler, G. R., Knapik, E. W., Deutsch, U., and Gruss, P. (1990). Spatially and temporally restricted expression of Pax2 during murine neurogenesis. *Development* **109**, 797–809.

Nutt, S. L., and Busslinger, M. (1999). Monoallelic expression of *Pax5*: A paradigm for the haploinsufficiency of mammalian *Pax* genes? *Biol. Chem.* **380**, 601–611.

Nutt, S. L., Vambrie, S., Steinlein, P., Kozmik, Z., Rolink, A., Weith, A., and Busslinger, M. (1999). Independent regulation of the two Pax5 alleles during B-cell development. *Nature Genet.* **21**, 390–395.

Oefelein, M., Grapey, D., Schaeffer, T., Chin-Chance, C., and Bushman, W. (1996). PAX-2: A developmental gene constitutively expressed in the mouse epididymis and ductus deferens. *J. Urol.* **156**, 1204–1207.

Okafuji, T., Funahashi, J., and Nakamura, H. (1999). Roles of Pax-2 in initiation of the chick tectal development. *Brain Res. Dev. Brain Res.* **116**, 41–49.

Otteson, D. C., Shelden, E., Jones, J., Kameoka, J., and Hitchcock, P. (1998). Pax2 expression and retinal morphogenesis in the normal and Krd mouse. *Dev. Biol.* **193**, 209–224.

Pagon, R. A. (1981). Ocular coloboma. *Surv. Ophthalmol.* **25**, 223–236.

Pfeffer, P. L., Bouchard, M., and Busslinger, M. (2000). Pax2 and homeodomain proteins cooperatively regulate a 435 bp enhancer of the mouse Pax5 gene at the midbrain-hindbrain boundary. *Development* **127**, 1017–1028.

Pfeffer, P. L., Gerster, T., Lun, K., Brand, M., and Busslinger, M. (1998). Characterization of three novel members of the zebrafish Pax2/5/8 family: Dependency of Pax5 and Pax8 expression on the Pax2.1 (noi) function. *Development* **125**, 3063–3074.

Porteous, S., Torban, E., Cho, N. P., Cunliffe, H., Chua, L., Mcnoe, L., Ward, T., Souza, C., Gus, P., Giugliani, R., Sato, T., Yun, K., Favor, J., Sicotte, M., Goodyer, P., and Eccles, M. (2000). Primary renal hypoplasia in humans and mice with PAX2 mutations: Evidence of increased apoptosis in fetal kidneys of Pax2(1Neu) +/– mutant mice. *Hum. Mol. Genet.* **9**, 1–11.

Puschel, A. W., Westerfield, M., and Dressler, G. R. (1992). Comparative analysis of Pax-2 protein distributions during neurulation in mice and zebrafish. *Mech. Dev.* **38**, 197–208.

Read, A. P. (1995). Pax genes-paired feet in three camps. *Nature Gene.* **9**, 333–334.

Reiger, G. (1977). krankheitsbil der Handmannschen sehnervenanomalie: windenbluten-(Morning Glory) syndrom? *Klin. Mbl. Augenheik.* **170**, 697–706.

Riley, B. B., Chiang, M., Farmer, L., and Heck, R. (1999). The deltaA gene of zebrafish mediates lateral inhibition of hair cells in the inner ear and is regulated by pax2.1. *Development* **126**, 5669–5678.

Ritz-Laser, B., Estreicher, A., Gauthier, B., and Philippe, J. (2000). The paired-homeodomain transcription factor Pax-2 is expressed in the endocrine pancreas and transactivates the glucagon gene promoter. *J. Biol. Chem.* **275**, 32708–32715.

Rothenpieler, U. W., and Dressler, G. R. (1993). Pax-2 is required for mesenchyme-to-epithelium conversion during kidney development. *Development* **119**, 711–720.

Rowitch, D. H., and Mcmahon, A. P. (1995). Pax-2 expression in the murine neural plate precedes and encompasses the expression domains of Wnt-1 and En-1. *Mech. Dev.* **52**, 3–8.

Royer, P., Habib, R., Courtecuisse, V., and Leclerc, F. (1967). Bilateral renal hypoplasia with oligonephronia. *Arch. Fr. Pediatr.* **24**, 249–268.

Salomon, R., Attie, T., Tellier, A. L., Antignac, C., Lyonnet, S., Vekemans, M., Munnich, A., Dureau, P., Broyer, M., Gubler, M. C., and Niaudet, P. (1999). PAX-2 mutations in oligomeganephronic renal hypoplasia. *J. Am. Soc. Nephrol.* **10**, 442A.

Sanyanusin, P., Mcnoe, L. A., Sullivan, M. J., Weaver, R. G., and Eccles, M. R. (1995a). Mutation of PAX2 in two siblings with renal-coloboma syndrome. *Hum. Mol. Genet.* **4**, 2183–2184.

Sanyanusin, P., Norrish, J. H., Ward, T. A., Nebel, A., Mcnoe, L. A., and Eccles, M. R. (1996). Genomic structure of the human PAX2 gene. *Genomics* **35**, 258–261.

Sanyanusin, P., Schimmenti, L. A., Mcnoe, L. A., Ward, T. A., Pierpont, M. E. M., Sullivan, M. J., Dobyns, W. B., and Eccles, M. R. (1995b). Mutation of the PAX2 gene in a family with optic nerve colobomas, renal anomalies and vesicoureteral reflux. *Nature Genet.* **9**, 358–363.

Schimmenti, L. A., Cunliffe, H. E., Mcnoe, L. A., Ward, T. A., French, M. C., Shim, H. H., Zhang, Y.-H., Proesmans, W., Leys, A., Byerly, K. A., Braddock, S. R., Masuno, M., Imaizumi, K., Devriendt, K., and Eccles, M. R. (1997). Further delineation of renal-coloboma syndrome in patients with extreme variability of phenotype and identical PAX2 mutations. *Am. J. Hum. Genet.* **60**, 869–878.

Schimmenti, L. A., Pierpont, M. E., Carpenter, B. L. M., Kashtan, C. E., Johnson, M. R., and Dobyns, W. B. (1995). Autosomal dominant optic nerve colobomas, vesicoureteral reflux, and renal anomalies. *Am. J. Med. Genet.* **59**, 204–208.

Schimmenti, L. A., Shim, H. H., Wirtschafter, J. D., Panzarino, V. A., Kashtan, C. E., Kirkpatrick, S. J., Wargowski, D. S., France, T. D., Michel, E., and Dobyns, W. B. (1999). Homonucleotide expansion and contraction mutations of PAX2 and inclusion of Chiari 1 malformation as part of renal-coloboma syndrome. *Hum. Mutat.* **14**, 369–76.

Schwarz, M., Alvarez-Bolado, G., Dressler, G., Urbanek, P., Busslinger, M., and Gruss, P. (1999). Pax2/5 and Pax6 subdivide the early neural tube into three domains. *Mech. Dev.* **82**, 29–39.

Shim, H. H., Nakamura, B. N., Cantor, R. M., and Schimmenti, L. A. (1999). Identification of two single nucleotide polymorphisms in exon 8 of PAX2. *Mol. Genet. Metab.* **68**, 507–510.

Stuart, E. T., Haffner, R., Oren, M., and Gruss, P. (1995). Loss of p53 function through PAX-mediated transcriptional repression. *EMBO J.* **14**, 5638–5645.

Tassabehji, M., Read, A. P., Newton, V. E., Harris, R., Balling, R., Gruss, P., and Strachan, T. (1992). Waardenburg's syndrome patients have mutations in the human homologue of the Pax-3 paired box gene. *Nature* **355**, 635–636.

Tassabehji, M., Read, A. P., Newton, V. E., Patton, M., Gruss, P., Harris, R., and Strachan, T. (1993). Mutations in the *PAX*3 gene causing Waardenburg syndrome type 1 and type 2. *Nature Gene.* **3**, 26–30.

Tellier, A. L., Amiel, J., Delezoide, A. L., Audollent, S., Auge, J., Esnault, D., Encha-Razavi, F., Munnich, A., Lyonnet, S., Vekemans, M., and Attie-Bitach, T. (2000). Expression of the PAX2 gene in human embryos and exclusion in the CHARGE syndrome. *Am. J. Med. Genet.* **93**, 85–88.

Terzic, J., Muller, C., Gajovic, S., and Saraga-Babic, M. (1998). Expression of PAX2 gene during human development. *Int. J. Dev. Biol.* **42**, 701–707.

Ton, C. C. T., Hirvonen, H., Miwa, H., Weil, M. M., Monaghan, P., Jordan, T., Van Heyningen, V., Hastie, N. D., Meijers-Heijboer, H., Drechsler, M., Royer-Pokora, B., Collins, F., Swaroop, A., Strong, L. C., and Saunders, G. F. (1991). Positional cloning and characterization of a paired box- and homeobox-containing gene from the Aniridia region. *Cell* **67**, 1059–1074.

Torban, E., Eccles, M., Favor, J., and Goodyer, P. (2000). PAX2 suppresses apoptosis in renal collecting duct cells. *Am. J. Pathol.* **157**, 833–842.

Torres, M., and Giraldez, F. (1998). The development of the vertebrate inner ear. *Mech. Dev.* **71**, 5–21.

Torres, M., Gomez-Pardo, E., Dressler, G. R., and Gruss, P. (1995). Pax-2 controls multiple steps of urogenital development. *Development* **121** 4057–4065.

Torres, M., Gomez-Pardo, E., and Gruss, P. (1996). Pax2 contributes to inner ear patterning and optic nerve trajectory. *Development* **122**, 3381–3391.

Treisman, J., Harris, E., and Desplan, C. (1991). The paired box encodes a second DNA-binding domain in the paired homeo domain protein. *Genes Dev.* **5**, 594–604.

Urbanek, P., Wang, S. Q., Fetka, I., Wagner, E. F., and Busslinger, M. (1994). Complete block of early B cell differentiation and altered patterning of the posterior midbrain in mice lacking Pax5/BSAP. *Cell* **79**, 901–912.

Ward, T. A., Nebel, A., Reeve, A. E., and Eccles, M. R. (1994). Alternative messenger RNA forms and open reading frames within an additional conserved region of the human PAX-2 gene. *Cell Growth Differ.* **5**, 1015–1021.

Weaver, R. G., Cashwell, L. F., Lorentz, W., Whiteman, D., Geisinger, K. R., and Ball, M. (1988). Optic nerve coloboma associated with renal disease. *Am. J. Med. Genet.* **29**, 597–605.

Wheat, W., Fitzsimmons, D., Lennox, H., Krautkramer, S. R., Gentile, L. N., Mcintosh, L. P., and Hagman, J. (1999). The highly conserved beta-hairpin of the paired DNA-binding domain is required for assembly of Pax-Ets ternary complexes. *Mol. Cell. Biol.* **19**, 2231–2241.

Wiggan, O., Taniguchi-Sidle, A., and Hamel, P. A. (1998). Interaction of the pRB-family proteins with factors containing paired-like homeodomains. *Oncogene* **16**, 227–236.

Winyard, P. J. D., Risdon, R. A., Sams, V. R., Dressler, G. R., and Woolf, A. S. (1996). The PAX2 transcription factor is expressed in cystic and hyperproliferative dysplastic epithelia in human kidney malformations. *J. Clin. Invest.* **98**, 451–459.

Xu, H. E., Rould, M. A., Xu, W., Epstein, J. A., Maas, R. L., and Pabo, C.

O. (1999). Crystal structure of the human Pax6 paired domain-DNA complex reveals specific roles for the linker region and carboxy-terminal subdomain in DNA binding. *Genes Deve.* **13**, 1263–1275.

Xu, W., Rould, M. A., Jun, S., Desplan, C., and Pabo, C. O. (1995). Crystal structure of a paired domain-DNA complex at 2.5 A resolution reveals structural basis for pax development mutations. *Cell* **80**, 639–650.

Zhang, X., Lin, E., and Yang, X. (2000). Sonic hedgehog-mediated ventralization disrupts formation of the midbrain-hindbrain junction in the chick embryo. *Dev. Neurosci.* **22**, 207–216.

Ziman, M. R., and Kay, P. H. (1998). A conserved TN_8TCCT motif in the octapeptide-encoding region of *Pax* genes which as the potential to direct cytosine methylation. *Gene* **223**, 303–308.

24

Cystic Renal Diseases

Sharon Mulroy, Cathy Boucher, Paul Winyard, and Richard Sandford

I. Human Clinical Disease Impact

II. Molecular Genetics of Human Renal Cystic
Diseases

III. Animal Models and the Pathogenesis of
Polycystic Kidney Diseases

IV. General Mechanisms Underlying Cystogenesis
and the Function of Proteins Causing
Polycystic Kidney Disease

V. Summary
References

I. Human Clinical Disease Impact

Renal cysts are common structural abnormalities seen in clinical practice. They may be detected as a specific or as an incidental finding by a wide variety of imaging techniques ranging from fetal ultrasonography to adult computed tomography (CT) imaging. They may vary in number from a few, that are a very common incidental finding at autopsy in the elderly to the many that are seen in the polycystic kidney diseases. Clinical sequelae resulting from renal cysts are therefore highly variable, with the most severe being the total loss of renal function seen in conditions such as bilateral renal cystic dysplasia and the polycystic kidney diseases.

Clinical conditions associated with renal cysts are listed in Table 24.1. They represent a broad range of congenital, familial, and noninherited conditions. Indeed, over 70 distinct syndromes associated with renal cysts are listed in On-line Mendelian Inheritance in Man (OMIM: www.ncbi. nlm.nih.gov/omim). They are also frequently associated with chromosomal aneuploidies. This classification system, coupled with age of presentation, forms a clinically useful system that can be of help in the differential diagnosis of this group of disorders.

The wide spectrum of disorders associated with renal cysts suggests that many different gene mutations and pathological processes result in cyst formation. It is likely that a greater understanding of the molecular pathogenesis of renal cyst formation will result in a new method of classification in the near future, as more genes are identified as being responsible for the familial cases and their function in normal renal physiology defined.

The pathogenesis of nonfamilial congenital disorders such as renal cystic dysplasia is discussed later. Other nonfamilial congenital cystic disorders may be secondary to renal tract obstruction and chromosomal aneuploidies, where the renal cysts are part of a wide spectrum of developmental abnormalities. Nonfamilial acquired cystic disease includes age-related simple cysts and the acquired cystic disease associated with chronic renal failure and dialysis therapy (Grantham, 1991; Levine *et al.*, 1991). Therefore, the precise clinical diagnosis of congenital and acquired renal cystic diseases requires an accurate family history, identification of associated clinical features, and a detailed assessment of renal structure and function. Familial renal cystic disease may occur without a family history, especially in autosomal-recessive disorders, but the characteristic clinical presentation, renal imaging, pathology, and increasing use

Table 24.1   Renal Cystic Diseases[a]

Congenital	Typical Presentation
Primary	
Cystic dysplasia	P/I
Chromosomal aneuploidies, especially trisomy 13	P
Secondary	
Renal tract obstruction	P/I
Vesicoureteric reflux	P/I/A
Familial	
Autosomal recessive	
Autosomal recessive polycystic kidney disease	P/I
Nephronophthisis (juvenile)	I
Meckel–Gruber syndrome	P/I
Bardet–Biedl syndrome	I
Zellweger syndrome	I
Autosomal dominant	
Autosomal dominant polycystic kidney disease	A
TSC2-PKD1 contiguous gene deletion syndrome	I
Von Hippel–Lindau syndrome	A
Oro-facial-digital syndrome type I	I
Glomerulocystic disease	I
Nephronophthisis–medullary cystic kidney disease	A
Branchio-oto-renal syndrome	I/A
Diabetes and renal cystic disease associated with HNF1β mutations	A
Acquired	
Simple cysts A	
Chronic renal failure	A
Hypokalemia I/A	

[a]A, adult; I, infantile; P, pre/neonatal.

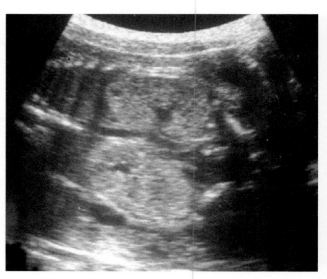

Figure 24.1 An antenatal ultrasound scan demonstrating bilateral enlarged hyperechoic kidneys typical of autosomal-recessive polycystic kidney disease. From Winyard *et al.*, 2001. Reproduced with kind permission from Wiley.

of molecular diagnostic test results make them relatively straightforward to diagnose (Hildebrandt *et al.* 2001).

Cystic renal diseases are the commonest inherited disorders that lead to endstage renal failure (ESRF). In the infant population, juvenile nephronophthisis and autosomal recessive polycystic kidney disease (ARPKD) predominate, and in the adult population, autosomal-dominant polycystic kidney disease (ADPKD) is the most common familial cause of ESRF. Juvenile nephronophthisis is often not considered a primary renal cystic disease, as cysts occur on the background of an atrophic kidney. However, the consistent appearance of cysts, a widespread functional and structural tubular defect, and the emerging details of the primary molecular defect suggest that it fits within the group of disorders discussed in this section (Hildebrandt and Omram, 2001).

A. Autosomal Recessive Polycystic Kidney Disease

This is a rare condition with a prevalence of 1:10,000 to 1:50,000 live births characterized by renal cysts and progressive hepatic fibrosis (Zerres *et al.*, 1996b). It may present *in utero* with enlarged hyperechoic kidneys on

ultrasound and oligohydramnios (Fig. 24.1); during the immediate postnatal period and the first year of life when it is associated with a high mortality; or later during childhood or even adulthood where symptoms of renal failure and portal hypertension become apparent.

Nearly half of affected individuals present in the first year of life with the remainder presenting beyond this age and occasionally in adulthood. The mortality in the first year is high, between 9 and 24%. The most frequent presenting features include abdominal distension and respiratory distress due to enlarged kidneys. For those surviving beyond the first year the prognosis is not as bleak as considered previously. There is a 50–80% survival at 15 years (Jamil *et al.*, 1999; Roy *et al.*, 1997; Zerres *et al.*, 1996a). During this period, hypertension and urinary tract infections are common and progressive renal insufficiency and hepatic fibrosis occur in up to 60% of cases. Rare survival into late adulthood with preserved renal function has been described associated with severe complications of portal hypertension (Fonck *et al.*, 2001).

The pathological features of ARPKD seen in the kidney comprise radially orientated collecting duct cysts (Fig. 24.2). The kidneys are generally enlarged and spongy with occasional macrocysts. Liver pathology comprises bile duct dilatation and proliferation with fibrosis surrounding enlarged portal areas.

The main differential diagnosis of ARPKD is ADPKD. ADPKD may rarely present in childhood but liver involvement is rare and renal imaging in parents in ARPKD is normal (Guay-Woodford *et al.*, 1996). Other conditions that must be distinguished include Meckel–Gruber syndrome, Bardet–Biedl syndrome, and the *TSC2-PKD1* contiguous

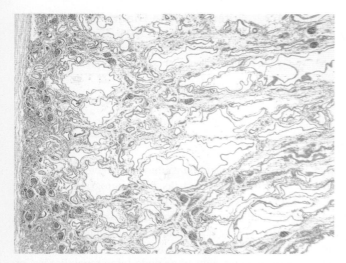

Figure 24.2 Typical pathological features of autosomal-recessive polycystic kidney disease with radially arrayed collecting duct cysts.

gene deletion syndrome. In these cases, other associated features of the condition should be actively sought.

Within families, variability of presentation and clinical severity is often seen and prenatal diagnosis is often sought. With no evidence of genetic heterogeneity, accurate prenatal diagnosis can be provided by linkage analysis (Zerres et al., 1998). The identification of the ARPKD gene, *PKDH1*, may also make direct mutation detection feasible (Ward et al., 2002).

B. Juvenile Nephronophthisis

The use of the term medullary cystic disease to describe the same condition has caused considerable confusion in the past. The medullary cystic disease–nephronophthisis complex is also used to describe the constellation of clinical, anatomical, and genetic abnormalities that make up this heterogeneous condition. However, the identification of a gene, *NPH1*, responsible for most cases of the autosomal-recessive juvenile disease means that a more rational basis for the classification of disease is imminent (Hildebrandt et al., 1997). Cases of the autosomal-dominant medullary cystic disease-nephronophthisis are, however, extremely rare (Fuchshuber et al., 1998; Scolari et al., 1998, 1999). Juvenile nephronophthisis is characterized by an autosomal-recessive mode of inheritance and a progressive tubulointerstitial nephritis with medullary cyst formation leading to ESRF in the first two decades of life (Gusmano et al., 1998). Clinically, patients present with polyuria and polydipsia secondary to a reduced urinary concentrating ability followed by a progressive loss of glomerular filtration rate leading to ESRF. Hematuria is absent, although mild proteinuria may be present. Hypertension develops late in the course of the disease. The disease may also rarely be associated with other abnormalities. The most common is tapetoretinal degeneration (Senior–Loken syndrome), which may result in early onset blindness.

The kidneys in juvenile nephronophthisis are typically normal or slightly reduced in size, and ultrasonography demonstrates the loss of corticomedullary differentiation and medullary cyst formation in the late stages of the disease (Blowey et al., 1996). The pathological features are of a noninflammatory tubulointerstitial nephritis and include tubular atrophy with a thickened basement membrane, interstitial fibrosis, occasional glomerular sclerosis, and glomerular cysts (Cohen and Hoyer, 1986). In later stages of the disease, medullary collecting duct cysts are prominent. Indeed, microdissection studies demonstrate extensive tubular diverticulae, suggesting that a very pleotropic defect in renal tubules occurs in juvenile nephronophthisis (McCredie and Baxter, 1976; Sherman et al., 1971).

C. Autosomal-Dominant Polycystic Kidney Disease

Unlike many of the other cystic renal diseases, ADPKD is common. With a prevalence of 1:800 of the population, it is one of the commonest single gene disorders of man, and the most common inherited disease leading to renal failure (Gabow, 1993). Between 5 and 10% of patients on renal replacement programs have a clinical diagnosis of ADPKD. With the cloning of the two genes responsible for virtually all cases of ADPKD, *PKD1* and *PKD2*, considerable advances have been made in our understanding of the disease phenotype and the mechanisms underlying cyst formation (Mochizuki et al., 1996; Ward et al., 1994).

The major clinical manifestations of ADPKD are due to the progressive increase in the number and size of renal cysts and the associated extensive interstitial fibrosis that lead to a progressive decline in renal function (Fig. 24.3). All affected individuals develop renal cysts by their fourth decade and the majority will have developed ESRF by their eighth decade. Disease associated with mutations in *PKD1* is more severe than that associated with mutations in *PKD2* with an average age of development of ESRF of 53 and 69 years, respectively (Hateboer et al., 1999). The development of cysts is usually accompanied by progressive nephromegaly that may lead to easily palpable kidneys and occasional abdominal distension. Other renal manifestations are secondary to cyst formation and include pain, hemorrhage, infection, and stone formation (Elzinga and Bennett, 1996). Cysts are also seen commonly in the liver and occasionally in the pancreas and other organs. Hepatic cysts are usually asymptomatic but can rarely cause massive hepatomegaly and predispose to hemorrhage and infection (Torres, 1996; Fig. 24.3). Congenital hepatic fibrosis is also a rare association with ADPKD (Mousson et al., 1997). A distinct clinical entity of polycystic liver disease without polycystic kidney

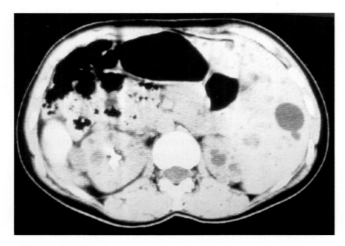

Figure 24.3 Abdominal CT scan of an individual with autosomal-dominant polycystic kidney disease. The kidneys and liver both contain multiple cysts.

Table 24.2   Criteria for the Diagnosis of ADPKD in an Individual at 50% Risk

Age	Minimum number of cysts
<30	Two cysts either uni- or bilateral
30–59	Two cysts in each kidney
>60	Four cysts in each kidney

disease also exists. This is not allelic with *PKD1* or *PKD2* based on genetic linkage studies. The gene located on human chromosome 19 remains to be identified (Reynolds *et al.*, 2000).

Hypertension is a common and early manifestation of ADPKD and may be the presenting feature (Chapman and Gabow, 1997). While current evidence suggests that it is secondary to the renal disease it may also represent a primary manifestation of the disease. Other cardiovascular abnormalities are well recognized. Mitral valve prolapse and other cardiac valvular abnormalities occur in up to 25% of affected individuals and intracerebral aneurysms in up to 10% (Pirson *et al.*, 2002; Lumiaho *et al.*, 2001).

ADPKD may also present *in utero*, in the neonatal period, and in childhood, providing evidence for the very diverse clinical presentation of disease due to mutations in the PKD genes. The development of cysts detectable *in utero* or in the neonatal period is associated with a high mortality and also predicts a very poor renal outcome (MacDermot *et al.*, 1998). Evidence supports a role for as yet unidentified disease-modifying genes in the pathogenesis of this very severe presentation of ADPKD. This presentation is very different from that in later childhood that follows a more slowly progressive course. Symptoms and signs of the disease tend to correlate with the severity of the renal cystic disease, including extrarenal manifestations such as hypertension (Fick-Brosnahan *et al.*, 2001).

Renal cysts in ADPKD are detected most usually with conventional transabdominal ultrasound where specific criteria for the diagnosis of the disease in a first-degree relative of an affected individual have been established (Table 24.2; Ravine *et al.*, 1994).

Other imaging techniques may be used in screening and diagnosis, including CT scanning (Fig. 24.3) and magnetic resonance imaging (MRI). Associated complications,

such as intracranial aneurysms, may be detected using more specialized techniques, such as MR angiography (Huston *et al.*, 1993). The very specific features of ADPKD detected by imaging make the use of renal biopsy as a diagnostic tool very rare. It may occasionally be indicated in pediatric practice where the typical cystic features have not developed, where there are clear signs of an underlying renal disease, or where a second renal diagnosis is suspected (Huston *et al.*, 1993).

The main conditions that may be confused with ADPKD are listed in Table 24.1. Imaging may form an important part of differentiating between these conditions, but molecular diagnostic testing may increasingly play a role in the management of these disorders as the genes for most of these conditions have been identified (Rossetti *et al.*, 2001, 2002).

The pathological features seen on biopsy and nephrectomy specimens in ADPKD vary depending on the stage of presentation. Biopsies from children and young adults with clinically mild or asymptomatic disease are uncommon but have demonstrated some of the early features of the disease that are absent from specimens taken from end stage kidneys (Fig. 24.4). In ADPKD, cysts may arise from the tubular epithelial cells from any region of the nephron and collecting ducts. Cysts are fluid-filled structures lined by a single layer of epithelium that may be flattened, cuboidal, or columnar. Rarely the cells appear hyperplastic and polypoid. This contrasts with the lesions seen in tuberous sclerosis complex (TSC) and von Hippel–Lindau syndrome (VHL) that also have malignant potential, a feature not seen in ADPKD. The basement membrane in ADPKD is usually thickened and abnormal in appearance. As the disease progresses, the pathological picture is also dominated by progressive interstitial fibrosis.

II. The Molecular Genetics of Human Renal Cystic Diseases

Mutations that perturb the normal function of a variety of genes and lead to human and murine renal cystic diseases have been characterized in recent years (Table 24.3, 24.4). This has provided the opportunity to define the precise molecular mechanisms underlying renal cyst formation in disease. As expected from the clinical diversity of renal cystic

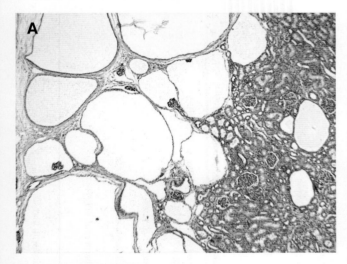

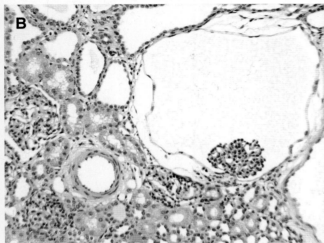

Figure 24.4 Pathological features of autosomal-dominant polycystic kidney disease seen in a childhood biopsy specimen. In the two sections, cysts can be seen to arise from all nephron segments and the glomerulus. The renal tissue architecture between cysts is well preserved.

diseases, there is clear genetic heterogeneity and the genes identified predict a range of proteins with very disparate cellular locations and functions. However, it is becoming apparent that many of these genes are classical growth or tumor suppressor genes involved in the regulation of cell proliferation. This is an observation predicted by the range of mouse mutations and transgenic models that produce a cystic phenotype (Table 24.3). A fully integrated functional map of the pathways that are involved in cyst formation therefore remains the goal of a large world wide research community.

A. Autosomal-Dominant Polycystic Kidney Disease

The mutational and functional characterization of the *PKD1* and *PKD2* genes has provided fundamental insights

into the potentially diverse and complex mechanisms involved in normal renal tubular epithelial cell function and cyst formation. Mutations in *PKD1* and *PKD2* cause virtually all cases of ADPKD (Peters *et al.*, 1993; Paterson and Pei, 1998). Both these genes have been cloned and the range of pathogenic mutations is being defined (Rossetti *et al.*, 2002). While the majority of *PKD2* mutations are nonsense and lead to a truncated and likely nonfunctional protein, the range of mutations in *PKD1* has been much harder to define due to its size and the presence of multiple transcribed copies (Thomas *et al.*, 1999; Rossetti *et al.*, 1997). Only the 3' end of the gene is single copy. Refinement of mutation screening techniques identifies the majority of *PKD1* mutations as truncating, with mutations at the 5' end of the gene also producing a more severe clinical phenotype (Rosetti *et al.*, 2002). This suggests that mechanisms such as haploinsufficiency and a dominant-negative effect, as well as loss of function as predicted by the two-hit hypothesis, are operating in cyst formation. The two-hit hypothesis predicted that a somatic mutation in 'PKD' genes was required for cyst formation in addition to an inherited germline mutation in a completely analagous manner to the Knudson two-hit hypothesis of tumor formation (Pei, 2001). The hypothesis was based on the slow rate of renal cyst formation and the very focal nature of the disease, with only 1–2% of nephrons undergoing cystic change. This has been confirmed in a small number of cysts by the identification of somatic mutations in *PKD1* and *PKD2* in renal and liver cyst-lining epithelial cells with defined germline mutations (Watnick *et al.*, 1996, 1998).

Polycystin-1

The predicted structure of the *PKD1* gene product, polycystin-1, suggests that the 460-kDa glycoprotein has a role in cell–cell or cell–matrix interactions (Fig. 24.5; Hughes, 1995). A large extracellular N-terminal portion contains multiple discrete domains that predict interactions with as yet unidentified extracellular ligands. These domains include a cysteine-rich flanked leucine-rich repeat (LRR); a C-type lectin-like domain; 16 copies of a novel PKD domain; an LDL-A domain; and a region with homology to a sea urchin protein called the REJ (receptor for egg jelly) domain (Sandford *et al.*, 1996). It also contains a conserved G-protein-coupled receptor proteolytic cleavage site (GPS) adjacent to the first predicted transmembrane domain, although it is not known if polycystin-1 is cleaved *in vivo* (Ponting *et al.*, 1999). In addition to this region, comparative sequence analysis predicts up to 11 transmembrane domains and a short intracellular C-terminal region (Sandford *et al.*, 1996). It is this latter region that has undergone extensive functional characterization, providing insights into the potential function of the whole molecule (Parnell *et al.*, 1998; Arnould *et al.*, 1998; Kim *et*

Table 24.3

Disease	Species	Model	Inheritance	Renal pathology	Chromosome Mouse/rat	Human	Gene(s)	Reference(s)
ARPKD	Rat	Pck[a]	AR	Collecting duct	9	6	Fibrocystin	Harris et al. (2002)
	Mouse	cpk[a]	AR	Collecting duct	12	—	cystin	Fry et al. (1985); Davisson et al. (1991); Hou et al. (2001)
	Mouse	bpk[a]	AR	Collecting duct	10[c]	—	?	Nauta et al. (1993)
	Mouse	orpk	AR	Collecting duct	14	—	TG737	Moyer et al. (1994)
	Mouse	$TG737^{\Delta2\text{-}3\beta gal}$	AR	Collecting duct	14	—	TG737	Murcia et al. (2000)
	Mouse	jck[a]	AR	Entire nephron	11	—	?	Atala et al. (1993)
	Mouse	inv	AR	?	4	—	inv	Yokoyama et al. (1993); Mochizuki et al. (1998)
	Mouse	CD1-Egfr^{mlCwr}	AR	Collecting duct cysts	11	7	Egfr	Treadgill et al. (1995)
ADPKD	Mouse	jcpk	AD/AR	Glomerulus	10[c]	—	?	Guay-Woodford et al. (1996)
	Mouse	Krd	AD	Not well defined	19	—	Del(19)TgN8052Mm	Keller et al. (1994)
	Mouse	kat[a]	AR	Entire nephron	8	4q	Nek1	Vogler et al. (1999); Upadhya et al. (2000)
	Mouse	hPKD1 transgenic	AD	Microcysts, predominantly glomerular	17	16	hPKD1 transgenic	Pritchard et al. (2000)
	Mouse	$Pkd1^{del34}$	AD	Entire nephron	17	16	Pkd1	Lu et al. (1997) Lu et al. (1999)
	Mouse	$Pkd1^L$	AD	Entire nephron	17	16	Pkd1	Kim et al. (2000)
	Mouse	$Pkd1^{del17\text{-}21\beta geo}$	AD	Entire nephron	17	16	Pkd1	Boulter et al. (2001)
	Mouse	$Pkd2^-$	AD	Entire nephron	5	4	Pkd2	Wu et al. (1998)
Nephronopthisis	Mouse	Tensin	AR	Triad[b], but PT cysts	1	2q35-q36	tensin	Lo et al. (1997)
	Mouse	α3β1 integrin	AR	Microcysts, Triad[b], but mostly glomerular changes and PT cysts	α3:11 β1 ?	α3:17 β1: 10p11.2	α3 and β1 integrin	Kreidberg et al. (1996)
	Mouse	Rho GDIα	AR	Triad[b], plus nephrotic syndrome	?	17q25.3	RhoGDIα	Togawa et al. (1999)
	Mouse	AP2β	AR	Triad[b], plus marked apoptosis	3	6p12	AP2β	Moser et al. (1997)
	Mouse	Ace	AR	Triad[b], plus marked apoptosis	11	17q23	Ace	Carpenter et al. (1996)
	Mouse	bcl-2	AR	Triad[b], plus marked apoptosis	2	18q21.3	Bcl-2	Veis et al. (1993); Sorenson et al. (1995)
	Mouse	pcy[a]	AR	Triad[b], entire nephrons, age of onset similar to NPH3	9	3q	?	Takahashi et al. (1991); Omran et al. (2001)
	Mouse	kd[a]	AR	Triad[b]	10	6q21	?	Lyon et al. (1971); Sibilac et al. (1998)
	Dog	Norwegian elkhound dog	AR		?	?	?	Finco et al. (1977)
Tuberous sclerosis	Mouse	Tsc2 ko mouse	AD	Cortical cystadenomas, renal cell carcinoma	17	16	tsc2	Onda et al. (1999)
	Mouse	Tsc1 ko mouse	AD	Cortical cystadenomas, renal cell carcinoma	2	9	tsc1	Kobayashi et al. (2001)
Von Hippel–Lindau	Mouse	Vhl^{lox1}	AD	Rare renal cysts only	6	3	vhl	Gnarra et al. (1997)

[a] Arose as a spontaneous mutation.

[b] Histological triad of NPH: tubular basement membrane disruption, tubular atrophy, and cyst development and tubulointerstitial infiltration with fibrosis.

[c] Mutations are allelic.

PT: proximal tubule

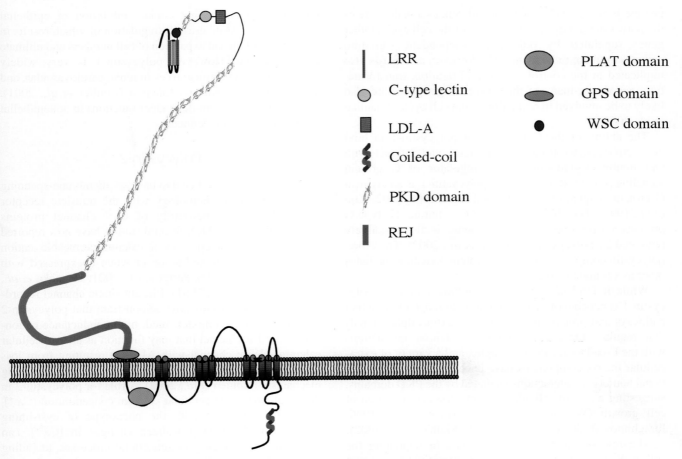

Figure 24.5 The domain architecture of polycystin-1.

al., 1999a, b; Vandorpe *et al.*, 2000; Sutters *et al.*, 2001; Parnell *et al.*, 2002; Nickel *et al.*, 2002).

The similarity between the clinical phenotypes caused by mutations in *PKD1* and *PKD2* (both in humans and in mouse models) suggests that polycystin-1 and polycystin-2 either interact directly with each other or function in different parts of the same signaling pathways. It has now been demonstrated that they form a polycystin complex via an interaction of polycystin-2 with a coiled-coil domain in the C terminus of polycystin-1 (Qian *et al.*, 1997; Newby *et al.*, 2002). The precise localization of this complex remains unclear, with polycystin-2 being localized to the plasma membrane or endoplasmic reticulum (ER), whereas polycystin-1 is localized to the cell surface (Cai *et al.*, 1999; Foggensteiner *et al.*, 2000; Ibraghimov-Beskrovnaya *et al.*, 1997).

Several models have been proposed to explain this observation. Either the two proteins interact across the "gap" between very closely apposed membrane compartments or polycystin-2 is resident in the ER and is trafficked to the plasma membrane only under certain conditions (Koulen *et al.*, 2002). The functional association of proteins in the plasma membrane and ER can occur with the interaction between cell surface L-type calcium channels and the ER RyR forming the basis of excitation–contraction coupling in muscle cells (Beam and Franzini-Armstrong, 1997). It therefore seems that this "conformational-coupling" model may be applicable to other cell types and be a mechanism for the regulation of intracellular calcium signaling by cell surface receptor activation (Kiselyov *et al.*, 1999).

The polycystin-1 C terminus also mediates a variety of additional interactions and cellular functions. These include interactions with G-proteins and related molecules, intermediate filaments, *Wnt* signaling pathways, and protein kinase C α (PKCα)-dependent and c-Jun N-terminal kinase-dependent activation of the transcription factor activator protein-1 (AP-1) (Arnould *et al.*, 1998; Kim *et al.*, 1999a, b; Parnell *et al.*, 1998, 2002). Cellular functions include a reduced growth rate, resistance to apoptosis and triggering of migration and branching morphogenesis in transfected renal tubular epithelial cells, and the induction of cell cycle arrest by the JAK-STAT-dependent upregulation of p21(WAF1) and inhibition of Cdk2 activity (Nickel *et al.*, 2002; Bhunia *et al.*, 2002; Boletta *et al.*, 2000). This latter process is also *PKD2* dependent and defines a major role of the polycystin complex in regulation of the cell cycle. Other genes regulated by polycystin-1-dependent signaling pathways remain to be identified. However, as AP-1 is also implicated in the control of cell proliferation, transformation, and death, many of these other target genes are also likely to be involved in regulation of the cell cycle (Shaulian *et al.*, 2002).

The ability of the C terminus of polycystin-1 to bind and activate heterotrimeric G proteins and interact with and inhibit degradation of the regulator of G protein signaling protein, RGS7, defines polycystin-1 as an atypical G-protein-coupled receptor (Parnell *et al.*, 1998, 2002; Kim *et al.*, 1999). The ability to activate $G_{i/o}$- but not G_q-type G proteins via the release of $G\beta\gamma$ subunits is also negatively regulated by polycystin-2 (Delmas *et al.*, 2002). Therefore, polycystin complex-regulated G protein signaling mediates other as yet undefined cellular functions.

While it is clear that the intracellular region of polycystin-1 is capable of modulating a wide range of signaling pathways and protein interactions, the extracellular signals that regulate these are unknown. Potential interactions with the E-cadherin/catenin complex and a variety of extracellular matrix components have been proposed with additional homotypic interactions mediated by the PKD domains, suggesting a role in cell adhesion and contact inhibition of cell growth (Weston *et al.*, 2001; Huan *et al.*, 1999; Ibraghimov-Beskrovnaya *et al.*, 2000; Malhas *et al.*, 2002).

The role of the polycystin complex in regulating the cell cycle and the observation that polycystin-1 forms part of cell–cell and cell–matrix adhesion complexes identify a potential role in the formation and maintenance of intact epithelial monolayers. Polycystin-1 has been localized to focal adhesions (the contact points between cells and extracellular matrix), adherens junctions (the contact points between cells), and desmosomes (Foggensteiner *et al.*, 2000; Ibraghimov-Beskrovnaya *et al.*, 1997; Huan *et al.*, 1999; Geng *et al.*, 1998; Scheffers *et al.*, 2000). Focal adhesions are dynamic structures that form a link between the extracellular matrix and the actin cytoskeleton (Geiger *et al.*, 2001). They regulate a number of processes, including cell–matrix interactions, cell motility, and signal transduction through complex phosphorylation cascades. Adherens junctions form cell–cell contacts and play an important role in the regulation of cellular proliferation and polarity. Desmosomes are intercellular adhesive complexes that attach to intermediate filaments and provide cells not only with mechanical strength, but also act as potential extracellular sensors (Green and Gaudry, 2000). Given that the phenotype of cyst-lining epithelia includes abnormalities of proliferation, differentiation, cell polarity, and apoptosis, components of all these complexes are attractive candidate ligands for polycystin-1. The polycystin complex may therefore be involved in the regulation of the correct positioning, growth, and contact inhibition of epithelial cells in tubular structures, dysregulation of which results in abnormal growth and expansion of cell numbers and ultimate cyst formation. However, polycystin-1 is very widely expressed and has critical roles in renal, cardiovascular, and skeletal development and function (Boulter *et al.*, 2001). Many other polycystin-dependent functions in nonepithelial tissues remain to be defined.

Polycystin-2

Polycystin-2 is predicted to have six membrane-spanning regions and shares homology with the transient receptor potential (TRP) superfamily of Ca^{2+} channel proteins (Mochizuki *et al.*, 1996). Several studies have now reported polycystin-2 to function as a calcium-permeable cation channel when expressed alone or when coexpressed with polycystin-1 (Gonzalez-Perret *et al.*, 2001; Hanaoka *et al.*, 2000; Vassilev *et al.*, 2001). Elegant single channel recordings of ER membranes now demonstrate that polycystin-2 functions as a calcium-activated, high conductance cation-permeable ER channel that may function as an intracellular calcium release channel with ADPKD resulting from loss of a regulated intracellular calcium release signaling mechanism (Koulen *et al.*, 2002). A role for polycystin-2 in the regulation of intracellular calcium concentration $[Ca^{2+}]_i$ would be consistent with the phenotype of cyst-lining epithelia in ADPKD. Localized changes in $[Ca^{2+}]_i$ can modulate a wide variety of subcellular processes, including cell proliferation, differentiation, and survival, all of which are abnormal in cyst-lining epithelia (Berridge *et al.*, 2000). The mechanism by which polycystin-1 regulates polycystin-2 channel formation and function remains to be identified, although a direct interaction and regulation via G protein signaling may both occur (Delmas *et al.*, 2002).

Therefore, a model of polycystin complex function in which diverse extracellular signals are transduced by polycystin-1 and initiate signaling events in the cell to regulate a wide variety of signaling cascades now has considerable experimental support. Some or all of these of these pathways will be polycystin-2 dependent through localized changes in $[Ca^{2+}]_i$—either through the movement of extracellular Ca^{2+} into the cell or by the local release of Ca^{2+} from intracellular stores. Whether polycystin-2 is an effector or regulator of polycystin-1 function and how and where polycystin-1 and polycystin-2 interact have not been defined. Their relationship is likely to be complex and it is probable that they function both together and independently in a temporal and tissue-specific fashion.

B. Nephronophthisis

Nephrocystin, the protein product of *NPHP1*, the gene responsible for juvenile nephronophthisis, may also be a component of focal adhesion signaling complexes. It is predicted to contain an *Src* homolgy 3 (SH3) domain, found most commonly in adaptor proteins involved in focal adhesion signaling complexes (Hildebrandt *et al.*, 1997). As discussed previously, focal adhesions are dynamic structures that form a link between the extracellular matrix and the actin cytoskeleton and regulate many important cellular functions, including cell migration and morphogenesis, cell proliferation, differentiation, and apoptosis.

Several lines of evidence support a role for nephrocystin in focal adhesion signaling. Nephrocystin has been shown to interact with known protein components of focal adhesion signaling complexes—Pyk2, p130cas, and tensin—and also colocalizes with both E-cadherin and p130cas at points of cell–cell contact in polarized epithelial cells where it interacts with members of the filamin family of actin binding proteins (Donaldson *et al.*, 2002; Benzing *et al.*, 2001). When targeted mutations are introduced into *tensin* (and *α3β1 integrin* or *Rho GDIα*, also components of focal adhesion signaling complexes), mice develop a nephronopthisis-like renal phenotype (Lo *et al.*, 1997; Wilson and Burrow, 1999).

Nephrocystin contains several other domains capable of mediating protein–protein interactions, and the identification of additional binding partners for nephrocystin and/or downstream targets of nephrocystin containing signaling complexes will help define cystogenic pathways that may overlap with those implicated in other cystic renal diseases (Donaldson *et al.*, 2002).

C. Other Cystic Renal Diseases

Two other conditions associated with renal cysts are tuberous sclerosis complex and von Hippel–Lindau disease. Both these conditions are caused by mutations in classic tumor suppressor genes, again suggesting that renal cysts arise due to the dysregulation of cell growth and proliferation.

Tuberous sclerosis complex is a hamartomatous condition caused by mutations in *TSC1* and *TSC2* (van Slegtenhorst *et al.*, 1997; Consortium TECTS, 1993). Patients with a contiguous gene deletion syndrome involving both *TSC2*, encoding tuberin, and *PKD1*, which are adjacent on 16p13.3, develop severe early onset polycystic kidney disease and usually progress to end stage renal failure in childhood or early adulthood (Brookcarter *et al.*, 1994). Tuberin has roles in protein sorting, as a GTPase-activating protein, cell cycle control, and interactions with several nuclear hormone receptor family members (Wienecke *et al.*, 1995, 1996; Soucek *et al.*, 1997; Xiao *et al.*, 1997; Henry *et al.*, 1998). It has also been shown to interact with hamartin, the protein product of the *TSC1* gene *in vivo*, but significant cystic disease has not been documented in affected members of *TSC1*-linked families (vanSlegtenhorst *et al.*, 1998). Hamartin has been shown to regulate cell adhesion via interactions with the ezrin-

radixin-moesin (ERM) family of actin-binding proteins and the GTPase Rho (Lamb *et al.*, 2000). Polycystin and tuberin containing complexes may therefore function in separate pathways, both regulating cell adhesion and proliferation, with defects in both producing a more severe renal disease than seen with defects in only one. Evidence for a direct link between these is also emerging with polycystin-1 mislocalized in tuberin-deficient cells (Kleymenova *et al.*, 2001).

von Hippel–Lindau disease is a familial tumor syndrome also associated with renal cyst formation caused by mutations in *VHL* (Latif *et al.*, 1993). Its protein product pVHL forms a multimeric protein complex and is important for the negative regulation of hypoxia response genes through the ubiquitination of the α subunits of hypoxia inducible factor (HIF) (Cockman *et al.*, 2000). Cells lacking pVHL overproduce products of HIF target genes, such as vascular endothelial growth factor (VEGF) and transforming growth factor (TGF) α. pVHL also has important roles in cell cycle control, differentiation, extracellular matrix formation and turnover, and angiogenesis (Clifford *et al.*, 2001).

The function of the *PKDH1* gene product fibrocystin is not known, but predictions from structural analysis suggest that fibrocystin is likely to function as a cell surface receptor molecule (Ward *et al.*, 2002).

Therefore, many of the abnormal cellular processes linked to renal cyst formation in human disease appear to be directly involved in the regulation of cell proliferation. This common theme is also continued in the discussion of mouse models of renal cystic disease. The combination of the functional characterization of human disease genes and the wide variety of cystic mouse models is demonstrating the very wide range of mechanisms that may lead to cyst formation.

III. Animal Models and the Pathogenesis of Polycystic Kidney Diseases

A large number of animal models with renal cysts as one of their major abnormalities have been described (Table 24.3). They share many characteristic clinical and histological features with human cystic kidney diseases and provide valuable tools in the investigation of the molecular pathogenesis of renal cystic disease. Many have arisen spontaneously and identify potential candidate genes for human cystic kidney disease. Cloning the genes associated with these diseases has provided insight into the molecules involved in cystogenesis and the pathways they regulate. Increasingly, targeted gene disruptions have been used to look at these mechanisms and their role in cystogenesis, and experimentally generated models are now available for many of the genes known to be involved in human cystic diseases.

Table 24.4

Disease[a]	Human disease gene	Human chromosome	Human accession No.	Mouse homologue	Mouse chromosome	Mouse accession No.
Meckel syndrome	MKS1	17q21-24				
Meckel syndrome	MKS2	11q13				
OFDI	OFD1	Xp22.2-p22.3	NM_003611	Ofd1	X	AA240611
Nephronophthisis 1	NPHP1	2q13	NM_000272	Nphp1		NM_016902
Nephronophthisis 2	NPHP2	9q22				
Nephronophthisis 3	NPHP3	3q22				
MCKD	MCKD1	1q21				
MCKD	MCKD2	16p12				
GCKD	TCF2	17q22	NM_000458	Tcf2	11	NM_009330
VHL	VHL	3p25-p26	NM_000551	Vhlh	6	NM_009507
TSC	TSC1	9q34	NM_000368	Tsc1		NM_022887
TSC	TSC2	16p13.3	NM_000548	Tsc2	17	NM_011647
ADPKD	PKD1	16p13.3	NM_000296	Pkd1	17	NM_013630
ADPKD	PKD2	4q21-q23	NM_000297	Pkd2	5	NM_008861
ARPKD	PKHD1	6p21.2-p12	AY074797	Pkdh1		

[a]OFD1, oro-facial digital syndrome type I; MCKD, medulllary cystic kidney disease; GCKD, glomerulocystic kidney disease; VHL, von Hippel–Lindau syndrome; TSC, tuberous sclerosis complex; ADPKD, autosomal dominant polycystic kidney disease; ARPKD, autosomal recessive polycystic kidney disease.

Many different rodent models of recessive PKD have been described. In the *pck* rat, animals develop collecting duct-derived renal cysts, ductal plate malformations, and hepatic cystic disease, similar to ARPKD (Sanzen *et al.*, 2001; Lager *et al.*, 2001). This model, with traditional positional cloning strategies, was used to identify the human ARPKD gene on chromosome 6, *PKHD1*, which encodes fibrocystin (Ward *et al.*, 2002).

The *cpk* mouse is, however, the best studied murine model of recessive PKD and arose spontaneously on a C57BL/6J background (Davisson *et al.*, 1991; Preminger *et al.*, 1982). Early renal changes involve the development of proximal tubular cysts. Later these regress with marked cystic change of the collecting ducts. Outcrossing the *cpk* mutation onto other genetic backgrounds (such as CD-1 and BALB/c) has resulted in cystic involvement of the liver and pancreas and ductal plate malformation (DPM) similar to the extrarenal manifestations of human ARPKD (Gattone *et al.*, 1996; Ricker *et al.*, 2000). The *cpk* gene has been identified and encodes a novel protein, cystin, that is localized to the apical cilia in polarized renal tubular epithelial cells (Hou *et al.*, 2002). The potential role of this protein in human disease is unknown, although it may be predicted to function in fibrocystin-associated pathways.

The *bpk* mouse also arose spontaneously, but the *bpk* gene has not been identified (Nauta *et al.*, 1993). Mice develop massive collecting duct cysts and biliary pathology, again similar to that seen in human ARPKD. Interestingly, the *bpk* locus on mouse chromosome 10 is allelic with *jcpk*, a phenotypically distinct murine cystic kidney disease (Guay-Woodford *et al.*, 1996).

In several murine models of ARPKD, polycystic kidney disease has been observed in association with left-right patterning abnormalities—mice lacking normal *Tg737* function and the *inv* mutant. *Inv/inv* mice develop *situs inversus* and polycystic kidney disease (Mochizuki *et al.*, 1998). The *orpk* mutant mouse was initially identified as a model for ARPKD (Moyer *et al.*, 1994). The *orpk* gene was subsequently shown to be a hypomorphic allele of the *Tg737* gene (*Tg737^orpk*) caused by an insertional mutation. A targeted null mutation, *Tg737^{Δ2-3βgal}*, is lethal in midgestation and embryos exhibit random left–right axis determination (Murcia *et al.*, 2000). Of interest is that murine polycystin-2 is also required for left–right axis determination (Pennekamp *et al.*, 2002). As the Tg737 protein, polaris, is required for the normal formation of primary cilia, which is also a novel site of expression of polycystin-2, this identifies a novel function for polycystin-2 (Pazour *et al.*, 2002; Yoder *et al.*, 2002).

Since the identification of *PKD1* and *PKD2*, many mouse models of ADPKD have been generated with different targeted mutations in *Pkd1* and *Pkd2* or transgenes (Boulter *et al.*, 2001; Bhunia *et al.*, 2002; Wu *et al.*, 1998; Herron *et al.*, 2002; Lu *et al.*, 2001; Pritchard *et al.*, 2000; Lu *et al.*, 1997).

Polycystin-1 and polycystin-2 are widely expressed and developmentally regulated (Boulter *et al.*, 2001; Guillaume *et al.*, 2000; Chauvet *et al.*, 2002). However, there has been considerable variation in the reported expression of polycystin-1 and polycystin-2 using a wide variety of different antisera, which has now been largely clarified by *Pkd1^{del17–21βgeo}* mice carrying a *LacZ* reporter gene and the previously mentioned RNA *in situ* hybridization studies

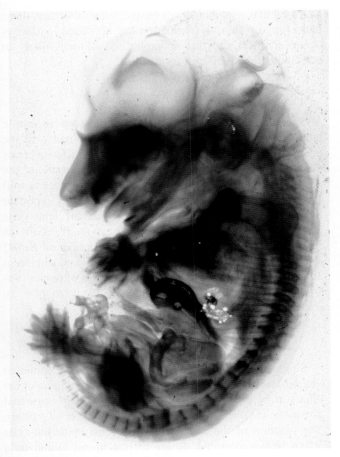

Figure 24.6 The expression of *Pkd1* demonstrated in the *Pkd1*[del17–21βgeo] mouse model of ADPKD. Widespread expression is seen in the cardiovascular and skeletal systems following X-gal staining at E13. 5 (Boulter *et al.*, 2001).

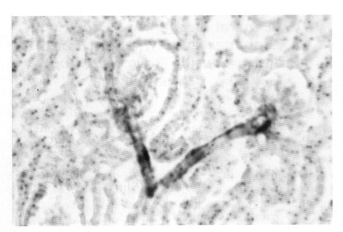

Figure 24.7 X-gal staining of frozen kidney sections from the *Pkd1*[del17–21βgeo] mouse demonstrates *Pkd1* expression in tubular epithelial cells with higher levels in the vasculature.

(Boulter *et al.*, 2001; Figs. 24.6 and 24.7). In the kidney, upregulation of *Pkd1* expression occurs in maturing tubule epithelial cells from E15.5. This is consistent with the development of renal cysts in all *Pkd1*–/– mice at E15.5, in which early nephrogenesis proceeds normally.

These data suggest that polycystin-1 and polycystin-2 are essential for the normal differentiation and maturation of kidney tubular epithelial cells. In addition, animal models have revealed important roles for the polycystins in cardiovascular (*Pkd1* and *Pkd2*) and skeletal development (*Pkd1*) not predicted from the pathology of human ADPKD (Boulter *et al.*, 2001; Lu *et al.*, 2001; Wu *et al.*, 2000). Further functions of polycystin-2 in left–right axis formation are intriguing (Pennekamp *et al.*, 2002). Renal cysts and laterality defects are also seen in mice mutants for *inversin* and *polaris*, which are required for the structural and functional integrity of monocilia (Mochizuki *et al.*, 1998; Yoder *et al.*, 2002). This defines a novel pathway involved in cyst formation.

Polycystin-1 and polycystin-2 are also expressed in the adult kidney where they are likely to have an important regulatory role, perhaps in the maintenance of differentiation. While loss of polycystin-1 or polycystin-2 function clearly leads to cyst formation, there is some evidence that altered levels of the protein may also cause cystogenesis (Pritchard *et al.*, 2000). Further, differences in the tissue and cellular expression of polycystin-1 and polycystin-2 also suggest novel tissue-specific functions for polycystin-1 that are independent of polycystin-2 (Foggensteiner *et al.*, 2000).

Several rodent models have clinical and histological features that resemble nephronopthisis (Table 24.4). Except for the *pcy* mouse, which shows synteny with the NPHP3 locus, none are syntenic with any form of human NPH-MCKD (Omram *et al.*, 2001). In *pcy* mice, renal cysts are found in all segments of the nephron and collecting duct and enlarge progressively with age, leading to renal failure. Cerebral vascular aneurysms have also been reported (Takahashi *et al.*, 1991).

Other models with a nephronopthisis-type renal phenotype include mice with targeted mutations in genes involved in focal adhesion signaling (*tensin*, *α3β1 integrin* and *Rho GDIα*) and genes regulating apoptosis (*Bcl-2*, *AP-2β*, and *Ace*). The *tensin* knockout mouse is of particular interest (Lo *et al.*, 1997). Tensin is an F actin-binding component of focal adhesions, which contains a src homology (SH2) domain, associates with p130[cas], and plays a central role in focal adhesion signaling. Tensin-deficient mice develop kidney disease, which is similar to nephronopthisis, although the cysts localize primarily to the proximal rather than the distal tubule.

A cystic phenotype has also been described in models overexpressing a transgene, the best described being the SBM transgenic mouse that has persistent expression of the *c-myc* protooncogene (Trudel *et al.*, 1991).

IV. General Mechanisms Underlying Cystogenesis and the Function of Proteins Causing Polycystic Kidney Disease

From this wealth of human and animal data, several general mechanisms implicated in renal cyst formation seem to be emerging. Cystic kidney diseases share many common features. Extensive characterization of the molecular and cellular defects in cyst-lining epithelial cells derived from human kidneys affected by polycystic kidney diseases and from a variety of rodent models of renal cystic disease has demonstrated generalized abnormalities in cell proliferation, differentiation, and apoptosis reflecting abnormalities in signaling pathways that regulate cell turnover.

A. Pathways Regulating Gene Transcription

Disruption of signaling pathways in polycystic kidney diseases results in abnormal gene transcription. Renal cyst formation is associated with the increased expression of many protooncogenes and growth factor receptors, such as c-*fos*, c-*myc*, c-K-*ras*, c-*erb* B2, and EGFR in human and animal models of cystic kidney disease (Wilson, 1996). Wnt- and AP-1-dependent pathways have been implicated in ADPKD (Arnould *et al.*, 1998; Kime *et al.*, 1999).

B. Apoptosis

The regulation of apoptosis is of particular interest in the development of cystic kidney disease. Elevated apoptosis has been reported in both human ADPKD and animal models (Woo 1995; Lanoix *et al.*, 1996; Winyard *et al.*, 1996). c-*myc* expression is upregulated, and there is evidence for a c-*myc* mediated apoptotic pathway that is *bcl-2/bax* independent (Trudel *et al.*, 1997). In addition, treatment with c-*myc* antisense RNA has been shown to ameliorate murine ARPKD (Ricker *et al.*, 2002). Upregulation of caspase activity has also been observed in cystic kidneys from *cpk* mice (Ali *et al.*, 2000). Targeted mutations in the antiapoptotic gene *Bcl-2* and in the *Ace* and AP2β genes (Sorenson *et al.*, 1996; Moser *et al.*, 1997) result in cystic kidney disease and are associated with a marked increase in apoptosis. In AP-2β deficient mice, downregulation of several antiapoptotic genes [bcl-X(L), bcl-w, and bcl-2] occurs at the end of embryonic kidney development in parallel to massive apoptotic death of collecting duct and distal tubular epithelia. Boletta *et al.* (2000) reported that the expression of full-length *PKD1* induces resistance to apoptosis and spontaneous tubulogenesis in Madin–Darby canine kidney (MDCK) cell lines, suggesting that *in vivo*, one function of polycystin-1 may be

as a negative regulator of apoptosis. Taken together, these data suggest that dysregulation of apoptotic pathways occurs in PKD and that apoptotic cell death is likely to contribute directly to cyst formation.

C. Cell Polarity

Specific defects in cell polarity are seen in cystic epithelial cells and have been implicated directly in the process of cyst formation (Wilson, 1997). The aberrant expression of selected basolateral proteins (such as the Na⁺-K⁺-ATPase or EGFR) on the apical cell surface is well documented in polycystic kidney disease. One important mechanism may be the failure to switch off fetal gene transcription. For example, mispolarization of EGFR in ADPKD is due to the persistent transcription of *ErbB2* (Nakanishi *et al.*, 2001). Adherens junctions and focal adhesion signaling complexes (potential sites of polycystin-1 and nephrocystin function) are also likely to play an important role.

D. Cilia Function

Single nonmotile cilia are expressed on the epithelial lining of much of the nephron as well as the epithelia of the biliary tract and the pancreatic ducts, sites of extrarenal disease in ADPKD and ARPKD (Wheatley *et al.*, 1996). The function of the apical cilium in renal epithelial cells is not well understood. Roles for the apical cilium as a mechanosensor or as an organelle responsible for sensing local environment have been proposed (Schwartz *et al.*, 1997). Increasing evidence from animal models links ciliary dysfunction, embryonic left–right patterning defects, and cystic kidney disease.

In *Tg737^{orpk}* mice, there is marked attenuation of the renal epithelial cilia. *Tg737^{Δ2-3βgal}* null mice die *in utero*, and exhibit random left–right axis determination and complete loss of embryonic nodal cilia (Murcia *et al.*, 2000). The *Tg737* protein product, polaris, is expressed in the basal bodies of both mono and multiciliated epithelial cells, including renal epithelia, and is required for assembly of the renal cilium (Taulman *et al.*, 2001; Yoder *et al.*, 2002). In the *inv* mutant, nodal cilia are present but dysfunctional and produce only weak leftward nodal flow (Okada *et al.*, 1999). Unfortunately, the function of renal cilia has not been fully characterized in this model. Cystin is also expressed in the apical cilia, localizing primarily to the axoneme (Hou *et al.*, 2002). Left–right patterning defects have not been identified in *cpk* mutants, and renal and biliary cilia appear structurally normal, suggesting that cystin is not involved in ciliogenesis (Ricker *et al.*, 2000). Instead, cystin may stabilize microtubule assembly within the ciliary axoneme. Paclitaxel (Taxol) and other related taxanes promote microtubule assembly. Taxane treatment in *cpk/cpk* mice produces a significant attenuation of renal disease progression (Woo *et al.*, 1997). In comparison, taxanes are ineffective in *Tg737^{orpk}* homozygotes, where a primary defect in ciliogenesis is likely (Sommardahl *et al.*, 1997).

That mutations in polaris, inversin, and cystin all produce polycystic kidney disease suggests a critical role for the cilium in tubular differentiation and epithelial cell polarity. This is also supported by the localisation of polycystin-1 and polycystin-2 to the primary cilium in cultured cells. Elucidating the biological role of these genes should provide important clues as to the function of this specialized structure and its role in cystogenesis.

E. Epidermal Growth Factor Signaling

Epidermal growth factor is of great interest in the development of polycystic kidney diseases. The EGF/TGF-α/EGFR axis is directly implicated in the progression of renal cyst formation (Richards *et al.*, 1998). Both EGF and TGF-α may be cystogenic *in vitro*, and transgenic overexpression of TGF-α in cystic mice accelerates the rate of cyst formation (Gattone *et al.*, 1996). Human and murine cyst fluid contains biologically active EGF, which can activate abnormally surface-expressed functional EGFRs in cystic epithelia. In murine models, blockade of the tyrosine kinase activity of the EGFR using novel chemical inhibitors has been shown to have beneficial effects on renal pathology, function, and overall animal survival (Sweeney *et al.*, 2000). Inhibition of TGF-α secretion also has a therapeutic effect in murine PKD (Dell *et al.*, 2001). These observations are paralleled by genetic studies that show animals homozygous for the cystic *orpk* mutation and a hypomorphic EGFR allele have a marked reduction in renal cyst formation directly correlated with reduced tyrosine kinase activity of the EGFR (Richards *et al.*, 1998).

V. Summary

The interaction of a cell with its neighbors and the extracellular matrix triggers numerous cellular responses that have essential roles in the regulation of its growth, differentiation, and fate. Specialized adhesion complexes and cytoskeletal organizations are required for both the development and the maintenance of normal tubular epithelial cell architecture and polarity and mediating signal transduction from the extracellular environement to the nucleus. There is increasing evidence that polycystin-1, polycystin-2, and nephrocystin are important components of these specialized signaling complexes in renal tubular epithelial cells and that their disruption leads to abnormal cell proliferation and differentiation and ultimate cyst formation. Other genes that are known to directly regulate cell proliferation are also implicated in cyst formation.

The targets of these signaling pathways are not well defined but are likely to be involved in regulation of the cell cycle and hence control proliferation, differentiation, and apoptosis of tubular epithelial cells.

Proteins in the apical cilia of epithelial cells may also be important in the regulation of these pathways or may cause cyst formation by an independent mechanism. The characterization of these pathways will define normal cellular physiology and offer further opportunities for therapeutic intervention.

References

Ali, S. M., Wong, V. Y., Kikly, K., Fredrickson, T. A., Keller, P. M., DeWolf, W. E., Jr., Lee, D., and Brooks, D. P. (2000). Apoptosis in polycystic kidney disease: Involvement of caspases. *Am. J. Physiol. Regul. Integr. Comp. Physiol.* **278**(3), R763–R769.

Arnould, T., Kim, E., Tsiokas, L., Jochimsen, F., Gruning, W., Chang, J. D., and Walz, G. (1998). The polycystic kidney disease 1 gene product mediates protein kinase C alpha-dependent and c-Jun N-terminal kinase-dependent activation of the transcription factor AP-1. *J. Biol. Chem.* **273**(11), 6013–6018.

Beam, K. G, and Franzini-Armstrong, C. (1997). Functional and structural approaches to the study of excitation-contraction coupling. *Methods Cell Biol.* **52**, 283–306.

Benzing, T., Gerke, P., Hopker, K., Hildebrandt, F., Kim, E., and Walz, G. (2001). Nephrocystin interacts with Pyk2, p130(Cas), and tensin and triggers phosphorylation of Pyk2. *Proc. Natl. Acad. Sci. USA* **98**(17), 9784–9789.

Berridge, M. J., Lipp, P., and Bootman, M. D. (2000). The versatility and universality of calcium signaling. *Nature. Rev. Mol. Cell. Biol.* **1**(1), 11–21.

Bhunia, A. K., Piontek, K., Boletta, A., Liu, L., Qian, F., Xu, P. N., Germino, F. J., and Germino, G. G. (2002). PKD1 induces p21(waf1) and regulation of the cell cycle via direct activation of the JAK-STAT signaling pathway in a process requiring PKD2. *Cell.* **109**(2), 157–168.

Blowey, D. L., Querfeld, U., Geary, D., Warady, B. A., and Alon, U. (1996). Ultrasound findings in juvenile nephronophthisis. *Pediatr. Nephrol.* **10**(1), 22–24.

Boletta, A., Qian, F., Onuchic, L. F., Bhunia, A. K., Phakdeekitcharoen, B., Hanaoka., K., Guggino, W., Monaco, L., and Germino, G. G. (2000). Polycystin-1, the gene product of PKD1, induces resistance to apoptosis and spontaneous tubulogenesis in MDCK cells. *Mol. Cell* **6**(5), 1267–1273.

Boulter, C., Mulroy, S., Webb, S., Fleming, S., Brindle, K., and Sandford, R. (2001). Cardiovascular, skeletal, and renal defects in mice with a targeted disruption of the Pkd1 gene. *Proc. Natl. Acad. Sci. USA* **98**(21), 12174–12179.

Brookcarter, P. T., Peral, B., Ward, C. J., Thompson, P., Hughes, J., Maheshwar, M. M., Nellist, M., Gamble, V., Harris, P. C., and Sampson, J. R. (1994). Deletion of the Tsc2 and Pkd1 genes associated with severe infantile polycystic kidney-disease: A contiguous gene syndrome. *Nature. Genet.* **8**(4), 328–332.

Cai, Y., Maeda, Y., Cedzich, A., Torres, V. E., Wu, G., Hayashi, T., Mochizuki, T., Park, J. H., Witzgall, R., and Somlo, S. (1999). Identification and characterization of polycystin-2, the PKD2 gene product. *J. Biol. Chem.* **274**(40), 28557–28565.

Chapman, A. B., and Gabow, P. A. (1997). Hypertension in autosomal dominant polycystic kidney disease. *Kidney Int. Suppl.* **61**, S71–S73.

Chauvet, V., Qian, F., Boute, N., Cai, Y., Phakdeekitacharoen, B., Onuchic, L. F., Attie-Bitach, T., Guicharnaud, L., Devuyst, O., Germino, G. G., and Gubler, M. C. (2002). Expression of PKD1 and

PKD2 transcripts and proteins in human embryo and during normal kidney development. *Am. J. Pathol.* **160**(3), 973–983.

Clifford, S. C., and Maher, E. R. (2001). Von Hippel-Lindau disease: Clinical and molecular perspectives. *Adv. Cancer. Res.* **82**, 85–105.

Cockman, M. E., Masson, N., Mole, D. R., Jaakkola, P., Chang, G. W., Clifford, S. C., Maher, E. R., Pugh, C. W., Ratcliffe, P. J., and Maxwell, P. H. (2000). Hypoxia inducible factor-alpha binding and ubiquitylation by the von Hippel-Lindau tumor suppressor protein. *J. Biol. Chem.* **275**(33), 25733–25741.

Cohen, A. H., and Hoyer, J. R. (1986). Nephronophthisis: A primary tubular basement membrane defect. *Lab. Invest.* **55**(5), 564–572.

Consortium TECTS. (1993). Identification and characterization of the tuberous sclerosis gene on chromosome 16. The European Chromosome 16 Tuberous Sclerosis Consortium. *Cell* **75**(7), 1305–1315.

Contreras, G., Mercado, A., Pardo, V., and Vaamonde, C. A. (1995). Nephrotic syndrome in autosomal dominant polycystic kidney disease. *J. Am. Soc. Nephrol.* **6**(5), 1354–1359.

Davisson, M. T., Guay-Woodford, L. M., Harris, H. W., and D'Eustachio, P. (1991). The mouse polycystic kidney disease mutation (cpk) is located on proximal chromosome 12. *Genomics* **9**(4), 778–781.

Dell, K. M., Nemo, R., Sweeney, W. E., Levin, J. I., Frost, P., and Avner, E. D. (2001). A novel inhibitor of tumor necrosis factor-alpha converting enzyme ameliorates polycystic kidney disease. *Kidney Int.* **60**(4), 1240–1248.

Delmas, P., Nomura, H., Li, X., Lakkis, M., Luo, Y., Segal, Y., Fernandez-Fernandez, J. M., Harris, P., Frischauf, A. M., Brown, D. A, and Zhou, J. (2002). Constitutive activation of G-proteins by polycystin-1 is antagonized by polycystin-2. *J. Biol. Chem.* **277**(13), 11276–11283.

Donaldson, J. C., Dempsey, P. J., Reddy, S., Bouton, A. H., Coffey, R. J., and Hanks, S. K. (2000). Crk-associated substrate p130(Cas) interacts with nephrocystin and both proteins localize to cell-cell contacts of polarized epithelial cells. *Exp. Cell. Res.* **256**(1), 168–178.

Donaldson, J. C., Dise, R. S., Ritchie, M. D., and Hanks, S. K. (2002). Nephrocystin conserved domains involved in targeting to epithelial cell-cell junctions, interaction with filamins, and establishing cell polarity. *J. Biol. Chem.* **10**, 10.

Elzinga, L. W., and Bennett, W. M. (1996). Miscellaneous renal and systemic complications of autosomal dominant polycystic kidney disease including infection. *In:* "Polycystic Kidney Disease." (M. L. Watson, V. E. Torres, eds). pp. 483–499. Oxford Univ. Press, Oxford.

Fick-Brosnahan, G. M., Tran, Z. V., Johnson, A. M., Strain, J. D., and Gabow, P. A. (2001). Progression of autosomal-dominant polycystic kidney disease in children. *Kidney Int.* **59**(5), 1654–1662.

Foggensteiner, L., Bevan, A. P., Thomas, R., Coleman, N., Boulter, C., Bradley, J., Ibraghimov-Beskrovnaya, O., Klinger, K., and Sandford, R. (2000). Cellular and subcellular distribution of polycystin-2, the protein product of the PKD2 gene. *J. Am. Soc. Nephrol.* **11**(5), 814–827.

Fonck, C., Chauveau, D., Gagnadoux, M. F., Pirson, Y., and Grunfeld, J. P. (2001). Autosomal recessive polycystic kidney disease in adulthood. *Nephrol Dial Transplant.* **16**(8), 1648–1652.

Fuchshuber, A., Deltas, C. C., Berthold, S., Stavrou, C., Vollmer, M., Burton, C., Feest, T., Krieter, D., Gal, A., Brandis, M., Pierides, A., and Hildebrandt, F. (1998). Autosomal dominant medullary cystic kidney disease: Evidence of gene locus heterogeneity. *Nephrol Dial Transplant.* **13**(8), 1955–1957.

Gabow, P. A. (1993). Autosomal dominant polycystic kidney disease. *N. Engl. J. Med.* **329**(5), 332–342.

Gattone, V. H., Kuenstler, K. A., Lindemann, G. W., Lu, X. J., Cowley, B. D., Rankin, C. A., and Calvet, J. P. (1996). Renal expression of a transforming growth-factor-alpha transgene accelerates the progression of inherited, slowly progressive polycystic kidney-disease in the mouse. *J. Lab. Clin. Med.* **127**(2), 214–222.

Gattone, V. H., 2nd, MacNaughton, K. A., and Kraybill, A. L. (1996). Murine autosomal recessive polycystic kidney disease with

multiorgan involvement induced by the cpk gene. *Anat. Rec.* **245**(3), 488–499.

Geng, L., Burrow, C., Bloswick, B., and Wilson, P. (1998). Polycystin-1 is associated with focal adhesion proteins and the actin cytoskeleton to form large multiprotein complexes. *J. Am. Soc. Nephrol.* **9**, A1906.

Geiger, B., Bershadsky, A., Pankov, R., and Yamada, K. M. (2001). Transmembrane crosstalk between the extracellular matrix–cytoskeleton crosstalk. *Nature. Rev. Mol. Cell. Biol.* **2**(11), 793–805.

Gonzalez-Perret, S., Kim, K., Ibarra, C., Damiano, A. E., Zotta, E., Batelli, M., Harris, P. C., Reisin, I. L., Arnaout, M. A., and Cantiello, H. F. (2001). Polycystin-2, the protein mutated in autosomal dominant polycystic kidney disease (ADPKD), is a Ca^{2+}-permeable nonselective cation channel. *Proc. Natl. Acad. Sci. USA* **98**(3), 1182–1187.

Grantham, J. J. (1991). Acquired cystic kidney disease. *Kidney Int.* **40**(1), 143–152.

Green, K. J., and Gaudry, C. A. (2000). Are desmosomes more than tethers for intermediate filaments? *Nature. Rev. Mol. Cell Biol.* **1**(3), 208–216.

Guay-Woodford, L. (1996). Autosomal recessive polycystic kidney disease: Clinical and genetic profiles. In: "Polycystic Kidney Disease." (V., Waton MaT, ed), pp. 237–266 Oxford Univ. Press, Oxford.

Guay-Woodford, L. M., Bryda, E. C., Lindsay, J. R., Avner, E. D., and Flaherty, L. (1996). The mouse Bpk mutation, a model of autosomal recessive polycystic kidney-disease (Arpkd) and Jcpk, a phenotypically distinct Pkd mutation, are allelic. *Pediatr. Res.* **39**(4 Pt2), 2151–2151.

Guillaume, R., and Trudel, M. (2000). Distinct and common developmental expression patterns of the murine pkd2 and pkd1 genes. *Mech. Dev.* **93**(1–2), 179–183.

Gusmano, R., Ghiggeri, G. M., and Caridi, G. (1998). Nephronophthisis-medullary cystic disease: Clinical and genetic aspects. *J. Nephrol.* **11**(5), 224–228.

Hanaoka, K., Qian, F., Boletta, A., Bhunia, A. K., Piontek, K., Tsiokas, L., Sukhatme, V. P., Guggino, W. B., and Germino, G. G. (2000). Co-assembly of polycystin-1 and -2 produces unique cation-permeable currents. *Nature* **408**(6815), 990–994.

Hateboer, N., v Dijk, M. A., Bogdanova, N., Coto, E., Saggar-Malik, A. K., San Millan, J. L., Torra, R., Breuning, M., and Ravine. D. (1999). Comparison of phenotypes of polycystic kidney disease types 1 and 2: European PKD1-PKD2 Study Group. *Lancet* **353**(9147), 103–107.

Henry, K. W., Yuan, X., Koszewski, N. J., Onda, H., Kwiatkowski, D. J., and Noonan, D. J. (1998). Tuberous sclerosis gene 2 product modulates transcription mediated by steroid hormone receptor family members. *J. Biol. Chem.* **273**(32), 20535–20539.

Herron, B. J., Lu, W., Rao, C., Liu, S., Peters, H., Bronson, R. T., Justice, M. J, McDonald, J. D. and Beier, D. R. (2002). Efficient generation and mapping of recessive developmental mutations using ENU mutagenesis. *Nature. Genet.* **30**(2), 185–189.

Hildebrandt, F., Rensing, C., Betz, R., Sommer, U., Birnbaum, S., Imm, A., Omram, H., Leipoldt, M., and Otto, E. (2001) Establishing an algorithm for molecular genetic diagnostics in 127 families with juvenile nephronophthisis. *Kidney Int.* **59**(2), 434–445.

Hildebrandt, F., Otto, E., Rensing, C., Nothwang, H. G., Vollmer, M., Adolphs, J., Hanusch, H., and Brandis, M. (1997) A novel gene encoding an SH3 domain protein is mutated in nephronophthisis type 1. *Nature Genet.* **17**(2), 149–153.

Hildebrandt, F., and Omram, H. (2001). New insights: Nephronophthisis-medullary cystic kidney disease. *Pediatr Nephrol.* **16**(2), 168–176.

Hou, X., Mrug, M., Yoder, B. K., Lefkowitz, E. J., Kremmidiotis, G., D'Eustachio, P., Beier, D. R., and Guay-Woodford, L. M. (2002). Cystin, a novel cilia-associated protein, is disrupted in the cpk mouse model of polycystic kidney disease. *J. Clin. Invest.* **109**(4), 533–540.

Huan, Y., and van Adelsberg, J. (1999). Polycystin-1, the PKD1 gene product, is in a complex containing E-cadherin and the catenins. *J. Clin. Invest.* **104**(10), 1459–1468.

Hughes, J., Ward, C. J., Peral, B., Aspinwall, R., Clark, K., Sanmillan, J. L., Gamble, V., and Harris, P. C. (1995). The polycystic kidney-disease-

1 (Pkd1) gene encodes a novel protein with multiple cell recognition domains. *Nature Genet.* **10**(2), 151–160.

Huston, J. D., Torres, V. E., Sulivan, P. P., Offord, K. P., and Wiebers, D. O. (1993). Value of magnetic resonance angiography for the detection of intracranial aneurysms in autosomal dominant polycystic kidney disease. *J. Am. Soc. Nephrol.* **3**(12), 1871–1877.

Ibraghimov-Beskrovnaya, O., Bukanov, N. O., Donohue, L. C., Dackowski, W. R., Klinger, K. W., and Landes, G. M. (2000). Strong homophilic interactions of the Ig-like domains of polycystin-1, the protein product of an autosomal dominant polycystic kidney disease gene, PKD1. *Hum. Mol. Genet.* **9**(11), 1641–1649.

Ibraghimov-Beskrovnaya, E., Dackowski, W. R., Foggensteiner, L., Coleman, N., Thiru, S., Petry, L. R., Burn, T. C., Connors, T. D., Van Raay, T., Bradley, J., Qian, F., Onuchic, L. F., Watnick, T. J., Piontek, K., Hakim, R. M., Landes, G. M., Germino, G. G., Sandford, R., and Klinger, K. W. (1997). Polycystin: *In vitro* synthesis, *in vivo* tissue expression, and subcellular localization identifies a large membrane-associated protein. *Proc. Natl. Acad. Sci. USA* **94**(12), 6397–6402.

Jamil, B., McMahon, L. P., Savige, J. A., Wang, Y. Y., and Walker, R. G. (1999b). A study of long-term morbidity associated with autosomal recessive polycystic kidney disease. *Nephrol Dial Transplant.* **14**(1), 205–209.

Kim, E., Arnould, T., Sellin, L., Benzing, T., Comella, N., Kocher, O., Tsiokas, L., Sukhatme, V. P., and Walz, G. (1999). Interaction between RGS7 and polycystin. *Proc. Natl. Acad. Sci. USA* **96**(11), 6371–6376.

Kim, E., Arnould, T., Sellin, L. K., Benzing, T., Fan, M. J., Gruning, W., Sokol, S. Y., Drummond, I., and Walz, G. (1999). The polycystic kidney disease 1 gene product modulates Wnt signaling. *J. Biol. Chem.* **274**(8), 4947–4953.

Kiselyov, K., Mignery, G. A., Zhu, M. X., and Muallem, S. (1999). The N-terminal domain of the IP3 receptor gates store-operated hTrp3 channels. *Mol. Cell* **4**(3), 423–429.

Kleymenova, E., Ibraghimov-Beskrovnaya, O., Kugoh, H., Everitt, J., Xu, H., Kiguchi, K., Landes, G., Harris, P., and Walker, C. (2001). Tuberin-dependent membrane localization of polycystin-1: A functional link between polycystic kidney disease and the TSC2 tumor suppressor gene. *Mol. Cell* **7**(4), 823–832.

Koulen, P., Cai, Y., Geng, L., Maeda, Y., Nishimura, S., Witzgall, R., Ehrlich, B. E., and Somlo, S. (2002). Polycystin-2 is an intracellular calcium release channel. *Nat. Cell Biol.* **4**(3), 191–197.

Kreidberg, J. A., Donovan, M. J., Goldstein, S. L., Rennke, H., Stepherd, K., Jones, R. C., and Jaenisch, R. (1996). Alpha-3 beta-1 integrin has a crucial role in kidney and lung organogenesis. *Development* **122**(11), 3533–3547.

Lager, D. J., Qian, Q., Bengal, R. J., Ishibashi, M., and Torres, V. E. (2001). The pck rat: A new model that resembles human autosomal dominant polycystic kidney and liver disease. *Kidney Int.* **59**(1), 126–136.

Lamb, R. F., Roy, C., Diefenbach, T. J., Vinters, H. V., Johnson, M. W., Jay, D. G., and Hall, A. (2000). The TSC1 tumour suppressor hamartin regulates cell adhesion through ERM proteins and the GTPase Rho. *Nature Cell Biol.* **2**(5), 281–287.

Lanoix, J., D'Agati, V., Szabolcs, M., and Trudel, M. (1996). Dysregulation of cellular proliferation and apoptosis mediates human autosomal dominant polycystic kidney disease (ADPKD). *Oncogene* **13**(6), 1153–1160.

Latif, F., Tory, K., Gnarra, J., Yao, M., Duh, F. M., Orcutt, M. L., Stackhouse, T., Kuzmin, I., Modi, W., Geil, L., *et al.* (1993). Identification of the von Hippel-Lindau disease tumor suppressor gene. *Science* **260**(5112), 1317–1320.

Levine, E., Slusher, S. L., Grantham, J. J., and Wetzel, L. H. (1991). Natural history of acquired renal cystic disease in dialysis patients: A prospective longitudinal CT study. *AJR Am. J. Roentgenol.* **156**(3), 501–506.

Lo, S. H., Yu, Q. C., Degenstein, L., Chen, L. B., and Fuchs, E. (1997).

Progressive kidney degeneration in mice lacking tensin. *J. Cell Biol.* **136**(6), 1349–1361.

Lu, W., Peissel, B., Babakhanlou, H., Pavlova, A., Geng, L., Fan, X., Larson, C., Brent, G., and Zhou, J. (1977). Perinatal lethality with kidney and pancreas defects in mice with a targetted Pkd1 mutation. *Nature Genet.* **17**(2), 179–181.

Lu, W., Shen, X., Pavlova, A., Lakkis, M., Ward, C. J., Pritchard, L., Harris, P. C., Genest, D. R., Perez-Atayde, A. R., and Zhou, J. (2001). Comparison of Pkd1-targeted mutants reveals that loss of polycystin-1 causes cystogenesis and bone defects. *Hum. Mol. Genet.* **10**(21), 2385–2396.

Lumiaho, A., Ikaheimo, R., Miettinen, R., Niemitukia, L., Laitinen, T., Rantala, A., Lampainen, E., Laakso, M., and Hartikainen, J. (2001). Mitral valve prolapse and mitral regurgitation are common in patients with polycystic kidney disease type 1. *Am. J. Kidney. Dis.* **38**(6), 1208–1216.

MacDermot, K. D., Saggar-Malik, A. K., Economides, D. L., and Jeffery, S. (1998). Prenatal diagnosis of autosomal dominant polycystic kidney disease (PKD1) presenting in utero and prognosis for very early onset disease. *J. Med. Genet.* **35**(1), 13–16.

Malhas, A. N., Abuknesha, R. A., and Price, R. G. (2002). Interaction of the leucine-rich repeats of polycystin-1 with extracellular matrix proteins: Possible role in cell proliferation. *J. Am. Soc. Nephrol.* **13**(1), 19–26.

McCredie, D. A., and Baxter, T. J. (1976). Familial juvenile nephronophthisis: Report of a case including microdissection studies. *Aust. Paediatr. J.* **12**(2), 118–127.

Mochizuki, T., Saijoh, Y., Tsuchiya, K., Shirayoshi, Y., Takai, S., Taya, C., Yonekawa, H., Yamada, K., Nihei, H., Nakatsuji, N., Overbeek, P. A., Hamada, H., and Yokoyama, T. (1998). Cloning of inv, a gene that controls left/right asymmetry and kidney development. *Nature* **395**(6698), 177–181.

Mochizuki, T., Wu, G., Hayashi, T., Xenophontos, S. L., Veldhuisen, B., Saris, J. J., Reynolds, D. M., Cai, Y., Gabow, P. A., Pierides, A., Kimberling, W. J., Breuning, M. H., Deltas, C. C., Peters, D. J., and Somlo, S. (1996). PKD2, a gene for polycystic kidney disease that encodes an integral membrane protein. *Science* **272**(5266), 1339–1342.

Mochizuki, T., Wu, G., Hayashi, T., Xenophontos, S. L., Veldhuisen, B., Saris, J. J., Reynolds, D. M., Cai, Y., Gabow, P. A., Pierides, A., Kimberling, W. J., Breuning, M. H., Deltas, C. C., Peters, D. J., and Somlo, S. (1996). PKD2, a gene for polycystic kidney disease that encodes an integral membrane protein. *Science.* **272**(5266), 1339–1342.

Moser, M., Pscherer, A., Roth, C., Becker, J., Mucher, G., Zerres, K., Dixkens, C., Weis, J., Guay-Woodford, L., Buettner, R., and Fassler, R. (1997). Enhanced apoptotic cell death of renal epithelial cells in mice lacking transcription factor AP-2beta. *Genes Dev.* **11**(15), 1938–1948.

Mousson, C., Rabec, M., Cercueil, J. P., Virot, J. S., Hillon, P., and Rifle, G. (1997). Caroli's disease and autosomal dominant polycystic kidney disease: A rare association? *Nephrol. Dial. Transplant.* **12**(7), 1481–1483.

Moyer, J. H., Lee-Tischler, M. J., Kwon, H. Y., Schrick, J. J., Avner, E. D., Sweeney, W. E., Godfrey, V. L., Cacheiro, N. L., Wilkinson, J. E., and Woychik, R. P. (1994). Candidate gene associated with a mutation causing recessive polycystic kidney disease in mice. *Science* **264**(5163), 1329–1333.

Murcia, N. S., Richards, W. G., Yoder, B. K., Mucenski, M. L., Dunlap, J. R., and Woychik, R. P. (2000). The Oak Ridge Polycystic Kidney (orpk) disease gene is required for left-right axis determination. *Development* **127**(11), 2347–2355.

Nakanishi, K., Sweeney, W., Jr., and Avner, E. D. (2001). Segment-specific c-ErbB2 expression in human autosomal recessive polycystic kidney disease. *J. Am. Soc. Nephrol.* **12**(2), 379–384.

Nauta, J., Ozawa, Y., Sweeney, W. E., Jr., Rutledge, J. C., and Avner ED. (1993). Renal and biliary abnormalities in a new murine model of autosomal recessive polycystic kidney disease. *Pediatr. Nephrol.* **7**(2), 163–172.

Newby, L. J., Streets, A. J., Zhao, Y., Harris, P. C., Ward, C. J., and Ong, A. C. (2002). Identification, characterization, and localization of a novel kidney polycystin-1–polycystin-2 complex. *J. Biol. Chem.* **277**(23), 20763–20773.

Nickel, C., Benzing, T., Sellin, L., Gerke, P., Karihaloo, A., Liu, Z. X., Cantley, L. G., and Walz, G. (2002) The polycystin-1 C-terminal fragment triggers branching morphogenesis and migration of tubular kidney epithelial cells. *J. Clin. Invest.* **109**(4), 481–489.

Okada, Y., Nonaka, S., Tanaka, Y., Saijoh, Y., Hamada, H., and Hirokawa, N. (1994). Abnormal nodal flow precedes situs inversus in iv and inv mice. *Mol. Cell.* **4**(4), 459–468.

Omran, H., Haffner, K., Burth, S., Fernandez, C., Fargier, B., Villaquiran, A., Nothwang, H. G., Schnittger, S., Lehrach, H., Woo, D., Brandis, M., Sudbrak, R., and Hildebrandt, F. (2001). Human adolescent nephronophthisis: Gene locus synteny with polycystic kidney disease in pcy mice. *J. Am. Soc. Nephrol.* **12**(1), 107–113.

Parnell, S. C., Magenheimer, B. S., Maser, R. L., Rankin, C. A., Smine, A., Okamoto, T., and Calvet, J. P. (1998). The polycystic kidney disease-1 protein, polycystin-1, binds and activates heterotrimeric G-proteins *in vitro*. *Biochem. Biophys. Res. Communi.* **251**(2), 625–631.

Parnell, S. C., Magenheimer, B. S., Maser, R. L., Zien, C. A., Frischauf, A. M., and Calvet, J. P. (2002). Polycystin-1 activation of c-Jun-N-terminal kinase and AP-1 is mediated by heterotrimeric G proteins. *J. Biol. Chem.* **23**, 23.

Parnell, S. C., Magenheimer, B. S., Maser, R. L., Zien, C. A., Frischauf, A. M., and Calvet, J. P. (2002). Polycystin-1 activation of c-Jun N-terminal kinase and AP-1 is mediated by heterotrimeric G proteins. *J. Biol. Chem.* **277**(22), 19566–19572.

Parnell, S., Magenheimer, B., Maser, R., Rankin, C., Smine, A., Okamoto, T., and Calvet, J. (1998). The polycystic kidney disease protein, polycystin-1, binds and activates heterotrimeric G-proteins. *J. Am. Soc. Nephrol.* **9**, A1939.

Paterson, A. D., and Pei, Y. (1998). Is there a third gene for autosomal dominant polycystic kidney disease? *Kidney Int.* **54**(5), 1759–1761.

Pazour, G. J., San Agustin, J. T., Follit, J. A., Rosenbaum, J. L., and Witman, G. B. (2002). Polycystin-2 localizes to kidney cilia and the ciliary level is elevated in orpk mice with polycystic kidney disease. *Curr. Biol.* **12**(11), R378–R380.

Pei, Y. (2001). A 'two-hit' model of cystogenesis in autosomal dominant polycystic kidney disease? *Trends. Mol. Med.* **7**(4), 151–156.

Pennekamp, P., Karcher, C., Fischer, A., Schweickert, A., Skryabin, B., Horst, J., Blum, M., and Dworniczak, B. (2002). The ion channel polycystin-2 is required for left-right axis determination in mice. *Curr. Biol.* **12**(11), 938–943.

Peters, D. J., Spruit, L., Saris, J. J., Ravine, D., Sandkuijl, L. A., Fossdal, R., Boersma, J., van Eijk, R., Norby, S., and Constantinou-Deltas, C. D., *et al.* (1993): Chromosome 4 localization of a second gene for autosomal dominant polycystic kidney disease. *Nature Genet.* **5**(4), 359–362.

Pirson, Y., Chauveau, D., and Torres, V. (2002). Management of cerebral aneurysms in autosomal dominant polycystic kidney disease. *J. Am. Soc. Nephro.* **13**(1), 269–276.

Ponting, C. P., Hofmann, K., and Bork, P. (1999). A latrophilin/CL-1-like GPS domain in polycystin-1. *Curr. Biol.* **9**(16), R585–R588.

Preminger, G. M., Koch, W. E., Fried, F. A., McFarland, E., Murphy, E. D., and Mandell, J. (1982). Murine congenital polycystic kidney disease: A model for studying development of cystic disease. *J. Urol.* **127**(3), 556–560.

Pritchard, L., Sloane-Stanley, J. A., Sharpe, J. A., Aspinwall, R., Lu, W., Buckle, V., Strmecki, L., Walker, D., Ward, C. J., Alpers, C. E., Zhou, J., Wood, W. G., and Harris, P. C. (2000). A human PKD1 transgene generates functional polycystin-1 in mice and is associated with a cystic phenotype. *Hum. Mol. Genet.* **9**(18), 2617–2627.

Qian, F., Germino, F. J., Cai, Y. Q., Zhang, X. B., Somlo, S., and Germino, G. G. (1997). PKD1 interacts with PKD2 through a probable coiled-coil domain. *Nature Genet.* **16**(2), 179–183.

Ravine, D., Gibson, R. N., Walker, R. G., Sheffield, L. J., Kincaidsmith, P., and Danks, D. M. (1994). Evaluation of ultrasonographic diagnostic-criteria for autosomal-dominant polycystic kidney disease-1. *lancet* **343**(8901), 824–827.

Reynolds, D. M., Falk, C. T., Li, A., King, B. F., Kamath, P. S., Huston, I. J., Shub, C., Iglesias, D. M., Martin, R. S., Pirson, Y., Torres, V. E., and Somlo, S. (2000). Identification of a locus for autosomal dominant polycystic liver disease, on Chromosome 19p13. 2–13. 1. *Am. J. Hum. Genet.* **67**(6).

Richards, W. G., Sweeney, W. E., Yoder, B. K., Wilkinson, J. E, Woychik, R. P., and Avner, E. D. (1998). Epidermal growth factor receptor activity mediates renal cyst formation in polycystic kidney disease. *J. Clin. Invest.* **101**(5), 935–939.

Ricker, J. L., Gattone, V. H., 2nd, Calvet, J. P., and Rankin, C. A. (2000). Development of autosomal recessive polycystic kidney disease in BALB/c-cpk/cpk mice. *J. Am. Soc. Nephrol.* **11**(10), 1837–1847.

Ricker, J. L., Mata, J. E., Iversen, P. L., and Gattone, V. H. (2002). c-myc antisense oligonucleotide treatment ameliorates murine ARPKD. *Kidney Int.* **61**(Suppl, 1), 125–131.

Rossetti, S., Burton, S., Strmecki, L., Pond, G. R., San, Millan, J. L., Zerres, K., Barratt, T. M., Ozen, S., Torres, V. E., Bergstralh, E. J., Winearls, C. G., and Harris, P. C. (2002). The position of the polycystic kidney disease 1 (PKD1) gene mutation correlates with the severity of renal disease. *J. Am. Soc. Nephrol.* **13**(5), 1230–1237.

Rossetti, S., Chauveau, D., Walker, D., Saggar-Malik, A., Winearls, C. G., Torres, V. E., and Harris, P. C. (2002). A complete mutation screen of the ADPKD genes by DHPLC. *Kidney Int.* **61**(5), 1588–1599.

Rossetti, S., Strmecki, L., Gamble, V., Burton, S., Sneddon, V., Peral, B., Roy, S., Bakkaloglu, A., Komel, R., Winearls, C. G., and Harris, P. C. (2001). Mutation analysis of the entire PKD1 gene: Genetic and diagnostic implications. *Am. J. Hum. Genet.* **68**(1), 46–63.

Rossetti, S., Ward, C., and Harris, P. C. (1997). A strategy for mutation screening in the duplicated region of the polycystic kidney disease 1 (PKD1) gene. *J. Soc. Nephrol.* **8**, A1756.

Roy, S., Dillon, M. J., Trompeter, R. S., and Barratt, T. M. (1997). Autosomal recessive polycystic kidney disease: Long-term outcome of neonatal survivors. *Pediatr Nephrol.* **11**(3), 302–306.

Sandford, R. N., Sgotto, B., Hughes, J., Harris, P. C., and Lockwood, M. C. (1996). Comparative analysis of the Pkd1 gene and its predicted protein, polycystin. *J. Am. Soc. Nephrol.* **7**(9), A1864–A1864.

Sanzen, T., Harada, K., Yasoshima, M., Kawamura, Y., Ishibashi, M., and Nakanuma, Y. (2001). Polycystic kidney rat is a novel animal model of Caroli's disease associated with congenital hepatic fibrosis. *Am. J. Pathol.* **158**(5), 1605–1612.

Scheffers, M. S., van der, Bent, P., Prins, F., Spruit, L., Breuning, M. H., Litvinov, S. V., de Heer, E., and Peters, D. J. (2000). Polycystin-1, the product of the polycystic kidney disease 1 gene, co-localizes with desmosomes in MDCK cells. *Hum. Mol. Genet.* **9**(18), 2743–2750.

Schwartz, E. A., Leonard, M. L., Bizios, R., and Bowser, S. S. (1997). Analysis and modeling of the primary cilium bending response to fluid shear. *Am. J. Physiol.* **272**(1 Pt 2), F132–F138.

Scolari, F., Puzzer, D., Amoroso, A., Caridi, G., Ghiggeri, G. M., Maiorca, R., Aridon, P., De Fusco, M., Ballabio, A., and Casari, G. (1999). Identification of a new locus for medullary cystic disease, on chromosome 16p12. *Am. J. Hum. Genet.* **64**(6), 1655–1660.

Scolari, F., Ghiggeri, G. M., Casari, G., Amoroso, A., Puzzer, D., Caridi, G. L., Valzorio, B., Tardanico, R., Vizzardi, V., Savoldi, S., Viola, B. F., Bossini, N., Prati, E., Gusmano, R., and Maiorca, R. (1998). Autosomal dominant medullary cystic disease: A disorder with variable clinical pictures and exclusion of linkage with the NPH1 locus. *Nephrol. Dial. Transplant.* **13**(10), 2536–2546.

Shaulian, E., and Karin, M. (2002). AP-1 as a regulator of cell life and death. *Nature Cell. Biol.* **4**(5), E131–E136.

Sherman, F. E., Studnicki, F. M., and Fetterman, G. (1971). Renal lesions of familial juvenile nephronophthisis examined by microdissection. *Am. J. Clin. Pathol.* **55**(4), 391–400.

Sommardahl, C. S., Woychik, R. P., Sweeney, W. E., Avner, E. D., and Wilkinson, J. E. (1997). Efficacy of taxol in the orpk mouse model of polycystic kidney disease. *Pediat. Nephrol.* **11**(6), 728–733.

Sorenson, C. M., Padanilam, B. J., and Hammerman, M. R. (1996). Abnormal postpartum renal development and cystogenesis in the bcl-2 (–/–) mouse. *Am. J. Physiol.* **271**(1 Pt 2), F184–F193.

Soucek, T., Pusch, O., Wienecke, R., DeClue, J. E., and Hengstschlager, M. (1997). Role of the tuberous sclerosis gene-2 product in cell cycle control. Loss of the tuberous sclerosis gene-2 induces quiescent cells to enter S phase. *J. Biol. Chem.* **272**(46), 29301–29308.

Sutters, M., Yamaguchi, T., Maser, R. L., Magenheimer, B. S., St. John, P. L., Abrahamson, D. R., Grantham, J. J., and Calvet, J. P. (2001). Polycystin-1 transforms the cAMP growth-responsive phenotype of M-1 cells. *Kidney Int.* **60**(2), 484–494.

Sweeney, W. E., Chen, Y., Nakanishi, K., Frost, P., and Avner, E. D. (2000). Treatment of polycystic kidney disease with a novel tyrosine kinase inhibitor. *Kidney Int.* **57**(1), 33–40.

Takahashi, H., Calvet, J. P., Dittemore-Hoover, D., Yoshida, K., Grantham, J. J., and Gattone, V. H., 2nd. (1991). A hereditary model of slowly progressive polycystic kidney disease in the mouse. *J. Am. Soc. Nephrol.* **1**(7), 980–989.

Taulman, P. D., Haycraft, C. J., Balkovetz, D. F., and Yoder, B. K. (2001). Polaris, a protein involved in left-right axis patterning, localizes to basal bodies and cilia. *Mol. Biol. Cell.* **12**(3), 589–599.

Thomas, R., McConnell, R., Whittacker, J., Kirkpatrick, P., Bradley, J., and Sandford, R. (1999). Identification of mutations in the repeated part of the autosomal dominant polycystic kidney disease type 1 gene, PKD1, by long-range PCR. *Am. J. Hum. Genet.* **65**(1), 39–49.

Torres, V. L. (1996). Polycystic liver disease. *In:* "Polycystic Kidney Disease" (M. L. Watson, V. E. Torres, eds.), pp. 500–529 Oxford Univ. Press, Oxford.

Trudel, M., D'Agati, V., and Costantini, F. (1991). C-myc as an inducer of polycystic kidney disease in transgenic mice. *Kidney Int.* **39**(4), 665–671.

Trudel, M., Lanoix, J., Barisoni, L., Blouin, M. J., Desforges, M., L'Italien, C., and D'Agati, V. (1997). C-myc-induced, apoptosis in polycystic kidney disease is Bcl-2 and p53 independent. *J. Exp. Med.* **186**(11), 1873–1884.

van Slegtenhorst, M., de Hoogt, R., Hermans, C., Nellist, M., Janssen, B., Verhoef, S., Lindhout, D., van den Ouweland, A., Halley, D., Young, J., Burley, M., Jeremiah, S., Woodward, K., Nahmias, J., Fox, M., Ekong, R., Osborne, J., Wolfe, J., Povey, S., Snell, R. G., Cheadle, J. P., Jones, A. C., Tachataki, M., Ravine, D., Kwiatkowski, D. J., *et al.* (1997). Identification of the tuberous sclerosis gene TSC1 on chromosome 9q34. *Science* **277**(5327), 805–808.

van Slegtenhorst, M., Nellist, M., Nagelkerken, B., Cheadle, J., Snell, R., vandenOuweland, A., Reuser, A., Sampson, J., Halley, D., and vanderSluijs, P. (1998). Interaction between hamartin and tuberin, the TSC1 and TSC2 gene products. *Hum. Mol. Genet.* **7**(6), 1053–1057.

Vandorpe, D. H., Chernova, M. N., Jiang, L., Sellin, L. K., Wilhelm, S., Stuart-Tilley, A. K., Walz, G., and Alper, S. L. (2000). The cytoplasmic carboxy-terminal fragment of polycystin-1 (PKD1) regulates a Ca^2+-permeable cation channel. *J. Biol. Chem.*

Vassilev, P. M., Guo, L., Chen, X. Z., Segal, Y., Peng, J. B., Basora, N., Babakhanlou, H., Cruger, G., Kanazirska, M., Ye, C., Brown, E. M., Hediger, M. A., and Zhou, J. (2001). Polycystin-2 is a novel cation channel implicated in defective intracellular Ca(2+) homeostasis in polycystic kidney disease. *Biochem. Biophys. Res. Commun.* **282**(1), 341–350.

Ward, C. J., Hogan, M. C., Rossetti, S., Walker, D., Sneddon, T., Wang, X.,

Kubly, V., Cunningham, J. M., Bacallao, R., Ishibashi, M., Milliner, D. S., Torres, V. E., and Harris, P. C. (2002). The gene mutated in autosomal recessive polycystic kidney disease encodes a large, receptor-like protein. *Nature Genet.* **30**(3), 259–269.

Ward, C. J., Peral, B., Hughes, J., Thomas, S., Gamble, V., Maccarthy, A. B., Sloanestanley, J., Buckle, V. J., Kearney, L., Higgs, D. R., Ratcliffe, P. J., Harris, P. C., Roelfsema, J. H., Spruit, L., Saris, J. J., Dauwerse, H. G., Peters, D. J. M., Breuning, M. H., Nellist, M., Brookcarter, P. T., Maheshwar, M. M., Cordeiro, I., Santos, H., Cabral, P., Sampson, J. R., Janssen, B., Hesselingjanssen, A. L. W., Vandenouweland, A. M. W., Eussen, B., Verhoef, S., Lindhout, D., and Halley, D. J. J. (1994). The polycystic kidney-disease-1 gene encodes a 14-kb transcript and lies within a duplicated region on chromosome-16. *Cell.* **77**(6), 881–894.

Watnick, T. J., Qian, F., Onuchic, L., and Germino, G. G. (1996). The molecular basis of focal cyst formation in Pkd1. *J. Am. Soc. Nephrol.* **7**(9), A1884–A1884.

Watnick, T. J., Torres, V. E., Gandolph, M. A., Qian, F., Onuchic, L. F., Klinger, K. W., Landes, G., and Germino, G. G. (1998). Somatic mutation in individual liver cysts supports a two-hit model of cystogenesis in autosomal dominant polycystic kidney disease. *Mol. Cell* **2**(2), 247–251.

Weston, B. S., Bagneris, C., Price, R. G., and Stirling, J. L. (2001). The polycystin-1 C-type lectin domain binds carbohydrate in a calcium-dependent manner, and interacts with extracellular matrix proteins *in vitro*. *Biochim. Biophys. Acta.* **1536**(2–3), 161–176.

Wheatley, D. N., Wang, A. M., and Strugnell, G. E. (1996). Expression of primary cilia in mammalian cells. *Cell Biol. Int.* **20**(1), 73–81.

Wienecke, R., Konig, A., and DeClue, J. E. (1995). Identification of tuberin, the tuberous sclerosis-2 product. Tuberin possesses specific Rap1GAP activity. *J. Biol. Chem.* **270**(27), 16409–16414.

Wienecke, R., Maize, J. C., Jr., Shoarinejad, F., Vass, W. C., Reed, J., Bonifacino, J. S., Resau, J. H., de Gunzburg, J., Yeung, R. S, and DeClue, J. E. (1996). Co-localization of the TSC2 product tuberin with its target Rap1 in the Golgi apparatus. *Oncogene* **13**(5), 913–923.

Wilson, P. D. (1996). Pathogenesis of polycystic kidney disease: Altered cellular function. *In:* "Oxford Clinical Nephrology Series" M. L. Watson and V. E. Torres eds., pp. 125–163, Oxford Univ. Press, Oxford.

Wilson, P. D. (1997). Epithelial cell polarity and disease. *Am. J. Physiol.–Renal Physiol.* **41**(4), F434–F442.

Wilson, P. D., and Burrow, C. R. (1999). Cystic diseases of the kidney: Role of adhesion molecules in normal and abnormal tubulogenesis. *Exp. Nephrol.* **7**(2), 114–124.

Winyard, P. J. D., and Chitty, L. (2001). Dysplastic and polycystic kidneys: Diagnosis, associations, and management. *Prenatal Diagnosis* **21**, 924–935.

Winyard, P. J. D., Nanta, J., Lirenman, D. S., Hardman, P., Sams, V. R., Risdon, R. A., and Woolf, A. S. (1996). Deregulation of cell survival in cystic and dysplastic renal development. *Kidney Int.* **49**(1), 135–146.

Woo, D. (1995). Apoptosis and loss of renal tissue in polycystic kidney diseases. *N. Engl. J. Med.* **333**(1), 18–25.

Woo, D. D., Tabancay, A. P., Jr., and Wang, C. J. (1997). Microtubule active taxanes inhibit polycystic kidney disease progression in cpk mice. *Kidney Int.* **51**(5), 1613–1618.

Wu, G., Markowitz, G. S., Li, L., D'Agati, V. D., Factor, S. M., Geng, L., Tibara, S., Tuchman, J., Cai, Y., Hoon Park, J., van Adelsberg, J., Hou, H., Jr., Kucherlapati, R., Edelmann, W., and Somlo, S. (2000). Cardiac defects and renal failure in mice with targeted mutations in pkd2. *Nature Genet.* **24**(1), 75–78.

Wu, G. Q., Dagati, V., Cai, Y. Q., Markowitz, G., Park, J. H., Reynolds, D. M., Maeda, Y., Le T. C., Hou, H., Kucherlapati, R., Edelmann, W., and Somlo, S. (1998). Somatic inactivation of Pkd2 results in polycystic kidney disease. *Cell* **93**(2), 177–188.

Xiao, G. H., Shoarinejad, F., Jin, F., Golemis, E. A., and Yeung, R. S. (1997). The tuberous sclerosis 2 gene product, tuberin, functions as a

Rab5 GTPase activating protein (GAP) in modulating endocytosis. *J. Biol. Chem.* **272**(10), 6097–6100.

Yoder, B. K., Tousson, A., Millican, L., Wu, J. H., Bugg, C. E., J. r., Schafer, J. A., and Balkovetz, D. F. (2002). Polaris, a protein disrupted in orpk mutant mice, is required for assembly of renal cilium. *Am. J. Physiol. Renal. Physiol.* **282**(3), F541–F552.

Zerres, K., Mucher, G., Becker, J., Steinkamm, C., Rudnik-Schoneborn, S., Heikkila, P., Rapola, J., Salonen, R., Germino, G. G., Onuchic, L., Somlo, S., Avner, ED., Harman, L. A., Stockwin, J. M., and Guay-Woodford, L. M. (1998). Prenatal diagnosis of autosomal recessive polycystic kidney disease (ARPKD): Molecular genetics, clinical experience, and fetal morphology. *Am. J. Med. Genet.* **76**(2), 137–144.

Zerres, K., Rudnik-Schoneborn, S., Deget, F., Holtkamp, U., Brodehl, J., Geisert, J., and Scharer, K. (1996a). Autosomal recessive polycystic kidney disease in 115 children: Clinical presentation, course and influence of gender. *Arbeitsgemeinschaft Padiatri. Nephrol. Acta Paediatr.* **85**(4), 437–445.

Zerres, K., Rudnik-Schoneborn, S., Steinkamm, C., and Mucher, G. (1996b). Autosomal recessive polycystic kidney disease. *Nephrol Dial Transplant.* **11**(Suppl. 6), 29–33.

25

Renal Cell Carcinoma: The Human Disease

Cheryl Walker

I. **Phenotypic Diversity of Renal Cell Carcinoma (RCC)**

II. **Molecular Genetics of RCC**

III. **The von Hippel–Lindau Tumor Suppressor Gene**

IV. **TSC-2 Tumor Suppressor Gene**

V. **c-met**

VI. **Other Genes Involved in RCC**

VII. **Animal Models for RCC**

References

Tumors of many types arise in the kidney: embryonic nephroblastoma (Wilms' tumor), renal mesenchymal tumors such as sarcomas, and tumors of epithelial origin broadly classified as renal cell carcinoma (RCC). RCCs arise from the epithelial cells of the renal nephron, primarily its proximal and distal segments, and are by far the predominant class of kidney tumors seen in adults (Figlin, 1999; Motzer et al., 1996). The renal nephron, the primary architectural unit of the kidney, is a very complex structure composed of an estimated 19 types of epithelial cells (Kriz and Bankir, 1988). Each of these cell types is highly specialized, displaying different patterns of gene expression and, in some cases, even having different embryological origins. These specialized epithelial cells have a precise spatial organization within the nephron, which allows them to accomplish the many important physiologic functions of the kidney, including fluid filtration, excretion of wastes, and regulation of salt and electrolyte content. It is therefore not surprising

that RCC exhibits an enormous amount of phenotypic diversity given the functional diversity of the renal epithelial cells from which these tumors arise.

RCC is an adult onset epithelial malignancy, accounting for approximately 85% of all kidney tumors and ~12,000 deaths annually in the United States (Figlin, 1999; Motzer et al., 1996). Most cases of RCC occur in the fourth to sixth decades of life, with the peak incidence in the sixth decade. The incidence of RCC has increased by over 30% in the past two decades, a reflection of both earlier diagnosis and the linkage between RCC and associated risk factors such as cigarette smoking (see later). These tumors have a very poor prognosis, due in large part to the fact that nearly 30% of all patients exhibit metastatic disease at the time of initial presentation. In patients with apparently localized disease, 50% ultimately develop distant metastases after removal of the primary tumor (Motzer et al., 1996).

RCC has an extremely variable clinical course. It is one of the few malignancies for which spontaneous regression is well documented (Vogelzang et al., 1992), and the 5-year survival for patients with solitary metastases has been shown in several series to range from 5 to 50% (Dineen et al., 1988; Golimbu et al., 1991; Maldazys and deKernion, 1986; Middleton, 1967; Wagle and Scal, 1970). Because of the variability in the treatment-free interval (even in association with metastatic disease), few clinical prognostic factors that can guide treatment have been identified. These are essentially limited to high performance status and lung-only metastatic disease (Figlin, 1999; Motzer and Vogelzang, 1997). There is presently no standard chemotherapeutic approach for RCC and no recognized systemic therapy is effective at

451

reducing the probability of relapse (Motzer and Vogelzang, 1997). Cytotoxic chemotherapeutic strategies, which are the standard treatment for other solid malignancies, only have been minimally efficacious for RCC (Yagoda *et al.*, 1995).

There are many well-documented risk factors for RCC (Table 25.1) (Ellis, 1997). These tumors display a strong sex bias, being nearly twice as frequent in men as women (Kosary and McLaughlin, 1993). Cigarette smoking is an exogenous risk factor, with a well-documented association with RCC, with an increased relative risk of 1.5 to 2.5 in smokers compared with nonsmokers (Tavani and Vecchia, 1997; McLaughlin *et al.*, 1995). Analgesic abuse is associated with a 2- to 3-fold increased relative risk for RCC, and a similar increased relative risk in women has been reported to be associated with the use of diuretics. Dietary factors such as high protein and fat consumption and obesity are also associated with an increased risk for RCC. Remarkably, almost 10% of patients with acquired renal cystic disease, usually those with end-stage renal disease undergoing dialysis, develop RCC. It has been estimated that the risk of RCC in these patients is 30-fold higher than in the general population (Brennan *et al.*, 1991). The molecular etiology of RCC may differ among tumors that develop as a result of different risk factors. For example, RCC that develop in patients with end-stage renal disease are predominantly not of the clear cell variety (see later). In one study of these tumors, they were found to lack mutations in the von Hippel–Lindau (VHL) tumor suppressor gene, consistent with these tumors having an etiology distinct from that of sporadic RCC (Hughson *et al.*, 1996). For hereditary RCC, individuals at high risk include members of certain cancer syndrome families, most notably von Hippel–Lindau disease and tuberous sclerosis (Lamiell *et al.*, 1989; Neumann, 1987; Washecka and Hanna, 1991).

I. Phenotypic Diversity of Renal Cell Carcinoma (RCC)

RCC is remarkable in the extent of phenotypic diversity displayed by these tumors. The most commonly used

Table 25.1   Risk Factors for Renal Cell Carcinoma

Hereditary predispositions	von Hippel–Lindau syndrome, tuberous sclerosis
Cigarettes	20–40% of RCC attributable to smoking
Analgesic use	Phenactin and acetaminophen
Diuretic use	Strongest in women when prescribed for weight loss
Occupational exposure	Cadmium
	Aromatic hydrocarbons
	Petroleum products
	Asbestos
Obesity	
Acquired polycystic kidney disease	30-fold higher risk than general population

Table 25.2   Classification of Renal Cell Carcinomas

Cytological variants	Histological variants
Clear cell	Solid
Chromophilic	Tubulopapillary
Chromophobe	Cystic
Spindle cell/pleiomorphic	
Oncocytic	

classification scheme, that of Thoenes (Thoenes *et al.*, 1986), recognizes no less than five different cytological and three different histological variants of this disease (Table 25.2). How much of this diversity is attributable to epiphenomena versus events involved in disease causation is an important question. Solid RCC, predominantly of the clear cell variety, constitute ~85% of all adult epithelial neoplasms of the kidney and are by far the predominant variant seen in the clinic (Fig. 25.1). Papillary tumors (Fig. 25.1), which are usually chromophilic, make up another 10% of RCC variants, and oncocytomas account for most of the remaining 5% of solid tumors that occur in adults (True and Grignon, 1997). Diagnosis and subsequent prognosis of RCC are often complicated by the range of phenotypic diversity displayed by these tumors (Motzer *et al.*, 1996). For example, in contrast to the poor prognosis associated with other types of RCC, the oncocytoma variant is considered benign and has good prognosis (Kovacs *et al.*, 1989; Lehman and Blessing, 1982; Pearse and Houghton, 1979).

An important but unresolved issue about RCC is the distinction between adenoma and carcinoma. At present, there is no way to accurately determine the prognosis of these small lesions. It is clear that size alone is not sufficiently prognostic, as tumors as small as 0.5 cm have been documented to metastasize (True and Grignan, 1997). This is compounded by the fact that small lesions are being detected more frequently because of the increased sensitivity of modern imaging techniques. Small lesions are found in 20 to 40% of patients (True and Grignon, 1997), and adenomas are found at an even higher frequency in patients with end-stage renal disease. Papillary RCC are also often associated with multiple papillary adenomas, and these early lesions have been reported to display many of the cytogenetic abnormalities found in this RCC variant (Kovacs, 1993).

II. Molecular Genetics of RCC

Specific RCC variants have distinct molecular etiologies, illustrated by their distinctive chromosome alterations (Fig. 25.2). In fact, it has been suggested by Kovacs and

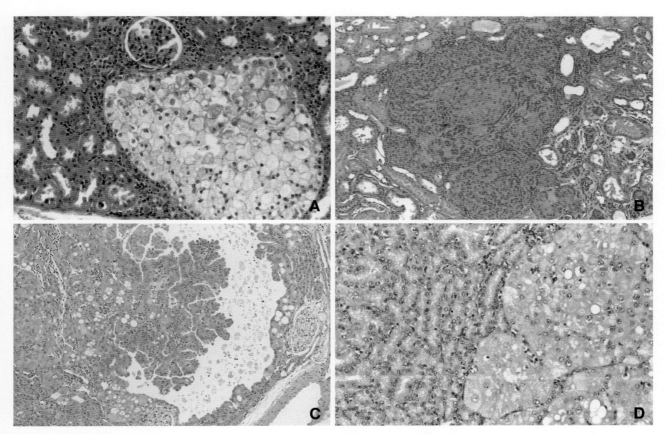

Figure 25.1 Histologic and cytologic RCC variants. H&E-stained sections of RCCs. (A) Typical solid, clear cell variant. (B) Solid chromophillic variant. (C) Papillary tumor. (D) Solid RCC exhibiting both eosinophillic and basophillic staining patterns.

others that a genetic approach to tumor diagnosis based on the pattern of chromosome alterations observed in particular types of tumors is a more appropriate and clinically relevant strategy than the currently used histological and cytological criteria (Kovacs, 1990; van den Berg *et al.*, 1997). For example, papillary tumors exhibit trisomy 7 and 17 and loss of the Y chromosome, whereas clear cell tumors are characterized by chromosome 3p deletions (Kovacs, 1993), suggesting that different molecular alterations are involved in the development of these two RCC variants. A distinct pathogenesis for papillary RCC is further supported by the observation that the kidneys of patients with papillary RCC often contain many focal areas of undifferentiated blastemal cells, so-called "metanephric rests," a finding specific to papillary RCC. This observation has been interpreted to suggest that papillary RCC may arise as a consequence of abortive nephrogenesis rather than from mature epithelial cells of the renal nephron (Kovacs, 1990). Several genes have been implicated in the development of different RCC variants, including the VHL and tuberous sclerosis 2 (TSC-2) tumor suppressor genes and the c-met protooncogene.

III. The von Hippel–Lindau Tumor Suppressor Gene

The VHL tumor suppressor gene is located on human chromosome 3p and is involved in both spontaneous and hereditary RCC (Decker *et al.*, 1997; Latif *et al.*, 1993). This gene is inactivated in tumors by several mechanisms, including mutation and silencing by DNA methylation (Gnarra *et al.*, 1994; Herman *et al.*, 1994; Shuin *et al.*, 1994; Whaley *et al.*, 1994). Even though the VHL tumor suppressor gene is expressed ubiquitously (Los *et al.*, 1996), its involvement in tumorigenesis appears to be restricted to RCC and some of the otherwise rare malignancies that occur with a high frequency in von Hippel-Lindau syndrome, such as hemangioblastoma (Kanno *et al.*, 1994; Oberstrass *et al.*, 1996), suggesting that its tumor suppressor activity is tissue specific. VHL is thought to be involved exclusively in the clear cell but not other RCC variants (Gnarra *et al.*, 1994; Kenck *et al.*, 1996), consistent with the hypothesis that different RCC variants may have different molecular etiologies. However, this conclusion may not be absolute, as loss

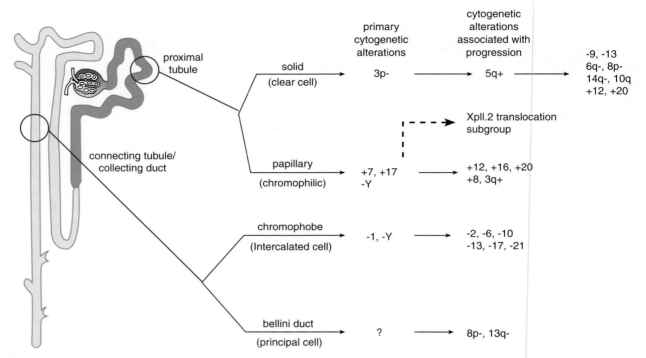

Figure 25.2 Cytogenetic alterations associated with RCC. Different histological and cytological RCC variants arise from different portions of the renal nephron and exhibit distinct cytogenetic alterations. Solid and papillary RCC both arise from the proximal tubule but contain different primary and secondary chromosome alterations. Chromophobe and Bellini duct tumors arise from intercalated and principal cells of the collecting duct, respectively, and also contain distinctive cytogenetic alterations. Primary cytogenetic alterations are those that occur early with a high frequency and are thought to be involved in the genesis of specific tumor subtypes, whereas additional chromosome alterations occur later during tumor development and are thought to be principally associated with tumor progression and/or metastasis.

of heterozygosity (LOH) at the VHL locus has been reported in collecting duct carcinomas (Fogt *et al.*, 1998).

Control of the presence or absence of many critical cellular regulatory proteins, such as those regulating the cell cycle, is often regulated at the level of protein stability. The primary function of the VHL tumor suppressor gene appears to be its ability to act as an E3 ubiquitin ligase, modulating protein stability by targeting proteins for ubiquitin-mediated degradation (Fig. 25.3). The VHL E3 ligase complex is composed of pVHL, which functions in substrate recognition, CUL-2, Elongins C and B, the RING finger protein Rbx, and an E2 ubiquitin ligase (Pause *et al.*, 1997; Kamura *et al.*, 1999; Iwai *et al.*, 1999; Gorospe *et al.*, 1999; Lisztwan *et al.*, 1999). This VBC (VHL/Elongin B,C/Cul-2 protein) complex has architectural homology with the SCF (Skp/Cul/F-box protein) family of E3 ligases (Stebbins *et al.*, 1999). Interestingly, mutations that inactivate the VHL tumor suppressor gene occur predominantly in the substrate recognition and Elongin B/C-binding regions of the protein, underscoring the importance of the E3 ligase activity of VHL to its function as a tumor suppressor (Tyers and Willems, 1999). Targets of the VBC ubiquitination complex include hypoxia-inducible factor 1α (HIF-1α) (Kamura *et al.*, 2000; Cockman *et al.*, 2000; Ohh *et al.*, 2000; Tanimoto

et al., 2000). HIF-1α is a transcription factor that mediates hypoxia-induced upregulation of genes such as vascular endothelial growth factor (VEGF), a potent inducer of angiogenesis. Inactivation of the VHL tumor suppressor gene results in loss of VBC ubiquitin ligase activity, stabilizing HIF-1α, and resulting in inappropriate expression of VEGF and other genes that can promote tumorigenesis.

IV. TSC-2 Tumor Suppressor Gene

The human tuberous sclerosis-2 (TSC-2) tumor suppressor gene located on chromosome 16 was cloned by the European Tuberous Sclerosis Consortium (1993), and its tumor suppressor function was first demonstrated for RCC in the Eker rat (see later). LOH of TSC-2 has also been shown to occur in renal hamartomas (Carbonara *et al.*, 1996; Green *et al.*, 1994; Henske *et al.*, 1995; Sepp *et al.*, 1996) and other tumors that occur frequently in tuberous sclerosis patients (Green *et al.*, 1994) who are also at increased risk for the development of RCC (Gomez *et al.*, 1999; Washecka and Hanna, 1991). Although less well studied than the VHL tumor suppressor gene, several cellular functions have been attributed to tuberin, the product of the TSC-2 tumor

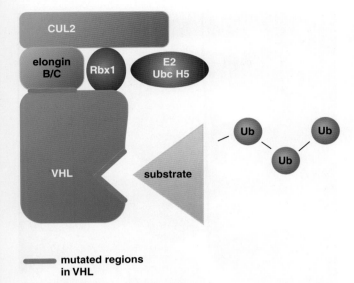

mutated regions in VHL

Figure 25.3 The VBC E3 ligase complex. One of the principal functions of the VHL tumor suppressor gene appears to be its ability to function as an E3 ligase. The VBC ubiquitin ligase complex is composed of several components, including VHL, which participates in substrate recognition, Elongins B and C, CUL2, the E2 UbcH5, and the RING protein Rbx1. Similar to the SCF family of ubiquitin ligases, the VBC complex targets protein substrates for degradation by the proteosome via ubiquitin conjugation.

suppressor gene. Tuberin shares some amino acid homology with GTPase-activating proteins (GAPs). Tuberin has been demonstrated to have GAP activity for Rab5 (Xiao *et al.*, 1997), which participates in endocytosis by regulating the fusion of early endosomes, and Rap1a (Wienecke *et al.*, 1995, 1996), which participates in signal transduction. A role for tuberin in regulating the cell cycle has also been suggested, where it may modulate the activity of p27 and regulate the G0 to G1 transition (Soucek *et al.*, 1997, 1998). Recently, tuberin has been shown to negatively regulate PBK signaling downstream of Akt and to be a target of this kinase (Manning *et al.*, 2001; Potter *et al.*, 2002; Dan *et al.*, 2002; Inoki *et al.*, 2002; Goncharova *et al.*, 2002; Liu *et al.*, 2002; Nellist *et al.*, 2002; Li *et al.*, 2002).

V. c-met

Hereditary papillary RCC has been shown to occur as a result of a germline mutation in the c-met protooncogene (Schmidt *et al.*, 1997). In epithelial cells, c-met functions as a receptor tyrosine kinase for hepatocyte growth factor (HGF), and germline mutations in c-met constitutively activate its kinase function (Fig. 25.4). Interestingly, the role of c-met in sporadic papillary RCC is quite variable. Schmidt *et al.* (1999) reported that only 17 of 129 sporadic papillary RCC contained c-met mutations. The differential frequency with which c-met becomes activated in

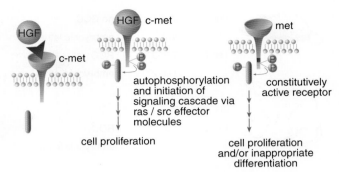

autophosphorylation and initiation of signaling cascade via ras / src effector molecules

cell proliferation

constitutively active receptor

cell proliferation and/or inappropriate differentiation

Figure 25.4 Constitutively active c-met participates in hereditary papillary RCC. Binding of HGF to its receptor c-met results in c-met auto-phosphorylation and intracellular signaling that can stimulate cell proliferation. Normally, HGF is produced by mesenchymal cells in a paracrine or endocrine fashion, signaling to c-met receptors located on adjacent or distant epithelial cells. Germline alterations in the c-met receptor have been identified in hereditary papillary RCC that constitutively activate the receptor, abrogating the need for HGF, stimulating cell proliferation inappropriately, and promoting tumorigenesis.

hereditary versus sporadic papillary RCC suggests that even within this one class of tumors, differences exist in the molecular etiology of hereditary versus sporadic forms of this disease.

VI. Other Genes Involved in RCC

Altered expression of transforming growth factor-α (TGF-α) and its cognate receptor, the EGF receptor (EGF-R), occurs with a high frequency in RCC (Fig. 26.5) (Mydlo *et al.*, 1989; Sargent *et al.*, 1989; Gomella *et al.*, 1989). TGF-α, which is normally considered a fetal mitogen, is not expressed to any significant degree by normal epithelial cells of the proximal or distal nephron in the adult kidney.

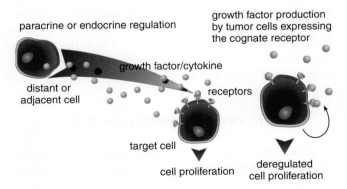

paracrine or endocrine regulation

growth factor production by tumor cells expressing the cognate receptor

growth factor/cytokine

distant or adjacent cell

receptors

target cell

cell proliferation

deregulated cell proliferation

Figure 25.5 TGF-α autocrine growth stimulation in RCC. Growth factors are normally secreted by cells in a paracrine or endocrine manner to stimulate cell proliferation of distant or adjacent cells that express receptors for these growth factors. Simultaneous expression of growth factors and their cognate receptors, e.g., TGF-α and EGF-R, can result in autocrine stimulation of cell proliferation and uncontrolled cell growth that can promote tumorigenesis.

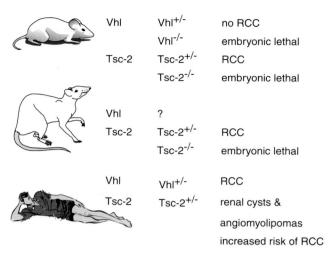

	Vhl	Vhl[+/-]	no RCC
		Vhl[-/-]	embryonic lethal
	Tsc-2	Tsc-2[+/-]	RCC
		Tsc-2[-/-]	embryonic lethal
	Vhl	?	
	Tsc-2	Tsc-2[+/-]	RCC
		Tsc-2[-/-]	embryonic lethal
	Vhl	Vhl[+/-]	RCC
	Tsc-2	Tsc-2[+/-]	renal cysts & angiomyolipomas increased risk of RCC

Figure 25.6 Species-specific differences in genetic factors predisposing to RCC. Germline alteration of a tumor suppressor gene is the strongest known risk factor for the development of cancer. Both VHL and TSC-2 tumor suppressor genes are involved in RCC, but they display distinct species-specific profiles in terms of their impact on the development of RCC. Germline alteration of the VHL tumor suppressor gene predisposes to the development of RCC in humans but not mice. Conversely, germline alterations in the Tsc-2 tumor suppressor gene predispose to RCC in mice and rats, but have a less pronounced role in the development of the human disease.

However, TGF-α is expressed abundantly in RCCs of the clear cell and papillary type derived from this region of the renal nephron, and these tumors also express high levels of EGF-R (Petrides *et al.*, 1990; Ishikawa *et al.*, 1990; Yao *et al.*, 1988). The concomitant expression of TGF-α and its receptor allows TGF-α to act as an autocrine growth factor in RCC, stimulating cell proliferation and promoting tumor growth (Atlas *et al.*, 1992). In addition to tumor suppressors (VHL, TSC-2) and the oncogene met, which have been clearly shown to be involved in the etiology of RCC due to their ability to predispose to these tumors via germline mutations, several other genes, such as phosphatase-tensin homologue (PTEN) and fragile histidine triad (FHIT), have been shown to be altered in RCC, but their association with specific subtypes of RCC has not been elucidated (Steck *et al.*, 1997; Teng *et al.*, 1997; Velickovic *et al.*, 1999; Hadaczek *et al.*, 1998).

VII. Animal Models for RCC

Two animal models for RCC have been described, which in addition to being useful experimental tools for studying

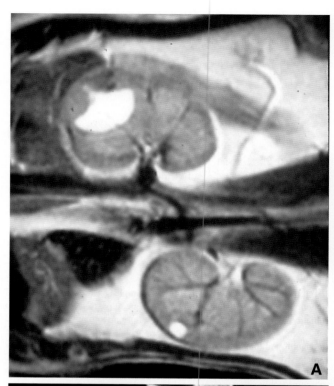
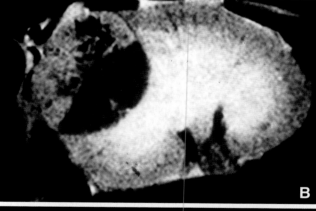
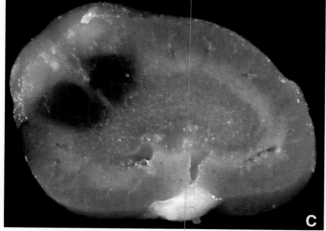

Figure 25.7 RCC development in the Eker rat. The Eker mutation in the Tsc-2 tumor suppressor gene predisposes to RCC in heterozygous rats carrying this mutation. Tumors that occur in these animals are multifocal and develop bilaterally. (A) MRI of Eker rat showing tumors on both kidneys. (B) MRI image of upper kidney shown in A. (C) Section through same kidney showing tumor at the upper left. Both cystic (black) and solid regions can be discriminated.

RCC, have served to highlight some interesting species-specific differences in the molecular etiology of this disease (Fig. 24.6). The Eker rat model for RCC was first described as an autosomal-dominant inherited form of RCC in the early 1960s (Eker and Mosige, 1962) and was later determined to be caused by a spontaneous mutation in the rat homologue of the Tsc-2 tumor suppressor gene (Yeung et al., 1994, Kobayashi et al., 1995). Rats heterozygous for the defective Tsc-2 allele in which the Eker mutation resides (Tsc-2$^{Ek/+}$) develop multifocal, bilateral RCC with 100% incidence by 12 months of age (Fig. 25.6) (Everitt et al., 1992). Tumors arise as a result of spontaneous or carcinogen-induced somatic alterations in the wild-type Tsc-2 allele, resulting in loss of function of this tumor suppressor gene (Walker et al., 1992; Yeung et al., 1995; Kobayashi et al., 1997). Tumors that arise in these rats are usually solid, chromophilic lesions (Everitt et al., 1992) (Figure 25.1), distinguishing them from the solid, clear cell variant that predominates in humans. Consistent with their histology, Eker rat RCC do not exhibit alterations in the Vhl tumor suppressor gene (Walker et al., 1996). Homozygosity of the Eker mutation (Tsc-2$^{Ek/Ek}$) is embryonic lethal, with mutant embryos dying between E11 and E13 of embryonic development (Rennebeck et al., 1998). Tsc-2 knockout mice have been described by two groups, which also develop spontaneous RCC with a high frequency (Onda et al., 1999; Kobayashi et al., 1999). Similar to the Eker rat model, tumors arise in heterozygous mice, and homozygous loss of Tsc-2 in these mice is embryonic lethal.

In contrast to the prominent role of Tsc-2 in the development of RCC in both mice and rats, inactivation of the mouse homologue of the VHL tumor suppressor gene does not result in susceptibility to RCC (Gnarra et al., 1997). Whereas germline inactivation of the VHL tumor suppressor gene predisposes humans to RCC, Vhl knockout mice fail to develop spontaneous RCC or precursor lesions such as dysplasias or adenomas associated with this disease. Heterozygous Vhl$^{+/-}$ mice are phenotypically normal, but homozygous mutant mice (Vhl$^{-/-}$) die between E10 and E13 of development due to defective placental vasculogenesis (Gnarra et al., 1997). Interestingly, while inactivation of the Vhl gene in mice may not contribute to the development of RCC, the case may be different in rats. Although observed only rarely in this species, RCC of the clear cell type have been noted to develop in rats in response to some nitroso-amine carcinogens, and these tumors have been shown to carry mutations in the rat Vhl gene (Shiao et al., 1998). Thus there is clearly species specificity in the molecular etiology of RCC, with the VHL tumor suppressor gene playing a prominent role in human RCC and the Tsc-2 tumor suppressor gene being the primary target for the development of these tumors in rodents. While the Vhl tumor suppressor gene appears able to participate in RCC in rats under very limited circumstances, the potential involvement of the TSC-2 tumor suppressor gene in human RCC remains an open question.

References

Atlas, I., Mendelsohn, J., Baselga, J., Fair, W. R., Masui, H., and Kumar, R. (1992). Growth regulation of human renal carcinoma cells: Role of transforming growth factor alpha. *Cancer Res.* **52**, 3335–3339.

Brennan, J. F., Stilmant, M. M., Babayan, R. K., and Siroky, M. B. (1991). Acquired renal cystic disease: Implications for the urologist. *Br. J. Urol.* **67**, 342–348.

Carbonara, C., Longa, L., Grosso, E., Mazzucco, G., Borrone, C., Garre, M. L., Brisigotti, M., Filippi, G., Scabar, A., Giannotti, A., Falzoni, P., Monga, G., Garini, G., Gabrielli, M., Riegler, P., Danesino, C., Ruggieri, M., Magro, G., and Migone, N. (1996). Apparent preferential loss of heterozygosity at TSC2 over TSC1 chromosomal region in tuberous sclerosis hamartomas. *Genes Chromo. Cancer* **15**, 18–25.

Cockman, M. E., Masson, N., Mole, D. R., Jaakkola, P., Chang, G. W., Clifford, S. C., Maher, E. R., Pugh, C. W., Ratcliffe, P. J., and Maxwell, P. H. (2000). Hypoxia inducible factor-alpha binding and ubiquitylation by the von Hippel-Lindau tumor suppressor protein. *J. Biol. Chem.* **275**, 25733–25741.

Dan, H. C., Sun, M., Yang, L., Feldman, R., Sui, X.-M., Yeung, R. S., Halley, D. J. J., Nicosia, S. V., Pledger, W. J., and Cheng, J. Q. (2002). PI3K/AKT pathway regulates TSC tumor suppressor complex by phosphorylation of tuberin. *J. Biol. Chem.* **277**, 35364–35370.

Decker, H. J., Weidt, E. J., and Brieger, J. (1997). The von Hippel-Lindau tumor suppressor gene: A rare and intriguing disease opening new insight into basic mechanisms of carcinogenesis. *Cancer Genet. Cytogenet.* **93**, 74–83.

Dineen, M. K., Pastore, R. D., Emrich, L. J., and Huben, R. P. (1988). Results of surgical treatment of renal cell carcinoma with solitary metastasis. *J. Urol.* **140**, 277–279.

Ellis, W. J. (1997). Epidemiology and etiology of renal cell carcinoma. In "Principles and Practice of Genitourinary Oncology" (D. Raghavan, H. I. Scher, S. A. Leibel, and P. H. Lange, eds), pp. 795–798. Lippincott-Raven, Philadelphia.

European Tuberous Sclerosis Consortium (1993). Identification and characterization of the tuberous sclerosis gene on chromosome 16. *Cell* **75**, 1305–1315.

Everitt, J. I., Goldsworthy, T. L., Wolf, D. C., and Walker, C. L. (1992). Hereditary renal cell carcinoma in the Eker rat: A rodent familial cancer syndrome. *J. Urol.* **148**, 1932–1936.

Figlin, R. A. (1999). Renal cell carcinoma: Management of advanced disease. *J. Urol.* **161**, 381–386; discussion 386–387.

Fogt, F., Zhuang, Z., Linehan, W. M., and Merino, M. J. (1998). Collecting duct carcinomas of the kidney: A comparative loss of heterozygosity study with clear cell renal cell carcinoma. *Oncol. Rep.* **5**, 923–926.

Gnarra, J. R., Tory, K., Weng, Y., Schmidt, L., Wei, M. H., Li, H., Latif, F., Liu, S., Chen, F., Duh, F. M., et al. (1994). Mutations of the VHL tumour suppressor gene in renal carcinoma. *Nature Genet.* **7**, 85–90.

Gnarra, J. R., Ward, J. M., Porter, F. D., Wagner, J. R., Devor, D. E., Grinberg, A., Emmert-Buck, M. R., Westphal, H., Klausner, R. D., and Linehan, W. M. (1997). Defective placental vasculogenesis causes embryonic lethality in VHL- deficient mice. *Proc. Natl. Acad. Sci. USA* **94**, 9102–9107.

Golimbu, M., Joshi, P., and Sperber, A. (1991). Renal cell carcinoma: Survival and prognostic factors. *Urology* **27**, 291.

Gomella, L. G., Sargent, E. R., Wade, T. P., Anglard, P., Linehan, W. M., and Kasid, A. (1989). Expression of transforming growth factor alpha in normal human adult kidney and enhanced expression of transforming growth factors alpha and beta 1 in renal cell carcinoma. *Cancer Res.* **49**, 6972–6975.

Gomez, M. R., Sampson, J. R. and Whittemore, V. H. (1999). "Tuberous Sclerosis Complex" Oxford Univ. Press, Oxford.

Goncharova, E. A., Goncharov, D. A., Eszterhas, A., Hunter, D. S., Glassberg, M. K., Yeung, R. S., Walker, C. L., Noonan, D., Kwiatkowski, D. J., Chou, M. M., Panettieri, R. A., Jr., and Krymskaya, V. P. (2002). Tuberin regulates p70 S6 kinase activation and ribosomal protein S6 phosphorylation: a role for the TSC2 tumor suppressor gene in pulmonary lymphangioleiomyomatosis (LAM). *J. Biol. Chem.* **277**, 30958–30967.

Gorospe, M., Egan, J. M., Zbar, B., Lerman, M., Geil, L., Kuzmin, I., and Holbrook, N. J. (1999). Protective function of von Hippel-Lindau protein against impaired protein processing in renal carcinoma cells. *Mol. Cell. Biol.* **19**, 1289–1300.

Green, A. J., Smith, M., and Yates, J. R. (1994). Loss of heterozygosity on chromosome 16p13.3 in hamartomas from tuberous sclerosis patients. *Nature Genet.* **6**, 193–196.

Hadaczek, P., Siprashvili, Z., Markiewski, M., Domagala, W., Druck, T., McCue, P. A., Pekarsky, Y., Ohta, M., Huebner, K., and Lubinski, J. (1998). Absence or reduction of Fhit expression in most clear cell renal carcinomas. *Cancer Res.* **58**, 2946–2951.

Henske, E. P., Neumann, H. P., Scheithauer, B. W., Herbst, E. W., Short, M. P., and Kwiatkowski, D. J. (1995). Loss of heterozygosity in the tuberous sclerosis (TSC2) region of chromosome band 16p13 occurs in sporadic as well as TSC-associated renal angiomyolipomas. *Genes Chromosomes Cancer* **13**, 295–298.

Herman, J. G., Latif, F., Weng, Y., Lerman, M. I., Zbar, B., Liu, S., Samid, D., Duan, D. S., Gnarra, J. R., Linehan, W. M., *et al.* (1994). Silencing of the VHL tumor-suppressor gene by DNA methylation in renal carcinoma. *Proc. Natl. Acad. Sci. USA* **91**, 9700–9704.

Hughson, M. D., Schmidt, L., Zbar, B., Daugherty, S., Meloni, A. M., Silva, F. G., and Sandberg, A. A. (1996). Renal cell carcinoma of end-stage renal disease: A histopathologic and molecular genetic study. *J. Am. Soc. Nephrol.* **7**, 2461–2468.

Inoki, K., Li, Y., Zhu, T., Wu, J., and Guan, K. L. (2002). TSC2 is phosphorylated and inhibited by Akt and suppressor mTOR signalling. *Nat. Cell Biol.* **4**, 648–657.

Ishikawa, J., Maeda, S., Umezu, K., Sugiyama, T., and Kamidono, S. (1990). Amplification and overexpression of the epidermal growth factor receptor gene in human renal-cell carcinoma. *Int. J. Cancer* **45**, 1018–1021.

Iwai, K., Yamanaka, K., Kamura, T., Minato, N., Conaway, R. C., Conaway, J. W., Klausner, R. D., and Pause, A. (1999). Identification of the von Hippel-lindau tumor-suppressor protein as part of an active E3 ubiquitin ligase complex. *Proc. Natl. Acad. Sci. USA* **96**, 12436–12441.

Kamura, T., Koepp, D. M., Conrad, M. N., Skowyra, D., Moreland, R. J., Iliopoulos, O., Lane, W. S., Kaelin, W. G., Jr., Elledge, S. J., Conaway, R. C., Harper, J. W., and Conaway, J. W. (1999). Rbx1, a component of the VHL tumor suppressor complex and SCF ubiquitin ligase. *Science* **284**, 657–661.

Kamura, T., Sato, S., Iwai, K., Czyzyk-Krzeska, M., Conaway, R. C., and Conaway, J. W. (2000). Activation of HIF1alpha ubiquitination by a reconstituted von Hippel-Lindau (VHL) tumor suppressor complex. *Proc. Natl. Acad. Sci. USA* **97**, 10430–10435.

Kanno, H., Kondo, K., Ito, S., Yamamoto, I., Fujii, S., Torigoe, S., Sakai, N., Hosaka, M., Shuin, T., and Yao, M. (1994). Somatic mutations of the von Hippel-Lindau tumor suppressor gene in sporadic central nervous system hemangioblastomas. *Cancer Res.* **54**, 4845–4847.

Kenck, C., Wilhelm, M., Bugert, P., Staehler, G. and Kovacs, G. (1996). Mutation of the VHL gene is associated exclusively with the development of non-papillary renal cell carcinomas. *J. Pathol.* **179**, 157–161.

Kobayashi, T., Hirayama, Y., Kobayashi, E., Kubo, Y., and Hino, O. (1995). A germline insertion in the tuberous sclerosis (Tsc2) gene gives rise to the Eker rat model of dominantly inherited cancer. *Nature Genet.* **9**, 70–74.

Kobayashi, T., Minowa, O., Kuno, J., Mitani, H., Hino, O., and Noda, T. (1999). Renal carcinogenesis, hepatic hemangiomatosis, and embryonic lethality caused by a germ-line *Tsc2* mutations in mice. *Cancer Res.* **59**, 1206–1211.

Kobayashi, T., Urakami, S., Hirayma, Y., Yamamoto, T., Nishizawa, M., Takahara, T., Kubo, Y., and Hino. O. (1997). Intragenic *Tsc2* somatic mutations as Knudson's second hit in spontaneous and chemically induced renal carcinomas in the Eker rat model. *Jpn. J. Cancer Res.* **88**, 245–261.

Kosary, C. L., and McLaughlin, J. K. (1993). Kidney and renal pelvis. In "SEER Cancer Statistics Review 1973–1990", (B. A. Miller, L. A. G. Ries, B. F. Hankey, *et al.*, eds.). National Cancer Institute, Bethesda, MD.).

Kovacs, G. (1990). Application of molecular cytogenetic techniques to the evaluation of renal parenchymal tumors. *J. Cancer Res. Clin. Oncol.* **116**, 318–323.

Kovacs, G. (1993). Molecular cytogenetics of renal cell tumors. *Adv. Cancer Res.* **62**, 89–124.

Kovacs, G., Welter, C., Wilkens, L., Blin, N., and Deriese, W. (1989). Renal oncocytoma. A phenotypic and genotypic entity of renal parenchymal tumors. *Am. J. Pathol.* **134**, 967–971.

Kriz, W., and Bankir, L. (1988). A standard nomenclature for structure of the kidney. The Renal Commission of the International Union of Physiological Sciences (IUPS). *Anat Embryol.* **178**, N1–N8.

Lamiell, J. M., Salazar, F. G., and Hsia, Y. E. (1989). von Hippel-Lindau disease affecting 43 members of a single kindred. *Medicine (Baltimore)* **68**, 1–29.

Latif, F., Tory, K., Gnarra, J., Yao, M., Duh, F. M., Orcutt, M. L., Stackhouse, T., Kuzmin, I., Modi, W., Geil, L., *et al.* (1993). Identification of the von Hippel-Lindau disease tumor suppressor gene. *Science* **260**, 1317–1320.

Lehman, H. D., and Blessing, M. H. (1982). Renal oncocytoma (pathology, preoperative diagnosis, therapy). *Prog. Clin. Biol. Res.* **100**, 589–596.

Li, Y., Inoki, K., Yeung, R., and Guan, K. L. (2002). Regulation of TSC2 by 14-3-3 binding. *J. Biol. Chem.* **2**, 2.

Lisztwan, J., Imbert, G., Wirbelauer, C., Gstaiger, M., and Krek, W. (1999). The von Hippel-Lindau tumor suppressor protein is a component of an E3 ubiquitin-protein ligase activity. *Genes Dev.* **13**, 1822–1833.

Liu, M. Y., Cai, S., Espejo, A., Bedford, M. T., and Walker, C. L. (2002). 14-3-3 interacts with the tumor suppressor tuberin at Akt phosphorylation site(s). *Cancer Res.* in press.

Los, M., Jansen, G. H., Kaelin, W. G., Lips, C. J., Blijham, G. H., and Voest, E. E. (1996). Expression pattern of the von Hippel-Lindau protein in human tissues. *Lab. Invest.* **75**, 231–238.

Maldazys, J. D., and deKernion, J. B. (1986). Prognostic factors in metastatic renal carcinoma. *J. Urol.* **136**, 376–379.

Manning, B. D., Tee, A. R., Logsdon, M. N., Blenis, J., and Cantley, L. C. (2002). Identification of the tuberous sclerosis complex-2 tumor suppressor gene product tuberin as a target of the phosphoinositide 3-kinase/akt pathway. *Mol. Cell.* **10**, 151–162.

McLaughlin, J. K., Lindblad, P., Mellemgaard, A., McCredie, M., Mandel, J. S., Schlehofer, B., Pommer, W., and Adami, H. O. (1995). International renal-cell cancer study. I. Tobacco use. *Int. J. Cancer* **60**, 194–198.

Middleton, R. G. (1967). Surgery for metastatic renal cell carcinoma. *J. Urol.* **97**, 973–977.

Motzer, R. J., Bander, N. H., and Nanus, D. M. (1996). Renal-cell carcinoma. *N. Engl. J. Med.* **335**, 865–875.

Motzer, R. J., and Vogelzang, N. J. (1997). Chemotherapy for renal cell carcinoma. In "Principles and Practice of Genitourinary Oncology," (D. Raghavan, H. I. Scher, S. A. Leibel, and P. H. Lange, eds.), pp. 885–896. Lippincott-Raven, Philadelphia.

Mydlo, J. H., Michaeli, J., Cordon-Cardo, C., Goldenberg, A. S., Heston, W. D., and Fair, W. R. (1989). Expression of transforming growth factor alpha and epidermal growth factor receptor messenger RNA in neoplastic and nonneoplastic human kidney tissue. *Cancer Res.* **49**, 3407–3411.

Nellist, M., Goedbloed, M. A., de Winter, C., Verhaaf, B., Jankie, A., Reuser, A. J. J., van den Ouweland, A. M. W., van der Sluijs, P., and Halley, D. J. J. (2002). Identification and characterization of the interaction between tuberin and 14-3-3 ζ. *J. Biol. Chem.* **277**, 39417–39424.

Neumann, H. P. (1987). Basic criteria for clinical diagnosis and genetic counselling in von Hippel-Lindau syndrome. *Vasa* **16**, 220–226.

Oberstrass, J., Reifenberger, G., Reifenberger, J., Wechsler, W., and Collins, V. P. (1996). Mutation of the Von Hippel-Lindau tumour suppressor gene in capillary haemangioblastomas of the central nervous system. *J. Pathol.* **179**, 151–156.

Ohh, M., Park, C. W., Ivan, M., Hoffman, M. A., Kim, T. Y., Huang, L. E., Pavletich, N., Chau, V., and Kaelin, W. G. (2000). Ubiquitination of hypoxia-inducible factor requires direct binding to the beta-domain of the von Hippel-Lindau protein. *Nature Cell Biol.* **2**, 423–427.

Onda, H., Lueck, A., Marks, P. W., Warren, H. B., and Kwiatkowski, D. J. (1999). Tsc2(+/–) mice develop tumors in multiple sites that express gelsolin and are influenced by genetic background. *J. Clin. Invest.* **104**, 687–695.

Pause, A., Lee, S., Worrell, R. A., Chen, D. Y., Burgess, W. H., Linehan, W. M., and Klausner, R. D. (1997). The von Hippel-Lindau tumor-suppressor gene product forms a stable complex with human CUL-2, a member of the Cdc53 family of proteins. *Proc. Natl. Acad. Sci. USA* **94**, 2156–2161.

Pearse, H. D., and Houghton, D. C. (1979). Renal oncocytoma. *Urology* **13**, 74–77.

Petrides, P. E., Bock, S., Bovens, J., Hofmann, R., and Jakse, G. (1990). Modulation of pro-epidermal growth factor, pro-transforming growth factor alpha and epidermal growth factor receptor gene expression in human renal carcinomas. *Cancer Res.* **50**, 3934–3939.

Potter, C. J., Pedraza, L. G., and Xu T. (2002). Akt regulates growth by directly phosphorylating Tsc2. *Nat. Cell Biol.* **4**, 658–665.

Rennebeck, G., Kleymenova, E. V., Anderson, R., Yeung, R. S., Artzt, K., and Walker, C. L. (1998). Loss of function of the tuberous sclerosis 2 tumor suppressor gene results in embryonic lethality characterized by disrupted neuroepithelial growth and development. *Proc. Natl. Acad. Sci. USA* **95**, 15629–15634.

Sargent, E. R., Gomella, L. G., Belldegrun, A., Linehan, W. M., and Kasid, A. (1989). Epidermal growth factor receptor gene expression in normal human kidney and renal cell carcinoma. *J. Urol.* **142**, 1364–1368.

Schmidt, L., Duh, F. M., Chen, F., Kishida, T., Glenn, G., Choyke, P., Scherer, S. W., Zhuang, Z., Lubensky, I., Dean, M., Allikmets, R., *et al.* (1997). Germline and somatic mutations in the tyrosine kinase domain of the MET proto-oncogene in papillary renal carcinomas. *Nature Genet.* **16**, 68–73.

Schmidt, L., Junker, K., Nakaigawa, N., Kinjerski, T., Weirich, G., Miller, M., Lubensky, I., Nuemann, H. P., Brauch, H., Decker, J., Vocke, C., Brown, J. A., Jenkins, R., Richard, S., Bergerheim, U., Gerrard, B., Dean, M., Linehan, W. M., and Zbar, B. (1999). Novel mutations of the MET proto-oncogene in papillary renal carcinomas. *Oncogene* **18**, 2343–2350.

Sepp, T., Yates, J. R., and Green, A. J. (1996). Loss of heterozygosity in tuberous sclerosis hamartomas. *J. Med. Genet.* **33**, 962–964.

Shiao, Y. H., Rice, J. M., Anderson, L. M., Diwan, B. A., and Hard, G. C. (1998). von Hippel-Lindau gene mutations in N-nitrosodimethylamine-induced rat renal epithelial tumors. *J. Natl. Cancer Inst.* **90**, 1720–1723.

Shuin, T., Kondo, K., Torigoe, S., Kishida, T., Kubota, Y., Hosaka, M., Nagashima, Y., Kitamura, H., Latif, F., Zbar, B., *et al.* (1994). Frequent somatic mutations and loss of heterozygosity of the von Hippel-Lindau tumor suppressor gene in primary human renal cell carcinomas. *Cancer Res.* **54**, 2852–2855.

Soucek, T., Pusch, O., Wienecke, R., DeClue, J. E., and Hengstschlager, M.

(1997). Role of the tuberous sclerosis gene-2 product in cell cycle control: Loss of the tuberous sclerosis gene-2 induces quiescent cells to enter S phase. *J. Biol. Chem.* **272**, 29301–29308.

Soucek, T., Yeung, R. S., and Hengstschlager, M. (1998). Inactivation of the cyclin-dependent kinase inhibitor p27 upon loss of the tuberous sclerosis complex gene-2. *Proc. Natl. Acad. Sci. USA* **95**, 15653–15658.

Stebbins, C. E., Kaelin, W. G. Jr., and Pavletich, N. P. (1999). Structure of the VHL-elongingC-elonginB complex: Implications for VHL tumor suppressor function. *Science* **284**, 455–461.

Steck, P. A., Pershouse, M. A., Jasser, S. A., Yung, W. K., Lin, H., Ligon, A. H., Langford, L. A., Baumgard, M. L., Hattier, T., Davis, T., Frye, C., Hu, R., Swedlund, B., Teng, D. H., and Tavtigian, S. V. (1997). Identification of a candidate tumour suppressor gene, MMAC1, at chromosome 10q23.3 that is mutated in multiple advanced cancers. *Nature Genet.* **15**, 356–362.

Tanimoto, K., Makino, Y., Pereira, T., and Poellinger, L. (2000). Mechanism of regulation of the hypoxia-inducible factor-1alpha by the von hippel-lindau tumor suppressor protein. *EMBO J.* **19**, 4298–4309.

Tavani, A., and La Vecchia, C. (1997). Epidemiology of renal-cell carcinoma. *J. Nephrol.* **10**, 93–106.

Teng, D. H., Hu, R., Lin, H., Davis, T., Iliev, D., Frye, C., Swedlund, B., Hansen, K. L., Vinson, V. L., Gumpper, K. L., Ellis, L., El-Naggar, A., Frazier, M., Jasser, S., Langford, L. A., Lee, J., Mills, G. B., Pershouse, M. A., Pollack, R. E., Tornos, C., Troncoso, P., Yung, W. K., Fujii, G., Berson, A., Steck, P. A., *et al.* (1997). MMAC1/PTEN mutations in primary tumor specimens and tumor cell lines. *Cancer Res.* **57**, 5221–5225.

Thoenes, W., Storkel, S., and Rumpelt, H. J. (1986). Histopathology and classification of renal cell tumors (adenomas, oncocytomas and carcinomas): The basic cytological and histopathological elements and their use for diagnostics. *Pathol. Res. Pract.* **181**, 125–143.

True, L. D., and Grignon, D. (1997). Pathology of renal cancers. *In* "Principles and Practice of Genitourinary Oncology," (D. Raghavan, H. I. Scher, S. A. Leibel, and P. H. Lange, eds.), pp. 799–811. Lippincott-Raven, Philadelphia.

Tyers, M., and Willems, A. R. (1999). One ring to rule a superfamily of E3 ubiquitin ligases. *Science* **284**, 603–604.

van den Berg, E., Dijkhuizen, T., Oosterhuis, J. W., Geurts van Kessel, A., de Jong, B., and Storkel, S. (1997). Cytogenetic classification of renal cell cancer. *Cancer Genet. Cytogenet.* **95**, 103–107.

Velickovic, M., Delahunt, B., and Grebe, S. K. (1999). Loss of heterozygosity at 3p14.2 in clear cell renal cell carcinoma is an early event and is highly localized to the FHIT gene locus. *Cancer Res.* **59**, 1323–1326.

Vogelzang, N. J., Priest, E. R., and Borden, L. (1992). Spontaneous regression of histologically proved pulmonary metastases from renal cell carcinoma: A case with 5-year follow-up. *J. Urol.* **148**, 1247–1248.

Wagle, D. G., and Scal, D. R. (1970). Renal cell carcinoma: A review of 256 cases. *J. Surg. Oncol.* **2**, 23–32.

Walker, C., Ahn, Y.-T., Everitt, J., and Yuan, X. (1996). Renal cell carcinoma development in the rat independent of alterations at the VHL gene locus. *Mol. Carcinogen.* **15**, 154–161.

Walker, C., Goldsworthy, T. L., Wolf, D. C., and Everitt, J. (1992). Predisposition to renal cell carcinoma due to alteration of a cancer susceptibility gene. *Science* **255**, 1693–1695.

Washecka, R., and Hanna, M. (1991). Malignant renal tumors in tuberous sclerosis. *Urology* **37**, 340–343.

Whaley, J. M., Naglich, J., Gelbert, L., Hsia, Y. E., Lamiell, J. M., Green, J. S., Collins, D., Neumann, H. P., Laidlaw, J., Li, F. P., *et al.* (1994). Germ-line mutations in the von Hippel-Lindau tumor-suppressor gene are similar to somatic von Hippel-Lindau aberrations in sporadic renal cell carcinoma. *Am. J. Hum. Genet.* **55**, 1092–102.

Wienecke, R., Konig, A., and DeClue, J. E. (1995). Identification of tuberin, the tuberous sclerosis-2 product. Tuberin possesses specific Rap1GAP activity. *J. Biol. Chem.* **270**, 16409–16414.

Wienecke, R., Maize, J. C., Jr., Shoarinejad, F., Vass, W. C., Reed, J., Bonifacino, J. S., Resau, J. H., de Gunzburg, J., Yeung, R. S., and DeClue, J. E. (1996). Co-localization of the TSC2 product tuberin with its target Rap1 in the Golgi apparatus. *Oncogene* **13**, 913–923.

Xiao, G. H., Shoarinejad, F., Jin, F., Golemis, E. A., and Yeung, R. S. (1997). The tuberous sclerosis 2 gene product, tuberin, functions as a Rab5 GTPase activating protein (GAP) in modulating endocytosis. *J. Biol. Chem.* **272**, 6097–6100.

Yagoda, A., Abi-Rached, B., and Petrylak, D. (1995). Chemotherapy for advanced renal-cell carcinoma: 1983–1993. *Semin. Oncol.* **22**, 42–60.

Yao, M., Shuin, T., Misaki, H., and Kubota, Y. (1988). Enhanced expression of c-myc and epidermal growth factor receptor (C- erbB-1) genes in primary human renal cancer. *Cancer Res.* **48**, 6753–6757.

Yeung, R. S., Xiao, G. H., Everitt, J. I., Jin, F., and Walker, C. L. (1995). Allelic loss at the tuberous sclerosis 2 locus in spontaneous tumors in the Eker rat. *Mol. Carcinogen.* **14**, 28–36.

Yeung, R. S., Xiao, G. H., Jin, F., Lee, W. C., Testa, J. R., and Knudson, A. G. (1994). Predisposition to renal carcinoma in the Eker rat is determined by germ- line mutation of the tuberous sclerosis 2 (TSC2) gene. *Proc. Natl. Acad. Sci. USA* **91**, 11413–11416.

26

The Tubule

William van't Hoff

I. Introduction

The renal tubule has the key role in the body's regulation of fluid, electrolyte, and acid–base balance. The vast majority of water and electrolytes in the glomerular ultrafiltrate are reabsorbed along the tubule by specialized transporters and channels, specifically localized in the tubular cell membranes, some in the luminal border and others in basolateral membrane. Any disturbance of tubular function, either congenital or acquired, can readily lead to profound electrolyte and volume disturbance. Much of the solute and water reabsorption occurs in the proximal segments, where 60% of the filtered sodium is taken up, along with water, potassium, bicarbonate, phosphate, amino acids, and low molecular weight proteins. Proximal tubule disorders may be isolated (confined to a single transporter) or generalized. In contrast, the distal tubule is responsible for the final modification of urine, utilizing specific transporters to regulate sodium and potassium reabsorption and proton secretion. Disorders of the distal tubule therefore tend to be isolated to a specific transporter. In the last few years, progress in molecular genetic research has revealed the structure, function, and effects of mutations in many of these transporters, which has led to major shifts in our knowledge of the function and dysfunction of the renal tubule. These advances apply not only to isolated renal tubulopathies, but also to several disorders of nephrolithiasis (e.g., cystinuria, Dent's disease) and to rare genetic causes of hypertension (e.g., Liddle's syndrome).

Most patients with genetic defects in tubular function present in the first years of life (some even antenatally). There are, however, several less severe disorders that present later or may be asymptomatic (e.g., Dent's and Gitelman's syndromes) and therefore the true incidence of some of these milder defects is uncertain. Children with a renal tubulopathy causing chronic dehydration, salt wasting, or acidosis will inevitably suffer poor growth, whereas excessive phosphate wasting will lead to rickets and retard bone development. Preliminary clinical assessment is based on the results of routine biochemical plasma and urine investigations. Further tests may include an assessment of urinary acidification and concentration, determination of specific proximal tubular markers, such as urinary levels of low molecular weight proteins (retinol binding protein, β_2-microglobulin) and enzymes (e.g., N-acetylglucosaminidase), amino acids, and phosphate reabsorptitive capacity. A renal ultrasound to determine the size and echogenicity (increased in nephrocalcinosis) of the kidneys should also be undertaken.

II. Proximal Tubulopathies

A. Proximal Renal Tubular Acidosis

The proximal tubule is the site of reabsorption of the majority of filtered bicarbonate, a process dependent on the formation of carbonic acid in the tubular cell, catalyzed by carbonic anhydrase type II. The carbonic acid dissociates and H^+ are secreted across the apical membrane into the urinary lumen, transported in exchange for sodium by the Na^+-H^+ exchanger (NHE-3). In the lumen, H^+ condenses with the filtered bicarbonate to form carbonic acid (catalyzed by apical carbonic anhydrase IV), which dissociates into carbon dioxide, which in turn diffuses readily across the apical membrane. The bicarbonate formed in the tubular cytoplasm is transported across the basolateral membrane by a Na^+ HCO_3^- cotransporter (NBC-1). Defective tubular bicarbonate absorption causes proximal renal tubular acidosis (RTA type II). In children, proximal RTA as an isolated defect is uncommon and it is seen more usually as part of the renal Fanconi syndrome (see later). Autosomal-dominant and -recessive forms of proximal RTA have been described, the latter in association with ocular defects (Brenes *et al.*, 1977; Igarashi *et al.*, 1994). Igarashi and colleagues (1999) have identified that patients with autosomal recessive RTA II associated with mental retardation, poor growth, glaucoma, and band keratopathy have loss of function mutations in the *SLC4A4* gene, which encodes the NBC-1 transporter. NBC-1 is expressed in both the kidney and the cornea so that loss of function causes both proximal tubular bicarbonate wasting and an increased corneal bicarbonate concentration, which in turn may lead to calcium deposition. Defects in NHE-3 have not been described in humans, but mice lacking NHE-3 develop a proximal RTA (Schulteis *et al.*, 1998).

B. Renal Fanconi Syndrome

The renal Fanconi syndrome comprises aminoaciduria, glycosuria, bicarbonaturia, phosphaturia, and rickets (or osteomalacia in adults). There are a number of genetic causes (see Table 26.1); most pediatric cases occur as part of a metabolic disorder and of these, cystinosis is the commonest and most severe.

1. Cystinosis

Nephropathic cystinosis is an autosomal-recessive disorder due to defective lysosomal cystine transport leading to excessive intracellular cystine accumulation, which, in turn, leads to multisystem damage (Gahl *et al.*, 1995). The proximal tubule is very sensitive to cystine storage, and patients (who usually present around 1–2 years of life) manifest the features of the Fanconi syndrome, with poor feeding, excessive thirst, delayed growth, weakness, and

Table 26.1 Inherited Causes of Renal Fanconi Syndrome

Disorder	Defective *gene*/protein
Cystinosis	*CTNS*/cystinosin
Tyrosinaemia Lowe's syndrome	Fumaryl acetoacetate hydrolase Inositol polyphosphate 5-phosphatase
Galactosemia	Galactose 1-phosphate uridyl transferase
Fructosemia	Fructose-1-phosphate aldolase B
Fanconi Bickel syndrome	*GLUT2*/Glut2 (facilitated glucose transporter)
Dent's disease	*CLC-5*/CLCN5 (voltage-gated chloride channel)
Mitochondrial disorders	Mitochondrial DNA
Wilson's disease	*Wc-1*/P-type copper transporting ATPase

rickets. All racial groups are affected but Caucasians commonly have blond hair and a fair complexion. Another site of cystine storage is the cornea, where cystine crystals can be seen on slit-lamp examination. In addition to general measures appropriate for any child with the Fanconi syndrome (rehydration, bicarbonate, electrolyte and vitamin D supplements, extra calories), patients with cystinosis require therapy with a cystine-depleting agent (cysteamine) to delay progressive glomerular damage, which, if untreated, leads to end-stage renal failure by 10 years (Markello *et al.*, 1993). Renal transplantation is successful and the disorder does not recur in the transplant but the grafted kidney does not correct the disorder and cystine continues to accumulate in nonrenal tissues, causing multisystem dysfunction (delayed puberty, hypothyroidism, diabetes mellitus, myopathy, and central nervous system involvement) (Gahl *et al.*, 1995). It is therefore essential for patients to continue their cysteamine therapy even after transplantation.

The proximal tubule is the first part of the nephron to be affected in cystinosis. Morphologically, proximal tubules develop a characteristic "swan neck" deformity with variability in epithelial size (Mahoney and Striker, 2000). The brush border may be attenuated and nuclei also appear heterogeneous. The glomeruli may appear normal but may also contain multinucleate giant podocytes. Cystine crystals are seen in interstitial cells and sometimes in podocytes but rarely in tubular cells, possibly because of rapid turnover (Gubler *et al.*, 1999).

For nearly 20 years, the biochemical defect in cystinosis has been known to involve the defective transport of cystine out of the lysosome (where it is produced as a result of protein hydrolysis). Classic countertransport studies by Gahl and colleagues (1982, 1995) demonstrated that lysosomal cystine transport was carrier mediated. However, it proved impossible to isolate the protein responsible for lysosomal cystine transport directly. A linkage study was therefore used to localize the cystinosis gene to the short arm of chromosome 17 with a Z_{max} of 9.87–10.89 for three markers (Cystinosis Group, 1995). A positional cloning approach was used to isolate the cystinosis gene, *CTNS*; the key advance being the discovery that a microsatellite marker, mapping to the cystinosis region, was deleted in 40% of cystinosis patients, but not in unrelated controls (Town *et al.*, 1998). The *CTNS* gene encodes a new class of integral membrane protein of 367 amino acids, cystinosin. This is predicted to have seven transmembrane domains, a specific lysosomal membrane targeting motif and heavy glycosylation toward the N-terminus (Town *et al.*, 1998).

The commonest mutation in the *CTNS* gene in cystinotic patients is a large deletion spanning exons 1–10 of the cystinosis gene and encompassing 57 kb of genomic DNA upstream (5′) of the gene. Forty percent of cystinosis patients have been found to be homozygous for this deletion and many others are deleted heterozygously (Forestier *et al.*, 1999; Anikster *et al.*, 1999). This larger deletion has only been identified in patients of northern European origin (found in 76% of such individuals) and can therefore be used as a diagnostic test. Patients who are heterozygous for the deletion or who do not have the major deletion can be screened using single-strand conformation polymorphism (SSCP) analysis (Orita *et al.*, 1989) covering the complete *CTNS* coding sequence and intron/exon boundaries (Town *et al.*, 1998). Overall, mutations have been detected in 90% of cystinotic patient samples using this strategy.

Several studies have shown that patients with the typical early onset form of cystinosis have truncating mutations on both alleles, likely to result in a complete loss of protein (Town *et al.*, 1998; Attard *et al.*, 1999; Shotelersuk *et al.*, 1998; McGowan-Jordan *et al.*, 1999). These mutations in the *CTNS* gene are consistent with the very severe phenotype of the disease and with the observations that patients have completely defective lysosomal cystine transport, whereas obligate heterozygotes exhibit 50% of normal transport activity (Gahl *et al.*, 1982). There are a small number of patients with milder forms of the disease, either presenting in adolescence with chronic renal impairment or an adult form with no nephropathy but manifesting only corneal cystine crystal deposition. Mutation analysis of these patients has usually shown that they have at least one nontruncating mutation in a nonconserved or functionally less important region of cystinosin (Attard *et al.*, 1999; Thoene *et al.*, 1999; Anikster *et al.*, 2000).

2. Pathogenesis of the Fanconi Syndrome

The exact mechanisms whereby molecular genetic defects such as those in cystinosis lead to the Fanconi syndrome are unknown. The biochemical abnormalities of phosphaturia, aminoaciduria, bicarbonaturia, glycosuria, and low molecular weight proteinuria suggest a widespread defect in reabsorption of these solutes in the proximal tubule. It is most unlikely that the underlying genetic defect affects every one of these solute cotransporters and it is more probable that there is interference with a common final step in the reabsorptive process. Equally feasible is a generalized increase in back flux of the solutes from the tubular cell into the urinary lumen. Investigators have made use of an animal model of the Fanconi syndrome, occurring spontaneously in the Basenji dog, and a toxin-induced system using maleate or cadmium. A Fanconi syndrome has been described in mice in whom the hepatocyte nuclear factor 1α gene has been "knocked out," but further work is required to understand the mechanism (Pontoglio *et al.*, 1996). Current evidence suggests three main pathogenetic mechanisms underlying the Fanconi syndrome: (1) alteration in membrane permeability, (2) reduction in sodium-coupled cotransport (mainly due to mitochondrial dysfunction), and (3) decreased transporter activity due to an effect on the endocytic recycling pathway (Bergeron *et al.*, 1995; Bergeron *et al.*, 1996). These mechanisms are shown in Fig. 26.1.

a. Toxin Models Administration of maleic acid to dogs or rats leads to a reversible Fanconi syndrome (see Bergeron *et al.*, 1995). Maleic acid is concentrated within mitochondria and decreases the activity of mitochondrial enzymes (e.g., 1α-hydroxylase and Krebs cycle enzymes). Maleate administration leads to a reduction in ATP concentration, NaK-ATPase activity, and intracellular phosphate (Eiam-Chong *et al.*, 1995). Maleate also reduces renal coenzyme A levels, possibly by trapping it as maleyl-CoA. In addition to these actions, maleate uptake by proximal tubular endosomes appears to be toxic to the membrane recycling pathway (thereby trapping transporters in endosomes) (McLeese *et al.*, 1996; Bergeron *et al.*, 1996). Endosomes are responsible in part for the recycling of apical membrane transporters. Megalin, a membrane glycoprotein (gp330) and a receptor for many ligands (e.g., albumin, low molecular weight proteins) is localized subapically at clathrin-coated pits. The megalin–ligand complex is internalized and processed through early and late endosomes where the ligand is released. Megalin can then be recycled to the cell surface. Maleate has been shown to partially inhibit the transport of lysozyme from endocytic vacuoles to lysosomes (Christensen and Maunsbach, 1980). Maleate may inhibit receptor binding to megalin by chelation of calcium, which mediates the interaction (Christensen *et al.*, 1992).

Cadmium toxicity in humans or animals produces a Fanconi syndrome. As with maleate, cadmium is concentrated in mitochondria, leading to enlargement and morphological changes. In animal work, there is evidence of toxic metabolite formation leading to mitochondrial dysfunction, a reduction in ATP, and decreased NaK-ATPase activity. 4-Pentenoate is a short chain fatty acid that causes a Fanconi syndrome in dogs, possibly by an abnormal phosphorylation

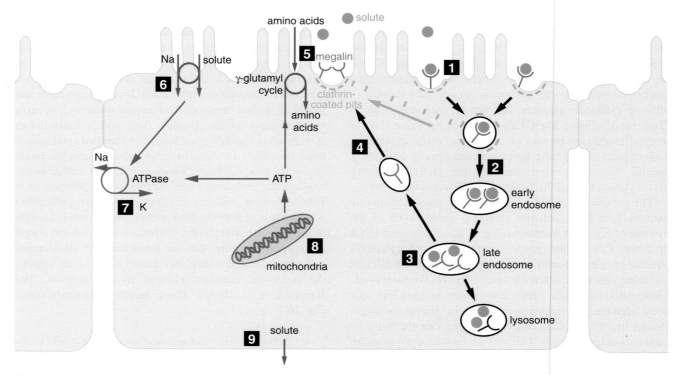

Figure 26.1 Schematic representation of solute and ligand entry into proximal tubular cell. (1) Ligands bind to megalin at clathrin-coated pits in the apical membrane and are internalized in early and late endosomes (2). The ligand dissociates (3) from megalin, allowing recycling of the megalin back to the apical membrane (4). Amino acids can be reabsorbed across the apical membrane by incorporation into the γ-glutamyl cycle (5), which is ATP dependent, or can, as with other solutes, undergo cotransport with sodium (6) into the cytosol. The gradient for this depends on sodium extrusion across the basolateral membrane (7), dependent on NaK ATPase. ATP, generated from mitochondrial oxidative phosphorylation (8), is required for adequate functioning of NaK-ATPase and the γ-glutamyl cycle. Solutes exit the basolateral membrane (9).

of brush border membrane proteins (Pouliot *et al.*, 1992), although accumulation of a toxic metabolite (3-keto-4-pentanoyl-CoA) may also inhibit mitochondrial enzymes and β-oxidation in the proximal tubule (Boulanger *et al.*, 1993).

b. Basenji Dog A spontaneous Fanconi syndrome develops in Basenji dogs, with evidence suggesting that the multiple transport abnormalities might be due to a membrane defect in the handling of sodium (Hsu *et al.*, 1992). Affected dogs have an alteration in brush border membrane fluidity, an increased cholesterol but normal phospholipid, and total free fatty acid content of the brush border (Hsu *et al.*, 1992, 1994).

c. Human Genetic Disorders Studies on the pathogenesis of the Fanconi syndrome resulting from metabolic disorders have focused on cystinosis, as it is the commonest inherited cause (reviewed in Haq and van't Hoff, 1999). Rats treated with cystine dimethyl ester (CDME), which leads to intralysosomal accumulation of cystine, exhibit features of the Fanconi syndrome with increased urine volume, excretion of phosphate, glucose, and amino acids compared

with untreated rats (Foreman *et al.*, 1987). Incubation of renal cortical tubule suspensions with CDME increases intracellular cystine concentrations to levels comparable to those found in renal tissue removed from patients with cystinosis. Cystine loading of isolated proximal tubules significantly reduces volume absorption, glucose, and bicarbonate transport without any change in the permeability of the tubules to mannitol or to bicarbonate. Oxygen consumption, oxidation of labeled glucose, lactate, butyrate, and especially succinate and ATP flux are also reduced significantly in cystine-loaded isolated renal cortical tubules (Foreman *et al.*, 1990; Coor *et al.*, 1991). Addition of exogenous ATP to the culture medium negates the reduction in volume absorption. The effect of cystine loading on phosphate concentration may be an important factor (Bajaj and Baum, 1996). Whether these *in vitro* studies, conducted in cells with intact lysosomal cystine transport (in contrast to cystinosis patients in whom lysosomal efflux is defective), provide an accurate model of the human condition remains unclear.

Marked intracellular phosphate depletion and a fall in ATP levels are also seen in hereditary fructose intolerance in which a rapid onset and reversible Fanconi syndrome is seen after the administration of fructose. Patients with hereditary tyrosinemia type 1 develop severe liver disease and a Fanconi syndrome as a result of a deficiency of fumaryl acetoacetate hydrolase (FAH), which therefore leads to the accumulation of a toxic metabolite, succinyl acetone (SA). Intraperitoneal injection of SA to rats leads to a Fanconi syndrome (Wyss *et al.*, 1992). *In vitro*, SA inhibits sodium-dependent phosphate transport by brush border membrane vesicles, decreases ATP production, and inhibits mitochondrial respiration (Roth *et al.*, 1991).

C. Dent's Disease

Dent's disease is an X-linked condition in which affected males develop aminoaciduria, hypercalciuria, low molecular weight proteinuria, nephrocalcinosis, and, in adult life, renal stones and chronic renal failure (Wrong *et al.*, 1994). The disorder, which is due to inactivating mutations in the *CLCN5* gene (Lloyd *et al.*, 1996), has been fully described in adults but increasingly, children with milder features have been reported. In Japan, where children have routinely undergone urine testing, a proportion with idiopathic low molecular weight proteinuria have been found to also have mutations in the same gene and are clearly manifesting a milder phenotype (possibly due to the younger age of ascertainment) (Lloyd *et al.*, 1997).

CLCN5 encodes a protein, CLC-5, one of a family of voltage-gated chloride channels, which is expressed in the kidney in the proximal tubule (in subapical endosomes and with vacuolar ATPase), thick ascending limb, and intercalated cells of the collecting duct. The vacuolar ATPase is involved in endosomal acidification, and defects in this process would be predicted to disrupt the endocytic recycling (e.g., of megalin and of apical transporters), thereby leading to generalized proximal tubular dysfunction. Such an effect has now been shown in mice in which the *clcn5* gene had been disrupted (Piwon *et al.*, 2000). In *clcn5*⁻ mice, apical proximal tubular endocytosis was disrupted, leading to internalization of NaP1-2 and NH3 transporters with consequent hyperphosphaturia (Piwon *et al.*, 2000).

D. Fanconi–Bickel Syndrome

The Fanconi–Bickel syndrome is a very rare autosomal-recessive disorder in which children present with hepatomegaly and failure to thrive. They have hepatic glycogenosis, fasting ketonuria, and hypoglycemia (features of decreased mobilization of glucose) and postprandial hyperglycemia, galactosemia, and galactosuria (features of decreased utilization of glucose and galactose) (Manz *et al.*, 1987). In addition, affected individuals have severe generalized proximal tubular dysfunction and rickets but, interestingly, do not suffer significant glomerular damage (in contrast to other forms of the Fanconi syndrome) (Santer *et al.*, 1998).

The abnormalities in carbohydrate metabolism led to speculation that these children had a defect in monosaccharide transport (Manz *et al.*, 1987). Santer and colleagues (1997) demonstrated that patients have mutations in a gene encoding GLUT2, one of four facilitated-diffusion glucose transporters, expressed in liver, small intestine, and proximal renal tubule. In the proximal tubule, the defect in GLUT2 at the basolateral membrane would explain the excessive urinary loss of glucose. Santer and colleagues (1997) speculated that this might lead to renal glycogen accumulation, which in turn might impair other renal tubular functions causing the characteristic Fanconi syndrome.

E. X-Linked Hypophosphatemic Rickets (XLHR)

XLHR is an X-linked dominant disorder and is the commonest inherited form of rickets with an incidence of approximately 1 in 20,000. Children (male or female) present usually between 1 and 3 years of age with short stature, poor motor development, and bow legs due to rickets. Biochemical investigations show normal plasma calcium, excess phosphaturia, hypophosphatemia, a normal parathyroid hormone level, and an elevated alkaline phosphatase concentration. Without treatment, patients suffer worsening growth and bony deformity requiring corrective osteotomies. The teeth are also affected with delayed dentition, excessive caries, and dental abscesses. However, the condition can be treated effectively by a combination of vitamin D (calcitriol or 1 α-calcidol) and phosphate supplements. There is a fine balance between undertreatment with consequent bone pain and deformity versus overtreatment resulting in nephrocalcinosis, hypercalcemia, abdominal pain, and diarrhea (secondary to the excessive phosphate doses).

The gene for XLHR was mapped to Xp22 and was subsequently isolated and named as *PHEX* (The HYP Consortium, 1995). This codes for a type II integral membrane protein, a member of the family of endopeptidases, which have a role in hormonal regulation. The pathogenesis of XLHR is complex, involving decreased reabsorption of phosphate in the proximal tubule, abnormalities of vitamin D metabolism, altered osteoblast function, and an effect of an as yet unknown circulating factor. Much information has been obtained from two murine models of the disorder. Hyp and Gy mice have skeletal and dental defects akin to the human condition, exhibit similar biochemical abnormalities, and have been shown to be due to defects in *Phex*, the murine homologue of *PHEX* (reviewed in Tenenhouse, 1999; Rowe, 2000). Hyp mice have a deletion of the 3′ end of the gene (Beck *et al.*, 1997), whereas Gy mice have a deletion involving the 5′ end but also extending to involve other genes (e.g., spermine synthase) (Strom *et al.*, 1997). This may explain additional features seen in Gy mice (hyperactivity, circling behavior, and male sterility).

The renal losses of phosphate are due to a defect in proximal tubular absorption. Hyp mice have a reduced transport of phosphate in proximal tubular preparations due to reduced expression of the high-affinity Npt2 (the type II renal sodium-phosphate cotransporter) at the brush border (for review, see Tenenhouse 1999). However, disruption of the mouse Npt2 gene, while resulting in phosphaturia and hypercalciuria, does not cause rickets (Beck *et al.*, 1998). This mouse model is more analogous to a very rare form of hypophosphatemic rickets associated with hypercalciuria.

As in patients with XLHR, the Hyp mouse has an inappropriately low $1,25(OH)_2$ vitamin D_3 level, resulting from increased catabolism due to upregulation of 24-hydroxylase (Roy *et al.*, 1994). The Hyp mouse also demonstrates abnormalities in bone mineralization and osteoblast function. Transplantation of Hyp mouse osteoblasts into normal mice causes abnormal bone formation (for review, see Rasmussen and Tenenhouse, 1995). Hyp mouse osteoblasts also have increased gluconeogenesis, abnormal responses to 1,25-dihydroxyvitamin D_3, reduced phosphorylation of osteopontin, but increased osteocalcin (Rasmussen and Tenenhouse, 1995). Clear evidence shows that a humoral factor may be implicated in XLHR. Parabiotic union of a Hyp mouse with a normal mouse causes hypophosphatemia and decreased renal phosphate reabsorption in the normal mouse. These effects were present even in a union in which both mice had had parathyroidectomies, suggesting that parathyroid hormone was not the humoral factor (Meyer *et al.*, 1989a,b). Nesbitt and colleagues (1992) transplanted a Hyp kidney into a normal mouse and demonstrated that it functions normally in a normal recipient, whereas a normal kidney leaks phosphate when placed in a Hyp recipient. Indirect evidence for a humoral factor comes from the observation that the reduction in plasma phosphate and renal phosphate reabsorption seen in individuals with oncogenic hypophosphataemic osteomalacia is negated by removal of the tumor. These studies suggest that XLHR and Hyp are caused by an alteration in the synthesis, catabolism, or effect of a humoral factor, which affects renal phosphate transport and vitamin D metabolism.

Identification of the *PHEX* gene as responsible for XLHR has begun to tie together these pathogenetic mechanisms (The HYP Consortium, 1995). *PHEX* is a zinc metalloendopeptidase with homologies to an M13 family of type II glycoproteins, some of which have a hormone regulatory function. *PHEX* or the mouse *Phex* gene are expressed in mouse bone and teeth and human fetal bone, lung, and ovary but, interestingly, it has not been identified in kidney (reviewed in Tenenhouse, 1999; Rowe, 2000). The gene is also expressed in tumors causing oncogenic hypophosphatemic rickets (OHO) (Lipman *et al.*, 1998). *PHEX* may regulate an as yet unidentified humoral factor, which in turn affects phosphate transport, osteoblast function, and mineralization. A glycoprotein, MEPE, has been isolated

from OHO tumors and has been considered as a candidate phosphaturic humoral factor (Rowe *et al.*, 2000).

III. Defects of the Thick Ascending Limb and Distal Tubule

A. Bartter's and Gitelman's Syndromes

These autosomal-recessive disorders, due to excessive tubular losses of sodium chloride, share features of hypokalemic alkalosis in the presence of normal blood pressure and an elevation of plasma renin and aldosterone. Within this group are a range of different phenotypes, which molecular advances have only recently begun to unravel. Children affected with the more severe form of Bartter's syndrome are often born prematurely after a pregnancy complicated by polyhydramnios and have very severe polyuria, electrolyte disturbance, and poor growth in the first weeks of life. Urinary calcium excretion is elevated markedly and they develop nephrocalcinosis (Rodriguez-Soriano, 1998). Occasionally, the biochemical presentation in the neonatal form is with metabolic acidosis and hyperkalemia and the more typical features evolve over time (Rodriguez-Soriano, 1998). There is a rare association with sensorineural deafness (Landau *et al.*, 1995).

The original case described by Bartter was of a boy with a milder variant of what is now known collectively as Bartter's syndrome (Bartter *et al.*, 1962). Such children typically present in the first few years of life with poor growth, polyuria, and polydipsia. Biochemical investigation reveals severe hypokalemia secondary to increased urinary losses of salt and potassium. In contrast to the more severe neonatal form, urinary calcium loss is normal or only slightly elevated and nephrocalcinosis is not observed. Patients with Gitelman's syndrome are often asymptomatic and discovered only during routine measurement of plasma electrolytes. Some present with marked weakness and/or tetany, due to profound hypomagnesemia, during an intercurrent illness (typically precipitated by vomiting or diarrhea). Symptoms of polyuria and poor growth are absent. Biochemically, there is hypomagnesemia and a hypokalaemic alkalosis, but in contrast to Bartter's syndrome, urinary calcium excretion is very low and nephrocalcinosis absent.

The treatment of both forms of Bartter's syndrome involves saline rehydration and then maintenance therapy with potassium supplements and indomethacin. Indomethacin has a profoundly beneficial effect in improving well-being, reducing polyuria, and increasing plasma potassium. Care is required, however, as there are a number of children who suffer gastroduodenal ulceration or other rare side effects (including benign intracranial hypertension) and it is inadvisable within the neonatal period. Patients with Gitelman's syndrome may not require treatment but, if necessary, should receive magnesium supplements (Rodriguez-Soriano, 1998; Bettinelli *et al.*, 1999).

In the 30 years after Bartter's case report, there were three principal theories to explain the pathophysiology of these syndromes: an abnormal pressor response to angiotensin II, excessive production of prostaglandins, or defects in renal tubular electrolyte transport. With time, the latter theory gained support, particularly with the recognition that therapy with frusemide (a loop diuretic) could mimic the Bartter phenotype, whereas thiazide treatment could reproduce the features of Gitelman's syndrome. Simon and colleagues (1996c) demonstrated linkage of Gitelman's syndrome to the locus *(SLC12A3)* encoding the thiazide-sensitive sodium chloride cotransporter (Na-Cl/TSC/NCCT) and identified mutations in affected families. This Na-Cl cotransporter is expressed in the distal tubule, and loss of function would be predicted to cause salt wasting, leading to hypokalemic alkalosis and hyperreninemia. Excessive urinary magnesium loss leading to hypomagnesemia may result in impaired distal tubular magnesium reabsorption, in turn secondary to the hypokalemic alkalosis (Quamme, 1997; Dai *et al.*, 1997). Because defects in the thiazide receptor had explained the pathogenesis of Gitelman's syndrome, defects in the frusemide-sensitive sodium–potassium–chloride cotransporter (Na-K-2Cl) were demonstrated to be the underlying cause of Bartter's syndrome in some children (Simon *et al.*, 1996a). The Na-K-2Cl cotransporter is expressed in the apical membrane of the thick ascending limb and is responsible for reabsorption of approximately 30% of the filtered sodium. Patients with mutations in the *NKCC2* gene coding for this transporter have severe salt wasting and demonstrate the phenotype of the neonatal form of Bartter's syndrome. However, many children with this form of Bartter's syndrome do not have mutations in this *NKCC2* gene, and most of these infants have been found to have mutations in a gene *(KCNJ1)* that encodes ROMK, one of a family of inwardly rectifying potassium channels (Simon *et al.*, 1996b). The ROMK channel serves to recycle potassium from the tubular cell back into the urinary lumen, thereby maintaining efficient functioning of the Na-K-2CL transporter (whose effect is rate limited by the luminal potassium concentration) (Hebert, 1995). In addition, ROMK serves to maintain the transcellular current flow in a basolateral to apical direction, balanced in turn by the transport of sodium across the paracellular pathway in a basolateral direction (Hebert, 1995). Loss of function of the ROMK channel leads to the severe phenotype of neonatal Bartter's syndrome (Simon *et al.*, 1996b). Calcium reabsorption in the thick ascending limb is linked to Na-K-2CL transport so that loss of function would be predicted to cause hypercalciuria and nephrocalcinosis, characteristic of this form of Bartter's syndrome. A locus for the severe neonatal Bartter's syndrome associated with

deafness has been mapped to chromosome 1p31 (Vollmer *et al.*, 2000).

The majority of patients with Bartter's syndrome, however, have a milder phenotype and do not have hypercalciuria or nephrocalcinosis. These patients do not have mutations affecting the Na-K-2Cl or ROMK channels. A number have mutations affecting a gene, *CLCNKB*, which encodes a renal chloride channel CLC-Kb, expressed in the basolateral membrane of the thick ascending limb and responsible for the transport of chloride from the tubular cell into the blood (Simon *et al.*, 1997). Dysfunction of this channel would be predicted to cause defective chloride reabsorption in the thick ascending limb, leading to the features of Bartter's syndrome, but further work is required to explain the difference in severity of the phenotype and the lack of overall effect on calcium reabsorption. In addition, there are several families with affected children in whom mutations have not been identified in any of the three genes described so far in Bartter's patients. It is tempting to speculate that dysfunction of other channels, involved in the regulation of sodium chloride transport in the thick ascending limb, may be implicated as the cause of the disorder in these remaining families.

B. Renal Hypomagnesemia Syndromes

In contrast to sodium and potassium, which are quantitatively mostly reabsorbed in the proximal convoluted tubule, the majority of filtered magnesium is reabsorbed in the thick ascending limb of Henle (TAL) and the final 1–5% is taken up in the distal convoluted tubule. The mechanism of magnesium reabsorption in the TAL is also different to that for other ions (e.g., sodium and chloride) and is dependent mainly on paracellular flux (reviewed in Quamme, 1997). This process is driven by the electrochemical gradient and is regulated carefully. Renal hypomagnesemia is most often due to drug therapy (e.g., thiazide diuretics), but can also occur in some tubulopathies (e.g., Gitelman's syndrome, see Section III,A). The molecular bases of two syndromes predominantly affecting renal magnesium reabsorption have been described.

The syndrome of hypomagnesemia, hypercalciuria leading to nephrocalcinosis, calculi, and chronic renal failure is inherited in an autosomal-recessive manner (for review, see Rodriguez-Soriano, 1987; Praga *et al.*, 1995). Affected individuals have polyuria but not salt wasting. A urinary acidification defect is usually observed, and some patients are initially diagnosed as suffering from distal RTA, but this is secondary to nephrocalcinosis (Rodriguez-Soriano and Vallo, 1994). A number of patients have ocular abnormalities. A genome-wide search and positional cloning strategy led to the identification of a gene, mutated in affected families, and coding for a protein, paracellin-1 (Simon *et al.*, 1999). Paracellin-1 is a protein of 305 amino acids, with four transmembrane domains, and is a member of the claudin protein family (Simon et al., 1999). It is expressed in the TAL, colocalizing with TAL-specific Tamm–Horsfall protein and with occludin, a marker of tight junctions (Simon et al., 1999). Paracellin-1 is clearly the principal mediator of magnesium reabsorption in the TAL and also affects the transport of calcium.

A second disorder of renal magnesium wasting associated with hypocalciuria is inherited in an autosomal-dominant pattern and presents with features of hypomagnesemia (tetany, seizures) (reviewed in Rodriguez-Soriano, 1987). This disorder has been mapped to chromosome 11q23 and a gene (FXYD2) encoding the Na^+, K^+-ATPase γ subunit found to be mutated in an affected family (Meij et al., 2000). This protein is localized to the basolateral membrane of distal convoluted tubular cells. A mutation, 123G→A, in the FXYD2 gene was identified in affected individuals and expressed in Sf9 insect cells. The mutant subunit accumulated in cytoplasm, whereas the wild type localized to the plasma membrane, colocalizing with other (α_1 and β_1) subunits (Meij et al., 2000). The effect of mutations in the gene coding for the γ subunit may be to reduce the expression and hence function of Na^+, K^+-ATPase at the plasma membrane.

C. Distal Renal Tubular Acidosis

Hydrogen ion secretion occurs predominantly from the α-intercalated cells of the cortical collecting duct. Carbonic acid formed intracellularly and dependently on carbonic anhydase II activity dissociates into H^+ and HCO_3^-. The H^+ ions are secreted into the lumen using a vacuolar H^+ ATPase, and the HCO_3^- is exchanged across the basolateral membrane by means of an electroneutral chloride-bicarbonate anion exchanger (AE1). There is also an exchange of H^+ ions for K^+ across the apical membrane of the α-intercalated cell, dependent on H^+ K^+ ATPase. Luminal secreted H^+ ions condense with HPO_4^- and NH_3.

Distal RTA is characterized by a failure of urinary acidification, hypokalemia, hypercalciuria leading to nephrocalcinosis, and, potentially, stone formation. In children, distal RTA usually occurs as an isolated tubulopathy (in contrast to proximal RTA, which is nearly always seen as part of a generalized Fanconi syndrome) and is commonly inherited in an autosomal-recessive manner. Affected children often present in the first years of life with poor growth, poor feeding, and vomiting. Rarely, distal RTA can be inherited in an autosomal-dominant manner. Therapy involves adequate and evenly spaced alkali supplements, which will lead to improvement in growth and will return the urinary calcium excretion to normal. Molecular defects have now been identified to explain the several forms of distal RTA. First, Bruce and colleagues (1997) identified mutations in the SLC4A1 gene, encoding the Cl^-/HCO_3^-

exchanger. Different mutations in the SLC4A1 gene have also been described in patients with several types of red cell fragility (e.g., hereditary spherocytosis and southeast Asian ovalocytosis), but urinary acidification in these patients is inconsistent with some individuals exhibiting distal RTA, whereas others have no defect (Rodriguez-Soriano, 2000). When one of the mutant SLC4A1 genes was expressed in Xenopus oocytes, chloride transport was reduced only slightly (Bruce et al., 1997). Other functional studies of mutant SLC4A1 genes demonstrated a failure of expression of AE1 at the oocyte surface and reduced chloride transport, both of which abnormalities could be negated by glycophorin A (Tanphaichitr et al., 1998). These findings may be explained by the mutation in SLC4A1 causing defective intracellular targeting of AE1, potentially into the apical rather than the basolateral membranes. Such an event would explain the observation of distal RTA and a high urine PCO_2 in a patient with southeast Asian ovalocytosis (Kaitwatcharachai et al., 1999).

A significant number of patients with autosomal-recessive distal RTA also have sensorineural deafness. The pathogenesis of this association has now been revealed by molecular studies (Karet et al., 1999). Mutations in ATP6B1, coding for the B1 subunit of apical H^+ATPases, are found in patients with recessive distal RTA with deafness, which would lead to failure of acid secretion (Karet et al., 1999). B1 subunit-containing H^+ATPases are also expressed in the inner ear epithelia so that it is possible that mutations in the gene coding for this unit could affect endolymphatic acid–base balance at this site and consequently impair hearing (Karet et al., 1999). Although alkali therapy corrects systemic pH, it does not affect progression of the hearing loss in these patients, as would be predicted because such treatment would not affect endolymphatic function (Karet et al., 1999). However, Karet and colleagues (1999) have found that a substantial number of families with distal RTA and deafness do not have mutations in ATP6B1.

Defects in ATP6N1B, another previously unidentified, subunit of the apical H^+ATPase, have been identified as a cause of autosomal-recessive distal RTA with normal hearing (Smith et al., 2000). ATP6N1B is expressed in the apical surface of α-intercalated cortical-collecting duct cells (colocalizing in immunoblot studies with ATP6B1) (Smith et al., 2000). The function of this subunit requires further investigation but it clearly has a role in complete functioning of the H^+ATPase proton pump, as loss of function mutations in the gene coding for it cause defective urinary acidification.

D. Carbonic Anhydrase II Deficiency

The carbonic anhydrase II deficiency syndrome is an autosomal-recessive disorder of osteopetrosis and hence increased bone fractures, renal tubular acidosis, cerebral calcification, mental retardation, and poor growth failure (Sly *et al.*, 1985). The renal tubular acidosis usually has features of proximal and distal dysfunction (Ohlsson *et al.*, 1986; Nagai *et al.*, 1997). Typically there is a hyperchloremic metabolic acidosis, a low urine ammonium excretion rate, and a low urine minus blood PCO_2 difference in alkaline urine. Other features of distal RTA, such as hypercalciuria and nephrocalcinosis, are uncommon but have been reported (Ismail *et al.*, 1997). Bicarbonate wasting, as seen in proximal RTA, is also observed, but other markers of proximal tubular function (e.g., amino acid and phosphate reabsorption) are normal. Osteopetrosis occurs as a result of defective acid secretion in osteoclasts, which is dependent on CAII and vacuolar H^+ATPases and which is necessary for bone dissolution.

The underlying defect is a deficiency of carbonic anhydrase II activity and immunoreactivity in erythrocytes (Sly *et al.*, 1983). Many reported cases are of Arabic extraction, and in these patients, one particular mutation of the CAII gene (a splice junction mutation at the 5′ end of intron 2) is very prevalent. A mouse model, deficient in the CAII gene, exhibits growth retardation and renal tubular acidosis. Lai and colleagues (1998) undertook "gene therapy" in this model by retrograde injection of the cationic liposome complexed with the CAII gene. Treated mice expressed the CAII gene and corresponding mRNA for up to 1 month and were able to acidify the urine after ammonium chloride loading.

IV. Disorders of the Amiloride-Sensitive Epithelial Sodium Channel

Pseudohypoaldosteronism type 1 (PHA1) is a disorder of massive urinary salt wasting, presenting in the neonatal period with weight loss, vomiting, and severe dehydration (Cheek and Perry, 1958). Investigations show hyponatremia, severe hyperkalemia, and raised plasma renin and aldosterone concentrations. There are two forms of inheritance, an autosomal-recessive type, which is more severe, affects several organ systems, and persists into adulthood, and an autosomal-dominant form, restricted to the kidney, which tends to improve with age (Hanukoglu *et al.*, 1991). Patients with the autosomal-dominant form have heterozygous mutations in the mineralocorticoid receptor gene *(MLR)*, predicted to cause loss of function leading to salt wasting (Geller *et al.*, 1998). A number of patients with this form of PHA 1 did not, however, have mutations in the *MLR* gene, suggesting the possibility that other genes may also be implicated. Mice in whom the *MLR* gene has been "knocked out" exhibit typical features of PHA1 soon after birth and die shortly afterward (Berger *et al.*, 1998).

The rarer autosomal-recessive form of PHA 1 is associated with mutations not in the *MLR* gene, but in α, β, or γ

subunits of the epithelial sodium channel (ENaC), which the MLR regulates (Chang *et al.*, 1996). Missense and frameshift mutations are predicted to cause complete loss of function, but others have more subtle effects on sodium transport. A G372S point mutation in the β subunit, in the intracellular side of the first transmembrane domain, affects a highly conserved residue and, in an oocyte expression system, reduces sodium transport by 50% (Grunder *et al.*, 1997). Similar studies have also shown that other mutations affecting the cysteine-rich regions of the extracellular domain reduce expression of the channel at oocyte surface, thereby reducing sodium transport (Firsov *et al.*, 1999).

Whereas these loss-of-function mutations lead to salt wasting and features of PHA 1, other mutations in the same subunits lead to gain of function with consequent excess sodium reabsorption and hypertension. This mechanism has been demonstrated as the cause of Liddle's syndrome, a rare autosomal-dominant disorder of hypertension and hypokalemia (Shimkets *et al.*, 1994; Hanson *et al.*, 1995). Pseudohypoaldosteronism type II is an autosomal-dominant disorder characterized by hypertension, hyperkalemia, metabolic acidosis, and a low plasma renin activity. It is not due to abnormalities of the MLR or EnaC, and the molecular genetic basis is unknown, although a linkage study demonstrated localization of PHA II genes to loci on chromosomes 1q31–42 and 17p11-q21 (Mansfield *et al.*, 1997).

V. Disorders of the Collecting Duct

A. Nephrogenic Diabetes Insipidus

Congenital nephrogenic diabetes insipidus (CNDI) is a disorder in which the kidney fails to respond to arginine vasopressin, leading to defective urinary concentration. It is usually inherited in an X-linked manner and affected males typically present in the newborn period with poor feeding and growth, irritability, recurrent vomiting, and constipation (Knoers and Monnens, 1999). The severe polyuria and polydipsia become evident within months but is usually not appreciated as abnormal in the neonatal period. In older children, potential problems include poorer growth, delayed bladder control, learning and behavior difficulties, and flow uropathy (megaureter and megacystis) (Hoekstra *et al.*, 1996; van Lieburg *et al.*, 1999).

Plasma biochemistry shows a raised plasma sodium and osmolality, a low urine sodium, and an inappropriately low urine osmolality. Confirmation of the diagnosis is provided by an inability to correct the urinary concentrating defect after administration of DDAVP (a synthetic form of arginine vasopressin). The basis of treatment of CNDI is to provide a high water intake and a feed restricted in solute load (in particular by sodium restriction). Pharmacological therapy to reduce urine output is also required and most children are treated with a combination of a diuretics (such as hydrochlorothiazide and amiloride) and indomethacin, a prostaglandin synthetase inhibitor (Knoers and Monnens, 1999).

CNDI occurs due to mutations in the gene encoding the vasopressin receptor in the collecting duct cells (V2R) (van den Ouweland *et al.*, 1992; Pan *et al.*, 1992). V2R is a member of the G-protein family of receptors and, when activated by binding of vasopressin, leads to the stimulation of adenylate cyclase to dephosphorylate ATP to cAMP. An increase in cAMP causes movement of intracellular vesicles containing aquaporin-2 (AQ-2) water channels to the apical membrane, thereby increasing water permeability (for review, see Knoers and Monnens 1999). The insertion of AQ-2-containing vesicles into the apical membrane is mediated by targeting proteins known as vesicle-associated membrane proteins (VAMPs) and syntaxins. Mutations in patients have been found throughout the V2R gene and prevent either vasopressin binding to the receptor or inhibit the signal transduction. Disruption of either of these processes prevents insertion of aquaporin-2 (AQP2) water channels to the apical membrane.

Female carriers of a V2R mutation in a CNDI family are generally asymptomatic. However, there are very rare instances of women with such a mutation who express the full renal phenotype and this variability can be explained by a variation in X inactivation (van Lieburg *et al.*, 1995). Females may also be affected if the disorder is inherited in an autosomal-recessive manner (which occurs in a small number of families). In these pedigrees, it has been found that affected individuals have heterozygous or homozygous mutations in the gene coding for the AQ-2 protein (Deen *et al.*, 1994; van Lieburg *et al.*, 1994). Autosomal-dominant inheritance of CNDI has also been described and alterations in the AQP2 sequence identified in two independent families (Bichet *et al.*, 1995).

Animal models of nephrogenic diabetes insipidus have now been created. Yun and colleagues (2000) generated mice deficient in the V2R gene and demonstrated that hemizygous male pups had a phenotype typical of CNDI. Likewise, mice in whom the AQ-2 gene has been disrupted also develop severe CNDI and die within days (Yang *et al.*, 2000). Other genes involved in the process of urinary concentration have also been identified. Mice deficient in other aquaporin water channels (AQP1,3,4) also all develop a phenotype of CNDI (Ma *et al.*, 2000; Yang *et al.*, 2001). Genes coding for proteins in other parts of the nephron have also been implicated in CNDI. Matsumara and colleagues (1999) investigated the role of sodium chloride reabsorption in the TAL of the mouse. These workers generated mice deficient in the *Clcnk1* gene (which codes for Clc-k1, a chloride channel involved in sodium chloride reabsorption in the TAL) and noted that they had polyuria (approximately five times more urine than wild-type or heterozygous mice)

and hyposthenuria (Matsumara *et al.*, 1999). Disruption of this channel affects the countercurrent multiplication system, thereby reducing the tonicity of the medullary interstitium and inhibiting final urinary concentration. These animal studies provide new insights into the physiology of renal urinary concentration and may offer new therapeutic targets for drugs aimed at reducing urine output.

VI. Conclusions

The rapid advances in molecular biology have provided major breakthroughs in our understanding of renal tubular function both in health and in disease. For many disorders in which mutations affect the function of a transporter or channel, the pathogenesis is now clear, although regulatory factors require further study. In contrast, there is still much to be learned about the mechanisms by which metabolic disorders affecting the proximal tubule (e.g., cystinosis) lead to generalized proximal tubular dysfunction. Molecular biology has, for some time, held the promise of gene therapy. This has now been accomplished in mice for one renal disorder, carbonic anhydrase II deficiency *(vide supra)* (Lai *et al.*, 1998). While many tubular disorders are amenable to simple electrolyte and pharmacological agents, others, such as CAII deficiency and some of the Fanconi syndromes, might surely benefit from these gene therapy advances.

References

Anikster, Y., Lucero, C., Touchman, J. W., Huizing, M., McDowell, G., Shotelersuk, V., Green, E. D., and Gahl, W. A. (1999). Identification and detection of the common 65-kb deletion breakpoint in the nephropathic cystinosis gene (*CTNS*). *Mol. Genet. Metab.* **66**, 111–116.

Anikster, Y., Lucero, C., Guo, J., Huizing, M., Shotelersuk, V., Bernadini, I., McDowell, G., Iwata, F., Kaiser-Kupfer, M. I., Jaffe, R., Thoene, J., Schneider, J., and Gahl, W. A. (2000). Ocular nonnephropathic cystinosis: Clinical, biochemical, and molecular correlations. *Pediatr. Res.* **47**, 17–23.

Attard, M., Jean, G., Forestier, L., Cherqui, S., van't Hoff, W., Broyer, M., Antignac, C., and Town, M. (1999). Severity of phenotype in cystinosis varies with mutations in the CTNS gene: Predicted effect on the model of cystinosin. *Hum. Mol. Genet.* **20**8, 2507–2514.

Bajaj, G., and Baum, M. (1996). Proximal tubule dysfunction in cystine-loaded tubules: Effect of phosphate and metabolic substrates. *Am. J. Physiol.* **271**, F717–F722.

Bartter, F. C., Pronove, P., Gill, J. R. Jr., and MacCardle, R. C. (1962). Hyperplasia of the juxtaglomerular complex with hyperaldosteronism and hypokalaemic alkalosis: A new syndrome. *Am. J. Med.* **33**, 811–828.

Beck, L., Soumounou, Y., Martel, J., Krishnamurthy, G., Gauthier, C., Goodyer, C. G., and Tenenhouse, H. S. (1997). Pex/PEX tissue distribution and evidence for a deletion in the 3′ end of the Pex gene in X-linked hypophosphataemic rickets. *J. Clin. Invest.* **99**, 1200–1209.

Beck, L., Karaplis, A. C., Amizuka, N., Hewson, A. S., Osawa, H., and Tenenhouse, H. S. (1998). Targeted inactivation of Npt2 in mice leads to severe renal phosphate wasting, hypercalciuria, and skeletal abnormalities. *Proc. Natl. Acad. Sci. USA* **95**, 5372–5377.

Berger, S., Bleich, M., Schmid, W., Cole, T. J., Peters, J., Watanabe, H., Kriz, W., Warth, R., Greger, R., and Schutz, G. (1998). Mineralocorticoid receptor knockout mice: Pathophysiology of Na⁺ metabolism. *Proc. Natl. Acad. Sci. USA* **95**, 9424–9429.

Bergeron, M., Gougoux, A., and Vinay, P. (1995). The renal Fanconi syndrome. *In* "The Metabolic and Molecular Bases of Inherited Disease," (C. R. Scriver, A. L., Beudet, W. S., Sly, D. Valle, eds.), 7th Ed., pp. 3691–3704. McGraw-Hill, New York.

Bergeron, M., Mayers, P., and Brown, D. (1996). Specific effect of maleate on an apical membrane glycoprotein (gp330) in proximal tubule of rat kidneys. *Am. J. Physiol.* **271**, F908–F916.

Bettinelli, A., Basilico, E., Metta, M. G., Borella, P., Jaeger, P., and Bianchetti, M. G. (1999). Magnesium supplementation in Gitelman syndrome. *Pediatr. Nephrol.* **13**, 311–314.

Bichet, D. G., Arthus, M., Lonergan, M., Balfe, W., Skorecki, K., Robertson, G., Oksche, A., Rosenthal, W., Fujiwara, M., Morgan, K., and Sasaki, S. (1995). Autosomal dominant and autosomal recessive nephrogenic diabetes insipidus: Novel mutations in the AQP2 gene. *Am. J. Soc. Nephrol.* **6**, 717. [Abstract]

Boulanger, Y., Wong, H., Noel, J., Senecal, J., Fleser, A., Gougoux, A., and Vinay, P. (1993). Heterogeneous metabolism of 4-pentenoate along the dog nephron. *Renal. Physiol. Biochem.* **16**, 182–202.

Brenes, L. G., Brenes, J. M., and Hernandez, M. M. (1977). Familial proximal renal tubular acidosis: A distinct disease entity. *Am. J. Med.* **63**, 244–252.

Bruce, L. J., Cope, D. L., Jones, G. K., Schonfield, A. E., Burley, M., Povey, S., Unwin, R. J., Wrong, O., and Tanner, M. J. (1997). Familial distal renal tubular acidosis is associated with mutations in the red cell exchanger (band 3, AE1) gene. *J. Clin. Invest.* **100**, 1693–1707.

Chang, S. S., Grunder, S., Hanukoglu, A., Rosler, A., Mathew, P. M., Hanukoglu, I., Schild, L., Lu, Y., Schimkets, R. A., Nelson-Williams, C., Rossier, B. C., and Lifton, R. P. (1996). Mutations in subunits of the epithelial sodium channel cause salt wasting with hyperkalaemic acidosis, pseudohypoaldosteronism type 1. *Nature Genet.* **12**, 248–253.

Cheek, D. B., and Perry, J. W. (1958). A salt wasting syndrome in infancy. *Arch. Dis. Child.* **33**, 252–256.

Christensen, E. I., and Maunsbach, A. B. (1980). Proteinuria induced by sodium maleate in rats: Effects on ultrastructure and protein handling in the renal proximal tubule. *Kidney Int.* **17**, 771–786.

Christensen, E. I., Gliemann, J., and Moestrup, S. K. (1992). Renal tubule gp330 is a calcium binding receptor for endocytic uptake of protein. *J. Histochem. Cytochem.* **40**, 1481–1490.

Coor, C., Salmon, R. F., Quigley, R., Marver, D., and Baum, D. (1991). Role of adenosine triphosphate and NaK ATPase in the inhibition of proximal tubule transport with intracellular cystine loading. *J. Clin. Invest.* **87**, 955–961.

The Cystinosis Collaborative Research Group (1995). Linkage of the gene for cystinosis to markers on the short arm of chromosome 17. *Nature Genet.* **10**, 246–248.

Dai, L.-J., Friedman, P. A., and Quamme, G. A. (1997). Acid-base changes alter Mg²⁺ uptake in mouse distal convoluted tubule cells. *Am. J. Physiol.* **272**, F759–F766.

Deen, P. M., Verdijk, M. A., Knoers, N. V., Wieringa, B., Monnens, L. A., van Os, C. H., and van Oost, B. A. (1994). Requirement of human renal channel aquaporin-2 for vasopressin-dependent concentration of urine. *Science* **264**, 92–95.

Eiam-Chong, S., Spohn, M., Kurtzman, N. A., and Sabatini, S. (1995). Insights into the biochemical mechanism of maleic acid-induced Fanconi syndrome. *Kidney Int.* **48**, 1542–1548.

Firsov, D., Robert-Nicoud, M., Gruender, S., Schild, L., and Rossier, B. C. (1999). Mutational analysis of the cystein-rich domains of the epithelial sodium channel (ENaC): Identification of cysteines essential for channel expression at the cell surface. *J. Biol. Chem.* **274**, 2743–2749.

Foreman, J. W., Bowring, M. A., Lee, J., States, B., and Segal, S. (1987). Effect of cystine dimethyl ester on renal solute handling and isolated renal tubule transport in the rat: A new model of the Fanconi syndrome. *Metabolism* **36**, 1185–1191.

Foreman, J. W., and Benson, J. L. (1990). Effect of cystine loading and cystine dimethylester on renal brush border membrane transport. *Pediatr. Nephrol.* **4**, 236–239.

Forestier, L., Jean, G., Attard, M., Cherqui, S., Lewis, C., van't Hoff, W., Broyer, M., Town, M., and Antignac, C. (1999). Molecular characterisation of CTNS deletions in nephropathic cystinosis: development of a PCR-based detection assay. *Am. J. Hum. Genet.* **65**, 353–359.

Gahl, W. A., Bashan, N., Tietze, F., Bernadini, I., and Schulman, J. D. (1982). Cystine transport is defective in isolated leukocyte lysosomes from patients with cystinosis. *Science* **217**, 1263–1265.

Gahl, W. A., Schneider, J. A., and Aula, P. (1995). Lysosomal transport disorders. *In* "The Metabolic and Molecular Bases of Inherited Disease." (C. R., Scriver, A. L., Beaudet, W. S. Sly, and D. Valle, eds.), 7th ed., pp 3763–3797. McGraw-Hill, New York.

Geller, D. S., Rodriguez Soriano, J., Boado, A. V., Schifter, S., Bayer, M., Chang, S. S., and Lifton, R. P. (1998). Mutations in the minerlacorticoid receptor gene cause autosomal dominant pseudohypoaldosteronism type 1. *Nature Genet.* **19**, 279–281.

Grunder, S., Firsov, D., Chang, S. S., Jaeger, N. F., Gautschi, I., Schild, L., Lifton, R. P., and Rossier, B. C. (1997). A mutation causing pseudo-hypoaldsteronism type 1 identifies a conserved glycine that is involved in the gating of the epithelial sodium channel. *EMBO J.* **16**, 899–907.

Gubler, M. C., Lacoste, M., Sich, M., and Broyer, M. (1999). The pathology of the kidney in cystinosis. *In* "Cystinosis" (M. Broyer, ed.), pp. 42–48. Elsevier, Paris.

Hanukoglu, A. (1999). Type I pseudohypoaldosteronism includes two clinically and genetically distinct entities with either renal or mulitple target organ defects. *J. Clin. Endocrin. Metab.* **73**, 936–944.

Haq, M. S., and van't Hoff, W. G. (1999). Cellular dysfunction in cystinosis. *In* "Cystinosis" (M. Broyer, ed.), pp. 14–19. Elsevier, Paris.

Hanson, J. H., Nelson-Williams, C., Suzuki, H., Schild, L., Shimkets, R., Lu, Y., Canessa, C., Iwasaki, T., Rossier, B., and Lifton, R. P. (1995). Hypertension caused by a truncated epithelial sodium channel γ subunit: Genetic heterogeneity of Liddle syndrome. *Nature Genet.* **11**, 76–82.

Hebert, S. C. (1995). An ATP-regulated inwardly rectifying potassium channel from rat kidney. *Kidney Int.* **48**, 1010–1016.

Hoekstra, J. A., van, Lieburg, A. F., Monnens, L. A. H., Hulstijn-Dirkmatt, G. M, and Knoers, V. V. (1996). Cognitive and psychometric functioning of patients with nephrogenic diabetes insipidus. *Am. J. Med. Genet.* **61**, 81–88.

Hsu, B. Y. L., McNamara, P. D., Mahoney, S. G., Fenstemacher, E. A., Rea, C. T., Bovee, K. C., and Segal, S. (1992). Membrane fluidity and sodium transport by renal membranes from dogs with spontaneous idiopathic Fanconi syndrome. *Metabolism* **41**, 253–259.

Hsu, B. Y., Wehrli, S. L., Yandrasitz, J. R., Fenstemacher, E. A., Rea, C. T., McNamara, P. D., Bovee, K. C., and Segal, S. (1994). Renal brush border membrane lipid composition in Basenji dogs with spontaneous idiopathic Fanoni syndrome. *Metabolism* **43**, 1073–1078.

The HYP Consortium (1995). A gene (PEX) with homologies to endopeptidases is mutated in patients with X-linked hypophosphataemic rickets. *Nature Genet.* **11**, 130–136.

Igarashi, T., Ishii, T., Watanabe, K., Hyakawa. H., Horio, K., Sone, Y., Ohga, K. (1994). Persistent isolated proximal renal tubular acidosis: A systemic disease with a distinct clinical entity. *Pediatr. Nephrol.* **8**, 70–71.

Igarashi, T., Inatomi, J., Sekine, T., Cha, S.-H., Kanai, Y., Kunimi, M., Tsukamoto, K., Satoh, H., Shimadzu, M., Tozawa, F., Mori, T., Shiobara, M., Seki, G., and Endou, H. (1999). Mutations in SLC4A4 cause isolated proximal renal tubular acidosis with ocular abnormalities. *Nature Genet.* **23**, 264–266.

Ismail, E. A. R., Abul Saad, S., and Sabry, M. A. (1997). Nephrocalcinosis and urolithiasis in carbonic anhydrase II deficiency syndrome. *Eur. J. Pediatr.* **156**, 957–962.

Kaitwatcharachai, C., Vasuvattakul, S., Yenchitsomanus, P. T., Thuwajit, P., Malasit, P., Chuawatana, D., Mingkum, S., Halperin, M. L., Wilairat, P., and Nimmannit, S. (1999). Distal renal tubular acidosis and a high urine carbon dioxide tension in a patient with southeast Asian ovalocytosis. *Am. J. Kidney Dis.* **33**, 1147–1152.

Karet, F. E., Finberg, K. E., Nelson, R. D., Nayir, A., Sanjad, S. A., Rodriguez-Soriano, J., Santos, F., Cremers, C. W. R. J., Di Pietro, A., Hoffbrand, B. I., Winiarski, J., Bakkaloglu, A., Ozen, S., Dusunsel, R., Goodyer, P., Hulton, S. A., Wu, D. K., Skvorak, A. B., Morton, C. C., Cunningham, M. J., Jha, V., and Lifton, R. P. (1999). Mutations in the ATP6B1 gene encoding the B1 subunit of the apical proton pump, H+-ATPase, cause renal tubular acidosis with sensorineural deafness. *Nature Genet.* **21**, 84–90.

Knoers, N. V. A. M., and Monnens, L. A. H. (1999). Nephrogenic diabetes insipidus. *In* "Pediatric Nephrology," (T. M. Barratt, E. D. Avner, and W. E., Harmon, 4th eds.), pp. 583–591. Lippincott, Williams and Wilkins, 1999.

Lai, L. W., Chan, D. M., Erickson, R. P., Hsu, S. J., and Lien, Y. H. (1998). Correction of renal tubular acidosis in carbonic anhydrase II-deficient mice with gene therapy. *J. Clin. Invest.* **101**, 1320–1325.

Landau, D., Shalev, H., Ohaly, M., and Carmi, R. (1995). Infantile variant of Bartter syndrome and sensorineural deafness: A new autosomal recessive disorder. *Am. J. Med. Genet.* **62**, 355–361

Lipman, M. L., Panda, D., Bennett, H. P., Henderson, J. E., Shane, E., Shen, Y., Goltzman, D., and Karaplis, A. C. (1998). Cloning of human PEX cDNA. Expression, subcellular localization, and endopeptidase activity. *J. Biol. Chem.* **273**, 13729–13737.

Lloyd, S. E., Pearce, S. H. S., Fisher, S. E., Steinmayer, K., Schwappach, B., Scheinman, S. J., Harding, B., Bolino, A., Devoto, M., Goodyer, P., Rigden, S. P. A., Wrong, O., Jentsch, T. J., Craig, I. W., and Thakker, R. V. (1996). A common molecular basis for three inherited kidney stone diseases. *Nature* **379**, 445–449.

Lloyd, S. E., Pearce, S. H. S., Gunther, W., Kawagushi, H., Igarashi, T., Jentsch, T. J., and Thakker, R. V. (1997). Idiopathic low molecular weight proteinuria associated with hypercalciuric nephrocalcinosis in Japanese children is due to mutations of the renal chloride channel (CLCN5). *J. Clin. Invest.* **99**, 967–974.

Ma, T., Song, Y., Yang, B., Gillespie, A., Carlson, E. J., Epstein, C. J., and Verkman, A. S. (2000). Nephrogenic diabetes insipidus in mice lacking aquaporin-3 water channels. *Proc. Natl. Acad. Sci. USA* **97**(8), 4386–4391.

Mahoney, C. P., and Striker, G. E. (2000). Early development of the renal lesions in infantile cystinosis. *Pediatr. Nephrol.* **15**, 50–56.

Mansfield, T. A., Simon, D. B., Farvel, Z., Bia, M., Tucci, J. R., Lebel, M., Gutkin, G., Vialettes, B., Christofilis, M. A., Kauppinen-Makelin, R., Mayan, H., Risch, N., and Lifton, R. P. (1997). Multilocus linkage of familial hyperkalemia and hypertension, pseudohypoaldosteronism type II, to chromosomes 1q31-42 and 17p11-q21 *Nature Genet.* **16**, 202–205.

Manz, F., Bickel, H., Brodehl, J., Feist, D., Gellisen, K., Gescholl-Bauer, B., Gilli, G., Harms, E., Helwig, H., Nutzenadel, W., and Waldherr, R. (1987). Fanconi-Bickel syndrome. *Pediatr. Nephrol.* **1**, 509–518.

Markello, T. C., Bernadini, I. M., and Gahl, W. A. (1993). Improved renal function in children with cystinosis treated with cysteamine. *N. Engl. J. Med.* **328**, 1157–1162.

Matsumara, Y., Uchida, S., Kondo, Y., Miyazaki, H., Ko, S. B. H., Hayama, A., Morimoto, T., Liu, W., Arisawa, M., Sasaki, S., and Marumo, F. (1999). Overt nephrogenic diabetes insipidus in mice lacking the CLC-K1 chloride channel. *Nature Genet.* **21**, 95–98.

McGowan-Jordan, J., Stoddard, K., Podolsky, L., Orrbine, E., McLaine, P., Town, M., Goodyer, P., MacKenzie, A., and Heick, H. (1999). Molecular analysis of cystinosis: Probable Irish origin of the most common french Canadian mutation. *Eur. J. Hum. Genet.* **7**, 671–678.

McLeese, J., Thiery, G., and Bergeron, M. (1996). Maleate modifies apical endocytosis and permeability of the endoplasmic reticulum membrane in kidney tubular cells. *Cell Tissue* **282**, 29–37.

Meij, I. C., Koenderink, J. B., van Bokhoven, H., Assink, K. F., Tiel Groenestege, W., de Pont, J. J., Bindels, R. J., Monnens, L. A., van Den Heuvel, L. P., and Knoers, N. V. (2000). Dominant isolated renal

magnesium loss is caused by misrouting of the Na+,K+-ATPase gamma-subunit. *Nature Genet.* **26**, 265–266.

Meyer, R. A., Jr., Meyer, M. H., and Gray, R. W. (1989a). Parabiosis suggests a humoral factor is involved in X-linked hypophosphatemia in mice. *J. Bone. Miner. Res.* **4**, 493–500.

Meyer, R. A., Jr., Tenenhouse, H. S., Meyer, M. H., and Klugerman, A. H. (1989). The renal phosphate transport defect in normal mice parabiosed to X-linked hypophosphatemic mice persists after parathyroidectomy. *J. Bone. Miner. Res.* **4**, 523–532.

Nagai, R., Kooh, S. W., Balfe, J. W., Fenton, T., and Halperin, M. L. (1997). Renal tubular acidosis and osteopetrosis with carbonic anhydrase II deficiency: Pathogenesis of impaired acidification. *Pediatr. Nephrol.* **11**, 633–636.

Nesbitt, T., Coffman, T. M., Griffiths, R., and Drezner, M. K. (1992). Crosstransplantion of kidneys in normal and Hyp mice: Evidence that the Hyp mouse phenotype is unrelated to an intrinsic renal defect. *J. Clin. Invest.* **89**, 1453–1459.

Ohlsson, A., Cumming, W. A., Paul, A., and Sly, W. S. (1986). Carbonic anhydrase II deficiency syndrome: Recessive osteopetrosis with renal tubular acidosis and cerebral calcification. *Pediatrics* **77**, 371–380.

Orita, M., Iwahana, H., Kanazawa, H., Hayashi, K., and Sekiya, T. (1989). Detection of polymorphisms of human DNA by gel electrophoresis as single-strand conformation polymorphisms. *Proc. Natl. Acad. Sci. USA* **86**, 2766–2770.

Pan, Y., Metzenberg, A., Das, S., Jing, B., and Gitschier, J. (1992). Mutations in the V2 vasopressin receptor gene are associated with X-linked nephrogenic diabetes insipidus. *Nature Genet.* **2**, 103–106.

Piwon, N., Gunther, W., Schwake, M., Bosl, M. R., and Jentsche, T. J. (2000). CLC-5 Cl⁻-channel disuption impairs endocytosis in a mouse model for Dent's disease. *Nature* **408**, 369–373.

Pontoglio, M., Barra, J., Hadchouel, M., Doyen, A., Kress, C., Poggi Bach, J., Babinet., and Yaniv, M. (1996). Hepatocyte nuclear factor 1 inactivation results in hepatic dysfunction, phenylketonuria, and renal Fanconi syndrome. *Cell* **84**, 575–585.

Pouliot, J. F., Gougoux, A., and Beliveau, R. (1992). Brush border membrane proteins in experimental Fanconi's syndrome induced by 4-pentenoate and maleate. *Can. J. Physiol. Pharmacol.* **70**, 1247–1253.

Praga, M., Vara, J., Gonzalez-Parra, E., Andres, A., Alamo, C., Araque, A., Ortiz, A., and Rodicio, J. L. (1995). Familial hypomagnesemia with hypercalciuria and nephrocalcinosis. *Kidney Int.* **47**, 1419–1425.

Quamme, G. A. (1997). Renal magnesium handling: New perspectives in understanding old problems. *Kidney Int.* **52**, 1180–1195.

Rasmussen, H., and Tenenhouse, H. S. (1995). Mendelian, hypophosphataemias. *In* "The Metabolic and Molecular Bases of Inherited Disease" (C. R. Scriver, A. L. Beudet, W. S. Sly, and D. Valle, eds.), 7th Ed. pp. 3717–3745. McGraw-Hill, New York.

Rodriguez-Soriano, J. (1987). Hypomagnesaemia of hereditary renal origin. *Pediatr. Nephrol.* **1**, 465–472.

Rodriguez-Soriano, J. (1998). Bartter and related syndromes: The puzzle is almost solved. *Pediatr. Nephrol.* **12**, 315–327.

Rodriguez-Soriano, J. (2000). New insights into the pathogenesis of renal tubular acidosis: From functional to molecular studies. *Pediatr. Nephrol.* **14**, 1121–1136.

Rodriguez-Soriano, J., and Vallo, A. (1994). Pathophysiology of the renal acidification defect present in the syndrome of familial hypomagnesaemia-hypercalciuria. *Pediatr. Nephrol.* **8**, 431–435.

Roth, K. S., Carter, B. E., and Higgins, E. S. (1991). Succinylacetone effects on renal tubular phosphate metabolism: A model for experimental renal Fanconi syndrome. *Proc. Soc. Exp. Biol. Med.* **196**, 428–431.

Rowe, P. S., de Zoysa, P. A., Dong, R., Wang, H. R., White, K. E., Econs, M. J., and Oudet, C. L. (2000). MEPE, a new gene expressed in bone marrow and tumors causing osteomalacia. *Genomics* **67**, 54–68.

Rowe, P. S. N. (2000). The molecular background to hypophosphataemic rickets. *Arch. Dis. Child.* **83**, 192–194.

Roy, S., Martel, J., Ma, S., and Tenenhouse, H. S. (1994). Increased renal

25-hydroxyvitamin D3-24-hydroxylase messenger ribonucleic acid and immunoreactive protein in phosphate-deprived Hyp mice: A mechanism for accelerated 1,25-dihydroxyvitamin D3 catabolism in X-linked hypophosphatemic rickets. *Endocrinology* **134**, 1761–1767.

Santer, R., Schneppenheim, R., Dombrowski, A., Gotze, H., Steinmann, B., and Schaub, J. (1997). Mutations in GLUT2, the gene for the liver-type glucose transporter, in patients with Fanconi-Bickel syndrome. *Nature Genet.* **17**, 324–326.

Santer, R., Schneppenheim, R., Suter, D., Schaub, J., and Steinmann, B. (1998). Fanconi-Bickel syndrome: The original patient and his natural history, historical steps leading to the primary defect, and a review of the literature. *Eur. J. Pediatr.* **157**, 783–797.

Schulteis, P. J., Carke, L. L., Meneton, P., Miller, M. L., Soleimani, M., Gawenis, L. R., Riddle, T. M., Duffy, J. J., Doetschman, T., Wang, T., Giebisch, G., Aronson, P. S., Lorenz, J. N., and Sull, G. E. (1998). Renal and intestinal absorptive defects in mice lacking the NHE-3 Na$^+$/H$^+$ exchanger. *Nature Genet.* **19**, 282–285.

Shimkets, R. A., Warnock, D. G., Bosits, C. M., Nelson-Williams, C., Hanson, J. H., Schambelan, M., Gill, J. R., Jr., Ulick, S., Milora, R. V., Findling, J. W., Canessa, C. M., Rossier, B. C., and Lifton, R. P. (1994). Liddle's syndrome: Hereditable human hypertension caused by mutations in the β subunit of the epithelial sodium channel. *Cell* **79**, 407–414.

Shotelersuk, V., Larson, D., Anikster, Y., McDowell, G., Lemons, R., Bernardini, I., Guo, J., Thoene, J., and Gahl, W. (1998). *CTNS* mutations in an American-based population of cystinosis patients. *Am. J. Hum. Genet.* **63**, 1352–1362.

Simon, D. B., Bindra, R. S., Mansfield, T. A., Nelson-Williams C., Mendonca, E., Stone, R., Schurman, S., Nayir, A., Alpay, H., Bakkaloglu, A., Rodriguez-Soriano, J., Morales, J. M., Sanjad, S. A., Taylor, C. M., Pilz, D., Brem, A., Trachtman, H., Griswold, W., Richard, G. A. John, E., and Lifton, R. J. (1997). Mutations in the chloride channel ClC-Kb cause Bartter's syndrome type III. *Nature Genet.* **17**, 171–178.

Simon, D. B., Karet, F. E., Hamdam, J. M., Di Pietro, A., Sanjad, S. A., and Lifton, R. P. (1996a). Bartter's syndrome, hypokalaemic alkalosis with hypercalciuria, is caused by mutations in the Na-K-2Cl cotransporter NKCC2. *Nature Genet.* **13**, 183–188.

Simon, D. B., Karet, F. E., Rodriguez-Soriano, J., Hamdan, J. H., DiPietro, A., Trachtman, H., Sanjad, S. A. and Lifton, R. J. (1996b). Genetic heterogeneity of Bartter's syndrome revealed by mutations in the K$^+$ channel, ROMK. *Nature Genet.* **14**, 152–156.

Simon, D. B., Lu, Y., Choate, K. A., Velazquez, H., Al-Sabban, E., Praga, M., Casari, G., Bettinelli, A., Colussi, G., Rodriguez-Soriano, J., McCredie, D., Milford, D., Sanjad, S., and Lifton, R. P. (1999). Paracellin-1, a renal tight junction protein required for paracellular Mg^{2+} reabsorption. *Science* **285**, 103–106.

Simon, D. B., Nelson-Williams, C., Bia, M. J., Ellison, D., Karet, F. E., Molina, A. M., Vaara, I., Iwata, F., Cushner, H. M., Koolen, M., Gainza, F. J., Gitelman, H. J., and Lifton, R. P. (1996c). Gitelman's variant of Bartter's syndrome, inherited hypokalaemic alkalosis, is caused by mutations in the thiazide-sensitive NaCl cotransporter. *Nature Genet.* **12**, 24–30.

Sly, W. S., Hewett-Emmett, D., Whyte, M. P., Yu, Y.-S. L., and Tashian, R. E. (1983). Carbonic anhydrase II deficiency identified as the primary defect in the autosomal recessive syndrome of osteopetrosis with renal tubular acidosis and cerebral calcification. *Proc. Natl. Acad. Sci. USA* **80**, 2752–2756.

Sly, W. S., Whyte, M. P., Sundaram, V., Tashian, R. E., Hewett-Emmett, D., Guibaud, P., Vainsel, M., Baluarte, H. J., Gruskin, A., Al-Mosawi, M., Sakati, N., and Ohlsson, A. (1985). Carbonic anhydrase II deficiency in 12 families with the autosomal recessive syndrome of osteopetrosis with renal tubular acidosis and cerebral calcification. *N. Engl. J. Med.* **313**, 139–145.

Smith, A. N., Skaug, J., Choate, K. A., Nayir, A., Bakkaloglu, A., Hulton,

S. A., Al-Sabban, E. A., Lifton, R. P., Scherer, S. W., and Karet, F. E. (2000). Mutations in ATP6N1B, encoding a new kidney vacuolar proton pump 116-kD subunit, cause recessive distal renal tubular acidosis with preserved hearing. *Nature Genet.* **26**, 71–75.

Strom, T. M., Francis, F., Lorenz, B., Boddrich, A., Econs, M. J., Lehrach, H., and Meitinger, T. (1997). Pex gene deletions in Gy and Hyp mice provide mouse models for X-linked hypophosphataemic rickets. *Hum. Mol. Genet.* **6**,165–171.

Tanphaichitr, V. S., Sumboonnanonda, A., Ideguchi, H., Shayakul, C., Brugnara, C., Takao, M., Veerakul, G., and Alper, S. L. (1998). Novel AE1 mutations in recessive distal renal tubular acidosis: Loss-of-function is rescued by glycophorin A. *J. Clin. Invest.* **102**, 2173–2179.

Tenenhouse, H. S. (1999). X-linked hypophosphataemia: A homologous disorder in humans and mice. *Nephrol. Dial. Transplant.* **14**, 333–341.

Thoene, J., Lemons, R., Anikster, Y., Mullet, J., Paelicke, K., Lucero, C., Gahl, W., Schneider, J., Shu, S. G., and Campbell, H. T. (1999). Mutations of CTNS causing intermediate cystinosis. *Mol. Gen. Metab.* **67**, 283–293.

Town, M., Jean, G., Cherqui, S., Attard, M., Forestier, L., Whitmore, S. A., Callen, D. F., Gribouval, O., Broyer, M., Bates, G. P., van't Hoff, W., and Antignac, C. (1998). A novel gene encoding an integral membrane protein is mutated in nephropathic cystinosis. *Nature Genet.* **18**, 319–324.

van Lieburg, A. F., Knoers, N. V. A. M., and Monnens, L. A. H. (1999). Clinical presentation and long term follow up of thirty patients with nephrogenic diabetes insipidus. *J. Am. Soc. Nephrol.* **10**, 1958–1964.

van den Ouweland, A. M., Dreesen, J. C., Verdijk, M., Knoers, N. V., Monnens, L. A., Rocchi, M., and van Oost, B. A. (1992). Mutations in the vasopressin type 2 receptor gene (AVPR2) associated with nephrogenic diabetes insipidus. *Nature Genet.* **2**, 99–102.

van Lieburg, A. F., Verdijk, M. A., Schoute, F., Ligtenberg, M. J., van Oost, B. A., Waldhauser, F., Dobner, M., Monnens, L. A., and Knoers, N. V. (1995). Clinical phenotype of nephrogenic diabetes insipidus in females heterozygous for a vasopressin type-2 receptor mutation. *Hum. Genet.* **96**, 70–78.

van Lieburg, A. F., Verdiijk, M., Knoers, N. V. A. M., van Essen, A. J., Proesmans, W., Mallman, R., Monnens, L. A. H., van Oost, B. A., van Os, C. H., and Dee, P. M. T. (1994). Patients with autosomal nephrogenic diabetes insipidus homozygous for mutations in the aquaporin-2 water-channel gene. *Am. J. Hum. Genet.* **55**, 648–652.

Vollmer, M., Jeck, N., Lemmink, H. L., Vargus, R., Feldmann, D., Konrad, M., Beekmann, F., van den Heuvel, L. P. W. J., Deschennes, G., Guay-Woodford, L. M., Antignac, C., Seyberth, H. W., Hildrebrandt, F., and Knoers, N. V. A. M. (2000). Antenatal Bartter syndrome with sensori-neural deafness: Refinement of the locus on chromosome 1p31. *Nephrol. Dial. Transplant.* **15**, 970–974.

Wrong, O. M., Norden, A. G. W., and Feest, T. G. (1994). Dent's disease: A familial proximal renal tubular syndrome with low-molecular weight proteinuria, hypercalciuria, nephrocalcinosis, metabolic bone disease, progressive renal failure and a marked male predominance. *Q. J. Med.* **87**, 473–493.

Wyss, P. A., Boynton, S. B., Chu, J., Spencer, R. F., and Roth, K. S. (1992). Physiological basis for an animal model of the renal Fanconi syndrome: Use of succinylacetone in the rat. *Clin. Sci.* **83**, 81–87.

Yang, B., Gillespie, A., Carlson, E. J., Epstein, C. J., and Verkman, A. S. (2000). Neonatal mortality in an aquaporin-2 knock-in mouse model of recessive nephrogenic diabetes insipidus. *J. Biol. Chem.*

Yang, B., Ma, T., and Verkman, A. S. (2001). Erythrocyte water permeablility and renal function in double knockout mice lacking aquaporin-1 and aquaporin-3. *J. Biol. Chem.* **276**, 624–628.

Yun, J., Schoneberg, T., Liu, J., Schulz, A., Ecelbarger, C. A., Promeneur, D., Nielsen, S., Sheng, H., Grinberg, A., Deng, C. X., and Wess, J. (2000). Generation and phenotype of mice harboring a nonsense mutation in the V2 vasopressin receptor gene. *J. Clin. Invest.* **106**, 1361–1371.

27

Diseases of the Glomerular Filtration Barrier: Alport Syndrome and Congenital Nephrosis (NPHS1)

Hannu Jalanko, Christer Holmberg, and Karl Tryggvason

The two best characterized genetic diseases of the glomerular capillary wall are Alport syndrome (AS) and congenital nephrosis of the Finnish type (NPHS1). AS is caused by mutations in genes coding for type IV collagen of the glomerular basement membrane (GBM). Cases of AS can be classified into X-linked and autosomal forms. Mutations in AS result in abnormal collagen chain composition and irregular ultrastructure of the GBM. The typical clinical features of AS are progressive nephritis, sensorineural hearing loss, and ocular changes. Several animal models of AS have been published and it is an attractive disorder for gene therapy. NPHS1 is a rare disease caused by mutations in *NPHS1*, encoding a cell adhesion protein nephrin. Nephrin is synthesized by podocytes and is localized at the slit diaphragm area of the glomerular capillary wall. Most mutations in the nephrin gene lead to massive proteinuria and nephrotic syndrome soon after birth, suggesting an essential role for the slit diaphragm in glomerular filtration. It is expected that our understanding of the molecular structure of the glomerular filtration barrier will increase dramatically in the near future.

In kidney glomerulus, primary urine is formed by ultra-filtration of plasma into urinary space through the glomerular capillary wall (Kanwar *et al.*, 1991). Normally, water and small plasma solutes pass through this filtration barrier easily, but the passage of proteins with the size of albumin and larger is almost completely restricted. The glomerular filtration barrier is composed of three layers: (a) fenestrated endothelium, (b) glomerular basement membrane of about 300–350 nm in thickness, and (c) podocyte-interdigitating processes that almost completely cover the external side of the GBM (Fig. 27.1). The adjacent processes are separated by 30- to 50-nm-wide slit pores that are covered by the slit diaphragm (Smoyer and Mundel, 1998). The flow of the glomerular filtrate is thought to follow the extracellular route, passing across the GBM and then across the slit diaphragm.

The GBM has been regarded as the most important component in the size and charge selective sieving of the kidney glomelurus. The main components of the GBM are type IV collagen, laminin, nidogen, and proteoglycans (Hudson *et al.*, 1993). Type IV collagen forms a three-dimensional network where the other components are attached. Proteoglycans (perlecan and agrin) are rich in anionic heparan sulfate moeties and contribute to an electric barrier for the negatively charged macromolecules, such as plasma proteins (Kanwar and Farquhar, 1979; Raats *et al.*, 2000). The slit diaphragm is a poorly characterized structure connecting the podocyte foot processes (Rodewald and Karnowsky, 1974). The molecular architecture and role of the slit diaphragm in glomerular sieving have largely been unknown. Recent findings, however, suggest that it is important, especially in restricting the passage of plasma proteins into urine (Tryggvason, 1999; Wickelgren, 1999; Somlo and Mundel, 2000).

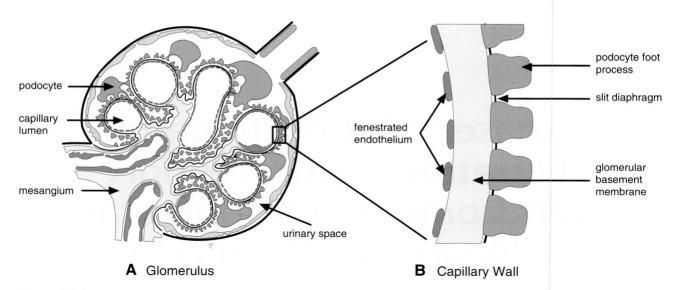

Figure 27.1 Schematic structure of the kidney glomerulus (A) and the glomerular filtration barrier (capillary wall) (B). Type IV collagen forms a three-dimensional network of the glomerular basement membrane, and nephrin is localized at the slit diaphragm Also see Chapter 11. (Fig. 27.1A is modified from the original of professor Wilhelm Kriz).

The glomerular filter is affected in a large number of acquired kidney diseases. Inherited defects, however, are quite rare as compared to kidney tubuli (Salomon *et al.*, 2000). The best known is Alport syndrome, which is caused by mutations in genes coding for type IV collagen of the GBM. The other recently characterized genetic disorder is congenital nephrosis of the Finnish type (NPHS1), which is caused by mutations in the *NPHS1* gene encoding a podocyte protein, nephrin. This protein most probably is a major component of the slit diaphragm. This chapter focuses on these two disease entities.

I. Alport Syndrome

Alport syndrome, also termed hereditary nephritis, was initially described in 1927 by A.C. Alport as an inherited kidney disease characterized by hematuria and sensorineuronal deafness. Later, ocular lesions were also associated with the syndrome. Today, diagnostic criteria used for differential diagnosis include a positive family history of hematuria with or without progression to end stage renal disease (ESRD); progressive sensorineural hearing loss; characteristic ocular changes (lenticonus and/or maculopathy); typical ultrastructural changes in the GBM; diffuse and abnormal GBM distribution of the α3, α4, and α5 chains of type IV collagen; and esophageal leiomyomatosis (Jais *et al.*, 2000)

The estimated prevalance of AS in the United States is 1:5000 (Atkin and Gregry, 1988) and accounts for 2.3% of the renal transplant population. In Europe, 1–2% of patients reaching ESRD have AS (Wing and Brunner, 1989).

A. Alport Syndrome and Type IV Collagen

AS arises from mutations in any one of the three genes coding for the α3, α4, and α5 chains of type IV collagen (Barker *et al.*, 1990; Mochizuki *et al.*, 1994; van der Loop *et al.*, 2000). Cases of AS can be classified into X-linked (COL4A5 mutations) and autosomal (COL4A3 and COL4A4 mutations) forms. The X-linked form is inherited as dominant, whereas autosomal forms exist both as recessive or as dominant. All these forms have similar clinical features.

Type IV collagen is a family of six α(IV) chains designated α1(IV) to α6(IV) chains (Hudson *et al.*, 1993). They are encoded by six distinct genes; COL4A1 to COL4A6. The two first are located pairwise in a head-to-head fashion on chromosome 13, COL4A3 and COL4A4 are located on chromosome 2, and COL4A5 and COL4A6 are located on the X chromosome. Mammalian type IV collagen genes are large (over 100 kb) and complex, containing as many as 52 exons (Zhou *et al.*, 1994). The primary structure of α chains is very similar, containing about 1600 amino acid residues. The amino terminus has a 25 residue noncollagenous domain, with a large Asn-linked oligosaccharide. A 230 residue noncollagenous (NC1) domain is located at the carboxyl terminus. The middle 1400 residue collagenous domain is composed of Gly-X-Y repeats (in which X is frequently proline and Y is frequently hydroxyproline) interrupted by about 20 short noncollagenous sequences.

The α chains assemble into triple-helical molecules, protomers, that self-assemble to form supramolecular networks by the formation of tetramers at the amino terminus and by dimerization at the carboxyl terminus through NC1 domains (Fig. 27.2). The α1 and α2 chains are ubiquitous in basement membranes. In contrast, the other α chains have more restricted tissue distributions, which presumably reflect specialized function. In the GBM, two networks with distinct chain composition have been identified: an α1–α2 (IV) network and an α3–α4–α5 (IV) network (Kalluri *et al.*, 1997; Gunwar *et al.*, 1998). In normal glomerulogenesis, the α1–α2(IV) network is assembled first in the embryonic glomerulus, but then there is a developmental switch to the synthesis of the α3–α4–α5(IV) network so that the adult GBM contains mostly the α3–α4–α5(IV) (Miner and Sanes, 1996; Harvey *et al.*, 1998; Heidet *et al.*, 2000).

The NC1 domain contains recognition sequences for the selection of chains and protomers that are sufficient to encode the assembly of the α1–α2 and α3–α4–α5 networks of the GBM (Boutaud *et al.*, 2000). The α3(IV)–α4(IV)–α5(IV) protomers contain disulfide cross-links between the triple helices through cysteine residues in the collagenous domain of the α3(IV) and α4(IV) chains. At eletron microscopy, α3(IV)–α4(IV)–α5(IV) protomers are characterized by loops and supercoiled triple helices not found in the α1–α2(IV)

protomers. The many cross-links in the α3–α4–α5(IV) network confer increased tensile strength and greater resistance to proteolysis compared to molecules composed of the α1 and α2 chains (Kalluri *et al.*, 1997). This is important to the GBM, which does not have a supportive extracellular matrix composed of collagen fibers.

B. X-linked Alport syndrome

The X-linked dominant transmission is the most frequent mode of inheritance (over 80% of AS patients). All mutations demonstrated so far in X-linked AS have affected the COL4A5 gene (Barker *et al.*, 1990; Knebelmann *et al.*, 1996; Jais *et al.*, 2000). Close to 300 mutations have been described in the X-linked AS differing from one family to the next. About 15% of the mutations are large gene rearrangements, such as deletions, insertions, inversions, or duplications. The rest are small mutations, with a large proportion of these single base changes. In general, there is a poor correlation between the nature of the mutation and the resulting phenotype. The mutations can result in complete

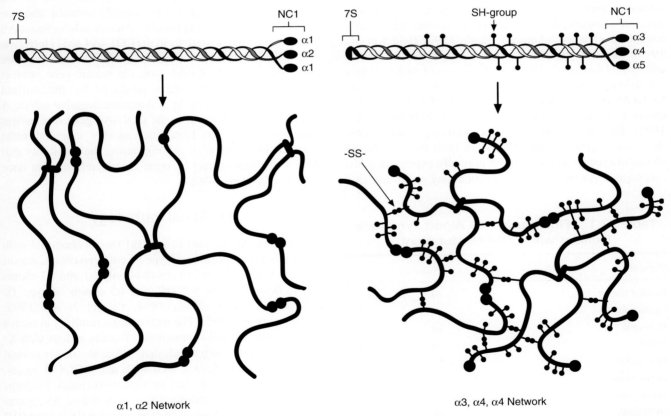

Figure 27.2 Diagram of the two distinct networks of type IV collagens in the GBM. (Top) The two protomers contain three α(IV) chains [either α–α2(IV) or α3–α4–α5(IV) composition]. Each chain contains a carboxy-terminal noncollagenous NC1 and an amino-terminal 7S domain. The protomers associate through NC1 domains and 7S domains to form suprastuctures (bottom). The many cross-links in the α3–α4–α5(IV) network confer increased tensile strength and greater resistance to proteolysis compared to molecules composed of α1(IV) and α2(IV) chains. The α3–α4–α5(IV) network is absent in Alport syndrome. Adapted and modified from Boutaud *et al.* (2000).

absence of the α5 chain or a truncated or malfunctioning protein.

The clinical picture of AS is quite heterogeneous. Affected males often develop renal failure anywhere from juvenile to adult age, whereas the clinical features in females are usually less severe and less progressive. The European Commity Alport Syndrome Concerted Action has delineated the AS phenotype and determined the genotype–phenotype correlations in a large number of European families (Jais *et al.*, 2000). Table 27.1 summarizes the findings in 195 familes with X-linked AS. All male patients were hematuric, and proteinuria was found in 95% of the patients. Progression to ESRD occurred in 78% of the patients. This was related to genotype so that large deletions and nonsense mutations changing the reading frame conferred to affected male patients a 90% probability of developing ESRD before 30 years of age, whereas the same risk was 50 and 70%, respectively, in patients with missense or splice mutations. The risk of developing hearing loss before 30 years of age was approixmately 60% in patients with missense mutations, contrary to 90% of the other types of mutations.

A small subset of patients with X-linked AS develops diffuse esophageal leiomyomatosis. Such patients may also have sensorineural deafness and ocular lesions, particularly congenital cataracts (Antignac *et al.*, 1992; Heidet *et al.*, 1995). The esophageal disease is a fully penetrant trait in females (Dahan *et al.*, 1995). In this group of patients, there are deletions involving the 5′ ends of the COL4A5 and COL4A6 genes, with the deletions in COL4A6 limited to exons 1, 1′, and 2 (Antignac *et al.*, 1992; Heidet *et al.*, 1995). This mutation may delete a common promoter in the intergenic region, leading to loss of both the α5 and α6 chains of type IV collagen (both normally expressed in the esophagus). Alternatively, the deletion may result in a gain

of function of COL4A6 with synthesis of a truncated α6 chain or involve an undiscovered gene in intron 2.

C. Autosomal Alport Syndrome

Autosomally inherited AS is about 10–15% of all AS cases. In the recessive form, hearing loss and eye lesions may be less common but otherwise the clinical picture is similar to that in X-linked AS (Atkins *et al.*, 1988; Mazzucco *et al.*, 1997). The patients develop hematuria, end-stage renal failure, and a disrupted GBM structure as in X-linked AS. Deafness and thrombocytopathy have been reported in some patients. Males and females are usually equally affected. The carriers are usually asymptomatic and there is often consanguinity. Mutations have been demonstrated in the COL4A3 and COL4A4 genes (Mochizuki *et al.*, 1994). Similar to the X-linked form, the mutations differ between families. The result of the mutations is also similar to X-linked AS with loss of the α3, α4, and α5 chains. The same possible pathogenic mechanisms described for X-linked AS could apply to the autosomal-recessive form of the disease, leading to a disrupted structure of the GBM.

Autosomal-dominant AS is a very rare form of the disease. Males and females are equally affected and one parent is likely to be symptomatic. Autosomal-dominant AS has been linked to mutations in the COL4A4 gene (Jefferson *et al.*, 1997; van der Loop *et al.*, 2000). Because there are two copies of the COL4A4 gene, the mutant gene product somehow negates the chain produced by the normal COL4A4 gene, possibly by a dominant-negative effect. A mutation has been found in the COL4A4 gene in a patient with benign familial hematuria, another inherited renal disorder (Lemmink *et al.*, 1996), raising the possibility that these two diseases may be more closely related than once thought (Kashtan, 1999).

D. Renal Pathology

Mutations in AS lead to a GBM that is abnormal with respect to its composition of type IV collagen chains. Alport GBM is in most cases composed of only α1 and α2 chains and is devoid of the α3, α4, and α5 chains of type IV collagen (Yoshika *et al.*, 1994; Peissel *et al.*, 1995; Ninomiya *et al.*, 1995). The mechanisms resulting in such a configuration are still controversial. Results in dogs with X-linked AS have suggested that there is transcriptional coregulation of COL4A3, COL4A4, and COL4A5 genes, which would explain the lack of all three α chains. Findings in human, however, indicated that in X-linked AS (mutations in the COL4A5 gene), the absence of α3(IV) to α5(IV) in the GBM resulted from events downstream of transcription, RNA processing, and protein synthesis (Heidet *et al.*, 2000). Intrestingly, the α3(IV), α4(IV), and α5(IV) chains were found to be synthesized in the podocytes,

Table 27.1   Findings in 195 Alport Syndrome Families with COL4A5 Mutation[a]

Variable	Frequency (%)
Familial history	89
Consanguinity	2.5
Hematuria	99
ESRD	76
Hearing loss	83
Ocular changes	44
Leiomyomatosis	5
GBM changes	98
Transplantation	54
Posttransplantation anti-GBM nephritis	4

[a] Adapted from Jais *et al.* (1999).

whereas the α1(IV) chain was produced by mesangial/endothelial cells.

In electron microscopy, the characteristic finding in AS is irregular thickening and multilamination of the GBM (Atkins *et al.*, 1988). These ultrastructural irregularities are perhaps associated with a high susceptibility to endoproteolysis of the GBM containing only α1(IV) and α2(IV) chains (Kalluri *et al.*, 1997). It is also possible that changes in the GBM composition may alter the homeostasis of GBM remodeling via changes in the interaction with cell adhesion molecules such as integrins (Sayers *et al.*, 1999).

The pathological lesions in AS are not restricted to the GBM. Typically, expansion of the mesangial matrix and podocyte foot process effacement occur quite early during the disease, and later, tubulointerstitial fibrosis is a prominent finding (Sayers *et al.*, 1999). The pathogenesis of these changes is not known. Using gene knockout mouse models, Cosgrove *et al.* (2000) demonstrated that two pathways, one mediated by transforming growth factor (TGF)-β1 and the other by integrin-α1β1, affect the pathogenesis in distinct ways. In AS mice, which were also null for integrin-α1, expansion of the mesangial matrix and podocyte foot process effacement were attenuated, whereas inhibition of the function of TGF-β1 prevented the focal thickening of the GBM. The authors suggest that integrin-α1β1 and TGF-β1 might provide useful targets for dual therapy aimed at slowing disease progression in AS (Cosgrove *et al.*, 2000).

E. Animal Models

Several animal models of AS have been published. Common features to all are that they resemble the human disease with delayed onset renal failure, hematuria, proteinuria, and lamellated GBM structure.

1. X-Linked AS

The best described animal model for AS is the Samoyed dog model of X-linked nephritis (Zheng *et al.*, 1994; Thorner *et al.*, 1996). Male affected dogs have proteinuria with or without hematuria by 2 months of age. Renal insufficiency usually develops by 5 months and progresses to uremia and death by 8–10 months (Jansen *et al.*, 1986). No visual or hearing defects have been detected in male- or female-affected dogs. Renal pathology includes a thickening and wrinkling of the GBM with multilamination. Up to 3 weeks of age, the GBM appears normal, but after 1 month, GBM abnormalities appear. By 4–5 months, all affected animals have abnormal GBM ultrastructures.

Available sequences for the dog α1–α6 chains of collagen type IV are >88% identical at the DNA level and >92% identical at the protein level to the respective human α(IV) chains (Zheng *et al.*, 1994; Thorner *et al.*, 1996). The genetic basis of canine X-linked nephritis is a single base mutation in the COL4A5 gene encoding the α5 chain of type IV collagen. The mutation introduces a stop codon causing a premature termination of transcription. This causes a 90% reduction of the α5 transcript. Unexpectedly, the α3 and α4 transcripts are reduced at least 77% in affected male dog kidney, whereas the α1 and α2 transcripts are expressed at comparable levels in normal and affected kidneys, suggesting a mechanism coordinating the expression of the α3, α4, and α5 collagen chains.

2. Autosomal AS

There are four animal models of autosomal-recessive AS reported. One is naturally occurring in English cocker spaniel dogs. Affected dogs present with proteinuria ± hematuria after 4 months of age and develop renal failure from 6 to 24 months of age (Lees *et al.*, 1997). The GBM shows the typical ultrastructural changes of AS. The α3 and α4 chains are absent from the GBM by immunohistochemistry, whereas the α5 chain is weakly expressed; staining for the α1 and α2 chains is increased. A similar observation has been noted in human AS (Kashtan and Kim, 1992). Carrier animals are asymptomatic. The molecular basis for the disease has not yet been determined.

The other three models for autosomal-recessive AS are all in transgenic mice, two involving a knockout of the COL4A3 gene (Miner and Sanes, 1996; Cosgrove *et al.*, 2000) and the third involving the COL4A4 gene (Lu *et al.*, 1997). In the COL4A3 knockout mouse, a deletion was created in the coding region for the NC1 domain of the α3 chain. Homozygous mice develop proteinuria at 2–3 months of age and die of renal failure by 3–4 months; no extra renal disease has been noted. The GBM shows typical Alport changes by 4 weeks of age. No staining for α3–α5 chains is detected in the kidney, whereas increased staining of GBM for the α1 and α2 chains is noted. Measurements of mRNA levels for the α(IV) chains have shown no detectable α3 message as expected and slightly elevated values for α1, α4, and α5 in mutant mice. This is in contrast to the Samoyed dog model in which the message levels of the α3, α4, and α5 chains were all reduced.

Taking the two sets of results together would mean that production of the α3 chain does not drive the α4 and α5 chains, but production of the α5 chain may drive the α3 and α4 chains. The α5 chain is detectable in the neuromuscular junction in mutant mice, suggesting that this chain does not require the presence of α3 and α4 chains in all sites in terms of collagen assembly. The newly described COL4A4 mouse model carries a tyrosinase minigene inserted into the COL4A4 gene (Lu *et al.*, 1997). Homozygous mice develop proteinuria at 2 weeks of age, followed by hematuria, azotemia, and death due to renal insufficiency between 8 and 12 weeks. Ultrastructurally, the GBM is thin and focally laminated. The mutation reduces the expression of both α3 and α4 chains at the mRNA level, but not α1 and α5 chains.

F. Gene Therapy

AS is an attractive disease for gene therapy. It almost solely affects the glomeruli, with the extrarenal complications not being life-threatening nor occurring in all patients. The kidneys have an isolated circulatory system, which lends itself well to organ-targeted gene transfer. Successful gene therapy of AS would require transfer of the corrected gene into the glomerular cells responsible for production of the GBM. We have been able to target gene transfer with high efficiency to glomeruli *in vivo* (Heikkilä *et al.*, 1996). This was achieved *in vivo* by perfusion of porcine kidney isolated from the systemic blood circulation using a separate oxygenated system and adenovirus as a vector. Using β-galactosidase as a reporter, the perfused kidneys showed activity in >85% of glomeruli. Furthermore, cells expressing in the glomeruli were podocytes and endothelial cells rather than mesangial cells or parietal epithelial cells (Heikkilä *et al.*, 1996; Tryggvason *et al.*, 1997).

Several problems remain to be considered if gene therapy is to be successful in AS. A transferred type IV collagen should be expressed in the correct cells, modified, and folded in the correct way into normal type IV collagen trimers. Additionally, these trimers should be able to incorporate into the GBM and restore its structure and function. It is not known yet if this kind of therapeutic response is possible to achieve. The second problem is that the adenovirus-mediated procedure provides only short-term expression of the transgene. The half-life of type IV collagen is estimated to be over a year and therefore the effects of transgene expression may last longer than the actual duration of expression. Nevertheless, the treatment would need to be repeated many times during life. Therefore, successful future gene therapy of this disease will depend largely on the development of a better transfer vector. Finally, the possible immunogenic effects of gene therapy in AS should be considered. Some patients with Alport syndrome after receiving a renal transplant form antibodies against the α3 chain or the α5 chain (Ding *et al.*, 1994; Rutgers *et al.*, 2000). An immune response may also result from the adenovirus itself that could also limit the success of transgene expression.

Extensive research still needs to be carried out before we can expect to be able to do successful gene therapy of AS in humans. Ultimately this will depend on the results obtained from animal model work, together with advances in gene transfer vectors to overcome the limitations that currently exist.

II. Congenital Nephrosis (NPHS1)

A. Epidemiology

The NPHS1 syndrome is a form of congenital nephrosis caused by mutations in the NPHS1 gene. It was first described by Hallman *et al.* (1956) in Finland. Since then, approximately 300 patients have been reported. The incidence in Finland is about 1 in 8000 (Huttunen, 1976) and about half of the published cases of NPHS1 come from Finland. However, patients have been reported from all over the world (Norio, 1966; Holmberg *et al.*, 1999), with most cases among whites, but also from other ethnic groups. Interestingly, a very high incidence of NPHS1 has been reported among the old order Mennonites in Lancaster County, Pennsylvania (Bolk *et al.*, 1999). In a subgroup of "Groffdale Conference" Mennonites, the incidence of NPHS1 was 1/500, which is almost 20 times greater than that observed in Finland.

B. Clinical Features

The clinical picture of severe NPHS1 [also called congenital nephrosis of the Finnish type (CNF)] varies only slightly (Huttunen, 1976; Holmberg *et al.*, 1999; Patrakka *et al.*, 2000). The majority (over 80 %) of NPHS1 children are born prematurely (<38th week) with a birth weight ranging between 1500 and 3500 g (Patrakka *et al.*, 2000). However, the newborns are rarely small for the gestational age (Table 27.2). Amniotic fluid is often meconium stained, but most neonates do not have major pulmonary problems. The index of placental weight/birth weight (ISP) is over 25% in practically all newborns. The reason for this is not known.

The basic problem in NPHS1 is a severe loss of plasma proteins, 90% of which is albumin. Proteinuria begins *in utero* and is thus detectable in the first urine sample tested. In one analysis (Patrakka *et al.*, 2000), nephrotic syndrome was diagnosed within the first week after birth in 82% of the Finnish patients and within 2 months in all. Microscopic hematuria and normal creatinine values during the first months were always observed. Heavy protein losses led to hypogammaglobulinemia and an increased risk for

Table 27.2 Features of 41 Finnish NPHS1 Infants with Fin-Major and Fin-Minor Mutations[a,b]

Variable	Frequency (%)
Born prematuraly (<38th week)	83
Placental weight/fetal weight >25%	98
Diagnosis of NS during the 1st week	81
Microscopic hematuria	96
Normal serum creatinine	100
Hypotonia	89
Cardiac hypertrophy/pulm stenosis	28

[a] Adapted from Patrakka *et al.* (2000).

[b] Both mutations lead to a truncated protein (Fig. 28.3) and total absence of nephrin molecule in the kidney glomerulus.

infections (Ljungberg *et al.*, 1997), as well as low anti-thrombin III levels and an increased risk for thrombotic complications (Holmberg *et al.*, 1999). Hyperlipidemia is also present as in other nephrotic syndromes (Holmberg *et al.*, 1999). Infants with NPHS1 do not have any major nonrenal malformations. However, minor functional disorders in the central nervous system and cardiac hypertrophy are common during the nephrotic stage. A small subgroup of the Finnish patients also has atethosis, whose etiology is still unknown.

Minor mutations in *NPHS1* lead to atypical and milder forms of nephrotic syndrome. So far, the reports on these patients are anecdotal (Lee *et al.*, 1999; Koziell *et al.*, 1999a,b; Beltcheva *et al.*, 1999). In some infants, the urinary protein concentration may be less than 20 g/liter, and the daily losses are modest. A systematic analysis of the milder forms of NPHS1 is clearly needed.

Children with classical NPHS1 are treated succesfully with active protein and nutritional support, followed by bilateral nephrectomy, dialysis, and renal transplantation (Holmberg *et al.*, 1999). The long-term kidney graft and patient survivals are generally good (Qvist *et al.*, 1999). Recurrency of nephrotic syndrome, however, may occur in patients with the Fin-major/Fin-major genotype (see later) (Laine *et al.*, 1994; Patrakka *et al.*, 2001). In cases with "mild" mutations, reduction of proteinuria may be achieved by antiproteinuric therapy (indomethacin and captopril), and this medication should be tried in non-Finnish patients with NPHS1 (Guez *et al.*, 1998; Heaton *et al.*, 1999; Patrakka *et al.*, 2000).

C. Renal Pathology

In typical cases of NPHS1, the kidneys are large compared to the weight of the patient (Huttunen *et al.*, 1980). The histologic alterations are polymorphic and progressive and no single finding is pathognomonic or necessary for the diagnosis. Glomeruli are present in nearly twice the normal number (Tryggvason and Kouvalainen, 1975). A proliferation of mesangial cells and an increase of PAS- and silver-positive matrix are characteristic. Dilations of the proximal, and sometimes also distal, tubules are the most characteristic findings. The tubular epithelium is tall in the beginning, but flat and atrophic in advanced cases. In the interstitium, round cell infiltration and fibrosis increase with age.

In electron microscopy the principal finding is the fusion and effacement of podocyte foot processes in glomeruli (Huttunen *et al.*, 1980). This is not specific for NPHS1 but is seen in many nephrotic kidney diseases. Kidneys from Finnish NPHS1 patients with Fin-major and Fin-minor mutations were studied carefully by electron microscopy (Patrakka *et al.*, 2000). These kidneys lack nephrin and, interestingly, no slit diaphragms were present between the podocyte foot processes.

D. Genetic and Molecular Bases of NPHS1

The gene responsible for NPHS1 was first linked by disequilibrium analysis to a critical 150-kb region on chromosome 19q13.1 (Kestilä *et al.*, 1994; Männikkö *et al.*, 1995). This region was then sequenced and a novel gene, named *NPHS1*, was identified (Kestilä *et al.*, 1998). The gene consists of 29 exons and has a size of 26 kb (Fig. 27.3). It codes for a novel protein, nephrin, which is a 1241 residue cell adhesion protein of the immunoglobulin family. The extracellular part of nephrin contains eight immunoglobulin-like modules and one type III fibronectin domain (Fig. 28.3). The immunoglobulin modules are of the type C2, which is found predominantly in proteins participating in cell–cell interactions. The intracellular domain has no significant homology with other known proteins, but it has nine tyrosine residues, some of which might become phosphorylated during ligand binding. *In situ* hybdrization for mRNA indicated that nephrin was synthesized by glomerular podocytes (Kestilä *et al.*, 1998), and nephrin was localized at the slit diaphragm area between the podocyte foot processes (Ruotsalainen *et al.*, 1999).

1. Mutations in Finnish Patients

Exon sequencing analyses of 49 Finnish patients revealed the presence of two important mutations in 94% of NPHS1 chromosomes (Kestilä *et al.*, 1998). The first mutation, a 2-bp deletion in exon 2, causes a frameshift, resulting in a stop codon within the same exon (Fin-major mutation). The second sequence variant was a nonsense mutation (CGA to TGA) in exon 26 (Fin minor) (Fig. 28.3). Fin major leads to a truncated 90 residue protein, and Fin minor leads to a truncated 1109 residue protein. In addition to Fin major and Fin minor, a few missense mutations in the Finnish patients were found (Lenkkeri *et al.*, 1999) The

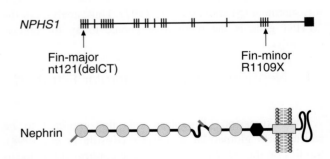

Figure 27.3 *NPHS1* gene and nephrin. *NPHS1* has a size of 26 kb and contains 29 exons. Fin major and Fin minor are the two most prevalent mutations found in Finnish NPHS1 patients. Both lead to truncated proteins of 90 and 1109 amino acid residues, respectively. Nephrin is a cell adhesion molecule containing 1241 amino acid residues. The extracellular part has eight Ig domains (circles) and one fibronectin type III motif (hexagon). The locations of three cystine residues are indicated by short bars.

uniform mutation pattern seen in the Finnish population can be explained by the founder effect.

2. Mutations in Non-Finnish Patients

Several reports on NPHS1 gene mutations in non-Finnish patients have been published (Lenkkeri *et al.*, 1999; Lee *et al.*, 1999; Koziell *et al.*, 1999a,b; Bolk *et al.*, 1999; Beltcheva *et al.*, 1999; Aya *et al.*, 2000). The patients come from Europe, North America, North Africa, Middle East, and Asia. In contrast to the Finnish patients, most non-Finns have individual mutations. These include deletions, insertions, nonsense, missense, and splicing mutations spanning over the whole gene. Also, mutations in the promoter area have been found. For the moment, over 60 mutations in *NPHS1* have been identified. Many of the missense mutations are located in exons coding for the Ig-like motifs (Lenkkeri *et al.*, 1999). Interestingly, the Fin-major and Fin-minor mutations are rare in non-Finnish patients. Enrichment of other mutations has been reported in different populations. In Mennonites, a 1481delC mutation is common and leads to a truncated protein of 547 residues (Bolk *et al.*, 1999). However, six out of nine patients from Malta were homozygous for a nonsense mutation R1160X in exon 27 (Koziell *et al.*, 1999b).

In addition, several sequence variants of the *NPHS1* locus have been detected in healthy individuals. The two most common polymorphisms are Glu117Lys and Asn1077Ser (Lenkkeri *et al.*, 1999). Findings indicate that this gene is quite susceptible to mutagenesis.

3. Prenatal Diagnostics

Proteinuria in NPHS1 starts already prenatally, which results in high levels of α-fetoprotein (AFP) in amnitotic fluid and maternal serum (Aula *et al.*, 1978). This finding is not specific for NPHS1 and may be caused by other fetal abnormalities, such as neural tube defects. In addition, carriers of the *NPHS1* mutations may show temporary fetal proteinuria and give false-positive results in the AFP measurement (Männikkö *et al.*, 1997). Thus, analysis of the NPHS1 gene is the method of choice today for precise diagnosis.

E. Nephrin and the Glomerular Filtration Barrier

The expression of nephrin was first reported to be restricted to glomerular podocytes (Kestilä *et al.*, 1998). Studies on the *NPHS1* promoter have confirmed this localization (Moeller *et al.*, 2000; Wong *et al.*, 2000). However, data obtained from mice suggest that nephrin is also expressed in some areas of the central nervous system and pancreas (Putaala *et al.*, 2000).

The localization of nephrin at the slit diaphragm in human kidney glomerulus (Ruotsalainen *et al.*, 1999; Holthöfer *et al.*, 1999) has been verified in mouse and rat (Putaala *et al.*, 2000; Holzman *et al.*, 1999; Ahola *et al.*, 1999). These data suggest an essential role for this structure in normal glomerular filtration (Wickelgren, 1999; Somlo and Mundel, 2000). We have proposed a hypothetical model where nephrin molecules from adjacent foot processes show a head-to-head assembly through homophilic interactions and form a backbone of the slit diaphragm (Ruotsalainen *et al.*, 1999). This architecture fits to the electron microscopic model of Rodewald and Karnowsky (1974), who suggested that the slit diaphragm has a zipper-like organization .

The central role of nephrin at the slit diaphragm is supported by the finding that in NPHS1 kidneys lacking nephrin, the filamentous image of the slit diaphragm is completely missing in electron microscopy (Fig. 27.4) (Patrakka *et al.*, 2000). Similarly, nephrin knockout mice lack a slit diaphragm and die of severe proteinuria soon after birth (Putaala *et al.*, 2001). Also, studies on human glomerulogenesis indicated that the early development and migration of junctional complexes between developing podocytes occur normally even in kidneys lacking nephrin, but the final maturation of slit diaphragms is defective (Ruotsalainen *et al.*, 2000). In normal nephrogenesis, nephrin is first expressed during the late S shape. The role of nephrin is also emphasized by the finding that antibodies against nephrin cause heavy proteinuria in kidney grafts transplanted to NPHS1 patients with the Fin-major/Fin-major genotype (Patrakka *et al.*, 2001). This recurrency of nephrotic syndrome in the kidney graft resembles the situation in Alport syndrome, wherein anti-GBM antibodies cause *de novo* nephritis in some transplanted Alport patients.

The actual molecular structure of the slit diaphragm is, as yet, unresolved. Nephrin has been found to be associated with the so-called CD2-associated protein in mouse (Shih *et al.*, 1999), and it is possible that this cytosolic protein anchors nephrin to the podocyte foot process. Other extra- and intracellular interactions of nephrin with podocyte molecules, however, are unknown.

III. Conclusions

Alport syndrome and NPHS1 are, so far, the best characterized disorders of the glomerular filtration barrier. Recent findings, however, have accelerated elucidation of the molecular structure of the glomerular filter. Boute and colleagues (2000) found that mutations in *NPHS2* coding for a podocyte protein, podocin, are associated with a hereditary form (autosomal recessive) of focal segmental glomerulosclerosis (FSGS). Similarly, Kaplan and co-workers (2000) showed that defective α-actinin 4 causes an autosomal-dominant form of FSGS. These proteins are synthesized and expressed by podocytes, and it seems that

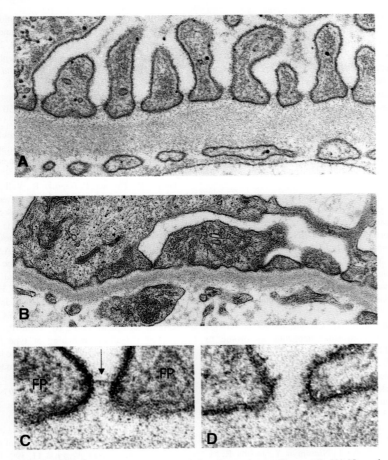

Figure 27.4 Electron microcopy of the glomerular capillary wall. (A) Normal kidney glomerulus showing regular podocyte foot processes, GBM, and fenestrated endothelium. Filamentous images of slit diaphragms can be seen in some slit pores. (B) NPHS1 kidney with effacement of foot processes. (C) Higher magnification of a normal slit pore with a slit diaphragm (arrow) between the foot processes (FP). (D) NPHS1 kidney with an "empty" slit pore (no slit diaphragm visible).

podocytes play a critical role in the pathogenesis of nephrotic syndrome. Molecules of the slit diaphragm and cytoskeleton are essential in regulating podocyte structure and function, and it is expected that our understanding of these molecular interactions will increase dramatically during the coming years.

Acknowledgments

This work was supported by The Sigrid Juselius Foundation, The Pediatric Research Foundation, The Finnish Academy, Helsinki University Central Hospital Research Fund, The Swedish Medical Research Council, Novo Nordisk Foundation, and NIH Grant DK 54724.

References

Ahola, H., Wang, S. W., Luimula, P., Solin, M.-L., Holzman, L., and Holthöfer, H. (1999). Cloning and expression of the rat nephrin homolog. *Am. J. Pathol.* **155**, 907–913.

Alport, A. C. (1927). Hereditary familial congenital haemorrhagic nephritis. *Br. Med. J.* **1927**, 1504–506.

Antignac, C., Zhou, J., Sanak, M., Cochat, P., Roussel, B., Deschenes, G., Knebelman, B., Hors, M. C., Tryggvason, K., and Gubler, M.-C. (1992). Alport syndrome and diffuse leiomyomatosis: Deletions in the 5′ end of the COL4A5 collagen gene. *Kidney Int.* **42**, 1178–1183.

Atkin, C. L., Gregory, M. C., and Border, W. A. (1988). Alport syndrome. *In* "Diseases of Kidney" (W. Schrier, and C. Gottschalk, eds.), pp. 617–641. Little-Brown, Boston.

Aula, P., Rapola, J., Karjalainen, O., and Seppälä, M. (1978). Prenatal diagnosis of congenital nephrosis in 23 high-risk families. *Am. J. Dis. Child.* **132**, 984–987

Aya, K., Tanaka, H., and Seino, Y. (2000). Novel mutation in the nephrin gene of a Japanese patient with congenital nephrotic syndrome of the Finnish type. *Kidney Int.* **57**, 401–404.

Barker, D. E., Hostikka, S. L., Zhou, J., Show, L. T., Oliphant, A. R., Gergen, S. C., Gregory, M. C., Skolnick, M. H., Atkin, C. L., and Tryggvason, K. (1990). Identification of mutations in the COL4A5 collagen gene in Alport syndrome. *Science* **248**, 1224–1227.

Beltcheva, O. J., Lenkkeri, U., Kestilä, M., Mannikko, M., and Tryggvason, K. (1999). Spectrum of nephrin gene mutations in congenital nephrotic syndrome *J. Am. Soc. Nephrol.* **10**, A2172.

Bolk, S., Puffenberger, E. G., Hudson, J., Morton, D. H., and Chakravarti, A. (1999). Elevated frequency and allelic heterogeneity of congenital nephrotic syndrome, Finnish type, in the old order Mennonites. *Am. J. Hum. Genet.* **65**, 1785–1790.

Boutaud, A., Borza, D.-B., Bondar, O., Gunwar, S., Netzer, K.-O., Singh, N., Ninomiya, Y., Sado, Y., Noelken, M., and Hudson, B. (2000). Type IV collagen of the glomerular basement membrane: Evidence that the chain specificity of network assembly is encoded by the noncollagenous NC1 domains. *J. Biol. Chem.* **275**, 30716–30724.

Boute, N., Gribouval, O., Roselli, S., Benessy, F., Lee, H., Fuchshuber, A., Dahan, K., Gubler, M.-C., Niaudet, P., and Antignac, C. (2000). NPHS2, encoding the glomerular protein podocin, is mutated in autosomal recessive steroid-resistant nephrotic syndrome. *Nature Genet.* **24**, 349–355.

Cosgrove, D., Rodgers, K., Meehan, D., Miller, C., Bovard, K., Gilroy, A., Gardner, H., Kotelianski, V., Gotwals, P., Amatucci, A., and Kalluri, R. (2000). Integrin $\alpha 1\beta 1$ and transforming growth factor-$\beta 1$ play distinct roles in Alport glomerular pathogenesis and serve as dual targets for metabolic therapy. *Am. J. Pathol.* **157**, 1649–1659.

Dahan, K., Heidet, L., Zhou, J., Mettler, G., Leppig, K. A., Proesman, W., David, A., Roussel, B., Mongeau, J. G., Gould, J., Grunfeld, J.-P., Gubler M.-C., and Antignac, C. (1995). Smooth muscle tumors associated with X-linked Alport syndrome: Carrier detection in females. *Kidney Int.* **48**, 1900–1906.

Guez, S., Giani, M., Melzi, M., Antignac, C., and Assael, B. (1998). Adequate clinical control of congenital nephrotic syndrome by enalapril. *Pediatr. Nephrol.* **12**, 130–132.

Gunwar, S., Ballester, F., Noelken, M. E., Sado, Y., Ninomiya, Y., and Hudson, B. G. (1998). Glomerular basement membrane. Identification of a novel disulfide-cross-linked network of $\alpha 3$, $\alpha 4$ and $\alpha 5$ chains of type IV collagen and its implications for the pathogenesis of Alport syndrome. *J. Biol. Chem.* **273**, 8767–8775.

Hallman, N., Hjelt, L., and Ahvenainen, E. K. (1956). Nephrotic syndrome in newborn and young infants. *Ann. Paediatr. Fenn.* **2**, 227–241.

Harvey, S. J., Zheng, K., Sado, Y., Naito, I., Ninomiya, Y., Jacobs, R. M., Hudson, B. G., and Thorner, P. S. (1998). Role of distinct type IV collagen networks in glomerular development and function. *Kidney Int.* **54**, 1857–1866.

Heaton, P. A., Smales, O., and Wong, W. (1999). Congenital nephrotic syndrome responsive to captopril and indomethacin. *Arch. Dis. Child.* **81**, 174–175.

Heidet, L., Cai, Y., Guicharnaud, L., Antignac, C., and Gubler, M.-C. (2000). Glomerular expression of type IV collagen chains in normal and X-linked Alport syndrome kidneys. *Am. J. Pathol.* **156**, 1901–1910.

Heidet, L., Dahan, K., Zhou, J., Xu, Z., Cochat, P., Gould, J. D., Leppig, K. A., Proesmans, W., Guyot, C., Guillot, M. *et al.* (1995). Deletions of both alpha 5(IV) and alpha 6(IV) collagen genes in Alport syndrome and in Alport syndrome associated with smooth muscle tumours. *Hum. Mol. Gen.* **4**, 99–108.

Heikkilä, P., Parpala, T., Lukkarinen, O., Weber, M., and Tryggvason, K. (1996). Adenovirus-mediated gene transfer into kidney glomeruli using an *ex vivo* and *in vivo* kidney perfusion system: First steps towards gene therapy of Alport syndrome. *Gene Ther.* **3**, 21–27.

Holmberg, C., Jalanko, H., Tryggvason, K., and Rapola, J. (1999). Congenital nephrotic syndrome. *In* "Pediatric Nephrology" (T. Barratt, E. Avner, and W. Harmon, eds.), 4th Ed., pp. 765–777. Lippincott Williams & Wilkins, Baltimore.

Holzman, L. B., St. John, P. L., Kovari, I. A., Verma, R., Holthöfer, H., and Abrahamson, D. R. (1999). Nephrin localizes to the slit pore of the glomerular epithelial cells. *Kidney Int.* **56**, 1481–1491.

Holthöfer, H., Ahola, H., Solin, M.-L., Wang, S., Palmen, T., Luimula, P., Miettinen, A., and Kerjaschki, D. (1999). Nephrin localizes at the podocyte filtration slit area and is characteristically spliced in the human kidney. *Am. J. Pathol.* **155**, 1681–1687.

Hudson, B. G., Reeders, S. T., and Tryggvason, K. (1993). Type IV

collagen: Structure, gene organization, and role in human diseases: Molecular basis of Goodpasture and Alport syndromes and diffuse leiomyomatosis. *J. Biol. Chem.* **268**, 26033–26036.

Huttunen, N.-P. (1976). Congenital nephrotic syndrome of Finnish type: Study of 75 patients. *Arch. Dis. Child.* **51**, 344–348.

Huttunen, N.-P., Rapola, J., Vilska, J., and Hallman, N. (1980). Renal pathology in congenital nephrotic syndrome of Finnish type: A quantitative light microscopic study on 50 patients. *Int. J. Pediatr. Nephrol.* **1**, 10–16.

Jais, J., Knebelman, B., Giatras, I., Marchi, M., Rizzoni, G., *et al.* (2000). X-linked Alport syndrome: Natural history in 195 families and genotype-phenotype correlations in males. *J. Am. Soc. Nephrol.* **11**, 649–657.

Jansen, B., Thorner, P, Baumal, R., Valli, V., Maxie, M. G., and Singh, A. (1986). Samoyed hereditary glomerulopathy (SHG): Evolution of splitting of glomerular capillary basement membranes. *Am. J. Pathol.* **125**, 536–545.

Jefferson, J. A., Lemmink, H. H., Hughes, A. E., Hill, C. M., Smeets, H. J. M., Doherty, C. C., and Maxwell, A. P. (1997). Autosomal dominant Alport syndrome linked to the type IV collagen $\alpha 3$ and $\alpha 4$ genes (COL4A3 and COL4A4). *Nephr. Dial. Transpl.* **12**, 1595–1599.

Kalluri, R., Gattone, V. H., Noelken, M. E., and Hudson, B. G. (1994). The alpha 3 chain of type IV collagen induces autoimmune Goodpasture syndrome. *Proc. Natl. Acad. Sci. USA* **91**, 6201–6205.

Kalluri, R., Shield, C. F., III, Todd, P., Hudson, B. G., and Neilson, C. (1997). Isoform switching of type IV collagen is developmentally arrested in X-linked Alport syndrome leading to increased susceptibility of renal basement membranes to endoproteolysis. *J. Clin. Invest.* **99**, 2470–2478.

Kanwar, Y. S., and Farquhar, M. G. (1979). Presence of heparan sulfate in the glomerular basement membrane. *Proc. Natl. Acad. Sci. USA* **76**, 1303–1307.

Kanwar, Y. S., Liu, Z. Z., Kashihara, N., and Wallner, E. I. (1991). Current status of the structural and functional basis of glomerular filtration and proteinuria. *Semin. Nephrol.* **11**, 390–413.

Kaplan, J. M., Kim, S. H., North, K. N., Rennke, H., Correia, L. A., Tong, H.-Q., Mathias, B. J., Rodriquez-Perez, J.-C., Allen, P. G., Beggs, A. H., and Pollak, M. R. (2000). Mutations in ACTN4, encoding α-actinin-4, cause familial focal segmental glomerulosclerosis. *Nature Genet.* **24**, 251–256.

Kashtan, C. (1999). Glomerular disease. *Semin. Nephrol.* **19**, 353–363.

Kashtan, C., and Kim, Y. (1992). Distribution of the $\alpha 1$ and $\alpha 2$ chains of collagen type IV and of collagens V and VI in Alport syndrome. *Kidney Int.* **42**, 115–126.

Kawachi, H., Koike, H., Kurihara, H., Yaoite, E., Orikasa, M., Shia, M. A., Sakai, T., Yamamoto, T., Salant, D. J., and Shimizu, F. (2000). Cloning of rat nephrin: Expression on developing glomeruli and in proteinuric states. *Kidney Int.* **57**, 1949–1961.

Kestilä, M., Lenkkeri, U., Männikkö, M., Lamerdin, J., McCready, P., Putaala, H., Ruotsalainen, V., Morita, T., Nissinen, M., Herva, R., Kashtan, C. E., Peltonen, L., Holmberg, C., Olsen, A., and Tryggvason, K. (1998). Positionally cloned gene for a novel glomerular protein-nephrin-is mutated in congenital nephrotic syndrome. *Mol. Cell* **1**, 575–582.

Kestilä, M., Männikkö, M., Holmberg, C., Gyapay, G., Weissenbach, J., Savolainen, E., Peltonen, L., and Tryggvason, K. (1994). Congenital nephrotic syndrome of the Finnish type maps to the long arm of chromosome 19. *Am. J. Hum. Genet.* **54**, 757–764.

Knebelmann, B., Breillat, C., Forestier, L., Arrondel, C., Jacassier, D., Giatras, I., Drouot, L., Deschnes, G., Grunfeld, J.-P., Broyer, M., Gubler, M.-C., and Antignac, C. (1996). Spectrum of mutations in the COL4A5 collagen gene in X-linked Alport syndrome. *Am. J. Hum. Genet.* **59**, 1221–1232.

Koziell, A. B., Lenkkeri, U., Grech, V., Trompeter, R. S., Barrat, T. M., and Tryggvason, K. (1999a). Nephrin mutations in congenital nephrotic syndrome: Further evidence for a critical role in the pathogenesis of proteinuria *J. Am. Soc. Nephrol.* **10**, A2205.

Koziell, A., Lenkkeri, U., Grech, U., and Tryggvason, K. (1999b). NPHS1 gene mutations in Finnish type congenital nephrotic syndrome patients of non-Finnish origin. *Pediatr. Nephrol.* **13**, C24.

Laine, J., Jalanko, H., Holthöfer, H., Krogerus, L., Rapola, J., von Willebrand, E., Lautenschlager, I., Salmela, K., and Holmberg, C. (1993). Post-transplantation nephrosis in congenital nephrotic syndrome of the Finnish type. *Kidney Int.* **44**, 867–874.

Lee, H. J., Gribouval, O., and Antignac, C. (1999). NPHS1 Mutations in non-Finnish CNF populations. *J. Am. Soc. Nephrol.* **10**, A2206

Lees, G. E., Wilson, P. D., Helman, R. G., Homco, L. D., and Frey, M. S. (1997). Glomerular ultrastructural findings similar to hereditary nephritis in 4 English cocker spaniels. *J. Vet. Int. Med.* **11**, 80–85.

Lemmink, H. H., Niilesen, W. N., Mochizuki, T., Schröder, C. H., Brunner, H. G., van Oost, B. A., Monnens, L. A. H., and Smeets, H. J. M. (1996). Benign familial hematuria due to mutation of the type IV collagen α4 gene. *J. Clin. Invest.* **98**, 1114–1118.

Lenkkeri, U., Männikkö, M., McCready, P., Lamerdin, J., Gribouval, O., Niaudet, P., Antignac, C., Kashtan, C., Holmberg, C., Olsen, A., Kestilä, M., and Tryggvason, K. (1999). Structure of the gene for congenital nephrotic syndrome of the Finnish type (NPHS1) and characterization of mkutations. *Am. J. Hum. Genet.* **64**, 51–61.

Ljungberg, P., Holmberg, C., and Jalanko, H. (1997). Infections in infants with congenital nephrosis of the Finnish type. *Pediatr. Nephrol.* **11**, 148–152.

Lu, W., Phillips, C. L., Pverbeekt, P., Meisler, M. H., and Killen, P. D. (1997). A new model of Alport's syndrome. *J. Am. Soc. Nephrol.* **8**, 1818A.

Mazzucco, G., Barsotti, P., Muda, A. O., Fortunato, M., Mihatsch, M., Torri-Tarelli, L., Renieri, A., Faraggiana, T., de Marchi, M., and Monga, G. (1998). Ultrastructural and immunohistochemical findings in Alport's syndrome: A study of 108 patients from 97 Italian families with particular emphasis on COL4A5 gene mutation correlations *J. Am. Soc. Nephrol.* **9**, 1023–1031.

Miner, J. H., and Sanes, J. R. (1996). Molecular and functional defects in kidneys of mice lacking collagen α3(IV): Implications for Alport syndrome. *J. Cell. Biol.* **135**, 1403–1413.

Mochizuki, T., Lemmink, H. H., Mariyama, M., Antignac, C., Gubler, M.-C., Pirson, Y., Verellen-Dumoulin, C., Chan, B., Schröder, C. H., Smeets, H. J., and Reeders, S. T. (1994). Identification of mutations in the α3(IV) and α4(IV) collagen genes in autosomal recessive Alport syndrome. *Nature Genet.* **8**, 77–82.

Moeller, M. J., Kovari, I. A., and Holzman, L. B. (2000). Evaluation of a new tool for exploring podocyte biology: Mouse NPHS1 5′ flanking region drives lacZ expression in podocytes. *J. Am. Soc. Nephrol.* **11**, 2306–2314.

Männikkö, M., Kestilä, M., Holmberg, C., Norio, R., Ryynänen, M., Olsen, A., Peltonen, L., and Tryggvason, K. (1995). Fine mapping and haplotype analysis of the locus for congenital nephrotic syndrome on chromosome 19 q 13.1. *Am. J. Hum. Genet.* **57**, 1377–1383.

Männikkö, M., Kestilä, M., Lenkkeri, U., Alakurtti, H., Holmberg, C., Leisti, J., Salonen, R., Aula, P., Mustonen, A., Peltonen, L., and Tryggvason, K. (1997). Improved prenatal diagnosis of the congenital nephrotic syndrome of the Finnish type based on DNA analysis. *Kidney Int.* **51**, 868–872.

Ninomiya, Y., Kagawa, M., Iyama, K., Naito, I., Kishiro, Y., Seyer, J. M., Sugimoto, M., Oohashi, T., and Sado, Y. (1995). Differential expression of two basement membrane collagen genes, COL4A6 and COL4A5, demonstrated by immunofluorescence staining using peptide-specific monoclonal antibodies. *J. Cell Biol.* **130**, 1219–1229.

Norio, R. (1966). Heredity in the congenital nephrotic syndrome: a genetic study of 57 Finnish families with a review of reported cases. *Ann. Paediatr. Fenn.* **12**(Suppl.),27–34.

Patrakka, J., Ruotsalainen, V., Qvist, E., Laine, J., Holmberg, C.,

Tryggvason, K., and Jalanko, H. (2001). Recurrence of nephrosis in kidney grafts of patients with congenital nephrosis (NPHS1): Role of nephrin. *Transplantation* **73**, 394–403.

Patrakka, J., Kestilä, M., Wartiovaara, J., Ruotsalainen, V., Tissari, P., Lenkkeri, U., Männikkö, M., Visapää, I., Holmberg, C., Rapola, J., Tryggvason, K., and Jalanko, H. (2000). Congenital nephrotic syndrome of the Finnish type (NPHS1): Features resulting from different mutations in Finnish patients. *Kidney Int.* **58**, 972–980.

Peissel, B., Geng, L., Kalluri, R., Kashtan, C., Rennke, H. G., Gallo, G. R., Yoshioka, K., Sun, M. J., Hudson, B. G., Neilson, E. G., Zhou, J. (1995). Comparative distribution of the α1(IV), α5(IV), and α6(IV) collagen chains in normal human adult and fetal tissues and in kidneys from X-linked Alport syndrome patients. *J. Clin. Invest.* **96**, 1948–1957.

Putaala, H., Soininen, R., Kilpeläinen, P., Wartiovaara, J., and Tryggvason, K. (2001). The murine nephrin gene is specifically expressed in kidney, brain, and pancreas: Inactivation of the gene leads to massive proteinuria and neonatal death. *Hum. Mol. Genet.* **10**, 1–8.

Putaala, H., Sainio, K., Sariola, H., and Tryggvason, K. (2000). Primary structure of mouse and rat nephrin cDNA and structure and expression of the mouse gene. *J. Am. Soc. Nephrol.* **11**, 991–1001.

Qvist, E., Laine, J., Rönnholm, K., Jalanko, H., Leijala, M., and Holmberg, C. (1999). Graft function 5–7 years after renal transplantation in early childhood. *Transplantation* **67**, 1043–1049.

Raats, C. J., van den Born, J., and Berden, J. H. M. (2000). Glomerular heparan sulfate alterations: Mechanisms and relevance for proteinuria. *Kidney Int.* **57**, 385–400.

Rodewald, R., and Karnoswky, M. J. (1974). Porous structure of the glomerular slit diaphragm in the rat and mouse. *J. Cell Biol.* **60**, 423–433.

Ruotsalainen, V., Ljungberg, P., Wartiovaara, J., Lenkkeri, U., Kestilä, M., Jalanko, H., Holmberg, C., and Tryggvason, K. (1999). Nephrin is specifically located at the slit diaphragm of glomerular podocytes. *Proc. Natl. Acad. Sci. USA* **96**, 7962–7967.

Ruotsalainen, V., Patrakka, J., Tissari, P., Reponen, P., Hess, M., Kestilä, M., Holmberg, C., Salonen, R., Heikinheimo, M., Wartiovaara, J., Tryggvason, K., and Jalanko, H. (2000). Role of nephrin in cell junction formation in human nephrogenesis. *Am. J. Pathol.* **157**, 1905–1916.

Rutgers, A., Meyers, K. E., Canziani, G., Kalluri, R., Lin, J., and Madaio, M. (2000). High affinity of anti-GBM antibodies from Goodpasture and transplanted Alport patients to α3(IV)NC1 collagen. *Kidney Int.* **58**, 115–122.

Salomon, R., Gubler, M. C., and Niaudet, P. (2000). Genetics of the nephrotic syndrome. *Curr. Opin. Pediatr.* **12**, 129–134.

Sayers, R., Kalluri, R., Rodgers, K., Shield, C., Meehan, D., and Cosgrove, D. (1999). Role for transforming growth factor-β1 in Alport renal disease progression. *Kidney Int.* **56**, 1662–1673.

Shih, N. Y., Karpitskii, V., Nguyen, A., Dustin, M. L., Kanagawa, O., Miner, J., and Shaw, A. S. (1999). Congenital nephrotic syndrome in mice lacking CD2-associated protein. *Science* **286**, 312–315.

Smoyer, W. E., and Mundel, P. (1998). Regulation of podocyte structure during the development nephrotic syndrome. *J. Mol. Med.* **76**, 172–183.

Somlo, S., and Mundel, P. (2000). Getting a foothold in nephrotic syndrome. *Nature Genet.* **24**, 333–335.

Thorner, P. S., Zheng, K., Kalluri, R., Jacobs, R., and Hudson, B. G. (1996). Coordinate gene expression of the α3, α4 and α5 chains of collagen type IV: Evidence from a canine model of X-linked nephritis with a COL4A5 gene mutation. *J. Biol. Chem.* **271**, 13821–13828.

Tryggvason, K. (1999). Unraveling the mechanism of glomerular ultrafiltration: Nephrin, a key component of the slit diaphragm. *J. Am. Soc. Nephrol.* **10**, 2440–2445.

Tryggvason, K., Heikkilä, P., Pettersson, E., Tibell, A., and Thorner, P. (1997). Can Alport syndrome be treated by gene therapy. *Kidney Int.* **51**, 1493–1499.

Tryggvason, K., and Kouvalainen, K. (1975). Number of nephrons in normal human kidneys and kidneys of patients with the congenital nephrotic syndrome: A study using a sieving method for counting of glomeruli. *Nephron* **15**, 62–68.

van der Loop, F., Heidet, L., Timmer, E., van den Bosch, B., Leinonen, A., Antignac, C., Jefferson, J., Maxwell, A., Monnens, L., Schröder, C., and Smeets, H. (2000). Autosomal dominant Alport syndrome caused by a COL4A3 splice site mutation. *Kidney Int.* **58**, 1870–1875.

Wickelgren, I. (1999). First components found for key kidney filter. *Science* **286**, 225–226.

Wing, A. J., and Brunner, F. P. (1989). Twenty-three years of dialysis and transplantation in Europe: Experiences of the EDTA registry. *Am. Kidney Dis.* **14**, 341–346.

Yoshioka, K., Hino, S., Takemura, T., Maki, S., Wieslander, J., Takekoshi, Y., Makino, H., Kagawa, M., Sado, Y., and Kashtan, C. E. (1994). Type IV collagen α5 chain: Normal distribution and abnormalities in X-linked Alport syndrome revealed by monoclonal antibody. *Am. J. Pathol.* **144**, 986–996.

Zheng, K., Thorner, P., Marrano, P., Baumal, R., and McInnes, R. R. (1994). Canine X chromosome-linked hereditary nephritis: A genetic model for human X-linked hereditary nephritis resulting from a single base mutation in the gene encoding the α5 chain of collagen type IV. *Proc. Natl. Acad. Sci. USA* **91**, 3989–3993.

Zhou, J., Leinonen, A., and Tryggvason, K. (1994). Structure of the human type IV collagen COL4A5 gene. *J. Biol. Chem.* **269**, 6608–6614.

28

Congenital Kidney Diseases: Prospects for New Therapies

Adrian S. Woolf

Human kidney and lower urinary tract malformations represent a major burden of disease and are sometimes life-threatening when renal failure is associated with bilateral disease. Fetal ultrasonography may allow prenatal detection of certain malformations, and genetic screening tests may also facilitate early diagnosis of several conditions, allowing the planning of either termination of pregnancy or "conventional" treatments, including fetal surgery and renal dialysis and transplantation in early life. Future therapeutic strategies, such as genetic engineering, renal precursor cell transplantation, and the use of novel drugs, may point the way forward in the treatment of otherwise intractable diseases, such as human kidney malformations and polycystic kidney diseases.

I. Introduction

As discussed in Chapter 21, human renal tract malformations account for up to 30% of all anomalies diagnosed prenatally. Severe renal failure will ensue when bilateral renal agenesis or dysplasia is present, the latter often accompanied by physical obstruction of the lower urinary tract. Indeed, such malformations account for most young children who require treatment with dialysis and renal transplantation (Lewis, 1999; Woolf and Thiruchelvam, 2001). In animal models, it is possible to perform surgical decompression of obstructed urinary tracts, which improves renal developmental potential (Glick et al., 1984; Edouga et al., 2001). In human fetuses, similar strategies are possible (Freedman et al., 1999), although such interventions have yet to be proven effective in randomized, prospective trials. In addition, animal studies have shown that administration of various growth factors may ameliorate damage to the developing kidney when associated with obstruction (Chevalier et al., 1998, 2000), which offers a potential therapy for future use in patients.

Other kidney diseases can be considered disorders of renal differentiation, e.g., polycystic kidney diseases, congenital nephrotic syndromes and genetic glomerular diseases, tubulopathies, and tumors—all aspects covered in this volume (Chapters 21–27). It is possible to conceive of novel therapies for such diseases based on our increasing basic knowledge of cell growth and differentiation. These include gene therapies, the use of renal precursor transplantation, and other novel strategies; these approaches are now reviewed briefly.

II. Gene Transfer Technologies

Diverse methodologies have been devised to introduce genes into mammalian cells; these include transfection of

DNA by physical means and transduction by viruses. Gene transfer is relatively easy to achieve and can be highly efficient in the artificial and controlled environment of cell culture. It is, however, more challenging to achieve gene transfer in the context of a solid organ in a living adult animal. The concept of gene transfer in the prenatal urinary tract *in vivo* might be even more difficult because of the relative inaccessibility of the target tissues. However, there are theoretical advantages of a gene transfer strategy into embryonic and fetal tissues. For example, some gene transfer strategies (e.g., retroviral transduction) are only feasible in replicating cells, and these are much more common during development than in the mature animal. Second, if a new gene becomes incorporated into the DNA of an embryonic target cell and that cell undergoes many rounds of division, there is a chance that a relatively large population of cells will express the gene. Beyond the feasibility of the gene transfer process itself, there are further technical challenges in the form of control of long-term regulation of gene expression.

Larson *et al.* (2000) used a fetal Rhesus monkey model to demonstrate that it was feasible to carry out adenoviral-mediated gene transfer from the amniotic fluid into lung and intestine; furthermore, treatment with a virus expressing the cystic fibrosis membrane conductance regulator resulted in accelerated lung differentiation. Mason *et al.* (1999) performed *in vivo* gene transfer in fetal sheep to alter morphogenesis of the developing cardiovascular system. Closure of the *ductus arteriosus*, a vessel directly connecting the fetal pulmonary artery with the aorta, requires prenatal formation of intimal cushions; furthermore, survival of newborns with severe congenital heart defects can sometimes be enhanced by an open *ductus arteriosus*. Normal closure of the *ductus* requires fibronectin-dependent smooth muscle migration. Mason and colleagues (1999) transfected the fetal *ductus arteriosus* with the hemagglutinating virus of Japan liposomes containing a plasmid encoding a "decoy" RNA, with the result of modulating fibronectin bioactivity; this resulted in delayed *ductus* closure. Another example, perhaps more relevant to the branching morphogenesis that occurs in normal metanephric development, is provided by Zhao *et al.* (1999). These authors demonstrated that the adenovirus-mediated transfer of decorin, a transforming growth factor-β (TGF-β)-binding protein, could inhibit TGFβ-induced inhibition of lung morphogenesis. This is interesting from the perspective of future modulation of abnormal nephrogenesis, as TGF-β is upregulated in kidneys in cases of human and experimental ovine fetal obstructive uropathy and may play a role in the abnormal growth of these dysplastic kidneys (Yang *et al.*, 2000, 2001).

With regard to genetic manipulation to affect kidney structure and/or function, there are several possible approaches. First, in experimental animals, very early embryos, at stages well before organogenesis, can be engineered so that kidney gene expression is modified. One example would be the replacement of a defective gene by homologous recombination in embryonic stem cells, precursors that then go on to form a whole embryo; in other words, a mutant gene would be replaced by a wild-type gene, which would come under the control of all the normal regulators of expression (Lai and Lien, 1999). This would offer the prospect of long-term normal regulation of corrected genes in the kidney and could be used for treating dominantly inherited diseases such as varieties of polycystic kidney disease. A variation on this theme is the prospect of long-term expression of an altered gene product that would protect against certain renal tract diseases. For example, the severity of renal failure after injury has been reduced in mice by the transgenic expression of a dominant-negative epidermal growth factor in renal tubules (Terzi *et al.*, 2000).

A less radical strategy for the treatment of congenital disease would require the direct introduction of new genes into the kidney. Several studies, in the early 1990s, showed that such direct gene transfer was indeed possible. Bosch *et al.* (1993) demonstrated retroviral transfer of a reporter gene into adult rat-regenerating tubular epithelial cells, albeit at low efficiency and with only transient expression. Woolf *et al.* (1993) reported that a similar vector resulted in gene expression in the mouse metanephros after *ex vivo* retroviral exposure; this resulted in limited reporter gene expression after microtransplantation of metanephroi into the renal cortex of neonatal host mice (see later). Koseki and colleagues (1991) also demonstrated the feasibility of expressing a reporter gene in the murine metanephros. After this early phase of research, other investigators attempted to optimize methodologies to make gene transfer into the postnatal kidney more efficient and also attempted to target specific populations of cells; e.g., Moullier *et al.* (1994) demonstrated high efficiency transfer of the LacZ reporter gene into rat tubular cells after retrograde infusion of adenovirus vectors into the lumen of the ureter; McDonald *et al.* (1999) achieved adenoviral gene transfer into kidney cortical vessels; and Tsujie *et al.* (2000) targeted renal interstitial fibroblasts by the artificial viral envelope-type hemagglutinating virus of Japan liposome method.

The studies just cited simply explored the possibility of transfer of reporter genes into the kidney. Other investigators have taken the technology a step further by introducing genes that alter the cell biology of the kidney. An early example is provided by Isaka *et al.* (1993), who induced fibrosis within glomeruli by *in vivo* transfection of TGF-β into the rat kidney. Others have begun to explore the possibility of gene transfer for therapeutic means. Two examples are relevant to the future treatment of humans with inherited kidney disease affecting the renal tubules and glomeruli. Lai *et al.* (1998) investigated mice with renal tubular acidosis secondary to carbonic anhydrase II deficiency due to a point mutation in this gene. Retrograde injection of the liposome-complexed carbonic anhydrase II gene into the urinary tract resulted in the gene being expressed in tubular cells and restored the ability of mice to acidify their urine, such treatment holds promise for patients with inherited renal tubular diseases (Chapter 26). Heikkila *et al.* (2001) achieved adenovirus-mediated transfer of the type IV collagen α5 chain cDNA into pig kidney, resulting in deposition of the protein in the glomerular basement membrane; such a strategy holds promise for treatment of patients with Alport syndrome (see Chapter 27).

A third strategy involves the production of soluble factors to ameliorate kidney disease by genetically engineered cells located outside the kidney. Isaka *et al.* (1996) expressed decorin in skeletal muscle in adult rats with immune-mediated glomerular inflammation; the decorin neutralized TGF-β-induced renal disease. A similar strategy has been used by Isaka *et al.* (1999) to suppress extracellular matrix accumulation in experimental glomerulonephritis. Celiker *et al.* (2001) performed systemic administration of a tissue inhibitor of matrix metalloprotienase DNA, and gene expression significantly inhibited growth of a Wilms' tumor model in nude mice.

III. Renal Precursor Cell Technology

An alternative to genetically engineering the kidney would be to rebuild a damaged or malformed kidney with new cells. In recent years, exciting advances have been made in unraveling cell lineages of diverse nonrenal tissues, including neural tissues (Steindler *et al.*, 2002) and bone marrow (Alison *et al.*, 2000; Reyes *et al.*, 2002)-derived cells; furthermore, the molecules that determine the survival, proliferation, and differentiation of these lineages are being explored. To an extent, similar advances are being made in renal precursor cell technology.

Humes *et al.* (1996) demonstrated that a single tubular cell from the adult mammalian kidney can proliferate and undergo morphogenesis to form a tubule in culture. The same group further demonstrated that such reconstituted epithelia could form a "bioartificial kidney" performing characteristic proximal tubule functions, including vectorial transport of electrolytes and glucose and 1α hydroxylation of vitamin D (Humes *et al.*, 1999b). Excitingly, such tubules could be used in an extracorporeal dialysis machine to form a "renal tubule assist device"; this technology was able to improve markers of outcome in an animal model of sepsis-related acute renal failure (Humes *et al.*, 1999a; Fissell *et al.*, 2002).

The aforementioned experiments used adult cells as a starting point to recreate new tubules. Can renal precursor cells be isolated and propagated for therapeutic uses? Various laboratories have demonstrated that it is possible

to harvest murine metanephroi, in the first days after the organs begin to form, and transplant them into sites in the postnatal animal where the embryonic organ will form mature structures, including vascular glomeruli, which filter blood to make urine. For example, one laboratory has used embryonic mouse kidneys transplanted into the cortex of neonatal mice, a site where nephrogenesis is ongoing (Woolf *et al.*, 1990, 1993; Loughna *et al.*, 1997). Similar results were obtained by Koseki *et al.* (1991). As discussed earlier, it is possible to retrovirally transduce such microtransplants before implantation to express reporter genes in mature structures, offering the prospect of genetically engineering kidney precursors, which then differentiate into functioning organs *in vivo*. The concept of using metanephric kidney transplants to replace the function of failing host kidneys has been investigated extensively in a murine model in which rudiments are transplanted into the omentum around the peritoneal cavity where they grow and connect with the host vascular system; after a period of growth, the ureter of the transplanted organ can be anastomosed surgically with the lower urinary tract of the host, and these transplants have a high enough glomerular filtration rate to maintain the life of the host when it is rendered anephric (Rogers *et al.*, 1998). The transplantation of fetal kidney cells may offer the additional advantage of rendering the host "tolerant" to immune attack from the host, as demonstrated by Rogers *et al.* (2001), who successfully transplanted embryonic kidneys that had a different major histocompatibility complex genotype to the host.

Two further animal experiments should be highlighted here. Ito *et al.* (2001) demonstrated that bone marrow constituted a reservoir for glomerular fibromusclular mesangial cells in the recovery period of a model of immune-mediated disease, and Poulsom *et al.* (2001) provided evidence that murine bone marrow cells can also contribute to renal tubule epithelia. Both studies followed the fate of genetically labeled cells in bone marrow transplants. These exciting studies raise the possibility that renal precursors may be present in the marrow cavity of mature animals, a very intriguing and potentially important discovery.

While the aforementioned studies used animal models, less is known about human kidney precursor cells. Certainly, the human metanephros, like the murine embryonic kidney, can be explanted into organ culture, where it will undergo differentiation (Matsell and Bennett, 1998). Indeed, human metanephroi can be transplanted into immune-deficient mice where they undergo considerable differentiation (Dekel *et al.*, 2002). Burrow and Wilson (1993) described the propagation and differentiation of human renal precursor cells from the cortex of fetal kidneys. Interestingly, Polusom *et al.* (2001) proved preliminary data that, in male patients who had received kidney transplants from female donors, some kidney tubule cells were found to contain markers for the Y chromosome; they interpreted these finding as consistent with the concept that circulating precursor cells, possibly from host bone marrow, could populate that adult kidney. Such presursors have yet to be isolated.

Other laboratories have begun to explore the *in vitro* biology of cultures of human cells from kidney malformations and tumors. For example, Yang *et al.* (2000) isolated a human dysplastic tubule line that expressed nephrogenic genes, such as PAX2 and BCL2; the proliferation of these cells, together with the expression of these markers, was inhibited by the addition of TGF-β, and this cytokine also shifted the phenotype of these cells from epithelial to stromal appearance. Gnarra and Dressler (1995) isolated PAX2-expressing human renal carcinoma cells and found that their growth was slowed by antisense oligonucleotides to this transcription factor. Such studies may, in the future, allow the isolation and *in vitro* manipulation of diverse normal and disease human kidney cell lineages.

IV. Experimental Treatments for Polycystic Kidney Diseases

Polycystic kidney diseases represent a varied genetic group of diseases in which terminal differentiation of renal tubule epithelia is abnormal, with resulting cyst formation and renal failure (Chapter 24). Woo *et al.* (1994) were able to recreate kidney cysts in a controlled culture milieu by using cells from an autosomal-recessive model called the *cpk/cpk* mouse. While the gene mutated in this animal codes for a ciliary protein (Hou *et al.*, 2002), the relationship of this structure to cyst formation has yet to be established. Woo *et al.* (1994) found that a drug called paclitaxel (Taxol), which stabilizes microtubules, was able to reduce cyst formation in culture; moreover, this drug, when administered to young *cpk/cpk* mice, reduced the growth of kidney cysts *in vivo* and prevented early death from renal failure. Thereafter, other investigators confirmed the efficacy of paclitaxel treatment in cultured cysts derived from a dominantly inherited rat model of polycystic kidney disease (Pey *et al.*, 1999).

Other lines of evidence point to abnormal growth factor signaling in the genesis of renal cysts; these observations have been exploited in order to design treatments for polycystic mice. Sweeney *et al.* (1998) demonstrated that the epidermal growth factor receptor was expressed basolaterally in normal mature renal epithelia but was expressed apically in autosomal-recessive mouse models of polycystic kidney disease; furthermore, these receptors are autophosphorylated and transmit mitogenic signals that lead to cyst growth. Epidermal growth factor can also be shown to stimulate renal cyst growth *in vitro* (Pey *et al.*, 1999), and a strain of mice with a mutated epidermal growth factor receptor, which leads to downregulated activity, is resistant to polycystic kidney disease (Richards *et al.*, 1998). Impressively,

when young polycystic mice were treated with a tyrosine kinase inhibitor specific for that growth factor receptor, cyst formation was slowed and renal function improved (Sweeney *et al.* 2000).

Could such a therapies be used for human polycystic kidney diseases? The answer is "perhaps." However, several words of caution are necessary. First, recessive human autosomal-recessive polycystic kidney disease, as opposed to the more common autosomal-dominant form, becomes established before birth; this would mean that treatment would have to theoretically start prenatally. Second, treatments such as paclitaxel are not effective in all types of murine polycystic models (Sommardahl *et al.*, 1997); any potential treatments should first be tested on animal models with the same genetic defect as human diseases, and this had yet to be done. Finally, the progression of human cystic kidney diseases is more variable than in mice, probably due to the varied genetic backgrounds of humans versus laboratory animals, and considerable thought would have to be given to which patients to treat. Nevertheless, strategies such as gene therapy, precursor cell transplantation, and the use of novel drugs may point the way forward in the treatment of otherwise intractable diseases, such as human kidney malformations and polycystic kidney diseases.

References

Alison, M. R., Poulsom, R., Jeffery, R., Dhillon, A. P., Quaglia, A., Jacob, J., Novelli, M., Prentice, G., Williamson, J., and Wright, N. A. (2000). Hepatocytes from non-hepatic adult stem cells. *Nature* **406**, 257.

Bosch, R. J., Woolf, A. S., and Fine, L. G. (1993). Gene transfer into the mammalian kidney: Direct retrovirus transduction of regenerating tubular epithelial cells. *Exp. Nephrol.* **1**, 49–54.

Burrow, C. W., and Wilson, P. D. (1993). A putative Wilms' tumor-secreted growth factor activity required for primary culture of human nephroblasts. *Proc. Natl. Acad. Sci. USA* **90**, 6066–6070.

Celiker, M. Y., Wang, M., Atsidaftos, E., Liu, X., Liu, Y. E., Jiang, Y., Valderrama, E., Goldberg, I. D., and Shi, Y. E. (2001). Inhibition of Wilms' tumor growth by intramuscular administration of tissue inhibitor of metalloproteinase-4 plamid DNA. *Oncogene* **20**, 4337–4343.

Chevalier, R. L., Goyal, S., Kim, A., Chang, A. Y., Landau, D., and LeRoith, D. (2000). Renal tubulointersitial injury from ureteral obstruction in the neonatal rat is attenuated by IGF-1. *Kidney Int.* **57**, 882–890.

Chevalier, R. L., Goyal, S., Wolstenholme, J. T., and Thornhill, B. A. (1998). Obstructive nephropathy in the neonatal rat is attenuated by epidermal growth factor. *Kidney Int.* **54**, 38–47.

Dekel, B., Amariglio, N., Kaminski, N., Schwartz, A., Goshen, E., Arditti, F. D., Tsarfaty, I., Passwell, J. H., Reisner, Y., and Rechavi, G. (2002). Engraftment and differentiation of human metanephroi into functional mature nephrons after transplantation into mice is accompanied by a profile of gene expression similar to normal human kidney development. *J. Am. Soc. Nephrol.* **13**, 977–990.

Edouga, D., Hugueny, B., Gasser, B., Bussieres, L., and Laborde, K. (2001). Recovery after relief of fetal urinary obstruction: Morphological, functional and molecular aspects. *Am. J. Physiol.* **281**, F26–F37.

Fissell, W. H., Dyke, D. B., Weitzel, W. F., Buffington, D. A., Westover, A. J., MacKay, S. M., Gutierrez, J. M., and Humes, H. D. (2002). Bioartificial kidney alters cytokine response and hemodynamics in endotoxin-challenged uremic animals. *Blood Purif.* **20**, 55–60.

Freedman, A. L., Johnson, M. P., Smith, C. A., Gonzalez, R., and Evans, M. I. (1999). Long-term outcome in children after antenatal intervention for obstructive uropathies. *Lancet* **354**, 374–377.

Glick, P. L., Harrison, M. R., Adzick, N. S., Noall, R. A., and Villa, R. L. (1984). Correction of congenital hydronephrosis in utero. IV. In utero decompression prevents renal dysplasia. *J. Pediatr. Surg.* **19**, 649–657.

Gnarra, J. R., and Dressler, G. R. (1995). Expression of Pax-2 in human renal cell carcinoma and growth inhibition by antisense oligonucleotides. *Cancer Res.* **55**, 4092–4098.

Heikkila, P., Tibell, A., Morita, T., Chen, Y., Wu, G., Sado, Y., Ninomiya, Y., Petterson, E., and Tryggvason, K. (2001). Adenovirus-mediated transfer of type IV collagen α5 chain cDNA into swine kidney *in vivo*: Deposition of the protein in the glomerular basement membrane. *Gene Ther.* **8**, 882–890.

Hou, X., Mrug, M., Yoder, B. K., Lefkowitz, E. J., Kremmidiotis, G., D'Eustachio, P., Beier, D. R., and Guay-Woodford, L. M., (2002). Cystin, a novel cilia-associated protein, is disrupted in the cpk model of polycystic kidney disease. *J. Clin. Invest.* **109**, 533–540.

Humes, H. D., Buffington, D. A., MacMay, S. M., Funke, A. J., and Weitzel, W. F. (1999a). Replacement of renal function in uremic animals with a tissue-engineered kidney. *Nature Biotechnol.* **17**, 451–455.

Humes, H. D., Krauss, J. C., Cielinski, D. A., and Funke, A. J. (1996). Tubulogenesis from isolated single cells of adult mammalian kidney: Clonal analysis with recombinant retrovirus. *Am. J. Physiol.* **271**, F42–F49.

Humes, H. D., MacKay, S. M., Funke, A. J., and Buffington, D. A. (1999b). Tissue engineering of a bioartificial renal tubule asist device: *In vitro* transport and metabolic characteristics. *Kidney Int.* **55**, 2502–2514.

Isaka, Y., Akagi, Y., Ando, Y., Tsujie, M., Sudo, T., Ohno, N., Border, W. A., Noble, N. A., Kaneda, Y., Hori, M., and Imai, E. (1999). Gene therapy by transforming growth factor-β receptor-IgG Fc chimera suppressed extracellular matrix accumulation in experimental glomerulonephritis. *Kidney Int.* **55**, 465–475.

Isaka, Y., Brees, D. K., Ikegaya, K., Kaneda, Y., Imai, E., Noble, N. A., and Border, W. A. S. (1996). Gene therapy by skeletal muscle expression of decorin prevents fibrotic disease in rat kidney. *Nature Med.* **2**, 418–423.

Isaka, Y., Fujiwara, Y., Ueda, N., Kaneda, Y., Kamada, T., and Imai, E. (1993). Glomerulosclerosis induced by *in vivo* transfection of transforming growth factor-β or platelet derived growth factor gene into the rat kidney. *J. Clin. Invest.* **92**, 2597–2601.

Ito, Y., Suzuki, A., Imai, E., Okabe, M., and Hori, M. (2001). Bone marrow is a reservoir of repopulating mesangial cells during glomerular remodeling. *J. Am. Soc. Nephrol.* **12**, 2625–2635.

Koseki, C., Herzlinger, S., and al-Awqati, Q. (1991). Integration of embryonic nephrogenic cells carrying a reporter gene into functioning nephrons. *Am. J. Physiol.* **261**, C550–C556.

Lai, L. W., Chan, D. M., Erickson, R. P., Hsu, S. J., and Lien, Y. H. (1998). Correction of renal tubular acidosis in carbonic anhydrase II-deficient mice with gene therapy. *J. Clin. Invest.* **101**, 1320–1325.

Lai, L. W., and Lien, Y. H. (1999). Homologous recombination based gene therapy. *Exp. Nephrol.* **7**, 11–14.

Larson, J. E., Morrow, S. L., Delcarpio, J. B., Bohm, R. P., Ratteree, M. S., Blanchard, J. L., and Cohen, J. C. (2000). Gene transfer into the fetal primate: Evidence for the secretion of transgene product. *Mol. Ther.* **2**, 631–639.

Lewis, M. (1999). Report of the pediatric renal registry. *In* "The UK Renal Registry: The Second Annual Report." (D. Ansell and T. Feest, eds.), pp. 175–187. The Renal Association, Bristol, UK.

Loughna, S., Hardman, P., Landels, E., Jussila, L., Alitalo, K., and Woolf, A. S. (1997). A molecular and genetic analysis of renal glomerular capillary development. *Angiogenesis* **1**, 84–101.

McDonald, G. A., Zhu, G., Li, Y., Kovesdi, I., Wickham, T. J., and Sukhatme, V. P. (1999). Efficient adenoviral gene transfer to kidney cortical vasculature utilizing a fiber modified vector. *J. Gene Med.* **1**, 103–110.

Mason, C. A., Bigras, J. L., O'Blenes, S. B., Zhou, B., McKintyre, B., Nakamura, N., Kaneda, Y., and Rabinovitch, M. (1999). Gene transfer in utero biologically engineers a patent ductus arteriosus in lambs by arresting fibronectin-dependent neointimal formation. *Nature Med.* **5**, 176–182.

Matsell, D. G., and Bennett, T. (1998). Evaluation of metanephric maturation in a human fetal kidney explant model. *In Vitro Cell Dev. Biol. Anim.* **34**, 138–148.

Moullier, P., Friedlander, G., Calise, D., Ronco, P., Perricaudet, M., and Ferry, N. (1994). Adenoviral-mediated gene transfer to renal tubular cells *in vivo*. *Kidney Int.* **45**, 1220–1225.

Pey, R., Bach, J., Schieren, G., Gretz, N., and Hafner, M. (1999). A new *in vitro* bioassay for cyst formation by renal cells from autosomal dominant rat model of polycystic kidney disease. *In Vitro Cell Dev. Biol. Anim.* **35**, 571–579.

Poulsom, R., Forbes, S. J., Hodivala-Dilke, K., Ryan, E., Wyles, S., Navaratnarasah, S., Jeffery, R., Hunt, T., Alison, M., Cook, T., Pusey, C., and Wright, N. A. (2001). Bone marrow contributes to renal parenchymal turnover and regeneration. *J. Pathol.* **195**, 229–235.

Reyes, M., Dudek, A., Jahagirdar, B., Koodie, L., Marker, P. H., and Verfaillie, C. M. (2002). Origin of endothelial progenitors in human postnatal bone marrow. *J. Clin. Invest.* **109**, 337–346.

Richards, W. G., Sweeney, W. E., Yoder, B. K., Wilkinson, J. E., Woychik, R. P., and Avner, E. D. (1998). Epidermal growth factor receptor activity mediates renal cyst formation in polycystic kidney disease. *J. Clin. Invest.* **101**, 935–939.

Rogers, S. A., Liapis, H., and Hammerman, M. R. (2001). Transplantation of metanephroi across the major histocompatibility complex in rats. *Am. J. Physiol. Regul. Inegr. Comp. Physiol.* **280**, R132–R136.

Rogers, S. A., Lowell, J. A., Hammerman, N. A., and Hammerman, M. R. (1998). Transplantation of developing metanephroi into adult rats. *Kidney Int.* **54**, 27–37.

Sommardahl, C. S., Woychik, R. P., Sweeney, W. E., Avner, E. D., and Wilkinson, J. E. (1997). Efficacy of taxol in the orpk mouse model of polycystic kidney disease. *Pediatr. Nephrol.* **11**, 728–733.

Steindler, D. A., and Pincus, D. W. (2002). Stem cells and neuropoiesis in the adult human brain. *Lancet* **359**, 1047–1054.

Sweeney, W. E., Jr., and Avner, E. D. (1998). Functional activity of epidermal growth factor receptors in autosomal recessive polycystic kidney disease. *Am. J. Physiol.* **275**, F387–F394.

Sweeney, W. E., Chen, Y., Nakanishi, K., Frost, P., and Avner, E. D. (2000). Treatment of polycystic kidney disease with a novel tyrosine kinase inhibitor. *Kidney Int.* **57**, 33–40.

Terzi, F., Burtin, M., Hekmati, M., Federici, P., Grimber, G., Briand, P., and Friedlander, G. (2000). Targeted expression of dominant-negative EGF-R in the kidney reduces tubulo-interstitial lesions after renal injury. *J. Clin. Invest.* **106**, 225–234.

Tsujie, M., Isaka, Y., Ando, Y., Agaki, Y., Kaneda, Y., Ueda, N., Imai, E., and Hori, M. (2000). Gene transfer targeting interstitial fibroblasts by the artificial viral envelope-type hemagglutinating virus of Japan liposome method. *Kidney Int.* **57**, 1973–1980.

Woo, D. D. L., Miao, S., Pelayo, J., and Woolf, A. S. (1994). Taxol inhibits progression of congenital polycystic kidney disease. *Nature* **368**, 750–753.

Woolf, A. S., Bosch, R. J., and Fine, L. G. (1993). Gene transfer into the mammalian kidney: Micro-transplantation of retrovirus-transduced metanephric tissue. *Exp. Nephrol.* **1**, 41–48.

Woolf, A. S., Palmer, S. J., Snow, M. L., and Fine, L. G. (1990). Creation of a functioning chimeric mammalian kidney. *Kidney Int.* **38**, 991–997.

Woolf, A. S., and Thiruchelvam, N. (2001). Congenital obstructive uropathy: Its origin and contribution to end-stage renal failure in children. *Adv. Ren. Replace. Ther.* **8**, 157–163.

Yang, S. P., Woolf, A. S., Quinn, F., and Winyard, P. J. D. (2001). Deregulation of renal transforming growth factor-β1 after experimental short-term ureteric obstruction in fetal sheep. *Am. J. Pathol.* **159**, 109–117.

Yang, S. P., Woolf, A. S., Yuan, H. T., Scott, R. J., Risdon, R. A., O'Hare, M. J., and Winyard, P. J. D. (2000). Potential biological role of transforming growth factor β1 in human congenital kidney malformations. *Am. J. Pathol.* **157**, 1633–1647.

Zhao, J., Sime, P. J., Bringas, P., Jr., Gauldie, J., and Wharburton, D. (1999). Adenovirus-mediated decorin gene transfer prevents TGF-β-induced inhibition of lung morphogenesis. *Am. J. Physiol.* **277**, L412–L422.

Index